D0811787

Pete Dunne's **ESSENTIAL FIELD GUIDE COMPANION**

Books by Pete Dunne

Tales of a Low-Rent Birder

Hawks in Flight

The Feather Quest

More Tales of a Low-Rent Birder

The Wind Masters

Before the Echo

Small-Headed Flycatcher. Seen Yesterday. He Didn't Leave His Name. And Other Stories

Golden Wings

Pete Dunne on Bird Watching

Pete Dunne's Essential Field Guide Companion

Pete Dunne's ESSENTIAL FIELD GUIDE COMPANION

Pete Dunne

Houghton Mifflin Company
Boston New York 2006

Visit our Web site: www.houghtonmifflinbooks.com.

Photographs by Linda Dunne and Pete Dunne

Library of Congress Cataloging-in-Publication Data

Dunne, Pete.
Pete Dunne's essential field guide companion / Pete Dunne.
p. cm.
ISBN-13: 978-0-618-23648-0
ISBN-10: 0-618-23648-1
1. Birds—North America—Identification. I. Title: Essential field guide companion. II. Title.
QL681.D882 2006
598'. 07'2347—dc22 2005021110

Printed in the United States of America

QUM 10 9 8 7 6 5 4 3 2 1

"I took them from a German soldier who didn't need them anymore," said my father, former Staff Sergeant Gerald W. Dunne, of the binoculars clasped in my hands.

Even at the age of seven, the import of this disclosure did not escape me, but as often happens in life, the urgency of the moment superseded due regard for the past.

"Can I use them to look at birds?" I asked.

"If you take care of them," he said, and I did, and I do, because those old 6x24 binoculars, the instrument that conferred my first intimacy with birds, remain in my care today.

This book is dedicated to the unknown German soldier.

And to my father,
who likewise doesn't need them anymore.

Contents

Acknowledgments

It has taken more than four years to move this project from concept to completion. During this time an unholy number of individuals, institutions, colleagues, and friends lent their counsel and support, and it is no exaggeration to say that this book could not have been completed without them.

First, and with due recognition, I wish to both acknowledge the contribution of and express my sincere gratitude to the many scientists, field guide authors and illustrators, and experts who have built the existing wealth of information relating to North America's birds. I have drawn all my life from this wellspring of knowledge, which makes my efforts here more nearly the work of a compiler and scribe than author.

But in particular I wish to recognize all the officers, directors, staff, editors, and, most of all, the contributing authors of the Birds of North America Series, volumes 1 through 716. This great work served as my primary reference for information relating to distribution, habitat, and migration patterns for the species contained in this book, and it represents, in my estimation, the most comprehensive summary of any region's bird life ever crafted. I suspect that few individuals have read as much of this work as I have. With no prompting I include here the Web site address (www.bna.birds.cornell.edu), and also without prompting I encourage all serious students of birds to avail themselves of this great body of work.

During the period of our travel and study, my wife, Linda, and I were helped by scores of individuals who gave freely of their time and talent—too many hundreds to enumerate here. My gratitude to all is nevertheless undiminished. But if singling out a few for special recognition does nothing to diminish the contribution of others, then I would like to express special thanks to Ann and Bob Ellis, proprietors of Camp Ellis—Home for Abandoned Dogs (and Linda's parents)—who not only served as our support base during assorted West Coast campaigns but in addition cared for our somewhat troublesome pets when Linda and I were on the road. Which, for months at a time, we often were.

In this same vein, I wish to thank the staff of New Jersey Audubon's Cape May Bird Observatory, who were obligated to shoulder extra work in support of my somewhat protracted "semi-sabbatical." In particular I want to thank my boss, Tom Gilmore, for granting an extended leave of absence. Then the extension. And then the extended extension. Tom, I owe you, and if the patience and latitude you accorded me could be converted (with no loss of mass or energy) into environmental activism, the world would have been saved yesterday. But the world hasn't been saved, and there is still work to do, so I'll be at my desk . . . tomorrow.

Speaking of saving the world (or at least big substantial chunks of it), during our travels Linda and I availed ourselves of the viewing opportunities afforded by many national wildlife refuges—the charge of a very special branch of the federal government, the U.S. Fish and Wildlife Service. Time and time again refuge personnel went out of their way to facilitate our efforts. Their professionalism and assistance were—and are—greatly appreciated. How they manage to do what they do with the funding provided is a miracle as big as the one they protect and preserve.

A number of respected friends and colleagues had their 2004 holiday season interrupted by the appearance of unsolicited manuscript that begged reviewing. These individuals not only provided useful insights regarding the species they were delegated but saved me considerable embarrassment. The only thanks I can offer is to commend them and their efforts to you: Cameron Cox, Richard Crossley, Jon Dunn, Shawneen Finnegan, Don Freiday, Tim Gallagher, Paul Gurris, Marshall Iliff, Greg Lasley, Michael O'Brien, Will Russell, Clay Sutton, Chris Vogel, Sheri Williamson, Chris Wood, and Kevin Zimmer. Their expertise is eclipsed only by their generosity, and my gratitude eclipses that.

I am particularly indebted to Paul Lehman, who in the capacity of editor and friend took upon himself the extraordinary task of reviewing this entire manuscript. There is no one whose knowledge base concerning North America's birds surpasses Paul's. His contribution to this book cannot be measured or gainsaid, but it may be called into question because of errors that will inevitably find their way into print. Accordingly, and in due deference to Paul, let it be known that no factual errors were missed by him. Rather, his corrections and well-considered suggestions were overlooked or stupidly dismissed by me.

In the same vein, I thank Wayne Petersen and Kevin Zimmer, who very generously agreed to review the page proofs and whose collective knowledge concerning the birds of North America brackets both coasts and all that lies between.

Thanks (in some cases, added thanks) are also extended to the following individuals who gave freely of their time and expertise in the field to make this a better book (and me a better birder): Jeff Bouton, Richard Crossley, Bob Dittrick, Mike Fritz, Steve Howell, Paul Lehman, Brian Patteson, Scott and Julie Roederer, Debbi Shearwater, Chris Vogel, Matt White, Sheri Williamson, and Tom Wood.

Grateful thanks are extended to my agent, Russell Galen, who facilitated the contract with Houghton Mifflin that resulted in this book; Terry Moore and Leica Sports Optics, for their generous support of our travels; and Bob Dittrick and Lisa Moorehead, for graciously allowing us to park the "Road Pig" in their drive in Eagle River, Alaska, for five weeks (while we went birding elsewhere).

Editors are a very special breed. It is their greatly underappreciated task to take pages of manuscript (in the case of this book, 1,800 of them) and convert them into something that will serve readers and save authors embarrassment. In this regard, Lisa White is both a credit to her profession and a candidate for sainthood. Thanks, too, to Shelley Berg for her production work.

Also in line for thanks is Anne Chalmers for her very compelling design of this book.

Finally, and without explanation, this acknowledgment would be sadly deficient without recognizing the contribution (and sacrifices) of Linda E. Dunne, whose many titles include Logistics Coordinator, Chief Navigator, Supply Master, RV Maintenance Chief, Dog Distracter, Proofreader, and Lost Binocular Finder. My wife is, in point of fact, the real birder in the family. Love you, sweetie. We can have our weekends back now. What do you say we take next Sunday off and go birding?

—Pete Dunne
Mauricetown, New Jersey
January 1, 2006

Pete Dunne's **ESSENTIAL FIELD GUIDE COMPANION**

Introduction

First, let's agree on what this book is not—it is *not* a field guide. Even though its focus is the identification of North America's birds, even though its ambition is to guide you to a correct identification, this is not a book intended to be carried in the field. It is a book that hopes to impart shortcuts to tricky identifications (the kind that expert birders use and beginners regard with mystified envy) and provide additional hints and clues to guide you to an identification when a review of your primary field guide has left that identification hanging in the balance.

It is, in short, a supplement, a field guide helper, and it distinguishes itself from other books dealing with bird identification in several ways.

First, and most notably, the species accounts are not supported by any illustrations. None. With such a wealth of illustration-driven field guides already available, it seemed silly to duplicate the efforts of so many fine artists and photographers. This book is designed to be used in concert (not to compete) with one or more illustrated guides.

Second, the text is lavish and stands on its own. In most field guides text is subordinate and supportive, used to draw the user's attention to or reinforce key points depicted by illustrations. The operative word here is *key*, because in most cases field guide authors are limited (by book size and competition for space with illustrations) in the amount of information they can impart. Most succeed in offering readers a summary of the best or most determining field marks but rarely are they able to provide all of them, and most descriptions are terse, imparting information in field guide shorthand.

This book has no such limitations. Here the text is the very centerpiece of each species account, and I think you will find that the style of this guide is livelier and more freewheeling than that of most books on bird identification. Birding, after all, is something that is supposed to be fun.

Finally, this book's identification process—the way in which an observer regards an unfamiliar bird and notes distinguishing characteristics—differs somewhat from that of most field guides. Wherever possible, I encourage birders to take a broad-brush approach to identification rather than focus on the fine details of bird parts or plumage. The approach here emphasizes overall color and shape, blatant patterns, behavioral traits, comparisons with similar species, habitat, and the other birds that share this habitat.

If you are familiar with one of my earlier books, *Hawks in Flight*, a field guide cowritten with David Sibley and Clay Sutton, then you no doubt see the similarities between the approach just described and the holistic method of identification pioneered in that field guide. Indeed, in many respects this is a *Hawks in Flight* that treats the rest of North America's regularly occurring species.

Where did the inspiration for this guide come from? Several places.

First, from Roger Tory Peterson, author of the first effective field guide to the birds. In 1998 Roger's widow, Virginia Marie Peterson, called on Noble Proctor and me to finish the fifth edition of Roger's *Eastern Birds* guide, which his death in 1997 had placed in limbo. The illustrations were mostly complete. All that remained was the balance of the text and an updating of the range maps (for which the considerable skills of Paul Lehman were enlisted).

The challenge, of course, was writing the unfinished and semifinished species accounts in Roger's style, which focused on one or two of a bird's key field marks (an approach that might be called identification shorthand). The problem was space and style. Time and again I was tempted to add to the account one or more field marks that would be useful to readers (and sometimes more useful, I felt, than the field marks accented by Roger in his illustration) but concluded that I could not. Adding information, even useful information, would undercut the very quality that generations of birders have found so compelling about the Peterson Field Guides: their simplicity.

It occurred to me that perhaps a supplement to *Eastern Birds*, offering readers more information than the guide itself could support, would solve the problem. Houghton Mifflin asked, "Why a supplement to just *Eastern Birds*? Why not a book treating all North American birds?"

And this is how I came to spend two and a half years traveling around North America, seeking out and studying the birds that enrich the continent, gathering many of the insights that are housed in this book. These travels were also the logistical footing that made it possible to do something that, to my knowledge, no field guide author has been able to do before: craft species accounts in real time, with the experience and insights from studying each species still fresh in my mind.

I wrote most of the descriptions in this book the same day I studied the species, using notes taken in the field. In all, fewer than twenty of the species accounts included here were written without the benefit of concerted, fresh, firsthand field study, and only four species are birds with which I have no experience: Himalayan Snowcock, Least Storm-Petrel, Ivory Gull, and Ivory-billed Woodpecker. Accounts for these species are based on the wealth of material available and the insights of colleagues.

Although 691 species and subspecies are treated in this book, some that have occurred in North America north of the Mexican border are not. What standards did I use to determine whether a species would be included?

The birds included here occur regularly (that is, annually or almost so) in the United States and Canada. Birds that are not here I deemed too rare, too geographically isolated, or (in the case of several introduced and recently established species) too biologically tenuous to make the cut.

Examples include Common Ringed Plover, an Old World species that breeds in the extreme northeastern arctic regions of Canada and Greenland (as well as on St. Lawrence Island, Alaska) and is sometimes encountered during migration on Alaska's western islands, and Cook's and Stejneger's Petrels, two deep-water ocean birds found well off the coast of California (and encountered only by an act of extreme volition).

Couldn't I have included all possible species so that the book would be complete? Certainly. But doing so would have undercut one of the guiding principles of this book and of my own approach to identification: simplify the identification process by first eliminating variables.

As anyone who knows me will tell you, I am not fond of details, and this lifelong antipathy for paying attention to minutiae has influenced my style of birding. As fortune had it, just about the time I was getting serious about birds, bird identification was focusing more and more on things like feather edges and feather wear and molt patterns. In short, birding was getting uncomfortably detailed for a broad-brush sort of guy.

I remember once attending an illustrated lecture on tern identification given by the late Claudia Wilds—a marvelous person and highly skilled birder who had forgotten more about tern plumages than I will ever know. I was a fair birder. Having spent many years studying birds at Cape May, New Jersey, I felt comfortable with my ability to distinguish Common and Forster's Terns. But after hearing Claudia's lecture, the focus of which was plumage patterns relating to molt and wear, I walked out convinced that I must have been deluding myself.

As it turns out, my ability to distinguish Common and Forster's Terns is fine. The fact is that the field marks I rely on to tell the two species apart have little to do with fine points of plumage. What I clue in on is overall shape (angular and compact for Common, gangly for Forster's), overall color (gray for Common, frosty white for Forster's), habitat (offshore for Common, tidal marsh creeks and ponds for Forster's), and, when it applies, time of year (in Cape May Forster's Terns arrive weeks earlier and depart weeks later than Common Terns).

Using several broad brush strokes (relating to size, shape, overall colors or pattern, and behavior), I find I can nail an identification, for most species, without resorting to fine plumage-related details.

Can all birds be identified by using a broad-brush, holistic approach? Possibly not (although I doubt that birds like Hammond's and Dusky Flycatchers tell each other apart by assessing each other's "primary projection"). But a greater reliance on manifest characteristics and traits can always advance the identification process, particularly when looks are fleeting and conditions less than optimal. And let's be honest—that's not the exception in birding, it's the norm.

Consider a standard field guide. What does it show? A broadside portrayal of a stationary bird showing key field marks to best effect. It is the ideal representation for the purposes of imparting information. But in the field a bird is often encountered at such a distance that distinguishing field marks are indiscernible, or the bird is backlit and reduced to a silhouette, or the bird is facing the viewer. Or even more commonly, the bird is in flight.

In situations like this beginning birders swallow their frustration, shrug their shoulders, and look for a more cooperative subject. Experienced birders, by contrast, start building an identification based on clues that have little to do with the field marks found in many basic field guides.

These same clues form the foundation of this book. If these tricks haven't gotten as much attention as they might have, it might be because they didn't fit neatly into the supportive role that text is designed to play in most traditional, illustration-driven field guides.

But not all guides. Which brings us to another inspirational thread leading to this work. Back when I was beginning to learn my birds, there were two very popular guides on the market. Roger Tory Peterson's *Birds of North America* and the "Golden Guide" written by Chandler Robbins. Both were similar in format and approach, and both relied on illustrations of birds with supporting text on the opposing page. As good as these guides were, there nevertheless were occasions when I would see a bird, note its field marks, turn to the guides, and still be unable to pin a name on it.

In most such cases I could narrow the possibilities down to two or three candidates. Northern or Louisiana Waterthrush, Common or Forster's Tern. But then I was stuck. That is when I reached for Pough.

I am dismayed by the number of birders who are not familiar with Richard Pough and his guides. Accounted among the most dedicated and accomplished conservationists of the twentieth century, he published *Audubon Land Bird Guide* and *Audubon Water Bird Guide* in 1946 and 1951, respectively (and a western version in 1957). What distinguishes Pough's guide is the organization. Illustrations and text are isolated, making the books not quite as utilitarian as Peterson or Robbins, but this presumed weakness is also the oblique source of the guides' strength. The text! The substance of the text fell under the heading "Habits" and dealt with the behavioral traits and habitat preferences of the species. It was often here, buried in Pough's text, that I found some catalytic nugget of information that propelled me (or at least nudged me) toward an identification.

My aim was to craft a book that compiles these catalytic and supportive clues and would allow birders to pin a name on a bird when diligent study and resort to a more classically arranged field guide have still left them short of an identification.

A word or two about the layout and text of this book. Even subtle and subjective information needs a structured carriage. Every species account is laid out in the same systematic manner with regard to age, sex, and subspecies. As mentioned, my approach to identification revolves around simplification; my sole objective when confronted by an identification challenge is to identify the species. One of simplicity's fundamental tenets is the reduction of variables. In this book, wherever possible, I ignore subtle distinctions and stress commonality—another way of saying that you will find minimal reference to subspecies, age classes, and sex except where such differences are manifest and affect the identification process.

This guiding principle went so far as to prompt me to exclude descriptions of plumages that are held briefly or are not commonly seen. An example is the transition plumage of juvenile Rock Sandpiper, which is worn for only a few weeks where the birds breed (in extreme western Alaska). It is commonly replaced by a nonbreeding plumage (technically first-winter plumage, which resembles adult nonbreeding plumage) before these young birds reach their winter range along more southern stretches of the West Coast (where they are most frequently seen). Because so few birders ever see this juvenile plumage, and many traditional guides treat this plumage (and there was nothing to add to their descriptions), it didn't warrant inclusion in this book.

Another example is European Starling. This species also has a juvenile plumage: newly fledged young are wholly drab grayish brown (in contrast to the glossy black plumage of breeding birds and the overall finely spotted plumage of birds in winter). But starlings are flocking species. When you see one, you commonly see many, and by size, shape, and association (not to mention the fact that if you wait long enough an adult starling will plop food into the young bird's mouth), most people can pin the right name to the bird, my account's terse reference to a juvenile plumage notwithstanding.

My efforts at simplification also extended to the terminology used in this book. To describe age classes and anatomy I tried not to use an esoteric term, even one that enjoys wide acceptance among birders, when a simple or more universally understood term would do. Describing the pale (or dark) line above a bird's eye as the "supercilium" is both accurate and precise, but "eyebrow" works as well.

For most species, the terms "first-year bird" or "first-winter" are generally avoided in favor of the broader and more generic age labels "immature" or "young." For some birds that go through successive plumages en route to adulthood, these differences affect the identification process (gulls, for example). In these cases I consolidated and simplified age classes where I thought I could and at times resorted to terms like "subadult" and "advanced immature."

I know that this is going to sound heretical to some people, but the fact is that you don't need to know whether an immature Bald Eagle is a two-year bird or a three-year bird to know that it is a Bald Eagle. All an observer need know is that advanced immatures (second- and third-year birds) are shabby-looking and their bodies are lavishly splattered with white.

When differences relating to plumage or structure do not radically diverge from the basic pattern between subspecies, I generally chose not to mention them. In several cases where differences between subspecies are marked (and the classification of the bird as to species or subspecies may be a matter of question or debate and future redeterminations likely), I chose to treat the identifiable forms as a full species (for example, Harlan's Hawk, Eastern and Western Willets, or Fox Sparrows). In one case, I lumped together two birds currently regarded as distinct species—Black-backed and White Wagtails—and classified them as "White Wagtail."

These exceptions aside, the ordering of species presented in this book follows the order newly established by the revised seventh edition of the American Ornithologists' Union's *Check-list of North American Birds.* This ordering of bird groups differs somewhat from earlier editions of the A.O.U. Checklist and, unfortunately, field guides currently in print (for instance, waterfowl not loons come first in the new order), but these changes are based upon the most recent, and widely accepted, evidence concerning the phylogeny and classification of birds. Future field guides and editions of guides already in print will almost certainly reflect these changes.

Indulge me for a moment as I regard the birding landscape and focus on the wealth of knowledge that constitutes all we collectively know about bird identification. I have already said that this holistic (or broad-brush) approach to bird identification is not my invention (although I am certainly among its practitioners), but I did not tell you who is responsible for these tricks and techniques.

The answer is—lots of people. Over the course of years and uncounted hours in the field, and since long before information-age birders were discussing and sharing identification information, hundreds of people—more nearly thousands—discovered many of the clues detailed in this book. There has never been any such thing as proprietary ownership of a field mark. Like a joke or a smile, or the whereabouts of a good bird, insight is something that is just shared and its value grows in the sharing.

Having said this, I wish to make it clear that I and I alone am responsible for any errors in this guide. One of the greatest disservices a birder or writer can commit is to present bad information that confounds rather than advances the identification process. Being mindful of this, I tried never to make observations that were not supportably true. As I came to discover in the review and editing process, however, I presented some information poorly or wrongly in the first drafts, and certainly not every last one of these errors can be caught and corrected.

Thus, I cordially and humbly invite readers who find fault with or take exception to the text to please let me know and, if you feel inclined, to share some of the useful tricks of the trade known to you. Future editions of this book will be richer for it—and birders everywhere better served.

I have already given due thanks and recognition to the host of individuals who directly or indirectly contributed to the information in this book. In closing, I would like to draw the reader's attention to a place that is a crucible for bird study and whose name has become synonymous with birding's frontier. This is Cape May, New Jersey. Since the pioneering days of Alexander Wilson ("the Father of American Ornithology") and John James Audubon, Cape May has attracted some of the best and brightest minds in bird study. They are drawn not only to the Cape's ornithological wealth and diversity but also to the energizing proximity of other keen minds. It is no mere happenstance that *Hawks in Flight* was written in Cape May, as well as *The Sibley Guide to Birds*,

The Shorebird Guide, and this book. These efforts and others are directly related to the opportunities for bird study found in Cape May.

Birding is a dynamic between birders and the opportunities and challenges presented by a region or place. Both shape the style and quality of birding. It is, perhaps, not an overstatement to say that there is a Cape May School of Birding and that this book is grounded in it.

A Guide to the Guide

HOW TO MAKE THIS BOOK WORK FOR YOU

There is nothing particularly complicated about a guide to bird identification. All it is (or hopes to be) is a book that explains *what* to look for to distinguish one species from another. In addition, both directly and indirectly it tells you *how* to go about doing so. For this book to work for you, you don't need to know any more than this. All you have to do is turn to the account of a species of interest. Read the text. Bring the information to bear in the field, or in the case of a bird you've already found and studied, compare the text to the details housed in your memory (or inscribed in your field notes) and see whether you have a match.

But if you want to maximize the potential of this book, and if you are the kind of person who is interested in the whys as much as the what and the how, then you are invited to keep reading. Certain principles govern the information provided here and the manner in which it is presented. If you understand these principles, this book will serve you better.

First, insofar as this book is designed to be a supplement, it is presumed that you already have one or more of the standard illustrated field guides to birds at your disposal. As they have been since the publication of the seminal Peterson field guide in 1934, a basic field guide is every birder's primary resource when confronting an identification challenge. This book is meant to augment these primary guides by offering more information. It also strives to present information as naturally as possible by replicating the identification process used by an experienced birder: looking at the big picture first and sleuthing for details later.

Inexperienced birders commonly use field marks to jump-start an identification. Experienced birders use field marks to confirm it. For very understandable reasons, standard field guides are thematically allied to the jump-start school. This guide is more wedded to process.

Don't Keep an Open Mind

Even before they sight a bird, experienced birders are bringing their experience to bear. They know that birds are creatures of habit and habitats and that the nature of a habitat encourages certain species to be there and discourages the presence of others. For example, you would expect to find a Carolina Wren in a suburban, coastal community in New Jersey. You would not expect a Rock Wren, a bird common to arid, rocky slopes.

Also, experienced birders know that different bird species have defining ranges (Rock Wrens are western birds that are not found east of the prairies, so they are not likely to be found in New Jersey) and that a bird's range is determined not only by geography but by seasons. The range of Rock Wren extends into southern British Columbia, southern Alberta, and southwestern Saskatchewan in the summer, but in the winter northern breeding members of this species retreat farther south. This species is not located in Canada in winter.

So when these birders go birding, their accumulated knowledge and experience enable them to predict which birds they are likely to encounter based on location, habitat, and time of year (among other clues). And because they are able to go into the field juggling fewer variables, the identification process is greatly simplified for them.

When a wrenlike bird pops up on a scree slope in June in the Rocky Mountain foothills just west of Calgary, Alberta, they can test a hypothesis—"Is it Rock Wren?" (the expected species)—rather than approach the problem by asking: "Now, which one of the nine species of wrens found in North America is this?"

But, you may be saying, I'm not an experienced birder, so I cannot apply such a search engine to filter what I see. That is exactly the function of the introductory paragraph in each species account.

Identification Right Think

The introductory paragraph for each species provides a biographical backdrop. The elements include STATUS, DISTRIBUTION, HABITAT, COHABITANTS, and MOVEMENTS/MIGRATION. STATUS relates to the bird's numeric abundance and condition of residency (whether it is a permanent resident, a summer or winter resident, a visitor, or a vagrant). You are likely to see birds that enjoy large populations and less likely to see those whose populations are small. The terms "common," "uncommon," and "rare" are most commonly used to describe a bird's status. A "common" bird is one you are very likely to encounter; "uncommon" refers to the bird you might see, but perhaps another, similar (and perhaps more common) bird should also be considered as a candidate. "Rare" birds are the ones you have only a slim chance of encountering. If you encounter a bird that resembles a rare species, your identification may well be correct, but you should approach the possibility with caution.

DISTRIBUTION defines the geographic area in which the bird is typically found. For some species, this remains fixed all year. For other species, distribution shifts seasonally. HABITAT describes the biological setting—climatic, topographical, and vegetative—that the species favors and offers examples of such settings. COHABITANTS are the other birds (or animals) that are also specialized for and likely to be found in a bird's preferred habitat. MOVEMENTS/MIGRATION provides the dates (and sometimes the routes and key staging areas) a species moves between its breeding and wintering areas; this passage sometimes carries the bird across regions that do not fall within that species' breeding or winter range.

Taken in sum, STATUS, DISTRIBUTION, HABITAT, COHABITANTS, and MOVEMENTS/MIGRATION constitute the biological framework that defines where a bird is likely to be and when it is likely to be there—and thus whether a species is likely to be what you believe it to be.

In a word, these elements of species' biographical backdrop define *probability*. Experienced birders use probability all the time, and inexperienced birders eventually come to appreciate it. They also come to understand that probability is not confining and in fact is empowering. It helps turn a complicated question ("Now, which one of the 800 species of birds found in North America is that?") into a simple one ("Is this the species I expect?").

You're in Cape May Point, New Jersey. You see a large wren in a suburban yard. The question you'll ask is: Is it Carolina Wren, the default large wren for the region? Almost always the answer is yes. But as salient a factor as *probability* is, it is not determining. It suggests, but it doesn't certify. Probability has a qualifying companion called *possibility*. Birds don't always follow the rules. They sometimes turn up outside their prescribed ranges and in marginal or ill-suited habitats or at odd times. Getting back to the aforementioned Rock Wren, it so happens that in December 1992 a Rock Wren was found in Cape May Point, New Jersey, rummaging around in the scattered debris of a house under construction.

So the last piece of information imparted in the opening paragraph, designated VI—short for VAGRANCY INDEX—is a conditional modifier. This index relates to the known vagrancy tendencies of a species or the possibility that it may turn up where it doesn't belong (in terms of its normal geographic distribution). There are five ratings.

0 No pattern of vagrancy. The chances of this species being seen outside its range are scant to nil.
1 Some slight tendency to wander, but such occurrences are regional, extending not far beyond the established borders of the species' range, or there are simply very few records of vagrancy.
2 The species shows some modest pattern of vagrancy. It is possible to encounter it outside its normal range but still not likely, and you should consider other, more likely possibilities first.
3 This species has demonstrated an established, widespread pattern of vagrancy. Ignore the range descriptions. This bird could be sighted almost anywhere.
4 The species is so widespread that there are few places left in North America for it to wander.

If you don't care to remember the particulars, just remember the rating system. The lower the number, the less likely a species is to wander.

Birds Are the Sum of Their Parts (and More), or, But How Did You Know It Was a Wren and Not a Swan?

The field marks used to differentiate birds relate most often to structure and plumage. Used in concert to make an identification, both are important. But a bird's structural characteristics are in many ways more fundamental and more determining. More than plumage, structural attributes (such as bill shape, neck length, body shape, leg length, or foot shape) link birds to closely related species; also, because these attributes vary less between the ages and sexes within a species, they are commonly not as variable or transitional as plumage. Accordingly, the *description* for every species looks first at structure and concludes with plumage, focusing first upon the most fundamental traits.

SIZE AND OVERALL SHAPE: Birders argue as to whether size or shape is a bird's most determining characteristic (the one experienced birders note first when making an identification). The fact is that most birders see and assess these qualities simultaneously, thus quickly simplifying the identification process.

STRUCTURAL CHARACTERISTICS: Bill size, shape, and length, head size and shape, the contours of a bird's neck, the shape of its body, the length of its legs, the shape of its feet—all constitute important, determining structural characteristics. These morphological traits divide birds into groups, such as sandpipers, hawks, gulls, warblers, or finches. Placing a bird in the right group is the next major step in the identification process. The rest comes down to details—those related to gradations in structure (like the small differences in the bill structure between Western, Semipalmated, and Least Sandpipers) and those relating to plumage.

PLUMAGE: Once size, shape, and structural components have been noted, identification is often a simple matter of making a determination between two (or three) similar species. This is the stage in the identification process where plumage is often most useful. Some plumage-related traits (field marks) are blatant (such as the all-red plumage of a breeding male Summer Tanager or the unique and colorful pattern of a male Harlequin Duck). Some are more subtle but no less determining (such as the lime green back of a nonbreeding Chestnut-sided Warbler). But the plumage characteristics of many birds are not so singular and are shared to some degree by closely related species (for example, the plumage of female Blue-winged, Cinnamon, and Green-winged Teals). In this case it is necessary to use a combination of plumage (and anatomical) traits to differentiate the birds.

And then there are birds that so very closely resemble related species that the similarities are commanding and the differences subtle. When dealing with species such as Dusky and Gray Flycatchers, you often have no alternative but to pay attention to fine details of plumage and structure. But there are additional clues that may build a case for one species and parry away the possibility of another. Many of these relate to behavior.

Before focusing on behavior, let's return to the beginning of this section, where I skipped a subjective but important element in the identification process. Birds are more than a bunch of isolated and idiosyncratic field marks. They are living entities. They are the sum of their parts, and they project qualities that are in concert with what they are. Birds often look different or behave differently because they *are* different.

It's easy to tell a wren from a swan. But telling Tundra Swan from Trumpeter Swan is more troublesome. You can look at the bill and try to catch a yellow spot on the base (a characteristic often seen on Tundra Swan) or the narrow, orange line that defines the gape of Trumpeter. Or you can look at the whole bird and determine whether it looks lithe (with its head erect), which is typical of Tundra Swan, or tired and slouched (with its neck couched or folded back onto the neck), which is the posture commonly adopted by Trumpeter Swan. Birders have a term for this projection of posture or sense of shape: GISS (General Impression of Size and Structure, pronounced *gizz* or *jizz*). *GISS* is a subjective clue, and its primary usefulness is to alert birders to the possibility that a bird is different or unexpected. It can also be very useful to the birder studying a bird at such a distance that classic field marks relating to plumage and shape are difficult to discern.

In the nickname or introductory tag line for each species, and often in a sentence introducing the species description, I have tried to capture something of the bird's essence or gestalt. It is not as determining as a field mark, but it is often suggestive and sometimes commanding.

BEHAVIOR: The inherent shortfall with most field guides is that they treat identification as a static process. They depict birds in a fixed posture, with distinguishing field marks shown to best effect. The reality is very different. Bird identification is in fact a dynamic process. Birds are animate. They move and assume different postures, often treating observers to views that are not replicated in guides and demonstrating mannerisms that are difficult or impossible to get across in a photo or illustration.

Behavior can be as determining as structure and plumage in differentiating birds, both between species groups and between individual species—even (and maybe particularly) in differentiating some that are very similar. Plovers walk, stop, and pick (like robins); sandpipers feed on the run. Semipalmated Sandpiper likes to keep its feet wet; Least Sandpiper more commonly forages on damp (even dry) ground, away from the water's edge (particularly when this smallest of sandpipers is feeding with other small sandpipers).

Band-rumped Storm-Petrels tend to be more skittish than Wilson's Storm-Petrels. When flocks sitting on the water are approached, the Band-rumped usually flushes first. Tennessee Warbler is usually a canopy species (particularly in spring); the similar Orange-crowned Warbler most commonly feeds lower (often in weedy tangles). Gray Flycatcher is a compulsive tail wagger; the tail of Dusky is given to the occasional jerk but is not habitually wagged with a downward pump.

FLIGHT: The bird behavior that our own species finds most captivating is also one that is most challenging to the birder. Flight is the characteristic that more than any other defines birds. The ability to move through the air is not unique to birds (and in fact is not even practiced by all birds), but it's the trait that garners both our attention and our envy. Thus, it is somewhat surprising that our ability to identify birds in flight and our means of describing them lag so far behind our ability to identify birds that are standing, swimming, or perching.

This is not true of all birds, of course. The flight profiles and styles of some birds (most notably hawks and seabirds) have been carefully studied for years, and field marks that work in this challenging arena have been codified. Each species account in this book describes the bird in flight. Some of these descriptions are cursory, and others are more detailed.

In attempting to describe birds in flight, I found that I needed to differentiate between terms that are often used interchangeably in everyday usage. As you read the species accounts, it will be helpful to understand how I use the following terms:

STRAIGHT FLIGHT: The bird moves forward without deviating from its course. *Example:* American Crow.

WANDERING OR TACKING: The bird angles left, then right, then left. *Example:* Northern Flicker or Say's Phoebe.

YAWING OR TWISTY-TURNY: The bird flies straight but not on an even keel—that is, it lists or favors one side, then leans to the other side. *Example:* American Woodcock.

UNDULATING: There is a regular and even rise and fall to the bird's flight. *Example:* many woodpeckers.

BOUNCY OR BOUNDING: There is a regular, mostly even, rise and fall to the bird's flight, with deep oscillations. *Example:* American Goldfinch.

JERKY: Bird flight characterized by abrupt, often irregular, bounce. *Example:* many warbler species.

RISE AND FALL: The bird changes its elevation—flying slightly higher, then dropping lower—but its flight is generally straight, and these altitudinal shifts do not show an even, undulating pattern. *Example:* Common Grackle or Baltimore and Bullock's Orioles.

REGULAR OR STEADY WING BEAT: The bird moves its wings in a steady, unbroken rhythm. *Example:* American Crow.

IRREGULAR WING BEAT: The bird moves its wings in a halting or broken rhythm: *Example:* Belted Kingfisher.

FLAP AND GLIDE, OR A SERIES OF FLAPS FOLLOWED BY A GLIDE: The bird's flight consists of a series of wingbeats punctuated by pauses lengthy enough to note that the bird is continuing to move forward. *Example:* Sharp-shinned Hawk.

OPEN-WINGED GLIDE: The bird glides with wings open and fanned. *Example:* Sharp-shinned Hawk.

CLOSE-WINGED GLIDE: The bird glides with wings closed and pressed to the side. *Example:* grackles.

SKIP/PAUSE: The bird flies with a momentary break in the rhythm (usually with wings closed) that is too terse to be called a glide. *Example:* many warblers.

I should point out that many birds alter their manner of flight according to conditions, objectives, and distance traveled. For example, sparrows (among many other species) flying short distances are frequently bouncy, but the same birds covering greater distance may be undulating or show a less energetic rise and fall. Birds flying downhill may glide extensively but then flap almost continuously, of necessity, when flying uphill. Birds heading into a wind will flap more (and glide less) than birds flying with a light tail wind. Courting birds exhibit all manner of energetic acrobatics that they never resort to when their objective is locomotion, not procreation.

Despite these variables, the flight of most species is fairly consistent, and while not necessarily defining, it can be an important aid to identification. The fact is that many birds are seen mostly in flight (such as many pelagic species), and some are most easily distinguished in flight. Flight identification is one of birding's frontiers. Here's another.

VOCALIZATIONS: Next to flight, song is the expression that best characterizes birds (and endears them to us). Many birds are easily and best identified by their songs, and not a few by their calls. And although most birds sing for only a portion of the year (just before and while they are nesting), some sing all year, and most make some identifying vocalizations or calls even in winter.

I was tempted to start each description of vocalizations with calls, the short utterances that both sexes make all year as opposed to songs, the more elaborate and lengthier vocalizations uttered mostly by males before and during nesting. In the end I sided with convention. I describe songs first (because they are, for the most part, more recognizable than calls), calls next, and then, if I was familiar with them or information was available, the flight calls of species.

PERTINENT PARTICULARS: Many of the species accounts conclude with "pertinent particulars," which sometimes summarize key points, sometimes compare similar species, and sometimes offer a tidbit related to the finding or identifying of a bird. Using this vehicle, there are two very pertinent particulars I want to bring to your attention. Both relate to how these species descriptions will serve you.

Since the capacity to describe sound is directly related to a person's ability to perceive it, I had my hearing tested at the beginning of this project and learned (as I have long suspected) that I am deficient in the upper middle range. My hearing is average for low tones, the lower-middle range, and very high tones, but at 4,000 Hz I am nearly 50 percent deficient in my left ear and 20 percent deficient in the right. What this means is that it is harder for me to hear sounds within this range than an average person, that very probably I miss (or mishear) notes in this frequency, and that my descriptions of bird songs will reflect this deficiency. (On the other hand, if you misspent your youth hanging around loud machinery and shooting trap without ear protection, as I did, then these descriptions are made to order.)

Second point: My sensitivity to color is acute. (Yes, I also had my vision tested.) As evidence, when I look through a wide assortment of binocular makes and models, I note on many a pink (or green or yellow) color bias, the byproduct of certain lens coatings. When I see the heads of male scaup, I have no difficulty perceiving purple or green (or determining which is dominant). When I see and describe white, I note a great difference between white and bright white. Here again, my senses (in this case a heightened one) affect not only my perceptions but my descriptions. Again, adjust accordingly.

Footnote on Field Guides

Insofar as this book is designed to be a companion to one or more of the very fine field guides that are available, in humility and deference I would like to offer readers my own thoughts about several popular guides. If you care to regard my comments as an endorsement, please feel free. If you conclude that my failure to discuss other guides constitutes a lack of regard for them, I wouldn't so presume.

As stated earlier, this project was initially conceived as a companion guide for Roger Tory Peterson's *Eastern Birds* (now also *Western Birds*, which is being revised). Both of Roger's guides have a proven track record. Both are celebrated for their compelling simplicity and the empathetic accord between the text and illustrations (because the author and the artist were one). Roger's approach to identification was anchored in the formative age of birding when everything relating to field identification was new. In the same way embryos pass through developmental stages that replicate the evolutionary advance of life on Earth (more accurately, life in the sea), all beginning birders must also begin with and pass through basic developmental stages before moving on to greater proficiency. Written on the most basic level, the Peterson Field Guides are easy for beginners to use.

The venerable "Golden Guide," or *Birds of North America,* by Chandler S. Robbins, Bertel Bruun, and Herbert S. Zim, remains a utilitarian masterpiece, best recognized for the simple brevity of its text and the illustrations by Arthur Singer. Although this guide seems to have fallen out of favor with birders in the inner circle, the partiality shown it by bird watchers who do not consider the label "birder" fundamental to their identity or other birders the cornerstone of their social network is impressive and can only be attributed to the book's continued merit.

The *National Geographic Field Guide to the Birds of North America,* now in its fourth edition, is a birding hallmark. It is celebrated for its thoroughness—in its scope, attention to detail, and focus on regional forms (subspecies). Virtually every species that has been recorded in North America is depicted (and thus many birders in Alaska, birders along the Mexican border, and pelagic birders on both coasts consider this guide their book of first resort). In addition, its studied focus on regional forms enhances the book's value especially in the West, where multiple subspecies are more often encountered. For the sake of accuracy, the text does use more technical and less generic terms than some guides (including this one), and presentation can rarely be seamless when multiple authors and artists are employed. Despite these minor concerns, this is a great field guide and one that I have found immensely useful for more than twenty years, both as a first and last resort.

The *Kaufman Field Guide to Birds of North America* is wonderful—intelligently conceived and carefully executed. What Kenn has done, through brilliance and conscious design, is to craft a Peterson Field Guide for this age. The text is not just simple and precise, but evocative. It doesn't just build an identification, but crafts an image of a living bird that can be carried into the field. It uses well-chosen photos organized in a way that facilitates comparison (an inherent shortfall of many other photo guides). I used this guide extensively during my travels (for pleasure as much as edification) and usually had it open to the appropriate page when writing my own descriptions in this book. Few birders (and certainly no beginning birders) should be without this guide. If you can own only two guides (because it's impossible to own just one), make this one of them.

In every generation there is someone who dominates his field. When it comes to knowledge concerning the identification of North American birds, few can stand on the same platform with David Sibley. But when it comes down to recasting the world of birds in a field guide, none have done it so brilliantly and completely as David has done in his *Sibley Guide to Birds.* This guide is a symbiotic fusion of illustrations that are unsurpassed in their accuracy and supportive text that is precise, groundbreaking, and spare—a layout that is comparative genius. The book may fall short in only one respect. Artists think visually. When they depict something, they assume that people will perceive it. Sometimes they do not. At times during my studies I discovered some characteristics relating to structure or plumage or posture that I thought served to distinguish a species, and when I turned to David's guide I found it depicted (David's eyes miss little and his eye is in direct communication with his brush) but unsanctified by supporting text.

Still and all, no book published in this century (and only one in the last) has been so catalytic a tool for bird identification. If you spend part of your life studying birds, you cannot be without the Sibley Guide.

And since you have read, to this point, a great deal of text dedicated to these ambitions, I must conclude that you are, like me, a serious birder who strives to be a more accomplished birder.

Species Accounts

WATERFOWL—GEESE, SWANS, AND DUCKS

Black-bellied Whistling-Duck, *Dendrocygna autumnalis*

Harlot-faced Squealer

STATUS AND DISTRIBUTION: Subtropical, mostly coastal species. In the U.S., mostly common to uncommon breeder in s. and e. Tex. and coastal La.; uncommon breeder in se. Ariz. north and west as far as Phoenix; recently established in cen. and s. Fla.; increasingly common wanderer outside of breeding areas. **HABITAT:** Open, shallow freshwater (but also tidal and brackish-water) ponds, wetlands, and streams. Forages in dry and submerged grain fields near water as well as on lawns; also found in richly vegetated marshes and lakes with shallow, vegetated shores. Roosts, mostly by day, on banks, sandbars, and tree branches. **COHABITANTS:** Fulvous Whistling-Duck, Mottled Duck, Common Moorhen. **MOVEMENT/MIGRATION:** Wholly migratory in the United States except for Florida birds and a few birds in southern and coastal Texas. Spring migration occurs in March and April; fall from August to October. VI: 3.

DESCRIPTION: Gooselike duck with an overpainted harlotlike face. Medium-sized (larger than Northern Shoveler; slightly larger than Fulvous Whistling-Duck; slightly smaller than Mallard), with a very long neck, long sturdy legs, a distinctly erect stance, and sinuous contours. On land, the body angles up, whereas the bodies of puddle ducks are horizontal. In the water, the profile is low.

The overall dark body (rufous brown back, neck, and breast; black belly), with a pale buff-colored slash along the side, provides a tasteful and conservative backdrop for the bird's most arresting feature—its overmade face showing too much gray pancake makeup, an oversized lipstick-colored bill, and an exaggerated pale ring around the eye. The sparse dark toupee on the crown is easily overlooked. The contrastingly pale head and pale slash running between the rufous back and black belly help identify adults (and grayish-billed, mostly all-brown young birds) at a distance.

BEHAVIOR: A social and noisy duck. Black-bellieds are most commonly seen in flight as they move between roosting and feeding areas at dawn and dusk, announcing their arrival with their distinctive squealing call. Commonly seen in flocks, but singles, pairs, and foursomes are frequently encountered (especially in flight). Among the least aquatic of waterfowl, this species often forages on dry fields (grain fields like rice and corn), and when in water, is normally found in depths shallow enough to allow it to stand. Feeds mostly at night. Roosts by day in pairs or flocks on some dry substrate next to the water (including in trees).

Very nimble on land (can even perch on cornstalks and reach otherwise unreachable morsels), and walks with a strut. Takes off with a lofting buoyancy; lands somewhat hesitantly with neck down and feet extended. Commonly associates with Fulvous Whistling-Duck.

FLIGHT: Distinctive wide-winged wilting profile. Wings are short, blunt, broad, and down-cupped; head and extending legs droop. Recalls a broad-winged ibis. At a distance, appears long-necked, short-tailed (legs disappear), and all-dark, except for the distinctive broad white stripe on the upperwing. Steady wingbeats are somewhat loose, somewhere between floppy and choppy, and shallow (they seem all down stroke); overall flight is buoyant, floating, and somewhat halfhearted or amateurish (again, recalls ibis). Flocks fly in what might generously be called a loose formation. Birds stay, for the most part, on the same horizontal plain but seem incapable of holding a V for long (and even flying single file looks like a struggle). Garrulous—if they are flying, they are almost certainly vocalizing.

VOCALIZATIONS: Very vocal, but not necessarily loud. Call is a squealing whistle followed by two (or more) high, sharp, stuttered peepings: "*OooeeE pee pee; OooeeE pee pee pee.*" The pattern and even the quality recall some of the vocalizations of Great-tailed Grackle (which will almost certainly be vocalizing nearby, particularly on the Texas coast). Also emits a more mellow whistled "*roo pear*" or "*roo pear pear*," a high chirping "*chit chit ch't ch't ch't*," and an excited, high, whistled trill, "*p'p'p'p'p'p'p*." When flushed, birds "*yip*" once or in a series. Many of the bird's vocalizations sound like shorebirds, including the "*keek*" note of Long-billed Dowitcher and the strident whistle of American Oystercatcher.

Fulvous Whistling-Duck, *Dendrocygna bicolor*

Rice Duck

STATUS AND DISTRIBUTION: Common Gulf Coast breeder from the Atchafalaya R. in La. to s. Tex. and coastal Mexico. Also found in se. and e.-cen. Fla. (absent along the Gulf Coast) and a regular visitor to the Salton Sea in Calif. Fla. birds are resident. Gulf Coast birds north of Corpus Christi and Salton Seabirds retreat south for the winter. **HABITAT:** Shallow freshwater wetlands, but in the United States virtually bonded to flooded rice fields. Also found in marshes and managed impoundments lush with floating vegetation—but it really, *really*, likes rice fields. **COHABITANTS:** Black-bellied Whistling-Duck, Mottled Duck, King Rail, Black-necked Stilt. **MOVEMENT/MIGRATION:** Spring migration from late February to late April; fall from late August to late October. Wanders in summer and fall, so may turn up well north of its normal range. VI: 3.

DESCRIPTION: A duck of a different color—the only native tawny orange duck in the United States. Medium-sized (larger than any teal; smaller than Mottled Duck and Black-bellied Whistling-Duck), with a large dark bill, a large and slightly peaked head, a sinuously long neck, a slightly plump body, long dark legs, and a short drooped tail. In the water, body profile is low and somewhat average-looking (but always seems large-headed, if not long-necked); when standing, posture is not as erect as Black-bellied (but more erect than a classic puddle duck).

Head, neck, and underparts are all tawny orange; back is blackish, with conspicuous orange tiger-stripe barring. The sides, back, and belly are knit with coarse white stitches. Sleeps with head tucked, making the neck look as though there is a dark ring around the base (actually it's the coiled dark hindneck). Plumage is essentially the same for all ages and sexes.

BEHAVIOR: This is an aquatic species: it doesn't graze on dry land or land in trees (both traits of Black-bellied Whistling-Duck). Feeds almost exclusively in shallow wet marshy or agricultural habitats, *particularly rice fields.* Wades and swims; forages by dunking its head, upending, and (in deeper water) diving. When shifting short distances (less than 5 ft.), thinks nothing of lofting into the air and landing with neck and legs pointed down. Feeds early and late in the day, but where persecuted for its addiction to rice, also feeds at night. Likes floating vegetation in shallow water, and commonly roosts (by standing) in a saladlike setting; also roosts on shorelines.

Very social (except during the breeding season). Flocks of several dozen birds are common; flocks in excess of several thousand are not unheard of where rice is abundant. This tame bird often allows close approach.

FLIGHT: Smaller, shorter-necked, and slimmer-winged than Black-bellied Whistling-Duck, but appears somewhat longer-tailed. (Actually the tail is short: the gray feet look like the tail at a distance, where the pink feet of Black-bellied Whistling Duck don't fool anyone.) Wholly black wings (no white patch) contrast with the tawny body. The white U at the base of the tail is visible at great distances. Appears overall gangly in flight, with paddle-shaped wings down-drooped. Flight is slow, with steady floppy wingbeats that are all down stroke; appears somewhat less ibis-like than does Black-bellied. Flies in a ragged V-formation, often very low over the water. Very vocal in flight.

VOCALIZATIONS: Common call is a high, thin, breathy, squeaky, descending two-note whistle: "*p'he/aeeer.*" Sounds very much like a squeak toy. Also makes a soft, high, accelerating laugh: "*he/ah he/ah he/ah he/ah he/ahea/hea/he/ah,*" and a low "*cup cup cup cup.*"

PERTINENT PARTICULARS: Fulvous Whistling-Duck does not remotely resemble any other native duck. It can only be confused with Ruddy Shelduck, an orange-bodied, pale-headed duck native to the Old World that is popular among exotic waterfowl collectors and does at times escape.

Greater White-fronted Goose, *Anser albifrons*

The Laughing Goose

STATUS: Common northern breeder and western and Gulf Coast winter resident. Rare in the East. **DISTRIBUTION:** Breeds in w. and n. Alaska, east across n. Yukon and Northwest Territories to the western shore of Hudson Bay. Winters primarily in the Central Valley of Calif. and e. Wash. in the West; also in s. and e. Tex., La., and Ark. and on the w. Miss. border; also occurs across much of Mexico. **HABITAT:** Nests on dry, well-vegetated tundra adjacent to wetlands, including rivers, streams, lakes, and ponds. Feeds in wet meadows and lagoons and on beaches. In winter, forages in crop fields and roosts in large bodies of open water or along open coastlines. **COHABITANTS:** In winter, found with Snow Goose, Ross's Goose, Cackling Goose, Canada Goose. **MOVEMENT/MIGRATION:** Wholly migratory. Spring migration from late February to early May; fall from late August to early January, with peak in October and November. VI: 3.

DESCRIPTION: A short-necked goose in a cryptic plain brown wrapper. If not for its pink or orange-pink bill and white-rimmed face, the bird would be invisible in open fields. Medium-sized (same size as Snow Goose), with short neck, compact body, and sturdy orange legs and feet.

Mostly plain dark grayish brown overall; distinguished by a pink bill with a white-rimmed base (looks like a pink and vanilla ice cream cone) and a scattering of black markings on the belly. The white undertail is often hard to see when birds are resting or feeding in corn stubble. Immatures are even plainer, showing a dull orange or pinkish bill but lacking the black markings on the belly and the white on the face. For identification, rely on the overall brown blandness. By comparison, immature dark-morph Snow Goose is cold dark gray.

BEHAVIOR: Highly social (like most geese); commonly found in small to large flocks, often in association with Snow Geese. Adept at feeding on land or in water. Forages by pecking, grazing, and upending (in water). Also digs for roots with its bill.

Quite wary; in mixed flocks, Greater White-fronted is often the first to flush and often shows great reluctance to land, circling several times. Also an early riser; among the first species to leave roost to forage. Vocal (particularly in flight).

FLIGHT: Has a very short neck and compact body, but also long, slender, tapered wings. Flight is more graceful and agile, less plodding, than Snow Goose. Overall plain pale brown, but shows two-toned upperwing (pale flash along the middle of the wing; darker trailing edge). Also shows a slight contrast between pale body and darker underwings. In flight, the white tip to the tail is not always easy to see, but is usually very apparent when the bird fans its tail when landing. The orange legs set against the white are often easy to see.

VOCALIZATIONS: Call is a hurried, high-pitched, squeaky two- or three-note laugh, "*kloyo leg leg*," that bears little resemblance to the honk of Snow or Canada Goose. The sound of a distant flock has a petulant, nagging quality. Feeding flocks emit a low harmonic buzz.

PERTINENT PARTICULARS: Along the Atlantic Coast, single or small groups of Greater White-fronted Geese from Greenland sometimes occur among flocks of migrating Canada Geese. Greenland birds have bills that are orange (adult) and yellow (immature), not pink, and the adults are commonly more heavily marked with black below and appear slightly darker-necked. Also, as a word of caution, Greater White-fronted Goose bears a more than passing likeness to a barnyard goose, the domesticated descendant of the Old World Graylag Goose—a gray-brown orange-billed goose.

Emperor Goose, *Chen canagica*

Goose of Tides

STATUS: Locally common but geographically restricted Alaskan resident. **DISTRIBUTION:** Breeds coastally in Siberia, n. Seward Peninsula, e. Nunivak I., and especially in the Yukon-Kuskokwim R. delta. Winters almost exclusively along the Aleutian Is., though some birds remain on the Alaska Peninsula east to Kodiak I., with stray birds reaching the Pacific Northwest south to Calif. **HABITAT:** This saltwater goose breeds and winters within reach of the tides. Breeds in coastal marshes usually no more than 10 mi. inland. Stages in tidal lagoons behind protective barrier islands, foraging on exposed mud or sand flats. In winter, forages in sheltered lagoons with extensive mud flats as well as sandy and rocky beach rich in marine vegetation exposed at low tide (for example, eelgrass and sea lettuce); roosts just above the reach of the tide. **COHABITANTS:** "Black" Brant, Sandhill Crane, Parasitic Jaeger; in winter, Glaucous and Glaucous-winged Gulls. **MOVEMENT/MIGRATION:** Spring migration from late March to mid-June; arrives on breeding grounds mid- to late May. Movement to molting/staging areas begins in midsummer; true migration begins in mid-August. By late September, birds are most concentrated on the northern end of the Alaska Peninsula. By late November, most have moved west into the Aleutians. VI: 2.

DESCRIPTION: Small, portly, white-headed, slate-colored marine goose. Small (smaller than Snow Goose; larger than Ross's; about the same size as Brant), with a petite and pale bill, short neck, plump body, and bright orange legs (yellow in immatures).

Adult plumage is simple, tasteful, and conservative, showing dark, shiny, blue-gray body and white or peach-colored head and white hindneck. Immatures (until early winter) are wholly dark—much like dark immature Snow Goose—but body feathers show pale tips, making the bird appear scaly or frosted. Immature Snow Goose is dull dark, bigger-billed, and dark-legged (immature Emperor has yellow legs).

In all plumages, Emperor's tail is white. A wink of it extends beyond the bird's shortish wings; Snow Goose's folded wings usually conceal the tail.

BEHAVIOR: Wholly coastal—Emperor Goose's life is defined by the reach and fall of the tide. In winter, feeds during the low tide cycle on beaches, rocky shores, and mud flats by reaching below the water's surface or walking in shallow water and picking. In spring, grazes on and roots for salt-tolerant plants in thawed estuaries. Usually found in pairs or small groups, but in fall, flocks may be large (numbering several thousand). Owing to its marine diet, this species does not commonly mix with nonmarine geese except for the odd, individual bird that turns up in the Pacific Northwest, where it commonly joins flocks of Snow, Ross's, and Cackling Geese. Flies little, preferring to walk or swim to and from feeding areas.

FLIGHT: Small, rotund, short-necked, and broad-winged profile. Overall plain, gray, and, at a distance, paler than expected. The white (or peach-colored) head/hindneck and U-shaped white tail are conspicuous. Wingbeats are quick, shallow, steady. Often flies just above the water.

VOCALIZATIONS: Call is a high, hoarse, petulant, and repeated "*klaha klaha.*" Also makes a quick, high, two- to four-note tooting yodel, "*reh ha ha,*" that sounds like a soprano Common Loon. Also emits a low nasal groan, "*eryah.*"

Snow Goose, *Chen caerulescens*

WA'WA

STATUS: Abundant, widespread, but somewhat geographically restricted both in summer and winter. **DISTRIBUTION:** Breeds as three separate populations—western, midcontinent, and eastern—in coastal arctic (and subarctic) regions from nw. Greenland to Wrangel I. (Russia). Winters in pockets from s. B.C. to cen. Calif., with the greatest concentrations found in the Sacramento Valley of Calif., in the Rio Grande Valley in N.M., in Tex., La., and Ark., up along the Mississippi R. drainage to Iowa and Neb., and coastally from N.J. to Ga. Other significant wintering areas include wetlands (and croplands) portions of Utah, s. Calif., the Playa Lakes region of Tex. and N.M., and Colo. Widespread migrant. In North America, rare only in the extreme Southeast. **HABITAT:** Breeds on arctic tundra. In winter and in migration, occurs most frequently on fresh and coastal marshes, slow-moving rivers, lakes, rice fields, and surrounding agricultural croplands. During spring and fall migration, Snow Geese put in to rest and feed at major staging areas along the four major waterfowl flyways (most commonly at refuges specifically managed for waterfowl). **COHABITANTS:** In winter and during migration, often found in mixed flocks with other geese. **MOVEMENT/MIGRATION:** Spring migration from early March to mid-May; fall from September to early December. In winter, flocks shift in response to food and freeze conditions. VI: 4.

DESCRIPTION: Medium-sized, fairly stocky, short-necked goose most often found in large flocks. Most adults are essentially pure white with black wingtips. The head and heavy pink bill give the face a longish, wedge-shaped look. The face is often stained yellow, orangy, or rust as a result of the bird's foraging in iron-rich earth, or it may be sullied with a muck mask if the bird has been foraging in coastal marshes. The upper and lower mandibles seem open or parted, as if the fit were improper or the bird were grimacing.

Other geese look cute or handsome. At close range, the face of Snow Goose looks ugly.

In all populations, but most commonly among birds wintering along the Gulf Coast and lower Mississippi Valley, a certain percentage of dark individuals occur—so-called Blue Geese. Blue Geese have slate-dark bodies and white heads. In flight, the upper and lower wing coverts are white or pale, and the trailing flight feathers are black.

Immature Snow Geese are dusky (particularly dusky-backed) versions of adults and have dark bills. Immature Blue Geese are overall dark gray (much, much darker than immature white morph birds) and lack the distinctive and contrasting white heads of adult dark-morph birds.

The Snow Goose expression is somewhere between a grimace and a grin—and not far from a leer.

BEHAVIOR: In winter and in migration, gathers in large flocks and forages on croplands or tidal drained marshes. On croplands, picks and plucks grain or shoots from the surface. The preferred technique is "grubbing" for rootlets in mud or shallow water, using the bill as a trowel. Snow Goose may forage with other geese, but the flocks tend to segregate out. Regularly associates only with the smaller, very similar, and in most places less common Ross's Goose; where their ranges overlap, also associates with Greater White-fronted Goose. As geese go, Snow Goose is not particularly wary (except where hunted).

FLIGHT: Overall long-winged and short-necked. White body and black wingtips are distinctive on white birds. (The white underwing coverts and dark trailing edge to the wing, as well as the white head on adults, distinguish Blue Goose.) Flies in classic V-formations, but in large flocks (numbering several hundreds) the Vs may overlap and morph into Ms or Ws . . . or both. Small flocks (fewer than 10) commonly organize themselves in a single offset line (half a V).

Wingbeats are quick and steady, but choppier than Canada Goose. Distant birds seem like shimmering white threads caught in the wind. If you hear high-flying geese and can't find them, chances are they are Snow Geese.

VOCALIZATIONS: Classic call is a rough-sounding single-note bugle "*Wha!*" sometimes given with a compressed two-note quality: "*W'rah!*" Call is higher-pitched, less distinctly two-noted, and not as sonorous as Canada Goose. The pitch and pattern of calls vary greatly between individuals in the flock—some are higher-pitched; some ascend while others are guttural and descending. The resulting cacophony is more varied, less musical, and more amateurish-sounding than the more harmonious honking of a flock of Canada Geese.

Foraging birds are noisy, making a nasal harmonic "*huh, huh, huh, huh-huh*" (this call is also used in flight).

PERTINENT PARTICULARS: Residents of New Jersey and Pennsylvania (as well as birders who travel there) are familiar with a great convenience store chain known as WAWA—the name that Native Americans ascribed to the geese that frequent the region. In keeping with the name, the company logo brandishes a goose—a Canada Goose. A great bird, but in this case possibly not the right bird. Canada Geese say "Ah-ronk," not "WAH"—a distinction a Lenape hunter might have made.

Ross's Goose, *Chen rossii*

Toy Snow Goose

STATUS: Local in the West, in winter, and in the Great Plains, the lower Mississippi Valley, and the Gulf Coast; common to fairly common but less numerous than Snow Goose; along the Atlantic Coast, uncommon to rare. **DISTRIBUTION:** Arctic and subarctic breeder characterized by isolated and somewhat limited nesting areas. Nests primarily in the cen. Canadian Arctic, especially at the Queen Maud Gulf Migratory Bird Sanctuary in the Northwest Territories. Also recorded on the North Slope of Alaska, on Baffin I., and along the west coast of Hudson Bay. Winters in the Sacramento and San Joaquin Valleys of Calif. and the Salton Sea; also in the Rio Grande Valley of N.M., the southern plains, the Gulf Coast, and the lower Mississippi Valley. Small numbers also winter among Snow Goose flocks along the Atlantic Coast from Del. to the Carolinas. In migration, common east to the Mississippi R. **HABITAT:** Breeds on low, flat, arctic tundra, often on islands in shallow lakes or along lakeshores. Vegetative ground cover seems optional. In winter, birds concentrate on agricultural lands, particularly grain fields and rice fields, and shallow wetlands (fresh or tidal). **COHABITANTS:** Often found in association with other geese, particularly Snow Goose and Cackling Goose. **MOVEMENT/MIGRATION:** Spring migration from late August to mid-December; fall from mid-February to mid-June. VI: 3.

DESCRIPTION: In most respects, a cute miniature or "toy" Snow Goose. In measured weight and overall bulk, about half the size of its larger cousin, but in comparison on the ground or in flight, seems about two-thirds the size. Overall less gangly, more appealingly compact than Snow Goose, with a distinctly rounder (sometimes slightly peaked) head, shorter neck, shorter body, shorter legs, and (most importantly) a cute wedge-shaped nub of a bill that shows little or no semblance of the grimace or "grin patch" conspicuous on the much longer (and uglier) Snow Goose bill. The slant on the bill is acute, almost 45°. On Snow Goose, it is doorstop-shallow. The bill on Ross's Goose meets the face at a near-right angle. On Snow Goose, the domed head and slanting forehead slope into the bill.

Hint: If you have to second-guess the bill structure, it is not a Ross's Goose. *Hint:* The oft-mentioned "warty protuberance" on the base of Ross's bill is not always present or noticeable, and it is unnecessary for the sake of identification. The color of the base of the bill, a blue-gray-purple, is more easily noted. Usually, however, bill size and shape alone suffice. Also, on Ross's Goose, the bill meets the face with a flush straight line. On Snow Goose, the joint is curved, indented toward the bill. Also, on Ross's Goose, the eye is set slightly forward, closer to the bill.

Adult plumage is like Snow Goose—this is a pure white bird with black wingtips. Rarely (if ever) do Ross's Geese show the orange- or yellowish-stained faces so often seen on Snow Geese. The dark or blue form is very rare and resembles Snow Goose, but the dark is darker and neater and more sharply edged, with pale edgings. Immatures are like adults but lightly washed with gray on the crown, neck, and upperparts; immature Snow Geese are overall much darker and easily distinguished from adult birds. Ross's expression is childlike and serene.

BEHAVIOR: Highly social. Where large numbers occur, Ross's Geese tend to segregate themselves from Snow Geese. Where Snow Geese are numerically superior, Ross's Geese integrate into the flocks but tend to cluster in pure Ross's subflocks within the group. Birds forage by picking or plucking and generally do not root in soil like Snow Geese.

FLIGHT: Flight silhouette is like a small short-necked Snow Goose. Flight is fast and nimble (for a goose). Wingbeats are quicker than Snow Goose, although differences are generally not apparent except in direct comparison.

VOCALIZATIONS: Not as vocal or garrulous as Snow Goose. Ross's call, "*keek*," is like Snow Goose, but higher-pitched, faster, and more abrupt—a cross between a tinhorn toot and a terrier yap.

PERTINENT PARTICULARS: It has been suggested that Ross's Goose lands more timidly than Snow Goose, braking with its wings right up to the point when its feet touch the ground. In mixed flocks, study the last birds flapping when flocks touch down. Also, Snow Geese and Ross's hybridize, producing young with intermediate traits.

Canada Goose, *Branta canadensis*

Common Canada Goose

STATUS: Common widespread northern breeder and southern winter resident; increasingly a ubiquitous and widespread resident. **DISTRIBUTION:** Breeds through virtually all of Canada, from the western and central interior of Alaska to Lab. and Nfld. A widespread, primarily northern breeder in the U.S., with the southern limit of its breeding range cutting across the middle of the country. Except for coastal areas, winters mostly south of Canada and se. Alaska, but occurs in se. B.C., sw. Alta., se. Ont., and se. Que. In the U.S., winters everywhere but arid (and forested) regions of e. and se. Calif., s. Nev., s. Ariz., sw. Tex., and parts of s. La., s. S.C., e. Ga., and cen. and s. Fla. **HABITAT:** Historically breeds on tundra marshes and ponds in the arctic; along rivers and ponds in forested regions; prairie pothole region to Nebraska and west to northern California. Now nests in a variety of human altered habitats, including managed wetlands, urban and suburban park ponds, and tidal marshes. In winter, found in a variety of coastal and interior wetland habitats, including agricultural lands (particularly corn and grain fields), corporate lawns, and golf courses. **COHABITANTS:** In winter and migration, commonly associates with other geese and waterfowl; also Killdeer and golfers. **MOVEMENT/MIGRATION:** Migration is protracted; timing and routes vary greatly between subspecies. In general, spring migration occurs from March to April; fall from September to November. VI: 2.

DESCRIPTION: Hardly necessary across most of North America. Large size, long neck, and general waterfowl characteristics define it as a goose. The contrast between the black head and neck and the uniform brown back and pale breast (on most subspecies) is unique. The white chin strap curving up on the cheeks is also a signature characteristic.

This species can only be confused with the similarly patterned, but much smaller, more compact, and petitely featured Cackling Goose. One subspecies of Canada Goose, the Lesser Canada, which breeds from central Alaska across northern Canada to Hudson Bay, migrates through the prairies to wintering areas in Colorado, New Mexico, Texas, Kansas, Oklahoma, and Louisiana and is overall smaller and smaller-billed than other Canadas, so it appears more like Cackling Goose. A considerably smaller Canada Goose among larger birds is therefore not automatically a Cackling Goose.

The Brant, a coastal marine goose, is smaller and more compact, with a black neck *and* black breast. (The limited amount of white on the Brant's throat can hardly be confused with Canada's wide chin strap.)

BEHAVIOR: Highly social. After the breeding season, gathers in large flocks that are semi-segregated by family groups. Flies out to forage during the day, often on harvested (or unharvested) grain fields, and returns to the protection of marshes, lakes, and ponds in the evening. Often forages with other waterfowl species. Flocks tend to segregate into satellite subgroups that apportion themselves fairly widely over an area. Very vocal—in fact, garrulous.

FLIGHT: Flight profile is long and slender, with a long extended neck. Wingbeats are strong, rhythmic, fluid, and constant, with the down stroke thrown forward (as if the bird were backpedaling or pushing away from some threat). Flies in well-regimented V-formations. Rises quickly and directly from land and water, sometimes with a preliminary step or two.

VOCALIZATIONS: The resonant "honk" of Canada Goose—the loud, uprising, distinctly two-noted "*ah-Ahnk*"—is so well known that it seems innate. Calling flocks sound uniform, with all birds more or less alike in volume, cadence, and pitch. The chorus of other geese, a medley of high, low, squeaky, and musical notes, is less homogenized.

Cackling Goose, *Branta hutchinsii*

Toy Canada Goose

STATUS AND DISTRIBUTION: Common northern arctic breeder; common but somewhat restricted winter resident. Breeds from w. coastal Alaska and the Aleutian Is., across the arctic coastal plain of n. Alaska, east to Baffin I. and the west coast of Hudson Bay. Winters primarily in the Willamette River Valley and the lower Columbia River Valley south into the Central Valley of Calif. Also occurs in sw. N.M., trans-Pecos Tex. (and adjacent Mexico), and along the Gulf of Mexico from w. La. to ne. Mexico. **HABITAT:** Nests amid tundra vegetation on islands and shores of ponds and along the edge of tidal wetlands. Winters on grassy meadows, rice fields, and agricultural land. **COHABITANTS:** In winter, Greater White-fronted Goose, Snow Goose, Ross's Goose. **MOVEMENT/MIGRATION:** Spring migration from early April to late May; fall from mid-September to late November. VI: 3.

DESCRIPTION: Cackling is a Ross's Goose dressed like a Canada Goose. Tiny and compact (smaller than the smallest Canada Goose; slightly smaller than Lesser Snow Goose (the smaller subspecies); slightly larger than Ross's, whose proportions it generally shares), this Canada-type goose has a rounded (or squared) head, a petite triangular nub of a bill, and a short thickish neck. The plumage pattern is similar to Canada Goose. Western birds are dark-breasted (like subspecies Dusky Canada Goose) and often show a white neck ring (which Dusky Canada lacks). Eastern birds are pale-breasted (like Canada Goose). Looks cute—a Canada Goose miniature.

BEHAVIOR: Like other Canada-type geese, a grazer. But whereas Canada Geese typically cluster in spaced family groups, Cackling grazes in tightly packed, massed flocks, as shorebirds do. Commonly grazes by itself, working an area thoroughly and turning grassy plains into close-clipped lawns. Outside of its normal winter range, mixes freely with Canada Geese, but often favors White-fronted, Snow, and Ross's Geese. Fairly tame (where hunting pressure is light).

FLIGHT: Profile appears chubby, tiny-billed, and short-necked. Flight is more agile and nimble than the larger Canada, and wingbeats are quicker.

VOCALIZATIONS: Call is a rapid, squeaky, high-pitched cackle. It doesn't honk.

PERTINENT PARTICULARS: The challenge presented by this species boils down to distinguishing the Richardson's Cackling Goose, which breeds in western coastal arctic Canada and winters primarily in western Texas, southeastern New Mexico, and coastal Texas, from Lesser Canada Goose, a taiga-breeding subspecies, found throughout much of interior Alaska and across northern Canada, whose breeding range abuts that of Richardson's and whose winter range it shares. The plumages of the two birds are almost identical. Size is a useful criterion, particularly if a Snow Goose offers a point of reference. Canada Goose is always larger than the Lesser Snow Goose wintering on the Gulf Coast; Cackling is very slightly smaller than Snow Goose or Greater White-fronted Goose. The petite, "toy," or Ross's Goose–like proportions of Cackling Goose (tiny nub of a bill; short thick neck), in conjunction with size, constitute the most reliable distinction. Also, Lesser Canada Goose honks like a higher-pitched Canada, while Cackling Goose gives a high-pitched, squeaky trill or chickenlike cackle.

Brant, *Branta bernicla*

The Marine Goose

STATUS: Common to very common, but restricted for the most part to arctic and coastal regions. **DISTRIBUTION:** Circumpolar in its distribution. Populations that winter in North America are found breeding from Baffin I. to Siberia. (A separate population breeds throughout the Queen Elizabeth Is. and n. Greenland and winters in Ireland.) Never found far from the coast (in the breeding season), Brant nests (often in colonies) in coastal estuaries and adjacent marshland as well as on gravel islands in deltas and lakes. Winters in marine waters along both coasts: the Atlantic subspecies from Mass. to N.C. on the Atlantic; the Pacific form ("Black" Brant) from Vancouver I. to Baja and along the Mexican side of the Sea of Cortez. **HABITAT:** In winter, found primarily in sheltered back bays, where it forages on exposed mud flats and in sheltered, shallow waters where eelgrass and sea lettuce flourish. Also sometimes forages on lawns and golf courses. **COHABITANTS:** In winter and in migration, does not commonly associate with other waterfowl except for a few rare interior migrants that may associate with other geese. **MOVEMENT/MIGRATION:** Migratory routes are narrowly defined. Atlantic birds move to and from wintering grounds on a route that carries them along the east coast of Hudson Bay. In fall, the migration is very constricted and wholly over land; in spring, some birds return via a coastal route to the Gulf of St. Lawrence, then fly over land. Occasionally birds turn up on large inland lakes. "Black" Brant uses the same coastal migration route in spring and fall. Spring migration from late January to early June, with peak between March and early May); fall from late August to early January, with peak in October and November. VI: 3.

DESCRIPTION: Small, dark, stocky marine goose whose likeness even to Cackling Goose is superficial. Overall compact, more angular than other geese, with a small head, a petite bill, and a short thick neck that seems at times to be reared slightly back. Overall darker and more contrastingly patterned than Cackling (and Canada), with more extensive black in the front (extends to the breast) and sides that are paler than the breast and the back (even on "Black" Brant). The white cheeks on Canada and Cackling Geese and the less distinct white necklace on Brant are very different. Atlantic immature birds are like adults but have barred upperparts. Young "Black" Brant is conspicuously darker and less contrastingly patterned than the adult but equally conspicuous in the dark uniformity of its plumage.

BEHAVIOR: Highly social. When not breeding, almost always found in large flocks in shallow marine environments. Forages, rests, and flies in compact, single-species flocks that may number in the hundreds. Feeds by tipping up (like a dabbling duck) or walking across exposed mud flats and rocks, neck extended, nibbling as it goes. Like most geese, garrulous. Unlike most geese, shifts locations frequently during the day, often in response to changing tidal conditions.

This tame species often allows very close approach and usually chooses to swim (or walk) if pressed.

FLIGHT: Flight is fast, nimble, and somewhat showoffish. Long, slender, acutely tapered wings seem elegantly incongruous on such a stocky bird. Wingbeats are quick and regular. Wings maintain a downturned configuration and an abbreviated upstroke that does not rise above the back. Rises directly from the water, without a running takeoff, and lands, wings braking, with little or no breasting glide.

Flies in compact flocks, tighter than other geese. Small groups fly in a single-wing (one side of a V) configuration. Larger groups attempt to form a U or boomerang-shaped configuration, but success is temporary. Flocks are sloppy and shifting. If you see geese that are in a nice, regimented V, be assured that they are not Brant. In winter, flies low, just above the water. In migration, commonly flies very high in flocks that seem to shift like smoke on the horizon.

Despite its structural incohesiveness, Brant is extremely nimble. Flocks fly, bank, and maneuver with a quick jaunty finesse that makes larger geese seem stodgy and plodding.

VOCALIZATIONS: The call is wonderful: a low, goose-like honk, with rolled r's, that sounds like a cross between a honk and a purr, "*r'r'r'r'ronk.*" Sometimes has a quizzical quality. Also makes a low, short, single-note, amphibian-like, croaking "*hawk.*"

PERTINENT PARTICULARS: Although this species tends not to mix with other waterfowl, individuals occasionally join flocks of scoter or Double-crested Cormorant (or other geese when inland).

Mute Swan, *Cygnus olor*

Nob-billed Swan

STATUS: Common but geographically localized, year-round resident. **DISTRIBUTION:** Found primarily from coastal s. Maine to the Delmarva Peninsula, much of Mich., and in pockets along the Great Lakes and in the Upper Midwest south to Iowa; also in and around Vancouver, B.C. Also established in w. Pa. and Ill. **HABITAT:** Breeds primarily on lakes and ponds, in freshwater or brackish marshes, and in slow-moving rivers and streams. In winter, generally remains on breeding sites unless forced by ice to relocate. Wintering concentrations occur in tidal bays and estuaries (even on open ocean during severe freezes), as well as in the open water at the mouths of rivers and streams, where moving water keeps waters open. **COHABITANTS:** Nothing wants to be close to this highly territorial bird. **MOVEMENT/MIGRATION:** Essentially nonmigratory. Individuals shift during post-breeding dispersal and to find open water in winter. VI: 1.

DESCRIPTION: Large, bulky, thick-necked, knob-billed, somewhat dowager-esque swan. More heavily proportioned, with more exaggerated curves, than the two native species. The body is long and sits high in the water, and the long neck is often held in a theatrical S-shaped curve, with the head and bill turned demurely down. Tundra and Trumpeter Swans commonly hold their necks more stiffly erect, with the head and bill meeting the neck at a right angle or only slightly angled down; the body language says alert, not demure. Mute Swan commonly cocks its long, pointed tail up, above the water. The tails of other swans are usually flush with the water. Also, in silhouette, the head of Mute Swan appears lumpy or has a prominent forehead; the heads of Tundra and Trumpeter are long, sloping, and cleanly tapered.

Adults are all white except for the distinctive and diagnostic bright orange bill, black party mask across the face, and swollen black nob at the base of the bill. Immatures are overall dingy or dusky, with grayish pink bills. The posture, the thick neck and head showing a prominent forehead, and the smaller bill help distinguish it from the more slender and clean-lined Tundra and Trumpeter Swans, with their straighter and thinner necks. Note also that immature Tundra Swan has a swollen, bubblegum pink bill. Mute's expression is disdainful and aloof.

BEHAVIOR: Tame—in fact, belligerent. Where habituated to people, Mute Swan is often reluctant to give ground on waterfront paths or park lawns. This is the swan most likely to be found on urban and suburban ponds, and the only swan likely to be encountered in such regions between May and October (but see "Pertinent Particulars"). Outside the breeding season, often found in small to large groups (in coastal bays may number several hundred individuals). Very territorial during the nesting season (and sometimes in winter). The threat posture—with wings raised theatrically over the back and the neck tucked in a tight S curve (the shape you see on merry-go-round swans) — is diagnostic. Movement on land is a slow ponderous waddle.

FLIGHT: Direct, strong, and impressive, with neck outstretched and wings that beat in a deep, slow, ponderous, even cadence. The throbbing, whistling hum of the wingbeats (a characteristic not shared by other swans) is audible at considerable distance. Birds can get aloft only by running across open surface. Flies in classic V-formation. Lands with effortful braking of the wings and often audible slapping of feet on water.

VOCALIZATIONS: Breathy, descending, trumpeting "*keeorrf*," often uttered in flight, sometimes with a rough ending "*keeorrf'f'f*." Also hisses and snorts.

PERTINENT PARTICULARS: A word of caution: Other native and non-native swans, in addition to Mute Swan, are sometimes introduced into landscaped ponds associated with cities, estates, and corporate parks. Escapes from zoos and aviaries are a perennial concern. Don't despair, but don't presume either.

Trumpeter Swan, *Cygnus buccinator*

Slouch-necked Swan

STATUS: Common in places, but enjoys a limited and spotty distribution and a relatively small population. **DISTRIBUTION:** Breeds in cen. and se. Alaska, s. Yukon, ne. B.C., and, in isolated populations, in s. B.C., Alta., Man., Ore., se. Idaho, sw. Mont., w. S.D., and nw. Neb. Also reintroduced into n. Minn., Wisc., Mich., and extreme se. Ont. Winters from coastal se. Alaska to n. Ore. and in isolated pockets in s. B.C., s. Ore., e. Nev., the junction of Idaho, Mont., and Wyo., se. S.D., and se. Ont. Birds from the reintroduced Great Lakes and Upper Midwest populations move south in winter to Kans., Iowa, Mo., and Ill. (and occasionally farther south). **HABITAT:** Breeds on clean freshwater lakes, ponds, and rivers in marshland and forests. Requires an area shallow enough for foraging, with emergent vegetation and some elevated habitat or structure (muskrat house, beaver dam, small island) for a nest site. In winter, concentrates in open freshwater streams, rivers, lakes, reservoirs, estuaries, and inlets. Forages in aquatic habitats as well as in grassy meadows and croplands. **COHABITANTS:** Nesting birds are often found on bodies of water occupied by Canada Goose and Barrow's Goldeneye; wintering birds are found with a variety of other waterfowl, but seem habitually accompanied by American Coot. **MOVEMENT/MIGRATION:** Spring migration of Alaskan and western Canadian populations occurs from late February to early April; fall from early October to mid-November. Other populations are mostly resident, although some consolidation of range occurs in winter. VI: 2.

DESCRIPTION: Big, hefty, black-billed, all-white swan — overall more robust than Tundra Swan. Huge (considerably larger than Tundra Swan; about the same size as Mute Swan), with a long, straight, wedge-shaped bill/head, a very long and variously contoured neck, and a large dome-backed body. Appears somewhat longer-faced, thicker-necked, and more dome-backed than Tundra Swan. Neck is straight and erect when alert, but otherwise the base is folded back atop the body in a lazy slouch, making Trumpeter seem more casual, bumpkinish, and less alert than Tundra.

Except for all-black bill (sometimes shows a flesh-colored tip) and black legs and feet, adults are all white. The head and upper neck of many individuals appear dirty or yellowish, stained by iron-rich water or soil (usually not the case with Tundra Swan). The flesh-colored gape is almost impossible to see at any appreciable distance. More easily noted is the absence of a yellow spot at the base of the bill (which is found on most, but not all, adult Tundra Swans). Immatures are dingy gray with pink bills that show black at the base (bills of young Tundras are just pink). Second-year birds are slightly grayer-backed than adults.

BEHAVIOR: Slow, deliberate, graceful feeder in the water. Feeds most often by upending, but may also submerge just its head and neck. Less graceful on land (in fact, somewhat plodding), but particularly in winter when foraging extensively on waste grain in agricultural fields. Feeding Trumpeter is often attended by other waterfowl (from Canada Goose to coot) that hope to benefit from the swan's dinner-table reach. Breeding Trumpeters are highly territorial (no more than one pair per one small lake), but at other times they are highly social and can be seen foraging in large flocks comprising many family groups. In migration, sometimes flies with geese and cranes.

Generally shy. Swims to avoid human intruders or, in tight confines or when pressed, flies.

FLIGHT: Swanlike profile, showing extremely long neck and rear-placed wings. Trumpeter's superior size and longer neck are difficult to assess unless there is opportunity to compare directly with Tundra Swan. Flight is direct and fast, with strong, regular, somewhat shallow wingbeats. Takeoffs require a great deal of running room.

VOCALIZATIONS: Loud, trumpeting honk or quavering bray, "*huh'huUh*." When approached (just before flight), may utter a low, soft, slightly resonant "*haw*" or "*huh*" (sometimes more than one), an utterance that has the same quality as a trumpet but lacks the energy or volume.

Tundra Swan, *Cygnus columbianus*

Slender-necked Swan

STATUS: Locally common but geographically restricted. **DISTRIBUTION:** In North America, breeds on tundra lakes and ponds primarily in arctic coastal wetlands (more particularly on river deltas) from Baffin I. to the west coast of Alaska; also found on the coasts of Hudson Bay. There are two distinct wintering populations: the eastern population winters coastally from s. N.J. to (rarely) Ga., with most birds found in N.C.; in the West, birds range from s. B.C. to the Central Valley of Calif.; several isolated interior groups are scattered west of the Rockies. Extremely rare south to the Salton Sea. **HABITAT:** Breeds on tundra lakes and ponds, commonly near seacoasts (that is, not in northern forest lakes and ponds, where Trumpeter Swan is found). In winter, forages in back bays, shallow estuaries, open lakes, freshwater rivers, and agricultural land. **COHABITANTS:** In summer, Pacific Loon, Long-tailed Duck; in winter, Canada Goose, other geese, Mallard, other puddle ducks. **MOVEMENT/MIGRATION:** Migrates across the interior of North America with staging stopovers in the Great Basin, eastern Great Lakes, the upper Mississippi Valley, eastern Pennsylvania, and southern New Jersey. Nonstop flights between wintering, breeding, and staging areas are the norm, but birds may appear anywhere along migratory routes. Spring migration from late February to May, with peak in late March to early May; fall from September to mid-December, with peak in October and November. VI: 3.

DESCRIPTION: Compared to Mute Swan and, to a lesser degree, Trumpeter Swan, a smaller, lither swan characterized by a slender neck and a slightly concave (adult) or slightly swollen (immature) bill. Adults are easily distinguished from Mute Swan by the all-black bill, and distinguishable from Trumpeter Swan at reasonable distances by the yellow teardrop at the base of the bill (sometimes lacking). With its slightly peaked head and slightly concave bill, Tundra looks happy or grinning. Adult Trumpeters, with longer, straighter, heavier bills, appear solemn or dour.

Their long, swollen, and straight bills (not upturned) make dusky young look more like Trumpeter Swan—but the bills of immature Tundra Swans are mostly pink. Bills of immature Trumpeter Swan are longer and less swollen, and they show a black base or just a pink saddle. Immature Tundra's expression is benign.

BEHAVIOR: Highly social: migrates, roosts, and forages together, sometimes in family groups, more often in large flocks. Feeds by reaching down with its neck (often upending) and pulling up submerged vegetation. Also forages by grazing on croplands. Subordinate to Mute Swan, which frequently drives Tundra Swans from its pond.

FLIGHT: Classically swanlike profile, with long neck extended and long wings set well back. Flight is strong, direct, and fast, with regular, steady wingbeats.

VOCALIZATIONS: One of the greatest sounds in nature. An eerie, haunting, winsome call that is part whoop, part sigh. Sometimes an abrupt "*whoo!*" Sometimes a sliding and sonorous "*w'hoo'oo?*" In chorus with other Tundra Swans, may sound clamorous and gooselike.

PERTINENT PARTICULARS: A last hint for separating Tundra and Trumpeter Swans. On Trumpeter, the black base of the bill seems to envelope the eye; on Tundra, the eye is tangent but distinct. If you can see the eye separate from the bill, it's Tundra.

Muscovy Duck, *Cairina moschata*

One Big, Black, Ugly Duck

STATUS AND DISTRIBUTION: Although domestic stock are widespread and common, wild Muscovies are found only along the lower Rio Grande from San Ygnacio to McAllen, Tex., where they remain scarce to rare. Range extends to South America. **HABITAT:** In the United States, occurs almost exclusively along the banks of the lower Rio Grande, but may also occur on nearby lakes and ponds. Outside the United States, occurs in brackish water. **COHABITANTS:** Double-crested Cormorant, Black-bellied Whistling-Duck, Ringed Kingfisher. **MOVEMENT/MIGRATION:** Permanent resident, but more common in summer. VI: 0.

DESCRIPTION: One big, black, ugly, and unmistakable duck. Huge (as large as some geese; almost as large as Double-crested Cormorant), with a large crested head, a fleshy, warty-based, pink-and-black-banded bill, a short neck, a robust body, a heavy long tail, and short legs set well forward. Females and immatures are less warty and show little or no crest.

Overall black (in good light shows an oily green sheen), with a wink of white peeking through the folded wing. Immatures are all black with no appreciable green sheen.

BEHAVIOR: Usually solitary or found in pairs; most commonly seen flying low over the Rio Grande early in the morning or in the evening. Perches on sturdy tree limbs. Feeds by upending, but may also graze on shorelines. Shy Muscovy flushes easily, flying out of sight. Usually silent.

FLIGHT: This big, bulky, short-necked, long-tailed, broad-winged duck is all black—really black—except for large white patches on the upper and lower surface of the wings. (Black-bellied Whistling-Duck has white only on the upperwing.) Immatures are uniformly black, with little or no white in the wings. Muscovy's neck is short and straight. (Double-crested Cormorant's neck is kinked, its head is elevated, and cormorants have no white in the wing.) Flight is fast and direct on steady, heavy, fairly rapid wingbeats.

VOCALIZATIONS: Basically silent.

PERTINENT PARTICULARS: Free-flying Muscovy Ducks whose ancestors were domesticated stock are common fixtures on duck ponds, and in some places (most notably Florida), feral populations have been established. Feral birds always show white patches on the head and body (and many are entirely white). Wild Muscovies show white only in the wings.

Wood Duck, *Aix sponsa*

Swamp Squealer

STATUS: It's a fortunate world in which a bird this beautiful can also be so common and widespread. **DISTRIBUTION:** Ranges over the e. U.S., s. Canada, and parts of the West and Northwest; absent as a breeding or wintering species only in the dry and tree-poor regions of the West. In the East, breeds from the Maritimes west to e. and cen. Sask., south through cen. N.D. and S.D., e. Neb., e. Kans., cen. Okla., and e. (and parts of n.) Tex. In the West, breeds from e. Mont. west to Wash., north to sw. Alta. and s. B.C., Idaho, Mont., parts of Colo., n. Utah, n. Nev., Ore., and Calif. In winter, retreats from all of Canada except extreme se. Ont. and sw. B.C., as well as most northern border states. Expands into s. Utah, e. Ariz., s. Colo., most of N.M., most of Neb., w. Okla., and s. Tex. **HABITAT:** At all times and seasons, Wood Duck seeks out mature hardwood swamps, slow-moving riparian watercourses, and thickly vegetated freshwater marshes (usually near woodlands or trees). A cavity-nesting species, Wood Duck is often found in woodlands, far from water. **COHABITANTS:** Generally does not socialize with other waterfowl—but you may see it foraging for acorns with squirrels and jays. **MOVEMENT/MIGRATION:** Spring migration from March to April; fall from September to November. Although virtually a freshwater obligate, during fall migration individuals and small groups sometimes join sea ducks (primarily scoter) migrating offshore, and during cold weather may retreat into brackish environments. VI: 2.

DESCRIPTION: Overall small duck that sits very high in the water. The stubby bill, crested head, and high stern (long-raised tail) make the Wood Duck silhouette unmistakable—a 15th-century galleon of a puddle duck, complete with ornate figurehead. The crest on males in breeding plumage suggests a sou'wester (rain hat) or a ponytail; on females, a slicked-back mane. In eclipse plumage, heads are bulbous and round.

Breeding adult males, with their stunning, almost improbable array of colors, render a description unnecessary. In eclipse (June to September), the multicolored bills, red eye, and white chin strap stand out against an overall subdued, slaty brown body. Females and immatures are overall muted—gray-brown, with subtle bluish and greenish highlights in the folded wings. The large white teardrop etched over the eye of adult females is unmistakable. Grayish flanks, dappled with white, suggest sunlight playing across the forest floor.

BEHAVIOR: A fairly antisocial Anser-phobic species, these birds are usually found, outside the breeding season, in small groups (6–20) and do not generally mix with other species. Part of Wood Duck's isolationism must relate to habitat. Most waterfowl prefer an open habitat, but this duck likes to be where it's dark and thick: heavily wooded swamps or sluggish streams with overhanging branches are ideal. It's active and animated on the water—bobbing its head like a barnyard chicken, picking food off the surface, or submerging its head. At dusk, post-breeding birds often gather in a sheltered spot in large communal roosts that may number several hundred individuals or more.

Wood Duck is shy, but not necessarily quick to flush. Startled, it may mill about nervously before lofting into the air. Often it is hugging a riverbank and must swim beyond overhanging vegetation before taking off. Females almost always call when flushed. Birds often perch on logs and branches on the water or high on tree limbs (along which they can walk nimbly).

FLIGHT: Long-bodied short-winged profile. In flight, shows a raised head, a distinctly long but somewhat blunt tail, and, except for a white belly patch, wholly plain dark plumage that is unique. Flight is usually direct, not particularly fast, and wingbeats are somewhat sluggish for so small a bird. Flocks fly in oval clusters with little or no shifting or maneuvering; birds flying through woodlands, however, maneuver very well through trees and branches.

VOCALIZATIONS: The female's distinctive alarm call—a wailing, high-pitched, two-part squeal, "*Ooo-EEEK!*"—is goose-bump-inducing eerie, and often repeated as the bird retreats. During courtship, females also make a winnowing call. Males are mostly silent.

PERTINENT PARTICULARS: Despite the adult male's unmistakable plumage, the birds can be amazingly cryptic when swimming among lily pads or pickerelweed or when hidden in the shadows. In flight, the all-dark, long-tailed silhouette can be confused only with American Wigeon (another longish-tailed duck), but the tail on wigeon is somewhat shorter and noticeably pointer; the head is not elevated as on Wood Duck.

Gadwall, *Anas strepera*

The Dapper Gray Dabbler

STATUS: Fairly common and conservative, in both attire and social commitments. Males make a fashion statement with tasteful gray; paired males and females remain together throughout the winter. **DISTRIBUTION:** Widespread nester with a spotty distribution across most of subarctic Canada and the U.S. Northeast and West. May nest anywhere from e. Que. and the Maritimes to the Alaska Peninsula, south to Del. and west to Iowa, cen. Kans., n. Tex., s. N.M., n. Ariz., and s. Calif. Primary nesting region is the prairie pothole region of Canada and the U.S. and throughout the n. Great Basin. Also widespread in winter, vacating most interior breeding areas and distributing itself across the U.S. East, South, Midwest, and Southwest, the Great Basin, and the West Coast (from Alaska to Baja). In general, if it's winter and you're next to open fresh water anywhere north of Central America, you are likely to find Gadwall. **HABITAT:** Nests along grass-lined wetlands, shallow ponds and watercourses, and wetlands studded with willows and shrubs, and in tidal estuaries. In winter and migration, found on reservoirs, impoundments, small ponds, and brackish estuaries and marshes. **COHABITANTS:** Associates with other dabbling ducks, primarily Mallard and wigeons. **MOVEMENT/MIGRATION:** Spring migration from February to April; fall from August to October. VI: 2.

DESCRIPTION: Medium-sized dabbling duck; in shape best likened to a compact gentrified Mallard. Adult male has a domed back and sits high on the water; female has a lower flatter-backed profile (in direct comparison, the difference is obvious). The head is relatively large and distinctively shaped—small-billed, and flat-faced peaked (or domed) forehead—unlike the long-billed sloping-head profile of Mallard. Gadwall's thin straight bill and flat forehead may recall the face profile of Hooded Merganser.

The plumage of the adult male is distinctive. The head and body are overall gray, shot through with subtle brown highlights and bracketed by a small black bill and a very obvious black butt. At extreme distances, the silver rump (actually the tertials of the folded wing) shines with reflected light. Females and immatures are cold brown, contrasting with a grayer head. Bills are black on top, orange along the sides. (Mallard has an all-orange bill with a dark saddle on top.) In all plumage, Gadwall shows a square bright white speculum that is obvious in flight and usually winks through the folded wing when the bird is sitting.

BEHAVIOR: A social duck; in winter, usually found in small (6–20) homogeneous groups (males and females are paired throughout the winter, so flocks are arranged in paired sets), but also mixes with other dabbling ducks. Prefers to stay close to the vegetated edge (where it retreats when pressed), and is not shy about occupying tighter confines, such as the smaller ponds and reed-enclosed watercourses shunned by many other species. Forages mostly on the water by picking food from the surface or swimming with head submerged. Appears to upend less than most dabblers.

Fairly shy, once flushed Gadwall rises rapidly (and usually silently), often in a twisting, acrobatic ascent. Groups often circle and do a flyover reconnaissance of their point of departure.

FLIGHT: Overall slightly more elegant than Mallard. The overall pale plainness of both males and females makes the white bellies and (especially) the bright white wing patch (absent on Mallard) stand out. Flight is fast, direct, and slightly more nimble than Mallard.

VOCALIZATIONS: Not very vocal. The female's quack is softer, more resonant than Mallard—a cross between a quack and a purr.

PERTINENT PARTICULARS: Males are not easily mistaken. Females and immatures are easily confused with the larger female Mallard and similarly sized wigeon. Mallard is overall warmer brown and lacks Gadwall's contrasting gray face. Also, Mallard has a large bill and sloping face, very unlike the flat-faced, domed-forehead, thin-billed profile of Gadwall. Both Mallard and Gadwall bills vary in the amount and distribution of orange, but the orange of Gadwall's bill is less prominent and is relegated to the bill's sides.

Like female Gadwall, American Wigeon has a gray head contrasting with a brown body, but also has an overall rounder, bulkier head, a blue (not black and orange) more petite bill, and unpatterned rufous orange sides, as compared to Gadwall's very obviously patterned sides.

Eurasian Wigeon, *Anas penelope*

Wigeon With and Without Contrast

STATUS: Uncommon West Coast winter resident; regular but rare along the Atlantic Coast; very rare inland. **DISTRIBUTION:** Winters coastally and modestly inland from B.C. to s. Calif. On the Atlantic, most common between coastal Mass. and N.C. Regular migrant and winter resident in w. Alaska. **HABITAT:** In North America, tidal estuaries, agricultural fields (flooded and unflooded), impoundments, shallow marshes and ponds, and park ponds. **COHABITANTS:** Almost invariably found among flocks of American Wigeon. **MOVEMENT/MIGRATION:** Not known to breed in North America. All birds are presumed to originate from Asia or Iceland. Arrives and departs with American Wigeon. VI: 3.

DESCRIPTION: Depending on age and sex, Eurasian Wigeon is either easily distinguished or easily overlooked among the ranks of American Wigeon. A compact medium-sized puddle duck, it shows, like the almost structurally identical American Wigeon, a large round head, a petite bill, and an acutely pointy tail.

The bright chestnut–colored head, pinkish breast, and silvery gray flanks and back of breeding males easily distinguish them from the ranks of gray-headed rusty-sided American Wigeon. The heads of eclipse and immature males are dark mahogany brown, showing little contrast with the now ruddy brown sides, *but again*, the heads are distinctly warm brown, not gray.

Female Eurasian and American Wigeons are very similar. Both have ruddy brown sides and breasts (tending toward orange or pink on American, chestnut on Eurasian). Both have heads and necks that range from warm brown (*clearly* Eurasian) to really gray (usually American but sometimes Eurasian). On gray-headed birds, note the degree of contrast between head/neck and breast. Female Eurasian Wigeon is grayer-breasted than American, resulting in little or no contrast between neck (or couched head) and breast. The ruddier-breasted female American Wigeon shows a slight (sometimes conspicuous) contrast between the gray head and ruddy breast. Also, female American Wigeon shows a paler forehead, which tends to make the shadowy eye patch (present on both species) stand out at a distance. At close range, note that American has a narrow black ring or weld encircling the base of the bill (absent on Eurasian).

BEHAVIOR: Like American Wigeon, Eurasian Wigeon is just as much at home foraging on dry land as in shallow water. Where multiple birds occur, they tend to cluster in the flock, making detection (particularly detection of females) easier.

FLIGHT: Round-headed, short-billed, slightly shorter-tailed profile. The bright chestnut heads of breeding males stand out, as do the white upperwing patch and the white belly (traits shared by American Wigeon). *Note:* The underwing of Eurasian Wigeon is uniformly gray, showing perhaps a trace of white; the underwing of American Wigeon males and females shows a distinct white core extending from the base to the bend of the wing.

VOCALIZATIONS: Males make a loud, emphatic, somewhat amusing, rising and falling, whistled "*WEEoo*." The kind of pronouncement a duck might make if it were being goosed, Eurasian Wigeon's call is very unlike the two-note whistle toot of American Wigeon. Females, in flight, make a low, repeated, croaking quack.

PERTINENT PARTICULARS: The fact is that adult male Eurasian Wigeons are found with regularity in the United States, and females rarely. This has led to speculation that males show more wanderlust than females and are thus more likely to occur in the United States, greatly simplifying identification. Of course, the disparity might relate to a sampling problem insofar as adult males are much easier to distinguish than females.

American Wigeon, *Anas americana*

The Happy Whistler

STATUS: Common puddle duck, particularly common in the West. **DISTRIBUTION:** Widespread northern breeder—from N.S. to Hudson Bay, north and west to the Beaufort Sea and the west coast of Alaska. Range extends south to the Great Lakes, the prairie pothole region of the U.S., the n. Rockies, and the Great Basin. In winter, found across much of North America south of Canada and south to Central America. Absent from interior New England, the Appalachians, all but the southernmost reaches of the Great Lakes, northern portions of the border states, portions of the s. Rockies, and arid regions of s. Calif. **HABITAT:** Breeds in freshwater marshes, ponds, and small lakes that lie adjacent to brushy and grassy cover. In winter, occupies freshwater marshes, shallow ponds, lakes, impoundments, tidal bays, parks, golf courses, and, especially, agricultural land. **COHABITANTS:** Assorted dabbling ducks, coots. **MOVEMENT/MIGRATION:** Spring migration from February to mid-May. Fall migration is both early and protracted, with some birds reaching wintering areas as early as mid-August and large influxes reaching southern wintering areas as late as December. VI: 4.

DESCRIPTION: Compact medium-sized puddle duck with a fairly distinctive profile, a very distinctive call, and a wanderlust nature. The body is fairly short, and the head large and roundish, but with a slightly peaked dome and a sloping forehead. The bill is unduckishly stubby—a short thick wedge the color of polished silver (tarnished in immatures). The tail is long and acutely pointed—a mock pintail. On the water, the tail extends well beyond the wings.

Breeding males, with their white or buffy crown (or "bald pate"), rakish green stripe behind the eye, and rusty-colored sides, are standouts among waterfowl. The scaly, almost reptilian gray heads of females and immatures contrast with the otherwise warm brown back and breast. Note also the stubby blue-gray wedge of a bill and sloping forehead (which help distinguish wigeons from the flat-foreheaded long- and thin-billed Gadwall) and the shadowy dark smudge over the eye that helps distinguish American Wigeon from Eurasian Wigeon.

BEHAVIOR: An active, amiable, and social duck that's nimble on land and mobile on water. Walks and runs with more balance, poise, and speed than most other ducks. In water, tends to feed through the ranks of other ducks, picking food from the surface as it goes and spending less time upending and working the bottom. Commonly grazes on short, cut grass. An artful klepto. Often attends larger waterfowl and snatches aquatic plants from their bills. Mixes freely with other waterfowl, but also spends much time on land, often foraging in tightly packed, single-species flocks (single, that is, except for the odd Eurasian Wigeon, which should be looked for, especially along the West Coast).

FLIGHT: Compact, with fairly slender wings and a distinctly long acutely pointed tail. Holds level in flight. Overall silhouette is more easily confused with Wood Duck than the more slender Northern Pintail. Wingbeats are quick, somewhat fluid, and steady. Flight is nimble, but not particularly fast. Flies in compact groups. (Wood Ducks are more strung out.) Males particularly, but females also, have a distinctive white patch on the leading upper surface of the wing (compare to Gadwall's square, white speculum on the trailing edge).

VOCALIZATIONS: Males make a quick, happy-sounding, two- or three-note whistle: "*We-whew*" or "*we-We-whew.*" Recalls a compressed "wolf whistle" that's more toot than whistle. Females make a low quack.

American Black Duck, *Anas rubripes*
Coastal Shadow Mallard

STATUS: Common Northeastern breeder and eastern wintering species; particularly common in coastal areas. **DISTRIBUTION:** Breeds from n. Lab. west to ne. Man. and south to coastal N.C., n. W.Va., n. Ind., cen. Wisc., and n. Minn. In winter, vacates extreme northern and interior breeding areas and redistributes widely east of the Mississippi R. (excluding s. Fla. and n. Maine) as well as in se. Minn., e. Iowa, n. Mo., se. Neb., extreme ne. Kans., and e. Ark. In Canada, winters north to coastal Nfld., along the St. Lawrence R., and west to the eastern shore of Lake Superior. **HABITAT:** In summer, breeds in northern wetlands and bogs (often surrounded by timber) and in tidal marshes; in winter, may be found in a variety of freshwater habitats, but shows a particular penchant for coastal estuaries, rivers, impoundments, and bays (will even go offshore to roost at night or escape heavy hunting pressure). **COHABITANTS:** Associates with other puddle ducks except in the extreme northern portions of its winter range, where it is likely to be the only puddle duck. Also found with Clapper Rail, Seaside Sparrow. **MOVEMENT/MIGRATION:** Spring migration from early February to mid-May; fall from mid-September to late December. VI: 2.

DESCRIPTION: A Mallard dressed for a funeral, except for the bill. Unlike most other waterfowl, males, females, and immatures are identical. Bodies are charcoal brown, with a contrasting paler, somewhat grayish head and neck. At a distance, when sunlit, the rump (actually the tertials overlying the rump) sometimes shines pale or coppery. Sides appear uniformly dark and unpatterned. (Mallard and Mottled Duck show a feather pattern.) The bill on males is yellow-green; female bills are duller and darker on the upper mandible. Feet, particularly on northern birds, are reddish. (On Mallard, feet are orange.) An iridescent patch on the wing (speculum) is purple without a white trim or border.

In flight, the contrast between the very white underwing linings and the otherwise all-dark duck is hard to mistake or overlook.

BEHAVIOR: This is one tough duck; able to tolerate bitterly cold conditions, it winters farther north than other puddle ducks. Somewhat warier than Mallard, and not quite as gregarious. Black ducks are inclined to keep to themselves—often in pairs or small groups (six or eight birds)—but in winter, or where heavily hunted, large numbers may gather in protected or sheltered areas. When black ducks do associate with other species, it is most often with Mallards. When gathered in larger numbers, the association seems casual, almost serendipitous. Unless startled and flushed, individuals and groups come and go from the flock without much sense of flock identity.

Prefers to forage on open mud flats and to loaf on banks, particularly along tidal creeks. Not so addicted to open spaces as many puddle ducks; individuals and small groups may tuck themselves away in small, tight confines out of sight. In migration, often seeks out protected coves on larger lakes to rest.

FLIGHT: The size and shape of a Mallard, but overall blackish brown, with a contrastingly pale head and very obvious white flashing underwings.

VOCALIZATIONS: Very Mallardlike, but lower-pitched and harsher.

PERTINENT PARTICULARS: See description of female Mallard for hints on how to distinguish from black duck. Also be aware that black ducks and Mallards hybridize freely. Hybrid birds usually show a white border on the speculum.

Mallard, *Anas platyrhynchos*

The Ubiquitous, Archetypal Duck

STATUS: Common, widespread, and hardly in need of an introduction. **DISTRIBUTION:** This cosmopolitan dabbling duck resides almost everyplace in North America at some point in the year. **HABITAT:** Tundra ponds, city parks, prairie sloughs, tule marshes, tidal flats, storm runoff catch basins, dry agricultural lands . . . the bird has even been known to place its nest amid desert cactus (but a stone's throw away from a cattle watering trough) and tucked into suburban shrubbery one six-lane interstate removed from the nearest water. Hardy, Mallard winters as far north as open water is found, retreating to fast-moving streams and coastal bays during freeze-ups, but unlike American Black Duck, tends to avoid salt water. In winter, many thousands gather in refuges in rice- and corn-growing regions of the country. **MOVEMENT/MIGRATION:** Where lakes or ponds do not freeze, a permanent resident. Spring migration from early February to mid-May; fall from September to late December. VI: 4.

DESCRIPTION: Large, long-billed, and classically shaped, *this* is the archetypal duck. The pale bodies of adult males support the distinctive iridescent green head, the bright yellow bill, the narrow white neck ring, and the plum-colored breast. (The short, curled tail feathers are a unique affectation.) Females and immatures are overall brown, with a slightly paler head and noticeably patterned back and sides; females are easily confused with several other female dabbling ducks.

The female's distinguishing traits include an impudently large orange bill bisected with a dark saddle, a white-bordered blue speculum, and whitish tail feathers and a whitish undertail that contrast with the brown body. The blue speculum may be hidden when the bird is sitting on the water, but a hint of the white border usually peeks through.

In flight, the female's underwing linings are pale, but the contrast between the body and underwings is not as dramatic as with black duck.

BEHAVIOR: Ridiculously tame (where they are fed); wary and hard to approach where heavily hunted. Mallards are a gregarious, social duck, commonly found in mixed flocks with other puddle or dabbling ducks. Forages easily on dry land. In summer, during molt, males lose much of their gloss and become sedentary, listless, and disheveled-looking.

FLIGHT: Mallards are strong, fast fliers, but the flight seems undramatic, routine. Wingbeats are strong, steady, regular—perfectly harmonious for a duck this size—but shallower than smaller dabblers.

VOCALIZATIONS: Given by females: a loud, resonant, soul-satisfying "*quack!*" When birds are nervous, the call is given in an evenly measured series, "*quack . . . quack . . . quack.*" The greeting call is a descending and accelerating "*QUAAACK . . . Quack, Quack, Quackquack.*" The feeding call is a gargled chuckling sound.

PERTINENT PARTICULARS: There are two important considerations for identification. First, along the Texas and New Mexico border with Mexico, Mexican Duck, a subspecies of Mallard (formerly regarded as a separate species), may occur. Males and females look much like female Mallard, but the tail and undertail are dark (not whitish). In the United States, hybrids between Mallard and Mexican Duck are the rule; pure Mexican Duck is a celebrated exception.

Second, while hybridization is common among many waterfowl species, Mallard seems to be particularly promiscuous. Where ranges overlap, hybridization with both American Black Duck and Mottled Duck often produces young with overlapping traits (such as a purple speculum with a white border). Hybridization with American Wigeon, Northern Pintail, and Gadwall also occurs. Of particular consternation to new birders are the birds found in parks and zoo ponds that result when Mallards mate with the large white domestic Long Island and Muscovy Ducks (sometimes producing "Mallards" that are twice the size of other puddle ducks) as well as other confusing mutts.

GENERAL PARTICULARS: Unlike the identification process as it applies to many other species, identifying waterfowl is rarely a matter of identifying a single bird. Because male and female waterfowl are often found together much of the year, identification is both, in one sense, easier and, in another, more dynamic.

Why easier? Because for most species, for most of the year, male ducks tend to be dramatically marked and easily identified. The more cryptic females suggest their identity by their close association with males, greatly simplifying the identification process. When you see a large brown duck swimming close to what is obviously a drake Mallard, the way to frame the identification challenge is not to look at the female and wonder which of the drab brown puddle ducks she is, but to check first to see whether the field marks are consistent with Mallard. If they are not, then go through the deck.

Why dynamic? Because the identification process is broadened and defined when two ducks show different field marks but can be presumed by their association to be the same species. For example, you see two ducks on the far side of the northern lake, over a mile away. All you can get on the drake is bright white sides and a dark head. It could be a Common or Barrow's Goldeneye or a Common Merganser, or possibly a scaup or Canvasback. Then the female turns and also shows a dark head and white breast. Identification: Common Merganser, since the females of the other possible species have dark breasts.

Mottled Duck, *Anas fulvigula*

Southern Black Duck

STATUS AND DISTRIBUTION: Common resident of the coastal marshes of n. Mexico, Tex., and La.; also occurs in reduced numbers in Miss. and Ala. Widespread across the Florida Peninsula. Also established near Savannah, S.C. Small numbers winter north to Kans., Okla., and Ark. **HABITAT:** Distinctly partial to fresh water, including freshwater prairie and brackish ponds; also ditches, rice fields, seasonally flooded wetlands, coastal marshes, and sheltered lagoons and bays. **MOVEMENT/MIGRATION:** Moves in response to water and food conditions, with no general seasonal pattern. VI: 2.

DESCRIPTION: Black Duck Lite—in plumage, intermediate between the darker American Black Duck and paler female Mallard (but almost exactly the same body plumage as Mexican Mallard). With its more slender head, thinner neck, longer bill, and slightly smaller body, Mottled Duck is slightly more sinuous and adolescent-like than Mallard and black duck. The bill on many individuals also seems slightly more robust or shoveler-like and shows a small, black spot at the base of the lower bill.

The overall dark brown body has a paler buffy head and neck but shows less contrast and less definition between the face/neck and body than American Black Duck. Black duck's silvery gray face contrasts markedly with the much darker body, and the line between the pale neck and the dark breast is sharply defined. On Mottled Duck, the buffy head and neck shade into the breast. At closer range, the sides and back of Mottled Duck are more patterned or mottled (even shabby). Black duck looks uniformly blackish, and you have to strain to see a feather pattern.

BEHAVIOR: Generally not as social as other puddle ducks. For much of the winter, pairs are the norm, although post-breeding flocks gathering in rice fields in late summer may number in the hundreds. Forages in shallow water, feeding from the surface or dabbling. Seems not to be as wary as American Black Duck, though its reactions are subject to habituation and hunting pressure.

FLIGHT: Whitish underwings contrast markedly with the darker body (like American Black Duck), but the contrast between face and body, again, is not as pronounced. Flight is direct and steady, and possibly not as fast or strong as American Black Duck.

VOCALIZATIONS: Like Mallard and black duck, but the female's "*quack*" seems not as loud or strident and is slightly higher-pitched.

PERTINENT PARTICULARS: There are several broad-brush distinctions to bear in mind when separating this species from other members of the Mallard Complex. Overall paler female Mallard (with an orange bill with dark saddle and whitish outer tails) and darker Mottled Duck (with all-yellow or greenish yellow bill with a dark spot at the base and all-dark tails) are the lesser identification challenges. Also, the geographic ranges of the Mexican form of Mallard and of Mottled Duck do not overlap except in extreme southern Texas.

Blue-winged Teal, *Anas discors*

Crescent-faced Teal

STATUS: Common, widespread, primarily northern breeder; common, primarily southern and coastal wintering species. Common and widespread migrant. **DISTRIBUTION:** Breeds across much of subarctic Canada, from se. Alaska to sw. Nfld. and the Maritimes. Also widespread in the U.S., but except for coastal N.C., coastal La., and much of Tex., absent as a breeder across the Southeast; also absent in early summer from large portions of Okla., Mo., Ark., cen. and s. Utah, Ariz., s. Nev., and all but ne. Calif. In winter, primarily coastal from N.C. and n. Calif. south to South America, but fair numbers also occur inland in Miss., La., Tex., and (in diminished numbers) s. Ariz. **HABITAT:** Breeds and winters in shallow freshwater marshes and ponds rimmed with vegetation (also in brackish estuaries). Particularly partial to small seasonal wetlands and ponds in spring and rice fields in winter. In migration, may also occur in small, heavily wooded ponds. **COHABITANTS:** American Coot, Cinnamon Teal. **MOVEMENT/MIGRATION:** Migrates very early (principally August and September) and returns late (March to early May). A regularly seen migrant over near-shore Atlantic waters and, especially, the Gulf of Mexico. VI: 2.

DESCRIPTION: Aptly named duck, but not above confusion with several other species. Structurally akin to a cross between Green-winged Teal and Northern Shoveler (but favoring Green-winged Teal).

Smallish (very slightly smaller than Cinnamon Teal; longer and leaner than the compact Green-winged Teal), with a long spatula-shaped bill, roundish head (showing a distinct forehead), and a long low-profiled body.

Males have a distinctive white crescent moon stamped at the base of the bill (and white flank patch). In comparison to female Green-winged Teal, female and immature Blue-wingeds are more contrastingly patterned, showing a slightly grayer (and more expressive) face, a slightly darker back, and paler sides. Female Blue-wingeds also have a suggestion of the male's white crescent at the base of the bill that is sometimes a prominent white patch and at other times more subtly akin to a worn area on the face. This pale patch contrasts with the bird's blackish bill. (Green-winged Teal's bill is about the same tone as the face so shows less contrast.) Highlighting white eye-arcs make the eyes of Blue-winged Teal easy to see, whereas on female and immature Green-winged Teals the eyes are lost.

BEHAVIOR: More inclined toward warmer temperatures and fresher water than Green-winged Teal. Forages, shoveler-like, in shallows. Often swims along with head submerged; infrequently turns tail up, like most other dabblers. Not particularly skittish, Blue-winged often sits on the water (bobbing its head nervously), considering its options before flushing (and often choosing to swim into the vegetation instead). Almost always found in shallow water, often right up against the bank. Associates with other puddle ducks (particularly other teal), but during migration, homogeneous flocks are the rule.

FLIGHT: Longer, more obviously front-heavy than Green-winged Teal. Flight is exceedingly fast, and more direct than Green-winged Teal; also appears slightly more fluid or languid. Flocks are more compact and less shifting.

VOCALIZATIONS: Less vocal than Green-winged Teal. Males make a high peeping squeak; females make a low, muffled, scraping quack.

PERTINENT PARTICULARS: Except in eclipse plumage, drake Blue-winged Teal is stunning. With its gray head shot through with blue and rose-colored hues and strikingly patterned sides, it must rank among the most beautiful birds in the world. Every birder should have the experience of training a spotting scope on the bird at close range and letting time suspend itself.

Perhaps half the time a sliver of blue winks through the folded wing; this wink of blue easily distinguishes Blue-winged Teal from Green-winged, but not from Cinnamon Teal or Northern Shoveler, which also have blue wing patches. Conversely, a wink of green through the folded wing does not automatically make the bird a Green-winged Teal. Blue-winged Teal, Cinnamon Teal, and shovelers also have greenish speculums.

Cinnamon Teal, *Anas cyanoptera*

Shoveler-billed Teal

STATUS: Common western breeder; winters in the Southwest. **DISTRIBUTION:** Breeds mostly in the Great Basin and intermountain West. Found from s. B.C., s. Alta., and sw. Sask. south to Mexico. Eastern border falls across e. Mont., Wyo., w. Neb., Colo., N.M., and extreme n. Tex. Absent as a breeder in extreme w. Wash., e. Idaho, s. Nev., and extreme sw. Ariz. Winters throughout most of coastal and cen. Calif., s. and cen. Ariz., s. N.M., s. Tex. along the Mexican border, on the Gulf Coast almost to La. (where it is very rare), and south into Mexico. **HABITAT:** Breeds in permanent and seasonal freshwater wetlands and alkaline bodies of water, preferring wetlands with emergent vegetation. In migration and winter, still selects shallow, well-vegetated, often seasonal bodies of fresh water, including flooded agricultural lands and tidal estuaries. **COHABITANTS:** When breeding, Eared Grebe, American Coot, Yellow-headed Blackbird. **MOVEMENT/MIGRATION:** Except in California and some southern portions of the bird's breeding range, the entire population vacates breeding areas in winter, many retreating into Mexico. Spring migration is very early and protracted, from late January into late May; fall migration is early, from early August into November. VI: 2.

DESCRIPTION: Structurally resembles a cross between a Green-winged Teal and a Northern Shoveler (but favors shoveler). Small, longish duck (larger than Green-winged Teal; smaller than Northern Shoveler; similar in size and shape to

Blue-winged Teal), with a large somewhat oval-shaped head, fitted with a long spatually flattened bill that recalls the outsized bill of Northern Shoveler and is slightly larger and more spatula-like than the bill of Blue-winged Teal. *Note:* The head of Blue-winged Teal is round (akin to Green-winged Teal) with a distinct forehead and a bill. Cinnamon Teal's forehead slopes down to the long exaggerated bill, giving the bird a very long-faced appearance and enhancing the shoveler-like appearance.

Adult male Cinnamon Teal is unmistakable. Even in eclipse plumage, the bird retains a signature warm cinnamon tinge (particularly about the face) and a red eye. (The eye of Blue-winged Teal is dark.) Females and immatures are overall brown and mottled—much like female and immature Blue-winged Teal. Their plainer faces and the warmer tones of their busy mottled sides, however, show considerable contrast and may remind you of the similarly plain-faced Mottled Duck.

BEHAVIOR: A social duck but not gregarious. Cinnamon Teal is commonly found in pairs from late winter until young are hatched (an association that makes the identification of paired females easy). In the fall and winter, the birds form small flocks (fewer than 50) and also associate with larger flocks of puddle ducks.

Cinnamon Teal forages in shallow, often very lushly vegetated (saladlike) water. Most commonly feeds by submerging its head and swimming (like a shoveler). Also picks at the surface. Likes to swim with its tail flush with the surface. (Tails of Blue-winged Teal may more often be elevated.) Nimble on land, often hauls out when not feeding. Not necessarily restricted to, or even attracted to, larger bodies of water. Often feeds in ditches or swales.

Less wary than some puddle ducks, but still not tolerant of close approach.

FLIGHT: Long-bodied profile, with a very projecting head and very slender (but not particularly long) wings. Flight is strong and nimble. Often flies in tight flocks that wheel and maneuver well, but is not as recklessly fast or erratic as Green-winged Teal—in fact, Cinnamon Teal seems to comport itself like a much larger duck than it really is.

VOCALIZATIONS: Fairly silent. Females make a short, low, flatulent quack.

Northern Shoveler, *Anas clypeata*

The Headless Duck

STATUS: Common, mostly northern and western breeder; common, primarily southern winter resident. Widespread migrant. **DISTRIBUTION:** Breeds throughout most of Alaska, the Yukon, s. Northwest Territories, e. and s. B.C., Alta., Sask., all but n. and ne. Man., n. and se. Ont., and along the St. Lawrence R. east to parts of N.B. In the U.S., breeds in Wash., w. and cen. Ore., ne. Calif. and the Sacramento Valley, parts of Idaho, n. Nev., much of Utah, parts of N.M., Mont., Wyo., N.D., S.D., nw. Neb., w. and s. Minn., n. Iowa, s. Wisc., se. Mich., and n. Ohio. Winters primarily in the s. and w. U.S., the Caribbean, and Mexico. In the U.S., occurs from coastal Mass. to Fla., up the Mississippi Valley to Ohio, west across the s. Great Plains to Utah, and north to s. B.C. In short, Northern Shoveler is absent where fresh and salt water are frozen for much of the winter. **HABITAT:** Breeds in shallow, mostly open wetlands. In winter and migration, prefers large shallow fresh- and saltwater lakes, wetlands, impoundments, and sewage treatment facilities. **COHABITANTS:** Often breeds and winters where muskrats thrive. In winter, associates with other puddle ducks, including Mallard, Gadwall, Northern Pintail. **MOVEMENT/MIGRATION:** In spring, migrates late by waterfowl standards, April into May, but early in fall, August to November. VI: 4.

DESCRIPTION: This medium-sized duck (smaller than Mallard; larger than a teal) is just too small for its bill, which is preposterously large, flat, and swollen-tipped, giving the bird a distinctly front-heavy appearance (a suggestion enhanced by the bird's high-riding stern).

A paint pallet on webbed feet, the male has a plumage pattern that is borderline gaudy—however, note the white chest! Only one other puddle duck, Northern Pintail, has a white chest. The pale brown females and immatures look like a paler, more contrastingly patterned female Mallard, but often the iridescent green speculum peeks through the folded wing.

In any event, the bill is all you really need to see.

BEHAVIOR: An aquatic vacuum cleaner. Feeds with head down, neck extended, and bill (and often much of the head) submerged for extended periods of time. Swims forward with head sweeping from side to side; only infrequently raises its head. Most other dabbling ducks habitually dunk their heads and bring them up again. Shovelers infrequently dabble and almost never forage on land; however, they often haul out on the shoreline to rest.

Generally social, Northern Shoveler in many places is found with many other puddle duck species. However, this species is partial to stagnant and polluted water, which is shunned by other puddle ducks, so don't be surprised to see them alone. Not quite as wary as other ducks (possibly because hunters disdain them).

FLIGHT: Good, strong, fast flier. Individuals in flocks stay tight. Flocks maneuver with near-teal-like finesse. In flight, birds look particularly front-heavy. The powder blue (male) or blue-gray (female) patch on the upper/inner portion of the wing is very distinctive in flight, but may be confused with the smaller Cinnamon and Blue-winged Teals.

Hint: Look at the bill!

VOCALIZATIONS: Females give a low, weak, muffled croak (or cough) and a weak hissing (and repeated) whistle when startled.

Northern Pintail, *Anas acuta*

A Duck Painted by El Greco

STATUS: Common and widespread breeder (more local in the East) and wintering species; occurs almost everywhere as a migrant. **DISTRIBUTION:** Northern and western breeder. Nests across virtually all of Canada and Alaska, except most of coastal B.C.; scarce or absent in most of Ont. and Que. Nests in n. N.Y., in states bordering the Great Lakes, and throughout the n. Great Plains, the n. Rockies, the Great Basin, most of Calif., and cen. Ariz. In winter, retreats from almost all of Canada (except coastal B.C.) and Alaska. Winters south of a line across s. Mass., n. N.J., cen. Md., e. Va., cen. N.C., w. S.C., n. Ga., sw. Tenn., w. Ky., s. Ill., s. Mo., s. Neb., n. Colo., w. Wyo., w. Mont., and w. Wash. **HABITAT:** Breeds extensively in the prairie pothole and tundra regions in shallow marshes, tundra ponds, and seasonal wetlands, showing a marked preference for open, unforested areas. In winter and migration, continues to favor shallow grassy wetlands, but is also common in agricultural grain fields and coastal wetlands and mud flats. **COHABITANTS:** In winter, found with assorted puddle ducks, including most notably Mallard. **MOVEMENT/MIGRATION:** Where they do not occur all year, pintails are among the earliest ducks to arrive in the fall (males in August), and birds continue south through November, with peak migration in September and October. Pintails are also among the earliest ducks to return north in spring: northbound birds have been recorded as early as late January, and lingering individuals are all but gone by May. VI: 4.

DESCRIPTION: Slim, elegant, even rakish-looking dabbling duck—with the elongated proportions of an El Greco painting. Overall slender, but with only a medium-sized body (smaller than Mallard), males *and* females should command instant recognition by their long-necked long-tailed profile alone. On the water, often rides low in the bow and high in the stern. At a distance, the neck and body seem almost to meet below the water line.

Breeding plumage males are conservatively colored (brown head, gray body, black butt), but conspicuously garbed. The white neck stripe and flank patch are visible at considerable distances, but the bright white breast distinguishes drake pintail from all other puddle ducks except the anything but elegantly proportioned male Northern Shoveler. Females and young can be distinguished from other puddle ducks by their warm buff brown bodies and even buffier heads (sometimes just faces). Nonbreeding males have grayer bodies but also have buff-colored heads. Easily overlooked and underappreciated (except at close range) is the long, slim, lead-colored bill of females and the silver-blue-sided dark-centered bill of males—elegant touches for a dapper bird.

BEHAVIOR: A social, even gregarious duck; in winter, may be found in large, homogeneous flocks but is just as comfortable in a mixed puddle duck crowd. Feeds by dabbling in open water and also by grazing through shallows or open fields, where it walks with neck extended and head down. Swims with its tail jauntily elevated, and reportedly dives for food on occasion. Comfortable on land, during daylight the bird often hauls out on shore to rest.

Northern Pintail is a very nervous duck, and often the first species in a mixed flock to spook. Suspicious birds raise their heads high, accentuating their long-necked appearance.

FLIGHT: Northern Pintail loses none of its slender elegance in flight, showing a very long neck, very long tail, and long, slender, angled-back wings. The contrasting dark head and "pin" tail make males hard to mistake. Females are distinguished by their slender profiles and plain, buffy uniformity. (Except for a narrow white trailing edge to the inner wing, females show no defining plumage characteristic.) The flight is direct, with the body held absolutely horizontal. Wingbeats seem somewhat slower and more elastic than other puddle ducks, but the flight is nevertheless fast. Northern Pintail maneuvers well, changing course with fluid control. When landing, often descends with a long, slow, droop-winged approach (body angled breast up) that is more characteristic of geese. In migration, flocks may form a V or be strung out in a long horizontal line.

VOCALIZATIONS: Females make a soft low quack. Males make a short piping trill that is very much like the call of male Green-winged Teal but lower-pitched and more mellow.

PERTINENT PARTICULARS: At a distance, and among mixed flocks, paired pintails (and Mallards) stand out as overall paler than the ducks around them.

Green-winged Teal, *Anas crecca*

Small, Dark, and Reckless

STATUS: Common northern breeder; common to abundant and widespread wintering species. **DISTRIBUTION:** In the U.S., breeds across virtually all of Canada, Alaska, and all northern border states, as well as Mass., Conn., n. and w. N.Y., n. Ohio, Wisc., n. Neb., Wyo., Colo., ne. Utah, e. Ore., n. Calif., n. Nev., and parts of Ariz. and N.M. In winter, vacates almost all northern noncoastal breeding areas. Winters widely throughout the South and West south of a line drawn from se. Alaska, across s. B.C., w. Mont., e. Wyo., ne. Colo., s. Neb., s. Mo., s. Ill., Ky., w. Va., Md., e. Pa., to coastal Mass. and N.S. **HABITAT:** Nests on shallow grassy edges in marshes and tundra ponds. In winter, found in freshwater marshes, shallow vegetated ponds, and swamps with standing dead timber north to the freeze-line and in tidal estuaries north to the Maritimes and southern Alaska. (Birds shift south and north as freeze conditions warrant.) **COHABITANTS:** Northern Pintail, Northern Harrier, Savannah Sparrow. **MOVEMENT/MIGRATION:** Birds migrate in bunched flocks (primarily at night), but also team up, singly or in small groups, with other species, including coastal migrants like scoter. Spring migration from February to early May; fall from August to November. VI: 4.

DESCRIPTION: A miniature dabbling duck (the smallest teal, and almost as small as Bufflehead), with a speculum so brilliant green that it challenges a folded wing to hide it and a flight so fast and maneuverable that it borders on reckless. Overall compact, Green-winged has a disproportionately large head, a short neck, an elfin-small bill, and a stocky body that seems somewhat dwarfish compared to a full-sized dabbling duck.

Adult males in breeding plumage are arresting. In almost any posture, at any reasonable distance, the vertical white slash between the breast and sides is easy to see. Swimming and in flight, the yellow rear taillights (a narrow stripe along the side of the tail) are conspicuous even at great distance. When backlit or from a distance, the dark head contrasting with a pale (but not white) breast distinguishes a sleeping Green-winged Teal from most waterfowl.

Females and immatures are overall brown (darker than female Blue-winged Teal) and fairly uniform, showing little contrast between the bill and face or between the back and sides. Compared to Blue-winged Teal, the bill is more petite and classically proportioned (not flattened and swollen at the tip). The green speculum sometimes winks through the folded wing of a sitting Green-winged, and the pale taillights (absent on Blue-winged) may be apparent.

BEHAVIOR: Often winters in large numbers, foraging in shallows. In tidal areas, birds feed by walking, bill down, across open mud. Green-winged Teal frequently rests out of the water on banks, logs, and even branches. Not particularly shy, Green-winged is among the last ducks to flush when approached. Once flushed, it may travel only a short distance before landing, even as other species leave the area.

FLIGHT: In flight, the bird is compact, almost stubby. The body appears up-angled, with the dark head elevated or cocked above the body. The steep forehead and petite bill make the head look blunt, square, and prominent. The wings are very pointy. At a distance, except for the pale bellies, both males and females appear overall dark. In both sexes, in poor light, and at a great distance, the bright green speculum may not be evident, but the absence of a pale blue patch on the upperwing distinguishes the bird from the longer-profiled Blue-winged and Cinnamon Teals.

As much as small size, it is the bird's distinctive flight that distinguishes it as a teal. Fast, agile, and flying in compact, synchronized flocks that change direction and altitude quickly, these teals easily recall a flock of shorebirds. Taking off, Green-winged leaps from the water and climbs faster and at a steeper angle than most ducks. In direct flight, it often jinks left or right or rocks to one side and then the other. Landing, the birds come in at what seems breakneck speed and stop with little or no glide on the surface.

VOCALIZATIONS: Males make a distinctive, high-pitched, chirping trill that recalls the sound of Spring Peepers. Females emit a hurried descending series of quacks that suggests a very small duck trying to imitate a big duck.

PERTINENT PARTICULARS: Putting it all together, what distinguishes Green-winged Teal is its very small compact size and the very obvious field marks on the males; females are distinguished by the very absence of obvious field marks. Most other small female ducks show something, including, in the case of the very similar Blue-winged and Cinnamon Teals, obvious blue patches flashing on the upper surface of the wing. The best way to tell a sitting Green-winged Teal from the other North American teal is bill size and shape. Looking at a female teal, pose this question for yourself: "Is there any conceivable way that I could turn this bird into a female shoveler (a bird with an outsized, spatula-shaped bill)?" If the answer is "yes," or even "possibly," the bird is not a Green-winged Teal.

Be aware that the Eurasian subspecies of Green-winged Teal (the Common Teal) is always possible where large concentrations of teal are found (especially along the East and West Coasts). Sometimes distinguished by the horizontal white slash running down the side, this bird can also (and sometimes more easily) be found by searching among male Green-winged Teals for the bird that doesn't show a vertical slash bordering the breast.

Although not impossible to separate, for all practical purposes, female Common and American Green-winged Teals are indistinguishable in the field.

Canvasback, *Aythya valisineria*
Speed

STATUS: Fairly common northwestern breeder and widespread but local winter resident. **DISTRIBUTION:** Breeds in n. and w. North America from cen. Alaska and across most of the Yukon, the w. Northwest Territories, e. B.C., most of Alta., w. and s. Sask., and s. Man. In the U.S., breeds in e. Wash., se. Ore., ne. Calif., parts of Idaho, ne. Utah, w. and n. Mont., parts of Wyo., w. Colo., much of N.D. and S.D., w. Neb., w. Minn., and nw. Iowa. Winters along both coasts from s. B.C. to cen. Mexico and from Mass. to the Yucatán Peninsula (except absent in s. Fla.). In the interior U.S., winters across the South and North, locally, to the s. Great Lakes, with a few birds present in coastal New England, throughout the Appalachians, and in the northern prairies and Rockies. **HABITAT:** Breeds in small fresh and alkaline lakes, large shallow or deep-water marshes, potholes, sloughs, and impoundments. At all times prefers habitat bordered by dense vegetation. In winter and migration, found in a variety of habitats, from deep freshwater lakes to coastal ponds, flooded fields, coastal bays and estuaries, and tidal mud flats. Prefers shallow water for foraging. **COHABITANTS:** In winter, often associates with Redhead. **MOVEMENT/MIGRATION:** Virtually the entire population retreats from nesting areas. Spring migration is early, from February to early May; fall is late, from October into December. VI: 2.

DESCRIPTION: Large (larger than Redhead; smaller than Mallard), distinctive diving duck whose low profile and elegant contours say *speed.* Head shape alone distinguishes both males and females. No other diving duck has a facial profile so slender and sloping (in a word, wedgelike). Redhead's plumage pattern resembles Canvasback's, but the head and bill shapes are very different (round with a flat forehead versus long and wedge-shaped).

Also unique among ducks is the male's black-bracketed white body—extensively white in the middle, dark fore and aft. Several eiders have white backs and show black sides; male Redheads have the same plumage pattern as Canvasback, but the body is distinctly gray, not bright white, and scaup show white sides but gray backs. Also, the head of male Canvasback is dark chestnut (going blackish about the bill). The head of Redhead is uniformly and brighter red. Female and immature Canvasbacks boast a shadow pattern of the male, showing gray bodies and pale brown heads, necks, and breasts. Female and immature Redheads are overall darker, browner, and more rumpled-looking, whereas the feathers of Canvasback seem smooth and even.

But shape, not plumage, is the first, middle, and last word in the identification of Canvasback. Plumage is just a footnote.

BEHAVIOR: A gregarious duck. Sometimes mixes with other waterfowl (most notably Redhead and scaup), but is commonly found in single-species flocks, often numbering in the hundreds. Dives vertically for food (primarily aquatic vegetation). Almost never walks on land, but in tidal areas will haul out and rest on banks until feeding areas are flooded on the next tide.

This is also a wary duck that, unless habituated to people (in a refuge or park), does not allow close approach. Requires a lengthy running takeoff to get airborne.

FLIGHT: Flight is swift, straight, and powerful. A long straight neck gives the bird an air of elegance. Males are mostly white, with bracketing black heads and butts. Females and immatures are very pale and plain and show very little contrast except for a sharp demarcation between the dark breast and white underparts. In migration, and often when birds are traveling a short distance, flocks form a well-regimented V.

VOCALIZATIONS: Females make a low, coarse, gargled growl, "*grrah grrah grrah,*" that sounds like a muffled and strangling American Crow. Males make descending, breathy, squeaky giggles.

Redhead, *Aythya americana*

False Canvasback

STATUS: Common but local, predominantly western breeder; widespread but unevenly apportioned winter resident. **DISTRIBUTION:** One of the prairie pothole breeders, nesting in Alaska, s. and e. B.C., s. and cen. Alta., s. and cen. Sask., s. Man., e. Wash., much of Ore., the Central Valley of Calif., parts of Idaho, n. Nev., most of Mont., w. and s. Wyo., n. Utah, much of Colo., n. and e. N.D., e. and s. S.D., parts of Neb., and w. Minn. Also breeds in parts of Ariz., N.M., Kans., the Texas Panhandle, Wisc., Mich., and n. Ohio. In winter, found in almost every state (as well as s. B.C. and Mexico) except for the northern prairie states, n. New England, and along the Appalachians to the Ky. border. **HABITAT:** Nests in a variety of freshwater wetlands, but predominantly uses seasonal wetlands bordered by cattails and bulrush. Also nests in agricultural ponds, reservoirs, impoundments, alkaline lakes, and sewage ponds. In winter, found primarily in coastal bays and lagoons where various sea-grasses flourish. Smaller numbers are found on inland lakes and reservoirs. **COHABITANTS:** Other diving ducks, including Canvasback,

Ruddy Duck, Lesser Scaup, Sora, Franklin's Gull, Black Tern. **MOVEMENT/MIGRATION:** Spring migration from March to May; fall from October to late December. Also makes a post-breeding/molting movement north of breeding areas in July and August. VI: 4.

DESCRIPTION: A bird whose similarity to Canvasback is highly exaggerated—not even feather-deep. Medium-sized (larger than Greater Scaup; smaller than American Wigeon; smaller than Canvasback), the overall compact and roundly contoured Redhead has a round head and a short dome-backed body (as compared to the lean, low, long-faced profile of Canvasback). Concentrate on the head shape: Canvasback's face, with its long sloping forehead and bill, looks like a doorstop; Redhead's steep (almost blunt!) forehead and curving bill give its face a ski-jump profile.

Plumage similarities between Canvasback and Redhead are superficial. Male Redhead is a dark gray-bodied duck with a truly red (rufous) head; drake Canvasback is a white-bodied duck with a dark chestnut head that often looks blackish at a distance or in poor light. The pale blue bill of male Redhead stands out at a distance; Canvasback's black bill merges with the dark crown.

Dingy brownish gray female and immature bodies are virtually patternless, with almost no contrast between head, breast, and body except for a slightly paler face (the contrast increases with distance). Female Ring-necked Duck is far more likely to be confused with Redhead than Canvasback. (Female scaup are easily separated from Redhead by the white ring around the base of the bill.) Female Redhead's plumage is uniform; female Ring-necked Duck has a dark back and paler sides. Head shapes differ as well: Redhead has a round head, and Ring-necked's is peaked.

BEHAVIOR: A social duck. Pairs usually feed together; in winter, flocks may number in the thousands. Where wintering numbers are low, often found in flocks of other diving ducks (most notably Canvasback). Dives with a curving, forward lurch, and in shallow water tips up like a dabbler. Usually not found on land. Takeoff requires a long running start across water.

Redhead is a shy duck that does not tolerate close approach.

FLIGHT: Overall compact profile in flight, appearing shorter-necked than Canvasback. Flight is direct, with steady but not particularly rapid wingbeats.

VOCALIZATIONS: A quiet duck. Females make a low, rough, gargled crow call (similar to female Canvasback). Males make a plaintive wail that recalls the cry of a distant gull.

Ring-necked Duck, *Aythya collaris*

The Ring-billed! Duck

STATUS: Common, fairly widespread northern breeder (at higher elevations in the West); common and widely distributed wintering species. **DISTRIBUTION:** Breeds across subarctic Canada, Alaska, and the northern border states from Nfld. and the Maritimes west to the Northwest Territories, the s. Yukon, and parts of B.C. Also breeds in e. Idaho, w. Wyo., ne. Nev., w. Colo., and cen. Ariz. Winters primarily in the U.S. and Mexico, mostly south of its northern breeding areas—that is, south of a line running through s. B.C., w. Mont., w. and s. Wyo., e. Colo., n. Okla., se. Neb., n. Mo., cen. Ill., s. Ind., n. Tenn., cen. Va., e. Pa., n. Conn., and e. Mass. **HABITAT:** Virtually a freshwater obligate. If it's winter and you're around open fresh water, chances are you'll see a Ring-necked Duck. Breeds in subarctic northern forests (taiga, boreal, and aspen parklands) in shallow but usually permanent well-vegetated marshes, bogs, and beaver ponds that also have open water nearby. Also found in prairie pothole habitat. In winter, small flocks (usually fewer than 50 birds) occur in a variety of still, sheltered, freshwater habitats (rarely brackish water), including smaller forest-rimmed ponds shunned by other divers. **COHABITANTS:** In winter, often associates with Ruddy Duck. **MOVEMENT/MIGRATION:** A late-fall and early-spring migrant. Spring migration from early February to early-May; fall from late September to early December. VI: 2.

DESCRIPTION: Small-bodied duck with an oversized conical head and a long narrow snout of a bill. Small diver (slightly larger than Lesser Scaup; slightly smaller than Greater Scaup), with a long narrow bill, a distinctive overlarge conical head (showing a slight indentation just behind the crown), and a compact body. Male Ring-necked's boldly patterned plumage is distinctive. Females and immatures are overall brown, less distinctly marked, somewhat less acutely proportioned, and easily confused with several other divers.

In adult males, the white wedge separating the ink-colored breast and gray sides is manifest at great distances. The tricolored bill (gray base, bracketing white rings, black tip) is diagnostic, but only the white ring near the tip of the bill is apparent at great distances. Silhouetted, the peaked head, with the dent in the rear, is idiosyncratic. Set high on a neck stretched thin when the bird is nervous, the head seems awkwardly oversized for such a compact well-proportioned body. The heads of females and immatures are less distinctly peaked than those of males (but always more so than Lesser Scaup). At reasonable distances, the combination of a pale eye-ring/eye-line, a shadowy impression of a wedge just behind the breast (not always visible), the narrow whitish ring between the black tip and gray bill, and the pale but not crisply defined base to the bill help distinguish Ring-necked from female scaup and Redhead. Some females show a pale scuff mark on the ear coverts (like scaup). On the water, Ring-necked Duck's wingtips are often elevated above the back when the bird is feeding or sleeping (not common on scaup). When feeding or swimming, the tail is usually not visible; on a sleeping bird, the tail is commonly up-cocked.

BEHAVIOR: In many places, this is the duck that heralds the end of winter; it often occupies lakes when they are still mostly covered in ice. Outside the breeding season, Ring-necked is a gregarious species and most commonly found in small homogeneous flocks, though it will associate with other divers and even dabbling ducks. It's a shy, nervous, somewhat schizophrenic duck. Feeding birds may be widely spaced, with every duck foraging independently, diving for submerged vegetation, and surfacing close to the point of entry. Or they may respond to one bird's initiative (and very likely your presence) by moving as a flock, every bird entering the water in the same direction and exiting facing the same way.

FLIGHT: Overall compact and large-headed profile. The gray trailing edge of the wing easily distinguishes this species from the white-trimmed wings of scaup. Flushed, their running takeoffs are short (and for a diving duck, abrupt!). Flight is fast, agile, and almost acrobatic; wingbeats are rapid, with wingtips reaching low. Landings seem recklessly fast: the bird brakes with a rapid flutter of wings and comes to a stop after a very short slide on the water. Clearly this is a bird designed to navigate in tighter confines than the average diving duck.

VOCALIZATIONS: Usually silent. Females make low, rough, purred snort.

PERTINENT PARTICULARS: The Eurasian Tufted Duck is a rare visitor to North America, and when present is most commonly found among scaup. Among scaup (or Ring-necked), Tufted Duck's all-white flanks contrasting with the blackish back of adult males draws the eye. Round, not peaked, heads, adorned with a dangling feathered queue, also distinguish adult male Tufted Ducks. Females and eclipse males bear a short tuft, or a vestige of one. Immatures are darker and browner-headed than immature Ring-necked. In all plumage, the trailing edge of Tufted Duck's wing in flight is telltale white; Ring-necked Duck's wing is uniform gray.

Greater Scaup, *Aythya marila*

Coastal Scaup with a Green Gloss

STATUS: Common but fairly restricted northern breeder; in winter, uncommon to very common along seacoasts; uncommon to rare inland (where Lesser Scaup dominates). **DISTRIBUTION:** Breeds primarily in Alaska, n. Yukon, and nw. Northwest Territories and along the southern and western shores of Hudson Bay and n. Que. and Lab.; also widely scattered in Alta., Sask., Man., and the Maritimes. In winter, mostly coastal, with primary concentrations found in coastal s. Alaska and from N.S. and Cape Cod to Sandy Hook, N.J. Ranges coastally from Nfld. to Ga., from s. Texas to the Florida Panhandle, and from the Aleutians to Baja; also found on the Great Lakes. Uncommon in both the U.S. and s. B.C. away from marine waters. **HABITAT:** Breeds on large, shallow, freshwater tundra lakes and ponds. In winter, most birds are found in marine waters, particularly large, protected, shallow coastal bays and river inlets, and in ice-free (and zebra mussel–rich) portions of the Great Lakes. In spring, on interior lakes, concentrates at ice-free river mouths. **COHABITANTS:** In winter, outside primary wintering areas, regularly found among flocks of Lesser Scaup. **MOVEMENT/MIGRATION:** A late-fall and early-spring migrant. Spring migration from late February to early June; fall from late September to early January. VI: 2.

DESCRIPTION: This husky, clannish, salt-water-loving diving duck is a street brawler of a scaup. Medium-sized (larger and overall bulkier than Lesser Scaup; smaller than Black Scoter; about the same size as Common Goldeneye), with a big, round, almost bulbous head that changes in shape from gently curved (almost flat) on top and bulging in the back to symmetrically ovate. (Much depends on the bird's posture and activity.) Most birds show a slight dimple on the back of the head, but it is difficult to call this a peak. The steel blue bill is long, heavy, and (viewed head on) proportionally broad.

Males are black in the front, pale in the middle, and black in the rear; females are basically brown, with one or two distinguishing adornments. The male's distinctive pattern and the bird's penchant for gathering in massive mixed-sex rafts in coastal areas distinguish it as a scaup. Now things get interesting.

Depending on the light and angle, the head of males appears black, with subtle green highlights, or unmistakably iridescent green, with an occasional momentary flash of purple. On most birds, finer barring on the back gives Greater a paler, cleaner gray mantle. Owing to the bird's greater body size and whiter flanks, the eye of an observer searching through a flock of Lesser Scaup for the odd Greater will be caught by the often sharply defined oval white sides.

Females and immatures are shaped like males. The heads of adult females are, if anything, more symmetrically round than those of males. A prominent white ring encircles the base of the bill, and nonbreeding females frequently show a prominent, pale scuff mark on the side of the face, sometimes described as a "crescent."

As a field mark, the larger black nib on the bill of Greater Scaup is difficult to see and functionally worthless.

BEHAVIOR: Scaup spend much of their time resting in large homogeneous rafts, in windy weather, often on the lee (more tranquil) side of the body of water. The difference in head shape is most apparent among sleeping birds. Feeding birds are often tightly massed. Feeding is orderly and regimented, with birds often diving in one direction, then emerging facing the other.

Scaup is a very unsuspicious duck, but prefers to swim when approached too closely (often with its head still tucked). If you are patient, it often returns. Flushed, scaup form a tight, clustered mass (recalling a ball of starlings when a hawk is present) that relocates out of harm's way.

FLIGHT: Overall compact profile. Strong, fast, agile flier. The trailing edge of the wing is mostly white, turning gray toward the tip. In Lesser Scaup, the white is restricted to the inner half of the wing (the secondaries). Half or more of the trailing edge on Greater Scaup is dingy or gray. Running takeoffs are long and labored. Wingbeats are rapid, and flight is direct and fast.

VOCALIZATIONS: Usually silent. Females make a low, short, gargled trill (sometimes more gulp than trill).

PERTINENT PARTICULARS: In winter, Greater Scaup is more common in northern regions (with males wintering farther north than females) and more tolerant of rougher, less protected water. In winter, Lesser Scaup is the dominant (and expected) scaup south of British Columbia, south of northern New Jersey, and throughout most of the interior—except in migration, when great flocks (some numbering in the tens of thousands) may be found on Lakes Erie and Ontario and along portions of the Mississippi River.

Lesser Scaup, *Aythya affinis*

Interior (and Southern) Scaup with a Purple Gloss

STATUS: Common and widespread northern breeder; common and in some times and places abundant and widespread wintering species (and migrant) across much of the United States. **DISTRIBUTION:** Breeds from central interior Alaska south and east across all but coastal Yukon and B.C., the w. Northwest Territories, all of Alta. and Sask., all but n. Man., n. and e. Ont., and w. Que. In the U.S., breeds at the junction of Wash., Idaho, and Mont.; in se. Ore. and ne. Calif.; and in se. Idaho, Colo., extreme se. Ont., and se. Que. Winters primarily and extensively across the U.S. and Mexico—in coastal areas, from s. B.C. and Cape Cod to Central America; in the interior, absent in the n. Rockies, much of the n. Great Basin, the northern plains, along the Appalachians, and in n. New England. **HABITAT:** Nests in seasonal and semipermanent (fresh or brackish) wetlands with tall, emergent vegetation (such as bulrush, cattail, or horsetail). In winter, concentrates along the coasts of lakes, reservoirs, coastal bays, and open water (for example, the Gulf of Mexico). In migration, found in large concentrations on the Great Lakes and on impounded portions of the Mississippi River, but also uses smaller marshes and wetlands (particularly in spring). **COHABITANTS:** In winter, may be found with Greater Scaup. In migration, sometimes joins flocks of scoter. **MOVEMENT/MIGRATION:** Lesser Scaup is a late-fall migrant (late September to mid-December) and late returnee in the spring (mid-February to late May). It is not uncommon to find wintering birds still present in mid-May. VI: 2.

DESCRIPTION: A dapper and gentrified diving duck. Smallish to medium-sized diver (smaller and more delicately proportioned than Greater Scaup; larger than Ruddy Duck; about the same size as Ring-necked Duck). The head is smaller and more interestingly shaped than Greater Scaup: the high forehead and modest peak high on the back of the head impart an air of jauntiness that Greater Scaup lacks. Lesser's neck is thinner, its bill is relatively thin, straight, and—seen head on—discernibly narrow, and its body is nicely contoured and compact.

The head on males is black, with an iridescent purple gloss that at certain angles and in certain lights offers flashes of green. This does not undermine the usefulness of this field mark. (Mallard heads likewise flash purple, but nobody disputes that the head of the "greenhead" is green.) Lesser's back is overlaid with coarse, dark barring that makes it appear darker than Greater Scaup. The white sides and flanks are often finely vermiculated, making the oval sides not quite so white. Overall adult male Lesser Scaup is slightly dingier and shows less white than Greater Scaup—but only in direct comparison is this apparent.

Females and immatures are overall brown and garbed in a shadow pattern of adult males. The white ring encircling the base of the bill is less extensive than on Greater Scaup, but still obvious. Compared to Greater Scaup, the pale scuff mark found on the side of the head of nonbreeding females is less apparent and in fact usually absent. (In summer, both species may show this mark.) Like males, the female's head is peaked top and rear.

BEHAVIOR: Similar to Greater Scaup, with which it readily mixes (although in mixed flocks the minority species tends to gather in clusters). As mentioned, Lesser Scaup is more prone to winter in interior freshwater habitats and is much more common than Greater the farther south you travel.

FLIGHT: The trailing edge of the wing is white only on the secondaries (restricted to the inner half of the wing). Greater shows more white—if you have to second-guess, it's probably Lesser.

VOCALIZATIONS: Usually silent. Females make a short nasal bark that is breathier, higher-pitched, and not as gargled as Greater Scaup.

PERTINENT PARTICULARS: Telling Greater from Lesser Scaup is challenging. As one very skilled birder of my acquaintance put it: "Why is it that after 20 years I still can't tell scaup apart?" The reason is that the differences between the species are slight and, to some degree, transitory. Head shape changes depending on what the bird is doing, and colors shift depending on light conditions and angle of view. Two pieces of advice may be helpful.

First, don't rely on one field mark. Build a case for an identification by looking at all pertinent field marks.

Second, use the advantages afforded by probability. Most observers see scaup during migration or as wintering birds. At such times, scaup are social and the two species mix. But for the most part, and depending on where you are, one species dominates and is thus the "default" scaup. The other species may be tucked into the ranks.

The way to approach the scaup identification challenge is make the assumption, based on habitat and range, that the birds you are looking at are predominantly the default species. In coastal New England in January, the default scaup is Greater; on a sewage lagoon in Hancock County, Mississippi, it is Lesser. Now test the hypothesis. Study the individuals in the flock and see whether the characteristics are consistent with those of the region's default scaup. Are you looking at the round head, green gloss, bulky proportions, and gleaming white sides of adult male Greater Scaup, or the peaked head with a purple gloss, elegant profile, and lightly tarnished white sides of Lesser? Or is it the round head, bulky profile, and prominent scuff mark on the face of female Greater versus the peaked head, slim profile, and little or no scuff mark of female Lesser?

Why take this approach rather than do a cold-start identification? Because most observers seeing scaup for the first time (or the first time in six months) don't lack knowledge or acumen—they lack confidence. Knowing the bird's *probable* identification imparts confidence and permits an on-the-spot refresher course for those who have not seen scaup for a while.

After several minutes' study—after you have ascertained what "well-rounded" or "peaked" looks like—go through the flock and search for the anomalous duck. The one whose size, head shape and color, bright white (or dull) sides, and well-scuffed (or unscuffed) face draw your eye.

Once you find a candidate, be warned—but be assured too—that if there are other minority scaup in the flock, chances are that the one next to your candidate is the same species.

Single birds or small single-species flocks are more difficult (although over most of North America, Lesser is much more likely). Head color and shape are very useful and, on true Greater Scaup, very distinctive. If you are having difficulty, you are probably looking at a Lesser Scaup that shares some Greater Scaup characteristics (a head that seems mostly round, for instance, and sometimes shows a hint of green).

Look at the overall profile. Is it elegant or bulky? Look closely at the head. My rule of thumb when viewing a distant scaup is to ask: "Is there any conceivable way to turn this bird into a female Ring-necked Duck (a bird with a very peaked head)?" If the answer is "no way," the bird is a Greater Scaup; if the answer is "yeah, maybe," it's Lesser.

Steller's Eider, *Polysticta stelleri*

The Dabblerlike Eider

STATUS: Uncommon to locally common, but restricted to Alaska. **DISTRIBUTION:** In North America, breeds on the arctic coastal plain between Icy Cape and Harrison Bay in Alaska, with Point Barrow situated at about midpoint. Also breeds just north of Nome on the Seward Peninsula and historically in the Yukon-Kuskokwim R. delta. In late summer, concentrates in the vicinity of Nunivak I., Cape Newenham, and western portions of the Alaska Peninsula. In winter, ranges along the Alaska Peninsula, east to Kodiak I. and Kamishak Bay, and along the Aleutians. **HABITAT:** Breeds on open, polygonal (honeycomb-patterned) tundra near freshwater ponds, usually within 20 mi. of the coast (occasionally up to 100 mi. inland), so is less tied to seacoasts than other eiders. Otherwise found in shallow marine habitats—coastlines, shallow lagoons, bays, and inlets, particularly where freshwater streams enter. COHABITANTS: When breeding, Snowy Owl, Red and Red-necked Phalarope, Pomarine Jaeger. In winter, usually avoids other waterfowl species, except individuals outside of the established winter range may join other eider species. In migration, reportedly flies with flocks of murres. MOVEMENT/MIGRATION: Coastal migrant. Spring migration from late April to early July; fall from late June to mid-October. VI: 1.

DESCRIPTION: A small, oblong-billed, stiff-tailed dabbling duck of an eider. A medium-sized diving duck but a small eider (much smaller than other eiders; about the same size as Harlequin Duck or a murre), with a big thickset Ruddy Duck bill, a distinctive square head (often showing a bump on the noggin and rear crown), a long body, and a long, stiff, up-angled tail. Overall shaped like an ungainly puddle duck.

Breeding males are garbed like circus clowns—mostly white (peach-blushed breast and sides) with a black collar, black-and-white-capped back, and black butt. The unique pattern and wealth of field marks notwithstanding, the bright white head alone is enough to distinguish this species. Females and eclipse males are overall unpatterned and dark ruddy brown, with bluish gray bills and a large, white-bordered, bluish speculum (more likely to be confused with a scoter or puddle duck than other paler brown and boldly barred female eiders). Immatures and subadult males are colder-toned, grayish brown, and lacking in the distinctive speculum pattern. Rely on size, head shape, and the heavy blue-gray bill.

BEHAVIOR: A *very* gregarious species. In late summer and winter, usually found in large, tight-packed, single-species flocks. Distant flocks, well populated with non-eclipse males, look white at a distance. In deeper water, flocks dive and surface synchronously (often sending up a cloud of spray when they submerge), but individuals and groups most commonly feed in shallow water close to shore, where birds dabble and tip up in the manner of puddle ducks.

Steller's is more agile on land than other eiders, walking with a waddling gait. Agile on the water too, this eider can lift off with hardly less effort than a puddle duck. Not very vocal.

FLIGHT: In shape, looks like a short-necked long-tailed puddle duck. The male's white head and large white patch on the upperwing show at great distances. The female's silvery white underwing contrasting with an all-dark body recalls American Black Duck (but the white-trimmed, blue speculum is more like female Mallard). Flight is direct, agile, and twisting and turning like a smaller diving duck (not lumbering or plodding like other eiders). Wingbeats are rapid, even, and steady and produce a whistle similar to goldeneye.

VOCALIZATIONS: Uncommonly silent. Females make a low croak.

Spectacled Eider, *Somateria fischeri*
Myopic Eider

STATUS: Uncommon to locally common, but pelagic most of the year and geographically restricted. DISTRIBUTION: There are two Alaskan breeding populations (and one Siberian): the northern population found along the Arctic Ocean between Wainwright and 50 mi. shy of the Canadian border, and the population nesting on the west coast of Alaska. In winter, all birds concentrate in the marine waters of the Bering Sea south of St. Lawrence I. Post-breeding/molt staging areas are found in the waters of Ledyard Bay (nw. Alaska) and Norton Sound (south of the Seward Peninsula) and on the south side of St. Lawrence I. HABITAT: Spends most of its time at sea. Adult males spend about one month at breeding areas, and adult females and young spend three to four months; subadult birds remain at sea until attaining sexual maturity (two to three years). Lowland breeder, occupying tidal estuaries, brackish and freshwater wetlands, ponds, rivers, and sedge meadows, all within 12 mi. of seacoasts. COHABITANTS: Brant, Pomarine Jaeger, Emperor Goose (breeding), other eiders; in winter, precious little. Nothing seems hardy enough to survive in the icy and ice-locked conditions where this bird seems to thrive. **MOVEMENT/MIGRATION:** Spring migration from early May to early June. In fall, males leave breeding area and head for molting areas from late May to late June; females leave from mid-July to late September. Birds leave molting areas from late September to late November. VI: 0.

DESCRIPTION: An eider that looks like a coal miner who just removed his protective goggles. Small eider or medium-sized sea duck (larger than Steller's Eider and Black Scoter; slightly smaller than White-winged Scoter; actual dimensions notwithstanding, appears about two-thirds the size of Common Eider in direct comparison), with a large profile-dominating head, thin neck (when head is elevated), and long low body profile. Head seems less sloping and more roundly contoured than Common Eider; the orange bill, while evident, is less commanding.

In all plumages, Spectacled Eider shows a unique pale "spectacle" surrounding the eye. Adult males are conspicuously black and white, patterned somewhat like adult male Common Eider, with a white back, black side, and white flank patch, *but* Spectacled shows a black breast and no black on the crown. (Common shows a black cap and a bright white breast.) Except for the obvious white spectacles, the head on males seems dirty or dingy. The green tones shown in many illustrations are difficult to see at a distance, so the bird appears to have a dirty face, with the dirt lodged between the bill and goggles and behind the eye. This dirty-headed appearance is evident even at distances where the spectacles cannot be discerned.

First-winter males and adult males in eclipse have dark bodies. Females are uniformly pale brown, with conspicuous narrow barring encasing the entire body (much like Common Eider), but whereas Common Eider shows a narrow dark line through the eye, Spectacled shows conspicuous pale goggles. The expression is myopic.

BEHAVIOR: Social with its own kind; antisocial around other waterfowl, although individuals may associate loosely with other eiders. At sea, feeds by diving (often to great depths). The bird is more surface-oriented inshore, where it feeds by dabbling, upending, or diving. Stays mostly in the water, walking only short distances over land. Except for breeding birds, this species is most commonly observed when migrating birds pass onshore sea-watch points.

FLIGHT: Looks like a small eider with a longish neck and broad, pointed wings set well rearward. The pattern on adult males resembles Common Eider except for the uncapped head and extensively black underparts. (Spectacled is black right to the throat; Common shows a white breast or seems white-fronted.) Females are mostly brown with blackish wings, resembling small female Common or King Eider except for all-dark underwings (no white wing pit). The small size is eye-catching, and the circular goggle patch (showing as a pale face) is fairly evident.

VOCALIZATIONS: Uncommonly silent. Females reportedly make a low croaking noise.

King Eider, *Somateria spectabilis*
Smiling Eider

STATUS: Common coastal arctic breeder; in places a spectacularly abundant migrant. Common northern, coastal winter resident, but much less common than Common Eider along the New England and mid-Atlantic coasts. **DISTRIBUTION:** Breeds in arctic coastal portions of Alaska, Canada, and Greenland. In winter, found along the coast of Alaska from the Pribilof Is. in the s. Bering Sea to Yakutat (rare visitor south to s. Calif.); in the Atlantic, from Lab. south to Va. (rarely farther south). Also, a few have been found on the Great Lakes. **HABITAT:** Breeds in a variety of tundra habitats, from low marshy areas adjacent to ponds, to bogs and dry, barren hillsides. Unlike the coastally centrist Common and Spectacled Eiders, King Eider commonly nests inland where lakes and ponds abound. **COHABITANTS:** Breeds in association with Snow Goose, Ross's Goose, Arctic Tern. In winter, associates with Common and Steller's Eiders and scoter (particularly White-winged Scoter). **MOVEMENT/MIGRATION:** Late-fall and early-spring migrant. Spring migration from early April to late June; fall from mid-October to early January. VI: 3.

DESCRIPTION: A compact solid-looking sea duck; more ducklike and less gooselike than Common Eider. Males are flamboyantly adorned (making females grin). Medium-sized (larger than Steller's Eider; smaller than Common Eider; about the same size as White-winged Scoter), with a large roundish head (lacking the jutting noggin of Common), a prominent bill, a short thick neck, and a compact dome-backed body showing (on most birds) two short, projecting, shark-tooth-shaped feather tufts (lacking in Common Eider). On females and immatures, the tufts are either missing or reduced to bumps.

Breeding males, with their bold black-and-white upperparts contrasting with their tangerine-blushed breasts and heads whose bulging orange bill and colorful pattern would not seem out of place at a Mardi Gras parade, are unmistakable. In winter, now-dark-breasted males retain a semblance of the bulging bright orange bill; bills of immature males (showing white breasts) lack the bulge but are likewise orange.

Females (adult and immature) are all brown and warmer- or sandier-toned than female Common Eider (especially on the head and neck), with a contrasting all-black bill. On Common Eider, it's hard to tell where the bill ends and the face begins. On King, it's obvious. The head shape is also key. Common Eider has a long, sloping face with an abrupt upward jut on the noggin and an expression that seems glum or dour. King Eider has a fuller, rounded head, a bill that shows a slight bulge where it meets the face, and an expression that seems grinning, sly, or happy. Most birds show a pale "wear" patch near the base of the bill. King Eider's sides show a scalloped, not barred, pattern.

BEHAVIOR: Outside of the breeding season, very social; feeds in large flocks and mixes with Common Eider and scoters south of the most concentrated wintering areas. In winter, feeds primarily by diving—sometimes in densely packed groups, sometimes in smaller spaced groups or alone. When several individuals occur in flocks of Common Eider, they tend to cluster and forage slightly away from the group (perhaps in deeper water).

Generally tame.

FLIGHT: Overall stockier and shorter-necked than Common Eider. Adult males show less white above than Common Eider. Flight is direct, fast, and low to the water. Wingbeats are quicker and movements more agile than Common Eider—in particular, King Eider seems able to lift off the water with less effort. Flies in V-formations, in horizontally strung lines, or in bunched flocks (less commonly in tandem like Common Eider).

VOCALIZATIONS: Male's call is a dovelike "*croo croo crooo.*" Females reportedly make a "muttering growl" as well as a grunting sound when threatened.

PERTINENT PARTICULARS: Historically, along the Atlantic Coast, King Eider was the eider most likely to be seen south of the breeding range of Common Eider. This is no longer the case. Common is now the more common of the two. Adjust your expectations accordingly.

Common Eider, *Somateria mollissima*

Goose Eider

STATUS: Very common to abundant breeder and resident along rocky northern coasts. Uncommon coastal winter resident south of its breeding range.

DISTRIBUTION: Breeds from w. Alaska across n. Canada and Greenland south to Mass. and (rarely) R.I. and Long Island. Winters in w. Alaska and Greenland and from Nfld. and e. Que. south to Long Island, with much reduced numbers occurring in Va., s. Fla. (rarely), and the e. Great Lakes (very rarely). **HABITAT:** Almost wholly marine. Nests in large colonies on coastal islands and forages in shallow, near-shore water (about 30 ft. deep) along unsheltered rocky coasts. On sandy coastlines, most commonly found near man-made seawalls, jetties, and bridges. **COHABITANTS:** In winter and migration, particularly south of breeding areas, associates with other sea ducks, most notably scoters. **MOVEMENT/MIGRATION:** Birds nesting in arctic Canada and Alaska move south of the ice pack in winter. Spring migration from mid-March to late May; fall from September to late December. VI: 2.

DESCRIPTION: Big, robust, wide-bodied sea duck with a wedge-shaped face and gooselike properties. Large (only slightly smaller than Brant; larger than White-winged Scoter), with a long sloping Canvasback-like face, a jutting bump on the forehead, a long but thick neck, and a long, broad-beamed, and somewhat humpbacked body. The bill is often angled demurely down. The tail is most commonly flush with the surface.

Black-sided white-backed adult males are unmistakable. (Even in eclipse, the back remains white.) Black caps make them appear eyeless. The bright yellow or lime green bill augmented by a lime green blush across the nape is also distinctive.

Adult females have grayish brown bodies with narrow black barring (which makes them look striped) and pale-tipped gray bills. Immatures are like females but overall warmer-toned, more finely barred, and showing only a slightly paler tip to the bill.

Note: Because the feathers at the base of the bill are as dark as the rest of the face on females and immatures, it is difficult to see the bill for the face. On female King Eider, the base of the bill shows a pale or worn spot on the side of the face, so the dark bill stands out. Also, on Common Eider, the slight forehead juts abruptly above the humongous bill. On King Eider, the bump on the bill is closer to midspan, shy of the forehead.

Because of the straight gape, Common Eider's look is aloof or bored, as well as dour or glum. King Eider, with an upturned culmen, looks sly or grinning.

BEHAVIOR: Gregarious. At all times of year, often found in tight-packed flocks; in winter, flocks may exceed 1,000 birds. Almost always found in salt water, often close to a rocky shoreline, where it dives for shellfish. Forages most often with other Common Eiders, but south of main wintering concentrations is usually found among flocks of scoter around jetties and seawalls.

Fairly tame; usually swims if pressed, but often returns if observers remain passive. In winter, rarely hauls out on land.

FLIGHT: Big, overall blocky, with a fairly long neck and short broad wings (recalls a compact goose or Muscovy Duck). Adult males look white in front, black behind. Larger females and immatures are obviously larger than scoter and somewhat peak-headed. (King Eider has a more rounded head.) Heavy, choppy, and struggling wingbeats are given with a clumsy up-and-down motion.

VOCALIZATIONS: Alarm call is a hoarse croaking that has been described as "*korr korr korr*"; also emits a trilled growl. Males make an eerie cooing call that recalls the amplified call of Rock Pigeon, "*w'hooo.*"

PERTINENT PARTICULARS: Several decades ago, King Eider was the expected wintering eider south of New England. Sightings of Common Eider were considerably fewer. This has changed, and Common Eider now lives up to its name, outnumbering King Eider along the mid-Atlantic coast.

Harlequin Duck, *Histrionicus histrionicus*

Beauty in the Rough

STATUS: Uncommon in the Northeast; more widespread in the Northwest; fairly restricted in its breeding and winter range, which is divided into Atlantic and Pacific populations. **DISTRIBUTION:** In e. North America, breeds close to freshwater rivers and streams in coastal Lab. west to Hudson Bay as well as in n. Nfld. and the Gaspe Peninsula. Winters in coastal waters from Nfld. to the mouth of Chesapeake Bay. In w. North America, breeds from the Seward Peninsula and southwest coast of Alaska to the Northwest Territories and south through B.C. and sw. Alta. into Wash., n. Ore., Idaho, Mont., and Wyo. Winters coastally from the Aleutians south to n. Calif. Also breeds in Siberia, Greenland, and Iceland. **HABITAT:** During the breeding season, breeds in fast-moving freshwater rivers and streams, which commonly cut through forests, right to the seacoasts. Also nests in treeless tundra or coastal habitat. Winters along rocky coasts, generally very close to shore. In the absence of natural rocky coast, accepts man-made jetties or seawalls. **COHABITANTS:** In winter, may be found with Common Eider, scoters, Long-tailed Duck, Bufflehead. **MOVEMENT/MIGRATION:** Migration is direct and brief, from breeding to wintering areas (which for many birds is close). In spring, both eastern and western populations migrate primarily April into May; fall migration, on the East Coast, is late September into November. In the West, fall migration is extremely protracted, with males first appearing in coastal wintering areas in early June and some females and immatures not reaching coastal areas until October. VI: 3.

DESCRIPTION: Small, colorful, and oddly configured duck (smaller than Black Scoter; about the same size as Long-tailed Duck). Overall compact, with a short, wide, and basically oval-shaped body. The backs of adult males are symmetrically domed; the bodies of females are more wedge-shaped, riding high in the front and low in the stern. (Both male and female birds appear front-heavy.) The oval-shaped head is flat on the forehead, round on top, and bulging behind—a configuration that in no small part can be attributed to Harlequin Duck's distinctly thick swollen neck. (The bird seems to be wearing a wig or a mane.) The oversized head/neck fusion makes the already small bill look cute and petite. The long tail is acutely pointed with ragged sides. Usually up-cocked and angled out of the water, it recalls Ruddy Duck.

Seen well and in good light, the plumage of adult males is unmistakable—a duck painted in the pattern and style of Pablo Picasso. Under the alchemy of distance, however, colors and patterns meld into a dull, eye-defeatingly uniform, cold brown (like females and immatures). The faces of females and immatures have a shadow pattern of the adult male—smudgy white at the base of the bill and a very distinctive, round or oval white spot over the ear. By comparison, female and immature Surf Scoters are overall long (not compact), with whopping big bills and smudgy white patches about the face. Female and immature Long-tailed Ducks are compact, like Harlequin, but have white sides (dark on Harlequin). Female and immature Buffleheads are blackish, not brown, and lack the white pattern at the base of the bill. Female Ruddy Duck has all-white cheeks, an unmistakably large bill, and a distinctly humped back.

BEHAVIOR: Tame, confiding, social (except at the southern limits of their winter range, where singles often occur). Allows approach to within a dozen feet. Commonly found in small tight-packed groups of 10 to 50. (In the Pacific, rafts numbering several hundred birds are possible.) Lone birds associate with other coastal birds, including Long-tailed Duck, Bufflehead, and scoter. Harlequin is a rugged duck that likes rough water, feeding in the turbulence of streams and in wave zones on rocky coasts. In summer, however, molting birds often seek out calm waters. Harlequin submerges with a quick, open-winged, upending dive. Leans forward when it swims. When not feeding, often hauls out onto rocks and gravel bars to rest. Moves with animated agility on land, often hopping rock to rock.

FLIGHT: Flight is rapid and low over the water, with quick, stiff, regular wingbeats. Flies angled up with head slightly raised. Profile seems small, blunt-headed (because of the width of the neck), slender-winged, and long-tailed. Recalls a slightly more compact Wood Duck. Adult males appear very ruddy to the rear. Flocks fly in tight shifting groups or horizontal lines. Landings are reckless, with no wing braking; after touching down at a shallow glide angle, the bird plows into the water on its belly and glides to a halt.

VOCALIZATIONS: Most commonly heard sound is a squeaky two-noted "*Yee-hee*" that is frequently repeated and often expressed when birds are anxious or nervous. Sound recalls a child's squeak toy and is the source of one of the bird's nicknames, "Sea Mouse." Also emits a soft murmured burp, "*whr*" or "*wh wh whr.*"

Surf Scoter, *Melanitta perspicillata*
The Skunkhead

STATUS: Common and widespread, but patchy nester across arctic and subarctic regions; in winter, in most coastal locations (found on all coasts but less common in the Gulf of Mexico), the most common of the three scoter species. **DISTRIBUTION:** Nests across n. Canada from Lab. to Alaska, although actual distribution and boundaries are somewhat speculative. Winters in shallow coastal waters: in the Atlantic, from N.S. south to cen. Fla.; on the Gulf, from e. Tex. to cen. Fla. (scarce); in the Pacific, from sw. Alaska to cen. Baja. Also occurs in zebra mussel–infested Great Lakes; scarce but widely scattered inland. **HABITAT:** Nests along small shallow lakes, often with rocky shores and little vegetation. In winter, occurs in shallow coastal water, usually within a mile of shore; also found in inlets and bays. Forages over rocky, pebbly, or sandy coast. **COHABITANTS:** In winter and migration, often found with other scoters, particularly Black Scoter. **MOVEMENT/MIGRATION:** Spring migration from mid-March to late May; fall from September to early December. Coastal movements are usually conducted offshore but within sight of land. Also migrates overland (presumably at high altitudes and at night), since some birds are found on interior lakes. VI: 3.

DESCRIPTION: Somewhat ungainly sea duck, distinguished by its robust, histrionic, doorstop-shaped bill. Medium-sized (slightly larger than Black Scoter; smaller than White-winged Scoter), with a bulbous bill that dominates the silhouette, making the bird seem very front-heavy and very long in the face. The orange-and-white bill pattern of adult males, coupled with the white forehead and white nape, renders the bird unmistakable at rest, on the water, or in flight at almost any distance. Females and immatures are pale brown with two whitish patches on the face—one behind the bill, one over the ear (a pattern shared with White-winged Scoter). At close range, the two spots are distinct; at a distance, female and immature faces look dark or dingy—in both cases, distinct from the pale face of female and immature Black Scoter. White-winged Scoter also has a substantial bill that is more upturned and less triangular and swollen than Surf Scoter. If undecided, wait. The bright white secondaries of White-winged Scoter are almost certain to peek through at some point. Also, many adult females and late-winter immature males show vestiges of the "skunkhead" white nape of breeding male Surfs.

BEHAVIOR: Very social. Feeds and rests in large fairly tight flocks. Rests on the water with bill angled slightly down. Dives by lunging forward with wings partially open. When exercising its wings, keeps the neck pointed forward, not angled down (like Black Scoter).

FLIGHT: Overall leaner, rangier, and pointier than Black Scoter, with a distinctly longer face/neck that makes the bird seem comparatively front-heavy. Underwings are uniformly dark, not as contrasting. Quick regular wingbeats appear angled back. Flies in long strings (sometimes wedges), generally just above the surface (as is typical of scoters). Strings shift and clump and string out again.

VOCALIZATIONS: Usually silent.

White-winged Scoter, *Melanitta fusca*

White-wedged Scoter

STATUS: Fairly common breeder in western Canada and Alaska; in winter, shows varied abundance in coastal waters; in more northerly areas, this is the dominant scoter species. **DISTRIBUTION:** Breeds largely in interior portions of Alaska, the Yukon, w. and s. Northwest Territories, n. and e. B.C., most of Alta., all but s. Sask. and Man., and nw. Ont. Winters along both seacoasts in bays and inshore waters, from Nfld. to the Carolinas (casual south to Fla. and in the Gulf of Mexico) and from the Aleutians to Baja (rare in s. Calif.). Also winters on the zebra mussel–infested s. Great Lakes. **HABITAT:** Nests on large freshwater lakes, ponds, rivers, and streams in northern forests; in coastal areas, also nests on brackish water. Generally not a tundra breeder. In winter, concentrates in shallow bays and inlets (also estuaries and shallow open coastal waters) over mollusk-bearing sandy or gravel bottoms. **COHABITANTS:** Mixes with other scoters in winter and in migration. **MOVEMENT/MIGRATION:** Because White-winged's migration routes are largely over land, over much of North America this is the scoter most likely to appear on inland lakes and rivers. Spring migration is from

March into May, with some nonbreeding birds lingering all summer; fall from September to January, with peak from October to late November. VI: 3.

DESCRIPTION: Large, dark, sleek sea duck and the largest of the three scoters (almost eider-sized, and somewhat eiderlike, owing to the sloping profile of the face). The body is long, with a largish wedge-shaped head, though the slope-billed profile diminishes somewhat at a distance. On the water, the neck seems particularly thick and stocky. The bill is heavily feathered at the base.

Males are overall black; females and young are rich blackish brown (warmer-toned than the other scoters), with one or two paler scuff patches on the face (similar to Surf Scoter, but on Surf Scoter the patch at the base of the bill forms a vertical slash; on White-winged, it is an oval). No matter what the plumage, the wings sport a large white patch on the inner trailing edge that is impossible to miss when the bird is in flight and that often peeks through as a narrow white wedge above the flank when the bird is sitting on the water. Adult males also have a white teardrop etched over the eye. The tip of the bill is orangy pink.

BEHAVIOR: Dives primarily for shellfish over sandy shoals or gravel, generally within sight of shore and beyond the wave zone. Gregarious; often gathers in tremendous flocks in more northern wintering areas; sometimes mixes with other scoter in southern portions of the wintering range and during migration. Exercise flaps on water are conducted with neck stretched and head raised.

FLIGHT: Profile is longer than other scoters, with a distinctly longer tail and, especially, a longer neck. The heavy-headed configuration of the perched bird notwithstanding, in flight White-winged's head seems slimmer, the distinction between the neck and head not as pronounced; the head shape seems to fall between the blunt-headed petite-billed Black Scoter and the long, sloping, wedge-shaped face of Surf Scoter. Wings are longer and appear narrower.

Unless the bird is severely backlit, none of this matters. The white wing patches are obvious at almost any imaginable distance. In a mixed scoter flock, White-winged appears manifestly larger.

Flight is like other scoters—direct and fast, with rapid and continuous wingbeats that are deeper than the other two scoter species (and more eiderlike). At a distance, you cannot see wings raised above the body on Black or Surf Scoter, but you can on White-winged. Like other scoters, White-wingeds fly low and single file but are less crowded and more widely spaced. (White-wingeds keep a body length behind the bird ahead, whereas other scoters fly nearly beak to tail.)

VOCALIZATIONS: Generally silent.

PERTINENT PARTICULARS: In flight, White-wingeds often betray themselves by staying farther behind the bird ahead than other members of the flock. Look for the break in the line. See whether the bird astern is larger and flashing white wing patches.

Black Scoter, *Melanitta nigra*

The Easy Scoter

STATUS: Common but isolated and restricted arctic breeder; common to uncommon wintering species along all three coasts. **DISTRIBUTION:** Eastern population breeds primarily east of Hudson Bay in Que. and Lab. Western population mantles the west coast and northern and interior portions of Alaska east to the northwestern corner of the Northwest Territories. Winters primarily from the Maritimes to n. Fla. and from the Aleutians to n. Baja (rare south of Calif.). Some birds also winter on the e. Great Lakes and the Gulf Coast from the Florida Panhandle to e. Tex. **HABITAT:** Breeds on fairly shallow freshwater interior lakes. In winter, primarily coastal, preferring shallow inshore waters within sight of land. Found along rocky, cobble, or sandy coastlines. **COHABITANTS:** In winter and migration, mixes with other scoter, most commonly Surf Scoter. **MOVEMENT/MIGRATION:** In migration, generally stays within a mile of coastlines. Individuals or small groups occasionally occur on interior lakes and reservoirs. Spring migration from mid-March to late May; fall from late September to mid-December. Nonbreeding summering birds are sometimes encountered along the coasts. VI: 3.

DESCRIPTION: Smallish sea duck with a compact body and a round head. Smallest of the scoters (larger than Common Goldeneye; smaller than Red-breasted Merganser), Black Scoter has a modest fairly narrow bill, a very round head, a short heavy neck, an average body, and an often up-cocked tail that add up to a more puddle duck–like appearance than the other two scoters.

Breeding males are overall black with an orange "golf ball" balanced on the base of the bill. On females and immatures, a pale face and well-defined dark cap stand out against the otherwise uniform, pale brown body. This pattern, combined with the bird's overall compactness and tendency to hold its tail up-cocked when on the water, recalls female Ruddy Duck.

BEHAVIOR: A social duck, Black Scoter is often found in small to medium-sized flocks that during migration may number up to several hundred birds. In migration, often seen with Surf Scoter. In winter, sometimes forages with Harlequin and Long-tailed Ducks. Feeding flocks are usually small (fewer than 50 birds) and compact. Feeds by diving in fairly shallow water (less than 30 ft. deep). Exercising wing flaps on the water are conducted by raising its body high and stretching its neck down, like a stiff, staged bow. (Surf and White-winged Scoters hold their necks straight.)

FLIGHT: Shape is overall plump and compact, but proportionally balanced. Face seems blunt. Body is pudgy or potbellied. Wings are a trifle shorter and blunter-tipped than Surf Scoter. Wingbeats are rapid and steady, and the flap is more vertical, not angled back. Paler flight feathers contrast with darker wing linings on the undersurface of the wings (looks like tarnished silver), but in poor light and at a distance this is difficult to see.

The adult male's orange-based bill and the white cheek on females and immatures vault the distance. Otherwise, use the blunt face and vertical wingbeat for distant identifications.

Black Scoter rises from the surface quicker and with less running takeoff than other scoters. Like all scoters, flies in long linear flocks that usually hug the surface and often coalesce into globular flocks before thinning out again. When flying higher (more than 100 ft.) over the water, a sloppy globular flock (with a trailing string of birds) is the norm.

VOCALIZATIONS: Usually silent.

PERTINENT PARTICULARS: At a distance, arguably the easiest scoter to identify—Black is the one that doesn't have a head shaped like a doorstop. The bright orange "golf ball" at the base of the adult (even second-year) male's bill is evident at great distances. The female's contrastingly paler face also stands out at distances, whereas the smaller, pale patches on the faces of female and immature White-winged and Surf Scoters disappear.

Long-tailed Duck, *Clangula hyemalis*

The Designer Duck

STATUS: Common arctic and subarctic breeder; common wintering species along northern seacoasts. **DISTRIBUTION:** Circumpolar breeder, found in North America from coastal Lab. to Hudson Bay and across the Northwest Territories (above the tree line); also in coastal areas in the Yukon and all but the southwest coast of Alaska. Winters in the Atlantic from Nfld. to S.C. (uncommon to Fla. and in the Gulf of Mexico; common in the Great Lakes), and in the Pacific from the Aleutians to cen. Calif. **HABITAT:** Breeds in freshwater tundra ponds, streams, and wetlands. After breeding, moves to deeper freshwater ponds. In winter, found in coastal marine waters and on large lakes. **COHABITANTS:** In summer, often found with Arctic Tern. In winter and migration, sometimes associates with other sea ducks, particularly scoter. **MOVEMENT/MIGRATION:** Migrates very early in spring, from late February to May, and very late in fall, from October to December. VI: 3.

DESCRIPTION: A dualistic duck. Males are showy and elegantly rakish, dressed to go out on the town—a designer duck. Females are stay-at-home frumpy overall (but not hard to identify). In plumage and silhouette, male and female look like different species.

Both sexes are small-bodied, with a large head, a short thick neck, and a low profile in the water. Heads of males are a large oval or almond shape; heads of females are equally large but more rectangular. Both have short bills. At all seasons, the female's is dark, and the male's has a bright pink saddle over the middle. The male's signature and namesake characteristic is the long, flagellum-like tail that often curls up out of the water (or may lie flush). Despite its compact body, the tail makes the male look elegantly long.

Breeding plumage males and females have fundamentally similar plumage. Both have mostly dark blackish or brownish black upperparts, dark breasts, and white bellies. Both have dark heads with large pale goggles over the eyes. Only the male has the rakishly long tail (actually elongated tail coverts). In winter, both birds become paler and more patterned—males lavishly so. Winter males are unmistakable. (Any illustration is worth 1,000 words.) Females are brown above with brownish breasts but otherwise whitish below—a description that does not rule out any one of several sea ducks. Happily, female Long-tailed Ducks have white faces with a dark bruise on the cheek. (Other female sea ducks have dark faces with white spots.) More than head pattern, the overall small, squat, unflattering squareness of the bird distinguishes it. Immatures are like females.

BEHAVIOR: Very social. Frequently found in small groups (often segregated by sex) or large flocks numbering in the hundreds. Seems disinclined to mix with other sea ducks. Occasionally found many miles from shore, but is also common in back bays and coastal waters. Often dives just beyond (even inside) the breakers on calm days. Courtship is spirited and frenzied (see "Flight"). Not particularly shy, but except in the breeding season or unless sick or injured, almost never found on land.

FLIGHT: Here again, the male's silhouette is unmistakable. Flies with head raised, plump-bellied body inclined (as though climbing instead of flying straight), and flagellum-like tail trailing behind. Wings are short but slender and curved back—increasing the graceful, almost greyhoundlike lines of the bird. The female's overall small, squat, squarish shape looks odd, whereas other small, fairly nondescript ducks look, at the very least, like the parts fit. The wings are short and appear blunter and wider than the wings of males.

Both males and females have black upper- and underwings that contrast with the paler (or patterned) back and very white underparts.

Flight is rapid, energetic, and twisting; the bird first lists to one side, then the other. Recalls the flight of a large alcid (particularly puffin). Wingbeats are constant and hurried on conspicuously down-drooped wings, giving the flight a fluttery or floppy sense. Flies in tight, compact, often shifting groups, low over the water. Lands with several braking wingbeats and then an ungraceful plopping crash to the surface.

VOCALIZATIONS: Noisy. The male's call, frequently heard in late winter, is a comical braying chant—as if a duck were singing opera: "*ah, ahna-lee*," or "*ah, ah, ahna-lee.*" May be repeated several times. Females make a low atonal grunt or quack, "*urk, urk urk.*"

Bufflehead, *Bucephala albeola*
Duckling in Black and White

STATUS: Common northern breeder; common and widespread winter resident. **DISTRIBUTION:** Breeds across most of boreal Canada and Alaska from w. Que. to B.C. and north into the Northwest Territories, s. Yukon, and interior Alaska. Southern nesting range extends into Minn., N.D., nw. Mont., and n. Wash. Also nests at higher elevations in e. Ore., n. Calif., nw. Wyo., and n.-cen. Colo. In winter, vacates breeding areas and reapportions itself along coastal areas in much of the U.S. and bordering Mexico. Coastally, it is found from N.S. south to n. Fla. and from the south coast of Alaska to n. Baja. Across interior portions of the U.S., it is absent only from New England to n. Ga. and across the n. Great Plains (north of Kans.) and the n. Rockies (ne. Wash., n. Idaho, and nw. Mont.). **HABITAT:** Nests along permanent freshwater ponds and small lakes in boreal forest and aspen parkland. In winter, largely coastal, seeking out sheltered bays, coves, and estuaries and avoiding open ocean. Also found in inland lakes, reservoirs, impoundments, and larger slow-moving rivers. **COHABITANTS:** In summer, Northern Flicker! (The primary excavator of Bufflehead nest cavities.) Also goldeneyes. In winter, occasionally Ruddy Duck, as well as other diving ducks. **MOVEMENT/ MIGRATION:** Spring migration from mid-February to late May, with peak in late March and April; fall from early October to mid-December. VI: 2.

DESCRIPTION: Tiny, compact, black-and-white duck shaped like a duckling. Small (smaller than Ruddy Duck; larger than Eared Grebe; about the same size as Pied-billed or Horned Grebe), with a disproportionately large, round head, a small, compact bill, and a plump body.

Where real ducklings are cryptically colored, the drake bufflehead, with its contrasting black-and-white pattern, is arresting. The identification centerpiece of the bird is its head, which is black with a white patch or "kerchief" over the nape or crown. At a distance, when Bufflehead has its head couched, the entire head appears white! (with a dark bordering face). The only bird that a male Bufflehead might be mistaken for is the slender, low-profiled, sliver-billed male Hooded Merganser. Drake Bufflehead hardly resembles this bird, however, except for the shared white on the head.

The plumage of the tiny, toylike female and immatures is overall dingy or dark, except for a distinct white ear patch. Distinct it is, but foolproof it is not. When female Bufflehead sits low in the water (as is often the case), the white ear patch and the head that is not quite so round and prominent as the male's make this bird a candidate for confusion with a distant, male Hooded Merganser with hood closed (or a female Ruddy Duck).

BEHAVIOR: The bird is animated, almost comical, particularly in late winter and early spring during courtship—a Keystone Cops routine on water. The norm is small groups, mostly loose and widely spaced, sometimes clustered (never tightly packed), always unregimented, with birds facing, submerging, and emerging in every direction. Dives are frequent and short. At a distance, feeding flocks suggest widely spaced chunks of ice bobbing in the chop—white chunks males, dirty chunks females. In late winter, birds may gather into large flocks of several hundred birds that are commonly subdivided into smaller groups. Bufflehead does not generally associate with other diving ducks.

The birds are moderately shy; if pressed, they usually choose to fly from observers instead of swim.

FLIGHT: Shows a small, compact, large-headed, and somewhat potbellied profile. Flies with the body angled up, head elevated. The white aviator's cap is eye-catching, but the bird's dramatic black-and-white pattern is the only field mark you need to see. The bird's bright pink feet are usually evident. Females and immatures are overall dark and dingy but show a tiny telltale white patch on the trailing edge of the inner wing that is whiter than that shown by female Hooded Merganser, another small, dark, fast-flying duck.

Takeoffs are short and more abbreviated than you'll see with any other diving duck. Landing with the body erect, the head up, and feet down, Bufflehead stops on a dime. The flight is fast and direct; wingbeats are extremely rapid (almost as fast as Hooded Merganser), with wingtips reaching well down. Wingbeats are audible, but not as shrill and not as far-reaching as Common Goldeneye. Small groups may fly in a loose string or cluster, usually close to the water.

VOCALIZATIONS: Mostly silent. Females produce a single, soft, questioning "*huh?*" that is sometimes given in sequence: "*huh? huh? huh? huh? huh?*" Males sometimes give a low soft bugle.

Common Goldeneye, *Bucephala clangula*

The Cub Head

STATUS: Common and widespread northern breeder. Common and widespread winter resident across most of the northern United States and southern Canada. **DISTRIBUTION:** Breeds from Nfld. to Maine and west to Alaska and B.C.; southern limits include the Adirondacks, n. Mich., Minn., Mont., and Idaho. Winters along all three coasts: in the Atlantic, from Nfld. to n. Fla.; along the Gulf of Mexico, from s. Fla. to n. Mexico; in the Pacific, from sw. Alaska to Baja. Also winters on large, open rivers, lakes, and reservoirs across s. Canada and most of the U.S., excluding much of the southeastern coastal plain, s. and w. Tex., and s. Ariz. **HABITAT:** Nests beside freshwater wetlands, lakes, and rivers flanked by mature forest. In winter, prefers shallow coastal bays, estuaries, and inlets and shuns deeper open coastline. In the interior, found on larger lakes and free-flowing rivers that are at least partially free of ice. **COHABITANTS:** Bufflehead, Barrow's Goldeneye, Common Merganser. In winter, does not commonly associate with other diving ducks except Barrow's Goldeneye. **MOVEMENT/MIGRATION:** Total population migrates just in advance of and on the heels of freezing temperatures. In spring, advances as early as late February and arrives at northern breeding sites in late April or early May. In fall, some birds reach southern wintering grounds as early as October, but peak movements often do not occur until November into early December. VI: 2.

DESCRIPTION: Sturdy diving duck with a peaked head and sloping profile for a bill. Medium-sized (larger than Ring-necked Duck; smaller than Black Scoter; about the same size as Greater Scaup and Barrow's Goldeneye), with a large distinctively shaped head, a short thick neck, and a low profile. The triangular or conically domed head peaks just bill-side of the center with a sloping forehead and curves, shoehorn fashion, into a fairly large, wedge-shaped bill. Barrow's Goldeneye has a domed head with a swollen nape and a near-vertical forehead that meets the bird's more petite, conical bill almost at a right angle. *Note:* When diving or feeding, Common Goldeneye often flattens its head, making the crown appear curved, not peaked; the overall head shape, however, is compressed, never swollen. Also, feeding ducks sit low on the water and thus have a low profile; resting birds sit higher.

Adult males appear all white with a black head. (The narrow black back disappears with distance.) In reasonable light, the green gloss of the head is easily seen. At almost any distance, the large, oval, white spot at the base of the bill is apparent. Females and immatures, on the water with head couched, are overall gray with all-dark heads (brown in good light). Heads raised, the necks are white (and the contrast with the head greater). The yellow on the bill of breeding females (seen in winter) is limited to the tip or across the middle. Often a hint of the white winks through the folded wing and appears as a white dash just above the flank. Also, the "golden" eye is apparent at surprising distances.

BEHAVIOR: A fairly gregarious but not particularly social diving duck. In winter, found in small flocks (3–50 birds) that do not mix with other divers (except Barrow's Goldeneye). In open water, birds may be spread out; in rivers, flocks tend to be more cohesive. Very shy: often rests and feeds well away from the shoreline, and does not tolerate close approach. Flushes with a thrashing, noisy, but short and foot-pattering takeoff.

Feeds by diving into the tide and the current and letting the water carry it back for another plunge. Flocks commonly dive and emerge together. Not commonly seen on land. Despite being a cavity-nesting species, does not commonly perch on limbs or branches.

FLIGHT: In shape, resembles a black-and-white (or dark-and-white) bowling pin on wings. The large head and thick, sturdy neck/breast project well ahead of the wings. The body angles slightly up. Flight is rapid, direct, and fast, with quick, steady, and somewhat choppy, effortful wingbeats. Flocks are tight. Flushed birds often circle back, passing close enough that you can hear the very audible whistling sound of the wings (sounds somewhat like a winnowing snipe).

VOCALIZATIONS: In winter, when flushed, females often make a short, low, huffing pant, "*arrh, arrh.*"

Barrow's Goldeneye, *Bucephala islandica*

The Duck That Swims into the Wind

STATUS: Common Rocky Mountain breeder and West Coast winter resident (but less common than

Common Goldeneye); uncommon and restricted in the Northeast. **DISTRIBUTION:** Breeds from s.-cen. Alaska to Kodiak I. and south through the interior across s. Yukon, e. B.C., sw. Alta., cen. Wash., cen. Ore., limited portions of Idaho, w. Mont., nw. Wyo., and nw. Colo.; winters coastally from Kodiak I. south to n. Wash.; less commonly to n. Calif. and in the interior north to s. B.C. and south to n. N.M., Utah, s. Idaho, and n. Calif. In the East, breeds west of the St. Lawrence R. in Que.; winters coastally from Nfld. south to Long Island (where it is rare); occurs less commonly on the St. Lawrence and on the Great Lakes. **HABITAT:** Freshwater and alkaline lakes in forested or parklike habitat; also beaver ponds, sloughs, and alpine and subalpine lakes. Prefers lakes that are mostly unvegetated, with trees surrounding a shallow, sloping shoreline, and not necessarily deep. Winters primarily in coastal water, including bays, inlets, and open coasts (often near a freshwater outlet). Also large, ice-free rivers and lakes. **COHABITANTS:** Red-necked Grebe, Trumpeter Swan, Common Merganser; in winter, associates with Common Goldeneye and scoters. **MOVEMENT/MIGRATION:** Spring migration from late February to late May; fall from early October to mid-December. VI: 3.

DESCRIPTION: Wedge-billed dark-headed diving duck whose flat-fronted, billowing, naped head shape makes it appear to lean forward as it swims. Medium-sized (larger than Bufflehead; smaller than Common Merganser; about the same size as Common Goldeneye), and shaped much like Common Goldeneye but with a smaller bill and a rounder, less crowned or peaked, oval-shaped head. Males in particular show a vertical forehead and full-maned appearance. Heads of females are rounder but still fuller and less peaked than female Common Goldeneye (a difference most apparent in direct comparison).

Black above and white below, adult males, when seen with Common Goldeneye, show overall less white. The white crescent on the face and the dark, crooked spur extending onto the bird's breast are evident at considerable distance. The purple cast to the head can be seen in flashes in good light. The head of female Barrow's is slightly darker and richer brown than Common (but direct comparison is helpful). In winter and spring, bills of females are more extensively (or wholly) rich orange-yellow. In winter, the bill of female Common Goldeneye shows a pale, yellowish tip or band. Immatures are like females, but show no yellow on the bill.

BEHAVIOR: Social (outside of the breeding season), but found in smaller flocks than Common Goldeneye. Feeds by diving in fairly shallow water, often (but not always) closer to shore and in shallower water than Common Goldeneye. In winter, prefers rocky coastlines (where mussels abound) and rivers. Like most diving ducks, does not usually leave the water, except to fly. Associates with Common Goldeneye and scoters.

Fairly timid, this bird flushes easily and takes off with an effortful running patter. Wings whistle in flight.

FLIGHT: In profile and manner of flight, like Common Goldeneye. Adult males show slightly less white above.

VOCALIZATIONS: A very quiet duck.

Hooded Merganser, *Lophodytes cucullatus*

A Dark Dart with Wings That Sizzle

STATUS: Uncommon but fairly widespread breeder, resident, and wintering species. **DISTRIBUTION:** Breeds across most of the e. U.S. and s. Canada, south of a line drawn from P.E.I. to e.-cen. Sask. and east of a line drawn from cen. N.D. to the Tex.–La. border; does not breed along the Gulf of Mexico or on the Florida Peninsula. A permanent resident in the Pacific Northwest across s. B.C., sw. Alta., most of Wash., n. Idaho, w. Mont., and w. and ne. Ore. In winter, eastern birds vacate most of Canada (except se. Ont.), n. Maine, n. Mich., n. Wisc., Minn., the Dakotas, w. Iowa, and n. Mo. and expand into Kans. and Okla., across most of Tex. and Fla., and along the Gulf Coast. In the West, wintering birds expand north along the southeast coast of Alaska and south across Ore., much of Calif., s. Idaho, nw. Utah, s. Nev., and w. and s.

Ariz. **HABITAT:** Nests in forested wetlands, swamps, beaver ponds, rivers, and creeks; also in man-made nest boxes in unforested marshes. In winter, birds are most heavily concentrated in shallow, brackish wetlands, tidal creeks, and ponds (less commonly in salt water) and avoid large, open bodies of water. Also occur in freshwater ponds, sluggish rivers, creeks, sloughs, and wetlands. **COHABITANTS:** Breeds with Wood Duck. In winter, usually keeps its own company, but may sometimes mix with other diving ducks. **MOVEMENT/MIGRATION:** Fall migration, in October and November, is late (or just ahead of advancing ice). In spring, birds begin leaving wintering areas in February, but individuals may linger into April. Arrival in breeding territories occurs almost as soon as frozen surfaces thaw (or even partially thaw). VI: 2.

DESCRIPTION: Long, slender, low-profiled dart of a duck with a histrionic headpiece. Medium-sized but petite (larger than Bufflehead; smaller than Common Goldeneye; about the same size as Wood Duck; much smaller than other mergansers), with a long, thin, popsicle-stick bill, an oversized and distinctly adorned head, and a low-profiled body that often sits so low in the water that it almost seems submerged.

The boldly patterned drake, adorned with a crest that opens and closes like a Chinese fan, is hard to overlook. With the hood closed, the head is anvil-shaped and adorned with a peanut-shaped white patch behind the eye. With the hood open, the head bulges into an oval fan with a pure white center and a black border that sets off a golden eye and a thin black bill. Distinctive plumage notwithstanding, at almost any distance, the color combination alone—black, white, and rusty brown sides—distinguishes male Hooded Merganser among North American waterfowl. Females and immatures are dark gray and drab overall, except for one eye-catching characteristic—a brownish wedge-shaped crest gilded with a buffy border. In poor light, or when folded, the border is merely pale-edged; when backlit, it glows with a translucent light that is unique among waterfowl—recalling the haloed bust of Nefertiti or, for old-horror-movie buffs, the hair of the Bride of Frankenstein.

When feeding or swimming strongly, Hooded Merganser keeps the tail submerged. On sleeping birds, and sometimes on swimming ones, the longish tail is raised at an angle.

BEHAVIOR: Extremely wary and not gregarious (singles and pairs are the rule and groups over a dozen the exception, except at roost sites), the birds flush and usually depart when surprised; they swim behind vegetation if approached. Hooded Merganser does not generally socialize with other ducks. Feeds by diving for fish, but also shovel-feeds in the shallows. Hooded is more comfortable on land than other mergansers; individuals may be seen standing on stumps or branches or hauled out onto banks (but remaining close to the water).

FLIGHT: A small sliver-winged dart of a bird. Profile is long, lean, and particularly long-tailed, showing a distinctively humpbacked (or droop-headed) silhouette that easily distinguishes it from other small ducks (especially Green-winged Teal and Bufflehead). Birds lifting off the water maneuver into a horizontal line that climbs slowly and evenly and sometimes doubles back, offering a second look. Flocks are tight, lenticular-shaped, and highly maneuverable. Flight is direct and very fast. Steady wingbeats are so rapid they *blur.* Wings produce a high-pitched trill.

VOCALIZATIONS: Usually silent (except at evening roost).

PERTINENT PARTICULARS: Green-winged Teal and Hooded Merganser occupy similar habitats in winter and are superficially similar in flight. Both are small, both have a rapid wingbeat, and both flock. Both are dark above and light below. Teal is compact and has a heads-up or angled-up profile. Hooded Merganser is overall slim and holds its body perfectly horizontal. Its wingbeat is also faster.

The best way to tell distant flocks apart is by behavior. Hooded Mergansers are regimented and orderly. Individual birds hold their position, and flocks maintain their shape. Green-winged Teals, often in large groups and constantly shifting places, are an undisciplined mob.

Common Merganser, *Mergus merganser*

Freshwater Merganser

STATUS: Common and widespread northern and western breeder; common and widespread winter resident. **DISTRIBUTION:** Breeds across forested Canada from Nfld. to coastal B.C., s.-cen. Alaska, and the Aleutians. In the e. U.S., breeds from n. Pa. and n. N.J. to Maine; also in n. Mich., Wisc., and Minn. In the w. U.S., a year-round resident in Ore., Wash., Idaho, most of Mont., n. Calif., nw. Nev., Wyo., w. Colo., parts of Utah, Ariz., and N.M. In winter, vacates northern and interior portions of its Canadian and Alaskan breeding range but remains in coastal areas. Winter range encompasses s. Canada, parts of n. Mexico, and most of the U.S., except the n. Great Plains, the Southeast from the interior of Va. to w. Tex., and sw. Ariz. **HABITAT:** In summer, common along northern woodland lakes, rivers, and creeks; prefers clear fresh water. In winter, large homogeneous flocks gather on deep, unfrozen freshwater lakes, reservoirs, and rivers, as well as tidal creeks and basins with lower salinity. **COHABITANTS:** Common Loon, goldeneye. In winter, may associate with Red-breasted Merganser and feeds in proximity to goldeneyes. **MOVEMENT/MIGRATION:** Among the earliest and latest of migrants: spring migration from late January to April; fall in November and December. In winter, shifts in response to freeze conditions, reoccupying frozen bodies of water as soon as they thaw. VI: 2.

DESCRIPTION: A large, elongated, wide-bodied, heavyset "saw-bill" (larger than any other duck except Common Eider) so strikingly plumaged it borders on poor taste.

The gleaming white sides, breast, and neck of adult males set off the prominent, iridescent green head (black at some angles or in poor light) and Day-Glo orange bill, which is long, heavy at the base, and acutely tapered. Coupled with the sloping forehead, the bill gives the bird a distinctly pointy-faced appearance.

Females and immatures, with their dark rich chestnut–colored heads, overly bright reddish orange bills, and prominent, seemingly painted-on white chin patches, look like overly made-up dowagers. The demarcation between the dark head and pale neck is abrupt and distinctive. At a distance, the female's head appears darker than the back. At extreme distances, the sloping forehead and back of the head make the female's head appear domed or crested. The expression is inscrutable. Even at close range, males appear to have no eyes.

BEHAVIOR: One tough duck. In winter, remains as far north as open fresh (and brackish) water allows. On the water, sits very high in the front and low in the stern—a wedge in the water. Swimming birds pull their heads down and appear neckless. In size, shape, and posture, Common Merganser recalls Common Loon. Diving birds surface away from the point of entry. Shy, they flush easily and often retreat to another body of water.

FLIGHT: The running takeoff is lengthy and labored; landing, the bird slows gradually on a long approach, brakes with the wings, and touches down feet first. In flight, the body is held horizontal. The long neck and wedge-shaped face, coupled with the long tail and severely tapered wings, make the bird look pointy in all directions. The white on the upper, inner half of the adult male's upperwing is hard to overlook; the white inner trailing edge of the female's wing is also prominent. Birds commonly fly low, just skimming the surface. Over water, they often fly in a line; flocks, traveling distances over land, form a V.

Flight is heavy, strong, direct, and fast; wingbeats are steady, shallow, and stiff, with much of the motion relegated to the wingtips. Common Merganser's wingbeats are somewhat like a loon's, but without the elastic quality.

VOCALIZATIONS: Usually silent. Females make a soft, atonal croak in flight.

PERTINENT PARTICULARS: Adult males may be confused with the larger Common Loon, but the *bright orange bill* is both distinctive and distinguishing. Females and immatures are easily confused with female and immature Red-breasted Merganser. Common Merganser's crest is thicker, fuller-bodied, and not as scraggly or unkempt as Red-breasted. It can be raised or lowered so that the bird sometimes looks thick-necked and large-headed or appears to be wearing a ponytail or Pileated Woodpecker–type crest. Or the crest can be erect and spiky when the bird is excited or courting, much like Red-breasted Merganser. Also, female Common Merganser shows a distinctive white chin; the chin on Red-breasted Merganser is dark or whitish, with no sharp contrast or demarcation.

Courting males (and sometimes females) raise their fanned tails out of the water. The white underparts on males sometimes glow with a pinkish wash.

Red-breasted Merganser, *Mergus serrator*

The Anadromous Merganser

STATUS: Common northern, primarily arctic breeder; common, primarily coastal winter resident. **DISTRIBUTION:** Breeds in the arctic and in boreal forest, nesting from Nfld. and Greenland south into Maine, Vt., s. Ont., n. Mich., n. Wisc., and e. Minn. and north of a line drawn from Lake Superior to the Yukon and the southeast coast of Alaska. In winter, found along all three coasts, from Nfld. to n. Mexico, and from the Aleutians to s. Baja. Also winters on the Great Lakes, the Great Salt Lake, and other, larger inland lakes and reservoirs. **HABITAT:** Nests along fish-bearing, brackish and freshwater wetlands on tundra or in boreal forest; also on lakes and rivers. In winter, occurs almost exclusively in marine environments. Selects bays, harbors, channels, rocky coastline, lagoons, tidal creeks, and large salt marsh ponds. Often forages in the surf or near rock jetties. In migration, small numbers are seen on inland lakes. **COHABITANTS:** Frequently nests on islands with gulls and terns. Forages alone in winter but is sometimes found in small, homogeneous groups. **MOVEMENT/MIGRATION:** Spring migration from late February to early June; fall from late September to December. VI: 2.

DESCRIPTION: A duck that looks like it was dressed out of a trunk found in the attic. Overall smaller, slimmer, and more spindly than Common Merganser. Adult male is striking, bordering on outrageous (easily distinguished from male Common Merganser by its dark breast and spiky "punk cut" crest). Females and immatures are similar to Common Merganser, but if Common Merganser can be likened to a dowager, Red-breasted Merganser recalls a lithe, casual, unkempt teen. Overall paler than Common Merganser, the orange-brown head is a shade lighter than the back and shows no sharp demarcation between head and the grayer neck. The crest is sparse, unkempt, and billowy (blowing freely in the wind). On birds emerging from the water, the head feathers hang limp. At extreme distances, the sparsely feathered crest disappears, making the head seem smaller, thinner, and gently curved on top, not bulky, domed, or crested. The orange bill (dark at a distance) is long, uniformly wide, and popsicle-stick thin; the bill combined with the high, prominent forehead makes the bird much less pointy-faced than Common Merganser. Males, with their red eyes, look crazed; females look alert.

BEHAVIOR: Red-breasted sits low in the water and seems not as front-heavy as Common Merganser. It often raises its spindly neck. More tolerant than Common, some Red-breasteds, habituated to people, allow very close approach. Feeds by diving, but in shallow water also drops its head below the surface (like a shoveler). Forages sometimes in a small group (up to 10 birds), but seems less inclined to gather in the large, tight flocks characteristic of Common Merganser.

If it's a large merganser, and it's in salt water, it is almost certainly Red-breasted Merganser. *Caveat:* Common Merganser may be common in brackish estuaries, particularly during prolonged freeze-ups.

FLIGHT: Profile is slender and long-necked, with long pointy wings — more spindly than Common Merganser. Male shows conspicuously less white overall (Common Mergansers seem mostly white); female is slightly paler than female Common (particularly paler-headed). Because its running takeoff is short, Red-breasted frequently forages in small tidal or tundra ponds. Flight is direct and strong, with the body held horizontal, and wingbeats are rapid and evenly measured.

VOCALIZATIONS: Mostly silent. In flight, and when startled, female has a call similar to Common Merganser, but higher-pitched.

PERTINENT PARTICULARS: The courting male engages in a stunning display. Stretching its neck up and forward, the bird abruptly bows, lowers its breast into the water, and opens its bill as if gasping for air—and looking as though an invisible foot has been planted on its back. It may repeat this display several times. Sleeping birds sometimes raise their normally submerged tails out of the water.

Ruddy Duck, *Oxyura jamaicensis*

Clearly the Bathtub Duck

STATUS: Common, primarily western breeder; common and widespread winter resident across the United States and Mexico. DISTRIBUTION: Breeds, sometimes in fragmented pockets, in every state and province from the central prairies to the Pacific Coast. Range extends north to the Northwest Territories and south to Central America, but is not found, as a breeder, across much of the Pacific Northwest or in arid portions of s. Calif. and Ariz. Also breeds in low densities in parts of the Midwest, around the Great Lakes, and in various mid-Atlantic states north to the Maritimes. In winter, vacates breeding areas across most of w. Canada and the northern prairie states. Distributes itself across much of the U.S., south of a line drawn from s. Maine, across the s. Great Lakes to s. Colo., and north to sw. B.C. HABITAT: Most partial to the prairie pothole region, nesting on marshes, managed impoundments, lakes, reservoirs, and stock ponds. Prefers habitats with extensive stands of aquatic vegetation and open areas for takeoffs and landings. In winter, uses same habitat for breeding but also breeds in brackish bays, large impoundments, tidal creeks, and estuaries. COHABITANTS: In summer, Horned and Eared Grebes, Black Tern, and a variety of other waterfowl; in winter, Bufflehead, Coot, and Ring-necked Duck.

MOVEMENT/MIGRATION: Spring migration from March to May; fall from September to early December. VI: 2.

DESCRIPTION: Compact comically proportioned diving duck that looks like it belongs in a bathtub. Small (larger than Bufflehead; smaller than Black Scoter; slightly smaller than Lesser Scaup), with a chunky, broad-beamed, and somewhat peaked or domed body, a disproportionately large, often slightly peaked head (at a distance, the bird seems all head), and a large slightly upturned bill—a Cyrano de Bergerac of a duck. The stiff long tail is cocked up at a slight angle and is usually raised when the duck is sleeping.

In breeding and nonbreeding plumage, the adult male has bright white cheeks that contrast with a blackish cap and an overall reddish (breeding) or gray-brown (winter) body. The turquoise blue bill is dull (tarnished silver) in winter. Females and immatures are like winter males, but their cheeks are dingier and bisected by a narrow, dusky stripe or chin strap.

All in all, Ruddy just looks cute—the perfect bathtub duck. Its expression is happy, grinning.

BEHAVIOR: Highly social, but most often found in homogeneous flocks that may number in the hundreds (sometimes found with American Coot or Ring-necked Duck). Sluggish and tame, Ruddy prefers to swim to safety or dive and flies as a last resort. Feeding birds are widely scattered, and diving is wholly unsynchronized. When foraging, generally does not raise the tail. Sleeping birds gather in tighter flocks, for the most part with tails raised. Ruddy Duck walks with great effort and stands erect. Its running takeoffs are lengthy and effortful, and the bird seems loath to fly.

FLIGHT: Overall small and compact, with an outsized head and up-angled body. Flight is rapid and direct. Wingbeats are rapid, regular, "buzzy."

VOCALIZATIONS: Generally silent. Females make a mellow peeping whistle, "*peur,*" uttered once or sometimes repeated several times. Females also make a short single-note laugh—a cross between a peep and a quack (a vibrant "*peur*" note).

PERTINENT PARTICULARS: In southern Texas and southern Florida, a somewhat similar stiff-tailed duck, Masked Duck, *Nomonyx dominicus*, is a rare visitor. Male Masked Duck is all ruddy with a black face and a bluish bill; females and immatures are overall buffier brown than Ruddy Duck and show two blackish lines across a buffy (not whitish) face. Also, Masked's bill shape is slightly different.

GAME BIRDS—CHACHALACA, QUAIL, PHEASANT, AND GROUSE

Plain Chachalaca, *Ortalis vetula*

Mexican Tree-Pheasant

STATUS AND DISTRIBUTION: Common resident species with a very limited distribution in the U.S.; a native population is limited to the lower Rio Grande Valley of Tex. Stable, introduced populations are found on Sapelo I. off the coast of Ga. and in San Patricio Co., Tex. **HABITAT:** Tall thorny scrub forest, thickets, and forest edge; well-vegetated residential areas. **COHABITANTS:** White-tipped Dove, Long-billed Thrasher, Olive Sparrow. **MOVEMENT/ MIGRATION:** Permanent resident. VI: 0.

DESCRIPTION: Plain, brown, raucous, and clannish tree-pheasant–like bird. Large (about the same size and shape as Greater Roadrunner or female pheasant), this sturdy bird resembles an emaciated long-tailed turkey (but with a feathered head).

Aptly named. Overall plain dull brownish olive with a white-tipped blackish tail. Ages and sexes are similar.

BEHAVIOR: A social forest bird. Never found far from a woodland edge. Commonly found in groups of 3–10 (and in fall as many as 20) and in pairs. Forages primarily on leaves and fruit in trees. Walks, nimbly and squirrel-like, along limbs, sometimes hopping and flying (more commonly gliding) only when it has to. Also forages on the ground by walking with a slinking crouch, running more often than flying short distances when disturbed.

To say the bird is noisy sells it short. Raucous is more accurate, and if you are so unfortunate as to be close when a flock erupts, a better description might be ear-splitting. Group calling begins at dawn and may continue off and on until midmorning. Calling also occurs at dusk and (particularly during a full moon) at night. Calling birds are usually perched high in trees; though they vocalize all year, they are most vociferous in spring.

Plain Chachalaca is shy and generally runs when approached or startled. Where habituated to people (like refuges and parks), may allow close approach. Readily attracted to seed and water at feeding stations.

FLIGHT: Looks big, long-tailed, long-necked, and brown; flashes warmer cinnamon tones from the underwings. Flies with long broad wings stretched to the sides and long tail widely fanned, showing a narrow white tip. Most flights are short; several strong rapid flaps are followed by a long glide. Commonly glides when descending from trees.

VOCALIZATIONS: Group chorus is a harsh, very loud, raucous clatter, "*(aw)RAW Ka Ka RAW Ka Ka RAW Ka Ka,*" that goes on and on, with some birds singing guttural bass, some shrill soprano. Also makes a low snorting purr, "*k'r'r'r,*" that recalls a stiff pack of cards being shuffled.

PERTINENT PARTICULARS: This is not a difficult bird to find or identify. Turkeys may be found in the same habitat but are bronze-colored, bare-headed, and much larger. Pheasants are not found in southern Texas.

Chukar, *Alectoris chukar*

Desert Partridge

STATUS: Fairly common introduced resident of the Great Basin. **DISTRIBUTION:** Found in s.-cen. B.C., e. Ore., e. Wash., s. Idaho, e. Calif., Nev., Utah, w. Colo., and parts of n. Ariz., w. Mont., sw. and n. Wyo.; also found in Hawaii. Outside of range, widely stocked for hunting. **HABITAT:** Arid rocky slopes and cliffs, often close to water. Habitat communities include desert shrub, pinyon-juniper, and montane brush. Found from below sea level to 10,000 ft. Does particularly well where introduced cheatgrass flourishes and where habitat has been overgrazed. **COHABITANTS:** Prairie Falcon, Rock Wren, Canyon Wren, Black-throated Sparrow. **MOVEMENT/MIGRATION:** Permanent resident; some down-slope movement in winter in response to snowfall. VI: 0.

DESCRIPTION: Boldly patterned rock-loving quail that says its name. Large (smaller than Ruffed Grouse; larger than California Quail), with a plump body, small head, short heavy neck, short tail, and short sturdy legs.

Boldly patterned adults are overall gray with a ruddy blush. Fortunately, the head pattern, featuring a black-bordered cream-colored throat and bright red bill, is distinctive, because in a typical rocky habitat this may be all you see. On a bird standing in the open, the zebra-striped black-and-chestnut bars on the buff sides are manifest, as are the reddish legs.

Immatures are overall brownish gray and boast a shadow pattern of the adult plumage, the most prominent aspects of which are a dark line through the eye, a pale throat, and red legs.

BEHAVIOR: Loves to be near or on rocks and outcroppings. Forages for most of the year in small flocks, walking along rocky, barely vegetated slopes, plucking seeds from the ground or reaching for them with head turned sidewise. Fairly shy. Crouches when approached or runs (normally uphill). When flushed, heads downhill, then circles and lands facing uphill.

FLIGHT: Overall stocky, with broad wings and a short square tail showing reddish sides. Explodes into flight with a rapid flurry of wingbeats, and often calls when it rises. Flight is heavy but strong. Over flat terrain, alternates flaps with glides. Over steep terrain, heads downhill flapping just enough to gain speed and altitude, then glides, banks, and circles as it lands.

VOCALIZATIONS: Says its name—but has to build up to it. Its territorial call is a series of low clucked notes that begin slow and gain in volume and tempo: "*ch chuh chuh Chuka Chuka Chuka ChuKAR, ChuKARA ChuKARA.*" When flushed, makes a loud whistled squeal, "*wheee!*" followed by the lower, more conversational note, "*whitoo, whitoo.*"

Himalayan Snowcock, *Tetraogallus himalayensis*

Ruby Mountain Snowcock

STATUS AND DISTRIBUTION: Asian resident introduced into the Ruby Mts. of ne. Nev., where the species became established in the 1980s. **HABITAT:** Found in alpine and subalpine meadows at elevations of 9,000–11,000 ft. surrounded by rocky ridges and protective escarpments. **COHABITANTS:** Golden Eagle, Black Rosy-Finch, Pica, Yellow-bellied Marmot, Mountain Goat. **MOVEMENT/MIGRATION:** Permanent resident. VI: 0.

DESCRIPTION: Large, shaggy, gray game bird with a maroon-trimmed white face found on the barren slopes of the Ruby Mountains. This bird isn't large—it's *huge* (smaller than Wild Turkey; about the same size as Greater Sage Grouse), with a grouselike head, a big shaggy body, short sturdy orange legs, and a short tail.

Mostly gray with a white face and throat and a mostly white breast embellished by narrow, maroon trim on the hindneck and throat. Undertail coverts are bright white. Immatures are less patterned (lack maroon) and overall paler gray.

BEHAVIOR: Small roosting flocks fly down-slope at daybreak and forage upward during the morning. After resting around midday, birds may return to lower foraging areas or continue up-slope to higher areas. Roosts (by walking or flying) just before dark. Because its size makes it so difficult to get airborne, avoids level ground. When disturbed, quickly waddles uphill with tail lifted, exposing bright white undertail and orange legs. Roosts on rock ledges. Often dust-bathes on Mountain Goat trails.

FLIGHT: Always downhill. Jumps off vertical points with flapping wings and settles into a breakneck power glide that doesn't deviate, unless it turns in a wide arc and lands facing up-slope. Flights are commonly less than 500 ft.

VOCALIZATIONS: The alarm call has been described as a single low whistle or a shrill rapid series of whistles similar to Long-billed Curlew. So if you think you hear Long-billed Curlew and you are high in the Ruby Mountains. . . .

PERTINENT PARTICULARS: You are not going to see this bird except by an act of volition. Identifying it once you do is the easy part.

Gray Partridge, *Perdix perdix*

Landed Quail (in the European Tradition)

STATUS: Fairly common but local resident species; less common at the western edge of its range. **DISTRIBUTION:** Old World native. In North America, most common in the Great Plains, from s. Alta., s.-cen. B.C., Sask., and s. Man. south into Wash., e. and cen. Ore., n. Nev., much of Mont., portions of Wyo., N.D., all but sw. S.D., e. and cen. Neb., w. and s. Minn., n. Iowa, s. Wisc., and nw. Ill. Also established in se. Ont. and bordering portions of N.Y., N.S, and P.E.I. Often released on game lands and semiwild shooting preserves, so may be encountered elsewhere. **HABITAT:** Primarily agricultural farmland but also prairie, grasslands, and small woodlands bordering fields and croplands. Much prefers grain fields with hedgerow borders, but can be found in plowed fields and even lawns. During heavy winter snows, often concentrates in sheltering woodlands. **COHABITANTS:** American Kestrel, Ring-necked Pheasant, Western (and Eastern) Meadowlark. **MOVEMENT/MIGRATION:** Permanent resident. VI: 0.

DESCRIPTION: Plump-bodied, grayish brown, quail-like bird with an orange face. Medium-small (larger than Northern Bobwhite or Gambel's Quail; smaller than Chukar), with a small head, a chicken-bill, a longish neck (tucked when resting), a plump, round body, short legs, and a short tail.

At any appreciable distance, appears pale grayish brown and featureless except for a contrastingly orange face. (Female's face is paler but contrasting enough to be noted.) At close range, the gray breast, barred flanks, and dark stain on the belly (males only) are apparent. Immatures are warm brown, with streaked (not barred) flanks.

BEHAVIOR: Shy, hair-triggered, flocking species—usually discovered as the entire flock explodes into cackling flight. Likes open flat or rolling agricultural land; in fact, in the best European tradition, thrives where the soil has been turned for grain crops, especially wheat. Among North American game birds, only Ring-necked Pheasant (like Gray Partridge, also introduced) is so much at home on farmland. Forages on the ground, feeding most actively early and late in the day (but may remain active all day in winter). Very social; almost never seen alone. For most of the year, found in coveys, except in spring (pairs) and in summer (family groups and coveys of males). Forages and often roosts in open stubble fields and is much more likely than most quail species to be found away from heavy cover; in winter, however, birds concentrate in bordering woodlands, particularly when snow covers the ground. A very hardy species, Gray Partridge can survive the bitter winters and snow cover of the northern prairies.

A shy bird that flicks its tail nervously if approached. Usually flushes when observers are 20 yds. away. (Most quail flush almost underfoot.)

FLIGHT: Stocky body with short round wings and a short fanned and reddish tail that stands out against the otherwise plain grayish brown upperparts. Flight is typical of grouse and quail, consisting of a rapid series of wingbeats followed by intervening short glides on slightly down-drooped wings. Flight is somewhat unsteady, and birds often veer when they land. Most flights are low, at eye-level, and cover less than 100 yds.

VOCALIZATIONS: Call is a weak, hoarse, emphatic, somewhat froglike "*sk'r'rEh?*" usually given two or three times. When flushed, makes a protesting series of pheasantlike cackles and raspy scraping sounds—"*skra ka ka ka kraaa kr'r'*"—accompanied by an audible flutter of wings.

PERTINENT PARTICULARS: The range of Gray Partridge overlaps with several species that tend to segregate themselves by habitat. Chukar prefers steep arid hillsides with sparse vegetation; Mountain Quail goes for dense cover; Northern Bobwhite is partial to grasslands, shrub lands, and bordering forest, but tends not to wander around in large open stubble fields (Gray Partridge's favorite haunt).

Ring-necked Pheasant, *Phasianus colchicus*

The Corn Rooster

STATUS: Variously common to uncommon widespread resident. The declining populations in many places are augmented through the stocking efforts of state game agencies and private hunting clubs. **DISTRIBUTION:** Scattered throughout much of the U.S. and bordering regions of Canada; absent only in colder interior portions of n. New England and N.Y. and most of the Southeast (south of a line drawn from n. Va. to trans-Pecos Tex.), but populations occur in se. Mo. and coastal Tex. Most heavily concentrated from Pa. west through the Midwest and into the southern and northern prairies. **HABITAT:** Classically associated with open agricultural land, especially cornfields, close to weedy ditches, overgrown fencerows, brushy borders, reed marshes, and woodlands. Also does well in abandoned industrial settings (often near wetlands), where its need for open weedy seed-bearing habitat and protective cover is met. **COHABITANTS:** American Kestrel, Vesper Sparrow, Eastern and Western Meadowlarks. **MOVEMENT/MIGRATION:** Nonmigratory; permanent resident. VI: 0.

DESCRIPTION: Distinctive and colorful (males) wild rooster. Large (smaller than Wild Turkey or Greater Sage-Grouse; larger than Sharp-tailed Grouse), with a small chickenlike head, a large body, short sturdy legs, and a rakishly long, thin, pointed tail that is as long or longer than the bird itself. Can be confused only with Sharp-tailed Grouse, sage-grouse, or Greater Roadrunner.

Adult males are unmistakable—a jaunty-looking bird clad in iridescent copper-and-gold chain mail and distinguished by a bright red face, a narrow white collar, and a brown, darkly banded, spiky plume of a tail. Females and immatures are all buffy brown, with patternless dark spotting and, like males, a long darkly barred plume of a tail. Female sage-grouse is shorter-tailed, more robust, and overall cold gray, not buffy. Sharp-tailed Grouse is shorter-tailed, grayer, and more lavishly and contrastingly patterned, with black-and-white spotting above and a heavily patterned breast contrasting with paler unmarked underparts.

BEHAVIOR: Often seen walking across open fields and along roadsides; when not seen, may explode from underfoot as you navigate a weedy ditch or approach the end of a cornfield. When flushed, males crow excitedly; females are silent.

Social except during breeding season. Females and males form small sexually segregated or mixed flocks numbering several individuals (usually males) to 20 or 30 (females). Where common, more than 100 birds may gather in sheltered roosting areas.

During breeding season, males patrol openly and vocalize frequently. Ground-nesting females locate their clutch and rear young in heavy vegetative cover (like fencerows and wetlands), but later in the season also use pastures, grasslands, and hay fields.

Spends most of its time on the ground, walking with a strut, pausing frequently. Prefers to walk or run for cover rather than fly, and often runs with tail cocked at an angle.

FLIGHT: Flies with body angled up, long tail projecting, short broad wings moving in a rapid blur. Explodes into flight, rising near-vertically or at a steep angle, heading straight away (often downhill). After an initial lengthy burst of wingbeats, commonly sets its wings and glides to a landing. Usually runs upon landing.

VOCALIZATIONS: Male call sounds like a truncated braying rooster, "*Korok-kok!*"—a sound perhaps aptly described as an old Model-A Ford car horn (though except for old-movie buffs, who's going to recognize that sound now?). When flushed, males make a loud, excited, two-noted crow given in a series: "*kutUck'kutUck'kutUck.*"

PERTINENT PARTICULARS: There are other exotic pheasant species. Most are ornamental, but some of those that have been introduced, like the green-bodied Japanese Green Pheasant, hybridize with Ring-necked Pheasant.

Ruffed Grouse, *Bonasa umbellus*

The Pahtridge (as They Say in New England)

STATUS: Variously common to uncommon permanent resident of primarily young, northern woodlands. **DISTRIBUTION:** Breeds across virtually all of forested Canada and across much of the n. U.S. Found from Nfld. to B.C. and north through the cen. Yukon and the s. Northwest Territories into cen. Alaska. In the East, found throughout New England, N.Y., N.J., Pa., Mich., Wisc., and Minn. and along the Appalachians to e. Tenn.; also found in s. Ind., s. Ohio, w. Ky., s. Ill., and portions of Mo., Ark., and Iowa. In the West, resident in much of Wash., Ore., nw. Calif., Idaho, w. and cen. Mont., and w. Wyo., along the Wyo.–S.D. border (Black Hills), on the N.D.–Canadian border, and in ne. Nev and ne. Utah. **HABITAT:** Over much of its range, almost a year-round aspen obligate. Also found in mixed deciduous-coniferous forests in early stages of succession and, in lower densities, in boreal forest. In all habitat types, prefers open understory. **COHABITANTS:** Northern Goshawk, Black-capped Chickadee, Hermit Thrush. **MOVEMENT/MIGRATION:** Permanent resident. VI: 0.

DESCRIPTION: The crested grouse of deciduous woodlands. Medium-sized (slightly larger than Spruce Grouse; considerably smaller than Blue Grouse or Ring-necked Pheasant), with a peaked head, plump body, and longish square-tipped tail.

Overall cryptic; either gray or slightly reddish above (particularly reddish on the tail), with a dark ruff of feathers at the base of the neck (often diffuse on females), blackish splotching on pale flanks, and a wide black band on the tip of the tail. Spruce Grouse and Blue Grouse have round, not peaked, heads. Male Spruce Grouse wears a distinctive black hood; the female is shorter-tailed, heavily barred below, and overall darker and more uniformly patterned. Blue Grouse is overall bigger, bulkier, and more darkly and uniformly patterned top to bottom, and it has a broadly pale gray–tipped (not black-tipped) tail. The expression is mean, serious.

BEHAVIOR: A bird equally at home on the forest floor and in the trees. Feeds on a mix of leaves, buds, fruits, and ground plants. Forages on buds of aspen, birch, and other trees, often in the upper canopy. Also forages on wild grape, smilax, mountain laurel, and assorted berries, acorns, and (in spring) flowers. At night, roosts in conifers or oak-leaf clusters, in stands of tightly packed saplings, or beneath snow cover. During the day, birds are never far from dense, protective cover, although they sometimes leave cover to dust-bathe and gather grit on roadsides.

Generally solitary and shy. In winter, some birds gather into small groups of four or five individuals to feed and roost together. Usually freezes when approached, and explodes into flight when distance closes to 10–30 yds. Generally does not fly far (less than 100 yds.). At times exhibits uncommon boldness, approaching and displaying before humans. Highly territorial birds have even been known to attack!

Does not respond to pishing. Territorial males, however, respond to imitations of their drumming, which can be done by thumping a closed fist against your chest.

FLIGHT: Flies with neck extended and head raised; shows a short round-winged and long-tailed profile—not that you'll see it. Ruffed Grouse in flight are usually birds that have been flushed, and when flushed, they generally go straight away showing at least the wide dark band on the tip of the tail. They explode into flight with a loud (and intentionally startling) whirring flutter of wings and rapid acceleration, then fly fairly low with rapid steady wingbeats and a somewhat wandering, veering, and rocking flight. Usually maneuvers to put the first tree it passes between you and it. Glides when landing.

VOCALIZATIONS: Generally silent. Males and females sometimes make peeping sounds just before flushing. The most commonly heard sound is an accelerating, percussive thumping (not a vocalization) made by the male's rapidly beating wings as it sits on a log or boulder: "*thump . . . thump . . . thump thumpthumppumppumppump.*" It's a bouncing-ball-accelerating-to-a-stop sound. You can approximate it by thumping on your chest with open cupped hands.

This extremely low-pitched territorial drumming is felt as much as heard.

PERTINENT PARTICULARS: Because of their wariness, and because they are often in dense cover, grouse are too often heard (flushing) and not seen. When approaching likely grouse cover (wild grape tangle, smilax fortress, stand of aspens), try approaching steadily and then stop about 30 yds. shy. Wait. This tactic seems to unnerve grouse, which, after 5–15 secs., lose their cool and flush—but now, you're ready for them.

Greater Sage-Grouse, *Centrocercus urophasianus*

Greater Sage-Pheasant

STATUS: Uncommon habitat-restricted resident of the intermountain West. **DISTRIBUTION:** Found in extreme se. Alta. and sw. Sask. (though very scarce now), se. Wash., e. Ore., ne. Calif., all but s. and w.-cen. Nev., all but nw. Mont., most of Wyo., w. S.D., sw. N.D., and scattered portions of n. Utah and w. and n. Colo. **HABITAT:** Sagebrush; in summer, also found in adjacent, open, grassy agricultural or shrubby habitats, often near water. **COHABITANTS:** Black-billed Magpie, Brewer's Sparrow, Sage Sparrow, Western Meadowlark. **MOVEMENT/MIGRATION:** Permanent resident. VI: 0.

DESCRIPTION: Large, sturdy, brownish gray, pheasant-like bird. Very large and robust (larger than Ring-necked Pheasant), with a round-topped head and a hawklike Roman nose of a bill, a short thick neck, a heavy body, and a long, narrow, pointed tail. Legs are very short and sturdy.

Males are brownish gray above, with scratchboard-fine pale lines etched throughout the plumage. The black throat, collar, and belly contrast with a white breast. Females are slightly shorter-tailed and slightly browner, with a brown throat and breast, but they share the male's distinctive black belly. Young are like adults.

Unlikely to be confused with any other bird except female Ring-necked Pheasant, which is tan (not gray) and lacks the belly patch. Female Blue Grouse (a bird found in sagebrush under pines but not in pure, open sagebrush habitat) appears longer-necked, shorter-tailed, and overall browner and has a whitish, not black, belly. Female Sharp-tailed Grouse is smaller, overall paler, and browner, with a slight wisp of a crest and a tail that is paler than the body (most evident in flight).

BEHAVIOR: During the nonbreeding season, forages in small flocks in sage or in open areas dominated by tall grass adjacent to sage flats. Most active at dawn and dusk, but also forages at midday. Flocks forage by walking and plucking the leaves, buds, and fruits of several species of sage, as well as insects in summer.

This shy short-legged bird crouches when approached; does not generally run. Reluctant to fly, but does so strongly, often traveling great distances.

FLIGHT: Distinctive silhouette, with an extended neck, long broad wings showing widely slotted primaries, and a long tail. The white underwing linings and black belly contrast with the bland uniformity of the upperparts; both are visible at great distances. Wingbeats are given in a powerful, heavy series punctuated by both short and long, somewhat unsteady, tipsy glides on down-drooped wings. In flight, flocks are loosely linear and fly fairly close to the ground.

VOCALIZATIONS: Displaying males combine two spaced and audible wing swishes with three low coos, followed by two loud pops. "*Swish . . . swish coo coo coo POP POP.*" Makes a low croaking "*rrk rrk rrk*" when disturbed, often just before or during flight.

Gunnison Sage-Grouse, *Centrocercus minimus*

Plumed Sage-Grouse

STATUS AND DISTRIBUTION: Uncommon and very geographically restricted resident of the Gunnison Basin in sw. Colo. and in adjoining se. Utah. **HABITAT:** Sage. **COHABITANTS:** Sage Thrasher, Sage Sparrow, but *not* Greater Sage-Grouse. **MOVEMENT/MIGRATION:** Permanent resident. VI: 0.

DESCRIPTION: A pheasant-sized sage-grouse found near Gunnison, Colorado. Looks much like Greater Sage-Grouse, but smaller, with a tail (on both males and females) that is paler than the back. The male's philoplumes (visible when displaying) truly look like a drooping feathered plume, whereas those of Greater Sage-Grouse look like a sparse, windblown cowlick. Other small distinctions between the two species are moot.

In Utah and Colorado, the two states where Gunnison Sage-Grouse is resident, Greater Sage-Grouse is restricted to ne. Utah and nw. Colorado; the ranges of the two sage-grouse do not overlap, and neither bird is prone to wander.

BEHAVIOR: Like Greater Sage-Grouse.

FLIGHT: Like Greater Sage-Grouse.

VOCALIZATIONS: Like Greater Sage-Grouse, makes a mix of loud "pop" sounds and audible wing swishes, but more complexly patterned. Gunnison begins with three pops, followed by three rapid wing swishes, three pops, a pause, then three more pops.

Spruce Grouse, *Falcipennis canadensis*

Forest Ptarmigan

STATUS: Common resident of northern coniferous forests. **DISTRIBUTION:** Ranges across boreal Canada and Alaska from w. Alaska to Lab. and Nfld. (absent in Canadian prairies and coastal portions of B.C. and w. Alaska). In the U.S., found in Maine, n. N.H., Vt., the Adirondacks of N.Y., n. Mich., ne. Minn., cen. and n. Wash., ne. Ore., much of n. and cen. Idaho, and w. Mont. Distribution is more restricted than Blue Grouse, which occurs throughout the Rockies and West Coast mountain ranges, south into Calif. and cen. Ariz. and N.M. **HABITAT:** Pine and spruce forests; prefers young (early to midsuccessional growth) and, particularly in winter, dense stands of trees. In summer, forages in more open habitats; in fall, occasionally occupies deciduous forest. **COHABITANTS:** Northern Goshawk, Boreal Chickadee, Gray Jay, Red Squirrel. **MOVEMENT/MIGRATION:** Most birds are sedentary. Some small percentage move from summer to winter territories, though for distances of no more than 5 mi. VI: 0.

DESCRIPTION: Stocky, grayish or brownish, ptarmigan-like grouse with a dark hood and white spotted underparts. Medium-sized (slightly smaller than Ruffed; considerably smaller than Blue; perhaps slightly larger than Willow Ptarmigan), with a small head, a small chickenlike bill, a shortish neck, a robust body, and a fairly short tail. Overall more compact than Blue Grouse (and much shorter-tailed).

More complexly patterned and white-spattered than Blue Grouse, particularly below. Adult males are bluish gray above (with a pale golden wash in the back) and white spotted below; underparts appear very scaly and suggest a girdle made of chain mail. The black head, neck, and breast give birds a dark, hooded appearance. Northern birds have a black patch on the belly. The scaly white band dividing the dark hood and belly shows as a narrow white belt.

Females have a red-and-gray color morph (like Ruffed Grouse), but show a shadow pattern of the male's pattern—darker hood and white scaled underparts. (Female Blue Grouse is grayer and more uniformly patterned below.) The tails of northern males and females show a warm brown band at the tip, as compared with a gray band on many Blue Grouse. Blue Grouse in the Rockies and Sierra Nevada show white spotting on the uppertail; the tails of interior birds are all dark.

BEHAVIOR: Much at home in trees and on the ground. In winter, spends much of its time aloft in pines and spruce, foraging on sturdier midlevel limbs. Walks along limbs or hops, reaching down to pluck needles. Reportedly prefers pine to spruce, and white spruce to black spruce. In summer, spends more time on the ground or on the forest floor beneath a canopy of branches, plucking buds and berries. Sometimes dust-bathes along roadsides.

Mostly solitary. In summer, females forage with chicks; in winter, birds may gather in small flocks of three to four birds (occasionally 10 or more). Very tame; allows such close approach that often goes undetected. Mostly silent.

FLIGHT: Typical grouse profile: neck extended, tail fanned, broad wings fully extended. The darker blackish tail contrasts with the paler body. A pale brown band at the tip of the tail is apparent in northern birds (both males and females). A rapid series of wingbeats is punctuated by a set-winged glide. Flights are normally short (less than 100 yds.). Flushed birds frequently fly to a nearby tree.

VOCALIZATIONS: Males give a series of low muffled barks, "*whuh whuh whuh/whuh,*" as well as a low, dry, grating chatter that recalls a Rock Ptarmigan: "*cow cow cacacaca.*" Males make a short, low, fluttering drum that starts fast and descends (rather than starting slow and accelerating, like Ruffed Grouse).

Willow Ptarmigan, *Lagopus lagopus*

The Grouse That Asks How? What? Where?

STATUS: Common northern arctic and subarctic resident; irregular winter resident in central and southern Canada. **DISTRIBUTION:** Breeds throughout most of Alaska, n. Canada, and the Canadian Rockies; the southern limits of the breeding range extend to s. and e. B.C., cen. and se. Northwest Territories, ne. Man., n. Ont., cen. Que., and Nfld. The southern limit of its irregular winter range cuts across cen. Alta., cen. Sask., n. Man., s. Ont., s. Que., and nw. N.B. (casually reaches the U.S.). **HABITAT:** Breeds in low, moist, shrubby or densely vegetated tundra; partial to willow thickets, marshy edge, and roadsides in forested areas. Usually found in level areas or gentle slopes and valley bottoms, but does occur at subalpine levels in the Canadian Rockies. Avoids the rocky, dry, elevated, and more barren habitats favored by Rock Ptarmigan—except in fall, when some birds go to higher elevations to molt. In winter, birds from higher elevations move to lower, more vegetated habitats, and northern birds (primarily females and immatures) move below the tree line to areas dominated by boreal forest (particularly in years of heavy snow cover). **COHABITANTS:** During breeding, Northern Harrier, Rock Ptarmigan, White-crowned Sparrow, Common and Hoary Redpolls. **MOVEMENT/MIGRATION:** Some populations move only short distances between breeding and winter areas; some move several hundred miles. Spring migration from early February to late April; fall from late September to late December. VI: 1.

DESCRIPTION: It's the ptarmigan in the willow thicket (or the boreal forest in winter). Large (larger and more robust than White-tailed; slightly smaller than Spruce Grouse; about the same size as Rock Ptarmigan), with a small head, a long neck, a heavy body, short legs, and a short tail. The bill is short but wide, and heavier and less petite than Rock Ptarmigan, with a more curved culmen.

All white in winter, except for the black tail (usually hidden). In summer, males are ruddy brown with white bellies, and the head and neck are distinctly chestnut brown. (This trait is evident in early spring, when the rest of the bird is still white.) Females are brown and lavishly flecked, barred, and spotted with black (wings remain white). Males are much ruddier and warmer-toned than Rock Ptarmigan. (In winter, males lack the male Rock Ptarmigan's black eye stripe.) Some females are slightly richer brown than female Rock Ptarmigan, but except for the latter's smaller bill, these two birds are very hard to separate in the field (or more accurately, the open tundra lying between a willow thicket and a rocky slope).

BEHAVIOR: In summer, found in pairs, family groups, or small flocks of unpaired males; in winter and migration, may form large flocks (usually fewer than 100 birds, but flocks in excess of 1,000 have been reported). Unlike other ptarmigans, males remain with females until chicks are reared. Willow does not engage in aerial displays; males fly to intercept females but display on the ground. Stands with neck elevated and walks in a crouch. Forages mostly on buds while standing on the ground or atop bushes. A tame bird that flushes reluctantly.

FLIGHT: In manner and appearance similar to Rock Ptarmigan. See description for that species.

VOCALIZATIONS: The male's low, croaking, clucking, chortling soliloquy sounds as if it were demanding (of an unsympathetic universe?): How? What? and Where? Example: "*(h)ow . . . (h)ow . . . wha Wha WHA Where?Where?Where? Ow whawhawha Where?*" The utterance starts out slow and philosophical but gains in speed, volume, and urgency. Overall much more varied (and theatrical) than Rock Ptarmigan's strangled call.

Rock Ptarmigan, *Lagopus mutus*

Barren Grounds Ptarmigan

STATUS: Common resident of arctic and alpine tundra throughout the Northern Hemisphere; both overall and in places where birders are likely to go, somewhat less common than Willow Ptarmigan. **DISTRIBUTION:** Widespread in Alaska, including the Aleutians, the Yukon, and across arctic Canada. Found at higher elevations south through B.C. and on heath lands in w. Nfld. In winter, range extends very irregularly south to n. Alta., n. Sask., n. Man., n. Ont., cen. Que., and s. Lab. **HABITAT:** Dry, open, rocky, hummocky or hilly arctic and alpine tundra; favors drier and more barren (less shrubby) areas than Willow Ptarmigan. In winter, *snow!*—and shrubby habitats above the tree line, but avoids the willow thickets and forest favored by Willow Ptarmigan. **COHABITANTS:** In summer, lichen, American Golden-Plover, Long-tailed Jaeger, American Pipit, Caribou. **MOVEMENT/MIGRATION:** Most populations are nomadic but nonmigratory; however, birds living at northern reaches of the breeding range move south to the tree line (and females to clearings below the tree line). Spring migration from late April to early June; fall from September to November. VI: 1.

DESCRIPTION: It's the ptarmigan living on the rocky, open, windswept slope or hilltop. A small grouse and medium-sized ptarmigan (very slightly larger than White-tailed Ptarmigan; very slightly smaller than Willow Ptarmigan), with a small head, a long neck, a plump body, a short tail, and short feathered feet. The bill is slightly smaller, narrower, and more petite than Willow Ptarmigan.

Males have three plumages—winter, spring (or transition), and late summer—and females have two—winter and summer (transition is rapid). In winter, both males and females are all white except for black tails (usually hidden), scarlet eyebrows (males only), and a small dark line through the eye (also seen on some females). In summer, males are grayish brown above and white below. Females are overall warm brown, with fine black spotting above and barring below. They look just like rocks and like female and young Willow Ptarmigans.

In spring (early April to late June), males are piebald, showing dark (grayish brown) heads and necks and a mix of dark and light feathers throughout the body (a pattern not seen in Willow Ptarmigan). In snowless terrain, the birds stand out.

BEHAVIOR: In summer, may be found alone, in pairs, or in groups of females with chicks or groups of single males. In winter and migration, sometimes gathers in flocks of several hundred birds. In summer, forages on vegetation on open tundra, and in winter, on vegetation poking through the snow (may be attracted to areas where snow has been removed by caribou, musk ox, and wind). Crouches when feeding, nipping buds with a sideways twist of the head. In spring, males engage in aerial flight displays during which they rise steeply into the air, stall, and parachute to the ground on open down-cupped wings (vocalizing begins at the onset of the stall). Very tame, allowing close approach. Flies reluctantly.

FLIGHT: Overall stocky, with a long neck, broad blunt wings, and a short wide tail. In winter, all white (except for black outer tail); in summer, shows dark upperparts and all-white wings. Flight is rapid, direct, and slightly tipsy, with wingbeats given in a hurried flurry followed by a short (or lengthy) open-winged glide with wings slightly down-drooped and turned back.

VOCALIZATIONS: Somewhat less vocal than Willow Ptarmigan. Most vocalizations sound like low croaking rattles—"*kara'a'ah*" or "*kah-ki-kah-ka'a'a'ar*"—as if making its last dying gasp. The sound is very unlike the guttural, varied, clucking, croaking, chortling (and comical) incantation of Willow Ptarmigan.

PERTINENT PARTICULARS: The question remains: How do you tell Rock and Willow Ptarmigans apart? In winter, male Rock has a black eye-line; Willow shows a black dot of an eye. In summer, male Rock is overall cold gray-brown; Willow has a rich deep chestnut–colored head and neck. Females are more difficult. Bill size is helpful; habitat, particularly in summer, is at best a clue.

White-tailed Ptarmigan, of course, always has a white tail. Rock and Willow always have black tails.

White-tailed Ptarmigan, *Lagopus leucurus*

Alpine Ptarmigan

STATUS: Uncommon and found only at high altitudes. **DISTRIBUTION:** At higher elevations from cen. Alaska and the e. Yukon and w. Northwest Territories south through B.C. and the southwest edge of Alta. into cen. Wash., extreme n. Idaho, and w. Mont. Also found in pockets in the Colo. Rockies and n. N.M. Reintroduced into the cen. Sierra Nevada in Calif. **HABITAT:** In summer, open alpine tundra at or above the tree line. Favors rocky habitat, low hilltops, wet areas adjacent to snow fields, and stream-side willow thickets. In winter, found predominantly in sheltered willow thickets and among stunted spruces bordering the alpine zone. In years with heavy snowfall, may occur well below the tree line. **COHABITANTS:** In summer, American Pipit, White-crowned Sparrow, rosy-finches. **MOVEMENT/MIGRATION:** Permanent resident, but moves to lower elevations in winter depending on the severity of the winter and the availability of food (leaves and buds). Spring migration from early April to early June; fall from late September to mid-November. VI: 0.

DESCRIPTION: A feathered oval rock with a small head and a short tail. A plump small grouse (larger than California Quail; smaller than Chukar), White-tailed is the smallest of the three ptarmigan species. When alert, stands with head and tail raised; when feeding, head and tail droop. When resting, looks just like a rock—granite in summer, white marble in winter.

More specifically, males in summer are brownish gray above, with blackish and dirty white mottling and a slightly grayer head. White underparts sport a distinctive black spotted breast. (Breasts on Rock and Willow are all dark to splotchy, but not spotted.) Females are overall grayish brown with fine black-and-yellow gilding to the feathers. Both male and female are overall colder and grayer than other ptarmigan species. Their white outer tail feathers easily distinguish them from Rock and Willow Ptarmigans (black tail feathers).

In winter, except for the black eye and black bill, the bird is utterly white. Other ptarmigans have black sides to the tails.

BEHAVIOR: Spends a great deal of time imitating a rock. Usually found (or not found) crouched amid professional rocks, and often atop a small knoll (for visibility). Sluggish feeder. Forages on the ground, head down, shuffling along at a painfully slow pace. Plucks leaves, buds, and flowers. In early summer, found in pairs; mid to late summer, males form small groups, as do females with chicks; in fall and winter, all birds form small to large flocks (50 birds or more). Moves by walking or running. Occasionally seeks cover in matted willows or ground-lying conifers. Flies strongly but rarely.

Extremely tame. You can approach to within inches of a White-tailed Ptarmigan without making it move, and if it's foraging in your direction, if you remain stationary it may simply walk by you. Frequently males can even be coaxed by poor imitations of their calls to walk about the feet of observers.

FLIGHT Strong but heavy flier. Stocky profile with very wide rounded wings. Wingbeats are rapid and steady, followed by lengthy, slightly down-drooped, and somewhat unsteady glide. When flushed, birds may fly great distances, downhill and out of sight.

VOCALIZATIONS: Male's call is a clucking chatter with unmistakable poultry yard overtones: "*wrk wrk wrk WHAAH!*" or "*wrk wrk wrk wrk wahha-haha*" (chuckled ending). Also makes a thin scream, "*reak!*" when provoked. Both sexes make a low clucking "*wrk . . . wrk*" or "*wrk'k.*"

Blue Grouse, *Dendragapus obscurus*

Meso-Grouse

STATUS: Fairly common resident of western mountains and foothills. **DISTRIBUTION:** Breeds from se. Alaska, se. Yukon, and sw. Northwest Territories through all but ne. B.C., sw. Alta., n. and w. Wash., w. and ne. Ore., n. and e. Calif. (Sierra Nevada), Idaho, w. Mont., mountains of Nev., Utah, Wyo., Colo., n. Ariz., and n. N.M. **HABITAT:** Breeds in a variety of habitat types, including coastal forest, open montane forests (young and old), shrub-steppe high desert (dominated mostly by sagebrush or bitterbrush), and alpine and subalpine tundra areas. In all cases, favors ground covered by a rich grass or shrub layer. Also partial to aspens. In winter, found almost exclusively in coniferous forests. **COHABITANTS:** Mountain Quail, Ruffed Grouse, Spruce Grouse, Steller's Jay, Dark-eyed Junco (in forests); Greater Sage- and Sharp-tailed Grouse (in shrub and grass); snowshoe hare. **MOVEMENT/MIGRATION:** In fall, moves from more open breeding areas to denser conifers, and thus is usually moving to higher elevations. Adult males' fall retreat into conifer forests begins in mid-June and extends to late October; shift to breeding areas occurs in April. Females and young often remain in subalpine meadow edges all summer. VI: 0.

DESCRIPTION: Big, dark, long-necked, long-tailed, slate blue grouse. Large (larger than Spruce or Ruffed Grouse; considerably smaller than Greater Sage-Grouse), with a smallish head, a long neck, a robust body, and a long broad tail. Distinctly longer-necked and longer-tailed than Spruce Grouse.

Overall darker and plainer than other grouse. Males are mostly bluish or sooty gray, showing almost no patterning below. Females and young are likewise mostly dark and unpatterned—sooty or gray-brown above, with a dark, bluish brown wash across the chest and touches of white on the belly (no barring as on Spruce or Ruffed Grouse). Most individuals show paler grayish band on the tip of the dark tail. Displaying males show a galaxy of white spread across the undertail coverts of otherwise dark, fanned tails.

BEHAVIOR: In summer, spends much of its time on the ground, but also feeds on buds of deciduous trees and shrubs; in winter, becomes mostly arboreal and sedentary, feeding almost exclusively on conifer needles from the vantage of elevated branches, hopping from branch to branch (often circling the tree in the process). Males on the West Coast usually vocalize from trees (and vocalize more often); interior birds vocalize from the ground (and not as often). Not a social species; in both spring and fall, birds are most commonly found alone, but in winter, groups of two to six birds are not uncommon. (Larger groups of up to a dozen birds are exceptional.)

A tame bird that, when approached, crouches and slinks to safety but at times runs and, particularly when surprised in the open, flies (if only to the closest elevated perch).

FLIGHT: Larger, longer-necked, and longer-tailed than Spruce Grouse. Males show an all-blackish tail; females have black sides with a brown core. Flight is strong, with wingbeats given in a rapid series followed by a glide. Birds may fly vertically to enter trees and glide long distances when flying from the tops of trees or downhill.

VOCALIZATIONS: Vocalizing males emit a series of five to six low hoots. Birds also make an assortment of low clucks.

PERTINENT PARTICULARS: Although not an uncommon bird, Blue Grouse's propensity to be in tree interiors or the heavy shrub layer often makes it difficult to find. In summer, look for Blue Grouse by driving roads that bisect good cover in the early morning. Females with chicks commonly feed along roadsides and meadow edges.

Sharp-tailed Grouse, *Tympanuchus phasianellus*

Grouse Where the Grass Meets the Trees

STATUS: Fairly common widespread, primarily northern and northwestern resident. **DISTRIBUTION:** There are two population strongholds: the northern prairies encompassing cen. and se. Alta., s. Sask., s. Man., sw. Ont., all but w. Mont., n. and e. Wyo., most of N.D., all but e. S.D., w. Neb., and n. Minn.; and n. and cen. interior of Alaska and w. Yukon. Disjunct populations are also found in s. and ne. B.C., ne. Wash., se. Idaho, parts of Colo., n. Wisc., n. Mich., and scattered locations in Ont. and w. Que. Also occurs (sparingly) across much of the Northwest Territories and the northern prairie provinces. **HABITAT:** Where grass and trees/shrubs meet. Primary habitats include prairie grasslands, broken forest with a grassland component, clear-cut boreal forest, and northern taiga/muskeg. Also found in agricultural land, particularly grain fields. In winter, concentrates in adjacent riparian, deciduous, and coniferous woodlands. Leks are most often situated on an elevated knoll with sparse vegetation.

COHABITANTS: Northern Harrier, Upland Sandpiper, and *lots* of other chickens, including Ring-necked Pheasant, Greater Sage-Grouse, Greater Prairie-Chicken. **MOVEMENT/MIGRATION:** Permanent resident. VI: 0.

DESCRIPTION: Pale spotted chicken with a pale pointy-tipped tail. Medium-sized (about the same size as Ruffed Grouse and Greater Prairie-Chicken), with a slightly peaked head, a fairly long neck, a robust body, and a longish wedge-shaped tail with projecting, blunt, central tail feathers. Sharp-tailed looks like it's in the terminal stage of its molt and the only feathers it hasn't dropped are the central pair.

Overall pale brown or brownish gray, with a galaxy of small white spots all over the upperparts and the breast, making the bird appear speckled or "frosted," depending on the distance. The whitish belly shows dark V-shaped feathers down the flank, but if you can see the pale belly, that is sufficient. Prairie-chickens are slightly warmer-toned and overall barred, not spotted; the barring is heaviest on the underparts, including the belly and flanks. The neck sacs of displaying males are purple, not reddish or orange, as on prairie-chickens. The longer, pointy tail shows pale sides that are often apparent even when the bird is on the ground (for example, at a lek).

BEHAVIOR: A social species. Found in flocks numbering several individuals to 20 or more for most of the year. In spring, males gather on communal leks, at times with other chicken species. Unlike other grassland chickens, Sharp-tailed is fond of shrubs, trees, and forest edge. In spring and winter, sometimes perches on treetops and in fact may fly from the top of one tree to land atop another. In winter, forages among tree branches, particularly when snowpack covers the ground, and spends much of its time in wooded cover; roosts in trees as well as on the ground. Despite its arboreal predilections, Sharp-tailed spends most of its time on the ground, particularly during warmer months. Forages actively at dawn and dusk, but all day in winter.

Fairly tolerant of approach. When flushed, flies strongly for about 100 yds. (or more).

FLIGHT: Overall robust, with short, blunt, broad wings and a narrow wedge-shaped tail showing distinct pale sides. (The tails of prairie-chickens are fanned, blunt, and dark.) Flight is straight, slightly unsteady, and often fairly high (for a chicken), consisting of a short series of rapid and somewhat floppy wingbeats followed by lengthier glides. Usually calls when flushed or when nervous.

VOCALIZATIONS: When flushed, makes a low "*kuh kuh kuh kuh kuh*" that sounds like a cross between a caw and a cluck. Displaying males make a gurgling gobble, "*ga'kagoop*," and a pigeonlike coo, "*w'hoo*"; they also make a cackle and a popping sound that has been likened to a cork pulled from a bottle.

Greater Prairie-Chicken, *Tympanuchus cupido*

Dances with Extinction

STATUS: Eastern population (Heath Hen) is extinct; Texas coastal population (Attwater's Prairie-Chicken) is nearly extinct; Greater Prairie-Chicken remains relatively common in the core breeding areas, but the population has declined in measure with the loss of native prairie habitat, and the bird now occupies only a fraction of its historic range. **DISTRIBUTION:** Now found in N.D., S.D., nw. Minn., Neb., ne. Colo., and n. Kans.; disjunct, remnant populations occur in Ill., Mo., and Okla. **HABITAT:** Native tall and mid-grass prairie. In spring, chooses slightly elevated or open areas (farmland, dirt roads, overgrazed pasture) for leks. **COHABITANTS:** Sharp-tailed Grouse, Upland Sandpiper, Horned Lark, Western (and Eastern) Meadowlark. **MOVEMENT/MIGRATION:** Some birds remain year-round in breeding areas; others relocate, sometimes more than 100 mi., with peak movements in June and in October and November. VI: 0.

DESCRIPTION: Robust prairie-chicken with encircling barred plumage (which helps to distinguish it from Lesser Prairie-Chicken) and a rounded tail (separating it from Sharp-tailed Grouse). Medium-sized (except for Attwater's race, slightly larger than Lesser Prairie-Chicken; smaller than Ring-necked Pheasant), with a small head, a short, sometimes flabby neck, a small chickenlike bill, a plump body, short sturdy legs, and a short blunt-tipped tail.

The mostly warm brown body is fully encircled by narrow dark and pale bands (the bird looks like it's been turned in a lathe), with a pale throat and dark-topped tail (white undertail is visible when males display). The male's long neck feathers (pinnae) make nondisplaying birds look like they are wearing dreadlocks; pinnae project to the rear (like Mercury's wings) or stand erect (like rabbit ears) when males display on leks. Eyebrows (eye-combs) are the same bright orange or yellow-orange color as the male's air sacs, which are manifest when the bird "booms" at leks (and which wink through the neck feathers at other times). Plumage of females and immatures are like adult male but lack the bright orange eye-comb.

BEHAVIOR: Very social bird year-round. In spring, a dozen or more males orbit around a single lek, and females often visit in groups. For the balance of the year, birds form mixed flocks that may number 100 or more. Greater Prairie-Chicken feeds mostly on the ground (less frequently in trees), foraging in the methodical manner of chickens on an assortment of leaves, seeds, buds, and insects. Agricultural grains are an important part of the winter diet; young require insects.

Most encounters with this species occur during the spring, when prairie-chickens engage in their celebrated courtship displays. Males boom and dance to entice females. When dancing, males stand horizontally, raise their pinnae and tails, and execute a foot-pattering jig. Be aware that other species use similar breeding strategies, including Sharp-tailed Grouse and sage-grouse.

FLIGHT: Stocky and blunt-winged, with a short, rounded tail. Flight is strong, fairly fast, usually low to the ground, and often covers great distances. Note the uniformly dark uppertail. (Sharp-tailed Grouse's tail shows pale sides.) Flight is a fairly lengthy flurry of rapid wingbeats followed by extensive and somewhat unsteady glides. Glides with wings cupped down.

VOCALIZATIONS: The male's boom (which may carry great distances on still mornings) has been likened to the sound of someone blowing into a beer bottle, as well as to a distant foghorn. The boom is three syllables, but don't be surprised if you hear only two: "*w'hoo'hoo*" or "*whoo doo dooh.*" Booming males often warm up by making a low, muffled, chuckling "*ruh ruh ruh ruh raa raa,*" the first part being faster than the second. Both sexes cluck and make a whoop: "*pwoik.*"

PERTINENT PARTICULARS: The challenge posed by the separation of Greater and Lesser Prairie-Chickens is mostly academic. The ranges are mostly distinct: very rarely do the birds overlap in western Kansas and western Oklahoma. Lesser Prairie-Chicken is overall slightly paler and grayer, and displaying males show reddish (not orange or yellow-orange) air sacs. The species can interbreed, and hybrid types are called, predictably but not inaccurately, "Guesser Prairie-Chickens."

Of more practical interest is the separation of Greater Prairie-Chicken and Sharp-tailed Grouse, a prairie grouse whose range overlaps with Greater's and which may use the same leks. Sharp-tailed Grouse is slightly colder-toned and slightly more crested and has a longer, pointier tail that is dark only at the tip. The tail of Greater Prairie-Chicken is short, round, and blackish above. Also, Sharp-tailed Grouse shows an array of white spots above (barred on Prairie-Chicken), as well as a dark chest and a pale lightly marked belly. (Prairie-Chicken is wholly barred below.)

Lesser Prairie-Chicken, *Tympanuchus pallidicinctus*

Chicken Garnished with a Slice of Mango and a Dollop of Raspberry Sherbet

STATUS: Uncommon and geographically restricted prairie resident; less common and widespread than Greater Prairie-Chicken. **DISTRIBUTION:** Found exclusively in se. Colo., sw. Kans., w. Okla. (along the Tex. border), the Texas Panhandle, and e. N.M. **HABITAT:** Mid-grass prairie community. Partial to habitats offering a mix of native low shrubs and mixed prairie grasses atop sandy substrate. Also found in shinnery oak/bluestem (grasses) habitat. Leks are typically located atop sparsely vegetated ridges or knolls. **COHABITANTS:** Horned Lark, Western Meadowlark, Cassin's Sparrow. In winter, also found on agricultural land, where it forages on grain. **MOVEMENT/MIGRATION:** Nonmigratory. VI: 0.

DESCRIPTION: A prairie-chicken with bright orange eyebrows (eye-comb) and a pale cadmium red air sac. Displaying males look like they have a slice of mango over their eye and a generous dollop of raspberry sherbet on the neck. Medium-sized (very slightly smaller than Sharp-tailed Grouse and Greater Prairie-Chicken, except for Attwater's race); structurally similar to Greater Prairie-Chicken, with a small head, a small chickenlike bill, a shortish neck, a plump potbellied body, short, sturdy legs, and a short rounded tail.

With brown-and-white plumage, Lesser Prairie-Chicken overall is slightly paler and grayer than Greater Prairie-Chicken. Underparts are distinctly barred, with narrow brown and white bands. Upperparts are slightly darker and more complexly patterned—overall Lesser is more finely flecked or white-spotted than crisply barred. (Greater appears more uniformly and evenly barred above and below.) Lesser's pointy, dark-edged, whitish pinnae usually drape down the sides of the neck but stand rabbit-ear-erect when the bird displays. Females and young are like breeding males, but lack the bright orange eye-comb.

BEHAVIOR: Like Greater Prairie-Chicken, Lesser is a very social bird. Males gather in communal leks of a dozen or more individuals in spring; females commonly attend leks in small groups. In fall and winter, birds gather in large flocks that may number more than 50 birds, although males are sometimes separate. All foraging is done on the ground, either in native grasses or in grain fields. Birds feed methodically, waddling like chickens.

Fairly shy, Lesser Prairie-Chicken flushes when approached.

FLIGHT: Robust body, broad blunt wings, and short rounded tail. Wings are slightly paler than the back; the tail is distinctly darker (blacker) than the body. Flight is direct, fairly fast, and usually low, but birds traveling to and from feeding or roosting areas may fly much higher. A rapid flurry of wingbeats is followed by a lengthy glide on down-bowed wings. Birds land with a fluttering, hovering flurry of braking wingbeats.

VOCALIZATIONS: On leks, the most frequently heard male vocalizations include a rolling gurgle that often begins with two spaced notes followed by a series, "*w'whulp w'whulp w'whulpw'hulpw'hulp-w'hulp,*" a sound that suggests the rolling-thunder ripple of thin sheet metal being shaken. Facing-off males often duet, alternating phrases. Lesser Prairie-Chicken also makes a loud, excited, rising-and-falling laughing sequence, "*rah rah RAH RAH rah'rah rarara,*" which recalls Laughing Gull; a descending grebelike chuckle, "*re'hehehe*"; and a moorhenlike bugle, "*re/uh, re/uh.*"

PERTINENT PARTICULARS: Be mindful that Lesser and Greater Prairie-Chickens are very rarely found together in Oklahoma and Kansas.

Wild Turkey, *Meleagris gallopavo*

The Great American Forest Fowl

STATUS: Common widespread forest resident now expanding its range in response to active management and, particularly in the East, forest regeneration and maturation. **DISTRIBUTION:** Exhibits a patchy distribution, but found in every U.S. state as well as extreme s. Canada and Mexico. Most heavily concentrated in the e. and cen. U.S. **HABITAT:** A mature-forest species bird that prefers oak-hickory woodlands in the Northeast, oak-pine forests in the South, cypress swamps in Florida, mesquite grasslands in the Southwest, and oak-savanna in California. Also frequents open agricultural land and has become increasingly suburbanized. **COHABITANTS:** Varied, but often found in same habitat as White-tailed Deer. **MOVEMENT/MIGRATION:** Nonmigratory. VI: 0.

DESCRIPTION: A very large, overall dark, and surprisingly slender version of the domestic turkey familiar to all. Can be confused only with Turkey Vulture, with which it shares a superficial likeness.

The body is long and humpbacked, but otherwise (except for displaying males) athletically lean. The neck is thin and long, the naked head comically small, the legs very long and sturdy, and the tail long and wide. Runs with head up. Forages with body held horizontal.

Overall dark and blackish, with bronze iridescence in the feathers except for the pale flight feathers that form a buff-colored triangle near the base of the tail. Displaying males, with tails fanned and body feathers fluffed, resemble the classic image emblazoned on Thanksgiving place mats and storefront decorations.

BEHAVIOR: Forages in flocks during the day, either in woodlands or in fields. Roosts at night high in trees. Birds flying from and (particularly) into roosts are extremely noisy, often breaking limbs and branches en route to a perch. The forest floor where Wild Turkeys have fed looks like it's been raked in patches.

A wary bird that generally walks (or runs) when approached. Flies, when startled or pressed, with a short running takeoff and a noisy flap of wings.

FLIGHT: Flight is strong and labored on steady wingbeats until the bird reaches altitude, at which point it glides on broad, planklike, down-cupped wings, with the tail fanned. Glides may cover long distances (particularly downhill), and Wild Turkeys have even been known to catch a thermal and do a soaring turn. Even very young turkeys (under two weeks old) are capable of flight. When startled, they fly (with great effort) to an elevated perch.

VOCALIZATIONS: The descending "gobble" of males in spring hardly warrants description. The female's scratchy yelp, which is usually given in a series of three or more, has a plaintive or curious quality. The alarm call is a sputtery "*putt.*" Individuals in flocks also make conversational clucks and purrs as they move.

PERTINENT PARTICULARS: Turkey Vulture, another all-dark bird with a naked head, often sits in trees during the day. After leaving their roosts, Wild Turkeys forage on the ground. Turkey Vultures may gather in a small group over a carcass and never graze or forage; they walk with a waddle (not a dignified strut) or hop and fly when approached (rather than run). Although Turkey Vultures commonly stand in open fields, only rarely will you encounter one on the forest floor.

Mountain Quail, *Oreortyx pictus*

Plumed Quail

STATUS: Fairly common to uncommon far-western resident, but restricted to higher elevations. **DISTRIBUTION:** At higher elevations in Wash. (very local), Ore., much of Calif., the western edge of Nev., and n. Baja. **HABITAT:** Hillsides dominated by dense shrubs and brush in an assortment of open or forested communities, from mountain forest to chaparral to desert scrub-brush. Found at elevations of 1,500–10,000 ft. Avoids habitat dominated by grasses. **COHABITANTS:** Varied, but includes Blue Grouse, Fox Sparrow (in mountain forest), California Quail, Wrentit, California Thrasher, Sage Sparrow (in desert scrub). **MOVEMENT/MIGRATION:** Moves to lower elevations (below snow level) in November and follows the retreating snow line north in the spring. Migration is believed to be mostly or wholly on foot. VI: 0.

DESCRIPTION: Large, tassel-plumed, short-tailed, and distinctively marked quail. A portly bird (slightly larger than California Quail), Mountain Quail is distinguished by a long neck and a long, thin, spiky plume that stands erect or, in high winds, billows behind the bird.

Often all you see of the bird is the head and breast, which in adults is mostly bluish gray, going olive brown on the back and tail. The maroon throat and flank patch are difficult to see at a distance or in shadow (where birds appear all dark), but the white border to the throat patch (which looks like the track of a tear down the bird's face) and the bold white slashes along the flanks (which make the bird look like it was finger-painted by a toddler) stand out. Immatures show the longish plume, the throat patch, and the pale outline of the adult's flank pattern.

BEHAVIOR: Generally secretive, confining itself to heavy brush and not venturing far from cover. Birds are most often seen early in the morning, when small flocks and family groups frequent (or at least cross) roadsides and in spring when males "crow." Crowing males commonly choose an elevated perch (often a boulder and less often a bush or tree, although Joshua trees are a favored perch in the Mojave Desert), but birds also crow on the run. Vocalizations begin at dawn and continue until midmorning; some crowing continues in the evening.

A shy bird that does not usually tolerate close approach and commonly retreats by running. Seems loath to fly. Flushed birds fly a very short distance, then run; even birds flushed from trees commonly fly directly to the ground, not away. Outside the breeding season, Mountain Quails form small flocks, sometimes numbering only a few individuals.

FLIGHT: Overall blocky, with broad square-cut wings and a very short tail (shorter than California Quail). Despite the bold flank pattern, birds often appear all dark. After an explosive takeoff, a short series of rapid wingbeats is followed by a slightly unsteady glide on down-bowed wings.

VOCALIZATIONS: The male's crow is a loud, plaintive, whistled yelp with a slight quaver, "*qwark . . . qwark.*" It sounds more like the echo of a whistle than the whistle itself and may recall the single warm-up notes of Northern Bobwhite. Females may respond with a low, somewhat Ospreylike whistle, "*yrk yrk yrk yrk.*" The bird's variable alarm call has been described as a "creeking" ("*cree-auk cree-auk cree-auk*") that varies in volume, pitch, and intensity.

PERTINENT PARTICULARS: Both Mountain Quail and the more widely distributed California Quail can occur in the same general habitat. Where both are found, Mountain Quail tends to be at higher elevations, occupies steeper slopes, and favors brushier and/or more forested canopied habitat.

Scaled Quail, *Callipepla squamata*

Cotton Top

STATUS: Fairly common southwestern quail, but mercurial and cyclical, with population increases and declines. **DISTRIBUTION:** The s. U.S. and Mexico. In the U.S., found primarily in w. Tex., most of N.M., and se. Ariz.; also found in sw. Kans., extreme w. Okla., and se. Colo. **HABITAT:** Arid grasslands with some shrubs for cover; also found in the transition zone between grasslands and desert scrub. Avoids both overgrazed habitats and the heavily vegetated scrub habitat favored by Gambel's Quail, but the two species occur together in the transition zone. Also frequents human habitations, being attracted to backyard feeding stations and piles of lumber and junk (for roosting). **COHABITANTS:** Swainson's Hawk, Greater Roadrunner, Loggerhead Shrike, Bendire's Thrasher. **MOVEMENT/MIGRATION:** Nonmigratory; permanent resident. VI: 0.

DESCRIPTION: Fairly uniform, all-gray (or brownish gray) quail with scaly-patterned underparts and a conspicuously white-tipped peaked or crested head. About the same size as Gambel's Quail, but slightly stockier, more angular, thicker-necked, shorter-tailed, and topped off with a peaked crest completely unlike the teardrop black plume of Gambel's. Seems more alert than Gambel's, more heads-up.

Plumage is almost devoid of color; except for the scaly or chain-mail underparts, however, the pattern is somewhat reminiscent of Gambel's.

Head on, the expression is slightly disapproving; from the side, the bird looks taken aback. But if you are close enough to gauge the expression, you've already seen all you need to identify this species.

BEHAVIOR: In winter, a flocking species. Roosts on the ground, under some protective cover (such as low-lying mesquite or an opportune brush pile). Forages on the ground in the morning. Hides on the ground in shady cover during the heat of the day. Forages again in the evening. Not as noisy or as tame as Gambel's. Generally harder to find and harder to approach. Prefers to run when pressed. Takes flight when startled, but doesn't fly far.

FLIGHT: Flight is low, fast, straight, and generally short. Stocky profile, with short wide wings and no outstanding plumage characteristics except one: the white cotton tuft of a crest is apparent in flight. The crest is flattened, but the frosty paleness can be seen. A rapid and audible series of wingbeats is followed by a prolonged and near-gravity-defeating glide in which the bird angles its wings slightly down and rocks unsteadily side to side.

VOCALIZATIONS: Most commonly heard call is a low scratchy-sounding "*chur-chur*," or "*chuch'er, chuch'er.*" In spring, males have a throaty assertive "shriek" call, "*Qwe/ur*," which sounds a bit like Mountain Quail but is more single-noted. The sound also recalls the single introductory notes that sometimes precede the classic namesake call of Northern Bobwhite.

California Quail, *Callipepla californica*
"Chi-ca-go"

STATUS: The most widespread and, in most places, most common of the western quail. (The very similar Gambel's is a close second.) **DISTRIBUTION:** Found from s. B.C. south into (mostly) w. Wash., sw. Idaho, most of Ore., Calif. south through Baja, w. Nev., and n. Utah. **HABITAT:** Frequents scrubby brush in lowlands and lower hilly slopes. Representative habitats include chaparral, sagebrush, oak parkland, riparian corridors, humid woodlands that have been opened by human activity, and suburbia. Requisite demands include low, brushy cover (for roosting, hiding, and loafing), water, and open, sparsely vegetated habitat for foraging. **COHABITANTS:** Mourning Dove, Bushtit, Bewick's Wren, Western Scrub-Jay. Range overlaps with Mountain Quail, particularly in winter, when Mountain Quail descends to lower elevations. **MOVEMENT/MIGRATION:** Nonmigratory; permanent resident. VI: 0.

DESCRIPTION: Handsome medium-sized quail well known for its jaunty forward-drooping plume and—owing to its unscripted intrusion into the audio portions of assorted Hollywood movies—its chanted call. Classically plump, but like Gambel's Quail, longer, leaner, and longer-tailed than many quail species. The teardrop-shaped plume (large and forward-falling on adult males, short and erect on females and immatures) distinguishes this species from all but the similar, desert-dwelling Gambel's. California Quail is overall slightly darker and less contrastingly patterned than the desert species. Also, in all plumages California Quail has a pale forehead, a richly white-flecked hindneck, and a distinctly scaled belly. (Gambel's belly is buff with black streaks.) The ranges of California and Gambel's abut (or almost so) but do not overlap.

BEHAVIOR: In fall and winter, found in large coveys that are most frequently seen walking or running across open areas. Roosts in trees, forages in the morning and evening, and spends much of the day in the shade beneath a protective ground-hugging tree or tangle. Walks with head up and bobbing—the movement is a cross between a strut and a waddle. Changes pace and pauses frequently. Runs with head up and body leaning forward. Flies when nervous, and takeoffs are accompanied by a noisy whir of wings. Feeds primarily on leaves and seeds from grasses and shrubs by scratching on the ground or leaping and plucking from above. Also sometimes forages in trees. Generally feeds in the open but never far from cover.

FLIGHT: Low, direct, usually just above the ground, and generally directed toward the nearest cover. Short round wings move in a blur. A rapid series of flaps is followed by lengthy wobbly glides during which, with the body slightly angled up, the bird seems not to lose altitude.

VOCALIZATIONS: Very chatty bird. Classic call is a theatrical laugh, "*ra Ha, her,*" whose pattern and enunciation sound to some as though the bird were saying, "*Chi-ca-go.*" The chant is usually repeated several times, as though the bird were reciting a lesson. Contact calls, usually given in a sequence, include a soft double-noted "*wr't-wr't, wr't-w'rt,*" which sounds like the last of the ketchup being squeezed from the bottle, and a rapid sputtering (part spit, part tick) "*sp't,-sp't, sp't-sp't,*" given when birds see a predator (or a birder coming too close to the flock).

Gambel's Quail, *Callipepla gambelii*
The Sonoran Quail

STATUS: Common warm-desert quail of the American Southwest and Sonoran Mexico. **DISTRIBUTION:** Found principally in the Sonoran Desert region of se. Calif., Ariz., and Mexico; also found in s. Utah, w. Colo., scattered locations in N.M. and extreme w. Tex. **HABITAT:** Desert areas dominated by cactus and brush, particularly honey mesquite; desert canyons, washes, and semidesert grasslands with scattered shrubs and brush; suburban areas where some natural habitat survives. Wherever found, depends on a standing water source. **COHABITANTS:** Greater Roadrunner, Verdin, Cactus Wren, Black-throated Sparrow. **MOVEMENT/MIGRATION:** Nonmigratory; permanent resident. VI: 0.

DESCRIPTION: A plump, longish-tailed, and distinctively patterned quail that is easily distinguished from Scaled Quail (whose range and habitat it overlaps), but less easily separated from California Quail (whose range it overlaps only slightly, if at all). Overall paler and slightly more contrastingly patterned than California Quail. Like California Quail, the round, domed heads of both males and females are adorned with a blackish teardrop-shaped plume that is large and drooping in males, small and mostly erect in females. Unlike California Quail, the flanks of Gambel's sport rust-colored pockets. This feature and/or the plain grayish hindneck is really all an observer needs to see to separate Gambel's and California (where geography has not done so already).

Scaled Quail is basically plain and gray with a distinctive and hard-to-overlook white-tipped crested head (the cotton top). Chukar, an introduced species that thrives in rocky desert canyons, is larger, more robust, and shorter-tailed than Gambel's. Plumage differences easily separate the species, and when you're observing flocks, those differences are fundamental and manifest. With Gambel's, male and female plumages differ—thus, flocks look mixed. With Chukars, males and females are identical and so flocks are uniform.

BEHAVIOR: Highly social, flocking quail. Fairly tame where habituated to people, Gambel's generally runs for cover (rather than fly) when threatened. At dawn, leaves its elevated roost in a tree or shrub and heads for water. Feeds on the ground on vegetation, staying fairly close to cover, then commonly sits out the heat of the day beneath a low-hanging tree or shrub. Forages again in the evening. Foraging birds move slowly, pecking as they go and vocalizing frequently.

Easily attracted to backyard feeding stations and especially water drips or trays placed on the ground. Most birds are seen in these situations or when they run (or sometimes fly) across open washes or highways. When flushed, may travel a fair distance, and commonly keeps running when it hits the ground.

FLIGHT: Explosive takeoffs. Flight is direct and fast. See "California Quail."

VOCALIZATIONS: Call is similar to California Quail's "*Chi-ca-go,*" but pronunciation is more emphatic and compressed—the last note is clipped short—and less distinctly three syllables. Also makes a loud descending wail that sounds somewhat peacocklike: "*Wha-eh-r?*" (drags "where?" out to three syllables). Often just before flushing, makes a rapid metallic spitting sound, "*pit* (pause) *p't, p't, p't,*" that is higher-pitched than California. Other flock sounds include low snorts, soft brays, and sputtery mutterings.

Northern Bobwhite, *Colinus virginianus*

Actually the Eastern *Bobwhite (Can't Abide Heavy Snow Cover)*

STATUS: Common to uncommon, predominantly eastern resident, but in many places declining. **DISTRIBUTION:** The eastern quail; range extends north to Mass., s. N.Y., se. Ont., Mich., s. Wisc., n. Iowa, se. S.D., cen. Neb., and se. Wyo. Western limits defined by e. Colo., extreme e. N.M., just east of Big Bend, Tex., and south into e. and s. Mexico. Attempts to reintroduce the Masked (southwestern) form to s. Ariz. are ongoing. **HABITAT:** Grassy or brushy fields, croplands, and early successional habitats (such as logged or burned forest); also parklike pine forest, mixed pine-hardwood forest, fallow land, and grass-brush rangeland. Optimal habitat offers a mix of fields, brush, and forest subject to periodic burns. **COHABITANTS:** Bachman's Sparrow, Field Sparrow, Eastern Meadowlark. **MOVEMENT/MIGRATION:** Permanent resident. Some seasonal shifts between higher and lower elevations noted in the Smoky Mountains. VI: 0.

DESCRIPTION: Plump, plumeless, pale-throated quail. Over most of its range, the only quail likely to be encountered. Small, robin-sized (but not robin-shaped!), roundly contoured quail with a round plumeless head, a short neck, an oval body, and a short tail. Forages in a supplicant-like crouch. Walks (and runs) with head elevated. When nervous, stretches its body and neck so that it looks like it's standing on tiptoe.

Overall dark ruddy brown with busy patterning above and a finely barred paler belly below. The overall dark intricacy of the bird makes the plain, white throat and white eyebrow of males (buffy on females and immature) stand out.

BEHAVIOR: Very social quail. For most of the year, found in coveys (3–25 birds) that feed together on the ground searching primarily for seeds and agricultural crops (such as corn, soybeans, and wheat). Coveys roost together, forming circles on the ground with tails pointed in and heads turned out; coveys also roost in trees.

From spring into late summer, males vocalize in the mornings and evenings, often from a moderately elevated perch (for example, within a bush, low in a tree, or on a fencepost). In breeding season, imitations of their distinctive call can stimulate birds to call at any time of day, and with persistence, they can also be coaxed to approach.

FLIGHT: Extremely compact, with short rounded wings. Explodes into flight with a rapid and audible whirring of wings, angling up and away, leveling off about eye level. Alternates rapid series of wingbeats and short glides. Flight is slightly twisty-turny.

When coveys flush, birds fan out, looking like a three-dimensional break on a pool table. Birds commonly do not fly far.

VOCALIZATIONS: Classic call is a loud whistled "*too-Wheet!*" ("*Bob White*"). Males sometimes add a prefix to this basic call consisting of several loud whistled warm-up notes: "*tooee . . . tooee . . . tooee . . . too-Wheet!*" Also emits a loud, harshly squealed "*Queeak!*" The contact call—used when scattered coveys reunite—includes a soft "*hoy,*" a louder "*hoy-poo,*" and a loud clear "*koy-lee.*"

Montezuma Quail, *Cyrtonyx montezumae*

Like a Rock or a Rolling Stone

STATUS: Uncommon, geographically restricted, and habitat-specific. **DISTRIBUTION:** A Mexican species; in the U.S., found in s. Ariz., sw. N.M., and scattered locations in trans-Pecos and sw. Tex. **HABITAT:** Thrives in the narrow habitat band between desert lowlands and forested elevations characterized by dry grassy hillsides with scattered live oaks, junipers, and occasionally mesquite or other trees (generally found at 4,500–6,000 ft.). Definitely not a desert species. Does occasionally come to backyard feeding stations where seed is scattered on the ground and water is offered. **COHABITANTS:** Acorn Woodpecker, Bridled Titmouse, Mexican Jay (but does not associate with these—or other—species). **MOVEMENT/MIGRATION:** Permanent resident. VI: 0.

DESCRIPTION: Males look like oval rocks dressed for Mardi Gras; females look like oval rocks dressed in fallen oak leaves. Small (smaller than Northern Bobwhite), plump, and round quail, with an exaggeratedly round head, an insignificant nub of a tail, and short but conspicuous legs.

Adult males are dark, with a distinctive, almost arresting harlequin-patterned party mask and white-spotted sides. Females, except for head shape and pattern (a toned-down version of the male's), are overall oak-leaf brown and distinctive in their lack of a conspicuous pattern. They look somewhat like dumpy dowagers with clown-sad expressions and their hair tied in buns.

BEHAVIOR: Spends most of its life not being seen. Found singly or in pairs and, in winter, in small flocks (commonly less than a dozen birds). Forages on the ground by scratching; moves with excruciating slowness. When approached, turns to stone. When flushed, explodes into flight, flies a short distance (30–100 yds.). Does not perch in trees. Does not vocalize to the degree most quail species do.

FLIGHT: In flight, looks like a stone with short round wings. Takes off with a loud "popping" of wings. Flaps for one or two seconds and then glides, often landing with a sharp banking turn and a tumbling roll (clearly not a bird designed to perch on limbs).

VOCALIZATIONS: Usually silent. Males make a loud, eerie, descending whistle aptly described as the sound of an aerial bomb falling or a circular saw cutting through a board: "*weerrrrrr.*" Females make an owl-like call consisting of a series of descending metallic-sounding notes. Flocks also make soft mewing sounds. Other sounds have been reported.

PERTINENT PARTICULARS: The challenge with this species is location, not identification. The most successful strategy appears to be to drive roads that bisect the birds' habitat for the first several hours after dawn, in the hope of surprising them on the road. Once located, birds may be amazingly compliant, allowing prolonged viewing and close approach, or they may slowly sneak away. If approached on foot, generally freezes.

LOONS

Red-throated Loon, *Gavia stellata*

The Smiling Loon

STATUS: Common, principally high arctic breeder, with a circumpolar distribution; in North America, heaviest concentrations are found near seacoasts. **DISTRIBUTION:** Breeds along the west coast of Canada, throughout most of Alaska, in arctic Canada south to the shores of Hudson Bay and east to coastal Que., in n. Nfld., and along the west coast of Greenland. Breeds south to the mouth of the Gulf of St. Lawrence, Que., and Vancouver I., B.C. Winters coastally from the Aleutians to s. Baja and, on the Atlantic, from the Maritimes south to n. Fla. **HABITAT:** Breeds primarily on small to large ponds in tundra, wetlands, bogs, and forests. Winters along seacoasts and in shallow inshore waters and bays. **COHABITANTS:** In winter, may be found near other seabirds (such as Pacific Loon, Common Murre, Horned Grebe), but the association seems serendipitous. **MOVEMENT/MIGRATION:** Generally coastal, but Red-throated Loon may be found in staging concentration on the Great Lakes and Delaware Bay; a few can also be found on smaller inland lakes. Spring migration from March to May, with a peak in April; fall from October to December, with peak in November (on both West and East Coasts). VI: 3.

DESCRIPTION: Pale, slender, diving bird with a rapier-thin and slightly upturned bill that makes it look happy or smiling. (The distinctive red throat patch is not usually present except in late spring and when birds are nesting.) Red-throated is a medium-sized seabird but a small loon (larger than a murre; smaller than Common Loon; about the same size as Pacific Loon). Swimming Red-throated sits lower on the water than other loons, with its bill angled up (a trait shared only with Yellow-billed Loon). In all plumages, the head is slender and smoothly contoured and seems to be about the same width as the neck, giving the bird a snaky-necked look. The forehead slopes, and the face tapers into the bill. On other loons, the forehead is steeper and the bill more distinctly a projecting appendage.

In breeding plumage, the back is charcoal gray, the head is pale gray, and the underparts are very white. In nonbreeding plumage, Red-throated is dark above, with fine, pale speckling that makes it appear battleship gray at a distance (and the overall palest of the loons). The face and neck on adults and many immatures show more white than on other loons, and the demarcation between the dark cap and hindneck and the white face and throat is sharp and stark. Other loons in winter show more restricted white on the face and neck or a dingy face and neck pattern (as do some immature Red-throated Loons in fall and early winter). In Red-throated Loon, the overall pale gray upperparts and slim, upturned bill are useful field marks.

BEHAVIOR: Generally solitary in midwinter, but more social in late winter and in migration. Migrates in small flocks numbering several birds to more than sixty. (Staging birds may gather in large aggregations numbering in the hundreds.) Dives for fish, generally close to shore and sometimes inside the breakers. Not usually found well offshore, as is the case with Pacific and Common

Loons. Often very tame. Where habituated to people, may forage within a few feet of fishermen or strollers. Prefers to swim when pressed, but unlike other loons is able to rise from the water with only a short and (for a loon) nimble takeoff. In winter and during coastal migration, does not generally fly over land if a marine alternative exists. In summer, often found on small, shallow ponds.

FLIGHT: Overall slender profile; cold battleship gray above, stark white below (and often on the face). Head and neck are long and slender, with the head drooped below the body. Very thin bill seems to disappear at any appreciable distance but, if seen, will always be closed. (Common Loon often flies with its bill agape.) Face seems cheeky or jowly. Head and forehead are smoothly contoured, with no discernible bump. Body is clean-lined, somewhat humpbacked, but devoid of a paunch in front of the legs. Feet project behind and are small and hard to see.

Wings are slender and angled sharply back, giving the impression that Red-throated throws the wings back when it flaps. (The wingbeat of other loons is more up and down.) Flight is swift, direct, and less lumbering than the larger Common Loon. Wingbeats are quicker, stiffer, and less rubbery than Common Loon's, and regular, without skip or glide except when landing.

In flight, Red-throated tends to lift its drooped head and tilt its bill up at an angle, as if swimming underwater and trying to see what is going on up above. In winter, it usually flies close to the surface (under ten feet) but in migration may fly several hundred feet above the water (much higher over land). Flocks are loosely bunched, never set in a formation, and birds are spaced and spread out both horizontally and vertically; often several or more birds are strung out well behind the main group. Red-throated does not migrate with any species except other loons.

VOCALIZATIONS: On territory, makes a long, descending, gull-like wail, "*rehhhhhr . . . rehhhhhr.*" Also on territory (and sometimes in winter, particularly when multiple birds are gathered), makes a series of low croaking quacks, "*quah quah quah quah,*" that recall a hoarse, excited barnyard duck. Also reported to make a "low growl."

PERTINENT PARTICULARS: In flight, Red-throated is slender, smooth-lined, jowly, and no-billed, and compared to Common Loon, paler, grayer, whiter-faced, and more slender-footed, with quicker, stiffer wingbeats. It flies in disorganized flocks that are strung out horizontally as well as vertically. In winter, Red-throated does not readily cross land and does not typically forage far out at sea.

Arctic Loon, *Gavia arctica*

White-flanked Loon

STATUS: Uncommon and highly restricted Alaskan breeder; rare vagrant along the West Coast. **DISTRIBUTION:** In North America, breeds coastally on Seward Peninsula and eastern portions of Kotzebue Sound. Rare elsewhere. **HABITAT:** In its limited North American range, seems most partial to brackish lakes or tidal bays and lagoons (but makes extensive use of freshwater lakes in Europe and Asia). In winter, often found in calm, sheltered bays devoid of currents and turbulence (which also seem less attractive to Pacific Loon). **COHABITANTS:** Pacific Loon, Common Eider, Bar-tailed Godwit. **MOVEMENT/MIGRATION:** Spring migration from late May to early June, with peak the first week of June; fall from September to October. VI: 1.

DESCRIPTION: Pacific-like loon showing a triangular or wedge-shaped white flank patch. Medium-sized (larger than White-winged Scoter; slightly larger than Pacific Loon, but with some overlap; smaller than Common Loon) and nearly identical to Pacific Loon in both shape and plumage, with the following exceptions.

In all plumages, Arctic Loon shows a triangular or wedge-shaped white flank patch bulging prominently above the water line just in front of the tail. All other considerations (slightly larger bill than Pacific; more Common Loon–like GISS; more prominent streaking on the neck; a darker hood that lacks the prominent, pale nape of Pacific in breeding plumage; greenish versus bluish throat patch, also in breeding plumage) are virtually insignificant. Just use the flank patch. It works.

In winter, however, Pacific Loon often shows a shadowy "chin strap." Arctic does not. While the presence of a chin strap does not certify that the bird is Pacific, it does preclude the possibility of Arctic.

BEHAVIOR: Like Pacific Loon, although there is some evidence that Arctic favors calmer water. Like Pacific, forages in deep lakes and in the open ocean (sometimes with Pacific). When Arctic occasionally flies in tandem with Pacific during migration, its greater size is readily apparent.

FLIGHT: Like Pacific Loon, but appears somewhat more lumbering owing to larger size. White flank patch is less apparent, but distinctive enough. *Note:* If the demarcation between the dark upperparts and white underparts is a straight, horizontal line, the bird is Pacific. If the white flank swells above and behind the wing, it's Arctic. If you're not sure, it's Pacific.

VOCALIZATIONS: Makes a harsh croaking "*karrr*"; also a gull-like crying.

PERTINENT PARTICULARS: Be aware that nonbreeding Red-throated Loon also shows a white flank patch.

Pacific Loon, *Gavia pacifica*

The Perfect Loon

STATUS: Common arctic and subarctic breeder; common West Coast winter resident. Abundant West Coast migrant. Nonbreeding birds regularly found in winter range. Regular migrant (in small numbers) in the interior West. **DISTRIBUTION:** Breeds throughout all but se. Alaska (also in w. Siberia); in sw. Yukon, across the Northwest Territories to extreme n. Sask., n. Man., n. Ont.; along the east coast of Hudson Bay and coastally in n. Que. Winters in coastal waters from s. Alaska to Baja and the Gulf of Mexico. **HABITAT:** Freshwater tundra lakes and ponds. Compared to Red-throated Loon, prefers larger, deeper, and less vegetated bodies. In winter, found along inshore coastal waters as well as bays and estuaries. Prefers deeper offshore waters and often concentrates along the kelp line. Generally less numerous in the shallow estuaries favored by Red-throated Loon. Also, Pacific Loon may favor or tolerate rougher water than Arctic Loon, which (like Common Loon) seems partial to back bays. **COHABITANTS:** While breeding, Tundra Swan, Long-tailed Duck, Long-tailed and Parasitic Jaegers. In winter, Red-throated Loon, Brandt's Cormorant, Glaucous-winged Gull. **MOVEMENT/MIGRATION:** Entire breeding population relocates to coastal waters for most of the year, staging impressive (mostly) spring and fall migrations just off Pacific Coast beaches. Spring migration from late March to early June, with peak in California in mid to late April; fall from October to December, with peak in California in early November. VI: 3.

DESCRIPTION: Compact, well-proportioned, dagger-billed loon. Pacific is a medium-sized water bird (larger than Surf Scoter; smaller than Common Loon or Brandt's Cormorant; the same size as Red-throated Loon and Western and Clark's Grebes) with a straight daggerlike bill *held horizontally,* domed head, thick neck, and domed back. In comparison to the balanced proportions and smooth contours of Pacific Loon, Red-throated looks spindly, and Common Loon awkwardly robust.

In breeding plumage, the puffy, shiny, gray head/nape is distinctive, and the combination of a black body with a black-and-white-striped saddle is unique (except for the very similar Arctic Loon). The blue throat patch appears black in most light. Because the sides are black to the water line, *Pacific Loon shows very little to no white in the flanks.* In nonbreeding plumage, it is mostly plain brownish gray above, with white cheeks, throat, and underparts. Pacific is slightly darker than Red-throated overall, and dark-hooded rather than capped like Red-throated (which shows a very white face and neck). Also, Pacific usually shows a dark chin strap. At a distance, in combination with the darker head, Pacific Loon appears white-chinned, whereas Red-throated looks white-faced. Immatures are like winter adults but paler.

BEHAVIOR: Commonly forages near shore in the company of Western Grebe, shearwaters, Brandt's Cormorant, Western Gulls, and California Gulls (also Rhinoceros Auklets) and in turbulent areas. During migration, gathers in large aggregations—in excess of 500 birds in some places such as Monterey Bay (and groups of several thousand are not unprecedented). Migrates in small to medium-sized flocks (10–35 birds); individuals are most commonly within 30 ft. of the water and well within sight of land (out to several miles).

FLIGHT: Slender horizontal profile with head level with the body (drooped in Red-throated Loon). Feet are horizontal (and thus difficult to see), not twisted vertically, like Common and Yellow-billed Loons. Plumage is dark above, with wings that appear affixed to the body with pale (barred) welds in breeding plumage. In breeding and nonbreeding plumage, dark upperparts and pale underparts are defined by a straight horizontal line, with *no white bulge intruding onto the rump behind the wing.* In winter, most Pacific Loons show a hooded head that frames a white throat patch visible at great distances. Wingbeats are regular, elastic, steady.

VOCALIZATIONS: On territory, makes a loud two- to three-part wail with the high, shrill, lonely quality of a gull (especially Mew Gull) and something of the pattern of Long-tailed Duck: "*er-Awh'rer-EEH . . . er-Awh'rerEEH.*" Does not sound like a yodel (that's Common Loon) or distant wolves howling. Also makes a low, descending, croaking growl, "*ow'h'h'h.*"

Common Loon, *Gavia immer*

A Big Bruising Longshoreman of a Loon

STATUS: Fairly common; while declining, remains the North American loon with the greatest breeding and wintering distribution and thus is the most likely loon species to be encountered away from the coasts. **DISTRIBUTION:** Breeds across almost all of Canada and Alaska as well as in n. New England and the states bordering the Great Lakes, a scattering of other mostly Canadian border states, such as Wash., Idaho, Mont., Wy., N.D., and n. Rockies. Winters along both seacoasts, from the Aleutians to Baja and from Nfld. to Mexico, and inland in the southern half of the U.S. **HABITAT:** Nests on larger, deeper, fish-bearing freshwater lakes in northern forests as well as on larger, deeper tundra ponds. Winters primarily in coastal marine habitat, including inshore waters, bays, inlets, and shoals, but may also be found well offshore (out to 40 mi.). In winter and during migration, also occurs on inland lakes, reservoirs, and slow-moving rivers. Many second-year birds apparently spend the summer in marine waters. **COHABITANTS:** Often nests on lakes that attract Bald Eagle and Osprey. **MOVEMENT/MIGRATION:** Migrates over land and along seacoasts. Spring migration from March to early June; fall from late August to December. VI: 4.

DESCRIPTION: Large, bulky, dagger-billed loon (larger than all but Yellow-billed Loon; same size as Double-crested Cormorant). On the water, rides somewhat high in the front, low in the back. The front-heaviness stems in part from Common's large broad head; big, formidable, daggerlike bill; and short, thick neck.

In breeding plumage, the head and bill are black, the back is checkerboard-patterned, and the neck is encircled by a diagnostic and (except for the wholly arctic breeding Yellow-billed Loon) near-idiosyncratic black ring. In winter, adults and immatures are dingy brown (not gray) above, with a poorly defined dark head and neck and a whitish throat that sometimes shows a shadowy trace of the neck ring or a pale collar. Also, in winter Common Loon shows a pale eyelid or eye-ring that is lacking in Pacific Loon, and the head has a bump or peak on the forehead. The bill is grayish (sometimes with a black tip) and just as formidable-looking. *Note:* Breeding plumage appears as early as March and may be retained when adults arrive on winter territory.

BEHAVIOR: Outside of the breeding season, generally solitary. Hunts by diving for prey but may remain on the surface, putting head underwater in an attempt to see prey before pursuit. Needs considerable room (25 yd. or more) to take off. After a long approach, lands feet first, followed by a breasting glide. Generally less tolerant of approach than Red-throated Loon, staying farther offshore, and moves away if approached.

FLIGHT: Very fast, direct, and the opposite of maneuverable. Must make wide sweeping turns (covering hundreds of yards) to turn or climb above obstacles.

The jumbo bill, bump on the forehead, distending paunch just fore of the vent, and conspicuously large, trailing feet (turned sideways!) give Common Loon a lumpy, large, and bulky profile (definitely not slender or clean-lined). Its wingbeats are rapid, rubbery, and regular, and the wingtips flick up and down, with as much up as down. The cadence and execution suggest both Peregrine Falcon and Double-crested Cormorant. Common Loon migrates alone or in small groups that often fly higher than Pacific and Red-throated and are usually spread out across a wide horizontal plain (that is, unlike Red-throated Loon, Common Loon flocks are not three-dimensional). Common Loon usually flies with its bill agape, and given the size of the bill, this trait is easily seen. Also, Common Loon most often migrates over land; except where they breed, Pacific and Red-throated Loons are rarely seen flying over land.

VOCALIZATIONS: Eerie haunting yodel that is as familiar as the howl of the Gray Wolf. Generally silent in winter, Common Loon frequently yodels during migration.

PERTINENT PARTICULARS: In migration, Common Loon takes off between dawn and two hours after sunrise. If birds are migrating at a speed of 60–70 m.p.h., observers within several hundred miles of a coastline can often project their arrival time by calculating the distance, dividing it by their speed, and adding the resulting hour figure to the time of sunrise.

Yellow-billed Loon, *Gavia adamsii*

Tanto-billed Loon

STATUS: Uncommon and remote arctic breeder; uncommon winter resident along the northern Pacific Coast (rare elsewhere). **DISTRIBUTION:** Breeds across n. Alaska from n. Seward Peninsula (also St. Lawrence I.) to the Canadian border (greatest numbers are found on the North Slope) and the northern portions and islands of the Northwest Territories. Winters along the Pacific Coast from Kodiak I. and the mouth of Cook Inlet south to Puget Sound (where they are rare). Very rare or casual farther south along the Pacific Coast; casual vagrant inland. **HABITAT:** Breeds on open treeless tundra, usually coastal, using large, clear, ice-free lakes. In winter, found in coastal marine habitats, often in protected inlets, bays, fiords. **COHABITANTS:** Pacific Loon, Long-tailed Duck, Glaucous Gulls. **MOVEMENT/MIGRATION:** Mostly coastal migrant (with passage noted in the Bering Sea), but in fall reportedly stages on large inland lakes (such as Great Slave Lake), mandating some overland route. Spring migration from early May to late June; fall from late August to late October. VI: 2.

DESCRIPTION: Big burly bird similar to Common Loon, with a long ivory-pale bill that is slightly upturned on the lower edge, like a tanto knife. Largest loon (slightly larger than Common Loon and Double-crested Cormorant). Compared to Common Loon, it shows a longer, paler bill (bright yellow in breeding adults; pale ivory in winter and immatures) that is fairly straight along the upper edge and, as mentioned, conspicuously up-curved along the lower edge. Common Loon's bill is mostly straight, daggerlike, and turned up only at the tip. Also, Yellow-billed's elevated bill, like Red-throated's (Common Loon holds the bill level), gives it a haughty look. Yellow-billed also shows a more prominent bump on the noggin and has a heavier neck, which appears about as thick as the head.

Yellow-billed's breeding plumage pattern is similar to Common's, but the bright yellow bill contrasting with the black head is obvious at almost any range. When a distant bird is silhouetted against a pale sky, however, the bill may disappear. Also, Yellow-billed appears whiter-backed and seems to have a more uniformly narrow collar compared with Common Loon's, which is elliptical (wider on the side of the neck).

In nonbreeding plumage, Yellow-billed becomes pale grayish brown above; overall it is slightly paler and browner-toned than Common, with a more uniformly plain, pale, pastel-shaded face and neck supporting a shadowy ear patch. Common's head is more contrastingly patterned, with a crown and hindneck and white throat showing no ear patch. Yellow-billed immatures are like adults but even paler and browner (putty-colored); the barred pattern on the back makes the body look like it was turned on a lathe. (Darker-backed immature Common Loons appear scalier.) In winter, the bill of Yellow-billed, among all age classes, remains more distinctly and uniformly pale than Common's—ivory as opposed to silvery gray, and dark only on the upper base rather than on the entire upper surface.

BEHAVIOR: For the most part a loner. Pairs occupy (and forcefully defend) lakes in summer, but the bird is mostly solitary in winter. Stages in coastal waters and large interior lakes in fall. Commonly migrates alone, though sometimes is seen in tandem in well-spaced pairs or trios. In summer, usually forages in deeper waters far from shore (less commonly in grassy areas near shore). Upon arrival, hunts at the edge of the ice (and is suspected of feeding under the ice). Submerges head to search for prey and often kicks up a spray of water when diving.

Rather shy, Yellow-billed does not tolerate close approach from shore or by boat and often swims long distances underwater.

FLIGHT: Long-necked, heavy-billed, heavy-bodied, slender-winged. Looks like a big, lumbering Common Loon. Bill is prominent and heavy; feet are turned 90° (like Common Loon), with flat side showing. The adult's bright yellow bill is distinctive in spring and summer; the head, bill, and face pale in winter. Flight is direct—straight as the loon flies—with slow, regular, steady wingbeats.

VOCALIZATIONS: Calls like Common Loon, but at slightly lower pitch and with slower delivery.

GREBES

Least Grebe, *Tachybaptus dominicus*

Toy Grebe

STATUS: Common resident, but geographically restricted to south Texas. **DISTRIBUTION:** Tropical and subtropical species found across much of Mexico and Central America. Most concentrated in the U.S. in s. Tex. along the Gulf of Mexico and Rio Grande Valley, but breeds north to San Antonio. **HABITAT:** A variety of freshwater and brackish-water habitats, ranging from large open ponds with little emergent vegetation to roadside ditches. Most commonly found in small, shallow, well-vegetated permanent or seasonal ponds. **COHABITANTS:** Pied-billed Grebe, Common Moorhen. **MOVEMENT/MIGRATION:** Permanent resident. VI: 1.

DESCRIPTION: Tiny, dark, yellow-eyed, petite-billed grebe. "Toy Grebe" is *tiny!* It appears two-thirds the size of Pied-billed Grebe, and half the size of Common Moorhen. Its compact, dome-backed body has a large head, fairly thick neck, and distinctly shaggy and cropped rear. The narrow, pointy, black bill is either straight or slightly upturned. Least Grebe most closely resembles Eared Grebe in shape.

Mostly dark, Least has traces of white to the rear; its overall cold grayish brown body contrasts only slightly with its grayer face and neck. Pied-billed Grebe is overall browner and warmer-toned, particularly on the head and neck. Least's bright yellow eye stands out and makes it look startled or angry. (Pied-billed looks happy or content.)

BEHAVIOR: Found singly or in pairs, often maneuvering or resting in vegetation. Leans forward while swimming and often drops its eyes below the water line to search for prey. Makes brief dives and doesn't travel far. Occasionally races across the water with rapid foot-pattering. Fairly secretive, Least Grebe prefers to forage in stands of emergent vegetation. It is also fairly lethargic, often sitting motionless for long periods.

FLIGHT: Small and slender profile, with head and neck outstretched, legs trailing. All dark except for contrastingly pale trailing edge of wing. Wingbeats frantically rapid. Commonly dives upon landing.

VOCALIZATIONS: Fairly loud, high, bugled "*eep!*" sometimes uttered with a compressed two-note quality: "*e'ehp*" or "*le'ehp.*" Calls also include a low, muttered "*mep.*" Chatter call is a rapid, high-pitched, slightly descending, whiny, and nasal trill, "*reh'h'h'h'h'h'h'h,*" with the notes all run together. Least Grebe's chatter is much faster and sharper than the low muffled chortle of Pied-billed Grebe and may suggest the giggle of fairies (if you are familiar with it).

Pied-billed Grebe, *Podilymbus podiceps*
The Hell Diver

STATUS: Common and widespread, but not necessarily apparent. **DISTRIBUTION:** Breeds throughout the U.S. south into Mexico and across much of s. Canada from N.S. across s. Que., cen. Ont., n. Man., n. Sask., n. Alta., to e. and s. B.C. and se. Alaska. **HABITAT:** Freshwater wetlands and reed-rimmed ponds with pockets of open water. Also breeds in brackish water. In winter and migration, also found in ponds, slow-moving rivers, and brackish impoundments that may lack emergent vegetation. Less commonly found in protected saltwater habitats such as bays and marinas. **COHABITANTS:** American Coot, Common Moorhen; also mixes with puddle ducks. **MOVEMENT/MIGRATION:** Protracted in both spring and fall. In spring, frequently appears as soon as lakes thaw. Spring migration from February to May; fall from late August into December as the freeze line plunges south and birds are squeezed out of the north. VI: 4.

DESCRIPTION: Small (smaller than Horned Grebe or American Coot; same size as Eared Grebe), brown, meatloaf-shaped water bird with a short thick neck and an oversized, chickenlike head. Pied-bill regulates buoyancy by compressing its feathers, changing shape in the process. Floating high, it has a distinctly humpbacked appearance; sitting low, it seems all head and neck and in fact may show only head and neck (or just the tip of the bill) above the surface. When partially submerged, Pied-bill can look like a miniature sea serpent.

In summer, the short, thickish, white, chickenlike bill is encircled by a dark ring. In winter, the bill is dull. At all seasons, Pied-billed is uniformly warm brown. In summer, its blackish forehead and throat are easily overlooked; in winter, the whitish throat does not stand out. A pale undertail is usually apparent on swimming birds. Juveniles have a striped facial pattern, but otherwise are similar to adults.

BEHAVIOR: Except in the breeding season, usually solitary and unsocial, although several resting birds may gather in close proximity. Feeds by diving. Submerges in one of several ways: diving in a headlong plunge; sinking without a ripple; or crash-diving—that is, bending in the middle and plunging, with bow and stern raised and only a column of water hanging in the air to show where a bird has been. The old gunner's term for the bird, "the Hell Diver," refers to its reported ability to react to the flash of a fowling piece and submerge before the shot reaches the water.

While often seen foraging in open water, Pied-billed Grebe, when not feeding, prefers to stay close to protective, aquatic vegetation and only occasionally hauls out onshore. This species almost never flies, preferring to dive and swim to safety. When taking flight, it runs with great effort across the surface. Ungainly on land, Pied-bill walks erect, penguinlike.

FLIGHT: Gangly front-heavy profile, with neck extended and feet trailing behind. Overall pale brown; white belly is often difficult to see. The uniformity of plumage and absence of a white flash or patch in the wing readily distinguish this species from other grebes, including Least Grebe, which shows a white trailing edge to the wing. Pied-billed's flight is weak, labored, and generally low over the water. Wingbeats are rapid and steady.

VOCALIZATIONS: Calls with a loud wild-sounding keening that incorporates bleating coos and mournful wails. Also makes a run on a series of notes that sounds like a rippling chuckle or someone blowing a satisfying series of toots into a handkerchief: "*ruhruhruhruhruh.*"

Horned Grebe, *Podiceps auritus*
The Cosmopolitan Grebe

STATUS: Common northern and western breeder; common coastal wintering species. **DISTRIBUTION:** Breeds across much of w. Canada and Alaska and bordering prairie states—from s. and w. Ont. across all of Man., Sask., Alta., and all but sw. B.C., north into the Northwest Territories, the Yukon, and across cen. Alaska. Also found in n. Man., N.D., and n. Mont. Winters along both seacoasts, from the Aleutians to Baja and from the Maritimes to Tex., and occasionally on inland lakes and reservoirs from the Carolinas to Fla. and into Okla. and ne. Tex. (also along the Colorado R. to Nev.). **HABITAT:** Breeds in small shallow freshwater ponds. Prefers shorelines or shallow areas rich in sedges, rushes, or cattails (for nesting) and large areas of open water (for feeding). In winter, strongly prefers the inlets, back bays, and other protected inshore habitats of coastal waters. Smaller numbers winter inland on large lakes and reservoirs. **MOVEMENT/MIGRATION:** All birds vacate breeding areas and fly across a broad front over inland areas to reach the coasts. Spring migration from March into early May; fall (more protracted) from July to December, with peak from September to November. VI: 2.

DESCRIPTION: Small (coot- or Bufflehead-sized), somewhat large-headed and heavy-necked grebe whose breeding and nonbreeding plumage may be confused with the more southern and western Eared Grebe. Horned's head is largish, flat-topped, peaked at the rear, and down-sloping to the tip of the short pencil-straight bill; at a distance, the head looks like a doorstop. (The pale bill tip is very hard to see, particularly in winter.) Its neck is thick compared to the scrawnier neck of Eared Grebe, and its body proportionally long, with a domed back that rides high in the middle and tapers to the rear. (Eared Grebe's body looks cropped and fluffy in the rear and rides high in the stern.)

Both males and females in breeding plumage are all dark, with a swollen face (the bird looks like it has mumps) and a bright yellow, tightly bound wreath atop the head. From behind, the golden plumes resemble lobes, or "horns." At closer range and in good light, the neck is clearly chestnut red.

In nonbreeding plumage, Horned Grebe is mostly gray and white. Bright white wraparound cheeks are standout features and contrast sharply with the crisply trimmed black cap. The throat and sides of the neck may also be bright white or slightly smudged with gray, but never so much so that the bird looks dirty-necked (like Red-necked Grebe) or dusky (like Eared). The remaining upperparts are dull, leaden gray, with streaky, paler sides.

BEHAVIOR: Typical grebe. Sits low in the water. Dives for fish. Seldom seen to fly, and getting airborne requires a long foot-pattering takeoff. Generally not social. Feeding birds spread out widely across feeding areas. Occasionally two or more feed in close proximity—perhaps for company, perhaps because the fishing is good—but unlike Eared Grebe, Horned Grebe does not gather in large numbers (several hundred birds).

FLIGHT: Very thin long-necked profile, with the body angled slightly up and showing a white wing patch on the inner trailing edge of the wing. Flight is straight but somewhat veering; wingbeats are rapid and steady. In migration, Horned Grebe sometimes briefly elevates its head (up periscope) and lowers it.

VOCALIZATIONS: A rapid, high, excited, twittering chatter, "*erEh'H'H'H'H'H'h*," that rises and falls and has both a breathy and bleating quality.

PERTINENT PARTICULARS: When Horned Grebe molts into breeding plumage in late winter and early spring, it often looks dirty-necked and may show a head pattern similar to Eared Grebe's.

Red-necked Grebe, *Podiceps grisegena*
The Dirty-necked Grebe

STATUS: Fairly common breeder; fairly common to uncommon along seacoasts and the Great Lakes in winter. **DISTRIBUTION:** Breeds from w. and cen. Alaska south through the Yukon and w. Northwest Territories to ne. Ore. and east across most of Alta., the southern half of Sask., s. Man., w. Ont., as well as n. and e. N.D. and the northern half of Minn.; also found in Idaho, w. Mont., and nw. Wyo. Winters along both seacoasts, from the Aleutians to n. Calif. and from Nfld. to New England (less commonly to the Carolinas). **HABITAT:** Nests on small, shallow, vegetated freshwater lakes and ponds in boreal forest, prairies, and tundra. Winters in marine habitat, especially shallow, near-shore waters, channels, inlets, and bays, but occasionally is found well offshore and on open freshwater lakes (including the Great Lakes). **COHABITANTS:** While breeding, assorted puddle and diving ducks, Wilson's and Red-necked Phalarope, Black Tern, Arctic Tern. **MOVEMENT/MIGRATION:** In both spring and fall, the Great Lakes constitute important staging areas; birds are more concentrated toward the east in spring. During winters severe enough to freeze large portions of the Great Lakes, large numbers of Red-necked Grebes may suddenly appear on the Atlantic Coast in January or February. Otherwise, spring migration from March to May, with peak in late April; fall from late July to November, with peak in August and September (and frequent stops on large inland lakes and reservoirs). VI: 3.

DESCRIPTION: A working-class grebe. A midsized, midproportioned diving bird, Red-necked is larger, lankier, and less compact and petite than Horned or Eared, and smaller, more robust, and less sinuously proportioned than Western or Clark's. Overall Red-necked has the body of a scoter and the head and neck of a loon.

The head and bill fuse into a long, sloping, wedge-shaped profile that starts with a peak at the back of the head and extends to the long, straight, pointy, distinctly yellow, and slightly down-angled bill. The neck seems heavy in breeding plumage but more slender in winter, with a hint of the grebish S-shaped curve. The compact body sometimes appears ragged or shaggy in the rear.

In breeding plumage, this grebe's red neck and white cheeks are distinctive. In nonbreeding plumage, it is overall dingy and uncontrastingly brown (not gray), with a whitish chin and dirty neck. In shape and overall plumage (and at a distance), Red-necked is more easily passed off as a female Red-breasted Merganser than a loon or a grebe, but the downward angle of the bill distinguishes it from mergansers, as does the dingy yellow-based bill.

BEHAVIOR: In winter, often solitary, but also feeds in pairs or loose groups. In migration, often found in large aggregations. Fairly sedentary. Forages for long periods in the same general area. Submerges with a graceful, arching dive. While breeding, often found along vegetated shorelines. In winter and migration, always found in open (often deep) water. Rarely flies; almost never hauls out on land (even in breeding season). Takes off with great effort.

FLIGHT: Slender and wilted profile, with both head/extended neck and legs drooping below the body. In winter, upperparts are uniformly dark and dingy, and underparts whitish. Slender wings, set well back, show conspicuous white patches on the leading and trailing edges of the inner wing. Wingbeats are steady, unhurried, limber, somewhat rubbery.

VOCALIZATIONS: Usually quiet, except when breeding. Calls include a loud, repeated *"krik . . . krik"* and an excited, squeaky stutter, *"eheheheheheeee"* (recalls a food-begging call); also has a gull-like wail, *"erAHH . . . erAHH."* In winter, sometimes makes a muffled grunt.

PERTINENT PARTICULARS: Many guides imply (and not a few anguished observers believe) that in winter it is difficult to distinguish Red-necked from the two smaller grebes. In fact, it's as simple as brown versus gray, overall dingy versus contrasting pattern, and big spiky yellow bill versus petite bill. In short, Red-necked Grebe is a very different sort of grebe.

Eared Grebe, *Podiceps nigricollis*

Crazed Grebe

STATUS: Common breeder of the western interior; abundant at post-breeding, premigratory staging areas. **DISTRIBUTION:** From e. B.C. across all but northern and some western portions of Alta. and into s. Sask. and s. Man.; south into w. Wash., se. Ore., and ne. Calif.; s. Idaho and into all but extreme w. Mont.; Wyo. and all but ne. N.D.; S.D., n. Neb., and w. Minn.; all but s. Nev.; Utah and all but e. Colo.; n. Arizona and nw. N.M. Winters coastally—along the Pacific from s. B.C. south, along the Gulf of Mexico from sw. La. south into Mexico. Also found away from coasts across s. Calif., extreme s. Nev., s. Ariz. and N.M., and all but n. Tex. Some wintering birds also found in s. B.C. and cen. Colo. A few winter in the East. **HABITAT:** Breeds on shallow freshwater (also saltwater) lakes, ponds with emergent vegetation, marshes, sloughs, and sewage ponds; prefers fresh water with no fish and no forest border. In winter and migration, prefers saltwater habitats but may winter on freshwater lakes. **COHABITANTS:** While breeding, assorted puddle and diving ducks, American Coot, Wilson's Phalarope, California

Gull, Black Tern. **MOVEMENT/MIGRATION:** Usually vacates interior breeding areas and relocates to coastal or more temperate latitudes after a lengthy post-breeding staging period, most commonly on the Great Salt Lake in Utah and Mono Lake in California. Spring migration from late February to late May; movement to staging areas from late June to early September; very protracted fall migration from September to late January, with a peak in December and January. VI: 3.

DESCRIPTION: Compact petite-featured grebe that likes company. Small (slightly smaller than Horned or Pied-billed), with a peaked head, a steep forehead, a slim, slightly upturned bill, a thin neck, a stocky dome-backed body, and a cropped rear. Its body is more compact and rides higher in the water than Horned Grebe's, exposing white undertail coverts. The slender bill disappears at a distance where the bill of Horned Grebe is still evident. In shape and proportions, Eared is more akin to Least Grebe than Horned Grebe.

In breeding plumage, Eared Grebe is all black with chestnut sides and a sparse wispy spray of bright yellow plumes behind crimson red eyes. At a distance or in poor light, it looks all black. Its nonbreeding plumage is darker overall than Horned Grebe's, with a dark face, a dirty neck, and two pale white spots bracketing the cheeks that are sometimes homogenized (not distinctly two). Eared's expression is crazed, demonic.

BEHAVIOR: Social in summer; breeds in colonies. Gregarious during post-breeding aggregations; in winter, found in large flocks that may number in the tens of thousands. Almost always seen on the water or, when breeding, on or near nests made of floating vegetation. Rarely struggles onto land (where it moves awkwardly in an upright posture). Almost as rarely seen to fly. Dives for food or may pick from the surface. Swims with head inclined forward, moving jerkily. When nervous or disturbed, rides lower in the water.

FLIGHT: Very slender and spindly, with a level profile. Takes off (or fails) with great, foot-pattering effort; flight is weak and labored. White wing patches are conspicuous. Wingbeat is quick, steady, and sputtery. Lands with a belly flop.

VOCALIZATIONS: High-pitched piping bugle. The short version, "*toowit,*" has a compressed two-note quality; the longer version, "*oooeet,*" recalls the two-note yelp of American Avocet. Eared Grebe also has a low "*wika, wika*" call and a low musical trill.

Western Grebe, *Aechmophorus occidentalis*

Serpent-necked Grebe (with the greenish bill)

STATUS: Common western breeder; common winter resident primarily on the West Coast. **DISTRIBUTION:** Breeds across most of w. U.S., cen. Mexico (permanent resident), and sw. Canada, including sw. Man., s. Sask., all but n. Alta., and s. and se. B.C. In the U.S., breeds in cen. Wisc., s. and w. Minn., all states west of and including N.D., S.D., n. and w. Neb., and Colo., and just east of the N.M.–Okla.–Tex. border (also s. Ariz.). In winter, found coastally from extreme sw. B.C. to Baja and the central coast of Mexico and in s. Nev., s. and cen. Ariz., N.M., w. Tex., and (rarely) coastal Tex. **HABITAT:** Large open lakes and marshes rimmed by emergent vegetation. Winters in coastal waters, bays, and estuaries, as well as in freshwater lakes. **COHABITANTS:** In winter, found with Common and Red-throated Loons and Clark's Grebe; occasionally found with Eared Grebe. **MOVEMENT/MIGRATION:** Spring migration from late April to mid-May; fall from late September to late November. VI: 3.

DESCRIPTION: Large (almost loon-sized) black-and-white water bird distinguished by a long, thin, swan-contoured neck and a greenish yellow marlin spike for a bill that is very long, straight, and pointy. Loons and Red-necked Grebe appear shorter and thicker-necked than Western Grebe, which, when roosting, draws its head back flush with the back and looks somewhat like a submarine.

Dark (blackish) above and bright white below, Western has a black cap/mask that extends below the eye and a wide blackish mane down the hindneck. Viewed head on, there is nothing but black above the bill; from the side, the black mane clearly shows against the white sides of the neck. The sides tend to be dark, with little or no white mottling toward the rear, but this characteristic is variable, making it difficult sometimes to distinguish Western Grebe from the overall paler and generally whiter-flanked Clark's Grebe.

The best field mark is the army drab, greenish yellow bill: it is darker and harder to see than the brighter orange-yellow bill of Clark's. If distance prevents you from being certain of what color the bill is, it's Western.

BEHAVIOR: Spends most of its life on (or below) the water, but in its aquatic environs moves with fluid finess (and seems graceful even when stationary). Sometimes submerges its head to search for fish. Dives either with a forward thrust of the head, the body just seeming to follow, or by springing out of the water. Flocks space themselves widely and dive independently. During courtship, the bird engages in spectacular displays in which it literally dances on water.

Western Grebe rarely flies. It roosts in groups on the water but generally does not associate with other birds (except Clark's Grebe) and commonly remains far from shore. Often observers are alerted to the bird's presence by a flash of white underparts as it turns on its side to preen. Barring this clue, the white neck stands out.

FLIGHT: Extremely long-necked and overall long and slender silhouette, with head held slightly lower than the body. The white wing stripe is slightly more obscure than Clark's. Flight is direct and generally low, with rapid, regular, steady wingbeats.

VOCALIZATIONS: High creaking or growling two-note trill, "*ker-kreek*," with the first part more trilled and the second part slightly gurgled. The sound may recall a harsh Green-winged Teal's trill; the pattern is like Marbled Godwit's "*godwit*."

Clark's Grebe, *Aechmophorus clarkii*

Serpent-necked Grebe (with the yellow bill)

STATUS: Common and widespread western breeder; common West Coast and interior winter resident. Less common than Western Grebe in northern and eastern portions of its range and, not coincidentally, less often encountered east of its range. **DISTRIBUTION:** Clark's Grebe breeds and winters in the same range as Western Grebe, but not on the Texas coast. **HABITAT, COHABITANTS, AND MOVEMENT/MIGRATION:** Same as Western Grebe. VI: 1.

DESCRIPTION: Much like Western Grebe, *except* for bright orange-yellow bill. Overall slightly paler above than Western Grebe, Clark's is less blackish and more overall grayish above, with more white spattering on the flanks and more restricted black on the head and hindneck. Clark's head looks classically capped (shows white above the eye); Western's looks like the black cap is pulled over its eyes. Viewed head on, the black cap looks like a badly fitting toupee (white in front of the eyes is visible). From the side, the narrower black hindneck stripe on Clark's Grebe is difficult to see, so the neck seems either whiter than Western Grebe's or all white (showing no apparent black border on the nape).

The bright orange-yellow bill stands out at great distances. If it's Clark's, you'll know. If you have to guess, it's the other grebe.

BEHAVIOR AND FLIGHT: Like Western, except that the white trailing edge to the upperwing is more extensively white—adding to the overall sense of paleness.

VOCALIZATIONS: Like Western, but more single-noted. A high, creeking or growling, amphibian-like trill: "*k'rrree.*"

PERTINENT PARTICULARS: Although bill color is most distinctive (and definitive), it's best to use a combination of traits when distinguishing between Western and Clark's Grebes. Check the amount of black around the eye, the amount of white on the flanks, and the width of the black neck mane. Since the birds feed and roost together outside the breeding season, finding (and identifying) individuals is sometimes only a matter of determining which species dominates a group, then searching for individuals that are overall paler (Clark's) or darker (Western). Be aware, however, that Clark's and Western Grebes do hybridize.

ALBATROSSES

Laysan Albatross, *Phoebastria immutabilis*

Slender White-headed Albatross

STATUS: Uncommon West Coast pelagic species; rare but regular within sight of land but more common well offshore. **DISTRIBUTION:** Breeds on several islands in the Pacific (the closest off Baja). Widespread wanderer across the North Pacific. In North America, found from the Aleutians and Gulf of Alaska (except in winter) and southern portions of the Bering Sea (summer) south to the waters off s. Mexico, commonly in deeper water (200 mi. from shore). Where ocean canyons are close to shore, birds may be found 50–75 mi. from land and, in the Aleutians, can be seen from shore. **HABITAT:** Open ocean waters, most commonly beyond the continental shelf. **COHABITANTS:** Other pelagic species, including Black-footed Albatross, Northern Fulmar, Sooty Shearwater, and Short-tailed Shearwater, but commonly associates only with other albatrosses. **MOVEMENT/MIGRATION:** Adults disperse away from breeding colonies July to November; most subadults are always at sea. This species has been recorded on the West Coast every month of the year, but most records (for California) fall between October and February. Regularly found in southwestern Alaskan waters, except in winter. VI: 1.

DESCRIPTION: Looks like an exceedingly long, slim-winged Western Gull and flies with heavy, fluid, time-suspended grace. As a large seabird but small albatross, Laysan Albatross dwarfs Western Gull and is about the same size as Black-footed Albatross. It is distinguished on the water by its size, white head contrasting with all blackish back, long pink bill, raised bump in the rump, and dark smudge across the eye.

BEHAVIOR: Spends most of its time at sea, in flight, but needs wind to get aloft. Feeds by sitting buoyantly on the surface, grabbing food with its bill. Attracted to ships and concentrations of birds.

FLIGHT: Exceedingly long-winged profile with white head and mostly white tail projecting about equal lengths. Arms are very long, and the hands short and angled back. Wings bow or droop along their length (particularly when the bird is banking aggressively). Like all albatrosses, Laysan coasts for extended periods without flapping its wings. Its wingbeats are slow, deep, and somewhat self-conscious-looking. Normal flight is casual but deliberate, heavy but fluid, and has a dreamlike quality. In high winds, flight is more dramatic, with tight, bounding or banking turns that make distant birds look like they are stitching the water in slow motion. Among its pelagic peers, Laysan recalls an awkward too-tall girl standing in the midst of a younger crowd.

VOCALIZATIONS: Silent at sea.

PERTINENT PARTICULARS: Rare (but increasingly more common) Short-tailed Albatross is the only other black-and-white albatross likely to be encountered off the West Coast (particularly in southwestern Alaskan waters). A big, broadly proportioned, lumbering bird with a massive pink bill, dusky cap, and white back, Short-tailed is not easily confused with the slender Laysan. For more information on Short-tailed, see "Pertinent Particulars" at **Black-footed Albatross.**

Black-footed Albatross, *Phoebastria nigripes*

Chocolate Albatross

STATUS: Uncommon (by seabird standards), but regularly encountered offshore. The most common albatross in waters along the West Coast from southern Alaska southward, this Pacific species is unlikely to be sighted from land. **DISTRIBUTION:** Breeds on a handful of islands and ranges across the North Pacific. In North America, found in coastal waters from the Aleutians to Mexico; range includes the Hawaiian Is. **HABITAT:** Found in marine environments, most commonly over continental shelf waters. Often found with fishing trawlers. **COHABITANTS:** Northern Fulmar, Buller's Shearwater, Sooty Shearwater, Pomarine Jaeger. **MOVEMENT/MIGRATION:** Breeds January to June. Spends the balance of its life at sea. Immatures (two- to four-year-olds) are wholly pelagic. Present along the Alaskan coast from May to October and off the West Coast year-round, though most common from mid-May to mid-August, with largest numbers found in northern waters. VI: 0.

DESCRIPTION: Conspicuously large, gangly, all-dark seabird with a whitish ring around the base of its bill and the fluid grace of a skater. A small albatross but a big seabird, Black-footed's wing span is more than twice the size of a shearwater's; slightly longer than Brown Pelican, the same size as Laysan Albatross. On the water, Black-footed looks long-necked, sturdy-billed, and hump-rumped and dwarfs most surrounding seabirds.

All dark brown except for a whitish face (which looks like a ring at a distance), Black-footed has a dark, slightly pinkish gray bill.

BEHAVIOR: Most commonly seen sitting on the water singly or in small groups (often amid concentrations of other seabirds or close to a fishing trawler) or flying at sea. Fairly active, flying even during calm conditions. Feeds by taking food from the surface—mostly squid and the discard from fishing boats. (Black-footed also seems to consume considerable amounts of plastic, reportedly with no ill effects.)

A tame bird, Black-footed flushes reluctantly from the water with a running takeoff. It follows ships and chums readily.

FLIGHT: A conspicuously all-dark albatross. Distinguished as an albatross by its size, incongruously long narrow wings, heavy bill, and the fluid time-suspended grace of its flight. Distinguished as Black-footed by its all-dark plumage. Over U.S. and Canadian waters, the only other albatross species likely to be encountered is the boldly dark-and-white-patterned Laysan Albatross. (See the discussion of the rare, immature Short-tailed Albatross at "Pertinent Particulars.") *Note:* Some aging adults have pale bellies and are paler about the face. They still look dark, not boldly black-and-white-patterned.

Overall Black-footed is more spindly-winged than Laysan and slightly front-heavy (the tail disappears at a distance). Flight is fairly active (for an albatross), with shallow wingbeats and frequent short glides. (Black-footed does not glide for endless, time-suspended periods, as many albatrosses do.)

VOCALIZATIONS: Silent at sea.

PERTINENT PARTICULARS: One other all-dark-brown albatross occurs in the Pacific: immature Short-tailed Albatross (*Phoebastria albatrus*), whose population (adults and immatures) probably does not exceed 5,000. Short-tailed Albatross is nevertheless regularly found in the waters off the Aleutians and in the Gulf of Alaska and has been recorded all along the west coast of Canada and the United States.

Short-tailed Albatross is a big, stocky, broad-winged, lumbering bird that is structurally more akin to Wandering or Galapagos Albatross and very unlike the slight spindly profile of Black-footed. Short-tailed's humongous, bubblegum pink bill is very distinctive, even at a distance. Although the bills of adult Black-footed may pale with age, they remain proportionally smaller and don't stand out as unique or different.

In keeping with its heavy structure, Short-tailed's flight is heavy, lumbering, and less dynamic.

PETRELS AND SHEARWATERS

Northern Fulmar, *Fulmarus glacialis*

Working-class Shearwater

STATUS: Common northern breeder and pelagic species along both coasts (but generally more numerous in the Pacific), especially in winter. **DISTRIBUTION:** Breeds in colonies on islands and isolated sea cliffs off Alaska, B.C., arctic Canada (including Baffin I.), and coastal Lab. and Nfld. Nonbreeding and post-breeding birds wander extensively at sea. While most common in northern waters, especially in winter, this species is irregular at all seasons south to s. Calif. and N.C.; occasionally occurs farther south. **HABITAT:** This cliff-nesting species prefers colder waters, at sea, generally well offshore; concentrates at the edge of the continental shelf and alongside fishing trawlers. **COHABITANTS:** At sea, shearwaters, skuas, jaegers, kittiwake, large whales, ships. **MOVEMENT/MIGRATION:** Along the Atlantic Coast (south of Canada), most common from December to March, with only small numbers present between May and August. In the Pacific, south of British Columbia, common from late August to April, with peak movements in November and March/April, but some birds are present year-round. VI: 2.

DESCRIPTION: Stocky thickset seabird that seems part shearwater, part gull. Medium-sized (smaller than Greater or Pink-footed Shearwater; slightly larger than Short-tailed or Sooty Shearwater or Black-legged Kittiwake; about the same size as Pomarine Jaeger), with a stubby thumb-sized bill, big head, thick neck, compact robust body, and short tail. Rides high on the water with bill angled demurely down, stern raised, recalling in shape and posture a horse on a carousel.

Plumage is variable; most birds are pale, silvery gray above and white below (very gull-like), but some are all dark, uniformly smoky gray, and more likely to be confused with a dark shearwater or petrel. Dark birds are much more common in western waters and very uncommon in the Atlantic. Northern Fulmar is distinguished on the water by its bulky shape, thick yellow bill (all plumages), and smudgy patch over the eye. Its expression is demure.

BEHAVIOR: A seagoing jack-of-all-trades. Northern Fulmar follows ships and feeds by sitting on the surface and tearing at carrion and offal, paddling on the surface and dipping for prey, or plunging to the surface and diving for prey. (It does everything apparently but snatch food from the air.) A highly gregarious and social species, it has been counted in the thousands where factory ships are operating. Often found with shearwaters and kittiwake, Northern Fulmar also attends feeding whales.

This bird comes readily to ships (and to chum) and often sits on the water near ships, anticipating a feeding opportunity.

FLIGHT: Overall heavyset, with wings held stiff and straight (not angled back like gulls or shearwaters). Shows a broad projecting head, broad untapered wings, and a short wide tail. Overall bigger-headed, broader, shorter-winged, and wider-tailed than shearwaters, Northern Fulmar is shaped like a small cargo plane, not a glider (that's a shearwater). Glides with wings straight out to the side or angled slightly but stiffly down. (Shearwaters droop their wings.) Light birds are contrastingly patterned, showing silvery gray upperparts, white heads and tails (Pacific birds show a dark tip), and conspicuous white patches bleeding through dark wingtips that are visible from above or below and indistinct or lacking on many dark birds.

Flight is tight, jerky, unglamorous, and more unsteady on the wing than a shearwater. Flies with a short, hurried, stiff, or batty series of wingbeats followed by a prolonged glide. Often flies close to the water but also habitually flies higher than shearwaters.

VOCALIZATIONS: Feeding groups make hoarse croaking or grunting sounds.

Black-capped Petrel, *Pterodroma hasitata*

Gulf Stream Petrel

STATUS: Common in the waters of the Gulf Stream, particularly in late spring and summer. Rare in the Gulf of Mexico and in waters north of Virginia. **DISTRIBUTION:** For a pelagic species, very restricted. Breeds in the West Indies between November and May. Disperses into the Caribbean and north, following the currents of the Gulf Stream north to at least the offshore waters of N.C., where it occurs year-round. **HABITAT:** Warm, tropical, deeper ocean waters (usually 100 fathoms or more). **COHABITANTS:** Audubon's Shearwater, flying fish, and assorted other pelagic birds. **MOVEMENT/MIGRATION:** Not well understood. VI: 2.

DESCRIPTION: Big, bounding, boldly patterned, and, by gadfly petrel standards, *stocky Pterodroma.* On the water, this large *Pterodroma* (larger than Audubon's Shearwater; slightly smaller than Greater Shearwater; about the same size as Sooty Shearwater) appears round-headed and thick-necked, with a stubby black bill.

But since it's a *Pterodroma,* you probably won't see it on the water, except on calm days.

Black-capped Petrel is dark gray-brown above with a white neck. Its black cap is restricted to the top of the head, and at close range this "cap" is revealed to be both a cap and a dark eye patch. The rare Bermuda Petrel and Fea's Petrel (both found in Gulf Stream waters) show a dark nape, but a few Black-cappeds also show a "dusky" collar. Black-capped's back is much darker than Fea's, and slightly darker than Bermuda's. Greater Shearwater has a long thin bill and black hat (not a skullcap or helmet) and a dark smudge on the belly.

BEHAVIOR: Usually seen singly (also in pairs and trios) flying in the classic arcing or bounding flight of *Pterodromas.* It is not uncommon to see multiple birds working a patch of ocean but remaining widely spaced, not clustered like shearwaters. Although this species approaches groups of birds working chum, it does not follow ships or mingle with other pelagic species.

FLIGHT: Silhouette like other *Pterodromas* but heavier—an amalgam of a short, bulky, neckless head (and stubby bill) with a longer tail, set on long, pointy, harshly angled back wings. Its dynamic flight reveals alternating flashes of upper- and underparts—a commanding combination of bright white unblemished underparts and a mostly white uppertail. This combination (not to mention the dynamic flight) easily distinguishes this species from the somewhat similarly patterned Greater Shearwater. Most Black-cappeds also show conspicuously white collars, and many have a molt pattern in summer that produces white patches in the wings.

If you are off North Carolina and see a pterodroma with a plumage pattern reminiscent of Crested Caracara, you are looking at a Black-capped.

Rare Fea's Petrel is overall paler above with a gray neck (showing no capped appearance) and usually shows all-dusky underwings compared to white on Black-capped. Fea's Petrel's all–pale gray tail lacks the black tip of Black-capped's, but more significantly, Fea's Petrel's tail base also lacks the gleaming, eye-catching white that is the signature characteristic of Black-capped. Overall darker Bermuda Petrel is much like Black-capped but (like Fea's) has a less contrasting pattern—no white collar and only a limited amount of white at the base of the tail.

Black-capped's dynamic, bounding, roller-coaster flight, on deeply bowed wings, seems reckless and aggressive. These birds give the impression that they are attacking the water, not covering distance. Wingbeats are somewhat stiff, shallow, and heavy—somewhere between choppy and floppy.

VOCALIZATIONS: Mostly silent at sea.

PERTINENT PARTICULARS: At great distances, what catches the eye is the flash of bright white underparts—could be Black-capped Petrel, could be Cory's Shearwater. If in the next instant you see a dark-backed bird showing a bright white head and tail (the black cap and tail tip are invisible), you are obviously not looking at a Cory's. Also, observers in waters where Black-capped is found should be mindful that other (but far less commonly seen) gadfly petrels may be in the neighborhood.

Cory's Shearwater, *Calonectris diomedea*

Mustard-shaded Shearwater

STATUS: Variously common to uncommon off the East Coast from late spring to early summer; more common in late summer and fall; rare in the Gulf of Mexico. **DISTRIBUTION:** Breeds in the Mediterranean (February to August). Disperses widely, with some portion of the population ranging west. Ranges—most commonly out of sight of land—from s. Fla. to the Maritimes. **HABITAT:** Warmer marine waters, usually well offshore. In spring, most common in Gulf Stream waters off the Carolinas. In late summer, more concentrated off southern New England, where summer-warmed waters occasionally bring the birds within sight of land, but large numbers also occur off the Carolinas. **COHABITANTS:** Black-capped Petrel, Band-rumped Storm-Petrel (spring); also Sooty and Greater Shearwaters. **MOVEMENT/MIGRATION:** Found in U.S. and (rarely) Canadian waters from May to early December. Spring birds are presumed to be nonbreeding subadults. VI: 2.

DESCRIPTION: Big, pale, pastel-shaded shearwater with a distinctly yellow bill. The largest shearwater (larger than Greater Shearwater; smaller than Herring Gull; about the same size as Northern Fulmar), Cory's is overall bulky on the water, showing a heavy yellow bill, large round head, short thick neck, and heavy body.

The color of its fairly featureless upperparts is diagnostic—a pale pastel-shaded yellow-brown (a Dijon mustard wash) or gray-brown that darkens toward the wingtips. Other eastern shearwaters are darker-backed; light Northern Fulmar is silvery gray above. Cory's Shearwater's large head appears hooded (Greater Shearwater looks black-capped), and its expression is gentle.

BEHAVIOR: So big, broadly proportioned, and sluggish that it reminds you of an albatross. Usually seen flying low over the water or sitting on the water in small groups (or among other shearwaters). Comes readily to chum, follows ships, and occasionally flies at considerable heights. This tame bird is easily approached on the water.

FLIGHT: Big, bulky, broad-winged, bullheaded, and humpbacked profile. The broad, somewhat blunt-tipped wings are conspicuously down-bowed, making the bird appear fatigued or wilted. Gleaming white underparts (including immaculate underwings) catch the eye at a distance. (Greater Shearwater is darker-backed and more patterned and overall usually shows less white below.) Cory's flies and banks sluggishly in several slow, measured, floppy wingbeats followed by a lengthy glide. In more dynamic flight, it rises abruptly and descends in a long banking glide. Its large size (projected by the bird's heavy movements), pale mustard-tinged upperparts showing no discernible pattern, and gleaming white underparts are easy to note at great distances.

VOCALIZATIONS: Usually silent at sea.

PERTINENT PARTICULARS: The identification of this large shearwater generally comes down to separating it from the similarly sized Greater Shearwater. Greater is darker-backed, dirtier-bellied, and contrastingly patterned, showing a well-defined dark cap and obvious white U at the base of the tail. Although Cory's Shearwater can have a pale U at the base of the tail, the head appears hooded, not capped.

Pink-footed Shearwater, *Puffinus creatopus*

A Big, Floppy, Western Shearwater with a Gentle Expression

STATUS: Common West Coast pelagic species. **DISTRIBUTION:** Waters of s. Alaska (rare) south to nesting areas in Chile. **HABITAT:** Coastal waters over the continental shelf; occasionally but not commonly seen from shore. **COHABITANTS:** Associates with other shearwaters, particularly Sooty. **MOVEMENT/MIGRATION:** Breeds between November and April (rare in U.S. waters in winter); in U.S. and Canadian waters, occurs April to November, with highest numbers from August to October. VI: 1.

DESCRIPTION: Big, plain-backed, white-bellied, pink-billed shearwater. Clearly the largest in any mixed flock of western shearwaters, Pink-footed has a big round head, a dark-tipped pink bill, and a long slightly humpbacked body.

Overall plain on the water, and grayish brown to pale mustard brown above. Dusky, pastel-shaded head and neck give way to whitish throat and breast without sharp contrasts. Patterned sides (the bird looks like it is smothered in pinkish hickeys) make Pink-footed look dark to the water line. (Cory's, which has been recorded in Pacific waters, has bright white sides.) If you can see Pink-footed's gentle expression, you can also see the near-diagnostic pink bill, shared only by all-dark, confusingly named, and probably conspecific Flesh-footed Shearwater.

BEHAVIOR: Fairly calm and genial. Usually found scattered throughout shearwater flocks (does not seem to form single-species flocks like Buller's Shearwater). Dives both to and beneath the surface to catch fish and squid. Occasionally follows ships and responds to chum.

FLIGHT: Overall large, bullheaded, and broadly proportioned shearwater with a robust body and long broad wings and tail (can seem gangly, in a bulky sort of way). In calm conditions, wings are held at a right angle to the body and are gently and evenly drooped in a glide. Pink-footed shows more crook to the wings than Buller's.

Flight is heavy and sluggish. Wingbeats are languid and floppy, given in a short or long series punctuated by often very lengthy glides. Flight is more inspired and wheeling in stronger winds, but seems accomplished rather than dramatic.

VOCALIZATIONS: Generally silent at sea.

Flesh-footed Shearwater, *Puffinus carneipes*

Dark-morph Pink-footed

STATUS: Rare West Coast pelagic visitor; a Pacific species that occurs chiefly in fall. **DISTRIBUTION:** Breeds in the sw. Pacific; a few birds wander to the Pacific Coast from s. Alaska to Baja. **HABITAT:** Coastal waters beyond the reach of onshore observers. **COHABITANTS:** Pink-footed and Sooty Shearwaters. **MOVEMENT/MIGRATION:** Spring migration in April and May; fall from August to early November. VI: 0.

DESCRIPTION: A dark-morph Pink-footed Shearwater with the species trademark pink bill (complete with a dark tip). This large, stocky, and robust shearwater is identical to Pink-footed in size and shape but, unlike pale Pink-footed, wholly blackish brown with a large contrastingly pink bill. (Smaller Sooty and Short-tailed Shearwaters have thinner blackish bills.)

BEHAVIOR: Very social; usually found among concentrations of other shearwaters. Readily attracted to ships and chum.

FLIGHT: Large, bulky, broad-winged shearwater (structurally identical to Pink-footed) with wings held at a right angle to the body and slightly drooped. Flesh-footed is wholly dark brown, like Sooty Shearwater, but lacks Sooty's whitish wing lining. In good light, most often when landing, the trailing edge of Flesh-footed's underwings flashes tarnished silver (like the wings of Turkey Vulture). Flight is heavy and sluggish; the bird's slow languid flapping is punctuated by lengthy stiff-winged glides (very unlike the near-frantic flight of Sooty Shearwater).

VOCALIZATIONS: Silent at sea.

PERTINENT PARTICULARS: Among rafts of Sooty Shearwaters, search for the one with the pink bill. Caution: Flesh-footed is easily passed off as first-year Heermann's Gull.

Greater Shearwater, *Puffinus gravis*

Shearwater with Contrast

STATUS: Common summer visitor over Atlantic continental shelf waters, but numbers vary between scarce and numerous. **DISTRIBUTION:** Ranges along the entire East Coast from Fla. to Nfld.; less common in the Gulf of Mexico. Most common south of Maine during spring migration. In summer, large numbers are found in the waters off Canada and n. New England, and in late fall, prior to migration, off New England. In most offshore waters, a few birds can be seen anytime between May and October. **HABITAT:** Colder waters over the continental shelf. Commonly stays beyond sight of land, but occasionally comes close enough to be seen from shore. **COHABITANTS:** Sooty Shearwater, Wilson's Storm-Petrel. **MOVEMENT/MIGRATION:** Spring movement north along the western Atlantic between May and July; common in northern waters from July to August; retreats south from September to November, but some birds may be present into December. VI: 1.

DESCRIPTION: Grayish brown shearwater most easily distinguished from other large Atlantic shearwaters by its more contrasting pattern. This large (larger than Sooty Shearwater and Black-capped Petrel; smaller than Cory's Shearwater) and sturdily proportioned shearwater has long, stiff wings (not as drooped or crooked as Cory's), a pointy head and fairly blunt tail that project about equal lengths, and a thin black bill. Overall proportions fall somewhere between the slender Sooty Shearwater and the broader, bulkier Cory's.

Mostly cold grayish brown above (darker toward the wingtips); white with spatters and patches below. The dark well-defined cap and dark-tipped tail are separated from the rest of the bird by an almost completely whitish collar and narrow tail band. (In flight, at a distance, it is the two white spots, fore and aft, that stand out.) The whitish underparts are marred by a broken dark collar, dark support bars near the base of the underwings, and a shadowy dark patch on the belly. Cory's Shearwater, by comparison, seems paler, browner, warmer-toned, and overall more homogenized and less patterned above, pure unblemished white below. Smaller, more angle-winged Black-capped Petrel shows more obvious white on the collar and rump, pure white underparts, and a white forehead (dark on Greater Shearwater).

BEHAVIOR: A very gregarious species, often seen in great numbers and with other large Atlantic shearwaters, particularly Sooty. (Cory's is partial to warmer waters.) The occasional Greater Shearwater appearing within sight of land often joins flocks of terns and gulls. Easily attracted to ships and to chum, the bird is aggressive and vocal where concentrations of feeding birds occur. It feeds by plunging to the surface as well as by diving (from the surface) for prey.

FLIGHT: Very stiff, sturdy profile with straight (not bowed) wings set at a right angle to the body. Flight is strong, combining a short series of quick stiff wingbeats with long glides. Greater's flight appears stiffer, more precise, and less floppy than Cory's. In high winds, joins other birds in bounding flight, arcing high, with wings near-vertical. (Other shearwaters behave similarly in these conditions.)

VOCALIZATIONS: The sound of squabbling Greater Shearwaters has been likened to a catfight, but individual birds sound more akin to a harshly bleating lamb.

PERTINENT PARTICULARS: In late fall and winter, Greater Shearwater loses much of its contrasting pattern; in fact, many birds have dark napes (show no contrast between the cap and mantle) and pale (not white) rumps. Confusion with the much smaller Manx Shearwater is likely, particularly when the birds serving as size references are Northern Gannet and Great Black-backed and Herring Gulls.

Buller's Shearwater, *Puffinus bulleri*
A Pterodroma Wannabe

STATUS: Uncommon to common West Coast pelagic species, with a narrow temporal range. **DISTRIBUTION:** Off North America, found in the waters off the southeast coast of Alaska to Baja, but considerably less common south of Point Conception. **HABITAT:** Generally found in warmer waters well away from shore, but occasionally seen from land. **COHABITANTS:** Other shearwaters; Pomarine Jaeger (but frequently found in single-species flocks). **MOVEMENT/MIGRATION:** Breeds in New Zealand between November and March. Dispersed birds reach the North Pacific (Alaskan waters) in June. Birds returning to New Zealand swing down the west coast of the United States and Canada from July to November, with peak movements in September and October. VI: 1.

DESCRIPTION: Pale elegant chimera of a shearwater. Distinctly but cryptically marked with a plumage and structure that hearken to one of the racy *Pterodromas*, but a flight that one authority likens to a "small albatross." Medium-sized shearwater (noticeably smaller than Pink-footed; about the same size as Sooty) but petitely proportioned with a smallish head and long slender bill and tail.

Overall pale gray above (in places somewhat silvery); gleaming white below (including the underwings). On the water, the dark cap contrasts with the white cheeks and throat; sides show white flanks. Pink-footeds have pale gray heads and pinkish gray hickeys along the sides that make them look dark to the water line.

BEHAVIOR: Buller's is a deeper-water shearwater often found in small homogeneous flocks, but it also mixes with other shearwater species (which it usually outmaneuvers and outclasses in competition). Buller's feeds by swimming and plucking or landing momentarily on the surface; it doesn't dive. It follows ships and responds well to chum.

FLIGHT: Somewhat racily built with a slender head and bill, long pointy tail, and long wings that have short arms and long hands (like *Pterodromas*) and are held in a stiff gull-like configuration (jutting up along the arms, angling down along the hands). Smaller, more compact, more angular, and decidedly smaller-headed than Pink-footed, whose long broad wings curve gently down along their length.

Buller's Shearwater's distinctive plumage makes structural peculiarities secondary. The silvery upperparts are branded with contrast—most conspicuously a dark M-pattern on the wing (like many of the pterodroma species). Underparts (body and underwings) are wholly and brilliantly white.

The pattern of the upperparts is not particularly stark (the point of it is to make the bird more cryptic against the shifting light/dark pattern of the sea), but a sense of dark-and-light contrast holds up even at great distances. By comparison, the upperparts of Pink-footed Shearwater look uniform whether the bird is near or far.

The flight of Buller's is fairly leisurely and buoyant, with wingbeats that are both stiff and spare and glides that are long and close to the surface in moderate winds and seas. The bird arcs and bounds in rougher conditions, with wings angled back and down.

VOCALIZATIONS: Silent at sea.

Sooty Shearwater, Puffinus griseus

Dark Shearwater with the Silver (wing) Lining

STATUS: Common and at times abundant Atlantic and Pacific pelagic species. In the Pacific, south of Alaska, generally the most common shearwater species; in the Atlantic, except in May and June, generally outnumbered by Greater and Cory's Shearwaters. **DISTRIBUTION:** Breeds in the Southern Hemisphere. In North America, ranges north in waters along both coasts from Lab. to Fla. in the Atlantic (rare along the Fla. coast and in the Gulf of Mexico); in the Pacific, found from the southeast and western coasts of Alaska south. Celebrated summer concentrations involving millions of birds occur off Calif. **HABITAT:** Ocean waters out to the edge of the continental shelf and commonly much closer to land; often clearly visible from shore. **COHABITANTS:** Offshore, assorted shearwater species; inshore, occasionally gulls. **MOVEMENT/MIGRATION:** Moves north along both coasts in spring and remains until fall. Peak numbers (multiple hundreds) occur between late May and June in the Atlantic; hundreds of thousands are present between May and October in California (common but less numerous in April). Very rare between November and April along the Atlantic Coast; rare from December to late March in California. VI: 2.

DESCRIPTION: Active, slender-billed, sooty brown shearwater with contrastingly silvery underwings. Medium-sized (larger than Black-vented and Manx; smaller than Greater and Pink-footed; about the same size as Buller's and Short-tailed) with a slightly squared head but a sloping forehead and longish bill. Overall very similar to Short-tailed Shearwater.

Appears *all* dark brown on the water, including bill and feet. Slightly paler throat is often indistinct or difficult to see.

BEHAVIOR: Often distinguished by its dark uniformity and the sheer weight of its numbers. Sits on the water in rafts of varying size (several birds to thousands). At times and places, pelagic birding is a matter of picking the odd bird out from among the multitudes of Sooties. Sooty commonly migrates and feeds within sight of land, so it is the shearwater species most likely to be seen by land-based observers along much of the Atlantic Coast, and except where Black-vented Shearwaters occur, along the Pacific Coast as well. Sooty often moves into protected near-shore waters and bays in fog.

This bird is an active feeder that dives for food and plunges beneath the surface while in flight. Although it attends fishing trawlers, it doesn't follow ships and often ignores chum slicks that draw other birds.

FLIGHT: Fairly heavy-bodied but very long-necked, -faced, and -billed. Shows exceedingly slender wings that are held stiffly to the sides and often seem more sharply pointed than those of most other shearwaters. Uniformly dark above; dark below except for silvery underwings. In late summer, some birds may show a pale bar on the upper wing. The extent and brightness of the paleness under the wing is variable, but it is usually extensive and always apparent. Flight is quick, direct, hurried. Wingbeats are choppy and slightly elastic, given in short rapid series followed by a lengthy glide. Sooty glides with wings angled stiffly and slightly down.

VOCALIZATIONS: Usually silent at sea, but occasionally makes a high, nasal, petulant, doll-like cry, "*haaa.*"

Short-tailed Shearwater, *Puffinus tenuirostris*

Short-billed Sooty Shearwater

STATUS AND DISTRIBUTION: In summer, a very common pelagic resident in the North Pacific (Alaskan waters) and the only shearwater found north of the Aleutians; uncommon to rare off the Pacific Coast (B.C. to Calif.) primarily from October (November in Calif.) to March (a few occur off Wash. and B.C. in summer). **HABITAT:** Cold, pelagic waters, commonly well offshore. **COHABITANTS:** Northern Fulmar, Mottled Petrel, Fork-tailed Storm-Petrel. **MOVEMENT/MIGRATION:** Breeds in Australia and migrates north to Alaskan waters, arriving in late May and retreating in October. Migrates down the U.S. West Coast primarily in fall and is mostly seen from October to March. VI: 0.

DESCRIPTION: Short-billed Sootylike shearwater. Medium-sized (larger than Mottled Petrel; smaller than Northern Fulmar; nearly identical in size and shape to Sooty Shearwater), but with a shorter bill, a rounded head, and a steeper, more abrupt forehead. Streamlined body ends with a blunt but not particularly short tail. Short-tailed seems blunter-headed, longer-tailed, and perhaps better-proportioned than Sooty.

Overall dark sooty brown. Some individuals show a pale chin. Underwings vary in degree of paleness.

BEHAVIOR: Found in immense numbers in the North Pacific and Bering Sea. Rare along the Pacific Coast in winter and usually outnumbered by the seasonally reduced numbers of Sooty Shearwater. An active species, Short-tailed flies with a sense of urgency. Often rests on the sea in great rafts during calm conditions, but makes the ocean surface boil with motion when foraging. Feeds mostly by diving (sometimes plunging) beneath the surface. Readily attracted to ships. (Sooty shows reluctance or only casual interest.) Very tame, Short-tailed allows ships to approach closely before flying. Also unlike Sooty, Short-tailed is mostly seen far from shore over colder upwellings, except in w. Alaska, where it is commonly seen from shore.

FLIGHT: Nicely proportioned, tapered, and slender-winged shearwater that does not show a particularly short tail. Bill is shorter than Sooty's, and forehead tends to be steeper. Also, head appears more compact and rounded.

Overall dark smoky brown with paler, grayer underwings (again, similar to Sooty). However, classic Short-tailed shows overall darker and more uniformly pale underwings; Sooty's underwing is usually lighter, with the pale (whitish) area more concentrated and more contrasting. Chins on Short-tailed sometimes appear whitish. Some individuals may appear darker-capped or even hooded.

Flight is close to the surface, stiff, and sometimes recklessly fast. Wingbeats are rapid, jittery (but not choppy), and stiff, with a slightly elastic quality. They are given in hurried series followed by a short glide on stiff, slightly curved back wings. Shape, movement, and wingbeats make the birds look like the Swift of the Sea.

VOCALIZATIONS: Silent at sea.

PERTINENT PARTICULARS: There is much discussion about the characteristics that distinguish (or unite) Sooty and Short-tailed Shearwaters. In the North Pacific, when searching for Sooty, try scanning through the Short-tailed ranks for the birds with lots of concentrated white under the wings and look for candidates that appear long-billed or long-headed. On the south coast of Alaska, finding Sooty shouldn't prove difficult. They are present from June through November. On the West Coast, in fall and winter, focus your study on birds that seem particularly attracted to your boat (their proximity alone commends them to study) and see whether your candidates seem short-billed and round-headed or have uniformly smoky gray or pale brown underwings (not a thick white band running down the center of the wing).

Manx Shearwater, *Puffinus puffinus*

Little Black Shearwater

STATUS AND DISTRIBUTION: Fairly common in North Atlantic waters from New England to Nfld., where it breeds; uncommon from Va. south in winter. Rare on the Pacific Coast from Alaska to Calif. in summer and fall (more likely in Wash. and Ore.) and in the Gulf of Mexico. Possible at any month off the mid-Atlantic Coast, but more so from March to May and in October and November. **HABITAT:** A cold-water shearwater (Audubon's prefers warm blue water),

Manx generally, but not exclusively, is found well offshore (in excess of 40 mi.) over the shallower waters of the continental shelf. **COHABITANTS:** tuna, Cory's Shearwater, Greater Shearwater, Pomarine Jaeger. **MOVEMENT/MIGRATION:** Spring migration from February to June; fall from August to December. VI: 2.

DESCRIPTION: Smallish, blackish shearwater with long pointy wings, a fairly short tail, and a masked face (often showing a thin, white curl behind the ear coverts). Medium-small (slightly larger than Audubon's Shearwater; much smaller than Greater), with a slender bill, round head, and wings extending beyond the tail when sitting.

Upperparts are black; underparts are conspicuously white, including the undertail. (Audubon's shows a dark undertail.) Dark crown extends well below and behind the eye, giving the bird a hooded or masked impression. (Browner Audubon's looks capped and paler-faced.)

BEHAVIOR: Somewhat antisocial by shearwater standards; commonly found as individuals among the ranks of more numerous shearwaters (such as Sooty and Greater Shearwaters) or in small single-species groups (fewer than 10). Non-storm-related sightings of birds from shore are usually of individual birds. Manx feeds by coursing over the surface and landing with a belly-plop to seize prey (often submerging its head or diving in the process). It is attracted to feeding fish (such as tuna) and concentrations of feeding birds and fishing boats, but doesn't follow ships. Fairly tame, Manx allows close approach by boats before flushing.

FLIGHT: Shows the balanced proportions of a large shearwater: long, narrow, pointy wings and head and tail projecting about equally. Audubon's appears more compact, broader-winged, longer-tailed, and overall blunter or more roundly contoured than Manx. Manx Shearwater is contrastingly patterned blackish above, extensively white below (including white undertail). Audubon's Shearwater shows dark undertail and more restricted white wing linings. Manx's choppy wingbeats are given in a short series followed by a modestly protracted glide. In high winds, its flight is energetic as it bounds with other birds, turning nearly perpendicular to the surface and giving scant (if any) wingbeats. Audubon's flight is more hurried, with quicker wingbeats and more frequent, less extensive glides.

VOCALIZATIONS: Silent at sea.

PERTINENT PARTICULARS: Worn individuals (in late summer and fall) may appear browner than spring birds.

Black-vented Shearwater, *Puffinus opisthomelas*

California Beach Shearwater

STATUS AND DISTRIBUTION: Breeds on islands off w. Mexico. Variably common to scarce in coastal waters off the southern and central coasts of Calif. from August to April (August to December off cen. Calif.). Most common in the fall. Rare from May to July and north of n. Calif. **HABITAT:** Warm, inshore waters usually within five mi. of shore, and well within the range of land-based observers. **COHABITANTS:** Heermann's Gull. **MOVEMENT/MIGRATION:** Disperses north along the California coast after the breeding season, which extends from February to July. VI: 1.

DESCRIPTION: Smallish, darkish, slender, Pink-footed–like shearwater seen from or within sight of the beaches of California. Medium-small shearwater, but the smallest shearwater commonly seen in California waters. With its slender black bill, round head, compact body, and short tail, Black-vented is much smaller than Pink-footed or Sooty Shearwater and slightly larger than the similar (and, in California, very rare) Manx.

Upperparts are dark, dingy brown (not blackish), and sides dingy white; underparts are white. Black-vented looks like a small, dark, dingy Pink-footed with an all-dark bill. The much larger Pink-footed is overall paler above with a pinkish bill; Manx is blackish above. Black-vented's whole face is dark or dingy, where Manx appears masked.

BEHAVIOR: Forages in warm, inshore waters in significant numbers in years when inshore currents are warm; in years when colder water temperatures reign, seen only in low numbers as most birds remain farther south. Flies low over the water, plopping to the surface to seize prey. Frequently seen alone or in loose single-species groups.

FLIGHT: Small and compact with shortish wings and a short tail. Brownish upperparts are apparent in good light. The face is variable; usually appears dark or dusky and lacking in definition or pattern. Undertail is dark. Flies fast, low, and direct with rapid, hurried, somewhat frantic, "clippy" or fluttery wingbeats and short glides. Pink-footed Shearwater is larger, bulkier, and overall paler, with a more leisurely flight.

VOCALIZATIONS: Silent at sea.

PERTINENT PARTICULARS: This species can be viewed from beaches or piers but, particularly during years of low occupancy, is more likely to be seen from the tips of the coastal projections lying between San Diego and San Luis Obispo Counties.

Audubon's Shearwater, *Puffinus lherminieri*

Little Brown Shearwater

STATUS: Tropical species. Common off the South Atlantic and Gulf Coasts from April to October. **DISTRIBUTION:** Found (normally well offshore) in the Gulf of Mexico and the Atlantic from Fla. north to the waters off s. New England, but irregular north of Va.; occurs in northern waters late in summer (primarily August and early September). Present year-round, in reduced numbers, south of Cape Hatteras and in the Gulf of Mexico. **HABITAT:** Gulf Stream and warm ocean waters well offshore. Often seen over blue waters with floating mats of sargasso weed. **COHABITANTS:** Cory's Shearwater, Bridled Tern. **MOVEMENT/MIGRATION:** Breeds in the Caribbean. Wanders north and west after breeding. VI: 2.

DESCRIPTION: Runty, brown-backed, blunt-winged, warm-water shearwater. Small, compact, somewhat roundly contoured (slightly smaller than Manx; distinctly smaller than Sooty).

Distinctly brownish above, and white below. The poorly defined dark cap stops just below the eye, so distant birds appear capped. Manx Shearwater is blackish above, with more extensive black on the head, so it appears dark-headed or masked.

BEHAVIOR: In U.S. waters, usually found in small flocks or as individuals mixed in with more common shearwaters. Often sits on the surface (usually in the company of other shearwaters) and submerges in pursuit of prey; also picks rapidly through sargasso weed. Does not generally follow ships.

FLIGHT: Small, overall compact, with short, fairly blunt-tipped wings and a fairly short tail. (Only in comparison with the stubby-tailed Manx does it appear longish.) Distinctly brownish above, with a face more white than dark. There is considerable variation in the amount of white showing on the underwings, but at a distance the underwings of most birds appear mostly white; the undertail is always dark (but not necessarily easy to see).

Flight is quick, playful, or casual, with wingbeats given in a hurried flutter followed by a short glide.

VOCALIZATIONS: Noisy around colonies; otherwise silent.

STORM-PETRELS

Wilson's Storm-Petrel, *Oceanites oceanicus*

Dances with Waves

STATUS AND DISTRIBUTION: Common in summer along the Atlantic Coast, where it is the most common storm-petrel and the one most likely to forage close to shore. Uncommon in the outer Gulf of Mexico, and rare on the Pacific Coast. **HABITAT:** Warm and cold ocean waters, often close to shore. **COHABITANTS:** Offshore: shearwaters and other storm-petrels; inshore: gulls and terns. **MOVEMENT/MIGRATION:** Found in Atlantic waters from late April to November, but not common before May or after September (October/November in southern waters). VI: 2.

DESCRIPTION: Small, blackish brown, curve-winged, clannish storm-petrel that habitually dances on the water. Small (smaller than Band-rumped or Leach's; slightly smaller than Purple Martin, which it resembles somewhat in shape). On the water, appears blackish brown with an obvious white patch on the sides of the rump.

BEHAVIOR: Where you find one Wilson's Storm-Petrel, you usually find others, and often enough they find you. This species is attracted to ships and comes readily to chum. Birds gather in fluttering clusters, hovering on raised wings and dancing on the surface with legs extended. Leach's rarely dances on the surface, Band-rumped sometimes dances, and Wilson's dances habitually. Wilson's also flies, just above the surface, moving in a direct flight (looking somewhat like a hunting harrier), and skips or bounces along the surface on outstretched wings.

FLIGHT: Small dark storm-petrel showing a gently curved leading edge of the wing (no sharp angles, no jutting elbow) and feet that project behind the blunt or round-tipped tail. Leach's and Band-rumped have longer, more angular, Merlin-shaped wings and show no projecting feet. Also, the fanned tails of Band-rumped and (especially) Leach's appear square-tipped or notched. The fanned tail of Wilson's is no-nonsense round.

Wilson's is overall more patterned and contrasting than either Leach's or Band-rumped, showing a dominating white rump, pale carpal bars, and slightly paler underwings. A buff or silvery carpal bar on the upperwing is prominent but compact (and does not extend to the leading edge of the wing). The U-shaped white rump commands nearly half the tail. On Band-rumped, the white rump is more trim-cut, less U-shaped, and is closer to one-third the length of the tail than half, while the usually bisected rump on Leach's is overall less eye-catching. Also, Wilson's is the only one of the three to show paler underwings.

Flight is slow, direct, delicate, somewhat butterfly-like, with shallow stiff wingbeats. Wilson's glides with stiff wings angled slightly down and, unlike Leach's, does not zigzag in flight (or if so, only slightly).

VOCALIZATIONS: Soft peeping notes at close range.

White-faced Storm-Petrel, *Pelagodroma marina*

Kangaroo Storm-Petrel

STATUS AND DISTRIBUTION: Breeds off the northwest coast of Africa. Rare off the Atlantic Coast in late summer. **HABITAT:** This pelagic species is found in deeper, offshore waters (usually in waters overlying submarine canyons), but also occurs (if rarely) in waters closer to shore (between 10 and 20 mi.). Because it seems comfortable with water temperatures in the mid to upper 70s, this species is not necessarily tied to the warmer waters of the Gulf Stream. **COHABITANTS:** Cory's and Audubon's Shearwaters, Wilson's Storm-Petrel. **MOVEMENT/MIGRATION:** Found from late July to early October off the Atlantic Coast, with peak in late August and early September. VI: 0.

DESCRIPTION: Small, pale storm-petrel that holds its wings outstretched and skips across the water like a stone—the only white-bellied storm-petrel found along the East Coast. Small (slightly larger than Wilson's Storm-Petrel; slightly smaller-bodied than Red Phalarope). On the water, all pale grayish brown, with a white face, small black bill, blackish cap, and eye patch. Resembles a storm-petrel in nonbreeding Red Phalarope clothing.

When feeding, holds its long, narrow, "paddle-shaped" wings open—frozen at a right angle to the body. Upperparts show a cryptic eye-defeating pattern of dark wings and tail, pale rump, and mocha-colored back. Underparts (sometimes hard to see) are all white except for contrastingly black trailing edge to the wings. At close range, the black eye mask and broken collar extending down the sides of the breast are easily noted. At a distance, White-faced appears mostly mocha-colored and distinctly paler than other storm-petrels, which look essentially blackish.

BEHAVIOR: White-faced's manner of feeding is unique. It forages by gliding and bounding, kangaroo-fashion, across the water on open, banking wings (with one wing angled higher than the other). Wilson's Storm-Petrel bounces and patters on the water with fluttering wingbeats and little forward momentum. White-faced sails across the water on motionless wings at foot-smacking intervals whose splashes are sometimes easier to discern than the bird itself. From a distance, it is as though a stone is being skipped across the water, but the splashes are regular and even (1–2 yd. apart). In U.S. waters, White-faced is always seen alone or in small flocks of Wilson's Storm-Petrels. It responds to (fish) oil slicks, but does not follow boats.

FLIGHT: In direct flight, holds its wings at a right angle to the body (not curved or angled back, like other storm-petrels) and angled slightly down. Long feet and legs extend well beyond the tail. Flies using a series of short, stiff, fluttering wingbeats and long glides (recalls Spotted Sandpiper). Flight is often erratic, with a rocking or banking motion.

VOCALIZATIONS: Silent at sea.

PERTINENT PARTICULARS: You will not mistake this bird for any other storm-petrel. You may confuse it with Red Phalarope. Be aware that Red Phalarope has been known to sit among groups of Wilson's Storm-Petrels on the water. Although not every pallid bird sitting among storm-petrels is automatically a White-faced, it is certainly a possibility worth investigating.

Fork-tailed Storm-Petrel, *Oceanodroma furcata*

Wave Fairy

STATUS: Common year-round in the colder waters of the northern Pacific. **DISTRIBUTION:** Greatest portion of the population breeds on southern coastal islands off Alaska, especially the Aleutians and the islands in the Gulf of Alaska and south to the Queen Charlotte Is. of B.C. Much less common farther south; colonies in Wash., Ore., and n. Calif. number in the hundreds, as opposed to the tens and hundreds of thousands farther north. In summer, forages over the waters of the continental shelf, though some birds, presumed to be nonbreeders, occur in the deeper waters of the North Pacific. In winter, most populations move out over deeper waters, with the greatest concentrations found in the North Pacific to the ice pack. In n. Calif., birds are present year-round in the waters not far from breeding colonies, and some birds range south, appearing off the central coast of Calif. and very rarely off s. Calif. **COHABITANTS:** At sea, found with assorted pelagic species, including Short-tailed Shearwater and Leach's Storm-Petrel (which it greatly outnumbers). **MOVEMENT/MIGRATION:** Disperses into deeper waters after breeding; most birds are found at sea between November and March, and birds are present in colonies from late March to November. VI: 1.

DESCRIPTION: Uniquely silver-hued, fork-tailed fairy of a bird. Medium-sized storm-petrel (slightly larger than Ashy; same size as Leach's) that sits buoyantly on the water and might easily pass for a phalarope in nonbreeding plumage (except for the tiny-billed, long-tailed profile).

On the water, appears pale silver-gray with darker wingtips and a dark eye mask.

BEHAVIOR: Arrives and departs from colonies in full darkness. Forages at sea by hovering (sometimes pattering), skimming, and (very commonly) sitting on the surface, jabbing like a feeding phalarope. On the water, can be very difficult to see; spends a great deal of time sitting on the surface, sometimes in groups. Fork-tailed commonly feeds in association with other Fork-taileds but is not as gregarious as many other storm-petrels. Approaches ships and is attracted to chum.

FLIGHT: Altogether distinctive: a pale, slender, silver-and-black-patterned storm-petrel with slender wings curved down and forked tail curled up at the tip. All other West Coast storm-petrels are dark brown; only Ashy has so deeply forked a tail.

At a distance, seems almost white. On closer view, the contrasting pattern caused by a dark bar on the upperwing and dark underwing catches the eye. Flies close to the surface with a direct, buoyant, unhurried, tentative flight. Appears overall somewhat frail. Wingbeats are quick, shallow, and fluttery. Fork-tailed's glides, with wings raised at the shoulder and down-curved along the length, are usually short, but this storm-petrel sometimes looks like Leach's, gliding shearwater-like with lots of banks and glides. The tail is not always flared, so does not always appear forked.

VOCALIZATIONS: Silent at sea.

PERTINENT PARTICULARS: In winter, Fork-tailed is more likely to be confused with the similarly sized and also pale gray Red Phalarope than another storm-petrel. Phalaropes fly fast and direct, however, with steady rapid wingbeats (nothing buoyant or tentative about it); moreover, they often fly in tight flocks, and they often call "*plink*" when lifting off the surface. They are, in short, very different.

Leach's Storm-Petrel, *Oceanodroma leucorhoa*

Drunken Nighthawk of the Sea

STATUS: Very common on breeding islands and surrounding waters; elsewhere, at other times, generally uncommon in offshore waters along both coasts; rare in the Gulf of Mexico. Where ranges overlap, usually much less common than Ashy Storm-Petrel (Pacific) and Wilson's Storm-Petrel (Atlantic). **DISTRIBUTION:** Breeds on coastal islands from the Aleutians to Baja, and from se. Lab. to Penikese I. in Mass. Forages out to and beyond 100 mi., but some (presumed) nonbreeding birds may be found in ocean waters well south and away from breeding colonies all summer. After breeding, Leach's disperses widely across both oceans, with the bulk of both populations wintering in tropical waters but some birds lingering in colder northern waters until winter. **HABITAT:** Breeds on offshore islands free of mammalian predators. Offshore, generally found well away from land (beyond 20 mi. and more commonly 30–40 mi.) foraging in both warm and cold waters, but appears to be most partial to the colder waters shunned by other species (or, at the very least, prefers to forage in waters shunned by the more gregarious storm-petrels). **COHABITANTS:** Loner tendencies notwithstanding, found among other pelagic birds, including Wilson's and Ashy Storm-Petrels. **MOVEMENT/MIGRATION:** Spring migration from early March to mid-June; fall from August to December. Probably absent from North American waters in January and February. In the Atlantic, greater numbers of birds migrate in spring, with peak numbers in late May and early June. VI: 2.

DESCRIPTION: Angular, long-winged, brown storm-petrel that flies like a drunken nighthawk or a bird fleeing the gates of hell. Medium-sized (larger than Wilson's; about the same size as Ashy and Band-rumped).

Leach's is overall brown and slightly paler than Band-rumped, but this is only evident in flight and in direct comparison. On the water, the bird shows only a wink of white at the base of the tail. (Band-rumped and Wilson's show more obvious white patches.)

BEHAVIOR: Usually solitary, but may feed in small groups. Most often seen flying through or by clusters of more gregarious species (like Wilson's). When feeding, flies slowly over the water, rarely dancing on the water (like Wilson's). Hovers occasionally or sits on the water, plucking food from the surface. When hovering, wings are barely raised above the horizontal. (Wilson's raises its wings high above the body.) Does not follow ships, as some other species do, but can be drawn to chum and resulting feeding concentrations (at least briefly).

FLIGHT: Very angular profile distinguished by long, slender wings that jut forward at the wrist and a long, conspicuously notched tail. If the softly contoured Wilson's Storm-Petrel can be likened to American Kestrel, Leach's is more akin to Merlin. Overall brown, Leach's has a pale carpal bar that extends all the way to the wrist and an oval rump patch usually divided by a narrow line. (It looks like the rump is undergoing mitosis.) Some birds show no bisecting line; some Pacific birds have an all-dark rump with no white at all. In spring and summer migration, most Leach's Storm-Petrels are in fresh plumage and therefore clean-cut. In the Atlantic, Wilson's is molting and appears ragged and shopworn.

The flight of Leach's is determined, energetic, erratic, and high-strung, combining the wandering jerkiness of a nighthawk and the deep pushing wingbeats of a jaeger. The bird seems to be fleeing from imminent danger, tacking left, then right, and often rising and falling in the dynamic fashion of a gadfly petrel. Leach's glides with an upward jut on the wing and in strong winds flies more shearwater-like.

Leach's is distinguished from Wilson's Storm-Petrel by its larger size, longer wings, and distinctly jutting elbows; its split and less conspicuous oval white rump patch; its notched tail with no trailing legs; and its flight style. It is distinguished from Band-rumped by its slightly paler color, longer-appearing and more slender wings, more obvious and more extensive carpal bar, forked tail, and flight style. Except in extreme southern California, where small numbers of all dark-rumped Leach's occur, it is distinguished from Ashy, Least, and Black Storm-Petrel by its white rump and less deeply forked tail. Finally, Leach's is distinguished from all other storm-petrels by its crazed, erratic flight.

VOCALIZATIONS: Vocal array includes purrs, chatters, screeches, and chips. Mostly silent at sea.

PERTINENT PARTICULARS: The easiest way to both pick out and identify Leach's Storm-Petrel is to let your eyes gravitate toward the bird that is flying like a drunken nighthawk—bounding and buoyant. By comparison, the flight of other storm-petrels looks tame or staid. Wilson's has a delicate, fluttery flight and glides with wings angled down. Band-rumped has a direct, slow, shearwater-like flight that combines several slow deep wingbeats (given with down-bowed wings) with long, often extensive glides.

Ashy Storm-Petrel, *Oceanodroma homochroa*

Ashy-winged Storm-Petrel

STATUS: Except at Monterey Bay, where it is common to abundant in fall, and the Farallon Islands, where most individuals nest, this species is uncommon to scarce, with a limited distribution and world population estimated not to exceed 10,000. **DISTRIBUTION:** Found almost exclusively in the waters over Calif.'s continental shelf from Cape Mendocino south to about Isla Cedros halfway down the Baja Peninsula. Breeds primarily on the Channel Is. and especially the Farallons. In winter, disperses offshore. **HABITAT:** Pelagic. Spends most of its time at sea within the waters of the California Current and, for the most part and except where it breeds, beyond sight of land. Breeds on the rocky slopes of small islands and inlets. **COHABITANTS:** At sea, Black and Least Storm-Petrels. **MOVEMENT/MIGRATION:** Nonmigratory but seasonally apportioned. Between April and September, found in the vicinity of breeding islands; between November and mid-February, found in offshore waters but still over the continental shelf. In September and October, large concentrations (estimated to constitute the bulk of the population) are found in the waters off Monterey. VI: 0.

DESCRIPTION: Dark long-winged storm-petrel with a tarnished silver (wing) lining. Medium-sized (slightly smaller than Black; larger than Least; about the same size as Leach's). On the water, slightly longer-tailed than Black.

Overall dark brown. Although Ashy is slightly paler and sometimes grayer than Black and Least, this difference is slight and is best noted through direct comparison.

BEHAVIOR: Nocturnal feeder. During the day, sits on the water in small groups or (in Monterey Bay and less commonly Morro Bay) in large rafts, often with other species. Plucks food from the surface while sitting on the water, or sometimes hovers low over the water into the wind.

Fairly tame. Allows close approach. Comes in to chum and oil slicks.

FLIGHT: Long-winged and long-tailed profile; overall larger than Least Storm-Petrel. Wings have slightly rounded contours along the leading edge; forked tail curls up at the tip (more noticeably than Black's; tail does not curl at all on Least or Leach's). Underwings show faintly paler linings, but the comparison is subtle, a matter of shades. Nevertheless, the underwings of the other all-dark storm-petrels are wholly dark. If you think you perceive a difference, you probably do.

Ashy's wingbeat is quick, fluttery, and mostly constant; it glides less often and less extensively than Black. The stroke is shallow and tentative, with wings barely rising above the body; Black Storm-Petrel has deep wingbeats. Ashy's flight is usually direct, occasionally wandering. It glides, with wings angled back and slightly down-drooped. In many respects, Ashy is kestrel-like.

VOCALIZATIONS: Silent at sea.

Band-rumped Storm-Petrel, *Oceanodroma castro*

Plain, Dark Storm-Petrel

STATUS: Uncommon summer resident off the Atlantic and Gulf Coasts; scarce in northern part of its range, but the most common storm-petrel in the Gulf of Mexico. **DISTRIBUTION:** Breeds off the coast of Spain and Africa. In the w. Atlantic, ranges from s. Fla. north to the waters off s. New England. **HABITAT:** Offshore marine waters. Usually found in deeper waters (beyond 100 fathoms; rarely closer than 20 mi. from shore). Selects warm, blue water, usually over upwellings, and seems most prevalent where floating sargasso weed flourishes. **COHABITANTS:** Audubon's Shearwater, Wilson's Storm-Petrel. **MOVEMENT/MIGRATION:** Found off the Atlantic and Gulf Coasts between late April and September; most common June to August. VI: 2.

DESCRIPTION: Large, dark, and refreshingly straightforward (but by no means easy to distinguish) storm-petrel. Large (in direct comparison, obviously larger than Wilson's; about the same size as Leach's), with structural characteristics that seem halfway between Wilson's and Leach's. On the water, in direct comparison, larger size is apparent.

Beneath overlying, closed wings shows white patches at the base of the tail (as does Wilson's, but not Leach's).

BEHAVIOR: Shy and standoffish. Does not follow ships. Does not linger long in chum slicks with Wilson's, choosing instead to work the edge of the group or just fly through. When sitting on the water with Wilson's, Band-rumped is the first to flush when approached by boats. Dances on the water (like Wilson's), and often settles on the water to feed.

FLIGHT: Overall darker and more robust than Wilson's or Leach's. Larger, *longer*-winged, and bulkier than Wilson's; broader-winged than Leach's, with a more rounded leading edge to the wing and a notch-tipped tail that is shorter and less deeply forked than Leach's. Unlike Wilson's, legs do not extend beyond the tail.

Overall blackish brown; darker, plainer, less patterned than either Wilson's or Leach's. Pale carpal bar on the upperwing is duller, less contrasting than the whitish bar of Wilson's or Leach's and more limited than Leach's. Underparts are uniformly dark; the underwings of Wilson's are contrastingly shiny. The squared-off white band on Band-rumped's rump covers less than half the tail. The white rump on Wilson's covers about half the tail and seems rounded. The rump patch on Leach's forms an oval or twin ovals.

Flight is mostly steady and direct or conservatively banking and tacking—stronger and less fluttery or tentative than Wilson's; calmer, more purposeful, and less wild than Leach's. Wingbeats are quick, snappy, shallow. Glides are extensive—sometimes direct, sometimes dynamic and tacking on down-bowed wings (not raised like Leach's or stiffly angled down like Wilson's). The pattern of quick wingbeats and prolonged glides makes Band-rumped appear more shearwater-like than other storm-petrels.

VOCALIZATIONS: Silent at sea.

PERTINENT PARTICULARS: The way to pick out Band-rumped is to forget about particulars and concentrate on manifest differences. The bird's larger, darker, plainer, and more deliberate manner catches the eye when set against the frenetic, dancing backdrop of a flock of Wilson's Storm-Petrels working a chum slick. Relax. Let your eye be drawn. Select a candidate. Now look for particulars.

Black Storm-Petrel, *Oceanodroma melania*

Biggest and Blackest Storm-Petrel of Them All

STATUS: Common pelagic visitor in summer and fall off southern California. **DISTRIBUTION:** Breeds off Baja; breeding range just reaches Santa Barbara I. in U.S. waters. Beginning sometime in August (after breeding), birds range north to Monterey Bay, Calif. (less commonly to the waters north of San Francisco). **HABITAT:** Warm pelagic waters, with concentrations over the continental shelf and drop-offs. At times, in southern California, can be viewed from shore. **COHABITANTS:** At sea, may be found with other storm-petrels, including Ashy and Least. **MOVEMENT/MIGRATION:** Disperses north along the California coast between August and October. VI: 1.

DESCRIPTION: Big, dark, long-winged, fork-tailed storm-petrel. Largest storm-petrel (larger than Ashy or Leach's; slightly larger than the pale Fork-tailed). In flight, shows long wings and long deeply forked tail. Pale carpal bar is more distinct than it is on Least, less obvious than on Leach's. Underwings are all dark (slightly paler on Ashy). Note all-dark rump—this and its flight distinguish Black Storm-Petrel from most Leach's Storm-Petrels.

BEHAVIOR: Usually feeds alone, but commonly roosts in single-species flocks and does join other feeding and rafting storm-petrels (particularly in years when few birds wander north). Responds to oil slicks, but is not particularly attracted to boats.

FLIGHT: Slow, deliberate, deep wingbeats interspaced with lengthy glides and a mostly direct flight. Has been compared to the flight of Black Tern. In size, shape, and dark blackish plumage, Black Storm-Petrel is most likely to be confused with a dark-rumped Leach's (whose numbers increase in late summer and fall off southern California). Leach's quick springy wingbeats and frantic zigzagging flight, however, are very different from the staid, direct flight and slow, stroking wingbeats of Black Storm-Petrel.

VOCALIZATIONS: Silent at sea.

Least Storm-Petrel, *Oceanodroma microsoma*

Batlike Storm-Petrel

STATUS: Late-summer and fall visitor to (primarily) southern and central coasts of California, with numbers varying—some years rare, some years common. **DISTRIBUTION:** Breeds off Baja. Commonly disperses north to s. Calif.; regularly (but not always) found north to Monterey (particularly in warm-water years); rarely occurs much farther north than Monterey. Storm-related records exist for se. Calif. and sw. Ariz. **HABITAT:** Coastal waters; casual from shore. **COHABITANTS:** Other storm-petrels, most commonly Black and Ashy. **MOVEMENT/MIGRATION:** Found in California waters between August and October. VI: 1.

DESCRIPTION: Small, all-dark (blackish), blunt-tailed bird with a batlike flight. Smallest storm-petrel (more than 2 in. smaller than Ashy Storm-Petrel; more than 3 in. shorter than Black), with a short, blunt, or modestly wedge-shaped tail. Equally dark Black Storm-Petrel is distinctly larger and lankier and has a long deeply forked tail. Browner Ashy Storm-Petrel is overall paler, with paler upperwing coverts, pale underwings, and a long deeply forked tail.

BEHAVIOR: A gregarious species, Least mixes freely with other storm-petrels, both while feeding and while sitting on the surface in large rafts.

FLIGHT: Considerably smaller and shorter-tailed than other storm-petrels, Least has a batlike shape. Its flight is erratic and indirect (thus batlike) on wingbeats that are deep, quick, and steady. Glides infrequently.

VOCALIZATIONS: Silent at sea.

TROPICBIRDS

White-tailed Tropicbird, *Phaethon lepturus*

Black-backed Streamer-tailed Tern

STATUS AND DISTRIBUTION: Regular summer and fall in warm deeper waters off the se. U.S. coast (less commonly in e. Gulf of Mexico). **HABITAT:** Warm, deep ocean waters far from land. In the southeast United States, most commonly found in the waters of the Gulf Stream. **COHABITANTS:** Audubon's Shearwater, Black-capped Petrel, Bridled Tern, flying fish. **MOVEMENT/MIGRATION:** Most common off North Carolina from May to September. VI: 2.

DESCRIPTION: Heavy-billed, boldly patterned, black-and-white ternlike bird with a flagellum for a tail. Medium-sized (smaller than Royal Tern or Red-billed Tropicbird; not counting tail, about the same size as Bridled or Sooty Tern), with a heavy orange-yellow bill, big head, stocky body, and long streamer tail.

On the water, adults are mostly white with a black saddle set well back (and a streamer tail raised like a drake Long-tailed Duck). Immatures (and some adults) lack the streamer tail and appear overall compact with well-spaced, coarse, blackish barring on the back. (Other young tropicbirds appear more uniformly gray-backed.) Black eye-line turns down to the rear (and does not meet across the nape).

BEHAVIOR: Spends most of its life over oceans but lives in the air. Breeds on cliffs (and coastal forts), and forages at sea (mostly out of sight of land). Pelagic outside the breeding season, spending most of its time aloft and usually very high above the surface. Feeds by plunging. Comes readily to ships, which it often circles or follows briefly, behaving like a kite under tow (not coincidentally, this species is reported to be attracted to kites towed behind ships) and often appearing "out of nowhere."

Despite its affection for kites, solitary. Seems not to be attracted to concentrations of birds to the degree that many other pelagic species are.

FLIGHT: Front-heavy streamer-tailed "tern" with slender curving wings set too far back. Overall smaller, more delicately proportioned than Red-billed Tropicbird. Long flagellum-like tail easily distinguishes this species from Royal Tern, whose narrow wings are harshly angled, not curved back.

Adults, showing boldly patterned upperparts (black hands and black braces across the upperwings) are distinctive. Immatures are overall paler and less contrastingly patterned above than Red-billed Tropicbird, with limited amounts of black on the hand (extending halfway to the bend in the wing) and an abbreviated darkish eye-line that does not meet at the nape.

Flight is direct, strong, and high, on wingbeats that are rapid, shallow, continuous, and snappy—faster and not as heavy as Red-billed.

VOCALIZATIONS: Mostly silent at sea.

PERTINENT PARTICULARS: This species is regarded as the tropicbird most likely to be encountered off the southeast U.S. coast, owing largely to the closer proximity of its breeding areas. Red-billed Tropicbird is the greater traveler, however, and is probably more numerous than prevailing wisdom might have it.

Red-billed Tropicbird, *Phaethon aethereus*

Red-billed Streamer-tailed Tern

STATUS AND DISTRIBUTION: Rare pelagic visitor to the waters off s. Calif., especially near the Channel Is. (where it is by far the most likely tropicbird to be encountered), and also to the South Atlantic, from Cape Hatteras south (where it is outnumbered by White-tailed Tropicbird). Has been recorded as far north as New Brunswick. **HABITAT:** Warm ocean waters, commonly more than 50 mi. from shore. **COHABITANTS:** A pelagic loner. But if you are in water that supports Black-capped Petrel, Audubon's Shearwater (in the Atlantic), or Least and Black Storm-Petrels (in the Pacific), you're in the ballpark. **MOVEMENT/MIGRATION:** Most likely to be found off both coasts in summer and fall. VI: 2.

DESCRIPTION: Large, white, streamer-tailed "tern" with a red bill, dusky back, and black trim along the leading edge of the outer wing. Largest of the tropicbirds (only slightly smaller than Royal Tern; White-tailed Tropicbird is closer in size to Sandwich and Bridled Terns or Long-tailed Jaeger). On the water (and despite long tail), overall robust, with a heavy red bill (yellow in immatures), a big, somewhat neckless head, a stocky body, and a long streamer tail that is lifted above the water (much like drake Long-tailed Duck).

Mostly bright white with shadowy gray back (shows as fine, dense gray barring at close range) and a thick black line through both eyes that curls up near the nape, nearly meeting. (The more limited eye-line on White-tailed Tropicbird turns down.)

BEHAVIOR: Flies very high, very fast, very directly, and very much all by itself. Although too often this bird is suddenly there and then just as suddenly gone, it is sometimes attracted to boats and may hang up, high over the stern, following for a short time. Red-billed plunge-feeds for fish, sometimes hovering before diving with half-folded wings (very gannet-like). On the surface, it sits high.

FLIGHT: Looks like a heavy (particularly front-heavy) tern with wings too narrow and set too far back for the body. At a distance, appears overall stocky and all white. Closer range reveals a large bill, a large head/neck projecting ahead of the overly narrow sickle-shaped wings, and a long, white, flagellum-like tail (lacking on immatures and some adults).

The pattern of dusky gray back and extensive black on the hands (not to mention the red bill) readily distinguishes adults from the more strikingly patterned adult White-tailed Tropicbird. Immatures are similar to young White-tailed, but show more extensive black on the leading edge of the wing (extending from the wingtips almost to the bend of the wing, where on White-tailed it stops halfway), full black eye stripes that meet at the nape (eye stripe on White-tailed is less distinct and stops short of the nape), and darker back that shows as finer and denser barring at close range (the back of young White-tailed is mostly white with coarse barring).

Flight is steady, direct, fast, and high. Wingbeats are steady, hurried, and heavy, with a powerful down stroke that is slightly hesitant (as in a young bird or a bird that hasn't flown in a while) and doesn't show the stiff crispness of most terns. The cadence is like that of a jaeger in pursuit flight. Red-billed Tropicbird gives the impression of a Royal Tern with the quicker wingbeat of a Least Tern.

VOCALIZATIONS: Mostly silent at sea.

PERTINENT PARTICULARS: Observers should be aware that a third species of tropicbird, Red-tailed Tropicbird, *Phaethon rubricauda*, is a very rare visitor to waters well off the coast of California. Slightly smaller but more heavily built than Red-billed, adult Red-taileds have almost pure white upperparts and a red streamer tail (which can be difficult to see against a blue sky). Immatures resemble young Red-billed but have black (not yellow) bills and no black in the wingtips.

SULIDS (BOOBIES)

Masked Booby, *Sula dactylatra*

Pelagic Gannet

STATUS AND DISTRIBUTION: Tropical species that breeds on the Dry Tortugas. Uncommon in summer in the Gulf of Mexico and rare in the Gulf Stream north to N.C. Occasionally wanders to s. Calif. **HABITAT:** Warm, deep ocean waters. In the United States, most sightings are well away from shore. **COHABITANTS:** Tuna, dolphin, and whales. **MOVEMENT/MIGRATION:** During the nonbreeding season, September to February, wanders hundreds of miles from colonies, but seems not to truly migrate. VI: 2.

DESCRIPTION: Gannetlike booby with a black face and yellow (or greenish yellow) bill. Large, sturdy, and proportionally balanced (slightly larger than Brown or Blue-footed Booby; distinctly larger than Red-footed Booby; smaller than Northern Gannet), with a tail that appears to project only slightly more than the head.

Adults with bright white bodies and black wings and tail are distinctive but not unique. Adult Northern Gannet is similar but has a bluish bill and yellow glaze to the head. (The head of Masked Booby is all white, and the bill bright yellow.)

Immature Masked Booby is mostly brown above, with a white collar, and, except for a brown neck, white below. The rump does not show when the bird is sitting on the water. Immature Brown Booby is wholly dark brown above, and paler brown below with no white collar (or rump). A bright yellow bill distinguishes it from the dark-billed, young Northern Gannet and Blue-footed Booby.

BEHAVIOR: Prefers to forage well offshore in deeper waters and, except near breeding colonies, is not commonly seen near shore. A social species, Masked Boobies are often seen foraging together in large groups. They sit on the water, attuned to the activity of predatory fish and mammals and the plunge-diving of other birds. They gather over fish being driven to the surface and execute dramatic and near-vertical plunge-dives from heights as great as 100 ft. Breeding birds commonly travel in strings, often low over the water, that show little rise and fall (as compared to the roller-coaster undulations of Northern Gannet). Masked Booby travels great distances to forage and is a tame, almost fearless, bird. On land, it most often stands on rocks or the ground; offshore, it sits on buoys and navigation towers.

FLIGHT: Sturdy booby that is less gangly and shorter-tailed than Northern Gannet. Adults are mostly white with a broad dark trailing edge to the wings, white rump, and black tail. (Adult Northern Gannet shows only black wingtips.) Immatures are mostly brown above but have a distinctive white collar separating the head and body; sometimes show a white rump.

Masked Booby flies with the body somewhat angled up and the bill held level and high, on slow measured wingbeats given in a series followed by a glide. Flight is commonly low over the water, with little change in elevation.

VOCALIZATIONS: Silent at sea.

Brown Booby, *Sula leucogaster*

Hooded Booby

STATUS AND DISTRIBUTION: Uncommon tropical pelagic bird found mostly in the waters off both Fla. coasts (regular near the Dry Tortugas). Rare in the Gulf of Mexico, north along the Atlantic Coast (with records to N.S.), and along the Pacific Coast north to Wash. (with some interior records in Ariz. and se. Calif.). **HABITAT:** Ocean waters, often close to land. **COHABITANTS:** In the tropics, other boobies. In the United States, don't expect to see one where you don't see Magnificent Frigatebird. **MOVEMENT/MIGRATION:** Possible off southern Florida any month of the year; sightings north of Florida are more common in late summer. VI: 2.

DESCRIPTION: Dark brown white-bellied booby (adults) with a pale yellowish bill (grayish in immatures). Medium-sized (much smaller than Northern Gannet; slightly smaller than Masked Booby; about the same size as Great Black-backed Gull), with a long, heavy, pointy bill/face, a long neck, a sturdy but slender body, and short yellowish legs and feet. On adults, the contrast between the bright yellow bill/face and the dark hood is manifest.

Adults are rich dark brown above, contrastingly white below, with a sharp demarcation separating the brown breast from the whiter underparts—the bird appears hooded. Immatures are wholly dull brown but show the hooded pattern of adults: a sharply defined dark breast contrasting with pale brown (not white) underparts. Young Masked Booby shows a dark head and neck, but white breast, belly, and collar. Young Northern Gannet is overall grayish (not brown), with pale spotting (in older birds, white mottling) above. Dark-morph Red-footed Booby is quite similar but has a pinkish purple base to the bill/face and orange legs; in flight, it shows a distinct blackish trailing edge to the upperwing.

BEHAVIOR: With the exception of Northern Gannet, this social, inshore booby is the *Sulid* most likely to be seen from shore. Nests and roosts in groups. Usually found feeding with other pelagic species (including gulls and Brown Pelicans, but mostly other *Sulid*), but also feeds and travels alone. Something of a low-altitude feeder: makes shallow, angled dives for prey and after lifting off may make another quick dive without bothering to gain altitude. Swims well and can spend long periods on the water. When out of the water, usually perches on the ground, but can also be found on rocks, buoys, ship superstructures, and dead tree branches (also favored perches of Red-footed Booby). When hunting, sometimes flies high; when commuting, commonly flies low over the water.

FLIGHT: Small *Sulid* with somewhat shorter wings and longer tail than Northern Gannet or Masked Booby. Wings appear less bowed and more angled down in a glide. Adults and immatures are all dark brown above; all other sulids (except dark-morph Red-footed Booby) show, at the very least, white at the base of the tail or a paler tail, if not a pale collar as well. The dark hood is very evident on adults and subadults, but less obvious on darker-bellied immatures.

Flight consists of strong, even series of flaps interspaced with relatively short glides. Brown Booby also glides along beaches and dunes. The cadence of its wingbeats is quicker and the flap deeper than Northern Gannet.

VOCALIZATIONS: Silent away from breeding areas.

Northern Gannet, *Morus bassanus*
The North Atlantic Booby

STATUS: Very common coastal breeder of northeastern Canada; in winter, an Atlantic and Gulf of Mexico pelagic species. **DISTRIBUTION:** Nests in colonies on sea cliffs found at the mouth of the St. Lawrence R. and along the coasts of Nfld., Lab., and Iceland. Winters along the Atlantic Coast from s. Maine to the Gulf of Mexico, south to the Tex.–Mexico border, frequenting the waters over the continental shelf. **HABITAT:** Ocean waters, often close to shore, but also found farther offshore, where it may concentrate near commercial fishing boats. In summer, some subadults remain within the winter range. **COHABITANTS:** Herring Gull, Great Black-backed Gull; farther from shore, shearwaters. **MOVEMENT/MIGRATION:** All birds vacate northern colonies for the winter. Spring migration from early February to late May; fall from late September to late December. VI: 1.

DESCRIPTION: Very large gangly seabird that executes spectacular near-vertical plunge-dives for fish. In flight, Northern Gannet is larger, stiffer, more slender, more angular, and more four-points-pointy (bill, tail, and the two wingtips) than the largest gull. On the water, it looks pointy-faced, thick-necked, and humpbacked.

Adults (at least four years old) are all white with all-black wingtips. The yellow wash on the head disappears at a distance, and the black wingtips may not be apparent. Except for a pale rump, first-year birds are overall smoky brown. Older subadults are variously piebald—younger birds are marked like white-headed and white-bellied juveniles, and older birds look more like splotchy adults.

BEHAVIOR: Feeds by flying into the wind (30–50 ft. up) with its body angled slightly up and bill angled down. Dives headfirst into the sea, sometimes with little or no splash. Large gulls don't dive (they plop to the surface), and they don't submerge. Northern Gannet is often seen feeding in large groups comprising mixed age classes. It also sits on the water, often in large groups. This bird does not commonly vocalize, except when clustered in a feeding frenzy, and does not generally haul out on land unless sick or injured (or breeding). Northern Gannet almost never flies over land (except when breeding).

FLIGHT: Very long, thin, gangly, overall pointy profile—with pointed head, pointed tail, and pointed wings that seem too slender for the rest of the bird. In hard glides or high winds, wings jut forward at the wrist and angle back at the hands, but in calmer conditions the wings are set at a stiff, right angle to the body. Flies with a coursing, undulating rise and fall—rises when executing a series of slow languid wingbeats; glides downhill. Often flies low over the water in a line with other gannets that recalls the rise and fall of roller-coaster cars on some invisible track. Wingbeat is slow and, for a seabird, somewhat stiff and effortful. Movements are dynamic and fluid, but also somewhat awkward.

VOCALIZATIONS: Usually silent at sea, except when competing for food (at a chum slick, for instance). Makes a raucous "*grrrrrou*" or "*grrah grrah grrah.*"

PERTINENT PARTICULARS: Distinguishing Northern Gannet from gulls is easy. They aren't shaped alike, they don't fly alike, and they don't feed alike. In southern waters, distinguishing distant adult and subadult gannets from distant adult and subadult Masked Booby, and juvenile gannet from Brown Booby, is more troublesome.

Both boobies are smaller, sturdier, and slightly more compact than Northern Gannet. Brown Booby is shorter-headed and longer-tailed; Masked Booby is longer-headed and shorter-tailed. Northern Gannet's head and tail are equally proportioned.

On both boobies, the wingbeats are quicker, deeper, and more crisply matter-of-fact. Although they fly with the same flap-and-glide pattern as gannets, boobies do not rise and fall in flight as theatrically as gannets—in fact, their flight is often low and straight with no undulation at all.

Gannets are commonly seen close to shore, and boobies—particularly Masked Booby—are more often found well at sea in deep water. Brown Boobies habitually perch on channel markers, whereas gannets rest on the water. And finally, Brown Boobies dive by angling into the water, whereas gannets (and Masked Booby) execute near-vertical dives.

PELICANS

American White Pelican, *Pelecanus erythrorhynchos*

The Herding Pelican

STATUS: Fairly common western and interior breeder and wintering species in the southern United States and Mexico. **DISTRIBUTION:** Breeds in scattered colonies in B.C., Alta., Sask., Man., and sw. Ont.; in Ore., n. Calif., the Idaho–Utah border, Mon., Wyo., Colo., N.D., S.D., and Minn. Winters in Calif., sw. Ariz., N.M., s. and e. Tex., La., s. Miss., s. Ala., Fla., and throughout much of Mexico to Central America. Permanent resident on the Tex. and La. coasts. **HABITAT:** Nests on islands in large freshwater lakes on the prairies and high plains (less commonly on lakes in forested habitat). Winters in coastal bays, inlets, and estuaries; less commonly inland. In migration, often follows large rivers and uses lakes, reservoirs, and impoundments as stopover sites. **COHABITANTS:** Double-crested Cormorant; gulls. **MOVEMENT/MIGRATION:** Spring migration from early March to late May; fall from mid-September to mid-November, east to about the Mississippi River. Individuals commonly wander outside their normal range. VI: 3.

DESCRIPTION: Huge all-white bird with a big, wedge-shaped, yellow-orange bill angled demurely down. Sometimes shows a narrow wink of black near the tail; sometimes doesn't. Immatures are like adults but slightly dusky on the head and back. Breeding adults show sparse blackish crowns that recall a badly fitted toupee.

BEHAVIOR: Except for vagrants wandering outside their normal range, American White Pelican commonly roosts, forages, and flies in flocks. Roosting birds sit in tight packs on islands and sandbars. Feeding birds, swimming very high in the water, form interception lines to concentrate fish. Birds dip their bills into the water to scoop out prey, and they feed both day and night. *Note:* American White Pelican does not plunge-dive like Brown Pelican (or Northern Gannet). It soars habitually, using thermals to gain lift and often reaching great heights, as flock members circle synchronously.

Away from breeding colonies, American White Pelican is often tame and generally silent.

FLIGHT: Despite its elevated head and projecting bill, the exceedingly long planklike wings make the bird appear oddly balanced in flight. The broad black trailing edge of the wing (visible above and below) is evident at heights where the rest of the bird melts into the sky.

Wingbeats are slow, measured, and ponderous, given in a lengthy series followed by a long gravity-canceling glide that seems not to lose altitude when flying over water. Sometimes flies in a rising and falling string of birds. When flying high, the birds often form a V.

VOCALIZATIONS: Call is a low, grunting "*raah*" or "*ruah.*"

PERTINENT PARTICULARS: Because of its great size and almost unique bill, this species can hardly be mistaken. Distant, nonbreeding adult Brown Pelicans, showing white heads and silvery gray bodies, can be confused with this species, but the bill of Brown Pelican is mostly dark and yellow only on the top. In flight, Wood Stork (in particular) and Snow Goose show a similar plumage pattern.

Brown Pelican, *Pelecanus occidentalis*

The Diving Pelican

STATUS: Common along southern U.S. seacoasts and somewhat common near coastal inland lakes; seasonal and less common along the northern coast of the United States. **DISTRIBUTION:** On the Atlantic, breeds from about the Va.–Md. border in the Chesapeake south to Fla., throughout the Caribbean, and along the Gulf of Mexico south to South America. Dispersing post-breeding birds and, in spring, prospecting birds range to s. New England. On the Pacific, breeds to the central coast of Calif. south; winters to San Francisco Bay and ranges, after the breeding season, as far north as

Vancouver I., B.C. Also found inland at the Salton Sea in Calif. and along the nearby Colorado R.; also at Lake Okechobee, Fla. Small numbers regularly winter across the interior s. U.S. **HABITAT:** Warm, coastal, marine environments, including near-shore waters, lagoons, beaches, sandbars, mangrove stands, fish docks, and pilings. **COHABITANTS:** Laughing Gull (in the East); Western Gull, Heermann's Gull, Royal Tern, Elegant Tern (in the West). **MOVEMENT/MIGRATION:** Mostly a resident species, but exhibits some post-breeding dispersal north to southern British Columbia between July and early January. On the Atlantic, nonbreeding birds wander north to Long Island in June (retreating in October), and many of the breeding birds in coastal Virginia and North Carolina retreat south in November and December, returning in February (but significant numbers remain in North Carolina). VI: 3.

DESCRIPTION: Very large, very familiar, grayish seabird whose signature feature is its massive bill. On dry ground or perched, stands very erect with outsized head raised and bill pressed demurely against the neck. On water, sits buoyantly high with large bill angled down. *Note:* Only the head and neck of nonbreeding adults is white. If the rest of the back is bright white (not silvery gray), it's the other pelican.

Immatures are like adults, but back and neck are brown, and underparts are whitish. Underparts of adults are blackish.

BEHAVIOR: Very social year-round. Rests and flies in groups ranging in size from several individuals to 100 or more (average 6–20). Flies over open water in search of fish. Executes a twisting wingover and plunges, near-vertically. Many other large seabirds plunge-dive (including gannets and boobies), but American White Pelican does not. Although social, Brown Pelican is mostly a solitary feeder and does not herd fish, as white pelican does.

On land, walks clumsily with a rolling sailor's gait. Perches on pilings and tree limbs (mangroves where found). Nests on the ground or in trees. Rises from the water without running.

FLIGHT: Profile is distinctive yet deceiving. Viewed wing on, the large body, large anvil-shaped head (raised above the body), and massive bill are manifest; the bird is decidedly front-heavy. From below, the bird seems curiously balanced—and all wing! Only a snub of a tail and the projecting tip of the bill extend beyond the long, broad, plank-like wings. The arm is exceedingly long, and the hand short. When soaring, wings are down and slightly bowed; when gliding, wingtips are angled down.

Flight is slow and heavy; Brown Pelican projects ponderous dignity. Wingbeats are shallow, heavy, and flowing and given in a series followed by a long—sometimes impossibly long, laws-of-physics-canceling—glide just over the surface of the water during which the bird seems not to lose speed or altitude.

Flocks fly in V-formation or in a string that rises and falls in roller-coaster fashion. Using updrafts off cliffs and dunes, Brown Pelican often soars aloft with flocks turning in synchronous fashion.

The Brown Pelican is so tame that where it becomes habituated to people, it may be a genial nuisance to dock owners and fishermen.

VOCALIZATIONS: Silent except during the breeding season.

CORMORANTS

Brandt's Cormorant, *Phalacrocorax penicillatus*

The Cormorant That Queues

STATUS: Common, strictly coastal West Coast resident and winter visitor. **DISTRIBUTION:** Breeds from B.C. (a few in se. Alaska) to s. Baja. After breeding, some birds disperse north to the southern coast of Alaska and south to the tip of Baja and throughout the Sea of Cortez. **HABITAT:** Inshore coastal waters, often near kelp beds; also large bays and, less commonly, estuaries and lagoons. Avoids fresh and brackish water. **COHABITANTS:** Forages with an assortment of coastal species, including loons, Western Grebe, Sooty Shearwater, and Brown Pelican, as well as other cormorants and assorted gulls. Often roosts with Pelagic and Double-crested Cormorants. Prefers rocky coasts. **MOVEMENT/MIGRATION:** Disperses north in July and August; retreats in September and October, but large numbers remain in British Columbia and Washington until late spring. VI: 1.

DESCRIPTION: Bulky black-billed West Coast cormorant that flies single file. Large (very slightly smaller than Double-crested; much larger than Pelagic), with a long straight black bill, a wedge-shaped face, a thickish neck, a long body, and a short tail. Pelagic Cormorant overall is slimmer and more Anhinga-like, with a smaller head and a thinner and proportionally shorter bill. Double-crested seems larger-headed and has a heavier (and distinctly bright orange) base to the bill and throat. Both Pelagic and Double-crested are longer-tailed.

Overall Brandt's is smoothly blackish with bronze or brown highlights in the folded wings, a faint purple gloss in the neck, and a pale buffy patch on the lower cheek/throat. The blue throat patch on breeding adults is hard to see and hardly necessary. Immature is like adult but browner, with paler throat and breast. Immature Brandt's is overall darker—particularly darker-breasted—than immature Double-crested.

Pelagic Cormorant is sleeker, with (at times) obvious purple gloss in the throat and green gloss on the upperparts (and no hint of brown in the plumage). Adult Double-crested Cormorant seems scaly-backed; immatures are brownish-backed and more extensively pale on the neck and breast.

BEHAVIOR: Social cormorant, often feeding in tight clusters with other Brandt's Cormorants or in mixed flocks with other fish-eating birds (and mammals). Pelagic Cormorant is more of a loner. Brandt's commonly feeds near kelp beds but is also found in open water away from shore. Rides very low in the water, often with only the head, neck, and part of the back showing. Avoids flying over land (even obstacles, like jetties), and commonly flies single file in long lines just above the water.

FLIGHT: Long, mostly straight neck (not kinked or up-raised like Double-crested); overall thicker-necked and shorter-tailed than Pelagic. All dark, but in good light shows pale brownish cast to the wings (Pelagic is just black). Wingbeats are steady and not as quick as Pelagic. Glides only when landing.

VOCALIZATIONS: Even by cormorant standards, uncommonly silent.

Neotropic Cormorant, *Phalacrocorax brasilianus*

The Grunting Duck

STATUS: Widespread New World cormorant. Common in the United States, but with a restricted southern breeding and wintering range; late summer post-breeding dispersal is fairly extensive but in limited numbers. **DISTRIBUTION:** In the U.S., breeds and winters primarily along the Gulf Coast from La. west to the Mexican border and a short distance up the Rio Grande. Also breeds, and occasionally winters, on inland reservoirs and refuges in e. Tex., N.M., and Ariz. **HABITAT:** In the United States, found in coastal marshes and swamps as well as on freshwater reservoirs and impoundments. **COHABITANTS:** Often found with Double-crested Cormorant. **MOVEMENT/MIGRATION:** Basically nonmigratory. Late summer and fall post-breeding dispersal, however, may find birds wandering as far north as the Great Lakes and the northern prairies and west to southern California. VI: 2.

DESCRIPTION: Small, slender cormorant with an Anhinga-esque air. Considerably smaller than Double-crested Cormorant (the species with which it is commonly found), with a slightly slimmer, less bulky head; a thinner, more petite bill; and a conspicuously longer, narrower tail. The bare skin around the base of the bill of adults and immature birds is more restricted and duller (dull yellowish or dull orange, as opposed to the bright orange of Double-crested). Distant Neotropics seem all head and dull, slender bill. On distant Double-cresteds, you can see that the birds have a face too (or, at the very least, a bill that is much brighter than the head). In breeding plumage, the wedge-shaped base of Neotropic's bill (the gular patch) shows a narrow white border. In winter, you have to look closely to see that the bare skin on Neotropic's face cuts a sharp wedge into the head (giving the bird the semblance of a grimace). On Double-crested, the weld between the bill and face is rounder and less pointy. Also, the more extensive orange on the face of Double-crested often projects just above the eye; you never see this on Neotropic.

Immature Neotropics and Double-cresteds are often easier to tell apart than adults. The classic young Neotropic is a rich brown below; the young Double-crested has a pale (buffy or cream-colored) throat and breast that contrast with a much darker belly.

BEHAVIOR: Social; often sits and feeds in groups (facilitating comparison with the much larger, bulkier, shorter-tailed Double-crested). Feeds by swimming and diving (like loons and other cormorants), but also plunge-dives (quite unlike any other cormorant). This is a nimble bird! Neotropic commonly perches on small, springy branches and narrow utility lines. When standing with Double-cresteds, frequently perches on the less stable structure. (If Double-crested is standing on a partially submerged log, Neotropic is on a limb.) Fairly tolerant of people (often nests close to human habitation), but does not like to be approached in tight quarters.

FLIGHT: In flight, silhouette is more slender and longer-tailed than Double-crested (with tail projecting about the same length as the head). The head on Neotropic is about the same width as the neck—giving the bird an Anhinga-like look; the head on Double-crested is bulkier. Like Double-crested, Neotropic flies with neck cricked and head elevated.

Flight is straight, strong, and fairly effortful. Wingbeats are steady and slightly quicker and choppier than Double-crested, with wings held in a downward curve. Neotropic often glides when coming in for a landing, circling to face into the wind, and brakes with wings and projecting feet. When a slender branch is the landing site, the maneuver requires considerable wing-flapping to maintain balance. When flying long distances in a group, Neotropics may form a V.

VOCALIZATIONS: Silent for most of the year. Breeding birds emit a series of loud, resonant, hippolike grunts, *"r'rauh"* or *"ruuh-aah."* Often one vocalizing bird sparks a chorus.

Double-crested Cormorant, *Phalacrocorax auritus*

The Abundant and Everywhere Cormorant

STATUS: Common—at times abundant—and widespread; found along both coasts and across most of interior North America. Except where Neotropic Cormorant also occurs, usually the only cormorant to be found in fresh water away from the coast. **DISTRIBUTION:** Breeds in scattered locations along both coasts, from sw. Alaska to Baja and from Nfld. to Cuba. Also scattered throughout the interior, particularly along the St. Lawrence R., around the Great Lakes, and throughout the northern interior West. In winter, birds vacate some northern coastal and interior regions and retreat to coastal and southern interior areas, from Mass. to Central America and from coastal Alaska to s. Baja, as well as to ice-free, freshwater lakes, rivers, and marshes throughout the South and parts of the West. In summer, nonbreeding subadults may occur almost anywhere there is open water. **HABITAT:** Breeds on lakes, reservoirs, impoundments, estuaries, and open coastlines, either on the ground or in stick nests located in large, dead (sometimes live) trees. Winters coastally or on inland bodies of water. Roosts on islands and sandbars and in hardwood swamps. **COHABITANTS:** Primarily Double-crested Cormorant, as well as other cormorant species. **MOVEMENT/MIGRATION:** Spectacular coastal migrations may number in the tens of thousands. Spring migration from late March to late May, with peak in the latter half of April; fall from early August to early November, with peak from September to mid-October. VI: 4.

DESCRIPTION: Over most of North America, particularly the interior, if you are looking at a cormorant, it's a Double-crested. Among other cormorants, only the Texas-Louisiana- and Arizona-New-Mexico-based Neotropic Cormorant is commonly found away from coastal areas.

Double-crested is a large, black, reptilian-looking bird distinguished as a cormorant by its shape and posture and as a Double-crested by a combination of size, head shape, bill color, and, in the case of immatures, a dirty white throat and chest contrasting with a darker belly.

Perched, adult Double-cresteds sit (often on a piling or elevated perch) with head drawn back, bill angled up, and tail angled down—suggesting a snake ready to strike. The head is proportionally undistinguished, the bill thin and hooked at the tip. The body language, more than the face, seems sinister. (At close range, the shocking blue eyes of the adult are more arresting than sinister.) The "double crests," suggestive of shaggy horns, are present only in the breeding season and are difficult to see.

Adults overall appear black (the iridescent green highlights are often difficult to see). Immatures are gray-brown above and two-toned below, pale on the neck and breast, and dark on the belly (but with no sharp demarcation between dark and light zones, just the opposite of immature Great Cormorant). The bill and skin at the base of the bill are bright orange (adults) or yellow-orange (immatures). At almost any distance, if the bird is black and you can't see orange, it's not Double-crested.

BEHAVIOR: Cormorants do a great deal of sitting on the ground or on elevated perches. After emerging from the water, they open their wings benediction-wide to dry their feathers. In the water, the bird sits very low—almost submerged—with the bill angled up. Takeoffs are long and labored, with a great deal of wing-thrashing on the water. (Loons also require long takeoffs but do less thrashing.) Fairly nimble, Double-crested is able to perch on heavier wire cables and tree limbs, but when found with Neotropic, is usually the one who has selected the sturdier perch.

FLIGHT: Big and sturdy-looking, with a distinctly raised head and cricked neck. The body seems up-angled or swaybacked, bent in a U (the product of the up-angled neck and low-hanging potbelly). Flight is stiff, effortful, somewhat lumbering—as if the R&D folks in the Creation Division left flight out of the original design specs, then forced through modifications just before production. The line of flight is direct, with steady and regular wingbeats; sometimes the bird makes short glides during which it loses no altitude. When not migrating, flying birds are usually solitary.

In migration, Double-cresteds fly in trios, small groups, and often in large sloppy flocks that aspire to a V but often settle for a W or a J (and sometimes several letters at once). Flocks are constantly shifting their formation (and creating new letter configurations). On days with good thermal production, Double-cresteds frequently soar, with wings at a right angle to the body, neck stretched, and tail fanned. The shape suggests a bulky Anhinga.

VOCALIZATIONS: Generally silent. When surprised and flushed, makes a muffled grunting sound.

PERTINENT PARTICULARS: Double-crested Cormorant can be confused with many different birds on many different levels. Perched or soaring, it suggests the longer-tailed, needle-billed, overall snakier Anhinga. In migration, large skeins might be mistaken for dark-bodied ibis or geese. Ibis are stick-figure birds whose shape suggests a wilted cormorant; Double-crested Cormorant is bulky, and the body bows up. Also, ibis glide frequently and in full or partial concert; when they glide, the line flows. Double-crested Cormorants glide infrequently, for shorter intervals, and not in concert.

Geese don't soar. Also, cormorant flocks often contain adults and immatures—birds with different plumage. For the most part, geese and ibis (except White Ibis) are plumage-static (all birds are similar).

Telling Double-crested from other cormorant species can be tricky (see species-specific discussions at the other cormorant sections). A flying Great Cormorant in a flock of Double-cresteds appears one-third again larger than the Double-crested (and

Great's bulk may make it appear even larger). In the West, Double-crested's range overlaps with those of Brandt's and Pelagic. Brandt's flies low over the water, single file, while Double-crested does not. Also, Brandt's and Pelagic fly with their necks straight; Double-crested shows a kink or elevated head.

Great Cormorant, *Phalacrocorax carbo*

A Neanderthal of a Cormorant

STATUS: Fairly common to uncommon; mostly coastal Atlantic cormorant. **DISTRIBUTION:** In North America, restricted as a breeder to n. Atlantic coastal areas (Nfld. to Maine), where it is usually a permanent resident. Winters south along the Atlantic Coast to S.C., but numbers fall off below New England (where it is the common wintering cormorant); scarce south of Va. A few birds also winter on the lower Hudson R. of N.Y. and the Delaware R. north to Trenton. **HABITAT:** Nests on rocky sea cliffs, often with Double-crested Cormorant. In winter, found in shallow coastal waters along rocky coasts or in waters marked by sandy beach, docks, wrecks, jetties, channel markers, seawalls, and sandbars. **COHABITANTS:** Double-crested Cormorant, Herring Gull, Great Black-backed Gull. **MOVEMENT/MIGRATION:** Some post-breeding birds disperse to nearby ice-free waters. Other birds, largely immatures, migrate south as individuals, pairs, or trios, often among flocks of Double-crested Cormorants. Spring migration from early February to mid-May; fall from late August to early December. A few nonbreeding subadults may linger south of breeding areas throughout the summer. VI: 2.

DESCRIPTION: Big, strapping, coastal cormorant; larger and bulkier than Double-crested Cormorant (although direct comparison helps). Overall proportions between the two species are similar, but Great is bigger and more square-headed, thicker-necked, and conspicuously heavier-billed. (The sloping forehead and heavy bill fuse, making the head look like one big wedge.) Double-crested's forehead and bill are more distinct. Great is also more humpbacked.

Breeding plumage is distinctive—Great is an all-black cormorant with a white throat patch and conspicuous white patch on its flank (seen January through, at least, May). Adult Double-crested is just blackish. Immature Greats and Double-cresteds are also easily distinguished: Double-crested's underparts are palest on the throat and breast and darkest on the belly; Great is just the opposite—the throat and upper chest are dark, and the belly is whitish. Overall, young Double-crested is warmer-toned below; Great Cormorant is colder and grayer, and the demarcation between upper- and lower parts can be seen at greater distance. Also, Great's bill is dull gray; the bill (and part of the face) of Double-crested is yellow-orange.

Adult Great Cormorant is most difficult to distinguish from Double-crested in summer and fall, when Great's white flank patch is absent. Great's bill is pale or horn-colored; though the white on the throat is more restricted, it is still evident on both adults and immatures. The bill, and even some of the face, on Double-crested is bright orange—a color that truly vaults distance.

BEHAVIOR: Like Double-crested, but seems to favor sturdier perches farther from shore. Often perches higher than Double-crested when the two are together.

FLIGHT: Greater's bulk translates into heavier flight. Whereas Double-crested seems slender (although not spindly), Great is somewhat humpbacked (recalling American Bittern) and never seems slender. In mixed flocks, Great Cormorant's greater size is immediately apparent. Great appears one-third again larger than Double-crested, although strictly speaking, the difference is slight and often hard to appreciate unless the birds are together.

VOCALIZATIONS: Silent away from nesting cliffs.

Red-faced Cormorant, *Phalacrocorax urile*

Aleutian Cormorant

STATUS AND DISTRIBUTION: A fairly common but localized and geographically restricted coastal resident found, for the most part, along a narrow band of ocean astride the Kuril and Aleutian Is. In the U.S., this range extends from Cordova west to outer Cook Inlet, Kodiak I., and both sides of the Alaska Peninsula (to about Cape Newnanham) and out along the Aleutians west to Attu. Also found on the Pribilof Is. in the Bering Sea. **HABITAT:** Breeds on rocky cliffs (primarily on islands); forages in marine waters along rocky shores, usually within sight of land. **COHABITANTS:** Pelagic Cormorant, Black-legged Kittiwake, Common and Thick-billed Murres. **MOVEMENT/MIGRATION:** Permanent resident. VI: 0.

DESCRIPTION: Slender, all-dark, marine cormorant with a pale bill and bright red face (adults). Medium-sized (slighter larger and bulkier than Pelagic; smaller and more slender than Double-crested), with a thin straight bill and slender head, neck, body, and tail. Very closely resembles Pelagic Cormorant (with which it often associates), but larger-headed, heavier-billed, thicker-necked, and overall not as sinuous or clean-lined. Also sometimes appears somewhat shabbier-looking than Pelagic.

Adults are overall blackish (with slightly green or purple highlights), except for the wings and tail, which are tinged dull brown. Adult Pelagic Cormorant shows no brown. In breeding plumage, the bird has a white flank patch and shows two feathered nobs on the head (absent on Pelagic). At all seasons, adult Red-faced shows *a bright red face!* and the red completely surrounds the eye. The very limited red on Pelagic's face is restricted to the base of the bill (and doesn't even appear to touch, much less encircle, the eye).

All-dark, immature Red-faced is much like Pelagic, but the bill is pale horn-colored. The bill on immature (as well as adult) Pelagic is black.

BEHAVIOR: Slightly more pelagic than the inaccurately named Pelagic Cormorant, but neither species wanders far from the coast. The distance this species wanders inland can very probably be measured in feet. The distance these birds forage offshore probably does not exceed 10 mi.; they usually stay within easy sight of land. Red-faced nests and roosts on cliffs and fishes in rocky coastal areas. Feeds with Pelagic Cormorant, but often forages in slightly deeper water.

Shy Red-faced breeds in small, clustered colonies (commonly numbering fewer than 50 pair) among Pelagic Cormorants and kittiwakes.

FLIGHT: Typical cormorant profile showing long outstretched neck, long tail, and wings set well back. Red-faced's overall profile approximates the very slender Pelagic Cormorant, with the head held horizontal to the body, but is less slender and streamlined. With its shorter, thicker neck that appears kinked and thin where it meets the body, ovate wings that appear structurally too narrow at the base, and somewhat bulbous tail, Red-faced seems like a Pelagic Cormorant that was built with ill-fitting parts. Also, Red-faced's belly sags, giving it a slightly potbellied appearance. Pelagic's underbody is straighter and leaner.

VOCALIZATIONS: Described as "groans, croaks, and hisses."

Pelagic Cormorant,
Phalacrocorax pelagicus
Greyhound Among Cormorants

STATUS: Common, widely distributed Pacific Coast resident. **DISTRIBUTION:** Resident from the western coast of Alaska to s. Calif. In winter, withdraws from much of its northernmost range in the Bering Sea and retreats south to central Baja. **HABITAT:** Marine, inshore waters (rarely out of sight of land); commonly frequents shallow bays and rocky coastlines (foraging closer to shore than Brandt's and Double-crested). Casually found on inland lakes and rivers (near seacoasts), particularly when salmon are running. Nests on cliffs, as well as in man-made structures such as buoys and derelict ships. **COHABITANTS:** In Alaska, nests on cliffs with Red-faced and Double-crested Cormorants and Black-legged Kittiwake. In more southern waters, associates with Double-crested and Brandt's Cormorants. **MOVEMENT/MIGRATION:** Mostly nonmigratory except for birds breeding in the northern Bering Sea, although some migration south to British Columbia is evident, and in winter dispersed birds range south to central Baja California. VI: 0.

DESCRIPTION: Small, sleek, serpent-necked, thin-billed cormorant that hugs the coasts. Small (obviously smaller than Brandt's or Double-crested; slightly smaller than Red-faced), with a small sleek head that seems hardly more than a swelling in the long thin neck. The popsicle-stick-thin bill is not appreciably wider at the base or conspicuous hooked at the tip. By comparison, the bulkier, more wedge-shaped heads of Brandt's and Double-crested Cormorant resemble ice cream cones. The body of Pelagic is slender, and the tail long. In winter, the head of the adult Pelagic sometimes shows a hint of the wispy crest found on breeding birds. Overall longer, leaner, more slender-headed, and longer-tailed than Brandt's or Double-crested, Pelagic is generally more spindly than those species, as well as darker-billed and longer-, more slender-necked than Red-faced.

Plumage is overall oily sleek and blackish, with a blackish face and a black bill. Adults have a greenish gloss to the body and purple gloss to the neck; immatures are dull black. Adult Double-crested is paler and shows scaly patterning on the back. Adult Brandt's shows less gloss and a buffy throat patch at the base of the bill. (Immatures are browner and paler-breasted and also show a pale throat patch.) Red-faced has a pale bill and red eye patch; in addition to its slightly slimmer contour, immature shows a pale, not black, bill.

In breeding plumage, both adult Pelagic and adult Red-faced show a white flank patch. In breeding plumage, the red patch at the base of Pelagic's bill is often difficult to see.

BEHAVIOR: The least social of cormorants. Frequently flies and feeds alone; roosts in small numbers; nests away from others of its species or in small widely spaced groups. When foraging in places frequented by Brandt's and Double-crested, often chooses to feed in rougher water close to a rocky shoreline.

Dives with higher-arching grace than Brandt's or Double-crested Cormorant.

FLIGHT: Shows a very slender, almost gaunt profile, with a straight, narrow, "broomstick" neck and short, squarish wings. Other cormorants appear heavier-headed and have thicker necks; Double-crested shows a pronounced bend or kink. Red-faced is more potbellied. Flight is straight, low, and fast. Wingbeats are stiff, hurried, and steady, with the cadence of a Common Merganser.

VOCALIZATIONS: Usually silent away from nest cliffs.

Anhinga, *Anhinga anhinga*

The Serpentine Cormorant

STATUS: Common and generally permanent resident of southern freshwater marshes; at times and in places (most notably south Texas), a very common migrant. **DISTRIBUTION:** In the U.S., breeds from coastal N.C. and S.C. across the southern half of Ga. and all of Fla.; also s. Ala., Miss., all of La., and portions of ne. and coastal Tex. south; also occurs in isolated pockets along large rivers in extreme w. Tenn. and s. Ark. In winter, extreme northern and interior birds retreat south into coastal Mexico and Central and South America. **HABITAT:** Still, shallow, sheltered freshwater swamps and marshes that offer elevated perches for birds to dry their feathers. Also found in mangrove lagoons in coastal areas, but generally not in large open areas of saltwater. **COHABITANTS:** Green Heron, White Ibis, American Alligator. **MOVEMENT/MIGRATION:** In most places, a permanent resident; large-scale movements occur, however, primarily in southern and coastal Texas in September and October, and again from March into May. Usually found in large, homogeneous flocks, although also mixes sometimes with migrating raptors (Swainson's and Broad-winged Hawk). VI: 3.

DESCRIPTION: Large, slender, serpentlike cormorant distinguished by its needlelike bill and long supple neck—which is all that's visible when the bird is fishing, since its body is almost always submerged. When perched, the overall slender body and very long tail help distinguish this species from Double-crested and Neotropic Cormorants. With its snaky neck and thin, pointy bill, Anhinga seems headless. It often perches erect with its long tail vertical, or nearly so; cormorants' shorter tails are usually slightly cocked.

Adult males are all black with silvery upper wings and back. (In breeding plumage, the neck is shaggy.) Females and immatures are like males, but their necks and breasts are buffy brown (making them look like they're wearing a ruff). Sleeping Anhinga often tucks the head so that it seems headless. When perched with neck extended, its posture and body language seem haughty.

BEHAVIOR: Generally found in colonies (with other Anhingas and with herons and egrets), but hunts alone. Spends most of its time perched on branches over water, often with wings spread (benediction fashion) for thermal regulation and drying (as do cormorants). Perches are not necessarily high and may be springy branches. Anhinga climbs and descends (by hopping) among the branches. It feeds by launching itself into the water and then hunting, submerged except for the head and neck, amid aquatic vegetation.

FLIGHT: Looks like a flat flying cross with a long straight (or slightly kinked or cocked) neck, straight, pointy, falconlike wings, and a long narrow tail that may be very broad when the bird is soaring. Exceedingly thin, with a body that seems compressed and one-dimensional. (Cormorants always show more bulk.) Adept at soaring and gliding, this bird can spend considerable time aloft without resorting to active flight. Wingbeats are given in a deep, stiff, regular series interspaced with long glides. After the final flap in a series, wings snap to a horizontal position. Anhinga soars with wings slightly down-drooped, straight-cut on the leading edge, and curved on the trailing edge. When gliding into a wind, Anhinga's flight seems too slow to bear such a gangly bird aloft.

In migration, flocks fly at extreme heights; birds circle in thermals by turning in the same direction (unlike hawks, but like cormorants). At high altitudes, the long thin neck disappears and the bird appears headless. The silvery upper surface of the wing flashes.

VOCALIZATIONS: Generally silent. When breeding, may make a dry, rattling chuckle or chattering that begins with several measured notes and merges into a rapid dry stutter before slowing to a stop: *"rah, rah, rah, rah, ac'ac'ac'ac'ac, hah, hah, hah."*

PERTINENT PARTICULARS: Soaring, single Double-crested Cormorants are often mistaken for this species. The cormorant can indeed soar well when conditions are optimal, but Anhinga is much more accomplished, soaring more frequently and for more extended periods.

Magnificent Frigatebird, *Fregata magnificens*

Gravity Suspended

STATUS AND DISTRIBUTION: Tropical and subtropical species, common in the waters of s. Fla. and the Florida Keys; nests on the Dry Tortugas. Uncommon to n. Fla. and rare along the Gulf Coast to Tex.; also rare in s. Calif. Casually wanders north to New England. May occur inland after hurricanes. **HABITAT:** Warm coastal and offshore waters. Frequently feeds in shallow lagoons, over reefs, and in deeper water out of sight of land. Nests on low scrub on islands away from people. **COHABITANTS:** Does not generally associate with other species, but pirates fish from a number of birds, including boobies, pelicans, gulls, and terns. **MOVEMENT/MIGRATION:** Usually disperses north between April and August. VI: 3.

DESCRIPTION: Very large dark stick figure of a bird. Gaunt with a small head, a long, heavy, and hook-tipped bill, exceedingly long wings, and an even longer tine-thin tail (which may be closed or splayed on perched birds).

Adult males are all black, with green iridescence. Females have bright white breasts; younger birds have white breasts and heads.

BEHAVIOR: Frigatebirds have two modes: perched and soaring. Favorite perches include tree limbs, channel markers, and the antennae or rigging of ships. In flight, they may remain aloft for hours, executing nimble and carefully controlled plunges to pluck prey from the water, or they may engage in spirited (and largely one-sided) pursuits to force other seabirds to surrender the contents of their crops.

Frigatebirds never dive into the water. They never sit or swim on the surface. They never walk. They are fairly social, however, frequently sitting in close proximity or soaring in small groups.

FLIGHT: Silhouette is unmistakable—a spindly, short-armed, long-handed scarecrow of a bird with a tined tail. At a distance, seems all wing and tail (the head disappears). At closer range, the body appears suspended from the wings. The overhead shape suggests an angular, contorted letter M on a stick, and this impression remains whether birds flap, glide, or soar. Head on, the birds show an Ospreylike crook in the wings. The tail may be open, forked, or closed like narrow forceps.

Rarely flaps; glides for hours. Forward momentum is often painfully slow, making it seem as though time and the laws of physics have been suspended. Frigatebird habitually micromanages the air by making slight adjustments in the set of the wings and tail. Executes astonishingly tight circles for so large a bird (seems able to reverse directions in place). Wingbeat is slow and dreamlike, with a deep pushing down stroke. The set of the wings makes it seem like the bird is braking with its wings, but the movement is forward.

Distant birds seem all black. Even the white heads and breasts of younger birds and females disappear.

VOCALIZATIONS: Call is a short chirpy trill.

PERTINENT PARTICULARS: Frigatebird does not represent a difficult identification challenge, but observers should be aware that not every long-winged fork-tailed bird observed flying over the Gulf of Mexico is inevitably this species. Swallow-tailed Kite (which superficially resembles immature Frigatebird) crosses the Gulf of Mexico during migration and has been seen soaring with frigatebirds over the Florida Keys. Also, there are very few records of other species of frigatebirds in North America.

HERONS, EGRETS, AND IBIS

American Bittern, *Botaurus lentiginosus*

Patience Cast in Feathers Imitating Reed Bed

STATUS: Uncommon and secretive but widespread northern breeder; uncommon wintering species along both coasts and across the southern United States. **DISTRIBUTION:** Breeds across most of subarctic Canada from Nfld. to cen. B.C. and north to the s. Northwest Territories. In the U.S., breeds in most northern states, with the southern border defined by coastal Va., s. Pa., w. W.Va., cen. Ky., cen. Mo., n. Kans., Colo. (also into n. and cen. N.M.), n. Utah, n. Nev., and cen. Calif. Winters coastally and into the coastal plain from s. N.J. to s. Tex. and from sw. B.C. south to Baja; also west across w. and s. Ariz., s. N.M., and throughout the Rio Grande Valley. Also throughout Mexico and much of Central America. **HABITAT:** Breeds in fresh (rarely tidal) marshes with tall wetlands vegetation (such as cattail); generally prefers larger to smaller wetlands and shallow areas. In migration and in winter, more expansive in its habitat selection, occupying freshwater wetlands dominated by tall reedy vegetation, phragmites marsh, tidal *Spartina* wetlands, lakeshores, sloughs, drainage ditches, and even upland fields and roadsides dominated by tall grass. **COHABITANTS:** Pied-billed Grebe, Sora, Black Tern. **MOVEMENT/MIGRATION:** Most birds (and virtually all interior birds) vacate breeding areas in advance of winter. Spring migration is early, generally March and April; fall is from late August to November, with peak in late September and October for much of the United States. VI: 1.

DESCRIPTION: Medium-large, vase-shaped, streak-necked heron, with a penchant for ridged immobility and a flair for not being seen. Shape and posture are changeable. When feeding, its long extended neck and near-vertical pointy bill (the classic stance) make it look like a snake that has swallowed a football. The long, thin, straight, and pointy bill seems an extension of the slender head, which looks like an extension of the neck, which seems like an extension of the body. When crouching, the bird may have its neck drawn partially back or fully retracted, making it appear neckless—all body and bill.

All ages and sexes show similar plumage. The body is overall rich warm brown; the neck is overlaid with reddish brown and buff, and the streaking, blackish whisker stripe bordering the face stands out. Expression is glum or dour.

BEHAVIOR: In keeping with the erect posture and reed-replicating neck pattern, the bird stands as immobile as the vegetation around it . . . or not. Sometimes swaying in a fashion that suggests wind-blown grass, American Bittern generally keeps to the edges of marsh, often partially (too often totally) concealed by reed. This solitary bird does not feed with other herons. During migration, two (rarely more) birds can be seen migrating together.

Sometimes extremely bold, American Bittern stands firm when approached, putting faith in its excellent camouflage (even, comically, at rare times when it is standing completely in the open). Also flushes when startled or, if scrutinized too long, slinks into the reed, often crouched and moving in a fashion that recalls one of the large rail. Movements are generally slow: it stalks prey or maneuvers for a strike with near-glacial speed.

Most active at dawn and dusk, when it is often seen flying over marshes. Also forages in daylight and at night. On breeding territory and in migration, American Bittern often vocalizes at night.

FLIGHT: Silhouette is distinctive. The bird is more spindly, angular, and pointy-winged than Black-crowned Night-Heron and Yellow-crowned Night-Heron. Back has a distinct hump; the underparts seem gently up-curved from the bill to the feet. Night-herons just look blocky—no hump, no curve. In flight, the trailing edge of the somewhat pointy wings is blackish, in contrast with the otherwise all-warm-brown bird. Flight is straight, somewhat stiff and rigid, and slightly jerky. Wingbeats are fairly quick and deep (quicker and deeper than a Black-crowned Night-Heron), with a slight hesitation on the down stroke.

VOCALIZATIONS: One of the classic sounds of the marsh—a resonant, imperfectly suppressed, three-note belch—"*gulp-G-gulp*"—sometimes likened to a stake being driven into the marsh; in tone and cadence nearly suggests a bassoon with a limp. When flushed, utters a low croaking bark, "*kok, kok, kok, kok,*" or "*rolf, rolf, rolf,*" that recalls night-heron but is more muffled and less raucous.

PERTINENT PARTICULARS: Immature night-herons are colder brown than American Bittern. Black-crowned Night-Heron, whose plumage pattern more nearly recalls the streaky bittern, is a stocky big-headed bird that lacks the sinuous qualities of a bittern. Immature Green Heron, another reed-hugging immobile hunter, has a streaked neck like a bittern's and a long neck and bill. But Green Heron is only slightly larger than half the size of American Bittern, habitually feeds in an angled-down crouch, and frequently perches on branches and logs. Bitterns rarely stand on anything but dry land or in the water.

Least Bittern, *Ixobrychus exilis*

Butterscotch Rail-Bittern

STATUS: Fairly common (but secretive) and widespread (but localized); primarily an eastern breeder, but also found in scattered western locations; more restricted as a winter resident. **DISTRIBUTION:** Breeds across the e. U.S. and extreme se. Canada. Western border cuts through e. N.D. and e. S.D., e. Neb., ne. Kans., w. Mo., w. Ark., and cen. Tex., north to the s. Okla. border, west to the Panhandle, and south into parts of Mexico and Central America; also breeds in extreme se. Man., se. Ont., and coastal N.B. Absent in interior Maine; occurs in all but s. N.H. and Vt. and in n. and cen. N.Y., cen. Pa., and along the Appalachians to ne. Ala. In the West, found in cen. Neb., e. N.M., s. and sw. Ariz., s. and cen. Calif., and s.-cen. Ore. Winters on the Florida Peninsula, along coastal Tex., in s. Calif. and s. Ariz., and south to Baja. **HABITAT:** Breeds in freshwater and brackish wetlands that offer both tall dense stands of emergent vegetation (such as cattail marsh, bulrush, saw grass, and phragmites) and open water. Occasionally found in salt marsh (especially in winter) and wooded swamps. **COHABITANTS:** Common Moorhen, Virginia Rail, Marsh Wren, Red-winged Blackbird. **MOVEMENT/MIGRATION:** Spring migration from early April to mid-June; fall from mid-August to early November. VI: 2.

DESCRIPTION: A scoop of butterscotch clinging to the reeds or crouched on a snag just above the water. *Tiny* heron (with neck folded, only slightly larger than Red-winged Blackbird and about the same size as Virginia Rail). With neck retracted, looks like a slug with a proportionally long, straight, pointy (heronlike) bill. With neck extended, looks like a snail with a long, straight, pointy bill. In shape, most closely resembles Green Heron, but much smaller.

Overall plumage is honey and golden brown above. Adult males show a black cap and back, but it's the buffy yellow body that catches the eye. Whitish underparts are generally not evident until the bird flies. Immatures are like adults but duller and more streaked on the throat. Green Heron is considerably larger and darker. Virginia Rail is much darker and rustier and has a down-curved (not straight) bill.

BEHAVIOR: Most commonly seen clinging to the reeds at the edge of an open body of water just above the water. Rarely stands in open water. When resting, the bill is horizontal (and often the only evidence that the golden mote is indeed a bird). When hunting, the bill, body, and (often) long extended neck are usually angled down. Remarkably quiescent. Usually sits for long periods without moving or maneuvers toward prey with the speed and manner of a snail. Very surreptitious. Feeds along open vegetative edges, but spends much of the day hidden. Very tame, often allowing close (and unobserved) approach, but flies (often across open water) when startled or walks to safety with a surprisingly quick rail-like gait, either on land or nimbly from reed to reed.

Generally solitary, Least Bittern is active both day and night. Though usually silent, males vocalize during the breeding season (often from a visible if not exactly exposed perch), and all birds quack throughout the year.

FLIGHT: Looks like a diminutive heron with legs trailing and neck tucked. Flies like a compact dark-backed rail, showing large, buffy, oval patches on the upperwing. Flight is struggling, straight, and just above (or below) reed-top height. Wingbeats are hurried, steady, effortful, deeply down-stroking.

VOCALIZATIONS: Song is a low, hurried, somewhat descending chuckle: "*kuhkuhkuhkuhkuh.*" In tone and pattern, recalls the cooing song of Black-billed Cuckoo. Also makes a loud, low, three-note quack: "*chack chack chack.*"

PERTINENT PARTICULARS: Most observers are not prepared for the diminutive size of this species, and the measurements given in field guides are at least partially to blame. Most measure the bird's length with its neck extended (as it appears in most specimen collections). The reality is different. With its neck retracted, a Least Bittern is about the size of a softball (Green Heron is about the size of a football).

Great Blue Heron, *Ardea herodias*

The Name Says It All

STATUS: Common, widespread, and, owing to its size and penchant for standing in open water, difficult to overlook (or misidentify). **DISTRIBUTION:** Breeds across most of the U.S., s. Canada, and much of Mexico. In Canada, breeds from the Maritimes to about cen. Alta., and on the coast from s. Alaska to coastal s. and se. B.C. In the U.S., absent as a breeder and wintering species only along the spine of the Rockies from e. Idaho to nw. N.M. In winter, retreats south to the limits of open water. **HABITAT:** Almost any open, still, or slow-moving fresh, brackish, or coastal fish-bearing water. Occasionally feeds in the surf, on mud flats with drought-trapped fish, or in dry fields in search of rodents. *Note:* Two distinct morphs, the all-white Great White Heron and the white-headed Wurdemann's Heron, are restricted to Florida and the Florida Keys, respectively, where they forage in shallow marine habitat. **COHABITANTS:** May associate with other large wading birds. **MOVEMENT/MIGRATION:** In spring, returns shortly after breakup—as early as February in some places, but most passage occurs in March and April. Post-breeding dispersal begins in late July. Fall migration from September to November, with peak in early October. VI: 2.

DESCRIPTION: The name says it all: the "Great" Blue Heron is North America's largest heron, standing nearly four feet tall. More accurately blue-gray than "Blue," except for a white face, dark cap, pale throat, and chestnut feathers on the thighs. But unquestionably a "Heron"—a long-billed, long-necked, long-legged wading bird most often seen standing in water. The bill is massive, and particularly heavy at the base. The stance is most commonly erect, with neck held straight and angled slightly forward—although the bird also holds its neck in an S-shaped curve or crouches with neck coiled and compressed against the body, making it look all body and bill.

Superior size notwithstanding, the combination of white head and black crown distinguishes Great Blue from all other herons. Immature Sandhill Crane (lacking a red cap) is overall gray (with rusty highlights), but shorter-necked and shorter- and thinner-billed, and its posterior is distinguished by a drooping feather duster of a tail. The short tail of Great Blue Heron is stiff and straight. Also, while Sandhill roosts in water, it habitually forages in dry fields. Great Blue Heron roosts in trees and infrequently forages in fields (but more so in the West).

The Great White morph is easily confused with Great Egret but is larger, structurally heavier, and particularly heavier-billed. Also, the legs of Great Egret are black; Great White's are pale.

BEHAVIOR: Stands tall, unmoving, or stalks slowly. Usually forages in deep water, away from the bank, but may wade up to its belly. Usually a solitary feeder, but will join other herons (including other Great Blues) in large, open, and productive waters. Drives other Great Blues away from smaller fishing holes.

Sometimes fishes from banks, docks, or logs. Roosts (and nests) in trees. Occasionally forages in dry fields or flooded fields for rodents, snakes, and other vertebrate prey (but not as much as the European Gray Heron).

FLIGHT: Large heron profile with head drawn back, neck bowed, and wings down-cupped, legs trailing. When taking off, landing, or chasing other herons out of its territory, Great Blue commonly extends its neck, which usually shows a bend or kink (the extended necks of cranes are perfectly straight). Steady wingbeats seem slow and ponderous. Does not usually glide, except to land. Does use thermals to soar, but often continues to flap while rising.

Because of their size, long-broad wings, and heavy wingbeat, distant Great Blues are habitually mistaken for eagles—particularly Bald Eagle. But eagles have straighter wings and a flap that describes a wide arch above and below the body. Great Blue Heron flies on severely down-bowed wings (seeming to cup the air), and the wings don't appear to rise above the body. The bird flies as if dipping its wingtips into an invisible cauldron, testing the air.

In migration, flocks of 8–12 birds are common. Flocks in excess of 50 birds are rare. Cranes often fly in much larger flocks.

VOCALIZATIONS: Call is a loud trumpeting "*wrreh*" or a harsher, lower-pitched, growling bray, "*wrrrah*," often given when the bird is flushed (or when you are standing in its place).

PERTINENT PARTICULARS: Great Blue Heron poses two major identification challenges—one for beginners, one for experts. The European Gray Heron is smaller (about the size of Great Egret), but its structure and plumage are very similar to Great Blue's. Luckily, Gray Heron has a grayer, paler neck, and its thighs are white (not chestnut). Gray Heron has occurred in the West Indies and is a species to watch for.

Ironically, the bird that beginning birders most often confuse with Great Blue is the infinitely smaller but also widespread Green Heron, which also has a dark cap, reddish neck, and bluish upperparts. Since many beginning birders focus on plumage and undervalue or misassess size, and because herons can extend and retract their necks, size becomes a mercurial standard. Let's make this easy: if it has a dark face, it's Green Heron; a whitish face belongs to Great Blue.

Great Egret, *Ardea alba*

The Stately Stalker

STATUS: Common and widespread across the United States and limited portions of Canada; as a breeder, widely spaced. **DISTRIBUTION:** Breeds coastally from s. Maine to South America on the Atlantic; north to cen. Wash. on the Pacific. Also breeds in fresh water throughout much of the interior of southern coastal states and in pockets throughout the Midwest, the eastern prairies, and western states. Wanders widely after breeding. In winter, retreats south from N.J. on the Atlantic and Gulf of Mexico (as well as from interior portions of the South) and expands to occupy all of the Pacific Coast south of Wash. while abandoning most northern interior breeding locations. **HABITAT:** Utilizes a variety of coastal and freshwater habitats, including tidal estuaries, tidal flats, beaches, freshwater marshes, tidal pools on rocky coasts, hardwood swamps, ponds, lakes, canals, ditches, impoundments, flooded farmland, and (occasionally) dry upland pasture and grasslands. Breeds in (often mixed) colonies, placing stick nests in woody vegetation, sometimes close to the ground but most often in the higher portions of trees. **COHABITANTS:** Other herons and egrets. **MOVEMENT/MIGRATION:** Wanders widely after breeding. Fall migration from early August to late November; spring return from late February to early May. VI: 2.

DESCRIPTION: Big, impressive, all-white heron with a bright yellow bill. Large (much larger than Snowy Egret; larger than Reddish Egret; smaller than Great Blue Heron or Wood Stork) and all white regardless of age or sex. In spring and early summer, adults sport long, feathery plumes that trail behind the tail when they are feeding or radiate from the back like a starburst of feathers when they are displaying.

Great Egret is easily distinguished from the equally common and widespread but smaller Snowy Egret and from first-year Little Blue Heron by its superior size and bright yellow bill. The white morph of Reddish Egret is intermediate in size between Snowy and Great, but has a bicolored bill (or all-dark bill if an immature).

More troublesome to identify, and more like Great Egret in size and shape, is the white morph of the Great Blue Heron (a.k.a. the Great White Heron), a form normally restricted to southern Florida. Great White Heron is larger, bulkier, and distinctly larger-headed and heavier-billed. Great Egret's legs are black, while Great White's are pale (flesh-colored to gray). Great Egret's expression is sad.

BEHAVIOR: Tame where habituated to people and may allow very close approach (in fact, forages on lawns in parts of Florida). Also social, often feeding with other herons. Owing to their larger size, Great Egrets are often found and cluster in deeper water. Great Egret almost always hunts in standing or flowing water, moving slowly with a deliberate stalking motion. (Both Snowy Egret and Reddish Egret are more typically active and high-strung.) Great Egret usually keeps its neck straight, or almost straight, and angled up and away from the body, but it may also retract its neck into a graceful S-shaped curve. Also hunts by remaining stationary, with neck extended, waiting for fish or other vertebrate or invertebrate prey. On dry land (more common in the West), captures lizards, small rodents, and large insects.

FLIGHT: Large all-white bird with head drawn back and neck coiled in a bulging S-shaped curve, suggesting an immense sagging goiter. By comparison, the necks of Snowy Egret and Reddish Egret are more squared and less distended and bulging. Great Egret's flight is graceful, stately, and steady; wingbeats are regular, somewhat caressing—not as stiff and quick as Snowy Egret, not as ponderous as Great Blue Heron; not as deep and angry as Reddish Egret. Glides, particularly when approaching for a landing, and also soars when thermals are strong.

VOCALIZATIONS: Usually silent, but when flushed or when chasing intruders from its fishing hole, makes a harsh low croak that is lower-pitched and more resonant than Snowy Egret.

PERTINENT PARTICULARS: In high breeding condition, the lores turn a shocking green, and the yellow bill deepens to yellow-orange.

Snowy Egret, *Egretta thula*
The Jack of All Herons

STATUS: Common coastal and widely distributed interior breeder; common, primarily coastal winter resident and widespread post-breeding wanderer in summer and early fall. **DISTRIBUTION:** Breeds coastally from Maine and n. Calif. south to Central America; widely scattered in the interior, with colonies in Pa., Ohio, Wisc., S.D., Ill., Ky., Ala., Ark., Neb., Okla., e. Tex., Mont., Wyo., Colo., N.M., Idaho, Utah, Ariz., Ore., Nev., and Calif. Winters coastally from s. N.J. and s. Ore. south. Also winters in cen. and s. Calif., sw. Ariz., and all of Fla. Disperses widely after breeding, with individuals, primarily young, ranging as far north as s. Canada. **HABITAT:** Breeds in mixed colonies with other heron species (usually near water) in barrier island woodlands, on dredge-spoil islands, in mangrove and freshwater swamps, and in phragmites marshes. Forages in tidal pools, salt marsh, tidal channels, tidal flats, sandy beaches with low wave action, wooded swamps, lakeshores, and drought-reduced ponds (and occasionally in dry open uplands). **COHABITANTS:** Other herons and egrets. **MOVEMENT/MIGRATION:** Except in parts of Texas and the Southwest, most interior breeders and birds north of Delaware vacate breeding areas before winter. Spring migration from March to May; fall from August to October. Post-breeding dispersal begins as early as April in southern populations. VI: 3.

DESCRIPTION: Distinctive, usually high-strung heron distinguished by its brilliant white plumage and array of feeding techniques—most of them active. Medium-sized (larger than Cattle Egret; about half the size of Great Egret; the same size but overall slighter than Little Blue Heron), with a rapier-thin black bill and long black legs. At the base of the bill are yellow lores (duller in immatures), and at the end of each leg are bright yellow feet (also duller in immatures). The rumps of breeding birds are adorned with an upturned, feathery cowlick of a plume (present from March to July/August). On immature birds, the bill and legs may be partially or wholly dull yellow-green, as are the lores. Even when legs are primarily yellow, the feet are a brighter yellow.

BEHAVIOR: Feeds primarily in standing water, using an array of techniques: stalking through pools, racing after prey, standing still with neck erect, or standing still and crouching with neck couched. Often feeds in large flocks (with or without other heron species). For the most part, feeding is active and energetic, with abrupt, jerky movements—a heron marionette with quicksilver reflexes. When feeding alone, often dabs its bill or tongue in the water to create fish-attracting ripples. Also pads its yellow feet in the water to create the same effect or stirs the bottom. When walking, most often draws the head and neck into an angular S-configuration, which stops short of being a Z, and steps high, often lifting its telltale yellow feet out of the water.

Feeds primarily at sunrise and before sunset. Commonly perches in trees or other elevated objects.

FLIGHT: Overall spindly and angular. Folded neck is squared in front (not roundly bowed). Long legs extend well beyond the tail, and the yellow feet are plainly evident. Wingbeats are direct, regular, and somewhat perfunctory (not lordly, ponderous, or graceful), as well as stiff, quick, choppy, and heavy on the down stroke. Snowy Egret does not commonly glide (except when descending for a landing), but does occasionally soar.

VOCALIZATIONS: Usually silent. When flushed or when engaged in an aggressive interaction, emits a harsh, grating, hawking croak, "*Ahr'aagh*" or "*hraagh*," that is higher-pitched and more nasal than Great Egret.

PERTINENT PARTICULARS: There are several other all-white herons. Snowy is easily distinguished from Great Egret by size and bill color (black in Snowy, yellow in Great); from Cattle Egret by shape (Cattle is stocky, Snowy is spindly) and bill color; and from immature Little Blue by a combination of traits (see discussion under "Little Blue Heron"). Great Egret is a slow stalking heron; Snowy is most often active and frantic. Cattle Egret is also a stalker, especially on dry land, a habitat normally ignored by Snowy.

Observers (particularly in New England and the Canadian Maritimes) have spotted a very similar Old World species, Little Egret, *Egretta garzetta*, a casual visitor to the Northeast. Most observations have been made in the spring, when twin, feathery nape plumes help distinguish this species from the shaggy-naped (but plumeless) Snowy.

Little Blue Heron, *Egretta caerulea*

A Heron of Poise and Bearing

STATUS: Common southeastern, south-central, and Atlantic Coast breeder; less common in the north central plains and the Southwest. Common in winter along the Atlantic and Gulf Coasts; rare in extreme southern California. An uncommon but regular and widespread post-breeding wanderer. **DISTRIBUTION:** Breeds and winters from coastal s. Maine to n. Mexico. Also breeds throughout s. Ga., s. Ala., most of Miss., e. Tenn., e. Ky., extreme s. Ind., s. Ill., most of Ark., s.-cen. Kans., e. Okla., and e. Tex. **HABITAT:** At all seasons, forages in a variety of fresh- and saltwater habitats, but seems more attracted to freshwater environments, particularly well-vegetated habitats. Favored haunts include ponds, lakeshores, swamps, tidal ponds, channels, and tidal flats. **COHABITANTS:** Other herons and egrets, Common Moorhen, Purple Gallinule. **MOVEMENT/MIGRATION:** Birds north of southern New Jersey and in the interior relocate to the southern Atlantic and Gulf Coasts for the winter. Spring migration from early February to late May; fall from late July to late November. After breeding, the young wander as far north as Newfoundland, across southern Canada west to the Rocky Mountain states, and north along the Pacific, irregularly, to southern Washington. VI: 3.

DESCRIPTION: Stiff, sturdy, slow-moving heron that likes to hunt methodically and keep a stiff extended neck. Medium-sized, Little Blue Heron is about the same size and shape as Snowy Egret, but the head is slightly larger, the neck thicker, the eye larger (somewhat bulging), and the bill heavier; Little Blue's bill is also distinctly bicolored (dark-tipped and blue-gray at the base). On most Little Blues, the bill is also very slightly downturned. Snowy Egrets have thinner, straighter bills that are mostly black (paler only at the base) and lores that are more distinctly yellow. Reddish Egret is larger and shaggier, with a heavier, straighter, and (when breeding) sharply defined bicolored bill (black-tipped and pink-based), and has a very different feeding style (more active and aggressive).

Adult Little Blue's plumage—all slate blue, with maroon highlights in the head and neck—is unique. At a distance, the bird simply appears all dark. Immatures (seen between June and the following May) are white, making confusion with the similarly sized but overall more spindly Snowy Egret inevitable, and confusion with the white morph of Reddish Egret possible. The legs of immature Little Blue are gray-green and sturdy; those of immature Snowy are yellow with a black-edged shin, and more slender. The dusky wingtips of immature Little Blue are often not apparent unless the bird is in flight.

The piebald plumage of second-year Little Blue Heron is also unique—white with patches of adult blue feathers intruding.

BEHAVIOR: At any age, Little Blue is a slow, careful, methodical hunter. Often stands for long periods with neck fully extended, angled slightly out, and head turned slightly askew, watching the water. Also stalks with easy, steady strides (somewhat recalling a large rail); does not raise its feet out of the water like Snowy Egret. Sometimes the head is erect; more often the body and extended neck are horizontal to the water and the bill is angled down. The head and neck sometimes sway, serpentlike, when Little Blue is stalking, making its movements appear sinister (but also suggesting the efforts of a severely myopic bird).

Like Tricolored Heron, Little Blue commonly hunts alone but will join mixed feeding flocks—particularly flocks of Snowy and Great Egrets. When feeding in a group, Little Blue often works the edge. Where birds are numerous and the fishing hole is small, most herons remain stationary, but Little Blue frequently forages through the pack in its steady, methodical stalking style.

FLIGHT: Little Blue's stockier proportions are an immature reflected in its flight, which is direct and unembellished. On young birds, wingtips have a dark, shadowy trim that is difficult to see at any appreciable distance. Flight is direct and somewhat sturdier than Snowy Egret. Wingbeats are medium-quick and choppy, but heavier and stronger than Snowy Egret (recalling a night-heron).

VOCALIZATIONS: A harsh, grating, squawk, "*raaaaah,*" much like Snowy Egret.

PERTINENT PARTICULARS: The easiest way to pick an immature Little Blue Heron out of a feeding flock of herons is behavior. Next to Snowy Egret, Little Blue seems stiff and sluggish; next to the mad thrashing of Reddish Egret, it's dead and embalmed. Also, Snowy most often holds its neck in an S-shaped curve or walks with its head and neck drawn back and cocked. Little Blue likes to extend its neck when it forages, showing only a slight curvature.

Tricolored Heron, *Egretta tricolor*

The Ninja Heron

STATUS: Common to fairly common, mostly coastal breeder and winter resident. **DISTRIBUTION:** Found year-round from coastal N.C. to ne. Mexico; breeds north to coastal s. Maine. In winter, also reaches coastal s. Calif. (rare). In spring and summer, overshooting adults and post-breeding birds may be propelled to the Maritimes, the Great Lakes, the n. and cen. Great Plains, and west across the American Southwest. **HABITAT:** Along the Eastern Seaboard, found in coastal marshes. In Florida and along the Gulf Coast, occurs in coastal marshes, mangroves, and shallow bays, but also on inland lakes and in freshwater marshes and rivers. **COHABITANTS:** Other herons and egrets. **MOVEMENT/MIGRATION:** Most birds are permanent residents. From Virginia north, spring migration is primarily in late April and May; fall movements are in August and September. VI: 3.

DESCRIPTION: Lithe, slender (almost emaciated) heron with serpentine contours, a stealthy demeanor, quicksilver reflexes, and plumage that is both distinctive and distinguished. Medium-sized (slightly taller than Snowy Egret and Little Blue Heron), with an exceedingly long, slender neck (as long as the body) and a long rapier-thin bill. Tricolored seems headless—just neck and bill. Little Blue Heron is shorter-necked, shorter-billed, and overall bulkier.

All ages and sexes have the same basic plumage pattern: dark, mostly bluish upperparts and neck, with a white belly and (best seen in flight) white underwing coverts. The contrast between the dark neck and the white underparts defines this bird. Some immatures show rusty necks (with pale throats) and rust suffused through the wings and upper back, but they usually look like washed-out (and somewhat scrawnier) adults.

In addition to the exceedingly long and slender neck, adults have two plumage traits that distinguish them at a distance or concealed in marsh grass or tidal creeks. In breeding plumage, a white plume long enough to blow in the breeze projects jauntily behind the head and distinguishes the bird from plumeless adult Little Blue. So too does the pale golden wash of plumes that covers the bird's rump. Viewed from behind and at a distance, the lower back shines pale. Adult Little Blue is all dark above.

BEHAVIOR: A solitary feeder that prefers to keep its own company in its own salt marsh pond. When it does associate with other herons, it often stays away from the main group. A quick, versatile, and often active feeder, Tricolored is adept at both stalking and still-hunting (or a combination of the two). Typically the bird moves forward, neck mostly extended, using exceptionally long strides; also crouches so low that the belly touches the water, with head and bill drawn back—cocked and ready.

More high-strung than most other herons. Often uses its wings for balance or extra propulsion when changing position (which it does often) or charging prey. Head moves in rapid, almost theatrical jerks. Does not usually fly in groups; more than three birds together at one time is a crowd.

FLIGHT: Maintains its slender profile in flight. The dark neck contrasting with the bright white underparts is evident at almost any distance. Flight is direct, and wingbeats are steady and regular but somewhat more buoyant, fluid, and casual than other herons. The bird seems less stable in a wind.

VOCALIZATIONS: A hawking croak, "*raahh*" (often repeated several times), but softer and less vehement than Snowy Egret; as much a purr as a growl.

PERTINENT PARTICULARS: Among all the herons, Tricolored's plumage most closely resembles that of the much larger, bulkier, heavier-billed Great Blue Heron. Plumage characteristics aside (and there are several that distinguish the two herons—not the least of which is Tricolored's dark head and Great Blue's white face), the feeding habits of the two birds differ dramatically. Great Blue is the slow, heads-up stalker, whereas Tricolored is the hyper, crouching ninja.

Reddish Egret, *Egretta rufescens*

Tyrannosaurus Rex of the Flats

STATUS: Uncommon year-round resident; geographically restricted and habitat-specific. **DISTRIBUTION:** A tropical coastal species ranging south to Central America. In the U.S., found primarily along the Atlantic and Gulf Coasts of Fla., La., and Tex. (also Miss. and Ala.). In winter, birds from the Pacific population may range north to s. Calif., where it is very rare. **HABITAT:** Almost always found in coastal areas foraging on tidal flats, lagoons, and salt ponds; rare inland or in fresh water. **COHABITANTS:** Laughing Gull, Least Tern, other herons and egrets. **MOVEMENT/MIGRATION:** Nonmigratory. In winter, shifts little outside established breeding range. Records of post-breeding dispersal (primarily coastal) extend to Rhode Island and Massachusetts, the Southwest, and the southern central plains. VI: 2.

DESCRIPTION: Resembles a larger, shaggier, rangier, more angular, and more frenzied Little Blue Heron. Largish heron (larger and bulkier than Little Blue; considerably smaller than Great Blue Heron), with a large head; straight, pointy, and (on breeding adults) distinctly bicolored bill (black-tipped, bubblegum pink–based); a long shaggy neck; a sturdy body; and long, sturdy legs. All in all, the bird just looks formidable and bad.

Reddish Egret has two color morphs — dark and white. Dark or reddish morph has pale slate blue body (like a sun-bleached Little Blue Heron) and a contrasting distinctly reddish neck. (Little Blue has reddish highlights that bleed through the blue feathers of the neck; Reddish Egret's neck is truly orange-red.) In good light, the entire bird is imbued with subtle reddish highlights that make the plumage more maroon than slate blue. Immature reddish morph is a pale pastel-shaded version of the adult and lacks the adult's shaggy neck; the bill is black. Usually less common, white-morph Reddish Egret (not found in the Pacific) is all white like a first-year Little Blue Heron, but its size, shape, and bright pink-based bill distinguish the adult. Immature Reddish has black legs and an all-black bill, whereas immature Little Blue shows greenish legs and a grayish pink base to the bill. Reddish Egret's expression is menacing, casually cruel, crazed.

BEHAVIOR: Forages alone in shallow water, away from other herons and egrets and is particularly intolerant of other Reddish. Moves like a mad linebacker, crouching and charging across the flats, leaping vertically with wings spread, turning, rearing its neck, jabbing left . . . right . . . then charging off again, with a bouncing, foot-pounding gait. Infrequently stalks, but does occasionally still-hunt (the classic hunting style of Little Blue Heron).

Put your binoculars down. The bird's angry animation alone is usually enough to distinguish it.

FLIGHT: Straight and steady, with a coiled neck (which makes the bird look big-headed) and disproportionately long wings. Wingbeats are deep and flowing, with a very heavy down stroke that closes in on the body. The bird looks like it's trying to gather or enfold the air. When feeding, it often lofts into the air or flies short distances with neck extended.

VOCALIZATIONS: Generally silent, but sometimes makes a low, protesting grunt or growl.

PERTINENT PARTICULARS: As noted, the angry, foot-pounding gait of the bird easily distinguishes it from other herons. Another subtle but equally distinguishing trait has to do with this aptly named bird's plumage. On dark-morphs, and in good light (particularly in flight), the entire bird is imbued with a reddish or maroon cast, a color that is limited to the neck of Little Blue Heron.

Cattle Egret, *Bubulcus ibis*

Egretta Terra (Earth Egret)

STATUS: Common to uncommon but widespread breeder with an even more expansive dispersal pattern; common southern resident. **DISTRIBUTION:** Breeds primarily across the se. U.S., but breeding has been confirmed in all but a handful of states, as well as two Canadian provinces. In summer and fall, may occur anywhere in the U.S. and across s. Canada. In winter, found in Fla., coastal Ala., Miss., La., Tex., and extreme s. Calif., as well as coastal Mexico and throughout Central America. Appears to be declining in some northern areas. **HABITAT:** Dry grassy upland habitats, particularly those occupied by grazing animals (such as cattle), but also roadsides and grassy highway center divides, airports, croplands, and city parks. Also forages in landfills and flooded agricultural fields. **COHABITANTS:** Cattle or other large grazing animals. Also follows mowers and tractors. **MOVEMENT/MIGRATION:** Northern and interior birds vacate to southern coasts or south of the United States in winter. Spring migration from late February to late May; fall from late August to January. Immatures disperse widely from July to November. A few birds wander up the West Coast to Washington in the fall. VI: 3.

DESCRIPTION: Small, compact, short-billed, thick-necked, short-legged egret most often found with cattle. Most closely resembles a yellow-billed Snowy Egret, but smaller, stockier, heavier-billed, and shorter-legged.

White in all plumages. Breeding birds ape the "punk" style, showing an orange-washed Mohawk crown, breast, and lower back and bright yellow-orange to red-orange bill and legs. In fall and winter, adults may hold traces of orange but have generally all-white plumage; bill is yellow, and legs are black. Immatures have dark bills and short black legs and feet.

BEHAVIOR: Where common, a gregarious species, found foraging in spaced flocks across grassy fields, often feeding among cattle (sometimes perching on their backs) or following tractors or mowing machines. Individuals and small flocks are also a common sight on lawns, roadsides, and garbage dumps. Unlike other egrets, does not wade in open water but may forage along the edge of lakes and sloughs and often gathers in numbers in fields flooded by irrigation (often with other herons and ibis).

Moves—usually walking, sometimes running—with a sinister, strutting, head-swaying grace. Stands with its neck lowered and extended (and body swaying while the head remains fixed). Flocks often feed by "leapfrogging," with birds at the rear flying over the flock. Flocks often make long morning and evening flights to and from feeding areas (flying in V-formation). In migration, may join with other herons and egrets.

FLIGHT: Blocky and compact profile with a squared-off front; the yellow bill is often hard to see at a distance. Flies and glides with down-cupped wings. Choppy and regular, wingbeats are slightly quicker than other egrets. Flight is direct. Recalls an oversized white Least Bittern.

VOCALIZATIONS: Usually silent—in fact, conspicuously so, almost never vocalizing while feeding or commuting. At colonies, utters an assortment of rough, nasal, unmusical calls.

PERTINENT PARTICULARS: A white egret feeding on dry land or among cattle is very probably, but not inevitably, this species. In some places, Great Egret often hunts for rodents in dry fields, and immature Little Blue Herons are reported to feed among Cattle Egrets.

Green Heron, *Butorides virescens*

Dart Gun in Feathers

STATUS: Common breeder across the eastern United States and along the southern Pacific Coast. Less common and more restricted in the interior West. **DISTRIBUTION:** In the U.S., breeds generally east of a line defined by cen. N.D. and cen. S.D., cen. Neb., w. Kans., w. Okla., and cen. Tex. Absent in n. Minn. In e. Canada, nests in se. Ont., extreme s. Que., s. N.B., and N.S. Winters coastally from S.C. to Central America and throughout the Florida Peninsula. In the West, nests from s. B.C. to n. Calif. and is a permanent resident south into Baja and east into s. Nev. and w. Ariz. Also occurs, but may not breed, across s. Ariz., N.M., and sw. Tex. **HABITAT:** Always associated with water (although may nest in dry woodlands not immediately adjacent to a body of water), Green Heron is primarily a bird of wooded and mangrove swamps, stream banks, drainage ditches, and all manner of shorelines (except sandy beach), fresh or marine. Prefers slow-moving or still water and dense vegetation. May feed in the open, usually on a log or branch that offers a strategic perch, but most commonly hugs the shore. **COHABITANTS:** Wood Duck, Belted Kingfisher, Bullfrogs. **MOVEMENT/MIGRATION:** All northern and interior birds vacate breeding areas for the winter. Migrates somewhat later in the spring than Great Blue Heron, with peak movement in April and May. Fall migration is somewhat uncertain and protracted owing to post-breeding dispersal, but August and September see the greatest numbers. By late October, Green Heron is rare north of its wintering range. VI: 2.

DESCRIPTION: Small, dark, somewhat improbably shaped heron that positions itself at gravity-defying angles and seems tethered to shore. Small (smaller than Black-crowned Night-Heron; larger than Least Bittern) and overall chunky (the exceedingly long neck is usually retracted), this football-shaped heron sports a disproportionately long, pointed bill on one end and is supported by disproportionately short, thick, yellowish legs on the other.

"Green" is something of a misnomer. The back is blue with greenish highlights, the breast and neck are chestnut, and the head is dark-capped. At a distance and in poor light, the bird looks all dark. Immatures are like adults, but the colors are muted; the wings are spangled with white, and the neck is covered with maroon streaks (not solid chestnut).

Green Heron has a crest that can be raised or lowered. At times, usually when perched on a tree or branch, the bird extends its neck so long that it exceeds the length of the body.

BEHAVIOR: Unlike most other herons, Green Heron habitually hugs the shallow waters of the shore. Crouched in the shadows or partially concealed in vegetation, with the neck coiled and the bill and body angled toward the water, the bird may stand motionless for minutes on end. When it sights its prey, the bird lunges with neck and body. This heron often hunts from a limb or stump, may stalk slowly in a crouch, and has been recorded to dive from a perch. Crepuscular, Green Heron feeds day or night.

Frequently seen perched in trees when away from water. Although solitary hunters, Green Herons often nest in loose colonies and during migration may be found in small flocks.

FLIGHT: For a heron, appears stocky and compact. Very front-heavy, long-billed profile (gives every indication that it is uncomfortably folded). Flight is straight and slow, with a deep down stroke, a halfhearted up stroke, and a crowlike cadence. Takeoff is hurried and jerky, with neck outstretched. Usually calls when flushed.

VOCALIZATIONS: Call is a loud, explosively gasped "*Skeow!*" Near nest sites, also makes a soft, muffled, conversational "*Kee-ow*" that sounds like a cross between a gulp and a growl. When disturbed, often when taking flight, utters a low, clucking "*kuk kuk kuk.*"

PERTINENT PARTICULARS: Green Heron is overall smaller and darker than American Bittern and much larger and darker than the butterscotch-colored Least Bittern — two other birds that like to hug the edge. Beginning birders' grave misjudgments of size often lead to confusion between Green Heron and Great Blue Heron — two species that are commonly encountered away from coastal areas and that both have generally bluish backs and ruddy necks. Forget size. Green Heron has a dark head; Great Blue Heron has a whitish face.

Black-crowned Night-Heron, *Nycticorax nycticorax*

Waterfront Thug

STATUS: Common widespread (but very local) breeder and fairly common wintering species across most of the United States and parts of Canada. **DISTRIBUTION:** Breeds across most of the U.S., but is conspicuously absent along the Appalachians and the n. Rockies. In Canada, breeds primarily in the prairie pothole regions of Alta., Man., and Sask. and along the St. Lawrence R. Southward breeding range extends into Mexico but bypasses much of the interior. In winter, range is considerably reduced; birds concentrate along the Atlantic coastal plain south of Mass. and in portions of the Mississippi and Ohio drainage, the coastal and some interior portions of Calif., the Great Basin, and the American South and Southwest all the way to Central and South America. **HABITAT:** At all times, associated with fish- or amphibian-bearing waters, either fresh or salt. Nesting and roosting colonies are usually found on vegetated islands in swamps or marshes, presumably because of the protection they afford. Most commonly roosts in trees or bushes but also, particularly in colder temperatures, in stands of phragmites offering a southern exposure. **COHABITANTS:** Nests and feeds with other herons and egrets. **MOVEMENT/MIGRATION:** Spring migration from March to mid-May. Post-breeding dispersal begins in July, but southbound migration probably waits until September, peaks in October, and continues into December. VI: 2.

DESCRIPTION: Stocky, thickset heron distinguished by mostly nocturnal habits, sedentary behavior, and malevolent red or orangy eye. With its slouching upright stance and head drawn down to its shoulders, the bird seems like a thug hanging out on a street corner. Medium-sized (larger than Green Heron; slightly smaller than American Bittern; about the same size as Yellow-crowned Night-Heron), with a slightly drooped pointy bill, a large head, a heavy neck, a chunky body, and stocky bright yellow legs. Typically the neck is retracted. Even leaning forward, with its neck extended, the bird remains chunky and sinister-looking.

The adult is mostly pale gray, with a white face and contrasting black cap and back. Immatures are overall grayish brown, with thick diffuse streaking below and lavish white spotting above. At all ages, legs are yellowish or greenish. The pale bill of immatures is particularly pale on the lower mandible and does not contrast sharply with the face (as does the blackish bill of Yellow-crowned Night-Heron).

BEHAVIOR: Black-crowned Night-Heron does a lot by moving very little. While mostly, but not wholly, nocturnal, this night-heron commonly roosts communally during the day. Leaving at dusk, often in globular flocks, it flies to aquatic (occasionally dry) habitats to forage alone, primarily by standing upright or crouching over a strategic fishing hole, waiting for prey to pass. Stalks prey by walking slowly, and feeds on a variety of animal prey—including snakes, mice, crayfish, crabs, and nestling terns—but primarily on fish.

FLIGHT: With head drawn back, looks overall stocky and solid, supported by broad, short, rounded, distinctly down-curved wings—looks like a milk jug on wings. The contrast between the black back and pale gray wings and body is readily apparent on adults; immatures appear uniformly gray-brown. Only the feet (not the legs) extend beyond the tail, enhancing the overall stockiness.

Flight is plodding and graceless, but smooth, not jerky. Steady perfunctory wingbeats are quick by heron standards but not particularly labored—this is a simple meat-and-potatoes flier going from point A to point B. Commonly glides when landing or executes a wide banking turn when coming in to roost. Flocks are generally loose, globular, or linear, recalling a flock of crows.

VOCALIZATIONS: The classic call, voiced often as birds leave or return to roost, is a loud "*WOK!*"

PERTINENT PARTICULARS: Young Black-crowned Night-Herons may be confused with American Bittern or young Yellow-crowned Night-Herons, which differ, not so subtly, in overall shape and color. Black-crowned Night-Heron is grayish brown and overall stocky. American Bittern is a rich golden brown with long slender (and distinctly streaked) necks. Immature Yellow-crowned Night-Heron is overall darker and grayer and has a bluish or purplish sheen; while still somewhat stocky, it's more classically heron-shaped. Also, the all-dark bill contrasts sharply with the paler face.

Yellow-crowned Night-Heron, *Nyctanassa violacea*

The Heronlike Night-Heron

STATUS: Common to fairly common across much of the species' primarily southeastern range. While less widely distributed than Black-crowned, where found, particularly inland, Yellow-crowned is often the more common night-heron. **DISTRIBUTION:** Breeds from coastal Conn. to coastal S.C., across s. Ga., all of Fla., all but ne. Ala., and north into the Mississippi R. drainage and w. Tenn., w. Ky., s. Ind., Ill., all but n. Mo., se. Kans., e. Okla., and Tex. Also found in parts of Minn., Iowa, Mich., Ohio, and e. Pa. In winter, primarily coastal. Found from coastal Ga., the Atlantic and Gulf Coasts of Fla., Ala., Miss., La., and south from the southern tip of Tex. Common year-round in the tropics. In spring and late summer, a few wander to cen. New England and the Southwest. **HABITAT:** Coastal, estuarine habitats, including barrier islands, coastal marshes, and mangroves; occurs inland in forested swamps and upland woodlands near rivers and lakes. Appears to have two primary requirements: a secluded wooded place to roost and nest, and crustaceans (such as crabs and crawfish) to eat. **COHABITANTS:** Sometimes found in colonies with other herons, but usually somewhat segregated (and located on or near the ground). **MOVEMENT/MIGRATION:** All but the southernmost interior breeding birds vacate breeding areas for the winter. Spring migration from February (early returns to the South) to late April (northern birds). Most fall migration occurs in August and September, but post-breeding dispersal begins as early as July. In spring adults, and in late summer and fall immatures, frequently appear outside the species' breeding range. VI: 3.

DESCRIPTION: Lanky dagger-billed night-heron—somewhere between Black-crowned Night-Heron and Little Blue Heron in structure and size. Medium-sized like Black-crowned, Yellow-crowned has a similar body shape but longer legs, a longer neck, and a larger, wider head fitted with a short, thick, straight, black, daggerlike bill. (The bill on Black-crowned is longer, thinner, paler, and slightly down-curved.) Structurally Yellow-crowned is more heronlike, and more commonly stands erect, than Black-crowned. Standing birds, though heavyset by egret standards, suggest kinship with their slender kin. Black-crowned Night-Heron, its head characteristically couched, just looks like a thickset thug.

Adults, with wholly slate gray bodies and striped heads, are beautiful and distinctive. Immature plumage is overall slightly darker than immature Black-crowned; though brownish, it has a distinct cold purple sheen. Forget about big white spots versus little white spots. *Just look at the color of the bird.* Is it somewhat bluish (like the adult Yellow-crowned) or flat dull brown? Note too that the all-black bill of Yellow-crowned Night-Heron shows greater contrast with the paler face than Black-crowned. Yellow-crowned's expression is startled.

BEHAVIOR: A communal nester but a solitary hunter. Several (or more) may hunt a crab-rich stretch of marsh, but they spread themselves out. The bird does a great deal of standing and waiting at the edge of tidal creeks (often with just the white top of its crown visible). Also stalks, in a slow half-crouch with neck partially drawn back, particularly in thicker (more concealing) cover. Feeds at all hours, even in the middle of the day when there are young to feed as well and when tidal conditions are favorable (as, in fact, does Black-crowned), but is most active at night.

Fairly tame. When roosting, often allows very close approach.

FLIGHT: Overall lankier, more classically heronlike profile than Black-crowned, showing a more coiled bulging neck and longer, more trailing legs. Flight is similar to Black-crowned, but slightly more graceful and flowing.

VOCALIZATIONS: Call is like Black-crowned Night-Heron, but higher-pitched.

White Ibis, *Eudocimus albus*

Pink-billed Wader

STATUS: Very common southern coastal species. **DISTRIBUTION:** In the U.S., breeds from extreme coastal s. Va. to coastal Tex. and Mexico. Outside the breeding season, commonly ranges to s. Del. and farther inland along the Atlantic coastal states; casual in the Southwest. **HABITAT:** Coastal marshes and estuaries, as well as mangrove swamps, freshwater marshes, and shallow ponds. Nests in colonies in trees, shrubs, mangroves on protected barrier islands, and interior swamps.

Sometimes forages on lawns or other dry, open upland habitat. **COHABITANTS:** Glossy Ibis, Greater Yellowlegs, Clapper Rail, Boat-tailed Grackle. **MOVEMENT/MIGRATION:** Somewhat nomadic, with populations shifting in response to water and drought conditions. Some post-breeding dispersal outside of normal range, but seemingly little true migration. VI: 3.

DESCRIPTION: Medium-sized, somewhat stocky heronlike bird with all-white plumage and a hot scarlet-pink distinctly down-curved bill. Usually it is the bright pink bill and face (juvenile's bill is pale pink) that catches an observer's eye.

Except for a small and easily overlooked black tip on the wings, adults are bright white. Unlike Glossy Ibis and White-faced Ibis, White Ibis young are two-toned piebald: pale and somewhat warm brown above, mostly dingy white below. In flight, white rump and lower back contrast with dark upperwings. Note that *underwings of young birds are white, not dark.* (One field guide shows this incorrectly.)

BEHAVIOR: Highly social. Often found in homogeneous flocks that may number 100 or more where birds are especially common, but also feeds with other large wading birds. When feeding, leans forward with head turned slightly down. Walks slowly, probing or picking food. Perches in the open or tucked in tree interiors.

FLIGHT: Flight silhouette is overall blockier (seems particularly to have square-cut wings) and slightly more front-heavy than the other two ibis species. Flight is a series of wingbeats that are perhaps more hurried than Glossy or White-faced and glides that are often shorter and more unsteady. Flies in flowing string or V-formation.

VOCALIZATIONS: A nasal moan with a compressed two-note quality (sometimes three): "*ra'ah*" or "*ra'ah'ah.*"

Glossy Ibis, *Plegadis falcinellus*

Black Curlew

STATUS: Common, primarily coastal breeder and resident along the Atlantic seaboard and portions of the Gulf Coast. **DISTRIBUTION:** In the U.S., breeds in coastal locations from s. Maine to Fla. and s. La. (rare in Tex.), but ranges north to N.B. Winters in breeding areas south of Va. and also in the vicinity of Mobile Bay. In spring and summer, wanders as far as the western prairies; casual to the far West. **HABITAT:** Fresh, saltwater, and brackish marshes; mangroves, mud flats, impoundment pools, and rice fields. Less commonly found in plowed fields and grassy fields and along roadsides. **COHABITANTS:** Snowy Egret, Eastern Willet, Forster's Tern. In Louisiana and Texas, range overlaps with White-faced Ibis. **MOVEMENT/MIGRATION:** Birds north of North Carolina are migratory. Spring migration from late February to May; fall from late June to late October, with peak in August and September. VI: 3.

DESCRIPTION: All-dark, gangly, droop-billed stick figure of a wading bird that resembles a black curlew. Medium-sized (larger than Green Heron; slightly smaller than White Ibis; same size as White-faced Ibis), with a long, drooping, pale reddish brown bill, a small round head, a long neck, a humpbacked body, and long legs.

In breeding plumage, neck and body are rich dark chestnut; blackish wings are shot through with iridescent green and rose-colored hues. Rose hues are more restricted than on White-faced, and the golden highlights seen on White-faced are mostly lacking on Glossy. The skin at the base of Glossy's bill is gray and narrowly edged in pale blue-white—a subtle pattern compared to the blatantly white-bordered red face of White-faced Ibis. Glossy's legs are dull red or just dark; White-faced's are usually brighter. The brown eye on Glossy is hard to see, whereas the red eye of White-faced Ibis can be noted over 200 yd. away through a spotting scope. In winter, Glossy's chestnut tones and facial pattern are subdued; the bird is overall dull blackish, with dull greenish highlights in the wings. Immatures are like adults but are overall smaller and duller, often show touches of white on the head and neck, and have a patterned (dark and pinkish gray) bill.

Winter adults may be separated by eye color (brown in Glossy, red in White-faced) and the presence of the narrow blue trim around the facial skin (absent on White-faced). Immature birds are probably not safely separable until late winter (as early as January), when the reddish eye develops on young White-faced, but immature Glossy shows the same bluish gray, white-trimmed facial skin pattern of adults.

BEHAVIOR: A social species, behaving more like a large shorebird than a heron. Birds forage together in fairly cohesive, single-species flocks, with heads down and humped backs raised. They sometimes gather into tight interception lines that move through the water, or they may feed individually, probing the water with their bills and raising their heads periodically to glance around. Glossy sometimes forages with other large wading birds (and shorebirds), but does not cluster tightly as it does in its own flocks. Among herons, found most often with Snowy Egret. Prefers open foraging areas, but also feeds (usually singly) in drainage ditches and small, tight, wet areas. Roosts and nests in trees. Adults are fairly shy and fly quickly when startled.

FLIGHT: All-dark, wilted stick figure of a bird with head and neck extended, legs trailing, and wings drooped and slightly angled back. Flight is direct, fairly slow, and often somewhat light and buoyant. Somewhat floppy, regular, steady, methodical wingbeats are given in a lengthy series followed by an open, droop-winged glide. Groups fly in single file or shifting V-formations that rise and fall like windblown banners. At a distance, flocks of Glossy Ibis and Double-crested Cormorants appear similar, but the wingbeats of cormorants are choppier and steadier, their glides are shorter, and the much heavier cormorants do not rise and fall.

VOCALIZATIONS: When flushed, emits a low nasal moan: "*raah, raah, raah. . . .*"

White-faced Ibis, *Plegadis chihi*
Golden-winged Ibis

STATUS: The common western ibis. **DISTRIBUTION:** Breeds widely across much of the w. U.S. and across s. Canada, from the Great Plains to s. Ore. and the coast of cen. Calif., south to n. Baja and ne. N.M. Winters in s. Calif., sw. Ariz., coastal Tex., and La. (a few to Miss. and Ala.) and coastal and s. Mexico to Central America. **HABITAT:** In interior locations, breeds in shallow fresh water with islands of emergent vegetation (primarily bulrush and cattails; also salt cedar). In Texas, California, and Louisiana, also found in coastal saltwater marshes as well as fresh. In migration and winter, may be found in a variety of wetlands habitats, including flooded agricultural (particularly rice) fields, wooded streams and swamps, sewage treatment facilities, mud flats, flooded pasture, and grassy fields. **COHABITANTS:** Assorted other wading species; Yellow-headed and Red-winged Blackbirds; along the Gulf Coast and outside its normal range, mixes freely with Glossy Ibis. **MOVEMENT/MIGRATION:** Arrives in spring between early March and early May; departs between mid-September and early November. VI: 3.

DESCRIPTION: Shortish, dark, gangly, curve-billed wading bird that looks nothing like herons and egrets and very much like the closely related Glossy Ibis. Smallish (same size as Snowy Egret) and fairly compact (by heron standards), with a distinctly long down-curved bill, a modestly long neck, a slender body, and sturdy but not particularly long legs. Overall profile is somewhat curvaceous; when feeding, looks very humpbacked.

Overall dark with flashing, shifting iridescence; chestnut and maroon head, neck, and body; and greenish wings that flash with rose and golden highlights. Breeding adults show pinkish facial skin with a full white border (which extends behind the eye) and reddish legs. (On some individuals, only the "joints" are obviously red.) Nonbreeding adults have red eyes (dull brown in Glossy Ibis) and golden iridescence in the wings (greenish in Glossy Ibis).

In adult plumage, White-faced Ibis may be overall slightly paler and redder than Glossy, including a paler or grayer bill. Immatures are probably not safely separable, however, until late winter, when the reddish eye color appears.

BEHAVIOR: Very social. Feeds in groups (among other wading birds and larger shorebirds) and more commonly in large flocks of 1,000 birds or more. Most commonly feeds in shallow water, picking and probing as it walks, sometimes with bill swishing.

Roosts in tall rank vegetation or brush over water. Flies out and returns to roost sites in waving, V-shaped flocks. Fairly shy and suggestive: when one bird flushes, usually they all flush.

FLIGHT: Long-necked profile overall. Slender, delicate, willowy or gangly, and slightly wilted, drooping at all points—drooped bill, down-cupped wings, and trailing legs held slightly below the body. Flight is mostly direct, mostly steady, not particularly fast. Slightly floppy wingbeats are given in an even, methodical series followed by a lengthy, buoyant, flowing, droop-winged glide. In flocks, birds flap and glide in semi-synchronous fashion. In even modest winds, flocks rise and fall like windblown banners. Given good thermal production, flocks also soar (briefly) on set wings, with all members of the flock turning the same way.

VOCALIZATIONS: Mostly silent. When flushed, makes a low "*raaah raaah raaah*" (sounds like the bird is clearing its throat).

Roseate Spoonbill, *Ajaia ajaja*

Pink Cloud on the Water

STATUS: Uncommon to fairly common, but distribution in the United States is limited. **DISTRIBUTION:** In the U.S., this Neotropical species breeds in s. Fla. and sparingly elsewhere along both Florida coasts; also breeds and winters from La. to Tex. south into Mexico. In winter, ranges up along the Florida Peninsula to St. Petersburg and Melbourne and occasionally as far north as Ga. Less common away from coastal areas. **HABITAT:** Mostly coastal, marine estuaries and shallow bays and mangrove-rimmed lagoons. Also freshwater marshes, rivers, and lakes. **COHABITANTS:** Reddish Egret, White Ibis, White Pelican (winter). **MOVEMENT/MIGRATION:** After breeding, birds expand their ranges along coastal areas and inland. Movement, particularly inland dispersal, is strongly linked to water conditions. During times of extreme drought, birds disperse more widely (and have occurred as vagrants along both seacoasts as far north as New York City and central California and inland to the Great Lakes and the Great Salt Lake in Utah). VI: 2.

DESCRIPTION: An unmistakable and improbable-looking bird. Medium-sized (about the size of Reddish Egret) pink-bodied, ibislike bird with a flat spatula-shaped bill. Overall compact (for a wading bird), with a stocky neck and shortish legs. Seen from the side, the compressed bill doesn't necessarily stand out at a distance, and when the bird is feeding, the tip is submerged for long periods.

Adult's body is overall pink and often spattered with deeper, redder patches and trim. (The all-pink flamingo is much larger and has extremely long legs and long neck and a markedly different posture both feeding and standing.) Young spoonbills are white with a faint pink blush on the body (and underwings) that deepens as they mature.

BEHAVIOR: A very social and gregarious species. Tends to feed, travel, and roost in homogeneous groups. When viewed in a distant mass of wading birds, spoonbills distinguish themselves as a pool of pink surrounded by white. The bird feeds with its body held horizontally, usually with the belly very close to the water, the head turned down, and the bill submerged (sometimes the head too). Feeds by sweeping the bill side to side (like some larger shorebirds)—an obvious and distinguishing trait at close range and (obliquely) at great distances too. Other wading birds feed by holding the bill out of the water (ready to jab) or by stitching the water with the head and neck rising and falling (ibis, for example). Spoonbills just keep foraging ahead in a mass of pink, horizontally aligned bodies with necks extended and angled into the water.

FLIGHT: Flies with neck extended and drooped below the body, resulting in a somewhat front-heavy and distinctly humpbacked appearance. Wingbeats are stiff, regular, even, steady, and strong without being heavy or ponderous. Glides are short. Occasionally soars. Flocks are typically small (less than half a dozen) and fly in a V or a line.

VOCALIZATIONS: Generally silent, but when disturbed makes a low, soft, froglike croaking: "*rurh*" or "*huh.*"

STORKS, VULTURES, AND FLAMINGOS

Wood Stork, *Mycteria americana*

The Ugly American

STATUS: North America's only stork. Locally common but restricted summer breeder and resident species. **DISTRIBUTION:** Breeds in se. S.C., several locations (primarily coastal) in Ga., and Fla., where it is a permanent resident. **HABITAT:** Nests and roosts in colonies in freshwater and marine forest habitats, sometimes with herons and egrets. Forages in a variety of freshwater and marine habitats, including swamps, fresh and tidal marshes, rice fields, sloughs, drainage ditches, farm ponds, tidal creeks, and pools—in short, almost any aquatic habitat that is shallow and (often) turbid and murky, even septic. **COHABITANTS:** Assorted herons and egrets, Anhinga, cattle, and (when soaring) Black and Turkey Vultures as well as other raptor species. **MOVEMENT/MIGRATION:** Most birds breeding north of Florida retreat south during the winter. After breeding season (May into summer), or if water conditions preclude breeding or result in breeding failure, birds disperse widely along the coasts and in the coastal plain, from North Carolina to Texas and into Mexico, as well as to the Salton Sea in California. This species often wanders even farther afield. VI: 3.

DESCRIPTION: Large, bulky, black-and-white wading bird with a penchant for murky water and a stance whose body language shouts "dejection." Dwarfs other white wading birds (except white form of Great Blue Heron). The body is bulky, the neck is fairly short, wrinkled, and gray, the head is skeletal, and the heavy bill is shaped like a Native American war club—all in all, at close range, this is an ungainly ugly bird.

Looks better from afar. All white with a narrow black racing stripe running along the side of the body. Immatures are like adults, but the neck is white-feathered, and the bill (on younger birds) is yellow or horn-colored. When perching, the bird stands with its body almost vertical and the bill turned downward, tucked against the neck, making it look dejected or contrite.

BEHAVIOR: Perches for long periods on trees or onshore. Forages for fish, often in turbid water, by walking slowly forward with the body horizontal to the water, the neck lowered, and the bill near-vertical and partially submerged. The bird "gropes" with a partially opened bill from side to side, sometimes rapidly opening and closing the bill with an audible clicking sound. Also stirs the bottom and pats the water with its feet, and frequently forages with wings partially open, shading the water.

In small water holes or canals, often forages alone. In large bodies of water, frequently found in groups (sometimes more than 100 birds), and also forages with other herons and egrets, particularly where drought-inspired concentrations of fish are found. Fairly tame; where habituated to people, allows close approach. When taking flight, runs one or two steps.

FLIGHT: Wood Stork is almost unmistakable in flight; might only be confused with American White Pelican and the very rare and geographically restricted Whooping Crane. Wood Stork is a large plank-winged bird with deeply slotted primaries; extends its head/neck and legs, which are all slightly drooped. The wings are likewise slightly drooped. The body seems suspended.

The bold black-and-white pattern (the bird looks white in front, black behind) is similar to American White Pelican, but soaring pelicans seem stiffly regimented in flight and all wing. (Just a stub of a tail and a bit of the drawn-back head/bill project in the front.) Storks look gangly and wilted.

Wingbeats are heavy, deep, ponderous, and given in a short series followed by a fairly short glide. Wood Stork lacks finesse and seems tired or dejected. Accomplished soaring birds, storks are commonly seen at altitudes beyond the limit of the unaided eye. When descending, they frequently spiral down on half-closed wings, with legs dangling.

VOCALIZATIONS: Silent except for bill clicking and infrequent hissing.

Black Vulture, *Cathartes atratus*

A Vulture Dressed for a Funeral

STATUS: Common, generally permanent resident throughout its extensive New World range. Most common coastally, but also found in Piedmont regions, where it avoids higher elevations. **DISTRIBUTION:** In the U.S., found south of a line drawn from se. N.Y. to the Rio Grande Valley of Tex. and s. Ariz. Also found across all but some northern portions of Mexico and throughout Central and South America. **HABITAT:** Prefers a mix of mature woodlands (for roosting and nesting) and open areas for foraging. Common along roadsides, in pastures, and on open range, where it searches for carrion. Roosts communally in mature deciduous and coniferous trees as well as on buildings, water, and communication towers. **COHABITANTS:** Turkey Vulture. **MOVEMENT/MIGRATION:** Nonmigratory throughout most of its range, although birds at higher altitudes and at northern extremes of the range relocate during colder months. VI: 2.

DESCRIPTION: Squat, squarish, truly black vulture—a vulture dressed for a funeral. Overall smaller and more broadly compact than Turkey Vulture, with a hooded gray head, blackish bill (adults at close range show a yellowish tip), and close-cropped triangle of a tail. Seems longer-legged than Turkey Vulture and not so shaggy or loosely feathered. Stands more erect, more alert than Turkey Vulture.

Overall black except for the bare gray skin of the adult's head and neck. Whitish wingtips are usually not visible on perched birds (but at close range the white shafts of the primaries sometimes peek through). Expression is casually cruel.

BEHAVIOR: Often seen perched on fenceposts or abandoned buildings or attending carrion on roadsides. Forages by flying at high altitudes and watching for the bellwether movements of other vultures (particularly Turkey Vulture). Very aggressive; chases Turkey Vultures and younger Black Vultures from kills. Walks, runs, and, when charging competitors at a kill, canters with opened wings. Feeds by grasping portions of the carcass (often inserting its head into the body cavity) and pushing back with both legs planted on the ground.

Often leaves roost sites early (particularly elevated roosts that afford a downhill glide to prospective foraging areas). Generally does not fly extensively until midmorning, when thermals are conducive for soaring.

FLIGHT: Silhouette is distinctly shorter, squatter, and blockier than Turkey Vulture; looks like a flying coffee table. The wedge-shaped head and triangular tail project equally ahead and behind squarish wings. Seen wing on, the bird appears very front-heavy, with the head and breast sagging below the wings. Whitish wingtips are visible from above and below. An observer's first impression of a distant Black Vulture is usually eagle, not vulture.

Wingbeats, given in a series, are quick, routine, choppy or flailing, and hurry-up. In most places, Turkey Vulture flaps very reluctantly. Black Vulture's glides are tighter, more controlled, and not as wobbly as Turkey Vulture, but still slightly unsteady. When soaring, Black Vulture's wings are flat or uplifted in a slight dihedral. In a glide, wings are slightly down-crooked (that is, angled down, not smoothly bowed).

VOCALIZATIONS: Silent except for hisses and grunts (if disturbed).

PERTINENT PARTICULARS: Turkey and Black Vultures forage and soar together. At a kill, Black Vultures are the ones at the carcass; Turkey Vultures are the ones standing around waiting. (If both vultures are standing around watching, it's because of the caracaras). In flight, Turkey Vultures are normally the birds in the lead. Trailing Black Vultures look like the runts in the litter, flapping hurriedly, trying to keep up with the bigger guys.

Turkey Vulture, *Cathartes aura*

Fence Walker in the Sky

STATUS: Common and widespread; the only vulture species likely to be encountered across much of North America. **DISTRIBUTION:** Breeds from s. Canada to South America. Winters south of a line drawn from Conn. to w. Tex. Also found in most of Calif. west of the Sierra Nevada as well as in extreme s. Ariz. **HABITAT:** Found in habitats offering a mix of woodlands and forest (for roosting and breeding) with adjacent open areas (for foraging). Also frequents elevated, rocky bluffs and buttes and roosts on man-made elevated structures (such as communication or water towers). Foraging habitats include grazing and croplands, beaches and lakeshores, roadsides. Sometimes makes fairly extensive crossings (1–2 mi.) over open water. **COHABITANTS:** May soar with Black Vulture, Wood Stork, other raptors. **MOVEMENT/MIGRATION:** Northern and interior western birds withdraw in the winter and migrate, often in large aggregations, through Texas, Mexico, and Panama. Spring migration is early and protracted—from early February into May, with peak in March and early April; fall migration is equally protracted, from late August into early December, with peak from late September to November. VI: 2.

DESCRIPTION: Large, loosely feathered, blackish bird that resembles an eagle with a shrunken, naked head that is red with a yellow bill on adults, gray with a dark bill on immatures. Perched profile appears long-bodied and humpbacked, giving the bird a dejected or brooding look. Legs are short, stocky, and whitish.

Overall blackish brown (not truly black), with pale scalloping on the folded wings. Expression is mournful.

BEHAVIOR: Most commonly seen standing on the ground (along roadsides and in pastures) or soaring overhead. Also perches on sturdy obstacles (fenceposts, tree limbs, buildings), and in this posture often fans and slightly droops its wings. Roosts communally, occasionally with several hundred birds.

Feeds on carrion on the ground. At large carcasses, multiple birds are the norm. Seems clumsy, almost oafish, on the ground. Walks with a waddling gait, but also runs, hops, and canters. Does not carry prey in its talons.

FLIGHT: Flight profile, with extremely long, broad, rectangular wings, small head, and long wide (but closed, rarely fanned) tail, is distinctive. Distant birds appear no-headed. Flies with wings uplifted in a pronounced V. Flight is tipsy and unsteady, as if the bird were walking an invisible rail fence in the sky, trying to maintain its balance. Sunlight on exposed silvery flight feathers makes the entire trailing edge of the underwings flash like mirrors catching the light.

A consummate soaring bird, in most locations Turkey Vulture rarely flaps. Wingbeats are slow, deep, heavy, and exaggerated. Gliding birds occasionally droop and theatrically straighten their wingtips.

In most places, birds commonly remain perched until thermals begin to rise. They often return to roost well before sunset (particularly in the winter), but in windy locations (such as coastal Texas), birds may be active at first light.

VOCALIZATIONS: Generally quiet. Sometimes hisses or makes low grunts.

PERTINENT PARTICULARS: On the ground or perched, Turkey Vulture may be mistaken for Wild Turkey, but turkeys have long stalking legs, longer necks, a more slender profile overall, and plumage that is more bronze-hued. If you see a flock of birds that are walking and foraging (not standing in a huddle), they aren't vultures. In flight, Turkey Vulture is habitually mistaken for adult Golden Eagle or immature Bald Eagle. Both eagles have larger heads and lack the silvery trailing edge of the wing, and unless buffeted by strong winds, eagles do not tip or rock in flight.

Although loath to resort to powered flight, Turkey Vultures make some exceptions. When just getting aloft in the morning, they often flap, and in very open, windy places (like coastal Texas), Turkey Vultures do a great deal of flapping.

California Condor, *Gymnogyps californianus*

Megavulture

STATUS: Rare resident species; though extremely localized, not particularly difficult to observe. **DISTRIBUTION:** Reintroduced in two locations: s.

Calif. north of Los Angeles, especially around Mt. Pinos in the Padre National Forest, and the Grand Canyon region in Ariz. and Utah, especially Grand Canyon Village along the South Rim. **HABITAT:** Montane forest and cliff faces (for roosting) and lower, sparsely vegetated foothills (for foraging). Specific habitat type seems less important than a reliable food source, expansive open space suitable for a large soaring raptor, and topography that offers good thermal production, updrafts, and launch points that permit birds to go aloft early with little effort. Flat open spaces, which do not meet these standards, are generally avoided. **COHABITANTS:** Turkey Vulture, Golden Eagle, ravens, biologists. **MOVEMENT/MIGRATION:** Individuals shift within their range in response to the availability of food and other factors; otherwise nonmigratory. VI: 1.

DESCRIPTION: Massive blocky soaring bird that is larger than an eagle and dwarfs a raven. Perched birds, with their small, naked heads, seem shaggy and hunchbacked. The adult's tangerine-colored head is distinctive; immatures have dark gray heads that become progressively more colorful over a five-year period.

Perched birds are all glossy black except for a whitish bar bisecting the wing at the base of the primaries (absent in immatures).

BEHAVIOR: Sits for extended periods, sometimes preening and stretching, usually on a cliff ledge or a sturdy tree limb near a cliff or drop. Generally late to go aloft in the morning (later than Turkey Vulture); returns early to roost sites. During the day, patrols, often at great heights, for large mammalian carcasses, which it locates by watching the movements of other condors, eagles, vultures, and ravens.

Highly social. Birds roost, forage, and feed together, but except when gathered around a carcass, tend to be in small groups or alone. At carcasses, California Condor generally dominates all species except Golden Eagle. Normally avoids road-killed animals along highways, but seems otherwise drawn to human activity, perhaps because it has become somewhat habituated to people as a result of the captive breeding and reintroduction program that is the foundation of today's wild population, but perhaps also because it regards any aggregation of large mammals as a potential food source. (Were it not for the guardrails, warning signs, admonishing rangers, and rescue teams found at the Grand Canyon National Park, the Condor's strategy to remain near the hordes of tourists would undoubtedly prove successful.)

FLIGHT: The bird's primary event. Overall massive, a trait observers can perceive both by direct comparison with other species circling nearby and obliquely by measuring how long it takes the bird to complete a circle (estimated at 16 secs. for a condor, 13 secs. for an eagle, and 8 secs. for a Red-tailed Hawk). Overall stocky and blocky, with a short, triangular tail, usually very little projecting head (when soaring, often has its head turned down as it searches the ground below), and broad wings showing widely splayed and upturned outer flight feathers when soaring. Wings are thrust slightly forward, and the silhouette is similar to the squat, wide-winged, short-tailed profile of a soaring Black Vulture.

When gliding, the bird has its splayed flight feathers closed and wingtips drawn down to acute points, but the rest of the wing, curiously, remains broad and perpendicular to the bird. Soars on flat wings or wings that are individually drooped (like Osprey, but not as severe).

Rarely flaps when aloft, except to play or to pursue an eagle that is too close to its nest ledge or cave. Wingbeats are slow and heavy—much in the manner and cadence of an eagle. Sometimes slowly droops and double-droops its wings (like Turkey Vulture).

Despite its size, California Condor often appears curiously unstable in flight—wobbling slightly, it uses its tail as a righting mechanism.

VOCALIZATIONS: Essentially silent.

PERTINENT PARTICULARS: While not always evident, all condors in the wild have numbered, plastic wing tags.

Greater Flamingo, *Phoenicopterus ruber*

Salt Flats Ornament

STATUS AND DISTRIBUTION: Tropical species, breeding in the West Indies. In the U.S., rare and localized to the southwest portion of Everglades National Park, where a small flock forms in fall and winter. The origin of this flock is a matter of debate; the prevailing opinion is that most, if not all, members of the flock are birds from breeding populations in the Bahamas and Cuba. Very rare vagrant elsewhere along the Gulf Coast. **HABITAT:** Expansive, shallow tidal lagoons and mud flats. **COHABITANTS:** Great White Heron, Roseate Spoonbill; in winter, American White Pelican. **MOVEMENT/MIGRATION:** Permanent resident. VI: 1.

DESCRIPTION: Greater Flamingo looks just like the lawn ornament, only larger. A large wading bird (much larger than Roseate Spoonbill; same size as Great White Heron), it has a boomerang-shaped bill, a very long S-shaped neck, an undersized body, and extremely long legs. Its great size and signature posture are very apparent.

Except for the black-tipped bill and sometimes a touch of black on the lower edge of the folded wing, adults are all bright pink. (Roseate Spoonbill has a white neck and pink body.) Subadults are slightly pink, and young birds have grayish bodies, legs, and bills.

BEHAVIOR: Social; birds feed in clustered flocks with heads dangling toward their feet (necks look wilted) and bills swishing back and forth beneath the water. By comparison, Roseate Spoonbill angles the bill down but in front of the body. Greater Flamingo runs to take off and land.

FLIGHT: Flies with exquisitely long neck and long legs fully extended. Shortish triangular wings show black trailing edge. Flight is rapid, direct, and low to the water (sometimes with a rise and fall); wingbeats are strong and steady. Flocks fly in strings or in a V-formation.

VOCALIZATIONS: Low gooselike honks.

PERTINENT PARTICULARS: Other flamingo species are kept in captivity and have at times escaped and turned up at a variety of sites across North America. These include Chilean Flamingo, *P. Chilensis*, Lesser Flamingo, *P. minor*, and the European subspecies of Greater Flamingo. As adults, none of these are as uniformly or intensely bright pink as Greater Flamingo. Any non-storm-related sighting of a flamingo in a place other than Florida Bay is almost certain to be a sighting of an escaped captive bird.

DIURNAL RAPTORS — KITES, HAWKS, EAGLES, AND FALCONS

Osprey, *Pandion haliaetus*

"It's the hawk that looks like a gull but it's not."
(Harry E. LeGrand)

STATUS: Common and widespread, primarily northern and coastal breeder, but breeds locally across large parts of the interior United States and Canada. Common to uncommon migrant throughout most of North America. Common to uncommon wintering species in southern coastal areas. **DISTRIBUTION:** Breeds across most of Canada and Alaska south of the tree line (absent in the southern prairie provinces). In the U.S., breeds coastally from Maine to La.; in the West, across w. Mont., most of Idaho, most of Wash., Ore., n. Calif., n. Minn., Wisc., and Mich.; also breeds in pockets in nw. Nev., Ariz., Utah, N.M., Wyo., Tenn., w. Ky., n. Ga., N.C., N.Y., and Pa. In the U.S., winters coastally from N.C. to Fla. west into cen. and s. Tex.; also along the Calif. coast and in se. Calif. and sw. Ariz. **HABITAT:** Varied habitat, from coastal habitat to coniferous forest to desert lagoons and man-made reservoirs. Whatever the habitat, however, nests are located within 15 mi. of shallow, fish-rich waters and situated on elevated or protected sites (usually near or surrounded by water) and at latitudes where water remains ice-free long enough for young to be raised. In migration, may occur almost anywhere, but seems most concentrated along seacoasts, ridge systems, and river corridors. In winter, found both in coastal locations and on ice-free inland lakes and reservoirs. **COHABITANTS:** Bald Eagle. **MOVEMENT/MIGRATION:** All Canadian, Alaskan, and interior and northern coastal breeders migrate; there is some evidence that Florida birds disperse north after the breeding season. Spring migration from late February to mid-May; fall from late August to late November. VI: 4.

DESCRIPTION: A masked, blackish-brown-and-white, fish-catching "eagle" most commonly seen perched on a sturdy snag (often near water). Large raptor (larger than Red-tailed Hawk; smaller than Bald Eagle), with a small head, a prominent down-hooked bill, and a longish and loosely feathered (shaggy-looking) body.

Except for a white cap, all dark brown above, including tail; except (sometimes) for a spotty necklace, all bright white below. A narrow dark mask extends from the bill to the nape, falling over the bright yellow eye. Immatures are much like adults, but upperparts show pale spotting.

Adult Bald Eagle is all blackish with an all-white head and tail. Juvenile Bald Eagle is all dark brown. More advanced immature eagles (which may show a dark mask through the eye reminiscent of Osprey) are very heavily mottled dark brown and dirty white all over the body and look ratty and unkempt.

BEHAVIOR: The only North American raptor that dives into the water, often submerging itself in the process. (Bald Eagle plucks fish from the surface; gulls plop to the surface but do not dive.) Hunts by circling and hovering and flying between 30 and 100 ft. over open water. Dives at an angle, headfirst. If successful, carries fish headfirst to an elevated perch, often some distance from the water, and often while being pursued and harassed by gulls and Bald Eagles. Unlike Bald Eagle, rarely feeds on dead fish.

Usually found alone or in pairs, but may nest in small colonies where nest sites are few and fish plentiful. Commonly perches on a high, sturdy perch. (Except when gathering nesting material, rarely perches on the ground, as eagles often do.) Fairly wary. Flushes when approached, but usually announces its displeasure by calling in a series of short, whistled notes before taking wing. May continue to call while retreating or while being pursued by pirating eagles.

In migration, birds often travel in pairs (sometimes in groups of three or four), with one bird following the other along the same flight path. Owing to their very light wing loading, Ospreys fly late in the day, sometimes continuing to migrate after the sun has set.

FLIGHT: Unmistakably gull-like in shape but more gangly—a rangy scarecrow of a bird with its long, narrow wings draped over a crossbar. In a glide, wings are cocked back and crooked down. Head on, the bird looks like a large droop-winged gull; from below, the bird's configuration conforms to the letter M. Even when gliding, the wings do not sharpen to a point, as with gulls; tips always show slotting primaries.

At a distance, birds seem headless and all crooked wing; a short broad tail is visible. The color is uniformly dark except for a bright white body that peeks from beneath the down-crooked wings. At closer range, the bird appears brown above (not blackish like eagles); the contrasting black-and-white pattern of the underwing is unmistakable.

Flight is commonly slow and buoyant. Wingbeats are shallow and stiff, with all the motion centered in the jutting elbow. In migration, Osprey rises quickly when thermaling and tends not to use its wings during the extensive glide. When flapping vigorously (as when pursued), the body jerks up and down.

VOCALIZATIONS: Call is a loud series of spaced piped whistles that climb the scale and accelerate toward the end: "*pew pew Pew Pew PEE PEE.*" Often punctuates the series with a higher-pitched, more strident double whistle, "*P'PEE; P'PEE.*"

PERTINENT PARTICULARS: Manifest dissimilarities notwithstanding, Swallow-tailed Kite is often mistaken for Osprey (and vice versa) because of similar size, plumage, and droop-winged configuration.

Hook-billed Kite, *Chondrohierax uncinatus*

Slothlike Kite

STATUS AND DISTRIBUTION: Tropical and subtropical raptor; rare U.S. resident: fewer than 60 individuals are estimated in its extremely limited range. Found exclusively in woodlands along the lower Rio Grande Valley in Tex. west to Falcon Dam. **HABITAT:** Riparian and scrub woodlands along the lower Rio Grande. Most reliably seen in Santa Ana National Wildlife Refuge (Texas), Anzalduas County Park (Texas), Bentsen–Rio Grande Valley State Park (Texas), Chapeno, Salineno, and Falcon Dam. **COHABITANTS:** Gray Hawk, White-tipped Dove, Altamira Oriole. **MOVEMENT/MIGRATION:** Permanent resident. VI: 0.

DESCRIPTION: One ungainly, unconventional, and unmistakable raptor. Medium-sized (slightly larger than Gray Hawk and Red-shouldered Hawk; smaller than Harris's Hawk), Hook-billed Kite is a somewhat buteo-like raptor, but overall elongated and uniquely proportioned. The elliptically shaped body (widest in the middle) and long wide tail give it the overall shape of an inverted bowling pin. Seems long-necked and small-headed, but is exceptionally large-billed and eaglelike in its proportions. Very short legs give perched birds a sawed-off appearance. A round pale green eye is the bird's most arresting feature; discernible at considerable distance, it gives the bird a crazed expression.

Adult males are gray—all gray above, and gray-breasted and gray-barred below. The two-toned adult female is even more distinctive: a brownish gray back contrasts with a reddish brown neck, breast, and barred belly (head has a dark toupee). Immatures are like females, but wholly barred below, with white cheeks. Tails of all birds have wide dark and pale bands.

A dark morph, common in the tropics, has only occurred once in the United States.

BEHAVIOR: A sloth with wings. Hook-billed Kite spends much of its time, whether alone, in pairs, or in family groups, in and below the canopy "hunting" tree snails. Typically forages by walking, parrot fashion, down limbs and plucking snails with its bill. Sometimes hangs upside down and thrashes and/or flies to a favored perch with prey carried in its bill. Also glides from perches, throwing itself against trunks, then pushing off with prey. Key feeding areas are marked with piles of shells beneath perches. (But don't count on a stakeout to see your bird—shells last a long time.)

Very shy. Generally flushes when spotted. Mostly silent.

FLIGHT: The silhouette is unique—a long narrow neck; large, exceedingly broad, paddle-shaped wings that are distinctly pinched at the base; and a long, relatively narrow, unfanned tail. In North America, only Harris's Hawk (with its obvious white basal half of a tail) comes close to approximating this shape. All birds show a translucent, white crescent window near the tip of the wing that is reminiscent of Red-shouldered Hawk. Wingbeats are slow and shallow. In level flight, a leisurely and short series of wingbeats is punctuated by a slow glide. In flight and when soaring, seems buoyant but ponderous—anything but nimble.

VOCALIZATIONS: Mostly silent. Call is a loud, rapid, mirthful chatter, *"k'hehehehehehe,"* that brings to mind some woodpecker species you just can't put your finger on.

Swallow-tailed Kite, *Elanoides forficatus*
The Wind Given Form

STATUS: Locally common but restricted southeastern breeder. Uncommon migrant in coastal Texas. Regular but rare wanderer outside its range. **DISTRIBUTION:** Breeds in coastal S.C. (Francis Marion National Forest), coastal Ga., over most of the Florida Peninsula and Panhandle, s. Ala., Miss., s.-cen. La., and e. Tex. **HABITAT:** Mature, often broken, southern pine and hardwood forests, particularly those flanking rivers. Forages in the skies over the canopy or over open country. **COHABITANTS:** dragonflies, Mississippi Kite, Red-shouldered Hawk, Barred Owl, Clouds. **MOVEMENT/MIGRATION:** Entire North American population retreats to South America in winter. Most presumably cross the Gulf of Mexico after staging in southern Florida in August. Spring migration from mid-February to mid-April, but some subadult birds may wander well north of their established range into May and June; fall migration from late July to early September. VI: 3.

DESCRIPTION: Hardly necessary— whether perched or in flight, the bird commands instant recognition. Large (much larger than Mississippi Kite; smaller than Osprey) and exquisitely slender, showing long, narrow wings and an even longer, deeply forked tail.

The white head and underparts contrast with the blue-black upperparts. Immatures are not quite so white and are shorter-tailed.

BEHAVIOR: A social kite (though commonly seen foraging alone) that glides with slow fluid grace over the treetops, swooping down and snatching prey from the branches or soaring overhead and plucking dragonflies from out of the air. Feeds on the wing by reaching down with its head to reach the food clutched in its talons. Swallow-tailed Kite is rarely seen perched, but when it does perch, it often chooses a high, exposed dead branch.

FLIGHT: The long, bowed wings and splayed tail are unique. Most observers are startled by the size (it's almost Osprey-sized and, except for the tail, not dissimilarly shaped), and occasionally by the fact that the tail is not longer. Swallow-tailed glides and soars, but rarely flaps. The tail twists and turns constantly, micromanaging the bird's flight. Wingbeats are stiff, shallow, and executed with a dreamy slow-motion quality.

VOCALIZATIONS: A high thin whistle.

PERTINENT PARTICULARS: If all birds were as easy to identify as this one, birding would hardly be a challenge.

White-tailed Kite, *Elanus leucurus*
Hover Kite

STATUS: Common resident and, although widespread in the New World, somewhat geographically restricted in the United States. **DISTRIBUTION:** Found throughout much of the Western Hemisphere. In the U.S., found in three distinct geographic areas: along the West Coast from w. Ore. (and a few in sw. Wash.) south to Baja and west of the Sierra Nevada; se. Tex.; and extreme s. Fla. (Dade Co.). In winter, also disperses across s. Calif., s. Nev., and s. Ariz. **HABITAT:** Open grasslands, most commonly at low elevations. Habitats include oak savanna, coastal grasslands, wetlands, agricultural and overgrown fields, and highway medians and edges. **COHABITANTS:** Red-tailed Hawk, Northern Harrier, White-tailed Hawk (Texas), American Kestrel. **MOVEMENT/MIGRATION:** Permanent resident showing some regular annual dispersal and irregular large-scale movements. VI: 2.

DESCRIPTION: Stocky, gray and white-and-black, falcon-shaped hawk that hovers, vigorously, over open grasslands. Medium-sized (larger than American Kestrel; smaller than Northern Harrier; about the same size as American Crow), with a largish head, a petite bill, a fairly robust body, very long wings, and a long tail. Appears somewhat broad-shouldered or top-heavy when perched.

Pearl gray above, with a bright white head and underparts—a simple pattern that showcases the bird's almond-shaped black eye patch and (usually conspicuous) black shoulder patch. When the bird is perched, the white tail is mostly hidden by overfolded wings. Immatures are like adults but show a warm rusty-buff wash on the head and breast. The expression is benignly aloof (but don't push it).

BEHAVIOR: An active, aerial predator that spends much of its time aloft. Hunts by hovering 20–75 ft. high, into the wind, with wings raised high and head and tail angled down. When prey is sighted, the bird parachutes on raised wings, closing the distance, checks its fall, then hovers again before making a final drop, talons first. If prey is not sighted, or the drop is aborted, the bird flies a short distance and hovers anew. Rarely perch-hunts (as many falcons do).

When not hunting, birds are commonly seen perched (often in pairs) on trees or shrubs (not wires). In winter, White-taileds may roost in numbers and be fairly communal as they head off at dawn to forage and return at dusk.

Individuals vary in their tolerance of approach, but in general this species seems calm.

FLIGHT: Falconlike profile, but appears top-heavy with disproportionately broad-handed (albeit pointy and long) wings and a tail that is too narrow. When flying, gliding, and soaring, wings are uplifted in a pronounced U-shaped dihedral. At a distance, looks headless. Seems overall pale, with a white head, white tail, and lots of black contrasts in the gray-and-white wing. When gliding, the black can be isolated to the shoulder and underwing tips. When birds are hovering, you see lots of black in the wings.

Flight is direct on slow, stiff, somewhat shallow, languid wingbeats. The wingbeats of a hovering bird seem deep and effortful viewed from the side, but more fluttery when viewed from behind.

VOCALIZATIONS: When provoked, makes a loud, low, raspy "*rrrrrch*" that has some of the qualities of Gray Catbird's call but is raspier and less whiny.

PERTINENT PARTICULARS: Not a bird easily confused with any other bird of prey, although it superficially resembles adult male Northern Harrier. Harriers, which also fly with their wings in a dihedral, hunt and hover close to the ground; they are rarely more than 10 ft. high. As such, a harrier's white rump patch (and the absence of black shoulder patches) should be only too apparent. When soaring overhead, harriers become blunter-winged and overall more accipiter-like. Soaring White-tailed Kite retains its falconlike lines.

Snail Kite, *Rostrhamus sociabilis*
Slender-billed Kite

STATUS: Uncommon, habitat-specific, and highly localized. **DISTRIBUTION:** A tropical raptor found from Mexico to n. Argentina. In North America, range is limited to the Florida Peninsula south of Orlando. **HABITAT:** Shallow, grassy, freshwater marshes that host Apple Snails, the bird's near-exclusive prey. **COHABITANTS:** American Alligator, Anhinga, Limpkin, airboat operators. **MOVEMENT/MIGRATION:** Permanent resident, but somewhat nomadic. Populations shift in response to water levels and the availability of snails. There is also some movement of northern birds to the southern end of the Florida Peninsula in winter. VI: 1.

DESCRIPTION: When this medium-sized raptor stands on a bush in the middle of the Everglades, its hooked bill looks so unnaturally long and slender that it seems to beg trimming. Snail Kite is buteo-shaped and -sized (same size as Red-shouldered Hawk), with an extremely long, slender, and acutely hooked bill and long wings that extend beyond the tail of the bird when perched.

Adult males are dark bluish gray with an orange bill and legs; females and immatures are brownish with boldly patterned faces, the signature slender bill, and thickly streaked underparts.

BEHAVIOR: Most commonly seen sitting conspicuously and upright on the springy tips of shrubs and trees, surrounded by expansive grassy marsh, or coursing low (less than 10 ft.) over the marsh in search of a snail, which it plucks from the surface (without getting more than its legs wet). Carries snail in its talons to a favored perch. May be seen feeding at any time of the day. Also soars, showing its distinctive tail pattern to best effect.

FLIGHT: A paddle-winged raptor with a short but boldly patterned triangle for a tail. Wide wings are acutely bowed. Each wing rises at the shoulder and curves down to the tip. The tail is bright white at the base (half the tail), with a black swath across the tip. The pattern is visible from above or below.

Flight is low and slow, coursing. Wing flap is floppy, and glides are floating. Hovers frequently. Drops and plucks snails and rises with deep wingbeats. Soaring birds hold their wings straight out to the sides and show a pinch-based paddle-winged configuration.

VOCALIZATIONS: Usually silent. Occasionally gives a harsh cackling "*ka,ka,ka,ka.*"

PERTINENT PARTICULARS: You don't happen upon Snail Kites. You search for them in the Kissimmee River Valley, St. John's River, Lake Okeechobee, Loxahatchee Slough, the Florida Everglades, and especially the adjoining Miccosukee Indian Restaurant astride the Tamiami Trail (route 41), the place where 90 percent of the birders in North America have seen this bird.

In its limited range, only three other raptors might be mistaken for Snail Kite: young Red-shouldered Hawk, which, like female and immature Snail Kites, has a streaked breast; Osprey, which is brownish-backed and has a distinct facial pattern; and Northern Harrier, whose low-coursing flight and white rump may make it the most likely candidate for confusion. Even with lavish amounts of wishful thinking applied, however, none of these species remotely resembles Snail Kite. Red-shouldered is a perch hunter, not a coursing raptor. Osprey needs sturdy perches; also, when hunting (over water), it circles or hovers much higher than the kite. Northern Harrier flies with its wings uplifted in a V. The kite's opened wings form a drooping, shallow M.

Mississippi Kite, *Ictinia mississippiensis*

The Colonial Kite

STATUS: Locally common, mostly southern breeder. Common to abundant migrant in southern Texas. Regular but rare spring and casual fall wanderer outside its normal range. **DISTRIBUTION:** Core breeding range covers e. and s. S.C., s. Ga., the Florida Panhandle, all but n. and ne. Ala., the periphery of Miss., all but sw. La., s. and e. Ark., and the Mo.–Ill. border; also breeds in sw. Kans., w. Okla., and n. Tex. Isolated populations are found in ne. Va., ne. N.C., sw. Ind., sw. Mo., cen. Iowa, w.-cen. Neb., se. Colo., cen. and e. N.M., and s. Ariz. **HABITAT:** In the Southeast, prefers large, unbroken, mature riparian and bottomland forests with adjacent areas for foraging. In the Great Plains, uses woodlands surrounded by prairie as well as riparian corridors, shelterbelts, and well-treed residential areas. Readily occupies woodlands in suburban settings (such as city parks and golf courses). **COHABITANTS:** dragonflies, Swallow-tailed Kite, Red-shouldered Hawk, American Crow. **MOVEMENT/MIGRATION:** The entire U.S. population migrates south in winter and is often seen in large flocks in Texas. Spring migration from early March to mid-June; fall from early August to early October. VI: 3.

DESCRIPTION: Social, nimble, buoyant, and very peregrine-like kite (in shape only, not in manner of flight). Medium-sized (smaller than a harrier; only slightly smaller than White-tailed Kite and Broad-winged Hawk) and fairly falconlike in overall shape, with a robust head and body and longish wings that are longer than the tail.

Perched adults are overall gray with a frosty pale head and a dark, overdone eye patch. Immatures have grayish upperparts and streaky, warm brown underparts (recalls adult male Merlin), but again, wings are longer than the tail (tail is longer on Merlin and Peregrine Falcon).

BEHAVIOR: Particularly in the Great Plains, virtually communal, nesting, roosting, and foraging in groups of 50 or more. In the Southeast, more solitary. Hunts almost wholly from the air, soaring and gliding at moderate heights, accelerating slightly to snatch an insect appearing in its path. Sometimes grabs prey from leaves or branches and less commonly hawks insects from a stationary perch; occasionally forages by walking or hopping on the ground. Usually consumes its prey on the wing.

Has become acclimated to (in fact, aggressive toward) people who get too close to nest sites.

FLIGHT: Long, pointy-winged silhouette; in size and shape (and fact!), very easily confused with Peregrine Falcon. There are two key differences. First, with a hand that is wider than the arm, Mississippi Kite has a Popeye look; the wing of Peregrine Falcon tapers evenly to a point. Also (and somewhat related), the outermost primary on the kite is shorter and even seems stunted. This may seem like a difficult thing to perceive, but it is often quite obvious in the field. Second, the fanned tail of Mississippi Kite flairs out toward the tip; Peregrine Falcon's does not.

In flight, the whitish head and whitish secondaries (upper surface only) contrast with the gray uniformity of the rest of the bird and are conspicuous at great distances. On immatures, the heavy-handed wing shape, short outer primary, and flared, conspicuously banded tail are all useful.

Flight is buoyant, nimble, and fairly swift. Feeding birds move nimbly, gracefully, and fluidly, using stiff, shallow wingbeats and floating glides.

VOCALIZATIONS: Call is a high-whistled "*peetee*" that recalls Broad-winged Hawk.

PERTINENT PARTICULARS: More than one very competent observer has mistaken a Mississippi Kite for a Peregrine Falcon. Two additional differences are worth nothing. Mississippi Kite's wingbeat is slow, measured, and stiff, whereas Peregrine Falcon's is just the opposite—rapid, continuous, fluid, and whippy. Also, peregrines don't hawk insects; kites do so for a living.

Bald Eagle, *Haliaeetus leucocephalus*

American Fish-eating Eagle

STATUS: Fairly common and widespread breeder; sometimes very common wintering species. **DISTRIBUTION:** Breeds across most of Canada and Alaska and scattered locations throughout the U.S. Winters coastally in Alaska and Canada and throughout the U.S. and n. Mexico, as well as border regions of Canada. **HABITAT:** Breeds in mature forests with tall, sturdy trees or on smaller isolated tracts away from human disturbance, commonly within a mile of seacoasts, rivers, reservoirs, or marshes. In winter, forages in the same habitat or in dry open upland, primarily where open fish-bearing water is found. In northern, icebound areas, Bald Eagle is partial to the open waters associated with dam spillways. **COHABITANTS:** Great Blue Heron, Osprey. **MOVEMENT/MIGRATION:** The pattern is complex. Most northern birds retreat south and many adult southern birds remain on territory year-round. In general, spring movements occur from late February to late May, and fall movements between mid-August and mid-December. First-year southern birds (and perhaps subadults and some adults) disperse north after breeding (May to August) and return August to October. Subadults are somewhat nomadic and move about in response to food availability; they may turn up anywhere, anytime. VI: 4.

DESCRIPTION: It's the bird on the back of the one-dollar bill with the olive branch in one talon and the arrows in the other. Huge (dwarfs large buteos like Red-tailed Hawk; much larger than Osprey and Turkey Vulture; about the same size as Golden Eagle, but with a different shape), with a big head, thick neck, very large bill, robust body, and short tail. At a distance, the silhouette of perched birds is somewhat elliptical.

Adults are unmistakable: all blackish brown except for an all-white head, yellow bill, and white tail. Juveniles are uniformly chocolate brown and dark-billed, with a bit of white showing at the base of the tail. Subadults (two- to three-year-olds) are ratty-looking; the body is lavishly spattered with white—particularly on the back and belly—and the bird may appear piebald, mocha-colored, or even tawny or creamy white with dark patches. Four-year-old birds have white heads and tails but commonly show a dark, Ospreylike slash across the eye and traces of black on the tip of the otherwise whitish tail.

BEHAVIOR: Spends most of its life sitting on sturdy branches on the banks of a watercourse. Also perches freely on the ground, ice, stumps, or low snags over open water. Very social but not necessarily amicable; multiple birds often sit or perch close together or try to feed from the same animal carcass (carp, goose, seal, deer). Also hunts and searches by cruising rivers (sometimes roadways) in search of carrion and live fish (which it plucks from the water with its talons without diving into the water, as does Osprey), and commonly pirates fish from Osprey. Although it feeds mostly on carrion, Bald Eagle (particularly the adult) is an accomplished hunter capable of capturing jackrabbits and flying down ducks and geese (usually by employing a combination stoop-and-pursuit). Commonly seen harassing ducks on the water. In Alaska, scavenges in landfills.

In migration, adults commonly fly in pairs but may be several minutes apart. In winter, often attended by crows or ravens. Adults and most immatures are fairly intolerant of human intrusion, but some birds may allow very close approach, particularly where habituated to people.

FLIGHT: Adults have long planklike wings and a white head and tail extending equal lengths. Immatures are usually bulkier, with long but somewhat broader, lumpier, or bulgingly contoured wings and a tail that is only modestly longer than the projecting head. First-year birds are mostly dark, with whitish underwing linings, dirty white feathers at the base of the tail, and little or no white showing in the flight feathers. Older birds (as early as first-winter) become ragged and have extensive white areas on the back, belly, and wings. At a distance, some birds look mocha-colored.

Bald Eagle soars on flat wings, but sometimes shows a slight down-droop (not as pronounced as Osprey) or a slight dihedral (not as pronounced as Golden Eagle). Glides on down-drooped or angled wings. Wingbeats are slow, measured, and ponderous, with a deep arch, and have about the same cadence (but not the shallow stroke) of a large gull.

Can only be mistaken for Golden Eagle.

VOCALIZATIONS: Call is a surprisingly weak, strangled, whistled, chirping stutter.

PERTINENT PARTICULARS: The image of the bird on the one-dollar bill is highly stylized, and while it may not help you identify a Bald Eagle, if you collect 20–30 of them, you can buy a decent field guide to the birds.

Northern Harrier, *Circus cyaneus*

The Coursing Hawk

STATUS: Common widespread raptor of open habitats. **DISTRIBUTION:** Breeds across virtually all of Canada and Alaska except for extreme northern regions. In the U.S., breeds in all the states north of coastal Va., much of the Midwest, and the western states west of Iowa and Mo. and north of n. Tex., n. N.M., s. Utah, and cen. Nev., into the Great Basin and intermountain West. Also breeds coastally in Wash., Ore., and Calif. south into Baja. Winters across most of the U.S. and Mexico except for interior New England, n. N.Y., n. Pa., n. Mich., n. Wisc., all but s. Minn., N.D., most of S.D., and all but w. Mont. **HABITAT:** A variety of open habitats, including dry and marshy meadow, coastal marsh, forest bog, tundra, alpine tundra, prairie, grazed pasture, and desert steppe. Commonly hunts at the edge of forests and on agricultural lands. Avoids dense forest and higher elevations. **COHABITANTS:** Varied depending on region and habitat, but includes White-tailed Kite, Rough-legged Hawk, Short-eared Owl. **MOVEMENT/MIGRATION:** Birds in much of Canada and the northern United States vacate breeding territories for winter. Extremely protracted fall migration occurs from late August to early January; spring from late February to late May. VI: 4.

DESCRIPTION: Slender, owl-faced, ground-hugging raptor with traits shared by both accipiters and falcons. Medium-sized (larger than Cooper's Hawk; smaller than Red-tailed), but overall lithe and slender, with a smallish head, a long tubular body, a very long tail, and long legs. Perched birds seem very accipiter-like.

Adult males are uniformly pale bluish gray above and bright white below, with a gray hood that extends to the breast. Females are brown above and creamy and streaked below (particularly heavy on the breast). Immatures are warm brown above and washed with cinnamon below. All birds show an owl-like facial disk.

BEHAVIOR: Northern Harrier's hunting flight is virtually idiosyncratic. Hunts low and slow over open ground, often following vegetative leading lines, frequently pulling up and hovering (often with long legs dangling). May execute a smart swift wingover as it fairly throws itself at prey. Usually hunts alone, but in winter, and even during breeding season, multiple birds may hunt in fairly close proximity. In winter, roosts communally (usually 5–30 birds) on the ground in tall grass or corn stubble. Birds leave roosts and begin hunting at dawn and often continue to hunt after sunset.

Perches are usually low and often on the ground, but on occasions (most often during the breeding season) birds sit high in trees (sometimes at the tips of springy twigs), their posture vertical. Prey is consumed on the ground. Generally silent, but calls as a warning to transgressing harriers, other raptors, and human intruders. Also very vocal during aerial courtship.

Not particularly aggressive toward most other raptors, but feuds often with Short-eared Owl. In migration, commonly flies very high. Usually solitary, it sometimes migrates in spaced or strung-out groups of two to six individuals. Not shy about crossing large bodies of water; sometimes encountered far out at sea.

FLIGHT: Buoyant grace. Silhouette is distinctly long-winged and long-tailed. When gliding, appears falcon-shaped; when soaring, is very accipiter-like (but shows longer wings). Males are overall stockier, females broader-winged. In all plumages, the base of the tail sports a distinctive, oval, white rump patch. Other birds of prey have pale rumps; this is a true patch.

Flies with a lazy, loping series of flaps punctuated by a floating glide during which the wings are lifted in a distinct V. Flight is buoyant but somewhat stiff and unsteady, rocking or tipsy. When gliding, pursing prey, or fighting strong winds, wings are drawn back and the bird looks very falcon- or kitelike.

VOCALIZATIONS: Call is a weak, breathy, somewhat petulant whistle, "*eeahhh.*" Courting males (and females) make a chattering "yicker"— "*keh-keh-keh-keh*"—when engaged in their spectacular sky-dancing displays.

PERTINENT PARTICULARS: The sky-dancing display involves both male and female birds climbing high above their territories and executing a series of roller coaster–like dives and climbs whose course describes a linked series of Us. Displaying birds commonly vocalize (yicker) and conclude by taking a perch on the ground.

Sharp-shinned Hawk, *Accipiter striatus*
The Artful Dodger

STATUS: Fairly common but secretive, and for the most part, northern nesting species; more widespread and more commonly seen in winter. **DISTRIBUTION:** Breeds from Nfld. and Lab. across all of Canada to interior Alaska; south throughout New England and along the Appalachians to n. Ala.; in the upper Midwest and throughout the Rockies and coastal forests south into Mexico and Central America. In winter, vacates all but coastal and s. Canada and s. Alaska and fans out across most of the U.S., Mexico, and Central America. **HABITAT:** Nests in a variety of forest habitats, but is most partial to conifers; more inclined toward younger and denser growth than Cooper's Hawk. In winter and migration, found in a variety of habitats, from forest to suburban yards where feeding stations concentrate small birds, the Sharp-shinned's principal prey. **COHABITANTS:** Almost any place where you find chickadees, juncos, and other small, woodland songbirds. **MOVEMENT/MIGRATION:** Highly migratory; though movements are broad-based, geographic features such as rivers, lakeshores, coastlines, and mountain ridges concentrate and direct the passage of birds. Spring migration from March to mid-May, with a peak in late April and early May; fall from late August to December, with peaks occurring in September (immatures) and October (adults). VI: 4.

DESCRIPTION: Small (jay- to flicker-sized), compact, bowling-pin-shaped hawk with virtually no neck and a head that is small and flat-topped but roundly contoured. The long slender tail has a blunt straight-cut or notched tip. Overall, the bird seems somewhat delicately proportioned.

Adults have slate blue backs and pale orange underparts that may look pale, even white, at a distance. Immatures are cold brown, with no orangy or warm tones about the neck (as on immature Cooper's Hawk); scattered whitish spots above; and cream below, with thick brown streaks or streaky bars that are heaviest on the breast (and make distant birds appear dirty-chested). The more lightly streaked immature Cooper's Hawk usually appears to have a pale chest, which contrasts with a dark hood.

On both adults and immatures, the whitish band on the tip of the tail is narrow and not crisply defined.

BEHAVIOR: Likes to hunt in the forest beneath the canopy, where a favorite strategy is to fly to a strategic, open perch 10–50 ft. up; wait, motionless, for a period of several minutes to half an hour; then move on. In more open, broken habitat, also sits low, often fully or partially concealed in low cover, and makes rapid direct forays toward prey. Also cruises low (often just above the ground) through bird-rich areas using obstructing cover (such as hedges or buildings) to shield its approach.

Kills prey by constriction (that is, does not use its bill). Plucks prey before eating, usually on the ground where the kill was made and less commonly from a stump or elevated perch.

At most times not social, even during breeding season (adults keep their distance). In migration, small groups of three to eight (mostly young) birds are often seen. Aggressive toward other birds of prey, particularly in migration. Frequent targets for harassment include Cooper's Hawk, Red-tailed Hawk, and perched owls (or owl decoys).

FLIGHT: The wings are short, stocky, and blunt, the tail exceedingly narrow, and the head short, projecting hardly at all ahead of the wings. Overall shape suggests a flying mallet. Cooper's Hawk, with its more projecting head and straight-cut leading edge to the wing, looks like a flying crucifix. The narrow whitish band on the tip of Sharp-shinned's tail is diffuse; that of Cooper's Hawk is wider, whiter, and more crisply defined. Immature Sharp-shinned appears dirty-chested; immature Cooper's Hawk shows whiter underparts contrasting with a well-defined hood.

Flight is generally straight. A series of wingbeats are interspaced with a glide. Wingbeats are quick and snappy, clipping instead of clopping. At a distance, beating wings recall a sputtering or flickering candle flame. Movements are abrupt and hair-trigger. Overall flight seems buoyant and bouncy.

When hunting, sometimes executes a woodpecker-like undulating flight.

VOCALIZATIONS: The alarm call, most often heard by observers as they enter the bird's territory, is a sharp, rapid, strident "*kek, kek, kek, kek.*" Very similar to Cooper's Hawk, but slightly higher-pitched.

PERTINENT PARTICULARS: Observers most commonly see Sharp-shinneds at their feeders. Most seed-eating birds fall within the prey range of both Sharp-shinned and Cooper's Hawk, but Sharp-shinned is most partial to smaller species (like chickadees and sparrows); Cooper's Hawk usually targets larger and more open-country prey, such as doves, pigeons, robins, and jays. This partiality is not absolute—Sharp-shinneds do regularly kill robins—but when separating accipiters, every little hint helps.

Cooper's Hawk, *Accipiter cooperii*

The Summer Accipiter

STATUS: Common and more commonly seen accipiter across much of the United States and southern Canada. **DISTRIBUTION:** Breeds farther south than Sharp-shinned—from w. N.B. across s. Canada and south over all but some extreme southern portions of the U.S. In winter, vacates most of Canada and some northern U.S. border states; expands across all of the U.S. and Mexico and into Central America. **HABITAT:** A woodland hawk that frequents a variety of forest and woodland habitats. Nesting habitat is frequently more open and more mature than Sharp-shinned's, with less predilection for conifers. Also readily adapts to broken forest habitat (including riparian corridors and city parks) and has become suburbanized in many places. In winter and migration, commonly hunts in open areas such as marshes, fields, and woodland edges—much more so than Sharp-shinned. **COHABITANTS:** Flickers, thrushes, jays; also smaller waterfowl and game birds. **MOVEMENT/MIGRATION:** The spring migration from March to May is earlier and peaks earlier than Sharp-shinned Hawk migration. Fall migration occurs from late August to late November, with peaks in early to mid-October (immatures) and mid to late October to November (adults). VI: 2.

DESCRIPTION: Large (crow-sized), lanky, bowling-pin-shaped hawk; overall more robust and broadly proportioned than Sharp-shinned. The head is larger and squarish. (*Note:* When threatened or when mantling prey, Cooper's Hawk's hackles are often raised, making the head look very square. Sharp-shinned never raises its hackles.) The neck is short, but Cooper's does not appear neckless, and it also frequently cranes its neck (something Sharpies cannot do). The Cooper's Hawk tail is long, straight, and narrow, most often showing a well-rounded tip (which may have a notch but is nevertheless round) or a lobed (but still overall round) tip.

Adults are slate blue above, orange below. The darker cap on the head is more evident and more defined than on Sharp-shinned. Immatures are warm brown above, often with orangy highlights about the face and neck. Underparts are creamy, with narrow crisp streaks that let a lot of the white show through, resulting in a paler, cleaner-chested appearance and a head that appears hooded. The white terminal band on the tail of adults and immatures is wide, more crisply defined, whiter, and just plain more conspicuous than on Sharp-shinned.

BEHAVIOR: Like Sharp-shinned, hunts from a perch, by low cruising, and by soaring (or flying) at some altitude and stooping. Unlike Sharp-shinned, commonly seen away from woodlands and in open areas, where it sits on exposed perches and might easily be confused with Northern Harrier. Hunts very early, often at first light. Forages primarily on medium-sized birds—thrushes, Mourning Dove, Blue Jay—but is also adept at taking small to medium-sized mammals. (Some individual birds become chipmunk or squirrel specialists.)

Solitary, except during the breeding season. Migrates alone, never in small groups (but will join other raptors in thermals). More vocal than Sharp-shinned, both in winter and in response to intruders during breeding season. Responds readily to squeal calls.

FLIGHT: Overall silhouette is lankier and more harrier-like than Sharp-shinned. The wings are longer, narrower, and more evenly tapered, the head more projecting, and the tail longer. Shape recalls a flying crucifix or Roman cross. Flight is like Sharp-shinned—a series of wingbeats followed by a glide. Wingbeats are stiffer and more "arthritic" (clopping instead of clipping) and are often given in a longer series. Glides are heavy, steady, and distinctly more prolonged than Sharp-shinned—the bird seems to slide or slither through the air rather than glide or bounce. Turns and movements are slower, smoother, and more deliberate than Sharp-shinned, not jerky or abrupt.

Cooper's Hawk soars more than Sharp-shinned and, during good soaring conditions, flaps less. In flight, the broad white tip to the tail becomes particularly evident, as does the clean-chested dark-hooded appearance of immatures.

When studying other raptors (or turning to regard human observers), Cooper's Hawk can swivel its head around without moving its body. Sharp-shinned must turn or drop a shoulder to get a view. Also, in flight, the very white undertail coverts of Cooper's Hawk, coupled with the overall lanky shape, lead many to the false conclusion that the bird has a white rump and must therefore be a harrier.

VOCALIZATIONS: Alarm call is like Sharp-shinned: a loud strident "kacking"—"*kak, kak, kak, kak, kak*"—that is lower-pitched than Sharp-shinned. Also makes a plaintive mewing sound, "*k'ehhh,*" that recalls but is louder than a sapsucker.

Northern Goshawk, *Accipiter gentilis*

A Street Brawler of a Raptor

STATUS: Uncommon but widespread and often retiring raptor of northern and higher-altitude forests. **DISTRIBUTION:** Breeds across boreal and most of arctic Canada and Alaska; absent only in open tundra and the southern prairie provinces. In the U.S., breeds in New England, N.Y., n. N.J., Pa., w. Va., n. Mich., n. Wisc., and n. Minn., through the Rockies south into Mexico and through the Cascades and Sierras south to cen. Calif. In winter, range extends farther south, and during periodic irruptions birds may occur well south of breeding range (but generally not as far south as the Carolinas and the states bordering the Gulf Coast). **HABITAT:** Breeds in a variety of forest types, from mature willow thickets in tundra regions to old-growth forest; prefers extensive, closed-canopy, mature, mixed hardwood and coniferous forest, but also found in pure coniferous forest. In winter, prefers mature forest but also hunts on the open edge (including marsh). **COHABITANTS:** Ruffed Grouse, Gray Jay, Ruby-crowned Kinglet, Yellow-rumped Warbler, Snowshoe Hare. **MOVEMENT/MIGRATIONS:** Primarily a permanent resident, but some birds withdraw from northern parts of the breeding range every year and some expand south of it. Spring migration from February to April; fall from October into December. During periodic (every 5–10 years) irruptions caused by declines in Goshawk's prey base, large numbers of birds move south, often reaching locations south of their normal wintering range. VI: 3.

DESCRIPTION: Forest accipiter the size of a buteo. Large, elongated raptor (considerably larger than Cooper's Hawk; smaller than Common Raven; about the same size as Red-tailed Hawk), with a smallish head, a long neck, a long tubular body, and a long broad tail. Overall, fairly sinuously lined.

Adults are overall gray—darker gray above, pale gray below—with a bold black mask across the eyes. Immatures are brown with pale spots above and coarsely streaked-on creamy underparts below. The face shows an owl-like circle pattern topped off by a pale eyebrow.

BEHAVIOR: This perch-hunting raptor has little patience. Hunts by alternating short flights and periods of perch-hunting that commonly do not last longer than 10–15 mins. Generally hunts low, taking perches in the understory level. Also makes extended hunting flights along forest edges, either by gliding or by using powered flight.

Using surprise, Goshawk may glide to a kill, but accelerates quickly if detected, often crashing through brush in the process. If unsuccessful, pursues prey tenaciously both in the air and not uncommonly on foot; pursuits have been known to last up to an hour.

A solitary hunter that is generally silent except when nests are approached. Can be attracted with a squeal call.

FLIGHT: A large raptor distinguished by very long tapered wings, a tubular body, and a long broad tail—overall quite buteo-like in size and proportions. Head does not project as much as Cooper's Hawk. Wings have the S-shaped contours of a Sharp-shinned but are considerably longer and more tapered at the hand and pointed at the tip. Because of the tubular body and wide tail (which looks like an extension of the body), the bird has sometimes been described as a "flying stovepipe."

Adults look pale below. Immatures are heavily streaked below, particularly on the chest (unlike immature Cooper's Hawk). They also lack the warm orange tones about the head and hindneck characteristic of immature Cooper's Hawk. Also, the streaking on the breast often assumes a checkerboard pattern, and the brown upperparts have a slightly bluish tinge that may give distant birds a faint purple cast. Wingbeats are deep, buteo-like.

VOCALIZATIONS: Provoked, the bird makes a harsh loud "cacking"— "*kah kah kah*"—that is angrier and higher-pitched than Cooper's Hawk.

PERTINENT PARTICULARS: Although Northern Goshawk and Gyrfalcon might seem unlikely candidates for confusion, they are of similar size, shape (long tail, long-tapered wings), and plumage. They may also breed in close proximity (although not in the same habitat), and both winter in open habitat dotted with broken forest or riparian woodlands. Similarities notwithstanding, there is one fundamental difference between the two species: Goshawk is a forest bird, and Gyrfalcon is an open-space obligate. So the sight of a possible Gyrfalcon heading into the trees should suggest another possibility.

Also, immature Northern Goshawk and immature Red-shouldered Hawk are about the same size and share similar plumage and even some flight characteristics. If you are hawk-watching and see a bird approaching that you initially identify as a Red-shouldered Hawk but that turns out to be a large accipiter, think Northern Goshawk. Conversely, if you see a bird that you first call an accipiter and then discover to be, in reality, a buteo, consider the possibility of Red-shouldered Hawk.

Gray Hawk, *Asturina nitida*

Mexican Goshawk

STATUS: Scarce and geographically restricted Neotropical species. **DISTRIBUTION:** In the U.S., found only in the lower Rio Grande Valley of Tex. (resident); also (rarely) in Big Bend and extreme se. Ariz. (breeder). **HABITAT:** Riparian woodlands dominated by mature cottonwoods or cottonwood groves (even individual trees) adjacent to mesquite woodlands, which it uses for hunting. **COHABITANTS:** Black-bellied Whistling-Duck, Brown-crested Flycatcher, Varied Bunting. **MOVEMENT/MIGRATION:** In Arizona, breeding birds depart between mid-September and early October; return from late February into early April. VI: 1.

DESCRIPTION: Pale, handsome, active buteo of Mexican border regions. Medium-sized (sized and shaped like a long-tailed Broad-winged Hawk), with a fairly large head, large bill, lean body (for a buteo), and longish-appearing tail. On perched birds, wings extend less than halfway down the tail. The particularly long-tailed immature has the bulkier body of a Broad-winged and the tail of a Cooper's Hawk.

Adults are overall pale gray except for a blackish tail creased with narrow white bands. Immatures are brown above and white below, with spare but heavy contrasting blackish brown streaking/spotting running in parallel rivulets down the breast and belly. The contrast between the white face and blackish eye-line and malar stripe is particularly striking, and much more starkly defined than on other young buteos (including Broad-winged Hawk) whose facial pattern is brown on buff (not blackish on white). Some observers find it easier, however, to focus on immature Gray Hawk's bolder and whiter eyebrow.

BEHAVIOR: Agile, nimble, and high-strung both perched and in flight. Perches often and easily on utility lines (a challenge for most buteos), but also commonly sits on limbs in the forest canopy, often revealing its presence by vocalizing. Typically perch-hunts, often low (less than 20 ft.), and intercepts prey by closing swiftly with rapidly beating wings. Does not usually pursue prey if the initial rush fails. Changes perches frequently (about every 15 mins.). Also hunts by soaring.

Usually solitary. Does not commonly tolerate close approach (often calls when flushed), but may nest close to houses that are adjacent to particularly attractive cottonwoods.

FLIGHT: Buteo-like silhouette, somewhat like Broad-winged, but with blunter wingtips and a longer tail. Both adults and immatures show extensive, pale, translucent wingtips. On soaring adults, the banding on the tail is very apparent. Soars and glides on flat wings and often soars at very low altitudes, making tight, banking turns. Looks high-strung and agitated, as though letting off steam.

Flight is direct and swift, with very rapid wingbeats that accelerate into a series followed by a glide.

VOCALIZATIONS: Emits a territorial call, a three-note whistle, that resembles the cry of a peacock: "*Ah-Waahah.*" Also a high, peevish, whistled "scream"— "*keeyah*"—given by adults and immatures.

Common Black-Hawk, *Buteogallus anthracinus*

The Riverside Hawk

STATUS: Uncommon and geographically restricted Neotropical species. **DISTRIBUTION:** Northern range extends through e. and cen. Ariz. to sw. Utah and east into cen. N.M. A small, disjunct population occurs in Jeff Davis Co., Tex., and birds are occasionally seen along the Rio Grande. **HABITAT:** In the United States, a riparian obligate. Nests and forages along perennial watercourses flanked by large mature trees (most notably cottonwoods). **COHABITANTS:** Assorted amphibians, Gray Hawk, Zone-tailed Hawk, Brown-crested Flycatcher. **MOVEMENT/MIGRATION:** Most withdraw from the United States for the winter, though some individuals may remain in the lower Rio Grande Valley. Departs from breeding areas from late August to late October; returns from early April to late June. VI: 1.

DESCRIPTION: A big, black, sedentary raptor commonly seen perched along, or flying over, a stream. Large (slightly larger than a Red-tailed; about the same length as Zone-tailed Hawk), but overall broad, bulky, big-billed, long-legged, and short-tailed, with wings that do not reach the tip of the tail. There is nothing petite or lithe about this couch potato of a bird.

Adults are overall blackish, with a hint of brown, and somewhat shaggy or loosely feathered, showing a conspicuously bright yellow cere and a single wide white band bisecting the tail. Some adults also show a white trim at the tip of the tail (Zone-tailed Hawk does not), but this mark abrades down as feathers wear.

Immatures are blackish brown above and heavily and richly patterned, particularly about the neck and face. The pale face is offset by a bold dark line through the eye and tawny streaking on the neck. Underparts are a rich tawny color overlaid with lavish streaking. Focus on the neck and tail: the sides of the neck boast large muttonchop sideburns, and the tail is a series of white and black bands ending in a wide dark terminal band.

BEHAVIOR: Fairly sedentary, Common Black-Hawk is almost invariably perched on a stout limb, log, or rock, over the water, passively waiting for prey. Once it sights a fish, frog, snake, or crayfish, it glides to intercept or may first fly closer. Also uses its long legs to advantage when walking through streams, sometimes using its wingtips to stir the water. When not hunting, commonly sits on a branch just below the upper reaches of the canopy. During courtship, adults may soar well above the trees.

Not social. Pairs space themselves out along watercourses. Generally shy, easily flushed.

FLIGHT: Very distinctive flat-winged flight silhouette showing exceedingly wide wings that reach forward and curve roundly on the trailing edge. All in all, most similar to Black Vulture. The head seems emaciated, thin. The short tail, with a single broad white band, is almost enveloped by the wing. (In fact, adults seem all-wing.) Immatures, only slightly less bulky, have a tail pattern that shows something of a black-and-white checkerboard pattern. Immatures also show a pale white patch near the tip of each wing (like immature Golden Eagle).

Wingbeats are slow and heavy. Movements in the air are stiff and somewhat unsteady, although courting birds can show uncommon aerial finesse.

VOCALIZATIONS: Call is an excited series of whistles that increase in volume, pitch, and tempo before trailing off.

Harris's Hawk, *Parabuteo unicinctus*

The Brush Hawk

STATUS AND DISTRIBUTION: Common to fairly common resident in s. Ariz., extreme s. N.M., along the Rio Grande Valley of trans-Pecos Tex., s. Tex., and in a narrow band across cen. Tex. to the southeast corner of N.M. **HABITAT:** Loves brush, particularly mesquite and desert scrub, but also found in more open desert habitats, savannah and coastal prairie, and marshes. A requisite component of all habitats is elevated perches (bushes, trees, saguaro cactus, utility poles) for hunting and nesting. **COHABITANTS:** Northern Bobwhite, Gambel's Quail, Scaled Quail, Ladder-backed Woodpecker, Ash-throated Flycatcher, Cactus Wren, Northern Mockingbird, rabbits. **MOVEMENT/MIGRATION:** Permanent resident. VI: 2.

DESCRIPTION: Slender, chocolate-colored, highly social, commonly roadside raptor. Medium-sized (larger than Northern Harrier and Cooper's Hawk; smaller than Red-tailed), with a roundish head, a large bill, a slender athletic body that seems to be going slightly soft, paunchy, conspicuously long legs, and a very long tail that extends well beyond the wings, which are so short that they barely conceal the bird's white uppertail. If the bird were to compete in an Olympics event, it would excel in the decathlon. May stand erect, with its tail plastered against the perch (very commonly a utility pole), or angle forward so that its shoulders seem to slouch and its body to sag. Perched birds are overall dark brown, showing a paler slash (which in good light shows chestnut) along the shoulder. The white undertail is usually concealed, and the white tip of the tail is difficult to see from the side (otherwise it is apparent). The bright yellow cere is usually very conspicuous, and sometimes the yellow legs and feet as well. Young birds show a darkish bib and heavily streaked underparts but are otherwise much like adults.

BEHAVIOR: Most often seen perched along roadsides, often on or near brush, or soaring over open or brushy country. If perched, look around. Where there is one Harris's Hawk, there are probably more. This species is highly social year-round. Birds roost, hunt, even nest cooperatively.

Hunts from high perches such as bushes, cacti, and trees, almost always from the highest point of the perch, moving its head to search the area below. Extremely nimble, Harris's Hawk can perch on springy branches or telephone lines and make vertical, spiraling, headfirst stoops onto prey from such perches. (Red-taileds are talon-firsters.) A tenacious hunter, Harris's runs and hops in pursuit of prey and throws itself into brush to flush or grab a meal.

Fairly tame. Often allows close approach, and when flushed, usually does not fly far—sometimes no farther than the next telephone pole.

FLIGHT: A compact, short, blunt, and paddle-winged buteo-like bird with wings that seem slightly pinched at the base. At a distance, appears all dark except for the white base of the fanned tail (visible above and below), which shows as a band (not a wedge, as with White-tailed Hawk). High overhead, the bird appears no-headed—all wing and tail—and the white band becomes invisible, making the dark tail seem separate from the all-dark bird. Immatures usually show crescent windows at the wingtips (like Red-shouldered Hawk and Hook-billed Kite). Normal flight—for example, when shifting perches—seems casual and buoyant, an unpatterned combination of slow, shallow, almost desultory flaps and slow lengthy glides. When gliding, soaring, or flapping, wings are slightly cupped or drooped. When pursuing prey, flight is rapid, reckless, and highly maneuverable. Also hovers.

VOCALIZATIONS: Seldom vocal. Call is a hoarse, harsh, nasal scream: "*Kraaaaaah.*"

Red-shouldered Hawk, *Buteo lineatus*

Crescent-winged Hawk

STATUS: Common resident in the South and in western California and southwest Oregon; uncommon northeastern breeder. **DISTRIBUTION:** Breeds in se. Canada from s. Ont. to s. Nfld. and in all the eastern states. Western border defined by cen. Minn., e. Iowa, e. Kans., cen. Okla., and e.-cen. Tex. to about Brownsville. In winter, vacates Canada, interior New England, all but s. N.Y., most of w. Pa., n. Mich., n. Wisc., and Minn. and extends its range into the e. Rio Grande Valley and ne. Mexico. In the West, a permanent resident from coastal sw. Ore. south through coastal Calif. and the Central Valley to n. Baja. Annual in small numbers in fall and winter in w. Nev. and Ariz. **HABITAT:** In the East, a forest raptor partial to large mature deciduous forest, particularly swamplands, but also found on wooded hillsides (often near water), suburban wooded lots, and in mixed deciduous and coniferous forest. In the West, found along riparian corridors and in oak woodlands, eucalyptus groves, and suburban and some urban areas with ample shade trees. **COHABITANTS:** Broad-winged Hawk, Barred Owl, Pileated Woodpecker; in the West, Nuttall's Woodpecker, Western Scrub-Jay, Oak Titmouse. **MOVEMENT/MIGRATION:** Partially migratory in the East. Spring migration from late February to early April; fall from mid-October to late December. VI: 2.

DESCRIPTION: Strikingly colorful (adult), fairly slender forest buteo with a smallish bill and a fairly long broad tail. Medium-large-sized (slightly larger than Broad-winged Hawk; smaller than Red-tailed), with a small bill and a tail that extends beyond the tips of the folded wings. On adults, eyes are dark (pale on immatures). The expression is somber.

Adults are strikingly patterned above—black with lavish white spotting and barring and a prominent rusty slash on the shoulder that is evident when the bird is perched. Head and underparts are rusty orange; tail is dark (blackish above; gray below), with several prominent narrow white bands that are less than half the width of the dark bands. The head pattern is distinctly plain, almost featureless. *Note:* Similarly patterned Florida birds are considerably paler, washed-out versions of most eastern birds. California birds are more richly colored.

Immature eastern birds are brown above, with pale spotting, and creamy below, with heavy, fairly even streaking; unlike Red-tailed and many immature Broad-winged Hawks, the breast is conspicuously streaked. Immature western Red-shouldereds are more akin to adults but coarsely patterned—a paint-by-number rendering of an adult.

BEHAVIOR: A perch-hunting raptor that usually perches below the canopy, on a sturdy limb, and often at the edge of a pond, stream, or swamp. Also hunts from perches along roadsides (fenceposts and utility poles) and forest edges, where it prefers to sit about midheight. Glides to capture prey or uses powered flight like an accipiter (particularly when attempting to capture birds). In winter, also haunts feeders.

Fairly tame in some places, such as suburban California and Florida, but shy in many other places. Flushes easily. Very vocal in late winter, spring, and summer and when nests are approached. At other times, moderately vocal.

In migration, does not flock (as Broad-winged does). On ridges, often flies higher than Red-tailed Hawk and on the less turbulent off-wind side.

FLIGHT: Adults are overall compact and clean-lined, with short paddlelike wings and a fairly short tail. Immatures, with longer planklike wings and longer tails, are lankier and more accipiter-like. Like adults, the lines on immatures are trim, lacking bumps or bulges. Red-shouldered glides with wings bowed down and soars with wings turned forward, as if reaching out to embrace something.

Adult Red-shouldered seems fairly dark overall; immatures are tawny, with a warmer cast than Red-tailed. In flight, the most obvious characteristic is the narrow crescent-shaped "window" near the tip of the wing. White (adult) or tawny (immature), it is visible above and below and evident at distances where things like "narrow white tail bands" on adults are invisible.

A more active and high-strung flier than Red-tailed Hawk, Red-shouldered often flaps in short quick series, even when gliding between thermals. Wingbeats are somewhat loose and floppy—not heavy like Red-tailed, and not choppy like Broad-winged.

VOCALIZATIONS: A series of loud, ringing, strident, two-noted screams: "*keeya, keeya, keeya.*"

Broad-winged Hawk, *Buteo platypterus*

Swarm Buzzard

STATUS: Common, primarily eastern breeder; at times and places, an abundant migrant. **DISTRIBUTION:** Breeds across most of the U.S. east of the prairies. The northern limit of its range extends from the Maritimes and e. Que. west across cen. Que., Ont., cen. Man., and cen. Sask. into cen. Alta., ne. B.C., and extreme se. Yukon (absent in s. Sask.). In the U.S., breeds east of a line cutting across w. Minn., w. Iowa, ne. Neb., e. Okla., and e. Tex.; also breeds in the Black Hills of S.D. and Wyo. Absent as a breeder only in coastal and s. Tex. and coastal La. and on the Florida Peninsula. Winters in Central and South America and s. Fla. Migrates primarily east of the Great Plains, but small to moderate numbers turn up at western hawk watches (for example, Goshutes, Grand Canyon, Golden Gate) every fall. **HABITAT:** Nests in large tracts of coniferous, mixed, and deciduous forest, often near water (usually younger forest than that favored by Red-shouldered). In migration, may be found in almost any woodland situation, including suburban neighborhoods. **COHABITANTS:** Cooper's Hawk, Ruffed Grouse, chipmunks. **MOVEMENT/MIGRATION:** A celebrated long-distance migrant. All birds vacate breeding areas after nesting, relocating primarily to the tropics, but a small number of mostly immatures winter in southern Florida, southern Texas, and, very rarely, California. Spring migration from late February to early July, with peak from mid-March to late April; fall from mid-August to late November, with peaks in early September and early October. VI: 2.

DESCRIPTION: Small, compact, perch-hunting forest buteo showing a broad white stripe through its tail. Small (slightly smaller than Red-shouldered Hawk; larger than Sharp-shinned Hawk; about the same size as, but bulkier than, Cooper's Hawk) and overall stocky, this barrel-chested buteo has a rounded head, a petite bill, and a short, narrow, almost accipiter-like tail.

Adults are cold brown above, ruddy-breasted with broken wavy bars below. Tail on, perched birds show a single broad white band. Immatures are brown above, with some pale spotting on the wings, and heavily but variably streaked below, with most individuals showing heavy, dribbled streaking down the sides of the throat and breast and on the belly, often leaving the breast wholly to relatively unstreaked. (Immature Red-shouldered, by comparison, is most heavily streaked on the breast.) The dingy tail is creased by many narrow dark bands and one distinctly wider and darker subterminal band.

Also has a rare dark morph that is most often seen in western portions of the bird's range. For a hawk, the brown-eyed expression is gentle.

BEHAVIOR: The consummate perch-hunting raptor. Commonly sits on an exposed perch below the canopy near forest clearings and edges and often near water. In migration and winter (in Florida), also perches on utility lines and poles. Hunts primarily for small mammals, reptiles and amphibians, and birds. In migration, also eats large insects.

Not particularly shy, and immature birds often allow close approach. In migration, commonly gathers in large flocks (called "kettles") whose swirling clusters within thermals recall water roiling in a pot. Celebrated concentrations numbering in the tens of thousands are most commonly associated with water barriers (like the Great Lakes). Prime viewing locations include Duluth, Minnesota; Detroit, Michigan; Corpus Christi, Texas; Vera Cruz, Mexico; and Panama in the fall; Santa Ana National Wildlife Refuge, Bentsen–Rio Grande Valley State Park (Texas), and Rochester, New York, in the spring.

FLIGHT: Silhouette is overall stubby. Wings are short and broad; shaped like a wide candle flame when soaring, and like a paring knife when gliding (curved along the leading edge, straight-cut along the trailing edge). In a glide, the tail may seem exceedingly narrow and accipiter-like. When soaring, wings are held straight out to the sides; when gliding, angles or bows wings slightly down.

Underwings of adults and immatures are whitish and relatively unmarked. Black borders make them look like stretched canvas surrounded by a black frame. Tails of adults commonly show a single broad white band (and less commonly, a second narrower broken band). Tails of immatures are mostly uniform grayish buff, but show a single dark subterminal band.

Wingbeats are quick and choppy. During migration, alternates periods of soaring with long glides. Early in the morning and late in the evening, when thermal lift is scant, birds may interspace their glides with bouts of flapping.

VOCALIZATIONS: Call is a loud, plaintive (or peevish), two-note whistle, with the second note higher and drawn out: "*pee-heeeee.*"

Short-tailed Hawk, *Buteo brachyurus*

Gravity's Master

STATUS: Scarce (fewer than 200 pairs in the United States) and geographically restricted tropical and subtropical species. **DISTRIBUTION:** In the U.S., breeds only on the Florida Peninsula and, to a very limited extent, in the Chiricahua and Huachuca Mts. of se. Ariz. In winter, Fla. birds retreat to the Everglades and Florida Keys. **HABITAT:** In Florida, nests in large closed-canopy forest and forages over habitats that offer a mix of woodlands and open space (savanna, marsh). In Arizona, nests in pine forests at higher elevations. **COHABITANTS:** Often hunts the same habitat as Red-tailed Hawk. **MOVEMENT/MIGRATION:** Short-distance migrant, flying to southern Florida and the Keys in October and early November, returning to breeding areas in February. VI: 1.

DESCRIPTION: Long-winged, Broad-winged Hawk–like buteo that flies and stoops with precise stop-and-go control. Small (about the same size as Broad-winged Hawk) and sturdy, Short-tailed perches in heavy cover and is rarely seen except aloft. Wingtips of perched adults reach the tip of the tail. (The tail of Broad-winged Hawk is longer than the wings.)

Has both a light and dark morph. Adult and immature light morphs are blackish brown above, all bright white below. A dark head gives birds a hooded appearance (like Red-tailed Hawk).

Adults show a chestnut patch on the side of the neck. Dark-morph adults are all blackish, with a bright yellow base to the bill. Immatures are all blackish brown above, with a dark chest and white spotting on the belly. *Note:* In Florida, most birds are dark morph.

BEHAVIOR: This is a consummate aerial predator—a buteo that specializes in plucking avian prey from trees and shrubs. Patrols the ecotone between woodlands and open habitat, often working the same area several times. Soars quickly aloft, turning tight circles in thermals or updrafts. Generally hunts from greater heights than other buteos (most notably Red-tailed). Holds into the wind, on set wings, remaining motionless in the air for extended periods, then advances forward (more than glides) with precise control. Does not hover. When prey is sighted, Short-tailed executes near-vertical stoops with reckless speed, sometimes checking its plunge en route, holding motionless, then dropping again.

Not social. Hunts alone or in pairs. The best time to find this species is midmorning, when birds are catching the first weak thermals of the day.

FLIGHT: Somewhat Broad-winged Hawk–like in overall shape, but wings are broader, less tapered, and blunter at the tip. The white underparts and underwing linings of light-morph birds contrast with dark gray flight feathers. Dark morph shows pale patches near the base of the outer flight feathers and a wide, dark band at the tip of a grayish tail. Soaring birds show a distinctive—and virtually diagnostic—flat-winged profile with the wingtips turned up. Wingbeats are hurried.

VOCALIZATIONS: Described as a high-pitched squeal that drops off; recalls a harsh Broad-winged Hawk or a sound somewhere between the piped whistle of a Broad-winged and the scream of a Red-tailed.

PERTINENT PARTICULARS: In Florida, the only other raptor that "kites" is Red-tailed Hawk—a bird whose white underparts fall short of pure white. Also, in the absence of other dark-morph buteos (except, rarely, the odd dark Swainson's Hawk), any dark-morph buteo seen in Florida is almost certainly Short-tailed Hawk.

Swainson's Hawk, *Buteo swainsoni*

The Harrier Hawk

STATUS: Common and widespread western breeder. Rare but regular vagrant in the East. **DISTRIBUTION:** Primarily a bird of the prairies and intermountain West. Breeds from s. B.C., cen. Alta., and s. and cen. Sask. south to n. Mexico. Absent as a breeder in e. Wash., Ore., most of Calif. (except parts of the Central Valley and the northeastern deserts), and w. Ariz. Eastern border cuts across w. and s. Minn., n. Iowa, e.-cen. Neb., cen. Kans., cen. Okla., and e.-cen. Tex. Disjunct populations also found in sw. Mo., n. Ill., and nw. B.C. **HABITAT:** Open grasslands and agricultural lands. For nesting, requires one bush or tree large enough to bear the weight of a flimsy stick nest (3–20 ft. high). **COHABITANTS:** Ferruginous Hawk, Western Kingbird, Western Meadowlark. **MOVEMENT/MIGRATION:** Almost the entire population engages in a spectacular mass exodus to and from Argentina. (A few birds winter in California and Florida.) Spring migration from late February to mid-April; fall from late August to early November. Strays regularly reach the East Coast and Florida between September and November. VI: 3.

DESCRIPTION: Slender long-winged prairie buteo that may be even more common—in summer, and in certain places—than the ubiquitous Red-tailed Hawk. Fairly large (smaller than Ferruginous Hawk; about the same size as Red-tailed), but overall longer and slimmer; its elongated teardrop shape is less chesty and robust than Red-tailed or Ferruginous Hawk. The head is round, the bill substantial (but not as dominating as Ferruginous Hawk), the body long and slender, and the tail long. Wingtips extend to (and sometimes beyond) the tip of the tail.

Shows light, dark, and intermediate morphs. Adult light morph (dominant plumage) is dapper, with unpatterned dark brown above and white below, a dark (gray or chestnut) bib, a white throat and white face, and a gray tail (which is often hidden by the wings—but the white face stands out like a headlight). Dark-morph adults have uniformly dark cinnamon-brown bodies and gray tails. Immature light and dark birds are highly variable, with mottled or dappled upperparts and streaked or dark splotchy underparts; the heaviest streaking is on the sides of the neck and the chest (a mock bib recalling adults). Second-year birds (returning from Argentina) are even more variable and mottled, and many appear white-headed. Immature birds also have a grayish cast to the tail.

BEHAVIOR: A versatile hunter, adept at soaring and stooping, perch-hunting, and foraging for prey on the ground—one of the few hawks to do so. Often stands and waits near gopher or prairie dog burrows or walks and runs through grass, pouncing on or plucking grasshoppers with their bill. Also attends farm equipment. When aerial hunting, commonly soars overhead or cruises low and slow (harrier-like). Despite its affinity for ground foraging, prefers an elevated perch. (Ferruginous Hawk more readily sits on some elevated patch of ground.)

At times, highly social. Migrates in great flocks that may number in the thousands and that often cluster, in their passage, in great towering tornadoes of birds (just as Broad-winged Hawks do). Nonbreeders as well as migrants often hunt communally in prey-rich fields. Fairly tolerant of humans.

FLIGHT: Overall slender with very long, slender, pointed wings that recall tapered candlesticks. In combination with the bird's fairly long tail, distinct dihedral, and fairly substantial white base to the tail, Swainson's Hawk is easily confused with Northern Harrier. (In fact, you would do well to think of Swainson's as the buteo that looks like a harrier.)

In all plumages, the dark trailing edge of the underwing (flight feathers) distinguishes the species from all but a few other hawks. Use caution in coastal Texas, where White-tailed Hawk occurs. In addition, be mindful that Osprey and, to a degree, Northern Harrier also show a dark, or partially dark, trailing edge.

Swainson's flight is nimble, buoyant, somewhat tipsy. When cruising, wingbeats are fairly slow, stiff, and deep. Soars with wings lifted in a stiff dihedral; glides with wings crooked in a shallow, somewhat gull-like configuration.

VOCALIZATIONS: A high thin scream that falls off at the end and recalls a weak Red-tailed Hawk: "*kreeee.*"

White-tailed Hawk, *Buteo albicaudatus*

Dances with Fires

STATUS: Common but geographically restricted Texas resident. **DISTRIBUTION:** Found from Tex. to Argentina; in the U.S., occurs only along the Tex. coast and Rio Grande Valley from Houston west almost to Laredo. Greatest concentrations occur in the Coastal Bend. **HABITAT:** Open and semi-open native grasslands; most U.S. birds breed in the coastal prairies and marshes. Avoids cultivated fields but is tolerant of pastures. **COHABITANTS:** White-tailed Kite, Aplomado Falcon, Crested Caracara, Eastern Meadowlark; in winter, Northern Harrier, Red-tailed Hawk. **MOVEMENT/MIGRATION:** Permanent resident. VI: 1.

DESCRIPTION: Stunning large buteo (slightly larger than Red-tailed Hawk; slightly smaller than Ferruginous Hawk). Large-headed, heavy-bodied, and long-legged. Most importantly from the standpoint of identification, the wings extend well beyond the short tail on perched birds. On other buteos, wingtips fall short of the tail (or meet the tip of the tail, as with Swainson's Hawk).

Adults appear all dark above, bright white below. (Only the rusty-backed Ferruginous Hawk shows as much white below.) The russet shoulders and wide dark band on the white tail should end any confusion. Immatures are all blackish brown, with a white patch on the breast and white mottling about the face. Subadults look like immatures from below (dark underparts, white breast), but much like adults from above (all dark, with cold gray upperparts except for a rufous slash along the shoulder). The tail on immatures and subadults is gray or brownish gray.

Descriptions relating to perched birds, however, are almost superfluous. Chances are you will see White-tailed Hawk in the air, where it seems to live.

BEHAVIOR: A bird tailor-made for a place as open and windy as the Texas coast. Commonly perch-hunts in the early morning hours (favorite perches include utility poles, trees, low shrubs, and utility lines); goes aloft as thermals perk and winds rise. Often flies very high during the day (so goes undiscovered), but hunts over open grasslands at lower altitudes, commonly under 100 yds. Searches for prey by hovering or kiting into the wind.

Attends grass fires the way moths seek candle flames, patrolling the advancing edge of the flames for fleeing prey. Twenty to thirty (or more) individuals may gather at a controlled or natural burn (accompanied by Swainson's Hawks, caracara, and Aplomado Falcon), and birds may continue to work a burned area a day or more after the fire. Except at times of fire, adults are found in pairs, but adults and young commonly remain together until January or February. During breeding season, nonbreeding subadults sometimes gather in small groups.

FLIGHT: Shape and plumage of adults is unique. Overall large, this bird has very broad wings that taper to fine points and combine with an excruciatingly short tail to craft a silhouette that seems all wing. (If you are familiar with the Bateleur of East Africa, then you are familiar with the contours of adult White-tailed Hawk.) The gleaming white body and white underwing lining contrast with the all-black trailing edge of the wings from below. All-dark upperparts cause the white rump and tail to gleam at almost any distance.

Immatures and subadults are more difficult. Considerably thinner-winged and longer-tailed than adults, they more closely approximate the flight profile of Swainson's or Ferruginous Hawk. But the white chest patch and white patch on the rump hearken to the adult pattern.

At any age, White-taileds glide and soar with wings held in an acute dihedral. They hover and remain motionless with a finesse that perhaps only Short-tailed Hawk can duplicate. When descending for prey, White-tailed may plummet or corkscrew and commonly checks its descent, holding motionless into the wind, before executing a final drop. Flight is buoyant and nimble, and wingbeats are typically slow and shallow.

VOCALIZATIONS: Call is a high harsh laughing series with the introductory note more protracted, distinctly different from the other, more hurried notes in the series, "*erEHH herEh herEh herEh.*"

Zone-tailed Hawk, *Buteo albonotatus*

Hawk in Turkey Vulture's Clothing

STATUS: Fairly common but geographically restricted southwestern breeder. A Neotropical species. **DISTRIBUTION:** The northern range extends to Ariz., N.M., and sw. Tex. Winters in Mexico, but a few occur every winter in s. Calif., s. Ariz., and s. Tex. **HABITAT:** Most commonly seen in riparian corridors, higher desert areas, and open coniferous (primarily ponderosa pine) forest at higher elevations. **COHABITANTS:** Turkey Vulture, Common Black-Hawk. **MOVEMENT/MIGRATION:** In the United States, birds depart breeding areas in late September and early October; return from mid-February to mid-April. VI: 2.

DESCRIPTION: Lean, athletic-looking buteo that looks and flies like a Turkey Vulture. Medium-large (smaller than Turkey Vulture or Common Black-Hawk; larger than Gray Hawk; about the same size as but more slender than Red-tailed Hawk), with a large round head, a proportionately large bill, a slender body, long and bright yellow legs, and long wings and tail. Common Black-Hawk is flatter-headed, larger-billed, overall stockier and more robust, longer-legged, and shorter-tailed. The bill of Zone-tailed is pale or yellowish at the base; bill of adult Common Black-Hawk is more extensively and conspicuously yellow. If you're in doubt, it's not Black-Hawk.

Overall grayish black (adult Common Black-Hawk is blacker and slightly brown-tinged), with a tail divided by one broad and one narrow white band (below); shows gray from above. *Note:* Only the broad white band usually shows when birds are perched or flying. Immatures are slightly browner-tinged and may show slight spare white speckling on underparts; otherwise like adults.

BEHAVIOR: In New Mexico and Texas, commonly nests at higher elevations in pine forests; in Arizona (and also New Mexico), often nests along riparian corridors (as does Common Black-Hawk) but hunts at higher elevations. Often perch-hunts early in the morning before vultures go aloft (again, like Common Black-Hawk), but thereafter hunts by cruising along ridgelines or soaring overhead, often with Turkey Vultures. Stoops on prey may be swift and dramatic, or may move away from intended prey and approach from behind some concealing obstruction.

Solitary as a rule, although breeding birds and adults and recently fledged young are seen together. Commonly associates only with Turkey Vulture, which plays the foil for this hawk in vulture's clothing.

FLIGHT: In shape (long, planklike, uplifted wings; long broad tail), plumage (all blackish with silvery flight feathers and a single, broad white band on the tail right where a vulture's excrement-covered feet would fall), and manner of flight (soars and banks with a wobbling, vulturelike unsteadiness), Zone-tailed Hawk resembles Turkey Vulture. There are, however, some key differences. Zone-tailed is overall slimmer and more symmetrically angular and has a wider, darker (and feathered) head that is very obvious when you look for it. Also, the lightly banded silvery flight feathers are not quite as contrastingly pale as on Turkey Vulture, and the white band on the tail is more conspicuous than the white feet of a vulture. *Note:* On some birds, a second narrower tail band may also be evident.

Adult Common Black-Hawk doesn't remotely resemble Zone-tailed (or Turkey Vulture) in flight: it's flat-winged and overall bulky, almost hulking, and looks like a short-tailed flying barn door (or a Black Vulture).

VOCALIZATIONS: Described as a squealing whistle and a high scream.

PERTINENT PARTICULARS: The trick to finding Zone-tailed Hawk is very simply to study each and every Turkey Vulture. Too often, observers become as indifferent to vultures as the Zone-tailed's prey. Zone-taileds go unseen and unidentified because observers don't look, or they look too casually.

Birds play a higher-stakes game than birders, and despite its accomplished mimicry, Zone-tailed is only able to fool some of the birds, some of the time. So if you see a pair of Western Kingbirds pulling feathers from the back of a presumed Turkey Vulture soaring overhead, give the presumed vulture the benefit of a second look. You'll probably be surprised.

Red-tailed Hawk, *Buteo jamaicensis*

The Real Roadside Hawk

STATUS: Common northern breeder and widespread resident species—across most of North America, the default buteo. **DISTRIBUTION:** Breeds across all of North America except n. Alaska and Canada north of the tree line. **HABITAT:** Extremely varied, but prefers open to fairly open country studded with elevated perches. Habitats include rangeland, farmland, grassland, mature forest with open understory, prairie, desert, urban parks, tidal marsh, and, especially, roadsides and interstates. **COHABITANTS:** Also varies, but if the habitat supports Great Horned Owl or American Kestrel, it probably supports Red-tailed. **MOVEMENT/MIGRATION:** Birds in Canada, Alaska, the northern prairies, and the northeastern United States (northern New England and New York State) vacate breeding areas. Spring migration from late February to early June, with peak in March and early April; fall from late July to early January, with peak in late October and early November. VI: 4.

DESCRIPTION: A big, stocky, broad-shouldered, short-tailed, dark-headed, rusty-tailed, and (usually) white-chested buteo perched at the edge of a field or along an interstate. Large hawk (much smaller than Golden Eagle; slightly larger than Red-shouldered Hawk; about the same size as Swainson's Hawk), with a fairly small, somewhat flattish head, a large bill, a barrel-like body, and a short wide tail. Overall stocky; the shape (particularly in adults) is somewhat elliptical and recalls a rugby ball (or a wide blunted football). *Note:* Immature Red-taileds share all these traits but are frequently rangier-looking and longer-tailed.

Red-shouldered Hawk is overall more slender than Red-tailed, with a round head, smaller bill, and longer tail. Rough-legged Hawk is longer-necked and has a pale head (except some dark morphs) and a petite bill. Ferruginous Hawk has a big flat head and a massive bill (head recalls a doorstop). Perched Swainson's Hawk is teardrop-shaped: broad at the shoulder, narrowing toward the tail.

Red-tailed plumage is variable, from all-black birds to birds showing rufous underparts to birds showing brown upperparts and white underparts that may be creased by a ragged, black "belly" band. No matter what the plumage, all adult Red-taileds have a reddish tail—ranging from a rosy-tinged blush on the tail of the pale prairie subspecies to a deep, rich rust or orange. The tails of immatures are brown (or blackish on young dark morphs) and narrowly banded. Also, the whitish scapulars on folded wings etch a pale V across the bird's back. This mark is not unique to Red-tailed (immature Red-shouldered shows a similar pattern), but it is very apparent.

East of the prairies, dark- and rufous-morph birds are rare. Eastern Red-taileds show a dark hooded head, an unmarked white breast, and (commonly and classically) a broad, black, coarsely marked belly band that divides the underparts in half. If the belly band is faint or lacking (as is often the case with some eastern adults and birds of the prairies and Southwest), then the white underparts are accentuated. The breasts of most other buteos are colored, patterned, or streaked. Exceptions are Short-tailed Hawk (Florida and Arizona), White-tailed Hawk (coastal Texas), and Ferruginous Hawk (western prairies).

Keep in mind that if it's a buteo and it's got an unmarked white (or rose-blushed) chest, chances are it's a Red-tailed Hawk.

In the West, dark-plumaged birds are common. Rufous-morph Red-tailed still shows a dark belly band. (Rufous Swainson's Hawk has a dark chest.) Wholly dark (blackish) immature Red-tailed is distinguished from Harlan's Hawk by the virtual absence of white spotting (particularly on the breast); from black Rough-legged by Rough-legged's more petite bill; and from dark Ferruginous by head shape.

BEHAVIOR: A conspicuous perch-hunting raptor that prefers to sit in the open on a high, sturdy perch—favorites include utility poles, cliff ledges, roadside fenceposts, highway billboards, and large (often dead) trees with sturdy limbs. Sits for long periods, surveying the ground, and often uses the same favorite perches day after day. Does not commonly perch on the ground (like Ferruginous Hawk) or on wires (like Broad-winged Hawk and Red-shouldered Hawk), except for very wide-gauge cables. Less inclined to sit on springy perches than Rough-legged Hawk.

Where birds remain on territory all year, mated pairs often sit together year-round, particularly from January on. In open woodlands, may hunt squirrels cooperatively. In open country (without updrafts), prefers to fly after thermals start perking (early midmorning) and spends long periods of time in the air. Generally shy; flushes if approached, or even if a parked vehicle remains too long. Often calls before leaving its perch or after taking flight, expressing, perhaps, irritation.

FLIGHT: Overall stocky, broad, and robust, with wide blunt-tipped wings showing somewhat muscular or bulging contours (bulging alula, bulging secondaries). Adults are very short-tailed and overall compact; immatures are longer-tailed and overall lankier. The red tail is very conspicuous on adults (shows pink from below). On immatures, the upperwing is two-toned pale on the outer half, slightly darker on the inner half. On light-morph birds, the white chest stands out at a considerable distance.

On soaring and gliding birds, wings are usually held in a smooth gently upsweeping U, with no abrupt jut at the shoulder and no sharp-angled V. Flight is steady, with little rocking, and wingbeat series are infrequent. Wingbeats are slow and heavy, but not ponderous or labored. When hunting, Red-tailed commonly faces into the wind, draws in its wings, and holds itself motionless and suspended in the air — a near-diagnostic aerial feat called kiting.

Among North American buteos, only Short-tailed Hawk and White-tailed Hawk habitually kite. The more buoyant Rough-legged Hawk and Swainson's Hawk kite sparingly and much prefer to hover (something Red-tailed does grudgingly).

VOCALIZATIONS: Call is a shrill, breathy, raspy, descending, compressed two-note whistle, "*rEE-ehhhhr.*" Sometimes the call is short and loud, and sometimes it's prolonged and slightly gurgled at the end.

PERTINENT PARTICULARS: Over most of North America, this is the default buteo — the first name you should try applying to a large, open-country raptor. If holding up the Red-tailed template to the bird in question produces no comfortable match, then consider other regionally tailored possibilities.

This species is habitually mobbed by crows, even though Red-taileds over most of their range target small mammals. Gulls and some waterfowl also lift off if a Red-tailed Hawk is soaring in menacing proximity.

Harlan's Hawk, *Buteo jamaicensis harlani**
A Red-tail of a Different Color

STATUS: Common but somewhat geographically restricted. **DISTRIBUTION:** Breeds in interior portions of s. and cen. Alaska, s. and cen. Yukon, se. Northwest Territories, and n. B.C. Winters primarily in the s. Great Plains, including se. Neb., w. Mo., e. and cen. Kans., w. Ark., and ne. Tex., but regularly occurs in other southwestern states, as well as e. Wash., s. Idaho, and n. Utah. **HABITAT:** Breeds in broken or open spruce forest or where dense forest abuts marshes or muskegs. Winters in open prairie, agricultural land, wetlands. **COHABITANTS:** In breeding season, Gray Jay, Boreal Chickadee, Bohemian Waxwing. In winter, often found in the same habitat as Rough-legged Hawk and Red-tailed Hawk. **MOVEMENT/MIGRATION:** Spring migration from late March to early May; fall from late August to early November. VI: 2.

DESCRIPTION: Blackish or starkly black-and-white northern buteo, with a whitish or grayish (not red) and variously mottled tail. Large, stocky buteo; structurally similar to Red-tailed Hawk, but perhaps, on average, slightly shorter-tailed.

Shows two color morphs, dark and light, with dark birds dominating; light birds, however, are probably not as uncommon as believed. Dark adults are overall cold black except for conspicuous whitish streaking or spotting on the chest and white on the face. Light adults are blackish above, with considerable white in the face and a mottled or marbled nape, and starkly white below, with a crisply marked black belly band. There is no brown in the plumage, and no pinkish wash on the breast. In both morphs, the tail is mostly gray or whitish, with a dark-banded or rufous-washed tip. Some tails are almost wholly white; some show longitudinal whitish streaks; and some show a pattern of narrow dark bands. In all birds, however, the tail is whitish, not bright red.

Immatures are mostly dark and overall colder and blacker-toned than immature Red-taileds, with lavish amounts of white spotting above and conspicuous white streaking below that is heaviest on the chest. The entire head may be mostly white (light morph), or the white may be limited to the face (dark morph). Tails show a pattern of dark zigzag or chevron-shaped bands; immature Red-tailed tail bands may be slightly wavy but not chevron-shaped.

BEHAVIOR: Very much an aerial predator; spends lots of time cruising overhead, often at great altitudes (much like Short-tailed Hawk in this regard). Also perch-hunts, often from utility poles, but also from fairly springy tops of trees — perches generally shunned by Red-tailed, particularly when sturdier perches can be found. Harlan's is easy to spook and less tolerant of approach than the average Red-tailed, so tends to flush at distances that most Red-taileds would find acceptable.

Usually solitary or in pairs, but in winter found in the same habitat as Red-tailed and Rough-legged Hawks.

*Harlan's Hawk is currently considered a subspecies of Red-tailed Hawk.

FLIGHT: Silhouette is similar to Red-tailed but overall slightly cleaner-lined, longer-headed, and straighter-edged along the leading edge of the wing. Adults are stocky and compact, with short broad wings (might recall an adult Red-shouldered). Immatures are rangier, more like Rough-legged. Glides with wings curved along the leading edge and mostly straight-cut along the trailing edge; Red-tailed's wing is more angled back. When gliding, wings are uplifted in a modest dihedral that sometimes shows more upward jut at the shoulder than Red-tailed.

On dark birds, pure white flight feathers contrast with the black body; the whitish (pale) tail is easily noted, but the white splotch on the chest may be hard to see at a distance. On light birds, underparts are bright white; the dark patterning is crisp, almost stark, in contrast. Immatures are like adults but more mottled with white, and they show dark (not whitish) tails. Like immature Red-tailed, the upperwings of immature Harlan's are two-toned, showing a paler hand and darker arm; while hard to see, the feathers on the underside of the wingtips are banded, not all black as on Red-tailed.

Very active, agile flier that spends hours aloft. Commonly hunts very high, covering lots of ground; in summer, often hunts over broken forest. Sometimes hovers but more often kites, with wings more extended than Red-tailed. Turns in tighter circles than Red-tailed. Upon sighting prey, often descends in tight tacking spirals before stooping. Wingbeats are quick and more fluid than Red-tailed, but when soaring, gliding, and turning, appears more tipsy or unsteady than Red-tailed.

VOCALIZATIONS: Call is a high whistled "*pe'e-arrh*" that falls somewhere between the shrill breathy "*kee yer*" of Red-tailed and the peevish whiny "*ehhhhh*" of Rough-legged Hawk.

Ferruginous Hawk, *Buteo regalis*

Russet-backed Prairie Eagle

STATUS: Uncommon to locally common western breeder and winter resident. **DISTRIBUTION:** Nests in s. Alta., s. Sask., e. Wash., e. Ore., n. and e. Nev., s. Idaho, most of Mont., most of Wyo., w. N.D., all but e. S.D., w. Neb., most of Utah, most of Colo., n. Ariz., and n. N.M. Winters across most of Calif., s. Nev., parts of Utah, Colo., sw. Neb., w. Kans., Ariz., N.M., w. Okla., and all but e. and extreme se. Tex., south into n. Mexico. **HABITAT:** An open-country obligate—grasslands, shrub-steppe and (to a lesser degree) pinyon-juniper, and deserts. Avoids higher elevations, forests, and narrow canyons. **COHABITANTS:** In summer, nests in same habitat as Swainson's Hawk, but the presence (or absence) of prairie dogs and ground squirrels is a better indicator of the bird's occurrence. In winter, commonly found where Golden Eagle, Red-tailed Hawk, American Kestrel, and Prairie Falcon occur. **MOVEMENT/MIGRATION:** Northern birds relocate to more southern and coastal habitat. Spring migration from late February to early April, with a peak in March; fall from late August to late November, with a peak in late October and early November. VI: 1.

DESCRIPTION: In most cases, a reddish-backed, pale-headed, and white-chested prairie eagle. Large long-profiled buteo (larger than Red-tailed Hawk; smaller than Golden Eagle), with a large-billed sloping doorstop–shaped head. Red-tailed is overall stockier, with a smaller, rounder head and smaller bill. At close range, the long yellowish gape (extending to the eye) and dark brow-line make Ferruginous look seriously unamused. Red-tailed looks more benign.

Found in both light and (much less commonly) dark morphs. Light adults and immatures have pale (streaky gray) heads and very white chests. Many Red-tailed Hawks have white chests, but almost all have a dark head and face. Rough-legged Hawks often have pale heads, but their bills are petite. Upperparts of adults are extensively russet, as are the feathered legs. Ferruginous tails range from white to rose. Immatures commonly show rusty traces in the wing, some spotting on the flanks and thighs, and pale gray tails (paler than the back).

Dark morphs are stunning—all dark (including the head), with chocolate and rust underparts and cinnamon-dusted backs. Tails are pale gray in adults, dark gray and banded in immatures.

BEHAVIOR: Primarily a perch-hunting raptor. Often takes advantage of strategic utility poles but also hunts prairie dogs and ground squirrels from slightly elevated ground (in winter, often in association with several other Ferruginous Hawks). Red-tailed may perch on elevated rocky outcroppings, but rarely on open ground. Perch-hunting birds may either glide to prey or fly over, hover, and descend. If unsuccessful, the bird usually flies to another perch. Also an accomplished bushwhacker, Ferruginous Hawk waits on the ground next to rodent burrows and grabs the occupants as they emerge. Other hunting techniques include cruising along slopes to surprise prey and, in open country and with an opportune wind, hovering in the manner of a Rough-legged Hawk.

Less tolerant of humans than many other buteos, and commonly hard to find—in part because of its penchant for sitting on the ground.

FLIGHT: Distinctive profile featuring a long wedge-shaped face, long, lanky, pointy wings, and a long tail. Light-morph birds are very white below; adults show dark leggings and rusty-tinged wing linings. Dark morphs show very white flight feathers and tail contrasting with dark underwings and body; at close range, the outer tips of the underwing linings show a white crescent. In all plumages, upperparts show three points of light—light patches near the tip of each wing and a tail that is lighter than the back. Pale, prairie Red-tailed also shows this pattern. Immature Rough-legged Hawk shows pale wingtips but either an all-dark or dark-banded tip to the tail.

Birds glide and soar with a dihedral that juts up at the shoulder and lifts gently along the length of the wing. Flight is direct and steady; wingbeats are slow, measured, shallow, and stiff.

VOCALIZATIONS: Whistled scream, "*kiiiiih*," is softer, shorter, and lower-pitched than Red-tailed or Swainson's Hawk.

Rough-legged Hawk, *Buteo lagopus*

The Petite-billed Buteo

STATUS: Common circumpolar arctic and high subarctic breeder; variously common to uncommon but widespread wintering species. **DISTRIBUTION:** In North America, breeds across w. and n. Alaska and east across n. Canada and the northern islands to ne. Que. Winters from the border regions of Canada south across the U.S., but short of the Mexican border. Absent in the southeastern states east of the Mississippi R. and south of Va. and cen. Ky. Also absent in most of La., s. Tex., and Calif. **HABITAT:** Breeds primarily in open wet and dry tundra regions, but when lemmings are abundant, range extends to the broken northern edges of forests. Nests primarily on steep cliffs. In winter, concentrates in open treeless areas, including prairies, pastures and agricultural lands, shrub- and sage-steppes, and salt and fresh marsh. Avoids areas with tall grass. **COHABITANTS:** In the arctic, Gyrfalcon, Peregrine Falcon, and ravens; in winter, Northern Harrier, Red-tailed Hawk, Prairie Falcon. **MOVEMENT/MIGRATION:** Entire population relocates south for winter. Spring migration from early February to late May, with peak from early March to early May; fall from late August or early September to late December (but there is little movement in the United States before October). VI: 3.

DESCRIPTION: A large, lanky, but broadly proportioned and boldly patterned buteo of open places. Perched birds appear to have slender necks and heads and petite bills. Except for these qualities and a slightly longer, less rotund profile, perched Rough-legged is proportioned much like Red-tailed Hawk.

Variable plumage includes light-morph and all-dark birds, but light morphs predominate and show boldly patterned underparts (paler chest and broad dark swath across the middle on immatures and females; dark bib and paler belly on adult males). Light-morph birds (and even immature dark morphs) appear pale-headed. In most cases, including light morphs, overall plumage appears white-spattered, richly spotted, even piebald. In all cases, the undertail is pale, with a single broad (or several narrow) terminal bands; in light-morph birds, the black-tipped white tail is also visible from above. Perched dark morphs are overall very black; they lack any white streaking on the breast (like Harlan's Hawk) and are smaller-billed than Zone-tailed and Red-tailed.

BEHAVIOR: This is a bird of open treeless places (where it can outcompete more perch-dependent raptors) and a buteo that is not shy about crossing open water. It hunts primarily on the wing, hovering habitually and often very high over open habitats. Hunting birds hover (less commonly kite) by holding into the wind and turning their heads side to side, studying the ground below. If they do not sight prey, the birds glide to another (usually nearby) location and try again. They often descend to lower altitudes (sometimes in several installments) before dropping onto prey. When perched, the bird very often sits on the springy tips of trees and shrubs (Red-tailed and Ferruginous Hawk prefer sturdier perches), but like other raptors, Rough-legged is quick to use utility poles. In early winter, Rough-legged is somewhat less timid than most Red-taileds. When breeding, commonly vocalizes when human intruders are still hundreds of yards away and before other raptors (most notably Peregrine Falcon) react, but remains silent in winter.

FLIGHT: Long, angular, broad- and plank-winged profile—overall more angular and lanky than Red-tailed Hawk. On light morph, the bold plumage pattern contrasts with pale underparts. Key marks include large black patches on the wrists, a broad black belly band (or, on adult males, a dark bib and pale collar), and a wide black or shadowy gray band on the tip of the tail (visible above and below). Like the light morph, many immature dark morphs also show black wrists and black belly bands, but these marks are set against a cryptic brown backdrop. The bodies of adult dark morphs are just black—except for the pale flight feathers, which are whitest in the hands, and the wide dark band at the tip of the tail. (This plumage pattern is not shared by any buteo that resembles Rough-legged Hawk or occurs where Rough-legged winters.)

When gliding and soaring, holds its wings in a slight dihedral, with wings that jut sharply up at the shoulder. From below, gliding birds show very long, broad arms but small, acutely tapered hands and long, broad tails.

Fairly buoyant in flight; more tippy than Red-tailed, but less than Northern Harrier. Flaps and glides habitually. (Red-tailed prefers to thermal aloft and glides without resort to powered flight.) Often hunts (and migrates) early and late in the day (when thermal production is nil). Wingbeats are slow, measured, regular, deep, and somewhat rowing.

VOCALIZATIONS: Call is a thin, petulant, descending, whiny protest: "*Ehhhhh.*"

Golden Eagle, *Aquila chrysaetos*
The Mountain Eagle

STATUS: Common and widespread breeder and resident in the West; uncommon to rare northern breeder and winter visitor in the East. **DISTRIBUTION:** Primarily a western and northern breeder in Canada found in e. and n. Que., Lab., and n. Ont. and west from ne. Man. north to Nunavut and west to cen. B.C., all of the Yukon, and all but extreme n. Alaska and parts of e. Alaska. In the U.S. and Mexico, breeds in all states west of the tall-grass prairies, that is, west of a line drawn from about w. N.D. to w. Tex. to cen. Mexico. In winter, most Alaskan and Canadian birds (except those breeding in s. B.C., Alta., and Sask.) retreat into s. Canada, the w. U.S. (west of and including the short-grass prairies), and Mexico. Smaller numbers fan out across the rest of the U.S., excluding the Florida Peninsula and many of the states bordering Canada. **HABITAT:** Primarily a bird of mountain canyons, buttes, rimrock, and grasslands, but also occurs in tundra, brush lands, and coniferous forests (but avoids unbroken forest). Often found along watercourses, perhaps because of the preponderance of cliffs they provide. In winter, in the West, Golden Eagle is most common in open, natural habitats (sagebrush, prairie, marshlands, oak savannah); in the East, most commonly associated with river systems situated in the southwestern Appalachian Plateau and coastal marshes. **COHABITANTS:** Common Raven. **MOVEMENT/MIGRATION:** Most Canadian and Alaskan birds vacate interior territories in advance of winter. Fall migration from early September to early January; spring from late February to early May. VI: 4.

DESCRIPTION: Big blackish (dark brown) raptor with a gold-burnished nape. Huge (dwarfs Red-tailed, Rough-legged, and Ferruginous Hawks) and overall broadly proportioned, with a somewhat smallish head, proportionately large bill, robust body, and shortish tail.

Adults are dark chocolate brown with a golden glaze on the hindneck and often a shiny paleness on the folded wings. Immatures are like adults, but more uniformly dark, and show bright white at the base of the tail.

BEHAVIOR: Primarily an aerial hunter. Soars high overhead or contour-hunts in hilly terrain along ridge tops in an effort to surprise prey. Also perch-hunts, most commonly in poor weather, using utility poles but also cliffs, ridge tops, and other elevated ground. Hunts alone or in pairs (sometimes cooperatively). At times, may fold up and stoop from a great height, glide along a descending angle, or pursue prey by active flight (even chasing prey on foot). Feeds on carrion where it is dominant over other birds of prey, including Bald Eagle but excluding California Condor.

Fairly sedentary; birds spend most of the day perched. Feeds most actively in the morning and late afternoon in summer, from midmorning to midafternoon in winter. Shy and mostly silent. Adults do not generally tolerate close approach.

FLIGHT: More buteo-like than Bald Eagle, with a relatively small head, long tail, and long but more contoured and tapered wings. The tail appears more than twice the length of the head. Adults are overall dark; they may show some slight pale or gray highlights in the wings and base of the tail, but the bodies are generally devoid of white. Golden head/nape may be so bright as to suggest a Bald Eagle. Immature birds show three points of white — one white patch near the end of each wing and a white-based dark-tipped tail. The amount of white varies. Some birds show large white patches on the upper and lower surface of the wings. Other birds show only a white border along the underwing coverts, but the white is confined to the base of the flight feathers; it does not spread onto the coverts. On immature Bald Eagle, the dingier white is mostly confined to the underwing linings (not the flight feathers), and most tails are not tipped with a broad crisply defined black swath of a border.

Soars and glides on wings that are slightly uplifted into a V (commonly leading to a mistaken Turkey Vulture identification). Eagles are steady on the wing; Turkey Vulture teeters and rocks. Wingbeats are slow, stiff, and heavy, but slightly more supple than Bald Eagle. Birds are slightly more agile than Bald Eagle.

VOCALIZATIONS: A high, weak, whistled, somewhat two-note "*cheeup . . . cheeup.*" Also emits a weak whistled twitter that is more chirped, less strangled, than Bald Eagle.

Crested Caracara, *Caracara cheriway*

The Ill-made Falcon

STATUS: Common (Texas) to uncommon (Florida and Arizona) resident with a restricted U.S. range. **DISTRIBUTION:** Widespread in Mexico, Central America, and South America, but in the U.S. found only in border regions of Ariz., s. and cen. Tex., and the prairies of interior cen. Fla. **HABITAT:** Dry open to semi-open grasslands, rangelands, and desert with scattered taller vegetation. Habitats include coastal savanna, pasture and cattle land, Sonoran Desert, brushy scrubland with an open-habitat component, and (in Florida) a mix of prairie and marshlands. Also frequents poultry yards, garbage dumps, and burned agricultural fields. **COHABITANTS:** Includes Turkey and Black Vultures, White-tailed Kite, American Kestrel, Cactus Wren, Eastern Meadowlark. **MOVEMENT/MIGRATION:** Nonmigratory. VI: 1–2.

DESCRIPTION: A bulbous-billed black-and-white roadside eagle wearing a badly fitting toupee. Medium-sized (larger than Red-tailed Hawk; slightly smaller than Turkey or Black Vulture), with a large squarish head, very heavy orange-based bill, a large blocky body, long, sturdy, pale legs, and a long tail.

Overall blackish or blackish brown, with an all-whitish head and neck (except for the dark cap of feathers that looks like a cheap toupee). The mostly white tail (with broad black-banded tip) is not always apparent on perched birds, but some hints of white from the tail (or white patches hidden in the folded wing) wink through. Immatures are browner and less contrasting, but patterned like adults.

BEHAVIOR: An accomplished scavenger as well as hunter, this very colorful bird is most often seen sitting on roadside perches, feeding on roadkill, or flying low over and along highways searching for roadkill (often very early in the morning). Commonly forages on the ground, stalking with head high, pausing frequently to survey the ground. Occasionally runs to secure food or prey with its bill, and also uses its foot to flip over objects that might conceal food.

Attends fires, walking behind the flames looking for victims (and often returns on subsequent days to search for additional victims). Also captures live prey by flying in pursuit.

Adults are not social, although young may remain with adults for several months after fledging. Nonbreeding young and subadults are more social and may gather in groups that number a dozen or more. Often found at large kills with vultures, where the caracara dominates.

FLIGHT: Large, sturdy, angular, and boldly patterned raptor. Planklike wings are set at a right angle to the body. Massive head/neck and tail project near equally ahead and behind. Overall very eaglelike; distinguished from eagles, however, by bold patterning.

Overall dark except for the white head, white (dark-banded) tail, and white patches on the wingtips (a great big cross in the sky with white at each terminus). This pattern shows above and below. Immatures are less contrasting but, again, similarly patterned.

Flight is overall heavy and lumbering but direct and fast, with lots of power but not much finesse—a meat-and-potatoes flier. Wingbeats are shallow and choppy on wings that are slightly down-drooped, both in flight and in a glide. Glides frequently, often for considerable distances, and with enough listing and correcting to seem a bit unsteady. Usually flies close to the ground. Does not commonly soar at high altitudes, but exceptions occur.

VOCALIZATIONS: Classic call is a loud rattle that has been likened to the sound of a stick run along a picket fence. Also "cackles" when threatened.

American Kestrel, *Falco sparverius*

Wire Falcon

STATUS: North America's most common and widespread falcon. **DISTRIBUTION:** Breeds across most of North America, from the arctic taiga to Central America. In the U.S. and Canada, absent as a breeder only along the West Coast north of s. B.C. and in most of s. and cen. Tex., s. and e. La., and s. Fla. In winter, vacates all of Alaska and Canada (except extreme s. B.C.), northern plains states, n. Mich., and n. New England; otherwise found across the entire U.S., Mexico, and Central America. **HABITAT:** At all seasons, a variety of open habitat, both barrens and short-grass, that offer hunting perches and, during the nesting season, trees (usually dead) with nesting cavities. Also nests on rocky cliffs, buildings and bridges with protective nooks, crannies, and crevices, and artificial nest boxes. Examples of habitat include meadows and grasslands, power-line cuts, farmland, cranberry bogs, prairie, desert, abandoned successional open land, and interstate medians in urban areas. **COHABITANTS:** Varies, but often found in the same habitat used by Red-tailed Hawk and Northern Harrier. **MOVEMENT/MIGRATION:** Spring migration from late March to early May; fall from early August to early December, with peaks in September and October. VI: 2.

DESCRIPTION: Small, slender, pale, but well-marked falcon most commonly seen perched on roadside utility lines. Small (same size and shape as Mourning Dove), with a larger head and slightly shorter tail than Mourning Dove. More slender and delicately proportioned than Merlin.

Colorful males are rufous above, with gunmetal blue wings, and pale and lightly spotted below, sometimes with a rosy blush across the chest.

Females are rufous brown and barred above, pale and lightly spotted below. Both birds show two vertical slashes across a pale face—a mustache and a sideburn. Immatures are like adults.

BEHAVIOR: Hunts from perches or utility lines and springy branches, often at or near the highest point of the tree. Appears to maintain its balance by habitually raising and lowering its tail, a trait that distinguishes American Kestrel from both Merlin and Mourning Dove. Where no perches are found, commonly hovers 20–30 ft. above the ground. Takes most prey (primarily insects and small mammals) on the ground (unlike Merlin, which takes virtually all its prey in flight), but also captures insects (primarily dragonflies) and sometimes small birds in flight. Sometimes consumes insects in the air, but usually consumes prey after taking it to an elevated perch.

More gregarious than Merlin. During migration, sometimes flies in small loose strings of two to ten birds. Also during migration, may sit within a utility pole length of other kestrels without aggression. Fairly timid. Flushes at distances where Merlin holds its perch, but when flushed, American Kestrel usually takes another perch quickly. Fairly vocal and defensive in the vicinity of its cavity nests.

FLIGHT: A slender, pale wisp of a falcon with long slender wings and a long narrow tail. Wings are more curved, more roundly contoured, and not as angled or acutely pointed as Merlin. Overall fairly pale (particularly pale below). The male's rufous tail and blue wings are easily noted; females, from below, seem uniformly pale and plain. When soaring, the trailing edge of both male's and female's wings shows a beaded string of white pearls.

Flight is light and buoyant, often tacking or wandering. Wingbeats are stiff, somewhat tentative, and given in a fluttery series followed by an open-winged glide on slightly down-drooped wings. American Kestrel slows but sinks little when it glides, and soars easily and well on long, narrow, and fairly blunt-tipped wings.

VOCALIZATIONS: Call is a loud, shrill, excited chatter: "*kleekleekleeklee.*"

PERTINENT PARTICULARS: Swallow species are fairly nonchalant when kestrels are present: in migration, they make only limited efforts to avoid them, and during the nesting season, they mob them. Not so Merlin. If a small distant falcon does not cause swallows to "ball up" (gather in tight defensive clusters), it's a kestrel.

Merlin, *Falco columbarius*

Falcon with Attitude

STATUS: Uncommon northern breeder and wintering species (generally less common than American Kestrel); particularly in coastal areas, may be quite common in migration. **DISTRIBUTION:** Breeds across all but extreme northern arctic regions of Canada and Alaska (absent in s. Ont., Que., N.B., and N.S.). In the U.S., breeds in n. Maine, the Upper Peninsula of Mich., ne. Minn., w. N.D., w. S.D., Mont., n. Wyo., n. and ne. Idaho, most of Wash., and cen. Ore. Winters along the Atlantic Coast and coastal plain from Mass. to Tex.; along the Pacific from coastal s. Alaska to Baja; and in the interior West from the e. Great Plains, including s. Man. and Alta., west to the coast; also found in the West Indies, Mexico, Central America, and n. South America. **HABITAT:** Open and semi-open habitat, including prairie, open or broken coniferous forest, bogs and rivers bisecting forest, tundra, alpine tundra, islands in large lakes, and coastal areas. In the prairies, has become somewhat urbanized and nests along tree-lined streets, cemeteries, and parks. In winter, found in many of these habitats, but frequently hunts in coastal areas where shorebirds abound and over agricultural land with a surfeit of longspurs and larks. **COHABITANTS:** Commonly uses old corvid nests to breed, so found in association with crows, ravens, and magpies. Nevertheless, very few creatures covered with feathers and smaller than a pigeon want to be near a Merlin. **MOVEMENT/MIGRATION:** Except for birds breeding along the coasts of western Canada and southern Alaska (the largely sedentary "black" Merlin) and in the northern prairie region of Canada and the Badlands region of the United States, all birds migrate south.

Spring migration from mid-February to mid-May; fall from late July to mid-November, with peak in mid-September to late October. VI: 4.

DESCRIPTION: Pigeon-sized, robust, fast, feisty, and mostly dark falcon. Fairly small (smaller individuals are only slightly larger than American Kestrel) and compact, with a large head, stocky body, and shortish tail—overall more broadly proportioned and robust than American Kestrel.

Adult males are slaty blue above, pale with ruddy brown streaks below. The helmeted head pattern is ill defined; a single dark mustache runs like a dark channel below the eye. Females and immatures are overall darker, browner, and more darkly streaked below—chocolate falcons. The expression is mean—real mean.

"Prairie" Merlin (which winters south into Texas, New Mexico, Arizona, Nevada, and California) is much paler than the more widespread taiga subspecies. Adult males are silvery blue above and orangy streaked below, with faces that may show no mustache at all. Females and immatures are pale brown—the color of Prairie Falcon. The Pacific Coast "black" subspecies is almost all dark, with an all-dark hooded face and underparts so heavily and darkly streaked that they appear mostly blackish.

BEHAVIOR: A perch-hunting raptor that commonly takes a sturdy (often low) perch looking out on a large open area. Unlike kestrel, rarely sits on wires. Attack is fast and direct. After an initial failure, often makes tenacious efforts to secure prey, pursuing it upward in a "ringing flight." Particularly fond of flocking species, like shorebirds and Horned Larks, which it most commonly secures in the air. Can also capture prey as agile as Tree Swallow. Young birds particularly target dragonflies. Often consumes insects on the wing; takes other prey to a perch for plucking and consuming.

Usually solitary (in fact, downright hostile toward other raptors), although pairs sometimes hunt together. In migration, immature birds "dogfight" with other birds of prey (and each other). Does not hover. Does not migrate in small groups. Merlin is less timid than American Kestrel and often allows closer approach before retreating; once airborne, however, Merlin may leave the area.

FLIGHT: Silhouette is stockier than kestrel, with wider, more acutely angled wings shaped more like isosceles triangles. Females and immatures are overall dark, with a hint of paleness about the throat. The undertail is half-dark, half-light—that is, pale from the base to midtail, and dark from midtail to the tip. Adult males are overall paler, with a rose or orange blush on the chest. (Adult male Merlin is more kestrel-like in plumage, but also much more compactly proportioned than kestrel.) Narrow white bands on the tail are often invisible unless the bird flairs its tail when breaking or soaring.

Wingbeats are rapid, steady, powerful, downward-flicking piston strokes. Flight is direct and fast. Migrating birds commonly fly low, just over the tops of vegetation. Hunting birds accelerate quickly and maneuver with reckless agility. When birds glide, they lose altitude quickly but slow little. Kestrels float and slow.

VOCALIZATIONS: A rapid shrill or squeaky chatter that is unmistakably strident and somewhat louder in the middle of the sequence: *"keh keh Keh Keh Keh keh keh."*

Aplomado Falcon, *Falco femoralis*

Yucca Falcon

STATUS: Uncommon Neotropical species reintroduced in Texas; a highly localized resident but apparently expanding; rare vagrant to the Southwest. **DISTRIBUTION:** Formerly found in se. Ariz., s. N.M., and trans-Pecos Tex. Reintroduced to Laguna Atascosa N.W.R. and Matagora I. in s. Tex., where breeding has been successful. **HABITAT:** In the United States, found in coastal and desert grasslands studded with yuccas and mesquite. In Mexico, favors woodland edge adjacent to expansive grasslands. **COHABITANTS:** White-tailed Kite, White-tailed Hawk, Eastern and Western Meadowlarks. **MOVEMENT/MIGRATION:** Permanent resident. VI: 1.

DESCRIPTION: A long, slender, colorful, and distinctly marked falcon of open grasslands. Medium-sized (considerably larger than American Kestrel and Merlin; same length as but more slender than Peregrine), with a small squarish head, a small bill, fairly wide shoulders, and an exceedingly long, untapered, round-tipped tail. *Note:* Aplomado's wingtips do not reach the end of the tail, as they do on Peregrine.

Adults and immatures are all dark above, with a very distinctive head pattern: a narrow black mask across the eyes and an abbreviated mustache stripe. From below, the white or pale breast and buffy or orange belly is separated by a dark vest. Looser feathers often billow or blow in the wind.

From a distance, what you see is an uncommonly slender all-dark or blackish falcon with an unaccountably and distinctly contrasting amount of white about the face, throat, and breast. Even slight and delicate kestrel looks stocky and robust compared to the elongated lines of Aplomado Falcon.

BEHAVIOR: A versatile eclectic hunter (even something of a brawler) that uses a mix of techniques and targets a variety of prey—from insects to lizards, rats to bats and, of course, birds. Searches for prey while perching, soaring, or cruising, often along watercourses. Captures insects by making kingbirdlike sallies from perches or by intercepting them on the wing, in the manner of Mississippi Kite. Flies down birds in tail chases, stoops onto grounded prey, hovers over prey that has taken cover in the scant protection afforded by thornbushes, and has been known to chase prey on foot (like Northern Goshawk). Paired adults even hunt cooperatively, with one bird flushing and one bird intercepting.

A nimble bird, able to perch on the springy twigs of low bushes and, like American Kestrel, utility lines. (Peregrine and Prairie Falcons prefer utility poles.) Fairly tame, often allowing closer approach than most raptors.

FLIGHT: Resembles a big kestrel. Wings are very long, overall narrow, and softly angled, with a backswept curve and blunted tips. The narrow untapered tail is exceedingly long (though the length is masked somewhat by the extreme length of the wings) and ends bluntly. The bird flies and glides with wings showing a slight droop along their length. At a distance, appears headless, all wing and tail.

Flight is very buoyant and somewhat kitelike. Wingbeats are stiff and fluttering (like American Kestrel), not whippy or elastic (like Peregrine). Sometimes hovers (even at high altitudes), with the tail fanned. When stooping on prey, may continue to flap well into the dive.

VOCALIZATIONS: A high, harsh, hurried series of "kek" notes: "*kek kek kek kek.*" Like Peregrine but faster. Also makes a single-note utterance.

Gyrfalcon, *Falco rusticolus*

What Is White, Gray, or Brown and Bad All Over?

STATUS AND DISTRIBUTION: Uncommon and restricted breeder of arctic and subarctic regions. In North America, breeds in arctic, coastal, and mountainous portions of Alaska (absent across most of the forested interior), the Yukon, extreme nw. B.C., and across most of the Canadian arctic coastal plain and arctic islands east to Lab. and Greenland. In winter, regularly wanders south to the Canadian border states, n. Wyo., and cen. S.D.; rare farther south. **HABITAT:** Breeds in regions dominated by arctic or alpine tundra. Nests are located on cliffs or isolated rock formations commonly flanking rivers and coasts and often near seabird colonies. In winter, mostly vacates higher altitudes, concentrating along seacoasts, in coastal wetlands, prairies, and high deserts, and on agricultural land. (There are also records of birds residing in and near cities, landfills, and rock quarries.) Winter habitat is typically open, flat, or rolling. **COHABITANTS:** Rock and Willow Ptarmigans, Common Raven. **MOVEMENT/MIGRATION:** Partial migrant: many adults remain in northern regions, and southern movements are conducted primarily by young and some females. Spring migration from early February to mid-April; fall from early September to late November, but this somewhat nomadic or far-ranging species may appear at any time in the winter. VI: 3.

DESCRIPTION: Large, long-tailed, broadly proportioned falcon that comes in white, gray, or brown and looks formidable and sinister. The largest of all falcons (but variable in size, with larger females the size of Rough-legged Hawk and small males closer to the size of female Peregrine), Gyrfalcon appears proportionally small-headed, heavy-bodied (somewhat potbellied), and long-tailed, with the tip extending well beyond the tip of the wings. The tail is variable. Sometimes it tapers to a candle-flame-shaped point; sometimes it appears broad and blunt. (Peregrine's tail is narrower and straight-edged and tapers modestly toward a blunt tip.) When resting on the ground, Gyrfalcon's body seems to melt into the substrate rather than perch on it.

There are three plumages: white (with upperparts patterned much like Snowy Owl), gray, or dark smoky brown. White birds are distinctive. Gray and brown birds are more monotoned and more uniformly dark (top to bottom) than Peregrine or Prairie Falcon, with a face pattern that is either indistinguishable from Merlin or darkly helmeted. In either case, the face lacks the distinct dark sideburn slash common in most Peregrines and, slightly less distinctly, in Prairie Falcon. Also, at close range, the nape on Gyrfalcon is usually touched with white. The expression is *bad*.

BEHAVIOR: Gyrfalcon is much at home on or near the ground. Walks with a sailor's gait and hops, even runs, in pursuit of prey. Perches on low hilltops and low-lying rocks, dunes, and even sandy beach (even when taller perches can be found), as well as on utility poles and towers. Commonly hunts low, just above the ground. Flies down prey in direct pursuit (Peregrine prefers to tower above prey and stoop) and readily snatches prey from the ground (something Peregrine rarely does). Ranges great distances. If you see a Gyrfalcon approaching, chances are you will be able to follow it to the horizon.

Mostly solitary; found alone or pairs. Generally shy and retiring, especially during the breeding season.

FLIGHT: Shaped much like big, broad, heavyset Peregrine, but showing broader blunter-tipped wings and a wide body and tail. Sometimes appears potbellied or humpbacked. White birds are almost unmistakable and can be confused only with a large albino raptor (such as Red-tailed Hawk). Gray birds might be confused with adult Goshawk. Brown birds could easily be confused with a dark immature Peregrine, but underwings of Peregrine are wholly dark, whereas underwings of Gyrfalcon are two-toned, with dark underwing linings flanked by pale flight feathers.

Flight is strong, direct, somewhat lofting. Wingbeats are steady, shallow, and slightly slower and not as elastic or whippy as the wingbeats of Peregrine.

VOCALIZATIONS: Call is a loud "*nyah, nyah, nyah*" that is lower-pitched than Peregrine.

Peregrine Falcon, *Falco peregrinus*

Crossbow in the Sky

STATUS: Uncommon localized but widespread falcon. **DISTRIBUTION:** Breeds widely across North America—from the Arctic to Mexico—but is found primarily in the arctic, the Rockies, and along the Pacific Coast. In the U.S., absent as a breeder in only a handful of states. In North America, winters primarily along seacoasts: in the Pacific, from the Aleutian Is. to Mexico (and farther south); in the East, from the Maritimes south along the Atlantic and Gulf Coasts. Also winters inland north to just above the U.S.–Canada border. **HABITAT:** Breeds in an array of habitats, from arctic tundra to mountain regions to deserts, as well as in many urban areas. Requires cliffs or clifflike structures (such as buildings or bridges) for nesting and open habitat for hunting (including the airspace over oceans, lakes, and rivers). In winter, may substitute tall man-made structures (such as water towers) for cliffs. **COHABITANTS:** Ravens, other cliff-nesting falcons. Often found near seabird colonies in breeding season, near pigeons in cities, and near shorebird and waterfowl concentrations in winter. **MOVEMENT/**

MIGRATION: Some populations (primarily birds breeding in the Pacific coastal Northwest and urban areas) are permanent residents. Most North American Peregrines migrate to the Neotropics, where they are primarily coastal. Spring migration from early March to early June, with peak from mid-April to mid-May; fall from late August to late November, with peak in late September to mid-October. Migration has a broad front, with concentrations found along the Great Lakes and mountain ridges and in coastal areas, particularly along the Atlantic Coast in fall and the Texas coast in spring. VI: 4.

DESCRIPTION: A conspicuously long-winged, long-tailed, bird-catching falcon that flies with the fluid grace of a professional skater. A medium-sized hawk and large falcon (larger than Merlin; smaller than Red-tailed; about the same size as Prairie Falcon or Northern Harrier), with a very large head, wide shoulders, and a long and tapered body (looks like an elongated teardrop). *Note:* There is great size variation between the smaller males and much larger females. Some males are almost as small as large Merlins, and some females larger than small Gyrfalcons.

Adults are blue-gray to blue-black, and paler and finely barred below. The pure white cheeks and chest contrast with a blackish hood whose single wide sideburn makes the bird look helmeted.

Immatures are overall darker and brown above, heavily streaked below. The helmeted facial pattern, although less contrasting, is manifest.

BEHAVIOR: Spends much of its time sitting on ledges or elevated perches (dunes, utility poles, fenceposts, driftwood). Flies out to intercept flying prey (birds or bats), or may hunt by flying low (or high) over open prey-rich areas or by "towering" high overhead, then stooping on prey, or by driving and pursuing prey aloft by circling below it. Often chases prey over water. Usually binds to prey, dispatches it by lowering its head and biting through the spinal cord, and then carries it to shore or a perch for plucking and consumption. On the breeding grounds (and sometimes in the case of young birds), hops on the ground in pursuit of avian and, less commonly, non-avian prey.

A solitary hunter outside the breeding season, but sometimes roosts with other Peregrines. Not as flighty as many birds of prey, but generally does not allow close approach. During migration, young birds are playful, engaging in aerial dogfights with other species of raptor. Adults are more reserved and are usually aggressive only toward Golden Eagle.

Has been known to respond to squeaking.

FLIGHT: Silhouette is very distinctive and can be confused only with Prairie Falcon, Gyrfalcon, and Mississippi Kite. Peregrine is a falcon with very long, slender, acutely tapered wings and a long tail (though the extreme length of the wings tends to mask the length of the tail). In level flight, wingtips are drawn back. In a glide, wings are curved along the leading edge and straight-cut along the trailing edge, conferring on the bird the shape of a "crossbow in the sky." When soaring, the widely fanned tail masks the extreme length of the wings, making the bird look much like a soaring Broad-winged Hawk (a buteo).

Flight is direct and may be fast or leisurely. In level flight, wingbeats are shallow, fluid, and regular; motion seems to ripple down the length of the wing, making the bird look whippy-winged. In execution and cadence, the wingbeat recalls Common Loon or, to a lesser degree, Double-crested Cormorant. When the bird is in rapid pursuit or just leaving a perch, wingbeats are quicker and deeper.

VOCALIZATIONS: Usually silent. When approached near its nest ledge, makes a loud, harsh, rapid "*kak kak kak kak kak.*"

PERTINENT PARTICULARS: Prairie Falcon is overall paler and sandier-colored than Peregrine, with blunter wingtips and obvious black "support struts" running along the inner lining of the underwing. Gyrfalcon is a broad, bulky falcon with contrastingly pale flight feathers below. (The underwing of Peregrine looks uniform.)

Merlin is overall more compact than Peregrine, with a weak mustache and a quicker, choppier wingbeat—the very antithesis of fluid. American

Kestrel, whose relative proportions are very similar to Peregrine, is overall paler and more delicately proportioned and has a quick, fluttery, tentative wingbeat.

Prairie Falcon, *Falco mexicanus*

Desert Falcon

STATUS: Fairly common and widespread western resident and winter resident. **DISTRIBUTION:** Breeds west of (and in) the short-grass prairies, from s. Canada (sw. Sask., s. Alta., and s.-cen. B.C.) south into Mexico. The eastern limits of its breeding range cut across w. N.D. and S.D., nw. Neb., e. Colo., and extreme w. Okla., south to trans-Pecos Tex. and a point just east of Big Bend. Does not breed in w. Wash., Ore., coastal Calif., and n. Idaho. Winter range is more extensive, encompassing all of the West and extending east to w. Minn., w. Iowa, w. Mo., cen. Okla., and all but eastern portions of Tex. **HABITAT:** Breeds in arid open habitats where intruding bluffs or cliffs provide nest sites, including prairie grasslands, alpine tundra, and shrub-steppe desert. In winter, found primarily on the dry grasslands of the Great Plains and the Great Basin, but also on croplands and wetlands. **COHABITANTS:** Breeds on cliffs used by Golden Eagle, Red-tailed Hawk, and Common Raven. In winter, found with Red-tailed and Ferruginous Hawks and Golden Eagle, but most importantly, with ground squirrels and Horned Lark. **MOVEMENT/MIGRATION:** As much a wanderer as a migrant. Post-breeding birds move east (and west), and then, possibly, south. Wandering birds may be encountered from late June to late April, with peak fall movement from late August to late October; most spring movement occurs in March. VI: 2.

DESCRIPTION: Large, pale, sparely plumaged desert falcon—overall paler and less contrastingly patterned than Peregrine. A large falcon and medium-sized raptor (larger than Merlin; smaller than Red-tailed; about the same size and shape as Peregrine), with a large squarish head and long wide shoulders, but overall lean body and long tail. The tail on perched birds may be either wide, partially fanned, and square-tipped or tightly closed, narrow, pointy-tipped, and shaped like a candle flame. The closed tail of a perched Peregrine is modestly tapered and shows a blunt tip (or would if Peregrine's longer wings didn't often conceal the tail).

Cryptically colored—desert colors for a desert raptor. Upperparts are plain, pale, and grayish brown; underparts are whitish with narrow streaks (immatures) or spare spotting (adults) that is heaviest on the flanks. The face is mostly whitish, scored by a single mustache slash that is usually the darkest point on the bird.

BEHAVIOR: Solitary, except in the breeding season. Most commonly seen perched (often on utility poles or high-tension towers), cruising low and fast just above the ground, or flying overhead (either soaring or utilizing powered flight). Whatever the bird's mode, it is almost certainly in the vicinity of a plenitude of ground squirrels. Barring ground squirrels, flocks of Horned Larks become favored prey, as do shorebirds, but almost any inattentive, open-country bird, from Mourning Dove to meadowlarks and rosy-finches, can earn a Prairie Falcon's favor.

Fairly shy. Does not tolerate close approach, and generally responds by leaving the area. Does not appear to respond to squeaking.

FLIGHT: Flight silhouette is very similar to Peregrine Falcon, but somewhat broader, blunter-winged, and longer-tailed. At times recalls an overgrown kestrel. In powered flight, may show an acutely pointed tail (unlike Peregrine, which never does). Overall pale except for the obvious dark underwing lining that remains visible even when the outline of a high-soaring bird melts into the sky.

Glides more than soars, with wings that are more curved than angled back; when gliding, appears to be no-headed. Wingbeats, when cruising, may be rapid and shallow; when targeting prey, wingbeats are deeper, with a more pronounced down stroke. Wingbeats are slightly stiffer, more mechanical, and not as fluid and rippling as Peregrine.

VOCALIZATIONS: Similar to Peregrine, but higher-pitched.

RAILS, COOTS, LIMPKIN, AND CRANES

Yellow Rail, *Coturnicops noveboracensis*

Ticking Rail

STATUS: Nocturnal and secretive, but probably more numerous than encounters suggest. Suffice it to say that this local, northern breeder and coastal winter resident is almost never encountered except by accident or concerted effort. **DISTRIBUTION:** Breeds locally in e. Alta., se. Northwest Territories, ne. B.C., Mont., Sask., Man., n. and cen. N.D., Que., cen. Minn., cen. Wisc., n. Mich., s. Que., n. Maine, and N.B.; a disjunct breeding population is found in s. Ore. and ne. Calif. Winters coastally from N.C. to Corpus Christi, Tex. (including much of the Florida Peninsula) and in small numbers in n. and cen. Calif. **HABITAT:** Breeds in open wet sedge meadows with either mud substrate or shallow standing water. Avoids tall rank vegetation (such as cattails) and meadows where brush, willows, or birch is established. Also found in high-marsh areas of brackish, coastal meadow in Quebec and northern Ontario. In winter, found primarily in upper portions of salt marsh meadows; favors areas with grasses that bunch and clump (leaving navigable channels) and with a shallow sheet of standing water; also found in rice fields. Avoids deep standing water, tall stands of emergent vegetation (favored by Virginia Rail and Sora), and the dense, unbroken stands of salt hay favored by Black Rail. In migration (less commonly as a breeder), also found in hay fields. **COHABITANTS:** In summer and winter, Wilson's Snipe, Sedge Wren, Nelson's Sharp-tailed Sparrow, Le Conte's Sparrow. Only in winter, Seaside Sparrow. **MOVEMENT/MIGRATION:** Wholly migratory. Spring migration from early April to mid-June; fall from late August to November. VI: 2.

DESCRIPTION: Looks like a small, yellowish, short-necked Sora or a small-headed mustard-colored Easter chick. Tiny rail (distinctly smaller than Sora; slightly larger than Black Rail—and House Sparrow!), with a small head, a small candy-corn-shaped bill, a plump and fairly tail-less body, and shortish legs (by rail standards). Overall shorter-billed, more compact, and shorter-legged than Sora. In all plumages, this darkly patterned mustard-colored bird has a front half that is mostly dirty yellow and a back half that is mostly blackish brown. (Most hopeful observers anticipate a yellower bird.) The back is mostly dark (sometimes it assumes a pattern of broken black stripes), with an intrusion of narrow buffy streaks that look like tongues of yellow flame licking down the back. The dirty yellow head has a diffuse dark cap and mask across the eye. The bill is yellow on breeding adults, but otherwise dark. The breast is buffy or dingy yellow. Flanks and undertail are immaterial—you'll never see them. Immatures are overall less yellow than adults (which are not all that yellow to begin with), but still show the narrow, buffy, streaked back pattern that, in addition to smaller size and a more compact and shorter-legged profile, serves to distinguish it from Sora.

BEHAVIOR: A secretive species that rarely forages along the edges of open water, as Soras commonly do. Spends most of its time on the marsh surface beneath a grassy canopy. Forages, mostly in the daytime, by walking; runs with body lowered and head extended. Males vocalize most often in full darkness, but may sometimes call during the day. Usually silent in winter, but wintering males may call prior to migrating (in late April and May). Flies reluctantly, most often when pressed (for example, when almost stepped on or run over by a rice-harvesting combine) or when provoked (either by another calling male or by use of tapes).

FLIGHT: Small and compact with short, broad, rounded wings and a head showing little projection. Legs dangle for short flights (several yards) but extend straight back on rare longer sojourns. Upperparts are plain, showing a yellowish cast, except for two obvious rectangular whitish wing patches on the inner trailing edge. (Some Soras show a pale trailing edge to the inner wing but not big white patches.) Flight is low and slow, usually just above the marsh and rarely as high as a person's head. Wingbeats are stiff, rapid, and steady, with wings raised as high on the up stroke. Overall flight is mostly direct, but weak and effortful. When flushed, does not fly as far as Sora.

VOCALIZATIONS: Call is an incessant, patterned "ticking" consisting of two measured notes followed by a brief hesitation, followed by three more rapid ticks: "*tic tic tictictic; tic tic tictictic.*" Tonal quality is sharp (recalls two quarters being struck together), but hollow-sounding at close range. Pattern is mostly steady, but the bird at times seems to lose its rhythm.

PERTINENT PARTICULARS: Birds can be enticed to approach by imitating or playing back their calls. Efforts to flush birds, either by employing groups to walk across the marsh in a parallel line or by dragging a weighted rope between two individuals, can be successful, but such tactics should never be used during breeding season, owing to the danger posed to the nests of this and other marsh species. Passively observing the area in front of combines when rice is being harvested may be the most productive (and least culpable) way of glimpsing this species.

Black Rail, *Laterallus jamaicensis*

A Kickydoo in the Night

STATUS: Uncommon to rare; habitat-specific; rarely seen (more commonly heard). **DISTRIBUTION:** Poorly understood and patchy even within its known range. Breeds coastally from coastal Long Island (sporadically to Conn.) south to s. Fla.; on the Gulf Coast along the western coast of Fla. to Ala. and in the marshes of se. Tex. In the West, found year-round in the San Francisco Bay area; also found in small numbers in the lower Colorado R. in Calif., Nev., and Ariz. Breeds rarely and irregularly in the Midwest. In winter, interior birds and birds north of Chesapeake Bay retreat south and establish themselves along the Atlantic and Gulf Coasts. **HABITAT:** Shallow marsh habitat (salt or fresh water) characterized by dense grass cover and shallow water. Habitats include higher portions of tidal marsh, salt hay farms, wet meadows, and flooded grassy fields. As a rule, favors habitats with shallower water more than other rails do, but in coastal California occupies marshes that are subject to high tides but also provide an adjacent belt of higher, drier, dense vegetation into which birds can retreat. **COHABITANTS:** Includes mosquitoes, Northern Harrier, other rail species, Saltmarsh Sharp-tailed Sparrow, Seaside Sparrow. **MOVEMENT/MIGRATION:** Spring migration from late March to mid-May; fall from mid-August to early November. VI: 2.

DESCRIPTION: A small, black, chicklike bird with oversized feet and legs—but you have to be exceptionally lucky to see the bird at all, much less the feet. Tiny rail (smaller than Yellow Rail or Sora; about the same size as Semipalmated Sandpiper), with a small, narrow, pointy bill, a small, fairly neckless head, a plump croissant-shaped body, and a short pinched stub of a tail.

Mostly charcoal gray with a galaxy of white flecks above. The red eye is easy to see at close range (and you will be close); the chestnut nape may be limited, or ruddy color may infuse into the upper back.

BEHAVIOR: Forages on wet (or moist) surfaces beneath a protective cover of grass, moving quickly along rodent runways. Contrary to popular belief, forages mostly during daylight hours, but very occasionally breaks into the open. Males vocalize mostly at night (usually beginning one to two hours after full dark, with another peak one to two hours before light), but birds may also vocalize during daylight (often persistently).

Not particularly shy—they don't have to be—but also almost impossible to find even when you are standing directly over them. Black Rail flushes reluctantly and flies short distances. Responds well to recorded calls. Appears to be less active (at least less vocal) on full-moon nights and during high tides.

FLIGHT: Tiny, all dark, with wide blunt wings. For short flights, legs dangle; for longer flights, legs trail. Flight appears weak when flying short distances, but strong and direct, with quick steady wingbeats, when covering greater distance.

VOCALIZATIONS: The male's song is a low, nasal, somewhat comical or flippant three-note chant: "*keekeeger . . . keekeeger,*" with the first two notes even and the final note descending in pitch. Also, when agitated, makes a low growl: "*grrr . . . grrr . . . grrr.*" Females make a soft cooing.

PERTINENT PARTICULARS: Let's be honest. Chances are you are not going to see this bird unless you resort to extraordinary (and perhaps ethically questionable) efforts. I have lived in Black Rail country for 30 years. I have heard scores of Black Rail and seen exactly one. The best way to gain a glimpse of this bird is to stake out a section of marsh at flood or storm tide and watch (or walk) the elevated edges where birds will be concentrated. Barring this, you can position yourself to scan down a narrow open lead or tidal creek (during low tide) close to where birds have been vocalizing and wait. Have a pure heart.

Clapper Rail, *Rallus longirostris*
Mud Hen

STATUS: Common and widespread along the Atlantic and Gulf Coasts. Four subspecies—two localized to coastal sections of California, one found in the interior Southwest, and one restricted to the Florida Keys—are all officially classified as endangered. **DISTRIBUTION:** Breeds coastally from Mass. (rarely s. Maine) to ne. Mexico. In the Southwest, the "Yuma" Clapper Rail occurs along the lower Colorado R. from s. Nev. southward and locally on other rivers and lakes east past Phoenix and around the Salton Sea. There are two coastal population centers in Calif.: one in the tidal marshes around San Francisco, the other extending from Santa Barbara south to Baja. **HABITAT:** In most coastal locations, found almost exclusively in salt- or brackish-water habitat dominated by cord grass (*Spartina*). Often seen foraging along tidal creeks and sloughs. In Florida, occupies mangrove swamps. The freshwater Yuma Clapper Rail occurs in a variety of marshy vegetation, including cattail and bulrush. **COHABITANTS:** Forster's Tern; in the East, Seaside Sparrow. In winter and migration, may occur with both Virginia and King Rails in brackish, tidal habitat. **MOVEMENT/MIGRATION:** Over most of its range, a permanent resident. Only birds on the Atlantic Coast north of the Carolinas are migratory, and except in southern New England marshes, some birds remain all year. Spring migration from March to May; fall from August to October. VI: 2.

DESCRIPTION: A large rail (larger than Virginia Rail, slightly smaller than King Rail, about the same size as Willet) that looks and behaves like a long-billed marsh chicken. The head and neck are somewhat serpentine, and the body is oversized and oddly contoured—bulgy on the breast and saggy in the rear. The long, slender, and somewhat down-curved bill seems like an extension of the head, not the face (the bird has little forehead), and the expression is mean. The tail may be drooped or (particularly when walking) up-cocked. Legs are thick, sturdy, and set well back.

Plumage differs markedly depending on location and subspecies. Atlantic Coast birds are overall grayish brown, with grayer cheeks, unpatterned upperparts, warmer (olive or buffy) underparts, and low-contrast gray-and-white barring on the flanks and belly. Gulf Coast birds are ruddier below and have more patterning above, but still have gray cheeks. California birds are overall more cinnamon and more nearly like King Rail—but King Rails aren't found west of Texas.

BEHAVIOR: A skulking species that spends much of its time out of sight. However, at dawn and dusk, during low tides and flood tides, and during the breeding season (March to May), the birds often forage, stand, bathe, and preen in the open along tidal creeks and on mud flats and matted wrack.

Forages by crouching low and walking with a movement that is part strut, part stalk. Movements are jerky. The head bobs forward and back, then jerks up (flashing a white undertail) and falls. The pace changes frequently, going from a fast strut to a creeping slink to a burst of rapid steps (sometimes aided by an energy-supplementing wing flap or two). Moves with predatory menace. Jabs its bill at prey sighted visually, probing only occasionally.

Usually solitary, but during breeding season pursuits (involving two or more birds) that break into the open are commonplace. Occasionally birds loft into the open for short flights or swim across open creeks, with bodies mostly submerged and heads and bills thrust forward. They occasionally climb onto roots or the low limbs of trees.

FLIGHT: Flight is like the Wright Brothers: low, weak, struggling, with much flapping on short wings and little gain. Neck is stretched forward; legs dangle behind. Lands with much wing braking, then a crash.

VOCALIZATIONS: May be oddly silent for hours on end, then a marsh erupts with rail vocalizations. Classic call sounds like a hoarse, raucous, descending belly laugh: "*RAH, RAH, Rah rah rah*" (five to ten notes). Also, in breeding season makes a loud, monotonous, and monotone series, "*Kek, kek, kek, kek*" (like King Rail), that may begin with slow notes, then gain tempo. (King Rail's tempo does not increase.) Also makes a cackled protest—"*Keh, Keh, Keh—kehkehkeh!*" or an explosive "*K'Keh!*"—and an angry "*Raaah!*" Many of these utterances recall the clucks and mutterings of chickens.

PERTINENT PARTICULARS: Because of its larger size and less contrasting pattern, Clapper is easily distinguished from Virginia Rail (which winters in coastal marshes and breeds in areas of lower salinity). Separation of Clapper from King Rail is more difficult (see "King Rail").

Clapper Rail can be coaxed to vocalize by even poor imitations of its call or by making a loud noise (for example, clapping your hands).

King Rail, *Rallus elegans*

Cattail Clapper

STATUS: Uncommon and very local eastern rail; reaches its greatest breeding and winter densities along the Atlantic and Gulf coastal plains. **DISTRIBUTION:** Breeds in scattered locations across much of the e. U.S. (also in extreme se. Ont.) north to Mass., Long Island, n. N.J., se. N.Y., s. Mich., cen. Wisc., cen. Minn., and se. N.D. Western border cuts across e. S.D., e. Neb., cen. Kans., cen. Okla., and cen. Tex. and south along the east coast of Mexico. Absent along the Appalachians to ne. Ga. In winter, northern interior birds vacate territories. Wintering birds are found in coastal N.J., Del., Md., Va., N.C., S.C., Ga., all of Fla., and Ala. Also winters in extreme s. Ill., w. Ky., w. Tenn., e. and s. Mo., Ala., La., s. Okla., and e. Tex. **HABITAT:** Breeds and winters in nontidal freshwater marshes with lush, emergent vegetation (particularly cattail, but also other tall marsh reed) as well as sedge and shrubs. Also found in brackish tidal marsh—where it is found in cord grass, phragmites, and rice fields—and in salt marsh in migration. **COHABITANTS:** American Bittern, Least Bittern, Sora, Marsh Wren, Swamp Sparrow, and sometimes Clapper Rail. **MOVEMENT/MIGRATION:** Uncertain. Spring migration probably from late March to mid-May; fall from early August to early October. VI: 2.

DESCRIPTION: A big, rusty, richly patterned rail of freshwater (less often brackish) marshes. Large (smaller than Green Heron; very slightly larger than Clapper Rail; much larger than Virginia Rail) and identical in shape and relative proportions to the Clapper Rail of coastal marshes, with a long slightly down-curved bill, a narrow head, a slender neck, a hefty body, and long sturdy legs.

King and Clapper Rails show the same basic plumage pattern but differ in several respects. King is overall no-nonsense ruddy; Atlantic Clapper is gray with warm highlights, particularly on the lower breast and sometimes on the very upper breast and throat. King shows reddish orange cheeks and breast (the gray is above and behind the eye); on Clapper, the cheeks and breast are gray or drab, showing just a wash of warm tones. King shows lavish black spotting on its upperparts (often arrayed as broken stripes) along a warm buff and reddish back. On Clapper, the spots are not neatly arrayed and are set against a gray backdrop. King shows a red patch or border on the lower border of the folded wing; Clapper shows a gray or brown border. King shows a crisp, contrasting array of black-and-white bars on the sides and flanks; on Clapper, the bars are more diffuse and gray and white. On

King, the uppertail is reddish; on Clapper, it's gray or brown.

Gulf Coast Clapper Rails are overall ruddier than Atlantic birds, but separable using a combination of the traits just described. More troublesome are hybrids, which are found in the brackish habitats where the ranges of King and Clapper overlap. Most hybrids (called "Kling Rails"), particularly those found on the Gulf Coast, are unidentifiable.

Immature King Rail has dark mottling on the breast and the same reddish trim on the wings seen on adults. Immature Clapper has an unpatterned pale wash on the breast and lacks the reddish trim on the wings.

BEHAVIOR: Like Clapper Rail, a skulker. Forages mostly behind a vegetative screen, but during the breeding season, males often stand in the open to vocalize, and individuals, pairs, and family groups can occasionally be seen standing in the open, usually amid emergent vegetation and not far from the protective cover of cattails. Sometimes seen in roadside ditches in the South.

Most partial to fresh water. If it's a large rail and it's in fresh water, it's King. If it's brackish water, Kings are most likely to be found in the higher, less saline stretches (for example, in the marsh area closest to the mouth of a freshwater stream, or in an area where phragmites and scattered willows grow, rather than salt marsh grass). Generally shy, but where habituated to people (like boardwalks over marshes), may be very tolerant. Most vocal (and perhaps most active) at dawn and dusk but also vocalizes and forages during the day.

FLIGHT: Like Clapper Rail, but shows bright reddish patches on the upperwings (wing coverts).

VOCALIZATIONS: Calls are similar to Clapper Rail, but King Rail's grunt call, "*Rah Rah rah rah rah,*" often sounds slightly lower-pitched, more resonant, and less hurried. The "kek" series is likewise lower-pitched, less hurried, and more even in tempo. (Clapper often quickens the tempo in the middle of the sequence.) Also, King Rail's "kek" series often goes monotonously on and on.

PERTINENT PARTICULARS: The Clapper Rails in California and the Southwest are much redder than Atlantic and Gulf Coast birds and, in fact, are sometimes considered a subspecies of King. However, the ranges of these Clappers and King do not overlap. Comparisons are more academic than practical.

Virginia Rail, *Rallus limicola*

Miniature King Rail

STATUS: Common widespread but fairly secretive marsh bird. **DISTRIBUTION:** Breeds across much of the n. U.S., s. Canada, and much of the West—from the Maritimes to coastal N.C., west across s. Canada and the Midwest to Alta. and s. B.C. as well as the central plains, the Rockies, and much of the U.S. west of the Rockies. Winters to the limit of open water (most commonly in coastal areas), from s. New England to cen. Mexico (including all of Fla.) and s. B.C. to cen. Mexico. **HABITAT:** Most partial to shallow freshwater marshes that offer a mix of tall, standing vegetation, open pools, and muddy edges. Also found in brackish water and in winter coastal marshes, which it shares with other rails but most notably Clapper. **COHABITANTS:** American Bittern, Sora, Marsh Wren, Red-winged Blackbird. **MOVEMENT/MIGRATION:** Spring migration from mid-March to mid-May; fall from mid-August to early November. VI: 2.

DESCRIPTION: Small, dark, long-billed rail. Shaped like King and Clapper Rail, but only half the size—making it about the same size as Sora, but with a long bill. Much is made of the bird's laterally compressed girth—an adaptation for maneuvering through tall, dense vegetation. Seen in the open, the birds often seem quite plump.

Seen quickly and in poor light, the bird is all dark. Under better conditions, overall ruddy tones can be seen to bleed through the upperparts. The throat and breast are burnt orange, and the contrasting gray face is blatant. (On Kings, the gray on the face is more restricted and ambiguous.) Set against the bird's generally all-dark plumage, the white undertail on the usually up-cocked tail, as well as the bright red-orange lower mandible, are standout features. Stubby-billed Sora is overall paler, grayer, and distinctly short-necked; Clapper is overall grayer; King has a yellow-orange bill and dark undertail. Virginia's expression is irritated, cross.

Immature plumage (held into September) is mostly blackish, with a spattering of white on the underparts and a touch of rust toward the rear. But the small size and long bill alone distinguish these birds.

BEHAVIOR: Fairly secretive, spending much of its time in thick reedy growth. At dawn and dusk, often observed on the muddy vegetative edge of ponds, where it stalks more than walks into and out of the open, frequently darting for cover. Movements are abrupt and jerky but also nimble and quick, with the tail nervously flicked. Also hunts visually, sometimes from shore, sometimes crouched (Green Heron–style) at a stick perch from which it may dive into water in pursuit of prey. Wades and swims across narrow spans of open water and also makes short, low flights that generally end with the bird dropping or running into cover. When suspicious, may stand concealed by vegetation with only its bright red-orange bill and gray face showing.

Though territorial, in winter multiple birds may occupy even small marshes, which erupt with vocalizations at dawn and dusk.

FLIGHT: Flight is weak and struggling. Flies with head forward and legs drooped (for short flights) or trailing behind when covering ground. Lofts into the air with a frantic fluttering of round wings and lands with a braking crash after a short, straight flight.

VOCALIZATIONS: Has three classic vocalizations. One is a descending series of grunting "*oinks*" that are richer, higher-pitched, more comically piglike than Clapper Rail, and sometimes given in a series of three, though more often in a longer series that loses volume and accelerates (in the classic "bouncing ball" way). A second vocalization is a brittle ticking tattoo that begins with single notes that soon double: "*kick, kick, kick, ka-dik, ka-dik, ka-dik, ka-dik.*" Finally, makes a higher-pitched scold that begins with a sharp stutter and ends in a police whistle–like trill: "*ch'ch'ch'cherrrr; ch'ch'ch'cherrr.*" Also makes short sharp twitters and squeals, frequently when disturbed or as it retreats.

PERTINENT PARTICULARS: This is a very hardy rail, capable of surviving prolonged freezes providing it can find a narrow lead of open water. While most vocal at dawn and dusk, also calls at night (occasionally in daylight) and responds to even poor imitations of its call (or the call of Sora). Making a loud noise (hand clapping, a squeal call, the plop of a rock thrown into the water) can also entice it to vocalize.

Sora, *Porzana carolina*
Masked Marsh Chicken

STATUS: Common, widespread, and, while somewhat secretive, the most likely freshwater rail to be seen. **DISTRIBUTION:** Breeds across subarctic and boreal Canada from w. Nfld. and N.S. west across s. Hudson Bay, n. Man., Sask., all of Alta. and into the Northwest Territories, all but n.-cen. and w. B.C., and sw. Yukon. In the U.S., breeds widely throughout the northern and western states, with the southern border defined by N.J., s. Pa., cen. Ohio, cen. Ind., cen. Ill., cen. Iowa, cen. Kans., e. Colo., e. N.M., s. Ariz., and s. Nev. The western border falls across e. Calif., w. Ore., and w. Wash., with disjunct populations found along the Ore. coast and the Central Valley of Calif. Winters in the extreme s. U.S. and in coastal states from Va. south to Central America and the West Indies, west through s. Tex., s. N.M., and s. and e. Ariz., north to s. Nev. and se. Utah, and along the West Coast from n. Calif. to Baja. **HABITAT:** Breeds in shallow freshwater wetlands dominated by tall emergent vegetation like cattail and bulrush. Also breeds in brackish, coastal marshes dominated by cord grass (*Spartina*). In winter and migration, may be found in fresh, brackish, and tidal marshes with emergent vegetation; also in drainage ditches, impoundments, rice fields, wet pastures, and (less commonly) dry, grassy meadows and fields. **COHABITANTS:** Least Bittern, Virginia Rail, Marsh

Wren, Swamp Sparrow, Red-winged and Yellow-headed Blackbirds. **MOVEMENT/MIGRATION:** Except for some southwestern and coastal areas, the entire population migrates. Spring migration from late March to late May; fall from mid-July to early November. VI: 4.

DESCRIPTION: Pear-shaped and pear-sized marsh bird with a candy-corn-shaped bill. Smallish, compact rail (larger than Yellow Rail; slightly smaller than Virginia Rail), with a smallish head, a conical bill, a heavy neck, an oval body, a pointy tail that is often up-cocked, and relatively long legs.

Overall cryptic—warm brown above; touches of gray on the face and breast; barred flanks (which are often hard to see). Standout features at a distance are the pale yellow bill, black face mask (adults), and white undertail. Immature birds are like adults but overall duller and less contrastingly patterned. The breast is dull yellowish, not gray, the bill is dingy, and they show little or no black on the face.

BEHAVIOR: A slow, methodical feeder, swimming with head raised or walking in a head-lowered crouch. Most often seen at the very edge of emergent vegetation (often in shadows), moving into and out of view or running across openings. Also a solitary feeder, although in winter and in migration, where there is one Sora there are often more. A shy bird that retreats quickly (by running) when startled, but often returns. Often swims, with a head-jerking and -bobbing motion. Sometimes dives when pursued. Responds well to tapes, squealing, and imitations of its calls. Flushes reluctantly, and flies like the Wright Brothers.

FLIGHT: Seems long-winged, long-legged, and frail in flight. Rises vertically from the marsh. Flight is mostly straight but often curving, with neck extended and legs dangling or trailing. Wingbeats are hurried and constant; flight is slow, effortful, and struggling; landings are clumsy.

VOCALIZATIONS: Classic call is a series of descending short notes that suggests a whistled whinny: "*We he he e e e e er.*" Also emits a mellow, plaintive, rising two-note whistle: "*tur-EEE . . . tur-EEE.*" Alarm is a high, sharp, surprised "*keek,*" sometimes given in multiples.

Purple Gallinule, *Porphyrula martinica*

The Southern Lily-Pad Chicken

STATUS: Fairly common, but often restricted and secretive resident of southern freshwater wetlands. Frequent vagrant well outside its normal range. **DISTRIBUTION:** In the U.S., limited to the Southeast, from coastal S.C. west across s. Ga., all of Fla., s. Ala., parts of Miss., s. and w. La., and e. and coastal Tex. (also localized in parts of Ark.). **HABITAT:** Lush, densely vegetated (matted and woody) marshes, ponds, lakeshores, and mature rice fields (most often near coastal locations). **COHABITANTS:** American Alligator, Anhinga, Least Bittern, Little Blue Heron. **MOVEMENT/MIGRATION:** Except in Florida, U.S. birds vacate breeding areas in winter. Spring migration from mid-March to early May; fall from August to late October. This species is a notorious vagrant (particularly in spring), with sightings particularly concentrated in the East north to Newfoundland. VI: 3.

DESCRIPTION: A medium-sized, serpentine-contoured, gaudily plumaged (and generally unmistakable) rail that likes to clamber about where it's thick. Slightly smaller than Common Moorhen, with a somewhat more slender and urnlike profile—the product of a long, heavy-based, but distinctly sinuous neck and a smallish head that slopes down to meet a prominent, outsized, and bright red bill.

Adult plumage features an iridescent green back and an undeniably purple head and neck; coupled with the bird's bright yellow legs, bright blue forehead shield, and bright red bill, this plumage renders it unmistakable. If second-guessing is your nature, note the absence of any white along the sides and the wholly bright white undertail, which will be prominent on the raised and repeatedly flicked tail. The greenish-backed immatures, with warm buffy brown on the head and throat and bright yellow legs, are also distinctive. The expression is predatory.

BEHAVIOR: Despite its plumage (or perhaps because of it), the bird likes to be where vegetation is tight and thick—often in and among rank weeds and branches overlying the edges of ponds and marshes. It does not tend to swim in open areas carpeted with surface weeds (as does Common Moorhen). It does habitually forage, by walking in a predatory crouch, across lily pads, and it spends a great deal of time clambering in rank vegetation and even climbing in the elevated branches of trees (something Common Moorhen does less frequently). Flicks its tail constantly. More fluid and more nimble than Common Moorhen. Moves its head and neck in a smooth, sinuous way (not with the spastic jerkiness of Common Moorhen). Also more inclined to fly than Moorhen (perhaps because it is less inclined to swim).

FLIGHT: Silhouette is a lanky and somewhat wilted rail showing long yellow legs and a short bill. All dark (except for the bright yellow legs). Flight is weak but steady; wingbeats are quick, regular, and somewhat floppy.

VOCALIZATIONS: Varied calls are generally more musical and pleasing than Common Moorhen. Call is a henlike chuckle, "*ha, ha, hah, hah,*" that is lower-pitched than Common Moorhen. Also emits low clucks that are commonly given in series that vary in tempo and pitch (sounds like a banjo being tuned). Also makes a higher-pitched, bugled "*heh.*"

Common Moorhen, *Gallinula chloropus*

Coot with a Racing Stripe

STATUS: Fairly common to uncommon and somewhat secretive freshwater marsh bird. **DISTRIBUTION:** Distribution is spotty. In the U.S., breeds across the East and s. Canada from the Maritimes to Fla.; west across extreme s. Que. and se. Ont. to w. Minn., e. Neb., cen. Kans., cen. Okla., and n. and e. Tex. Also found in isolated pockets in Utah, Nev., Calif., N.M., and Ariz. Most eastern birds depart in winter. Winters from coastal N.C. to s. and sw. Tex. Year-round in Cuba, Mexico, and Central America. **HABITAT:** Typically freshwater lakes, marshes, impoundments, ponds, and ditches with extensive thick, emergent vegetation (for cover) and open water rich with floating or submerged vegetation (for feeding). Classic plant types include cattail, phragmites, cord grass, and (in the South) water hyacinth. **COHABITANTS:** Pied-billed Grebe, Least Bittern, American Coot, Red-winged Blackbird, Muskrat. **MOVEMENT/MIGRATION:** All northern (and many southern) birds appear to migrate, although wintering numbers may be expansive in parts of the South. Spring migration from March to May; fall from August to November. VI: 2.

DESCRIPTION: A cootlike water bird best and most easily distinguished from American Coot by the narrow white racing stripe running along its sides (and in breeding adults, the bright red bill and face or shield). Smaller, less dumpy, and more angular and elegantly contoured than a coot, with a longer, more slender neck, a more petite bill, a flat forehead, and, when swimming, a distinctly pointy and up-angled posterior with wingtips that jut above the tail. The expression is alert, intense—but the dark eye against the dark face is difficult to see.

Adult plumage is basically dark, with a narrow white line (sometimes broken into feathery brush strokes) etched along the bird's sides and a white undertail bisected with a black center (all white in Purple Gallinule). The folded wings of adults have a subtle brownish cast. Mostly gray and drab, immature Common Moorhen is similar to an immature American Coot, but slighter and slimmer (more suggestive of a Sora), with a darker, more petite bill, overall browner plumage above, and usually a discernible white racing stripe down the sides.

BEHAVIOR: Not as gregarious as American Coot, and not as prone to be in the open, away from cover. Commonly hugs lushly vegetated shorelines (or, more maddeningly, lurks just behind the vegetation). Forages in the open, however, where surface vegetation is lush on lakes and ponds, well away from shore and tall stands of reeds. Feeds by walking along in a crouch or atop lush vegetation, or swimming with neck straight and angled up and forward. Most commonly picks food from the surface; less commonly dabbles, and much less commonly dives. Not as prone to climb in tall or woody vegetation as Purple Gallinule, but may do so.

Moves its head with exaggerated jerks as it swims. Jerks its tail when it walks. Movements are generally more deft and elegant than American Coot, with which it regularly associates.

FLIGHT: Flies with even more reluctance than Coot! Much prefers to swim away from danger, or even dive. In flight, overall slimmer and longer-legged than coot (with no trailing white edge to the wing). Flight is weak, effortful, and straight, on steady wingbeats, but often veers before landing.

VOCALIZATIONS: Like American Coot, makes a variety of single- and multiple-note bugled cluckings, but generally higher-pitched and more cluck than bugle. Typical call is a strangled, two-part, bugled ensemble—the first part is rapid, and the last three (or so) notes are slowed: "*reh, reh, reh, reh, reh; reh reh reh.*"

PERTINENT PARTICULARS: Observers tend to try to turn this species into Purple Gallinule—in part because the less common species is celebrated for its tendency to wander well outside of its normal range, and in part because the birds are structurally similar (though not identical), and in good light the front of Common Moorhen shows iridescent purple hues. It is easy to turn an adult Common Moorhen into a Purple Gallinule. It is almost impossible to misidentify a real adult Purple Gallinule. If you are struggling with the identification, it isn't Purple Gallinule.

American Coot, *Fulica americana*

Blackened Teapot with a White Spout

STATUS: Common breeding species across much of North America; more common in the West. Sometimes abundant in winter. **DISTRIBUTION:** Widespread but complex. In the U.S., basically a northern and western breeding species; largely absent as a breeder east of the Mississippi R., except for isolated pockets in the Maritimes, w. N.Y., w. Pa., coastal N.C. and S.C., Fla., coastal Miss., and wrapped around the s. Great Lakes. In the West, breeds essentially everywhere (including Mexico, but excluding nw. B.C.). In winter, withdraws from northern regions and expands east; found from coastal s. New England to Fla. and west (except for colder, elevated portions of the Appalachians) and south of a line drawn from w. N.Y. to s. Idaho, then north into s. B.C. and north again along the coast to extreme s. Alaska. **HABITAT:** Breeds in freshwater bodies of water that offer both stands of emergent vegetation and open water for foraging—extreme examples include large lakes with reed beds along the shore and shallow, reed-choked marshes with open pockets. Size is not particularly important. Birds have been known to breed in sloughs, canals, sewage, and small farm ponds. In migration and winter, also found in large open bodies of water without emergent vegetation, coastal bays, and brackish and tidal marshes. Also in winter forages freely on shortly cut grass and lawns near ponds (such as city parks and golf courses). **COHABITANTS:** Ruddy Duck, Red-winged Blackbird. In winter, Eared Grebe, assorted puddle ducks, geese, and swans. **MOVEMENT/MIGRATION:** Although some birds are permanent residents, others are short- to medium-distance migrants; spring migration from March to early May; fall from August to early November. VI: 2–3.

DESCRIPTION: Small (Green-winged Teal–sized), compact, all-blackish water bird with a contrastingly white chickenlike bill. Looks like a charcoal gray teapot with a white spout. Overall plump (a sumo wrestler of a rail), with a dome-backed, broad-beamed, and roundly contoured body that sits high on the water. The head is large, the neck short and thick, and the rear (at a distance) looks blunt—wingtips are flush with the back (never elevated like Common Moorhen). The very abbreviated up-cocked tail is hardly noticeable. On the water, head and neck are curved like a bullnecked swan. On land, looks dumpy and stumpy-legged.

Adults, except for the white bill and barely discernible white outer trim on the tail, are all charcoal gray (no brown in the plumage). Immatures are dingy brownish gray and darker above; the heavy bill is pale gray.

BEHAVIOR: Highly gregarious—in winter, may be found in flocks or rafts numbering in the hundreds and thousands. Also found with assorted ducks and water birds. While often feeds or rests in vegetated areas, not shy about being in open water, and also commonly forages on land (particularly fond of short cut grass). Feeds by plucking from the surface and dabbling and diving (something moorhen rarely does). A slower and more methodical feeder than Common Moorhen. Doesn't like to reach for food (plucks what's right in front of its breast). Often attends larger waterfowl (like geese and swans), genially pilfering aquatic plants from their bills. Loath to fly, and becomes airborne only after a lengthy, wing-flailing, foot-pattering takeoff that as often as not results in the bird settling once more on the water. The head jerks as it swims. Vocal both day and night.

FLIGHT: Posture is like a somewhat potbellied rail with an arched back, slightly drooped head, and trailing legs. Flight is usually low and labored, with narrow flailing wings beating steadily. Course is usually straight, but birds sometimes veer (as if too much "English" was applied). The narrow white trailing edge to the inner wing is usually visible on birds flying away.

VOCALIZATIONS: Many and varied, but almost invariably comical. Repertoire includes low clucks, growled mutterings, and bugled grunts, trills, and guffaws that are like Common Moorhen but generally lower-toned. Most commonly heard call is a bugled two-noted "*ree-uh, ree-uh, ree-uh,*" whose pattern and cadence recall a saw moving through lumber.

Limpkin, *Aramus guarauna*

The Wailing Rail

STATUS: Uncommon to locally common resident species. **DISTRIBUTION:** Widespread in New World tropics. In the U.S., occurs only in Fla. (rare in extreme se. Ga.), except absent in the western panhandle region and the extreme northeastern coast (Jacksonville/St. Augustine). **HABITAT:** All manner of clean freshwater swamps, particularly those with slow-moving rivers and spring-fed waters; open, grassy wetlands; and lakeshores and canals with a lush cap of floating vegetation. The bird's presence in any habitat is contingent upon the presence of apple snails—its favored prey. **COHABITANTS:** American Alligator, Little Blue Heron, Snail Kite, Black-necked Stilt. **MOVEMENT/MIGRATION:** Permanent resident. Some northern birds, however, are believed to retreat south after the breeding season (May to August), with returning birds noted from January to March. VI: 1.

DESCRIPTION: Looks like a large, dark, white-paint-spattered ibis. Slightly larger than ibis, but similar in structure. Limpkin has a stouter bill that is heavier throughout its length and curved mostly at the tip, a comically small head, a shorter, heavier, less S-curved neck, and longer legs. Feeding birds, leaning forward with bills angled down, are slightly less hunchbacked than ibis, and the neck is not as sinuously curved. When alert, the bird's posture is almost erect, and it resembles a bowling pin with an acutely down-angled bill. (Ibis holds its bill horizontal.)

Overall olive brown, with a head and neck finely flecked with pale feathers (appears hoary at a distance) and a body (particularly the front half) spattered with white.

BEHAVIOR: Not particularly secretive, just very difficult to see in dark wooded swamps and (particularly) Everglades-type grasslands. More often heard than seen (and heard day and night). Feeds by walking along banks with a slow striding gait, wading through water (generally not up to its belly) or striding across heavy mats of vegetation, probing with its bill. Bill thrusts are driving, almost angry, and describe an arch in front of the bird. Ibis, by comparison, stitches the water with rapid movements of the bill. Ibis also feeds on dry land—something Limpkin does not do.

Often forms loose breeding colonies, but forages alone or close to a mate or young (never in a grouped flock). When disturbed, stands erect, bill angled down (somewhat storklike). When calling, leans forward with body and neck horizontal. Flies well (for a rail), lofting from the water with little effort. Perches on tree limbs and walks along branches with ease. The expression is harried, worried.

FLIGHT: Profile, again, is ibislike, but a stocky, very broad-winged ibis (structured somewhat like Hadada Ibis of Africa). Unlike Glossy Ibis, Limpkin's head and bill are angled down. Also, when flying short distances, the bird dangles its feet; for longer distances, feet are raised, trailing behind. Flight is buoyant; wingbeats are slow, regular, and executed with a stiff, snapping, cranelike upstroke. Short glides are steady and buoyant. Lands with wings raised, over the body.

VOCALIZATIONS: One of the greatest sounds in nature (and arguably one of the best known, since Hollywood dubbed this New World species call into the soundtracks of assorted old jungle movies). Classic call is a wailed scream: "*keowww,*" that sometimes begins with a stutter, "*k'k'kke-owww.*" Suggests a musical Sandhill Crane. Also makes a low clucking: "*klok, klok.*"

Sandhill Crane, *Grus canadensis*

The Gray Crane

STATUS: Common widespread breeder across much of central and western Canada and Alaska and in isolated and widely scattered pockets in the Lower 48. In winter, a somewhat restricted but often abundant species across portions of the southern United States and Mexico and, as a migrant, at key staging areas along migratory routes. **DISTRIBUTION:** Breeds from Mich., Wisc., and Minn. north to James Bay, Canada, then north and west across virtually all of mainland North America and the islands of the Northwest Territories (mostly absent as a breeder in the southern Canadian prairies), nw. B.C., s. Yukon. and extreme n. Alaska. Also breeds in the Great Basin, along the Idaho–Wyo. border, and in n. Utah, nw. Colo.; coastal Miss., se. Ga., and n. Fla., as well as Cuba. A few birds now breed in Ohio, w. Pa., and upstate N.Y., and Maine. A small resident flock in s. N.J. probably contains some pure Sandhills, but many birds are hybrids, showing characteristics of Common Crane (*Grus grus*). All birds except the Miss., Fla., and Cuban populations relocate south for the winter, with various populations and subspecies found in Calif., s. Ariz., N.M., n. Mexico, much of Tex., sw. Okla., coastal La., and Fla. **HABITAT:** Breeds in a variety of open and generally wet habitats, including marshes, forest bogs, open grasslands, beaver meadows, bulrush and cattail marsh, wet meadows, and transition-zone wetlands with woody growth. In winter and migration, concentrates in large marshes and grain fields of all sorts, but particularly corn stubble, cattle range and dairy farms, intertidal marshes and mud flats, salt lakes, rivers, and deltas. **COHABITANTS:** In winter, often associates with geese, particularly Snow Geese. **MOVEMENT/MIGRATION:** Migratory populations follow different routes and have slightly different timetables. In the East, birds fly from eastern Canada and the Great Lakes to Florida, a route that carries them west of the coastal states and east of the Mississippi. In central and western North America, migrating cranes may be encountered from the eastern prairies to the Pacific, but singular in its significance and numbers is the concentration of staging birds found along the Platte River Valley in central Nebraska in March. Spring migration from mid-February to late May, with peaks from March to early May; for northern birds, fall migration from September to December, with peaks in October and November. VI: 3.

DESCRIPTION: There are only two cranes in North America. This is the gray one, distinguished from most other long-necked heronlike birds by its somewhat inelegant, bottom-heavy body type, its drooping, feather-duster tail, and an array of behavioral traits. Sandhill Cranes range in size from the larger, more southern-breeding "Greater Sandhill" to the smaller "Lesser Sandhill," which breeds in the arctic. In terms of plumage and structure, they are very similar.

The body seems somewhat vaselike when the bird is standing upright; the neck is short (compared to herons and egrets), and the head embarrassingly small, but the long, straight, daggerlike bill is formidable. When foraging, birds lean forward with the body horizontal and the neck curved up—a posture that makes distant birds look like two-legged, one-humped camels.

In winter, adults are uniformly pale gray except for a paler face, a distinctive red crown, and a smattering of rusty splotches on the wings and back. In breeding plumage, the wings and back are overall rustier. Immatures are like adults, but they are overall slightly browner and lack the red cap.

BEHAVIOR: It is rare to see a single crane. Small family groups are typical; in winter and migration, great concentrations involving hundreds, even thousands, of feeding and (particularly) roosting birds are the norm. Forages across open country by day, often in dry, open, upland habitat. In winter and migration, roosts in marshes or roosts by standing in rivers.

A foraging Sandhill Crane both picks from the surface and probes. Moves steadily, head slightly elevated or angled down, with a halting stalk. Where acclimated to people (as on refuges), can be quite tame, allowing close approach. More wary where hunted.

Cranes are usually heard before they are seen—in part because they are extremely noisy and their vocalizations carry, but also in part because, despite their large size, their gray plumage makes them cryptic. When flying between roosting and feeding areas, they are very vocal.

FLIGHT: Flight is distinctive. In active flight, Sandhill Crane is both angular and balanced. Straight planklike wings are set at a right angle to the body, the neck stretches out in front, and the legs trail behind (except in cold weather, when legs may be drawn up). Wingbeats are steady, slow, regular, and distinguished by a crisp, smart upstroke. It seems as though the bird is caressing the air.

When gliding or soaring in a thermal (which it does often), Sandhill seems to wilt or slouch. Wings turn back at the tips, making the bird appear pointy and slightly down-drooped. The legs and neck droop as well, although the head remains slightly elevated.

When soaring in a thermal, flocks turn in the same direction. In small groups or when flying short distances, birds form a line. When migrating over greater distances, they fly in a V-formation (like geese) and occasionally migrate in formation with geese.

VOCALIZATIONS: One of the greatest sounds in nature. Adult's call is a loud, resonant, bugled trill. In migration, a flock sounds like an orchestra of ungreased creaking wooden hinges. Also makes a low growl or a rumbling purr. Young cranes make a weak, pathetic-sounding, chirping trill that sounds very much like the amplified food-begging call of a nestling songbird.

PERTINENT PARTICULARS: Great Blue Heron and Sandhill Crane are similar in size and shape, but very different in flight. The heron usually flies with its neck coiled and on down-bowed wings. The wingbeat, which looks as though the bird is dipping its wingtips into an invisible cauldron, seems hardly to reach above the body. Cranes fly with necks outstretched and on straight stiff wings that rise and fall smartly and crisply above and below the body. Also, cranes fly in V-shaped, often large flocks (like geese). Great Blue Herons fly in small mobs, and migrating flocks rarely exceed 20 birds.

Whooping Crane, *Grus americana*

Great White Crane

STATUS: Rare and geographically restricted. **DISTRIBUTION:** There are two wild populations in North America. The migratory population breeds in Wood Buffalo National Park in s. Northwest Territories and n. Alta. and winters on the cen. Tex. Gulf Coast in and around Aransas N.W.R. In migration, birds pass through ne. Alta., s. Sask., ne. Mont., w. N.D. and S.D., cen. Neb. and cen. Kans., w. and cen. Okla., and e. and cen. Tex. The nonmigratory population has been established in the Kissimmee Prairie in Fla. A third migratory population has recently been introduced to cen. Wisc. (Horicon Marsh N.W.R.), with a wintering territory on the Fla. Gulf Coast. An earlier reintroduction effort in the Idaho Rocky Mts. failed. **HABITAT:** Wild migratory population breeds in extensive, shallow, taiga forest wetlands. In winter, birds establish territories in brackish coastal wetlands, foraging on tidal flats, ponds, marshes, and open farmland. During migration, birds forage on open cropland and roost in freshwater wetlands. The nonmigratory Florida population is found in saw palmetto prairie dotted with lakes and wetlands. **COHABITANTS:** In Wood Buffalo National Park, found in the same habitat as Pacific Loon and Sora. In winter, associates with assorted egrets, including Reddish Egret, American White Pelican, Sora, White-tailed Hawk, Laughing Gull. **MOVEMENT/MIGRATION:** Spring migration from Aransas from late March to early May, with breeders arriving in Canada as early as late April. Fall migration from mid-September to early October, though many birds spend several weeks in southern Saskatchewan and reach Texas between late October and mid-November. VI: 1.

DESCRIPTION: An impressively large, stately, white crane—North America's tallest bird. About 25% larger than Greater Sandhill Crane, Great Blue Heron, and Wood Stork; dwarfs Great Egret (which occupies the same habitat in Florida and coastal Texas). Overall bulkier than an egret, with a hefty-based, pointy, and straight bill, a long thickish neck, a robust body, and a fluffy, drooped feather-duster of a tail. (The posteriors of egrets are more streamlined.) Legs are long and black.

Overall bright white (black wingtips are usually concealed unless the bird is in flight.) The red on the adult's crown and around the base of the bill shows up at great distances. Immatures have a rusty-buff-colored wash over the head, neck, and upperparts that becomes patchy over the course of the winter.

BEHAVIOR: At Aransas N.W.R. (the place to go to see this bird), look for the large, white, long-necked birds feeding in pairs, with or without one or two young. Cranes are monogamous and largely antisocial, maintaining their bond, proximity, and territory throughout the winter. Egrets are single, not paired, and normally widely spaced, but they are also concentrated among other bird species where fish are abundant. American White Pelicans roost and feed in flocks.

When feeding, Whooping Crane rarely stands still for long. Walks with the body horizontal, the neck coiled in an S configuration, and the head lowered below the bulk of the body with the bill pointed down. (Herons feed with their necks elevated above the back.) A walking Whooping Crane suggests an all-white, two-legged camel; a standing bird looks like a long-necked vase (or a big, white, alert Limpkin). Overall movements are steady, stately, flowing. When not feeding, walks with head erect.

FLIGHT: Flies with neck and legs extended (though in colder temperatures may draw legs in) and long rectangular wings set at a right angle to the body. Overall white, with black wingtips—the same basic pattern as Wood Stork, White Pelican, and Snow Goose. Snow Goose is much smaller, however, as well as shorter-necked, and shows no trailing legs. Both White Pelican and Wood Stork show black along the trailing edge of the wing, not just in the wingtips. Herons and egrets keep their necks coiled in flight, not extended. At a distance and in some lights, Sandhill Crane may seem very pale and silvery, but not truly white.

VOCALIZATIONS: Call is a loud trumpeting bugle that is higher, clearer, and less croaking than Sandhill Crane.

SHOREBIRDS — PLOVERS AND SANDPIPERS

Black-bellied Plover, *Pluvialis squatarola*

Gray Plover with the Haunting Whistle

STATUS: Common high arctic breeder; common wintering species along all three coasts; uncommon migrant away from coastal areas. **DISTRIBUTION:** Breeds in w. and n. Alaska and along the north coast and islands of arctic Canada. Winters coastally from Mass. (locally in s. N.S.) and se. B.C. south to South America (also inland at the Salton Sea). Some nonbreeding birds remain in coastal areas through the summer. **HABITAT:** Breeds on open tundra well above the tree line—often, but not always, in wetter areas than American Golden-Plover. In winter, forages over tidal mud flats and along sand beaches. In migration, flocks forage freely on tidal marsh, newly plowed fields, flooded agricultural land, airports, and sod farms where it sometimes associates with golden-plovers. **COHABITANTS:** In winter, often found with Dunlin. In migration, may associate with golden-plovers and may migrate with sea ducks. **MOVEMENT/MIGRATION:** Spring migration from early March to early June, with peak in late April and May; fall from early July to early November. VI: 3.

DESCRIPTION: A bulky, squarish, grayish street brawler of a plover that likes to keep its feet damp. Large and sturdy (larger than American Golden-Plover, Red Knot, and Surfbird; smaller than Willet), with an overall stocky and robust body, more potbellied than chesty, and showing more girth than golden-plovers (most evident head on and from the rear); a large, somewhat squared head; and a short, thick neck. The bill is heavy and long (nearly as long as the head is wide), and often noticeably swollen at the tip. Legs are long, but appear proportionally shorter than those of golden-plovers. There is nothing delicate about the bird.

Breeding plumage is distinctive. Males show white crowns, silvery gray backs (mottled black and white at close range), and jet-black underparts (except for a white undertail). From a distance, Black-bellied Plover seems whitish or silvery above and black below—just the opposite of most birds. Females have the same pattern but are slightly browner above and somewhat mottled below, and the definition is not so crisp or clear-cut.

Nonbreeding plumage is basic gray with some white speckling above, dirty white below. Juveniles are like nonbreeding adults but have considerably more white speckling above and pale gray streaking on the underparts (particularly the breast). In some juveniles, the upperparts are washed with gold that is reminiscent of juvenile American Golden-Plover. In all birds, the expression is formidable.

The bird's posture seems hunched, or slouched, when foraging, as though dejected. The large head, heavy bill, and legs that seem a tad too far back give the bird a front-heavy appearance.

BEHAVIOR: Black-bellied Plover is a predator. It crouches as it moves and stalks more than walks, taking a few steps followed by an alert pause. Another step follows, and the bird leans forward, head turned slightly askew, studying, analyzing. Then a rush of steps. A downward jab. A savoring moment. And then the stalk begins anew. Everything about the bird's movements seems menacing. In conjunction with the overlarge head, no-necked appearance, large body, and forward-leaning posture, those movements easily conjure the image of a Tyrannosaurus Rex of the mud flats.

Mud and sand are the bird's preferred elements. Black-bellieds think nothing of mucking it up, wading in ooze that any self-respecting golden-plover would shun. It feeds among other shorebirds, even foraging in shallow water (something other large plovers don't do), but it commonly forages away from other Black-bellieds. Predators need their space. When the tide shifts, birds gather into small flocks that fly to high-ground roosts, where they often stand as a segregated unit, flanked by lesser species. In migration, often flies with other shorebird species and even, on occasions, with waterfowl.

Predator or not, Black-bellieds do not tolerate close approach and are among the first birds to fly when pressed, calling as they climb.

FLIGHT: Large, bulky, angular profile, with a large projecting head, big chest, and very angular and triangular wings that appear shorter than they are. In all plumages, the black axillaries (wing pits) and pale tail distinguish it from any of the golden-plovers. The prominent white wing stripe (buffy and less distinct on a golden-plover) is easily noted. A strong, forceful flier once airborne, but needs a step or two to get aloft. The line of flight is usually direct—no nonsense, no frills. Wingbeat is regular, powerful on the upstroke and down stroke, and somewhat stiff and arthritic, almost choppy. For short distances, often flies in a fairly well-regimented string. For greater distances, birds fly in a V or a line.

VOCALIZATIONS: Call is one of the most recognizable calls in nature. A long, loud, trisyllabic whistle, "*Pee-ur-Eee*," that has a lonely, haunting quality. Heard once, it is easily remembered (and imitated). Also makes a truncated, softer, and hurried "*Pee-ur,*" often when flushed.

PERTINENT PARTICULARS: In winter, the face pattern recalls Mountain Plover (ranges overlap in Texas and California), but Black-bellied is heavy-billed and shows a whitish (not black-tipped) tail.

American Golden-Plover, *Pluvialis dominica*

Buffy Plover That Yelps

STATUS: Common arctic and subarctic breeder; common spring migrant through Great Plains; uncommon fall migrant along the Atlantic and Pacific Coasts (rare in spring). By far the more common and widespread of the two North American golden-plovers. **DISTRIBUTION:** Breeds in western coastal Alaska and across the northern two-thirds of the state east across most of the Yukon to extreme n. B.C., as well as arctic portions of the Northwest Territories to the western and southern shores of Hudson Bay. Winters in South America. **HABITAT:** Nests on dry rocky tundra, most often on elevated slopes; also lower, wetter tundra. During migration, favors open, short-grass habitat, including prairie, pasture, golf courses, airports, and sod farms, but also uses mud flats, estuaries, and beaches. Not commonly seen standing in water like Black-bellied Plover. **COHABITANTS:** In summer, Long-tailed Jaeger, Lapland Longspur. In migration, Black-bellied Plover, Killdeer, Upland Sandpiper. **MOVEMENT/MIGRATION:** In fall, most birds pass over eastern Canada and migrate off the Atlantic Coast, flying nonstop to South America, but some take a course that carries them over the eastern half of the continent; a few others migrate along the Pacific Coast, stopping and foraging along the way. Spring migration occurs from late March through May and is concentrated through the eastern prairies (although smaller numbers appear farther east, and fewer still farther west). In spring, along the primary migration route in Texas, American Golden-Plovers can occur in large numbers. Adults begin fall migration in July, with peaks in late August and early September; juveniles migrate from September into November. A few birds linger into December, and a very few winter in North America, primarily on the southern coasts. VI: 3.

DESCRIPTION: In all plumages, overall smaller, uniformly darker, and more delicately proportioned than Black-bellied Plover. If Black-bellied Plover is a street brawler, golden-plover is gentry. The body is slimmer, sleeker, and longer than Black-bellied, and better proportioned—more chesty than portly, more pear-shaped than squared. The head is small and round and not neckless. The short bill is slim, almost petite, and tapered to a point, not bulbous and blunt. The legs are longish, but not overtly so. The primaries clearly extend beyond the tip of the tail, enhancing the lean lines of the bird and offering perhaps the best way to distinguish it from the very similar Pacific Golden-Plover (and perhaps a way to distinguish Black-bellied, whose wings also extend just to the tip of the tail).

Breeding plumage is distinctive and distinguished: brown with inlaid golden flecks above; uniformly black below from face to tail; a white sash on the sides of the neck (but not the sides). Females are like males but often show a broken white border between the folded wing and the all-black underparts (including a black undertail), and the black is not as solid or intense, especially on the face.

Nonbreeding plumage is overall warm gray with slight contrast between the upperparts and slightly paler underparts. Black-bellied Plover is colder and grayer above, with contrastingly paler underparts. Juveniles are overall buffier, warmer, and washed with gold above, and grayer with little patterning below. All birds show a distinct whitish eyebrow that kicks up at the rear and, in combination with the darker crown, gives them a capped appearance. In Pacific Golden-Plover, the eyebrow is usually buffier or golden (on juveniles) and often turns down at the rear.

BEHAVIOR: Forages in single-species flocks, usually in short grass or tilled fields, or may mix with other species, most notably other plovers, like

Killdeer and Black-bellied Plover, and Buff-breasted Sandpiper, and Upland Sandpiper. Golden-Plover's smaller groups (5–20 birds) tend to group tighter than Black-bellied and seem more cohesive or coordinated. Posture is erect, not slouched, and birds are alert, almost haughty. Golden-Plover's movements are deliberate, and with its head held up, it seems less myopic in its efforts to find prey than Black-bellied. Often reaches for food with a theatrical bow, and sometimes runs ahead to snatch prey. Fairly wary, but perhaps less skittish than Black-bellied.

FLIGHT: Snub-billed, with a very long and slender scimitar-winged profile; less bulky than Black-bellied. In flight, even allowing for the white neck of breeding-plumage birds, appears uniformly dark (or uniformly buffy) above and below. Rump/tail is as brown as the back; wing pits are pale, not black. Flight is fast, fluid, direct, and more buoyant, less choppy, and less effortful than Black-bellied. Wingbeats are rapid, *regular,* and cleaving, with a deep down stroke. Fluidity also characterizes flocks, which move with an air of coordinated casualness; groups are cohesive, but not regimented.

VOCALIZATIONS: Song is a monotonously repeated, shrill, bisyllabic, ringing, whistled yelp, "*per'chee*" or "*ker'lee,*" and sometimes a more single-noted "*p'reee.*" In flight, makes a more liquid, mellow, bisyllabic yelp, "*KLEE-up*" or "*KLEE-yeep*" (accent on the first syllable), that is more urgent and less haunting than Black-bellied's whistle. Also gives a single-note variation.

Pacific Golden-Plover, *Pluvialis fulva*

Plover with a Racing Stripe

STATUS AND DISTRIBUTION: Common coastal breeder on the western coast of Alaska, breeding contiguously from about Point Hope to, at least, the mouth of the Kuskokwim R. (possibly farther east). Range encompasses the entire Seward Peninsula and extends perhaps 100 mi. inland in sw. Alaska. Most birds winter outside the continent, but a few regularly occur in coastal Calif. **HABITAT:** Breeds in arctic and subarctic tundra. Where range overlaps with American Golden-Plover, Pacific favors flatter, lower, wetter, less rocky, and more vegetated habitats. Where the species do not overlap, Pacific is found in the same higher, drier stony slopes occupied by American Golden-Plover. In winter and migration, found in a variety of open habitats, including coastal marshes, mud flats, tilled fields, beaches, sod farms, and golf courses; also found in drier, sandier habitats, such as beaches. **COHABITANTS:** During breeding, Whimbrel, Bar-tailed Godwit, Long-tailed Jaeger. **MOVEMENT/MIGRATION:** Most migration is conducted over the Pacific or along the coast of Asia. In spring, birds reach Alaska in late April and continue through May; in fall, adults depart in late July, and juveniles from late August to early October. VI: 2.

DESCRIPTION: A golden-plover with a white racing stripe running the length of its medium-sized body, from forehead to tail. *Very similar in size and shape to American Golden-Plover,* with these differences: the body appears slightly more compact and less streamlined, and the bill and legs average slightly longer. Wings extend ever so slightly beyond the tip of the tail (a quibbling distance). On American Golden-Plover, longer wingtips project unequivocally beyond the tail. Also (and this is sometimes easier to note), on Pacific Golden-Plover the tertials almost reach the tip of the tail; the tertials on American Golden-Plover are distinctly shorter—shorter than the tip of the tail, and shorter than the primaries.

In breeding plumage, male Pacific is slightly paler and more golden-washed above and overall shabbier and unkempt-looking. (American looks dark and dapper.) The white border separating the golden upperparts and black underparts is narrow at the shoulder and runs the length of the body to the mostly white undertail. American's fuller white sash stops at the shoulder, and the undertail is black. There are two important considerations. First, the white racing stripe is often ragged and broken, and sometimes indistinct. At a distance, such poorly marked birds, showing obvious white only on the neck and sides of the breast, might easily be mistaken for American Golden-Plover — so pay close attention to the undertail: white equals Pacific, and black equals American. Second, the underparts of female American may show a pattern similar to Pacific, but again, there should be at least some black feathers on the undertail. Also, female Pacific shows a dull, grayish border on the lower edge of the folded wing.

In nonbreeding plumage, adult Pacific is more golden-washed above, with a buffy face and eye stripe. American is grayer and shows a whitish eye stripe. Juvenile Pacific is even more conspicuously golden, with the golden wash extending onto (at least) the breast.

BEHAVIOR: Similar to American Golden-Plover.

FLIGHT: Shape is like American Golden-Plover, but toes can sometimes be seen extending beyond the tail. In breeding plumage, the white undertail coverts and the white border between the gray-brown underwings and black underparts help distinguish this species. In nonbreeding plumage, the overall golden wash of Pacific distinguishes it from the grayer American.

VOCALIZATIONS: Song is a clear, unhurried, mellow two-note whistle: "*Pee perEE . . . Pee perEE.*" Also makes a high, clear, rising, whistled "*peeEE.*" Flight call is a whistled "*clu-EE*" (rising on the second syllable).

Snowy Plover, *Charadrius alexandrinus*

Beach Pixie

STATUS: Uncommon breeder; locally common to uncommon winter resident along the Gulf and West Coasts. **DISTRIBUTION:** Coastal and inland breeder. Nests along the Gulf Coast from the southwestern tip of Fla. to the Yucatán; along the Pacific from s. Wash. (rare) to Central America. Interior populations are found primarily in Utah (Great Salt Lake region), cen. Colo., se. N.M., w. Neb., w. and nw. Tex., w. Okla., and Kans., and from s. Ore. to w. Nev. and s.-cen. Calif. and the Salton Sea. In winter, most birds withdraw to coastal areas, but some winter in s.-cen. Calif. and the Salton Sea. **HABITAT:** Dry, sandy, sparsely vegetated habitat adjacent to a large or small brackish or freshwater source, including beaches, tidal flats, salt flats, saline lakes, dry mud flats, river bars, salt evaporation ponds, riverbanks, and shorelines. **COHABITANTS:** Wilson's Plover (East), Sanderling, Least Tern. **MOVEMENT/MIGRATION:** Most coastal birds are resident, although some, if not most, northern breeders move south. The migratory period is protracted: spring migration from mid-February to mid-May; fall from mid-June to early November. VI: 2.

DESCRIPTION: A small, pale, sand-colored plover that likes it hot and likes it dry. Smaller, trimmer, and more elfin-featured or pixielike than Piping Plover, except for the largish, squarish head, which makes the bird somewhat front-heavy. The bill is sandpiper long and narrow, and always black (unlike Piping Plover's stub of a bill). Eyes are almond-shaped (round in Piping Plover), and legs are long and blackish (always yellow-orange in Piping Plover).

Gulf Coast and, particularly, western birds are slightly darker above than Piping Plover and putty-colored, with warm buffy highlights, particularly about the head and neck. In breeding plumage, the accentuated black facial pattern and broken collar are enough to distinguish Snowy. In winter, with the plumage pattern muted, rely on the bill shape, small body, long legs, and leg color to clinch the identification — even though the bird's overall slighter proportions are very apparent. Piping Plover, by comparison, looks plump and dumpy. Snowy's expression is coy.

BEHAVIOR: A social bird that gathers in winter in flocks with other Snowy Plovers. Though it also feeds (and sometimes roosts) among other shorebirds, seems more inclined to keep to itself than other plovers. Along seacoasts, frequently forages in dry areas well above the tide line but also forages in the wet sand. Inland, most often feeds in damp or wet areas. Forages in typical plover fashion: run . . . stop . . . search . . . run and stop to grasp prey. Also probes into dry and wet sand and mud and charges insects around carrion, flycatching from the ground. Tame and reluctant to fly. Generally runs when approached.

When not feeding, usually crouches (often in slight indentations) on the higher, drier portions of the beach (and away from human traffic). Individuals in flocks are in close proximity but not massed (like sandpipers), a configuration that may make them even more difficult to detect than they already are.

FLIGHT: Overall slender and mostly pale, showing narrow white sides to the tail (and a less distinctive white stripe in the wing than Piping Plover). Flight is direct, fast, and nimble, on rapid wingbeats.

VOCALIZATIONS: Flight call is a soft "*kerWee*" that has the same pattern as Semipalmated Plover but is more nasal. Also gives a low, purring, slightly rough single-syllable call.

Wilson's Plover, *Charadrius wilsonia*

High Beach Plover

STATUS: Fairly common coastal breeder; more restricted permanent resident. **DISTRIBUTION:** Breeds coastally from s. Va. to South America; also along the west coast of Mexico (but not the U.S.). Winters along both coasts of Fla. and in Mexico, south. **HABITAT:** Sparsely vegetated coastal habitats, including sand beach, sand dunes, salt flats, mud flats, and spoil islands. Commonly roosts and nests in the drier upper beach, often in small patches of vegetation (such as saltwort and glasswort) or in open sandy breaks in vegetation; also seems partial to driftwood and trash. Forages on the upper beach in the drier areas shunned by other shorebirds, but also forages in the intertidal zone (wet sand and mud flats). **COHABITANTS:** Piping Plover, Snowy Plover, Least Sandpiper, Least Tern. **MOVEMENT/MIGRATION:** Most northern birds are migratory. Spring migration from mid-February to April; fall from mid-July to mid-October. VI: 2.

DESCRIPTION: Large-billed, long-legged Semipalmated-like plover. Medium-sized (larger than Semipalmated Plover; much smaller and more compact than Killdeer), with an oversized squarish head, a prominent bill, a proportionally undersized body, and long flesh-colored legs. The long, thick, pointy, jet-black bill stands out against the pale brown-and-white face. Usually looks neckless, but when alert, stands very erect and neck may be hyperextended.

Grayish brown above and bright white below, with a single dark band across the chest. Underparts are more extensively white and brighter white than Semipalmated Plover; the dark upperparts seem one size too small, exposing more white up the sides. As a result, Wilson's Plover is not just an overall paler bird but more extensively white — a color that vaults distances. The dark breast band is variable, ranging in width from a wide collar to a narrow ring, and in color from black in breeding males to pale brown or orangy brown in females. The face pattern — a broad white eye stripe and forehead — is less contrasting but more extensively white than Semipalmated Plover, and many birds show a touch of orange behind the eye or on the nape — a color Semipalmated doesn't show. Females, juveniles, and winter males are paler versions of the breeding male, and their paleness enhances the contrasting prominence of the large black bill.

BEHAVIOR: A bird that would rather run than fly. Forages on the high beach and dry salt and mud flats (less commonly on wetter areas near the water's edge). Feeds in the stop-and-go fashion of plovers, but pauses longer and runs greater distances between pauses than Semipalmated Plover, sometimes charging five or more yards to snatch prey. Semipalmateds seem tame and programmed: run several steps, stop, run several steps, stop, pick, run several steps. . . . Pausing at irregular intervals, and for varying amounts of time, the more predatory Wilson's Plover runs for varying distances, with long strides and in a slinking crouch, accelerating when prey is sighted and stretching its neck as it closes with its quarry. Semipalmated commonly reaches down to pluck prey; Wilson's Plover reaches forward. In spring and summer, Wilson's Plovers are often found in pairs. In winter, they form flocks of 10–30 birds and often roost with Piping and Snowy Plovers.

Wilson's is tame . . . to a point. Seems to allow approach to about 20 yds., then runs, *fast.* Even when pursued, very reluctant to fly, preferring at all times to run from danger.

FLIGHT: Except for the bill, fairly compact. Long slender-winged profile shows overall slightly less contrast than Semipalmated Plover—but a whiter face and contrastingly black bill. Flight is strong and mostly direct, sometimes banking or tacking, with quick, stiff, regular wingbeats. When flying short distances, occasionally angles the body up and flies with choppier wingbeats (recalling the flight of a phalarope).

VOCALIZATIONS: Common calls include a high, slurred, whistled "*peep*" (given by males) and a low, mellow, liquid "*churp*" or "*ch'churp.*" Also emits a short, high "*cheep*" (often doubled or trebled) when approached. Song is a two- or three-part, slurred, liquid gurgle and trill: "*skleeu'r'r*" or "*chik-wr'r-wheet.*"

Semipalmated Plover, *Charadrius semipalmatus*

Mud-colored Plover

STATUS: Common and widespread arctic and subarctic breeder; common winter resident in southern coastal areas; very common coastal migrant but also widespread across the interior in both spring and fall. **DISTRIBUTION:** Breeds across Alaska, the Yukon, and the Northwest Territories to and around the shores of Hudson Bay, then east along the north coast of Que., Lab. to Nfld. Winters along all three coasts, from Va. in the East and n. Calif. in the West, south through Mexico, the Caribbean, Central America, and almost all of coastal South America. **HABITAT:** Breeds primarily on gravel, loose shale, or sand and gravel shorelines of rivers and seacoasts. Also nests on tundra with an assortment of short vegetation. In migration and winter, forages on mud flats, beaches, salt marshes, riverbanks and sandbars, freshly plowed agricultural fields, alkaline ponds, and very shallow standing water. **COHABITANTS:** In migration, may associate with Sanderling on beaches and other small peep on mud flats. In winter, sometimes feeds and roosts with Piping and Snowy Plovers. **MOVEMENT/MIGRATION:** Spring migration from mid-April to early June; fall from early July to late October. VI: 3/4.

DESCRIPTION: A small handsomely marked plover distinguished by its size, shape, bold plumage pattern, and probability. Except for Killdeer, the larger two-banded land lubber of the plover clan, Semipalmated is the small banded plover most likely to be seen across most of North America.

Small and neckless (slightly smaller than Sanderling, with which it commonly associates on beaches), with a large domed head with a flat pushed-in forehead, a stubby (sometimes very slightly upturned) bill, and yellow-orange legs. In breeding plumage, the bill is the same color as the legs. Nonbreeding and juvenile birds have mostly black bills with some orange near the base of the lower mandible. Even from a considerable distance, the large round eye, highlighted by a narrow orange ring, stands out.

Adult upperparts are smooth, unpatterned brown (mud-colored!), with a prominent black mask over the eyes. White underparts are bisected by a single bold, black, complete breast-band. On slightly paler juveniles, upperparts are faintly scalloped, and the head pattern and (usually) broken breast-band are brown, not black.

BEHAVIOR: Feeds with other shorebirds, especially Sanderling, turnstones, and other small plovers, but rigidly maintains a proper distance from other feeding Semipalmated Plovers, defending its feeding territory from transgressors. Spreads out across mud flats or in linear fashion along the tide line, where it forages on wet sand, just above the turbulent zone favored by Sanderling but below the higher, drier portion of the beach frequented by Piping Plover (where they occur). Like a proper plover, walks, stops, picks. The walk is somewhat halting: advancing, Semipalmated brings its head down and forward, giving the impression that it is tugging against a halter or leaning into the wind. Sights prey visually. Picks (rarely probes) prey and draws it from the substrate. (With worms, a veritable tug-of-war often ensues.) The birds also foot-patter to either draw prey to the surface or disclose their presence. After feeding, birds often cluster in homogeneous flocks.

FLIGHT: Flight is rapid and direct, on very long, slender, swept-back wings set well forward. The domed head and small bill make the bird look very blunt-headed.

VOCALIZATIONS: Very distinctive, given frequently in flight. A whistled two-noted "*Chew-EE.*" During aggressive interactions, gives a three-note warning followed by an angry, descending, shorebird whinny: "*rrh, rrh, rrh, rhe'h'h'h'h'h'eh.*"

Piping Plover, *Charadrius melodus*

The Sand Wraith

STATUS: Uncommon and declining: population estimates do not exceed 2,500 pairs. **DISTRIBUTION:** Breeds primarily along Atlantic Coast beaches from e. Nfld. to N.C. and along the rivers and sandy lakeshores of the n. Great Plains: s.-cen. Alta., s. Sask., s. Man., e. Mont., most of N.D., cen. and e. S.D., most of Neb., w. Iowa, n. Kans., e. Colo., n. Tex., and w. Okla.; also found locally on the western end of Lake Superior. Winters on Atlantic Coast beaches from N.C. to cen. Fla. and on the Gulf Coast from s. Fla. to n. Mexico. **HABITAT:** Almost always associated with sandy or cobble beach and alkaline flats with sparse vegetation. Less frequently found on mud flats. **COHABITANTS:** Least Tern, Sanderling, other small plover species. **MOVEMENT/MIGRATION:** A hardy and early-spring migrant; some returning coastal birds arrive in late February, and the balance by mid-April. Most interior birds arrive in the later half of April and early May. In fall, departure begins in early July and extends into November, but most birds depart from Atlantic Coast beaches by late August. VI: 2.

DESCRIPTION: A small pale plover of pale landscapes. When it moves, you see it. When it doesn't, you don't. In size, shape, plumage, and behavior, suggests a plump bleached-out Semipalmated Plover with a narrow (usually broken) breast-band, but ghostly pale upperparts easily distinguish Piping from the mud-colored Semipalmated. In poor light and at great distances, Piping is stockier and more compact, with a larger head, a thicker, stubbier-appearing bill, and proportionally shorter legs. Piping Plover also has a more horizontal profile. (A foraging Semipalmated usually keeps its head more elevated and seems also to show less clear definition and less cleft between the head and the back.)

At all times, Piping Plover has bright yellow-orange legs—a trait that easily distinguishes it from the smaller, slimmer, more angular, slightly darker-backed, and gray-legged Snowy Plover. In winter, dark-billed juveniles and adults are most easily distinguished from Snowy Plovers by their larger size, plumper, more compact shape, paler upperparts, stubby (now black) bills, and, when all else fails, yellow legs. Piping's expression is gentle and serene.

BEHAVIOR: Active when feeding, running in short spurts above the water's edge, picking with near-sandpiperish rapidity. Sometimes feeds in small groups of other Piping Plovers (generally fewer than a dozen), but birds maintain their distance from one another. Often found with Sanderling, Western Sandpiper, and Semipalmated Plover. In such aggregations, most often stays higher up, on the drier (paler) portions of the beach, where it is better camouflaged.

Extremely tame. When approached, prefers to run rather than fly. When not foraging, roosts on the upper beach (sometimes with other shorebirds), where it is very difficult to detect. Responds quickly to imitation of its alarm call—which, of course, you shouldn't do (endangered species). Because of their cryptic plumage, Piping Plovers are often heard before they are seen.

FLIGHT: In flight, displays a surprisingly white and contrasting wing stripe and pale rump (actually uppertail coverts) and a dark-tipped tail. Flight is fast, direct, and slightly more fluttery than Semipalmated Plover. Exaggerated display flight is slow, stiff-winged, batlike.

VOCALIZATIONS: A mellow, winsome, somewhat ventriloquial two-note whistle: "*Pee-po.*" When disturbed, emits a series of low-pitched ascending notes that sounds like a whistled keening, "*peo'rp, peo'rp, peo'rp,*" and often ends with the "*Pee-po*" note.

PERTINENT PARTICULARS: Piping Plover is an endangered species. Don't press it. If the bird seems agitated or is approaching you (with the obvious objective of attracting your attention), back off. There are young or a nest nearby.

Finding Piping Plover on a beach is often complicated by the profusion of winter-plumage Sanderling. The trick is to look (even without binoculars) for the pale bird that isn't constantly in motion. Sanderlings are always charging after the retreating surf and jabbing their bills into the sand as they go. Piping Plovers stay higher on the beach (except when feeding), feed more methodically and less aggressively, and pause frequently while foraging.

Killdeer, *Charadrius vociferus*

The Noisy Plover

STATUS: Common and widespread breeder and winter resident. **DISTRIBUTION:** Found as a breeder across virtually all of North America except arctic Canada, Alaska, and s. Mexico. In winter, vacates all but the west coast of Canada and northern interior portions of the U.S. subject to heavy and lasting snowfall, occupying much of the e., w., and s. U.S., all of Mexico, Central America, and South America to Chile and Peru. **HABITAT:** At all times of year, found in open areas, both wet and dry, that are covered by short vegetation, bare earth, or mud. Typical habitats include heavily grazed pasture or prairie, mud flats, sandbars, cultivated fields, sod farms, golf courses, corporate lawns, airports, parking lots. Often found near water (usually fresh, sometimes salt), but, like Mountain Plover, is highly terrestrial, foraging freely on dry ground well away from water. **COHABITANTS:** In summer, American Kestrel, Horned Lark; in wetter areas, migrating shorebirds. **MOVEMENT/ MIGRATION:** Migration is early and protracted. Birds return to southern breeding areas in late February, with a peak in March. Northern birds arrive by mid-April. Fall migration is extremely protracted: birds depart from some northern areas in June, and there is evidence that birds may continue to migrate through the southern United States in December and January. (Late movements are likely to be triggered by harsh winter conditions.) Fall migration peaks from September to early November. VI: 4.

DESCRIPTION: Gangly medium-sized plover (slightly longer but more slender than American Golden-Plover), with a distinct shape—a long, slender, horizontal posture. Killdeer is distinctly marked as well: its uniform dark (brown) back, bright orange rump (visible in flight), and double-black breast-bands are etched across white underparts. Killdeer is also distinctly vocal, saying its name, "*Killdee,*" loudly and often. Adults and grown young are similarly patterned. Very young birds (commonly found with adults) show a single breast-band, like the smaller banded-plover species.

BEHAVIOR: Feeds with the classic, halting, walk-stop-pick plover M.O., but movements are more abrupt and jerky. Often found close to humans (nesting in driveways, for example), but generally shy when approached, if vocal—calls loudly when approached, bobs its head, and runs, haltingly, to increase the distance. Stressed birds do a broken-wing imitation that exposes their bright orange rump.

In migration and winter, a fairly social flocking plover. Flocks number 6–30 birds. Associates with other plovers, particularly American Golden-Plover and Mountain Plover, but does not typically form a mixed feeding flock.

In migration, often flies in flattened and widely spaced V-formation.

FLIGHT: Overall slender, with long, slender, pointy, angled-back wings and a long narrow tail. Killdeer appears falconlike and is about the size and shape of American Kestrel. Normal flight is strong and mostly direct, with regular but stiff wingbeats that appear tired or somewhat tentative and heavy on the down stroke. If flushed or pursued, flight is twisty-turny, but not erratic. In migration, flight is strong and direct, with steady wingbeats.

The bird's shape and white underparts are very distinctive.

VOCALIZATIONS: Classic call is a loud, strident "*Keeldee'e!*" (killdeer), voiced when the bird is nervous or alarmed. Males in aerial display render an easily recognized variation of this call, repeated over and over. Distress call is a rolling, twittering trill that sounds like the "deer" portion of the song with a rolled *r*: "*de'r'r'r'r'r'r'r.*" The trill rises and falls, increases and decreases in volume and pitch, and is often punctuated by a sharp "*dee'e.*"

Mountain Plover, *Charadrius montanus*

More Accurately, Plains Plover

STATUS: Uncommon prairie breeder; uncommon southwestern winter resident. **DISTRIBUTION:** Breeds in the high short-grass prairies, from cen. and s. Mont., all but nw. Wyo., e.-cen. Colo., ne. Utah, n. N.M., extreme sw. and s. Okla., and extreme nw. Tex.; also in the Davis Mts. of Tex. Winters in the Central Valley of Calif., the central coast (very local), and se. Calif.; also in s. Ariz. and s. Tex. south to n. Mexico. **HABITAT:** Breeds in open, short, dry, flat, grazed (and overgrazed) grasslands and prairies. Feeds and winters on plowed fields, disturbed fields (such as prairie dog towns), and grazed and burned grasslands; also on coastal prairie and alkaline flats. **COHABITANTS:** During breeding, Horned Lark, Lark Bunting, McCown's Longspur, prairie dogs; in winter, Killdeer, Long-billed Curlew, Horned Lark. **MOVEMENT/MIGRATION:** Vacates breeding areas in winter; relocates in the Southwest and northern Mexico. Migrates in small flocks (fewer than 50 birds). Rarely seen during migration except in a very few locations. Spring migration from later half of February to mid-April (early migrants presumably stop over somewhere en route); protracted fall migration from late June to late November. VI: 2.

DESCRIPTION: An elegant, cryptically colored, almost patternless prairie-plover. The species has little to do with mountains (except to vault them during migration). Medium-sized (smaller than Black-bellied Plover or golden-plover; about the same length as Killdeer), but akin to golden-plover in stance and structure. Large and round-headed, with a plover-short but unploverlike slender and pointy bill, longish neck, full but contoured body, longish tail, and longish pale legs. The posture on an alert bird is long-necked-erect. Seems haughty. When feeding, is more hunched and horizontal (more Killdeerlike).

Overall uncontrasting. Soft, seamless, pinkish tan brown above and whitish below, with a fawn-colored blush across the breast (grayish buff in winter). The pale whitish face (dingier in winter) stands out and sets off the black eyes, black bill, and, in summer, black forehead and black lores. The expression is demure and baleful—and the facial pattern bears an eerie similarity to the face pattern of Pronghorn Antelope.

BEHAVIOR: Prefers habitat that is so dry, scabby, and vegetatively pauperized that most other birds wouldn't give it a second glance. Moves in the halting stop-and-go pattern of plovers, but is more fluid and less jerky than Killdeer. Walks three to five steps, head down, then pauses, with head raised or cocked slightly askew. When sights prey, rushes forward, reaches down, and plucks. Also pries into soil and foot-patters.

Feeds alone or in small groups (up to 250 birds in some winter flocks, but more commonly a dozen). Associates with other shorebirds only when they're using the same plowed field. When approached, usually faces away and crouches, using its neutral back color to melt into the ground.

Loves prairie dog towns, unfurrowed plowed fields, and very short grass. Dislikes tall thick grass and wet areas (although forages in irrigated plowed fields, often with wading birds like yellowlegs and Pectoral Sandpiper, as well as in drier mud flat edges). Fairly shy. Does not allow close approach, and flies, even if only for a short distance, when pressed.

FLIGHT: Silhouette is very symmetrical, with head and tail projecting equal lengths. Flight is fast, direct, and very slightly twisty-turny. Wings are distinctly down-bowed (somewhat like Killdeer), but movement is more elastic. Wingbeats are given with a quick regular cadence (like one of the larger plovers), but also have some of Killdeer's tentativeness. In winter, often seen flying in tight flocks that melt into the landscape as soon as they touch down.

VOCALIZATIONS: For a plover, uncommonly silent. In courtship, birds give a rolling, drawled, and repeated whistle: "*wee wee.*" In flocks, birds give a conversational "*kip*" call. When anxious or disturbed (flushed), birds make a low, short, liquid, rattling "*royk-k-k*" and a low throaty protest: "*reaah.*"

American Oystercatcher, *Haematopus palliatus*

Beach Chisel-Bill

STATUS: Common and very conspicuous coastal shorebird—in part because it has distinctive traits, and in no small part because it is highly vocal. **DISTRIBUTION:** In the U.S., this New World Oystercatcher is restricted to the coastal areas from s. Maine (also N.S.) to Tex. A resident of both Mexican coasts (and most coastal areas of Central and South America), its Pacific range extends only to the Sea of Cortez. In the Pacific U.S., the species occurs only as a vagrant to s. Calif. **HABITAT:** Forages primarily on sand or mud in the intertidal zone as well as on mussel and oyster shoals and rocky coasts. At high tide, found roosting on beaches, sandbars, open salt marsh, dredge spoil, rock jetties, and rocky coast above the tide. **COHABITANTS:** Willet, Marbled Godwit, Laughing Gull, Black Skimmer. **MOVEMENT/MIGRATION:** Largely a permanent resident. In winter, vacates breeding areas north of New Jersey for a short time. Spring migrants arrive as early as late February; the majority of birds move in March, but some continue until late April. Fall movements probably begin in late August, with peaks in November, though some birds may winter as far north as Long Island. VI: 2.

DESCRIPTION: A large, chunky, black-hooded, brown-backed, white-bellied shorebird—but all anyone needs to see is the long, chisel-shaped, orange-red bill. The striking red-rimmed yellow eye makes the bird look goat-eyed crazy and distinguishes this species, in conjunction with the white underparts, from all other oystercatcher species. Young birds show dark eyes and dark-tipped bills.

BEHAVIOR: A fairly social bird, often found foraging and roosting in small flocks comprising several individuals to a dozen; in winter and migration, may gather in flocks numbering several hundred. Freely associates with other wading shorebirds (including Willet), but such proximity may be serendipitous.

Forages by walking or running on tidally exposed sand or in shallow water. A slow, methodical, studied feeder that probes using deep or jabbing thrusts. Sometimes extracts mollusks from the substrate and carries them to shore to open them by "hammering" the shell.

Extremely noisy during courtship. Couples and trios (and sometimes more) engage in spirited shallow wing batting, flights that show off plumage to best effect, and vocal animated "piping" displays on the ground.

Very wary. Normal stance is usually heads-up alert, but when approached, the head becomes even more elevated. Usually attempts to walk away from approaching danger, but if pressed, flushes at distances from which most other shorebirds show no concern.

FLIGHT: Big, heavy, broad-winged, long-billed shorebird. In flight, American Oystercatcher's pied wing and tail pattern is blatant; could be confused with Willet or Hudsonian Godwit, but is more striking and more contrasting. Coupled with the bird's broad, somewhat rounded (and unshorebirdlike) wings, overall stocky structure, and big red bill, this pattern makes confusion with other species unlikely. Flight is fast, direct, and unglamorous. Wingbeats are quick, deep, choppy, and steady.

VOCALIZATIONS: Makes loud, shrill, strident (even maniacal-sounding) descending (or piped) whistles. Call begins with two to three loud, well-spaced descending whistles that increase in tempo, becoming a loud, piping, prolonged giggle that varies in volume and pitch and trails off at the end. "*Wheer. Wheer. Wh'Heh'Heh'Heh'Heh'Heh,Heh'h'-h'h.*" The "*wheer*" call is also used singly and often has a rising-and-falling or slurred two-note quality: "*wh-'EEr.*"

Black Oystercatcher, *Haematopus bachmani*

Rock Chisel-Bill

STATUS: Common resident on the West Coast of the United States. **DISTRIBUTION:** Strictly coastal from the Aleutians to Baja. **HABITAT:** Primarily rocky coastlines, but also cobble beach and, infrequently, sandy beach and mud flats. **COHABITANTS:** Pelagic and Brandt's Cormorants, Surfbird, Black Turnstone. **MOVEMENT/MIGRATION:** Over most of its range, relocates locally after breeding rather than migrate. Many but not all northern birds migrate to more southern locations. Spring migration from March to May; fall from late August to mid-November. VI: 0.

DESCRIPTION: An animate chunk of black rock with a bright red bill and flesh-colored legs. Large, stocky, all-blackish shorebird (much larger than Surfbird and Black Turnstone; about the same size as Whimbrel), with a bright red chisel-like bill and short, thick, contrastingly pale legs. Young birds show dark-tipped bills.

BEHAVIOR: Usually seen alone or in pairs; in winter, roosts in well-spaced flocks. Stalks slowly but purposefully up and down wet mollusk-studded rocks in the intertidal zone, periodically driving its bill into partially opened mussels or chiseling off limpets or barnacles. Hops and flutters to higher locations to avoid waves. Roosts alone, with other oystercatchers, or amid other shorebirds, with head tucked (and red bill hidden).

Not particularly shy; allows close approach. Frequently calls before flushing and taking flight.

FLIGHT: A sturdy meat-and-potatoes flier. Overall stocky and black, with broad, somewhat blunt-tipped wings, a short tail, a slightly elevated head, and a conspicuous bright red, slightly down-angled bill. Wings are down-curved. Wingbeats are steady, hurried, stiff, and choppy. Flight is generally straight and low over the water.

VOCALIZATIONS: A loud whistled "*wheer*" or "*wheep*" that is generally indistinguishable from American Oystercatcher.

PERTINENT PARTICULARS: Identifying Black Oystercatchers is less challenging than finding them. When roosting with bills tucked, the birds' stocky forms meld with their rocky substrate. The pale flesh-colored legs help distinguish bird from rock. Also, the red-rimmed golden eyes that turn your way are distinctly unrocklike. At high tide, scan around the edges of flocks of more contrasting clusters of rock-roosting shorebirds, or wait until low tide, when exposed mollusks signal feeding time for Black Oystercatcher.

Observers in southern California should be aware that Black Oystercatcher and American Oystercatcher interbreed where their ranges overlap in Mexico and that hybrids have occurred in northern Baja and southern California.

Black-necked Stilt, *Himantopus mexicanus*

Coral-legged Water Strider or "Marsh Poodle"

STATUS: Common but widely scattered breeder and wintering species; found only as a vagrant across a good part of North America. **DISTRIBUTION:** Primarily a western species, but also found along all coasts. Breeds mostly in the Great Basin, scattered locations in Calif., and the southern plains; found in e. Wash., e. Ore., s. Idaho, s. Alta., Mont., n. Nev., Utah, Wyo., w. N.D., w. S.D., Colo., Ariz., N.M., and w. and cen. Tex. Also breeds coastally from Del. to Fla., on the Ala.–Miss. border, in La. and Tex., and up the Mississippi River Valley to Mo. and Ill. Winters in Calif., coastal and proximal interior portions of Tex. and La., the southwest coast of Fla., and throughout much of Mexico and coastal Central America. **HABITAT:** Breeds on salt ponds, sewage ponds, rice fields, impoundments, and freshwater marshes with emergent vegetation. Winters in rice fields, tidal marshes and flats, mangrove swamps, and flooded fields (in addition to breeding areas). **COHABITANTS:** Eared Grebe, Least Grebe, American Avocet, Wilson's Phalarope. **MOVEMENT/MIGRATION:** Most birds from interior locations vacate breeding areas in advance of winter and spend considerable time staging or loafing in places short of their ultimate wintering site. Spring migration from late March to late May; fall from early July to October. VI: 3.

DESCRIPTION: A stiffly and formally attired shorebird perched on long red circus stilts. Large (Greater Yellowlegs–sized) anorexic shorebird with a small round head, a needle-fine bill, a long neck, and a slender body—all set on impossibly long coral red legs.

Upperparts are all black (female is slightly browner), and underparts all white. No other species remotely resembles this bird.

BEHAVIOR: Feeds singly or in widely spaced groups (not a flocking bird like avocet). Prefers water with some vegetation, but also feeds on open flats. Moves with long, striding, delicate, and precise movements. Frequently turns and circles. Often angles its body sharply down in front when it feeds. Sometimes runs, but usually stalks. Jabs at prey on the surface or, in murky water, may submerge and swish its bill. Loath to swim; when foraging, does not like to get its belly wet.

Fairly tame, often allowing close approach. Intolerant of trespass while nesting or when young are present. If you don't know the call, you soon will.

FLIGHT: Profile is exceedingly long, slender, and delicately angular. The bill angles down, and wings are straight, narrow, stiffly triangular, and held at a near-right angle to the bird. Its exquisitely long legs dangle slightly and seem like a trailing red string. Wingbeats are quick, stiff, downflicking, and tentative or hesitant—a quick down stroke is followed by a slight hesitation on the upstroke.

Wings are entirely black, both above and below.

VOCALIZATIONS: Call is a loud, persistent, terrier-like yap: "*rrap*" or "*jaap*."

American Avocet, *Recurvirostra americana*

Pied Scythebill

STATUS: Common widespread western breeder and wintering species in the southern United States; primarily coastal. **DISTRIBUTION:** Breeds primarily in the prairies and the intermountain West. In Canada, found in se. B.C., s. Alta., s. Sask., and sw. Man.; in the U.S., e. Wash., e. Ore., Calif., Nev., Idaho, most of Mont. and Wyo., and w. and cen. N.D., S.D., and Neb.; Utah, Colo., Ariz., N.M., and Tex. Regularly but not always found breeding east of its core range. Winters coastally from n. Calif. to Baja and east to sw. Ariz.; on the Tex. and La. coasts; in N.C. to and around the Florida Peninsula to the Florida Panhandle; and in virtually all of coastal Mexico and much of the interior. **HABITAT:** Breeds on ponds, potholes, shallow alkaline wetlands, freshwater lakes, impoundments, and water treatment facilities. Edges may be vegetated, but prefers open unvegetated shores. Winters on tidal mud flats, shallow bays, and impoundments. **COHABITANTS:** Commonly found with Wilson's Phalarope, Black-necked Stilt, and other long-legged waders. Occasionally loafs with smaller gulls. **MOVEMENT/MIGRATION:** Except in California, breeding birds vacate interior nesting areas and move to coastal locations. Spring migration from mid-January to May; fall from early July to mid-November. VI: 3.

DESCRIPTION: Elegance with an upturned darning-needle-slender bill. Large (godwit-sized) long-legged wader with a round head, sinuously contoured neck, pear-shaped body, and long sturdy legs.

But the bird's signature characteristic is a long slender bill that turns sharply but demurely up near the tip. The body is as boldly striped as a barber pole—black-white-black-white from back to belly. In breeding plumage, the head and neck are blushed with burnt orange, and in winter with grayish white. Long pale blue legs complete the package. The expression is aloof and calmly serene.

Fairly tame and highly confrontational when breeding areas are approached.

BEHAVIOR: Semicolonial. Breeds in small colonies on inland lakes. Often winters in flocks numbering in the hundreds, usually with other waders, most notably and commonly its elegant sidekick, Black-necked Stilt. Forages by walking through water sometimes picking at the surface for prey, but more commonly by leaning forward, dropping its bill (or entire head) under the surface, and sweeping from side to side like an aquatic scythe. On land, walks with mincing steps, as if its feet hurt. In deeper water, swims more adroitly and readily than other large waders.

Rests either by standing (on one leg) in water or by crouching on dry land. Fairly quiet; more vocal when disturbed.

FLIGHT: Distinctive, gangly, balanced profile—a long drooping neck, long wings, and long trailing legs; the bird looks slightly drooped or wilting on all points. Broad wings are angled back and down, with an acute bend (not curve) to the wing. The wings are so boldly and conspicuously patterned black and white that any more description is superfluous. Wingbeats are quick, regular, down-pushing, and slightly delicate to tentative.

VOCALIZATIONS: Call is a loud, ringing, somewhat shrill "*oooeet*" or "*r'leep,*" given with a compressed two-note quality. Sometimes makes a repeated "*leap leap leap*" that sounds very much like the call of Greater Yellowlegs (particularly at a distance when the two-notedness attenuates to one), or perhaps a shrill Willet.

Greater Yellowlegs, *Tringa melanoleuca*

The Velica-Tringa

STATUS: Common and widespread shorebird whose protracted migration makes it as familiar a species to most North American birders as it is to those who reside in the taiga forests where Greater Yellowlegs nest or those at coastal locations where the species winters. **DISTRIBUTION:** Breeds in a broad but well-defined geographic band that runs from Nfld. to w. (but not coastal) B.C. and north, after a several-hundred-mile gap, into s. and cen. Alaska. Winters (somewhat farther north than Lesser Yellowlegs) in coastal regions from s. New England and s. Ore. south to Mexico and beyond. Wintering birds extend inland to the Central Valley of Calif., across s. Ariz., s. Nev., and much of Tex., the s. Great Plains, the lower Mississippi River Valley, the interior South, and throughout the Florida Peninsula. **HABITAT:** Breeds in muskeg, open forests (usually with water nearby), and bogs. In winter, makes extensive use of wetlands, primarily coastal wetlands, but is also common on tidal flats, pond edges, wild-rice plantations, and flooded agricultural land. In migration, uses these habitats as well as vernal ponds, sewage lagoons, lakeshores, and river bottoms—essentially any fairly substantial wet area that supports fish, amphibians, or aquatic insects. **COHABITANTS:** In summer, Short-billed Dowitcher, Palm Warbler; in winter and migration, other shorebirds, smaller herons, and egrets. **MOVEMENT/MIGRATION:** Spring migration begins very early across much of North America, extending from March to early June, with a peak in late April; fall migration extends from late June or early July (approximately a week later than Lesser Yellowlegs) to November (several weeks later than Lesser Yellowlegs). VI: 4.

DESCRIPTION: There is little delicacy to this large, gangly, sturdily proportioned shorebird distinguished by long bright yellow legs and a long, slightly upturned bill. Resemblance to Lesser Yellowlegs is only plumage-deep.

Medium-large (larger than Lesser Yellowlegs; slightly smaller and less stockily proportioned than Willet; only slightly shorter than Hudsonian Godwit), with long legs and a long, slightly upturned bill that is heavy at the base and about twice the width of the head. Although the sight of a straight-billed Greater Yellowlegs is not unlikely, in most birds the bill is at least slightly upturned, and in many birds it's unequivocally upturned—to the point that confusion with a distant Hudsonian Godwit is a distinct possibility.

Plumage in breeding adults is brown-gray above, marbled with lavish amounts of black, and spangled with white flecks. Head and breast are likewise heavily marked. Underparts are white, but flanks are extensively barred (generally more so than Lesser Yellowlegs).

The plumage of nonbreeding birds is plainer: gray above (including neck and breast), but flecked with white; whitish below (lower breast and belly). Juveniles largely resemble nonbreeding adults, but upperparts are frosted with small pale freckles.

BEHAVIOR: Less gregarious than Lesser Yellowlegs. Mixes with other shorebirds (most often Lesser Yellowlegs, Willet, and Hudsonian Godwit), but amid medium-sized shorebirds, often feeds apart, in deeper water, and is frequently found feeding among herons and egrets (particularly Snowy Egret).

A more active, angry, and aggressive feeder than Lesser Yellowlegs. Walks with longer strides—a Tyrannosaurus Rex of a shorebird. Greater Yellowlegs usually hunts with the head up, visually searching for prey. It runs to catch fish and other aquatic life, stabbing and jabbing its bill into the water. (Lesser Yellowlegs is more apt to pick.) Also bill-sweeps with head down and bill submerged, charging ahead with bill sweeping side to side. (While using this technique, several Greater Yellowlegs may move abreast, forming an interception line; though Lesser Yellowlegs may also bill-sweep, they appear more methodical than aggressive.)

Noisy and alert, Greater Yellowlegs becomes vocal when approached, raising its head and neck, leaning forward, and yapping constantly.

FLIGHT: Overall angular and gangly, with a long pointed bill and (usually) long trailing legs. Wings are uniformly dark above (no wing stripe), and tails are mostly white with a grayish tip. The heavy bill makes the bird appear front-heavy. Flight is overall heavier and less delicate than Lesser Yellowlegs (which has a short upstroke and a powerful wing-drooping down stroke), but is otherwise similar. The legs usually trail behind the tail, but in cold temperatures the legs may be drawn up against the body, altering the flight profile enough to suggest a dowitcher. During long migratory flights, generally flies in larger V-shaped flocks (often in excess of 100 birds) than Lesser Yellowlegs. When traveling locally, often mixes with Lesser Yellowlegs, dowitchers, Willet, and other shorebirds.

VOCALIZATIONS: A distinctive call. A loud, ringing three- to five-whistle "*TEW, TEW, TEW.*" When agitated, the "*Tew*" note is held longer—"*Tu'ee, Tu'ee, Tu'ee*"—and is repeated ad nauseam until you leave or the bird flies. In spring, also a yodeling call.

Lesser Yellowlegs, *Tringa flavipes*

The Delicate Yellowlegs

STATUS: Common widespread northern nester; common winter resident primarily along southern coasts; best known as a spring and fall migrant across most of North America. **DISTRIBUTION:** Breeds somewhat farther north than Greater Yellowlegs—from the eastern side of James Bay, Que., across n. Ont., n. Man., n. Sask., n. Alta., ne. B.C., the s. Northwest Territories, and all but n. Yukon, into s. and w.-cen. Alaska. Winters along both coasts, from N.J. and San Francisco south, as well as throughout most of Fla., the lower Mississippi River Valley, the interior South, and e. Tex., but most birds winter outside the continental U.S., from the West Indies and Mexico south to s. Chile and Argentina. **HABITAT:** Open or semi-open boreal forest and taiga habitat (not commonly found on open tundra) in drier and more vegetated habitat than Greater Yellowlegs favors, and generally adjacent to muskegs, ponds, bogs, and sedgy pools. During migration and winter, found in a variety of wetland habitats, including fresh- and saltwater wetlands, mud flats, shorelines, rice fields, saltwater lagoons, sewage treatment facilities, and rain puddles—almost anyplace land and water meet. **COHABITANTS:** In migration, often associates with dowitchers, Stilt Sandpiper, Greater Yellowlegs; also Pectoral Sandpiper. **MOVEMENT/MIGRATION:** Spring migration takes place from early March to May, with a peak across most of North America in April. This species is among the earliest of fall migrants. Failed breeders appear as early as the third week in June, peak movement occurs in late July through August, and migration ends in October. VI: 4.

DESCRIPTION: A tall, slender, elegant shorebird distinguished by its delicate features, its refined manners, and, most of all, its very long bright yellow legs—the feature that quickly distinguishes this species from all North American shorebird species except Greater Yellowlegs. Greater differs from Lesser primarily in size, structure, and voice. Confusion is also possible with Solitary Sandpiper, Stilt Sandpiper, and Wilson's Phalarope, but the dissimilarities with these species are hard to ignore.

Everything about this bird—bill, neck, body, legs—is slender: a shorebird to inspire El Greco. Medium-sized (stands moderately taller than dowitchers and Stilt Sandpiper, with which it commonly feeds) and finely proportioned, with a bill that is thin, pointy, straight, uniformly dark, and about one and a half times the width of the head (two times greater in Greater Yellowlegs); some sources suggest that the bill is only slightly longer than the head is wide. The problem in the field is noting where the head ends and the bill begins.

The upperparts of breeding adults, including the breast, are brownish gray and richly spangled with black spattering and white flecking. The belly is white, and the sides marked sparingly with black feathers.

In winter, birds shed the black spangles and become basically gray above (with fine white speckling) and, except for the gray breast, white below. Juveniles resemble nonbreeding adults but have slightly browner upperparts and are more lavishly patterned with small pale freckles. The expression is mildly curious, alert.

BEHAVIOR: In migration and winter, a gregarious species often found in small flocks of 3–25 individuals. (Feeding Greater Yellowlegs is more solitary.) Mixes freely with other shorebirds, most commonly dowitchers, Stilt Sandpiper, and Greater Yellowlegs. Uses a number of foraging techniques, but is primarily a wading bird that leans well forward when feeding, with neck extended and bill pointed toward the water, and picks more than jabs at prey on the surface or below in a quick but deliberate fashion. Greater Yellowlegs, by comparison, is a more active and aggressive feeder, chasing through the shallows and stabbing at the water. Dowitchers are probers, not visual hunters—they keep their bill tips down in the water.

Both yellowlegs commonly perch (display) in trees. They also swim and, when foraging in deeper water, choose to swim rather than fly to shallower water.

FLIGHT: Slender, delicate, long-winged, and long-legged profile—better proportioned and balanced than front-heavy Greater Yellowlegs. The yellow legs commonly trail behind the tail, but in cold temperatures birds may draw their legs in, altering the flight profile. Flight is direct, light, buoyant, and more casual and nimble than Greater Yellowlegs—as if the bird could dance on air. Wings are down-drooped and do not appear to rise above the body. Wingbeats are regular and more perfunctory, even casual, than crisp; a split-second hesitation seems to follow the pronounced down stroke.

Often flies in flocks that form a loose, shifting, and rarely symmetrical V. Birds vocalize in flight, but perhaps less frequently than Greater Yellowlegs.

VOCALIZATIONS: A softly uttered two- to three-note call, "*tew*" or "*tew, tew,*" that lacks the loud ringing quality of Greater Yellowlegs. The display song, often given when descending during spring migration, is a lively, rolling, yodel: "*tewdileeoodlyoodlyoodlyoo.*"

Solitary Sandpiper, *Tringa solitaria*

Puddlepiper

STATUS: Fairly common and widespread northern breeder, nesting for the most part in remote forested areas. Common and widespread migrant. Except in southern Texas and southern Florida, winters south of the United States and Canada. **DISTRIBUTION:** Breeds in boreal forests across all but eastern- and westernmost portions of Canada and through most of Alaska. In the U.S., nests only in extreme ne. Minn. Winters from s. Tex. south. **HABITAT:** Nests in freshwater bogs and ponds in spruce and coniferous forests. In migration, virtually a freshwater obligate (rarely seen in salt or brackish water). Seems most at home in vegetatively confined woodland pools, vernal ponds, swamps, sodden manure pits, ditches, and puddles. When found on larger bodies of water, is almost always hugging the edge or areas clogged with low vegetation or emergent stumps in drought-stricken reservoirs. **COHABITANTS:** As the name implies, Solitary Sandpiper frequently travels alone, does not join shorebird flocks, and is hardly ever encountered in groups exceeding half a dozen individuals. Occasionally found with Spotted

Sandpiper, individual Least Sandpipers or Pectoral Sandpipers, or farm-pond ducks. **MOVEMENT/MIGRATION:** The bulk of spring migration across North America takes place mid-April to mid-May; fall migration of southbound birds occurs from late June to October, with peaks from mid-July to mid-September (rare after October). VI: 3.

DESCRIPTION: A small dark Tringa that likes to stay in the shallows and the shadows. Smaller, more compactly proportioned (particularly shorter-necked), and more angular than Lesser Yellowlegs. The straight bill is shorter and thinner, and the legs are shorter and greener (and seem set too far forward), but it is the overall profile more than the individual parts that distinguishes Solitary.

Upperparts (including the head and breast) of adults and juveniles alike are dark—olive brown in adults, browner in juveniles. Seen closely, back and wings are spangled with pale flecking. Also evident in all plumage are the white "spectacles" around the eye that make Solitary Sandpiper seem alert.

The tail's boldly patterned black-and-white bands radiating from a dark center are visible whether the bird is offering a side profile or heading away in flight.

BEHAVIOR: Solitary Sandpiper is a loner, foraging regularly in small pools and tight aquatic edges that other shorebirds ignore. Moves (and often stands) in a crouch, with tail and body teetered up in the rear; head raised slightly, each step seems calculated, and all movements somewhat stiff. Yellowlegs moves steadily ahead, covering ground twice as fast in an easy loose-jointed fashion. Solitary Sandpiper is methodical and precise, working slowly and meticulously—a bird that could spend the entire day in one small puddle. It sometimes bobs its body, stiffly, as it moves (and more so when nervous). Flushes easily when startled, rising almost vertically in tight surroundings, calling as it rises. In more open areas, may fly only a short distance or even return to the same spot if not hard-pressed. If approached casually, often allows close approach.

FLIGHT: The silhouette alone is unique. Overall compact but angular, with narrow pointy wings angled acutely back so that the primaries are almost parallel to the body. Birds flying overhead show a great contrast between the uniformly dark underwings and the whitish underparts. Flight is rapid, direct, and even-keeled; wingbeats are stiff, somewhat irregular, and arrhythmic and seem all flicking down stroke.

Descent is rapid, almost breakneck. The bird simply stops flapping, draws its wings in tight, and angles sharply toward the earth, slowing with a few short braking bounces and landing with wings held dramatically aloft, as if inviting the approval of the judges.

When the bird is shifting from one feeding spot to the next, flight is stiff and jittery—but also somewhat light and buoyant, as if the bird were dancing on the air. Usually calls when flushed.

VOCALIZATIONS: Most often heard call is a high, sharp, clear, whistled "*Pee-Peet*" or "*Pit-Weet*" (sometimes trebled) that recalls Spotted Sandpiper but is louder, sharper, and more assertive. Most often vocalizes when flying overhead; not surprisingly, this is how most birders discover that there is a Solitary Sandpiper in the vicinity.

Eastern Willet, *Catoptrophorus semipalmatus semipalmatus*

The Calico

STATUS: Common breeder on the Atlantic and Gulf Coasts; distribution in winter is not clear, but birds probably winter entirely south of the United States. **DISTRIBUTION:** Breeds coastally from sw. Nfld. (rare) to n. Mexico (except s. Fla.); winters south of the U.S. **HABITAT:** Breeds in tidal wetlands and beaches; also pastures in N.S. Forages on open marsh, tidal pools and creeks, sandy beaches, sand and mud flats, and (rarely) rocky shores and rock jetties. Avoids fresh water. **COHABITANTS:** Clapper Rail, American Oystercatcher, Seaside Sparrow. In winter and migration, may roost with Greater Yellowlegs and godwits. **MOVEMENT/MIGRATION:** Birds north of Virginia vacate breeding areas, and many are presumed to fly by coastal and over-water routes to South America. More southern breeders are thought to move shorter distances. Northern birds depart as early as late June through August. In spring, birds return to Virginia in late March and reach Nova Scotia by early to mid-May. VI: 3.

DESCRIPTION: A sturdy, inelegant, straight-billed, coastal shorebird in a plain brown wrapper whose vivid wing pattern explodes in flight. Fairly large (larger and more robust than Greater Yellowlegs; smaller than Whimbrel or Hudsonian Godwit), with a long, heavy, straight, and relatively untapered bill, a thick neck, a compact body, and long, thick, grayish legs. Compared to the slender elegance of yellowlegs, Willet seems thickset and clunky.

Overall bland. Breeding adults are brownish, and coarsely and heavily patterned above and below (except for whitish bellies). Nonbreeding adults and juveniles are plain brownish gray or dun above (showing little or no patterning), with unbroken putty-colored breasts (again, only the bellies show white). With their plain faces, birds have a dull or vacuous expression. By comparison, yellowlegs look pert and alert.

For comparison with the western subspecies, see "Description" section under "Western Willet."

BEHAVIOR: A bird of generalized tendencies. Tends to be solitary when feeding, but may also gather in small spaced groups. Tends to forage on shorelines and exposed substrate more than other wading birds, but also wades up to its belly like a godwit. For the most part, Willet is a visual hunter, moving sluggishly, methodically, and somewhat ungracefully along shorelines and exposed mud flats—walking, stopping, and picking or probing. Also stitches the water, godwit-fashion; runs along the edge of receding surf, pauses, and drives its bill into the unsettled sand.

Except when taking flight, fairly quiet most of the year. When breeding, Willets are positively noisome, giving their territorial song day and night, from the ground, from elevated perches, or in flight. Fairly tame and easily approached; when trespassers cause nesting adults to become frenzied, birds hover and stoop until the intruder retreats (to the scant safety of the neighboring Willets' territory). In airborne displays, birds hover and flutter high overhead, calling attention to themselves with their incessant calling and very dramatic wing pattern.

FLIGHT: When it lifts its wings, the wrapping comes off, exposing a bold black-and-white zigzag pattern on both the upper and lower wings that earned the bird its colloquial name, "the Calico." Both American Oystercatcher and Hudsonian Godwit have similar but weaker patterns. In flight, Willet seems somewhat stocky and balanced—wings are broad, shortish, and blunt, and the head/bill and legs/tail project equally. Head on, the wings show two white headlights on the leading edge; going away, birds show lots of white on the wings and tail.

The wingbeat is stiff, measured, even, somewhat choppy, with about as much upstroke as down stroke. In migration, birds fly in a messy wedge formation, and sometimes in flocks numbering 200–300 individuals. But the flamboyant wing pattern renders all other characteristics secondary.

VOCALIZATIONS: Territorial song is a loud, monotonous, whistled chant: "*pih-will-willet; pih-will-willet.*" In flight and when approached, often gives a loud, clear, yelping two-note bray: "*kyah-yah!*" When excited or agitated, breeding birds emit a monotonous yapping: "*kleep, kleep, kleep.*" Birds defending territories from intruders emit a shrill, trilling, somewhat breathy shriek, "*kre'e'e'e'eh . . . kre'e'e'e'eh,*" but if you are close enough to hear it, you hardly need to rely on the call to identify the frenzied calico-patterned bird trying to scalp you.

PERTINENT PARTICULARS: Eastern Willet is currently considered a subspecies.

Western Willet, *Catoptrophorus semipalmatus inornatus*

The Inelegant Godwit

STATUS: Common breeder in the northern prairies and the Great Basin; common winter resident on all three coasts. **DISTRIBUTION:** Breeds in se. Alta., s. Sask., sw. Man., se. Ore., ne. Calif., n. Ariz., s. Idaho, cen. and ne. Mont., n. Utah, w. and cen. Wyo., nw. Colo., much of N.D., parts of S.D., and nw. Neb. Winters from n. Calif. (locally to sw.

Wash.) to South America, along the Gulf Coast, and along the Atlantic Coast north to Va. and south into South America. Some nonbreeders remain along the Calif. coast in summer and also winter inland at the Salton Sea. **HABITAT:** Breeds in sparse grasslands near seasonal and permanent wetlands (fresh or brackish). In winter, found on beaches, bays, estuaries, and (rarely) rocky coasts. **COHABITANTS:** In winter, often associates with Marbled Godwit, Long-billed Curlew, and Whimbrel. **MOVEMENT/MIGRATION:** Spring migration from late March to late May; fall from late June to late November. *Note:* On the Atlantic Coast, most Willets observed after August are Western Willets. VI: 3.

DESCRIPTION: Like Eastern Willet (see "Description" at "Eastern Willet") but slightly larger, lankier, grayer, paler, and all in all more godwitlike. Compared to Eastern Willet, overall slightly taller and longer-legged; also, the bill is longer, thinner-based, and often slightly upturned.

Breeding adults are grayer and more sparsely and less coarsely marked above and below. In winter, adults and juveniles are paler and grayer (less dun-colored), with whiter underparts. (The gray-washed breast of young birds is commonly broken, showing a white center.)

BEHAVIOR: More landlocked than Eastern Willet. Forages, godwitlike, in deeper water, but is still much more shoreline-oriented than yellowlegs, godwits, and avocets.

FLIGHT: Like Eastern Willet.

VOCALIZATIONS: Similar to Eastern Willet, but somewhat lower-pitched.

PERTINENT PARTICULARS: Western Willet is currently considered a subspecies.

Wandering Tattler, *Heteroscelus incanus*

Slate-colored Rockpiper

STATUS: Generally uncommon, a status directly related to the bird's relatively small population, its mostly remote breeding area, and the ecologically restricted but geographically expansive scope of its wintering habitat. **DISTRIBUTION:** Breeds in ne. Siberia, Alaska, the Yukon, and nw. B.C. In the Western Hemisphere, winters along seacoasts, from n. Calif. to Peru. Most individuals winter on the islands of the s.-cen. Pacific. **HABITAT:** In breeding season, found mostly in rocky montane tundra, most commonly (but not always) close to a water source, such as rocky moraines, kettle lakes, and braided creeks and rivers; also breeds adjacent to tidal coasts. In winter, almost always found on rocky coastlines and rarely inland or on mud flats or sandy beaches. **COHABITANTS:** Surfbirds (both summer and winter); Black and Ruddy Turnstones (winter). **MOVEMENT/MIGRATION:** Entire breeding population relocates in winter. Spring migration from mid-March to early June; fall from mid-July to early November. VI: 1.

DESCRIPTION: A longish, all-gray shorebird with short yellow legs that feeds where surf and rock meet. Looks like a plump, overgrown Spotted Sandpiper or a short-legged Willet. Medium-sized (slightly larger than Surfbird), with a fairly long straight bill, a short neck, a long body, and long wings that extend beyond the tail when folded. Legs are short and yellow, making the bird appear crouched.

In all plumages, uniformly gray above. Breeding birds have heavily and narrowly barred underparts (look all gray at a distance; sometimes gray-and-white-spangled). Nonbreeding and juveniles have uniformly gray breasts and white bellies (which may or may not be evident). The face has an indistinct dark eye-line that gives it a Spotted Sandpiper–like pattern.

BEHAVIOR: Generally solitary or found in well-spaced twos and threes that feed away from groups of other shorebirds. Forages along rocky streams and in the high-energy surf zone of rocky shorelines. Walks rapidly, and not very gracefully, in a half-crouch, taking long steps and covering ground quickly. By contrast, dances nimbly and delicately to avoid the surf, with fluttering wings raised high above the body. Very commonly feeds on rocky surfaces that are the same gray tone as the bird. When roosting (commonly away from other rock-foraging birds), Wandering Tattler is very difficult to pick out. Jerks its head when it feeds, and bobs its tail when it stands (but not as habitually as Spotted Sandpiper).

FLIGHT: Very long-winged, fairly long-billed profile. Plain, unpatterned, gray upperparts are unique. Flight is straight, steady, and usually low, on steady, choppy, stiff wingbeats.

VOCALIZATIONS: A series of hollow whistled notes all on the same pitch.

Spotted Sandpiper, *Actitis macularia*

Teeter-ass Sandpiper

STATUS: Common, widespread, versatile, and somewhat anomalous-looking. **DISTRIBUTION:** Enjoys the most extensive nesting range of any North American sandpiper—a range that encompasses virtually all of Canada and Alaska and the northern two-thirds of the U.S. Winters from coastal S.C. and s. Ga. to Calif.; north along the Pacific Coast to sw. B.C.; and south throughout the Caribbean, Mexico, Central America, and most of South America. **HABITAT:** Occupies an array of aquatic (and semi-aquatic) habitats, from fast-moving streams to upper ocean beaches, rocky coasts, riverbanks, wet and dry mud flats, stock ponds, swimming pools, and irrigation ditches. Also forages along roadsides and other dry habitats. Nests, on the ground, in or under vegetative cover that is fairly close to fresh water (but occasionally 100 yds. or more away). **COHABITANTS:** Likes to be alone. **MOVEMENT/MIGRATION:** Migrates alone or in small groups (fewer than a dozen individuals). Spring migration from mid-March to early June, with a peak in May; fall from late June to November, with a peak in July and August. VI: 4.

DESCRIPTION: A small, somewhat roundly contoured sandpiper, constructed with parts that seem not quite to fit. It is easily distinguished in summer by its uniquely spotted underparts, in winter by its plain brown wrapping, and at all seasons by its very curious shape, posture, and antics. The body is plump in the front, pinched thin in the rear. The head is somewhat squarish and neckless. The bill seems a bit too heavy and too straight (not to mention, in breeding plumage, too orange), the tail (projecting beyond the wings) seems too long, and the pale legs may be a bit too short.

The upperparts are dun-colored and, except for a black eye-line and pale eyebrow, about as plain as plain gets. In breeding plumage, the breast and to a lesser extent the flanks and belly are richly spotted in a manner that recalls Wood Thrush. In nonbreeding plumage, the spots are replaced by a shadowy half-collar that curls down the sides of the breast (looks like a thumbprint, as if someone with dirty hands had pinched the neck). Juveniles are like nonbreeding adults.

BEHAVIOR: Very much a *shore* bird. While Spotted Sandpiper does occasionally stand in water, it much prefers to keep its feet dry, feeding just at the water's edge, atop a mat of vegetation or well away from the water (on jetties or raised dikes adjacent to wetlands). Very active and nervous, Spotted walks with a distinctive and near-perpetual "bobbing" or "teetering" motion, with the body horizontal and the head slightly raised. Once prey is sighted, the bill darts forward, or, if not in range, the bird crouches and runs with head lowered and body angled up toward the rear.

The bird's walking gait often seems halting or wandering, with an occasional sidestep. Pursuit of prey (such as flies on a mud flat) may be very energetic; the bird may even leap into the air to pick prey off of vegetation and flying insects out of the air.

At high tide or on the breeding ground, often sits on exposed perches—treetops, fenceposts, driftwood, even snow fence. Commonly solitary, but several individuals may forage together or sit out a tide together.

FLIGHT: Flight is distinctive. With the body suspended between wings as straight and narrow as tongue depressors, the bird seems to skip across the surface of the water on wings that do not flap so much as intermittently vibrate or jitter, then freezes into a droop-winged glide. The long tail is broadly fanned and forms a uniquely shaped blunt wedge. The wings are bisected by a pencil-thin white line running down their length. The head is slightly raised, as if the bird were swimming (instead of flying) and trying to keep its head above water. Flight may be direct or wandering, but it is usually of short duration and low.

VOCALIZATIONS: Utters a soft, three-penny whistle of a call, "*pee, pee, pee,*" that is similar to Solitary Sandpiper but not so sharp, loud, or emphatic. Also, when flushed, utters a soft, whistled "*pee-ury-hurry,*" in tones that would serve for a bedtime story.

PERTINENT PARTICULARS: This very distinctive, and in so many respects unique, sandpiper shares some of the same habitats, mannerisms, and calls of the larger, darker, and more angular Solitary Sandpiper, but the differences are also manifest. Bear in mind that Spotted is hyperactive and likes being in the open; Solitary is deliberate and likes tight places and shadowy confines. Also, Spotted's habit of leaning forward and chasing prey across open ground is mimicked by several other sandpipers, most notably Wilson's Phalarope.

Upland Sandpiper, *Bartramia longicauda*

Bug-eyed Sandpiper

STATUS: Common breeder in central portions of its range; less common to scarce and declining in the East and far West. **DISTRIBUTION:** There are two distinct breeding populations, southern and northern. The southern range extends from the Maritimes south through New England to Del.; west across s. Canada (skirting the northern rim of the Great Lakes) and through Pa., Ohio, n. Ind., and n. Ill. Farther west, the range expands north across Minn., into the lower half of Man., Sask., and Alta.; south into w. Mo., all but w. Kans., n. Okla., e. and n. Colo., all of N.D. and S.D., and all but w. Mont. Also found very locally in e. Ore. and Wash. Northern population breeds in the w. Northwest Territories, much of the Yukon, and e. and n. Alaska to the north slopes of the Brooks Range. **HABITAT:** Native prairie and grassy tundra; also hay fields, grazed pasture, commercial blueberry farms, airports, and plowed fields (for foraging). In migration, any open grassy expanse, including agricultural lands and sod farms. **COHABITANTS:** Horned Lark, Bobolink, Eastern and Western Meadowlarks, Savannah Sparrow. **MOVEMENT/MIGRATION:** Entire population vacates North America in winter. Spring migration from mid-March to mid-May; fall from late June to mid-October, with a peak in August. VI: 2.

DESCRIPTION: A uniquely shaped upland shorebird. Medium-sized (between Lesser and Greater Yellowlegs in size), with a tiny head and a bug-eyed expression. This curiously shaped bird is both plump and lanky, with parts that seem fitted for two different birds. The head is small and round, with a large black eye, and the straight yellowish bill is too short for a bird that probes, but too long for a bird that picks. The neck is plucked-chicken scrawny, but the body is chesty and plump. The tail is long—particularly for a shorebird. The yellow legs are fairly long—just about right for a bird that stalks through grass.

In all plumages, overall warm brown above, with blackish and whitish spangling. Face is pale and plain (makes the eye stand out). Except for the darkly streaked neck and chevron patterning down the flanks, underparts are pale.

BEHAVIOR: It's an upland shorebird and a loner. Avoids shorebird concentrations, standing water, and mud flats. Feeds on short dry grass or, barring this, broken expanses of taller grass (including the untilled edges of agricultural land and coastal dunes). Breeds in taller, denser vegetation, but relocates to shorter grass soon after chicks hatch. In migration, found in fields with American Golden-Plover, Killdeer, Buff-breasted Sandpiper, and Baird's Sandpiper, but does not really associate with these species.

Feeds by stalking in a halting glide. Moves slowly forward, half-crouched, head moving rhythmically back and forth. Elevates its head when it pauses. Rushes forward to seize prey or returns to the stalk. Movements appear somewhat pigeonlike—but like a pigeon in slow motion.

When not feeding, often seen perched on fence-posts or rocks in fields. When it alights, it often raises its wings above the body. Fairly tame. Responds to imitations of its call.

FLIGHT: Silhouette is unique. Short-headed and plump-bodied (sags somewhat at the neck/chest), with an extremely long tapered tail and long triangular wings. Bird seems all wings and tail. Flight is straight. Wingbeats are steady and seem to quiver, not beat (recalls the jittery flight of Spotted Sandpiper, but there is no pause and glide). Usually flies alone; calls in flight.

VOCALIZATIONS: Song is a plaintive, rising-and-falling, slurred whistle: "*wooolEE WEElurrr.*" The call often heard when the bird is flying overhead is a brusque, low, liquid "*quidyquit,*" which is not a harsh or loud sound, but it carries incredibly far.

Whimbrel, *Numenius phaeopus*

Coastal Curlew

STATUS: Common subarctic nester; common southern coastal winter resident and mostly coastal migrant. The most widespread North American curlew, and the only one commonly found in the eastern half of the United States and Canada (north of Georgia). Rare migrant in most of the interior. **DISTRIBUTION:** Breeds along the west coast and interior portions of Alaska, the Yukon, and the nw. Northwest Territories; also west of Hudson Bay from the se. Northwest Territories and n. Man. east to the mouth of James Bay. Winters coastally from San Francisco Bay (less commonly north to s. B.C.) and Cape Hatteras, N.C., south to South America. **HABITAT:** Nests on subarctic and alpine tundra ranging from dry heath to grass-sedge lowlands and hummocky taiga bogs. In winter and migration, found in tidal wetlands, sandy beaches, mud flats, rocky shores, upper beaches and dunes, fields, short grass, meadows, and mangroves. **COHABITANTS:** When nesting, golden-plovers; in winter, Long-billed Curlew, Willet, godwits. **MOVEMENT/MIGRATION:** All birds retreat south after nesting. Spring migration from March to early June, with a peak in April and May; fall from late June to late October, with peaks in July (adults) and late August (juveniles). VI: 3.

DESCRIPTION: A large, grayish brown, stripe-headed curlew of coastal habitats. Large solid-looking shorebird (larger than Willet; smaller than Long-billed Curlew; about the same size as oystercatcher), with a long downturned bill, a small flattish head, a long neck, a football-shaped body, and moderately long legs. Long-billed Curlew is considerably larger (in direct comparison, seems almost twice as large); the adult's bill is so long that it makes Whimbrel's look stunted. Long-billed also has a rounder head and a plumper, more humpbacked body.

In all plumages, overall dull grayish brown, with a slightly darker patch on the back, barring down the sides, and black stripes through the face and crown. (Sleeping birds, with heads tucked, look dark-capped.) Long-billed Curlew is overall warmer brown (blushed with pinkish tones), with plain unbarred sides, a pattern of broken black rivulets running off the bird's back, and a plain face and unpatterned head. (The slightly darker crown is the same color and shade as the hindneck.)

BEHAVIOR: A visual hunter that feeds primarily in daylight and works alone or in small spaced flocks. Usually found on dry or muddy substrate (less commonly in water). Whimbrel strides purposefully along, picking, pecking, probing, and sometimes lunging at prey. Favored prey includes fiddler crab, which the bird captures by plunging its bill down the animal's burrow. Also feeds on insects and berries, which are engulfed with a backward toss of the head.

In migration, may be found in flocks of several hundred (but foraging birds spread out). Roosts and sometimes forages with Long-billed Curlew and Marbled Godwit. On rocky coasts, commonly sits out tides on rocks above the splash zone (Long-billed Curlew is a beach- and mud-flat-firster), and in mangroves roosts in trees.

On the Atlantic Coast, fairly wary, flushing readily when approached (usually calls). On the West Coast, can be very tame.

FLIGHT: Distinctive, long-winged, long-billed, chesty, and heavy-bodied profile. Overall grayish brown with no discernible pattern or pinkish blush in the wings. Flight is strong, direct, and steady on down-curved wings. Wingbeats are stiff, slow, and regular, with a deliberate pushing down stroke. Flocks fly in a loose shifting V that flows like a banner blowing in the wind, or, when shifting short distances, they may just fly as a mob.

VOCALIZATIONS: Call is a sharp, loud, whistled stutter, "*kikikikikikiki,*" often given several times when the bird flushes.

Bristle-thighed Curlew, *Numenius tahitiensis*

The Pale-rumped Curlew

STATUS AND DISTRIBUTION: Rare (fewer than 3,500 breeding pairs) and geographically restricted breeder. Found only in two locations: a southern population is located in the s. Nulato Hills on the n. Yukon delta, and a northern population is situated in the cen. Seward Peninsula (northeast of Nome). In late summer, stages at the mouth of the Yukon-Kuskokwim R. **HABITAT:** Dwarf-shrub (tundra) meadows situated on rolling hills with gentle slopes. Prefers to forage in expansive areas of extremely short dry vegetation, but territories also encompass wetter tussocky lowlands and taller willow-shrub mats. **COHABITANTS:** During breeding, American Golden-Plover, Whimbrel, Bar-tailed Godwit, Long-tailed Jaeger. **MOVEMENT/MIGRATION:** Arrives in the second and third weeks of May. Stages from June to August, with adults leaving breeding areas from mid-June to mid-July; juveniles leave breeding areas by early August and depart staging areas in late August. VI: 1.

DESCRIPTION: A warm-toned whimbrel with a pale rump situated on the rolling hills north of mile 72 of the Kougarok Road outside of Nome. Hails you with a whistle. Large shorebird (about the same size and shape as Whimbrel; slightly larger than Bar-tailed Godwit), with a long down-curved bill, a smallish head, a long neck, an elliptically shaped body, and sturdy gray legs. Compared to Whimbrel, the bill appears somewhat shorter and more curved throughout, with the downturn on the curlew beginning just past midpoint. On Whimbrel, the downturn seems relegated to the last third of the bill (that is, closer to the tip).

Most adults are overall warm-toned (Whimbrel is colder gray-brown), with a lavish array of large pale spots on dark upperparts. At a distance, the back seems frosted. Underparts show darker patterning on the neck and upper chest and a cleaner belly. At a distance and in flight, Bristle-thighed Curlew seems bibbed. Whimbrel, by comparison, is overall grayer and plainer, unpatterned above, and more extensively marked below—more pale-bellied as opposed to bib-chested. Juveniles are similar to adults. *Note:* The dark upperparts of juvenile Whimbrel show fine pale speckling. The back of Bristle-thighed is spotted, making it appear paler and buffier.

BEHAVIOR: Most often seen in display flight over breeding areas, where birds call attention to themselves with their eerie calls and flutter-and-glide displays. Also seen foraging in dry close-cropped portions of tundra, where it walks with slow methodical steps when feeding on berries and moves faster when pursuing insects. Very tolerant of neighbors (including American Golden-Plover, Whimbrel, and Long-tailed Jaeger), and seems to associate with Bar-tailed Godwit. Vocalizes at all hours, from the ground and especially in the air, but seems most vocal in the early hours of the morning (5:00–10:00). Shy and not easily approached.

FLIGHT: Stocky curlew with long wings and a head and tail projecting about equal lengths. Uniformly dark above except for a conspicuous pale (blond to cinnamon) rump that easily distinguishes this species from the dark-rumped Whimbrel. Underparts are mostly pale, with a bibbed chest.

Straight flight is strong, direct, and very fast, with steady wingbeats. In display, flutters its wings in a quivering series (much like Upland Sandpiper) followed by prolonged parachuting glides on down-bowed wings showing a very short arm and very long hand. These straight, curving, or circling glides may last a minute or more and cover distances in excess of one mile. Whimbrel has a similar display, but the wingbeats appear slower and stiffer (less quivering), and vocalizations differ.

VOCALIZATIONS: The whistled call, given in the air or on the ground, "*cheeooeet,*" sounds like the human hailing whistle or like Black-bellied Plover's call, but it's quicker, softer, and less mournfully plaintive. Song is a prolonged, eerie, rising-and-falling, whistled keening, "*oo-ree-chah oo-ree-chah oo-ree-chah,*" that usually begins with several warm-up notes. Both the quality and the rising-and-falling pattern of the song recall the sounds of an old ham radio when the operator is turning the dial to find a faint signal. By comparison, Whimbrel's song is a monotonous gargled whistle: "*wr'wr'wr'wr'wr'wr'wr.*"

Long-billed Curlew, *Numenius americanus*

Prairie "Cur-lee"

STATUS: Fairly uncommon western breeder; locally common wintering species along the California and Texas coasts, the Salton Sea, and a few locations in southern Arizona; rare winter resident from coastal North Carolina to Florida. **DISTRIBUTION:** Breeds primarily in the northern short-grass prairies and the n. Great Basin. Found in s. B.C., s. Alta., and sw. Sask., e. Wash., eastern two-thirds of Ore., ne. Calif., n. Nev., s. Idaho, n. Utah, most of Mont., w. N.D., w. S.D., western two-thirds of Wyo., e. Colo., and nw. N.M. Winters along the Calif. coast south to Baja and in se. Calif., coastal Tex., and sw. La. south into Mexico. Low numbers also winter along the Atlantic and Gulf Coasts from N.C. to Ala. **HABITAT:** Nests in prairie-brush steppe and short-grass prairie; less commonly in agricultural fields and fallow fields. Avoids tall-grass prairie and areas with ample shrubs. Winters in tidal estuaries, dry and flooded agricultural land, wet and dry pastures, and sandy beaches. In migration, may be found in sloughs, holding ponds, and sewage treatment ponds. **COHABITANTS:** In summer, Horned Lark, Chestnut-collared Longspur. In winter, Great Egret, Reddish Egret (Texas), Mountain Plover, Willet, Marbled Godwit. **MOVEMENT/MIGRATION:** Entire breeding population relocates to southern and coastal wintering areas after breeding. Spring migration from late February to early April; fall from early June to mid-August. VI: 3.

DESCRIPTION: An impressively large shorebird with an even more impressive bill. Large and puddle duck–sized (larger than Whimbrel and Marbled Godwit), with a round head, a long neck, a long football-shaped body, long sturdy legs, and a very, very long thin bill that is straight for most of its length, then droops markedly at the tip. Whimbrel's bill is shorter (usually considerably shorter), sturdier, and seems more curved throughout its length.

In all plumages, overall brown with a warm cinnamon or pinkish cast and no overt plumage pattern. The head and face are very plain — sometimes with a shadowy eye-line, sometimes with a pale center to the slightly darker crown, but basically plain. The back is uniformly spattered with black, pale, and ruddy flecks. Buff or salmon underparts are plain and lightly patterned, showing little contrast with upperparts.

BEHAVIOR: When breeding, fairly vocal. At other times, fairly quiet, but usually calls when taking flight. In winter and migration, found singly, in pairs, or in small groups (usually fewer than 50 birds; occasionally considerably more). Sometimes forages with herons and egrets, other times in segregated flocks. Forages most commonly when falling tides expose hard mud flats; feeds less commonly on sandy beach (which is favored by Marbled Godwit and Willet). After feeding, commonly loafs, often standing in a linear array along the shoreline.

In summer, primarily feeds by pecking. In winter, feeds mostly by probing for marine crustaceans and invertebrates with a jittery motion and its head turned slightly askew — not a smooth up-and-down probing motion. Walks (and often runs) with long strides across mud flats, beach, or shallow water. Fairly shy. Does not tolerate close approach.

FLIGHT: Pale brown and very long-winged, with a very obvious long drooped bill. Underwings show a pinkish or cinnamon wash; the trailing edge of the upperwing flashes warm ruddy brown. Overall flight is slow, direct, and heavy, but also elegant, with steady shallow wingbeats and down-drooped wings. (Long-billed Curlew's cadence and heaviness recall a goose, its flight is slightly battier than Marbled Godwit, and its wing shape looks more like a gull.) Usually makes a long gliding landing, sometimes circling on set wings as it lands. Flocks are often loose and strung out.

VOCALIZATIONS: Call is a loud, clear, whistled, bisyllabic yelp, "*cur-lee!*" that rises at the end.

Hudsonian Godwit, *Limosa haemastica*

Ring-tailed Marlin

STATUS: Not uncommon, but geographically restricted in both its breeding range and migratory pathways. **DISTRIBUTION:** Breeds in scattered locations in coastal w. and s. Alaska, the nw. corner of the Northwest Territories, nw. B.C., the southern shore of Hudson Bay—and probably elsewhere. In spring, migrates through the prairies (e. Tex., sw. La., Okla., Kans., w. Mo., e. Neb., e. S.D., and cen. Sask.). In fall, the migration is more eastern; from major staging areas in the Quill Lakes region of Sask. (presumably Alaskan birds) and James Bay, Canada, the birds appear to migrate nonstop to South America. Some birds (numbering in tens to as many as 200) turn up mostly in coastal areas in New England and the Maritime Provinces, with a few in s. N.C. **HABITAT:** Breeds in the ecotone where tundra and boreal forest meet—typically large open muskeg with bogs and ponds that have islands of spruce or bordering forest. In migration, uses a variety of open shallow wetlands (both salt and fresh), including tidal estuaries, mud flats, sandy beaches (along seacoasts, lakes, and rivers), freshwater marshes, tidal pools, flooded crops (particularly rice fields), sewage lagoons, and, less commonly, dry open uplands (such as pastures). **COHABITANTS:** During breeding, Whimbrel, Short-billed Dowitcher, American Golden-Plover, Bar-tailed Godwit (Yukon delta). In migration, may feed and roost with Willet, Greater Yellowlegs, Marbled Godwit, dowitchers. **MOVEMENT/MIGRATION:** Arrives in Texas in late March and (mostly) April; in Saskatchewan and coastal Alaska in late April; at breeding grounds in late April to early June. Birds begin gathering in staging areas in late June; most adults depart by mid-August, and juveniles in early September. Some birds linger along the Atlantic Coast until early November. VI: 3.

DESCRIPTION: A large humpbacked shorebird with a broad white band across the base of the tail and a long modestly upturned marlin-spike for a bill. Large (larger than Greater Yellowlegs; smaller than Marbled Godwit and Whimbrel; about the same size as Willet), but generally smaller than most observers anticipate, and individuals vary in size. Has a very long, slender, pointy, upturned bill, a round head, a long neck, a robust body, and fairly short sturdy legs. Overall, has somewhat curvaceous lines; feeding birds seem distinctly humpbacked.

Breeding birds are dark or blackish above, with a grayish head/neck contrasting with dark, rich chestnut underparts (males) that usually (but not always) show blackish barring (particularly on the belly). Paler females show more distinctly barred underparts. Larger Marbled Godwit is more monotone, all barred, and buff brown. Bar-tailed Godwit, in breeding plumage, is paler and more orange-red below, *including the head and neck;* underparts show little or no barring. Also, the bill of Hudsonian Godwit is distinctly two-toned—half-pink (to midbill), half-black. Bar-tailed's bill shows pink just at the base.

In nonbreeding plumage, adult Hudsonian Godwit is just plain gray, above and below, but this plumage is not commonly seen in North America, where most adults are seen in fall in transition, with mostly pale underparts mottled or spotted with black and red (looks like an exuberantly applied crop of hickeys). Winter Marbled Godwit retains an overall warm buffy cast. Winter Bar-tailed Godwit shows lightly streaked gray upperparts and a gray breast contrasting with a whitish belly.

Juveniles are mostly plain and slightly two-toned—brownish gray above (and on the breast) and paler on the belly—and most closely resemble Willet rather than Marbled Godwit, which is overall buffier and warmer-toned and shows less contrast between upper- and lowerparts, or juvenile Bar-tailed, which shows a mottled pattern above.

BEHAVIOR: Looks and feeds like a big long-billed dowitcher. Feeds rapidly, but moves forward with deliberate grace, thrusting its long bill deep into the water and mud, often in a series of jabs. Often feeds up to its belly and submerges its head. By comparison, Greater Yellowlegs looks frantic and crazed, and Willet looks sluggish. On the breeding ground, Hudsonian Godwits commonly perch on the tops of trees. In spring, commonly found in small flocks (several birds to several dozen). In fall, where the bird occurs over North America (mostly in the East), it is usually solitary but may occur in pairs, trios, or flocks of up to a dozen birds. Solitary birds frequently roost with Greater Yellowlegs, Willet, and gulls, but often feed with dowitchers. Often roosts standing in water.

Response to humans varies—some birds are wary, some are tame. Calls during migration. Flocks fly in V-formation.

FLIGHT: Large, distinctive, long and slender-winged profile. Flies with neck extended, long bill angled slightly down, and legs trailing behind the tail. The narrow white wing-bar is easy to note against the dark wing, but the bright white basal half of the tail, which appears to be a white rump, cannot be missed. Marbled Godwit has a dark rump and salmon flash in the wings; Bar-tailed Godwit appears uniformly gray above, with no ringed tail and no white wing stripe.

Flight is strong, fast, and direct, with deep, steady, down-pushing wingbeats.

VOCALIZATIONS: Call is a loud, two-note, whistled yelp: "*kowEE*" or "*err-Yeh*" (phonetically rendered: "*Godwit*"). Notes are higher than other godwit.

PERTINENT PARTICULARS: When roosting among other medium-large grayish shorebirds (like Greater Yellowlegs and Willet), this species does not particularly stand out. Be prepared to search for only a slightly taller and bulkier bird. The best time to see this species along the East Coast is between mid-July and late September, when strong coastal storms (particularly hurricanes) block the bird's offshore passage and birds reroute to coastal locations. In coastal New England and the Maritimes, Hudsonian Godwit is a regular no matter what the weather.

Bar-tailed Godwit, *Limosa lapponica*

Plain-tailed Godwit

STATUS: Common breeder in western and northern Alaska. Rare vagrant along both coasts. **DISTRIBUTION:** Breeds within 100 mi. of the coast from Bristol Bay (southern limit) north and east to just shy of the Canadian border; generally scarce east of Prudhoe Bay. **HABITAT:** Breeds on open treeless tundra—both wet and dry—on the coast and in hilly interior. In many places, seems partial to wet sedge meadows. Where range overlaps with Hudsonian Godwit, Hudsonian prefers wet bogs closer to forest edges. In migration, found on tidal sand and mud flats and estuaries; in fall, occurs in sedge meadows. **COHABITANTS:** Pacific and American Golden-Plovers, Whimbrel, Bristle-thighed Curlew. In migration, assorted shorebirds, especially Hudsonian Godwit and dowitchers. **MOVEMENT/MIGRATION:** Among the world's most celebrated migrants, believed to travel nonstop from Alaska to New Zealand and Australia. Arrives in Alaska between early May and early June; departs early July to September, with most birds vacating breeding areas and moving to staging areas in the Yukon-Kuskokwim delta and the Alaska Peninsula in July and August. At staging areas, some birds may remain into October. VI: 2.

DESCRIPTION: Overgrown Alaskan dowitcher with a tail and rump the same color as the back. Large shorebird (much larger than a dowitcher; slightly larger than Hudsonian Godwit; slightly smaller than Whimbrel), with a long, exquisitely slender, slightly upturned bill, a small head, a longish neck, a robust body, and fairly short sturdy legs. Proportionally similar to Hudsonian Godwit, but slightly shorter-legged.

Adult male breeding plumage shows dark brown upperparts and an orange face and underparts—orange right to the tail. Upperparts are flecked with gold; underparts show scant barring (limited mostly to the flanks). The color of the underparts recalls Red Knot or the prairie-form of Short-billed Dowitcher. Hudsonian Godwit's underparts are brick red (closer to Long-billed Dowitcher), with conspicuous blackish barring on the belly and a whitish or broken white undertail. Breeding females are grayish-backed, with a buffy face and breast and whitish unbarred belly. (Female Hudsonian is gray-faced, with conspicuous, narrow, reddish and blackish barring below.)

In winter, Bar-tailed is overall grayish (like Hudsonian) but more patterned, showing dark streaks on the back and breast. Hudsonian is plain unpatterned gray. Juvenile Bar-tailed is warm-toned and darkly marbled above, with a slightly streaked breast; juvenile Hudsonian is overall plainer, paler, and grayer, with just a warm wash across the breast (no streaking or spotting).

If all this is too hard to sort out, just keep in mind that Bar-tailed is more plainly patterned than Hudsonian in summer, and more streak-backed (adults) or mottled (juveniles) in winter.

BEHAVIOR: A social bird that forages with other shorebirds and, particularly in fall, in large single-species groups that may number in the thousands (spring flocks are small—fewer than a dozen). When breeding, appears to nest close to other defense-minded shorebirds, which form an "attack umbrella" over the godwit's nest. In summer, plucks berries and insects from the tundra and probes wetter areas. Almost wholly coastal the rest of the year, working tidal flats (both dry and inundated) and beaches, where it probes for food much like a dowitcher. Very calm, and often allows close approach during breeding season. Near human habitation, often seen along roadsides and on roads.

FLIGHT: Overall slender, long-billed, long-winged profile. Upperparts are uniformly plain, with no obvious distinction between the back, wings, and tail. There is no bold pale wing stripe, and no bold white-based, black-banded tail (that's Hudsonian). The narrowly barred tail that gives the bird its name is visible at close and moderate ranges, but for the purposes of identification, all you need to see is that the tail pattern is not boldly bisected—half-black, half-white.

Wingbeats are strong and deep. Flies and glides with wings severely drooped, showing a short arm and long hand.

VOCALIZATIONS: Flight song is a hurried, nasal, excited, and even series of two-note utterances, "*kerWir kerWir kerWir,*" whose cadence and pattern might suggest an impatient Northern Cardinal or Carolina Wren and in quality is richer and slightly lower pitched than Hudsonian Godwit. The song is sometimes proceeded by a nasal introductory note. The "Godwit" call, "*Ki'Whik,*" is similar to Hudsonian Godwit.

Marbled Godwit, *Limosa fedoa*

Tawny Sword-bill

STATUS: Common breeder; very common winter visitor along the California coast and the Salton Sea; fairly common on the Texas coast; uncommon to rare farther east on the Gulf Coast, in coastal Florida, and on southern Atlantic beaches north to Virginia. **DISTRIBUTION:** Breeds primarily on the prairie grasslands of s. and cen. Alta., Sask., Man., cen. and n. Mont., most of N.D., nw. Minn., n. and cen. S.D., the west coast of James Bay, Canada, and the base of the Alaska Peninsula. Winters coastally in nw. Calif. and the Salton Sea south to Central America; along the Gulf Coast from w. La. to the Yucatán Peninsula; and along the west coast of Fla. Also found in reduced numbers on the Atlantic from N.J. to Miami. **HABITAT:** Nests in short, usually native grasslands near wetlands and ephemeral ponds. Avoids tall, dense vegetative cover. In winter, commonly found in shallow back bays, coastal mud flats, sandy beaches, estuaries, and salt ponds. In migration, also found along the edges of drawn-down reservoirs, grasslands, hay fields, and golf courses. **COHABITANTS:** Willet, Whimbrel, Long-billed Curlew. **MOVEMENT/MIGRATION:** Entire population vacates breeding areas and relocates to coastal wintering areas by crossing over the interior. Greatest movement occurs from the Great Plains west; regular but considerably less significant movements occur throughout the Midwest and along the East Coast. Spring migration from early April to late May; fall from late June to mid-December. VI: 3.

DESCRIPTION: A large, overall tawny shorebird wielding an arrestingly long and slightly upturned bill that just seems too big for it. A *big* bird (obviously larger than Eastern Willet; only slightly larger than Western Willet; smaller than Long-billed Curlew; about the same size as Whimbrel), with a small head, a long neck, a large body, long black legs—and an almost grotesquely long, slightly upturned, mostly pinkish bill. When standing erect and when walking, the bird angles its long bill down. When foraging, appears very humpbacked.

In all plumages, fairly plain, uniformly buffy brown, with no bold patterning and no sharp contrast between upper- and lowerparts. On breeding adults, upperparts are finely speckled with black and cinnamon, and underparts are lightly barred (net effect: pretty uniform). The tawny underparts of nonbreeding adults and juveniles are unbarred. The contrast between upperparts and underparts is slightly greater, but birds are still basically overall warm pale brown. Also, in all plumages, the folded wings and the tail show salmon undertones that bloom when the bird takes flight.

Very plain-faced, except for a smudgy shadow in front of the eye. The expression is serene, or a little tired.

Winter plumage Hudsonian Godwit and Willet are gray above, whitish below. Whimbrel is colder gray and lacks any hint of warm buff below. The larger Long-billed Curlew shows much the same plumage as Marbled Godwit, but the bill curves down, not up.

BEHAVIOR: Very social except when breeding. Feeds in large flocks that spread out across tidal flats, in shallow water along shorelines, or, less commonly, in high dry beach. Feeds by probing in the mud (or sand), somewhat like a dowitcher, but commonly takes two or three long strides, then probes the water. Also stitches as it walks. (Dowitchers usually take one step, probe one to three times, then take another step.) Moves directly without wandering. Covers ground quickly. Really likes to get into the wet. Frequently forages up to its belly in water, occasionally swims, and dunks its head as it probes. Willet is normally much more reserved, working the shallower edges and probing without plunging.

Roosts on high ground, often with other large waders; in the absence of large waders, may roost with gulls. Fairly tame where acclimated to people. Moderately vocal and often calls in flight.

FLIGHT: Distinctively large, slightly front-heavy, and very pointy (pointy bill, pointy wings) profile, with long trailing legs that (unlike Whimbrel and Long-billed Curlew) extend well beyond the tail. The (upturned) bill is angled slightly down. Upperwings and tail show salmon-colored flashes; underwings flash cinnamon and salmon. Flight is direct and fast, on steady, fluid, and fairly shallow wingbeats. Flocks traveling any distance commonly fly in a V.

VOCALIZATIONS: The classic (and namesake) call, often given in flight, is a trumpeting nasal bray or yelp, "*reeh*" or "*ereehor*" or "*kerWEK!*" (*Godwit!*). The pattern and tonal quality recall the Gull-billed Tern or Laughing Gull. Also has a longer elaboration of the basic "Godwit" call—a loud, hurried, rolling "*reh karehah karehah karehah heh heh heh heh heh*" that sounds like the bird is laughing at the end. Also, a softer, nasal, and more conversational rendering of the "Godwit" call, "*wah*" or "*wahah,*" is heard among feeding flocks.

Ruddy Turnstone, *Arenaria interpres*

Horsefoot Snipe

STATUS: Common coastal species with a circumpolar breeding distribution and a cosmopolitan view of the planet. **DISTRIBUTION:** Accounted among the world's northernmost breeding species, in North America breeds on the north and west coasts of Alaska, the arctic islands north of the Canadian mainland, and the north coast of Greenland. Winters exclusively along seacoasts, in the New World from Mass. (a few in s. N.S.) and s. Wash. south to the tip of South America. **HABITAT:** Breeds on rocky coasts and tundra, almost always in proximity to a marsh, stream, or pond. In winter and migration, found on almost any coastline (including inland lakes during migration), but is attracted to rocky coastlines, rock jetties and breakwaters, mud flats and upper beaches (particularly those festooned with shells, cobble, kelp, or seaweed). Also sometimes found in parking lots, on sidewalks (with lots of debris), or even at garbage dumps. **COHABITANTS:** In winter, often associates with Purple Sandpiper (East Coast), Rock Sandpiper, Black Turnstone, surfbirds, and Rock Sandpiper (West Coast), and Dunlin and Sanderling (both coasts). In migration, found with Red Knot on sandy beach. **MOVEMENT/MIGRATION:** Primarily a coastal migrant, in spring Ruddy Turnstone does not commonly reach breeding grounds until late May or June. In May, staging groups numbering from ten into the thousands gather on coastal beaches before departing for the arctic. In fall, returning birds appear in late July, but numbers peak in August, September, and October. During migration, may occur inland, most often on the shores of large lakes, especially the Great Lakes, where high counts of several hundred birds are possible. VI: 3.

DESCRIPTION: This colorful, comical, harlequin-patterned shorebird is hard to mistake—its feeding habits make it hard to ignore. A medium-sized sandpiper (larger than Purple or Rock Sandpiper; smaller than Surfbird or Red Knot), with an unconventional shape. The body is pointy and ovate, shaped somewhat like a croissant. The head appears too small for the body; the bill is short, pointy, and seems very slightly upturned; and the bright orange legs are stumpy, comically short. The posture seems to sag as the bird stands. Except for Black Turnstone of the Pacific Coast, the silhouette is unique.

The plumage is likewise difficult to mistake, and even harder to take seriously. Harlequin-patterned breeding birds, with their rich black, white, and chestnut pattern, are distinctive. Brownish-backed, black-bibbed nonbreeding adults retain a shadow pattern of the distinctive head and breast pattern and have orange legs (young are similar). Any of these traits distinguishes this species from the distinctly black-and-white-patterned Black Turnstone of the West Coast.

BEHAVIOR: Highly social, Ruddy Turnstone mixes freely and in close quarters with assorted other small and medium-sized shorebirds. This species is also something of a ruffian in the flock, the uncouth cousin at the family picnic. Turnstone waddles along, sometimes running, with amazing single-mindedness; shouldering its way through smaller sandpipers, it stays focused on the next bit of flotsam on the beach, which it deftly flips with its wedge-shaped bill, disclosing whatever marine animal might be hiding below. Captures prey by pecking, jabbing, probing. Adept at capturing prey on the run, and also feeds on carrion and the fly larvae associated with it.

An accomplished excavator, one or several turnstones working in concert root like feathered pigs, tossing geysers of sand aside. These excavations may be so large that they mask the bird's body from view. On rocky coasts and jetties, the bird searches methodically in crannies and uses its bill to hammer barnacles into submission.

FLIGHT: Short but slender-bodied; slender, angular, and pointy-winged. Bold dark-and-white-patterned upperparts are distinctive and can be confused only with Black Turnstone, whose upperparts are more distinctly black and white; this species is brown or reddish brown, black, and white. Flight is strong and direct, with stiff hurried wingbeats. Not particularly nimble, Ruddy Turnstone seems to require a braking effort or premeditated care when landing. Flocks are tight-packed and globular.

VOCALIZATIONS: A short, sharp, protesting "*chur*" or "*chees.*" Also makes a low, broken, or stuttering rattle in flight.

Black Turnstone, *Arenaria melanocephala*

Rock Turnstone

STATUS: Common coastal breeder in Alaska; common winter resident of western coastlines. **DISTRIBUTION:** Breeds along the west coast of mainland Alaska. Winters from the south coast of Alaska and Kodiak I. to Baja and the west coast of Mexico. Some nonbreeding individuals remain on the winter range through the breeding period (late May to early July). **HABITAT:** Nests on coastal sedge meadows. Winters coastally (almost never found inland), primarily along rocky high-energy coastlines, but may also be found on gravel beaches, mud flats, and sandy beach. **COHABITANTS:** In winter and migration, most commonly associated with Surfbird and Ruddy Turnstone. **MOVEMENT/MIGRATION:** Entire population vacates breeding areas, though some fly only to the south coast of Alaska. Spring migration from March to early

June, with a peak in late March to early May; fall from late June to early November. VI: 1.

DESCRIPTION: A croissant-shaped, nutpick-billed study of a shorebird done in charcoal; a non-ruddy Ruddy Turnstone. Medium-sized (smaller than Surfbird; larger than Sanderling or Rock Sandpiper), with a somewhat elongated and curiously shaped body featuring a small neckless head, a small pointy bill that appears slightly upturned, a plump body, and a pointy rear. Overall seems front-heavy, short-legged, and somewhat stockier than Ruddy Turnstone.

Plumage appears coarse, ragged-edged, and distinctly dark—charcoal gray onto black above, contrasting with a white unmarked belly (distinctly blacker and darker-bibbed than other rock-hopping shorebirds like Surfbird, Wandering Tattler, and Rock Sandpiper). Overall blacker and less patterned than Ruddy Turnstone, showing much less white about the head (even in breeding plumage), a fuller, more extensive, ragged-edged bib, and blackish or dull legs that sometimes show a hint of orange but not the bright orange legs that are so conspicuous on Ruddy Turnstone. All ages and sexes are essentially similar.

BEHAVIOR: A busy, industrious little rockpiper that waddles up and down slippery rocks within the wave and spray zone, jabbing at mussels or barnacles with its nutpick-like bill or using the flattened top to flip over shells or roll back algae to expose prey. Also probes on the breeding ground. Commonly found foraging in small flocks (usually less than a dozen birds), often in the company of Surfbird and Ruddy Turnstone. Roosts (again, with Surfbird and Ruddy) above the splash zone but close to feeding areas. Fairly tame and generally calls when it flies.

FLIGHT: Overall angular and pointy, with long narrow wings, and overall slightly more compact than Ruddy Turnstone. Flight is direct and fast, with strong shallow wingbeats. Lands with much fluttering and not much finesse. Boldly patterned black-and-white upperparts are arresting. (Surfbird shows only an exaggerated white wing stripe and a white-based tail.)

VOCALIZATIONS: Call is a broken, breathy, medium-high-pitched, rattling trill: "*ch'chrrrrr.*" Recalls Belted Kingfish or a ground squirrel.

Surfbird, *Aphriza virgata*

Blunt-billed Rockpiper

STATUS: Uncommon to fairly common tundra breeder; fairly common West Coast winter resident. DISTRIBUTION: Breeds in the interior of Alaska and n. and w. Yukon. Winters from sw. Alaska coast south to Tierra del Fuego and points between. HABITAT: Breeds in dry montane or ridge-top tundra; winters exclusively on rocky coastlines, but occasionally feeds on sandy beach or mud flats (most commonly during migration). COHABITANTS: (Nonbreeding) Wandering Tattler, Black Turnstone, Rock Sandpiper. MOVEMENT/MIGRATION: Wholly and highly migratory. Breeds May to July. Found in coastal locations from early July to mid-May. Seen away from coastal locations only when breeding. VI: 1.

DESCRIPTION: A robustly plump snub-billed sandpiper. Medium-sized shorebird (larger than Black Turnstone; about the same size as Red Knot) and roundly compact, with a small, round, neckless head, a blunt nub of a bill, a rotund body, and short, sturdy, yellowish legs. Much larger, plumper, chestier, and paler than the Black Turnstones with which it will almost certainly be associating.

Breeding adults are coarsely marbled gray and black above, with a large rufous orange swath riding above the wing, and coarsely spotted below, with a bibbed breast that gives way to a dark-spotted white belly. In winter, conspicuously plain with slate gray upperparts, a gray-bibbed breast, a white belly, and lightly spotted flanks. The orange or yellow-orange base to the bill is often difficult to see at a distance; the yellowish legs stand out against dark rocks. Juveniles are similar to nonbreeding adults.

Shape and bill structure seem ploverlike to some, but Surfbird, while sturdy and compact, is plumper and more roundly contoured than the larger, square-headed, and longer-legged Black-bellied Plover (which may also be found on rocky coasts).

BEHAVIOR: Highly social, almost clannish. Commonly found roosting above the splash zone with Black Turnstone, Rock Sandpiper, and Black Oystercatcher, as well as other species. Commonly feeds with other "rockpipers," but also forages in tight segregated groups away from other species (most often during its rare appearances on mud flats, where it may associate with Red Knot). Feeds in the intertidal zone by prying mussels and limpets off rocks. When foraging in shallow water, submerges its head and feeds with a jittery motion; on mud flats, picks.

Fairly tame, allowing close approach. Generally silent, but sometimes calls when flushed.

FLIGHT: Compact and wide-bodied in flight, with nominally long wings. You'll probably overlook the shape. Upperparts show a Hudsonian Godwit–like pattern, with a white, black-tipped tail and prominent white wing stripes—a unique pattern among small and medium-sized shorebirds. Flight is direct and strong, with steady wingbeats. Flocks are cohesive, and in mixed flocks with turnstones and Rock Sandpiper, Surfbird's superior size is distinctive.

VOCALIZATIONS: Usually silent. Flushed birds sometimes make a turnstonelike "*churt.*"

PERTINENT PARTICULARS: Among the rockpipers, Surfbird is clearly paler than Black Turnstone (and more patterned below), and shorter-billed than Rock Sandpiper (and less spotted below). Surfbird's yellow legs are also distinctive. Wandering Tattler and Surfbird are similar in color, basic pattern, and size, but differ radically in shape (tattlers are long, Surfbird is stocky), feeding behavior, and social proclivities. Surfbirds are clannish; tattlers are loners.

Red Knot, *Calidris canutus*

The Robin Snipe

STATUS: Almost wholly coastal; at times numerically common in primary staging areas; relatively uncommon away from staging areas. **DISTRIBUTION:** A circumpolar breeder, in North America nests on dry elevated tundra on the North Slope of Alaska, Canada's arctic islands, and n. and e. Greenland. Most North American knots winter in extreme s. South America, but also in the intertidal zones of both seacoasts from n. Calif. and Mass. south. **HABITAT:** In winter and migration, forages primarily below the high-tide line on sandy beach, exposed sod banks, tidal flats, and occasionally tidal salt marsh. Prefers areas near the mouths of rivers, bays, and beaches subject to high-energy wave action. **COHABITANTS:** In winter, often forages with Sanderling or Dunlin; in migration, forages with dowitchers, Dunlin, and Ruddy Turnstone. **MOVEMENT/MIGRATION:** One of the earth's most celebrated long-distance migrants. The entire population vacates northern breeding areas in a series of long migratory jumps that ferry a portion of the population to the tip of South America. Fall staging areas in North America include Hudson and James Bays, select Atlantic beaches in Canada, Massachusetts, and New Jersey, and San Francisco Bay. In spring, major concentrations are found in Delaware Bay, the New Jersey–Delaware border, and the Copper River delta and Yukon-Kuskokwim River delta in Alaska. Spring migration from late April to early June; fall from mid-July to November, with a peak in late July to mid-August in the northeastern north, and in late August to early October in Florida and on the West Coast. Small numbers are found scattered in the interior, mostly in the fall. VI: 3.

DESCRIPTION: A ruddy-breasted (in winter, dingy) Sanderling on steroids. A large sandpiper (dowitcher-sized) proportionately similar to Sanderling, but much larger, slightly more humpbacked, and a little paunchier. The head appears a little too small for the body, the eye a little too small for the head, and the legs a little too short for the bird. Short-necked, the bird can seem neckless in cold temperatures. The straight bill is slightly longer than the head is wide, thick at the base, and tapered at the tip (resembles a golf tee).

Breeding plumage is unmistakable—a silver-gray back, upperparts that are often spangled with orange, and a pale to deep orange face and underparts (hence the old market gunner's name, the Robin Snipe). In nonbreeding plumage, birds are dingy gray above (halfway between Western Sandpiper gray and Dunlin brown), dull white below, with a gray-washed breast and lavish amounts of gray speckling along the sides (right to the tail). Particularly when the bird is feeding, the barred rump and tail often peek through the wingtips and stand out against an otherwise bland backdrop.

Sanderling is considerably paler. Dunlin is browner-backed, with whiter (unmarked) flanks and a proportionally longer and distinctly down-drooped bill. Dowitchers are the same size as knots and likewise gray-toned, but longer-legged and distinctly longer-billed. Also, dowitcher upperparts appear slightly patterned or scaly, not plain. (Juvenile knots have scaly backs, but most are in nonbreeding plumage by October.) The expression is myopic and piggy-eyed. Red Knot appears slightly more intelligent than Dunlin, but not by much.

BEHAVIOR: Less frantic and energetic than smaller sandpipers; prefers walking over running. Often stays close to the wave action (when feeding on sod banks), and is not timid about foraging below tide line in the wave surge. Where Sanderling retreats before the oncoming wave, Red Knot commonly lets the wash roll up to its belly (even in winter). Feeds by quick probing in the sand or tidal wash, often submerging its head. Also picks for food on the surface.

Highly social. Most often feeds in tight homogeneous groups (knots?), whether by themselves or on mud flats and beaches crowded with other shorebird species. On beaches, found with Black-bellied Plover, Sanderling, Dunlin, turnstones; on tidal flats and sod banks, found with dowitchers.

FLIGHT: Sheds most of its sense of stockiness in the air. Body is elliptical and held horizontal, making a distinctly humpbacked profile. Wings are very long, pointy, and angled back; the tail appears short and blunt. The pale gray tail and rump contrasting with the darker back is distinctive. Wingbeats are rapid and steady on swept-back wings; flight is direct and fast. When traveling short distances, flocks are often globular (packed tight in the front and looser in the back). In migration or when shifting from feeding to roosting areas, birds order themselves in a loose V.

VOCALIZATIONS: Generally silent. Sometimes makes a fairly loud godwitlike "*kowek*" that is unlike any call of the birds it associates with.

Sanderling, *Calidris alba*

The Bird That Plays Tag with the Waves

STATUS: Common along U.S. seacoasts except from mid-June to mid-July; in migration, locally common around the Great Lakes and alkaline lakes of the northern Great Plains; uncommon to rare elsewhere. **DISTRIBUTION:** A circumpolar species that breeds, for the most part, north of the Arctic Circle. In North America, nests almost exclusively on the arctic islands north of the Canadian mainland. Winters from New England and s. N.S., and from s. B.C. and Kodiak I., Alaska, south to Tierra Del Fuego. **HABITAT:** Among the planet's northernmost breeding birds, Sanderling nests on barren or sparsely vegetated ground. Winters on sandy coastal beaches, particularly beaches with a shallow profile and dynamic wave action, but also occasionally found on rocky shores, jetties, and (less commonly) mud flats. In migration, occurs in small numbers on the shores of large lakes and reservoirs. **COHABITANTS:** In winter and migration, may associate with Western Willet, Dunlin, Western Sandpiper, Red Knot, other sandpipers. **MOVEMENT/MIGRATION:** Spring migration from late April to mid-June; fall from early July to early November. In many coastal areas, absent only during midsummer. VI: 3.

DESCRIPTION: This one is easy. A chunky, fist-sized, ultra-pale peep (distinctly paler than other sandpipers) that keeps to the beach (doesn't like to get its belly wet) and plays tag with the waves. Over much of coastal North America, this defines Sanderling, the default sandpiper of sandy beaches. Gain familiarity with this bird, and you can use it as a measure for all the rest.

Sanderling is a large compact peep (larger than Western or Semipalmated Sandpiper; smaller than Red Knot; same size as Dunlin and Rock and Purple Sandpipers), with a straight moderately short bill, a round neckless head, a humped back (particularly when feeding), and wings that barely extend beyond the tip of the tail.

In nonbreeding plumage (carried for much of the year), birds are pale gray above and very white below. Against this pale backdrop, the straight black bill, the black eye, the short black legs, and the signature smudgy black shoulder patch (lacking on some individuals) stand out. In breeding plumage (assumed quickly in May, discarded by August), the male head and breast turn rusty red, and the back is suffused with reddish feathers. Female upperparts and breasts become more patterned and reddish- or saffron-washed. Juveniles (seen in late summer into fall) are like winter adults, but upperparts are boldly spangled with black-and-white feathers, and the birds wear a smudgy dark cap.

BEHAVIOR: This is the flocking sandpiper that plays tag with the waves. Other sandpipers (notably Dunlin, Red Knot, and Western Sandpiper) may ape this behavior, but Sanderlings play the game for keeps. These birds *run,* seemingly all the time, on legs that blur. Up the incline of the beach and down. Chasing retreating waves. Angling their tails sharply up. Driving their bills into the wet sand with a brief, rapid, stitching motion. Turning and retreating, en masse, as the next wave approaches. If they misjudge the intensity of a wave's rush, they take wing in a synchronized kaleidoscope of black and white, often calling in protest. Landing, they resume the game.

Feeding birds gather in tight groups that are always in motion. Heavily used beaches seem to pulse with the charge and retreat of birds whose movements are wedded to the waves (unless they are chasing each other, as they often do). At high tide, the birds gather on higher ground in tightly packed flocks. A tame bird, Sanderling can be approached very closely and often tries to put distance between itself and an intruder by hopping away on one foot.

FLIGHT: Fairly compact, humpbacked, and long-winged profile. In all plumages, blackish upperwings bracket and accent a broad white wing stripe that is hands down more striking than the wing stripe on other small sandpipers. Flight is hurried, direct, unwavering, not twisty-turny, and somewhat effortful. Wingbeats are quick and describe a moderately deep arc, with wingtips rising equally above and below the body. Flocks shifting position on a beach are linear and globular, and the flight is generally carried out just offshore.

VOCALIZATIONS: Most often heard sound is a sharp, distinctive, metallic "*plink.*"

Semipalmated Sandpiper, *Calidris pusilla*

Blunt-billed Sandpiper

STATUS: Common and at times abundant migrant at migratory staging areas. **DISTRIBUTION:** Semipalmated nests in the arctic and high arctic regions of North America, from Lab. to the west coast of Alaska; winters coastally from the Antilles to Argentina and through Central America and the coasts of Ecuador to n. Chile. *This species does not winter in the continental U.S. or Canada. That's Western Sandpiper.* **HABITAT:** Nests on open tundra and amid low-lying vegetation near water. In migration, favors shallow fresh and salt water, mud flats, lakeshores, and impoundments; prefers places with muddy bottoms. Less commonly found on sandy beaches. **COHABITANTS:** Mixes freely with other peep during migration, particularly Western, Least, and White-rumped Sandpipers. **MOVEMENT/MIGRATION:** Migrates primarily east of the Rockies in the United States (where, in most places, it outnumbers the similar Western Sandpiper), using a strategy that incorporates long nonstop flights of several thousand miles to staging areas. Migration is fairly compressed: spring from late April to early June; fall from July to October. VI: 3.

DESCRIPTION: A small, compact, nicely proportioned, grayish peep that likes to keep its feet wet. Small sandpiper (larger than Least; slightly smaller than Western), with a straight, stout, usually blunt, sometimes swollen-tipped bill of varying length, a round, somewhat neckless head (showing a steep forehead), a compact body, and black legs. Very short-billed birds are easy to identify. Longer-billed individuals are easily confused with Western Sandpiper.

Plumage is at the grayer end of the scale. Breeding adults have gray-brown upperparts with lightly (but variably) streaked and generally pale breasts and bright white underparts. The gray head shows traces of rufous on the crown and ear patch; the shoulders (scapulars) show a faint touch (not a band) of rufous-tinged feathers. Nonbreeding adults are uniformly gray above and very white below, with a shadowy gray necklace. Juveniles are the brownest of the lot; pale feather edges make them appear scaly-backed.

BEHAVIOR: Semipalmated likes crowds (numbering in the hundreds and thousands) and mixes freely with other peep. Semipalmated also likes to keep its feet wet. Even on drawn-down mud flats, this bird prefers the wetter portions or forages in residual puddles.

A thorough, methodical, and somewhat single-minded feeder. When feeding, picks more than probes. Western is more often a prober. On mud flats and dry land, Semipalmated likes to forage with its tail angled up—sometimes comically so. Least Sandpipers commonly maintain a more horizontal profile; Westerns (particularly shorter-billed Westerns) vary.

When foraging through shallow water, Semipalmated holds its head and body horizontal. Western (particularly longer-billed Westerns) holds its head slightly elevated (like White-rumped).

FLIGHT: Small, long-winged, and not as spindly as Least Sandpiper. An excellent flier, Semiplamated's flight is stronger, more direct, and less wandering than Least, and Semipalmated flocks are more tightly packed.

VOCALIZATIONS: Classic call is a short assertive burp: "*krip*" or "*jrrt.*" Also makes a short, sharp, thin "*jeet*" or "*chi*" that recalls Western but is not so screechy. In addition to these calls, feeding birds, which are noisome, make a low, petulant shorebird whinny: "*weh-eh-eh-eh-eh-eh.*"

PERTINENT PARTICULARS: It's important to bear in mind that in all plumages, Least Sandpiper is browner and has a streakier, dirty, or darker chest. Because Semipalmated has whiter underparts and more ventral surface area showing, it stands out, at a distance and in a mixed flock, as showing lots of white (as does Western Sandpiper); Least, by contrast, disappears into the backdrop. In distinguishing Western Sandpiper and Semipalmated, remember that both breeding and juvenile Western Sandpipers are more distinctively marked than Semipalmated of the same age class and show richer and redder plumage.

The problem is that nonbreeding Westerns and Semipalmateds are similar in plumage. But from November through March, virtually all Semipalmated/Western-type sandpipers are, by default, Westerns, which winter in the coastal United States. Also, in fall, adult Westerns molt earlier than Semipalmateds. In July and August, the ratty-looking sandpipers (some birds look like chunks have been bitten out of them) are Westerns. Come September and October, Westerns, now in fresh nonbreeding plumage, look neat and clean; the worn-looking birds are probably Semipalmateds.

One last point about molt: Semipalmated Sandpiper molts its flight feathers on the wintering grounds. In North America, in fall migration, birds missing primaries are Western Sandpipers.

Western Sandpiper, *Calidris mauri*
The Gnomish Sandpiper

STATUS: A numerically superior species that breeds on the west coast of Alaska. Common coastal (and near-coastal) winter resident. Very common, at times abundant, in coastal staging areas in the West; far less common, but not necessarily uncommon, east of the Rockies (particularly in fall). **DISTRIBUTION:** Breeding range is restricted to the west coast of Alaska (and ne. Siberia). Found inland during migration, particularly in fall. Winters along all coasts: in the Pacific, from s. B.C. to Peru; in the Atlantic and Gulf, from N.J. to Venezuela. **HABITAT:** Nests in drier portions of the tundra in close proximity to wetter feeding areas. In migration and winter, prefers mud flats, particularly tidal mud flats, but also forages on sandy beach, lakeshore, and shallow impoundments.

COHABITANTS: Mixes with assorted other peep.
MOVEMENT/MIGRATION: (Brief) spring migration from second week of April to May; (protracted) fall migration from late June to late October, with peaks in July (earlier migrating adults) and late August into early September (juveniles). VI: 3.
DESCRIPTION: A slightly front-heavy, chestier, droopy-billed, gnomishly proportioned version of Semipalmated Sandpiper. Many individuals are easily distinguished on the basis of plumage and shape. The identification of some generically proportioned birds in nonbreeding plumage is more problematic.

Smallish sandpiper (larger than Least; considerably smaller than Sanderling; similar in size and shape to Semipalmated), with a longish bill that is pointy and slightly drooped at the tip, a round head that may seem raised uncomfortably high above the back, a portly chest, a short body, and shortish legs (slightly longer than Semipalmated). Next to the trim, proportionally balanced lines of Semipalmated, Western appears a bit ungainly and gnomishly front-heavy. Some Westerns (females) have bills that recall Dunlin's—exceedingly long, drooping, tapered to a point. If you see this bill, the case is closed—the bird's a Western. Other birds have bills that are shorter and straighter, overlap in size, and approximate the shape of a longer-billed Semipalmated. These bills suggest, but they don't certify.

Structurally, a feeding Semipalmated often has a no-necked appearance. A feeding Western probes and keeps its round head raised above the body. Also, as is often apparent in direct comparison, the legs of Western are longer. In water of the same depth, Western stands slightly taller.

In breeding plumage, Western is more brightly and distinctly patterned, with a reddish wash on the crown, a reddish ear patch, and reddish shoulders contrasting with mostly paler and grayer upperparts. In addition, Western's face is distinctly whiter and contrasts with a chest and flanks that are more heavily spotted (even somewhat narrowly streaked), with chevron-shaped spots along the sides.

In summer (fall migration), vestiges of breeding plumage (a red scapular, a chevron on the flanks) may be evident in Westerns into August, and molting adults typically appear ratty or moth-eaten, whereas later-molting Semipalmateds still look neat. Clean gray-backed juvenile Westerns show the red scapulars of adults, and like adults seem pale-headed. Juvenile Semipalmateds, by comparison, are dirty-faced and dirty-breasted and show no distinctive red slash on the shoulder.

In nonbreeding plumage, Western is gray above, very white below, with little or no streaking on the breast, and very similar to Semipalmated, which is very slightly browner and darker across the breast.
BEHAVIOR: A gregarious and social sandpiper that, even more than Semipalmated, likes to keep its feet wet. During migration in coastal staging areas in the West, may be found in homogeneous flocks that number in the tens of thousands, but also mixes with assorted other peep. In winter, often found amid flocks of Dunlin. Feeds on mud flats and in shallow standing water, primarily by probing. Western is a prober first and a picker second—just the opposite of Semipalmated, *but both species pick and probe.* Western is able to forage in deeper water than Semipalmated, so may stand farther out or apart. When feeding, Semipalmated is usually dominant and in a challenge will drive Western away.
FLIGHT: Direct, fast, and no different than Semipalmated except for obvious characteristics related to bill size, plumage, and molt, which can often be seen in flight.
VOCALIZATIONS: Classic call is a high-pitched squeaky "*cheep*" or "*g'reet*" (recalls White-rumped Sandpiper).
PERTINENT PARTICULARS: One of the keys to locating and identifying Western and Semipalmated is to know that in migration Western is the dominant "gray" peep throughout most of the West and Semipalmated is the default sandpiper in the East. Particularly helpful is the knowledge that Western winters in the United States and Semipalmated does not. Any Semipalmated/Western-type sandpiper seen in the United States between mid-

November and late March is all but certain to be Western.

Least Sandpiper, *Calidris minutilla*

Peep with Dry Feet

STATUS: As a migrant, common, widespread, and tolerant of marginally suited wetland habitat—making it the most likely peep to be encountered inland. **DISTRIBUTION:** Breeds throughout most of the arctic and subarctic from Nfld. to Alaska. Winters along both coasts, from s. N.J. to Fla., and from s. B.C. to Calif. and across much of the South, north to Mo. and Okla. **HABITAT:** Breeds on open tundra and in wet boggy areas, most often in proximity to taiga habitat. In migration and winter, found on mud flats, shorelines, short grassy meadows, tidal estuaries, riverbanks, farm ponds, irrigation ditches, sod farms, puddles, and, less commonly, sandy beach and plowed fields. On the coasts, frequently occupies the tight vegetated or drier edges of estuaries and impoundments shunned by most sandpipers. In migration, occupies shorelines and drier areas, including areas well away from water. **COHABITANTS:** Associates with other peep. **MOVEMENT/MIGRATION:** Spring migration from April to June; fall from late June to November, with peaks in May and August. VI: 4.

DESCRIPTION: A small, dark, pointy-faced sandpiper that likes to keeps its feet dry. When Least is standing in a mixed flock, its status as the smallest and darkest of the peep is manifest. Least is overall petite, spindly (compact without being stocky), and pointy—pointy-faced and pointy-tailed. The head is small, the face plain, the eye black and beady. The forehead slopes down to the bill, which is overall thin, pointy, and conspicuously drooped at the tip.

In breeding plumage, adults are warm brown above; in nonbreeding plumage, gray with distinct brown overtones. Juveniles are eye-catchingly ruddy or rufous above. In all plumages, Least is darker and browner above than the other small peep and usually more heavily streaked on the chest; Least can even have a bibbed appearance that recalls the much larger Pectoral Sandpiper. Least's underparts are white, but not brilliant white. Legs are yellowish on adults, greenish on juveniles.

BEHAVIOR: Found in small flocks (5–30 on average) and in mixed flocks, but also and often found alone. Any lone peep, particularly one in marginal habitat (for example, baked mud away from the water's edge, or tight watery leads flanked by rank, tiny puddles), immediately suggests Least Sandpiper. When feeding with Semipalmateds and Westerns, the darker-backed Least usually stays on the mud; it likes to keep its feet dry. The grayer sandpipers are most often found standing in the water.

An active feeder, Least seems to crouch as it moves. Longer-billed and somewhat shorter-legged than Semipalmated, it holds its body more horizontal. Semipalmated commonly feeds with its tail cocked higher (and almost vertical when making aggressive displays to other sandpipers). Generally well spaced as they feed, Leasts are also bullied by other sandpipers and chased from one bird's feeding area to another (one more reason it is often off by itself).

Least is tame and often allows very close approach.

FLIGHT: Overall pointy and spindly, with wings that are acutely angled back at what seems an awkward angle. The petite and pointy bill and overall dark brown upperparts are surprisingly easy to note. Flight is somewhat weaker and more fluttery, shifting, and batlike than the other peeps. Flocks, which tend to be small, are not tightly packed.

VOCALIZATIONS: Classic call, often given in flight, is a reedy, trilling "*brEEEp?*"—posed as a question. In aggressive confrontations, makes a musical stutter—"*eh-eh-eh-eh.*"

PERTINENT PARTICULARS: More than other sandpipers, Least can be attracted to your imitations of its squeal call. Individuals and flocks respond from 100 yds. away, circle, and may land at your feet.

Least Sandpiper might be confused with Pectoral Sandpiper, which also is very brown, also has yellow legs, and also favors grassy edges. Pectoral is a much larger sandpiper, however, and the streaking on the breast (the bib) is heavier and more crisply truncated.

Insofar as migrating Least Sandpiper is common, widespread, and often found in small and marginally suited habitats, it is typically the peep most likely to be encountered across interior North America. Observers should know, however, that at times and places encounters with other species may be just as likely or even more so—for example, Baird's Sandpiper, whose primary migration route cuts through the center of the continent.

White-rumped Sandpiper, *Calidris fuscicollis*

Streaky-flanked Sandpiper

STATUS: Northern breeder. Locally common spring migrant on the Great Plains; uncommon farther east. Casual west of the Rockies. **DISTRIBUTION:** Breeds in the high arctic from Alaska's North Slope across coastal n. Canada and the arctic islands east to Baffin I. Winters in s. South America. In migration, employs a strategy of long nonstop flights and few staging areas. In spring, migration is concentrated in the central prairies (Cheyenne Bottoms, Kans., is a major staging area), although birds may also be found east to the Atlantic Coast. In fall, migrates primarily down or off the Atlantic Coast, where large numbers are found in e. Canada and lesser numbers along the Eastern Seaboard. Then flies, presumably nonstop, to South America. **HABITAT:** Breeds in open tussocky tundra and wet meadows characterized by low vegetation, often near ponds or the coast. In migration, found in both fresh- and saltwater marshes, impoundments, mud flats, and lakeshores. Also sometimes found on sandy beaches and dry short-grass habitat. **COHABITANTS:** In migration, found with assorted other peep, particularly Semipalmated, Baird's, and (in deeper waters) Stilt Sandpipers. **MOVEMENT/MIGRATION:** A late-spring migrant, White-rumped is not commonly found before May 1 (late April in Texas); birds routinely, and suddenly, appear during the first ten days of June. The peak period for spring migration in North America is late May. In fall, migrates later than most peep. Adults do not commonly appear before the last week of July or early August. Juveniles migrate from mid-September to early November, and may linger longer. VI: 2.

DESCRIPTION: A sturdy long-lined peep with a signature field mark. Except for Curlew Sandpiper, White-rumped is the only small sandpiper with a white rump. A large peep but a small sandpiper (larger than Semipalmated; smaller than Sanderling; same size as Baird's), with a longish, slightly drooped bill, a round head, a long, sturdy, slightly chesty or tubular body, and long wings that extend well beyond the tail, giving the bird an overall long profile. Only Baird's shows such long lines. Next to a White-rumped, most other peep look stocky.

The black legs are seated precisely in the middle. The bill, which seems a little short, is slightly drooped at the tip. (The bill on Baird's appears straighter.) At close range, the base of the bill is dull orange.

Breeding plumage most recalls the overall stockier, longer-billed (and paler-faced) Western Sandpiper. Like Western, White-rumped is overall gray with lots of symmetrical streaking, particularly on the breast and down the flanks—with chevrons extending all the way to the base of the tail. The crown, ear patch, and scapulars are touched or blushed with rufous. In nonbreeding plumage, birds are overall gray but distinguished by a very white eyebrow, a shadowy gray bib across the chest, and (often) vestigial pencil-fine dark streaks down the flanks. Also, because of their length, the black wingtips contrast with the gray back (like Baird's).

Juvenile plumage is warmer and somewhat ruddy, with white-edged back feathers that give the upperparts a scaly-backed appearance similar to juvenile Baird's. Juvenile Baird's is overall buffy, not ruddy, with a buffy wash across the chest and no streaking down the flanks. Also, Baird's does not show the very pronounced white eyebrow that juvenile White-rumped shares with the adult.

BEHAVIOR: A very active, almost aggressive feeder that walks quickly and directly across mud flats. Stopping after several steps, it makes several rapid half-bill-length jabs into the mud, then continues on. White-rumped also picks, but less commonly. A heads-up stalker in water, particularly the shallow tidal puddles it favors. Moves several steps with head raised, then drives bill forward in an exploring stitching probe. Often stands with its belly to the water, and sometimes submerges its head.

May occur in homogeneous groups, but often mixes with other peep, particularly Semipalmated and Baird's Sandpipers. Defends feeding territory aggressively, chasing other peep away while vocalizing angrily.

FLIGHT: For a peep, has a large long-winged profile, with a broad white rump. A strong swift flier with strong wingbeats. Flight is often swerving or tacking; the body inclines to one side and then the other, alternately presenting the upper and lower surfaces of the body (so the white rump winks). Often calls in flight.

VOCALIZATIONS: Classic call is a very high sharp "*jeet*" or "*j-jeet*" that may recall the sound of two coins being struck together. Challenge call is a very high-pitched screechy chatter that has the same quality as the "*jeet*" call.

PERTINENT PARTICULARS: Finding White-rumped often means sorting through lots of Semipalmated and Baird's Sandpipers. White-rumped is obviously larger and longer than Semipalmated, and in spring more eye-catchingly colorful above and more streaked on the breast. In fall, the long profile, chesty appearance, and white supercilium are easily noted.

Baird's is most like White-rumped in shape, and juveniles have a similar scaly-backed appearance. But Baird's is a frequenter of dry ground, often short grass, and feeds by picking. White-rumped is more often a wet habitat prober. Also, White-rumped is chestier and more robust; Baird's is more trimly elegant.

In spring, both White-rumped and Baird's migrate primarily through the central plains and commonly forage on the mud flats. In breeding plumage, White-rumped shows touches of red and is more lavishly streaked below. Baird's is cold grayish brown above and shows no red highlights, and the streaking below is limited to the chest; the flanks are clean and white.

Baird's Sandpiper, *Calidris bairdii*

The Gentleman Peep

STATUS: Uncommon arctic breeder; fairly common migrant through the prairies; uncommon coastal migrant in fall; very rare coastal migrant in spring. **DISTRIBUTION:** Mostly high arctic breeder found from the west coast of Alaska and ne. Siberia, across the northern coastal plain of Canada and northern island groups to the northwest coast of Greenland, the northern two-thirds of Baffin I., and the northwest edge of Hudson Bay. Also breeds in interior Alaska in the Alaska and Brooks Ranges. In spring, migrates primarily through the plains states—west of the Mississippi, east of the Rockies, and, in Canada, west of Hudson Bay. Fall migration is mostly concentrated through the prairies but spans the continent owing to the more broad-based movements of juveniles. **HABITAT:** Breeds on dry tundra (often adjacent to ponds), sandy beach, gravel bars, and bare soil. In migration, partial to dry barren or short-grass habitat, but also found in mud flats, flooded fields, river banks and bars, lakeshores, plowed fields, pastures, golf courses, sod farms, and high sandy beach; less commonly found in tidal flats and estuaries. Far less coastal than other peep. **COHABITANTS:** In migration, other peep, most notably Least Sandpiper (which also forages freely in drier habitats), but also Killdeer, Buff-breasted and Pectoral Sandpiper; in spring, White-rumped Sandpiper. **MOVEMENT/MIGRATION:** A champion long-distance migrant that winters in southern South America. Spring migration from mid-March to mid-June, with a peak in April to early May; in fall, failed breeders reach the continental United States by late June, most adults pass in late July to August, and juveniles from late August to early October. VI: 3.

DESCRIPTION: A trim, slender, elegant, and conservatively garbed peep. A small shorebird but medium-sized peep (1 in. larger than Western Sandpiper; 1 in. (or more) smaller than Pectoral;

same size as White-rumped Sandpiper), with a straight pointy bill, a round head, a short but slender neck, and a long and slender but somewhat portly and slightly humpbacked and potbellied body—attributes mostly eclipsed by the bird's overall long slender profile. (These attributes do, however, help distinguish Baird's from the similarly sized and also long-profiled White-rumped Sandpiper.) Long wings, which extend beyond the tail, are usually crossed and elevated above the tail when the bird feeds. Short centrally positioned legs give the bird a balanced profile and horizontal posture.

Adult plumage is overall warm, fairly uniform (herringbone bland), and neat, trim, and uncontrasting. In all plumages, upperparts are brownish gray, but transfused with a subtle warm buffy wash that is particularly evident on the very plain face and breast. Underparts show a well-defined buffy wash across the breast but are otherwise white. There is no hint of red in the plumage of Baird's. Almost every other peep, at some point in the plumage cycle, shows at least a bit of red somewhere.

In breeding plumages, a buffy breast band shows fine darkly flecked streaks, and the back is spattered with widely spaced, black-centered, white-edged feathers that make the birds look black-spotted. Juveniles are extremely buffy, almost straw-colored, above, with a very buffy face and breast and pale-edged back feathers that make upperparts appear very scaly.

BEHAVIOR: Social, but not fond of multitudinous crowds. Most commonly found in small flocks (fewer than a dozen birds), and often off by themselves. Frequently forages on dry land while other peep are standing in water nearby, but also wades like other peep. Posture is horizontal. Movements are quick but deliberate; Baird's picks more than it probes. Often aggressive with other peep, driving smaller shorebirds away.

Tame even by peep standards. Usually calls when flushed.

FLIGHT: Short plump body, but very long broad-based wings. Upperparts are dark and bland (the pattern almost homogenized); the pale wingbar is narrow and weak, and the grayish dark-centered tail doesn't stand out as on other peep. The breast seems dark, like Pectoral Sandpiper, but shows no white ovals at the base of the tail. Flight is straight, nimble, and fast, with rapid steady wingbeats (not appreciably different from other peep).

VOCALIZATIONS: Call is a soft, low, descending, trilled "*prrrrt.*" Most closely resembles the "*brup*" call of Pectoral Sandpiper, but is slightly softer and less abrupt.

PERTINENT PARTICULARS: Like Baird's, Least Sandpiper also tends to feed on dry land, away from feeding flocks, but is easily distinguished from Baird's by its size, its pinch-faced, its droop-billed look, and its short tail and wings (among other dissimilarities). White-rumped is most like Baird's in size and shape. White-rumped is overall beefier (big-chested as opposed to slightly humpbacked and potbellied), with a longer down-drooped bill, a bolder face pattern with a whiter eyebrow, and a more heavily streaked breast with streaks down the flanks (which Baird's never shows). Also, breeding-plumage adult White-rumped shows a reddish wash about the head, face, and back; juveniles have rust on the crown and back.

I am compelled to point out that one highly qualified reviewer disagreed that breeding birds show any warm tones in their plumage. The disparity is not a matter of right or wrong, but of perception. You, like the reviewer, may see a colder plumage, one devoid of warm tones. Calibrate my description accordingly.

Pectoral Sandpiper, *Calidris melanotos*

Super Peep with the Corduroy Bib

STATUS: Common to uncommon migrant east of the Rockies; scarce but regular west of the Rockies (mostly in the fall). **DISTRIBUTION:** Breeds on the arctic coastal plain of Russia, across n. Alaska to the Northwest Territories. Winters in South America from Peru to s. Brazil south to Patagonia.

Most ne. Asian birds also travel to South America, but limited numbers winter in Australia and New Zealand. **HABITAT:** Nests on wet grassy tundra encompassing small ponds and elevated hummocks. In migration, favors wet grassy areas, including freshwater marshes, flooded fields, shallow prairie potholes, large puddles, and tidal wetlands. May also be found on dry short grass (including prairie, sod farms, and airports). On impoundments or mud flats, most often forages around edges. **COHABITANTS:** In migration, Lesser Yellowlegs and Solitary, Least, and Baird's Sandpipers, among other shorebirds. **MOVEMENT/ MIGRATION:** In spring, migrates through North America from mid-March to May on a route that bisects the middle of the continent, but measurable numbers also pass east of the Mississippi to the Atlantic Coast. In fall, most birds continue to favor the same midcontinent route, but greater numbers also occur in the East, and smaller but regular numbers are found in the West. Fall migrants appear in early to mid-July. Number of adults peaks in late July to mid-August; juveniles in early September to October. Most birds are gone by mid-November. VI: 3.

DESCRIPTION: A big beefy peep or a medium-sized sandpiper (larger than White-rumped or Buff-breasted; about the same size as Solitary Sandpiper) distinguished by its variable but largish size, well-patterned upperparts, idiosyncratic shape, and distinctive corduroy bib. The overall structure seems ungainly. The distinctly large and sturdy body is somewhat humpbacked, big-chested, and occasionally potbellied. The neck is unpeeplike long. The head seems a little too small, and the sloping forehead, combined with the slightly down-drooped bill, makes the bird appear pinch-faced. Folded wings project beyond the tail, giving the bird a longish, somewhat White-rumped-like profile, and the legs, bright yellow to yellow-green, seem a trifle short.

Size varies greatly between males and larger females. Flocks of Pectorals (particularly in flight) can be identified by noting the size array between otherwise similarly proportioned and plumaged birds.

Breeding-plumage adults are heavily scallop-patterned above, with feathers that are somewhat gilded or rufous-edged. In nonbreeding plumage, upperparts are less patterned and more uniformly gray-brown; juveniles are darker, with chestnut-edged feathers on the wing (and a more rufous crown) that make them appear ruddier above. In all plumages (but particularly spring), Pectoral's breast is fitted with a distinctive pattern of heavy streaks: it ends with a crisply cut lower edge that makes the bird look as if it were wearing a corduroy bib. The balance of the underparts is creamy white.

The bill is two-toned—reddish or fleshy yellow at the base, and dark toward the tip. Among mud-flat loving peep, only the diminutive Least Sandpiper also has yellow legs.

BEHAVIOR: Pectoral seems to have its own rules. Sometimes found in small flocks (commonly no more than 50 birds and frequently fewer than a dozen), "Pecs" also spread out among other peep (often Least Sandpiper) and are not uncomfortable about foraging alone in a sheltered pocket, even when numbers of other shorebirds, including other Pectorals, are nearby.

Likes short-grass vegetation, likes the edge, and likes to feed methodically and slowly—waddling or shuffling, stopping to pick or probe, shuffling on again. The profile is horizontal. When found with only one other shorebird, it is very commonly a Least—another bird of the edge.

A tame bird. When alert, stands very tall. When flushed, seems to be going full speed almost from the moment of liftoff (and full speed is very fast).

FLIGHT: Overall broad, heavy, and sturdy profile, with long broad wings. Distinctly brownish above; shows virtually no pale wing stripe but does show distinctive white oval tail patches along the flanks that recall Ruff. The dark bib is usually very apparent. When looking at flocks, note the marked difference in sizes between males and females. Flight is fast, and when flushed, often very erratic and twisty-turny (like a somewhat tempered snipe); nevertheless,

Pectorals fly in tight synchronized flocks. Regular flight is more direct, but includes some tacking left and right. Usually silent when feeding; flushed birds vocalize continuously (until they land).

VOCALIZATIONS: A soft, flat, atonal grunt or burp: "*djjjrt*," or "*brrup*."

PERTINENT PARTICULARS: Sharp-tailed Sandpiper, *Calidris acuminata*, is an Asian species that is similar to Pectoral Sandpiper in size, structure, and plumage, is regular in western Alaska in fall, and has a widespread pattern of vagrancy across North America, especially along the West Coast. Most sightings are of juvenile birds, which, when foraging with Pectoral Sandpipers (which they commonly do), appear overall ruddier and richer-colored, showing a reddish back, a brighter reddish crown, and a faintly streaked (mostly limited to the sides of the chest), warm buffy or orange-buff wash across the chest. Breeding adults are heavily spattered below with dark chevronlike spots that extend down the flanks (thus no sharply cut bib) and have a reddish cap. Nonbreeding adults have a pale orangy-buff wash across the chest (also no sharply cut streaked bib). Sharp-tailed's call is not as rough as Pectoral.

Purple Sandpiper, *Calidris maritima*

Eastern Rockpiper

STATUS: Fairly common, but isolated and restricted in both its breeding and winter range. **DISTRIBUTION:** A high arctic breeder with an interrupted but essentially circumpolar distribution. In the arctic, breeds on open tundra on the west coast of Greenland and on the islands lying between Greenland and mainland Canada. Winters on the Atlantic Coast from Nfld. to S.C. Very rare inland (mostly on the e. Great Lakes) and south to the Gulf. **HABITAT:** In winter, almost exclusively rocky coasts, jetties, and seawalls. **COHABITANTS:** In winter, Dunlin, Ruddy Turnstone. **MOVEMENT/MIGRATION:** Migration is almost wholly coastal. Spring migration in April and May; fall from late September to December. VI: 2.

DESCRIPTION: In winter, resembles a plump, dark, stubby-legged (but yellowish-legged) Dunlin wearing a featureless slate-colored cowl. Forages exclusively on wave-washed rocks. Closely resembles Rock Sandpiper of the Pacific Coast, but ranges do not overlap. Ignoring Rock Sandpiper, Purple Sandpiper does not constitute a difficult identification problem. The near-idiosyncratic bluish brown cast of the plumage, coupled with the bird's restricted range and its near-slavish devotion to rocky substrates, distinguishes it. The bright yellowish legs and orange base to the slightly downcurved bill are distinctive but hardly necessary.

In breeding plumage (seen in late spring), loses the blue-brown uniformity, becoming streaky on the face, spotted on the breast, and adds an infusion of rusty feathers to the back; legs turn dark green. Juveniles resemble breeding adults, but commonly reach the Atlantic after molting into winter plumage.

BEHAVIOR: Extremely tame, foraging at the feet of observers. Very industrious. Waddles and clambers up and down slippery algae-covered rocks picking at tiny mollusks with quick jackhammer-like jabs. Forages in unsynchronized groups below the splash zone, often fluttering to high ground to avoid waves. Generally silent. Often seen in the company of the larger Ruddy Turnstone, the slightly smaller Sanderling, and occasionally Dunlin, which is easily distinguished from Purple Sandpiper in winter by its overall paler plumage, unstreaked underparts, black legs, and all-black bill.

FLIGHT: Straight and fast, with rapid unbroken wingbeats. Flies in fairly cohesive, linear, species-specific flocks (that land on rocks!). In flight, appears uniquely dark, and particularly dark-chested (Sanderlings are distinctly black and white; Dunlin is undistinguished contrast-shedding brown). Purple Sandpiper's exceedingly narrow white wing stripe stands out because of the darkness of the wing. Purple Sandpiper is not unique in pattern, but the contrast between the white belly and white rump (bisected with a black center) and what is, on balance, its all-dark plumage makes it an eye-catching bird.

VOCALIZATIONS: Unusually silent for a shorebird. Feeding birds make soft chattering notes.

Rock Sandpiper, *Calidris ptilocnemis*

Intertidal Sandpiper

STATUS: Common Alaskan breeder; uncommon and local West Coast winter resident. **DISTRIBUTION:** Breeds coastally from the Seward Peninsula (and Siberia) south to the Aleutians, the n. Alaska Peninsula, and w. Kodiak I. Winters from the Aleutians and Prince William Sound south to the south coast of B.C. An uncommon winter visitor in n. Calif. **HABITAT:** Not limited to rocky coasts, but partial to them. In summer, found on open, dense, coastal tundra not far from the sea. In winter, strictly coastal, foraging entirely in the intertidal zone. Substrates include rocky coast, gravel beach, peat banks, mud and sand flats, and snow and ice. On rocky coasts, forages on algae-covered or mussel- and barnacle-encrusted substrate. Roosts above the high tide on rocks, beaches, piers, and shipwrecks. **COHABITANTS:** Breeds near Black-bellied Plover, Bar-tailed Godwit, Western Sandpiper, Dunlin. Winters with Dunlin, Ruddy and Black Turnstones, Surfbird, Wandering Tattler. **MOVEMENT/MIGRATION:** Rock Sandpiper's pattern of movement is complex, complicated by variables relating to subspecies and post-breeding movements. In general, spring migration from late March to early June; fall from late June to mid-November. South of the breeding area, birds are generally present September to May; peak fall movements are later than for other shorebirds. VI: 1.

DESCRIPTION: A chunky, short-legged, colorful (breeding) or dark, sooty gray sandpiper that likes rocks. A small shorebird but a large peep (slightly larger than Dunlin; smaller than Black or Ruddy Turnstone or Surfbird), with a longish, slightly downturned bill, a round head, a short neck, a chunky body, and short greenish legs. In shape, looks like a short-legged, short-winged, and slightly shorter-billed Dunlin.

Breeding plumage varies in tone and pattern but is basically Dunlinlike, with a reddish back and tarry-looking black patch on the lower chest that is less extensive and set higher in the front than Dunlin (does not extend behind the legs). Lighter birds are the most Dunlinlike, but they show a dark ear patch that Dunlin lacks. Darker races look like richer red versions of winter birds and have very dark faces. (Dunlin's face is whitish.)

In winter, uniformly sooty gray above and dark-chested with heavily spotted white underparts below. Legs are dull greenish yellow. The bill is black except for a dull orange base. The larger Surfbird has a stubby bill and bright yellow legs. Wandering Tattler has a long straight bill and bright yellow legs. Dunlin is pale brown (dun-colored).

BEHAVIOR: Preoccupied shorebird that moves nimbly but slowly on rocks, beaches, or mud flats; in winter, particularly in southern portions of its range, almost always found on rocks. Semisocial; usually found feeding and roosting in small flocks, but individual birds, mixed among other rock-pipers, are not uncommon. A tame bird that allows close approach.

FLIGHT: Stocky and short-winged profile; all-dark above (much darker than Dunlin), with a contrasting and distinctive white wing stripe and a white tail with a black stripe down the center. (Surfbirds and turnstones have all-white bases to the tail; tattler is uniformly gray above.) Flight is strong, fast, and direct on rapid steady wingbeats.

VOCALIZATIONS: Like Purple Sandpiper, mostly silent in winter. Flight call is low, scratchy, grating.

PERTINENT PARTICULARS: Rock and Purple Sandpipers, while similar in plumage, are as different as east and west (coast). Neither species has ever been recorded on the coast or breeding areas of the other, and interior records of Rock Sandpiper are nonexistent. With this in mind, discussions relating to plumage differences between the two species are more academic than practical. Given this species' more expansive approach to habitat, however, not to mention Purple Sandpiper's slavish devotion to rocks, any "Purple" Sandpiper feeding on an Atlantic Coast beach or sod bank probably rates a second studied look.

Dunlin, *Calidris alpina*

Mud and Multitudes

STATUS: Common and, in winter and in northern coastal marshes, the most common sandpiper (in places, numerically abundant). Except in the Mississippi Flyway, an uncommon interior migrant. **DISTRIBUTION:** Breeds in arctic and subarctic regions across the Northern Hemisphere; in North America, found from the west coast of Alaska to the western and southern coasts of Hudson Bay, and east to the mouth of James Bay. Winters coastally in the Pacific from the south coast of Alaska to Baja and cen. Mexico, and on the Atlantic from Mass. (a few to s. N.S.) to cen. Mexico; also winters on the Yucatán Peninsula. **HABITAT:** Nests on wet tundra and sedge meadows. Winters in coastal tidal estuaries. In migration and winter, found on seasonal marshes, lakeshores, agricultural fields, and farm ponds. **COHABITANTS:** Mixes with other sandpipers, including Black-bellied Plover and dowitchers. **MOVEMENT/MIGRATION:** Migration is heaviest along both seacoasts, but passage across inland regions, particularly locations that lie between the Gulf Coast and breeding areas west of Hudson Bay, also support birds in both spring and fall. Both spring and fall migration, during which birds make use of key staging areas, is protracted. Spring migration from March to early June, with a peak in late April and early May; fall from late August to December, with a peak from September to December. VI: 3.

DESCRIPTION: Large, somewhat comically proportioned peep. Distinctly larger than Western Sandpiper (with which it often associates). The body is stocky, the head exceedingly round, the bill long, pointy, and wilted at the tip, and the legs sturdy and black. Breeding plumage is distinctive, almost histrionic—the bird's white face and red back are ballasted by a sharply defined black patch on the belly that extends behind the legs. (The patch on Rock Sandpiper stops short of the legs.)

In winter, becomes a nearly featureless dun-backed bird with dingy underparts; wearing an expressionless face and a shadowy hood that cuts across the breast, Dunlin is distinctive in its uniform plainness. Western Sandpiper is whiter (particularly on the breast) and grayer. Juvenile plumage is rarely seen outside of breeding areas.

The expression is somewhat vacuous. Duh!

BEHAVIOR: The bird loves mud and company. Winter flocks often number in the hundreds and thousands. Roosting birds gather in tightly packed dun-colored masses. Foraging birds apportion themselves across tidal flats. Dunlin feeds by walking several steps, then executing an exploratory probe into the muck. If it likes what it feels, it drives its bill in, frequently clear to the face. Dunlin forages on exposed mud or in shallow water. Other foraging substrates include sandy beaches, grassy tidal estuaries, and shores of freshwater ponds and lakes. Flocks are conversationally noisy and tame around humans. Massed concentrations cannot help but draw satellite predators. Dunlin takes wing frequently, coalescing and maneuvering in tightly packed flocks.

FLIGHT: Overall stocky, with a long bill and long slender wings. Flight is direct and sizzlingly fast, literally. The wings of passing flocks make a hissing sound. Shifting flocks are linear; birds are cohesive but not tightly packed. When evading falcons, birds fuse into tightly packed flocks that move as a single entity and change shape like pooled mercury at the mercy of a child's finger. As the flocks climb, wheel, split, and re-form, they flash light (underparts) and dark (upperparts). These dancing distraction displays may go on for several minutes.

VOCALIZATIONS: Call is a short, high-pitched, breathy trill. Feeding birds make a sound that recalls the soft "*chur*" of bluebirds. Also makes a mellow and high-pitched "*peep*."

PERTINENT PARTICULARS: Nonbreeding Curlew Sandpiper is similar to winter Dunlin but is overall more elegant and elongated, as well as grayer and less dun-colored, with a longer, finer-tipped bill, a longer neck, and a more tubular body (with extending wingtips). It's also slightly longer-legged, so look for Curlew Sandpiper to be feeding taller than Dunlin or slightly away (in deeper water) from it.

Curlew Sandpiper, *Calidris ferruginea*

Brick-breasted Sandpiper

STATUS: Rare migrant primarily along the Atlantic Coast (casual elsewhere). **DISTRIBUTION:** Has bred in Alaska and may breed elsewhere in arctic North America, but is primarily an Old World species, breeding in northern Russia and wintering in

Africa, Asia, and Australia. **HABITAT:** Nests on open tundra. In migration, favors mud flats or muddy shorelines and shallow wet areas that are totally or mostly free of vegetation; also occurs on sandy beaches. Found in both fresh and salt water. **COHABITANTS:** In migration, forages with other small sandpipers and often roosts with Dunlin and Red Knot. **MOVEMENT/MIGRATION:** Primarily coastal. In North America, most spring migrants are found in May; in fall, July to September. VI: 3.

DESCRIPTION: A larger-than-peep-sized and somewhat elongated sandpiper with a conspicuously down-bowed bill. The head is distinctly round. (Stilt Sandpiper has a flatter head and a sloping forehead.) The neck is short and thick (but compared to the neckless look of Dunlin, Curlew Sandpiper shows a neck). The body is somewhat tubular and more roundly pudgy than chesty. Its long tapered profile results from long wings that reach well beyond the tail. Curlew recalls a more tubular White-rumped Sandpiper (most illustrations show the bird too stocky).

The bill is more bowed than downturned, and curved throughout its length; it even seems to rise and fall along its length, not just bend down at the tip. The black, fairly long legs seem set just a little too far back.

Breeding adults are stunning. The head, breast, and belly are brick red, and the back red-spangled (more richly colored than breeding-plumage Sanderling). Most Curlews are located and identified because they are in this distinctive plumage or, in summer, have traces of it. Winter adults are gray above, white below. Juveniles are buffy on the face and breast and scaly-backed.

In flight, the underwings are starkly white (like Dunlin). The tails are gray, but the rump is white (like Stilt Sandpiper).

BEHAVIOR: A very active single-minded feeder that on mud walks and probes rapidly, like a small sandpiper; in the water, it gets up to its belly and probes the bottom, submerging its head as a matter of course. Mixes with other shorebirds, particularly peep, and Dunlin and behaves (as does Dunlin) like a large peep.

Its penchant for foraging in deeper water—because of its longer legs and cavalier attitude about going right up to the gunwales—often physically segregates it from the shorter-legged Dunlin. Conversely, the Curlew's comparatively shorter legs may force it to stand apart (in shallower water) from Stilt Sandpiper.

When feeding, the body is distinctly horizontal, and the bill almost vertical.

FLIGHT: Recalls Stilt Sandpiper (which also has a white rump), but Curlew shows a broad white wing stripe (which Stilt lacks) and does not show long trailing legs, as Stilt does. Flight is fast and direct, with the body slightly angled up.

VOCALIZATIONS: Call is a soft, short, low, slightly trilled "*chirrup*" that is more abrupt than Dunlin's flight call and not as breathy or shrill.

Stilt Sandpiper, *Calidris himantopus*

Stitching-billed Sandpiper

STATUS: Generally less numerous than the medium-sized shorebird species it feeds with; nevertheless, a fairly common spring and fall migrant through the prairies, and fairly common along the Atlantic Coast in fall. Rare along the Pacific Coast. **DISTRIBUTION:** An arctic breeder located on the coastal plain from Barrow, Alaska, east to the Northwest Territories and in several other disjunct northern locations (including portions of coastal Hudson Bay). Winters principally in interior South America, but some birds occur in parts of North America, including the southeast coast of Fla., the Gulf Coast from La. to the Yucatán Peninsula, the Salton Sea of Calif., and portions of Mexico and Central America. Spring migration passes mainly through the middle interior of North America, although a few birds appear on the East Coast. Fall migration mirrors the spring route except that substantially greater numbers appear on the Atlantic Coast and smaller numbers on the Pacific. **HABITAT:** Nests on wet to semi-wet tundra meadow dominated by sedge and assorted dwarf shrubs. In migration and winter, forages in shallow pools, small ponds, and flooded fields. Prefers fresh water but may also occur in estuary ponds and brackish impoundments. **COHABITANTS:** In migration, most commonly associates with dowitchers and Lesser Yellowlegs. **MOVEMENT/MIGRATION:** Spring migration from late March to early June, with a peak in early May; fall from late June to late October, with peaks in early August (adults) and mid-August to mid-September (juveniles). VI: 3.

DESCRIPTION: A tall, slender, delicate, but only medium-sized shorebird that seems a structural and behavioral hybrid between a Lesser Yellowlegs and a dowitcher. The body is somewhat curvaceously lined and appealingly proportioned: the head is somewhat round, the neck long, and the body ovate—dowitcher-shaped but not dowitcher plump. The bill is Lesser Yellowlegs long, but not so thin and tapered and decidedly down-drooped near the tip. The legs are very slender and very long, but olive green or yellow-green, not bright yellow. In short, the bird looks like a shorter, slightly plumper, curvaceous Lesser Yellowlegs with a wilting bill.

Breeding plumage suggests a boldly patterned dowitcher but is overall darker, more strikingly patterned, and distinguished by blackish-feathered mosaic above, symmetrical narrow black barring below, and a splash of chestnut on the cheeks and crown. The expression is gentle.

Nonbreeding-plumage adults are a soft, pale, mostly unpatterned gray—overall paler than either dowitchers or yellowlegs. Even into late summer, many adults retain vestiges of their breeding plumage, particularly vestigial barring on the sides, which readily distinguishes them. Juveniles are warmer buffy gray with scaly upperparts. This scaly pattern is quickly lost on the back but retained into the fall on the wings (like Wilson's Phalarope).

BEHAVIOR: More like a dowitcher than a yellowlegs. Forages almost exclusively in water, often wading up to the belly. Often feeds with dowitchers (also Lesser Yellowlegs). Homogeneous flocks tend to be small—half a dozen to a dozen individuals (rarely more than 30, except in May on the Great Plains). Dowitcher flocks are frequently larger. When mixed with dowitchers, Stilts tend to cluster and may, because of their longer legs, stand slightly off by themselves in deeper water.

In deeper water, forages by stalking forward with its body horizontal and head angled up. (Yellowlegs keep their tails up and heads down; dowitchers keep their bodies and heads on a horizontal plain.) In shallow water, angles its head down (more like yellowlegs).

Like dowitcher, Stilt Sandpiper is a prober. Unlike dowitcher, Stilt doesn't just forage blindly ahead. Moves deliberately and then dabs or eases its bill into the water with a series of jerky mini-jabs (seems as though it's stitching the water). At almost any distance, this motion is distinctive and diagnostic. Stilt habitually puts its head under water (sometimes the neck too). Dowitchers dunk their heads more reluctantly. Stilt is mostly silent, tame, and not particularly prone to flush.

FLIGHT: Overall slender profile; smaller and slighter more delicately proportioned than Lesser Yellowlegs. Stilt is flatter-bodied and not as dome-backed as dowitcher and has narrow wings and legs that trail behind the tail. Wings and back are dark; the tail is gray, becoming white near the base. Flight is direct and fairly fast, but more buoyant than dowitcher and not as stiff as yellowlegs. Wingbeats are more shallow than dowitcher and slightly quicker.

VOCALIZATIONS: A short, soft, abrupt "*jrrt.*"

PERTINENT PARTICULARS: Stilt Sandpiper is often compared to Dunlin and Curlew Sandpiper, largely because of the shared curvature of the bill. Dunlin is stockier, chunkier, noticeably shorter-necked, and much shorter-legged. Dunlin, in winter, is also browner (dun-colored not gray) and has a shorter and darker eyebrow and a dark rump. It feeds in large flocks on open mud flats—Stilt prefers quieter, more sheltered pools. Curlew is a differently shaped bird—more tubular, more slender, with long projecting wings. Its feeding behavior is peeplike. Stilt is more dowitcher-like.

Buff-breasted Sandpiper, *Tryngites subruficollis*

Dovelike Sandpiper

STATUS: Uncommon spring and fall migrant through the central plains east of the Rockies; scarce along the Atlantic Coast in the fall; very rare on the Pacific Coast. **DISTRIBUTION:** Breeds on dry tundra ridges in the high arctic from Alaska's North Slope and the islands and coastal mainland of Canada's Northwest Territories. Winters in the Pampas region of Argentina. **HABITAT:** At all seasons, prefers short dry grass, drier mud edges to wetlands, or tundra. In migration, occurs on prairie, tilled farmland, rice fields, golf courses, airports, sod farms, and drought-enhanced lakeshore and sandy upper beaches with sparse vegetation. Generally shuns very wet areas. **COHABITANTS:** In migration, American Golden-Plover, Killdeer, Baird's Sandpiper. **MOVEMENT/MIGRATION:** Spring migration to coastal Texas from late March to early May; birds arrives at the northernmost reaches of the breeding range by mid-June; fall migration from late July to late September, with most birds (juveniles) sighted in August and September. VI: 3.

DESCRIPTION: A smallish dovelike shorebird (larger than Baird's Sandpiper; about the same size as or slightly smaller than Pectoral Sandpiper) whose shape, color, and mannerisms make it distinctive, almost winsome. The body is long, fairly lean, and streamlined, like one of the long-profiled peep, but the yellow legs are too long and the posture too erect. The head is distinctly round, the face plain, the eye large and black, and the bill plover-short, phalarope-straight, and overall petite. The expression is gentle, baleful.

Adults and juveniles are fundamentally similar. The scaly-backed buff-colored plumage is eye-catching and almost unique (can only realistically be confused with juvenile Baird's Sandpiper, or perhaps juvenile Ruff). In flight, the upperparts are uniformly dark (golden warm); underparts (including underwings) are contrastingly white. In shape, recalls golden-plover but a golden-plover with contrasting upper- and lower parts.

BEHAVIOR: A social species, Buff-breasted is often found in the company of American Golden-Plover, but also associates with other upland foraging shorebirds, including Least Sandpiper. Always active, always moving, the bird feeds by walking in a halting, weaving, wandering manner (suggesting a tipsy drunk and mimicking a starling's waddle). The raised head jerks pigeonlike, and the tail swings side to side. When it sights prey, the bird leans forward, often securing the prey with a horizontal peck. Buff-breasted is a tame bird that often allows close approach and prefers to walk away from danger rather than fly. Frequently freezes when other birds depart and remains behind. In spring, sometimes executes its raised wing display, flashing white underwings.

FLIGHT: Body is ploverishly compact, and wings are long and tapered. Flight is strong and fast; twists and turns in flight, exposing white underwings in flashes, but the flight does not seem erratic. Executed with finesse, the twists and turns seem like polished maneuvers, not panicked flight.

VOCALIZATIONS: Usually silent. Call is a low "*brrrt*," recalling Baird's Sandpiper.

Ruff, *Philomachus pugnax*
Anomalous Shorebird

STATUS: Rare but regular, primarily coastal vagrant (or migrant?). A few are found in California in winter. **DISTRIBUTION:** Old World breeder, although annual and in some locations regular sightings suggest that low numbers may breed in arctic regions of the New World. **HABITAT:** Breeds in taiga bogs, marshes, and grassy edges of ponds and lakes. In migration, most often found in small, shallow, usually grass-rimmed ponds; also on flooded agricultural lands, lakeshores, and more open coastal marshes with some areas vegetated and others mud. Much prefers fresh to salt water. **COHABITANTS:** Pectoral Sandpiper, Lesser and Greater Yellowlegs, Least Sandpiper. **MOVEMENT/MIGRATION:** In spring, most sightings fall between late March and mid-May, with a peak in mid-April; in fall, late June to early October. VI: 3.

DESCRIPTION: A curiously shaped shorebird that seems pieced together with parts of other species, but with a twist or two of its own. In plumage and shape, Ruff is both distinctive and unique; structurally, it's like a cross between Pectoral Sandpiper and a yellowlegs. Medium-sized to medium-large shorebird, with great variation in size between males and females—some are as large as Greater Yellowlegs and some as small as dowitcher. Ruff has a short, slightly downturned (Pectoral Sandpiper–like) bill, a small round (Dunlin-shaped) head, a slender neck that is neither long nor short, a robust Pectoral Sandpiper–shaped body, and longish yellow legs that are nevertheless shorter than those of yellowlegs. Because it often stands with its neck retracted, Ruff can seem unnaturally bulky or plump-bodied and hunchbacked, and its already-too-small head can seem even smaller. When foraging (with head lowered), some of the back feathers stick up like an unruly cowlick.

Breeding males are as variable as they are unique (and some perhaps are even grotesque), showing a skeletal-thin face adorned with head plumes and encircled by a billowing ruff or feathered collar. Birds may be all black, white, red, or a combination of colors. Breeding females are more subdued, showing mostly blackish upperparts and splotchy black spotting on the breast. The base of the bill is orange. Legs are usually orange, but are also often green. Adults in winter are grayish-backed, with whitish throats and a gray wash across the breast and a distinctive white base to the bill. Juveniles show dark scaly backs and a buffy face, neck, and breast; they resemble a large, awkward, long-legged Buff-breasted Sandpiper. The expression is sleepy or not too bright.

BEHAVIOR: Likes to be in shallow standing water amid broken or open vegetation. Flooded harvested cornfields are perfect, and mud flats festooned with clumps of vegetation are great. When vegetation becomes dominating or confining in places that Ruffs have frequented for years, the birds do not return. Ruff is most frequently found feeding among Greater and (particularly) Lesser Yellowlegs as well as Pectoral Sandpipers. In water, moves somewhat sluggishly, with head down; on dry ground, often moves fast, with head up or leaning forward. Often feeds up to its belly in water, and frequently submerges its head. Fairly tame.

FLIGHT: Breeding males, of course, are unique—bulky and distinctly colored. Otherwise, the overall shape recalls a gangly long-winged Pectoral Sandpiper. Shows, in all plumages, white underwings, a white U on the rump, and feet projecting beyond the tail. Flight is aptly described as lazy: wingbeats are sluggish, deep, and often irregular, with occasional glides that may be brief or (particularly when landing) protracted.

VOCALIZATIONS: Usually silent. Flushed birds sometimes make a low muffled grunt.

PERTINENT PARTICULARS: When Ruffs are standing among Pectoral Sandpipers, their larger size is apparent. When feeding among yellowlegs, Ruff is bulkier, smaller-headed, shorter-legged, and usually in more shallow water.

Short-billed Dowitcher, *Limnodromus griseus*

Lesser Dowitcher

STATUS: Common, generally subarctic (taiga) breeder and common migrant, especially along seacoasts; less common inland. Common coastal winter resident. **DISTRIBUTION:** Breeds in three disjunct geographic regions, each relating to a different subspecies. The Atlantic form (*L. g. griseus*) breeds in n. Canada from Lab. to e. Hudson Bay. The prairie form (*L. g. hendersoni*) nests in northern portions of Man., Sask., and Alta. The Pacific form (*L. g. caurinus*) is restricted to nw. B.C., sw. Yukon Territory, and the south-central coast of Alaska. This species winters coastally, from n. Calif. to cen. Peru, and from Va. to Brazil. The Pacific form is relegated to the West Coast, the prairie form occupies coastal sections along the Atlantic and Gulf Coasts, and the Atlantic form presumably is found south of the U.S. in the West Indies and South America. **HABITAT:** Breeds in the transition zone between forest and tundra where tamarack and spruce are broken up by sedge meadows and bogs. Except when breeding, Short-billed shows a marked preference for coastal marine habitats, particularly open unvegetated mud flats next to shallow bays, lagoons, estuaries, salt marshes, and sandy beach. During migration, in interior areas, also found in freshwater marshes and mud-rimmed ponds. **COHABITANTS:** When breeding, Wilson's Snipe, Solitary Sandpiper, Greater Yellowlegs. In migration, Lesser Yellowlegs, Red Knot, Stilt Sandpiper,

Long-billed Dowitcher. **MOVEMENT/MIGRATION:** Spring migration from mid-April to late May (Pacific form migrates on a slightly more accelerated timetable). In fall, first southbound dowitchers arrive in the last week of June, peak in July (adults), end in August (juveniles), and linger into November. In most coastal saltwater locations, where great numbers of dowitchers are found, Short-billed vastly outnumbers Long-billed Dowitcher. On the Gulf Coast and in California, both species occur in numbers, and Long-billed may dominate. In migration, in interior portions of the eastern half of the United States, both Long-billed and Short-billed can be expected. Beginning in the eastern prairies, Long-billed dominates, and farther west, Short-billed is rare in the interior. VI: 3.

DESCRIPTION: A somewhat comically proportioned medium-sized shorebird (1 in. larger than a snipe; slightly smaller than Long-billed Dowitcher), with a plump oval-shaped body and (despite the shortcomings of the name) a distinctly long bill. The resulting profile is distinctive—suggesting a small meatloaf on a stick—but not unique: Long-billed Dowitcher is similar in structure and plumage. Differences between the two species are more of degree than of kind. The degree of difference between Long-billed and Shore-billed also varies between subspecies. The Atlantic form of Short-billed is most unlike Long-billed in size, shape, and breeding plumage; the prairie form most closely resembles Long-billed in breeding plumage; and the Pacific form comes closest to Long-billed in size.

The head is round, and the bill long (but not unwieldy), straight, and sometimes slightly drooped at the tip. The legs are reasonably long but seem shorter than they should be for a bird with such a long bill. The body is plump, but the lines are trim and unexaggerated, not mesomorphic. When feeding, the bird has a gently domed but not humped back. The short neck easily distinguishes dowitchers from other "long-billed" shorebirds (such as yellowlegs). The dowitcher expression is vacuous or befuddled.

Bill length varies between subspecies of Short-billed Dowitcher, and between sexes of both Long-billed and Short-billed (females have longer bills), resulting in much overlap. Where only Atlantic-form Short-billed Dowitcher occurs, bill length can frequently distinguish the two species. But where the two longer-billed subspecies of Short-billed occur, bill length is best used only to suggest or support identifications, not certify them.

In breeding plumage, Short-billed is dark above, with a mosaic of pale and orange-edged black feathers on the back. The face and underparts are orange, with profuse amounts of black speckling and barring on the sides of the neck, the sides, and the flanks. Atlantic and Pacific birds show white bellies, particularly white between the legs. The prairie race is more colorful: more orange shows in the back, and there is brighter and more extensive orange below (and frequently no white showing between the legs).

In all plumages, tails are marked with narrow black-and-white bands of near-equal width that are obvious in flight but can only be seen otherwise beneath the folded wings when birds are at rest or feeding.

Nonbreeding plumage is undistinguished gray above, with a gray-flecked chest and gray barring coursing down the flanks of the underparts. Juveniles are similar to breeding adults, but show overall more bright orange above (in almost Day-Glo intensity in some fresh-plumage birds) and just a wash of color below (no dark flecking or barring except on the upper sides of the breast). Most important on juveniles, the tertial feathers, which overlie the tail (actually the folded wingtips) when the bird is standing, are tiger-striped—barred black and orange. On Long-billed Dowitcher, the feathers are wholly dark, with a narrow pale or orangy border.

BEHAVIOR: Somewhat desultory and cowlike (just moving along, doing what the bird in front is doing), dowitchers like crowds, and they particularly like crowds of dowitchers. Feeding birds cluster among other feeding shorebirds, coalescing into dowitcher pockets. Dowitcher feeds primarily by standing in water, up to and sometimes above the belly, and probing. Standing with the body horizontal and the bill pointed vertically, the bird plumbs the depths, either dropping and raising its bill in a smooth, rhythmic, mechanical fashion (like a slow-moving sewing machine) or, driving the bill down to the surface, making two or three mini-probes before drawing the bill up, taking a step, and trying the next spot. (Godwits feed similarly.) The tip of the bill is commonly not raised out of the water.

Also feeds on exposed mud by mini-probing (rather than picking off the surface). Swims well (for a shorebird); when entering water too deep to forage, may elect to swim to shallows. Fairly tame, but may be slightly more quick to flush than Long-billed Dowitcher.

FLIGHT: Fairly compact, with slender wings angled back and long bill angled slightly down. The body is distinctly dome-backed and hang-bellied; feet extend just beyond the tail. In all plumages, both this species and Long-billed Dowitcher show a white stripe running up the rump and lower back—as though a finger-painting toddler had slapped a finger of white paint up the bird's back. The black-and-white barred tail blurs to pale gray. Flight is strong, direct, and fast. Wingbeats are steady, regular, quick, and moderately shallow, with wingtips falling below the body only slightly more than they are raised above. Flies in a globular string.

VOCALIZATIONS: Classic call is a rapid musical stutter: "*Tu, tu tu*" or "*Tchu, tchu, tchu.*" Also emits a more rapid "*d'd'd*" or "*'d' dee,*" whose quality recalls a mallet strummed quickly down the keys of a xylophone.

Long-billed Dowitcher, *Limnodromus scolopaceus*

Buffalo Bill

STATUS: Common northwestern arctic breeder; common western migrant, but generally less numerous than Short-billed Dowitcher along the East Coast. Away from coastal areas, the more likely dowitcher to be encountered. Common winter resident. **DISTRIBUTION:** Breeds in ne. Siberia, w. and n. Alaska, and east to the Northwest Territories. Winters coastally (but farther inland than Short-billed Dowitcher) from Vancouver south to Central America, from N.J. to the Yucatán Peninsula, and in interior portions of California, s. Ariz., N.M., s. Tex. (and Mexico), and the Florida Peninsula. **HABITAT:** Breeds on arctic tundra. At all times prefers freshwater environments, particularly those characterized by short vegetation, including marshes, wet grassy meadows, flooded rice fields, agricultural land, lakeshores, saline lakes, and open mud flats. Also occurs in assorted coastal marine environments, often with the mud flat–loving Short-billed Dowitcher. **COHABITANTS:** During breeding, Pomarine Jaeger, Glaucous Gull; in winter, assorted puddle ducks and shorebirds. **MOVEMENT/MIGRATION:** Spring migration is protracted—from late February to late May across most of North America (but very rare in spring along much of the East Coast). Fall migration is protracted and fairly late, from early July to November, with peaks from late July to early September (adults) and in September and October (juveniles). VI: 3.

DESCRIPTION: Like Short-billed, a dumpy, ungainly, somewhat slothful shorebird, but one often distinguished by an amazingly long, almost (at times) godwitlike bill and a humpbacked (godwitlike) profile. Medium-sized (slightly larger than Short-billed Dowitcher; much smaller than Hudsonian Godwit) and shaped much like Short-billed Dowitcher but overall bulkier. Long-billed Dowitcher stands taller, has a wider body, and displays a distinctly humpbacked appearance that is most pronounced when it is feeding (recalls the posture of a godwit or, if you are really imaginative, a bison).

The bill is sometimes so long that it appears unwieldy, and it commonly droops at the tip. Although the bill length of male Long-billed approximates that of female Short-billed, the bill of female Long-billed is almost always longer than the longest-billed Short-billed. In other words, given a 50–50 sex ratio, about half of all Long-billed Dowitchers (the females) have a bill length that hints at the bird's identity, and perhaps half of these (one out of every four birds) have a bill length that distinguishes them. Truly long-billed dowitchers—those with unwieldy godwit-long bills—are Long-billed Dowitchers.

Long-billed Dowitcher is also longer-legged. In a mixed crowd, they stand slightly taller than Short-billeds. In water, they can forage in deeper water, and if they are feeding among Short-billeds, a space can be seen between the water and their bellies, whereas the Short-billeds' bellies are touching the water.

In all plumages, Long-billed is overall darker than Short-billed — in direct comparison, often distinctly so. In breeding plumage, the bird seems dressed for Halloween. Upperparts are a mosaic of orange-centered dark-trimmed feathers. Darker and more richly colored than Short-billed (though prairie-form Short-billeds are very close), Long-billed's face, neck, and underparts to the tail are dark orange to brick red; the bird is heavily spotted on the sides of the neck and bedecked with wedge-shaped bars down the sides and flank. The hindneck (very visible when birds are feeding and facing observers) is brightly spangled with orange — giving the appearance of an orange yoke, which Short-billed does not show.

Note: Molting Long-billed often loses color on the belly early, replicating the pale-bellied plumage pattern seen on some Short-billeds. Look for unmolted feathers between the legs, which will be darker and redder than Short-billed.

In nonbreeding plumage, Long-billed is overall slightly darker, duller gray, with less patterning on the back; it has a darker, more uniform gray breast (Short-billed's breast seems frosted with white or slightly mottled) and dark flanks that seem ribbed rather than banded or patterned.

Juveniles show the most distinctive plumage. Backs are dark warm brown (not orange-spangled); breasts have a gray or buffy gray (not orange) wash; the tertials overlying the tail have all-dark centers with a narrow orange border (tertials of Short-billed are banded, orange and black). In all plumages, Long-billed's tail is basically dark, etched with narrow white (or orange) lines. In flight (especially in direct comparison in mixed flocks), appears slightly darker-tailed than Short-billed — more charcoal gray than pale gray.

BEHAVIOR: Similar to Short-billed. Probes for food using smooth, deep, mechanical plunges that bring the bill to the surface of the water; may submerge the head. Prefers fresh water (but certainly uses brackish and salt water) and grassy or vegetated wetlands. (Short-billed loves tidal mud.) When found in mixed flocks, tends to cluster in species-specific groups rather than mix randomly with Short-billeds.

May be tamer than Short-billed. If birds flush, those remaining are likely Long-billeds. Also more vocal than Short-billed. When flushed, Long-billed almost always calls; Short-billed often does not.

FLIGHT: Silhouette is similar to Short-billed but somewhat bulkier. Bills of longer-billed females appear incongruously long. In full breeding plumage, shows entirely dark underparts (no white belly or white between the legs). Long-billed's darker tail (charcoal gray, not plain gray) is usually apparent only in direct comparison. Flight is like Short-billed: direct and fast, with rapid wingbeats describing an arc that brings tips above the body and slightly more below the body.

VOCALIZATIONS: An abrupt sharp "*keek*" that is manifestly different from Short-billed's musical stutter, "*tu, tu, tu.*" A word of caution: When flushed, the "*keek*" is sometimes trebled but the quality remains the same.

Wilson's Snipe, *Gallinago delicata*

The Pucker-faced Sandpiper

STATUS: Common and widespread breeder and winter resident. **DISTRIBUTION:** Breeds across most of arctic and subarctic Canada and Alaska. In the continental U.S., found in all states bordering Canada and the Great Lakes as well as at northern and higher elevations in the West (south to Colo., Utah, Nev., and n. Calif.). In winter, vacates all but some southern and coastal portions of its breeding range, distributing itself across much of the U.S., Mexico, and Central America south of a line drawn from Mass. to the south shore of the Great Lakes (also se. Ont. and along the St. Lawrence R.) to n. Iowa, cen. Neb., e. Wyo., sw. Mont., and s. B.C., and north along the coast to se. Alaska. **HABITAT:** Breeds in open, shallow, sedgy, marshy areas with no tall vegetation and in wet meadows and alder and willow swamps. In winter, occurs in marshes, pastures and farmland with shallow standing water, drainage ditches, and (particularly during prolonged freezes) along river and stream banks, in springs, and in tidal estuaries. **COHABITANTS:** Owing to its large distribution, many and varied. In winter and migration, often found in the same habitat as Least Sandpiper and Pectoral Sandpiper. **MOVEMENT/MIGRATION:** Spring migration from March to May; fall from August to November, with peaks from April to early May and from October to early November. VI: 4.

DESCRIPTION: Looks like a streaked gaunt dowitcher that's been sucking on a lemon. Medium-sized shorebird (larger than Solitary Sandpiper; smaller than a dowitcher), with an extremely long, straight, slender bill, and a full-breasted body that tapers toward the rear. The head seems embarrassingly small for the neck and bill. Fairly short legs are made even shorter by the bird's tendency to crouch.

At all times, plumage is overall grayish brown and boldly, if cryptically, patterned. Emerging from the cryptic clutter is a pattern of longitudinal lines that cut horizontally across the face and run along the back. The pattern both sleekens the bird and imparts to the face an expression of smugness or the suggestion of a pucker.

Birds are often partially hidden, with only the head and a portion of the fully extended neck showing. Useful points include the smallness of the head, the high and somewhat rearward placement of the eye, and the bird's tendency to angle the bill acutely (almost contritely) down. (Dowitchers hold the bill closer to horizontal.)

BEHAVIOR: Feeds in wet, grassy, or vegetated places, usually with shallow standing water (rarely found on drained mud flats). Probes deeply with the bill, and often pauses after two or three thrusts, with the bill stuck in right to the hilt. Any association between the usually solitary Wilson's Snipe and other shorebirds is serendipitous, but during migration Wilson's may occur in loose groups (up to 50) that flush and fly in small widely spaced squadrons (not the cohesive flocks of dowitchers). Almost never do all birds flush together; usually one or more remain crouched and hidden. A shy bird, Wilson's does not tolerate close approach. Usually flies just at the point when it might be seen, but sometimes walks (or runs) to better cover.

On territory, commonly perches on the tops of trees, fenceposts, or utility wires.

FLIGHT: Like dowitcher, but more spindly and angular and longer-billed. Plumage is all dark except for a white belly (no white stripe up the back). Flight is rapid and erratic, with many a twist and turn. Takeoff is explosive, near-horizontal, and usually vocal. Birds gain altitude at a shallow rising trajectory, commonly with a twisting dodging pattern. Wingbeats are rapid and down-flicking on wings angled sharply back; bills are angled acutely down. (Dowitchers hold their bills closer to horizontal.) Flushed birds often circle (sometimes more than once), but usually land somewhere else with a braking flutter.

Displaying birds spiral aloft, then execute a series of dives and climbs.

VOCALIZATIONS: Flushed birds make a loud scraping protest, "*yrrrch,*" frequently repeating this call as they fly. During courtship, flight displays are accompanied by a winnowing sound caused by air passing through the outer tail feathers as the birds execute pendulum-like dives. The winnowing sounds like a series of low whistled notes that seemto climb the scale or accelerate in tempo. Then a pause. Then another series: "*hoo hoo hoohooHooHooHooHoo Hoo.*" The eerie quality of this call recalls the tremolo call of Eastern Screech-Owl (more accurately, Boreal Owl). On territory, often from an elevated perch (such as a bush, tree, or fencepost), birds also make a loud, repetitive, screechy "*yert . . . ch yert . . . ch yert . . . ch*" that, in cadence and quality, recalls the yelp of a female Wild Turkey.

American Woodcock, *Scolopax minor*

Meatloaf on a Stick

STATUS: Fairly common denizen of eastern second-growth forests and adjacent open areas, but largely nocturnal and almost invisible until flushed, with great reluctance, from its woodland confines. **DISTRIBUTION:** Breeds locally east of the prairies, east of a line drawn from s.-cen. Man. south to e. Tex., and north across s. Ont. and Que. to s. Nfld. (but absent as a breeder along the Gulf Coast and

s. Fla.). In winter, withdraws from northern regions, residing south of a line extending from Long Island to se. Pa., w. Va., w. N.C., cen. Tenn., s. Mo., and ne. Okla. south across e. Tex. almost to the Rio Grande Valley. **HABITAT:** Breeds in forests, particularly young, regenerating, deciduous forest, with openings or proximal edge. Shows a particular fondness for alders, aspens, and birch. In winter and migration, occupies almost any woodland habitat but favor areas with fairly heavy but broken understory. At night, at all seasons, usually feeds in fields, pastures, agricultural land, lawns, the edges of freshwater and slightly brackish marshes, and moist woodlands—anywhere earthworms, its principal prey, are found. **COHABITANTS:** Does not commonly associate with other species; does not become active before Whip-poor-will and Chuck-will's-widow begin to vocalize. **MOVEMENT/MIGRATION:** Spring migration (at night) from January to April, with a peak in March; fall from September to December, with a peak in October and November. VI: 1.

DESCRIPTION: A plump, softball-sized, oval-shaped, long-billed, woodland shorebird—the only shorebird that inhabits the forest floor. Appears neckless, tailless, and short-legged (the head seems like a bump on the body), and overall resembles a meatloaf on a stick. In all plumages, upperparts are cloaked in an eye-defeating gray, black, and buff pattern that makes the bird melt into the forest floor. Underparts are orange buff. Overall the plumage is warm-toned, soft and cuddly, and devoid of any definable pattern. Standout features include the extremely long flesh-colored bill and the large black eyes set improbably high and rearward on the head. The expression is gentle.

BEHAVIOR: For the most part, nocturnal. Sallies out from woodlands to forage in open areas at dusk; returns at dawn to woodlands, where it roosts on the forest floor. Generally goes unseen until flushed—often from underfoot. In summer, also forages during the day, but often in damp woodlands (not open fields), and when approached, crouches and freezes (like a roosting bird). Not found in standing water, in salt water, or on open exposed mud flats, and generally not seen in open areas during daylight.

Gait is a slow shuffle, often with the body bobbing or undulating rhythmically but the head remaining stable and fixed. Probes for earthworms using long thrusts followed by a pause. Generally solitary except during migration or when freezing temperatures concentrate birds in unfrozen places. Never flocks. Never joins with other shorebirds. Never perches on anything but the ground.

FLIGHT: A plump long-billed bird struggling to keep aloft on short triangular wings that move in a fluttering blur. When flushed, rises nearly vertically with an audible twittering of wings, body acutely angled up, legs often dangling. After clearing the understory, lines out, flying drunkenly (but silently), listing first to one side and then the other, often zigzagging between trees and branches. Generally does not fly far, sometimes circles and returns close to the same location, and often glides, unsteadily, just before landing.

Aerial courtship display is arresting and easily observed at dawn and dusk. Birds spiral aloft, calling as they climb, then descend rapidly to earth like broken kites. After an interval, the display is repeated.

VOCALIZATIONS: Silent except during courtship display (does not call when flushed). During the grounded portion of the courtship display, males make a low-pitched reedy buzz, "*peent,*" that is repeated every 5–10 secs. As birds go aloft, they emit a frenzied twittering whistle that spirals or fades in and out. When birds descend, the twittering notes become spaced, slowed, and more mellow, and they are usually given in a repeating series of three to five (suggests to some the sound of water dripping). After a brief intermission, the "*peent*" buzz begins again.

Wilson's Phalarope, *Phalaropus tricolor*

A Gazelle Among Sandpipers

STATUS: Common, primarily western breeder; abundant at post-breeding and premigratory staging areas. Because of its breeding range and nonpelagic nature, this is the most likely phalarope to be encountered on land outside of the Arctic. **DISTRIBUTION:** Breeds from the s. Yukon across e. B.C., all of Alta., s. Sask., and sw. Man., south across e. Wash. and Ore., ne. Calif., n. Nev., all of Idaho, Mont.,

Wyo., N.D., S.D., w. Minn., n. Utah, cen. Colo., and n. and w. Neb. Also breeds in Wisc. and Mich., se. Ont., and sw. Que. Winters in South America. **HABITAT:** Breeds in shallow freshwater marshes. Stages in highly concentrated saline lakes. In migration, may be found in either fresh or salt water. **COHABITANTS:** During breeding, Eared Grebe, Sora, Wilson's Snipe, Black Tern, assorted puddle and diving ducks. In migration, may be found among other shorebirds, including Lesser Yellowlegs, Stilt Sandpiper, dowitchers. **MOVEMENT/MIGRATION:** All birds relocate to South America for the winter. Many relocate to premigratory staging areas located in several hypersaline lakes in the western United States prior to migrating, including Mono Lake in California and the Great Salt Lake in Utah. Spring migration from late March to late May; fall dispersal from mid-June to late August; fall migration to South America from late July to early October. VI: 3.

DESCRIPTION: A slender, delicately featured, aquatic sandpiper. Smallish shorebird with a small head, a long neck, a long, straight, needlelike bill, a longish slender body, and fairly long yellowish legs. Most closely resembles the larger Lesser Yellowlegs and similarly sized Stilt Sandpiper—but is eminently more finely featured (makes a yellowlegs look oafish).

Breeding females (more colorful than males) are stunning—soft gray above with a black and maroon racing stripe that curves sinuously from the eye to the sides of the neck to the back, then fans out across the back and folded wing. Bright white underparts are washed with orange about the neck. Breeding males are dull gray above and white below, with a muted suggestion of the female pattern that manifests itself as a blotch of warm rusty tones about the neck.

In nonbreeding plumage, adults are contrastingly pale gray above and white below (no black through the eye), with yellow legs. Similarly patterned juveniles are darker, browner, and scalier above, but still paler and less blackish and streaked above than juvenile Red-necked Phalarope.

BEHAVIOR: It floats! And by choice, not because it wandered into deep water. On the water, sits high in the rear with tail elevated, which, in combination with its long-necked head-elevated posture, makes the bird seem swaybacked. Swims with quick jerky head movements and abrupt downward jabs of the bill. Also crouches low in the water with head extended (in shallow water also stalks in this manner), snatching insects with deft jabs of the bill or lowering the bill (occasionally the head) into the water and bill-sweeping like an avocet. Feeds on mud flats in much the same way—with tail raised and long neck stretched out and low. Like Spotted Sandpiper, runs after insects. Wilson's Phalarope is an active angry feeder that seems to be moving twice as fast as all the other shorebirds. If you want to find the Wilson's Phalarope among all the other shorebirds feeding in shallow water, look to see where all the ripples are.

When alone or in small numbers, frequently feeds with dowitchers, yellowlegs, and Stilt Sandpiper. Not shy, and often allows close approach. Not vocal. Rarely vocalizes during migration.

FLIGHT: Long and elegant profile with projecting head and tail, slight chestiness, and long, slender, slightly angled-back wings (like a slender, paler, softer-lined Solitary Sandpiper). The white face and white tail (actually white rump and gray tail but looks all white) contrasts with the darker upperparts.

Flight is stiff, jerky, buoyant, and slightly tacking or wandering. Wingbeats are stiff, with an emphatic down stroke and a slight hesitation on the lift. Glides more frequently and with more grace and finesse than Red-necked. Lands with a snipelike braking flutter; on water, takes slightly longer to settle than Red-necked Phalarope, with more flutter and foot patter.

VOCALIZATIONS: Call is a terse, low, soft, muttered "*rrrnk*" that sounds like a muffled Black Skimmer's "*yap.*"

Red-necked Phalarope, *Phalaropus lobatus*

Dark Water-elf

STATUS: Common arctic breeder; fairly common offshore migrant; uncommon to rare migrant in most inland locations outside of its breeding range, but may be abundant in select hypersaline lakes in the Great Basin. **DISTRIBUTION:** Circumpolar breeder; in North America, breeds from Alaska across all of the Yukon, east to the southern edge of Hudson Bay, n. Que., all but s. Nfld., s. Baffin I., and Greenland. Winters at sea, mostly in tropical waters. **HABITAT:** Breeds on dry tundra habitat, primarily at the transitional edges of lakes, ponds, bogs, and marshes. In winter, wholly aquatic; found in tropical seas. In migration, found offshore along both seacoasts, but much more numerous in the Pacific. Concentrations found at upwellings; also may occur on almost any manner of open fresh- or saltwater environment, from lakes to estuaries to (especially) sewage treatment pools, but may be especially numerous on salt lakes. **COHABITANTS:** In migration, mixes with both Wilson's (inland) and Red (at sea) Phalaropes; sometimes swims among longer-legged wading shorebirds. **MOVEMENT/MIGRATIONS:** Whole population vacates breeding areas in winter. Spring migration from mid-April to late May; fall from late June to early November, with birds concentrating in post-breeding staging areas. VI: 3.

DESCRIPTION: A small, darkish, elfin-featured aquatic sandpiper with a short needlelike bill. Small (smaller than Sanderling or Stilt Sandpiper; larger than Western Sandpiper), with a round head, a long neck, an extremely thin pointy bill, and a longish, somewhat angular body. Overall smaller, shorter-billed, more stockily proportioned, more angular, and less sinuously lined than Wilson's Phalarope; slightly smaller, thinner-billed, and less chubby than Red Phalarope.

On the water, sits higher than Wilson's Phalarope and is more front-heavy and flatter-backed. Does not flatten itself against the water. Keeps its head elevated and picks downward.

Overall, at all seasons, plumage is darker and more contrasting than Wilson's Phalarope. Breeding males and females are charcoal gray, with rusty highlights about the head and neck and a collar that encircles a bright white throat and two to four jagged buffy streaks down the back. Nonbreeding plumage is gray above, with obvious dark streaking on the back, and white below. The white face and neck contrast with a dark cap, a black patch across the eye, and a black bill. Juveniles are like adults but darker still, almost blackish above, with narrow golden streaks on the back and a warm wash across the breast (but not the face and neck).

The key to identifying Red-necked Phalarope in winter or at sea is the darkly streaked back. Larger, plumper Red Phalarope has plainer and paler (unstreaked, silvery gray) upperparts.

BEHAVIOR: A highly social aquatic sandpiper; mostly pelagic when not breeding. Forages in flocks (numbering a handful to many thousands), feeding on plankton and aquatic insects. Picks prey from the surface and does not commonly submerge its head (as does Wilson's Phalarope). Swims, with jerky movements, in a linear search pattern. Also spins, as do all phalaropes, presumably to suck prey to the surface in the resulting vortex, and during the breeding season plucks insects from emergent and overhanging vegetation. Does not commonly venture onto land, except during the breeding season. On land, its short legs limit the bird to a halting clumsy waddle.

Calm and very tame. Allows close approach. Does not usually forage with other shorebirds except other phalarope species; and in migration, when large concentrations of Wilson's and Red-necked occur, tend to segregate themselves into different flocks.

FLIGHT: Small, angular, compact in flight—more sandpiper-like. Elevated head and short tail project little. Slender wings are acutely angled back. Flight, in flocks, is direct, fast, often twisty-turny, and low over the water. When alone, when landing, and when flying at higher altitudes, flight seems more erratic, more wandering, than Wilson's. Wingbeats are quick and steady; glides are short and perfunctory (usually executed when birds bank and land).

Lands more abruptly than Wilson's Phalarope. Braking is fluttering and hurried, the foot pattering is perfunctory, and the bird drops with a plop. In flight, the white wing stripe contrasting with otherwise dark upperparts is very distinctive.

VOCALIZATIONS: Call is a low flat "*spik*" (sometimes rendered with a compressed two-note quality: "*spi/unk*") that most recalls Sanderling's "*plik*" but is lower-pitched, not as sharp or metallic-sounding, and drags slightly at the beginning.

Red Phalarope, *Phalaropus fulicaria*

The Pelagic Sanderling

STATUS: Common but restricted, mostly arctic coastal breeder; outside the breeding season, wholly pelagic. Inland, even coastal sightings of this species are rare, and most occur during fall migration. Much more likely to be seen on and from shore on the Pacific Coast than the Atlantic. **DISTRIBUTION:** Circumpolar; in North America, breeds along the coastal plain from w. Alaska (Yukon-Kuskokwim delta) north along the Bering and Beaufort Seas across n. Canada and the arctic islands (east to n. Que.) and the west coast of Greenland. In winter, in North America, some birds winter off Calif. (occasionally Ore.) and n. Baja and off the Carolinas and Ga. **HABITAT:** Breeds in low, wet, flat tundra or sedges dotted with an abundance of small shallow ponds. In winter, stays well offshore (commonly farther than Red-necked), with concentrations found along oceanfronts where water temperatures and density change. Also attracted to areas of turbulence, upwellings, oil slicks, and foraging whales. Inland, often forages along the shallow grassy edges of lakes, ponds, and sewage pools (as opposed to sitting out in open water). **COHABITANTS:** In summer, Arctic Tern, Sabine's Gull, caribou. In winter, Red-necked Phalarope, whales. **MOVEMENT/MIGRATION:** Spring migration from early April to early June, with a peak in mid-May. Fall migration is protracted, from July to December, with large staging concentrations in the Bay of Fundy in August and September and peak numbers off the southern Atlantic Coast in October and December; in the Pacific, peak movements occur in late September. VI: 3.

DESCRIPTION: In nonbreeding plumage, a swimming Sanderling. Small shorebird (larger, heftier, more compact, and less spindly than Red-necked Phalarope; only slightly larger than and shaped somewhat like Sanderling), with a straight, thickish (not needlelike), yellow-based bill, a small round head, a shortish neck (for a phalarope), a robust chesty body, and short legs.

Breeding females (and less colorful but similar males) are stunning—black above, etched with fine gold lines, and burnt orange below. Large white cheeks and mostly yellow bill dominate the head. In winter, smooth pale gray above, with no streaks, and white below. The head shows a black eye patch and a black bill. Juvenile plumage is similar to Red-necked juvenile, but less overtly streaked above, and may show a warm buffy wash on the face, neck, and breast. Note the slightly thicker bill of Red Phalarope, in contrast to the needle-fine bill of Red-necked, and in winter the absence of black on the shoulder (characteristic of Sanderling).

If you see a Sanderling-like shorebird floating on the water far out at sea, you can be confident that it is not a Sanderling.

BEHAVIOR: Highly social; usually seen in well-spaced flocks well offshore. (But at times and in places, flocks may number in the hundreds and may be mixed with Red-necked Phalarope.) Swims buoyantly on the surface, riding high in the front and low in the rear. Feeds by turning or spinning in the water and picking at the surrounding water (also sometimes upends like a dabbling duck). Inland, this species is more likely to forage along the edges of ponds by wading. (Red-necked Phalarope mostly swims.) In summer, in the arctic, also forages on dry land, where it runs, somewhat clumsily, after insects.

Very tame. Often allows close approach, on both land and sea.

FLIGHT: In winter, chunky shape and pale gray plumage recall Sanderling. (Red-necked appears more spindly.) Overall pale and smooth gray above. (Red-necked appears darker.) Flight is low, strong, steady, direct, and not as erratic or twisty-turny as Red-necked. Birds lift off the water easily.

VOCALIZATIONS: Call is a sharp "*plik*" that is higher-pitched than Sanderling.

SKUAS AND JAEGERS

Great Skua, *Stercorarius skua*

Winter Skua

STATUS AND DISTRIBUTION: Breeds in Greenland and Iceland. Rare to uncommon in winter off the

Atlantic Coast from Greenland south to N.C. **HABITAT:** In winter, found in deeper ocean waters well away from land. Sightings by land-based observers are extremely rare and almost always associated with winter storms. Very rarely recorded in summer off New England and the Maritimes. Almost never recorded inland. **COHABITANTS:** Northern Fulmar, Black-legged Kittiwake, large gulls. **MOVEMENT/MIGRATION:** Most commonly sighted off the Atlantic Coast between October and March, but possible late August to mid-April. VI: 1.

DESCRIPTION: A big, bad, coarsely plumaged seabird that gets what it wants. Large (larger than Pomarine Jaeger; slightly larger than South Polar Skua; about the same size as a small Herring Gull), with a short, heavy, hooked bill, a large round head, a thick neck, a beer keg of a body, a short tail, and sturdy legs. Somewhat resembles a dark, squat, immature gull, but distinctly more thickset, barrel-chested, and more menacing-looking than any gull.

Adult plumage ranges from dark to pale. Most birds are dark brown, with lavish yellowish or rufous streaks on the neck that give darker birds a neat streaky quality and paler birds a bleached-out or ragged quality. Both light and dark adults usually show tinges of cinnamon (wholly lacking in South Polar Skua). Most individuals look dark-capped.

Immatures are overall smoother and cinnamon brown, with a dark brown head and wings. At close range, upperparts are etched with narrow, wavy, cinnamon lines. Note the absence of barring on the underparts (characteristic of immature Pomarine Jaeger).

BEHAVIOR: Usually solitary at sea, but often attends commercial fishing vessels and naturally occurring concentrations of feeding seabirds. Usually found in colder waters, but not always. Feeds by scavenging, fishing, and pirating food from other seabirds, and despite its robust size, is agile enough in flight to pursue and kill small birds at sea as well. Sits high on the water. Often fearless and aggressive on land, but may be hard to approach on the water.

FLIGHT: Big, robust, heavyset seabird with a beer keg–shaped body, long, broad, pointy wings, a large projecting head, and an embarrassingly short triangular tail. Overall larger, more broadly proportioned, and conspicuously shorter-tailed than Pomarine Jaeger. Slightly heavier-billed and more potbellied than South Polar Skua.

Plumage is overall dark, with large obvious white patches on the upper- and lower wings (more conspicuous than the patches on Pomarine Jaeger). Underwing coverts are dark but brownish-tinged (black on South Polar). Many adults appear shaggy or heavily streaked and spotted above. (Darker South Polar is more cleanly uniform.) Juveniles are smooth reddish brown with darker heads. (South Polar is cold and gray-brown.)

Flight is strong and direct, with strong steady wingbeats and infrequent glides. Often flies high and frequently soars.

VOCALIZATIONS: Silent at sea.

PERTINENT PARTICULARS: Both adult and subadult Great Skuas and South Polar Skuas molt after breeding, but as with the birds' breeding schedules, the timing differs. Adult Great Skua molts its primaries between August and March, adult South Polar Skua between March and August. In April and May, Atlantic skuas showing primary molt should be South Polar; in fall, skuas showing primary molt should be Great. Also, Great Skua drops its primaries one or two at a time, showing few gaps in the wing. South Polar Skua's molt is more rapid than other skuas, and more feathers (two to five per wing) may be missing or being replaced at one time; as a result, the wings of molting South Polar Skuas appear very ragged.

South Polar Skua, *Stercorarius maccormicki*

Summer Skua

STATUS AND DISTRIBUTION: Breeds in Antarctica. Uncommon off both coasts in North America between May and October; most common in late spring (along both coasts) and fall (West Coast), reaching s. Alaskan waters and Greenland. **HABITAT:** Well offshore, often near concentrations of gulls and shearwaters. Only casually seen from land (most sightings are storm-related), and almost never recorded inland. **COHABITANTS:** Shearwaters, Pomarine Jaeger, large gulls. **MOVEMENT/MIGRATION:** Present in waters off North America from late April to late October, with peaks from mid-May to mid-July on both coasts and from late August to late October on the West Coast. VI: 1.

DESCRIPTION: A big, bloated, crop-tailed jaeger with a blond nape. Large seabird (larger than Pomarine Jaeger; slightly smaller than Herring Gull or Western Gull, but overall stockier and more robust), with a short, thick, hooked bill, a rounded head, a bullneck, a beer keg–shaped body, a short tail, and sturdy black legs. Pomarine Jaeger is smaller-headed, slender-necked, longer-tailed, and overall slighter in build, with a bill that's slimmer, more pencillike than skua.

Plumage varies, ranging from light to intermediate to dark. Most birds are smooth cold gray-brown (lacking any trace of warm brown or cinnamon tones). Wings and back are usually darker than the head and lower parts. (Many Great Skuas appear monotoned.) On most adults, the nape is paler than the body (showing, at the very least, a golden sheen), and often the entire head seems encircled by a broad pale or golden collar. The head may be uniformly dark, or the face (particularly the area in front of the eyes) may be darkened, making the bird look masked but not capped. Very pale birds show smooth blond heads and underparts contrasting with dark grayish brown wings and tail touched with light pale streaks (but their paleness alone distinguishes them). Great Skuas, in all adult plumages, are more coarsely plumaged, with streaked necks, ragged or mottled upperparts, and pale streaking on the breast.

Juveniles are uniformly smooth cold grayish brown, with no pale nape; paler birds show darker wings. Great Skua, by comparison, is dark-headed and overall warmer-toned; many individuals are distinctly cinnamon brown overall.

The expression is bad to the bone.

BEHAVIOR: Usually solitary at sea, though sometimes seen in pairs. Often seen close to large aggregations of other seabirds (commonly near commercial fishing vessels), which they parasitize like the playground bully. Also found alone feeding by scavenging and active fishing. Spends a great deal of time sitting (very high) on the water. Very aggressive and intimidating, South Polar Skua brutalizes other seabirds more than it steals from them, swarming all over them rather than merely chasing them. Birds sometimes use troughs for concealment when approaching feeding birds. A suspicious bird, South Polar is not easily approached by boat, but does follow ships and respond to chum.

FLIGHT: A large, humpbacked, barrel-bodied bird with wings almost as wide as the body and a short broad tail. Makes the longer-tailed Pomarine Jaeger look delicate. The pale nape/collar (when present) is usually easy to see, as are the large white patches on both the upper and lower surface of the wing. In all plumages, underwing coverts are blackish (Great Skua's are brownish).

Flight is fast and direct, with slow, heavy, steady wingbeats. Usually flies low when closing on other birds but quite high when traveling distances.

VOCALIZATIONS: Silent at sea.

PERTINENT PARTICULARS: Separation of this species from Great Skua is complicated by variables relating to plumage but somewhat simplified by geography and the calendar. South Polar is the only skua found in the Pacific. In the Atlantic, where both skuas are found, South Polar is present from April to the end of September, and Great from the middle of September to late May. Theoretically the species overlap during April and May and during September and October, but the actual overlap in any given location is less because these dates represent the extreme dates for the entire coastal region between North Carolina and the Maritimes. In fact, by the time South Polar arrives off North Carolina in mid-April, Great, which usually departs by this time, is mostly gone from Carolina waters. When South Polars retreat from North Carolina in late September, Great's arrival in North Carolina, the southern limit of its winter range, is still a month and a half away.

Some nonbreeding Great Skuas may be present off Massachusetts, the Gulf of Maine, and the Maritimes all summer (where South Polar Skua is present), but thus far there are no winter records of South Polar Skua.

Pomarine Jaeger, *Stercorarius pomarinus*

The Skualike Jaeger

STATUS: Uncommon to fairly common and somewhat nomadic high arctic breeder; in summer, far more coastal and more geographically restricted than either Parasitic or Long-tailed Jaeger. Uncommon to sometimes common spring and fall migrant along both coasts; very rare inland, but on smaller bodies of water more likely than Parasitic Jaeger. **DISTRIBUTION:** Breeds along Alaska's North Slope, and in Canada, found on assorted arctic islands of the Northwest Territories east to the northeast shore of Hudson Bay. Winters well at sea from Calif. and s. N.C. south (occasionally farther north). **HABITAT:** Breeds in low-lying wet tundra and marsh with an abundance of lakes. Breeds when and where lemmings reach their cyclic peak. In migration and winter, found offshore, often near concentrations of shearwaters, other seabirds, gulls, and (sometimes) fishing trawlers (out to about 40 mi. from shore). Parasitic Jaeger is usually closer to shore; Long-tailed is even farther offshore. However, during migration this species is more likely to turn up on smaller inland bodies of water than Parasitic Jaeger. **COHABITANTS:** In migration, usually associates with shearwaters. **MOVEMENT/MIGRATION:** Almost wholly coastal; often in small strings of 3–10 birds (occasionally more). Spring migration from mid-April to early June; fall from late July to December. VI: 3.

DESCRIPTION: A large, dark, overall robust, skua-like jaeger. Large and broadly proportioned (larger than Sooty Shearwater or Parasitic Jaeger; smaller than Western Gull or South Polar Skua), with a large head, a heavy bicolored bill, a fat body, long tapered wings with a broader base than Parasitic, and a short, thick, blunt tail that shows long spoon-tipped central tail feathers in adults and blunt-tipped central tail feathers in immatures and subadults. Overall more heavyset and skua-like than Parasitic or Long-tailed Jaeger.

Like other jaegers, plumage is highly variable, with dark, light, and intermediate morphs as well as immature and subadult plumages. In general terms, Pomarine appears all dark—darker than Parasitic. Light-morph adults (showing whitish bellies) are also often overall more coarsely marked and scruffier-looking than the more pastel-shaded trimmer-cut Parasitic. The dark cap on light morphs is more extensive than on Parasitic—more helmet or mask than cap—and it extends to the base of the upper bill. (Parasitics show a white spot at the base of the bill.)

Upperwings show a prominent white flash near the tip (as does Parasitic). Underwings show pale double bars that are evident at close range on most individuals and can be seen even at a considerable range on some very well-marked individuals. Immatures and subadults have broad dark-and-white bands on the rump (uppertail) that give them a pale-rumped look at a distance. Overall paler immature Parasitic Jaeger shows less contrast between the back and rump.

BEHAVIOR: A big aggressive bully of a jaeger, but at sea less piratic than Parasitic Jaeger. Forages primarily for lemmings when breeding; rarely chases small birds. At sea, often found attending fishing trawlers, around which it simply outmuscles shearwaters and gulls for discarded catch. (Pomarine can also dive beneath the surface.) Steals food from other seabirds with a rapid all-out bum's rush, although rarely engages in the long tail chases characteristic of Parasitic Jaeger and terns, and instead commonly picks on larger, slower species. Terns easily outmaneuver Pomarine—which accounts for Pomarine's near-absence in nearshore waters, where terns and Parasitic Jaeger are commonly found.

Like Parasitic Jaeger, Pomarine often sits on the water. In migration, commonly flies high above the surface in small groups (of 10 or more birds) strung out single file. Other jaegers commonly fly in smaller, less regimented and spaced groups.

FLIGHT: Flight profile is overall heavier, more gull-like than Parasitic Jaeger. A bulkier head, broader-based wings, chestier body, and heavy, short, blunt tail all help distinguish this bird from the slighter, trimmer Parasitic. When present, the long spoon-tipped central tail feathers on adults (which do not undulate in flight) are very apparent.

Flight is powerful, with deep gull-like wingbeats that are steady, mechanical, and almost relentless, with little gliding and little changeup in pattern or cadence.

VOCALIZATIONS: Silent at sea.

PERTINENT PARTICULARS: Even very experienced seabirders often mistake Pomarine Jaeger for skuas at first glance. Skuas are larger, bulkier, and overall more compact, with even heavier bodies, shorter tails, and extremely wide wings.

Parasitic Jaeger, *Stercorarius parasiticus*

Falconlike Jaeger

STATUS: Arctic breeder; uncommon marine and rare but regular Great Lakes and Salton Sea migrant, but at times and under certain conditions may be common along the coast. Except for the Gulf Coast (where Pomarine Jaeger may predominate), the most likely jaeger to be seen from both coastal shores and the shores of large inland lakes (such as the Great Lakes). Casual inland. **DISTRIBUTION:** Breeds along the southern and western coasts of Alaska east across arctic Canada to the south shore of Hudson Bay; also in extreme ne. Canada east to Greenland. Winters offshore primarily in the Southern Hemisphere; also regularly found in small numbers in s. U.S. waters off s. Calif., Tex., and Fla. **HABITAT:** Coastal tundra and inland foothills, often near lakes or fiords and commonly in marshier habitat than Long-tailed Jaeger. In winter and migration, becomes pelagic, but in migration commonly occurs within sight of land, often entering bays and inlets, and is regularly seen on large inland bodies of water (most notably the Great Lakes). Also migrates well offshore. **COHABITANTS:** In migration, often found in concentrations of feeding terns. **MOVEMENT/MIGRATION:** Spring migration from mid-April to early June; extremely protracted fall migration occurs from August to early December, with a peak in early September to early November. VI: 3.

DESCRIPTION: A dark, nimble, aggressive seabird that looks and flies like a cross between a small gull and Peregrine Falcon. A medium-sized jaeger (larger than Forster's or Common Tern; smaller than Heermann's or Ring-billed Gull; about the same size as Laughing Gull, kittiwake, or Elegant Tern), Parasitic is an intermediate size between the usually larger, bulkier Pomarine and the usually slighter smaller and more ternlike Long-tailed.

Overall slender and falconlike, with a proportionally small head and bill, a slender body, long, narrow, and acutely pointed sweptback wings, and a long projecting tail. The profile says "menacing speed." Gulls, by comparison, look bulky and even-keeled, balanced fore and aft; terns look speedy, but stiff and frail. Parasitic appears slimmer and more streamlined than Pomarine and has a proportionally smaller head, more slender wings (particular the arm), a trimmer body, and a slimmer and more cleanly attenuated tail. Adults have projecting, narrow, pointy central tail feathers that may appear tined or single-bladed. They vary in length but are never so long that they appear flagellum-like (as with Long-tailed Jaeger) or undulate or ripple in flight. The central tail feathers of immatures form short twin spikes.

Pomarine looks bulky and skualike, with a big head and bill, broad wings, a chestier body, and a heavy bulky tail with long twisted tail feathers that thicken toward the tip. The tails of immatures have blunt central tail feathers.

Parasitic plumage is variable, confusing, and very similar to other jaegers. Whether light, dark, or intermediate morph, both Pomarine and Parasitic Jaegers look dark at a distance. On light- and intermediate-morph adults, the dark cap is less extensive than Pomarine (and slightly more extensive than Long-tailed). The cap stops short of the bill, so the bird shows a white spot between the eyes that is particularly obvious when it's standing on open tundra.

In flight, Parasitic shows a white flash on both the upper- and underwings. The flash on the upperwings may be bold or subtle, but it is almost always more extensive than the white-trimmed leading edge to the wing that Long-tailed Jaeger shows. The underwings show a single unbroken white flash. Pomarine's underwing pattern usually shows a pale double bar — made up of one large pale flash near the tip of the underwing and a second paler band near the base of the primaries.

Also, immature Parasitics are commonly paler, particularly paler-headed; immature Pomarine is commonly very dark. At a distance, the finely barred uppertail (rump) pattern found on immature and subadult light- and intermediate-morph Parasitics shows little contrast with the rest of the bird, whereas on Pomarine the broadly barred uppertail makes distant birds appear more pale-rumped.

BEHAVIOR: When breeding, hunts open tundra for birds and mammals by flying low and hovering frequently. Often hunts in pairs: one bird flushes prey by walking on the ground (sometimes with wings open), while the other waits above. In migration, becomes piratic, attending and harassing flocks of terns (also parasitizes smaller gulls and alcids) in nearshore waters. Commonly sits on the water until it spots a fish-laden tern, then closes rapidly, flying close to the surface. Parasitic's pursuit of terns usually involves a vertical component, which catches an observer's eye, since most bird movement is horizontal. It is not uncommon for two or more jaegers to team up against a single tern.

In migration, over water, may be close to the surface or moderately high (less than 50 ft. above the water).

FLIGHT: Overall very angular and falconlike, with long, narrow, pointed, and sweptback wings and a long-tailed profile (both adults and immatures). In flight, appears fluid, determined, and somewhat sinister. When flapping, the wings are distinctly bowed and gull-like. When the bird is gliding, the bow on the wings may be severe (extra gull-like). In normal flight, wingbeats are fairly shallow, with a gull-like stiffness along the arm and a Peregrine-like fluid flex at the wingtip. In pursuit flight, wingbeats are faster, deeper, and pushing, but with a shifting cadence that is sometimes faster, sometimes slower.

Pomarine is more sluggish and gull-like in flight; adult Long-tailed Jaeger is buoyant and serenely languid.

VOCALIZATIONS: On territory, gives both a long call and a yelp. Generally silent in migration.

PERTINENT PARTICULARS: Cutting to the chase (so to speak), Parasitic Jaeger is the falconlike dark bird chasing the pale bird high above the concentration of feeding terns. On the West Coast, be aware that the mostly or all-dark Heermann's Gull also pirates fish from small gulls and terns and is about the same size as Parasitic (or slightly larger). Heermann's Gull is not as agile as a jaeger, and usually not as tenacious in its pursuit.

Long-tailed Jaeger, *Stercorarius longicaudus*

Gray Jaeger

STATUS: Common arctic breeder. Uncommon pelagic migrant well offshore in the West; very uncommon (mostly fall migrant) off the East Coast. Rare inland (and mostly in fall). **DISTRIBUTION:** Widespread breeder across arctic and subarctic regions. In North America, found across most of Alaska (excluding the Aleutians, the southern coast, and cen. Alaska), most of the Yukon, the n. Northwest Territories, and the arctic islands east to the northeast shore of Hudson Bay, Baffin I., and the west coast of Greenland. Winters in the oceans of the Southern Hemisphere. Migrates well offshore, commonly beyond the drop-off of the continental shelf (and almost never within sight of land). **HABITAT:** Breeds on arctic and alpine tundra, often far from seacoasts. In most places, prefers higher, drier, less vegetated tundra, but may also occupy lower, wetter, tussocky tundra. In migration, found over marine waters. Though very rare inland during migration, Long-tailed is the most likely jaeger to be seen away from bodies of water. **COHABITANTS:** During breeding, Willow and Rock Ptarmigans, American Golden-Plover, lemmings. In migration, found in the same waters as Arctic Tern and Sabine's Gull. **MOVEMENT/MIGRATION:** Mostly well offshore. Spring migration from May to early June; fall from late July to early October. Nonbreeders may be present off both U.S. coasts in summer, but are absent between mid-November and late April to early May. VI: 3.

DESCRIPTION: The smallest, sleekest, lightest, and grayest of the jaegers. A small jaeger (slightly smaller than Parasitic Jaeger) but a medium-sized seabird (excluding its long central tail feathers, about the same size as Mew and Laughing Gulls), with a short thickish bill, a small rounded head, a long, slender, tapering, athletic-looking body, and a tail that, with or without the long central tail feathers of breeding adults, appears long.

Overall smoothly plumaged bird. Breeding adults show a trim limited black cap, pale grayish to grayish brown upperparts, a creamy white breast and upper belly (with no breast band), and a long, narrow, flagellum-like tail that extends half the body length beyond the end of the wings.

There is, it is believed, no dark morph.

By comparison, adult light-morph Parasitic Jaeger is slightly darker and browner above, with a more extensive cap, a (usually) dark breast band, a longer, narrower bill, and a pale white spot above the base of the bill.

The plumage of adult Long-tailed looks trim, crisp, expensive, and tastefully coordinated. Parasitic Jaeger looks like a Long-tailed that spent ten hours at the office and three hours in traffic. The dark and coarsely garbed Pomarine Jaeger looks like it slept in its suit in the park. Winter-plumage Long-taileds are not seen in North America.

The flagellum-tailed breeding adult is distinctive, but immature birds are challenging. Immature Long-tailed, like other jaegers, may be light, intermediate, or dark. In all plumages, birds are slightly colder and grayer than other jaegers and devoid of any orange or reddish tones. Like adults, immatures have a short thick bill. The projecting central tail feathers are usually blunt. Light birds have pale heads and pale bellies. The pale band on the lower chest of intermediates makes the bird look as though its darker feathers were worn away. Dark birds show distinct narrow whitish edgings to the back feathers (orange on Parasitic) and distinctive white barred undertail coverts (which Parasitic often lacks).

BEHAVIOR: On territory, Long-tailed sits conspicuously high on the open tundra, displaying its white chest. Hunts by coursing over the tundra, sometimes swooping, sometimes hovering, often with other Long-tailed Jaegers, but most often hunts alone. Also hunts on foot while on land and along the edges of ponds.

At sea, most often seen flying very high and directly (50–200 ft. high), either alone or in small well-spaced groups. Foraging birds cruise low and slow, swoop to pluck prey from the surface, and are reported to plunge-dive. Long-tailed often settles on the water, and swims well. Sometimes associates with other seabirds (shearwaters and terns), and pursues them less often than does Parasitic Jaeger. Inland, migrating immature birds occasionally land on plowed fields and roadsides and forage for insects and earthworms. Usually tame, Long-tailed is easily approached on land and by boat, but normally ignores boats at sea.

FLIGHT: Sinuously slender, with an aerial finesse enhanced by its fluttering flagellum-like tail (which may sometimes, however, be missing or broken). Adults are short-headed, slender-bodied, long-winged, and long-tailed. Birds with broken or missing tail feathers show a long body projection with more body projecting behind the wing than the wing is wide.

The pale grayish upperparts (showing contrastingly darker hands and a trailing edge to the wing) are apparent at great distances; the unbanded white breast is even more obvious. At closer range, upperwings show a narrow tracing of white on the outer leading edge and no conspicuous white patch, as on other jaegers. Underwings are always all black, though some subadults show a bit of fine white barring.

Flight is light, leisurely, fluid, buoyant, and dreamlike. Wingbeats on migrating birds are slow and measured: the body rises and falls with each wingbeat, and you cannot help but recall the time-suspending flight of frigatebird (an image conjured by no other jaeger). Birds often resort to shearwater-like glides, at times (particularly in strong winds) gliding for several hundred yards.

Immatures are often as difficult to identify as adults are easy. It helps to not think of them as the same creature. Concentrate on what makes immature Long-tailed different from immature Parasitic Jaeger (even Pomarine). Young Long-tailed appears shorter in the front, with longer, more slender wings and more bird coming out the rear than Parasitic.

The projecting central tail feathers are blunt, not pointed. The upper surface of the wing shows a narrow white trim along the leading edge, not a white flash (as on Parasitic). The white flash on the underwing is more restricted to the base of the primaries.

Above all, remember that immature Long-tailed is overall a colder gray tone than Parasitic and Pomarine—it has no orange or rufous tones. Light-morph birds appear pale, almost white-headed and white-bellied. Intermediate birds show a pale tie-dyed collar on the lower chest. Dark birds usually show conspicuous white barring on the undertail. (Dark Parasitic has a dark undertail or a pattern that is hard to see.)

VOCALIZATIONS: Call is a high yap, "*k'yap*" or "*k'lap*," given on the breeding grounds. Silent at sea.

GULLS

Laughing Gull, *Larus atricilla*

Ichabod Gull

STATUS: Common to very common on the Atlantic and Gulf Coasts and locally on the outer coastal plain. **DISTRIBUTION:** Breeds from Nova Scotia to s. Tex.; winters coastally from s. Va. to South America and across interior portions of the Florida Peninsula and s. Tex. **HABITAT:** Breeds on coastal marshes, rocky islands, and sandy beaches. Forages on beaches, inshore waters, and, in the interior, plowed fields, airports, parking lots, and landfills. Roosts on lakes and reservoirs. **COHABITANTS:** Clapper Rail, Forster's Tern, Seaside Sparrow, French fry–bearing tourists. **MOVEMENT/MIGRATION:** In July and August, Atlantic Coast birds disperse north and south along the coast and inland rivers. Individuals turn up between March and October at scattered locations in the U.S. interior. In the West, Mexican breeding birds disperse north as far as the Salton Sea. From late August to late December, all Atlantic birds north of Virginia (and many farther south) migrate south. Northward migration begins in early March and continues into May. VI: 3.

DESCRIPTION: A gangly, long-snouted, darkish-backed gull of beach, marsh, and inshore waters—an Ichabod Crane of a gull. Smallish (smaller than Ring-billed; larger than Bonaparte's; larger and longer-profiled than Franklin's), with a small flattish head, a long droop-tipped bill, a long lean body, long wings that extend well beyond tail, and long legs.

Breeding adults (March to August) are distinguished by their all-black heads, reddish bills, and dark gray backs, whose shade falls midway between Ring-billed Gull and Great Black-backed Gull, is slightly darker than Franklin's Gull, and is slightly lighter than Lesser Black-backed Gull. In winter, the bill and legs turn black, and the black head is replaced by a broken shadow of a dirty gray cap. (The vestigial cap of Franklin's Gull is fuller and darker; Bonaparte's Gull shows a single band and a single black ear spot; Black-headed Gull has two bands across the head—more sharply defined than Laughing Gull—and a black ear patch.)

Juvenile birds are mostly warm brown and scallop-patterned above, contrasting with a plain, patternless, grayish brown head, breast, and sides. First-winter birds (plumage is often assumed in the fall) have the dark grayish back of adults but the warm brown wings of juveniles. The head is often more darkly and extensively capped but remains shadowy (that is, it's never black and sharply defined like immature Franklin's Gull). Laughing Gull's expression is sleepy and dreamy.

BEHAVIOR: A bold, noisy, omnipresent, opportunistic gull—at home whether loafing on the beach, pilfering food from sunbathers, following tractors tilling soil, floating in a current picking morsels out of the chop, pirating fish from smaller terns and other Laughing Gulls, or snapping up flying insects (such as emerging greenhead flies) in midair. Forages by walking and picking along shorelines as tides fall, but may skim food from the water surface while flying.

Highly social, nesting in large noisy colonies and feeding and roosting in flocks. Swims buoyantly, riding high in the water. Able to perch on utility lines, even marsh grasses.

FLIGHT: A long and narrow-winged profile gives birds a gangly appearance. Adult birds show extensively blackish wingtips (visible above and especially below) that blend into the gray (or brown) of the upperwing. Juveniles are dark above, with a contrasting white tail tipped with a wide black band. The brownish chest makes them look hooded.

Flight seems stiff and somewhat effortful (for a gull). Wingbeats are shallow, regular, fairly quick, and perfunctory. Soars well. Able to hover and move nimbly enough (if not very gracefully) to snap up prey in the air.

VOCALIZATIONS: Call is a loud, strident, nasal, two-syllable belly laugh: "*KeeYa, KeeYa KeeYa.*"

PERTINENT PARTICULARS: Juvenile Laughing Gull is sometimes mistaken for light-morph Parasitic Jaeger. About the same size, both show dark upperparts and pale underparts, and when Laughing Gull is flying low and fast or, especially, pursuing terns, its shape is not that dissimilar from Parasitic Jaeger's. The similarities, however, are superficial; mindfulness is the best antidote.

Franklin's Gull, *Larus pipixcan*

Prairie Black-headed Gull

STATUS: Common breeder in northern prairies; common and, in places, abundant migrant through the central and interior western United States; rare on both coasts. **DISTRIBUTION:** Breeds across e. and cen. Alta., all but ne. Sask., w. Man., n. and cen. Mont., all but sw. N.D., ne. S.D., and disjunct pockets in Minn., s. Idaho, Nev., and n. Utah. Migrates regularly farther west across the Great Basin and the Rockies. Winters primarily along the west coast of South America. **HABITAT:** Nests on mats of floating vegetation on freshwater lakes and marshes. Also found in marshes with tall rank growth (such as cattails, bulrushes, and phragmites). In migration, found on lakes, ponds, estuaries, flooded fields, pastures, cropland, prairies, salt ponds, salt lakes, mud flats, and sewage ponds. **COHABITANTS:** During breeding, Eared Grebe, Wilson's Phalarope, Black Tern, Yellow-headed Blackbird. **MOVEMENT/MIGRATION:** Spring migration from early March to late May, with a peak in late April and first half of May; fall movement is protracted (late July to mid-December), with birds wandering widely across the West and the Great Plains after breeding; migration begins in earnest in late September and goes through early November. VI: 3.

DESCRIPTION: Compact, charcoal gray–backed, black-headed gull. Small (closer to Bonaparte's Gull than Laughing Gull in size), with a round head, a short straight bill (not thin and petite like Bonaparte's; shorter and not droop-tipped like Laughing), and short legs. Overall compact without being petite, and showing a nicely proportioned head—not the flat-headed, long, and droop-snouted profile of Laughing Gull.

Adults in breeding plumage, with their black heads and charcoal gray upperparts, are similar to Laughing Gull, but their broken white eye-ring is more prominent (you'll note it before you know to look for it), and the closed wingtips show a black-and-white pattern (not just tiny white spots on a black background). Most adults show a faint pink blush on white underparts.

The head of adult winter birds is covered by a blackish well-defined pilot's helmet, or "took," that makes the broken eye-ring stand out. A poorly defined shadowy gray stripe or two runs over the back of winter Laughing Gull's white head, making it look like a balding bird trying to make the most of the hair (or hood) it has left.

Juveniles and immatures are fundamentally black, gray, and white. The black "pilot's helmet" is well defined and full, and the broken white eye-ring prominent. The back and the folded wings are smooth brownish gray (you have to look very close to see the brown). Except for a shadowy suggestion of a collar or hood extending along the sides of the neck/breast, underparts are white, and the white tail is tipped with a narrow black band that doesn't quite reach the outer edge. By comparison, immature Laughing Gull has conspicuously brown wings, a scaly back, a dirty wash across the breast, and a black banded tail that covers one-third of the tail and goes edge to outer edge.

BEHAVIOR: Very social gull. Nests in colonies. Forages with other Franklin's Gulls (also with Bonaparte's, Ring-billed, and Laughing) after breeding and during migration. In migration, forages by the hundreds and thousands in search of earthworms in plowed fields or waste grain. During breeding season, hawks insects over marshes; after breeding, concentrates on midge- and fly-infested lakes, where they bob and turn (phalarope-like), plucking insects from the surface or snapping insects out of the air. Apparently does not pirate food (unlike Laughing Gull).

Nimble, agile, and able to hop over obstacles and perch on cattail stems. Not particularly shy, but not indifferent to approach. When pressed, flies (en masse).

FLIGHT: Adult Franklin's is overall stocky, with a short body and wide, short, blunt-tipped wings—recalls Little Gull more than Laughing Gull. Juveniles have a different shape, with longer acutely pointed wings (much more like Laughing Gull). Wings of adults are all gray above and grayish-washed below, with a small black tip cut off from the rest of the wing by a white band. The pattern is visible above and below. Subadults have the small black tip but lack the white band and thus more closely resemble Laughing Gull. Wings of juveniles have a simple two-toned pattern with no overt brown: above, gray on the arm and blackish on the hand; below, all white with dark gray tips. The white tails of adults have a faint pale gray center; juveniles' white tails have a narrow black tip that doesn't quite reach the outer edge of the tail.

Note: In midsummer, when adults are molting, black wingtips may be absent, and the entire wingtip appears white. (The upper surface of the wing is often piebald-patterned as well.)

Flight is direct and somewhat heavy; wingbeats are hurried, stiff, choppy. In migration, birds frequently fly very high.

VOCALIZATIONS: Franklin's call, "*eeeyah, eeah eeah eeah eeah,*" is quicker and higher-pitched but not as loud or piercing as Laughing Gull. Vocalizations have the pattern of Laughing Gull but the pitch and muted petulance of Ring-billed Gull.

Little Gull, *Larus minutus*

Gull with the Dark Mittened Wing

STATUS AND DISTRIBUTION: Rare breeder (very local and irregular on the shores of the Great Lakes and of Hudson and James Bays); uncommon but regular on the e. Great Lakes in fall and early winter and in late winter and spring along the Atlantic Coast, where groups of a dozen birds or more are possible. Very rare elsewhere in the U.S. **HABITAT:** Breeds in shallow fresh or brackish wetlands. In winter and migration, usually coastal, foraging at the mouths of inlets, bays, and tidal estuaries; also forages on inland lakes and reservoirs, sewage lagoons, and the warm-water outfalls of power plants. A few birds winter along the Niagara River between New York and Canada, and some unknown number of birds may winter well offshore. **COHABITANTS:** In summer, Common, Forster's, and Black Terns. In winter, almost always found with Bonaparte's Gull. **MOVEMENT/MIGRATION:** In fall, two movements occur on the Great Lakes—one from mid-July to late August, and a second from September to November. On the Atlantic Coast, most fall migration occurs later, in November and December, and in spring from late February to mid-May. VI: 3.

DESCRIPTION: A baby Bonaparte's Gull—tiny, roundly proportioned, with pigeonlike compactness, and overall pale except for a dark head (cap in winter) and underwings that flash black in flight (adults). The smallest gull (smaller than Bonaparte's Gull; smaller than a pigeon!), Little Gull has a petite bill, a round head, a fairly slender neck, a plump compact body, short wings, and a short square-tipped tail. Bonaparte's looks gangly by comparison. Juvenile Littles have pointier wings, so are more akin to Bonaparte's.

Adults in breeding plumage have pale gray upperparts (and white wingtips), a fully hooded black head, bright white (or pinkish) underparts, and reddish legs. In winter, the head shows a dark skullcap and dark ear patch. Immatures are like adults and similar to immature Bonaparte's, but young Little Gull has a dark skullcap with attached ear muffs (Bonaparte's has an ear patch and no skullcap), and its folded wings show more black (particularly at the shoulder).

BEHAVIOR: In summer, breeds in pairs or in small colonies, but commonly seen feeding alone. Otherwise, usually found among flocks of Bonaparte's Gulls (also at times, terns). Commonly feeds by flying into the wind 10–20 ft. over the water, swooping, and plucking prey from the surface. Often lands on the water, where it floats high and buoyantly, turning phalarope-like and sometimes dunking its head. Also hovers and reportedly plunge-dives like a tern. Also roosts with Bonaparte's Gulls, frequently situated at the end or edge of a group. (If there are multiple birds, they usually sit together.) Sometimes persecuted by Bonaparte's Gulls, so may sit slightly (a few inches) away from the larger gulls. Little Gull is tame and allows close approach.

FLIGHT: Looks like a small, compact, short-tailed, blunt-winged or mitten-winged gull. Immatures are pointier-winged but still shorter and proportionally broader than Bonaparte's. Adults are overall pale except for all blackish underwings. Immatures are conspicuously patterned above, showing a dark M pattern on the upper surface of the wing so bold that the wing seems mostly blackish and shows no hint of the whitish flash seen in young (and adult) Bonaparte's. *Note:* Unlike adult Little Gull, the immature's underwing is pale, not black. The square-cut tail is tipped with a black band. Flight is light, buoyant, nimble; wingbeats are quick, deep, and somewhat fluttery or choppy.

VOCALIZATIONS: Call is a terse nasal *"kek,"* infrequently given.

PERTINENT PARTICULARS: In mixed flocks, the blackish underwings of adult Little Gull truly stand out. Although not quite as apparent, the darker upperwing pattern of immatures also draws the eye. Adult Black-headed Gull, which is conspicuously larger and bulkier than Bonaparte's Gull in direct comparison (and simply dwarfs Little Gull), also shows dark under the wings, but the black is limited to the outer portion of the wing (the hand); it doesn't cover the entire underwing, as is the case with adult Little Gull.

Black-headed Gull, *Larus ridibundus*
Meso-Bonaparte's Gull

STATUS AND DISTRIBUTION: Old World species, breeding in Iceland and s. Greenland (and sparingly in Nfld.). Common winter visitor in Nfld.; uncommon winter visitor to coastal waters of the Maritimes and locally in New England; rare in the mid-Atlantic states. Very rare elsewhere except s. Bering Sea and, in spring, the Aleutians, where small numbers are regular. **HABITAT:** Coastal waters, including inlets, back bays, estuaries, sheltered harbors, and fresh- and saltwater ponds; also found on fishing docks, alongside trawlers at sea, at sewage outlets, and on large inland lakes. **COHABITANTS:** Where winter ranges overlap, often found with Bonaparte's Gull (also Ring-billed Gull). **MOVEMENT/MIGRATION:** Generally present in North American waters from September or October into May, but nonbreeding birds may be encountered at any time. VI: 3.

DESCRIPTION: A big, pale, stocky Bonaparte's-like gull with a brown (not black) hood and a reddish or orangish (not black) bill. Medium-sized (slightly larger than Bonaparte's; smaller than Ring-billed; about the same size as Black-legged Kittiwake), with a short, straight, reddish (or yellowish in immatures) bill, a smallish head, a longish neck, and a compact body. Overall bulkier and less elegant than Bonaparte's Gull.

Slightly paler above than Bonaparte's. Breeding adults wear a charcoal brown fencer's mask that is browner and less extensive than the hood of Bonaparte's Gull, leaving the hindneck wholly white. (Looks like a "black-headed gull" whose hood has shrunk.) In winter, most birds (adults and immatures) show two dark, shadowy, vertical slashes on the head—one above the eye, one behind the eye. Most Bonaparte's Gulls show a round black ear spot (and perhaps the suggestion of a shadowy smudge above the eye).

In comparison to the petite black bill of Bonaparte's Gull, Black-headed's heavier, distinctly reddish bill (or yellow-orange to orangy-red in immatures) stands out. Legs of adult Black-headed are brighter red; legs of immatures are a rich yellow-orange (the same color as the bill!), whereas the legs of immature Bonaparte's are pale pink.

BEHAVIOR: Much like Bonaparte's Gull. Often found where people are found; often seen in association with other gulls. Feeds on and over water, along coastlines, and on agricultural fields. This species seems more terrestrial than Bonaparte's, and more like Ring-billed Gull, and it often visits inland freshwater ponds and parking lots. A tame bird that comes readily to food (whether offered or not).

FLIGHT: Looks like a big, bulky, somewhat gangly Bonaparte's Gull. Like Bonaparte's, Black-headed shows a triangular white flash along the leading edge of the wing, but the underwings show shadowy dark hands (recalling Little Gull). The underwings of Bonaparte's are pale, almost translucent. In a mixed flock, in a melee of white and silver, it is the dark underwings that stand out.

Immature Black-headed shows a more limited white flash on the upperwing than adults, and less dark shadow on the underwing, but the pattern is still evident. The upperwing pattern of immature Black-headed is more extensive and diffuse than Bonaparte's (and perhaps not so contrasting). These characteristics in combination with Black-headed's larger size and dark banded tail make distant young Black-headed look more like immature Ring-billed Gull. If you see an immature Bonaparte's-like gull that keeps reminding you of Ring-billed, look again.

Flight is graceful and buoyant, but more wheeling, not as quick, nimble, and ternlike as Bonaparte's. Wingbeats are slower and not as snappy.

VOCALIZATIONS: Call is a descending, gargled, crowlike "*ahr'r'r*" that is more protracted and not as flat as Bonaparte's.

Bonaparte's Gull, *Larus philadelphia*
Ternlike Gull

STATUS: Common and widespread arctic and subarctic breeder; common along all three coasts in winter, and inland in migration and winter. **DISTRIBUTION:** Breeds across s. Alaska and most of interior Canada south to within about 200 mi. of the U.S. border and east to cen. Que. Except for the southeast coast of Alaska and the south shore of Hudson Bay, does not breed near seacoasts. Winters coastally from se. B.C. to cen. Mexico, n. Mexico to Fla., and N.S. to Fla. Also found inland across the se. U.S. and N.M. (in mild years, also on Lakes Erie and Ontario). In migration, may occur anywhere in the U.S. and s. Canada. **HABITAT:** Nests most commonly in trees along lakes and marshes lying at the borders of the boreal forest. Winters in coastal bays, harbors, inlets, and beaches, concentrating in the vicinity of currents and upwellings. Also found along unfrozen interior lakes, rivers, reservoirs, sewage lagoons, and sewage and power plant outfalls. Does not commonly forage on landfills. **COHABITANTS:** In winter, Red-throated Loon, Clark's and Western Grebes, Red-breasted Merganser, other gulls, Forster's Tern. **MOVEMENT/MIGRATION:** Wholly migratory, concentrating along seacoasts and inland rivers. Spring migration from late February to late May; more protracted fall migration begins in late July and ends in late January, with many birds making interim long-term stopovers in places such as the Great Lakes, the Bay of Fundy, and other large interior lakes. VI: 2.

DESCRIPTION: A small, pale gray, ternlike gull with a flashing white wedge along the leading edge of the outer wing. Small and fairly compact (smaller than Franklin's or Mew Gull; about the same size as Forster's Tern), with a small head, a petite ternlike black bill, a slender neck (when extended), and a compact body. On land, looks plump-chested and short-legged. On the water, sits high, phalarope-like.

Overall pale—pale gray above (paler than Laughing and Franklin's Gulls; about the same color as Common Tern) and bright white below. Often the gray extends down the neck onto the breast, giving the bird a pale collar at a distance. (Adult Black-headed Gull has a white neck and breast and no collar.) Breeding Bonaparte's has a black head (with a white nape) and a rosy blush on the breast. In winter, shows a well-defined black dot over the ear and a narrow shadow of a band extending eye to eye over the crown. Immatures are like adults but show a broken blackish brown slash along the folded wing. Legs are pink and pale.

BEHAVIOR: A social gull, often seen foraging in groups that may number from a half-dozen to several thousands. Often forages in flowing and choppy water. Nimble and active, Bonaparte's often hovers just above the chop or flies to the head of a confluent, where it lands and lets the current sweep it through food-rich water. On the water, it sits high and spins and picks at the surface phalarope-like. Commonly forages close to shore, but concentrations are also found well offshore, where upwellings occur. Immature birds, in particular, find sewage treatment facilities irresistible. Walks quickly on land, with a slight waddle. Perches easily on slender branches when nesting. Often roosts in tight flocks on land or water during migration and in winter. Fairly quiet (in winter), but not mute. A tame bird that often allows close approach.

FLIGHT: Short-bodied with long, fairly straight, triangular wings. On adults, the large, white, narrow triangular wedge along the leading edge of the outer wing flashes conspicuously. On immatures, which also show a blackish slash on the upperwing and a black-tipped tail, the white wedge is muted and more constricted but still shows.

The patterned upperparts of immatures are conspicuous but not overwhelming. (In other words, don't confuse this plumage with the bolder, blacker patterns of immature Black-legged Kittiwake or Little Gull.) Keep two key points in mind: The bill is petite and entirely black, and the underwings (except for the narrow black trim on the trailing edge) are wholly whitish.

VOCALIZATIONS: Call is a short, flat, nasal protest: "*yaah.*"

PERTINENT PARTICULARS: Flocks of Bonaparte's Gulls are the fertile ground from which sightings of Little Gull and Black-headed Gull spring. The key to finding these much less common species among Bonaparte's is mindfulness. When birds are grouped on land, scrutinize individuals that are standing slightly apart, often at the ends or edges of a flock. Black-headed appears bulkier and stands taller than Bonaparte's. Little Gull shows a darker, fuller black cap (in winter).

Heermann's Gull, *Larus heermanni*

Lipstick-billed Gull

STATUS: Common nonbreeding West Coast species. **DISTRIBUTION:** Breeds south of the U.S. in coastal Mexican waters. Disperses north after breeding (nests March to July), beginning in June along the Calif. coast and regularly north to s. B.C. **HABITAT:** Almost wholly found in coastal marine environments—beaches, rocky coasts, harbors, bays, kelp beds—but also ranges well offshore. Very rare inland. **COHABITANTS:** Inshore, forages and loafs with other gulls, particularly Western and Glaucous-winged Gulls; offshore, joins Brown Pelican, shearwaters, and gulls in pursuit of schooling herring. **MOVEMENT/MIGRATION:** Found in coastal California nearly year-round (uncommon March to late May). Most common off the coast of British Columbia in July and August, with some individuals lingering into November. VI: 2.

DESCRIPTION: A striking dark-bodied gull with an oversized (and garishly applied) lipstick-colored bill. Slender and medium-sized (larger than Ring-billed; smaller than California; about the same size as Laughing Gull), with a long, somewhat droopy, bright red-orange bill (adults only) that appears too large for the small head and gives the bird a long-faced appearance.

In all plumages, distinctly dark-bodied. Adults are a distinguished gray—dark gray above, slightly paler gray below. Breeding birds show brilliant white heads. Immatures are all dark brown, with darker heads; they look as though they were cast from pure, smooth, dark chocolate. The darker plumage, dark legs, and lack of conspicuous patterning distinguish Heermann's immatures from other immature gulls. The darker head and long, drooping, pink-based bill distinguish them from jaegers and shearwaters. Subadults are like immatures but dark gray-brown, with a red bill.

BEHAVIOR: An active, nimble, aggressive feeder. Commonly forages well offshore in flocks hovering over schools of baitfish. Also concentrates its feeding efforts around kelp beds and shorelines. Swims and plucks food from the surface or lofts into the air and dives to capture submerged prey. Also plunge-dives while flying or hovering, and snatches fish from cresting breakers. Commonly pirates prey, jaeger fashion, from other birds, and attends concentrations of cormorants, pelicans, and marine mammals to snatch fish.

A very social gull, Heermann's gathers in roosts that may number several hundred birds, though more commonly 20–30. Tame, and where habituated to people who like to feed gulls, very tame.

FLIGHT: Overall stocky profile with broad pointy wings and a short tail. Adults are mostly gray (except the white head in breeding plumage), with a distinctly black, white-tipped tail. Immatures (and subadults) are all dark, showing no white patches in the wings. Flies with wings held in an exaggerated tucked bow, with fairly heavy wingbeats.

VOCALIZATIONS: Away from breeding grounds, generally less vocal than other gulls.

PERTINENT PARTICULARS: It might seem hardly worth mentioning on so distinctive a bird, but the legs in all age classes of Heermann's Gull are black, a characteristic it shares only with Black-legged Kittiwake. The legs of juvenile jaegers, which may show wholly dark bodies and may sometimes stand on beaches, are pale gray.

Also, the large white patch on the outer wing of a very small percentage of adult Heermann's Gulls may cause observers to conclude that they are looking at a jaeger.

Mew Gull, *Larus canus*

A Gentle, Dark-eyed Gull

STATUS: Common northwestern breeder; common West Coast winter resident. **DISTRIBUTION:** Breeds across all but n. Alaska, the Yukon, n. B.C., and into the Northwest Territories and n. Sask. Coastal year-round resident from Kodiak I. south to s. B.C., and a winter resident south to cen. Baja (somewhat scarce south of Los Angeles). Away from the coast in winter, occurs locally (and rarely) to w. Mont., e. Ore., and interior portions of Calif. Very rare or casual visitor farther east. **HABITAT:** Eclectic in both breeding and wintering. Nests in marine and freshwater habitats, including tundra, marshes, ponds, lakes, streams, rivers, cliffs, and meadows. Post-breeding and in winter, found in nearshore beaches, open ocean, estuaries, lagoons, tidal flats, harbors, sewage treatment facilities, pastures, plowed fields, rivers, lakes, and garbage dumps. Not generally found out of sight of land. **COHABITANTS:** In summer, often found with Arctic Tern; in winter, most often associates with California and Heermann's Gulls and terns. **MOVEMENT/MIGRATION:** Nearly complete and primarily coastal migrant. Spring migration from early March to mid-May; fall from early August to early December. VI: 3.

DESCRIPTION: A pleasingly proportioned, kittiwake-like gull with a gentle brown-eyed expression. Medium-sized (smaller than Ring-billed or California; larger than Bonaparte's), but small for a "white-headed" gull, with a proportionally large round head, a pointy petite, almost thrushlike bill, a plump chestiness, and long wings. Overall profile is long, slender, and horizontal, with legs set well forward. Ring-billed Gull is slightly larger, stockier, and more erect, with legs set closer to the middle of the bird. Also, Ring-billed's bill is conspicuously thicker, stubbier, and more blunt-tipped, and its yellows eyes seem to squint, making it look sly, not gentle. California Gull is larger and more angular, with a longer pencil stub for a bill.

Breeding adults show a dark gray back (darker than Ring-billed) and a wide white-feathered gap (or weld) between the gray of the folded wing and the black wingtips. The white weld on Ring-billed's wing is narrow to nearly absent. In winter, Mew Gull is dirtier-headed, showing, in particular, a dark hindneck or a shadowy collar.

Juvenile and first-winter Mew Gulls are overall browner, plainer (particularly plain-faced), and less contrastingly patterned than Ring-billed, and are also conspicuously lacking in black and white—what is black and white in Ring-billed is brown and tan in Mew. First-winter birds (even subadults) show a dark smudge on the belly not found on Ring-billed. First-summer birds show much pale wear but retain their warm brown tones and show no black (particularly in the folded wingtips). Subadults (second-winter) are like adults but may show a ring around the bill and a dark tip to the tail.

BEHAVIOR: A very social bird, often loafing on beaches or foraging in fields together. Uses several foraging techniques. Sometimes "flutters" low and slow over the water, with head down and legs dangling, plucking food from the surface. Also sits high on the water, buoyantly spinning in the manner of a Bonaparte's Gull, or swims against the current and snaps up edibles as they pass. Commonly flies low along cresting waves, gaining lift, and loves kelp beds. In summer in the northern interior, often seen perched at the tops of spruce trees.

Fairly tame, often allowing close approach. Roosting birds often segregate themselves apart from larger gulls, but mix with California and Ring-billed Gulls.

FLIGHT: In overall proportions, somewhere between Ring-billed Gull and kittiwake. Appears long-tailed, with moderately slender wings. Adults show lavish amounts of white spotting in the black wingtips. Winter adults and subadults appear hooded, owing to the extensive tan-brown mottling; immatures are overall brownish gray, with a subdued pattern on the upperparts (not the contrasting black, white, and gray of immature Ring-billed). Flight is buoyant, nimble, playful; wingbeats are quick but light and caressing.

VOCALIZATIONS: Cry is a high petulant-sounding yip that is breathier and more squealing than Ring-billed.

Ring-billed Gull, *Larus delawarensis*

Parking Lot Gull

STATUS: Common to abundant and widespread northern and principally inland breeder; common, primarily coastal winter resident (but also found inland). Across much of North America, this is the white-headed gull most likely to be encountered inland. **DISTRIBUTION:** Breeds across s. and cen. Canada from cen. B.C. and the s. Northwest Territories east to James Bay, then south along the Ont.–Que. border and north along coastal Que. to e. Lab. and Nfld. In the U.S., breeds in Wash., Ore., ne. Calif., Idaho, n. Nev., Mont., Wyo., N.D., n. S.D., all but s. Minn., Wisc., Mich., and along the southern border of the Great Lakes and L. Champlain. In winter, except for the Great Lakes and along the Columbia R., vacates breeding areas. Winters along seacoasts and through the coastal plain from s. Maine to n. Mexico; from B.C. to cen. Mexico; and up the Colorado R. to Utah, N.M., and much of the s.-cen. and se. U.S.—s. Ky., Ill., Tenn., Mo., s. Neb., e. Colo., and the Great Salt Lake—to just south of the Mexican border. A few wander north in late summer to se. Alaska. **HABITAT:** Breeds primarily inland on low sparsely vegetated islands or peninsulas extending into lakes, sometimes along rivers, and also along the coast of northeastern Canada. Winters mostly along seacoasts, but uncommon beyond sight of land. Also found on freshwater lakes, rivers, reservoirs, and landfills and in mall parking lots and cities. **COHABITANTS:** Associates with other gull species, particularly California, Herring, and Laughing. **MOVEMENT/MIGRATION:** Migrates along both seacoasts and inland across all of the United States and southern Canada, often following rivers. (Early-) spring migration from February to April; (very protracted) fall migration from late June to late December. VI: 2.

DESCRIPTION: Handsome, medium-sized, white-headed gull. The body is sturdy and chesty but not portly, recalling Herring Gull, but conspicuously smaller and more elegantly proportioned. The head is proportionally large and round, and the bill straight and fairly stout, with no bulge at the tip. Compared to Mew Gull, Ring-billed's bill seems thick and blunt. The legs are set slightly forward, giving the bird an upright posture and making it appear modestly long in the rear. Does not appear angular or gangly.

Adult and second-winter plumages are essentially similar on this three-year gull, and recall adult Herring Gull: a pale gray back and wings, a white head, white underparts, and black wingtips bedecked with small white spots. Unlike Herring Gull, Ring-billed's legs and bill are yellow or slightly greenish (not pink). The namesake black ring encircling the bill is narrow and well defined. (The black ring on the bill of third-winter Herring Gull is wider.) In winter, the brownish streaking on the head and hindneck is generally sparse and sprinkled, not splotchy or coarsely applied (like Herring and Mew Gulls). Overall, feathers appear loose or layered. The expression is sometimes nefarious, sometimes shy or sleepy.

First-winter plumage seems half-adult, half-immature—and wholly uncoordinated. The combination of a mostly pale grayish back, large patch-patterned wings, and dirty streaking on the head closely resembles the plumage of a second-winter Herring Gull, but the smaller bill is pinker and the wing coverts are darker. (And of course, the head and bill are shaped different.) The underparts are white with some sparse dark barring or spotting below the wing. See Mew Gull account for comparison with this species.

BEHAVIOR: Commonly found inland, far from coastal beaches, particularly beaches dominated by larger wintering gulls. Often seen foraging in plowed fields and golf courses or standing atop streetlights illuminating parking lots for shopping malls and fast-food restaurants. Also found in reservoirs, landfills, and parks with lakes. In coastal areas, forages in marshes, mud flats, tidal creeks, coastal lakes, and beaches.

Social and not quite as quarrelsome as many other gulls. Often gathers in large numbers to exploit an opportune food source (like a recently plowed field rich in earthworms). Active and mobile. Searches for food by walking with quick mincing steps. Also sits on the ground or atop an elevated perch, scans for a food opportunity (for example, an immature bird with a smelt or a dropped bag of French fries), then flies to grab it.

FLIGHT: Balanced profile with equally projecting head and tail. A smallish gull with a big gull shape. Flight is light, buoyant, and more nimble than larger white-headed gulls. Wingbeats are quicker and choppier. Rises easily from the water and lands with only a few running steps, if any.

VOCALIZATIONS: Cry is like Herring Gull but much higher-pitched and more petulant-sounding. Begins with a breathing upsliding squeal followed by a short series of notes: "*eeYAH yah yah yah yah.*"

California Gull, *Larus californicus*

Teenage Herring Gull

STATUS: Common breeder in the intermountain West and northern Great Plains; in winter, common along the West Coast and parts of its U.S. breeding range. **DISTRIBUTION:** Breeds from interior portions of the Northwest Territories south through e. Alta., all but ne. Sask., sw. Man., e. Wash., cen. Ore., ne. Calif., w. Nev., most of Idaho, Mont., N.D., ne. S.D., n. Utah, and portions of Wyo. and Colo. Winters coastally and in the adjacent interior from extreme s. B.C. to the central coast of Mexico; also in w. Wash., s. Idaho, and pockets elsewhere in its breeding range. Regular migrant in se. Alaska, n. N.M., and w. Ariz. Rare to casual visitor farther east. **HABITAT:** Nests on islands in rivers or fresh or salt lakes; forages in a variety of dry, wet, and generally open habitats, often well away from colonies. In winter, concentrated in marine habitats such as beaches, estuaries, and rocky coasts, as well as rivers, streams, and the habitats it frequents in summer—such as farmland, garbage dumps, marshes, and sewage treatment facilities. **COHABITANTS:** Where ranges overlap, commonly nests with Ring-billed Gull; on seacoasts, associates with Mew, Ring-billed, Heermann's, Western, Glaucous-winged, and Herring Gulls. **MOVEMENT/MIGRATION:** Most birds leave interior breeding areas in early to mid-July and migrate to the Pacific Coast, returning between late February and early to mid-May (with peak movements between March and late April). During the nonbreeding season there is a widespread, coastal movement involving large numbers of birds that first fly northward in early fall then south in late fall and winter. VI: 3.

DESCRIPTION: Looks like a long, slender, angular, somewhat awkwardly proportioned teenage Herring Gull with a narrow straight bill that seems too long and too thin for the face. Medium-sized (larger than Ring-billed or Heermann's Gull; smaller than Herring and Western Gull), with an undersized, round-topped (or slightly peaked), and slender head, a conspicuously long, narrow, straight bill that is barely swollen at the tip, a long neck, a slender body, and a long-winged profile.

Adult and subadult plumage is like Herring or Ring-billed, but with slightly darker upperparts. In winter, shows a very dark mottled hood and a smudgy patch/line through and behind the eye. Adults' greenish yellow legs and oversized red/black spot near the tip of the bill are easy to see at close range. The dark eye distinguishes adults and subadults from yellow-eyed Herring or Ring-billed. Also, the legs and base of the bill of subadult California Gull are a unique blue-gray-green. The legs of comparably plumaged Herring and Ring-billed Gulls are pink.

Juveniles and immatures are dark brown with paler heads and faces and a more contrasting overall pattern than Herring Gull. They show a pink base to the black bill, but the bill and head shape—a round crowned head with a long, narrow, straight bill versus a long-snouted sloping face profile and bill that thickens at the tip (Herring) or a very round head with a short thick stub of a bill (Ring-billed)—is often easier to note.

BEHAVIOR: Like other white-headed gulls. Very social. Nests in colonies, and forages alone or in large well-spaced groups. Roosts with other gulls (especially Mew Gull on seacoasts and Heermann's Gull in California). Commonly flies great distances from nest sites to forage. Uses a variety of foraging strategies and is adept at snatching prey from the air. Also runs with head down and bill agape through clouds of brine flies, inhaling mouthfuls.

FLIGHT: Appears short-tailed, long-faced, and slender-winged. Flight is stiff but nimble, with somewhat deep wingbeats.

VOCALIZATIONS: Cry is a rapid shrill bleat that is higher-pitched and squeakier than Herring Gull. Also emits a rough, low "*ah ah ah ah,*" like a distant or muffled Great Black-backed Gull.

PERTINENT PARTICULARS: Although California Gull is not the only gull that nests on interior lakes, owing to its range (which encompasses many of North America's popular summer retreats), California and Ring-billed are the gulls most likely to be encountered.

Herring Gull, *Larus argentatus*
The Protogull

STATUS: Circumpolar breeder. In North America, common and widespread, primarily northern breeder. In winter, common coastal resident, particularly in the East, where its numbers greatly exceed all other large gulls; also winters in ice-free portions of the interior United States. **DISTRIBUTION:** In North America, breeds across most of Canada north of a line drawn from s. Maine across n. N.Y. and along the south shore of the Great Lakes, then north and west to the Yukon and northern and central portions of B.C., then west into interior and western coastal portions of Alaska. Also breeds along the Atlantic Coast south to N.C., and locally even farther south. In winter, vacates virtually all northern and interior portions of its breeding range. Winters coastally from Nfld. to Central America, and from the Aleutians to the central coast of Mexico. In the interior, winters on the Great Lakes, through much of the Mississippi R. basin, on large lakes of the Great Plains and interior West, and through the Columbia R. drainage to Idaho. On the East Coast and the Great Lakes, large numbers of nonbreeding subadults remain south of breeding areas throughout the summer. **HABITAT:** Nests on islands and in coastal and freshwater habitats, particularly rocky islands but also well-drained marsh hummocks, sandy beaches, and artificial spoil islands. In winter, most common near large bodies of water, including lakes and rivers; also found on open agricultural land, parking lots, and (particularly) open landfills. Also winters at sea, but generally not more than 50 mi. from coasts. **COHABITANTS:** In winter, associates with other large gulls, including California, Great Black-backed, Glaucous-winged, and Western. **MOVEMENT/MIGRATION:** Absent from northern and interior breeding areas from November to mid-March. In the East, large offshore movements are noted in November and December. VI: 4.

DESCRIPTION: Large gull that is highly variable in structure, plumage, and size (but smaller than Great Black-backed Gull and most Glaucous Gulls). A four-year gull whose first and second years are fundamentally similar except for a dark ring around the tip of the pink bill in the second year. (The all-dark first-year bill becomes pink-based in spring.) The third-year bird is much like the adult—or like a big pink-legged Ring-billed Gull insofar as it still has a ring near the tip of the bill.

In sum, there are only two plumages to deal with—immature and adult.

Overall somewhat gangly and angular, particularly about the head and face. Although variable, Herring heads are generally long and narrow (wedge-faced or long-snouted), with a peak to the rear and a long sloping profile that tapers into the longish and fairly narrow bill. The gonydeal angle is large and acute, and the bill is long and almost raptorially hooked. The heads of most Herring Gulls, especially males and immatures, look gaunt, shrunken, even skeletal—and if any gull deserves to be called ugly, this one is it.

Heads are often couched, but when extended, the neck too is constricted near the head. Legs are positioned just about the middle of the body.

Breeding adults are classically plumaged: pale gray above, with a white head/neck and white underparts featuring a prominent row of white dots on the black wingtips that project beyond the white tail. In winter, heads, necks, and sometimes upper breasts become diffusely mottled—often heavily so—without symmetry or pattern, and the birds take on a disheveled and unkempt look. The shabbiest-looking gull on a winter beach is usually Herring Gull.

Immature (first- and second-year) birds are basically all brown—some are darker (coffee), and some are lighter (mocha). At close range, underparts may seem somewhat patchy or patterned but not streaked. In the East, a large entirely dark brown gull is almost certainly a Herring Gull. In the West, first-winter Western Gull, California Gull, and perhaps a darker Thayer's Gull share this pattern.

Herring Gull's bill is long, moderately narrow, and thickened at the tip. First-winter birds have black bills, but usually with a pale pinkish base (particularly later in the winter). Second-winter birds have a two-toned bill—pink with a dark tip. By comparison, the bill of first-winter Thayer's Gull is all black, straight, and more petite, and its head is rounder. The bill of first-winter Western Gull is short, heavy, and black, with pink only on the base of the lower mandible. (Second-winter Thayer's and Western, with a mix of adult and immature plumage, cannot be confused with all-brown first- and second-winter Herring Gull.)

First-winter California Gull is smaller, more slender, slimmer-billed, and overall more contrastingly, less uniformly patterned, showing a whiter head and face, fading underparts, bold contrast on the folded wing between the paler shoulder and the much darker greater coverts, and even darker tertials.

Adult Herring's expression is angry, crazed; immature's is benign.

BEHAVIOR: Differs little from other large gulls. Highly social. Forages and loafs with other Herring Gulls of differing age groups as well as other gull species.

FLIGHT: Straight and direct. Wings are fairly narrow. Wingbeats are steady, regular, shallow, and languid. In winter, more heavily mottled adults often seem hooded. The upperwings of first- and (particularly) second-winter birds flash pale inner primaries. Birds moving to and from feeding areas often fly in V-formations. When flushed, birds often cluster and soar in large kettles (often at extremely high altitudes). These kettles are easily distinguished from kettles of hawks by the classic, sweptback, down-crooked, "gull-winged" configuration, as well as by the mosaic of plumages (brown birds mixed with pale birds).

VOCALIZATIONS: Cry is a loud, shrill, fairly high-pitched, down-scale keening—"*Keeya-ya-ya-ya-ya-ya.*" Also makes a shrill gargled "*Te-eee'r,*" sometimes with a trilled ending.

Thayer's Gull, *Larus thayeri*

Iceland West

STATUS: Canadian arctic nester. Uncommon winter resident along the West Coast; very uncommon to rare in winter on the Great Lakes and in the interior West and Great Plains; very rare in the East. **DISTRIBUTION:** Breeds in scattered colonies in the arctic islands of the Nunqvut and nw. Greenland. Range abuts and slightly overlaps with more eastern Iceland Gull. Winters primarily along the West Coast and shelf waters from se. Alaska to n. Baja, but more common from n. Calif. north. **HABITAT:** Breeds on coastal rocky cliffs. In winter, found mostly in coastal areas and (less commonly) on rivers, reservoirs, and large lakes, as well as landfills, fish-processing plants, dams, and (occasionally) lawns and agricultural lands. Particularly attracted to landfills and pig farms. **COHABITANTS:** Other gulls, including California, Herring, Western, and Glaucous-winged. **MOVEMENT/MIGRATION:** Wintering birds are present on the West

Coast from September (in the north) through April. VI: 3.

DESCRIPTION: A medium-sized gull (slightly larger than California; slightly smaller than Herring; about the same size as Iceland), with the structure of Iceland and the plumage of Herring.

Adults are gray above (very slightly darker than Herring and Iceland Gulls) and white below, with black wingtips on the folded wing. Smaller than Herring, Thayer's also has a rounder head, a more petite bill, deeper pink legs, and (most of the time) a dark eye. In winter, the head and hindneck are usually heavily streaked (like Herring, but usually more so).

First-winter birds appear patently uniform and virtually devoid of contrasting pattern—almost a larval gull. Plumage ranges in color from uniformly tan to uniformly dark gray-brown, but the wingtips and tertials are always slightly darker than the rest of the bird. On first-winter Iceland, Glaucous, and Glaucous-winged Gulls, the wingtips and tertials are the same pale color as the body. On close examination, the upperparts on Thayer's have a very fine marbled pattern. The head, neck, and underparts are clean, patternless, and creamy smooth. By comparison, first-winter California, Herring, and Western Gulls are overall darker (often much darker) and more coarsely patterned, with wingtips and tertials that are much darker (not somewhat darker) than the body. Also, Thayer's Gull looks overall neater, trimmer, cleaner, even preppier—not as rough or coarsely plumaged as most large gulls on western beaches.

Second-winter Thayer's is, again, like second-winter Herring, but wingtips are tan or brown (not black, as on Herring) and, again, note the finer Iceland Gull-like features.

BEHAVIOR: Like other gulls.

FLIGHT: Wingtips on adults appear blackish from above, pale below. On Herring Gull, wingtips are black above and below; on Iceland, wingtips are mostly white above and below, although many northeastern birds show a limited touch of dark gray (not black) near the tip that is less extensive than Thayer's. First-winter Thayer's is overall pale and fairly uniform except for the broad dark outer half of the tail.

VOCALIZATIONS: A very quiet gull (unlike Herring); vocalizations are similar but lower-pitched, more petulant sounding than lonely or keening.

Iceland Gull, *Larus glaucoides*

The Petite White-winged

STATUS: Common but restricted breeding gull of the northeastern Canadian arctic. Uncommon wintering species south of Canada. **DISTRIBUTION:** Breeds on coastal cliffs in Greenland and extreme n. Canada. Winters along the Atlantic Coast, the Great Lakes, and larger rivers from Lab. to New England; also found in reduced numbers to N.C. and inland to the Ohio and Tennessee River Valleys. **HABITAT:** In winter, commonly found in harbors and inlets; also on open ocean waters and landfills, reservoirs, and dams. **COHABITANTS:** Associates with other large gulls. **MOVEMENT/MIGRATION:** Spring migration from March to early May; fall from October to December (but birds continue to shift south as harsh weather conditions and food availability dictate). VI: 3.

DESCRIPTION: A medium-sized white-headed gull (between Herring and Ring-billed in size), with compact features but overall a nicely proportioned, even delicate shape. Decidedly cub-headed (that is, head is small, round, and wide), with a small, thin, straight bill and a short thick neck. Folded wings are long (projecting the length of the bird's bill beyond the tip of the tail), giving the bird a sense of slender elegance. When standing, seems to be leaning back on its short bright pink legs. (Herring Gull legs are longer, not so pink, and the stance is more vertical.)

Iceland is a four-year gull, but adults and third-winters are similar; juveniles and first-winters are distinctive; second-winter birds combine adult and juvenile traits but are also distinctive.

Adults are overall pale with a white head, neck, and underparts and pale gray wings and back. Whitish folded wingtips are variably barred or banded—in light to medium gray (grayer than the mantle but not black). In winter, head is finely mottled with gray but not streaked. The bill is bright yellow, and the eye dull yellow (often appears dark at a distance). The expression is sleepy, coy, serene.

Juveniles and first-winter birds are overall pale, with little contrast between upper- and lower parts. Most are creamy buff with darker brownish speckling or pale barring above; some are browner but still show little overall contrast—particularly between the pale wingtips and the body. By summer, bodies are almost pure white. The bills of first-winter birds are mostly all black—a standout feature against the pale body. The eye is dark brown, and the expression is gentle.

Second-winter birds have whitish bodies (like a very pale first-winter bird), pale gray mantles (like adults), darkish primaries contrasting with the paler wing (like Thayer's), and a pinkish, gray-based, dark-tipped bill (suggestive of first- and second-winter Glaucous Gull).

BEHAVIOR: Calm, not particularly aggressive gull; generally subordinate to larger gulls in the flock.

FLIGHT: Overall compact (relative to Herring Gull), with a shorter head and wider wings (particularly at the base), but a longish tail. Young birds are mostly uniformly pale. Adults are uniformly white below (no black wingtips) and above show only touches of dark gray near the tips. Flight is more nimble than Herring Gull, with quicker wingbeats, but since wingbeat tempo relates to weather conditions and a bird's objectives, the difference is often only apparent in direct comparison.

VOCALIZATIONS: Like Herring Gull, but shriller. In winter, virtually mute.

PERTINENT PARTICULARS: Among winter gull concentrations, first-year and adult birds far outnumber other age groups. Second-winter Iceland is relatively uncommon but easily detected among gull flocks dominated by Herring Gull by its overall paleness, and distinguished from Glaucous Gull by its smaller size and petite bill, which is dull pink only on the basal half (not bubblegum pink with only a dark tip, like first- and second-winter Glaucous).

Lesser Black-backed Gull, *Larus fuscus*

Dirty-necked Black-backed Gull

STATUS: Scarce to uncommon, mostly East Coast winter resident; largest numbers occur in the mid-Atlantic region. Rare to very rare inland to the Rockies and along the Gulf Coast and in summer along the East Coast. Casual in the West. **DISTRIBUTION:** Breeds in northern Europe and Iceland; winters in increasing numbers along the Atlantic Coast from the Maritimes (few) to n. Mexico; also occurs on the Great Lakes and (somewhat pelagic) is often found well offshore. **HABITAT:** Like other large white-headed gulls, winters primarily along coastlines as well as on large inland lakes and rivers where a surfeit of food is found (fishing harbors, landfills). **COHABITANTS:** Other large gulls, particularly Herring Gull. **MOVEMENT/MIGRATION:** Clearly migratory, but this aspect is little studied. Although some second- and third-year birds (and a few adults) summer in North America, most wintering adults are gone from May through August. In North America, greatest numbers occur from November through February. VI: 3.

DESCRIPTION: A well-proportioned but decidedly long-winged white-headed gull. Small (much smaller than Great Black-backed; slightly smaller than Herring Gull), with a fairly sleek body, clearly not as robust as Great Black-backed. Wings extend well beyond the tail, giving the bird a long profile. A somewhat domed or rounded head is fitted with a short thick-tipped bill that is not as heavy and massive as Great Black-backed.

Adults look like paler miniature Great Black-backeds; first-winter birds resemble streaky colder brown Herring Gulls. Adult and third-winter Lesser Black-backeds have a slate-colored (not black) back that may be only slightly darker than adult Herring Gull but usually falls midway between Herring Gull gray and Great Black-backed black. The legs are bright yellow on summer adults, duller in winter (pink on Great Black-backed), and in winter the streaking on the head, face, and even the neck and sides of the breast is extensive. Great Black-backed is distinctly white-headed in winter, showing scant, if any, streaking. The streaking on

Herring Gull is variable but normally coarser than Lesser Black-backed.

If it's winter, you're in Massachusetts, and you see a Black-backed Gull wearing a dirty disheveled hood, it's a Lesser.

First-winter Lesser Black-backed is overall a colder, grayer brown than Herring, and more mottled and streaked, particularly on the underparts. The head is slightly paler than the body, and an obvious dark shadow through the eye makes the bird look like it's in need of sleep. (Herring's face is uniform and the same color as the rest of the bird.) The bill on Lesser is all black. By late winter and in summer, Herring's bill gets pink at the base. Lesser Black-backed's bill remains wholly black into the second winter.

Second-winter Lesser Black-backed is already donning dark gray adult feathers on the mantle. Second-winter Herring still shows a brown back; third-winter Herring shows the pale gray (not dark gray) mantle of the adult bird.

Lesser Black-backed's expression, marked in all winter plumages by the shadow around the eyes, is heavy, tired, cross, out-of-sorts.

BEHAVIOR: Like other large gulls. Spends much time loafing on shores, on ice floes, or around food sources. May tend to loaf among Herrings rather than Great Black-backeds, an association that makes detection of adults easy and enables Lesser Black-backeds to cluster in close proximity.

FLIGHT: Usually appear longer and more slender-winged than Herring or (particularly) Great Black-backed. Very dirty-headed adults stand out in flocks. First- and (particularly) second-winter birds have very pale, almost white rumps that contrast markedly with the dark back and fairly uniformly dark wings.

VOCALIZATIONS: Calls are similar to Herring Gull's, but a Herring Gull with a head cold.

Yellow-footed Gull, *Larus livens*

Salton Sea Black-backed Gull

STATUS AND DISTRIBUTION: Resident in the Sea of Cortez in Mexico. Uncommon to fairly common visitor to the Salton Sea in e. Calif. between June and October, with numbers peaking in late summer. Rare but regular in winter. **HABITAT:** The barren shoreline and rocky outcroppings of the Salton Sea. Occasionally forages on agricultural fields. **COHABITANTS:** Ring-billed Gull, California Gull, Herring Gull. VI: 1

DESCRIPTION: Why labor the obvious? This is the only large blackish-backed gull likely to be encountered on the Salton Sea. (The very similar Western Gull is rarely seen there.) The bulkily proportioned adult looks like the heavier-billed Western with bright yellow legs (pink on Western). First-winter birds resemble second-winter Westerns, being dark gray above (already!), but they show a whiter head, underparts, and rump than the all-brown first-winter Western and Herring Gulls (the only other large gull species commonly found on the Salton Sea). By first summer, birds show yellow on the legs.

BEHAVIOR: Much like other large gulls.

FLIGHT: Like Western Gull. First-winter birds show a very white rump. Second-winter birds are whiter-headed and more cleanly contrasting than subadult Herring and Western Gulls.

VOCALIZATIONS: Like Western Gull, but lower-pitched.

PERTINENT PARTICULARS: While it is possible to find this species in winter, you are much more likely to be successful in August and September, when numbers peak. Be warned: Summer daytime temperatures well in excess of 100° and high humidity conspire to make a visit to the Salton Sea a miserable experience. Bird early in the day, then seek shelter in an air-conditioned room.

Western Gull, *Larus occidentalis*

Pacific Black-backed Gull

STATUS: Common Pacific coastal resident. **DISTRIBUTION:** Breeds from the central coast of Wash. to cen. Baja. In winter, ranges north along the coast to se. B.C. and south to the tip of Baja; nonbreeding birds are found throughout the summer north to B.C. Rare but annual in the Sea of Cortez. **HABITAT:** Almost exclusively coastal, ranging seaward out to 25 mi. and, along rivers, several miles inland. Nests on islands and artificial structures (such as abandoned piers); roosts and feeds on beaches, landfills, and open fields. **COHABITANTS:** Nests in association with other coastal species, including cormorants, pelicans, and (in Oregon) Glaucous-winged Gull. In winter, mixes with other gull species. **MOVEMENT/MIGRATION:** Permanent resident throughout most of its range, but coastal expansion north and south occurs from October to April. VI: 1.

DESCRIPTION: The big, bulky, and only dark-backed gull found on the beaches of California and Oregon; virtually a fixture dockside and along beaches. A large, stocky, bordering on squat white-headed gull (same approximate size as Herring and Glaucous-winged Gulls), with a thick swollen-tipped bill, a rounded or flat-topped head, a long neck, a barrel-chested body, and fairly short wings and tail. Overall more robust, larger-headed, and thicker-billed than Herring Gull.

Adult upperparts are dark slate gray—not black, but distinctly darker than other West Coast gulls. In summer and winter, head remains white—no hood, no streaks. First-winter birds are overall smudgy, ugly, and grayish brown, showing a cold grayish cast to the smooth, plain, and unpatterned head and breast and thick all-black bill. Immature Herring Gull is overall warmer brown (lacks the gray cast to the head and breast), paler-headed, thinner-billed, and more mottled/marbled in body plumage. Second-summer Herring Gull becomes patchily white (it's particularly pale about the head) and has a two-toned bill—pale with a dark tip. Second-summer Western Gull remains overall darker, including the bill, which becomes pale only at the base. Adult slate gray back feathers are showing by the second winter and easily distinguish the bird from the much paler gray plumage of subadult Herring Gull.

BEHAVIOR: Like other large gulls—at home loafing on beaches or floating on ocean surfaces far from shore. Consumes a range of food items from carrion to live prey. In intertidal areas, commonly forages alone. At sea, gathers in widely spaced groups over and near prospective food sources (such as schooling fish or trawlers), and is often the first species to react to a promising food event.

FLIGHT: Large and compact profile with distinctly broad and shortish wings and a head and tail that project equally. Herring Gull is overall more gangly and slender-winged. Western is a strong flier, with fairly heavy and slow wingbeats.

VOCALIZATIONS: Similar to Herring Gull, but rougher and lower-pitched.

PERTINENT PARTICULARS: No other dark-backed gull species is likely to occur within the range of Western Gull. The simplified challenges presented by this species become separating immature Western from Herring Gull and from Western/Glaucous-winged hybrids, which are very common where breeding ranges overlap in southern British Columbia, Washington, and northern Oregon and are also scattered throughout wintering populations. Immature California Gull is easily distinguished from Western by its smaller size, slender build, and conspicuously narrow (not thick) bill.

Adult Western/Glaucous-winged hybrids are slightly darker gray above than adult Herring Gull but have the heavy bill and body structure of the parents. In winter, the heads are darkly streaked. (Western Gull has a white head.) Immature hybrids are overall a pale brown mocha color, with wingtips a shade or two darker than the body. Immature Glaucous-winged is overall pale gray-brown, with wingtips the same color as the body—a feature that also helps distinguish hybrids from the overall more slender and delicately featured Thayer's Gull.

Glaucous-winged Gull, *Larus glaucescens*

Silver-backed Gull

STATUS: Common West Coast breeder and winter resident. In summer, across most of the breeding range, this is the default large gull. **DISTRIBUTION:** Breeds coastally from the west coast of Alaska (south side of Norton Sound and St. Matthews I.), including offshore islands, to ne. Ore. In sw. Alaska, also breeds away from coasts on inland lakes. Winters coastally from sw. Alaska to Baja and the northern Sea of Cortez. **HABITAT:** Year-round, found on or near coastal waters (rarely occurs on fresh water, but regular inland up the Columbia River). Favored habitats include bays, inlets, harbors, estuaries, beaches, mud flats, and (particularly in winter) offshore waters (commonly around commercial fishing boats). Also found increasingly in coastal cities, parks, and landfills. **COHABITANTS:** During breeding, Bald Eagle, Black-legged Kittiwake, Northwestern Crow. In winter, Glaucous Gull in Alaska and Canada; in the lower 48 states, Western, California, and Herring Gulls; Thayer's Gull. **MOVEMENT/MIGRATION:** Although many northern birds are resident, others move south. Spring migration from late February to early May; fall from late August to late November. Birds are present in significant numbers in California from October to March. VI: 2.

DESCRIPTION: The big silver-gray gull with wingtips the same color as the back. Large and overall stocky (slightly larger than Herring Gull; slightly smaller than Glaucous Gull; the same size and much the same shape as Western Gull), with a big wide head, a fairly short thick-tipped bill, a heavy, chesty, compact body, and pink legs (dull gray in younger birds). Distinctly more robust and thicker-billed than Herring Gull and more compact than Glaucous Gull.

Upperparts of adults are silver gray (about the same tone as adult Herring Gull; much darker than Glaucous Gull). First-winter birds are wholly and smoothly plain tan or brownish gray (tarnished silver), showing little or no pattern, and are overall a colder gray tone than first-winter Glaucous. In both cases, *the wingtips are the same color as the upperparts*. In other large gulls, the wingtips contrast. Glaucous Gull shows white wingtips; Herring, Western, and Thayer's show black or (in immatures) darker wingtips. The eye of adults is dark (light on adult Herring Gull and Glaucous Gull). The bill of first-winter Glaucous-winged Gull is black, and the contrast with the pale head is marked. First-winter Glaucous Gull has a pink bill with a dark tip. First-winter Herring Gull has a dark bill (becoming pale-based by midwinter), but the diminished contrast between the bill and the darker head/body won't catch your eye. Also, the expression of first-winter Glaucous-wingeds is dull or dazed (in other words, not too bright). Second-winter birds look like first-winter birds but show some silver-gray in the back. Third-winter birds look like adults with black-tipped bills.

BEHAVIOR: Like other large gulls. Adults are commonly found in pairs year-round, but birds also forage alone and among other large gulls.

FLIGHT: Overall fairly compact, with wide-based wings that taper acutely to a pointed tip (adult); young birds appear rangier and longer-winged. Overall pale gray above, with a white head and tail. The adult's dusky wingtip (or darker trailing edge to the tip) is evident at close range but not at a distance—wings appear uniformly pale gray above. However, the trailing edge of the underwing is gray, and the leading edge contrastingly white (a pattern like a pale adult Western Gull). On Glaucous and Herring Gulls, the underwing shows little contrast, and what contrast there is shows just a opposite pattern—gray on the leading edge, white on the trailing. Flight is languid but strong, with slow shallow wingbeats.

VOCALIZATIONS: Similar to other large gulls. Lower-pitched and less strident than Herring Gull; more sonorous and not as coarse as Western Gull.

PERTINENT PARTICULARS: This species commonly hybridizes with Western Gull where their ranges overlap (also with Herring and Glaucous Gulls), and young can have traits of both parents. For example, first-winter Glaucous-winged/Glaucous hybrids have blocky bodies but pink-based bills. Birds may also have characteristics that fall between (such as medium-gray bodies with slightly darker gray wingtips for adult Western/Glaucous-winged hybrids). If you find a gull with a mixed or confusing array of traits, consider the very real possibility that it's a hybrid. If you like challenges, sleuthing a confusing gull's parentage is the game for you. If not, there is no harm or shame in just moving your spotting scope to the next gull.

Glaucous Gull, *Larus hyperboreus*

The Great White Gull

STATUS: Common arctic nesting species; the only large breeding gull with a circumpolar distribution; in winter, uncommon to rare south of Canada. **DISTRIBUTION:** In North America, breeds close to and above the Arctic Circle from Greenland and Lab. (also the east side of Hudson Bay) across northern coastal Canada to the northern and western coasts of Alaska. Winters for the most part south of its breeding areas; found from s. Greenland and the Hudson Straight south to Va. (occasionally to Fla.); in the Pacific, from cen. Bering Sea to n. Calif. (occasionally farther south). **HABITAT:** Most often found in marine environments. Breeds on rocky cliffs, islands, or open tundra not far from water. Winters along seacoasts, large freshwater lakes and rivers, and urban landfills, as well as offshore to the limit of the continental shelf. **COHABITANTS:** During breeding, Common Eider, Polar Bear; in winter, other large gulls. **MOVEMENT/MIGRATION:** Clearly migratory, but little information is available. In northern areas, movements are observed in September and October; the return to arctic breeding areas occurs in April and May. In southern reaches of the wintering range, most movement, involving mostly non-adult birds, occurs from November to March. VI: 3.

DESCRIPTION: A big, bulky, white-winged gull that, except for juveniles, is overall pale onto white. Proportions are overall robust and awkwardly front-heavy—too big and heavy in the front, too cropped in the rear. The head is large, somewhat oval (definitely not round!), and sloping along the forehead. The bill is long, straight, and narrow without being thin, and without much bulge at the tip; the neck is heavy but long, the chest robust, and the legs long. On most birds, the rear seems comically abbreviated, the combination of a tail that's too short and wings that extend less than the length of the bill past the tip.

Glaucous is a four-year gull. Adults and third-winters are similar; first- and second-winters differ in color shade, not pattern, so are also similar.

Adults are exceedingly pale, with bright white heads, necks, and underparts and exceedingly pale blue-onto-white upperparts—noticeably paler than adult Iceland and appreciably paler than Glaucous-winged. At a distance, birds simply appear white—as many second-winter birds and some first-winters truly are. In winter, however, many adults have some dark streaking on the head and neck. The expression is haughty, guileful, predatory, nefarious.

First-winter birds range from overall creamy buff to nearly all white. In all cases, the back, wings (except the tips), and tail are usually etched with fine dark golden brown vermiculations (similar to, but usually not as pronounced as, first-winter Iceland Gull). Glaucous-winged Gull usually looks smudgier. First-winter Glaucous's expression is closeted, secretive, piggy-eyed.

The bill and the wingtips provide two keys to identification. In first- and second-winter birds, the bill is bubblegum pink, with a crisply defined black tip. Iceland's bill is all black in the first winter and bicolored in the second winter, but the pink is dull and the black "tip" may command half the bill. Glaucous-winged's bill is also black in the first and second winters, with just a small irregular pale area appearing at the base. In all plumages (except very fresh juvenile plumage), Glaucous's folded wingtips are not just pale and unpatterned—they are pure white! All Glaucous-wingeds (and almost all Icelands) have some patterning or darker shading on the wingtips.

BEHAVIOR: In winter, flocks with other larger gull species in places where food is available. In the north, also winters at sea, where birds are found either individually or, given an auspicious food source, in numbers that may exceed several thousand. In mixed flocks, often seems somewhat testy or dominant; perhaps as a result, birds often have more space around them than do their neighbors.

An efficient predator as well as forager, in summer Glaucous haunts nest ledges and predates upon chicks and smaller adult alcids. Hunts on foot and from the air.

FLIGHT: Flight silhouette retains the sense of bulk and front-heaviness. Also seems to have very broad wings with rounded or blunter tips. Flight is slow and heavy; wingbeats are fairly shallow on wings that may appear flatter and not as down-crooked as other large gulls.

VOCALIZATIONS: Like Herring Gull, but with slightly deeper, coarser notes.

PERTINENT PARTICULARS: The similarity between Glaucous Gull and Iceland Gull is only feather-deep. The birds may have similar plumage characteristics, but their shapes are very different—big and blocky (Glaucous) versus pigeon-headed and slender. Compare the shapes of Great Black-backed Gull and Mew Gull. Distinctive, right? The difference between Glaucous and Iceland is only slightly less pronounced.

Great Black-backed Gull, *Larus marinus*

The Beach Master

STATUS: Common, primarily northeastern gull (does not occur on the West Coast and is rare in the Gulf of Mexico). Recorded annually west to the western Great Plains; casual farther west. **DISTRIBUTION:** Breeds coastally from Lab. to Va. (with a few to N.C.) and along the St. Lawrence R. to Lake Ontario (and a few pairs west to Lake Michigan). Winters coastally from Nfld. to n. Fla., but also found away from the coasts of New England and the mid-Atlantic states and around the e. Great Lakes. Nonbreeding subadults remain in the winter range. **HABITAT:** Breeds on coastal rocky inlets, dredge-spoil islands in salt marsh, and barrier beach, often with Herring Gulls but prefers more elevated sites. Vacates extreme northern portions of its range during colder months; otherwise a permanent resident. Winters in coastal and inland habitats adjacent to open water, usually near towns, dams, or landfills. Also very common at sea, but generally not beyond the coastal shelf. **COHABITANTS:** Associates with other large northeastern gulls, but smaller gulls generally give it a wide berth. **MOVEMENT/MIGRATION:** Coastal movements noted from October to March. VI: 2.

DESCRIPTION: Large and hulking, North America's largest gull, easily distinguished by its superior size, thickset proportions, and (in adults) all-blackish back and wings. Except for the relatively uncommon Lesser Black-backed Gull, the only large dark-backed gull in the East.

Adults and immatures are larger, distinctly bulkier, and more curvaceous than Herring Gull, with large square or domed heads, heavy swollen-tipped bills, and long, thick, but sinuous necks. Feeding birds seem humpbacked. Great Black-backed never looks gaunt or reptilian, but it does seem formidable, and its movements have a slow slinking quality.

A four-year gull, but essentially only three plumages to deal with. Adults and third-winter birds have gleaming white heads and bodies contrasting with black backs and wings. In winter, heads remain white (or only slightly streaked). First-winter birds are striking and beautiful, with crisp and contrasting blackish and white marbling (or a salt-and-pepper pattern) above. (Herring Gull is mostly brown and hardly striking.) The distinctly whitish head makes the heavy black bill stand out. Second-winter birds are like first-winters except for a pink-based dark-tipped bill and a charcoal gray mantle that hearkens to the dark backs of adults. The expression is mean.

BEHAVIOR: More coastal and pelagic than Herring Gull; also more solitary and more predatory. Swims (or at low tide walks) across intertidal zones, dipping and sometimes diving or picking up marine organisms (such as crabs, clams, and starfish). Also kills water birds (to the size of coots and small ducks), alone or with other Great Black-backeds, by harassing them to exhaustion and then grasping and shaking them with its bill. Scavenges on refuse fish at landfills and other opportune food concentrations (for example, natural fish kills or storm-deposited clams on beaches), where it clearly dominates over Herring Gull. Parasitizes food from a number of birds, including other gulls, Osprey, and Bald Eagle.

FLIGHT: Flight has a gull-like grace but is heavier and more lumbering than Herring Gull, with distinctly broader wings and a broader, heavier, more projecting head. The combination of the black back and wings and the all-white head and tail on adults is hard to mistake. From below, the broad dark trailing edge to the white wings and body is also distinctive. On first- and second-winter birds, a mostly whitish body easily distinguishes them from uniformly brown Herring Gull.

VOCALIZATIONS: Cry is a deep, hoarse, resonant, and somewhat sinister "*haw, haw, haw, haw, haw.*" Slower than Herring Gull, and never shrill-sounding. Also makes a low gruff growl, "*glowl,*" that sounds like the bird is clearing its throat.

Sabine's Gull, *Xema sabini*
Tricolored Gull

STATUS: Coastal arctic and subarctic breeder; fairly common pelagic migrant off the West Coast; very rare along the East Coast and inland. **DISTRIBUTION:** Breeds coastally from the north side of the Alaska Peninsula north along coastal Alaska to the nw. Northwest Territories and a number of islands in the arctic island group, east to Greenland. **HABITAT:** Nests on swampy low-lying tundra, often near freshwater lakes or pools; also found in tidal sloughs and meadows and extensive wetlands dotted with ponds and lakes. **COHABITANTS:** Often nests with Arctic Tern, as well as Red-throated Loon, Tundra Swan, Red-necked Phalarope. In migration, may be found with Long-tailed Jaeger, Arctic Tern, and Bonaparte's Gull, and often attends seals, porpoises, and whales. **MOVEMENT/MIGRATION:** Whole population vacates breeding areas and winters well south of the United States. Most migration is conducted offshore, but a few birds (mostly juveniles) migrate over land in fall, across Canada and the United States (very few in spring). Spring migration from late April to early June; fall from late July to early November, with a peak between late August and early October. VI: 3.

DESCRIPTION: A small trim gull with a bold tricolored wing pattern (seen in flight). Adults appear overall crisply patterned and handsome; juveniles look gentle and delicate. Small (slightly larger than Bonaparte's; smaller than kittiwake or Mew Gull), with a petite bill (adult bills have a yellow tip that can be seen at close range), a small head, a long neck, and a long-bodied, long-winged profile. On land, appears very slender, with a horizontal posture.

Breeding adults have a dark gray hood with a narrow black collar, a gray back (slightly darker than Black-legged Kittiwake), and blackish wingtips showing white spots so conspicuous some individuals appear barred. Nonbreeding adults have a partial hood that covers the rear crown and nape (paler than the blackish hood of Franklin's Gull; more like Laughing Gull). The very attractive juveniles are uniformly gray-brown above (slightly mustard-colored) and white below. A brownish shawl-like hood is the same color as the back, and despite fine pale scalloping, the birds appear very smooth above. First-winter plumage (almost never seen in North America) resembles first-winter Laughing Gull, but the breast of Sabine's is white; Laughing Gull's is gray.

BEHAVIOR: A moderately social gull. In migration, usually seen in small groups (5–15), but flocks of several hundred birds have been reported. On the East Coast and in the interior, singles are more common. Normally seen well away from land; infrequently joins other small gulls on the beach. A versatile feeder. Forages mostly by swimming buoyantly, snatching prey from the surface. Also hovers and plucks prey, plunges (from heights of about 15 ft.), and scavenges food on beaches, walking nimbly. In migration, birds resting on the water often gather in tight clusters. Fairly tame, allowing close approach when sitting on the water. Less attracted to chum (or ships) than are kittiwakes.

FLIGHT: Appears short-bodied, with dominatingly long, broad, angular wings (immatures appear more slender). The bright white underwings catch your eye, but it is the bird's distinctive tricolored upperwing pattern that holds it (and clinches the identification). *Note:* Immature Black-legged Kittiwake has a wing (and tail) pattern that can be confused with Sabine's, but the kittiwake has a mostly white head; Sabine's is dark-headed (immature) or dark-hooded (adult). Also, the kittiwake has narrow, curving, boomerang-shaped wings, whereas Sabine's Gull's wings are broader and more angular.

Wingbeats are stiff, shallow, and steady, but curiously slow and heavy for the bird's size. Flight is overall lofting, buoyant, and ternlike, but not as nimble and ternlike as Bonaparte's.

VOCALIZATIONS: Call is a harsh, grating, somewhat ternlike "*keeeyr.*" Young make a descending chattering trill.

Black-legged Kittiwake, *Rissa tridactyla*

The Seagull

STATUS: Abundant but somewhat geographically restricted northern coastal breeder; common but almost exclusively pelagic winter resident along both coasts. Rare inland. **DISTRIBUTION:** In the East, breeds coastally in Nfld., Lab., and ne. Que.; in the West, along the south and west coasts of Alaska. Some scattered colonies also occur in the Northwest Territories. In winter, found from the Aleutians south to Baja, and from e. Lab. to N.C. (very rarely Fla. and the Gulf Coast). **HABITAT:** Nests on coastal cliffs, steep slopes, and boulders. In winter, found in marine waters from inshore out to about 150 mi. In northern portion of its winter range, commonly forages close to shore; in the coastal United States, somewhat less common but still regular from shore; southern birds are usually far from land. **COHABITANTS:** Breeds (and in winter feeds) with murres and other alcids, Northern Fulmar, cormorants. **MOVEMENT/MIGRATION:** Birds leave nesting colonies as early as August and appear off the East and West Coasts between October and April. VI: 3.

DESCRIPTION: A pale, slender-winged, seagoing gull that recalls a bulky and roundly contoured tern in flight. Small and fairly compact (larger than Bonaparte's Gull; smaller than Northern Fulmar; about the same size as Mew or Laughing Gull), with a large head, a straight greenish yellow bill (black in immatures), a compact body, a short tail, and an upright posture.

Adults in breeding plumage are very plain and pale, with a white head and underparts, pale gray upperparts (slightly darker than Herring Gull), and jet-black wingtips devoid of any white spotting. The greenish cast to the bill (more yellow in summer) is eye-catchingly anomalous, and the gentle dark-eyed expression winsome. In winter, shows a gray nape and blackish ear patch (which makes the head more prominent). Immatures are much like immature Bonaparte's Gull, showing a black ear patch, a broken black line running the length of the folded wing, and a narrow black-tipped tail, but in addition to being overall bulkier, kittiwakes differ from Bonaparte's in having a distinct narrow black collar.

BEHAVIOR: A pelagic gull most commonly seen cruising high above the ocean surface. Often the first species to arrive when surface food (or a boat) appears; often the first to leave when things get crowded. Commonly feeds by swimming on the surface and plunging and plucking for prey (also dives), and is adept at stealing food from other seabirds. Very social; nests in colonies with other kittiwakes and feeds in flocks. In summer, does not generally forage more than a few miles from the colony, but in winter commonly occurs well beyond sight of land. Bold bordering on indifferent to observers.

FLIGHT: A pale, fairly compact, rectangular-bodied gull with long, narrow, boomerang-shaped wings. At times suggests a bulky tern. The head (particularly the dark-collared heads of immatures and gray-naped winter adults) seems bulky, domed, overly prominent, and the tail fairly long and narrow. Wings are conspicuously long and slender (particularly near the tip), and though angled back, seem bowed or curvaceously contoured (not harshly and sharply angled like Bonaparte's Gull).

At a distance, adults appear very pale, almost white. At closer range, the upper surface of the wing is two-toned—darker gray on the arm, paler gray on the hand (below the black wingtip). This contrast is evident even at distances where the all-black triangular wingtip is hard to see. Immatures show a black M pattern on the upperwing that contrasts with their otherwise white flight feathers. The black collar distinguishes it from similarly patterned immature gulls. Flight is buoyant and nimble, with tight cornering and frequent banking that favors one side over the other (like terns). Wingbeats are stiff and steady, with an exaggerated upstroke and shallow down stroke.

VOCALIZATIONS: Noisy at colonies, enunciating a quick nasal rendering of its name: "*kit-y-Waak,*" often given in series. At sea, gives a short, assertive, single-note "*cah!*" or "*eeyah!*" (with a compressed two-note quality) that recalls a Snow Goose yelp or bark.

Red-legged Kittiwake, *Rissa brevirostris*

Pug-faced Kittiwake

STATUS AND DISTRIBUTION: Uncommon and geographically restricted, breeding on four island groups in the Bering Sea (Commander Is., Buldir I., Bogoslof I., and Pribilof Is.); in winter, forages in the deeper waters of the North Pacific, beyond the rim of the continental shelf. Even where it breeds and feeds, it is outnumbered by its close relative, Black-legged Kittiwake. **HABITAT:** Nests on cliffs; feeds over deep ocean waters. **COHABITANTS:** Northern Fulmar, Black-legged Kittiwake. **MOVEMENT/MIGRATION:** In winter, apparently disperses across the deeper waters of the North Pacific, mostly away from the waters of the Bering Sea (but a few remain in the southern Bering Sea in winter). Returns to nesting cliffs in April and early May. VI: 1.

DESCRIPTION: A small, compact, dark-winged, pug-faced pelagic gull. Small (larger than Sabine's; slightly smaller than Black-legged Kittiwake or Ivory Gull), with a stubby bill, a large rounded head, a somewhat compact body, a short tail, and short reddish legs (red also on immatures). With its steeper forehead and short bill, appears pug-faced next to Black-legged Kittiwake. At close range, note the bird's large balefully dark eye.

Adults have white heads, neck, and underparts and dark gray upperparts—a shade or two darker than Black-legged Kittiwake, and about the same shade as Sabine's Gull. In winter, shows a dark ear patch (like Black-legged Kittiwake). Immatures are like winter adults but have a narrow dark collar.

BEHAVIOR: Spends most of its life at sea and appears to be somewhat solitary. Where birds are feeding, Red-legged Kittiwakes usually arrive singly. Flies high (30–40 ft.) over the sea. Drops to the surface to pursue prey, sometimes plunging below the surface. Also sits on the surface and dips. Feeds day and (particularly) night, often with Black-legged Kittiwake.

FLIGHT: Overall stockier and more pigeonlike than Black-legged Kittiwake, but also appears slightly longer and slimmer-winged. Because its upperparts (and underwings) are distinctly darker than Black-legged, its black wingtips do not stand out quite as much as on Black-legged, and the contrast between the dark upperparts and the white head and body is stark. Immatures lack the bold black M pattern seen on young Black-legged, so appear much like adults (and adult Sabine's Gull). Flight is strong, direct, and mostly steady, with deep backstroking wingbeats (punctuated by periodic half-skips) that are quicker than Black-legged Kittiwake.

VOCALIZATIONS: Silent at sea; noisy at colonies. Calls are like Black-legged Kittiwake, but higher-pitched.

PERTINENT PARTICULARS: Although it's possible to see the red legs on birds in flight, you'll have more success focusing on more obvious characteristics—the darker upperparts, the pug face, the slender wings, the minimal contrast between the upperwing and the wingtip.

Ross's Gull, *Rhodostethia rosea*

The Pink Seagull

STATUS: Rare and extremely localized arctic breeder; uncommon at the edge of ice packs in winter. Very infrequent winter visitor outside arctic regions, with most sightings on the East Coast south to Maryland and on the Great Lakes (also scattered elsewhere in the interior and the Pacific Northwest). **DISTRIBUTION:** Breeds primarily in coastal Siberia, but also breeds on several Canadian arctic islands and Greenland and (irregularly) at Churchill, Man. **HABITAT:** Breeds on open tundra; forages over ponds. Outside the breeding season, forages at the edge of ice packs (perhaps more widely at sea). **COHABITANTS:** In summer, Arctic Tern; in winter, Ivory Gull. As a vagrant, often found among Bonaparte's Gulls. **MOVEMENT/MIGRATION:** Spring migration in May and early June; fall from late September to December. VI: 2.

DESCRIPTION: A small, slender-winged, wedge-tailed darling of a gull. Small (slightly larger than Little Gull; smaller than Black-legged Kittiwake; about the same size as Bonaparte's Gull), with a petite bill, a round head (sometimes with a bump or peak), a short neck, a plump breast, long wings (extending well beyond the pointy tail, whose wedge shape is apparent in flight), and short legs. Overall appears very dovelike.

Adults are overall pale, with a pale gray back and variably pink-blushed underparts. Breeding birds show a unique narrow black collar (reduced to only a tiny ear patch in winter); in winter, pink tones are usually diminished or absent or mixed with gray. Immatures are shaped like adults but patterned like young Bonaparte's or Little Gull, showing a pale body with a small dark ear patch, a wide dark racing stripe along the folded wing, black wingtips, and a black tip to the tail. Young Ross's is overall more delicately featured than Bonaparte's (particularly smaller-billed) and more akin to a long-winged Little Gull (except that Little Gull has a dark cap). In winter, the absence of obvious dark patterning on the heads of adults and immatures makes the dark eye stand out.

BEHAVIOR: Spends most of its life at sea. Feeds by flying and hovering 10–15 ft. over the water, then swooping to snatch prey, plop to the surface, or plunge beneath the surface. Also floats and turns, phalarope fashion, on the water, and in summer walks pigeonlike along the shallow edges of ponds searching for food (bobbing its head as it walks). When found in the United States and southern Canada (both on land and at sea), this tame gull is usually in the company of Bonaparte's Gulls.

FLIGHT: Overall small, slender, delicate, and racy-looking, with a small head, long pointy wings, and a long wedge-shaped tail. Adults are overall pale, with shadowy slate gray underwings. Like immature Little Gull or immature Kittiwake, immature Ross's sports a dark M-shaped upperwing pattern and a dark-tipped tail; it also shows gray underwings. Note that immature Ross's has a pale head (no dark cap, no pronounced ear patch, and no black collar) and only a touch of black on the tip of the tail, not the black terminal bar seen on other species. Flight is buoyant, nimble, and somewhat ternlike when feeding; fast and direct, with deep wingbeats, when the bird is going places.

VOCALIZATIONS: Silent away from breeding areas.

PERTINENT PARTICULARS: The pink underparts of adults are not unique to this species. Other small gulls (including Little and Bonaparte's Gulls) may show a faint pink wash, and some species (most notably Franklin's) may be very pink. Don't be confused, and don't presume. Also, in flight, the shadowy underwing of adult Ross's may be confused with the even blacker underwing of adult Little Gull.

Ivory Gull, *Pagophila eburnea*

Ice Pigeon

STATUS AND DISTRIBUTION: Rare. Breeds in arctic Canada; winters mostly offshore, at the edge of the ice pack, in the Bering Sea and from the Labrador Sea to the Davis Strait; rarely occurs south to the Maritime Provinces of Canada and very rarely to New England and the Great Lakes. **HABITAT:** In winter, gathers at the edge of the ice pack; in summer, frequents water with high concentrations of floating ice. **COHABITANTS:** In summer, Brant, Thayer's Gull; in summer and winter, Polar Bear, seals (live and dead). **MOVEMENT/MIGRATION:** Breeds between mid-June and early September; otherwise at sea. VI: 2.

DESCRIPTION: A smallish, pigeonlike, pure white gull with a round black eye and short black legs. Medium-sized (larger than Ross's and Sabine's Gulls; smaller than Thayer's Gull; about the same size as Black-legged Kittiwake and Mew Gull), with a small pale-tipped bill, a round neckless head, a plump-breasted body, and stubby black legs.

Ivory Gull is all white—not ivory white, but white-white, with an obvious black eye and legs (eliminating the possibility of albinism). Immatures (the bulk of southern wanderers) are shaped like adults and, although white, have dirty faces (making the face appear deformed) and spare dark speckling on the upperparts and wings. The tip of the tail, hidden by the long wings, has a narrow black band.

BEHAVIOR: During the breeding season, colonial; in winter, found alone or in small flocks. Behaves like other gulls, but is not often found in mixed-species flocks. Forages at sea, mainly at night, searching the open water at the edge of the ice pack for food or prey. Also feeds on carrion and offal and attends carcasses left by feeding Polar Bear and human hunters; also attends feeding whales. When feeding aloft, dives ternlike into the water, but when temperatures allow, often sits on the water for extended periods. Reported to investigate anything red left on the ice or snow. Waddles like a pigeon.

FLIGHT: Appears stocky, broad-winged, and bright white (even immatures, which also show a dark face, a narrow dark tip to the tail, and a dark edge to the trailing edge of the wing). Flight is fast (reportedly faster than Kittiwake), buoyant, and graceful, with quick wingbeats.

VOCALIZATIONS: Described as a harsh, descending, ternlike "*kreearr.*"

TERNS AND SKIMMER

Gull-billed Tern, *Sterna nilotica*

Laughing Tern

STATUS: Uncommon and largely restricted to coastal regions in the United States. **DISTRIBUTION:** In the U.S., breeds along the Atlantic and Gulf Coasts from Long Island to s. Mexico. Except for disjunct populations at the Salton Sea and San Diego, West Coast birds are restricted to Mexico. In winter, absent from the Atlantic north of Fla. Winters primarily along the Gulf Coast and in s. Fla. south through Central America. **HABITAT:** Preferred substrate is sandy beach, but may also nest in salt marsh. Most commonly seen foraging over salt marsh and along the edges of bays, but also hunts over dry fields, airports, and freshwater ponds (sometimes many miles from the coast). **COHABITANTS:** Laughing Gull. Nests in small numbers among terns and other beach-nesting birds (for example, Common Tern, Black Skimmer). **MOVEMENT/MIGRATION:** Returns to Atlantic Coast nesting areas in April and May. Departs by the end of August. VI: 2.

DESCRIPTION: A medium-sized marsh tern that is larger, bulkier, blockier, and more dapper than Forster's Tern. The body is compact, angular, and somewhat wedge-shaped, particularly in flight, when the flat underparts and upperparts angle up from the tail to the blunt roundly contoured head. The bill is stubby and black (recalls Fish Crow), the legs are long and black, the tail is short, and the wings, when folded, are very long, extending well beyond the tail. While aptly named, the whole bird seems somewhat gull-like.

Adult plumage is a study in black and white. The upperparts are pale gray, but this is a technicality. Overall the bird simply looks white; against this backdrop, the black cap, black bill, and dark trailing edge of the primaries jump out. In a mixed flock, Gull-billed and Sandwich Terns are pale standouts—first because of their color, and second because Gull-billed's longer legs elevate the bird above its medium-sized tern cousins. In winter, adults wear an opaque eye patch that is not as dark or distinct as Forster's Tern.

Juveniles are slightly thinner-billed (and pinkish-billed), and somewhat buffy above, with a shadowy eye patch.

BEHAVIOR: Despite its relative scarcity, Gull-billed's feeding behavior makes it an easy bird to find. The bird flies 10–20 ft. high, swoops down to snatch prey with its bill, then loops back aloft. The swoops are fluid and generally steep—sometimes even vertical when the bird overshoots its prey and must reverse direction, executing a turnaround swoop. Does not plunge-dive, as do most other terns, but rather plucks prey from the surface. Does hunt over dry land, snapping prey from the ground and insects out of the air. Both of these activities automatically distinguish Gull-billed Tern from any other tern (except for the very dissimilar Black Tern). Hunting Gull-billeds may work a small area for many minutes or use a linear search pattern, covering great distances over open marsh and along shorelines.

Usually a solitary hunter that is not found among feeding flocks of terns, but multiple Gull-billeds may forage over the same area. Even when sitting among a group of terns, Gull-billeds frequently sit just a little apart.

FLIGHT: Blocky and angular, with long broad wings and a front-heavy short-tailed profile. Direct flight is fast and buoyant, but perfunctory. Wings are straighter and stiffer than Forster's Tern, and wingbeats are more shallow, less hurried. The grace of a working bird can be seen in the loping cadence and directness of the flight, which seems smart, professional, and precise, not languid and easy.

VOCALIZATIONS: Call is a loud distinctly two-noted yap, "*ka-Wek!*" that is often repeated several times. Recalls Black Skimmer, but is more clipped and less nasal. Sounds as if the bird is giving a short "Ah, hah! I told you so!" laugh.

Caspian Tern, *Sterna caspia*

The Big Red (-billed) One

STATUS: Fairly common to uncommon northern breeder and winter resident along southern coasts, with a widespread but patchy distribution. By far the most likely of the large orange-billed terns to be found away from coastal areas. **DISTRIBUTION:** Almost cosmopolitan, breeding on every continent but Antarctica. In North America, breeding colonies are geographically localized and widely scattered on both coasts and also in interior locations, primarily around the Great Lakes, w. Canada, and the w. U.S.; also in coastal Nfld. and N.C. and scattered along the Gulf Coast and West Coast, in sw. Wash., and along coastal Calif. from San Francisco south to Baja, most often in association with large freshwater rivers or bodies of water (including the Great Lakes). In winter, interior and northern coastal breeders retreat to southern coastal waters—from N.C. and cen. Calif. south to Central America. Also winters inland in Fla. and s. Tex. and in parts of Mexico and Central America. In migration, can appear virtually anywhere in the U.S., as well as in the eastern and prairie provinces of Canada. **HABITAT:** Nests on fairly flat, open, sparsely vegetated sandy or pebbly beaches, rocky shores, and dredge spoil in a variety of habitats, including estuaries, barrier islands, freshwater islands, rivers, and salt lakes. In winter, found primarily along coastal beaches, large lakes, rivers, canals, and marshes. **COHABITANTS:** Nests in colonies with gulls and terns, including Herring, California, Ring-billed, and Western Gulls. Also associates with large gulls and terns in winter. **MOVEMENT/MIGRATION:** Spring migration from March to mid-May; in fall, birds appear outside breeding areas as early as late June and may linger in coastal areas north of established winter range until early November. In fall, adults are often accompanied by one or two young. VI: 3.

DESCRIPTION: A large formidable-looking tern with a crewcut and a dark-tipped red to orange-red bill. This very large tern is halfway between Ring-billed and Herring Gulls in wing length and overall proportions and can be confused only with the slightly smaller and more coastal Royal and Elegant Terns. Caspian is overall larger, less angular, and more broadly proportioned than the other large terns. The head is very large, the tail is short with a shallow V, and the legs are long, black, and sturdy. Like other large terns, Caspian shows pale gray upperparts and white underparts. It is distinguished, however, by a crewcut-cropped (not shaggy) crest that becomes sprinkled with white on the forehead in winter but remains full or mostly full all year; blackish (not grayish) wingtips; and, most of all, a long daggerlike red-orange bill that is straighter, heavier, and usually redder than either Royal or Elegant Tern.

Looking close, you can see that the bill is darkened just shy of the paler tip. Royal's bill never shows this. Seen from behind, Caspian's crest extends like a dorsal widow's peak down the nape of the neck. Royal's crest is sharply truncated behind the head. Caspian's tail is short, and its dark wingtips extend well beyond the tip (more so than on Royal). Standing amid Royals, longer-legged Caspian towers above them.

Juveniles are like adults except their bills are slightly shorter and less red (more orange), and their folded upperwings and back are dusky gray at a distance, finely scalloped with brown up close.

The expression is formidable and ill-tempered.

BEHAVIOR: More solitary than most terns. Likes to roost at the end of spits and often chooses to roost with medium-sized to large gulls rather than smaller terns. In parts of North America, migrates in flocks. For most of the year, Caspians prefer to forage alone or, as is often the case, in the company of one or more young, which remain with parents well into the fall.

Hunts higher (15–50 ft.) than other terns. Rarely forages far offshore and instead cruises shorelines and the edges of vegetation, often returning to work the same stretch of shoreline again and again. On sighting prey, birds wheel over and plunge.

FLIGHT: In flight, the body is robust and portly, the large head and bill projecting. Wings are exceedingly long, broad (much broader than Royal), and symmetrically tapered, not segmented (that is, the wing shows a distinct arm and hand, as does Royal Tern). Overall a very front-heavy bird and more roundly proportioned (more gull-like and less ternlike) than the comparatively spindly Royal.

Seen from below, dark wingtips make Caspian look as though it's wearing black mittens. Above, the upperparts are distinctly white on adults—whiter than Royal Tern. The dark trailing edge to the outer primaries is often seen as a wedge of black intruding into the white of the wing.

Point-to-point flight is direct and fast, but birds sometimes wander a bit, as if they haven't quite made up their mind about their destination. Hunting birds, flying into the wind, may work their way very slowly up a shoreline. Wingbeats are heavy, smooth, and deeper and more flowing than the light, crisp, stiff, cleanly executed stroke of Royal Tern. Caspian's flight is graceful, but there is something joyless or ponderous about it.

Often seen flying exceedingly high, but it's debatable whether it simply chooses to fly high or its vocalizations call attention to it when it's high overhead.

VOCALIZATIONS: Very vocal. The bird's presence is often heralded by a harsh loud call that has been described as heronlike but more nearly approximates the sound of a cat being stepped on: "*aieeeYOW!*" Call is often repeated at regular intervals. Also makes a short, gruff, sequential "*rah rah rah,*" as well as a low rough chortle, "*ra'ha'ha'ha'ha'ah,*" that recalls the feeding call of Mallard. Food-begging young make a whiny, quavering, peeping "*pe'he'he'heep.*"

Royal Tern, *Sterna maxima*

Friar Tern

STATUS: Common breeder and winter resident along the southern Atlantic and Gulf Coasts; summer post-breeding wanderer to New England and, in fall and winter, along the southern coast of California. **DISTRIBUTION:** This predominantly tropical tern with populations in East Africa and North and South America is almost never found away from coastal areas. On the Atlantic Coast of North America, breeds from Va. to Fla. and throughout the Caribbean and Gulf of Mexico. On the West Coast, breeds no farther north than San Diego Bay. On the Atlantic, nonbreeding adults are regular north of the nesting areas in spring. In late summer (June through September), many adults and young disperse northward, some going as far north as New England on the Atlantic Coast. In s. Calif., north of breeding areas, the greatest numbers of Royals occur in fall and winter and are found north to Morro Bay. **HABITAT:** Nests in large colonies on open sandy beaches and protected islands. Forages in bays, inlets, and nearshore and offshore waters (sometimes well offshore). Frequently seen sitting and roosting in large groups, sometimes with smaller terns and gulls. **COHABITANTS:** Sandwich, Elegant, and Forster's Terns. **MOVEMENT/MIGRATION:** On the Atlantic, northern breeders reach colonies in March and April; prospecting adults move north in May and June, followed by a general post-breeding dispersal from July to November. In California, birds are most common north of breeding areas from early October to March. VI: 2.

DESCRIPTION: A large tern—among North American terns, second only to Caspian in size. Similarities in plumage, size, and bills notwithstanding, Royal and Caspian differ markedly in breadth and heft. Caspian is a broad, bruising, heavy-billed dockworker of a tern; Royal is a slender-winged diver.

Conspicuously larger than medium-sized terns (like Common), sitting and in flight Royal is distinctly pale (pale gray above, white below) and easily distinguished by its long, pointy, completely orange bill and narrow shaggy crest cut in the likeness of a Friar Tuck haircut. Only for a short time (April and May) do adult birds have an all-black crest like Caspian.

When Royal is perched, the tail is considerably shorter than the wings. In flight, the tail is deeply and sharply forked (not just dented in the middle like Caspian, and not flanked by long outer tines like Common or Forster's). When resting on the water, Royal cocks up its tail and wingtips at an angle.

Juveniles have yellow bills, an upperwing pattern consisting of longitudinal bars and sooty wingtips, patterned backs, and the limited Friar Tuck crest of adults.

BEHAVIOR: A social tern. Often seen loafing in flocks or hunting over baitfish with other terns (particularly Sandwich Tern) and gulls. Young birds (that will be calling) follow and get food from adults until well into the fall. Royal usually hunts away from the surf (in bays or offshore) and commonly hunts and flies higher (20–30 ft. above the surface) than the smaller terns. Hunts visually. Dives straight down or at an angle to secure prey. Seems not to like flying over land, making only short crossings for convenience.

FLIGHT: Front-heavy in flight, the body is shaped like a wedge or a door stop—flat below, angling up from the tip of the tail to the broad back and head. When Royal is in direct flight, the bill is pointed straight ahead and its color makes it obvious. When hunting, the bill is angled down and the silhouette of the bird becomes blunt. Wings are exceedingly long, narrow, angular, and gangly. The arm and hand are somewhat segmented. Where most tern wings have a seamless symmetry, the wings of Royal Tern seem distinctly two-parted—the arm and the hand. Flight is direct, unhurried, and somewhat gull-like. Unhurried wingbeats are light, crisp, and shallow, with wings held slightly down-bowed.

VOCALIZATIONS: A distinctive, high, loud, somewhat trilled and two-noted "*Khr'r'r'r'rEp!*" ("Cheer-up") that is more musical and less grating than the harsh call of Caspian Tern. Young's call is like adults, but higher-pitched and less two-noted, "*Kreeeek*"; young also emit a piping whistle, usually given in a series of three. When breeding, adults make a short muffled "*reh.*"

PERTINENT PARTICULARS: A group of birds to which Royal Tern bears more than a passing likeness are the tropicbirds, particularly Red-billed Tropicbird, whose plumage approximates that of a juvenile Royal Tern. Standing, tropicbirds are compact, whereas Royal Tern is overall slender. In flight, tropicbirds have hurried, choppy, steady wingbeats; Royal Tern's are slower, measured, and cleaving.

Elegant Tern, *Sterna elegans*

Slender-billed Tern

STATUS: Common at times and places, but otherwise seasonal and geographically restricted. **DISTRIBUTION:** Approximately 95% of the world population (estimated at fewer than 30,000 pairs) breeds on Isla Rasa in the n. Gulf of California. Most of the balance are found in three isolated colonies in coastal s. Calif.: one in Los Angeles, one at Bolsa Chica Ecological Reserve in Orange County, and one on San Diego Bay. Post-breeding birds disperse north along the coast to about Humboldt Bay, Calif., but some birds wander farther (especially in El Nino years) to Ore., Wash., and (rarely) s. B.C. **HABITAT:** Strictly coastal. Forages in nearshore waters; roosts on beaches and coastal mud flats. **COHABITANTS:** Nests and roosts with Caspian, Royal, and Forster's Terns. **MOVEMENT/MIGRATION:** Returns to breeding colonies from mid-March to late May; dispersal north along the coast begins in June, with a peak between July and September; most birds migrate south of California before mid-November. VI: 1.

DESCRIPTION: Royal Tern Lite. Medium-sized (about 3 in. longer than Forster's Tern; 3 in. shorter than Royal Tern; dwarfed by Caspian Tern), and in structure and plumage, looks like a smaller, slender, shorter-legged, and more *elegant* Royal Tern. The bill of adult Elegant, however, is slightly but distinctly longer, thinner, and more drooped than Royal. Other structural differences between Elegant and Royal are nearly impossible to assess unless the birds are side by side, but Elegant's more elegant bill shape is easy to see. Also, Elegant's bill color ranges from coral to orange to yellow, and the tip is often yellower or more translucent. Royal's bill is orange or yellow (no reddish tones) and uniform to the tip.

In breeding plumage, the black crest of breeding adult Elegant Tern is commonly longer, shaggier, and held longer into the summer than the crest on Royal Tern (whose forehead may be white by May). Royal Tern's cap resembles the crewcut of Caspian Tern; Elegant looks like a bird badly in need of a trim. From behind (even in nonbreeding and immature plumage), Elegant retains a fuller slicked-back crest—a tern with a D.A. haircut. In front, outside of the breeding season, Elegants show a receding hairline as compared to the advanced balding of Royal's Friar Tuck cut. In addition, in nonbreeding plumage, Elegant's fuller crest usually enfolds the bird's black eye, rendering it invisible. On Royal Tern, you can see the eye.

Also, in breeding plumage, Elegant's breast commonly shows a pinkish blush (which may be retained into the winter); the breast on Royal is never pinkish.

Except for shorter bills and pale (yellow) legs, juveniles are much like nonbreeding adults, so much less patterned above than young Royal Terns.

BEHAVIOR: Highly social. In the United States, breeds with other tern species. During the summer northern dispersal, may gather in the hundreds at traditional sites (like Elkhorn Slough and Bolinas Lagoon). Forages for fish in nearshore waters by hovering near the surface, swooping, and plucking prey (much like Gull-billed Tern) or landing momentarily on the surface and snatching prey. Infrequently plunge-dives like Royal Tern. Gathers in large loafing flocks after feeding, typically in semi-segregated flocks that may incorporate Caspian and Royal Terns and other species. Elegants are a flighty species—flocks often erupt into spontaneous flight, then resettle, while Caspians continue to lounge on the beach.

Very vocal. Calls when feeding and loafing. Also, young birds accompany and are fed by adults into the fall. If you see a mystery immature tern in the fall that you would like to turn into a Royal, just wait and see who comes in to feed it.

FLIGHT: Long and slender-winged, long-faced, and short-tailed. Overall slimmer, lankier, more angular, and more front-heavy than Royal Tern. Flight is quick, nimble, buoyant, and somewhat stiffer than Royal (recalls Sandwich Tern), with stiff regular wingbeats of varying quickness.

VOCALIZATIONS: Common call is a harsh, distinctly two-noted "*kir-eek*" that sounds like a baritone Sandwich Tern. Also emits a longer, rising, rolled-r's "*cher'r'reek*" that sounds like a quicker and creakier rendering of Royal Tern's "cheer-up."

PERTINENT PARTICULARS: The problem of separating Royal and Elegant is greatly simplified by geography and the calendar. Elegant is regularly found the length of the California coast south of Los Angeles (and northward) from March to November, and from May to early November across the balance of the state. Royal Tern does not breed in California, except in very small numbers south of Los Angeles, and does not commonly occur in California waters until fall and winter, when Elegant Tern is absent, and even then Royal occurs no farther north than Morro Bay.

Unless your sighting lies south of Morro Bay and falls between the months of October and March, chances are that you are looking at an Elegant Tern. The greatest overlap between the two species occurs in October and November, when identification can commonly be facilitated by side-by-side comparison.

Sandwich Tern, *Sterna sandvicensis*
With Mustard on Its Bill

STATUS: Common but localized Atlantic and Gulf Coast breeder; common and more widespread winter resident along southern coasts. **DISTRIBUTION:** Breeds from Cape Charles, Va. (rarely Md.) to n. Fla. and, in the Gulf, from cen. Fla. to Mexico and south. Distribution is spotty along the Atlantic, with most birds breeding along the Tex. and La. coasts. In winter, found from about Daytona Beach, Fla., south and throughout the Gulf of Mexico. **HABITAT:** Strictly coastal. Nests and roosts on flat sandy beaches, barrier islands, and dredge-spoil islands. Feeds offshore but generally within 2 mi. of the coast (also in bays, harbors, and, particularly, inlets). **COHABITANTS:** Laughing Gull, Royal Tern, Black Skimmer. **MOVEMENT/MIGRATION:** Birds in the Atlantic wander north to

New Jersey after breeding and retreat south before the onset of winter. Spring migration from mid-March to early June; post-breeding dispersal begins in July; final retreat south by early November. VI: 3.

DESCRIPTION: A slender, pale, shaggy-crested, black-billed tern with a dab of mustard on the tip. Medium-sized (larger than Forster's Tern; smaller than Laughing Gull or Royal Tern; about the same size as Gull-billed Tern); when standing, crest just reaches the lower mandible of the Royal Terns that are almost certainly nearby. Overall somewhat angular, with a large flat-topped head, an exceedingly long, thin, black bill (with a surprisingly inconspicuous yellow tip), a long thick neck, and a wedge-shaped body showing a blocky chest and slender tapered rear (folded wings extend beyond the shortish tail). Shape most closely resembles Elegant Tern, and to a lesser degree (ignoring the thick bill) Gull-billed Tern.

Very pale gray above (paler than Royal Tern; about as pale as Gull-billed Tern), and white below. In breeding plumage, the black crest is shaggy and spiky at the rear. By midsummer, adults show white foreheads, but the crest is still shaggy above the nape. Juveniles resemble adults but are lightly patterned above (but less patterned than juvenile Forster's, Common, or Royal Tern) and have crestless black caps (no shagginess) and long, thin, blackish bills with pale (but not yellow) tips.

BEHAVIOR: A very social tern; feeds and loafs with other Sandwich Terns and seems equally at home with Royal Terns. Feeds by hovering 15–20 ft. above the water and plunging, vertically, for fish (often submerging). Also swoops and plucks prey from the surface in the manner of Gull-billed Tern. When not feeding, loafs on shorelines near the water's edge or on pilings.

Moderately tame, but when approached, jumps when all the other terns jump.

FLIGHT: A distinctive profile combines the compact wedge-shaped front-heaviness of Gull-billed Tern with the very long slender wings of Forster's Tern. The short tail, usually closed to a narrow tine, is difficult to see. Moreover, the bill's yellow tip is often invisible, making the bill appear shorter than it is. In contrast to the compactness of the body, the wings are long and slender throughout their length and sharply angled.

Overall pale (very much like Gull-billed Tern), usually showing a dusky wedge intruding into the tip of the upperwing and a narrow dark trailing edge to the tip of the underwing. Upperwing may also show a paler rendering of the two-toned pattern of Forster's Tern—gray on the arm, paler on the hand.

Wingbeats are regular, forceful, slightly fluid or flexing, and quicker than Royal. Flight is strong, direct, no-nonsense; movements are nimble without being delicate.

VOCALIZATIONS: Common call is a high, harsh, slightly trilled, two-noted "*ke'h'h-yeh!*" or "*kee-ak!*" or "*ch'r'r-eek!*" that is more screechy and less musical than Royal Tern's "cheer-up." Also makes a soft nasal "*re'uh*"; a squeaky "*ye'ah*"; a low, nasal, Forster's Tern–like wheeze, "*ra'a'a'h*"; a short nasal scrape, "*jeh*"; and a Whimbrel-like stutter, "*reh'he-h'heh'heh'heh.*"

PERTINENT PARTICULARS: Perhaps because Sandwich Tern and Royal Tern both have crests, have somewhat similar calls, and commonly associate with each other, many descriptions focus on separating the two. In terms of size, shape, and plumage, Sandwich is more likely to be mistaken for the equally pallid and also black-billed Gull-billed Tern (when sitting) or the overall pale and slender-winged Forster's (in flight). Packed in a flock, Gull-billeds have conspicuously round heads, whereas the head of Sandwich Tern is squared and flat-topped. In flight, Sandwich is more front-heavy and angular than Forster's, with a longer bill and shorter tail. Gull-billed, in flight, is broader-winged than Sandwich.

Roseate Tern, *Sterna dougallii*

The Pallid Streamer-tailed Tern

STATUS: Uncommon and restricted breeder along the Northeast coast and the Florida Keys. Very rare migrant along the Atlantic Coast. **DISTRIBUTION:** Breeds from coastal Que. and N.S. south to Long

Island. A southern population is found in the Caribbean, and individuals breed in the Florida Keys. Winters coastally off South America. HABITAT: Nests most commonly on high beach areas in and amid rocks and vegetation. Feeds in shallow marine waters (often near shore or over bars) or over schools of predatory fish. COHABITANTS: Other terns and gulls, especially Common and Forster's Terns in the North; Sandwich Tern and Brown Noddy in the South. MOVEMENT/MIGRATION: Migrates offshore; rarely encountered far south of nesting areas (although regular in spring along the New Jersey coast). Northern birds arrive, amid flocks of Common Terns, in May; begin departing as early as mid-July, with peak movements seen in late July and August. From late summer until early September, birds gather at staging sites, especially Cape Cod. VI: 1.

DESCRIPTION: A pallid, elfin-featured, medium-sized tern so elegantly slender that it makes other *Sterna* terns seem angular and ungainly. Streamer tail notwithstanding, birds are overall smaller and slighter than Common and Forster's Terns—both resting and in flight.

Distinctly whiter than Common, Forster's, and (to a lesser degree) Arctic Terns—a distinction easily noted whether birds are sitting together or in flight. The pinkish blush on the underparts of breeding adults is usually apparent but not eye-catching. Sitting birds show pale wingtips and an all-white tail that projects well beyond the wingtips. Except in June and July, the adult bill is all black (or red just at the base). The overall paleness of the bird makes the bill and black crown pop out.

Juveniles (rarely seen outside of breeding areas) have scaly backs that may be warm or buffy but not brownish. They most closely resemble miniature Sandwich Tern.

BEHAVIOR: Nests on higher portions of the beach than Common Tern, often amid rocks or vegetation. Feeds offshore, often with other northern and tropical terns and often over sandbars, rips, or areas where predatory fish are driving small bait-fish to the surface. Very active, almost aggressive feeder. Plunges for prey more often than Common or Forster's, and hovers less. Roseate often dives with accelerating wingbeats and commonly immerses itself completely in the water. Although Roseates flock-feed among Common and Forster's Terns, they tend to maneuver themselves slightly away from the main body of the flock.

FLIGHT: Overall smaller, slighter, shorter-winged, and longer-tailed than Common and Forster's Terns. The body is somewhat wedge-shaped: wide at the head/chest and tapered toward the tail—more like Least and Gull-billed Terns than the tubular body shape of Forster's and Common. The long white streamer tail is often invisible at a distance. When visible, it appears flagellum-like and even ripples as the bird flies.

Compared to other terns, Roseate's wings appear shorter, more slender throughout their length (wings of other terns are distinctly broader at the base and more triangular), more curved, and less acutely angled along the leading edge. The black line along the leading edge of the wingtip is often distinct against the otherwise white backdrop of the bird. On the wings of Common and Forster's, the black is on the trailing edge.

Flight is rapid, almost reckless. Wingbeats are rapid, more shallow, stiffer, and more batty than Common and Forster's.

Taken in sum—the front-heavy profile, the overall paleness, the short, slender, curved wings with shallow rapid wingbeats, the long streamer tail feather—Roseate Tern conjures the image or impression of tropicbird, an impression nobody gets when watching Common or Forster's Tern.

VOCALIZATIONS: Call is a terse, harsh, shorebirdlike, two-note "*chivik.*" Also makes a harsh, raspy, angry, and ascending "*zraaaaAP!*"

Common Tern, *Sterna hirundo*

Gray-breasted Tern

STATUS: Common breeding bird in northern regions and along the Atlantic Coast; rare in winter along the Gulf Coast and southern Florida south to Argentina and Chile. DISTRIBUTION: Breeds from the sw. Northwest Territories south to Mont., east across Canada and N.D. to Lab. and coastal Nfld. Also breeds in scattered colonies around the Great Lakes, along the St. Lawrence R., in n. Maine, and coastally from Nfld. south to S.C. In the U.S., winters in very small numbers from Tex. east to the Florida Keys. In migration, occurs routinely across the e. U.S. and in states bordering the Pacific Ocean, but may occur almost anywhere. **HABITAT:** In coastal areas, nests on sandy and cobble beaches or rocky islands free of tall vegetation; forages in both nearshore and offshore waters. In the interior, prefers treeless rocky or gravel-covered islands in lakes (occasionally rivers). Also nests on spoil, piers, gravel roofs, and matted wrack in tidal estuaries (as does Forster's Tern). **COHABITANTS:** Other terns, including Arctic, Forster's, Least, and Black Skimmer; also smaller gulls. **MOVEMENT/ MIGRATION:** Spring migration from late April to June; fall from July to early November. VI: 3.

DESCRIPTION: A handsome, trim, fairly compact tern that is darker and more angular than the similar Forster's Tern. The bill is usually shorter and not so drooping as Forster's. The tail does not extend beyond the wings, and the legs are shorter. In sum, a more compact bird.

Adults in breeding plumage are pale gray above, with a glossy jet-black cap. Folded wingtips are dark gray to almost black (darker, in all cases, than the back). Underparts are also pale gray (Forster's Tern's are white), leaving the cheeks white. Bill is bright red-orange with a black tip. Legs are the same color as the bill (though, when standing, usually only the tarsus is visible).

Many Common Terns retain vestiges of the gray underparts into September, but by then most birds are changing into nonbreeding plumage—black bills and legs, pale underparts, and a dark carpal bar on the bend of the wing. The forehead becomes white, turning the black cap into a skullcap that wraps completely around the nape.

By fall, juveniles are gray (brown-backed when newly fledged) and show the black skullcap and dark carpal bar of winter adults.

BEHAVIOR: Spends much of its time (in fact, much of its life) in the air—foraging, traveling to feeding areas, or migrating. Gathers in large—sometimes very large—flocks, often with other terns and gulls. Feeds fairly close to shore (within 1 mi.), but may travel many miles to reach prime foraging areas. Often feeds in large numbers, frequently over areas of tidal turbulence, in shoal water, or where predatory fish are forcing smaller fish to the surface. Fishes by cruising over the surface at altitudes of 3–15 ft. and plunges when prey is sighted, often submerging itself. Often hovers over prey before diving, and also picks prey from the surface without diving.

Sometimes sits on floating objects offshore, but does not commonly rest on the water. A consummate klepto-parasite, Common Tern pirates prey from other terns.

FLIGHT: Overall slightly more compact than Forster's Tern, with wings that appear wider, shorter, and more symmetrically triangular. Uniformly gray upperwing tips are bisected by a distinct dark wedge. Forster's Tern may also show darker worn flight feathers, but the entire wingtip becomes opaquely shaded and there is no impression of a narrow well-defined wedge. Common's flight is graceful, direct, and slightly stiffer, crisper, and more perfunctory than Forster's. Wingbeats are steady, with a slight skipping or halting quality—a quick shallow upstroke followed by a deeper pushing down stroke that hesitates momentarily before the next beat.

VOCALIZATIONS: Classic call is a harsh, angry, drawn-out, and descending "*KEE-uuur*"—held longer than Forster's Tern's more nasal, less angry-sounding "*zurrrr.*" Also makes a harsh "*keh, keh, keh, keh*" that often precedes the "*KEE-uuur.*" Attack call is an angry chattering prelude followed by a louder, rising, nasal, two-noted scream: "*Kak-Kak-Kak-Kak—Ky-AAK.*" When you hear this call, it is in your interest to be elsewhere.

Arctic Tern, *Sterna paradisaea*

Mirror-winged Tern

STATUS: Common arctic and subarctic breeder; uncommon pelagic migrant; rarely seen from shore away from breeding areas; rare inland. **DISTRIBUTION:** Circumpolar breeder. Nests throughout

Alaska and across n. Canada from nw. B.C., the Yukon, the Northwest Territories, nw. Sask., n. Man., n. Ont., cen. Que., and Lab. Also breeds coastally in Nfld. and south along the Atlantic Coast to Mass. A very few isolated pairs nest in Wash. and Mont. **HABITAT:** Nests on open sparsely vegetated habitat near water, including lake and ocean shores, islands, sandy beaches, sand and gravel bars, tundra, marshes, bogs, and short-grass meadows. Forages over streams, ponds, marshes, and coastal waters. Migrates over coastal marine waters, primarily well offshore. **COHABITANTS:** During breeding, Long-tailed Duck, eiders, Red Phalarope, jaegers, Common Tern, Aleutian Tern. In migration, found in mixed flocks with other terns, especially Common, Forster's, and Roseate. **MOVEMENT/MIGRATION:** The torch-bearer among long-distance migrants, Arctic flies round-trip from the Arctic to the Antarctic. Spring migration from March to early June, with a peak in the coastal United States in May and early June; fall from July to November, with a peak from August to mid-September. VI: 3.

DESCRIPTION: A slender, elegant, pixie-featured tern with wings the color of a mirror's surface. Medium-sized (larger than Least; smaller than Common), with a petite straight bill, a smallish rounded head, a slender body, and a long streamer tail that extends beyond the tips of folded wings. Legs are stubby—shorter than Common Tern—and set well forward so that the bird appears to be resting on its feet (or standing in a hole). In spring and summer, the bill is mostly or wholly deep rich red, but particularly early in the nesting season, individuals may show varying amounts of black toward the tip.

Overall paler and more uniformly gray than Common Tern. The narrow white cheek patch sandwiched between the black cap and gray underparts is very conspicuous. Juveniles lack the long streamer tail feathers of adults (wings are longer than the tail), but like adults, they appear short-billed and short-legged. Also on juveniles, the folded wing shows a dusky trace of a carpal bar. (On young and nonbreeding adult Common Tern, the carpal bar is bold and blackish.)

BEHAVIOR: Arctic Terns have a dualistic nature: birds breed in colonies (often with Common or Aleutian Tern), but across most of the arctic (where Arctic is the only tern) nest as single pairs on open tundra. Feeds with other terns and gulls in ocean waters, but frequently forages alone on tundra lakes, rivers, and marshes. Feeds by plucking food from the surface (like Black Tern) and by diving (like Common Tern). Walks with a mincing shuffle. Commonly perches on rocks, emergent portions of submerged branches, road signs, utility lines, logs, boards, and other floating debris at sea, but at sea also rests on the water. Very noisy around colonies and very aggressive toward intruders. Usually tame, allowing close approach.

FLIGHT: A slight, slender wisp of a tern that makes Common Tern appear clunky. Appears small-headed (at closer range, note the straight petite bill) and slight-bodied, but with very long, very slender, sweptback wings and a long slender tail that flutters when the bird is hovering and ripples slightly and seems upturned at the tip in level flight. Arctic's sweptback wings are more curved or acutely angled compared to Common and Aleutian Terns. At a distance, Arctic seems no-headed and all long slender wings.

Overall gray but more pallid, less contrasting, and more monotoned than Common Tern. (Common Tern shows darker upperparts and paler underparts.) Wingtips show virtually no black at the tips. The entire upperwing is uniformly silver gray.

Flight is light, fluid, and slightly bouncy, with slow, deep, caressing wingbeats and a quick upstroke. Overall more buoyant, less plodding, than Common Tern. If Arctic Tern waltzes, Common Tern does a polka.

VOCALIZATIONS: Call is a single-note "*Eeeeeh*" and a rougher, ascending, two-noted "*eeeeah*" that is higher-pitched and not as drawn out and angry-sounding as Common Tern. Also emits a high, short, mellow "*pip pip*," as well as a sharper "*kip*" or "*chih*." Arctic Tern makes many of the sounds that Common Tern makes, but softer, higher-pitched, and more mellow.

PERTINENT PARTICULARS: Forget what you may have heard about a narrow black trailing edge to the primaries. Narrow is relative. Concentrate on the upperwing. It is all silver gray—like a mirror. Common Tern's upperwing is darker gray with a black wedge at the tip. Forster's Tern's is two-toned—gray on the arm, frosty on the hand. Roseate Tern's is white with an obvious blackish leading edge. Aleutian Tern is overall darker with a contrasting white tail.

Forster's Tern, *Sterna forsteri*

White-breasted Tern

STATUS: Common breeder and winter resident. A medium-sized tern seen in the interior of the United States in summer or anywhere in the United States in winter is likely this species. **DISTRIBUTION:** Primary breeding areas lie in the Great Basin, the northern prairies, the Great Lakes, and coastal portions of N.J., Del., Md., Va., N.C., Ala., Miss., La., Tex., and Calif. (San Francisco Bay and the Salton Sea). Winters coastally from Va. to Mexico and throughout Fla., Ala., Miss., La., and e. Tex., and on the Pacific from n. Calif. south to Central America. Widespread in migration. **HABITAT:** A marsh tern, Forster's breeds in fresh-, brackish, and saltwater marshes, including the narrowly vegetated edges of lakes and ponds. In winter, occupies beaches and forages in nearshore waters, bays, and inlets, particularly around jetties. Infrequently encountered out of sight of land. **COHABITANTS:** Associates with other terns. **MOVEMENT/MIGRATION:** Spring migration begins in March (in coastal areas), but northern interior breeders may not arrive until May. Generally a late-fall migrant; after peak movements in October, some birds linger into early December in coastal areas north of the winter range. Regular post-breeding dispersal occurs (in small numbers) north to New England. VI: 2.

DESCRIPTION: A rangy whitish marsh tern with white underparts and a two-toned upperwing. Medium-sized (slightly larger than Common; smaller than Sandwich), with a longish bill that is slightly drooped, a slender tubular body, and a long tail that extends well beyond the folded wings.

Overall distinctly paler than Common Tern, and more shabbily or loosely feathered. From April to early July (sometimes later), the bill is orange (Common Tern's is orange-red), the blackish cap is full, the upperparts are pale gray, and underparts are white. (Breeding Common and Arctic Terns show grayish underparts.) Folded wingtips are pale gray (not blackish), and the cap is not quite as jet-black as Common Tern.

Forster's Tern loses its cap earlier than Common. By August, most adults (and all young) have a black bill and a black patch or teardrop etched over the eye that easily distinguishes them from winter Common Tern (which wears a dark skullcap that curls completely around the nape). Some nonbreeding Forster's Terns show a grayish wash across the nape, but the black eye patch contrasts markedly with it. Juveniles are like winter adults but have buffy upperparts (and no carpal bar, as do juvenile and nonbreeding Common Terns).

BEHAVIOR: Forster's Terns spend a great deal of time sitting on piers, beaches, and spoil islands in flocks, often mixed with other terns, gulls, and other birds. When foraging, Forster's Tern cruises less than 20 ft. above the surface of the water with its bill angled down to dive for fish swimming near the surface; also hovers in place, then dives. Prefers more sheltered areas (would choose a lagoon over open ocean waters), and though more coastal in winter, not commonly seen far offshore. Nests on accumulated dead vegetation (such as a muskrat house or tidal wrack). Noisy and gregarious, Forster's frequently vocalizes while feeding. Does not commonly rest on the water. Does not steal prey from other terns.

FLIGHT: Adult Forster's is overall whitish, slender, and gangly, with wings that appear long and narrow for the body. (Common is compact, angular, trim-lined, well proportioned, and overall darker and grayer.) In addition to being overall paler, the upperwing of Forster's is also two-toned—light gray from the bend of the wing to the body, frosty white from the bend of the wing to the tip. *Note:* This contrast is often more evident when you don't use binoculars. Distance tightens contrasts. In late summer and fall, the black eye patch on adults and young stands out at a considerable distance. The leading edge of the arm is pale on Forster's, shadowy or dark on Common Tern. Forster's flight is fluid and looser, with deeper, somewhat more languid wingbeats than Common.

VOCALIZATIONS: Classic call is a soft, descending, nasal buzz, "*zurrrr*," that is lower-pitched and not as harsh and two-noted as Common. Also makes a sharp "*kick*" call, which may be repeated several times and often follows the "*zurrrr*" call. Also, during aerial courtship, chants rapidly and excitedly, "*kan-katakan, katakan, katakan, katakan.*"

Least Tern, *Sterna antillarum*

Easily Piqued Beach Pixie

STATUS: Often common, widespread, highly localized coastal, riverside, and reservoir breeder. **DISTRIBUTION:** Breeds locally from s. Maine to n. Mexico and in coastal Calif. north to San Francisco Bay. In the interior, breeds locally along the Mississippi R. north to Ill. and along the Red R., the Arkansas R., the Platte R., and the Missouri R. north and west to Mont.; also breeds in scattered inland locations in S.D., Neb., Okla., Colo., N.M., and Tex. Winters south of the U.S. (casual in extreme s. Tex.). Rare spring and summer visitor to se. Calif. and s. Ariz. **HABITAT:** Most often found nesting on sandy coasts, rivers rich in stable sandbars or islands, and reservoirs with sandy shores; also on alkaline flats. Has also acclimated itself to dredge spoil, gravel roofs, and open sand mines situated near waters rich in small fish. Nests above the high-tide line where vegetation meets open beach. Feeds in shallow waters, and prefers sheltered inlets, bays, and tidal creeks to rougher open water. **COHABITANTS:** Snowy and Piping Plovers often breed in and near Least Tern colonies; also Common Tern and Black Skimmer. **MOVEMENT/MIGRATION:** Migration is compressed considering the geographic span. Birds appear on the Gulf Coast in late March and reach New England by early May. Birds in the northern interior are on territory by mid to late May. In fall, birds depart from nesting areas not long after chicks have fledged: June and July in the South, later in more northern parts of the range. Stragglers very rarely persist into October; the majority depart by late August and early September. VI: 2.

DESCRIPTION: A high-strung pixie of a tern best known for its diminutive size and intolerance for intruders. Least Tern is short-bodied, short-tailed, and long-winged (folded wingtips reach beyond the tail) — all angles and all energy. It appears only half the size of medium-sized terns.

Least Tern is very pale gray above, very white below. In summer, it is distinguished by its white forehead, thin bright yellow bill, and yellow legs. Post-breeding adults have black bills and a head pattern resembling Common Tern or Arctic Tern. Black-billed juveniles are buffy above, their backs are finely vermiculated, and their head pattern (dark eye patch and grayish wash) is most like Forster's Tern.

BEHAVIOR: Very gregarious, spending much of its time in and around breeding colonies. After the nesting season, birds gather in largely homogenous flocks on the beach, but may also mix with other tern species. As noisy as they are social, Least Tern colonies are distinguished by the birds' active and vocal courtship.

A very active feeder, with the head angled down, Least Tern moves back and forth 10–20 ft. above the surface, constantly hovering, swooping low (like a Gull-billed Tern), and diving frequently. Dives almost twice as much as other terns, and its recovery from a failed dive is almost as fast as the descent. Walks quickly and well, with a bit of a waddle. Intolerant of intruders, including humans, and hovers and stoops at them until they quit the area. Nervous and high-strung. Flushes easily when sitting, frequently jerks its head from side to side.

FLIGHT: In flight, shows a compact, front-heavy, long-winged profile. The head/bill appears long and projecting; the tail is short, more notched than forked, and lacks long outer tail feathers. Wings are exceedingly long and narrow for the bird, and very angular, jutting forward at the wrist. This forward jut of the wing, combined with the projecting head and short tail, gives the bird a front-heavy appearance.

Direct flight is rapid, straight, buoyant, and slightly bouncy. On the wing, birds seem hurried and high-strung. Rapid, deep, and stiff wingbeats dominate the silhouette and control the flight of the bird—a tern inspired by a swift. Very acrobatic—the bird can fly one way, turn, and fly another way without skipping a beat. Even micro-adjustments in angle and in ascents and descents seem controlled entirely by the wings—the body is just going along for the ride.

VOCALIZATIONS: A rapid, excited, high-pitched, and screechy chant, "*Ir-ree ch'dik*" or "*rrrE-kidik*," repeated five to six times. Also makes an excited high-pitched "*p'cheet*." When annoyed, emits a high-pitched and often repeated "*pwik, pwik, pwik*" or "*preek, preek, preek*." Attack call is a lower-pitched, raspy, attention-getting snarl: "*Ehrrrch!*"

Aleutian Tern, *Sterna aleutica*
Darktic (dark + arctic) Tern

STATUS: Fairly common but geographically restricted and coastal. **DISTRIBUTION:** In North America, found in scattered colonies along the west and south coasts of Alaska, including the Aleutian Is. **HABITAT:** Coastal tern. Breeds on flat islands and on the protected inside of barrier beaches. Substrate is usually vegetated with dwarf shrub, grasses, or a mix. Also reported to breed on freshwater lakes and in coastal marshes. **COHABITANTS:** Arctic Tern. Less commonly Mew Gull. **MOVEMENT/MIGRATION:** Wholly migratory. Arrives in Alaska between early May and early June; departs during August. VI: 0.

DESCRIPTION: A shadowy gray tern with a contrasting short white tail seen on the Alaskan coast. Overall more compact than Arctic (the only other tern found within this species' North American range), with an all-black bill and wings that extend well beyond the relatively short forked tail. (On Arctic, the tail extends beyond the wings.)

Uniformly dark gray in breeding plumage—except for the whitish face and white tail—and *distinctly* darker than Arctic Tern (almost as dark as the back and wings of Black Tern). The white forehead and black line through Aleutian's eye are easily noted. Has a black bill and legs (compared to red in breeding Arctic). In winter, shows white underparts. Juvenile is heavily marked with rufous orange above and has a gray tail that is paler than the upperparts.

BEHAVIOR: A social tern that breeds in colonies (often with Arctic Terns) but is not particularly gregarious. Usually feeds by itself or with other Aleutian Terns. Where the two species are working the same stretch of water, one species-group works through, then the other. Forages almost exclusively by swooping down and plucking prey from the surface (like Black or Gull-billed Tern). Appears to favor turbulent water (near inlets and tidal rips or behind breakers), where food is carried to the surface, but may also be found up to 5 mi. offshore. Sits on floating debris. At times, travels high over land in large noisy groups.

FLIGHT: A long-winged short-bodied tern. Longer-headed, shorter-tailed, and wider-winged than Arctic. The white forehead seems abrupt, and the black bill longish and, in level flight, snobbishly elevated. Aleutian is not so much a slender tern, but it is certainly an angular one.

Uniformly dark gray with a bright white tail. Arctic Tern shows more contrast below between the gray belly and whitish (almost translucent) underwing. The narrow dark trailing edge of Aleutian's secondaries is easy to note except at a distance. Also, the upperwing of Aleutian is slightly two-toned—darker at the base, paler toward the tip. (Arctic's upperwing is uniformly pale gray.) Flight is strong, direct, and often high; wingbeats are quick, crisp, deep, and angrier than Arctic Tern, with an emphatic down stroke and a slight hesitation on the upstroke.

VOCALIZATIONS: Call is an excited or anxious, descending, twittering chuckle. Also makes a high squeaky "*wheetoo*" that recalls a sharper version of a House Sparrow chirp.

PERTINENT PARTICULARS: With so many other ways to tell Arctic from Aleutian, this one seems unnecessary, but Arctic hunts with its bill pointed down, almost vertical. Aleutian holds its bill more horizontal.

Bridled Tern, *Sterna anaethetus*

Pelagic Flotsam Tern

STATUS AND DISTRIBUTION: Breeds throughout the Caribbean (very rare breeder in the Florida Keys). Fairly common pelagic species off the s. Atlantic Coast and in the Gulf of Mexico. Uncommon to rare north to Md. (very rare to the waters off s. New England), mostly in late summer. Rare vagrant inland (following tropical storms). **HABITAT:** Forages exclusively in warm marine waters, especially the Gulf Stream. Most commonly found along sargasso weed-lines and habitually seen perched on floating debris. **COHABITANTS:** Cory's Shearwater, Audubon's Shearwater, Sooty Tern. **MOVEMENT/MIGRATION:** Found in southern waters March through November; off North Carolina May through October; off Maryland June through September. VI: 2.

DESCRIPTION: A dark smoky gray (not black), black-capped and black-masked tern often seen perched, with its hefty-looking tail and wings slightly elevated, on a board in warm marine waters. Medium-sized (larger than Common; slightly smaller than Sooty; about the same size as Brown Noddy), with a longish bill, a long-bodied profile that appears inelegantly thick and heavy-tailed, and longish legs set well forward.

Upperparts are dark grayish to brownish gray, and underparts are bright white. Contrastingly blacker head pattern shows a narrow black eye-line with the white eyebrow extending past the eye. The white-sided tail is sometimes hard to see on perched birds. Shorter-tailed juveniles are paler gray above, with only a shadowy rendering of the dark (brownish) cap and dark line through the eye. Adult Sooty Tern is truly black above; juvenile Sooty Tern shows sooty above with a sooty breast (much more like the smaller Black Tern than Bridled Tern).

BEHAVIOR: Habitually sits on floating debris. (Sooty Tern almost never perches while at sea.) Forages by flying close to the ocean surface (under 30 ft.), swooping toward the surface, and snatching prey (in the manner of Gull-billed and Black Terns). Sometimes dives, but does not totally submerge. Mostly solitary or in pairs, but sometimes forms small opportunistic flocks over baitfish. Also joins other seabirds, including Cory's and Audubon's Shearwaters, Wilson's Storm-Petrel, and Sooty Tern. Moderately shy. Flies when crowded by boats, but often lands on the next closest bit of floating debris.

FLIGHT: Very long narrow-winged profile, with a long deeply forked tail. Upperparts show a tricolored pattern—a black cap, a pale gray nape (sometimes not easy to see), and a charcoal back. Underwings appear white below with darker tips. (The undersides of Sooty Tern's primaries are more extensively dark.) The white sides of the tail are very conspicuous when the bird flairs the tail (as it does when actively feeding). Flight is smooth and buoyant; wingbeats are slow, elastic, deliberate, and shallow, on wings that are slightly down-bowed. Feeding birds are graceful and acrobatic, incorporating wheeling, banking, gliding, hovering, and dipping as part of their routine.

VOCALIZATIONS: Calls include a soft rising whistle and an emphatic repeated barking when disturbed.

Sooty Tern, *Sterna fuscata*

Pelagic Air-Tern

STATUS AND DISTRIBUTION: Tropical tern. In the U.S., a few breed on islands off the coasts of Tex. and La., and quite a few breed off the Dry Tortugas in the Florida Keys. Ranges north to the waters off N.C. (mostly in the Gulf Stream). Hurricanes occasionally ferry birds inland to the Great Lakes and as far north as the Maritimes. Regular visitor to offshore Gulf of Mexico. Casual to s. Calif. **HABITAT:** Air! Except for hurricane-spawned inland occurrences, this strictly marine species is found over (not commonly on) warm, deeper, offshore waters. **COHABITANTS:** Brown Noddy, Bridled Tern, Tuna. **MOVEMENT/**

MIGRATION: Present in the Gulf of Mexico and southern Florida from March to November; off the Carolinas from June to October. VI: 3.

DESCRIPTION: A distinctive and conservative-looking black-and-white tropical marine tern. Medium-sized (slightly larger than Common or Bridled Tern; smaller than Royal Tern or Black Skimmer; about the same size as Brown Noddy), with a long straight bill and a long, somewhat front-heavy and short-legged profile.

All black above (no pale collar) and white below. A crisp face pattern shows a white forehead and black crown. (Bridled Tern shows a black crown and a white eyebrow that extends over the eye.) Juvenile Sooties are mostly brownish black (beaded with white spots) and have sooty heads and breasts. (Juvenile Bridled Tern has a mostly white head and all-white breast.) Sooty plumage somewhat resembles adult breeding adult Black Tern, but juvenile Sooty appears almost twice the size of Black Tern.

BEHAVIOR: Outside the breeding season, spends its life aloft, day or night. Almost never recorded sitting on the water. Rarely lands on debris (as does Bridled Tern). Hunts from 3–60 ft. over the ocean surface, swooping and plucking prey from the water (without submerging) or snatching small fleeing fish in midair. Often flies and soars at considerable heights.

As social as Bridled Tern is not. Although it is sometimes seen alone or in pairs, Sooty is more often found in the large aggregations of mixed seabirds that gather over schools of predatory fish or mammals (such as tuna and whales).

FLIGHT: Large, long-winged, long-tailed tern that appears broader-winged, shorter-tailed, and overall slightly more robust than Bridled Tern. Distinctly black above (no contrast, no pattern, and in especially good light may be tinged brown), and white below except for the black trailing edge of the wing. The narrow white trim on the outer tail feathers is very hard to see. Mostly all-dark juveniles show a white belly and undertail and white underwing linings. Flight is strong and powerful, with deep, purposeful, jaegerlike wingbeats. Soars extensively and glides for extended periods.

VOCALIZATIONS: A quick high-pitched "nasal laugh," usually with two to three syllables: "*wida-a*" or "*wide-a-wake.*" May be repeated ("*wideawakewideawake*").

Black Tern, *Chilidonias niger*

Shadow of a Tern

STATUS: Common to uncommon widespread but localized breeder across the northern United States and southern Canada; uncommon to fairly common migrant; nonbreeding birds are uncommon to common in summer along the Gulf Coast and at the Salton Sea in California. **DISTRIBUTION:** Primary breeding area is the northern prairies. Range includes widespread portions of B.C., Alta., sw. Northwest Territories, s. and cen. Sask., s. and cen. Man., all but n. Ont., s. Que., and s. N.B. In the U.S., breeds locally in n. Calif., Ore., e. Wash., Idaho, n. Nev., n. Utah, Mont., Wyo., Colo., N.D., S.D., w. Neb., cen. Kans., Minn., n. Iowa, Wisc., n. Ill., Mich., n. and s. Ind., n. Ohio, ne. Pa., w. and n. N.Y., and n. and cen. Maine. **HABITAT:** Breeds in shallow and well-vegetated freshwater marshes and in lakes and ponds rimmed by emergent vegetation. In migration, found in all manner of fresh, brackish, and marine habitats, and often seen well offshore, as well as on dry fields and tilled cropland. **COHABITANTS:** During breeding, Lesser Scaup, Sora, Marsh Wren, Red-winged Blackbird; in migration, often found amid groups of other terns. **MOVEMENT/MIGRATION:** Spring migration from late April to early June; fall migration (beginning very early) from early July to early October. Nonbreeding birds may appear throughout the summer. VI: 2.

DESCRIPTION: A small shadow-colored tern that picks instead of plunges. Small and compact (smaller than Common or Forster's Tern; slightly larger and plumper than Least), with a short bill, a round head, short legs, and relatively long wings that extend well beyond (in fact eclipse!) the short square-tipped tail. Profile is long and slender, with legs that seem set too far forward.

In breeding plumage, shows a black head and body and shiny gray upperparts (you'll probably never see the white vent). No other extra-tropical

North American tern has an all-black head. In late summer, loses the black except for a solid dark cap, but the upperparts remain a smoky gray that is several shades darker than the other terns with which Black Tern may associate (except for southern pelagic species like Bridled and Sooty).

BEHAVIOR: Very social. Breeds in small loose colonies. Most often seen feeding in loose flocks of birds coursing slowly, low over the water (3–4 ft. when windy; 10–12 ft. when still), all moving in the same direction (into the wind). Each bird, in turn, stalls, rises abruptly, and swoops toward the surface, often with a nimble banking turn. Prey is most often plucked from the water. Sometimes birds plop momentarily to the surface. Almost never do they plunge-dive. Black Tern also flycatches (over grassy areas of the marsh or over land) and tail-chases larger insects.

Perched birds are often seen on an elevated perch. In migration, in coastal locations, they sit and feed among other tern species, where their darker color betrays them. During the summer (also in coastal locations), nonbreeding birds sometimes associate with Least Terns.

Fairly tame, but very defensive when breeding.

FLIGHT: An overall compact bobtailed body wedded to very long, broad-based, pointed wings. Feeds with the bill turned down. The tail is short, slightly forked, and innocuous when bird is in level flight, but roundly fanned when it wheels and maneuvers. Flaps with wings slightly down-drooped. Overall seems more roundly contoured, not quite as angular, as other terns.

Breeding birds are distinctive—a frosty smoky gray with a beefy black chest. The pattern and colors (even the shape somewhat) recall a breeding-plumage Black-bellied Plover. Underwings (and tail) are paler still—silver gray—and it is this flash of silver that catches the eye when distant birds bank and wheel. Head on, birds show a pale leading edge to the wing that contrasts with the black head. In winter, Black Tern looks like a small shadow-colored tern with a narrow white collar. The black cap and a dark blotch on the side of the breast are easily noted.

When feeding, flight is buoyant and languid, with mostly regular shallow wingbeats that seem both batty and floppy—looks like a slow-flying Tree Swallow. In normal flight, wingbeats become fast and deep, and the flight erratic, wheeling, nighthawklike. From below, wings seem longer and more slender. High-flying birds often vocalize.

VOCALIZATIONS: Calls include a high, harsh, nasal squeak, "*ehk*" or "*eek*" or "*wee'uk*," uttered when the bird is agitated. Also emits a high, squeaky, twittering laugh, "*reh'heheh*," that at a distance may recall Least Tern. Also makes a soft nasal "*ke/uh*" or "*ye/ah*."

Brown Noddy, *Anous stolidus*

Upside-down Tern

STATUS AND DISTRIBUTION: Tropical marine tern. In the U.S., breeds only on the Dry Tortugas sw. of the Florida Peninsula. Ranges north to s. Fla. and adjacent waters. Casual in the Gulf Stream north to N.C. Extra-Florida appearances are hurricane-related. Winters at sea in warm tropical waters. **HABITAT:** Warm tropical seas. When breeding, commonly forages within sight of nesting colonies; is not as pelagic as Sooty Tern. **COHABITANTS:** Breeds in association with Sooty Tern and Black Noddy (although the latter species does not breed in the United States). **MOVEMENT/MIGRATION:** Absent from the Dry Tortugas between November and mid-January. VI: 1.

DESCRIPTION: An all-brown jaegerlike tern with a silver white crown—just the opposite of the pale-bodied, dark-capped pattern of most terns. Medium-sized (slightly smaller than Laughing Gull and Parasitic Jaeger; slightly larger than Black Noddy; about the same size as Sooty Tern), with a longish, slightly down-drooped bill, a very long, slender, long-winged body, and short legs.

All dark brown except for the white crown and (at close range) grayish face. Immatures are just all brown. Except for the very similar Black Noddy, no other tern is all dark.

BEHAVIOR: Colonial nester. At sea, often found foraging over concentrated baitfish with other pelagic species. Usually flies within a few yards of the surface. Does not dive. Hovers over the surface, swooping to snatch prey while remaining airborne or belly-plops to the surface and grabs. Also settles on the surface and swims (like a gull). Often found concentrated over predatory fish (like tuna) in tightly packed flocks (often mixed with other species).

A tame bird that commonly lands on ships, floating debris, and the backs of sea turtles. Also nimble enough to perch on the branches of trees.

FLIGHT: Looks like a dark, floppy, scraggly-tailed tern. Overall slender with long slender wings. The bill is angled slightly down; the tail is usually closed, so appears long, stiff, narrow, and blunt. When fanned, the tail is curiously shaped—somewhat lobed or wedge-shaped, it often looks scraggly near the tip (very unternlike). Overall brown except for the pale forehead on adults, which is visible at great distances. (Head on, the bird appears completely white-headed.) Upperparts are slightly paler than underparts. (Jaegers are paler below or uniformly dark.) Smaller, darker Black Noddy appears uniformly blackish.

Flight is direct and commonly below 10 ft., with rapid wingbeats that are looser, floppier, and more languid than similarly sized terns. The upstroke is unnaturally high (somewhat like Black Skimmer, but not so exaggerated). The bird tilts from side to side, favoring one side and then the other, and seems to lean forward in flight, giving the impression that it's always flying into the wind.

Occasionally glides short distances but rarely soars.

VOCALIZATIONS: Quiet at sea—and curiously, also at nesting colonies. Adult's call is a crowlike caw.

PERTINENT PARTICULARS: The very similar Black Noddy, *Anous minutus* (VI: 1), is an irregular visitor to the Dry Tortugas and is otherwise exceedingly rare in U.S. waters. (It has been recorded in Texas, for example.) Black Noddy is smaller and slimmer, with a longer, thinner bill, and entirely blackish or brownish black, except for a white cape that is often more extensive and crisply defined than on Brown Noddy and a tail that may appear slightly paler than the rest of the bird. In spite of their generally darker plumage, Black Noddies showing worn plumage are slightly paler and browner—particularly on the wing—so may look more like Brown Noddy.

Black Skimmer, *Rynchops niger*

Beach Scissor-bill

STATUS: Common coastal species. **DISTRIBUTION:** Along the Atlantic, breeds from coastal Mass. south to coastal Central America; winters from S.C. south. In Calif., found as a resident restricted breeder and nonbreeder from about Santa Barbara south (including the Salton Sea); also suspected of breeding in San Francisco Bay. **HABITAT:** Nests on sand or cobble beach as well as man-made dredge-spoil islands. Forages primarily in sheltered estuaries, tidal creeks, bays, lagoons, rivers, salt marsh ponds, harbors, and beaches with low-energy surf. **COHABITANTS:** Piping and Snowy Plovers, Laughing Gull, Royal, Common, and Least Terns. **MOVEMENT/MIGRATION:** On the Atlantic, birds breeding north of South Carolina vacate breeding areas in advance of winter. Spring migration from mid-March to early June; fall from late September to early December. VI: 1.

DESCRIPTION: A distinctive black-and-white beach bird with an outlandishly proportioned and structured bill. Largish and ternlike (larger than a Common Tern; smaller than a Royal Tern; about the same size as a Laughing Gull), with an overall low profile, a large flattish head, a long heavy-based bill with an extending lower mandible, a very long narrow body, and even longer wings that extend beyond the tail.

Adults are black above, white below. The long straight bill is black with a bright red base.

This bird is unmistakable, so don't labor over the identification. It hardly seems worth mentioning that in nonbreeding birds the hindneck is white. Young birds are structurally similar to adults but mottled brownish gray above, white below. The red on the bill is not so bright.

BEHAVIOR: Most commonly seen flying low over shallow nearshore water with the body angled down and the bill open, the lower mandible cutting through the water. Young birds sometimes run through shallow water, skimming. Black Skimmer rests in large tightly packed flocks on beaches, islands, shorelines, gravel roofs, and parking lots. When approached, flocks rise in long flowing waves that wheel in a synchronized manner and commonly return to alight in the same spot.

Feeds primarily at low tide and dusk, but is also active at night. Walks well. Often calls when feeding. Fairly social; several birds work up and down the same stretch of water together. Young birds in colonies are often seen crouched or sprawled in the sand (looking half-dead).

FLIGHT: Almost always low, just above the surface of the water. Silhouette is unmistakable—a slender bird that flies with its outlandishly long bill (and often entire body) angled down and long slender wings raised theatrically high above the body. Wingbeats are slow, measured, and slightly elastic, each stroke raised high above the body but not below. Flocks, flying with unsynchronized wingbeats, seem to flail the air. Wheeling flocks, turning as one on set wings, move in a long flowing line that flashes all black, then all white. Birds land with several braking steps.

VOCALIZATIONS: Call is a nasal terrier-like yap: "*yaah, yaah, yaah, yaah.*"

ALCIDS—AUKS, MURRES, AND PUFFINS

Dovekie, *Alle alle*

Bumblebee Auklet

STATUS AND DISTRIBUTION: Abundant breeder in the eastern high arctic, nesting primarily on the coast of Greenland. (Small numbers are also found on St. Lawrence I. in the Bering Sea.) Pelagic in winter; common in northern coastal and offshore waters south to coastal Nfld. and the Grand Banks and usually found south to Georges Bank. Irregular (but at times abundant) off the New England coast; occurs annually south to off N.C. (very rarely farther). After December, rarely sighted from shore. Occurs casually inland in the Northeast, mostly in November and December after storms. **HABITAT:** Nests on talus slope along seacoasts in the high arctic. Pelagic in winter, concentrating (in large flocks) along the shelf edges of underwater drop-offs (most notably the Grand Banks). Also seems concentrated along temperature breaks where warmer water supports a plankton bloom. **COHABITANTS:** Other alcids, but tends to breed and forage in single-species flocks. Large-scale influxes of this species often coincide with, or are heralded by, onshore "crashes" of Leach's Storm-Petrel. **MOVEMENT/MIGRATION:** Spring migration from April to early June; fall movement from late August to December, with birds appearing off the eastern United States from late October through December. This species engages in infrequent, spectacular, massed irruptions (or crashes) during which many thousands of birds are seen along seacoasts and, in rare historic instances, hundreds of birds turn up inland. These irruptions appear to be related to food, not storms. VI: 2.

DESCRIPTION: A tiny, chunky, black-and-white alcid whose shape and whirring wings recall a bumblebee at sea. Very small and starling-sized (appears half the size of Atlantic Puffin and Black Guillemot; slightly smaller than Parakeet Auklet; almost twice the size of Least Auklet), with an abbreviated finchlike bill, a large turtlelike head, a short thick neck, and a compact body. At sea, when feeding, often crouches on the water, with wings drooped to the side. At a distance, appears to have no bill, almost no head, and almost no tail—looks like a black-and-white Nerf football.

Distinctly black-and-white alcid with an all-black head, throat, and breast in summer and a black cap with goggles in winter. At a distance, however, it's easier to note the extensive amount of white behind the eye. In the Atlantic, in winter, Dovekie is most likely to be confused with an immature puffin (which shows a larger bill and a dark grayish face) and perhaps with Razorbill (whose long pointed tail is usually up-cocked). Of course, both puffins and Razorbill are much larger. In summer, Razorbill chicks (fresh at sea) look deceptively like Dovekie. In the Bering Sea, Dovekie is more distinctly black and white than other small alcids except in late summer and fall, when juvenile Least Auklets (which are also black above and white below) abound.

BEHAVIOR: Gregarious at all times. In winter, found in flocks (well away from land), but in extreme southern portions of the bird's winter range, individuals are the norm. Not particularly shy; often allows close approach by ships. Takes off quickly with no start-up bounce on the water.

FLIGHT: Small, plump, blunt-faced, black-and-white bumblebee of a bird that might be mistaken for a shorebird. Whirring wingbeats are too rapid to see. Flight is low, fast, and often darting or tacking.

VOCALIZATIONS: Silent at sea.

Common Murre, *Uria aalge*

Brown-headed Murre

STATUS: Abundant but somewhat localized, primarily northern coastal breeder. Common pelagic species in winter; in the Pacific, south of Alaska, much more likely to be encountered than Thick-billed Murre; in the Atlantic, south of Canada, *less* common than Thick-billed Murre close to shore. **DISTRIBUTION:** Circumpolar. In w. North America, scattered breeding colonies are found on the mainland and offshore islands from nw. Alaska south to Monterey, Calif. In the Atlantic, breeds from the sw. coast of Greenland to Lab. south to the northern coast of the Gulf of St. Lawrence (Que.) and the Bay of Fundy between N.B. and N.S. and possibly soon at Machias Seal I. in s. N.S. In winter, found offshore over deep, cold, continental shelf waters from the limit of sea ice south to s. Calif. (occasionally n. Baja) and, in the Atlantic, to Cape Cod (occasionally to waters off Va.). **HABITAT:** Pelagic species. Breeds on rocky coastal cliffs or flat rocky areas atop isolated headlands and islands. In winter, found in coastal waters over the continental shelf (much more numerous from shore in the Pacific than the Atlantic), but may also occur up inlets and bays devoid of estuaries. **COHABITANTS:** Shares nesting cliffs with many species, including other alcids, cormorants, gulls. Forages with shearwaters, cormorants, gulls, jaegers, other alcids. In winter, often found with Razorbill and Rhinoceros Auklet. **MOVEMENT/MIGRATION:** Varies between populations in degree (from no migration at all by some birds breeding in the Farallon Islands of California to 100% migration of birds in the northern Bering Sea), timing, and distance traveled, but in general spring migration occurs from March to late May; fall from late August to December. VI: 2.

DESCRIPTION: A large slender-billed alcid with a distinctly brownish cast, particularly on the head and nape. Large (slightly larger than Razorbill; slightly sleeker than Thick-billed Murre; much smaller than any loon), with a long narrow bill (relative to Thick-billed Murre), a narrow head, a slender neck, and a long slender body. On land, looks very penguinlike; on the water, the bill held slightly elevated recalls a short-necked Red-throated Loon. (Thick-billed Murre holds its bill more horizontally.)

Overall black, but with a distinctly brownish cast above; white below. Dusky streaks on the flanks are invisible at any distance. In summer, the narrow white eye-ring and pencil-fine line trailing behind the eye of many breeding Atlantic birds may be hard to see. In breeding plumage, the demarcation between the all-dark head/neck and the white breast forms a modest inverted V; on Thick-billed, the white breast intrudes, forming an acute inverted V. In winter, most of Common Murre's face is white; on Thick-billed, only the throat is white.

Although the narrower bill shape and posture are often useful field marks, the best way to tell Common Murre from Thick-billed (or Razorbill) is color. Common Murre is overall browner, particularly on the head, and also slightly paler; thus, the darker brown head contrasts with the slightly paler back, a contrast most evident when birds are in flight.

BEHAVIOR: A highly social as well as numerous bird; estimates place the North American population as high as 10 million. Habitually seen in flocks numbering anywhere from several birds to over 100. Forages in fairly tight groups on the water (but be careful—Pacific Loon may also cluster). Dives for fish from the surface and remains submerged for extended periods.

Most often seen on the water in groups whose presence is usually signaled by the flash of white sides as birds bob in the waves or lie on their sides to preen. Also seen commuting to and from feeding areas. Flies in tight-packed V-formations or lines, skimming just above the water.

FLIGHT: Symmetrically elliptical body with narrow wings that are slender throughout their length. More pointy-faced, less potbellied, and more slender than Thick-billed Murre. Dark above and white below, but overall slightly paler and, when seen in good light, distinctly browner than Thick-billed, with a darker brown head contrasting with a slightly paler back. Wingbeats are rapid, constant, stiff, and somewhat flailing; flight is direct and fast, but hardly nimble or maneuverable. Birds yaw somewhat (exposing underwings) and sometimes elevate their heads slightly.

VOCALIZATIONS: Silent at sea.

Thick-billed Murre, *Uria lomvia*

Blackish Murre

STATUS: A bird of northern seas whose numbers approach staggering abundance in breeding colonies; less common in winter south of breeding areas, where it is seen as individuals or in small groups. Along the west coast of North America from British Columbia south, this species is far less likely to be found than the more southerly breeding Common Murre. Along the Atlantic, from New England south, Thick-billed appears to winter closer to shore; despite the superior numbers of breeding Common Murre, Thick-billed is the murre most likely to be seen by coastal observers. **DISTRIBUTION:** Breeds on coastal mainland and island cliffs in the West from the Arctic Ocean south to coastal Alaska (and, sparingly, south to coastal B.C.); in the East, from Ellesmere I. and Greenland south to the n. Gulf of St. Lawrence. In winter, birds retreat to the southern edge of the ice pack and range south to s. B.C. (where it is very rare) and coastal New England (less commonly farther south). **HABITAT:** Except during infrequent incursions, almost always associated with coastal waters, spending much of its life offshore, foraging over the continental shelf waters for fish and other marine animals, sometimes diving to depths in excess of 200 ft. **COHABITANTS:** Black-legged Kittiwake, Common Murre, other alcids. **MOVEMENT/MIGRATION:** South of winter ice, appears not to range far from breeding areas. The relocation or migration that does occur is accomplished by swimming. In fall, in southernmost portions of the winter range, small numbers of passage birds are seen between November and January (and yes, these are flying). Spring migration is somewhat speculative, but most birds appear to have departed the southern limit of their range by April. This species is known to stage irruptive movements that may be related to the extent of the ice pack or to coastal storms blocking offshore passage. Although these rare conditions may cause birds to occur inland, this has not happened for many decades. VI: 2.

DESCRIPTION: A large thickset alcid that is overall sturdier and more compact than Common Murre. The bill is shorter, thicker, and heavier than Common Murre, the upper mandible curves down, and a pencil-fine white line tracing the gape turns down at the lower face. These bill features make Thick-billed look like it's frowning or grimacing. When sitting on the water, holds its bill relatively straight, not up-angled as does Common Murre. The tail is short and flush with the tips of the wings.

All ages and sexes are blackish above and white below. On the water, adults in breeding plumage are blacker-headed and blacker-backed than Common Murre. In summer, the white underparts form a sharp white wedge on the fore neck (more restricted and rounded in Common). In winter, the white on the head is limited to the throat (no white behind the eyes, as seen on nonbreeding Common).

At close range, the sides and flanks are white, with no smudgy streaking.

BEHAVIOR: Unless standing upright on a nest ledge, this species is almost always seen floating on or flying above salt water. Most commonly seen foraging alone, but may also occur in small groups. On the water, holds its bill horizontal or only slightly raised. Dives for food, and may remain submerged for up to three minutes.

Groups (numbering several individuals to several dozen birds) usually fly in a string just over the surface of the water. When traveling greater distances, birds may fly higher (40–50 ft. above the water) and form shifting, not-very-well-organized V-formations. Birds sometimes fly in mixed flocks—most commonly with Common Murre, but also with other alcids and even with some waterfowl.

FLIGHT: Like Common Murre, the bird is elliptical—humpbacked and potbellied—but overall slightly stockier and paunchier than Common. The bill is straight, not raised. Upperparts are uniformly dark, with little contrast between the head and back. The white flanks extend farther up the sides than on Common. This characteristic and the absence of streaking are particularly useful aids to identification when birds are heading away. Flight is fast and unwavering (somewhat loonlike), with rapid, continuous, effortful, unvarying, and somewhat stiff wingbeats. Takeoffs are labored—the bird uses its wings to pull itself across the surface to gain flight speed. In flight, wings are used primarily for propulsion. Navigation is controlled by the feet.

VOCALIZATIONS: Most commonly heard call is a loud, harsh, croaking "*aH-Rahh,*" which is also uttered in a descending sequence: "*RAH-rah-rah-rah-rah.*"

PERTINENT PARTICULARS: It has been suggested that in some places where Common and Thick-billed Murres nest together, the species can be distinguished by the color of the droppings under their ledge—pink for Thick-billed, white for Common. Insofar as the two species sometimes forage at different depths and target, to some degree, different prey, this observation may not lack for . . . substance?

Razorbill, *Alca torda*

The Last Auk

STATUS: Common but restricted breeder; generally uncommon offshore winter resident, but may be very common at times, such as after a storm. Except for Black Guillemot, the most likely alcid species to be seen from shore, particularly in southern portions of its winter range. **DISTRIBUTION:** Breeds from coastal Lab. to the east coasts of Nfld. and N.S., in scattered locations in coastal Que. and the islands in the Gulf of St. Lawrence, as well as on Machias Seal I. in N.B. Also, small numbers now breed in coastal Maine. Winters in coastal waters from Nfld. south to N.C.; very rarely farther south. **HABITAT:** Breeds on rocky islands and sea cliffs in colder northern waters. Winters in open continental and coastal waters. Most commonly found over sandy bottom in waters less than 200 ft. deep. **COHABITANTS:** At all times of year, may be found with murres, Atlantic Puffin, Black-legged Kittiwake. **MOVEMENT/MIGRATION:** Coastal migrant. Spring migration from early March to late May; fall from September to early January, with a peak in November and December. VI: 2.

DESCRIPTION: Razorbill looks like a blunt-billed penguin. A large thickset alcid (larger than female Long-tailed Duck or Atlantic Puffin; slightly smaller than Common Murre), with a large, anvil-shaped, heavy, blunt bill, a very short stocky neck, a long-bodied profile, and a long pointy tail. Unlike murres and puffins, Razorbill's pointy tail is often raised above the water. With its shorter bill and pointier-faced look, immature Razorbill is more like Thick-billed Murre.

Extremely black above and very white below; on the water, shows lots of white on the sides. In winter, adults and immatures show white behind the eye. Of the two murres, the blacker, heavier-billed Thick-billed is most like Razorbill, but in winter the latter's head is all dark with a white throat patch. Common Murre has white behind the eye like Razorbill, but also has a thinner, pointier bill and is charcoal brown, not black. Atlantic Puffin is smaller, more compact, shows less white on the sides, and has a black chest band.

BEHAVIOR: Usually seen singly or in small groups sitting on (or flying low over) the water. Dives with wings open (as do all alcids) and pointy tail conspicuously displayed. Flies and swims single-file. Groups dive synchronously.

FLIGHT: Looks like a cross between a football and a food processor. Body shape is symetrically elliptical. (Murres appear slightly more lumpy or imbalanced.) Feet are tucked beneath the long tail. (Murres' feet trail, adding to the sense of imbalance.) Razorbills going away show lavish amounts of white on the sides (flank/rump). Underwings are paler than puffin. Wingbeats are rapid, whirring, and slightly slower than puffin. Flight is rapid, direct, and perhaps more shifting and agile than murres.

VOCALIZATIONS: Silent in winter.

Black Guillemot, *Cepphus grylle*

Inshore Atlantic Alcid

STATUS: Common, mostly inshore resident of northern rocky coasts; Beaufort and Bering Sea population is small but expanding. In the Atlantic, the most likely alcid to be seen inshore. **DISTRIBUTION:** In North America, there are two breeding populations. One occurs on the north coast of Alaska and the Yukon. The second, more extensive population ranges from Greenland and many of the eastern arctic islands south into Hudson Bay, the north coasts of Que., Lab., Nfld., and the Gulf of St. Lawrence, south through the Maritimes to the southern coast of Maine. In winter, insofar as the ice pack allows, remains near breeding colonies; some Alaskan birds retreat into the n. Bering Sea, and Atlantic birds concentrate at the edge of the ice pack, showing little expansion south of the breeding range and moving south only to Long Island, N.Y. (where they are scarce). **HABITAT:** Breeds on rocky coasts; forages in shallow waters close to shore. **COHABITANTS:** In winter, Horned Grebe, Eared Grebe, Common Eider, Red-breasted Merganser. **MOVEMENT/MIGRATION:** Varies depending on timing and extent of ice pack. In Massachusetts (south of the breeding area), birds are present late October to early May. VI: 1.

DESCRIPTION: Pointy-billed auk—distinctly black in summer, and distinctly whitish in winter. Medium-sized alcid (slightly larger than Atlantic Puffin; distinctly smaller than Red-necked Grebe; same size as Horned Grebe and Pigeon Guillemot), with a straight, pointy, golf-tee-shaped bill, a slender tapered head, a fairly long (by alcid standards) but thickish neck, and a compact body. Bill is held horizontally. Birds ride high in the water and show a rise or bump toward the rear. In winter, suggests a fat phalarope; at a distance, might be mistaken for Horned Grebe (though grebes are overall darker, particularly darker-capped, and have thinner necks).

In summer, all black except for a large white oval patch on the folded wing. In winter, mostly pale—gray and white—with all-white head/neck and a dusky crown and back. No other alcid is mostly white except adult Pigeon Guillemot, whose range barely overlaps with Black Guillemot in the northern Bering Sea.

At all times, the bright red legs and feet can be seen when the bird dives.

BEHAVIOR: Unlike most alcids, sticks close to shore—even in rocky back bays—and is not ultra-gregarious. Colonies are small; in winter and summer, solitary birds are not uncommon. Unlike grebes, opens wings as it dives. Tame.

FLIGHT: In flight, appears plump and potbellied, small-headed and neckless, with slender, somewhat paddlelike wings and a slightly up-angled body. The bright red feet trail behind the bird. Underwings are bright and unequivocally white (wings of Pigeon Guillemot are silvery gray). Wingbeats are quick and choppy. Flight is low and direct, but slightly twisty-turny, so that at a distance white wing patches wink and disappear.

VOCALIZATIONS: Call is a thin high-pitched whistle: "*eeeee.*"

PERTINENT PARTICULARS: In winter, among diving birds off the New England coast, this species is the palest bird on the water.

Pigeon Guillemot, *Cepphus columba*

Pacific Guillemot

STATUS: Common coastal seabird of inshore waters; by and large the most likely alcid to be seen from shore. (Common Murre is a close second.) **DISTRIBUTION:** In North America, resident from nw. Alaska to s. Calif.'s Channel Is. In fall and winter, disappears from most of the Calif. (except Monterey) and Ore. coasts. Over most of its range, the only guillemot; in the n. Bering Sea, range overlaps with Black Guillemot. **HABITAT:** Rocky coastline. Breeds on cliffs, among boulders (also piled driftwood, bridges, wharfs, old buildings, and beached ships), and on islands and rocky mainland coasts. Feeds inshore in water 30–100 ft. deep; less commonly to the edge of the continental shelf. In winter, seeks protected inshore waters. Avoids sandy coast. **COHABITANTS:** Loons, cormorants, scoters and other sea ducks, murres and other alcids. **MOVEMENT/MIGRATION:** Nonmigratory over most of its range; however, extreme northern birds withdraw south to the edge of the ice pack, and some southern birds fly north (some as far as southern Canada) after breeding. VI: 0.

DESCRIPTION: A plump-bodied, small-headed, pointy-faced seabird with an all-black plumage pattern that may (or may not) show a big white wing patch. A medium-sized alcid (larger than Dovekie; smaller than a murre; about half the size of Surf Scoter), with a conspicuously small head, a slender, pointy, very slightly upturned bill, a fairly long neck (for an alcid), and a short pointy tail that's slightly up-angled. Sits high on the water. Sometimes seems humpbacked, sometimes flat-backed, with the head angled up. (Thick-billed Murre looks slouched and dejected.)

In breeding plumage, the body and bill are all black (murres show white sides), and the legs bright red (as is the mouth lining, which is visible when the bird is calling). The white wing patch is usually conspicuous, but the intruding black wedge jutting up along the lower edge may be very difficult to see; some birds sitting on the water show no white until they fly. In winter plumage, birds become pale and somewhat piebald-patterned—fundamentally white with a dark crown, a pale gray-brown back, and dark wings broken with a white patch. Immatures are overall dingier and slightly darker brown, but still overall paler and showing more white than a murre. In all plumages, feet remain bright red.

BEHAVIOR: Forages in shallow, usually protected inshore waters (rarely more than 1 mi. from shore), but is also often seen sitting or standing on rocks, where its red legs are plainly visible. Dives by opening its wings and diving, using its wings to swim. Breeds in small colonies; in winter, may gather in larger flocks, but most commonly feeds alone (that is, unpaired).

FLIGHT: Somewhat small-headed, long-necked, potbellied, and rear-heavy profile, with short thin lens or paddle-shaped wings (narrow at the base and tip). The head is somewhat raised, and the red trailing legs give the impression that the bird has a red tail. The white wing patches are plainly evident (though the bisecting black wedge is often hard to see), and in breeding plumage, the wings look slightly browner than the black body. The silvery grayish underwings distinguish it from Black Guillemot (whose underwings are unequivocally and entirely white except for a very thin dark border). Flies somewhat weakly, low over the water, with quick, steady, slightly elastic wingbeats.

VOCALIZATIONS: Call is an excited, high-pitched, whistled trill, "*Ti ti ti-ti-ti-ti-ti-ti-i-i-iii,*" that accelerates near the end. Pattern recalls Whimbrel. Also emits a high "*peep peep peep*" that recalls the chip of Northern Cardinal.

PERTINENT PARTICULARS: The bird most commonly confused with Pigeon Guillemot is the much smaller Marbled Murrelet, another common inshore feeder whose immature plumage is also somewhat piebald: dusky brown above, dingy white below, and showing a whitish patch in the wing. Murrelets are usually found, however, in pairs or small groups; guillemots feed alone. Also, the murrelet is more compact, shorter-necked, and distinctly shorter-billed.

Marbled Murrelet, *Brachyramphus marmoratus*

Wedge-faced Murrelet

STATUS: Uncommon to fairly common West Coast resident. **DISTRIBUTION:** Breeds from the Aleutians and the north coast of Bristol Bay in Alaska to just south of San Francisco. In winter, ranges very rarely south to the Channel Is. off Calif. (less commonly to n. Baja). **HABITAT:** Nests in coniferous coastal forests—primarily on moss-covered tree limbs but also in burrows on scree slopes and cliffs—most commonly within 1 mi. of the shore but sometimes as far as 4 mi. inland. Forages in shallow waters, primarily in bays, inlets, and harbors, or within several miles of shore; is often visible from shore. Seems attracted to areas with stronger currents, upwellings, and, in particular, the waters at the mouths of rivers and glacial streams. In Alaska, may be found out to 30 mi. (or more) from shore; unlike other alcids, this species sometimes occurs on freshwater lakes. **COHABITANTS:** Pelagic Cormorant, Glaucous-winged Gull, Common Murre, Pigeon Guillemot, Kittlitz's Murrelet. **MOVEMENT/MIGRATION:** Poorly understood. In Alaska, some movement is evident in late April and May, but some birds can be found in the waters near breeding areas all year, provided that water remains open. VI: 0.

DESCRIPTION: Small inshore alcid with a pointy face and a dark cap that eclipses the eyes. Small (considerably smaller than Pigeon Guillemot; slightly smaller than Ancient Murrelet; very slightly larger than the similar Kittlitz's Murrelet), with a fairly long pointy bill, a large flat head that makes the bird look long or wedge-faced, a thick neck (when relaxed), a compact body, and a short tail. On the water, sits with bill slightly elevated and tail conspicuously up-cocked.

In summer, mostly dark blackish brown with a darker cap, but many birds continue to show patchy white on the folded wing (scapulars) and the sides of the rump all summer. In good light, upperparts project a ruddy cast. Undertail coverts are conspicuously white. In winter, all white below and black above, with a horizontal white slash above the folded wing and a white patch near the tail. The head is distinctly black-capped with the bottom pulled over (below) the eye. Immature is a dingier, browner version of winter adult.

In all plumages, eyes are enveloped by the extensive black cap and difficult to see.

BEHAVIOR: Usually not a gregarious species. On the water, seen singly (the norm in winter) or in pairs, threes, or fours during the breeding season, but in Alaska concentrations in excess of 50 birds in protected back bay areas are reported. Does not usually associate with other alcids, except Kittlitz's, whose range overlaps. Forages by diving, usually in fairly shallow inshore water. Pairs often dive together, and may call to each other when surfaced.

A shy bird that dives repeatedly when approached by boats (flashing a white undertail), always surfacing farther away. Flies when pressed.

FLIGHT: Plump but overall pointy—pointy face, pointy tail, and (especially) exceptionally long, thin, sweptback, pointed wings (resembles somewhat a plump-bodied Chimney or Vaux's Swift). In summer, all dark, but in good light shows a ruddy cast to the upperparts. In winter, blackish upperparts sport white at the top of the wings and white flanks, but show little or no white in the outer tail. Going away, the pale cheeks wrapping around the neck show as two pale spots on either side of the neck.

Flight is fast and low, with rapid steady wingbeats. Constant twisting or yawing in flight gives the bird a drunken or unsteady appearance. Lands with a rapid braking helicopter-flutter of wings.

VOCALIZATIONS: Call, often given on the water, sounds like a musical high-pitched gull: "*keer.*" It might remind you of a single note excerpted from American Robin's song.

PERTINENT PARTICULARS: Long-billed Murrelet, *Brachyramphus perdix*, breeds in Siberian waters but shows a widespread pattern of vagrancy (VI: 2). It is most like Marbled Murrelet, but the dark brown breeding adults show a distinctly pale throat and appear more capped than masked (or having a cap that's not drawn below the eye).

Nonbreeding birds and immatures (the plumage most likely to be seen in North America) show dark napes (no white collar), white eye arcs (like MacGillivray's Warbler), and small white patches on the sides of the nape. The bill is slightly longer, but not something to go by in the field. Given Long-billed Murrelet's pattern of vagrancy and the lack thereof with Marbled and Kittlitz's, this species is by far the most likely of the three to be found inland, particularly far away from the West Coast. But a word of caution: Ancient Murrelet also has a pattern of inland vagrancy.

Kittlitz's Murrelet, *Brachyramphus brevirostris*

Pinch-faced Murrelet

STATUS: Uncommon and geographically restricted coastal breeder. **DISTRIBUTION:** Alaskan species breeds coastally along the west, south, and southeast coasts as well as the Aleutians and Kodiak I. The breeding stronghold for the species is Glacier Bay. In winter, most birds are presumed to winter over the coastal shelf, but some remain in sheltered bays and coves, usually among Marbled Murrelets. **HABITAT:** Breeds on scree slopes or rocky cliffs as far as 50 mi. inland. In summer, forages in protected bays, inlets, and fiords, particularly those that feature the glacial icebergs and gray water associated with glacial melt from rivers and streams. Kittlitz's Murrelet is far less likely to be found in open coastal waters than Marbled, but both species forage in glacial runoff. **COHABITANTS:** Pelagic Cormorant, Black-legged Kittiwake, Pigeon Guillemot, Marbled Murrelet, Rhinoceros Auklet. **MOVEMENT/MIGRATION:** Poorly understood. Spring movements occur between late March and early June; fall departures are noted as early as mid-July in some places, and some birds linger in northern breeding areas until forced out by winter ice in late October. VI: 0.

DESCRIPTION: A small pinch-faced alcid that looks carved out of gray granite. Small (smaller than Pigeon Guillemot; in direct comparison, only slightly smaller than Marbled Murrelet), with a round head and a very short bill that gives the bird a pinched-face look, in contrast to the longer wedge-faced look of longer-billed Marbled Murrelet. With its thick neck, compact body, and short tail, Kittlitz's is overall slightly smaller and more compact than Marbled. The short bill appears less elevated than on Marbled, and the shorter tail seems, perhaps, not so up-cocked.

Breeding plumage is somewhat variable; most birds are overall dark gray, flecked with white. In good light, golden highlights turn upperparts on some birds golden brown. Overall paler than Marbled, particularly paler-headed, and more uniformly plain. Kittlitz's lacks any semblance of a dark cap or the piebald patches seen on many breeding Marbleds.

In winter, like most alcids, Kittlitz's is black above, white below. Upperparts show a whitish bar above the wings and white near the flanks (like Marbled). Unlike Marbled, the black cap is restricted, making the face white and the black eye easy to see. Immatures are like winter adults.

BEHAVIOR: An inshore feeder (usually within sight of land) that is often found in the milky water associated with glacial melt. Seen singly, in pairs, or in small groups (larger groups have been noted); commonly forages with Marbled Murrelet, as well as other alcids and gulls.

A shy bird that does not tolerate close approach by boats. Reacts by diving and swimming to a more acceptable distance, and flushes when pressed.

FLIGHT: A small elliptical torpedo set on narrow flailing wings. In direct comparison, appears slightly shorter-winged than Marbled, with a slightly quicker wingbeat. Appears to lift off the water with greater ease than Marbled, but birds that have fed heavily often bounce a time or two like Marbled.

In flight, shows dark upperparts and pale underparts (in winter, black above and white below), and at all times a narrow dark ring encircling the throat. Going away, shows white patches at the sides of the head (which Marbled shares) and white sides to the tail (which Marbled lacks).

VOCALIZATIONS: Less vocal than Marbled. Call is described as a low groan. Sounds like gull with a head cold saying "Ah" for the doctor: *"aaah,"* or *"raahr."*

Xantus's Murrelet, *Synthliboramphus hypoleucus*

White-wedged Murrelet

STATUS: Uncommon Pacific Coast breeder and resident; rare off Oregon, Washington, and British Columbia. **DISTRIBUTION:** Breeds on the Channel Is. off s. Calif. and several islands off Baja (late February to June). In winter, simply not seen anywhere much at all. **HABITAT:** Nests on steep slopes and cliffs with sparse vegetation. In winter (July to February), found well offshore in the warm waters of the Davidson Countercurrent (avoiding colder inshore waters). **COHABITANTS:** Does not commonly associate with other pelagic species. **MOVEMENT/MIGRATION:** Birds disperse from nesting areas as early as March and as late as July, apparently making their way gradually to wintering areas, and return to nesting islands in February and March (possibly as early as December or January). VI: 2.

DESCRIPTION: Small, clean-lined, crisply patterned blackish and white alcid; easily distinguished from most small alcids, but virtually identical to Craveri's Murrelet. A small, streamlined, and nicely proportioned alcid (slightly larger than Cassin's Auklet and Red Phalarope, but much more slender), Xantus's Murrelet has a large flattish head, a pointy bill, a longish neck, and a low flat profile on the water. Craveri's is almost identical, but its bill is very slightly longer and thinner, and its tail is slightly longer and appears more overtly up-cocked (making the white undertail more conspicuous).

Upperparts on Xantus's are all blackish (but bluish black in winter and early spring, and faint gray with sometimes a touch of brown in late summer on worn birds), and underparts are all white. Because birds sit so low on the water, they appear mostly black except for the white chin and lower half of the face, the white throat, and the white breast.

Keep in mind that Xantus's has a small white cleft or wedge that juts up into the black face directly in front of the eye. On Craveri's, there is no wedge. On the southern race of Xantus's, this white cleft is so extensive that it curls around and over the eye. That's the good news. The bad news is that this subspecies is a very rare visitor to U.S. and British Columbia waters in late summer and fall.

Also note that, on the water, Xantus's usually shows more extensive white on the sides of the breast; Craveri's is more likely to look just black along the water line. Because the tail on Xantus's is not commonly up-cocked, the white undertail coverts are difficult to see. On Craveri's, the white undertail is more often apparent.

Also, Xantus's shows bright white underwings; Craveri's underwings are a mottled dingy gray. While this characteristic is of no use when wings are closed, sitting birds sometimes lift their wings momentarily. Watch for a white flash.

BEHAVIOR: Almost always seen in pairs away from other pelagic birds. Swimming birds elevate their heads and lean forward (as do Craveri's Murrelet). Flushes easily, springing directly from the water, and often travels a long way (usually out of sight) before setting down again. When pressed, may call (keep your fingers crossed).

FLIGHT: Profile is slender and pointy-faced, with slim curved-back wings. Flight is low and direct, with a slightly rolling or unsteady quality. The projecting narrow (broken) black collar jutting onto the breast of Craveri's is hardly a blunt suggestion on Xantus's. If the bird is truly Xantus's, you'll see nothing but an even transition between black and white (that is, no bulge onto the breast at the shoulder). Underwings are bright white, not mottled and dingy, but this mark is almost impossible to see when birds are flying directly away (which is usually the case).

VOCALIZATIONS: Usually silent, but occasionally makes a high squeaky peeping, *"zee, zee, zee,"* which is distinctly different from Craveri's trill. A word of caution: If phalaropes are in the vicinity, don't mistakenly ascribe their sharp *"plik"* note to a mystery murrelet.

PERTINENT PARTICULARS: First of all, know that Xantus's Murrelet is generally more common and more likely than Craveri's. Rather than approach the identification by juggling two variables, try using one. Assume the bird is Xantus's, and test to see whether field marks support this assumption. Do the upperparts look plain black with a slightly bluish or grayish (not brown) tinge? Does "fairly short" seem to describe the bill? Does the black-and-white border on the face show the expected white cleft in front of the eye?

Don't rely on a single characteristic. Build a case using several. And don't take your eye off the bird. When Xantus's deigns to raise and flutter its wings, the opportunity is gained or lost in a flash.

Craveri's Murrelet, *Synthliboramphus craveri*

Collared Murrelet

STATUS: Mexican breeder; uncommon and irregular late-summer to fall visitor off the California coast. Generally less common than Xantus's. **DISTRIBUTION:** Breeds in the Sea of Cortez; winters approximately 60 mi. off the coast of Mexico. Found regularly in coastal waters off the s. Calif. coast. Occurs almost annually to about Monterey; very rarely farther north. **HABITAT:** In the United States, found in the warm waters of the Davidson Countercurrent (especially in El Niño years), where sharp drop-offs in the ocean floor occur. **COHABITANTS:** Does not join other seabirds but is often found in the same waters as Buller's Shearwater. **MOVEMENT/MIGRATION:** Found in California waters between mid-July and late October. VI: 1.

DESCRIPTION: An ever so slightly browner version of Xantus's Murrelet. A small alcid, Craveri's has a flat head, a slender bill, a thick neck, and a flat-backed body; it sits low on the water with its head raised and its tail often distinctly up-cocked, flaunting white undertail coverts. On shorter-tailed Xantus's Murrelet, the white undertail coverts are usually less conspicuous. The bill on Craveri's is slightly longer and more slender than on Xantus's (just as Black-and-White Warbler's bill is longer and more slender than Black-throated Gray's).

Craveri's is blackish above and white below, but in good light, upperparts are duller and browner-tinged than Xantus's (which is equivocally, not obviously, black, with a bluish or grayish sheen). The lower border of the black face on Craveri's is sharp and clean to the edge, with no hint of a cleft in front of the eye. Black upperparts bulge at the shoulder, giving the bird a slightly more collared look (if it's sitting high) or, more often, a more dark-to-the-water-line look.

BEHAVIOR: Usually found in pairs in deeper water, Craveri's is difficult to see when waters are not calm. The bird flushes first, flying out of sight before stopping to ask questions. Swims with head raised. Leaps from the water with minimal (if any) foot-pattering in the takeoff. Dives, calls, and may flutter its wings before departing.

FLIGHT: Like Xantus's, Craveri's has a slender, long, and sweptback-winged profile with rapid and steady wingbeats. Watch for the dark collar bulging onto the sides of the breast and for mottled and dingy underwings.

VOCALIZATIONS: Makes a weak chirping trill unlike the sharp single-note peep of Xantus's.

PERTINENT PARTICULARS: This is a warm-water species that moves closer to shore in late summer when warmer currents press closer to the shoreline.

Ancient Murrelet, *Synthliboramphus antiquus*

Wandering Murrelet

STATUS: Common breeder and resident in the colder coastal waters of the North Pacific; variously common to scarce winter resident along the U.S. Pacific Coast. **DISTRIBUTION:** Breeds coastally from the Aleutians (possibly the Pribilof Is.) east and south to the Queen Charlotte Is., B.C. A few wander north to the Bering Sea. In winter, range extends south to the central coast of Calif. and very rarely to s. Calif. More than any other alcid species (except Long-billed Murrelet), Ancient Murrelet has a habit of wandering and has turned up on inland bodies of water all the way to the East Coast of the U.S. (One has even been recorded for England.) **HABITAT:** Breeds on vegetated islands (often forested). In summer and winter, forages out over the waters of the continental shelf, but may also be found in inshore waters, where it may be seen from land. **COHABITANTS:** Black-legged Kittiwake, Rhinoceros Auklet, Marbled Murrelet. **MOVEMENT/MIGRATION:** Disperses away from colonies from August to March; returns in March and April. Most vagrancy occurs in late October and November. VI: 2.

DESCRIPTION: A long-profiled, squarish-bodied, well-patterned alcid that can commonly be seen from shore. Small (slightly larger than Marbled Murrelet; smaller than Crested Auklet; considerably smaller than Pigeon Guillemot,) with a short bill, a blocky head, a long, flat-backed, and low-profiled body, and a short tail. Looks about the shape of an 8-in. piece of two-by-four floating in the water, with a head that seems to lean forward. Auklets appear stubbier and more compact; other murrelets appear more slender and delicate.

On the water, overall gray above, with a black face and throat (just capped in winter) and a prominent white collar extending from throat to nape. At close range, the pale gray of the back is apparent (other alcids look black or brown); the black face/throat pattern of breeding adults easily distinguishes this species from other alcids, and breeding plumage is often assumed in December. At a distance, appears dark to the water line, with bracketing patches of white set fore and aft—white collar and white undertail. Other small black-and-white alcids usually show more white along the water. In good light, Ancient's pale bill stands out, as does the contrast between the black cap and gray back.

BEHAVIOR: Unlike most murrelets, which are commonly seen singly or in pairs, Ancient Murrelet forages in small groups and (wayward birds excepted) is not inclined to feed alone. Not particularly skittish, Ancient reportedly feeds around the bows of slow-moving vessels. Like other alcids, feeds by diving, and flocks often dive in unison. Lifts off from the water with little effort.

FLIGHT: Fairly blocky in flight, appearing long-tailed or with wings set too far forward. There is a great deal of alcid trailing behind the wing; other small alcids appear more symmetrically balanced. The white collar makes the head look unattached to the body. Flight is fast and maneuverable, with rapid, steady, stiff wingbeats; the body twists and turns in flight.

VOCALIZATIONS: Calls, uttered at breeding colonies, include a variety of high squeaky chirps.

Cassin's Auklet, *Ptychoramphus aleuticus*

Auklet with the Quizzical Expression

STATUS: Common and widespread West Coast auklet. **DISTRIBUTION:** Found in coastal waters from the Aleutians to Baja. **HABITAT:** Coastal waters out to a distance of 100 mi. or more, with concentrations found over the upwellings associated with the drop-off of the continental shelf. Nests on coastal islands in natural clefts or in excavated burrows in soft soil to which it returns nightly. **COHABITANTS:** At sea, Sooty Shearwater, Pomarine Jaeger, Rhinoceros Auklet. **MOVEMENT/MIGRATION:** Northern birds are apparently migratory, but little information is available. Nonbreeding southern birds remain near nesting colonies or forage farther offshore. VI: 0.

DESCRIPTION: Small, plump, lead-colored auklet—in flight, resembles a dark salt shaker with whirring wings. South of Alaska and away from inshore waters, the most likely small alcid to be encountered. Small and overall compact (same size as Red Phalarope, but slightly plumper; appears less than half the size of Rhinoceros Auklet). Head shape is comically unique—round and prominently crowned with a steep flat forehead; the short, pointy, grebelike bill looks like it's set too low on the face. The body is stocky, with no apparent neck or tail. In many respects, recalls a short-necked Least Grebe.

On the water, overall dingy brownish gray. (In summer, molting birds appear browner than depicted in most illustrations.) The small white eyebrow and touch of yellow at the base of the bill are visible only at close range.

BEHAVIOR: Commonly encountered offshore singly, in pairs, or in small groups (fewer than 10 birds) at the drop-off of the continental shelf; in late fall and winter, however, large concentrations in excess of 10,000 birds (presumably birds from Alaska and British Columbia) may be found well offshore. Fairly easy to approach. Flees from approaching boats by diving, thrashing across the surface water on rowing wings, or flying with a takeoff that is less labored than many alcids.

FLIGHT: A short, plump, somewhat front-heavy profile shows a large round head distinct from the body. Short rounded wings move in a fluttery blur. Overall dark (often showing a brownish cast), with a touch of paleness on the underwing linings and lower belly and two pale lines flanking the tail (often hard to see). Lifts off from the water with little effort. Flies in fairly compact but sloppy flocks just above the surface. Flight is mostly straight, but slightly twisty-turny (particularly just after takeoff).

VOCALIZATIONS: Silent at sea. Birds in colonies make a series of low harsh hisses.

PERTINENT PARTICULARS: Cassin's Auklet's range largely overlaps with the similarly plumaged but distinctly flat-headed and puffin-sized Rhinoceros Auklet and the overall more slender, pinch-faced, and slender-winged Marbled Murrelet, which is mostly dark brown in summer and contrastingly black above and white below in winter. In western Alaska, Cassin's Auklet might also be confused with the all-gray Crested Auklet or (the rare) Whiskered Auklet, but is shorter, rounder-winged, and distinctly paler on the underwing lining and belly than either of these species.

Parakeet Auklet, *Aethia psittacula*

The Auklet with the Mona Lisa Smile

STATUS: Common Alaskan pelagic species, but less numerous than Least and Crested Auklets where these two species breed. **DISTRIBUTION:** Nests in scattered locations throughout the Bering Sea, the Aleutians, and the south-central coast and southeastern coasts of Alaska. In winter, ranges into deeper water, mostly south of the Aleutians, but very rarely leaves Alaskan waters; casual south to Calif. **HABITAT:** Breeds on rocky islands and the coastal mainland, using rock cliffs, talus slopes, boulder beaches, and grassy slopes. When foraging, reportedly avoids areas with turbulence and seas with scattered ice. **COHABITANTS:** Least Auklet, Crested Auklet, murres, puffins. **MOVEMENT/MIGRATION:** After breeding, most of the population appears to retreat to deeper ocean waters immediately south of the Aleutians. Spring migration from early March to late May; fall from mid-August to November. VI: 2.

DESCRIPTION: A chunky black-and-white auklet with a stubby bright orange bill. Medium-sized (larger than Whiskered; smaller than Pigeon Guillemot; slightly larger than Crested Auklet), with a thick, blunt, orange bill, a large head, and a stocky body.

Parakeet Auklet is charcoal gray above and white below. Its large size and orange bill distinguish it from Least Auklet; white underparts (showing as white gunnels all around the bird as it sits on the water) distinguish it from the also orange-billed but all-dark Crested Auklet (as well as from the smaller and more restricted Whiskered Auklet). Bills of immatures are blackish, not orange; otherwise, they're like winter adults. Parakeet Auklet's expression is secretly happy.

BEHAVIOR: An auklet that shuns crowds, preferring to forage in small groups away from areas of massed concentration; tends to nest in species-segregated groups, avoiding areas dominated by other species. When moving to and from feeding areas, flocks are small (fewer than 10 individuals); one or two birds may join flocks of murres, Crested and Least Auklets, puffins, and other alcids.

FLIGHT: Stocky, roundly proportioned, somewhat parrot-shaped alcid with a large head, a potbellied body, and fairly broad and blunt-tipped wings. Overall more robust than Crested Auklet, with which it is commonly seen and from which it is easily distinguished by plumage.

Blackish above and white below; the orange bill of adults is evident at considerable distance (but less so than Crested Auklet). Flight is straight, rapid, and usually low over the water. Wingbeats are fairly sluggish (for an alcid) and choppy.

VOCALIZATIONS: Generally silent at sea. Birds in colonies make a staccato horse whinny.

Least Auklet, *Aethia pusilla*

Feathered Ping-Pong Ball

STATUS: Abundant but geographically restricted coastal and pelagic species with little inclination to wander outside its breeding range. **DISTRIBUTION:** Breeds in remote coastal islands in the Bering Sea and along the Aleutian Chain. Winters in the s. Bering Sea and the waters south of the Aleutians to Kodiak I. **HABITAT:** Breeds on rocky cliffs, talus slopes, and boulder-encrusted fields—habitats rich in crevices. Forages in open ocean and close to shore. Concentrates where upwellings carry plankton to the surface. **COHABITANTS:** Breeds in association with other alcids, including Parakeet and, especially, Crested Auklets. **MOVEMENT/MIGRATION:** Where waters remain ice-free, birds remain year-round, but there is a sizable migration from the northern and central Bering Sea to Aleutian waters. VI: 0.

DESCRIPTION: A stubby diminutive Ping-Pong ball or rubber ducky of an alcid; Least Auklet matches the size and affects some of the manner of the smaller sandpipers. Tiny and compact (same size as a plump Semipalmated Sandpiper; appears half the size of Crested Auklet, with which it often associates), with a small sparrowish nub of a bill, a round neckless head, and a small stocky body.

Breeding adults appear moth-eaten: grayish black above with spare white spotting, and whitish underparts have varying amounts of dark mottling. Plumage is variable: some individuals show dark breasts and patchy flanks, while others are mostly white below with sporadic darker feathers. Even the lightest birds have enough patterning below to frame a conspicuous white throat. In winter, birds are grayish black above, with a pale white line etched over the folded wing, and extensively white below. The pale tip of the bill on breeding birds is difficult to see except at close range.

Crested Auklet, with which Least often associates, is all dark, all year, and appear about twice the size of Least.

BEHAVIOR: A very gregarious species. Breeds in large, often multi-species colonies (which may number in the hundreds of thousands). In colonies, often gathers in large tight wheeling groups (suggesting a flock of shorebirds in the presence of a falcon). Usually forages in single-species flocks, but often flies with Crested Auklets. Forages both inshore and well offshore, sometimes singly but more often in groups.

FLIGHT: Appears diminutive in flight, showing a stubby plumply compact body and narrow slightly sweptback wings. In summer, dark-fronted birds show contrasting white faces and white bellies. (The white is on the throat but looks like it surrounds the bill.) In winter, birds are black above, bright white below, with a narrow white line that welds the wing to the body. Birds twist and yaw in flight, showing flashes of white underparts.

Birds lift easily from the water and fly low. Flocks form tight elongated bundles that are clotted at the front and strung out behind. Size and configuration suggest a flock of small shorebirds, as do the bird's wingbeats, which are rapid and steady, with wings moving so rapidly they blur.

VOCALIZATIONS: Call, heard in colonies, is a descending series of chips that sound somewhat parakeet-like and recall the food-begging call of a young Royal Tern (only more musical): "*eh-ehehh-ehh-ehh-ehh-ehh.*" Also makes a harsh scraping trill, "*reh'h'h,*" repeated sequentially.

Whiskered Auklet, *Aethia pygmaea*

Aleutian Auklet

STATUS: Locally common but geographically restricted; found exclusively on the islands of the Aleutian, Commander, and Kuril Chains. **DISTRIBUTION:** In North America, breeds from Unimak I. west to Buldir I. Winters in the nearshore waters of the Aleutians, with the highest concentrations near breeding colonies. **HABITAT:** Breeds on rocky cliffs and talus slopes. Feeds in nearshore marine waters, often close to colonies, and in strong tidal rip and upwellings; very often feeds in channels between islands and does not commonly feed more than about 5 mi. from shore in summer, 10 mi. in winter. **COHABITANTS:** Least Auklet, Crested Auklet, "Peale's" Peregrine Falcon. **MOVEMENT/MIGRATION:** Permanent resident. VI: 0.

DESCRIPTION: A small blackish gray or brownish black alcid that resembles a small Crested Auklet, showing dull dingy white below and toward the rear (in flight). Small and compact (larger than Least; appears about two-thirds the size of Crested), Whiskered looks like Crested (round head, thick neck, stocky body), but has a smaller bill and pointier face. On Cassin's, the large bright orange-red bill is easy to see. On Whiskered, the bill is smaller, duller, and hard to see.

On the water, appears blackish or blackish gray (with some brown showing in the wings at times). At close range, under ideal conditions, note the wispy sprig of a dark plume, which is otherwise difficult to see (but obvious on Cassin's). At a distance (even in flight), the fine white V between the pale eye and dark red bill is more easily noted (especially in breeding condition). The narrow plume behind the eye is harder to see.

BEHAVIOR: Highly social. Likes crowds and is almost never seen alone. Likes to forage in strong rip currents in tightly packed groups. Its disinclination to wander away from its breeding islands explains why it ranks among the most difficult of North American birds to see. Where it occurs, concentrations in the thousands may be expected. Also feeds (and habitually flies) with Crested Auklet—allowing direct comparison and easy detection: not only are Whiskereds smaller, but they are the ones in the rear of the flock.

FLIGHT: Looks like a bumblebee with a face that is distinctly more pointed than Crested and a smaller, darker bill that is much harder to see. All dark above, but pale on the aft portion of the underparts (appears whitish from midbelly to the undertail). Larger Crested Auklet is all dark below; larger Parakeet Auklet is wholly bright white below; small Least Auklet shows lots of white mottling on the body (breeding plumage) or is all white below (winter). Like Crested Auklet, Whiskered has upperparts that often appear two-toned in flight—blackish toward the front, brownish toward the rear—but unlike Crested, Whiskered never appears all cinnamon or brown-tinged as Crested's may in certain lights.

Flight is direct and fast, with the body angled slightly up and wingbeats frenzied. Flies low over the water in tight, globular strings, often between 10–20 birds.

VOCALIZATIONS: Most often described call is a single high-pitched mew. Birds in colonies often make a chattering series of mew notes. Also squeak when disturbed.

Crested Auklet, *Aethia cristatella*

Dark Chocolate Auklet

STATUS: Common coastal breeder and resident of western Alaskan ocean waters. **DISTRIBUTION:** Coasts and waters of the Bering Sea and the Aleutians. In winter, ranges as far east as Kodiak I. **HABITAT:** Breeds on remote islands and coastal rock cliffs free of vegetation. Forages offshore in deeper water, usually over upwellings and convergences, but also commonly feeds close to shore, often just beyond breaking waves. **COHABITANTS:** Murres, Least Auklet, Whiskered Auklet, Parakeet Auklet. **MOVEMENT/MIGRATION:** Birds that breed where winter seas are encased in ice retreat south in the waters of the southern Bering Sea to off Kodiak Island. Little else is known. VI: 0.

DESCRIPTION: An entirely brownish black alcid with a bright orange bill and a jaunty plume. Medium-sized (larger than very similar Whiskered; smaller than guillemots; about the same length as Parakeet Auklet, but not as plump), with a stubby bright orange bill, a smallish head, and a compact body.

Uniformly charcoal brown. In some lights, the front half of the bird appears blacker than the rear; in bright, shallow light, the bird (particularly the underparts) looks like cinnamon-dusted chocolate. The feathered plume normally bends forward (in front of the bird's eyes), but when birds are feeding, it may be slicked back and momentarily invisible.

Easily distinguished from Parakeet Auklet by its dark underparts. Not as easily distinguished from the smaller, rarer, more restricted Whiskered Auklet. Immatures and second-year birds have shorter, more scraggly, and less plumelike crests. Bills of immatures are dark.

BEHAVIOR: A very gregarious alcid. Commonly seen flying and foraging (year-round) in large flocks. Nests in colonies that may number one million individuals or more. Nesting birds commonly segregate themselves from other alcids but are reportedly more accepting of Least Auklets. V-shaped flocks are tight—one bird length apart—and disorganized, but less disorganized or clotted at the front than Least Auklet flocks.

FLIGHT: Overall compact, with fairly slender wings. In flight, very slightly smaller than Parakeet Auklet (with which it often is seen) and less robust. The orange bill of adults stands out at great distances. In flight, the plume is flattened against the head. More steady in flight than Least Auklet (not so twisty-turny). Wingbeats are rapid and steady, with an angry down stroke.

VOCALIZATIONS: In breeding season, away from colonies, makes a high chihuahua-like yap, "*re/ah*," which it sometimes gives a compressed two-note quality, "*eeyow*," and often utters in a monotonous series that may accelerate in times of high excitement into a yapping chortle.

Rhinoceros Auklet, *Aethia monocerata*

One Big Ugly Lump of an Auklet

STATUS: Common and widespread West Coast seabird. **DISTRIBUTION:** Occurs from the Aleutians to s. Calif. In summer, population densities vary widely, with distribution tied to breeding colonies (nearly three-fourths of the population is found off B.C.). In winter, found from cen. B.C. to s. Baja, with most of the population found off the coast of Calif. **HABITAT:** Continental shelf waters. Nests on islands and mainland cliffs. **COHABITANTS:** In summer, Brandt's Cormorant, Common Murre, Cassin's Auklet; in winter, Pacific Loon, cormorants, Black-legged Kittiwake, Common Murre, Cassin's Auklet. **MOVEMENT/MIGRATION:** Some birds remain near the nesting colonies year-round; others disperse south after breeding. Spring migration March to May; fall (presumably) from August to December. VI: 0.

DESCRIPTION: A large, stocky, anvil-headed alcid. Puffin-sized (much larger than Cassin's Auklet; smaller than a murre), with a big, flat-topped, anvil-shaped (or wedge-shaped) head, a straight heavy bill, a very thick neck, and a robust body.

Overall dark and dingy—dark brownish above, with paler sides and a conspicuously yellow to orange bill (dark in immatures). Rhinoceros's "horn," wispy white whiskers, and eye stripe (found on breeding birds) are hardly necessary for identification. The expression is formidable and predatory. The bird looks like a thug: everything about it says, "Don't mess with me."

BEHAVIOR: Often forages within sight of land, singly or in small groups (4–40). Commonly joins mixed flocks of birds over small fish, and when not feeding, loafs on the surface in concentrated rafts (which may number several hundred individuals). Fairly tame, and easily approached by boats. Struggles to leave the water.

FLIGHT: Heavy, potbellied profile, with an up-angled body, a raised head (puffin keeps head level), and short triangular wings. Flight is fast, direct, and heavy, with steady wingbeats slow enough so that individual wingbeats may be noted. All dark above, with a dirty chest and an obviously paler (whitish) belly. The yellow bill is usually conspicuous.

VOCALIZATIONS: Generally silent.

PERTINENT PARTICULARS: On the water, most likely to be confused with the similarly sized and likewise dark and dingy immature Tufted Puffin. The puffin's head is round (not flat-topped), with a larger, heavier (and in winter), more orange-colored bill. Cassin's Auklet is about half the size of Rhinoceros Auklet and seems overall round and petite (anything but formidably proportioned).

Atlantic Puffin, *Fratercula arctica*

Rainbow-billed Alcid

STATUS: Locally common coastal breeder in the North Atlantic; uncommon and widespread in winter in northern ocean waters; very rarely sighted from land. **DISTRIBUTION:** Breeds locally and coastally from Greenland to Lab., e. Que., Nfld., the Maritimes, and Maine. In winter, occurs mainly well offshore, south of Greenland to Mass. Less common but regular, also well offshore, to Va. **HABITAT:** Pelagic and dispersed outside of the breeding season; most often found over the deeper waters of the continental shelf. Breeding colonies are usually situated on soil-covered islands, but also on rocky talus slopes. **COHABITANTS:** In summer, in nesting colonies, Leach's Storm-Petrel, Black Guillemot, Razorbill, murres, Arctic Tern. In winter, Great Skua, Black-legged Kittiwake, Dovekie. **MOVEMENT/MIGRATION:** Present in colonies from late April to mid-August (otherwise at sea). South of breeding areas, birds appear from October to May. VI: 1.

DESCRIPTION: A portly penguin with the face of a sad-eyed clown. Stocky, medium-sized, black-and-white alcid (considerably shorter than Razorbill or murre; about the same size as guillemot, but plumper), with a big head and thick neck, an all–grayish white face, a prominent, if not massive, multicolored triangular bill, and bright orange feet. In colonies, stands mostly upright. On the water, sits high and appears front-heavy. When alert, the head is raised and the neck appears tapered (like a smokestack).

In winter, the adult's bill is slightly smaller and not so bright, and the face becomes darker gray (making the head appear dark at a distance). The immature's bill is darker, smaller, and somewhat lemon-shaped, but it still appears thick and conspicuous compared to other alcids. Winter puffins most closely resemble winter Razorbill but are overall smaller and darker (except for the white breast, dark to the water line). Also, Razorbill shows white on the throat and behind the eye and a long pointy tail; puffins appear squared to the rear and tailless.

BEHAVIOR: Atlantic Puffin tends to be found singly or in very small groups (often no more than two). Most other alcids (including Razorbill) are more commonly found in larger assemblies. This species is pelagic; birds spend most of their lives far from shore and, outside the breeding season, are rarely seen from shore except during storms. Atlantic Puffin is fairly tame on the water. Lifts off with ease, but seems most inclined to swim and dive when approached on the water.

FLIGHT: In flight, stocky and, although elliptically shaped (like Razorbill and murres), more rotundly blunt fore and aft, with a slightly downturned bill. Wings too are blunter and show dark underwings. (Murres and, particularly, Razorbill show white underwings.) Atlantic Puffin's bright orange feet are easy to note, whereas the black feet of murres and Razorbill merge with the upperparts. In winter, puffin feet are a duller yellow-orange, but usually still apparent. Also shows a black breast band and no white trailing edge to the wing. Overall flight is like other alcids.

VOCALIZATIONS: Silent at sea. Birds in colonies make a groaning or moaning sound near or from burrows.

Horned Puffin, *Fratercula corniculata*

Collared Puffin

STATUS: Common northern coastal breeder and Pacific pelagic species. **DISTRIBUTION:** Breeds coastally in Siberia and northwestern and southern coasts of Alaska, including the Aleutians, south locally to islands in B.C. (where it is rare). In winter, ranges widely over the North Pacific, most commonly in deeper waters far away from shore. Nonbreeding birds spend all year in deep water. Regular in winter (and perhaps summer) well off the coasts of Wash., Ore., and Calif. Rare years bring numbers of birds closer to shore from Wash. to Calif., with many recorded from spring and early summer. **HABITAT:** Breeds on rocky cliffs. Forages in waters close to the colony and out to distances in excess of 20 mi. In winter, pelagic. **COHABITANTS:** Pelagic Cormorant, Black-legged Kittiwake, murres, Least and Crested Auklets, Tufted Puffin. In winter, may be found among assorted pelagic species. **MOVEMENT/MIGRATION:** Most individuals (all birds in ice-covered waters) vacate nesting areas after breeding. Winters from the southern Bering Sea south. Spring migration from early March to mid-June; fall from early September to late December. VI: 2.

DESCRIPTION: A big, barrel-bodied, blunt-faced, black-and-white alcid. Large (smaller than either murre; about the same size as or slightly smaller than Tufted Puffin), with a conspicuously broad and colorful bill, a domed head, a short neck, and a stocky body.

Distinctly black and white. The white face is prominent, whereas the white on Tufted Puffin's face seems to be just part of the multicolored bill pattern. On the water, the white showing on Horned's breast and flanks is diagnostic. Tufted is black to (and below) the water line. In winter, Horned's bill is less colorful (in fact, at a distance, it looks all dark), but remains bulbously prominent. The white cheek become dusky gray. Birds stop looking collared and appear more hooded. Immatures resemble winter adults.

BEHAVIOR: More social than Tufted Puffin. Breeds in colonies numbering up to tens of thousands of individuals. Although it may be seen alone or in pairs, this species commonly forages in groups as large as a dozen individuals and also joins multi-species flocks that include kittiwakes and Glaucous and Glaucous-winged Gulls. May feed close to colonies or range out beyond 20 mi.

FLIGHT: Distinctive stocky barrel-shaped profile—overall more compact and much more blunt-headed than murres. Also blunter-headed than Tufted Puffin, which shows a projecting bill. (For some reason, at a distance, the bill on breeding Horned Puffin disappears, giving the bird a square-faced look.) Of course, Horned can always be distinguished from the all-black Tufted by its white underparts. The trick, then, becomes separating Horned Puffin from murres.

Horned Puffin is a compact bird—overall smaller, stockier, heavier-bodied, and blunter-faced than the elliptically shaped murres. Murres hold their bills straight. Puffins appear to angle their heads down in flight. In breeding plumage, Horned Puffin looks bisected or collared, with the bright white face and white belly separated by a black collar. In winter, Horned appears dark-hooded. (Winter-plumage murres show white cheeks/throats.)

Flight is rapid, direct, and often high above the water. Wingbeats are constant, rapid, even, and somewhat rubbery.

VOCALIZATIONS: Usually silent at sea.

PERTINENT PARTICULARS: If you are familiar with Long-tailed Duck (a.k.a. "the poor man's alcid"), you may be surprised by how much a distant approaching Horned Puffin resembles a winter male Long-tailed: both have a white head and body, a dark breast band, and all-blackish wings, and even the cadence of their wingbeats is similar.

Tufted Puffin, *Fratercula cirrhata*

Black Puffin

STATUS: Common and widespread Pacific breeder; rare in winter in coastal areas; more common well offshore. **DISTRIBUTION:** The more widely distributed of the Pacific puffins. Breeds coastally from w. Alaska (and Siberia) south to (possibly) San Miguel I. in s. Calif. In winter, moves away from breeding areas and coastal waters, wintering in the deeper waters of the Pacific south to n. Baja (very rare). **HABITAT:** Breeds on grassy slopes of islands and protected mainland cliffs. Feeds well offshore. **COHABITANTS:** During breeding, Common Murre, Thick-billed Murre, Horned Puffin; in winter, Cassin's Auklet, shearwaters, storm-petrels. **MOVEMENT/MIGRATION:** Some birds may remain near nesting colonies north to the Aleutians, but most move well offshore after breeding. Spring migration from mid-February to late May; fall from early September to late December. VI: 0.

DESCRIPTION: An outlandish-looking all-blackish puffin with a humongous bill and a golden mane. Large, sturdy alcid (smaller than either murre; slightly larger than Horned Puffin and Rhinoceros Auklet), with a heavy heart-shaped bill, a large domed head, a thick neck, a stocky body, and short orange legs.

Wholly black body with an arresting head pattern—a bright orange, yellow, and white bill/face capped with a golden mane. Tufted's bill really stands out as a bill, whereas Horned Puffin often appears more blunt-faced than heavy-billed. In winter, Tufted is more subdued (lacks white in the face) but still wholly black, with an orange heart-shaped bill and traces of gold on the nape. Immatures are like winter adults but browner and lack gold on the nape.

On the water, never shows a white chest. That's the other Pacific puffin.

BEHAVIOR: A pelagic alcid that spends its nonbreeding life (and prematuration years) in the deeper open waters of the Pacific (and is thus scarce in coastal areas in winter). Inshore, most commonly seen standing on nest cliffs or flying out to sea to forage, usually alone or in pairs, less commonly in small strings or loose bunches. Usually solitary, but may forage with other alcids. Commonly flies higher than most other alcids (offering ventral views!), but also flies close to the water (sometimes with other alcids, particularly murres and Horned Puffins).

FLIGHT: Large and compact, somewhat barrel-shaped, long-faced. Posture is slightly elevated. Profile shows a heavy bill and a robust, hump-backed, neckless body. The bill appears angled down, and the tail (actually the feet) turned up. Wings are narrow at the base, broader and blunt at the tip (the bird looks like it's wearing mittens).

All-black body. Horned Puffin has conspicuous white underparts; duller, grayer Rhinoceros Auklet shows a large white oval on the belly.

Flight is strong, fast, and direct, with rapid, steady, rubbery wingbeats. Cadence is about the same as Black or Surf Scoter. Looks anxious, hurried—a bird late for an appointment. Flight may be low over the water, but also high.

VOCALIZATIONS: Silent at sea.

PIGEONS AND DOVES

Rock Pigeon, *Columba livia*

Bronx Petrel

STATUS: The common city (and suburban and rural) pigeon almost anywhere you travel in the world. **DISTRIBUTION:** In North America, common and widespread across s. Canada (including Nfld.), the continental U.S., Mexico, and Central America. Local in urban areas in Alaska and n. Canada. Distribution everywhere is spotty; birds are heavily concentrated in urban areas, on rural agricultural lands, and along interstate highways. **HABITAT:** Typical habitats includes city parks, streets, and buildings, highway overpasses, factories and warehouses, barns, and agricultural lands (particularly grain-producing). Also occasionally found on seacoasts and interior cliffs, like the wild stock native to Europe. Not found in forests or heavily vegetated habitats. **COHABITANTS:** Starlings, House Sparrows, humans. **MOVEMENT/MIGRATION:** Permanent resident. VI: 4.

DESCRIPTION: A medium-sized, plump, and stocky pigeon with a small head, short neck, and short tail. A bluish gray bird with a broad-brush plumage pattern—a slate blue front (head, neck, and breast), a pale blue, almost whitish body, and a slate blue rear (rump and tail). The plumage of other native and wild North American pigeons is more uniform.

The hindneck shines with green iridescence. Two blackish wing-bars bisect the wing. No other North American pigeon (or dove) has a wing-bar.

BEHAVIOR: Spends much of its time walking in a wandering, somewhat aimless fashion on the ground (or concrete) foraging for food. Also perches, usually in small to medium-sized flocks, on buildings, statues, roofs, highway overpasses, and utility lines. Where alternatives exist, generally does not perch in trees.

Tame to the point of being underfoot. Particularly in urban areas, attracted to anyone eating anything.

FLIGHT: In flight, big chest and pointy wings make the bird appear very falconlike. Flight is fast, powerful, and frequent. Flocks habitually and spontaneously take wing, coalescing into a tightly packed mass that sallies out, circles, and commonly returns to the launch point. Wingbeats are usually quick and fairly regular, hard-pumping and clipped. Often glides in a rocky unsteady manner, with wings raised in a severe dihedral. Takes off with a noisy snapping and slapping of wings. Lands with a braking flurry of wings. Despite its speed and tight flock formations, not a particularly agile or graceful flier.

VOCALIZATIONS: Most frequent vocalization is a soft gurgling coo.

PERTINENT PARTICULARS: Often scattered among the ranks of Rock Pigeons are feral pigeons—domesticated birds from wild Rock Pigeon stock that have lapsed back into a more native lifestyle. Structurally identical to Rock Pigeon, feral pigeons manifest an array of plumages—from blackish to white to ruddy to black and white . . . red and white . . . piebald . . . white with a black head. . . .

The multicolored element these birds add to Rock Pigeon flocks quickly identifies such flocks as Rock Pigeons (and their allies). Also, though all-dark feral pigeons superficially resemble North America's two dark native pigeon species, Red-billed Pigeon (in North America found only along the Rio Grande) and White-crowned Pigeon (found only in extreme southern Florida), these two species, unlike feral pigeons, are highly arboreal and very wary.

White-crowned Pigeon, *Columba leucocephala*

Florida Tree-Pigeon

STATUS: Common but very restricted breeder and resident tropical species. **DISTRIBUTION:** Found throughout the Antilles and coastal Central America. In the U.S., found only in extreme s. Fla. and the Keys. **HABITAT:** Nests in mangrove islands. Forages on fruit-bearing trees in wooded hammocks and the tropical hardwood forests of the Keys and the Florida mainland. Favored trees include poisonwood and strangler fig. **COHABITANTS:** Pileated Woodpecker, Mangrove Cuckoo, White-eyed Vireo, Black-whiskered Vireo. **MOVEMENT/MIGRATION:** Some U.S. birds may be found all winter, but most appear to move south. On a daily basis, birds leave roosting and nesting sites and fly to feeding areas (as far as 50 mi. away). VI: 0.

DESCRIPTION: An all-dark (slate blue) pigeon with a white cap seen in extreme southern Florida and the Keys. Slightly larger than Rock Pigeon, and distinctly rangier, with a longer neck, a longer and distinctly wedge-shaped tail (straight-cut across the tip; exceedingly narrow at the base), and blunt-tipped wings. Immatures are like adults, but the crown is grayish (not white).

BEHAVIOR: A treetop species that often forages for berries on the crowns of the tallest trees or perches atop the tallest dead snags. Nimble and acrobatic, White-crowned sometimes hangs upside down while plucking fruit. Infrequently seen on the ground. A social bird that usually feeds in small groups, although it's often seen flying alone. During the nonbreeding season, leaves roost at dawn; forages during the day; returns to roost at dusk. Shy and does not tolerate close approach. (Too often your first indication that the birds are present is the sound of them exploding from the canopy above.) Also frequently circles several times before landing to feed if other birds are not present. Commonly flies over water.

FLIGHT: Profile is classically pigeonlike, with a slightly up-angled body and the head slightly raised above the body. A strong fast flier, but with a hesitant or tentative quality. Wingbeats are regular, with a deep emphatic down stroke but a slight hesitation on the upstroke.

VOCALIZATIONS: Call is a low, deep, three-part coo that is repeated five to eight times, the middle phrase slightly hurried: "*croo croo'uh croo.*" Also makes a low purring growl.

Red-billed Pigeon, *Columba flavirostris*

Yellow-billed Pigeon

STATUS: Tropical pigeon of lowlands in Mexico and Central America. Very uncommon breeder in the southern Rio Grande Valley in Texas. **DISTRIBUTION:** In the U.S., along the Rio Grande from Bentsen–Rio Grande Valley State Park west to, at least, San Ygnacio. **HABITAT:** Tropical deciduous forest in lowlands and foothills; also broken forest, forest edge, and forest with open areas (particularly partial to fig trees). In Texas, associated with the remnant forest and tall scrub along the Rio Grande, where it is partial to tall trees. **COHABITANTS:** Brown-crested Flycatcher, Green Jay, Audubon's Oriole. **MOVEMENT/MIGRATION:** Permanent resident over most of its range. In the United States, reduced numbers occur within the Rio Grande Valley from October to January, but normally a few can be found. VI: 0.

DESCRIPTION: A large all-dark pigeon with a yellow-tipped bill (the red base is visible only at close range). Immatures show less yellow on the bill. The head is large, the forehead bulging, the face drooping. The neck seems extremely long and thin when extended, but when the head is tucked, the coiled neck accentuates the bulging-headed appearance. The tail is short.

Plumage is mostly dark. In good light, shows maroon highlights and a mauve-colored head.

BEHAVIOR: A bird of the canopy and tree interior, where it walks slowly and somewhat unsteadily (on very thin branches), plucking fruit. Most often seen in small flocks, flying over treetops or perched in dead or leafless vegetation. When feeding in the canopy, generally flushes before being seen. Rarely seen on the ground, but does land to drink. Most active early in the morning and late in the evening.

Extraordinarily shy and exasperatingly difficult to detect (at least north of Mexico).

FLIGHT: Fairly stocky and robust profile; recalls White-winged Dove, but the head projects more. Overall dark, including dark underwings; in good light, front half of the body seems mauve-tinged. Flight is strong, direct, and very fast; wingbeats are raised high on the upstroke and thrown low on a pushing down stroke that is held for a fraction of a second, making the cadence seem not irregular but slightly out of sync.

VOCALIZATIONS: A low, coarse, cooed pattern that begins with one long growling "*cooo*" followed by "*up-cup-a-coo; up-cup-a-coo; up-cup-a-coo*" (repeated up to five times).

PERTINENT PARTICULARS: This is not an identification challenge. In the Rio Grande Valley, Red-billed is the only forest pigeon. Smaller feral pigeons are birds of buildings and bare earth and are most common in cities (where Red-billed does not occur). Western Band-tailed Pigeon does not occur in the Rio Grande Valley.

Band-tailed Pigeon, *Columba fasciata*

The Last Forest Pigeon

STATUS: Common western breeder; more restricted winter resident. **DISTRIBUTION:** There are two distinct populations. The coastal population ranges from southern coastal B.C. to n. Baja and winters from cen. Calif. to Baja. The interior population breeds from cen. and s. Utah to cen. Colo. south into cen. and e. Ariz., all but e. N.M., and trans-Pecos Tex. to Big Bend; winters in extreme se. Ariz., sw. N.M., and Big Bend south into Mexico and Central America. **HABITAT:** Coastal birds in the Pacific Northwest inhabit temperate coniferous forests to altitudes of 1,000 ft.; in California, found in mixed conifer forests, primarily Douglas fir, as well as oak and pine-oak woodlands and suburban areas rich in oaks. Interior birds frequent montane coniferous woodlands at 5,000–8,000 ft. (occasionally lower). Most commonly found in pine–Douglas fir; also favors Gambel's Oak, oak-juniper and pinyon-juniper, Ponderosa Pine, and White Fir. **COHABITANTS:** Hairy Woodpecker, Hutton's Vireo, Steller's Jay, Chestnut-backed Chickadee (coastal), Mountain Chickadee (interior), Western Tanager. **MOVEMENT/MIGRATION:** Most northern coastal birds and most interior birds vacate breeding areas in winter (although some birds remain in urban centers where food is available). Spring migration from late February to mid-June; fall from mid-August to late November. Even in areas where birds are present year-round, flocks may move great distances in response to food needs, so numbers vary. VI: 3.

DESCRIPTION: A large, robust, woodland pigeon that resembles a stretched-out, pale, and monotoned Rock Pigeon. Larger than Rock Pigeon with a domed head and long face, a long slender neck, and a full body that tapers to a long tail.

Uniformly slate gray with a mauve blush (particularly on the breast). The narrow white collar on the hindneck (absent in immatures) is surprisingly easy to see, as is the mostly bright yellow bill. (Greenish iridescence on the necks of adults is inconspicuous.)

BEHAVIOR: Usually seen in flight, in small groups, or flying over treetops as it travels to and from feeding and roosting areas. Also commonly seen perched high atop trees where (unless approached) birds may be studied at length. Rock Pigeon almost never perches in trees and even less frequently perches at the very tops of trees.

Band-tailed Pigeon also sits low, in the understory, particularly when feeding on berries, and navigates by walking along limbs. In flight, weaves a path through the branches with reckless speed. While commonly found at higher altitudes, does descend in winter and in response to drought.

Feeds on the ground, foraging on waste seed in agricultural areas as well as on natural (and cultivated) fruit in trees (and orchards). In winter, acorns constitute a staple. Also comes to feeders. Generally quiet, often sedentary, and, unlike Rock Pigeon, shy. An observer's first indication that the birds are about is often the loud sound of their wings as they take flight. Unlike Rock Pigeon, Band-tailed rarely circles and returns.

FLIGHT: Usually flies in small loosely strung-out flocks (but single birds are not uncommon). Overall longer-profiled than Rock Pigeon—longer-tailed, longer and rounder-tipped wings—but unmistakably large and robust. The white neck collar is easy to see because everything else about the bird is so uniformly pale (bluish gray) and plain. The paler gray wide band on the outer half of the tail is often visible when birds flair their tails and land. Flight is direct and strong. Wingbeats are powerful, regular, and steady, with a particularly forceful down stroke.

VOCALIZATIONS: Uncommonly silent. Call is a soft, low, two-noted "*(uh) whooo*" (pause) "*(uh) whooo.*" Sounds owl-like (also Blue Grouse–like) in pattern, as if the bird is drawing in a breath (*uh*) and letting it out (*whooo*). End usually sounds strained, as though the bird is running out of breath.

Eurasian Collared-Dove, *Streptopelia decaocto*

Megadove

STATUS: Eurasian species first found in North America in the late 1970s. Now common and spreading. **DISTRIBUTION:** Formerly concentrated in the states bordering the Gulf of Mexico as well as s. Ga. and the coastal Carolinas. Now abundant on the Great Plains and found to cen. Calif., s. B.C., and s. Sask. May appear almost anywhere in North America. **HABITAT:** A resident of small to medium-sized towns (usually shuns large cities) and suburban areas; also found on agricultural lands, particularly in grain-growing areas. **COHABITANTS:** Mourning Dove, White-winged Dove. **MOVEMENT/MIGRATION:** Permanent resident. VI: 3.

DESCRIPTION: A big, burly, pigeonlike dove. Slightly larger than Mourning Dove but bulkier; while broadly proportioned, more athletically trim and distinctly longer-tailed than Rock Pigeon. Tail is shorter than Mourning Dove, as well as broader and square-cut at the tip. Overall pale (buffy gray), particularly on the head. (Heads of Mourning and White-winged Doves are the same tone and color as the body.) Most closely resembles the smaller, paler, more petite, and also collared Ringed Turtle-Dove, an introduced species that at present has no viable population in the United States. Eurasian Collared-Dove is very easily identified in flight.

BEHAVIOR: Conspicuous and vocal. Most commonly found perched on roofs, grain elevators, or utility lines (often with Mourning Dove) or foraging on the ground in yards, along roadsides, and in parking lots for grain or grit. Seems to prefer unkempt yards over well-tended ones. Perches almost horizontally. (Mourning Doves sits with its tail angled down.) Walks with a head-bobbing motion. Flies frequently but generally not far. Wings do not whistle.

FLIGHT: Very broadly proportioned in flight, with wide fairly blunt wings. If Mourning Dove is falconlike, Eurasian Collared Dove is more accipiter-like. The overall sense of paleness is retained in flight, but a distinctive contrasting pattern emerges. Wings are pale along the arm, dark on the hand. The widely fanned tail shows a broad whitish terminal band above and a two-toned pattern below—white on the distal half, blackish closer to the body. Flight is strong, stiff, and somewhat pigeonlike; while less nimble than Mourning Dove, given to frequent histrionic swoops and dives.

VOCALIZATIONS: Makes a three-note coo, "*hoo Hoo, cook*," with the last note more abbreviated, slightly lower-pitched, uttered after a slight hesitation, and then repeated. Similar in quality to Mourning Dove, but quicker, lower-pitched, and not as mournful-sounding. Also emits a soft, descending, nasal bray that recalls a catbird's scold (or the call of King Penguin). *Note:* There is also a domesticated breed of this species that utters a two-note call.

Spotted Dove, *Streptopelia chinensis*

Lace-necked Dove

STATUS: Uncommon to common (and declining and retracting) but localized and geographically restricted resident of southern California. **DISTRIBUTION:** In the continental U.S., found primarily in and around the Los Angeles Basin. Locally, range includes Los Angeles and Orange Counties and scattered populations found in urbanized portions of the s. San Joaquin Valley (particularly Bakersfield). **HABITAT:** Urban and suburban habitats. Prefers older established neighborhoods that feature a mix of mature evergreen trees, an understory of shrubs, open grassy areas (for feeding), and a dependable water source. Also found on ranches and in industrial areas (where habitat requirements are met). **COHABITANTS:** Mourning Dove, Northern Mockingbird, House Finch. **MOVEMENT/MIGRATION:** No seasonal movements. VI: 0.

DESCRIPTION: A large, robust, ruddy-toned dove wearing a lacy shawl. About the same size as Mourning Dove, but chestier and overall more robust, with a long but distinctly wider, heftier, square-cut tail. (Spotted Dove looks like you could easily lift it by the tail; if you tried to do this with Mourning Dove, the spiky tail would snap.) When bird is leaning forward while foraging, wingtips show as little triangles that jut above the back. (On Mourning Dove, wingtips are less elevated and more horizontal.)

Overall slightly darker and ruddier (particularly ruddier-breasted) than Mourning Dove, with no black spots on the wings. At close range, adults show a distinctive, broad, dark collar/shawl showing clustered white spots on a black background. (The collar is faint or lacking on immatures.) Eurasian Collared Dove is slightly larger and paler, with a narrow black band on the hindneck.

BEHAVIOR: Usually seen perched on utility lines or roofs or foraging in the open on the ground, usually in lusher, more heavily vegetated habitats than those favored by Mourning Dove; also visits feeders. Not particularly social. Usually seen singly, in pairs, or in small spaced groups, and generally does not associate with Mourning Doves except at feeders. Fairly tame. If flushed while feeding, often flies to the nearest tree or rooftop and shortly returns. Upon landing, elevates its tail.

FLIGHT: Overall dark, stocky, and distinctly heavy-tailed. When flushed, shows a broken broad white band on the tip of the dark tail (not along the sides of the tail, like Mourning Dove). Flight is direct, fast, and generally low, with stiff irregular wingbeats. Wings flutter audibly when birds take off but do not whistle in flight.

VOCALIZATIONS: The cadence and quality of the three-part cooing call recalls Inca Dove's "*no hope*" (but with an extra note): "*wha h'o'o'o'o pope.*" Middle note is trilled or gurgled.

PERTINENT PARTICULARS: Despite differences in plumage and shape, Spotted Dove is not so manifestly darker, redder, and shorter-tailed that distinguishing it from Mourning Dove is automatic. Some illustrations show Spotted Doves that are darker, redder, and shorter-tailed than they appear in reality. Observers searching for the bird need to study, not just glance at, all doves. The long heavy (not spiky) tail and unspotted wings are more readily noted than the namesake spotted collar.

White-winged Dove, *Zenaida asiatica*

White-trimmed Dove

STATUS: Common; permanent resident in some areas, but very migratory in others. **DISTRIBUTION:** Except for s. Florida, primarily a bird of drier southwestern regions breeding from se. California and extreme s. Nevada (migratory) east across Arizona (partially migratory), New Mexico (mostly resident), se. Colorado, sw. Kansas, Oklahoma, Texas, sw. Louisiana, south into Mexico, Central America, and the West Indies. In winter, birds from w. Arizona west and from the c. Southern Plains vacate breeding areas while some birds expand east along the Gulf and Atlantic Coasts north to North Carolina. **HABITAT:** Breeds in dense thorny woodlands (typical trees ebony, mesquite), deserts where cactus and palo verde dominate, and riparian woodlands; also orchards, and residential areas where stands of older trees and shrub are found. Feeds in grain fields, desert, and backyard feeding stations. Optimal habitat is one where woodlands or desert habitat is contiguous to feeding areas. **COHABITANTS:** Mourning Dove, Inca Dove, Phainopepla, Verdin, Curve-billed Thrasher. **MOVEMENT/MIGRATION:** A large portion of the population moves to southern portions of the range in fall, returning in spring. Some flocks may make short flights with frequent feeding stopovers; others are believed to fly nonstop to wintering areas. Spring migration from March through May (peaks in April and early May); fall August into December (peaks from September through November. VI: 3.

DESCRIPTION: A large, sturdy, pigeonlike dove—slightly larger but distinctly bulkier than Mourning Dove whose overall measurements are exaggerated by its long tail. Head is flatter and more oblong (Mourning Dove's head is round) with a longer down-drooped bill (makes White-winged Dove seem comparatively long in the face). Tail is shorter, broader, and blunter than the tine-tail of Mourning Dove (but does not appear short except in comparison).

Slightly grayer, paler, and (ignoring for the moment the distinctive and diagnostic white racing stripe that defines the lower edge of the folded wing) overall plainer than Mourning Dove (that is, no black spots on the wings). Shows a blue-ringed reddish eye (black in Mourning Dove). Immature is paler faced but otherwise similar to adult.

Expression: Resigned; forlorn.

BEHAVIOR: Social dove, feeding in small to large flocks with this and other dove species. Eats primarily seed, and while preferring to feed in larger open areas, will readily adapt to suburban or even urban feeders in yards bracketed by vegetation. Forages on the ground by picking up grain and larger seeds, but (and unlike Mourning Dove) is adept at perching on the stalks of sturdy seed-bearing plants (example: corn and sorghum). This dexterity permits it to utilize hanging bird feeders, which may account for its growing numbers in towns and residential areas.

FLIGHT: In flight, stocky, pigeonlike, and here, the tail does appear short. Shows a distinctive and contrasting wing pattern—blackish wingtips; white crescent center; gray base. Tail is darker than the body and shows a broad pale terminal band. The white in the wing distinguishes it from all North American doves and pigeons, and only Eurasian Collared-Dove shares the broad whitish terminal tail band.

Flight is swift, direct, usually at medium height (within 200 feet of the ground) and commonly covers great distances. May fly singly, in small groups, or in large flocks. Wingbeats are quick, regular, and emphatic. Overall flight is not as twisty-turny as Mourning Dove.

PERTINENT PARTICULARS: This species regularly wanders north of its normal range, even as far north as s. Canada.

VOCALIZATIONS: A distinctive cooed "*Who cooks for you?*" Also a five-note variation: "*La-coo-kla-coo-kla.*"

Mourning Dove, *Zenaida macroura*

Teardrop with a Tail

STATUS: North America's most common and widespread dove and, except for southern and western portions of the United States and Mexico (and where Eurasian Collared-Dove has not become established), North America's *only* dove. **DISTRIBUTION:** Breeds across much of s. Canada (from B.C. to the Maritimes) and the entire continental U.S., the West Indies, and south into Mexico. In winter, vacates breeding areas in the n. Great Plains north of a line drawn across Wyo., s. S.D., cen. Minn., and n. Wisc., as well as parts of Ont. and Que. **HABITAT:** At all seasons, prefers open and semi-open habitats (shuns large unbroken forest). Particularly prevalent in agricultural lands, especially where grain is raised, but also common in weedy fields, prairies, and deserts where water and trees (including planted trees) are found. Highly suburbanized, Mourning Dove frequently nests in rosebushes, ornamentals, and planted pines. A common bird at bird-feeding stations, where it favors millet, and at cattle feedlots. **COHABITANTS:** Forages with other doves. **MOVEMENT/MIGRATION:** Birds living in the interior at the northern limit of the species range (primarily the northern Great Plains) vacate breeding areas in winter. Otherwise a permanent resident, but many birds retreat south to the Southwest and Mexico for the winter. Spring migration from March to May; fall from September to November. VI: 2.

DESCRIPTION: A long, slender, medium-sized dove (from bill to tail, slightly longer than White-winged Dove; smaller than Eurasian Collared-Dove; about the same size as Common Grackle) distinguished by its distinctly long, stiff, acutely pointed tail. The head is small (almost ridiculously so), the neck slender, and the body plump—these features, in conjunction with the tail, make the bird look like a teardrop with a tail or a pear on a stick. Overall grayish tan, with a rose-blushed breast and a cluster of black spots on the folded wings. The turquoise-ringed eye is balefully black.

No other dove seems so overall plain, so microcephalic, so tine-tailed.

BEHAVIOR: Dawn and dusk, Mourning Dove is flying in flocks (and is often the first bird seen or heard in the predawn gloom); mornings and afternoons, it is feeding on the ground. The rest of the time it sits on utility lines or any tall perch that accommodates at least a few birds. Very social. Feeds in pairs when breeding; otherwise in small (several birds) to large flocks (hundreds of birds).

Feeds on open, bare, or lightly vegetated ground, moving in the aimless, wandering, head-bobbing manner of many pigeons, pecking at scattered seeds. Walks and sometimes runs (somewhat). Body and tail are held horizontal. Usually tolerant of humans, but sometimes flighty, particularly where hunted. When Mourning Dove takes off, it explodes, rising almost vertically with the white-edged tail fanned, wings flailing, and wind whistling audibly through its wings as it rises. Likes water, particularly in desert areas.

FLIGHT: Direct and extremely fast. Birds have a very streamlined silhouette, with wings that are always sharply angled back, never extended. Wingbeats may be regular or choppy and irregular. Flight is twisting and turning: the bird lists first to one side, then the other. Lands clumsily and tentatively, with braking flaps and a light touchdown (often needs more than one try when landing on a branch). Courtship flight is dramatic: a series of deep choppy flaps followed by a prolonged droop-winged glide that curves, tacks, and wobbles like a poorly thrown Frisbee. An observer's first impression is always, "It's a hawk!"—but it's not.

VOCALIZATIONS: One of nature's most familiar songs—even if many people attribute it to an owl, not a dove. A mellow, plaintive, whistled query that begins with a three-part cooing phrase followed by three spaced coos (the last sometimes slightly hurried): "*WhoOOoo? Who? who'Who?*"

PERTINENT PARTICULARS: Mourning Doves are roughly similar in size and shape to American Kestrel, another bird that habitually perches on utility lines. Kestrels on wires, however, are almost always alone; doves are often in groups. Also, a dove on a wire keeps its tail rigid, whereas American Kestrel habitually bobs its tail.

Inca Dove, *Columbina inca*

The Dove with "No Hope"

STATUS: Common suburban dove of the Southwest. **DISTRIBUTION:** Found across the s. U.S. from se.

Calif. to extreme sw. Nev., w. and s. Ariz., cen. N.M., and Tex., as well as se. Colo., s. Okla., sw. Ark., and s. La. Also occurs in all of Mexico, into Central America. **HABITAT:** At all times of year, favors human habitations (cities, villages, towns, farm buildings), open areas (pastures, parks), and floodplains adjacent to riparian woodlands. **COHABITANTS:** Common Ground-Dove. **MOVEMENT/MIGRATION:** Mostly a nonmigratory, permanent resident, but a few wanderers turn up well outside of the bird's normal range. VI: 2.

DESCRIPTION: Small (hardly more than the size of a large sparrow), pale, slender, long-tailed dove. An observer's first impression is that it's a baby Mourning Dove, but Inca is overall paler than Mourning, with a distinctly scaly appearance, *particularly* on the back, and no hint of iridescence in the feathers. Upperparts are buffy gray (and scaly); head and underparts are pale gray (and scaly).

BEHAVIOR: A tame bird that allows very close approach; observers often become aware of its presence only when it explodes into flight from almost underfoot. Take another step, and one or more birds will take flight. Flushed birds usually take a perch in a nearby tree. (Ground Dove is more likely to land on the ground, then run beneath some concealing cover.)

Roosts and nests in shrubs and trees. Forages in open areas with short or no vegetation (such as lawns and backyard feeding stations) and open areas adjacent to streams or riparian woodlands. Does not commonly forage in woodlands or under brushy cover. When not feeding, Inca Doves sit in trees, usually close to the ground or at mid-height, often clustered together with bodies touching. Inca walks with a mincing shuffle, the head bobbing back and forth. Very vocal; calls regularly year-round.

FLIGHT: Very fast and usually short. Takeoffs are explosive and noisy (wings make a rattling/shuffling sound)—the bird seems to be going full speed right out of the blocks. In flight, note that tops and undersides of wings flash red, like Common and Ruddy Ground-Doves, but that the shape is long-tailed, like Mourning Dove. Also, Inca shows flashing white sides to the tail, whereas Common Ground-Dove has a black-trimmed tail with white that is limited to the outer tail tips (and is hard to see).

VOCALIZATIONS: Song is a lonesome, repeated, cooed, two-note chant, "*pol-pah*"—or as it is more often phonetically rendered, "*no hope.*" Also makes a low trilling "*coo.*"

Common Ground-Dove, *Columbina passerina*

Tiny Ground-Dove

STATUS: Common year-round southern species. More common in the Southwest than the Southeast. **DISTRIBUTION:** Ranges widely throughout the Caribbean, Mexico, Central America, and n. South America. In the U.S., found across much of the South from s. S.C. across the southern two-thirds of Ga., s. Ala., Miss., La. (rare), s. Tex., sw. N.M., s. Ariz., and s. Calif. **HABITAT:** Dry, often sandy areas with short and open vegetation, including open pine woodlands, mesquite thickets, woodland edges, coastal dunes, riverine thickets, agricultural fields, suburban areas, orchards, successional fields, and roadsides. **COHABITANTS:** Inca Dove. **MOVEMENT/MIGRATION:** For the most part, a permanent resident. Some birds disperse in late summer and fall, however, and there is evidence that some northern birds retreat toward warmer, coastal, or more southern portions of the range. VI: 2.

DESCRIPTION: A tiny dove (only slightly larger, but plumper, than House Sparrow), with a short neck and a very short tail. Overall plump and compact but dove-shaped, complete with a small, thin, mostly bright red-orange bill (pale pink in immatures). The rump has a jutting bump. A short, stiff, down-cocked tail seems like a structural afterthought.

Overall pale grayish brown, with rosy highlights on the face, breast, and wings (but not the back) and some spare dark spotting on the wings. The scaly head, neck, and breast make the bird look like it's wearing a chain-mail mantle (but this characteristic is not easily seen at a distance). The scaly blue feathers on the crown contrasting with the ruddy face makes the heads of males look capped. Inca Dove is overall paler (particularly on the face), longer-necked, distinctly longer-tailed, lacking in any rosy glow, and coarsely scaly all over.

BEHAVIOR: A distinctly terrestrial dove, this busy little bird spends most of its time on the ground, moving with short mincing steps as it forages. Movements are animated, jerky (the head bobs and the breast seems to jiggle and shimmer as it walks)—and quick! Common Ground-Dove seems to do everything with flickering quickness. Forages in sparsely vegetated areas (including bare earth and close-clipped lawns), but prefers to be close to brushy cover. When pressed, flies to the closest cover (such as the woods' edge). Nevertheless, birds are generally tolerant of humans, allowing close approach.

Most commonly found in pairs or in small flocks (and, where ranges overlap, mixes freely with Ruddy Ground-Dove and Inca Dove). At times, also forages with other dove species.

FLIGHT: Tiny and compact in flight—seems more like a tiny quail than a dove. Short, stocky, roundly pointed wings flash with rufous. The short tail is fanned and shows a black trim (views are brief, so white corners are often not seen). Takeoffs are explosive (quail-like). Flight is direct, quick, jerky, and normally short, with stiff wingbeats given in a rapid series that is broken or punctuated by a halting rhythm-breaking skip or a half-pause. When birds take off, wings make a fluttering sound—like a deck of cards being shuffled—that is not as loud or clattery as Inca Dove.

VOCALIZATIONS: A simple, uncomplicated, and somewhat monotonous "*cooah,*" given in series.

PERTINENT PARTICULARS: Chances are you won't even notice the bird before it flushes, and if you do, you may dismiss it as a sparrow—but a very active head-bobbing sparrow that walks instead of hops. Once it flushes, showing a black tail and flashing red in the wings, all resemblance to a sparrow ends.

Ruddy Ground-Dove, *Columbina talpacoti*

Blue-headed Ground-Dove

STATUS: Scarce and restricted fall and winter resident; occasional summer resident. **DISTRIBUTION:** Tropical species that breeds as far north as Mexico. Ranges annually into s. Calif., s. Ariz., s. N.M., and casually into Tex. **HABITAT:** Dry desert scrub bordering cropland and roadsides. Commonly feeds on dry earth. **COHABITANTS:** Associates with both Common Ground-Dove and (often) Inca Dove. **MOVEMENT/MIGRATION:** Some birds wander north after breeding. VI: 2.

DESCRIPTION: A tiny smoothly plumaged dove with a ruddy body and a pale head. Small (almost half the size of Mourning Dove; slightly longer than the very similar Common Ground-Dove), with a round head, a short slightly down-drooped bill, a plump body, and a shortish tapered tail. Slightly less compact and longer-tailed than Common Ground-Dove.

Adult male's body is totally ruddy, contrasting with a pale bluish gray head and an all-dark bill. Plumage is smooth, with no hint of scaling on the nape, head, or breast. The head does not look capped, and black spotting on the wings extends above the folded wings and onto the shoulders (coverts). Females and immatures (which are more commonly seen in the United States) are slightly paler, slightly more brown-gray, overall plainer, and more smoothly plumaged than Common Ground-Dove, with no hint of scaly pattern on the nape, head, or breast. Their heads are slightly paler than their bodies—brownish gray as compared to the male's bluish gray. Like the male, females and immatures have blackish bills (often dark gray going darker toward the tip, with no pink or red at the base), and more extensive black marks on the wing extend up onto the back. Also, on many females the pale edges of the wing coverts give wings a frosted or feathery look.

In part because of their longer tails, slightly longer neck structure, and smoother, more brown-gray plumage, female Ruddy Ground-Doves more nearly recall small or baby Mourning

Doves than does Common Ground-Dove, which is more like a plump baby Inca Dove.

BEHAVIOR: A tame dove that sometimes feeds toward observers. In the United States, often found among flocks of Inca Doves, but also forages with Common Ground-Dove. When flushed, may retreat into bordering vegetation.

FLIGHT: Like Common Ground-Dove in shape, but shows overall more black (on the tail and underwings). Flight is fast and direct but halting, with rapid sputtery series of wingbeats punctuated by a skip or a pause (see "Flight" section of "Common Ground-Dove").

VOCALIZATIONS: Call is a monotonous series of low bisyllabic coos, "*per-woo, per-woo, per-woo,*" that are lower-pitched and slightly faster than Common Ground-Dove's single-note call.

White-tipped Dove, *Leptotila verreauxi*

Moaning Dove

STATUS AND DISTRIBUTION: Common but highly restricted resident in the U.S. found only in the lower Rio Grande Valley of Tex.; widespread through much of Mexico and Central and South America. **HABITAT:** Dense riparian woodlands dominated by Texas ebony, cedar elm, and mesquite; often near resacas. Has expanded into citrus groves and well-vegetated suburban areas. **COHABITANTS:** Plain Chachalaca, Green Jay, Olive Sparrow, Altamira Oriole. **MOVEMENT/MIGRATION:** Permanent resident. VI: 0.

DESCRIPTION: A pale, busy, stocky, ground- and thicket-loving dove whose call is a quavering moan. Medium-sized (shorter, but bulkier, than Mourning Dove; about the same size as White-winged Dove), with a round head, a short neck, a plump body, and a blunt too-short tail (that is, for those who take the long narrow tine-tail of Mourning Dove as the norm).

Overall pale and plain but more contrasting than Mourning and White-winged Doves; shows dull pale brown upperparts, a paler grayish and rose-blushed head and breast, and a whitish belly. The white corners on the tip of the tail are often not visible until the bird takes flight. Wings lack the black spotting of Mourning Dove and the white wing border of White-winged. Expression slightly crazed, in a preoccupied sort of way.

BEHAVIOR: A busy, somewhat mechanical-moving dove. Walks quickly across the forest floor (a habitat shunned by Mourning and White-winged Doves). Walks into the open to feed (particularly where food is placed), but never strays far from dense cover. Usually seen singly or in pairs (sometimes several individuals gather at bird-feeding stations); though sometimes seen with other doves at watering holes or feeding stations, does not join mixed-species flocks. For the most part, stays under cover, where only its low cooing, given from a perch near the ground, betrays its presence.

Shy, flushes easily, and heads directly for the nearest cover. Wings make a twittering whistle.

FLIGHT: Overall stocky, exceedingly chesty, short-tailed profile. The contrast between darker upperparts and paler underparts is very apparent, as are the reddish flashes winking from the bird's underwings and blackish undertail. (The white tip to tail is easily overlooked, however, when glimpses are brief.) Flight is direct and recklessly fast, not twisty-turny (like Mourning Dove). Birds rocket across open areas and penetrate the forest edge without slowing.

VOCALIZATIONS: Song is a low quavering moan: "*wh'whoo'oo.*" Because the first part, which sounds like a quick muffled intake of breath, is often not audible at a distance, the song has a run-on, quavering, single-note quality (very different from the more complex moan-and-coo-coo-coo pattern of Mourning Dove). To many, the song suggests the sound of blowing into an empty bottle.

PARROTS AND PARAKEETS

Monk Parakeet, *Myiopsitta monachus*

The Widespread Parakeet

STATUS: Introduced species; restricted but locally common; more hardy and widespread than other introduced parakeet species that thrive in more temperate regions. **DISTRIBUTION:** Found across portions of the Florida Peninsula. Colonies are also found in R.I., Conn., N.Y., N.J., Va., Chicago, Ala., La., Tex., and Ore. **HABITAT:** In the United States, primarily associated with urban and suburban areas, including residential areas, city parks, power-line towers (where they frequently nest), and baseball fields (where light towers serve as nest sites). Also nests in trees, including palms. **COHABITANTS:** Assorted urban and suburban bird species. **MOVEMENT/MIGRATION:** Nonmigratory; permanent resident. VI: 1.

DESCRIPTION: Looks looks like a bright green American Kestrel (and sounds like a screechy one), with a longer pointier tail. The pale gray forehead and gray breast quickly distinguish it from other introduced green-bodied parakeets. There's no reason to labor this point: it's a parakeet, and where it is found, it is neither particularly difficult to locate nor easily confused with any native species.

BEHAVIOR: Noisy, both in flight and when perched. Commonly seen in small flocks that have already announced their arrival or presence with their chatter. Feeds on the ground and in trees. Moves slowly and methodically by walking with a slow sailor's gait and clambering, using feet and beak.

Tame and habituated to people. Its large sloppy stick nests (found in clusters in trees and on utility poles and platforms) are very conspicuous.

FLIGHT: Often flies in noisy groups. The very long narrow tail is conspicuous. In flight, pointy wings may show blue above and below, but in general, it's an all-greenish bird with a pale face and breast. Flight is rapid and shifting, with frequent course changes. Wingbeats are rapid and steady, on down-curved wings that do not rise above the horizontal.

VOCALIZATIONS: A loud screechy rattle with a slight upward inflection: "*rrrrrch.*" Sounds like someone running fingers up a (screechy) washboard.

PERTINENT PARTICULARS: Non-native Parakeets have become established in North America, primarily in California, Texas, and Florida. These include two South American species, White-winged Parakeet, *Brotogeris versicolurus*, found primarily in the Miami area, and Yellow-chevroned Parakeet, *Brotogeris chiriri*, which is numerous in the Los Angeles area (also in Miami). The populations of these and other introduced species wax and wane, but some are now common. Wherever you encounter a parakeet species, but especially in states where large populations of exotic parakeets thrive, be mindful of other possibilities besides Monk Parakeet.

Green Parakeet, *Aratinga holochlora*

The 10th-and-Zinia-keet

STATUS AND DISTRIBUTION: Mexican resident whose northern range approaches the U.S.–Mexico border. A small population has established itself in the Rio Grande Valley between McAllen and Brownsville, Tex., some of which are presumed to be of natural origin. **HABITAT:** Suburban and urban areas as well as outlying woodlands and orchards. Most commonly seen in the evening when birds gather to roost. **COHABITANTS:** Mourning Dove, Great-tailed Grackle, House Sparrow. **MOVEMENT/MIGRATION:** Permanent resident, but roosting sites shift frequently. VI: 0.

DESCRIPTION: A screechy, mostly green parakeet seen in and around McAllen, Texas—often at the intersection of 10th and Zinia Sts., but there are several sizable McAllen flocks and flocks in other towns in the area, including Brownsville. Medium-sized (slightly larger than both Mourning Dove and Red-crowned Parrot, the other indigenous parrot found in the lower Rio Grande Valley), and easily distinguished from Red-crowned Parrot by its more slender build and distinctly long pointy tail. (Red-crowned's tail is short and blunt.)

It's all green! Bright green above, and ever so slightly yellowish green below. Legs sometimes show a yellowish tinge. The bill is horn-colored or pinkish, and on some individuals the head is garnished with small orange flecks. The gray eye-ring is hard to see. (Many other introduced parakeets have whiter and/or more prominent eye rings.)

BEHAVIOR: Roosts in trees around suburban neighborhoods—very commonly palms. In winter (when most people visit the Rio Grande Valley), flies out from roosts in small groups (6–30) to forage during the day; returns about an hour before sunset. Highly social and very noisy. Birds announce their arrival with a loud screechy call.

Like all parakeets, very nimble and acrobatic. Walks along limbs (and wires) and hangs upside down, using its bill for stability and mobility.

FLIGHT: Looks long-tailed, sickle-winged, and blunt-headed—much like a large-headed, pointy-tailed, somewhat front-heavy American Kestrel. Appears overall green except for the pale bill and grayish tail and trailing edge to the wings.

Flies on down-curved, somewhat blunt-tipped wings. Flight is direct and fast or wandering and tacking, with rapid, steady, somewhat batty wingbeats. When landing, often glides toward perches and brakes with fluttering wings and a tail fanned into a splayed and spiky array of feathers. Flocks are noisy and fairly tight, but shifting and usually showing a horizontal spread.

VOCALIZATIONS: In flight and when perched, makes a loud, gargled, screechy chatter: "*re'h'h, re'h'h.*"

Red-crowned Parrot, *Amazona viridigenalis*

The New American Parrot

STATUS AND DISTRIBUTION: Mexican species with an established population, presumably of natural origin, in the lower Rio Grande Valley of Tex. between McAllen and Brownsville. Also occurs in other southern cities, most notably in the Los Angeles Basin, but these populations are thought to originate from escaped caged birds. **HABITAT:** In the United States, found primarily in wooded urban and suburban habitats. In Mexico, inhabits broken forest and open habitats dotted with trees. **COHABITANTS:** In Texas, Great Kiskadee, Green Parakeet, Northern Mockingbird, Great-tailed Grackle. **MOVEMENT/MIGRATION:** Permanent resident. VI: 0.

DESCRIPTION: Stocky, red-crowned, green parrot seen flying around Brownsville, San Benito, Harlingen, Weslaco, and McAllen, Texas, and eastern Los Angeles. Medium-sized (slightly smaller but conspicuously bulkier than Green Parakeet; about the same size as White-winged Dove), with a big blunt head, a pale bill, a chunky body, and a short tail.

All green except for the pale bill and red crown. Folded wings usually show a wink of red and a touch of blue near the tip. The yellow tip to the tail is not conspicuous. Red-lored Parrot, another Mexican species with established but suspect populations in the United States, is very similar to Red-crowned but shows a duller bill, yellow cheeks (Red-crowned has no yellow in the cheeks), and red on the head that is restricted to the forehead, not covering the entire crown.

BEHAVIOR: Most often seen in flight, flying to or from roost sites, but also seen at the roost sites themselves. (Locations are monitored by local birders and birding organizations.) Begins to arrive at roost sites about one hour before sunset; some winter roosts may number 100 or more birds. Red-crowned is a social bird found mostly in small (noisy) flocks or in pairs. Moves sluggishly and methodically and is fairly tame, unperturbed by locals tending to their lawns or birders trying to get a closer look.

FLIGHT: Overall stocky, compact, and proportionally balanced profile with a large head and a short tail projecting equally ahead and behind short, blunt, broad wings. In flight, wings are downcurved, and head appears downturned. Overall green below (with a yellowish tip to the tail). From above, shows a red crown and a red patch on the mid–trailing edge of the wing.

Wingbeats are hurried, choppy, and steady. Flight is fast but somewhat labored.

VOCALIZATIONS: Call is a loud, harsh, grating, screechy, gargled "*Ree'ahr'r'hah'ah.*" Also emits a more musical "*ree-ic,*" as well as softer twittery and chirpy notes.

PERTINENT PARTICULARS: Other parrots, primarily escaped cage birds, may be encountered, principally in southern Florida and southern California, except one species, Thick-billed Parrot, *Rhynchopsitta pachyrhyncha*, a native of the mountains in northern

Mexico, once bred in the mountains of southeastern Arizona and southwestern New Mexico. This large, green, black-billed, long-tailed parrot shows red on the crown and on the shoulder.

As a matter of course, know that encounters with escaped exotic parrots of assorted species are possible anywhere. Be particularly mindful when birding the Chiricahua Mountains of Arizona and the Animas Mountains of New Mexico that any large green parrot is worth your studied attention.

CUCKOOS, ROADRUNNER, AND ANIS

Black-billed Cuckoo, *Coccyzus erythropthalmus*

Black-billed "Coocoocoo"

STATUS: Widespread breeder and migrant, but uncommon, irregular, secretive, and to some degree nocturnal. **DISTRIBUTION:** Overall a more northern species than Yellow-billed Cuckoo. Breeds extensively across n. and e. U.S. and s. Canada. Ranges from s. Alta., cen. Mont., and e. Wyo. east across s. Canada and the n. U.S. to the Atlantic seaboard, excluding n. Maine and w. N.B. but including w. N.S. and P.E.I. Southern border cuts across s. Wyo., ne. Colo., se. Neb., cen. Kans., s. Mo., s. Ill., s. Ind., e. Ky., e. Tenn., w. N.C., and w. and n. Va. to the Delmarva Peninsula. **HABITAT:** Small isolated woodlands, groves, forest edges, and brush and willow thickets (often near water). Prefers deciduous habitat, but also found in mixed coniferous-deciduous woodlands. More inclined to occupy forest, and less suburbanized than Yellow-billed Cuckoo. **COHABITANTS:** Yellow-billed Cuckoo, Eastern Kingbird, Willow Flycatcher, Gray Catbird, Brown Thrasher. **MOVEMENT/MIGRATION:** Wholly migratory; winters in South America. Spring migration from mid-April to mid-June, with a peak from mid to late May; fall from late August to late October. VI: 2.

DESCRIPTION: Slightly proportioned secretive cuckoo with a red-rimmed eye. Slender, long-tailed, and rather fine-featured, Black-billed is a smaller and less robust and tubular cuckoo than Yellow-billed, with a small, more slender, and all-black bill and a narrower tail.

Bland and cold-toned, Black-billed is a visually unengaging cuckoo. In all plumages, the bird is grayish brown above; adults are white below, and immatures slightly buffy. Upperparts are plainer and colder-toned than Yellow-billed, with only a hint of rufous, if any, in the flight feathers. The undertail pattern is very subdued—grayish brown with narrow white stripes. (Yellow-billed has a white undertail with narrow black bands.)

Given the secretive nature of the bird, however, chances are that you'll see no more than the bird's face peaking at you through the foliage. Happily, this view may simplify identification. You'll probably be able to focus on a small, slender, all-blackish bill (grayer on immatures) and a red-ringed eye (yellow on Yellow-billed). The eye-ring of immatures is indistinct; look for a plainer, less contrasting head pattern.

BEHAVIOR: A low-down skulker. Black-billed generally feeds lower than Yellow-billed. Perch-hunts from within foliage, often remaining stationary for long periods then running and hopping to grasp prey. Also adept at tearing apart tent caterpillar tents from a strategic perch that it can reach without effort (or overt motion).

Males vocalize day and night from within foliage and (at night) in flight. Black-billed does not commonly perch in the open. Flies reluctantly. Normally freezes if approached. When flushed, commonly heads for the nearest cover.

Responds reluctantly to pishing, but may be incited to call (try imitating a screech-owl). Often seems sluggish and somnambulant during the day, raising speculation that the species may be at least partially nocturnal.

FLIGHT: Overall slender, long-tailed, and somewhat short-winged. Overall slighter and colder-toned than Yellow-billed Cuckoo. The chestnut flash in the wings is equivocal (not blatant), and the undertail pattern difficult to see. Flight is swift and direct, usually fairly close to the ground, and slightly twisty-turny (like Yellow-billed Cuckoo).

VOCALIZATIONS: Coo notes are hurried, almost stuttered: "*coocoocoo . . . coocoocoocoo.*" In quality, pattern, and cadence, recalls Least Bittern. Also makes a slow "*k'awp k'awp k'awp*" that is similar to Yellow-billed but faster, higher-pitched, and less coarse.

Yellow-billed Cuckoo, *Coccyzus americanus*

Cinnamon-winged Cuckoo

STATUS: Common and widespread, but an irregular secretive breeder, with a patchy distribution in the West. **DISTRIBUTION:** Breeds throughout the e. U.S., generally east of a line drawn from e. N.D. and n. S.D. to e. Wyo., e. Colo., and e. and s. N.M. south into Mexico. Scattered populations are found in Mont., Idaho, se. Nev., e. Calif., and Ariz. Absent in n. Minn., n. Maine, n. Vt., and n. N.H. In Canada, found in extreme se. Ont. and extreme s. Que. **HABITAT:** Deciduous woodland, woodland edge, and woodland clearings, usually with a low dense (or broken) scrub component; also found in shrubby late successional fields and thickets. Likes rivers and other watercourses (in much of the West requires riparian woodlands with large cottonwood trees) and marshes; avoids unbroken mature forest. Forages in shrubs and mature trees. **COHABITANTS:** Great Crested Flycatcher, Gray Catbird, Yellow Warbler, Summer Tanager. **MOVEMENT/MIGRATION:** Spring migration from mid-April to mid-June; fall from August to October. Somewhat later in the West than in the East. VI: 3.

DESCRIPTION: A large, tube-shaped, brown-backed, white-bellied bird with a short curved, thrasher-like bill that is yellow on the bottom. Large and slender (same size as Mourning Dove and Black-billed Cuckoo; very slightly smaller than Mangrove Cuckoo), with a short down-curved bill, a flattish head that seems like an extension of the long tubular body, and a very long, straight, untapered tail. Overall slightly more robust than Black-billed Cuckoo.

Mostly plain grayish brown above, white below, but overall more colorful and visually engaging than Black-billed. The tail is plain above, but distinctly patterned black-and-white below: large white spots on a black background or narrow black bars on a white background. The dark-topped bill shows a rich yellow lower mandible. The face bears the suggestion of a dark mask through the eye (absent on Black-billed) and has no red eye-ring (as on Black-billed). Folded wings usually have a reddish tinge along the lower border, but it may be faint. Immatures have reduced amounts of yellow on the bill (sometimes limited to the base of the lower mandible).

BEHAVIOR: A sedentary, slothlike, stop-and-go skulker that may spend minutes on end on the same perch, moving only its head in slow, wandering turns to study the foliage around it before flying to another perch. May perch in plain sight, sometimes high in trees (seeming to assume that its hyper-erect sticklike posture will conceal it), or may perch partially or wholly hidden. When moving among branches, commonly crouches. When flying, snakes its way deftly through the foliage. Captures prey by hopping, walking, or running along limbs and reaching forward. Sometimes hovers and plucks.

Overall movements seem clandestine, sinister, predatory. Responds somewhat to pishing, but usually approaches and remains hidden from view (except for its head poking through the foliage). Usually seen alone or in pairs. In migration, may be found in small groups.

FLIGHT: Overall long and slender with a small head, a long body, an extremely long tail, and relatively short pointy wings. Most birds show obvious rufous flashes on the outer wings, but now and again you'll encounter birds whose wings are hardly ruddier than Black-billed. The uppertail shows patches or touches of white that you may attribute to the tip. (Black-billed's tail seems all-dark above, with small white spots that are hard to see.) Flight is fast, direct, and fluid, on hurried wingbeats that may be steady or executed with a slightly shifting rhythm. Birds sometimes twist and turn in flight (moving like skaters, or as if maneuvering around unseen obstacles). Although flight is generally smooth, it conveys a sense of stiffness and a sense of hurry or anxiety—as if the bird is uncomfortable being in the open.

VOCALIZATIONS: Several call types. Common is a low loud clucking or knocking that starts fast and insistent and loses speed and interest at the end: "*kluk luclucluclucluc luc luc k'lowp k'lowp k'lowp.*" Also makes a hard, rapid, even knocking series, "*tok tok tok tok tok,*" that is commonly repeated several times. Also makes a series of 5–10 soft spaced coos that slow and weaken toward the end, "*coo coo coo coo . . . ,*" and can be confused with Black-billed Cuckoo.

Mangrove Cuckoo, *Coccyzus minor*

Buff-bellied Cuckoo

STATUS: Uncommon, elusive, and, in the United States, highly restricted. **DISTRIBUTION:** Primarily tropical coastal species; in the United States, found only in extreme s. Fla., from Miami to the Keys, and north along the Gulf to just north of St. Petersburg. **HABITAT:** In the United States, restricted to fairly mature stands of Red and Black Mangrove and adjacent coastal hardwood habitats. **COHABITANTS:** Black-whiskered Vireo, Prairie Warbler. **MOVEMENT/MIGRATION:** Some birds may migrate, but others remain in coastal mangroves throughout the winter; given their secretive nature, their seasonal silence, and the near-impenetrable nature of their habitat, they may be very difficult to locate. VI: 1.

DESCRIPTION: A darkish, attractive, and very habitat-specific cuckoo. In size and shape, resembles a snub-billed Yellow-billed, which has yellow on its bill and large white spots on its undertail. Mangrove, however, is more contrastingly and richly plumaged: the yellow on its bill is restricted to the lower mandible (an evident but not blatant feature), and the upper mandible is all black.

Olive brown upperparts are slightly darker and richer than Yellow-billed, with no hint of red in the wings. White on the throat and upper breast (like Yellow-billed) gives way to warm buff or peach-colored underparts. The head pattern is distinctive: a dark mask across the black eye sets off the bird's gray crown and hides the eye. (On Yellow-billed, the eye is easy to see.)

BEHAVIOR: Solitary or paired. Spends most of its time foraging in the canopy of mangroves and other hardwoods, hopping onto or ascending branches. Pauses, usually for less than a minute, to study the surrounding vegetation with slow contortionist head movements. Flies a short distance to another tree (but generally not the closest tree). Most often seen perched conspicuously on dead branches above mangrove enclaves (and sometimes roadside utility lines) or flying across the road. In spring and summer (less frequently in winter), draws attention to itself by calling.

Tame. When foraging, often flies toward observers. Responds poorly (if at all) to pishing.

FLIGHT: Mangrove looks much like Yellow-billed, with a tail bracketed by large white spots, but has no rufous in the wings. Flight is cuckoo-direct, with rapid wingbeats and a fast fluid passage through the air (like Yellow-billed). Lands at full speed, braking with a fanned tail.

VOCALIZATIONS: Has two calls, sometimes given together: a low gargled rattle, "*rar'r'r'r'r'r*" (sounds very much like Anhinga), and a slow, evenly spaced, growled sequence, "*rah, rah, rah, rah, rah*" (much like the "engine cranking call" of Cactus Wren). The latter may be followed by a lower-pitched "*ough, ough, ough,*" a call that suggests Elegant Trogan. The "*rah, rah, rah, rah, rah . . . ough, ough, ough*" sequence is similar in pattern to Yellow-billed Cuckoo, but overall slower, rougher, and less musical.

Greater Roadrunner, *Geococcyx californianus*

Great Ground Cuckoo

STATUS: Common permanent resident across most of its largely southwestern range; less common in peripheral portions of its range. **DISTRIBUTION:** From sw. Mo., cen. Ark., and w. La. west across extreme s. Kans., Okla., Tex., se. Colo., all but nw. N.M., all but ne. Ariz., sw. Utah, s. Nev., and all of s. Calif. north through the Sacramento Valley. Ranges south into Mexico. **HABITAT:** Arid and semiarid areas with sparse low-lying vegetation. Classic habitats include the Sonoran Desert, mesquite thickets, creosote bush flats, pinyon-juniper, chaparral, and cactus-grasslands. Also occupies oak savanna and farmlands. Not found in thickly vegetated woodlands, urban areas, expansive prairie, or unvegetated desert. Has become suburbanized. **COHABITANTS:** Scaled Quail, White-winged Dove, Verdin, Cactus Wren. **MOVEMENT/MIGRATION:** Permanent resident. VI: 0.

DESCRIPTION: A large, loping, long-tailed, ground-loving cuckoo—virtually a caricature of itself. Large—about pheasant-sized—and overall gray and gangly, with a bushy-crested head, a large heavy bill, sturdy legs, and a long animate tail that constitutes half the length of the bird.

It's unmistakable.

Heavily streaked and spotted plumage looks grayish at a distance. Crest and tail are brownish black.

BEHAVIOR: A slinker. Moves with stealth and grace along the open edges of vegetation (usually avoids dense vegetation), dry streambeds, gullies, roadsides, field edges, and railroad tracks. Moves forward in a crouch, head down. Stops and elevates its head; raises and lowers its tail. Mostly solitary, but sometimes seen in pairs.

Most commonly seen crossing roadways. Also sometimes seen perched on rocks, fenceposts, or even trees, a vantage it most commonly gains by hopping up limbs. On cold mornings, may be found sunning itself with wings drooped and back feathers spread (exposing bare skin). Prefers to run if surprised, but if flushed, lofts itself vertically into the air, then lines out in level flight.

FLIGHT: The large, long-tailed, paddle-winged profile is unmistakable. Short flight is effortful and labored, but also buoyant. Wings move in a short heavy flutter followed by either a short or a surprisingly lengthy glide. Usually hits the ground running.

VOCALIZATIONS: Usually silent. Makes a series of low cooing sounds, "*who'o who'o who'o who'o who'o*," that sounds like the whimpering cries of a heartbroken dog. Also makes a rattling chatter (sometimes sequentially) that sounds like stiff playing cards being shuffled.

Smooth-billed Ani, *Crotophaga ani*

Florida "Ah-nee"

STATUS AND DISTRIBUTION: Widespread tropical and subtropical resident. In the U.S., increasingly scarce; found only in Fla. between the lower keys north to the south end of Lake Okechobee. **HABITAT:** Open savanna, scrub, disturbed areas, and low vegetation in rural and suburban areas. Partial to airport edges, roadsides, brushy ditches, sugar cane, cow pastures, and tall dense stands of reed or grass bordering lakes and ponds. **COHABITANTS:** Cattle Egret, Mourning Dove, Boat-tailed Grackle, cattle, cane field workers, landscapers on mowers. **MOVEMENT/MIGRATION:** Permanent resident. VI: 1.

DESCRIPTION: A bird with the body of a grackle and the head of a turtle. Large, gangly, primitive-looking blackbird (slightly larger than Groove-billed Ani; about the same size as Boat-tailed Grackle), with a short, blunt, bulbous bill, a large head, a long loosely feathered body, and a long, narrow, loosely unkempt tail. The swollen silvery black bill has a bony ridge on top that is often hard to see; the bill may simply appear larger and more dominating than Groove-billed's.

Overall black, showing some purple and bronze highlights (but less iridescent than grackles). Because of the light and the way their feathers lie, birds often appear hooded or maned—unkempt. Immatures are smaller-billed and browner-toned in the wings.

BEHAVIOR: Highly social; a pack animal. Roosts, forages, travels, and breeds in small groups (usually fewer than 10) or pairs, but in winter sometimes gathers in larger flocks. Forages in bushes and dense grasses and reportedly catches large insects in flight. On the ground, hops, walks, and makes short flights. In bushes, clambers about using the beak as well as the feet. Because of its attraction to stirred-up insects, often forages in close proximity to cattle and lawn mowers. When foraging, posts sentinels. Birds roost touching shoulder to shoulder on the same branch; on cold mornings (or after rain), they sun themselves with wings spread wide. Tame.

FLIGHT: Distinctive, slender, long-headed, long-tailed (wedge-shaped when fanned) profile with short rounded wings. Boat-tailed Grackle has longer and pointier wings. Smooth-billed Ani's flight is slow, straight, and normally low. Wingbeats are given in a short, hurried, choppy series followed by a lengthy glide with wings held stiffly to the sides. Birds frequently fly single file and call as they fly.

VOCALIZATIONS: Namesake call is a slurred, rising, quavering, whistled "*oooereet*" (*Ani!*). The quality and pattern may recall female Wood Duck (or curlew). Also reportedly makes short "growls, coughs, and barks" during the predeparture period at the roost site.

Groove-billed Ani, *Crotophaga sulcirostris*
Big Floppy Texas Blackbird

STATUS AND DISTRIBUTION: Tropical species and s. Tex. resident; fairly common in summer; scarce and local in winter and found mostly close to the coast. Breeds from Big Bend, along the Rio Grande Valley, to the coast and north to Corpus Christi. Winters north and east of its breeding range along the Gulf Coast to La. and Miss. (casually farther).

HABITAT: Open and semi-open country, but most partial to habitats where thickets and thorny brush lie adjacent to grassy marshes, fields, pastures, and roadsides. **COHABITANTS:** Northern Bobwhite, Red-winged Blackbird, Great-tailed Grackle. **MOVEMENT/MIGRATION:** Most Texas breeders apparently vacate nesting territories between September and April. During this period, birds appear north of their breeding range as far north and east as Mississippi and western Florida, and casually all the way to Ontario and New Jersey and west to Arizona and California. VI: 3.

DESCRIPTION: A big, unkempt, loosely feathered, blackbirdlike cuckoo with a short swollen bill. Large and gangly (larger than Common Grackle; smaller than Boat-tailed or Great-tailed Grackle), with a big head and bulbous bill (head looks like a bill with a well-groomed mane), a long narrow body, short, often drooped wings, and an exceedingly long loosely arrayed tail that is often cocked askew of the body.

Adults are all dull black and shabbily or loosely feathered. (The lead-colored bill is the palest part of the body.) At close range, plumage may show some greenish or bronze highlights, but the species lacks the iridescence of grackles. Immatures are smaller-billed and browner-toned.

BEHAVIOR: A pack animal with platoon instincts. Where true blackbirds move together—flushing and flying as a unit—anis move with coordinated stealth. First one bird moves across an opening, then another . . . and another. . . . Groups may number several individuals or two dozen or more. Forages along brushy edges, frequently burrowing into the brush or hopping clumsily branch to branch; also forages in tall and short grass. Great and Boat-tailed Grackles also forage on short grass, but they are rarely found in tall grass, and they never maneuver through dense vegetation. Anis are also attracted to feeding cattle and the insects they flush.

Anis like to stay close to the ground. Their movements are slow and clumsy: they clamber and hop in bushes, walk slowly on the ground, and run or hop when prey is sighted. They seem not so much to perch as to cling to branches or festoon themselves on bushes. When foraging in brush, they spend most of their time out of sight. Grackles stay in the open. Roosting anis commonly occupy the same perch with bodies touching. On cold and wet mornings, birds often array themselves, placidly and en masse, on the sunny side of hedges, looking disheveled and dejected.

FLIGHT: Extremely long and gangly, with a long projecting head/neck, short round wings, and a long partially fanned tail. Flight is direct, low, and somewhat unsteady, with birds tending to yaw a bit. Although flight is also fairly slow and floating, wingbeats are hurried and somewhat floppy, given in a regular series followed by a prolonged glide on wings that *snap* on the upstroke to a horizontal position. Dips and raises its fanned wedge-shaped tail when it lands.

VOCALIZATIONS: Call is a high, squeaky, whistled "*skee-er'Ree skee-er'Ree.*" Also makes a weak baby-bird-like sucking "*peeps*" and a low, mellow, keening, somewhat hawklike, two-note whistle, "*pee'er pee'er pee'er*." Also emits a low "*chrk*."

PERTINENT PARTICULARS: This is the ani most likely to appear as a vagrant across most of North America. Smooth-billed Ani, whose declining population is found only in southern Florida, has only appeared, as a vagrant, along the Atlantic Seaboard, where records of Groove-billed Ani are as numerous as those for Smooth-billed.

OWLS

Barn Owl, *Tyto alba*

Golden Monkey-faced Owl

STATUS: Widespread and fairly common to uncommon permanent resident in southern portions of its range; less common in colder northern areas. **DISTRIBUTION:** Cosmopolitan owl. In North America, found in every state but Maine and Vt.; in states bordering Canada generally occurs only in southern portions. In Canada, found in extreme s. B.C. and just north of Lake Erie. **HABITAT:** Generally open habitats at lower elevations, including farmland, grassland, deserts, and fresh and coastal marshes. **COHABITANTS:** Occasionally roosts in groves with Long-eared Owl; otherwise solitary. **MOVEMENT/MIGRATION:** Through most of its range, a permanent resident. Young may disperse hundreds of miles after fledging (peak movement from mid-September through October). Most northern adults vacate buildings where breeding occurs during colder months but may merely relocate to more protected or better-insulated roost sites. VI: 1.

DESCRIPTION: A pale, slender, long-legged, monkey-faced owl that shrieks. Medium-sized (slightly larger than Short-eared or Long-eared Owl; considerably smaller than Snowy Owl), with a large, round, "earless" head, a long slender body, a short tail, and distinctly long legs.

Overall pale; only Snowy Owl, the whitest owl, is paler. Tawny or golden brown above with grayish highlights, and tawny or ghostly white below. The always white heart-shaped face framing all-dark eyes is unique among North American owls. Long stiltlike legs are shared by Burrowing Owl, but Barn Owl is twice the size.

BEHAVIOR: Most commonly glimpsed as it flushes from the structure into which you just intruded or when "caught" at the edge of your headlight beam as it flies in open country. A golden-backed owl that turns an all-white face your way as it angles away. Almost wholly nocturnal. Birds do not commonly leave roost sites until it is almost dark. Roost and nest sites include some natural structures (large tree cavities, cuts or holes in dirt banks, caves) as well as many man-made structures, including church steeples, abandoned wells, trap houses, elevated deer (hunting) blinds, bridges, mine shafts, silos, stacked hay bales, and, of course, barns. Also roosts in trees, usually well hidden and commonly near the top (as compared to Long-eared Owl, which often roosts quite low, occasionally only 3–4 ft. from the ground). Hunts by coursing over the landscape (like Short-eared Owl), but also perch-hunts (like Great Horned Owl).

At times, very vocal. Calls in flight, and during migration and dispersal. Usually shy, flushing easily, but when roosting in conifers, may sit very tight. Can be attracted by squeaking, even before full dark.

FLIGHT: Large-headed short-tailed silhouette with wings that are broader and more tapered toward the tip than Long-eared or Short-eared. Overall very pale, showing no apparent pattern. Flight is buoyant and mothlike, with slow stiff wingbeats. Hunts by flying slowly 5–10 ft. above the ground, often hovering before dropping onto prey.

VOCALIZATIONS: Classic call is a long, loud, hissing shriek: "*shreeeeh*." Also makes a series of twittering clicks and quieter hisses.

Flammulated Owl, *Otus flammeolus*

Diminutive Dark-eyed Screech-Owl

STATUS: Western breeder; not uncommon, but not easily seen. **DISTRIBUTION:** Widespread in the mountain forests of the West. Range includes s. B.C.,

e. and cen. Wash., much of Idaho, w. Mont., parts of Ore., Calif., Nev., Utah, Ariz., s.-cen. Wyo., w. and cen. Colo., w. and cen. N.M., south into Mexico. Winters south of the U.S. **HABITAT:** Classically a bird of ponderosa pine, yellow pine, and fir forests, but comfortable in other open, dry, mature, montane pine forests with a dense or bushy understory and a surfeit of insect prey. Also, locally, found in aspens. **COHABITANTS:** Pygmy Nuthatch, Steller's Jay, Western Tanager, Red Crossbill. **MOVEMENT/MIGRATION:** Entire population is presumed to vacate U.S. and Canadian breeding areas for winter, with the route conforming to the bird's breeding range and largely confined to higher elevations. Fall migration poorly understood but believed to fall between August and November; spring return in April and May. VI: 2.

DESCRIPTION: A small, cryptically dark, somber-eyed owl of western pine forests. Small (smaller than Whiskered Screech-Owl; larger than Elf Owl; same size as Northern Pygmy-Owl), with a large mostly tuftless head, a compact body, and a short tail. Overall appears more roundly contoured than screech-owls.

Mostly sooty gray with touches of rust on the face and traces of rust in the wings; mottled underparts are scored by darker blotchy streaking. The bird's most distinguishing characteristic is its large, round, arrestingly black eyes. All other small owls have yellow eyes. The only other dark-eyed owls (Barn, Barred, and Spotted) are two to four times larger than the diminutive Flammulated.

BEHAVIOR: Wholly nocturnal, with peak activity just before dawn and just after dusk. Secretive and security-conscious. Roosts on a branch next to the trunk of the tree, where it very closely resembles a broken branch. Also vocalizes from positions close to the trunk, and if attracted by imitations of its call, generally stays away from forest openings as it returns the vocal challenge. Occasionally roosts in young pines, choosing perches that offer a protective canopy, and at times found in pairs. Hunts from perches, often along open edge. May be semicolonial, with several breeding pairs localized in the same area.

FLIGHT: Disproportionately small-bodied and long-winged. Flight has been described as less maneuverable than other woodland owls. Often hovers to pluck prey from branches.

VOCALIZATIONS: Male's song is a series of low, even, well-spaced, slightly rising hoots, "*poop . . . poop . . . poop,*" repeated ever two to three seconds. Notes are sometimes doubled: "*po-poot or p'do poot.*" Female's hoot is higher-pitched, longer in duration, and quavering.

PERTINENT PARTICULARS: Flammulated Owl has short catlike ear tufts that lie close to the head but are raised when the bird is disturbed.

Western Screech-Owl, *Otus kennicottii*

Hair-streak Screech

STATUS: Common widespread western resident. **DISTRIBUTION:** Coastal Alaska (north to Cordova), w. and s. B.C., Wash., Ore., Idaho, w. Mont., Calif., Nev. (very local), Utah, w. Wyo., w. and se. Colo., all but e. Ariz., and trans-Pecos and cen. Texas south into Mexico. Range abuts Eastern Screech-Owl in Colo. along the Arkansas R. and in Tex.; overlaps with Whiskered Screech-Owl in se. Ariz. **HABITAT:** Found in almost any woodland habitat, including suburbia, but most common in deciduous and riparian woodlands. Also found in hemlock and fir forest, oak savanna, palms (California), parks, and (in desert regions) well-watered areas that mimic riparian habitats. Where range abuts Eastern Screech-Owl, Western favors drier areas; where range overlaps with Whiskered Screech-Owl, more common at lower elevations. **COHABITANTS:** ("Red-shafted") Northern Flicker, Gilded Flicker, Pileated Woodpecker, Western Wood-Pewee, and nuthatches. **MOVEMENT/MIGRATION:** Nonmigratory. VI: 0.

DESCRIPTION: A trim gray screech-owl with pencil-fine black streaks down a cleaner breast. Small (slightly larger than Whiskered Screech-Owl; same size as Eastern Screech-Owl and European Starling), with pointy ear tufts, a squarish head, a stocky body, and a short tail. In size and shape, identical to Eastern Screech-Owl.

The always gray Western Screech-Owl has no true red morph (although birds in the Pacific Northwest are somewhat brownish); otherwise, much like Eastern Screech-Owl, except the longitudinal blackish streaking on the underparts is crisper and finer. By comparison, the streaking on the underparts of gray Eastern Screech-Owl and Whiskered Screech-Owl is thicker, heavier, and blotchier and commonly overlaid with short dark bars. In other words, the dark streaks on Western's underparts look like fine brush strokes, and the streaks on Eastern and Whiskered look like the birds' underparts were pieced together with crude surgical stitches. Also, Western's bill is blackish, whereas the bill on the other two species is pale.

BEHAVIOR: Mostly nocturnal perch-hunter that becomes active 15–30 mins. after sunset and normally returns to roost half an hour before sunrise. Sometimes hawks insects, but usually captures prey on the ground or in the water (crayfish). In warmer months, commonly roosts in trees—either coniferous or deciduous—and usually on a branch close to the trunk; in winter, more commonly selects a cavity. Males begin vocalizing in January and February and are most vocal at dusk (particularly later in the breeding season).

FLIGHT: Like Eastern Screech-Owl.

VOCALIZATIONS: Western has two primary songs. The "bouncing ball" call is 12 or so short, low, whistled toots that begin slow and accelerate: "*too too tootootootootootoo.*" This is the call most often heard at dusk. In the "double trill" (sung by both sexes), the first note is short, the second prolonged.

Eastern Screech-Owl, *Otus asio*

The Owl Next Door

STATUS: Common and widespread resident. If you live in the United States east of the Rockies and have a modest-sized woodlot nearby, chances are you live near Eastern Screech-Owl. DISTRIBUTION: Found from w. Tex. through e. Colo. to e. Wyo., into the Badlands of Mont., north into extreme se. Man., and east to the coast. Absent in n. Minn., n. Mich., and n. New England, but present along the St. Lawrence R. in Que. HABITAT: Found in open woodlands of almost any tree size, forest age, total acreage, and species composition. Readily accepts parks, orchards, and suburban neighborhoods (with trees). Eastern's only requirements seems to be a relatively open understory and trees with cavities suitable for roosting and nesting, but birds will adopt nest boxes. COHABITANTS: If you've got habitat to meet the needs of Eastern Gray Squirrel, chances are it can sustain Eastern Screech-Owl. MOVEMENT/MIGRATION: Nonmigratory; permanent resident. VI: 0.

DESCRIPTION: Small, large-headed, somewhat blocky or stocky owl with conspicuous and widely spaced ear tufts and features that are somewhat harsh and angular, particularly when the bird is standing alert and erect. Comes in both a rufous and a gray form; gray birds are numerically dominant across most of the bird's range (particularly toward the West), but in some mountainous portions of the South, and east of the Appalachians, rufous birds outnumber gray.

Rufous birds are unmistakable. Gray birds seem crafted from craggy gray tree bark and thus look extremely cryptic. Very similar to Western and Whiskered Screech-Owls—both western species—except for a paler greenish bill (black in Western).

This is not a particularly cute or cuddly owl. Its expression ranges from Oriental inscrutable (when sleepy and eyes are drawn down to slits) to really ticked off (such as when someone imitates its call).

BEHAVIOR: A solitary nocturnal perch-hunter that becomes most active just at dark. Generally confines its activity to below the canopy, at an average height of about 15 ft. Captures prey on the ground (sometimes forages on the ground). Commonly roosts in tree cavities during daylight, but may also roost in conifers and thick vegetation. In winter, frequently seen framed in cavity openings (sometimes partially emerged) in direct sunlight.

Tame and easily approached; easily excited and attracted by imitations of its call (although less responsive in late spring and early summer). Responds by calling back, flying close, or landing nearby but remaining silent.

FLIGHT: Flight is direct, silent, not particularly fast, and frequently close to the ground. Flight silhouette is overall stocky, with distinctly short round wings. Wingbeats are quick and steady. Frequently drops as it leaves the perch, flies close to the ground, and rises abruptly before taking the next perch.

VOCALIZATIONS: Eastern has two easily recognized (and imitated calls). The contact call sounds like a low, mellow, gargled or trembling whistle, lasting two to four seconds, all on the same pitch; it may be loud or softly uttered, giving the impression that a bird no more than 10–20 ft. away is much farther away. Also makes a descending nasal "whinny" that it often (but not exclusively) uses to assert territoriality. In courtship, both calls are given by both sexes.

PERTINENT PARTICULARS: While almost wholly nocturnal, Eastern Screech-Owl is most vocal at dawn and dusk, can be enticed to respond to whistled imitations of its call in full daylight, and can even be persuaded (particularly in May and June, when young have just fledged) to approach in daylight. Also responds to imitations of other, smaller owls (such as Northern Saw-whet) and can be drawn in by squeaking. A word of caution: Although a rare occurrence, screech-owls and other owls have struck and injured people trying to entice them with calls.

Scolding chickadees and jays often alert observers to roosting Eastern Screech-Owls, although a mobbing cluster of birds around a cavity does not guarantee that a bird is present.

Whiskered Screech-Owl, *Otus trichopsis*

Screech Owl in Dots and Dashes

STATUS: A subtropical species; fairly common within its very restricted U.S. range. **DISTRIBUTION:** Resident from extreme se. Ariz. (barely sw. N.M.) to Central America. Well-known and easily accessed U.S. concentration points include the Santa Catalina, Santa Rita, Huachuca, and Chiricahua Mts. of Ariz. **HABITAT:** Montane pine-oak forest, commonly at higher elevations than Western Screech-Owl (in excess of 5,000 ft.). Also frequents upper portions of riparian woodlands dominated by sycamore. **COHABITANTS:** Greater Pewee, Mexican Jay, Painted Redstart. Habitat overlaps with Western Screech-Owl (at lower elevations) and Flammulated Owl. **MOVEMENT/MIGRATION:** Permanent resident. VI: 0.

DESCRIPTION: A small, grayish, structurally abbreviated, and diffusely patterned screech-owl (smaller than Western; larger than Flammulated). Overall much like Western (or gray Eastern), but more compact, with shorter ear tufts (often not showing) and smaller feet that make perched birds look clubfooted.

Plumage is overall gray, sometimes touched with brown, and somewhat diffusely or shabbily patterned (not as crisply and contrastingly marked as Western). Eyes show an orange tinge (yellow on Western; dark on Flammulated). A paler bill (black on Western) usually shows a greenish cast.

But since it's night, you probably won't be able to note any of these field marks and will be identifying the bird by call, as most observers do.

BEHAVIOR: A nocturnal hunter that becomes active just before dark. Roosts close to the trunks of trees by day or in protective foliage. An active aerial hunter, Whiskered Screech-Owl appears to take most prey in the air by sallying to and from branches beneath the canopy. Also, but less commonly, takes prey on the ground.

Both males and females vocalize, frequently calling at dusk. Whiskered responds well to imitations of its song, as well as the tremolo warble of Eastern Screech-Owl.

FLIGHT: Appears slightly more compact than Western Screech-Owl and shorter-winged. Flight is like other small owls, with stiff rapid wingbeats, but movements are slightly more nimble.

VOCALIZATIONS: Typical song is a short, hurried, slightly rising-and-falling series of mellow evenly spaced toots: "*poo poo poo poo poo poo poo.*" After a pause, the song is repeated. Another very distinctive song consists of a persistent series of toots given in a broken or halting pattern that recalls the dots and dashes of Morse code: "*pudo po po, pudo po po, pudo po po.*"

Great Horned Owl, *Bubo virginianus*

The Five (or Four to Seven) Hooter

STATUS: Common resident. **DISTRIBUTION:** Essentially everywhere in North America—forests, woodlands, deserts, grasslands, suburbia, city parks. Absent only in the treeless arctic regions and true urban centers. **HABITAT:** Shows a preference for open woodlands, particularly edge habitat. **COHABITANTS:** Red-tailed Hawk, Great Horned's diurnal counterpart. **MOVEMENT/MIGRATION:** Nonmigratory. Present year-round in well-defined territories. VI: 1.

DESCRIPTION: A neckless broad-faced beer keg of a bird topped with the devil's own horns. The great size, rotund proportions, wide girth, and ear tufts that recall feline ears (or devil horns) readily separate this bird from all other owls. The broad head and no-tailed appearance distinguish distant birds from perched hawks. Generally stands erect; calling birds become horizontal, with head and tail raised, creating the impression of a plump Chinese fortune cookie.

The white throat (very apparent at dawn and dusk) helps distinguish the bird from the superficially similar but smaller, slimmer, and more angular Long-eared Owl. Great Horned is generally brown or gray, but color varies by geography, ranging from warm brown in the East to blackish brown in the Northwest, to cold gray in desert regions, to ghostly pale silver and white in northern forests.

One of the few birds that can be identified by smell, owing to a partiality for skunk as prey. The expression is sleepy malice—or just plain malice.

BEHAVIOR: A solitary perch-hunter; except when attending young, almost always seen alone. Never roosts communally. Not strictly nocturnal. Active shortly after sunset and just before sunrise; when young are in the nest, also hunts by day. At twilight, comes out to woodland edges, sits conspicuously high, and calls to mate (who responds). Likes sturdy perches, including rocky outcroppings. Roosts in large trees, often very high and close to the trunk; also roosts on cliffs.

Generally shy and retiring. Flies readily when approached. Sometimes sits on the ground. Walks with a lumbering sailor's gait.

FLIGHT: Flight is generally low and hurried. Profile is a blunt-headed beer keg of a bird set on broad wings. May flap continuously or flap in a series and enter into an extended gravity-defying glide. Flight is direct, steady, point to point. Wingbeats are heavy, choppy, and hurried (anything but buoyant), with wings angled down at the wrist. Glides on slightly down-curved wings and seems not to lose altitude. In the open, generally drops low, hugging the ground. In woodlands, may fly at any height.

VOCALIZATIONS: The archetypical owl call. Ranked among the world's most familiar sounds (in fact, nicknamed the "five hooter"). Call is a loud, low-toned, solemn, two-part hooted incantation. The female's (?) call is faster, higher-pitched, and six- to seven-noted, with the second, third, and forth notes quicker and compressed: "*Who, whowhowho?* (pause) *W'hoo, who.*" Sounds like the bird is asking a question, then answering itself. The male's (?) call is lower-pitched, slower, and more solemn, given in two parts, each two-noted: "*Who'who* (pause) *who, who.*" Adults (particularly females) and young also communicate with a loud shrieked "*Ehr-reeeh,*" which combines a shrill breathiness with a raspy timbre (not as shrill as Barn Owl's scream). Also, females (?) make a loud brassy braying sound. Heard in winter is a series of low, slightly accelerating "*who who whowhowhowhowho.*"

PERTINENT PARTICULARS: The nocturnal equivalent of the equally widespread Red-tailed Hawk. (In fact, in many places Great Horneds appropriate Red-tailed nests.) If you have Red-taileds, you almost certainly have Great Horneds in the neighborhood. Absolutely hated by crows, who amuse themselves by gathering around roosting owls and haranguing them with a gritty vehemence they inflict on no other enemy. A cautionary note: Ear tufts are erectile. When threatened—such as when a bird is approached on the nest—the ear tufts may be flattened against the skull and invisible.

Snowy Owl, *Nyctea scandiaca*
Owl as Melting Snowman

STATUS: Uncommon and nomadic arctic breeder; uncommon winter resident in subarctic central Canada and in the northern United States from the Pacific Northwest to the Great Lakes and New England. Irregular in winter across eastern Canada (including Newfoundland), the mid-Atlantic states, and the Midwest. Rare in the Central Plains south to Colorado and Oklahoma. Casual farther south. **DISTRIBUTION:** Breeds on the coastal plain from w. Alaska across the northern portions and islands of the Northwest Territories to n. Que. and Lab. Winters south of its breeding range, as described under "Status." **HABITAT:** Breeds from just north of the tree line to the coast on open tundra, most commonly at lower elevations (below 1,000 ft.). Prefers open rolling tundra with a surfeit of promontories that serve as perches. In winter, favors open habitats, including prairies, farmland, beaches, salt marshes, and partially frozen arctic ocean waters. In migration, may turn up anywhere, including downtown buildings in urban areas. **COHABITANTS:** Northern Harrier, Rough-legged Hawk. **MOVEMENT/MIGRATION:** Regular migrant to the northern Great Plains, southeastern Canada, and New England—places with snow cover much of the winter; irruptive and unpredictable elsewhere. In irruptive years, owls appear in greater numbers and usually penetrate farther south, with some nomadic movement throughout the winter. Wintering birds may be found between November and April, but peak movement occurs between December and early March. VI: 3.

DESCRIPTION: Looks like a small, soot-flecked, partially melted snowman with bright yellow eyes. Large owl (same size as Great Horned Owl), with a large, round, tuftless head and a stocky body. Seen head on, seems very erect. More horizontal-looking from the side, with a heavy body that seems to have melted and collapsed into the surrounding snow (or beach).

Plumage ranges from almost pure white (adult males) to the color of old soot-covered snow (immature females) on white bodies darkened by wavy black lines (only the face remains pure white). Most birds fall in between, showing conspicuously white bodies flecked with gray spots or broken barring.

BEHAVIOR: An owl of open treeless places that likes to sit conspicuously atop sturdy elevated (and often man-made) structures but is equally at home sitting on the ground, normally where a slope or rise offers a vantage point. Common perches include fenceposts, utility poles, roofs, duck blinds, water towers, ice floes, dune ridges, and driftwood. Does not commonly perch in trees. Hunts during daylight hours, but is most active at dawn and dusk. In winter, spends much of its time conspicuously perched, often on the same perch for hours; turns its head in swiveling jerks and leans forward when something catches its eye.

In winter, not social (you will never see two Snowy Owls sitting side by side). Tame, often allowing close approach. Flies, for the most part, very close to the ground, and (unlike Short-eared Owl) rarely hovers.

FLIGHT: Large owl with a fairly pointy (bullet-shaped) face and long, broad, buteo-like wings showing tips uncommonly tapered for an owl. Most birds are conspicuously white or whitish; others are very dark, but with whiter heads and faces. Flight is direct and strong, but somewhat jerky. Wingbeats are stiff and slow (somewhat buteo-like), with a heavy down stroke that causes the body to appear jerky. Wingbeats may be constant and more hurried (when approaching prey) or punctuated by long glides.

VOCALIZATIONS: In winter, generally silent.

PERTINENT PARTICULARS: An all-white owl can be challenging to locate in a snowy landscape. Scan all elevated points and look twice at any abrupt bumps or projections; look three times if the bump is darker than the surrounding snow. Perched Snowy Owls draw the ire of assorted birds. Be mindful of harriers and kestrels (or other raptors) stooping on snowy lumps; Herring Gulls frequently circle over owls sitting on the beach, calling attention to themselves and the owl by their incessant calling. The absence of gulls may also pinpoint the location of Snowy Owls known to be wintering in the area. Gulls tend to apportion themselves fairly evenly along a beach. A stretch of beach showing a sizable gap in the larid ranks is a stretch worth scanning.

Northern Hawk Owl, *Surnia ulula*

Spruce Ornament

STATUS: Fairly common northern resident; rare and irruptive winter wanderer. DISTRIBUTION: Resident across all of Alaska and Canada north to the tree line. Southern range limit cuts across cen. B.C., sw. and cen. Alta., cen. Sask., sw. Man., and cen. Ont. west of the St. Lawrence R. in Nfld. In winter, may wander south to the northern U.S. border states. HABITAT: Coniferous and mixed coniferous forest bordering open or marshy places. Most partial to the less vegetated northern and altitudinal edges of the boreal forest and muskeg festooned with small and widely spaced spruce. Also found in stands of aspen. Avoids dense woodlands; partial to open areas offering an ample number of fairly low hunting perches. In winter, particularly during irruptions, may be found in almost any open or semi-open habitat, including farmland and city parks. COHABITANTS: Northern Shrike, Gray Jay, White-crowned Sparrow. MOVEMENT/MIGRATION: A nomadic species that shifts within its defined range in response to prey availability (primarily rodents), so that places with high numbers one year may be avoided the next. At intervals of three to six years, this species initiates irruptive migrations that propel numbers of individuals well south of their normal range. These irruptions vary in both magnitude and geographic scope—they may involve the northern Great Lakes one year, southeast Canada and northern New England another. In addition, an irruptive year is sometimes followed by a second "echo" movement the following year. VI: 3.

DESCRIPTION: A distinctive long-tailed owl with a black-framed face and dark-patterned head that sits, like a Christmas tree ornament, at the very tip of stunted spruce trees. Medium-sized (considerably larger than Boreal; much smaller than Great Gray; about the same size as Cooper's Hawk or Peregrine), with a broadly oval-shaped head and body and a long accipiter-like tail. The round head often appears square because of the squared facial pattern. From the side, the face appears flat, with no bill showing. The body may be erect when relaxed, but hunting birds commonly lean forward. The tail may point down or be cocked at an angle. Small eyes make the bird look somewhat myopic.

Shows a distinctive pattern of large whitish spots bordering a solid dark back above, dense brown barring below, a white freckled forehead, and a black-framed squarish face. From a distance, appears two-toned, with a slightly brownish (at close range, olive brown) body and a grayish head. The bold blackish head pattern is not particularly contrasting, but it is evident. Face on, the black-framed face can be seen; from the side, lamb-chop sideburns; from behind, a round skullcap. At a distance, generally paler underparts and the spotted pattern above the wing make the folded wing look like a dark gray racing stripe running the length of the bird.

BEHAVIOR: A daytime hunter. Seems to spend most of its life imitating the wind-sculpted top of stunted spruce trees between 4 and 30 ft. in height. Also perches on utility poles and other man-made structures. Hunts mainly by sight, swiveling its head to search for prey. Though the bird rarely shifts perches—sitting for half an hour or more on the same branch—it may move to a closer perch when prey is sighted before making a final approach. When prey is captured, often flies to a strategic tree and caches the catch before returning to the hunt (not uncommonly to the same perch). Solitary, hunts alone, and seems tolerant of other raptors in its territory, except when young are recently fledged.

Uncommonly tame and reluctant to flush; usually allows close approach, even as close as right beneath the tree in which it is perched.

FLIGHT: Distinctive profile shows a large blunt head, a long narrow tail, and long wide-based wings that taper toward a blunt tip. Recalls a heavy-bodied falcon. Flight is direct and often low. Birds leave perches and drop toward the ground when approaching prey, and ascend abruptly when taking a perch. Wingbeats are hurried, stiff, and mostly regular, with a pronounced down stroke. (Most owls have evenly measured wingbeats.) May flap continuously (especially when crossing large open areas), but commonly interspaces long series of wingbeats with lengthy glides. Very nimble and able to maneuver well in tight confines; commonly hovers.

VOCALIZATIONS: Calls include a lengthy tooting whinny that increases in volume: "*wh'huhhuhhuh uhhuhHuhHuhHuh.*" Also emits a shrill, rising, breathy shriek, "*wra/eeik,*" and a rapid shrill chatter, "*ki kikikikikiki.*"

PERTINENT PARTICULARS: Often found along roadsides, both summer and winter, but most likely to be seen in spruce forest. Drive slowly through such a forest, being particularly attentive at the edges of clearings and bogs. Search the tops of dead and live trees for a solid rounded mass or swollen top. Also watch (and listen) for agitated American Robins and Gray Jays, which frequently harass hawk owls.

Northern Pygmy-Owl, *Glaucidium gnoma*

Long-tailed Pygmy Tooter

STATUS: Generally uncommon, but also unobtrusive and easily overlooked. **DISTRIBUTION:** This western owl has a fairly extensive but spotty distribution. Found from extreme se. Alaska and the northern two-thirds of B.C. to sw. Alta., then south, generally in mountain forests, to Wash., Ore., Calif., Idaho, w. Mont., w. Wyo., Utah, and bordering Nev.; occurs extensively through mountainous regions of Colo., Ariz., and N.M., south through elevated portions of the Mexican interior to Central America. Shows some plumage variation across its range. Broken distribution may relate to two (or more) pygmy-owl species. **HABITAT:** Found in a variety of forest habitats, from coniferous (assorted pines, spruce, hemlocks, and yellow pine) to mixed coniferous-deciduous, to aspen-cottonwood bottomlands, oak forests, and oak-riparian canyons (in Mexico, pine-oak and scrub). In winter, uses same habitats, but may also occupy towns, suburbs, parks, and wooded cemeteries or haunt backyard bird-feeding stations. Although nonspecific in terms of forest type, this species does prefer more open, broken, or edged forest—habitat well suited for a sight-guided perch-hunter. **COHABITANTS:** Steller's Jay, Mexican Jay, Mountain Chickadee, Brown Creeper. **MOVEMENT/MIGRATION:** Nonmigratory; permanent resident, but makes some altitudinal shifts in winter. VI: 0.

DESCRIPTION: A small, stocky or plump, earless, long-tailed owl with a head that looks like it's been stepped on. Tiny! Without its longish tail, Northern Pygmy-Owl is barely fist-sized. Smaller (but blockier) than the woodpecker whose holes it often uses to nest; smaller even than some of the birds it preys upon. The head seems flat and compressed, little more than an elevated portion of the body. In cold temperatures, with feathers fluffed, the bird is the same size and shape as a softball; the head seems like a bulging aneurysm slightly off to the side. The yellow eyes are close-set, giving the bird a somewhat myopic appearance. The tail is often animate, twitching, and the bill is yellow-green.

Comes in two colors—reddish (Pacific Coast) and grayish (Rocky Mountains). Both forms are mostly dark with whitish spotting—a clustered galaxy of tinny white dots on the head, with larger spots on the back and wings. The whitish belly is overlaid with blackish dribbling streaks. The long tail is conspicuously branded by narrow white bands.

The expression is indifferent, myopic. *Note:* Unlike Northern Saw-whet, Flammulated, and Elf Owls, this is not a cute owl.

BEHAVIOR: A visual perch-hunter. Hunts in daylight, with peak activity at dawn and dusk (possibly into the evening). With a few branches breaking up its outline, sits in the interior of a strategic tree or shrub. The stance, like a sprinter relaxing in the blocks, is between vertical and horizontal. Turns its head in quick jerks to study prey possibilities. Explodes into flight with little warning beyond an occasional quick sideways twitching of the tail. At close range, and unlike owls who rely on their ears more than their eyes to pinpoint prey, wingbeats are audible. Also stalks prey by dropping limb to limb.

Solitary. Even members of pairs hunt alone. Frequently mobbed by smaller birds whose harangue can be used to locate perched owls. Responds readily to imitations of its call. Can often be coaxed to cross narrow valleys and to within less than 20 ft.

FLIGHT: Surprisingly small, with short wings and a long narrow tail. Flight is rapid, direct, and distinctly undulating. Wingbeats are given in a rapid angry flurry interspaced with a terse closed-wing glide. Recalls the flight of a shrike (another visual predator of small birds and mammals).

VOCALIZATIONS: Across much of its range, gives a fairly slow (slower than the cadence of Northern Saw-whet Owl), two-noted, tooting whistle, "*poot-poot . . . poot-poot . . . poot-poot,*" that often has an irregular quality. Pacific Coast birds give a single-noted call. When agitated, may make a short, rapid, undoubled series of toots: "*pootpootpootpoo.*"

PERTINENT PARTICULARS: In southern Arizona, the ranges of Northern Pygmy-Owl and Ferruginous Pygmy-Owl overlap—but not the habitats. Ferruginous is a bird of low desert areas; Northern is a bird of forested higher elevations.

A key point in identifying a pygmy-owl is to remember that because of its small size (and woodpecker-like flight), most observers dismiss the bird in flight. Small birds, which have less latitude to make any such misidentification, are very attuned to the flight of this owl. Train yourself to be alert to the alarm calls, not just the scolding harangues, of small birds.

Ferruginous Pygmy-Owl, *Glaucidium brasilianum*

Ever So Slightly Reddish Pygmy-Owl

STATUS AND DISTRIBUTION: Rare resident in s. Ariz.; uncommon in extreme s. Tex. west to Zapata and north to Falfurrias. **HABITAT:** Dry woodlands and desert. In Arizona, occupies riparian woodlands dominated by cottonwoods and mesquite; also upland palo verde–mesquite–acacia scrub and desert washes with saguaro–palo verde. In Texas, found in dry, upland, oak-mesquite forest and mesquite-ebony riparian forest. **COHABITANTS:** In Arizona, Gila Woodpecker; in Texas, Golden-fronted Woodpecker. **MOVEMENT/MIGRATION:** Permanent resident. VI: 0.

DESCRIPTION: A tiny, plump, pale brown owl with a narrow orange tail. Tiny (smaller than a starling; larger than Elf Owl; same size as Northern Pygmy-Owl), with a small compressed-looking head, a plump body, and a relatively long narrow tail. When agitated or alert, appears more slender and the head is rounder and more elevated. Looks beady-eyed and dimly hostile.

The implications of its name notwithstanding, Ferruginous Pygmy-Owl is not distinctly rufous. Upperparts are pale brown—rufous brown about the head and collar, with an olive brown back and a contrastingly dull orange and dark barred tail. The back of the head shows two black "eyes" highlighted by a white eyebrow (like Northern Pygmy-Owl). The facial pattern shows more white than Northern Pygmy-Owl. The blurry grayish brown streaking on the white underparts, which is often more patterned than Northern Pygmy, coalesces into two lines along either side of the breast (suggesting a set of ribs). From a distance, underparts are divided by three pale vertical lines—one central line (between the ribs) and one on either side of the ribs. When the bird is perched in vegetation, it is often these pale lines that catch an observer's eye.

At close range (which is often easily accomplished), the crown is showered with a galaxy of tiny white dashes, not spots as on Northern Pygmy-Owl.

BEHAVIOR: Active during daylight (particularly during the breeding season) as well as at night, but most hunting (and vocalizing) occurs around dawn and dusk. Generally solitary or found in pairs. Hunts and roosts beneath the woodland canopy, commonly 10–20 ft. off the ground and often on an outer limb. Launches itself from perches with noisy wingbeats (noisy for an owl), targeting a variety of prey but very commonly birds (including birds larger than itself). Very quick. Often shifts positions, hopping to a new perch or launching itself before the movement can be comprehended.

Very tame and reluctant to flush; often allows close approach and more often than not goes unseen. Responds well to imitations of its call (except in heavily birded areas) and calls during the day (mostly early and late). Because it also often sits conspicuously and shifts perches during the day, it is frequently and noisily mobbed by small songbirds. Be alert.

FLIGHT: Tiny, compact, short-winged, long-tailed owl that might easily pass for a songbird. Flight is very rapid, with wingbeats too rapid to count. Birds often drop when coming off perches and fly with a deeply undulating flight.

VOCALIZATIONS: Call is a rapid continuous series of high, slightly rising, whistled toots, "*pwih pwih pwih pwih,*" that are faster and higher-pitched than Northern Pygmy-Owl and lack the doubled or two-noted quality of that relative species.

Elf Owl, *Micrathene whitneyi*

Burrowing Owl Lite

STATUS: Common but geographically restricted southwestern owl. **DISTRIBUTION:** Breeds in s. Ariz., sw. N.M., trans-Pecos Tex., the Big Bend Region of Tex., and the lower Rio Grande Valley in Tex. **HABITAT:** Breeds in saguaro cacti desert, riparian forest, thorn woodlands, and montane evergreen woodlands. Has become somewhat suburbanized, and commonly breeds along roadsides. **COHABITANTS:** In saguaro desert, Gilded Flicker and Gila Woodpecker (whose nest cavities it uses); in other riparian and forest habitats, may nest in the same tree (but not the same hole!) as Western or Whiskered Screech-Owl and Northern Pygmy-Owl. **MOVEMENT/MIGRATION:** In winter, birds appear to retreat into Mexico, where insect activity is greater. Spring migration from late February to late April; fall from late August to early October. VI: 0.

DESCRIPTION: A minuscule, "yappy," cavity-nesting owl that looks like a miniature Burrowing Owl. Smaller than any other owl, and smaller in overall length than many sparrows; however, Elf Owl *always appears larger than it is because of its stocky build.* Shaped like a small Idaho potato, with its large, rounded, tuftless head, compact body, very short tail, and long bare legs, Elf Owl looks like a slightly shorter-legged Burrowing Owl.

Overall brownish gray or grayish brown; plumage is an amalgam of gray, rufous, warm buff, and whitish spots—but overall more homogenized and less patterned than other small owls. Paler underparts are lavishly overlaid with blended gray and rufous streaks. The pale cross-looking eyebrows and bracketing white upper and lower borders on the folded wing often show in a beam of light.

BEHAVIOR: Almost strictly nocturnal owl that is most active before dawn and just after dusk. Males generally make their initial appearance at a nest hole when it's too dark to perceive colors. Birds are found at nest holes from March to early July. Thereafter, adults and young roost in shrubs, evergreens (such as Alligator Juniper), and deciduous trees and clumps of mistletoe. Nest holes may be as low as 4 ft. from the ground and are usually old woodpecker holes, constructed by various species, located in cactus, trees, fenceposts, utility poles, and bluebird boxes. Nest sites are often used over and over again (but birds commonly use different holes).

Early in the breeding season, highly vocal. The male's high-pitched terrier-like "yips" are distinctive and unique among owls. Usually perch-hunts, but also captures insects on the wing (often insects drawn to light beams); may also hop or run after prey on the ground.

FLIGHT: Overall tiny and compact. Wingbeats are given in a rapid frenzy. Rapid undulating flight borders on bouncy.

VOCALIZATIONS: A short descending nasal chuckle: "*ur'chur chur chur chur chur.*" Also a high descending whistle, "*eeeur eeeur.*"

Burrowing Owl, *Athene cunicularia*

Potato on Stilts

STATUS: Widespread in the West and Florida, where it is uncommon to scarce. **DISTRIBUTION:** Except for coastal forests of w. Wash., Ore., and n. Calif.,

n. Idaho, and w. Mont., breeds across most of the West. Northern range extends to se. Alta., s. Sask., and s. Man. Eastern border is defined by w. Minn., w. Iowa, se. Neb., e. Kans., e. Okla., and ne. Tex. Breeding and wintering range extends south into Mexico. In the East, found year-round on the Florida Peninsula. Most western birds retreat south in fall. The winter range lies south of a line drawn from San Francisco to the southern tip of Nev., across the lower middle of Ariz. and N.M. to Okla., southeast to Houston (also along the coast into w. La.). **HABITAT:** Open, dry, grassy or sparsely vegetated desert habitat, generally devoid of trees. Often found close to prairie dog towns. Also inhabits airports, golf courses, cemeteries, irrigation ditches bordering farmland and roadsides, vacant lots, and residential areas. Except in Florida, where birds excavate their own burrows, the presence of burrow-excavating animals or artificial burrows (constructed or accidental) is mandatory. **COHABITANTS:** Horned Lark, Prairie Dog. **MOVEMENT/MIGRATION:** Birds breeding in Canada and all but the southern United States migrate. Spring migration from March to early May; fall in September and October. VI: 2.

DESCRIPTION: A small, brown, dirt-loving owl that resembles a potato on stilts. Overall small (slightly larger than screech-owl, smaller than a prairie dog, less than half the size of Barn Owl), compact, and gracelessly contoured like an Idaho potato or a loaf of bread. Head looks squashed and flat (like it was stepped on), and legs are uncommonly long (stilt-like) and unfeathered. Because of the long legs, the bird seems larger than it really is. Seen close, most people are shocked at the owl's small size.

Overall pale brown. Lavishly flecked with white spots above, and dark-chested and darkly barred below (young birds have pale buffy underparts). Eyes are uncommonly bright yellow, and the expression is fierce.

BEHAVIOR: Stands somewhat tensely, on the ground, in broad daylight, usually atop a mound of dirt at the mouth of a burrow and often surrounded by prairie dogs. Leans forward when standing "erect" and seems menacingly hunched (as if standing with its wings in its pockets). Also perches on rocks, short bushes, fenceposts, and road signs, but rarely higher—the bird likes to be near the ground.

Active day and night. Perch-hunts, stalks prey on the ground, and sallies out to flycatch insects during the day. Somewhat colonial. Multiple adults frequently breed in close proximity (given a sufficient quantity of burrows). Can be coaxed to use artificial burrows (PVC pipe works well), and also uses water culverts and drains. Fairly tame. Where acclimated to people (such as around golf courses), may allow close approach. When nervous, bounces jerkily up and down. When pressed, flies or disappears down its burrow.

FLIGHT: Silhouette is broad and proportioned somewhat like a longer but mittened or blunt-winged screech-owl. Feet are sometimes visible trailing behind the short tail. Flight is straight and direct. Wingbeats are steady, even, and quick, but not hurried. Usually glides when coming in for a landing (or at intervals during lengthy flights), with wings jutting up sharply at the shoulder and flattening out along the hand.

VOCALIZATIONS: Burrowing Owl has varied calls but is not an overly vocal species. Males make an evocative two-note cooing, "*cuh-huh,*" that sounds somewhat quail-like. Call is a muffled bark: "*whoof*" or "*whew.*" Birds also do a rapid chatter, "*whe'h'h"h'h'h*" or "*chrchachacha,*" and a loud "*reah,*" followed by a descending chortle: "*reh'h'h'h.*"

Spotted Owl, *Strix occidentalis*

Dappled Owl

STATUS: Uncommon to rare western resident; widespread but localized. **DISTRIBUTION:** Found from s. B.C. to n. Baja; also found in isolated locations in s. Utah, Wyo., s. Colo., Ariz., and N.M. south into n. and cen. Mexico. **HABITAT:** Varies according to region. "Northern" Spotted Owl largely depends on mature and old-growth coniferous forests. Roost sites have a high canopy and multiple layers. In California, birds use oaks and riparian woodlands at lower elevations, mixed to all conifers at higher elevations. They also favor mature trees with high canopy and a "complex" forest structure (trees of multiple sizes and diameters). In the Southwest, "Mexican" Spotted Owl is found in Douglas fir and mixed conifers or pine-oak forest as well as in steep-sided canyons with conifers or riparian woodlands where birds may roost in trees or on rock cliff walls. **COHABITANTS:** Band-tailed Pigeon, Steller's Jay, Swainson's Thrush, Hermit Warbler; in the Southwest, Black-headed Grosbeak, Hepatic Tanager. **MOVEMENT/MIGRATION:** Permanent resident. Some altitudinal shifting occurs among some members of the population. VI: 0.

DESCRIPTION: Dark somber-eyed forest owl with a sunlight-dappled breast. Medium-sized (smaller than Barred or Great Horned; much larger than Western Screech-Owl; slightly larger than Barn Owl), with a big head, no ear tufts, and a bulky oval or teardrop shape. On roosting birds, the head and body are seamlessly fused.

Overall grayish brown above with pale spotting (most significantly, twin galaxies of white dots on top of the head). Mostly pale below with broken horizontal barring that makes the underparts look rough and scaly and replicates the pattern of dappled light shining through the canopy. *Note:* Northern Spotted Owl is overall darker, with underpart feathering that is 50–50 dark and light. The larger, browner Barred Owl (whose range overlaps with Northern Spotted Owl), with its horizontally spotted neck ruff and vertically streaked underparts, looks disheveled. Spotted Owl's dark brown eyes seem shrunken, the bill is dull greenish, and the expression is one of sullen smoldering anger.

BEHAVIOR: Almost strictly nocturnal, foraging most frequently for the first two hours of darkness and again just before dawn. Roosts during the day—in a cool shady location (often away from tree trunks) during warm weather; high in the canopy and closer to trunks in colder weather. Selects a site with a protective overhang (branches or cliff face) when raining.

Consummate perch-hunter that avoids hunting in open areas (although may hunt forest edge). Tame, often allowing very close approach.

FLIGHT: Overall stocky, with short broad wings and a short tail. Several quick wingbeats are followed by a lengthy glide (but you'll be lucky to see this much).

VOCALIZATIONS: Sounds like a hesitant Barred Owl with a cold. Call is a series of haltingly spaced, low, muffled hoots: "*wuh. huh/huh. whuh.*" Or "*wuh. huh/huh. whuhwhuh.*" Females make a rising squealing whistle that may recall Boat-tailed or Great-tailed Grackle.

Barred Owl, *Strix varia*

Maniacal Forest Owl

STATUS: Common widespread resident of mature forest. **DISTRIBUTION:** Found across the entire e. U.S. and s. Canada, as well as nw. North America. Eastern U.S. range extends west to Minn., se. N.D., e. S.D., e. Neb., e. and s.-cen. Kans., nw and cen. Okla., and cen. Tex. (reaching the s. Edwards Plateau). In Canada, found from the Maritimes across s. Que., s. Ont., s. Man., cen. Sask., cen. and w. Alta., B.C., s. Yukon, and sw. Northwest Territories (also se. Alaska). In the w. U.S., found in Wash., Ore., n. Calif., Idaho, and nw. Mont. Also breeds in Mexico. **HABITAT:** Mature deciduous and mixed deciduous-coniferous forests. Most partial to extensive closed canopy and mature or old-growth forest near water (including narrow riparian woodlands), but is not limited to this habitat. Avoids young timber, but at times occupies small woodlots, including city parks. **COHABITANTS:** Red-shouldered Hawk, Spotted Owl, Pileated Woodpecker. **MOVEMENT/MIGRATION:** Nonmigratory. VI: 2.

DESCRIPTION: A big, imposing, barrel-shaped, somber-eyed owl that looks like it's wearing a shabby stain-streaked coat with a closed fur collar. Large and stocky (considerably smaller than Great Gray Owl; slightly smaller than Great Horned; slightly larger than Spotted), with a large, round, tuftless head, a barrel-shaped body, and a short tail.

Grayish brown upperparts are dappled with pale spots; creamy underparts are lavishly streaked with pale brown brush strokes that give the bird a disheveled look. The dark brown eyes and bright yellow bill are framed by large gray disks. A fluffy barred ruff looks like a fur collar or a shabby scarf wrapped several times around the bird's neck. The expression ranges from haunted to really ticked off. Spotted Owl is similarly shaped but overall less disheveled-looking and shows pale dappled or scaly (not streaked) underparts.

BEHAVIOR: Likes big trees and sturdy limbs beneath a shadowy canopy; unlike Great Horned or Great Gray Owl, rarely sits at the forest edge. Usually perches high in trees close to the trunk; hunts primarily from perches that may be at times less than 10 ft. from the forest floor. The bird has been known to hunt on foot and in water, and is also fond of bathing. Mostly nocturnal, but at times hunts and vocalizes during the day (again, within forest). Responds well to tapes and even poor imitations of its call, but persistence may be required to elicit the bird's response (after it has approached unseen). At times very vocal, usually on still nights; at other times, inexplicably silent. Usually found in widely spaced pairs.

FLIGHT: Looks like a grayish brown barrel with broad blunt wings. Flight is quick and surprisingly nimble. Wingbeats seem hurried; glides are extensive and agile, with birds commonly maneuvering around brush.

VOCALIZATIONS: Classic call is a two-part series consisting of eight deep, loud, measured hoots (phonetically rendered): "*who cooks for you; who cooks for YOU All?*" More emphatic, less muffled than Great Horned. Distant birds may initially be mistaken for a barking dog. Birds also (and commonly) emit a varied and maniacal cacophony of barks, laughs, cackles, and hooted guffaws. Such recitals often engage multiple pairs.

PERTINENT PARTICULARS: Where ranges overlap, Red-shouldered Hawk is found with Barred Owl, and owls respond (often aggressively) to Red-shouldered calls. Also, Barred Owl can be attracted during daylight with recordings and imitations, but persistence (up to a half-hour) is important.

Great Gray Owl, *Strix nebulosa*

Myopic Owl

STATUS: Uncommon resident of northern boreal forests; irregularly irruptive. **DISTRIBUTION:** Resident across much of boreal Canada and Alaska, including cen. Alaska, all but n. Yukon, the Northwest Territories (below the tree line), all except sw. B.C., most of forested Alta., cen. and s. Sask., cen. Man., all but se. Ont., and sw. Que. In the U.S., breeds in e. Wash. and along the Cascade and Sierra Mts. through Ore. and cen. Calif.; also found in n. Idaho, w. Mont., nw. Wyo., ne. Minn., nw. Wisc., and the Upper Peninsula of Mich. During periodic winter irruptions, may wander to n. Minn., n. Wisc., se. Ont., s. Que., n. N.Y., and New England.

HABITAT: Northern forests, especially taiga, but also coniferous, deciduous, or mixed northern forest habitats. Most often found near bogs, along forest edges overlooking fields, in mountain meadows, and on cultivated land, airports, and roadsides. Nests in raptor nests (frequently Northern Goshawk nests) or snapped tree trunks. Avoids hunting in areas with heavy shrub layer. **COHABITANTS:** Northern Goshawk, Boreal Owl, Great Horned Owl, Gray Jay, voles. **MOVEMENT/MIGRATION:** In most years, sedentary or, at least, loath to range beyond the limits of its breeding range (except for some shift to lower elevations or to areas with less snow cover). However, this species also engages in periodic irregular irruptions that may be either widespread or more regional and that involve fewer than 100 birds or several hundred. VI: 3.

DESCRIPTION: A huge flat-faced gray owl with small yellow eyes surrounded by exaggerated concentric-ringed lenses. Arrestingly large bird (distinctly larger than Great Horned or Snowy Owl; dwarfs Barred Owl), with an imposing round head, a long thickset body, and a long broad tail.

Overall sooty gray with coarse, muted, and diffuse patterning that makes the bird's concentrically ringed facial disk and the white "bow tie" stand out. (The white trim recalls a sloppy cat caught lapping milk out of a saucer.) Note the small yellow eyes (not large and dark brown) and the darkly mottled but not streaked underparts (that's Barred Owl). Great Gray's expression is sometimes myopic, sometimes regally aloof, but not solemn or haunted (that's Barred Owl).

BEHAVIOR: An imposing owl most often seen perched upright on the edges of open fields or along roadsides. Sometimes sits high (on utility poles, for instance) but more often perches low, at midheight, among the trees and not uncommonly just above the ground, with only the head showing above the vegetation. For the most part a perch-hunter. Makes long glides to prey (which is often hidden beneath the snow). Also courses low over fields, sometimes hovering before plunging, sometimes just upending and crashing headfirst.

In summer (and in winter when food-stressed), hunts during daylight. Otherwise, most active at dawn and dusk and at night. Many individuals are very tame, allowing close approach. Responds to squeaking.

FLIGHT: Appears huge, with a big head, a barrel-like body, a longish tail, and profile-dominating long, wide, round-tipped wings with pale buffy tips. Looks like a broad-winged log with a sawed-off face. Wingbeats are deep, slow, and ponderous; mothlike flight is buoyant and regal. Great Horned Owl, by comparison, has hurried wingbeats and a flight that looks heavy and harried.

VOCALIZATIONS: Call (of both sexes) is a series of 5–10 very low (almost inaudible) evenly spaced hoots uttered at a rate of about one per second. The male's call reportedly is lower-pitched. Other described calls include a "deep growling" that may be short or drawn-out.

Long-eared Owl, *Asio otus*

A Great Horned Drawn by El Greco

STATUS: Widespread northern and western resident; very widespread but somewhat irregular winter resident. Generally more numerous than reflected in reported observations. Mostly nocturnal; when seen in flight, frequently misidentified as Short-eared Owl, which it closely resembles in size, shape, manner of flight, and habitat preference. **DISTRIBUTION:** Breeds in open and semi-forested areas of Canada and the n. U.S. from N.B. south to Pa., west across forested Canada and the bordering northern states to the southeastern Yukon, and across the n. Great Plains south to Iowa and nw. Neb., then south through the Rockies, the Great Basin, and the Cascades and Sierra Nevada south to Calif., Ariz., and N.M. A disjunct breeding population is also found in w. Ore. and coastal Calif. south into Baja, as well as in western portions of Va. In winter, may occur anywhere in the U.S. and Mexico except coastal portions of the Southeast (from coastal Ga. to La.). **HABITAT:** At all seasons, commonly roosts in dense stands of vegetation adjacent to open and grassy marshy or desert areas used for hunting. In desert areas, less dense vegetation is acceptable. Where available, often prefers stands of young (not mature) conifers for roosting as well as breeding. In the West, breeds in willow thickets, cottonwoods, and tamarisk; in the East, sometimes roosts beneath the skirt of pin oak branches, and in some coastal locations roosts in bayberry and wax myrtle thickets. **COHABITANTS:** Varied, but frequently nests where Cooper's Hawk nests; in parts of its range, forages where Short-eared Owl hunts. **MOVEMENT/MIGRATION:** A permanent resident where it breeds; some birds migrate south almost to the Gulf Coast, then through all of Texas into Mexico. Spring migration (early) from February to early April; fall (protracted) from early September to early December. VI: 2.

DESCRIPTION: Medium-sized owl (longer than Screech; much smaller than Great Horned) that superficially resembles a dark and less broad-beamed Great Horned Owl, with a bright orange face and wide staring yellow eyes—a Great Horned Owl drawn by El Greco. Most frequently seen in roosts, where birds respond to intrusion by assuming an alert, elongated-onto-gaunt posture, with ear tufts standing up at attention—or they flush! At other times, the bird appears more compact (and surprisingly small). Ear tufts may be depressed and hardly visible, but are usually conspicuously erect. In extremely cold (subzero) temperatures, Long-eared can fluff up and approach the beer-barrel contours of Great Horned.

Overall dark but richly patterned, with dark gray upperparts and sloppily streaked and barred underparts (the bird looks as though it was crudely stitched together). The facial disk may be orangey or buff and often upper- and lower parts are suffused with touches of orange. The plumage of Great Horned Owl is more blandly homogenized; its underparts are somewhat splotchy on the breast, but otherwise tidily barred and vermiculated, not richly and thickly streaked the length of the bird. Also, the face of Eastern Great-horned Owl is more ochre-colored and not as orange—that is, not nearly so arresting. Finally, the ear tufts of Great-horned Owl are set wide, to either side of the head, and resemble the ears of a cat; Long-eared's ear tufts are set closer together and resemble licking candle flames.

BEHAVIOR: Roosts in thick growth (often in young conifer stands), but like Short-eared Owl, hunts and perches in open areas. Nests in the abandoned nests of other birds, most notably Black-billed Magpie, American Crow, and Cooper's Hawk. Prefers nests situated in stands of trees rather than isolated trees. Unlike most other owls, often roosts communally in the winter, with 2–20 individuals in close proximity, sometimes separated only by 2–3 ft. Also roosts alone. Favored roost perches are close to the tree trunk and 2–15 ft. above the ground; birds frequently return to the same perch night after night.

Often sits tight, allowing close approach. Sallies out to hunt as dusk deepens (less commonly at sunset, as is common with Short-eared and Great Horned Owls). At dawn (one hour to 45 min. before sunrise), birds sometimes perch at the edges of open areas before going to roost. In small trees or bushes, favored perches are unencumbered, stiff and sturdy (not springy), often dead or leafless and exposed branches.

FLIGHT: Classic owl with a large head and a short tail; the posture appears slightly humpbacked. Broad wings appear shorter, broader, and wider-handed than Short-eared Owl. Overall darker, particularly on the underparts, so does not have Short-eared's hooded appearance. In the open, alternates a series of wingbeats with a glide on down-bowed wings (more drooped than Short-eared). With slightly drooped wings, birds fly in a leisurely way on wingbeats that are looser and floppier but not as stiff and mothlike or harrier-like as Short-eared. In woodlands, very nimble; generally flies low and away when flushed, turning quickly and landing in the first place it finds with obstructing cover. In daylight hours, very reluctant to leave its roost grove. In short, Long-eared Owl's flight suggests an accipiter or Red-shouldered Hawk; Short-eared Owl's recalls Northern Harrier.

In migration, sometimes flies in small strung-out groups of two to four birds (like Sharp-shinned Hawk).

VOCALIZATIONS: Male's call is a series of long, low, well-spaced hoots: "*whooh . . . whooh . . . whooh . . . .*" When alarmed or irritated makes a loud, but somewhat muffled three-note bark, "*whuh, whuh, whuh.*" Also (particularly in winters), a loud, quavering, catlike yowl: "*yeEEEEeEEoh.*"

PERTINENT PARTICULARS: In winter, roosting birds seek sheltered places that provide cover, easy access (and escape), and comfort. Prime locations have a southern exposure that blocks northerly or westerly winds and catches the warming sun's rays and are easy to reach from the side or from above. Small open areas surrounded by heavy cover are ideal, as are young stands of evergreens close to open feeding areas. Birds usually abandon such stands when trees reach about 20 ft. and the stand becomes more open, making birds more vulnerable to Northern Goshawk and Great Horned Owl.

Short-eared Owl, *Asio flammeus*

The Mothlike Owl

STATUS: Fairly common and widespread owl of open grassy places. **DISTRIBUTION:** Breeds across virtually all of Canada and Alaska as well as much of the n. U.S. south to n. N.J., Pa., n. Ohio and Ind., Ill., Mo., Neb., Colo., Utah, Nev., and n. Calif. In winter, vacates most of its breeding range north of the U.S.–Canada border except sw. B.C. and extreme se. Ont. and Que. Winters throughout the U.S. and n. Mexico except n. Maine, n. Mich., n. Minn., and n. N.D. **HABITAT:** Tundra, prairies, marshes, coastal grasslands, shrub-steppe, airports. **COHABITANTS:** Northern Harrier. **MOVEMENT/**

MIGRATION: Among the most migratory of the owls (single individuals are occasionally seen migrating in daylight). Also somewhat nomadic. Spring migration from February to May; fall from late August to December. VI: 3.

DESCRIPTION: Medium-sized owl, somewhat roundly contoured (definitely not angular or slender), with a large, round, hornless head; the tiny spikelike ear tufts are usually not raised and appear inconsequentially small when they are. Overall pale and streaky (males) or buffy-cinnamon brown and streaky (females). The streaking is particularly prominent around the head/hindneck and forms a collar or bib on the throat and chest that sets off the pale oval face. Yellow eyes are flanked by dark bags, making the bird look tired or haggard.

BEHAVIOR: The most likely owl to be encountered in open grassy habitat, and among the most likely to be encountered hunting in daylight. Commonly roosts on the ground in tall grass, often communally. Also roosts in trees, sometimes with Long-eared Owls, particularly when snow is deep. Active at all hours of the day during the breeding season. More crepuscular in winter, and although most often seen at dawn and dusk, often active in daylight (particularly on dark cloudy days). Hunts by coursing low over marshes, prairies, meadows, and tundra. Perches in the open, usually low to the ground, and often on the ground.

Commonly engages in aerial dogfights with Northern Harrier and other Short-eared Owls in which Short-eared's superior maneuverability and buoyancy easily distinguish it. Short-eared will be the bird flying above the harrier.

FLIGHT: Looks like a pale beer keg on wings. The head is heavy and blunt, the tail short and rounded, and the wings broad, long, somewhat tapered, and blunt (the bird seems to be wearing mittens). In active flight, wings are stiff and straight. When gliding, wings jut up along the short arm and flatten or droop along the very long hand (that is, there's no acute dihedral, as with Northern Harrier). The wings also jut forward at the wrist, making the head less prominent and making the bird seem all wing.

Males are paler than females, with whitish underparts, a pale trailing upper surface of the wing, and a distinctly pale face. On darker females, the upper surface of the wing shows only a pale window panel just short of the dark tip. On both males and females, the underwings are distinctly pale. This is usually what draws the eye of an observer scanning in low light conditions. In both sexes (but particularly males), the bibbed chest contrasting with the paler (less streaked) belly makes the bird appear hooded.

Flight is buoyant, floating, nimble, and aptly described as mothlike (except of course that moths don't commonly glide). Wingbeats are stiff but caressing (not floppy like Northern Harrier), and project a slow-motion quality, even though the tempo and arc described by the wings are frequently variable—a quick, short, and deep sequence may be followed by one that is more shallow and slower. By contrast, harriers plod along with the same rhythm and cadence.

Hunting birds quarter low over the ground but may vary their altitude, climbing then descending. (Hunting harriers generally hug the ground.) The movements of Short-eared are quicker, more abrupt, and overall more nimble than a harrier. Short-eared Owl also soars, and when scrapping with a harrier, easily outclimbs it.

VOCALIZATIONS: A terrier-like yap with an angry warm-up growl: "*yeeeeAp.*"

PERTINENT PARTICULARS: An observer can easily entice hunting (and roosting) birds to approach very close by making squeaked imitations of mice; however, it is rare for a Short-eared Owl to be snookered twice.

Boreal Owl, *Aegolius funereus*

The Astonished Owl

STATUS: Wide-ranging but uncommon nocturnal rarely vocal (and damned difficult to find) northern resident. **DISTRIBUTION:** Resident from cen. Alaska north to the south slope of the Brooks Range, across boreal Canada to Lab. and n. N.B. (absent in the Canadian prairies). In the U.S., found in mountain forests in cen. and e. Wash., n.

Ore., n. and cen. Idaho, w. Mont., portions of Wyo., ne. Utah, Colo., n. N.M., and n. Minn. **HABITAT:** In northern regions, typically found in black and white spruce and subalpine fir forest with elements of aspens, poplar, and birch. In southern mountains, found in subalpine forest dominated by fir and Englemann spruce. Generally prefers unbroken mature forests, in part because of reduced ground clutter, and in part because larger older trees support the woodpecker cavities Boreal Owl uses for nesting. Also hunts open edges, particularly in spring. **COHABITANTS:** Spruce Grouse, Boreal Chickadee, Gray Jay. **MOVEMENT/MIGRATION:** Permanent resident (often maintaining small territories), but periodic irruptions cause incursions south of the bird's range into se. Canada and the n. U.S. (into Maine, Mass., N.Y., Mich., Wisc., and Minn.) that often coincides with movements of other northern owls, including Northern Saw-whet Owl, a potential bellwether. Peak movements have been noted in October and November as well as in February. VI: 3.

DESCRIPTION: A small square-faced boreal forest owl whose face shines with astonishment (then it looks cross). Fairly small but broad-beamed (larger than Northern Saw-whet or screech-owls; smaller than Northern Hawk Owl or Long-eared Owl), with a large, square, tuftless head, a stocky body, and a short tail. Appears bulkier and more square-headed and square-faced than Northern Saw-whet. If the overall shape of Northern Saw-whet can be likened to a soup can, Boreal Owl is proportioned more like a coffee can.

Olive brown with pale spots above; whitish with heavy streaks below. Overall colder and grayer-toned than Northern Saw-whet, particularly about the head and face. The mostly pale facial disk is framed by a black rectangular border (accentuating the square-headed appearance) and lacks the distinctive white V pattern emblazoned between the eyes of saw-whets. Juveniles (until September, then plumage is like adults) are uniformly blackish brown with narrow whitish eyebrows that do not meet across the forehead. Juvenile Northern Saw-whets are distinctly two-toned, with chocolate heads and buffy underparts, and they show a broad white X stamped between the eyes, extending below the bill.

BEHAVIOR: Solitary and all too silent. Males vocalize between January and April (also in October) but become mostly silent after birds are paired. Birds vocalize at three peak times—just after dark, just before midnight, and before dawn. Despite long hours of daylight in the northern summer, very nocturnal. In winter, does not become active until full dusk. In summer, commonly brings prey to nestlings during dim twilight hours. Hunts from perches, making short flights to new locations every 5–10 mins. Except when with newly fledged young, almost always alone.

Nests primarily in woodpecker cavities (often in deciduous trees), but roosts on branches of (usually) coniferous trees, close to the trunk, and usually chooses a different perch every night. Readily adopts nest boxes.

FLIGHT: Overall broad, stocky, and shaped much like saw-whet. Wingbeats are hurried and steady. Glides only when approaching prey.

VOCALIZATIONS: Song is a series of low, hurried, whistled toots that recall the winnowing of snipe (but end abruptly without falling off at the end): "*poo poo Poo Poo Poo Poo Poo Poo.*" The series may begin softly and gain in volume; on still nights, the song may be heard more than a mile away. Each series consists of six to eight notes followed by a pause, then another series. Males also make a low soft trill when approaching the nest site and emit a loud harsh "*skew*" when provoked. Young birds make a sharp chirp.

Northern Saw-whet Owl, *Aegolius acadicus*

The Winsome Tiger or "You just want to put it in your shirt pocket and take it home with you" (Anon. first-time observer)

STATUS: Fairly common, widespread, but unobtrusive woodland owl. Far less frequently encountered (or attracted) than a screech-owl. Nevertheless, over most of North America this is the only small tuftless owl likely to be encountered. **DISTRIBUTION:** A year-round resident across s. Canada and the n. U.S. as well as in higher elevations. The northern limit of its range runs from the Maritimes west across cen. Que., Ont., Man., Sask., Alta., n. B.C., and coastal Alaska to Anchorage. The southern limit is defined by a line drawn from N.J. west across cen. Pa., cen. Ohio, n. Ind., n. Ill., ne. Iowa, sw. Minn., ne. N.D., and s. Sask. Also occurs at higher elevations in Md., Va., W. Va., N.C., and Tenn., and in the West at higher elevations through the western mountains into Mexico, with populations found in all western states. In winter and migration, found throughout the U.S. except extreme s. Ariz., s. and e. Tex., and the Southeast. **HABITAT:** Breeds in a variety of open and closed forested habitats but is most commonly found in coniferous woodlands (various spruce, fir, and pines) and frequently in habitats that are low and wet. Winters in a variety of woodland habitats, many of them deciduous, and particularly in migration may occur in suburban and urban areas, including apartment building balconies. **COHABITANTS:** Varied. Nests in holes excavated by flickers and Pileated Woodpecker. **MOVEMENT/MIGRATION:** Somewhat irruptive (mostly east of the Rockies), with some birds migrating every fall and larger incursions (comprising mostly first-year birds) occurring every several years. Spring migration from late February to May; fall from late September to December, with a peak in October and early November. VI: 3.

DESCRIPTION: A small, winsome, and (in the estimate of birds up to the size of Northern Flicker and American Robin) ferocious owl. The roundly contoured, ovular, somewhat heart-shaped, and tuftless head is wide, and indeed wider than the body when the bird elongates in a threat posture. In combination with the chunky body and stub of a tail, the overall shape and size recalls an avocado (narrow end down). The facial disk is pale brown, with a white V stamped between the yellowish eyes. Upperparts are slightly olive brown and basically unspotted. (Beaded white spots are concentrated on and toward the wings.) The pale breast is boldly streaked with chestnut brown and much warmer-toned than the back. Juvenile plumage (held into September) is dark chocolate above, cinnamon below, and the white V on the all-dark face is a standout feature. The expression shows umbrage when alert, and appears shrewd when relaxed.

BEHAVIOR: Solitary and exceedingly tame. Hunts only at night; roosts by day (which is when most observers see them). Prefers to roost in conifers where available, and in honeysuckle tangles or hollies if not. Typical roosts are in denser younger growth or thickly tangled vegetation that has a protective dome or canopy; does roost, however, particularly during migration, in the open (such as on porch furniture on elevated apartment building patios). Most often roosts away from trunks, usually not more than 12 ft. above the ground and occasionally less than 3 ft. from the ground. Frequently roosts with prey in its talons, and commonly uses the same perch night after night (or perhaps multiple owls use the same favored perches every night).

When approached too closely, becomes elongated and thin. Also turns away from observers and occasionally moves its wing across the breast as if trying to shield its paler underparts.

FLIGHT: Overall small and stocky-bodied, with short rounded wings. In size, shape, and manner of flight, suggests American Woodcock, but the flight is more direct and less tipsy, and the wings are silent, not twittering. Perch-hunts and flies silently toward prey, generally low, less than 6 ft. above the ground. May either fly directly or undulate when approaching prey. Wingbeats are quick and constant.

VOCALIZATIONS: The male's typical vocalization is a monotonous and unvarying series of whistled toots. The female's call is softer and more irregular. Also makes a shrill, ascending, nasal whine: "*eeeep!*"

PERTINENT PARTICULARS: Saw-whet is smaller, rounder, and overall more petite and cuter than a screech-owl. After one look, you want to put it in a pocket and take it home with you. Boreal Owl is bigger, stockier, square-headed, and mean-looking. Elf Owl, a diminutive bird with a small head, generally roosts in woodpecker holes. (Saw-whets roost in the open.) Pygmy-owls have long tails and are active daytime hunters.

Saw-whet is most vocal just at dawn. Although not as overtly responsive to imitations of its call as a screech-owl, saw-whet is frequently attracted to imitations and, remaining silent, lands nearby. Sometimes it takes 20–30 mins. before it can be incited to respond. It pays to be attentive to the sound of wings striking branches (evidence that a bird has come in) and to play a flashlight over branches (even over the ground) after 10–15 mins. of trying to initiate a response.

NIGHTHAWKS AND NIGHTJARS

Lesser Nighthawk, *Chordeiles acutipennis*

The Trilling Nighthawk

STATUS: Common breeder in the Southwest. **DISTRIBUTION:** The Central Valley of Calif. south to include most of se. Calif., s. Nev., the southwest corner of Utah, w. and s. Ariz., s. N.M., and s. Tex. to Corpus Christi; south into Mexico and Central and South America. **HABITAT:** Hot, arid, and desert lowlands covered by sparse vegetation or low brush. Often found around dry desert washes, sandy flats, or agricultural fields. Also attracted to more lushly vegetated habitats close to ponds or human dwellings; such locations are insect-rich owing to the habitat and/or the presence of artificial light sources. **COHABITANTS:** Elf Owl, assorted bat species. **MOVEMENT/MIGRATION:** All birds vacate the United States in winter. Spring migration from early March to early May, with a peak in April; fall from early August to late October, with a peak in late August to mid-September. VI: 2.

DESCRIPTION: A medium-sized, small-headed, slender-winged nightjar of dry open places. Larger and more slender than short, plump, big-headed Common Poorwill; smaller, more elegant, and shorter-tailed than long-tailed Common Pauraque; about the same size and shape as Common Nighthawk.

That's the rub.

On perched birds, the wingtips reach the tips of the tail. On Common Nighthawk, wings extend beyond the tips of the tail.

Brownish (males) or brownish gray (females and immatures) plumage is overall plainer and less patterned than Common Nighthawk, showing overall more whitish spots and patches and fewer blackish ones. Browner males have a white cut across the throat; grayer (and more Common Nighthawk–like) females and immatures have a buffy throat.

BEHAVIOR: Spends much of the day mantling a stout branch or tree limb or roosting on the ground. Gets up at dusk to hunt, but may also hunt during the day (especially late into the morning), particularly when there are young to feed. Forages singly, in pairs, or in groups. Often flies very low (just over the ground or tops of vegetation), but during daylight hours tends to forage much higher.

Captures most prey on the wing. Also drinks by skimming low over the surface of ponds.

FLIGHT: Looks and flies like a tipsy falcon. Overall slender, shaped somewhat like a notched-tailed Mississippi Kite. The smallish head does not project appreciably, wings are slender and angled back, and the tail is long but seems short in relation to the wings. In relation to Common Nighthawk, the Lesser Nighthawk's contours are rounder and overall more compact, less acutely angular, and shorter- and blunter-winged.

Overall browner, warmer-toned, and more uniform below than Common Nighthawk, but with more apparent barring on the tail. White or buffy oval patches cutting across the tips of the wings are closer to the wingtip than on Common Nighthawk and make the wingtip look like a perfect triangle (all sides equal). Common Nighthawk's wingtip is an isosceles triangle (the sides longer than the base). A white chin strap and white points on the tips of the tail are apparent on males, but absent on females.

Flight is erratic, mixing extended periods of flapping with equally extensive glides and much banking and maneuvering. Wings, held in a pronounced V above the body, jut up at the shoulder and then out and up. Wingbeats are stiff, given in a jerky irregular sequence or a quivering flutter when birds are climbing to intercept prey. Most of the stroke occurs above the body, with wings slightly down-bowed. Glides may be steady or tipsy. Compared to Common Nighthawk, Lesser's flight is a little slower, looser, more floating, and more fluttery, not as stiffly erratic, and often much closer to the ground.

VOCALIZATIONS: Refreshingly simple. A long, low, whistled, toadlike trill. Common Nighthawk has an abrupt "*peent*" similar to American Woodcock.

Common Nighthawk, *Chordeiles minor*

Common Duskhawk

STATUS: Common to fairly common and widespread breeder. **DISTRIBUTION:** Breeds across most of North America south of the tree line. In Canada, absent in extreme w. B.C. and the sw. Yukon and, in the East, extreme e. Que., all but s. Lab. and Nfld. In the U.S., absent in Alaska, all but n. and e.-cen. Calif., s. Nev., w. and s. Ariz., and sw. Tex. **HABITAT:** Almost everywhere, but most often found in open farm- and rangeland, grasslands, sagebrush, open burned or logged forest, coastal dunes, and the roofs and skies above cities. **COHABITANTS:** Killdeer, Mourning Dove, Chimney and Vaux's Swifts. **MOVEMENT/MIGRATION:** Spring migration from late April to early June; fall from early August to mid-October. VI: 2.

DESCRIPTION: A wheeling, drunken-looking, falconlike bird feeding high overhead at dawn and dusk. Medium-sized nightjar (slightly larger than American Kestrel; much smaller than American Crow; very slightly larger than Lesser Nighthawk and Antillian Nighthawk), showing a tiny bill, a small head, big white patches near the tips of the wings, and a long body that is big in the chest and tapers acutely toward the tail. On perched birds, wings extend beyond the tip of the tail. (On Antillean and Lesser Nighthawks, the wings and tail are equal length.)

Overall mottled and mostly gray with dark, pale, and some brownish patches. Perched birds usually show a touch of white between the head and chest, a touch of white at the bend of the wing, and sometimes a touch of white on the wing (just above the rump).

BEHAVIOR: By day, perches length-wise along tree limbs or atop fenceposts, open ground, and rooftops. Becomes most active just after sunset (and before sunrise), but on territory displays throughout the day and in migration may be seen in daylight (often at great heights). Also active to a lesser extent at night where birds forage near street or parking lot and stadium lights. Sometimes hunts low over water, but more commonly high. In migration, often seen in wide, spaced, horizontal waves, but in daylight has been seen to gather in large clusters or kettles (much like Broad-winged Hawk).

A tame bird that allows close approach.

FLIGHT: Falconlike shape showing long, very pointy-angled back wings and a long, narrow, slightly notched tail. The white rectangular patch bisecting the outer wing (positioned slightly more than halfway between the wrist and the tip) gives the dark wingtips the shape of an isosceles triangle (sides longer than the base); the wingtip on Lesser Nighthawk forms a perfect triangle (all sides equal). Also, Common's underparts are more contrasting and show a two-toned underwing—pale on the arm, dark on the hand—as well as a slightly paler undertail. Lesser Nighthawk is more uniform below.

Flight is wheeling, darting, and tipsy. Wingbeats are given in a quick or slow series followed by an unsteady raised-wing glide.

VOCALIZATIONS: Call is a buzzy, reedy "*beeeez't*" uttered after the bird executes a rapid wing flutter. On territory, at the bottom of their dive display, males make a booming sound.

PERTINENT PARTICULARS: Except in southern Florida and the American Southwest, this is the default nighthawk. Lesser Nighthawk may be separated from this species by vocalization and careful attention to field marks. Antillean Nighthawk is safely separable form Common Nighthawk only by call.

Antillean Nighthawk, *Chordeiles gundlachii*

"Rickery-dick"

STATUS AND DISTRIBUTION: Uncommon and geographically restricted to the lower Florida Keys, although it may occur as far north as Homestead, Fla. Most common on Key West and Big Pine Key. **HABITAT:** Nests in denuded open areas, including bare ground, beach, and gravel roofs. Hunts in the skies over towns, airports, cane fields, and forests. **COHABITANTS:** Gray Kingbird, Black-whiskered Vireo, Common Nighthawk. **MOVEMENT/MIGRATION:** Arrives on the Keys in early April; departs by mid-September. VI: 0.

DESCRIPTION: A buffy nighthawk hunting over the Key West airport that says "*Rickery-dick*" (as its call is phonetically transcribed on the Cayman Islands). A small nightjar that is slightly smaller than Common Nighthawk (which occupies the same habitat and range) and Lesser Nighthawk (which Antillean also closely resembles). On perched birds, wingtips reach the tip of the tail. (On Common Nighthawk, wingtips extend slightly beyond the tail.) Otherwise, Antillean is essentially similar to Common Nighthawk and best separated by call.

BEHAVIOR: Much like Common Nighthawk—a mostly crepuscular bird that hawks insects at various heights, often in association with other nighthawks.

FLIGHT: Slightly smaller, shorter- and rounded-winged, and overall buffier than Common Nighthawk, but very similarly proportioned. The white patch on the wings of females is smaller, dingier, and less distinct. Flight is also similar, but wingbeats are, perhaps, somewhat stiffer.

VOCALIZATIONS: Call is a distinctive, buzzy, froglike, two-note hiccup: "*wheebik.*" Multisyllable variations have been rendered as "*killi-kadick-dick-dick-dick*" and "*picky-pick-pick.*"

Common Pauraque, *Nyctidromus albicollis*

Long-tailed Nightjar

STATUS AND DISTRIBUTION: Common resident but in the U.S. restricted to s. Tex. south of Laredo, San Antonio, and Rockport. Range extends south into South America. **HABITAT:** In Texas, found in dense scrub woodlands dominated by mesquite and Texas ebony; also occurs in mesquite-oak savannas. Usually vocalizes from and roosts in woods near forest openings and edges, including fields, roads, and roadsides. **COHABITANTS:** Ferruginous Pygmy-Owl, Green Jay, Long-billed Thrasher. **MOVEMENT/MIGRATION:** Permanent resident. VI: 0.

DESCRIPTION: A noisy distinctly long-tailed nightjar, sitting on the road. Large (larger than Lesser Nighthawk; slightly smaller and less bulky than Chuck-will's-widow), with a largish head, a plump breast, and a conspicuously long straight tail extending well beyond the wingtips (the bird looks as though you could lift it by the tail).

Mostly mottled gray with a brownish eye patch and a fine, pale, zigzag line pattern where the folded wing and back meet.

BEHAVIOR: Roosts on the ground by day, generally in leaves and brush; calls and forages by night. Hunts by crouching—most commonly on the ground in an open place, less commonly from a low elevated perch—and throwing itself aloft, then returning to the launch point, sometimes with a banking open-winged flourish. If flushed during the day, usually flies a short distance before landing—again, on the ground. If flushed at night, normally flies only a short distance and may return to the same point.

Usually found alone, though sometimes in pairs or small family groups. A tame bird; during the day flushes from underfoot, and at night usually allows approach to about 20 ft. Vocalizes all year, so easy to find.

FLIGHT: Distinctive shape and pattern make this one of the easiest nightjars to identify. Shows long, wide, fairly blunt wings with conspicuous white crescents (buffy in females) near the wingtips and a long tail with a white outer edge (males) or outer tips (females). *Note:* The white tips on the female's tail may be difficult to see, the white sides on the male's tail are hard to miss. Wingbeats are deep, stiff, rapid, and lofting, given in a series followed by a glide, with wings steeply raised and thrown slightly forward and the tail partially fanned. Flight is usually close to the ground and erratic, darting, nimble, weaving, and highly maneuverable. Birds move easily through and around brush (usually not above it). Silent in flight.

VOCALIZATIONS: Call is loud and distinctive. Usually begins with a low, stuttering, whistled warm-up followed by a rising-and-falling nasal yowl: "*what-wr't, what-wr't, what, what'w'REEo*" (repeated). The loud last part of the song ("*what'w'REEo*") is the part phonetically transcribed "*Pauraque.*" Also makes a shorter, slightly more musical rising-and-falling nasal growl, "*we'er,*" which projects some of the sound quality produced by an Australian didgeridoo. Other calls include a low muttered "*wurt, wurt*" and a low tremolo trill that recalls Eastern Screech-Owl.

Common Poorwill, *Phalaenoptilus nuttallii*

Tiny Nightjar

STATUS: Fairly common widespread western breeder; rare to uncommon wintering species in the Southwest. **DISTRIBUTION:** Found primarily in the U.S. and n. Mexico; breeds from the prairies west. Eastern limits of range cut across cen. N.D. and S.D., e. Neb. and Kans., w. Okla., and w. Tex. Range extends north into sw. Sask. and s.-cen. B.C. Absent in n. Idaho and w. Mont., coastal Wash., Ore., coastal n. Calif., and the Central Valley. In winter, found in s. Calif., s. Ariz., and sw. N.M. and along some western portions of the Rio Grande in Tex. **HABITAT:** Dry open grasslands with or without scattered brush; also found on hillsides and rocky plateaus with scattered or moderately heavy brush. **COHABITANTS:** Somewhat varied, but includes Common and Lesser Nighthawks, Whip-poor-will, Rock Wren, Rufous-crowned Sparrow, Western Meadowlark. **MOVEMENT/MIGRATION:** More questions than answers here. Most or all northern birds apparently migrate to more temperate portions of the range (the southern United States and Mexico). In response to cold stress, this species hibernates, but how commonly or widely remains unknown. Spring migration in late April and early May; fall in September. VI: 0.

DESCRIPTION: A small, compact, big-headed, short-tailed nightjar. Overall small (starling-sized), with an exceptionally large neckless head and a stubby plump body that ends abruptly—no tail! On other nightjars (except nighthawks), the tail extends beyond the tail.

Overall gray with a blackish face and a narrow white collar. If you are lucky enough to find a roosting bird (without flushing it), it looks like a round-eyed lump of gray marble.

BEHAVIOR: A mostly nocturnal bird that also feeds and vocalizes at dawn and dusk. Forages by sitting on the ground or low perches and fluttering quickly aloft to catch insects, then gliding to the ground; frequently returns to the same spot. Uses paved and dirt roads as launch pads. Unlike nighthawks, does not course over open areas.

During the day, commonly roosts on the ground, facing downhill (also roosts in low limbs of trees). Vocalizes most ardently at dawn and dusk, but also during the night. Responds readily to tapes and generally easy to approach. If flushed during the day, does not fly far (commonly lands within 50 ft.).

FLIGHT: A stubby body with fairly short wide wings. Silhouette looks more owl-like than many other nightjars. Usually low flight is lofting and moth-like, with deep, silent wingbeats.

VOCALIZATIONS: Easily recognized song is a low, mellow, distinctly two-noted whistle, "*poor-Will,*" repeated over and over. More plaintive and less rollicking than the three-note chant of Whip-poor-will; not as loud or raucously shrill as the incantation of Common Pauraque.

Chuck-will's-widow, *Caprimulgus carolinensis*

Ruddy Nightjar

STATUS: Common southeastern breeder. **DISTRIBUTION:** Found across most of the se. U.S., though absent at higher elevations of the Appalachians. Breeds north to Long Island (less commonly Mass.), coastal N.J., s. Del., e. Md., all but w. Va., s. W. Va., s. Ohio, s. Ind., se. Iowa, n. Mo., e. Kans., w. Okla., and cen. Tex. south to the boundaries of the

Edwards Plateau. Absent in coastal Tex. and La. Winters in Central and South America and the e. Caribbean as well as s. Fla. **HABITAT:** A variety of woodland, including deciduous, pine, dune holly-hackberry, and live oak. Openings and edges (roads, agricultural fields, pastures, low scrub, broken woodland) are essential habitat components. Where ranges overlap with Whip-poor-will, Chuck commonly occupies more open areas, Whip more heavily forested areas. **COHABITANTS:** Wild Turkey, Great Crested Flycatcher, Carolina Chickadee. **MOVEMENT/MIGRATION:** Except birds in southern Florida, all birds retreat from the United States for the winter. Spring migration from mid-March to early June; fall from late August to early October. VI: 2.

DESCRIPTION: A large, long-tailed, almost uniformly ruddy brown nightjar. Large (slightly larger than Pauraque; distinctly larger than Whip-poor-will), with a tiny bill, a massive flat neckless head, and a long tapering body and tail, with the tail extending well beyond the wingtips. (Wings of Common Nighthawk extend slightly beyond the tip of the tail.) Appears distinctly larger and flatter-headed than Whip-poor-will.

Overall ruddy brown with a throat that is slightly paler than, or the same color as, the breast. The pale line separating the throat and breast is indistinct or lacking. The paler braces over the folded wing are not as contrastingly gray or eye-catching as on Whip-poor-will.

BEHAVIOR: Commonly roosts on limbs (usually close to the ground), crouching parallel with the branch. Also roosts on the ground. Forages actively at dusk and dawn, less commonly at night, except on bright moonlit nights or in places brightened by artificial light. Also occasionally forages in daylight, particularly on dark rainy days. Primarily an aerial hunter, flying low along the edges of fields and roadsides and in gaps through the forest. Also hunts by sallying up from the ground or bushes to snap up flying prey and on the ground (particularly in late July and early August) when molting.

Fairly tame and reluctant to flush. Once flushed, often flies a short distance before settling again.

FLIGHT: Long-winged profile with a long broad tail—larger and longer-winged than Whip-poor-will. Overall dark, except for pale sides to the tail on males. Strong, buoyant, agile flier. Flies with a varied but often lengthy series of wingbeats punctuated by lengthy glides on wings that are uplifted in a slight dihedral. Flights are generally low, often weaving around vegetation with tight banking turns, but birds may also fly (and forage) high.

VOCALIZATIONS: Song is a loud, rolling, four-note rendering of the bird's name: "*Chuck wills WI-dow.*" The "*chuck*" sounds like a note struck on a wooden block and is often inaudible at a distance. Emphasis is on the penultimate note, "*WI.*" Whip-poor-will makes the final note louder and stronger. Chuck also makes a croaking growl and a clucking sound (much like the opening note of the bird's song). Flushed birds emit a series of low rough barks: "*qwuh qwuh qwuh.*"

Buff-collared Nightjar,
Caprimulgus ridgwayi
Desert Washjar

STATUS: Resident of Mexico and Central America. In the United States, rare and geographically restricted breeder. **DISTRIBUTION:** In the U.S., found in se. Ariz. and extreme sw. N.M. Most observations are from Guadalupe Canyon, N.M.; Guadalupe Canyon, Ariz.; McCleary (a.k.a. Florida) Wash, below Madera Canyon; and the Santa Rita Mts. of se. Ariz., principally near California Gulch. **HABITAT:** In the United States, found mostly in heavily vegetated dry desert washes and desert canyons. Common trees include mesquite, acacia, and hackberry. **COHABITANTS:** Western Screech-Owl, Common Poorwill, Varied Bunting, coatis. **MOVEMENT/MIGRATION:** U.S. birds presumably retreat south outside the breeding season. Birds vocalize from late April to late August, but vocalizations drop off after June. VI: 0.

DESCRIPTION: It resembles two orange eyes seated atop a bush, glowing in the beam of a powerful flashlight. Medium-sized nightjar (slightly larger than Common Poorwill; slightly smaller than Whip-poor-will), with proportions more akin to the longer-tailed Whip-poor-will. Poorwill, by comparison, looks compact and tailless.

Except for the narrow buffy collar that completely encircles the neck, brownish gray and showing less overall patterning than Whip-poor-will.

BEHAVIOR: Crepuscular and nocturnal. Roosts beneath a bush by day; forages mostly at first light and late dusk. Perch-hunts from atop bushes and short trees; sallies out to snatch flying insect prey, and commonly returns to the same perch, using favored perches night after night. Common Poorwills, by comparison, hunt from the ground and commonly do not return to the same perch.

FLIGHT: Like Whip-poor-will.

VOCALIZATIONS: A rapid, rising, accelerating series of sharp notes, beginning with "clucks," progressing into "chirps," and ending with hurried flourish: *"tk tk tk tk tkchercherchercherlee!"*

Whip-poor-will, *Caprimulgus vociferus*

Forest Nightjar

STATUS: Widespread and locally common woodland breeder; winter resident along the Gulf Coast. **DISTRIBUTION:** Primarily the Northeast, but also found in portions of Ariz., N.M., s. Calif., s. Nev., and trans-Pecos Tex. In the Northeast, found north to N.S., e. N.B., cen. Maine, s. Que., s. Ont., s. Man., and cen. Sask. and west to w. Minn., n. and e. Neb., e. Kans., and e. Okla. Southern range limit falls across sw. and cen. Ark., se. Mo., w. Ky., ne. Miss., cen. Ala., and cen. Ga. Absent in coastal portions of S.C., N.C., n. Ind., and nw. and cen. Ohio. Winters in Fla. (less commonly north to coastal Ga.) and along the Gulf Coast to Mexico and Central America. **HABITAT:** A variety of dry, mostly extensive, deciduous or mixed-forest habitat characterized by breaks, openings, and/or open canopy and little underbrush. In the East, seems very partial to oak and pine-oak woodlands. In the Southwest, associated with wooded canyons. Nighthawks and Poorwill prefer open, often arid, habitat; Pauraque inhabits brush. Where range overlaps with Chuck-will's-widow, Whip-poor-will favors higher elevations and drier and more interior portions of forest. In coastal areas, Chucks are more common near the coast, and Whips occupy the interior. **COHABITANTS:** Yellow-throated Vireo, Wood Thrush, Ovenbird, Scarlet Tanager; also Mexican Jay and Painted Redstart. **MOVEMENT/MIGRATION:** Wholly migratory. Spring migration from late March to late May; fall from early September to early November. VI: 2.

DESCRIPTION: A round-headed, long-tailed, grayish nightjar showing no white in the wing. Medium-sized (smaller than Chuck-will's-widow; slightly larger than Common Nighthawk or screech-owl; about the same size as American Robin), with a tiny bill, a largish neckless head, a long tapering body, and a tail that extends well beyond the folded wingtips. (On Common Poorwill and nighthawks, the wingtips and tail are about equal length.)

Cryptically dapple-patterned, but overall grayish, showing touches of warm brown in the cheeks and wings (and on females the tip of the tail), a throat that is darker than the breast, and a gray black-bordered scar just above the wing (the bird looks like something took a bite out of it). Chuck-will's-widow is uniformly ruddy brown, with a throat that is paler than the breast. Whip-poor-will's expression is aloof.

BEHAVIOR: Roosts, length-wise, on low limbs or on the ground, usually in heavy cover. When flushed, commonly travels only a short distance before landing. Mostly crepuscular; active when bats are active, hunting in the dim light between full darkness and half an hour before and after sunset (but longer on moonlit nights). Hunts by coursing through open areas and woodlands, but perhaps even more often by sallying from perches near forest openings, often using favored perches night after night. Seems particularly fond of poorly traveled roads: using the road surface as the launch point, birds leap vertically to snatch prey from the air. Vocalizing males are commonly low, sometimes perched on a fallen log or a low limb, and usually close to the forest edge.

FLIGHT: Smaller, with shorter wings and a shorter tail than Chuck-will's-widow. Whip-poor-will's silhouette suggests a robust Cooper's Hawk. Tails of males flash bright white corners. Flight is light, nimble, and buoyant—recalls an erratic cuckoo. Stiff wingbeats are given in a hurried series punctuated by prolonged, mostly flat-winged glides. During aerial hunting, birds are active and wheeling, changing altitude and direction quickly and often and landing with a rapid vertical descent.

VOCALIZATIONS: Celebrated song is a loud, clear, whistled rendering of the bird's name, "*whip-per-WILL*," with the accent on the last note. (A barely audible "cluck" note sometimes introduces the song but can be heard only at close range.) The same phrase is sung repeatedly at one-second intervals for periods of up to 10 mins. or more. Birds are most vocal at dawn and dusk and on moonlit nights. Whip-poor-will may continue to vocalize (at dawn or dusk) until late in the season (into September), and reportedly vocalizes on occasion in winter. Also makes a growl similar to Chuck-will's-widow.

SWIFTS

Black Swift, *Cypseloides niger*

Sky-waltzing Swift

STATUS: Scarce and highly localized western breeder. **DISTRIBUTION:** Occurs locally in B.C., extreme se. Alaska, w. Alta., Wash., n. Idaho, nw. Mont., Ore., Calif., Utah, Colo., Ariz., and N.M. Also found in Mexico and Central America. **HABITAT:** Breeds in caves or deep ledges near or behind waterfalls, most commonly at high altitude; also breeds on sea cliffs. Forages over open montane or forested habitat, often at great heights. *Note:* Fewer than 100 breeding sites are known for this species in North America. **COHABITANTS:** Essentially none. At times, may forage in association with other swifts or swallows. Streams below waterfalls may host American Dipper. **MOVEMENT/MIGRATION:** Migrates in single-species flocks that can number several hundred or more; also seen migrating with flocks of other swifts and swallows. Spring migration from late April to late June; fall from late August to late October. VI: 1.

DESCRIPTION: Large all-dark swift seen almost exclusively in flight. Overall larger, longer-tailed, and longer-winged than Vaux's and Chimney Swifts. The tail and head projection on the smaller swifts seems fairly even, with the tail only slightly longer than the head. On Black Swift, the tail is unmistakably longer, less tapered, and more square (or notch-tipped). The shorter wings on the smaller swifts bulge or widen along the trailing edge and seem blunted at the tip. On Black Swift, the exceedingly long and narrow wings are uniformly tapered to a fine point.

Vaux's and Chimney Swifts are sooty brown with distinctly pale throats. You won't need to second-guess how very black a Black Swift is.

BEHAVIOR: Leaves nest cliffs at first light and returns just before sunset. Spends the day aloft, foraging at high altitudes for flying insects; also concentrates over lakes. In cloudy, foggy, or rainy conditions, may hunt lower (possibly at lower altitudes farther from the nest ledge). While never seen in large numbers (except in migration), does not necessarily hunt alone. Insects concentrated enough to attract one Black Swift often attract several, and groups may grow to 20–30 birds.

FLIGHT: While foraging, moves with the fluid grace of a skater—not as jerky and jittery as Vaux's and Chimney Swifts, and not as stiffly erratic as White-throated Swift. A long shallow series of wingbeats is followed by long tacking glides on slightly down-drooped wings. Wingbeats are slower and more fluid than other swifts, with more rippling flex seen in the wing. While foraging, covers great distances as a matter of course, often moving up and down the length of a valley.

VOCALIZATIONS: Except for occasional low chips, generally silent.

Chimney Swift, *Chaetura pelagica*

A Cigar Butt on Wings

STATUS: Common summer resident. **DISTRIBUTION:** Breeds across the U.S. and s. Canada east of the Rockies (except for s. Fla.), although generally less common in extreme western portions of its range. Rare but almost regular summer visitor locally in

s. Calif. Casual elsewhere in the far West. Winters in South America. **HABITAT:** Nests, for the most part, in chimneys, so most common over cities, towns, and villages. **COHABITANTS:** Sometimes forages with swallows, most notably Barn Swallow and Purple Martin. **MOVEMENT/MIGRATION:** Spring migration from mid-April to mid-May; fall migration from late August to late October. VI: 2.

DESCRIPTION: In the immortal words of Roger Tory Peterson, "a cigar with wings." Almost never seen except in flight. The head is small, the tail is longer, and the wings are extremely long, curved back, and thin. Tails can be blunt or fanned. At high altitudes, the bird looks headless — a sickle with a tail. At close range, seen from below, the throat and flight feathers are paler. Otherwise, the bird is all dark chocolate. Its western counterpart, Vaux's Swift, is slightly smaller, paler-throated, and paler-rumped, soars less, and nests in trees. Separation of the two species is difficult.

BEHAVIOR: Spends virtually all its life aloft in open obstacle-free air. May fly at very high altitudes or sweep insects out of the air over lake surfaces. Rarely seen alone; frequently seen in pairs or trios. May feed with swallows, among whom its shape, dark underparts, quivering wingbeat, and superior speed distinguish it. In migration, in flocks of 6–20, birds are widely spaced, and formations are loose and generally linear. Birds going to roost may number in the hundreds and thousands; their spiraling descent into chimneys is a spectacle. Very vocal.

FLIGHT: Is variable. Sputtery and direct; erratic (with many a jink left and right) and recklessly fast; even tipsy (with birds rocking unsteadily from side to side). Sweptback wings retain their slender sickle shape no matter what the bird is doing. Wings are stiff and quiver more than flap in rapid series punctuated by long sweeping glides on stiff down-angled wings. In courtship, birds briefly soar using a raised and fixed-wing glide.

VOCALIZATIONS: A high-pitched, excited, rising-and-falling twittering or chittering — a cross between a chatter and a twitter. Recalls the song of House Wren played on a Tin Pan Alley piano.

Vaux's Swift, *Chaetura vauxi*
Tree Swift

STATUS: Fairly common northwestern breeder; fairly common migrant west of the Rockies, particularly in Pacific Coast states. A few birds may winter in southern California and along the Gulf Coast east to northern Florida. **DISTRIBUTION:** Breeds from coastal s. Alaska across w. and s. B.C., sw. Alta. (barely), Wash., Ore., cen. Calif., n. Idaho, and w. Mont. **HABITAT:** The air above mature coniferous and mixed coniferous-deciduous forests; shows a preference for old-growth forest. Generally found at lower altitudes than Black Swift, but anticipate a fair amount of overlap. **COHABITANTS:** Pileated Woodpecker, Cassin's Vireo, Violet-green Swallow, Hermit Warbler. **MOVEMENT/MIGRATION:** Spring migration from mid-April to late May; fall from mid-August to early October. VI: 2.

DESCRIPTION: North America's smallest swift (in flight, about 20% smaller than Chimney Swift; about the same size, in flight, as Cliff Swallow). A dark swift, but slightly paler (and eminently more western) than Chimney Swift. Like Chimney Swift in flight — a thumb-shaped bird flanked by narrow, pointy, sickle-shaped wings. Appears slightly shorter-winged and more compact than Chimney Swift; distinctly less angular and stubbier-tailed than White-throated Swift; much smaller (and paler) than Black Swift.

In flight, appears all dark with a pale sheen on the throat/breast and rump that is slightly paler and more extensive than that seen on Chimney Swift. But differences are subtle, and the conspicuous dearth of sightings east of the Rockies (where Chimney Swift breeds) almost certainly relates to the difficulty of separating the two species.

BEHAVIOR: A forest species that nests most commonly in hollow trees and old Pileated Woodpecker excavations (and much less commonly in chimneys). Usually seen cruising over forests — as well as grasslands, lakes and rivers, and cities and towns near breeding habitat — hunting high-flying insects with speed but stiff and somewhat unsteady movements. Often seen with slower-flying swallows, but seems to fly through more than feed with them.

A gregarious species. Birds nest in relative proximity within the same forest area (usually in different trees) and roost communally, with hundreds, even thousands, of birds gathering into tight flocks that circle over roost trees (and, in migration, chimneys) before spiraling down in a tornado of birds.

FLIGHT: Shaped like Chimney Swift—"a cigar butt with wings"—but slightly smaller, more compact, and overall slightly paler, with more extensively pale throat/breast and rump. Flight is rapid and mostly straight, sometimes swooping or climbing, and more than a little unsteady. Feeds through swallows. Wings move in rapid, shallow, quivering series of wingbeats followed by short or extensive glides on slightly down-angled wings. May fly with a little more recklessness and a little less stability than Chimney Swift.

VOCALIZATIONS: A rapid series of sharp twittering chips, "*chee chee chee,*" that are higher-pitched and somewhat weaker than Chimney Swift's.

White-throated Swift, *Aeronautes saxatalis*

North American Needle-tailed Swift

STATUS: Common summer resident of western mountains; fairly common wintering species in southern portions of the United States and south. **DISTRIBUTION:** Breeds throughout much of the intermountain West from s. B.C. east to e. S.D. and south into Mexico and Central America. Absent in western portions of Wash. and Ore. and portions of n. Calif. In winter, withdraws from much of its breeding range; found from n. Calif. across s. Nev., s. Ariz., s. N.M., and trans-Pecos Tex. south into Mexico and Central America. **HABITAT:** Cliffs, rock formations, and the skies above adjacent forests, meadows, foothills, and prairies. Nests on cliffs and clifflike man-made structures (for example, highway overpasses and buildings). Often feeds at extreme heights (beyond the range of the unaided eye), but also forages low over lakes, particularly during cold or inclement weather. **COHABITANTS:** Prairie Falcon, Canyon Wren. **MOVEMENT/MIGRATION:** Most birds in the United States migrate. Spring migration in April and May; fall from late August to late October. VI: 1.

DESCRIPTION: A slender-winged, needle-tailed, black-and-white dart—North America's only boldly patterned black-and-white swift. (Other species are more uniformly dark.) Overall trim and slender, with a long tail that tapers to a fine point. (Tails of other swifts are short and blunt.) Occasionally White-throated fans its tail, showing a notch or forked tip, a characteristic that is just as distinguishing.

Boldly patterned black and white. White saddlebags over the rump are standout features when birds pass at eye level. The white stripe running from the throat to the belly distinguishes the bird when viewed from below. *Note:* At eye level, the white underparts are usually not visible, and the face looks dusky, not white. Also, at extreme distances and in poor light, the plumage pattern is hard to see, so White-throated may be confused with Black Swift. Rely on shape.

BEHAVIOR: Almost always seen in flight, usually in flocks of 10–50 birds. (Greater numbers occur when birds are migrating, concentrating over an unusually productive food source, or going to roost.) Sometimes feeds with swallows and other swift species. Leaves roost in the morning and returns in the evening (in winter returns en masse at sunset), but unlike Black Swift, White-throated may be found around roosting areas all day. Spends the day on the wing, hunting along ridge lines or high in thermal columns or updrafts, which carry and concentrate insects aloft. Very vocal around nest cliffs and roosts. At other times, generally silent.

FLIGHT: Swift, darting, and somewhat stiff or uptight. A rapid series of stiff wingbeats is followed by a protracted, somewhat unsteady and tipsy glide executed on slightly down-angled wings. Glides more than Vaux's and Chimney Swifts, and seems more recklessly fast. More agile than Black Swift, but not as maneuverable as swallows.

VOCALIZATIONS: One of nature's most ethereal sounds. A long, shrill, descending trill. Beginning notes are distinct but soon compress into a breathy twittering trill. The notes themselves recall the tinkling of ceramic wind chimes. Birds in a flock often call together, creating a surreal medley of sound.

PERTINENT PARTICULARS: Observers whose impulse is to assign the name "swallow" to any darting aerial hunter and who focus on the white patches bracketing White-throated Swift's rump are apt to confuse this species with Violet-green Swallow. Although shape alone distinguishes the two species, Violet-green's brilliant all-white face and underparts are standout features, and visible from almost any conceivable distance. The bisecting white stripe notwithstanding, the body of White-throated Swift appears mostly dark, and the face dirty or dusky.

HUMMINGBIRDS

Broad-billed Hummingbird, *Cynanthus latirostris*

Flutter-tailed Emerald

STATUS: Primarily a Mexican species; in the United States, a common breeder within its restricted range: DISTRIBUTION: Extreme se. Ariz. and sw. N.M.; vagrant in Tex. and Calif., and has occurred in such far-flung locations as Mich. and N.B. HABITAT: A woodland species found most commonly in oak woodlands and sycamore-cottonwood riparian corridors within mountain canyons at altitudes between 3,000 and 6,000 ft. Also occurs in desert foothills residential areas. COHABITANTS: Rufous-crowned Sparrow, Varied Bunting. MOVEMENT/MIGRATION: Over most of its overall range, a resident. In the United States and northern Mexico, present from March to mid-October; a few birds remain in desert lowlands (particularly at feeders) through the winter. VI: 2/3.

DESCRIPTION: A blackish green red-billed beauty with a tail that flutters like a windblown cape. Medium small, Broad-billed is about the same length as Black-chinned, but more robust, with a round head, a long neck, a compact body, and an animate blackish tail that is forked when closed, heart-shaped when fanned. The long, mostly reddish, slightly down-curved bill is distinctive.

In all plumages, but especially males, overall darker, and more uniformly so, than most other small hummingbirds of the same age and sex. Adult males are dark green above (the head may appear all black in poor light), and dark green with bluish iridescence below (the white undertail is easily overlooked); the tail is distinctly blackish (and gray tips are easily ignored). Adult females and immatures are emerald green above and uniformly dingy gray below (immature males have some of the adult male's blue-green iridescence on the throat), with a whitish throat, white undertail coverts, and white outer tips to the broad, blackish, distinctly wedge-shaped tail. A broad dark mask through the eye and a bracketing white eye stripe give the bird a mournful expression.

BEHAVIOR: The tail moves like a rippling banner. Opening, closing, fluttering . . . even when the bird lands, the tail seems difficult to tame. It quivers to a halt. Unlike other smaller hummingbirds, Broad-billed's wingbeats are slower and discernible, not just a gray blur. Generally less aggressive than other hummingbirds. Multiple Broad-billeds seem content to share the same feeder; aggressive displays seem limited to face-offs and vocalizations. Birds call regularly.

Forages on nectaring flowers and insects. A tame bird that responds well to imitations of Northern Pygmy-Owl.

FLIGHT: A little hummingbird with wingbeats and movements that seem as slow as one of the larger hummingbirds. Flight is nevertheless fast (although not quite as fast as some other hummingbird species) and direct, with some altitudinal rise and fall.

VOCALIZATIONS: Song is a high buzzy sizzle, "*zeega zeega zeega*," which has a mindless, monotonous, machine-shop cadence and repetition; this sequence is interspaced with a stuttering "*chichidit*." Call is a stuttering/chattering "*chidid*" that recalls a rough-sounding Ruby-crowned Kinglet; also emits a scraping "*skrrree*." Chase call is a short, dry, descending chatter: "*chh cheh ch'cheh; ch'trr chtrr*."

White-eared Hummingbird, *Hylocharis leucotis*

Green Thumb with a Short Bill and a White Slash

STATUS: Rare southwestern breeder and post-breeding wanderer from Mexico; geographically restricted. **DISTRIBUTION:** In the U.S., this Mexican species has bred in the Huachuca and Chiricahua Mts. of se. Ariz. and the Animas Mts. of N.M. Vagrant from June to August in s. Ariz., N.M., and w. Tex. in the vicinity of the Davis, Guadalupe, and Chisos Mts. Other records include one each from Miss., Colo., Mich. **HABITAT:** Pine and oak-pine mountain forests at 4,500–10,000 ft. In the United States, most commonly seen at feeding stations in proper habitat. **COHABITANTS:** Magnificent Hummingbird, Mexican Jay, Painted Redstart, Yellow-eyed Junco. **MOVEMENT/MIGRATION:** Arrives in Arizona, New Mexico, and Texas in April, retreats south into Mexico in September. VI: 1.

DESCRIPTION: Small, dark green, thumb-sized, thumb-shaped hummingbird with a conspicuous and diagnostic big white slash behind the eye. Overall stocky and compact (slightly smaller than Broad-billed; the same length as, but shorter-billed than, Black-chinned), with a large peaked head and high forehead, a short, thin, mostly straight, orange-based bill, a short body, and a short squared-off tail. By comparison, the bill of Broad-billed Hummingbird is longer and more curved along its length. Wings just reach the tip of the tail.

Overall dark green with a blackish head brandishing a strikingly broad white stripe that broadens behind the eye. Females and immatures (the birds most often seen) are patterned like males, but with paler grayish green underparts and a head pattern that highlights a broad black mask across the face in addition to the contrastingly white slash behind the eye. Somewhat resembles the much more common Broad-billed Hummingbird (particularly female Broad-billed), but White-eared's bill is shorter, its head is bulkier, and its white eyebrow is much more prominent and usually broadens near its terminus. (Broad-billed's overall less distinct eye stripe narrows and grows indistinct.)

BEHAVIOR: Aggressively defends nectar sources. Tail is less fluttering and animate than Broad-billed.

FLIGHT: Like other hummingbirds.

VOCALIZATIONS: Call is a sharp single- or double-noted "*ch'chit*" that recalls Anna's and Broad-tailed Hummingbirds and is distinctly unlike the chattering stutter of Broad-billed.

Berylline Hummingbird, *Amazilia beryllina*

"Cinnemerald" Hummingbird

STATUS AND DISTRIBUTION: Resident in the foothills and highlands of Mexico. Rare but near-annual visitor to se. Ariz.'s "sky islands"—most notably the Huachuca and Chiricahua Mts.—and the Davis Mts. in Tex. **HABITAT:** In the United States, open broadleaf (especially oak) and pine-oak woodlands and edge as well as riparian woodlands dominated by maple and sycamore. Most birds attend feeding stations near appropriate habitat. **COHABITANTS:** Broad-billed Hummingbird, Magnificent Hummingbird, Painted Redstart. **MOVEMENT/MIGRATION:** Permanent resident in Mexico; absent in the winter in the United States. VI: 1.

DESCRIPTION: A distinctive half-cinnamon, half-emerald ("cinnemerald") hummingbird that looks like the upper half (the head and upper body) is shining with a green finish and the bottom half (the lower back, wings, belly, and tail) is still wearing the rusty undercoating. Medium-sized (slightly larger than Broad-billed; distinctly smaller than Magnificent), with a modest, mostly straight bill, a round head, a portly body, and a square-tipped (or slightly notched) tail about the same length as the wings. Overall somewhat plump and dumpy, recalling a short-billed Buff-bellied Hummingbird.

The head, back, and breast are brilliant green—the bird looks as though it's wearing a glittering chain-mail cowl. The cinnamon tail flashes purple on the top side in good light (dark center in poor light). Dark wings show traces of cinnamon that are most concentrated at the shoulder and along the lower edge. The bill is dark and reddish at the base of the lower mandible. The belly is grayish or cinnamon gray. Immatures are patterned like adults, but are more subdued, sometimes showing a green-encased buffy throat and chest.

No other hummingbird seen in Arizona is so green on the head, back, and breast. No other hummingbird shows a cinnamon red tail or cinnamon wings. Most closely resembles Buff-bellied Hummingbird (of southern Texas), but Buff-bellied shows a distinctly reddish bill with a dark tip, a very warm buffy belly, and no red in the wings.

BEHAVIOR: Territorial hummingbird that defends its patch from all but the larger hummingbird species. Feeds on a variety of flowers as well as insects. Although males vocalize from conspicuous perches, birds spend most of their time in the shadowy understory.

FLIGHT: Like other hummingbirds.

VOCALIZATIONS: Call is a short, dry, low, buzzy "*drzzzr.*" The "song" is described as a long series of "high-pitched hisses, twitters, and chips."

PERTINENT PARTICULARS: Although hybridization between hummingbirds is not uncommon, Berylline/Magnificent hybrids seem disproportionately common in Arizona given the low number of Berylline Hummingbirds there (or is this why?).

Buff-bellied Hummingbird, *Amazilia yucatanensis*

Texas Copper-tail

STATUS AND DISTRIBUTION: Common but restricted summer breeder in se. Tex.; scarce winter resident in coastal Tex.; rare in coastal La., Miss., and east. Range extends from Tex. south through e. Mexico to the Yucatán, Belize, and n. Guatemala. **HABITAT:** In the United States, found mostly in brush woodlands dominated by mesquite and acacias and in oak islands in grasslands. Also attracted to flower gardens and hummingbird feeders (particularly in winter). **COHABITANTS:** Green Jay, Black-crested Titmouse, Olive Sparrow. **MOVEMENT/MIGRATION:** Between October and March, small numbers of birds appear along the Gulf Coast east to Florida. Most Texas breeders are presumed to retreat south into Mexico. VI: 2.

DESCRIPTION: A plump, dark, curve-billed Texas hummingbird. Medium-sized (larger and dowdier than Ruby-throated or Black-chinned), with a no-nonsense down-curved bill and distinctive emerald and copper plumage.

The dark body seems bisected—half green, half red. The upper half (head, breast, back) is greenish, and the lower half (belly, lower back, tail) is mostly brownish red. It's the only hummingbird in Texas that shows this half-and-half pattern, but if you're a stickler for namesake characteristics, the buff-colored belly is easily noted. The black-tipped reddish bill is also conspicuous, and this characteristic (as well as dark, not orangy, wings) distinguishes Buff-bellied from the similar Berylline Hummingbird (a rare visitor to southeastern Arizona and casual in west Texas). Immatures are like adults but duller.

BEHAVIOR: A hummingbird adept at maneuvering in brushy woodlands in short halting flights. Feeds on a variety of flowers (including tree blossoms), primarily in shaded woodland areas but in winter often found in more open habitats (including suburban yards with trees, bushes, and feeders). Between feeding bouts, commonly perches near the nectar source, often quite low and not uncommonly amid flowers. Also captures insects, mostly by hover-gleaning and less commonly by flycatching.

Generally solitary. Usually dominant over Ruby-throated and Black-chinned at feeders. Fairly quiet although males can be noisy and aggressive at times; but wings make a low hum.

FLIGHT: Plump contours and a distinctive reddish-lobed tail. Flight is like other hummingbirds, but wingbeats appear slower—barely a blur. Feeds fairly methodically. Wings show no trace of red or orange.

VOCALIZATIONS: Makes a ticking sound (somewhat like Rufous Hummingbird) and a low, somewhat sputtered "*tzeet . . . tzeet.*"

Violet-crowned Hummingbird, *Amazilia violiceps*

Snow White

STATUS: A Mexican species; in the United States, uncommon and restricted. **DISTRIBUTION:** Extreme se. Ariz. and bordering sw. N.M. Casual in w. Tex. and Calif. Strongholds in the U.S. include Guadalupe Canyon, Sonoita Creek (Patagonia), and the Mule Mts. near Bisbee. **HABITAT:** Breeds at 3,500–5,500 ft. in riparian corridors dominated by Arizona Sycamore. **COHABITANTS:** Black-bellied Whistling-Duck, Broad-billed Hummingbird, Thick-billed Kingbird, Rose-throated Becard. **MOVEMENT/MIGRATION:** Migratory status is uncertain. Frequently absent from breeding areas in winter, but a few birds winter at well-maintained feeders. VI: 1.

DESCRIPTION: A slender, blue-capped, coral-billed hummingbird. Medium-large (larger than Black-chinned; smaller than Blue-throated), with a slender round head, a gazelle neck, and a long bill. Easily distinguished by the combination of a reddish bill, a bluish crown, and brilliant white underparts—all features that are unaccountably accompanied by plain, dull, greenish bronze upperparts. In truth, and for the purposes of identification, the red bill and violet crown are superfluous. No other North American hummingbird is so immaculately white below. Immatures have less blue on the head and darker-tipped bills; otherwise, they are unmistakably similar.

BEHAVIOR: Likes to forage high, in the middle and upper strata of the canopy, but descends to forage on low-lying flowers and, in winter, becomes an altitudinal obligate (forages at whatever height a feeder is hung). Also forages on insects by gleaning them from foliage or snapping them up in midair—moving directly from point to point to point in a geometric three-dimensional game of connect-the-dots. Flips its tail while feeding, but not habitually. Aggressively defends feeders from other hummingbirds.

FLIGHT: Longish and slender profile with a long bill and distinctly long slender wings. Flight is very fast and direct. Forays are lengthy and somewhat stiff in execution.

VOCALIZATIONS: Chase call is a high-pitched squeaky squeal that degenerates into a sputtering descending stutter: "*sque'ekk'k'k'k'k*" (sometimes abbreviated to a short high-pitched trilling squeak). Also makes an unmusical stutter: "*t'ch.*"

Blue-throated Hummingbird, *Lampornis clemenciae*

Riparian Missile That Goes "Seet"

STATUS: Uncommon breeder; geographically restricted and localized; habitat-specific. **DISTRIBUTION:** Largely a Mexican species whose range extends into se. and e.-cen. Ariz., sw. N.M. (casual), trans-Pecos Tex. (a few), where it is a regular breeder in the Chisos and Guadalupe Mts. **HABITAT:** In the United States, inhabits riparian (sycamore-maple) canyon woodlands, often adjacent to mixed evergreen woodlands. **COHABITANTS:** Sulphur-bellied Flycatcher, Mexican Jay, Painted Redstart. **MOVEMENT/MIGRATION:** In the United States, most retreat south for winter. Spring migration from early February to late May; fall from late August to early November. A few winter at feeders in Arizona; can withstand subfreezing temperatures. VI: 2.

DESCRIPTION: A dark gray-bellied giant of a hummingbird. In the United States, can be confused only with the equally large Magnificent (and the very rare Plain-capped Starthroat). Dwarfing and dominating most other hummingbirds, Blue-throated has a proportionally small, mostly straight bill, a small head, a long neck, and a long body profile that is mostly long wings and long tail. The profile of perched birds seems sinuously curvaceous.

In all plumages, overall dark (but not black), with smooth gray underparts, a double white slash bracketing a dark face, and a super-sized black tail that is conspicuously and broadly tipped with white. Magnificent Hummingbird is jet-black below (adult males) or mottled green below (females and immature males), and its duller greenish tail lacks white in males and shows less flashy, more restricted gray tips in females.

In good light, Blue-throated's rump has a dull bronze sheen. On Magnificent, the rump is iridescent green and its call notes are less shrill.

BEHAVIOR: Flies swiftly up and down canyons with a loud warning "*seet!*" When you hear it, you have just enough time to turn and watch it pass. Sips nectar from larger flowers (such as mountain sage, columbine, and penstemon). Mostly insectivorous before the rainy season. Flycatches from perches, and during prolonged foraging flights maneuvers around trees and bushes, gleaning insects from foliage (also pilfers insects from spider webs). Dominates hummingbird feeders, and interactions between two Blue-throats using the same sets of feeders—not just the same feeder—can be physical, involving, in extreme cases, stabbing and stomping. Perched males begin calling near first light.

Less shy than Magnificent, and generally not shy at all.

FLIGHT: Seems huge in flight, with very long wings and wingbeats that, relative to smaller hummingbirds, seem to move in slow motion. When hovering, opens and fans its tail frequently, flashing white tips. Flight is fast, often spanning great distances, and distinctly undulating, with an abrupt upward rise at the end of each ascent. Males frequently vocalize when they fly. Birds also sometimes glide on set wings.

VOCALIZATIONS: Male's conspicuous song is a loud, whistled, somewhat monotonous "*soot soot seet*" or "*soot seet seet.*" Also makes a faint series of raspy chips, twitters, and hisses. Flight call is a single, loud, piercing "*seet!*" or "*seet seet.*" Also makes loud, sometimes rapid chips.

Magnificent Hummingbird, *Eugenes fulgens*

Black Beauty

STATUS: Fairly common but geographically and habitat-restricted; for several reasons, not as common at feeding stations as many other species. **DISTRIBUTION:** One of the "sky islands" inhabitants, breeding in the Chisos and Guadalupe Mts. of Tex.; the Chiricahua, Huachuca, Santa Rita, Santa Catalina, White, and Pinaleno Mts. of Ariz.; and the Animas, Peloncillo, Pinos Altos, and Sacramento Mts. of sw. and s.-cen. N.M., south into Mexico. **HABITAT:** Mixed oak-pine woodlands at 4,500–8,000 ft.; riparian forest in canyon bottoms. **COHABITANTS:** Blue-throated Hummingbird, Painted Redstart, Hepatic Tanager. **MOVEMENT/MIGRATION:** Most individuals vacate the United States in winter, but a few may be found at Arizona feeders. Spring migration from mid-February to early May; fall from mid-September to late November. VI: 2.

DESCRIPTION: A dark and beautiful giant among U.S. hummingbirds; can be confused only with Blue-throated Hummingbird. Large (only slightly smaller than Blue-throated), with a large head, a distinctly long straight (or mostly straight) bill, a long neck and body, and a moderately long tail. The head is distinctly and curiously shaped: sometimes shaggy with a raised crest, and sometimes peaked with a protruding bump to the rear (the bird looks hydrocephalic). The bill is key: longest in females, it appears half again longer than the bill of Blue-throated.

Adult males are dark green above, black below, with a purple crown and a bright green gorget. In most lights, however, except for the tiny white tear behind the eye, the bird appears all black—except for the very rare Green Violet-ear, the only all-black-looking hummingbird. Immature males have a shaggy semblance of the adult's plumage, including a dark chest and sparse greenish gorget. Females and immature females are green to golden green above, scaly grayish green below; a greenish tail has narrow white-tipped corners. They look very much like female and immature Blue-throated.

There are some key distinctions between female Magnificent and Blue-throated. Magnificent has an obviously long bill; the bill of Blue-throated seems short for the size of the bird. Magnificent has scaly greenish gray underparts; Blue-throated is plain uniform gray. Magnificent usually has a distinct white triangular teardrop behind the eye (and sometimes a pale trailing eye-line); Blue-throated has an obvious white eye-line and an obvious white mustache. The corners of Magnificent's tail are narrowly tipped with gray; the corners of Blue-throated's tail are broadly tipped with white (a contrast heightened because

Blue-throated's tail is black and female Magnificent's is green). On perched birds, the lower third of a Blue-throated's undertail is white; just the tip on female Magnificent's tail is white.

You can also play probability. Female Magnificents are considerably less common than males (estimates are eight males for every female). Female Blue-throats seem only slightly less common than males.

BEHAVIOR: A "trapline" hummingbird, Magnificent runs a circuit between nectar sources (as opposed to defending a single flower patch or feeder). Also highly insectivorous and less bound to nectar (and feeders) than many other hummingbird species. At feeding stations, most commonly seen early in the morning and late in the evening. Visits during the day may be very infrequent (more regular in winter).

Dominant at feeders among smaller hummingbirds except Rufous; subordinate to Blue-throated. Often announces its arrival with a short distinctive call. Fairly shy. Often reluctant to feed when observers crowd a feeder, and leaves at the slightest provocation.

FLIGHT: Magnificent's great size is apparent not only in its proportions but in the relatively slow-motion wings. Flight is strong, fast, and slightly undulating, with shallow and long waves.

VOCALIZATIONS: Call is a distinct, terse, clear, high-pitched "*chip*" or "*chee*"—sometimes uttered with a compressed two-note quality: "*t'chee.*" Not particularly loud, and not as shrill as the "*seep*" call of Blue-throated, it nevertheless carries. Chase call is a high-pitched stutter: "*r'eechchchch.*"

Plain-capped Starthroat, *Heliomaster constantii*

Lance-billed Hummingbird in a Miniskirt

STATUS: Rare but near-annual vagrant from Mexico. **DISTRIBUTION:** Breeds from the s. Sonoran Desert to Costa Rica. In summer, wanders into se. Ariz. **HABITAT:** Desert scrub, tropical woodlands, riparian corridors, oak forest. Partial to streamsides. **CO-HABITANTS:** Common Black-Hawk, Violet-crowned Hummingbird, Green Kingfisher. **MOVEMENT/MIGRATION:** Permanent resident in Mexico. Some post-breeding wandering. Most U.S. recorded occurrences were between June and September. VI: 1.

DESCRIPTION: A large, gangly, overall dull, and somewhat moth-eaten-looking hummingbird. Large (about the same size as Blue-throated), with an exceedingly long, slightly curved, lancelike bill, a large oddly shaped head (the bird looks like it has a hydrocephalic bulge on the rear crown), a long neck, a slender body, and a shortish tail with folded wings extending slightly beyond the tip.

Overall dull green or bronzy gray, but very conspicuously marked here and there with white—a broad white malar fluffy white tufts that sometimes protrude above the folded wing, a rumpled white rump, and a white longitudinal stripe parting the grayish underparts. A small reddish gorget is indistinct. Plumage for all ages and sexes is similar although immatures may lack red in the gorget.

BEHAVIOR: To the dismay of those who haunt hummingbird feeders in hopes of finding rarities, this bird much prefers insects to sugar-water. Often perches on the tops of trees. Hunts by making extended hovering forays (often over streams), sallying back and forth in shallow pendulum swings to snatch insect prey.

Does not frequent feeders as often as many other hummingbirds, and does not defend feeders. Owing to the extreme length of their bills and the design of most feeders, starthroats stand distinctly and uncomfortably erect at feeders, trying to get their long bills into the opening.

FLIGHT: Aptly described as swiftlike, with long, slender, sweptback wings and a disproportionately short tail. Wingbeats seem slow or sluggish. The bird habitually wags and fans a short miniskirt-like tail with a gray-tipped border. In flight, a series of wingbeats is often followed by a long, slightly unstable, swiftlike glide.

VOCALIZATIONS: Mostly silent. Song said to recall Magnificent Hummingbird. The call, which has been described as a flycatcher-like "*tsip*" or "*teep,*" is heard when the bird is out to snatch prey—hovering, turning, searching, and sallying out to intercept another target.

PERTINENT PARTICULARS: The size and very long bill (never mind all the white spattering) make this species fairly easy to identify. Blue-throated Hummingbird is as large as Starthroat, but has a short bill and long tail (Starthroat has the opposite combination) and, except for the face and outer tail tips, has no white on the body.

Lucifer Hummingbird, *Calothorax lucifer*

Fork-tailed Agave Sipper

STATUS: Rare and highly restricted breeder and wanderer. **DISTRIBUTION:** In the U.S., this Mexican species is found primarily in Big Bend National Park in Tex. and in extreme se. Ariz. and sw. N.M. **HABITAT:** Steep rocky desert canyons and hillsides dotted with ocotillo, sotol, century plant, and other hardy desert species. **COHABITANTS:** Black-chinned Hummingbird, Rock Wren, Varied Bunting, Rufous-crowned Sparrow. **MOVEMENT/MIGRATION:** Breeder only. Returns to the United States from mid-March to mid-April; departs late August to late September (extreme arrival and departure dates are uncertain). Birds ascend to higher altitudes to forage in July, then descend in August. VI: 1.

DESCRIPTION: A small, curve-billed, fork-tailed desert hummingbird with a unique shape that seems to have been forged from ill-fitting parts. Small (slightly smaller than Black-chinned; slightly larger than Calliope) exceedingly top-heavy hummingbird with a large head, a long, fairly heavy, down-curved bill, a small body, and a long, exceedingly thin, scraggly, fork-tipped, or just pointy tail that gives the bird a spindly, pointy posterior. The tails of females and immatures are shorter but still extend well beyond the wingtips when perched. Perched birds appear to slouch and look dejected. Even in flight, the tail is usually closed.

Adult males are dingy green above and greenish with a white collar and rusty touches on the flanks below. The gorget looks like a large, bushy, purple beard. The tail is dark. Females and immatures are dull greenish above, and lavishly washed with salmon or cinnamon below and on the face and neck. (Young males get spotty touches of purple gorget late in the summer.)

Adult males most closely resemble Black-chinned Hummingbird (which also have long and down-curved bills), but the bill on Lucifer is no-nonsense curved; Black-chinned's bill is equivocally curved. Black-chinned also has a trim clean-cut gorget; Lucifer's gorget looks like it was slept in.

Another bird Lucifer might be confused with is the similarly sized Costa's Hummingbird, which also has a shaggy purple gorget. Costa's is a compact, short-billed, short-tailed bird (not top-heavy and scraggy-looking like Lucifer's). Also, the billowing gorget on Costa's overflows onto the head and face. Lucifer just has a purple beard.

Female and immature Lucifers are not easily mistaken for anything else. The salmon-tinged face, coupled with a distinctly down-curved bill, is unique. Nevertheless, the lavish extent of the color on the underparts will probably distinguish the bird as "something different" before particulars (like the bill and the long thin tail) are noted.

BEHAVIOR: When in bloom, favors agave above all other flowers. In early spring, targets Mexican buckeye. Later, when flowers emerge, birds are drawn to penstemon or woolly paintbrush. But when the agaves bloom in May, that's the nectar source—in fact, Lucifers are so dedicated to the flower that when blooms appear at higher elevations in July, they follow (accounting for the birds' altitudinal migration and their appearance at feeders at higher elevations where they did not occur earlier in the season).

Comes readily to hummingbird feeders, but is fairly timid and generally bullied by other species (especially Black-chinned Hummingbird).

FLIGHT: Distinctly front-heavy, short-winged, long-tailed profile. Rarely flares its tail.

VOCALIZATIONS: Call is dry and high-pitched, recalling Broad-tailed. When excited, call is given in a rapid series.

Ruby-throated Hummingbird, *Archilochus colubris*

The Default Eastern Hummingbird

STATUS: Common nesting species—the only hummingbird species native to the eastern United

States and Canada; rare to uncommon from southern Florida to North Carolina and west along the Gulf Coast to east Texas in winter. **DISTRIBUTION:** Breeds across s. Canada from the Maritimes across s. Que., Ont., Man., and Sask., north across cen. Alta. to the B.C. border. In the U.S., breeds in all states east of a line drawn across the e. Dakotas, e. Neb., e. and s.-cen. Kans., cen. Okla., and e. Tex. to the Mexican border (does not breed in extreme s. Fla.). Casual in the West. **HABITAT:** Mixed and deciduous woodlands, especially edge, openings, stream borders, and roadways. Forages widely in suburban gardens. **COHABITANTS:** Many and varied, but includes Eastern Wood-Pewee, Red-eyed Vireo, Blue Jay. **MOVEMENT/MIGRATION:** Except for a few lingering individuals in the South, the entire population retreats to southern Mexico and Central America (also extreme southern Florida) after breeding. Spring migration from early March to late May; fall from early August to early November, with a peak in late August and September. VI: 2.

DESCRIPTION: It's the only hummingbird commonly found in the East. Smallish and stocky (same size as Black-chinned, Rufous, and Allen's Hummingbirds), with a straight medium-length bill, either a long deeply forked tail (adult males) or a shortish tail (females and immatures) that is slightly longer than the folded wingtips. Folded wings are narrow, saberlike, and pinched or constricted at the tips.

Adult males are greenish above and white-chested, with grayish green flanks and belly below. A sharply defined black gorget is confined to the throat and flashes a metallic ruby red in sunlight. The plumage of adult females and immatures is very similar to Black-chinned Hummingbird (bronzy green above; whitish below), but shows more contrast between the dark head and face and the white throat and has generally whiter underparts and a darker (greener) crown. The species are best separated by the shape of the wingtips. The gorget of immature male Ruby-throateds shows reddish spotting by fall (purple spots appear on immature Black-chinneds).

Note the absence of any red in the tail.

BEHAVIOR: An inveterate flower kisser that is especially partial to red and orange tubular flowers but also snaps up insects in the air and from spider webs and is reported to drink sap from sapsucker wells. Commonly perches high on an exposed (often dead) branch between feeding trips, but in rainy weather may huddle low in vegetation near flowers or feeders.

Aggressively defends its nectar source from other Ruby-throateds, but subordinate to Rufous and Allen's where both species occur together. Attracted easily to artificial feeders, but somewhat shy about coming to feeders when people are present until acclimated to their presence.

FLIGHT: Like other hummingbirds. While hovering, flutters its tail sparingly or not at all (in other words, not a compulsive tail flutterer like Black-chinned).

VOCALIZATIONS: Call is a terse, soft, muttered spit or "*cht*" (sounds like a cross between a squeak and a skuff). Chase call is a sharp snarl followed by a rapid wheezy stutter: "*zzzzz chikachikachika.*"

PERTINENT PARTICULARS: If you are watching a hummingbird east of the Mississippi River and the calendar says it's sometime between April and July, you are almost certainly looking at Ruby-throated. From August into December and throughout the winter in states bordering the Gulf Coast, vagrant western hummingbird species (most notably but not exclusively Rufous, Allen's, and Black-chinned) may turn up. Rufous, Allen's, Calliope, and Broad-tailed show red-orange in the tail (although it may be limited in some individuals). Black-chinneds are generally slimmer and slightly longer-necked than Ruby-throated, and have longer and slightly down-curved bills. After September, any slender long-billed hummingbird that turns up in the East should be scrutinized closely. The shape of the wingtips is key. Narrow, fairly straight wingtips that are pinched at the tip are classic Ruby-throated. The sight of a wide wing that is conspicuously up-turned toward the wide or bulbous tip is grounds for a call to a local birding expert to help confirm a possible Black-chinned.

Black-chinned Hummingbird, *Archilochus alexandri*

The Dapper Hummer

STATUS: Common widespread western hummingbird; particularly common at lower elevations in the Southwest. **DISTRIBUTION:** Breeds from s. B.C., e. Wash. and Ore., most of Idaho, extreme w. Mont., most of Utah, Nev., sw. Wyo., and w. and s. Colo., with a scattered distribution throughout Calif., all but s.-cen. Ariz., N.M., all but northern and eastern portions of Tex., and south into cen. Mexico. Winters in small numbers along the Gulf Coast from Tex. to Fla. Casual elsewhere in the East. **HABITAT:** Fairly generalized. In arid regions, prefers riparian woodlands (canyons or flood plains) dominated by sycamore and cottonwoods. Also favors pinyon-juniper, oak shrub lands, salt cedar, orchards, and urban and suburban areas with tall trees. Readily attracted to hummingbird feeders and flower gardens. **COHABITANTS:** Varied. **MOVEMENT/MIGRATION:** Most Black-chinneds winter in western Mexico. Spring migration from March through mid-May; fall from late July to early October, with a peak in August and early September. Some post-breeding dispersal to higher elevations has been noted prior to departure to Mexico. VI: 3.

DESCRIPTION: A slender, slightly proportioned, long-billed hummingbird—very similar in size and proportions to the eastern Ruby-throated Hummingbird. Medium-sized (slightly smaller than Anna's and Broad-tailed; slightly larger than Costa's and Calliope), with a small flat-topped head, a thin neck, a well-proportioned body, and a tail that extends just beyond the wingtips. The disproportionately long bill is slightly to moderately down-curved. The primaries of the folded wings are wider than on Ruby-throated, more clublike than saber-like, and roundly blunt, not pinched, at the tip.

Green-backed adult males are exceptionally dapper and easily distinguished by the narrow, crisply defined, bright white collar separating the all-black head and the dappled greenish-vested underparts. The amethyst throat-band is narrow, confined (does not flair or bulge to the sides), and only visible in good light.

Females have greenish backs, a duller, grayer forehead and midcrown, a whitish or grayish face, a white throat, and pale grayish underparts—overall a very plain uncontrastingly plumaged hummingbird. Overall uncontrasting plainness, in combination with size, slight proportions, and, especially, a long down-curved bill helps distinguish this species.

Immatures are much like females. Males show a heavy "five o'clock shadow" on the face and usually show incidental or clustered purple spotting low on the throat; immature females have very gray crowns and are usually very white below.

BEHAVIOR: A habitual tail wagger—wags (and flares) tail when hovering, and sometimes continues to wag when perched at a feeder. Forages on flowers from ground to canopy; also hawks insects. Pugnacious, but usually subordinate to similar-sized and larger species.

FLIGHT: Nimble and active; when hovering, habitually wags its tail. Tail is all black on adult males; tails of females and immatures have white outer tips (no rufous panels near the base).

VOCALIZATIONS: Call is a soft, terse, low, dry, muttered spit or "*tsk*" or "*ch*" (similar to Ruby-throated). Chase call is a low wheezy snarl that degenerates into a wheezy stutter: "*sneechchcheechchch.*"

Anna's Hummingbird, *Calypte anna*

Raspberry-headed Hummingbird

STATUS: Common far-western breeder and resident; the most common and familiar hummingbird species along the California coast. **DISTRIBUTION:** Found primarily along the West Coast and in the Southwest. Breeds from w. B.C. to Baja. Ranges east to the borders of the western coastal states and expands eastward into w. and s. Nev., across all but ne. Ariz., s. and cen. N.M., and extreme w. Tex., south into n. Mexico. In winter, slightly more widespread; birds expand into nw. Mexico. Casual north to Alaska and the e. U.S. **HABITAT:** Historically riparian habitats, chaparral, savanna, and coastal scrubland, but has become increasingly urban and suburbanized; attracted to gardens and hummingbird feeders year-round.

COHABITANTS: Varied, but include Rufous and Allen's Hummingbirds, Black Phoebe, Bushtit. **MOVEMENT/MIGRATION:** Some birds disperse to higher elevations and into the Southwest after breeding (December to May); return to breeding areas after leading a semi-nomadic existence (but some stay put year-round). Nomadic period begins in May and continues into early January. VI: 2.

DESCRIPTION: A darkish, portly, somewhat shabby hummingbird—but the entire head of the adult male blooms at the touch of sunlight. Medium-sized (slightly longer than Allen's or Black-chinned; the same size as Broad-tailed), with a large squarish head, a short to medium-length straight bill, a portly, even potbellied body, and a somewhat variable-length tail—long on adult males (extending well beyond the wingtips), shorter on females (extending just beyond the wingtips).

In all plumages, greenish above, grayish green below (many individuals appear drab or dingy). The male's throat, face, and crown are entirely iridescent in good light; otherwise, the entire head looks dark and loosely hooded. The iridescence shines rose, copper, or gold, depending on the light. Adult females and immature males have darkly beaded throats and a smattering of iridescent rose-colored feathers (usually clustered in the center of the throat). Immature females are more smoothly grayish below and lightly beaded on the throat, with no iridescent feathers. Except for a stouter build, a shorter, straighter bill, and a slightly grungier appearance (especially showing a more mottled or marbled pattern on the sides), female Anna's is most similar to female Black-chinned.

BEHAVIOR: Unlike Black-chinned, Costa's, and (to a lesser degree) Broad-tailed, does not flutter its tail when hovering. Sips nectar from flowers and also picks and hawks insects out of the air, most commonly in winter. Vigorously defends its own flower patch, but at feeders sometimes appears less aggressive than some other species. Prefers open sunny areas.

FLIGHT: Fast and direct.

VOCALIZATIONS: Song consists of two (sometimes three) distinct parts—an introductory series of evenly spaced, low, raspy, squeaky stutters, followed by a series of zinging pendulum-swinging notes, one down-sliding, the next two up-sliding: "*z'uh z'uh z'uh ZEEer zeEEH zeEEH ZEEer zeEEH zeEEH.*" Overall voice quality sounds like a cross between a squeak and a sizzle. Call is a high sharp "*chih*" or "*chit*" (higher-pitched than Black-chinned or Broad-tailed); also gives a high-pitched "*seep*" that recalls Blue-throated Hummingbird. Chase call is a rapid, squeaky, raspy chatter with a compressed two-note quality: "*tchi tchitchi tchitchi tchitchi tchitchi.*" Wings make a fluttery sound like cards being shuffled. At the bottom of their dive, displaying males utter a squeaky "*pop.*"

Costa's Hummingbird, *Calypte costae*

Desert Dumpling with a Fu

STATUS: Common southwestern resident and breeder. **DISTRIBUTION:** Summer resident across s. Calif. (north to Lompoc and the San Joaquin Valley), Baja, and s. Ariz. south into n. Mexico. Breeding range extends up along the Calif.–Nev. border to Death Valley National Park, across s. Nev., extreme sw. Utah, and w. and s. Ariz. Small numbers winter in s. Calif., s. Ariz., and s. Nev. Casual north to Alaska and the Rocky Mountain states. **HABITAT:** Dry and desert habitats, in particular desert scrub in the Sonoran and Mojave Deserts and, in coastal California, chaparral and coastal scrub. In the Sonoran Desert, breeds along desert washes lined with palo verde and acacias and on rocky hillsides rich in ocotillo and palo verde; in the Mojave Desert, partial to riparian areas. **COHABITANTS:** Black-chinned Hummingbird, Verdin, Wrentit, California and Black-tailed Gnatcatchers. **MOVEMENT/MIGRATION:** A late-winter and spring breeder (mid-January to mid-June). Where not a permanent resident, vacates breeding areas from April to mid-June (some apparently go to higher elevations); returns November to late March. VI: 2.

DESCRIPTION: Tiny hummingbird with a gorget so oversized that it droops like a purple "Fu Manchu" mustache. Small (slightly smaller than Black-chinned; larger than Calliope; about the same size as or slightly larger than Lucifer), compact, and dumpy hummingbird with a large, round, neckless head, a short, fine, straight or slightly drooped bill, a plump slightly humpbacked body, and a tiny nub of a tail. Looks like a short-billed, short-tailed dumpling with wings that extend just beyond the tip of the tail.

Dull green or green-gold above. Males have a dark green vest; females and immatures are immaculately gray or white below (no buff!). The purple gorget of the adult male hangs like an exaggerated purple mustache. It's so eye-catching that many observers fail to notice that the crown is likewise purple (black in poor light). Anna's is the only other small hummingbird with iridescence extending beyond the throat and onto the crown. Half of adult female Costa's Hummingbirds show a cluster of purple spots on the center of the throat; by late summer, immature males already show a sparse "Fu."

Adult males can be confused only with adult Lucifer, whose full purple gorget also billows to the side. But Lucifer is long-billed, long-headed, long-necked, and long-tailed—anything but dumpy. Female and immature Black-chinneds are longer-billed, and Anna's has dingy, not clean, underparts. Calliope males have a sparse beard, and females have rufous flanks.

BEHAVIOR: A fairly reticent, almost retiring, hummingbird. Seems more deliberate, not as high-strung, and certainly not as aggressive as Black-chinned. At feeding stations, picks a favorite feeder (often away from the rest) and waits until it's vacant—does not aggressively defend. In natural settings, feeds heavily upon ocotillo, chuparosa, and other desert flowers. Considerably suburbanized (although not as much as Anna's), particularly in California, where it comes readily to hummingbird feeders as well as non-native flowers. Also flycatches.

An intermittent tail pumper. Often feeds with the tail unmoving and angled down. When pumped, tail moves in a regular rhythmic manner.

FLIGHT: Like other hummingbirds. Often perches only a short distance from a nectar source.

VOCALIZATIONS: Call is a high, sharp, abrupt "*tik*" that is often repeated in a spaced series that recalls a Bushtit's chatter. The call is very different from the soft spit of Black-chinned.

Calliope Hummingbird, *Stellula calliope*

Thimbelina with a Sparse Beard

STATUS: Fairly common and fairly widespread western breeder. **DISTRIBUTION:** From cen. and e. B.C. to w. Alta., south through cen. and e. Wash. and Ore., n. and e. Calif. (also found in scattered elevated parts of s. Calif. and n. Baja), most of Idaho, w. Mont., nw. Wyo., Nev., and Utah. Winters in sw. Mexico. Casual east, mostly at feeders in fall and winter. **HABITAT:** A montane breeder, ranging from 3,500 ft. to the tree line. Breeds in young regenerating growth following fires or logging; also in aspen and willow thickets along streams. In spring migration, appears in desert washes, at lower elevations, and (in California in the mountains) along the coast. Fall migration is conducted largely at higher elevations. **COHABITANTS:** Broad-tailed Hummingbird, Rufous Hummingbird, Mountain Bluebird, Lincoln's Sparrow. **MOVEMENT/MIGRATION:** Wholly migratory. Spring migration from March to May; fall from early July to late September. VI: 3.

DESCRIPTION: Someone has to be the smallest. Meet North America's tiniest breeding bird, a compact water droplet–shaped hummingbird with an abbreviated straight bill and a short tail. Short wings barely extend beyond the tip of the tail. In comparison to other hummers at a feeder, the bird's small size helps distinguish it, and the short thin bill confirms the identification. (On some immatures, the bill is only slightly longer than the length of the head!) The bird even slouches when it perches—projected dejection in response to its Lilliputian size.

Adult males, with greenish backs, green-vested, whitish underparts, and scraggly reddish beards streaking across both sides of the white face, are distinctive and unique. Females and immatures are golden green above, with a warm buffy or rufous wash across the breast and down the flanks. This pattern, coupled with the short tail, is unique. (All other buffy-flanked hummers, such as Broad-tailed, Allen's, and Rufous, have tails that project beyond the wings.) Also, the throat on Calliope is neatly and symmetrically beaded.

BEHAVIOR: Likes to stay low. Often perches on low branches, even where higher perches abound; frequently forages on ground-hugging nontubular flowers that other hummingbirds ignore. On territory, however, commonly sits high, often on dead twigs near the tops of willows or alders. Hawks insects and also visits the sap-wells of sapsuckers. Very defensive and territorial during breeding season; more retiring during migration, although adult males may defend favorite feeders vigorously and successfully from larger species. When feeding on flowers, does not wag its tail, but often cocks its tail up, at a right angle to the body. When perched, sometimes jerks wings and tail. In migration, generally silent.

FLIGHT: Overall tiny and compact, with short wings and a short, somewhat cropped or squarish tail. Flight is like other hummingbirds—fast, slightly swerving, with steady wingbeats. Perhaps because of its minute size, Calliope seems slower and more bumblebee-like.

VOCALIZATIONS: Generally silent. Call is a high soft "*chip*," also described as a thin "*tsip*."

Broad-tailed Hummingbird, *Selasphorus platycercus*

The Mountaineer That "Zings"

STATUS: Common widespread breeder in the southern and central Rockies and the mountains of the Great Basin. DISTRIBUTION: Found in extreme se. Ore., s. Idaho, sw. Mont., Wyo., Colo., Utah, Nev. (and bordering portions of e.-cen. Calif.), n. and e. Ariz., all but e. N.M., and trans-Pecos Tex. Winters in Mexico and Central America. Casual in the East, mostly at feeders. HABITAT: Breeds most commonly at higher elevations, including mountain meadows and foothills in northern portions of the range and boreal zone forests on higher mountains farther south. Habitats include pinyon-juniper, ponderosa pine, Englemann spruce, subalpine fir, streamside willows, pine-oak woodlands, and aspen groves. Males (occasionally females) are seen foraging above the tree line. COHABITANTS: Common Raven, Mountain Chickadee, Pygmy Nuthatch, Western Tanager. MOVEMENT/MIGRATION: Virtually the entire population vacates the United States in winter (a few winter in the Southeast). Spring migration from early March to late May; fall from late July to early October. VI: 2.

DESCRIPTION: A robust hummingbird best known for (and easily recognized by) the metallic trill given by the wings of adult males in flight (except when in molt). Medium-sized (larger than Black-chinned; much larger than Calliope) and overall stocky, with a big head, a wide body, and a long, pointy, candle flame–shaped tail that extends beyond the wings when the bird is perched (as do the tails of Rufous and Allen's). An average-looking, almost prototypically proportioned hummingbird, if a little husky. The bill is moderately long (one and a half times the width of the head) and usually very slightly curved but sometimes straight. In overall proportions, Broad-tailed appears most like Anna's, but is not so big-headed or potbellied and is much longer-tailed.

In all plumages, upperparts are bright green. The male's red gorget is large, confined to the throat, and unique in its rose magenta color. Seems metallically brighter than other hummingbird gorgets (perhaps because of the bracketing white breast). Even when the bird is turned away from the sun, the deep reddish hues are evident (unlike male Ruby-throated Hummingbird, whose gorget turns black away from light). Females and immatures have a shadow pattern of the male's gorget—outlined in neatly beaded green dots and a buffy wash down the flanks. Throats may be so spotted that the face appears darkened by a "five-o'clock shadow." The white tips to the tail are often displayed as birds flash their tails. The orange base to the outer tail feathers is more limited than in Rufous or Allen's, and barely evident when birds are perched.

BEHAVIOR: Forages on a variety of flowers, frequently in mountain meadows (at lower altitudes during migration). Fairly mild-mannered and less aggressive than Rufous or Allen's, which commonly dominate at feeders. Not a compulsive tail wagger, although birds open and flash their tails when approaching a feeder (displaying white outer tail tips).

FLIGHT: Fast, straight, and slightly undulating, with an abrupt rise at the end of the descent.

VOCALIZATIONS: Call is a high-pitched "*tchih*" that is slightly higher-pitched than Black-chinned, Rufous, and Allen's. Also makes a somewhat soft and reserved "*chihchihchih.*" The zinging electric trill made by the male's wings in flight is almost unique.

Rufous Hummingbird, *Selasphorus rufus*

The Devil in Feathers

STATUS: Common western breeder and migrant; uncommon and restricted in winter; rare but known vagrant elsewhere in North America. **DISTRIBUTION:** Breeds in the Northwest from s. and se. Alaska across the s. Yukon, all but ne. B.C., sw. to s.-cen. Alta., Wash., Ore. to the Calif. border, the northern and western two-thirds of Idaho, and w. Mont. Winters primarily in Mexico, with northern limits reaching s. Calif. Also, increasingly found along the Gulf Coast from Tex. to Fla.; more rarely at feeders in the Northeast. **HABITAT:** Breeds in brushy or second-growth openings in coastal and interior forests. In fall migration (beginning in early July), favors alpine meadows. In winter in the United States, most commonly found in suburban gardens supplemented by hummingbird feeders that offer protectively dense vegetation for roosting. **COHABITANTS:** Often breeds in the same habitat as Calliope Hummingbird. **MOVEMENT/MIGRATION:** The entire breeding population vacates breeding areas in winter, and birds move south along two routes: the Pacific Coast and through the Rockies. In spring, most birds follow the coastal route north. Spring migration from early February to late May; fall from late June to mid-October. VI: 3.

DESCRIPTION: A hummingbird celebrated (and reviled) as much for its fiery demeanor as for its red-hot raiment. Moderately small (larger than Calliope; the same size as Black-chinned) and somewhat top-heavy, with a stocky upper body (neckless head, short straight bill, compact body, shortish wings) and a long, pointy, withered-looking tail that projects beyond the wingtips. Calliope is overall stubby; Broad-tailed is robust, with a long, broad, hefty tail.

In all plumages, red-orange is lavishly applied as a warning to other hummingbirds to back off. Rufous shows more extensive and brighter tones than any other hummingbird species except Allen's. The classic adult male is copperplated above and below with dark greenish wings, a lavish, almost billowing, fiery red-orange gorget (sometimes flashes green and gold), and a bracketing narrow white band across the chest. (*Note:* A small percentage of adult males have a mostly green back and look very much like adult male Allen's.) Adult females and immatures are green above and nominally white below, but the upperparts (particularly the rump) are infused with coppery touches; the flanks and undertail are heavily and broadly burnished with copper orange; the usually heavily spotted gorget shows a cluster of iridescent red spots in the center of the lower border (absent on some birds); and the flared tail is dominated by an extensive hot orange core with a black-and-white border.

In short, even in nonbreeding adult plumage, Rufous burns hotter and brighter than other hummingbirds (except Allen's).

BEHAVIOR: All ages and sexes aggressively defend flower patches and feeders from other, often larger, hummingbirds, both during and outside the breeding season. When facing off against rivals or approaching an occupied nectar source, Rufous frequently hovers and fans its tail, displaying a warning red flag, and vocalizes in "chips and chitters." When feeding and hovering, both wags and fans its tail. When sitting, often jerks its tail (and sometimes wings) nervously.

Forages primarily on tubular flowers. Also captures insects by snapping them up in midair or plucking them from leaves and grasses. Where territories overlap with Calliope Hummingbirds, Calliope frequently forages on ground-hugging blossoms.

FLIGHT: In adults, the fanned tail (most easily studied when fanned in a confrontational posture) shows feathers that are somewhat broader and less spindly than Allen's—particularly the outermost tail feather. In reality, these characteristics are often evident only in photos (or with birds in the hand). Also, in adults, the tail feathers just left and right of the two central tail feathers have a pinched tip, like the point of a fountain pen. On Allen's, the tips are evenly tapered (structurally no different from the rest). Immature birds, especially immature females, are indistinguishable in the field. Flight is like other hummingbirds, but more extensive reddish plumage is evident in flight. The wings of adult males make a high trilling sound, similar to the ringing sound made by the wings of male Broad-tailed but more muted, less musical and pleasing—it sounds in fact like an angry bee and makes you want to duck.

VOCALIZATIONS: Call is a sharp hard "*tchip.*" Battle cry is a shrill "*eeee,*" followed by a hissing, stuttering, sharp-edged "*zikity zikity zeek*" or "*zee chip-pitty chippity*" (sounds like a cross between a Black-chinned battle cry and a dentist's drill).

PERTINENT PARTICULARS: Although Allen's Hummingbird's breeding range is more coastally restricted and most birds follow a coastal migration route to wintering areas in south-central Mexico, some unknown percentage of the population does pass inland, where they mix with Rufous Hummingbirds. It has been estimated that 5–10% of Allen's and Rufous–type hummingbirds in southeastern Arizona in midsummer are Allen's.

Allen's Hummingbird, *Selasphorus sasin*

Selasphorus with Green and Red Sauce

STATUS: Fairly common breeder, but geographically restricted—virtually endemic to California. Uncommon fall migrant in the Southwest (July to September). Casual vagrant in the East (but owing to the likelihood of confusion with Rufous, perhaps more common than is recognized). **DISTRIBUTION:** Breeds along the Pacific Coast from s. Ore. (Curry Co.) to n. San Diego Co. and the Channel Is. (where it is a year-round resident). **HABITAT:** Moist coastal scrub and riparian forest (including introduced eucalyptus) between sea level and 1,000 ft., but within the reach of summer sea fog. **COHABITANTS:** Pacific-slope Flycatcher, Steller's Jay, Wrentit, Hooded Oriole. **MOVEMENT/MIGRATION:** Among the earliest of migrants—both spring and fall. Spring migration from early January to mid-March; fall from mid-May to late September. *Note:* Allen's and Rufous migration periods overlap, but Allen's is earlier in each period. The movement of Allen's to and from wintering grounds in south-central Mexico is thought to be almost entirely coastal in spring, but a certain percentage of fall migrants cut the corner, flying across the interior Southwest. VI: 2.

DESCRIPTION: Hot copper tempered by a cool green mantle—but in shape, plumage, and temperament, differs from Rufous Hummingbird only to feather-splitting degrees. Smallish (larger than Costa's; smaller than Anna's; the same size as Rufous) and top-heavy, with a large neckless head, a stocky body, and a long, thin, spindly tail that projects beyond the tips of the shortish wings.

Adult male is hot reddish orange with a green crown and back and a white-trimmed billowing orange-red gorget (sometimes flashes green and gold). *Note:* Some adult male Rufous Hummingbirds heralding from Oregon—where (wouldn't you know it) the ranges of Rufous and Allen's overlap—also have green backs.

Females and immatures are greenish above, paler below, and lavishly and conspicuously washed with buffy orange on the sides, tail, and undertail (also and variously elsewhere). The gorget often shows a smattering of bright reddish orange feathers, commonly clustered in the lower middle. The extent and brightness of its warm plumage usually distinguishes Allen's from other hummingbirds except Rufous. Calliope has a pale peachy wash on the flanks. Female and immatures Broad-taileds may have ruddy-washed flanks, but the red in the tail is segregated into flanking twin spots at the base—it never appears as a hot red core to the entire tail.

BEHAVIOR: Territorial during breeding season and in migration, but seems not quite as aggressive as Rufous. (Rufous/Allen's–type hummingbirds that deign to share a hummingbird feeder with another hummer—as opposed to driving it off—are often suspected of being Allen's.) Feeds by hovering at flowers and feeders; also pursues insects by hawking and gleaning from foliage. Defends fairly small flower-rich areas and feeds frequently.

FLIGHT: Silhouette seems short-tailed. Flight is like other hummingbirds. In face-offs with other hummingbirds, and when hovering, frequently fans its tail, showing an extensive (but variable) reddish core (with a black-and-white border in females and immatures). At close range, in adult males, the tail feathers flanking two central tail feathers are symmetrically pointed, not notched or pinch-tipped, as seen on male and female Rufous. Also, Allen's outer two tail feathers are slightly narrower and more spindly than on Rufous (but this is very difficult to perceive in the field). Immature Allen's and Rufous Hummingbirds are indistinguishable in the field based on present knowledge.

VOCALIZATIONS: Same as Rufous. Wings of males also make a shrill trilling.

TROGONS

Elegant Trogon, *Trogon elegans*

Sycamore Trogan

STATUS: Uncommon tropical species whose breeding range just brushes the United States. **DISTRIBUTION:** In the U.S., se. corner of Ariz. and extreme sw. corner of N.M. Extends south through Mexico; also occurs in Central America. Casual in w. and s. Tex. **HABITAT:** Occupies a range of habitat in Mexico, but in the United States found almost exclusively in riparian canyon woodlands dominated by sycamores in association with pinyon-pine and juniper. **COHABITANTS:** Arizona Woodpecker, Dusky-capped Flycatcher, Sulphur-bellied Flycatcher, Painted Redstart. **MOVEMENT/MIGRATION:** Over most of its range, a permanent resident. Migratory in the United States. Arrives in April and May; departs mid-September to November. Has been known to overwinter in Arizona (and south Texas). VI: 2.

DESCRIPTION: Uniquely shaped bird with resplendent plumage. A large (slightly larger than a jay), slender, long-tailed bird. The head is large and round, the bill conspicuously short, wide, and bright yellow, the body plump, and the tail long (constitutes half the bird's length), untapered, and squared and flared at the tip.

The adult male looks like a bird painted by a color-struck four-year-old. The head, breast, and back are iridescent green. The face is black, the wings bluish, the belly red, and the tail oxidized copper (greenish and rufous) above, broadly banded gray-and-white below. The red ring encircling the eye and the white seam that knits the greenish chest and red underparts hardly warrant mention (except to point out that in poor light the white line across the chest is often the most visible field mark). Females are more subdued but patterned like males. Overall browner, females have a yellow bill, a white patch behind the eye that stands out against the black face, a white band across the chest, a reddish belly (not as bright as the male's), and a copper-colored tail above, pale gray-and-white bands below. Immatures are like females, but grayish below (still showing a narrow white chest band), with lavish white spotting on the wings.

BEHAVIOR: Forages like it has brain damage. Sits upright and immobile for extended periods except for the head and up-turned eye, which moves with smooth and excruciatingly slow turns to survey the surrounding foliage. The bird then suddenly goes berserk, throwing itself against the foliage in a fluttering frenzy of wings and fanned tail, grabs a large insect (or plucks fruit), and takes a perch, once again composed. Does most of its foraging below the canopy, often below 20 ft. Vocalizing males may be much higher in the trees (and males vocalize frequently).

Fairly tame; can often be approached to within 50 ft. Commonly perches close to the trunk and, given its sedentary feeding style, is easy to overlook despite the amazing plumage. Responds to pishing (when you happen near a nest) and, at all times of year, to imitations of Northern Pygmy-Owl.

FLIGHT: A big, long-tailed, wing-fluttering bird flashing white outer tail feathers. Flight is loose and deeply undulating (bounding for short flights); despite the hurried, almost frenzied beating of the wings, the overall passage is smooth, almost buoyant.

VOCALIZATIONS: Classic call is a measured series of loud, low, throaty croaks: "*raah, raah, raah, raah, raah raah.*" Sometimes sounds like a distant dog barking. Also gives a higher-pitched version that has a compressed two-note quality: "*wh'renk, wh'renk, wh'renk.*" Alarm call is a descending chuckle that sounds somewhat amphibian-like or like an accelerated Gray Squirrel alarm.

Eared Quetzal, *Euptilotis noexenus*

Mountain Quetzal

STATUS AND DISTRIBUTION: Endemic to the mountains of w. Mexico; ranges north to within 100 mi. of the U.S.–Mexico border. A few wander irregularly to the mountains of se. Ariz. where they have nested (and wintered). **HABITAT:** Pine forests on mountain slopes at elevations of 6,000–10,000 ft. Occasionally visits the riparian canyons more typical of Elegant Trogon. **COHABITANTS:** Thick-billed Parrot, Steller's Jay, Hepatic Tanager. **MOVEMENT/MIGRATION:** Nonmigratory, but if birds can reach the Huachuca, Chiricahua, and White Mountains, they are clearly capable of more than local movement. VI: 0/1.

DESCRIPTION: A small-headed, wide-bodied quetzal with a dark bill and no white on the breast. Large trogon (larger than Elegant Trogan; smaller than Cooper's Hawk; about the same size as Band-tailed Pigeon), with a small black bill, an undersized head, a big wide body, and a long wide tail that tapers toward the square tip. The shape may recall a wide-bodied neckless pigeon, but one that perches as though its back is broken. Elegant Trogon is overall more slender and has a large head, a yellow bill, and a long narrow tail that flares toward the tip (rather than tapers).

Overall dark—dark green and blue above and on the head and breast, with a red belly; the white underside of the tail has a narrow dark center. Note the absence of a ring around the eye (the dark eye is very hard to see) and the lack of a narrow white band between the dark green breast and red belly. Females and immatures are like males but show gray heads and breasts, not green.

BEHAVIOR: Routinely crosses canyons in pine forests to forage singly, in pairs, or in small family groups. Feeds on large insects and fruits (especially madrone berries), usually at mid and upper levels of trees. Feeds by making short, sallying flights to pluck prey from branches (sometimes insects in flight). Responds to imitations of its skree-chuck (distress?) call.

FLIGHT: Profile shows a long broadly tubular body and long broad wings. Entirely dark above, red and white below. Flight is undulating, with a slight rocking or listing, and often more protracted than the short-distance flight common for Elegant Trogon.

VOCALIZATIONS: Song is a lengthy series of "quavering whistles" that begin softly and swell in volume (very unlike the low croaking of Elegant Trogan). Distress call is a loud ascending squeal ending with an abrupt chuck note: "*skreeCHuk!*" (often repeated).

KINGFISHERS

Ringed Kingfisher, *Ceryle torquata*

Red-bellied Kingfisher

STATUS AND DISTRIBUTION: Common resident along the Rio Grande downriver from Laredo, Tex. Less common away from the river valley, but at times wanders north to cen. Tex. **HABITAT:** Primarily along wooded banks of the Rio Grande and nearby ponds and resacas. Also commutes overland between aquatic habitats (often very high).

COHABITANTS: Green Kingfisher, Great Kiskadee, and (in winter) Belted Kingfisher. **MOVEMENT/MIGRATION:** Permanent resident. VI: 1.

DESCRIPTION: A shaggy-crested, robust, red-bellied kingfisher almost the size of a Pileated Woodpecker. Ringed Kingfisher closely resembles Belted Kingfisher but is considerably larger. Ringed also has a distinctly heavier bill and more girth to the body.

Upperparts are all blue (like Belted); underparts are mostly rufous (very unlike Belted). The female's blue breast-band pales to spectral insignificance. The white collar is very conspicuous and is often what draws your eye when birds are perched against a riverbank. At a distance, the pale base of the bill projects well, making birds seem pale-billed. (Belted Kingfisher is always dark-billed.)

BEHAVIOR: A noisy kingfisher. Usually vocalizes (rattles) as soon as you step into view. Commonly sits conspicuously on a sturdy branch along a riverbank. Resting birds may be only 3–4 ft. above the water. Hunting birds are often higher; more commonly 15–35 ft. Often remains on the same perch for extended periods. Unlike Belted Kingfisher, does not commonly hover. Usually solitary or seen in pairs.

A shy bird, Ringed often flushes when observers are distant but does not commonly travel far (only a couple of hundred yards or until out of sight). Usually flies within 50 ft. of the surface, but when crossing land, may be much higher.

FLIGHT: Large balanced profile, showing broad wings and a projecting heavy bill/head and tail. The size (and perhaps the shape and colors) may recall Green Heron. The rufous underparts (belly on males, belly and underwings on females) are easily noted. All-blue upperparts usually flash white patches near the tips of the wings, but this mark may be very hard to see on some individuals. From below, the trailing edge of the wing appears translucent.

Flight is strong, straight, and overall smoother and less jerky, high-strung, and erratic than Belted Kingfisher. Wingbeats are deep and pushing—slower than Belted Kingfisher, but given in a similarly halting or irregular pattern of two or three measured wingbeats followed by one hurried beat.

VOCALIZATIONS: Ringed has a brittle stuttering rattle call that is similar to Belted Kingfisher but slower, lower-pitched, and more prolonged. Ringed also makes a chattering stutter, "*ch'lak*," a heronlike "*ah ah ah*," and in flight a low rough, gracklelike mutter, "*gr'ick*," or a low gruff snort, "*k'chak*."

Belted Kingfisher, *Ceryle alcyon*

The Aqua Kestrel

STATUS: Common and widespread near water; ranked among North America's most widely distributed birds and the only kingfisher found outside of south-central Texas and southeastern Arizona. **DISTRIBUTION:** Breeds across most of the U.S. and s. Canada. Absent as a breeder only in arctic regions, s. Fla., s. La., s. Tex., and the arid Southwest (se. Calif., s. Nev., s. Utah, Ariz., and all but n. N.M.). In winter, vacates northern interior regions of Canada, Alaska, and colder northern portions of the continental U.S. and ranges south across most of the U.S., Mexico, the Caribbean, and n. South America. **HABITAT:** Open fish-bearing water with or near dirt or clay banks for nesting. In winter, the availability of open, nonturbid, fish-bearing water (fresh or salt) largely determines its presence or absence. **COHABITANTS:** Where it breeds, often found near Bank Swallow colonies. In both winter and summer, Great Blue Heron. **MOVEMENT/MIGRATION:** Fall migration begins in August or early September. In spring, birds appear as soon as iced-over lakes and rivers thaw; thus, across vacated regions, spring migration spans from March to May, with a peak in April. Movements seem concentrated along coastlines and major water courses. VI: 4.

DESCRIPTION: A dagger-billed, shaggy-crested bird perched over water. Medium-sized (between Blue Jay and a crow), Belted Kingfisher is distinctive and easily distinguished by its oversized head, shaggy double-peaked crest, and oversized daggerlike bill. In all plumages, birds are slate blue above, white below. The head and back are separated by a white collar (which makes the bird's head seem disembodied at a distance); the breasts of adult males and immatures are creased by a single prominent gray-blue breast-band. Adult females have two bands—blue above, rust below.

BEHAVIOR: Except during the breeding season, solitary. Perches conspicuously on branches, utility lines, Osprey platforms, and bridges, usually over fresh or salt water. Dives for fish by plunging vertically or at an angle. Also hovers, like a stocky aquatic kestrel, 10–40 ft. above the surface. Nests in burrows it excavates in riverbanks or lakeside bluffs as well as in the sand or dirt banks associated with sand mining operations, sometimes a mile or more from water.

Very territorial and very aggressive toward other kingfishers encroaching on its territory. Also very vocal and very wary. Even distant kingfishers (and approaching observers) elicit a vocal response.

FLIGHT: The overall blocky profile is almost idiosyncratic, showing a massive head and bill, a short tail, and short, wide, blunt wings. Prominent white wing patches flash in flight, making birds hard to misidentify. The bird's flight is almost equally idiosyncratic—an aerial fingerprint. A series of slow, casual, down-stroked wingbeats are alternated with rapid sputters, making flight direct and not particularly hurried, but somewhat schizophrenic. In its slow mode, the wingtips almost touch below the bird; in the accelerated series, the wings blur. At great distances, the flight seems jerky, hesitant, stop-and-go. The birds also have a steadier unvaried wingbeat that in cadence and execution recalls the rowing motion of Pileated Woodpecker or American Crow.

In migration, birds often fly very high. Leaving a perch, they may drop toward the water and fly just above the surface. Often call in flight.

VOCALIZATIONS: Call is a scratchy chattering rattle or a stuttering uneven snare-drum tattoo. The call can be loud and protracted (when the bird is really peeved) or short and muttered (if the bird is trying not to draw attention to itself). The louder call carries great distances and often alerts birders to the arrival of high-flying or distant birds.

PERTINENT PARTICULARS: You want to see Belted Kingfisher? Just add water. At the icy edge of its winter range, birds perch over (and defend) any patch of open water.

Green Kingfisher, *Chloroceryle americana*
River Dart

STATUS AND DISTRIBUTION: Fairly common in the lower Rio Grande Valley of Tex. and in s.-cen. Tex. from Del Rio east to the s. Edwards Plateau. Scarce in se. Ariz. Also found throughout much of Mexico and Central and South America. **HABITAT:** Shorelines of clear, slow-moving, freshwater rivers, streams, and ponds rimmed with brush, trees, reeds, or exposed roots for hunting perches and soil banks for nest cavities. Streams may be very small—even eddies off the main course. Also uses water culverts and ditches. **COHABITANTS:** Ringed Kingfisher, Great Kiskadee (Texas), Gray Hawk, Tropical Kingbird (Arizona). **MOVEMENT/MIGRATION:** Permanent resident. VI: 1.

DESCRIPTION: A feathered dart. *Tiny* sliver of a kingfisher (much smaller than Belted Kingfisher; slightly smaller than Golden-fronted or Gila Woodpecker), with a large head, a stiletto-like bill, an undersized body, no legs (just feet), and a short stub of a tail. Appears about half head and bill—a loon's head fitted to a woodpecker's body.

The dark (in good light, dark green) head and upperparts are bisected by a narrow white collar. Mostly white underparts are broken by a band across the chest—chestnut in males, broken green in females.

BEHAVIOR: Most commonly seen perched or flying, singly, just above the water. In the history of the world, nobody has ever been heard to utter these words: "Look at that flock of Green Kingfishers way up there." Commonly perches on a bare twig, root, or reed about 3 ft. (or less) above or slightly back from the water. "Dives" at an angle, and skips or splashes across the surface of the water; rarely submerges. (Belted Kingfisher usually makes vertical dives and usually submerges.) If Green aborts the dive, may return to the same perch. Whether successful or unsuccessful, Green

Kingfisher usually takes another perch after completing a dive. If flushed, usually moves out of sight.

Often very tame, allowing close approach; since it often sits up against the bank and low against the water, it often goes unseen. Usually quiet, but "ticks" when approached too closely. Also raises its head and jerks its tail upward.

FLIGHT: A dark dart with wings, showing a small, front-heavy, slender-bodied, slender-winged, short-tailed profile. Because the bird will probably be going dead away, don't be dismayed if you can't see white underparts or a white collar, but you may notice white outer tail feathers. Flight is direct and fast, mere inches above the surface. Wingbeats are given in a rapid, deep, down-stroking series in which every third or so wingbeat seems to freeze momentarily on the down stroke, giving the flight a distinctive jerkiness—*like a stone skipped across water.*

VOCALIZATIONS: Most commonly heard call is a hard-sounding (but softly uttered) "tick" that recalls the sound of two stones being struck together. Given sequentially, it sounds like the ticking of a Geiger counter. Also makes a low, terse, sharp trill, "*greerk*," a nasal buzz, "*re'h'h'h*," and a low crackling rattle, "*ja'a'ah*."

PERTINENT PARTICULARS: Perched birds are often very difficult to pick out. The soft "pebble-clicking" notes will alert you to its presence, and the white collar (not the green back, not the chestnut breast-band) is what will catch your eye.

WOODPECKERS

Lewis's Woodpecker, *Melanerpes lewis*

Black Sally

STATUS: Fairly common western woodpecker with a spotty and somewhat irruptive distribution. **DISTRIBUTION:** Breeds from cen. B.C. into s. and e. Ore., n. Calif., most of Idaho, w. and s. Mont., n. Nev., w. and e. Wyo., eastern and southern portions of Utah, the western two-thirds of Colo., n. Ariz., and nw. and cen. N.M. Winters, in varying numbers, in the U.S. Southwest north to s. Ore. and n. Colo., east to the Okla. and Tex. Panhandles, and south to Mexico. **HABITAT:** Breeds in mostly semi-open habitat, most commonly open ponderosa pine forest, open riparian woodlands, and recently logged or burned coniferous forest. Also found in oak woodlands, orchards, pinyon-juniper, and ranch land, but seems particularly attracted to burned-over forest. In winter, most commonly found in open oak woodlands, oak savanna, and orchards. **COHABITANTS:** Acorn Woodpecker, Western Wood-Pewee, Mountain Bluebird, Townsend's Solitaire. **MOVEMENT/MIGRATION:** Most northern birds vacate breeding areas, but many birds in southern areas are present all year. In late summer, somewhat nomadic; flocks head for higher elevations or areas with ample mast or fruit (such as orchards). Spring migration from late March to late May; fall movements and migration from early August to early December. VI: 3.

DESCRIPTION: A big, strikingly patterned, but distinctly dark, broad-winged woodpecker that likes to sit high and sally out in pursuit of insect prey. Large (almost flicker-sized) and bulky, Lewis's is long-tailed and curiously shaped—the pale neck makes the somewhat smallish head seem unattached to the body.

Overall dark; blackish above, with an all-dark head (the red face often cannot be seen at a distance), paler collar, and paler gray-and-rose-colored underparts that are often surprisingly easy to see when birds are perched high atop twigs or branches. Immatures are overall darker with only a hint of a pale collar and rosy-tinged underparts.

BEHAVIOR: Maneuvers up and down trunks with the hitching aplomb of other woodpeckers, but rarely bores for insects. Also forages on the ground, but doesn't dig in the soil. This is an open-country woodpecker, one that excels, like no other woodpecker, in the art of hawking insects, kingbird-fashion, from elevated perches. Commonly sits at the tops of trees—often on springy twigs—from which it sallies out like a feathered boomerang (except the blocky shape more nearly resembles a Frisbee). After snapping the insect out of the air in a swooping intercept, commonly banks with a flourishing half-soar (recalling one of the Old

World rollers) and glides or flaps and glides back to a perch. Also eats acorns that it pries from branches, then carries to a stout limb to remove shells.

Forages alone, in pairs, or in small flocks in the fall. Occasionally engages in antagonistic disputes with Acorn Woodpeckers over mast rights and disputed claims to stored acorns. Fairly shy; generally shifts perches when approached.

FLIGHT: Both the shape and the flight have been aptly described as crowlike. The long broad wings dominate the flight profile (the bird looks virtually rectangular), and the head and tail project equidistant. Wingbeats are languid, heavy, measured, regular (or sometimes slightly irregular), and strongly down-stroking (crowlike!). Flight is straight and buoyant, with a slow-motion, floating, dreamlike quality and no undulations. Glides frequently.

VOCALIZATIONS: Generally silent. Despite its large size, vocalizations have a higher-pitched quality that would seem to belong to a smaller woodpecker. Rattle call is like flicker, but higher-pitched. Also makes a squeaky metallic "*che/uh*" (sometimes doubled) that recalls a ground squirrel alarm.

Red-headed Woodpecker, *Melanerpes erythrocephalus*

Tricolored Woodpecker

STATUS: Variously common to scarce resident, breeder, and winter resident across its mostly eastern range. **DISTRIBUTION:** Found from the prairies east. Northern limits cut through s. Sask., s. Man., extreme se. Ont., and Que. Western limits of the breeding range reach cen. Mon., n.-cen. and se. Wyo., e. Colo., and e. N.M. Does not breed in New England (except w. Vt.), portions of N.J., Del., the s. Appalachians, extreme s. Fla., and coastal La. In winter, vacates northern and northwestern portions of its breeding range and expands south into n.-cen. Tex. and coastal La. **HABITAT:** Shows a strong preference for dry, mature, open, parklike forests or woodlands with a high canopy and little or no understory, including beech or oak woodlands, mixed pine-hardwoods and pine savanna, riparian woodlands, beaver swamps (with standing dead timber), and utility poles in prairie habitat. Commonly occurs in older suburban neighborhoods where mature trees abound. Colonizes recently burned areas or areas where flooding or insect damage has left large stands of standing dead timber. In winter, most common in woodlands where acorns and other mast are abundant. **COHABITANTS:** Red-bellied Woodpecker, Northern Flicker, Eastern Bluebird. **MOVEMENT/MIGRATION:** Partial migrant. The level and extent of the bird's northern evacuation depend very much on the abundance or scarcity of acorns. Spring migration from late March to late May; fall from late August to early November. Ridge tops and coastlines appear to be migratory leading lines. VI: 2.

DESCRIPTION: A distinctive, boldly patterned, red-white-and-black banner of a woodpecker. Medium-sized and somewhat compact (smaller than a flicker; larger than Downy; about the same size as Red-bellied or Hairy), with a round head and a fairly long straight bill.

Adults are unmistakable. No other North American woodpecker is so distinctly partitioned red, black, and white. The red on Red-bellied Woodpecker is limited to the top of the head, and it has dingy underparts and no white showing on the folded wing. Red-breasted Sapsucker (whose range does not overlap with Red-headed) has an all-red head but otherwise bears little resemblance to Red-headed. Immature Red-headed has a brown to reddish brown head and a browner back (than adults), but overall shares the bold pattern of adults.

BEHAVIOR: Something of a cross between Lewis's Woodpecker (the sally-master) and Acorn Woodpecker (the nut-stasher). Red-headed commonly forages low (below 30 ft.), in mostly open areas. In summer and on warm days in winter, often perch-hunts (or, more accurately, trunk-hunts, insofar as the launch point is usually the trunk of a tree, not a limb) for insects that it snatches out of the air. Sometimes flies straight up and back to its perch; sometimes makes floating sallies and returns to the same perch (much like Lewis's Woodpecker) or travels on to another. Red-headed, one of the few ground-foraging woodpeckers, makes gliding stoops to land and pluck insects from the ground. In winter, commonly forages higher, often near the tops of dead trees, where it pries at bark in search of insects, chisels into wood using slow deliberate strokes, or excavates cached acorns, which it then carries to another perch to open and consume. Changes perches with deliberation and is often sedentary for long periods.

Not particularly social, although, in places, birds appear to breed in small colonies. Pairs often winter together, and Red-headed often shares the same habitat with Red-bellied and other woodpecker species. Not particularly responsive to pishing, although birds sometimes vocalize if piqued by a screech-owl imitation. Fairly tame, often allowing close approach.

FLIGHT: Fairly stocky, with broad shortish wings. The bold white lower back patch seen on perched adults and immatures metamorphoses into a unique broad white rectangular swath that extends across the trailing edge of both wings and across the rump. Flight is buoyant and straight on steady, quick, somewhat rowing wingbeats that are given in long series punctuated by intermittent skips and pauses. While foraging, often glides short distances. When landing, often uses a series of braking open-winged glides.

VOCALIZATIONS: Call is a sharp "*queerp*" that recalls Red-bellied but is shriller and less rolling. Also makes a short, soft, low, dry, descending rattle, "*t'r'r'r'r*," that recalls Acorn Woodpecker and is often given with a backward toss of the head. The drum is short, soft, and slow.

Acorn Woodpecker, *Melanerpes formicivorus*

Acorn Clown

STATUS: Common but (un)fairly restricted western woodpecker. **DISTRIBUTION:** In the U.S., extreme s.-cen. Wash. south to Baja, most (but not all) of Ariz., N.M., and trans-Pecos Tex. south to Big Bend; a few groups occur in s. Utah and s. Colo. Range extends south through cen. Mexico into Central America. **HABITAT:** Coastal woodlands, montane woodlands, riparian corridors, suburbanized areas, and grassy park habitat with a plenitude of oak trees. **COHABITANTS:** Western Scrub-Jay, White-breasted Nuthatch, titmice. **MOVEMENT/MIGRATION:** Generally sedentary, although family groups relocate if the acorn crop in their territory fails. VI: 1.

DESCRIPTION: A medium-sized, somewhat stocky woodpecker (slightly larger than a starling) whose histrionic plumage, animation, and noisome vocalizations demand attention. The conservatively patterned black back, dark-bibbed and darkly streaked breast, and white-bellied body lay a staid foundation for the bird's outlandishly patterned "clown face." From the golf tee–shaped bill to the white face, crazed yellow eye, and red beanie, the bird is the caricature of a clown. No other woodpecker looks remotely similar; no further description is necessary.

BEHAVIOR: Active, animate, acrobatic, noisome, excitable, and highly social. Family groups of up to a dozen or so birds hold down the center ring in the area surrounding their "granary trees"—the trees used to store the clan's stash of acorns. Family groups commonly gather in the tree—an avian Keystone Cops routine that incorporates much body bobbing, position turning, wing displaying, and even, occasionally, just quiet sitting. Like all woodpeckers, Acorn Woodpecker clings to and climbs vertically along trunks. But the bird is adept at sitting astride branches and hopping lightly and nimbly along limbs and up branches, shifting perches frequently. Does not commonly go to ground (unless an acorn is dropped).

Feeds primarily on the acorns it plucks from trees and stores in nooks, crannies, and acorn-sized holes excavated in the granary trees for that purpose. Also flycatches, sallying out from perches to snatch insects in the air and landing with an acrobatic swoop. In summer, drinks sap from trees, as well as sugar-water from hummingbird feeders.

FLIGHT: In flight, the bright white oval wing patches and white rump are hard to overlook. Overall stocky (seems all short wide wing). Wingbeats are somewhat slow, fluttery, and batlike or butterfly-like. Flight is straight and undulating—a series of wingbeats followed by a descending glide with wings partially opened.

VOCALIZATIONS: A raucous gargled cacophony. Most commonly heard call is a low, strangled, rising-and-falling "*r'r'rA-Ha*" that is often repeated three to four times and has a cadence and grating quality that recall lumber being sawed. Also makes a gargled, protesting, ascending "*r'r'aaah!*"

Gila Woodpecker, *Melanerpes uropygialis*

Boisterous Woodpecker with a Cherub's Face

STATUS: Common bordering on omnipresent, but restricted to arid regions of the Southwest. **DISTRIBUTION:** Permanent resident, primarily of s. Ariz. deserts, extending west to the Salton Sea and east to extreme sw. N.M. **HABITAT:** Very common in deserts supporting saguaro cactus, but also found in riparian woodlands and residential areas with trees suitable for nesting. **COHABITANTS:** Gilded Flicker (in saguaro deserts), Ladder-backed Woodpecker (in riparian and suburban habitats). **MOVEMENT/MIGRATION:** Nonmigratory, but shows some seasonal shifts in response to food needs. VI: 0.

DESCRIPTION: An active, noisy, unadorned, plain-looking desert woodpecker. Medium-sized (larger than Ladder-backed; smaller than Gilded Flicker; about the same size as Hairy Woodpecker, Red-bellied Woodpecker, and Golden-fronted Woodpecker, which it closely resembles), with a long straight bill, a smooth slender profile, and a forked (at the very least, notched) tail.

The head and underparts are a plain, pale, putty brown; upperparts are black-and-white barred. The male has a tiny red skullcap; the unmarked heads on all other Gilas make them the plainest of the "zebra-backed" woodpeckers. The expression is plain and benevolent.

BEHAVIOR: Active—when feeding, changes position and perches frequently. Noisy—with an array of vocalizations and a love of making itself heard. Omnivorous—forages on both plant and animal matter and uses a variety of foraging techniques. In trees, forages mostly on trunks and heavier branches, moving both up and down, probing for and gleaning insects. In summer, feeds heavily on the fruits and flowers of saguaro. Also flycatches, forages on the ground, forages in small bushes, and comes readily to backyard feeding stations, where it prefers suet and peanut butter mixes. Also eats small lizards and the eggs and nestlings of birds.

Found alone or in pairs. Fairly tame. Responds readily to pishing.

FLIGHT: Fairly compact profile, with broad rounded wings and a short tail. Shows obvious round whitish windows on the wingtips and a darkly barred rump. Flight is deeply undulating (a pattern of regularly spaced U's). Wingbeats are given in a hurried series followed by a closed-wing glide.

VOCALIZATIONS: Many and varied, including a loud protesting "*yak yak yak*"; a short, descending, rolling trill, "*chr'r'r'r'r'r*"; a high-pitched accipiter-like protest, "*reh reh reh*" (sounds like a cross between a mew and a yap); and a loud terse "*mew.*" If you are at all familiar with the calls and vocalizations of Red-bellied Woodpecker, you can fairly ascribe them to Gila.

Golden-fronted Woodpecker, *Melanerpes aurifrons*

Saffron-naped Woodpecker

STATUS: Common but geographically restricted resident. **DISTRIBUTION:** In the U.S., found only in s. and cen. Tex. and sw. Okla. Also occurs in e. Mexico and Central America. **HABITAT:** Fairly eclectic; found in a variety of open and wooded habitats, but seems most partial to dry woodlands, especially where range overlaps with Red-bellied Woodpecker. Typical habitats include mesquite brush land, riparian woodlands, cottonwood groves, oak-juniper savanna, second-growth habitat, suburban yards, and urban parks. **COHABITANTS:** Ladder-backed Woodpecker, Black-crested Titmouse. **MOVEMENT/MIGRATION:** Permanent resident. VI: 0.

DESCRIPTION: The "zebra-backed" woodpecker with a rich golden or orange-yellow nape. Slender, straight-billed, and medium-sized (larger than Ladder-backed Woodpecker; smaller than Northern Flicker; the same size and shape as Red-bellied Woodpecker, whose range overlaps the northern and eastern portions of Golden-fronted's range).

Conspicuously striped above, and uniformly pale brown below. Usually easily distinguished from Red-bellied by the golden nape (red on Red-bellied) and the dab of yellow at the base of the bill (which is harder to see). The tail is wholly black (no white center), and the rump is wholly white (no black spotting). Except for a small red patch on the crown of adult males, sexes and ages are similar.

At great distances or in harsh light, the yellow nape may be hard to see. Fall back on the absence of red, which is usually easier to note. Also, the white rump often winks through the folded wings when birds are perched.

BEHAVIOR: A foraging more than a boring woodpecker. Spends much of its time on sturdy branches, working both the upper and lower sides, picking and probing for food more than pecking (drilling). Also forages on smaller branches, forages freely on the ground (like flicker), flycatches, and is easily attracted to fruit (oranges, apples) and sugar-water in hummingbird feeders.

Usually solitary or in pairs. Usually vocal. Often very tame.

FLIGHT: Fairly slender and overall pale—at a distance appears gray and yellow-headed—but shows three distinctive white patches: one toward the tip of both wings and one on the rump. The white rump is particularly conspicuous (even more so than on Red-bellied) owing to the absence of dark spotting and contrasting all-black tail. Wingbeats are rapid and given in a short series followed by a closed-wing glide. Flight is flowing and deeply undulating.

VOCALIZATIONS: Calls, which are much like Red-bellied but somewhat louder and raspier, include a dry "*che/uh*" and a softer "*chur.*" Also emits a descending and accelerating "*chuh chuh chuhchuhchuhchuh*" that is louder and harsher than a similar call by Red-bellied.

Red-bellied Woodpecker, *Melanerpes carolinus*

The Eclectic Woodpecker

STATUS: Common, largely eastern U.S. resident of woodlands and suburban habitat. **DISTRIBUTION:** Found south of a line drawn from e. Mass. across extreme se. Ont. west to se. N.D. Western limit reaches se. S.D., much of Neb., Kans., Okla., and cen. Tex. (where the range of the very similar Golden-fronted Woodpecker begins). **HABITAT:** Essentially a woodpecker of eastern hardwood and mixed forests. Its partiality for wetter habitats—such as bottomlands and riparian corridors—accounts for its minor western territorial incursions along the Missouri, Platte, and other rivers that support substantial woodlands. **COHABITANTS:** Hairy Woodpecker, Tufted Titmouse. If you see Black-crested Titmouse, start considering Golden-fronted Woodpecker, but note that the ranges of Red-bellied Woodpecker and Black-crested Titmouse do slightly overlap. **MOVEMENT/MIGRATION:** Mostly resident, but some birds wander in the nonbreeding season, including north to northern New England and southern Canada. VI: 3.

DESCRIPTION: A pale, slender, and distinctly marked woodpecker—both perched and in flight. Medium-sized (smaller than a flicker; larger than Hairy Woodpecker or a sapsucker), with a long straight bill and a long body—seems less stocky, more nubile, and less blocky or angular than other eastern woodpeckers.

Despite the crisp black-and-white "zebra-backed" barring on the upperparts, Red-bellied often appears overall pale, uncontrastingly patterned, and generally buffy-toned, distinguished by its crimson helmet and very plain face. The red on the male's crown runs from the bill to the nape; the female's helmet is parted, with red limited to the base of the bill and the back of the head (crown is buffy). Immatures have orange-toned napes; otherwise the ages and sexes are similar.

Difficult to confuse with any other woodpecker—except in central Texas, where the ranges of Red-bellied and Golden-fronted abut.

BEHAVIOR: An active nimble woodpecker that seems very much at home anywhere on a tree—the main trunk, heavy branches, twigs; high in the canopy or low at the base; on a live tree or a dead one. Gleans larvae and insects, pries in crevices, bores for insects, flycatches, gathers acorns and nuts, plucks berries. . . . What nonperishable items it doesn't eat it often stores for later. Hitches itself up trunks in classic halting woodpecker fashion. Frequently angles itself astride the trunk when descending. Frequently perches at the tops of trees (like a flicker), clinging to springy branches.

Generally solitary outside of breeding season. Somewhat shy except where it has become habituated to people around feeding stations—as it does readily, being particularly fond of sunflower seeds, peanuts, suet, and peanut butter.

FLIGHT: Fairly slender, overall pale woodpecker. The white oval wing patches near the tips of the wings and the white rump are standout features. Flight is deeply undulating, much more so than a flicker. Wingbeats are given in a rapid, deep, angry series followed by a closed-wing glide. Often flies for long periods over open treeless terrain. Often calls in flight.

VOCALIZATIONS: Call is a loud rolling "*kweer*" (sounds like a musical flicker rolling its r's). Also makes a soft muffled cough: "*chuh, chuh, chuh.*" Drum is even, lengthy, fairly typical sounding.

PERTINENT PARTICULARS: In southern Florida, the crimson helmet of adult males thins on the forehead, creating a head pattern that lies somewhere between more northern males and females.

Williamson's Sapsucker, *Sphyrapicus thyroideus*

The Beautiful Sapsucker

STATUS: Uncommon woodpecker of the western mountains with populations that are localized and disjunct. **DISTRIBUTION:** Breeds in broken pockets and narrowly linked corridors in all the western states in and west of the Rockies as well as in s. B.C. In winter, withdraws from northern portions of its range, where it is found locally in s. Ore., Calif., Ariz., and the western half of N.M. (occasionally east to w. Tex.) as well as in w. Mexico. **HABITAT:** Breeds in higher to middle elevations in forests dominated by conifers (spruce-fir, Douglas fir, ponderosa pine, lodgepole pine) as well as in mixed deciduous-coniferous forests, especially those that include quaking aspen (an important nest tree). In winter, found in middle to lower altitudes, concentrated in mixed forest habitat, primarily oak-juniper and pine-oak. **COHABITANTS:** Mountain Chickadee, Red-naped Sapsucker, Dusky Flycatcher, Pygmy Nuthatch. **MOVEMENT/MIGRATION:** Northern populations are migratory; birds at higher altitudes descend to lower elevations often in advance of winter. VI: 2.

DESCRIPTION: A medium-sized woodpecker (about the size of Hairy Woodpecker) distinguished among woodpeckers by the striking plumage of adult (and immature) males and the more cryptic garb of females. In other woodpeckers, the plumage of males and females is similar (if not identical).

The classic sapsucker body is complete with a strong, straight, pointy sapsucker bill. The male is unmistakable. Shiny and uniformly black above (head, back, tail), it has a head creased by two white stripes and a bold white wing patch. The prominent white rump is visible only in flight. Underparts are black on the chest, yellow on the belly. Immature males are like adults but lack the red throat and yellow belly. Females are overall pale and, in spectral defiance of the wraparound pattern of narrow blackish and grayish barring, have a brownish cast. Their heads are warmer and buffier, and their faces conspicuously plain. The blackish breast-band and yellow belly (not usually visible, since birds forage with their bellies pressed against tree trunks) are absent on immature females. Although female Williamson's lacks a white patch in the wing, a very conspicuously white rump (visible in flight) distinguishes this species from other female sapsuckers.

BEHAVIOR: For most of the year, feeds on the sap of conifers it gets by drilling rows of shallow holes in the bark. In summer, switches to insects. Sometimes works up from the bottom of one tree and then flies to the base of another when the canopy is gained (much like a creeper). These birds seem just as inclined, however, to switch trees as to climb and will forage horizontally through a woodland—landing, searching a trunk for ants or other insects, then flying on to another tree. Because of this active feeding pattern, Williamson's is usually conspicuous when breeding—an advantage from a birding standpoint because the species also tends to be quiet.

Most commonly found on the trunks of trees. Usually nests in aspens, both live and dead, and in snags as well as the main body of the tree.

FLIGHT: In flight, the white wing patches and rump of males (and the pale rump of female) are standout features. Wings are broad. Flight is light, buoyant, and casually undulating. Lengthy and leisurely series of wingbeats are punctuated by terse closed-wing glides. Individuals commonly move great distances through trees (passing up any number of possible foraging sites on the way).

VOCALIZATIONS: Fairly quiet. Call is a loud, harsh, slightly slurred "*reeah reeah reeah*" that is louder, lower, and harsher than Red-naped Sapsucker (might recall a hoarse Red-shouldered Hawk scream). Males and females together also make a low muttered "*chuh, chuh, chuh*" (given in series), as well as a low nasal purr/trill, "*r'r'r'r'r'rr,*" that peters out at the end. Drum is a series of short multiple (stuttering) taps; the first is longer, the balance shorter, and given in a halting rhythm that slows to a stop.

Yellow-bellied Sapsucker, *Sphyrapicus varius*

Quiet Taps for a Cryptic Woodpecker

STATUS: Fairly common and widespread northern breeder and southeastern winter resident, but somewhat cryptic and secretive. **DISTRIBUTION:** Breeds in northern forests, across Canada from Nfld. to Alta. and ne. B.C. and the Yukon. In the U.S., breeds in Maine, interior New England, N.Y., n. N.J., n. and cen. Pa., n. Mich., n. and w. Wisc., Minn., and the e. Dakotas. Winters across the mid-Atlantic and Southeast from Cape Cod south and west across n. Kans., south through Okla., most of Tex., and e. and cen. Mexico. Very rare in the w. U.S. **HABITAT:** Breeds in early successional deciduous and mixed forests and woodlots, often near water (such as beaver ponds). In mixed woodlands, this species is particularly partial to aspens and birch. In winter, found in a variety of woodlands but favors hardwoods and avoids pure conifers. Birds are most common in open forest (including large city parks) or at the forest edge; also found in bottomlands. **COHABITANTS:** In summer, Least Flycatcher, American Redstart. In winter, Carolina Chickadee, Brown Creeper, Eastern Bluebird, Yellow-rumped Warbler. **MOVEMENT/MIGRATION:** Highly migratory, wintering almost entirely south of its breeding range. Spring migration from late March to early May; fall from late August to late November. VI: 2.

DESCRIPTION: A retiring, somewhat disheveled, cryptically plumaged woodpecker. Small to medium-sized (slightly smaller and more compact than Hairy), with a fairly long straight bill and a slightly smaller and more peaked head.

Yellow-bellied Sapsucker's upperparts are less contrasting than Hairy's—a blurrier black and a not-quite-white as opposed to pure black and white. The pattern replicates the 3-D dappling of sunlight on branches. *Note:* Most field guides show the birds with more contrast and crisper definition than is evident in the field. Shabbily patterned underparts are tarnished yellow with shadowy barring. All in all, and unlike most woodpeckers, the basic plumage of adults is muted—a backdrop that makes the distinctive red-black-and-white pattern of the face and the very white racing stripe defining the edge of the folded wing stand out.

Males have red crowns and red throats; females have red crowns and white throats. Immatures lack red and, while otherwise patterned like adults, are overall more brownish, muted, and disheveled looking. *Note:* The white racing stripe is not always conspicuous. Learn not to depend upon it.

In flight, the white bar bisecting the upperwing and the white rump are distinctive. When the bird is perched, the center of the tail shows broken white. Other black-and-white woodpeckers (except the other sapsuckers) have dark centers.

BEHAVIOR: A reclusive sedentary woodpecker. In winter, tends to be solitary; in migration, several birds may occupy a sap-rich tree. Feeds by tapping shallow holes into tree trunks and branches, drinking the sap, and consuming the insects drawn to the sap. Also eats fruits and buds. Target trees include poplar, willow, birch, maple, hickory, alder, and (to a lesser degree) conifer.

A somewhat bipolar feeder — active and passive. At times, forages methodically, often remaining in the same spot for several minutes or in the same tree for hours. At other times, very energetic, moving frequently from tree to tree in floating open-winged sallies.

A shy bird, Yellow-bellied often moves to the far side of a limb to avoid study. Despite its energetic drilling (as evidenced by the plenitude of horizontal rows of holes etched in the trunks of favored trees), observers are often unaware of the bird's presence. Dislodged shards of bark are few, and the bird's tapping is surprisingly soft, almost soundless. Often birders are alerted to the bird's presence by its call.

FLIGHT: Overall fairly compact with broad round wings and little projection of the head and tail; compared to Red-bellied and flickers, appears all wing. Overall fairly pale and undifferentiated except for the white bar on the wings and the whitish rump. Underparts often show a yellowish cast. Flight is evenly undulating and fairly buoyant; wingbeats are given in a rapid series followed by a closed-wing glide. Flight of flickers is straighter, less undulating. Red-bellied Woodpecker is slimmer and has faster and angrier wingbeats.

VOCALIZATIONS: Call is similar to Red-naped and Red-breasted Sapsuckers — a loud, nasal, squealing, somewhat two-noted "*reeuh reeuh reeuh*." Also heard year-round is Yellow-bellied's soft, descending, plaintive, catlike (or catbirdlike) mewing sound. Drum begins with a rapid series of four or five taps that slow and lose their rhythm, often ending with one or two double taps.

Red-naped Sapsucker, *Sphyrapicus nuchalis*

Aspen Sapsucker

STATUS: Common breeding sapsucker across most of the interior West (replaced by Red-breasted Sapsucker along the Pacific Coast). **DISTRIBUTION:** Breeds from se. B.C. and sw. Alta. south into cen. and e. Wash., cen. and e. Ore. (also the Calif.–Nev. border), Idaho, w. Mont., most of Wyo., w. S.D., the eastern two-thirds of Nev., Utah, the western two-thirds of Colo., ne. Ariz., and n. N.M. Winters in se. Calif., extreme s. Nev., Ariz., sw. N.M., and trans-Pecos Tex., south into Mexico. **HABITAT:** Breeds in deciduous or mixed pine-deciduous forests; shows a marked partiality for stands of aspen and poplars and for open ponderosa pine and dead trees in beaver swamps; also favors montane coniferous forests and birch groves. Avoids breeding near forest edges (but breeds in poplar stands surrounded by open areas) and oak-pine woodlands. In winter, more eclectic in its choice of woodlands, including oaks and orchards. **COHABITANTS:** In summer, Western Wood-Pewee, Hammond's Flycatcher, Western Tanager. **MOVEMENT/MIGRATION:** Vacates all but southernmost portions of the breeding range in winter. Spring migration from mid-March to early May; fall from late August to November. VI: 1.

DESCRIPTION: A distinctive and distinguished-looking black-white-and-red woodpecker. Very similar to the eastern Yellow-bellied in all but head pattern, but except in winter in west Texas, the ranges do not overlap.

Compact small to medium-sized woodpecker (smaller than Hairy), with a moderately long straight bill. Overall plumage is somewhat shabby — blackish above with grayish mottling on the back, dirty white with dirty spattering below. Muted contrasts and dappled patterning make the bright white stripe running along the folded wing a standout feature.

The head pattern is distinctive — a black-and-white face pattern bracketed by a red throat that is not completely framed by black, a red cap, and a red nape. The red on Yellow-bellied is more restricted on the throat and completely framed by black; most significantly, red is entirely absent on Yellow-bellied's nape. The more coastal Red-breasted Sapsucker has an extensively red head, throat, and breast.

Immature Red-napeds show the same pattern as adults (including the white wing stripe) but have dingy brown heads and breasts and lack any red on the head. They are somewhat in between the plumage of paler brown immature Yellow-bellied and darker brown Red-breasted and share other similarities with both of these birds. *Note:* The immature plumage of Red-naped is held only into the fall; immature plumage of Yellow-bellied is held through the winter.

BEHAVIOR: Drills, maintains, and defends sap-wells in trees year-round. In summer, fairly active and more insect-driven. Hops along trunks and branches, and changes trees frequently. Often forages low on trunks and in stands of willows. Plucks insects from branches. Flycatches with finesse, often making long lofting sallies to snap insects out of the air.

Most often found in poplars and aspens. Where ranges overlap, may nest in close proximity to Red-breasted Sapsucker (and occasionally hybridizes with it).

FLIGHT: Fairly short-bodied, with long wide wings. The white bar cutting across the inner middle of the wing, the pale rump, and the bright red facial pattern are distinctive. Flight is nimble, buoyant, and variably undulating. Wingbeats are given in a lengthy series followed by a glide.

VOCALIZATIONS: Similar to Yellow-bellied. On territory, makes a loud, full-bodied, yelping "*reeuh reeuh reeuh*" that is higher-pitched and more squealing than Williamson's Sapsucker (and may recall the cry of Ring-billed Gull). Call is like Yellow-bellied—a soft, descending, catlike, mewing whine. Also makes a softer but somewhat flickerlike "*rehh rehh rehh rehh rehh*" and a "*wicka wicka.*" Drum, usually given from the top of a tree or a well-elevated branch, is a series of single taps that begin steadily and slow to an irregular tapping halt, punctuated by a double tap. Sounds somewhat like a windblown wooden gate creaking to a halt.

Red-breasted Sapsucker, *Sphyrapicus ruber*

The Red-hooded Woodpecker

STATUS: Fairly common West Coast breeder, resident, and winter resident. **DISTRIBUTION:** Range is more limited and coastal than Red-naped. Breeds from se. Alaska and B.C. (absent in the northeast and southeast corners of the province), south through w. Wash., w. Ore., n. Calif., and along the Sierras south to the higher mountains of s. Calif. Winters, not far inland, from sw. B.C. south to n. Baja. A few winter in w. Nev. Casual in Ariz. **HABITAT:** Breeds in a range of coniferous forests; prefers old-growth. Also found in mixed deciduous-coniferous habitats and riparian woodlands of aspen and cottonwood. In winter, also found in a variety of forest habitats—from open cottonwoods (and suburban neighborhoods with exotic plantings) to dense coastal cedars as well as other woodland habitats that do not support breeding. **COHABITANTS:** In summer, Hairy Woodpecker, Hammond's Flycatcher, Steller's Jay. **MOVEMENT/MIGRATION:** Birds in interior British Columbia withdraw in winter. Spring migration from mid-March to early May; fall from late August to November. VI: 1.

DESCRIPTION: A western red-hooded woodpecker with a white racing stripe down the side. Small to medium-sized (larger than Downy or Nuttall's; smaller than Hairy; the same size as Red-naped Sapsucker), with mostly blackish upperparts (showing limited white on the back), dingy grayish or yellow-tinged underparts, and a distinctive all-red head whose red extends down to and over the upper breast. (The red on Red-naped stops at the throat.) Blackish upperparts show two dappled and poorly defined white stripes running down the back and a bold white racing stripe running down the sides (along the folded wing). Some birds show a whitish line running below the eye, but this shouldn't detract from the overall sense of red-headedness. Immatures show an all-dark grayish brown hood (where adults show red) that is darker, richer, plainer, and more hoodlike than the head patterns of immature Red-naped and Yellow-bellied Sapsuckers.

BEHAVIOR: Like other sapsuckers, a driller and custodian of sap-wells. In summer, becomes insect-oriented, pursuing them actively on the branches (as well as needle tips) of assorted conifers and deciduous trees. Seems most partial to conifers and very partial to snags and branches (as opposed to main trunks), often perching on, drumming from, and excavating snags for nest sites. In winter, far less active, more trunk-oriented, often working quite low and remaining quiescent on the same trunk for lengthy periods.

Not very social. Found in pairs and family groups in summer; in winter, often solitary.

FLIGHT: Like other sapsuckers—overall fairly compact, showing a short body and broad wings. The white slash on the wings and the (particularly) white rump are easily noted. The all-red head is easy to see in good light. Flight is variously undulating—sometimes shallow, sometimes pronounced. Wingbeats are given in a regular, fairly lengthy series followed by a closed-wing glide.

VOCALIZATIONS: Like Red-naped and Yellow-bellied. In winter, the only call given is the mew call.

PERTINENT PARTICULARS: This species hybridizes with Red-naped Sapsucker, and possibly with Yellow-bellied.

Ladder-backed Woodpecker, *Picoides scalaris*

The Cactus Woodpecker

STATUS: Common desert woodpecker of the American Southwest. **DISTRIBUTION:** In the U.S., occurs from se. Calif. (also Baja) across s. Nev., extreme sw. Utah, Ariz., s. and e. N.M., the junction of Colo., Kans., and Okla., and all but easternmost Tex. Extends south through Mexico. **HABITAT:** Desert, particularly deserts rich in cactus, mesquite woodlands, and riparian woodlands; also partial to pinyon-juniper and live oak. **COHABITANTS:** Juniper Titmouse, Verdin, Cactus Wren, Curve-billed Thrasher. **MOVEMENT/MIGRATION:** Nonmigratory; permanent resident. VI: 0.

DESCRIPTION: The default small black-and-white desert woodpecker (larger than Downy; smaller than Hairy; the same size as Nuttall's). Specializes in habitats so marginally suited for woodpeckers that encounters are often something of a surprise. Can be confused only with Nuttall's (ranges overlap slightly in southeastern California). Classically shaped and proportioned woodpecker with a fairly long, straight, or ever so slightly downturned bill.

Crisply patterned above, with a uniform black-on-white barred pattern. (Downy and Hairy are distinctly blackish above, with white spotting on the wings and a white patch/stripe centered on the back.) Underparts are dingy, with sparse fine spotting.

Overall paler and buffier than Nuttall's, particularly on the face and upper back. Ladder-backed has a black-on-white face pattern and uniform barring on the back right to the nape. The darker Nuttall's shows a white-on-black face pattern and has a wide black shawl across the upper back. On male Ladder-backeds, the red extends across the crown. On Nuttall's, it is restricted to the back of the head.

BEHAVIOR: Industrious and fairly active woodpecker. Shifts positions often, with many side movements and much perch-changing, but not as acrobatic as Nuttall's. Forages extensively on cactus (particularly Joshua trees and cholla) as well as mesquite and (in riparian areas) cottonwoods. Often forages low (occasionally on weed stalks), but does not commonly land on the ground. Males and females forage in close proximity. Males favor larger limbs and trunks, females the narrower branches. Forages mostly by probing, but also bores for insects. Association with other bird species seems incidental. Fairly tame. Allows close approach, but is not very responsive to pishing. When shifting to nearby perches, birds often glide or parachute on short wide wings.

FLIGHT: Typical woodpecker flight. Rapid series of wingbeats followed by a closed-wing glide, resulting in a bouncy, deeply undulating flight.

VOCALIZATIONS: Rattle call is like Downy Woodpecker's, but it's lower-pitched, the notes are not as sharp, and they drag, noticeably at the end, becoming stuttering and harsh. Male call is a sharp "*peek*" (flatter than Downy and less strident); females give a lower-pitched "*prk.*"

PERTINENT PARTICULARS: Habitat alone defines this species. Hairy and Downy Woodpeckers are forest and woodlands species, not desert. Nuttall's Woodpecker is an oak and riparian specialist and, where ranges overlap, is most often found foraging on the trunks or heavy branches of trees, not exploring the dietary possibilities hidden in scrub or cactus. There is one exception to the rule: in coastal southeastern Texas, where Hairy, Downy, and Nuttall's do not occur, Ladder-backed fills the vacuum by foraging in live oaks.

Nuttall's Woodpecker, *Picoides nuttallii*

Woodpecker in Three Dimensions

STATUS AND DISTRIBUTION: Fairly common but restricted; found only in Calif. and n. Baja, generally west of the deserts. **HABITAT:** Most often found in oak woodlands, but also frequents riparian corridors dominated by cottonwoods and willows, suburban and rural homes surrounded by assorted mature trees, mixed montane oak-conifer woodlands, and, less commonly, purer pine forests. **COHABITANTS:** Acorn Woodpecker, Oak Titmouse. **MOVEMENT/MIGRATION:** Nonmigratory; permanent resident. VI: 0.

DESCRIPTION: A small, black-and-white, ladder-backed woodpecker with a fairly long straight bill. Overall slender and somewhat angular; not as compact as the smaller Downy or as rangy as the larger Hairy—two species from which it is easily distinguished by its more homogenized spotted/barred plumage pattern. On Downy and Hairy, the black-and-white plumage is more segregated and more contrastingly apportioned (note particularly the all-white backs). Nuttall's is very similar in size, shape, and plumage to the more widespread Ladder-backed Woodpecker, whose range overlaps slightly with Nuttall's in southeastern California and Baja. Nuttall's, however, is distinctly black and white; similarly patterned Ladder-backed is buffy, not creamy, below. Also, Nuttall's barred back pattern is topped off by a broad black collar. On Ladder-backed, the barring is uniform right up the back.

Adult male Nuttall's is distinguished from females by a red patch on the back half of the crown. Male Ladder-backed has a much more extensive red cap that covers most of the crown.

BEHAVIOR: A busy active woodpecker that frequently shifts positions, perches, and even trees. Often circumnavigates limbs and narrow branches rather than hitch itself along vertically. Other woodpeckers move slowly, methodically, directing their attention to whatever lies under their bills. Nuttall's sits and looks around, left . . . right . . . turning its head over a shoulder, investigating even the branches behind it. Unlike most woodpeckers, does not bore for insects with its bill held at a right angle to the tree. Prefers to move with the bill pointed ahead. Probes nooks and crannies and flakes off bark obliquely, its head turned askew, often reaching left or right to secure prey. Forages on narrow branches that boring woodpeckers would avoid.

FLIGHT: Classic roller-coaster flight—a rapid series of wingbeats, a closed-wing glide. May cover great distances.

VOCALIZATIONS: Call is a short metallic stutter: "*Trrit.*" Often given in a sequence, with the last call lengthened into a rapid, stuttering, staccato, musical trill: "*trrit, trrit, trrit, t'r'r'r'r'r't.*" Pattern recalls Hairy Woodpecker, but the rendering is more rapid and higher-pitched.

Downy Woodpecker, *Picoides pubescens*

Tiny Backyard Woodpecker

STATUS: Common and widespread. **DISTRIBUTION:** A permanent resident across most of North America. Avoids only treeless tundra regions and the desert Southwest. **HABITAT:** Found in a variety of habitats—forests, woodlots, orchards, city parks, abandoned lots, and suburban yards. Markedly prefers deciduous trees. Also forages, woodpecker-like, on weedstalks, cornstalks, and phragmites, and is readily attracted to backyard feeding stations, where it favors suet. **COHABITANTS:** In winter, joins mixed-species flocks that include chickadees and titmice. **MOVEMENT/MIGRATION:** Nonmigratory; permanent resident. VI: 1.

DESCRIPTION: A petite, stubby, active, black-and-white woodpecker with a small bill. Our smallest woodpecker (slightly larger than Tufted Titmouse; slightly smaller than Ladder-backed and Nuttall's Woodpeckers; more than 2 in. shorter than the similar Hairy Woodpecker), Downy has a round head, a petite wedge-shaped bill, a compact body shaped like a salt shaker, and distinctive black-and-white plumage.

Except for the larger, rangier, longer-billed, and similarly plumaged Hairy Woodpecker, Downy is easily separated from all other North American woodpeckers by the combination of its small size, white unmarked back, and all-white underparts. Birds of the Pacific Northwest are dirty white but are otherwise similar. Adult males have a small red patch on the back of the head. The expression seems benign, nonplused.

BEHAVIOR: Not surprisingly for so small a bird, Downy is more versatile and nimble than most woodpeckers. Feeds by clinging to tree trunks and branches down to the size of twigs. Hitches itself up the trunk in a rapid, jerky, stop-and-go fashion, using its short stiff tail for a prop, sometimes hanging upside down. Forages for insects and larvae by pecking on limbs or weed stalks; also flakes bark off of dead trees. Frequently solitary or in the company of a mate, but in winter also accompanies foraging flocks of chickadees and titmice.

Downy is tame, particularly in urban and suburban areas where it habituates to people. When approached, frequently moves around the tree rather than fly.

FLIGHT: Appears overall stubby in flight: stubby head, stubby tail, and short round wings that beat rapidly and audibly. Flight is bounding, and undulations are deep and tight.

VOCALIZATIONS: Call is a sharp high-pitched "*pik!*" or "*peek*" that varies somewhat in pitch and volume. Rattle call often begins with one to three measured high-pitched warm-up notes (which resemble the "*peek*" call), then degenerates into a descending, high-pitched, descending stutter: "*peek . . . peek . . . peek/peek/peekpeekpeekpeek.*" Sounds peeved.

Drum is short (less than 2 secs.); individual taps are discernible but too rapid to count. Drums persistently, with 10–15-sec. pauses between sequences.

PERTINENT PARTICULARS: Downy Woodpecker has several plumage characteristics that help distinguish it from Hairy Woodpecker, including black spotting on the outer tail feathers and a less prominently curled mustache stripe that doesn't extend onto the sides of the breast. These distinctions, however, are subtle. Downy's small size and small bill (less than the width of the head), coupled with its less specialized habitat preferences, are usually sufficient to distinguish the two species.

Responds to pishing and is particularly irked by imitations of screech-owl, pygmy-owl, or saw-whet owl.

Hairy Woodpecker, *Picoides villosus*

Forest Woodpecker

STATUS: Fairly common and widespread permanent resident. **DISTRIBUTION:** Found over much of North America south of the tree line. Absent only in extreme s. Fla., portions of the Great Basin, parts of s. Calif., s. Nev., w. Ariz., e. N.M., and all but e. Tex. and parts of trans-Pecos Tex. **HABITAT:** A forest bird that is partial to unbroken mature deciduous and coniferous forests but, like Downy Woodpecker, occupies a variety of habitats, including smaller woodlots, mixed deciduous-coniferous woodlands, riparian corridors, parks, cemeteries, and suburban habitats. In the West, partial to pines. In all habitats, prefers larger mature trees and robust limbs. Not likely to occur in newer suburban neighborhoods devoid of woodlands, and unlike Downy, does not forage on weed stalks, cattail, or phragmites. **COHABITANTS:** Varied. Where they occur, often found in the same habitat as Pileated Woodpecker. **MOVEMENT/MIGRATION:** Nonmigratory; permanent resident, but shows a small movement into the Great Plains in fall and winter. VI: 1.

DESCRIPTION: A medium-sized, sturdy, boldly patterned black-and-white woodpecker whose plumage resembles the smaller Downy Woodpecker (with a white back and white unbarred sides), but whose size and bill distinguish it. The head is large and somewhat oblong or squarish. The bill is heavy and long—as long as the head is wide. Its head and bill give Hairy a wedge-faced appearance, quite unlike the round-headed petite-billed look of Downy.

The unbarred white outer tail feathers and projecting black wedge onto the breast (the handlebar to the mustache) are helpful field marks, but not necessarily apparent in the field. The red on the back of the head of adult males is broken and often not as prominent or bright as on Downy. Hairy, while best distinguished by size, is sometimes more disheveled-looking and has an expression that seems formidable, even mean.

Pacific birds are overall darker and soiled-looking, with less white spotting in the wings and some dark spattering on the flanks and back.

BEHAVIOR: A typical woodpecker with a penchant for large limbs and mature trees. Solitary or paired, birds forage at all heights. Hairy needs more room than Downy, ranges more widely, and is prone to making long flights to other foraging locations (within woodlands or between woodlots). More retiring than Downy, when pushed Hairy often quits the area. Also more solitary; not commonly associated with foraging woodland flocks in winter.

FLIGHT: Like Downy, bounding, but not so steeply undulating. Gives every impression of being a big woodpecker, not runty or stubby like Downy.

VOCALIZATIONS: Call is a loud, sharp, high-pitched "*Peek*" or "*Speak!*"; held longer than Downy's call, Hairy's call sometimes has a compressed two-note quality ("*S'Peek*"). Although it varies in pitch, duration, and volume (explaining why after years of birding you still have trouble telling Downy from Hairy!), it is almost always louder, sharper, and more sustained than Downy. Rattle call is a loud, high-pitched, even stutter that gets right into it (no warm-up notes); sounds truly angry, not just peeved. Drum is rapid, fairly lengthy series of run-on taps with long pauses between drums (commonly four to five drums per minute).

PERTINENT PARTICULARS: People have difficulty distinguishing Downy from Hairy because they give too much weight to plumage and not enough to the powers of human perception. Downy is a tiny compact woodpecker that would have to stretch to see over a coffee mug. Hairy is a full-sized big-billed street brawler of a woodpecker that would have no trouble peering over a beer mug.

Arizona Woodpecker, *Picoides arizonae*

Plain Brown-backed Woodpecker

STATUS: Common but geographically restricted in the United States. **DISTRIBUTION:** In the U.S., found only in se. Ariz. and the extreme southwest corner of N.M. Range extends south into w. Mexico. **HABITAT:** Montane oak or pine-oak woodlands with alligator juniper and riparian woodlands rich in sycamores and walnut. **COHABITANTS:** Blue-throated Hummingbird, Elegant Trogon, Sulphur-bellied Flycatcher, Painted Redstart. **MOVEMENT/MIGRATION:** Permanent resident. VI: 0.

DESCRIPTION: A uniquely plain brown-backed woodpecker showing no barring above and lavish spotting below. Small to medium-sized, slightly larger than Ladder-backed, but overall stocky, almost chubby, with a long, straight, sturdy bill and a short tail.

Overall dark and muted—plain smoky brown above, grayish and richly spotted with brown below. The bold face pattern is a unique mix of brown, blackish brown, and cream. From behind, the neck shows lavish amounts of white with a narrow blackish border down the nape. Ages and sexes are similar. Males have a small red patch on the rear crown.

The white eyebrow gives the bird a serious expression.

BEHAVIOR: A woodpecker that loves oaks and junipers (although also found in pines). Generally forages low, mostly on trunks and sturdy branches (in summer, also twigs). Moves methodically and deliberately, but nimbly, working up the trunk in slow spirals or hitching its way down with the body slightly angled or astride the tree (in other words, not vertical). Unlike Hairy, does not drill for insects. Pokes into nooks and crannies and flakes off bark with soft chiseling blows. Often targets soft damaged portions of trees (rotting knots and bowls). Usually flies only a short distance to a nearby tree.

Works alone or with its mate. Also joins winter flocks. A tame bird that responds fairly well to pishing.

FLIGHT: Overall compact, and except for the white neck, dark and plain. Flies with a series of wingbeats punctuated by a lengthy closed-wing glide. Flight is deeply undulating and lofting; seems more buoyant than Hairy.

VOCALIZATIONS: Call is a soft "*tchew tchew tchew.*" Rattle call is harsh and descending. Also makes a soft "*pcheep*" that sounds like Hairy but is softer. Taps on trees are soft.

Red-cockaded Woodpecker, *Picoides borealis*

The Colonial Woodpecker

STATUS: Rare to locally common southeastern species. In prime habitat, birds tend to cluster in limited areas. **DISTRIBUTION:** The coastal plain from se. Va. to cen. Fla. west across Ga., Ala., Miss., s. and cen. Ark., se. Okla., La., and e. Tex. Distribution within this range is fragmented. **HABITAT:** Large, contiguous, open, mature pine forest, particularly longleaf and loblolly pine. **COHABITANTS:** Brown-headed Nuthatch; Bachman's Sparrow (in park grasslands), Eastern Meadowlark. **MOVEMENT/MIGRATION:** Sedentary year-round resident. VI: 0.

DESCRIPTION: A black-and-white woodpecker with prominent white cheeks and a penchant for mature pines—the only woodpecker within its range with white cheeks. Medium-sized (slightly smaller than Hairy Woodpecker), with a bill proportioned like a male Hairy but overall less robust and conspicuously longer-winged and -tailed than Hairy. Wingtips cross over the back of feeding birds and are often conspicuously elevated or lifted, not flush with the back.

Head is distinguished by white cheeks and a black crown. The minute red tuft set back and above the eye of males is rarely visible. Back and wings are wholly blackish, with beaded white spots; appears barred at close range, uniformly dark gray at a distance. (Hairy and much smaller Downy have white backs.) Underparts are whitish with flecking down the sides, blurring the contrast between the upper- and lower parts and making underparts appear dirty white, not bright white. (The contrast between the largely black upperparts and starkly white underparts of Hairy and Downy is crisp and stark.)

BEHAVIOR: Hitches itself along the trunks and branches of pines, flaking off shards of bark with a prying bill. Bark scales falling to the forest floor often herald the presence of a bird feeding above; the smooth red-scarred surface of well-worked trees also attests to the presence of birds in the area. A colonial bird, Red-cockadeds nest in family groups and sometimes forage together, particularly early in the day. Several cavity trees, identified by a telltale varnish of pine sap surrounding the nest holes, may be clustered within several hundred feet of each other.

An active feeder that moves up (and down and sideways) on a tree's main trunk and branches. Shifts position (and trees) frequently; prefers larger trees. Fairly tame. Very vocal, almost chatty. Infrequently lands on the ground, where it moves with short hops. Not commonly attracted to suet (if at all).

FLIGHT: Appears slightly longer-winged and -tailed than Hairy or Downy. Upperparts are uniformly dark—there is no white back, and the white outer tail feathers often cannot be seen. Flight is classic undulating woodpecker flight, with rapid wingbeats followed by a closed-wing glide. Undulations are pronounced, acute.

VOCALIZATIONS: Most common call is a soft, high, squeaky/raspy "*chee*" or "*chu,*" sometimes uttered with a slight vibrancy: "*ch'r'r'r.*" Recalls the "*keek*" call of Long-billed Dowitcher. Drum is a soft, short, rapid sequence given frequently (intervals of 5–10 secs.). Commonly more than one male in a group is drumming.

PERTINENT PARTICULARS: This is not a difficult identification challenge. The only other black-and-white woodpecker in Red-cockaded's habitat is the much smaller Downy. (The larger Hairy Woodpecker is normally absent where Red-cockadeds nest.) Also, as an endangered species, Red-cockaded is intensely studied. Many nest sites are festooned with flagging or otherwise labeled.

Also, this species is commonly found in the same habitat favored by Bachman's Sparrow, whose song is so far-reaching and easily recognized (a single clear note followed by a musical, varied, warbled trill) that it might serve as an audio marker for Red-cockaded. Nevertheless, there are certainly places that support one of these species but not both.

White-headed Woodpecker, *Picoides albolarvatus*

The Trunk-vaulting Woodpecker

STATUS: Fairly common western woodpecker, but restricted in both habitat and range. **DISTRIBUTION:** Extreme s. B.C. (very rare), e. Wash., e. Ore., Idaho, and Calif., with populations somewhat disjunct. **HABITAT:** A mountain species associated with tall, open-canopy, mixed coniferous-deciduous forest dominated by large-cone pines, especially ponderosa pine but also sugar pine, white pine, Coulter pine, Jeffrey pine, and Douglas fir. **COHABITANTS:** Williamson's Sapsucker, Western Wood-Pewee, Dusky Flycatcher, Pygmy Nuthatch. **MOVEMENT/MIGRATION:** Permanent resident. Some evidence of irregular late summer and fall relocation and concentration to areas of good pinecone production. Also sometimes moves down-slope in winter. VI: 1.

DESCRIPTION: A stunning, distinctive, unique woodpecker—in both appearance and behavior. Hairy Woodpecker–sized and overall glossy black, with a wholly white face and white-laced primaries that form a white wedge on the folded wing and, in flight, a bold white patch near the wingtips (visible from above and below).

Males have a red patch on the back of the head; otherwise, the sexes are similar. Immatures are browner but similarly patterned.

BEHAVIOR: An active nimble woodpecker that habitually uses its wings for foraging. Forages in large part but not exclusively on tree trunks (aspen is a favorite), often very close to the ground. Moves frequently and routinely between trees, often at first light feeding high in the canopy, where insect activity is greatest and, in late summer, where seed-bearing cones are found. Hitches up tree trunks like other woodpeckers, but often vaults to large sections of trunk in vertical wing-flapping ascents—from trunk to trunk and limb to limb. Specifically targets and investigates places where limbs meet the trunk and areas of rough or indented bark where insects might hide. Sometimes flakes bark from a tree; rarely bores into the tree. When feeding on large cones, clings acrobatically to the sides and sometimes hangs beneath the cone. Flies nimbly between trees and also flies high over the canopy when crossing larger distances.

FLIGHT: Broad-winged and fairly long-tailed profile. The white head and white wing patches are distinctive. When foraging, flight is strong, direct, and undulating, with wingbeats that seem theatrically slow. When ascending trunks, wingbeats are audible. When flying great distances, flies with a flourish (has a cape-in-the-wind-like quality); wingbeats are rapid, perfunctory, and given in a quick series followed by a closed-wing glide. Overall flight is smoothly undulating.

VOCALIZATIONS: Call is a sharp, metallic, two-noted "*chi'dik*" whose pitch and quality are somewhat like Nuttall's or Hairy. Rattle call is rapid, strident, sharp, and usually short: "*ch'di'di'di'dik.*" Has a small woodpecker's high pitch but a large woodpecker's volume.

American Three-toed Woodpecker, *Picoides dorsalis*

The Shabby Three-toed

STATUS: Uncommon and localized, with shifting nomadic populations that tend not to wander greatly outside of their established range. **DISTRIBUTION:** A permanent resident across northern regions. Found from Nfld. and Lab. south to n. New England, where it is rare (also found in the Adirondack Mts. in N.Y.), west along the northern shores of the Great Lakes, across boreal Canada to Alaska and B.C., and south, on the east slopes of the Cascade Mts. of Wash. and Ore. and the Rockies into w. Mont., Idaho, Wyo., Colo., n. and e. Ariz., and n. N.M.; also occurs in the Black Hills of S.D. and Wyo. **HABITAT:** Boreal forests, particularly spruce, subalpine fir, and lodgepole pine forests. Concentrations occur in recently burned areas (within about the past three years) and forests heavily damaged by bark beetles. **COHABITANTS:** Black-backed Woodpecker, Western Wood-Pewee, Olive-sided Flycatcher, Gray Jay, Boreal Chickadee. **MOVEMENT/MIGRATION:** Nonmigratory; permanent resident. Some post-breeding dispersal of young is seen in late fall and early winter (November and December). Birds shift to and concentrate in forests affected by fire or insect infestations. VI: 2.

DESCRIPTION: A medium-sized darkish woodpecker more easily confused with the slightly larger Hairy Woodpecker than the even larger and blacker Black-backed Woodpecker. Shaped like Hairy, but has a more oval (not square) head and a slightly larger bill.

Overall rather unkempt, even shabby. The head and tail are jet-black, but the back and wings are flatter, duller, and slightly browner. The face is creased by two white lines (one extending behind the bill and the other, sometimes fainter, behind the eye), and the back is spattered with grayish white; the pattern resembles a Christmas tree. (*Note:* Birds found in the Rockies have much more white on the back and so more closely resemble Hairy Woodpecker.) Grayish white underparts are spattered along the flanks with dark barring—enhancing the overall sense of shabbiness and the lack of contrast between upper- and lower parts, which is more distinctive on the clean-cut Hairy.

Males have a yellow forehead, but both males and females show white flecking on the crown. (The crowns of Hairy and Black-backed Woodpeckers show no pale speckling.) Neither sex shows red on the head. Immatures are similar to adults.

BEHAVIOR: Generally quiet and unobtrusive. Foraging birds work trees tenaciously hammering and flaking off bark, pausing at intervals. Birds often return to the same trees until the bark is stripped. Forages mostly on trunks—both high and low. Where found with Black-backed Woodpecker, generally works higher up on the trunk. Also reported to favor trees that are moderately fire-damaged (not wholly blackened), which makes sense. The bird's not quite jet-black plumage and pale back patch more nearly replicate the bark pattern of fire-scored—not scorched and blackened—trees. Not shy; often allows close approach. Does not respond well to pishing.

FLIGHT: Quick, nimble, and somewhat wandering as it weaves a path through burned woodlands. Wingbeats are somewhat crisp and stiff. Undulates like other woodpeckers.

VOCALIZATIONS: Call is a sharp "*wik*" that is higher-pitched than Black-backed and Hairy (closer to Downy). Rattle call is a rapid string of raspy "*wiks.*" Drum is variable, usually short, and infrequently given—a single regularly spaced series (slightly slower than Hairy) that accelerates and falls off at the end and lasts slightly more than a second. Occasionally birds execute shorter sequential drums (two to four of equal duration given with a pause in between of less than a second) or, after a drummed series, pause and punctuate the set with two quick taps.

Black-backed Woodpecker, *Picoides arcticus*

The Sharp-dressed Three-toed

STATUS: Uncommon, localized, and somewhat nomadic, concentrating in burned-over and beetle-infested forest areas. **DISTRIBUTION:** Permanent resident of northern regions. Found across forested regions of Canada, from Nfld. and s. Lab. south to n. Maine and n. N.H. (also in the Adirondacks of N.Y.), west across the northern boundary of the Great Lakes (and both shores of Lake Superior and n. Lake Michigan), west to the southern and eastern interior of Alaska, the s. Yukon, and the mountainous regions of B.C., south along the Cascade and Rocky Mts. into Wash., Ore., n. Calif., Idaho, w. Mont., extreme nw. Wyo., and the Black Hills of Wyo. and S.D. **HABITAT:** Boreal and montane coniferous forests comprising spruce, tamarack, red fir, hemlock, ponderosa pine, and lodgepole pine. Uncommon in unburned forest; also selects insect-damaged timber. Burned areas are colonized as quickly as three months after burning and remain attractive to woodpeckers for a period of approximately four years. **COHABITANTS:** American Three-toed Woodpecker, Olive-sided Flycatcher, Tree Swallow, Gray Jay, Mountain Bluebird. **MOVEMENT/MIGRATION:** Nonmigratory, but occasionally irruptive, staging periodic but very infrequent movements south of its normal range. VI: 2.

DESCRIPTION: A medium-sized woodpecker (the same size as Hairy Woodpecker) distinguished by its all-black upperparts—head, back, wings, and tail are all glossy jet-black. Slightly larger, blacker, and distinctly more crisply and contrastingly patterned than three-toed. The face is divided by a single white line (no white streak behind the eye). Sides are more crisply and distinctly barred (less haphazardly spattered) than American Three-toed.

Males have a yellow patch on the forehead that is lacking in females and reduced or absent in immature males. Except for the yellow patch, crowns are jet-black (no speckling).

BEHAVIOR: A slow, methodical, somewhat awkward feeder. Working with apparent effort and a tail flipped outward for balance, the bird moves up the tree in a spiraling search pattern. Bores forcefully, throwing its whole body into the tapping action. Black-backed workings look like bullet impact sites—round holes surrounded by pale or reddish halos where the bark has been blasted away.

Holes often have a vertical track (one atop the other). Black-backed concentrates its efforts on lower trunks. Occasionally forages on logs or larger limbs on the ground.

FLIGHT: The white outer tail feathers are evident; otherwise black above, whitish below (more contrasting than American Three-toed). Flight is undulating and somewhat more casual or leisurely than three-toed, with a short series of wingbeats followed by a closed-wing glide.

VOCALIZATIONS: Noisier than three-toed. Call is a low "*chrk*" that recalls Red-winged Blackbird and is lower-pitched than three-toed. Makes a dry rattle, "*rha-ah-ah-ah*," that sounds like a stick being dragged over a concrete sidewalk; also emits a high-pitched single note repeated rapidly: "*whikwikwikwik*." The call has a pattern and cadence that recalls a thumb being run down a comb, and it sounds somewhat amphibian-like. The drum is like American Three-toed, but longer.

Northern Flicker, *Colaptes auratus*

Peterson's Woodpecker

STATUS: Common and widespread; at some point in the calendar year found virtually everywhere in North America. **DISTRIBUTION:** Except for se. Calif., se. N.M., and s. Tex., breeds everywhere in North America south of the tree line. In winter, found almost everywhere in the continental U.S. and some border regions of Canada (coastally north to s. Nfld. and se. Alaska). **HABITAT:** Found almost everywhere trees are found. Most partial to forest edge, small woodlots surrounded by open country, open forests, and park habitats. Habitat types include riparian woodlands, flooded swamps or marsh edge with dead standing timber, burned woodlands, shelterbelts, city parks and suburbia, open pine forest (such as ponderosa pine), aspen stands, and cottonwoods. Shuns only deep woodlands with thick understory. **COHABITANTS:** Varied, but where they occur, may forage in the same habitat as Red-headed, Red-bellied, Acorn, and Lewis's Woodpeckers; also American Kestrel and bluebirds (which use old flicker nest holes). **MOVEMENT/MIGRATION:** Vacates northern portions of its range in winter. Spring migration from late March through May; fall from September to November. In some coastal and peninsula concentration points, the number of migrants may reach into the hundreds. VI: 2.

DESCRIPTION: A *large*, distinctly patterned, long-billed, and pinheaded woodpecker that confounds nonbirders by foraging on suburban lawns and having the temerity not to be a robin. Except for a gray crown, the bird is overall pale brown, with lots of black barring above, spotting below. The black bib across the chest is distinctive. The "Yellow-shafted" (eastern) form has a red triangle on the nape (lacking in "Red-shafted"). The male "Yellow-shafted" form has a black mustache (red in "Red-shafted"). Perched, the yellow, or red, shafts of the flight feathers often peek through the upper surface of the wing. So too does a sliver of the white rump.

BEHAVIOR: A woodpecker that apes a measure of passerine traits. Northern Flicker spends a great deal of time on the ground, foraging for ants and grubs, sometimes in the company of blackbirds. Moves short distances with a series of hops; also shuffles forward. Fairly casual about human presence. In taller grass, raises its head to gauge your approach, returns to foraging, then raise its head after a few seconds to reevaluate the situation.

Sits propped against trunks in typical woodpecker fashion, but when flushed, commonly flies to a tree and sits high, very erect, and very conspicuously on springy twigs. Even on sturdier limbs, it often sits astride the branch, with its tail dangling down (not planted, woodpecker fashion, against the bark). Bows head and body nervously, particularly upon landing, and yelps, protesting the intrusion.

FLIGHT: Overall lanky and somewhat rear-heavy, with a slim projecting head, a long straggly tail, and long broad wings. In flight, easily recognized by the flashing yellow or red underwings and big white rump. Flies with a series of heavy flaps and a closed-wing glide, in typical woodpecker fashion, but the undulations are long and shallow (almost absent) and not as rolling as many woodpeckers. Flight is generally direct, but when flushed, birds often tack to the left and right before lining out. Sometimes their course wanders slightly in level flight as well—recalling a skater moving across the ice. Lands with an amateurish series of breaking flaps that are audible at close range (as are the wingbeats upon takeoff).

VOCALIZATIONS: A very vocal bird with several distinctive calls, including a loud piercing yelp: "*YE-Ur!*" Also utters a sometimes ringing, sometimes muttered, plaintive-sounding "*wicKa, wicKa, wicKa, wicKa, wicKa, wicKa.*" Song is a loud, prolonged, ringing stutter, "*Keh/Keh KEH, KEH, KEH, KEH, KEH,*" that has an angry or defiant quality. Recalls Pileated Woodpecker, but sounds more strident and less rich and resonant than the larger woodpecker. When flushed or startled, gives a muffled exclamation: "*Wha-who-who.*" Drum is loud and even—sometimes slow, sometimes more rapid.

PERTINENT PARTICULARS: On the Great Plains, where the ranges of the "Yellow-shafted" and "Red-shafted" forms abut, intergrades that show a mix of characters occur.

Gilded Flicker, *Colaptes chrysoides*

Saguaro Flicker

STATUS: Common but restricted geographically and by habitat. **DISTRIBUTION:** Found in s. Ariz. and bordering portions of extreme e. Calif. and s. Nev. (eastern limit falls short of N.M.) south into nw. Mexico and Baja. **HABITAT:** Most closely linked to Sonoran Desert scrub habitat, where it nests in saguaro cactus, but also found in riparian desert woodlands. **COHABITANTS:** Gila Woodpecker, Ash-throated Flycatcher, Cactus Wren, Black-throated Sparrow. **MOVEMENT/MIGRATION:** Permanent resident. VI: 0.

DESCRIPTION: A "Yellow-shafted" flicker with the red mustache of a "Red-shafted" seen clinging to a saguaro cactus. A large woodpecker (almost 2 in. larger than Gila, but slightly smaller than Northern Flicker), Gilded Flicker is overall slightly thinner, rangier, and perhaps longer-billed than Northern.

Plumage is similar to the "Yellow-shafted" Northern Flicker—complete with a black chest patch (more a blotch than a chevron), yellow underwings, and a prominent white rump—but overall Gilded is paler, grayer (particularly on the back), and less contrastingly patterned. The red malar on males is usually conspicuous. The brownish or mustard-colored cap is usually hard to see in the harsh desert light. The undertail is mostly black (just yellow at the base), like "Yellow-shafted."

BEHAVIOR: Much like Northern Flicker. Spends a great deal of time foraging on the ground for ants and other insects. Also relishes the fruits of various cacti. Generally found alone or in pairs. Does not associate with other species.

Fairly tame, allowing close approach. When flushed, often takes a high perch some distance away and calls in protest. Does not respond well to pishing.

FLIGHT: Fairly long-winged and long-tailed woodpecker. Shows a conspicuous white rump and yellow underwing linings in flight; the yellow on the undertail seems less conspicuous. Flight is mostly straight (slight to no undulation). Wingbeats are given in a lengthy series followed by a closed-wing glide.

VOCALIZATIONS: Like Northern Flicker.

Pileated Woodpecker, *Dryocopus pileatus*
The Log Cock

STATUS: Uncommon to common permanent resident of eastern, northern, and coastal western mature forests. **DISTRIBUTION:** Found across much of North America, with greatest densities in the Southeast. Ranges across most of e. North America, from N.S. south to Fla. and west to the eastern borders of the Great Plains—e. N.D., ne. S.D., e. Neb., se. Kans., e. Okla., and e. Tex. (absent in coastal portions of s. New England and the mid-Atlantic states, as well as in the agricultural regions of Ohio, Mich., Ill., and Wisc.). Northern range extends across forested regions of Canada to coastal B.C., then south into w. Mont., n. Idaho, Wash., Ore., and n. Calif. **HABITAT:** Mature deciduous or coniferous forests; prefers old-growth conifers, cypress swamps, bottomlands, mature hardwood forests, or younger growth with at least some larger mature trees standing. Also occurs in some suburban areas with large trees. Standing or fallen dead timber and logs are key components of Pileated habitat. **COHABITANTS:** Varied but include Broad-winged Hawk, Barred Owl, Hairy Woodpecker, nuthatches. **MOVEMENT/MIGRATION:** Nonmigratory; permanent resident. VI:1.

DESCRIPTION: With one notable and exceedingly rare exception, North America's only crow-sized, crested, and long-necked woodpecker. (Other woodpeckers appear neckless.) While Pileated's great size, overall ranginess, and conspicuous crest distinguish it, the distinctive plumage pattern also renders it unmistakable.

Bodies of perched birds appear uniformly black—the perfect backdrop to show off the bird's striking black-and-white head and fiery red crest. Much smaller Red-headed Woodpecker is tricolored—red head, black back, white wings and back. The white in Pileated's wings is generally visible only in flight.

BEHAVIOR: The bird doesn't drill holes. It excavates cavities, pounding into the trunks of dead and decaying trees (and live ones as well), crafting signature rectangular or oblong-shaped cavities that may be several inches deep—so deep that the holes in backlit trees are transfused with light, suggesting cathedral windows. Birds may be seen perched high, near the upper trunks of trees, but generally below the canopy.

Unlike many woodpeckers, the "Log Cock" also likes to work low, frequently foraging on branches or logs lying on the forest floor, stumps, and the bottom portions of trees. Most commonly seen alone or in pairs. Forages by flying through the forest. Explores trees, sturdy limbs, stumps, and fallen logs for signs of boring insects. Digs for insects with slow heavy "thumps" given in short series. The sound of working birds and the large shards of wood falling to the forest floor alert observers to the bird's presence. Fairly shy; puts the trunk between itself and observers and flies when pressed, frequently leaving the area. Usually calls when flushed.

Territories are large. Birds frequently shift locations and are often seen flying long distances, from one hillside to the next. Fairly vocal. Calls routinely while foraging, particularly if suspicious or disturbed. Almost always calls when flushed or in response to the call of, or an encounter with, an owl. Often responds quickly and vocally to the imitation of Barred or Spotted Owl (even screech-owl) and attacks owl decoys.

FLIGHT: Flashy and distinctive in flight—a large broad-winged black bird with a reddish crest and flashing white patches on the upperwings and white underwings. Flight is usually straight, not undulating, with a slight bounce or jerk. Wingbeats are smooth, regular, rowing, and somewhat crowlike in cadence and execution, but the flight is overall more buoyant and more flowing than a crow. Pileated looks like a swimmer doing the breast stoke. When birds are descending (for example, flying downslope), they do sometimes undulate.

VOCALIZATIONS: Call is a loud ringing fanfare that recalls an amplified flicker or wild laughter and has an echo quality. Usually begins with one or two spaced notes that accelerate into a series: "*Kuk, kuk, kukkukkukkuklkuk.*" Often trails off at the end. Also makes a more conversational, clucking call, "*wuk, wuk,*" that lacks the ringing quality of the fanfare and has a halting, irregular cadence. Drum (given infrequently) is short, slow, loud, and heavy, given at the rate of one to two per minute, and often slows and softens at the end.

Ivory-billed Woodpecker, *Campephilus principalis*

The Second Coming of the Lord God Bird

STATUS/DISTRIBUTION: Exceedingly rare resident of mature southern riverside forests and for nearly half a century mistakenly believed extinct. In 2004, at least one individual was rediscovered in se. Ark.

HABITAT: Extensive bottomland, swamp forests dominated by mature hardwoods (oaks, sweet gum, cypress-tupelo); historically pine forest. Large numbers of standing dead trees infested with beetle larvae is an essential component. **COHABITANTS:** Wood Duck, Pileated Woodpecker, Lazarus.

MOVEMENT/MIGRATION: Nonmigratory, permanent resident, but this species is known to shift its territory in response to food availability.

DESCRIPTION: An impressively large, crested, black, white, and (males) red woodpecker whose size and splendor was enough to provoke observers to exclaim: "Lawd, Gahd!" Very large woodpecker, approximately three inches longer than Pileated Woodpecker, which it most closely resembles and is the only other North American bird with which it can be confused.

In all plumages, distinguishes itself from Pileated by showing limited white on the face (white does not reach the bill), having a black chin (white in Pileated), and when perched, showing one or two, depending on your angle, white stripes running down the back, merging into an extensive white patch across the lower back (folded wings). This back pattern is most similar to that shown by the much smaller Red-headed Woodpecker. Except for the neck, face, and sometimes one or two small patches of white winking through the folded wings, the body of Pileated Woodpecker is wholly black. Crest of Ivory-billed males is bright red; females black (on Pileated, both males and females show red crests). Young are browner than adults and have shorter crests. The ivory-colored bill, while distinctive, may be confused with the sometimes pale bill of Pileated.

BEHAVIOR: Forages over great distances, sometimes moving several miles in a day across a territory of between three and six square miles. Usually feeds high, above midheight in dead or dying trees, but there are accounts of the bird foraging on the ground. Feeds by flaking bark from trees, hammering with sideways blows and levering loose bark away with the bill. As with American Three-toed Woodpecker, bark-stripped trees may show where Ivory-billed has been foraging. Also excavates borings like Pileated. Historically, birds were commonly seen in pairs.

A late riser, Ivory-billed often doesn't leave roost holes until after sunrise. It then flies to feed—sometimes less than a mile, sometimes three or more miles. Generally works its way back to the roost site during the day, arriving around sunset. Communicates by making double taps, the second sounding like an "immediate echo" (*fide* Tanner) of the first.

FLIGHT: Strong, direct, fast, and straight, without undulations. In flight the bird is reported to look "surprisingly like a [Northern] Pintail" (*fide* Tanner), showing a long and slender neck, long tapering tail, and rather narrow wings. By comparison, Pileated looks more compact. The bold, broad, white trailing edge of Ivory-bill's upperwings renders it almost unmistakable. The upperwings of Pileated show conspicuous but not extensive white windows limited to the base of the outer flight feathers, and the underwings, while extensively white, have a black trailing edge. Ivory-billed shows a white trailing edge above and below.

VOCALIZATIONS: Call is a nasal yap or "*kent*" that has been described as sounding like the call of a loud Red-breasted Nuthatch, and in tone and quality sounds like a note blown through a clarinet or a tin horn. The call is sometimes repeated in a rapid series or doubled with the second note lower-pitched.

FLYCATCHERS

Northern Beardless-Tyrannulet, *Camptostoma imberbe*

The Drab Bush-Mite

STATUS: Uncommon breeder in southeastern Arizona; rare in Texas. **DISTRIBUTION:** A Mexican and Central American species that reaches the northern limits of its range in se. Ariz., occasionally Guadalupe Canyon, N.M., and the tip of the Rio Grande Valley in Tex. **HABITAT:** In Arizona, the riparian deciduous forests and riverside mesquite thickets that cut through arid and semi-arid country; also found along dry washes. In Texas, deciduous forests—most commonly including mesquite, acacia, hackberry, willow, and elm—that have epiphytic moss as a vegetative component. **COHABITANTS:** Black-chinned Hummingbird, Cassin's Kingbird, Rose-throated Becard, Blue Grosbeak, Lesser Goldfinch; in Texas, Ferruginous Pygmy-Owl, Olive Sparrow, Altamira Oriole. **MOVEMENT/MIGRATION:** Resident species throughout most of its range, but in the United States mostly migratory, arriving between early March and late April and departing late August to late September; a few individuals are present October through February. VI: 1.

DESCRIPTION: One small homely-looking bird—almost singular in its drab uniformity. Tiny (smaller than any other flycatcher and most warbler species) and incongruously shaped, with a large head, a short swollen bill, a long body, and a short narrow tail. Looks (and behaves) like an odd vireo more than a flycatcher. Head shape is variable, but most often peaked with a back-turned topknot (recalls a titmouse). When the bird is agitated, the crest is erect and bristly, making the head incongruously large and round. The bill appears very narrow from below, and short, blunt, and slightly downturned from the side.

Overall bland. Brownish gray above, with slightly browner wings and a slightly darker cap; paler and dingy below. Pale wing-bars are indistinct bordering on nonexistent. (The indistinct pale eyebrow and the fine dark line through the eye are sometimes easier to see.) Don't despair if you don't see any recognizable field marks. This is the hallmark of Northern Beardless-Tyrannulet. Aside from overall size and shape, the only characteristic that stands out is the orange-based bill.

BEHAVIOR: A flycatcher in name only. Tyrannulets spend most of their time gleaning insects from bark, branches, and, to a lesser degree, foliage; they rarely flycatch. Northern Beardless-Tyrannulet likes to forage high inside the canopy, but is also found low. Moves through the foliage in short hops, changing perches frequently (much like an active vireo), picking at insects it sights visually from a stationary perch. When foraging, moves its tail with an upward jerk.

Males sing near the tops of trees (but often not the peak), frequently from dead twigs. Singing tyrannulets often remain on the same perch for extended periods, but also makes frequent 180° shifts. Sings most commonly from dawn (sometimes before dawn) to early morning, but also intermittently throughout the day.

Fairly tame. Responds poorly to pishing.

FLIGHT: Small, thin, stubby silhouette. Flight is quick and shrewlike, bouncy and nimble, with no pause or glide. Many of the bird's movements are too abrupt to apprehend. Wingbeats are mostly steady, rapid, and fluttery. Flight sometimes wanders or changes direction abruptly; birds are able to fly vertically, when changing limbs, and hover.

VOCALIZATIONS: Two songs are common; though similar, each is distinctive. The first is a descending series of short, clear, plaintive, whistled notes: "*Peeur peeur peeur peeur peeur peeur.*" The second song has two parts and is higher-pitched, more assertive, and slightly squeakier: "*Chibu chee ch'chee cheeher.*" This song has some of the quality, even the intonations, of Lesser Goldfinch. The call is a plaintive, squeaky, ascending "*eeah?*" sometimes "*eeah p'leu*" (second part descending).

Olive-sided Flycatcher, *Contopus cooperi*

The Portly Flycatcher with an Open Vest

STATUS: Uncommon to fairly common northern and western montane breeder; surprisingly uncommon migrant. **DISTRIBUTION:** Breeds across boreal Canada from Nfld., the Maritimes, and s. Lab. west to all but extreme w. and n. Alaska (absent in the southern prairies and se. Ont.). In the U.S., breeds in n. New England, n. N.Y., n. Mich., n. Wisc., and n. Minn., in the Rocky Mts. south to cen. Ariz. and n. N.M. and in the Guadalupe Mts. in Tex., and along the Cascade and Sierra Mts. south to n. Baja. Winters in Central and South America. **HABITAT:** Montane and northern coniferous forests. Typically found near forest edges or openings, including lakeshores, beaver ponds, bogs, marshes, meadows, canyons, and logged areas; this species is particularly common in recently burned forest. Prefers dead snags and the very tops of trees (above the surrounding canopy). **COHABITANTS:** Black-backed Woodpecker, Tree Swallow, Violet-green Swallow, Steller's Jay. **MOVEMENT/MIGRATION:** Entire population vacates North America after breeding. In spring, arrives late (late April to early June) and departs early (late July to early October). Movements seem somewhat concentrated in mountainous regions. VI: 2.

DESCRIPTION: A large robust flycatcher distinguished by its big head and barroom voice. Overall stocky, even portly, with a bull head, wide body, and short cropped tail. The head is conspicuously peaked, and the bill fairly long and heavy. The eye is large. (Pewees are beady-eyed.) Perched birds—with their heads high and shoulders thrown back—seem to be standing at attention or about to give a speech.

Mostly dark and somewhat shabby-looking, Olive-sided is a small-town politician in an old dark suit that doesn't fit anymore. Dingy gray upperparts and extremely blurry streaked sides give the bird an open-vested appearance. (Eastern, Western, and Greater Pewees button the top button on their "vests.") The "white shirt" peeking through the open vest is a standout feature and is usually easier to distinguish at a distance than the vest itself. Also, in good light, the pale lower mandible can be surprisingly conspicuous. Conversely, the fluffy white tufts protruding above the folded wings are often not visible. The expression is fierce, compared to the more benign expression of pewees.

BEHAVIOR: Most commonly perches conspicuously above the canopy or at the forest edge atop a high bare snag. Feeds almost exclusively by sallying out for flying insects (which it pursues tenaciously), and almost always returns to the same strategic perch. In open burned-over areas, may forage lower (below what would have been the canopy). Very aggressive toward potential predators and harbors a special dislike of jays.

On territory and in spring migration, males (and sometimes females) are very vocal. Sings early and late and often during the day, particularly early in the nesting season.

FLIGHT: With its heavy body, short tail, and short triangular wings, Olive-sided recalls Cedar Waxwing in shape (but not in manner of flight). Sally flight is fast and direct, with rapid wingbeats. Pursuit may be acrobatic. Return flight or long-distance flights (generally over the canopy) are more leisurely; wingbeats are fluttery, but flight is straight and unwavering, with no undulations.

VOCALIZATIONS: Both song and call are distinctive. The two-part song is phonetically rendered "*Hic, three beers*," but the loud plaintively whistled beverage order is more accurately rendered "*whip*" or "*pip reebeer*," with the song rising on "*ree*" and falling on "*beer*." *Note:* At a distance, the first note is often inaudible. Call is a mellow "*pip*" frequently given in a series of three, "*per pih, pih*," whose cadence and, at a distance, attenuated quality recall the "*beep beep beep*" call of American Robin.

Greater Pewee, *Contopus pertinax*

José Maria

STATUS: Common but geographically restricted breeder. **DISTRIBUTION:** At higher altitudes in Mexico and Central America and in the U.S., found only in the mountains of cen. and se. Ariz. and sw. N.M. Very rare in the Davis Mts. of Tex. Casual vagrant in Calif. and s. Tex. **HABITAT:** Open mature pine woodlands (often with a low oak understory) and higher riparian woodlands dominated by Arizona sycamore and Arizona walnut. **COHABITANTS:** Western Wood-Pewee, Steller's Jay, Olive Warbler, Grace's Warbler, Hepatic Tanager. **MOVEMENT/MIGRATION:** Wholly migratory in the United States. Arrives between mid-March and mid-April; departs late August to late October. In drought years or after breeding, birds exhibit some limited down-slope movement. VI: 2.

DESCRIPTION: A large lanky flycatcher that sits high atop a tall snag and calls attention to itself in all seasons with its plaintively whistled song, phonetically rendered "José Maria." Overall slender, almost gaunt (not portly like Olive-sided); also somewhat angular. The head seems small for such a long body (a pewee trait), but the peaked and slightly wispy crest is often conspicuous (easily distinguishing the bird from Western Wood-Pewee). The long, straight, and, from below, very wide bill shows a bright yellow-orange lower mandible. In the absence of any other color on the bird, it can be seen through binoculars at a considerable distance.

Dingy gray above, paler gray below. Plumage is not as smooth and cleanly tailored as Western Wood-Pewee, but also not contrasting and shabby like Olive-sided. Sexes are similar; immatures are washed with yellow on the belly and have buffy wing-bars.

BEHAVIOR: Likes to sit high, bolt upright (or nearly so) on top of a tall pine. Sings tenaciously, but not incessantly (long pauses between sets); although this species vocalizes most often in the morning and evening, it may sing during the day. When perched, moves the head in exaggerated jerks. Sallies out for flying insect prey, often covering considerable distances to intercept prey. Very aggressive toward Western Wood-Pewee; tail-chases are commonplace.

Appears more sedentary than a wood-pewee, changes perches infrequently, and commonly returns to the same perch after a hunting sally.

FLIGHT: Flight profile is long—long face, long tail—with broad pointy wings, though not so acutely pointed as Olive-sided or (especially) Western Wood-Pewee. Flight is rapid and direct; down-stroking wingbeats are strong but slightly jerky, with intermittent hesitations in the otherwise regular cadence.

VOCALIZATIONS: Song is simple, sonorous, and somewhat plaintive: "*soo Say* (pause) *so Reeo.*" Call is a low "*prrp*" or "*pip*" that is similar to Olive-sided Flycatcher.

Western Wood-Pewee, *Contopus sordidulus*

Perch-changing Pewee

STATUS: Common breeding bird in a variety of western woodlands. **DISTRIBUTION:** In North America, found from the western edge of the prairies to the West Coast. Eastern limits are w. Man., cen. Dakotas, w. Neb., nw. Kans., e. Colo., e. N.M., and trans-Pecos Tex. Northern limit of range extends into s.-cen. Alaska, and southern limit to Baja (and south into Mexico). Absent as a breeder in coastal regions of w. Canada and Alaska and portions of Ore. and Wash., as well as in the Sacramento Valley of Calif. and the southwestern deserts. **HABITAT:** Not choosy. Almost any open woodland or woodland adjacent to edge will serve. Favored habitats include ponderosa pine woodlands, riparian corridors dominated by sycamores and/or cottonwoods, open mountain coniferous forests, and pinyon pine. Generally not found in dense woodlands, and favors drier habitats than Eastern Wood-Pewee. **COHABITANTS:** Olive-sided Flycatcher, Plumbeous Vireo, Cassin's Vireo, Mountain Chickadee, Yellow-rumped Warbler, Pacific-slope Flycatcher, Purple Finch, Bridled Titmouse, White-breasted Nuthatch. **MOVEMENT/MIGRATION:** Wholly migratory; casually wanders east to the Gulf Coast in fall. Spring migration from early April to early June; fall from early August to October. VI: 2.

DESCRIPTION: A slender generally dark flycatcher (about 1 in. longer than most *Empidonax* flycatchers; 1 in. smaller than Olive-sided Flycatcher), with a thin bill, a small slightly peaked head, a long body, longish wings, and a fairly long tail.

Overall brownish gray, trim, and clean-looking (not unkempt or disheveled), with a smooth and slightly opened vest. Face is plain, almost expressionless. Wing-bars are narrow, indistinct, silver-edged slivers (often only a single wing-bar is evident). Distinguished from grayer *Empidonax* flycatchers by its larger size, expressionless dark face (with only the barest hint of an eye-ring), barely discernible wing-bars (adults), and dirty undertail coverts. Olive-sided Flycatcher is bigger-headed, stockier, and shabbier, with a big bill and a big eye. Western is overall slightly darker and dingier than Eastern Wood-Pewee, with narrower, less conspicuous wing-bars and a slightly more open-vested appearance, but the two species are best distinguished by range and voice.

BEHAVIOR: A consummate perch-hunter. May sit high, above the canopy, or low (but generally more than 10 ft. above the ground). Hunts from snags or living branches. Specializes in lengthy sallies; covers more ground than most *Empidonax* flycatchers. Most prey is captured in the air after a direct approach and a short set-winged glide. Less commonly plucks prey from vegetation (hovering may be involved). Frequently does not return to the same perch; either continues on to another location or returns to the point of departure but takes another perch. Does not flick its tail like many smaller flycatchers.

FLIGHT: Appears long-headed with short triangular wings. Flight is straight, direct, and unwavering, with rapid and steady wingbeats. Seems to be moving slowly for the amount of energy being expended. When covering large distances, sometimes flies with some slight hesitation, as if uncomfortable being out in the open for so long, or with some jerkiness, as though running out of fuel.

VOCALIZATIONS: Song is a burry petulant-sounding whistle, "*wheeer?*" or "*vee-eeurr?*" that rises on the first syllable and descends on the second, sometimes with a compressed three-note quality. Recalls somewhat Common Nighthawk. A variation is a more wistful and melancholy-sounding ascending "*err-eee*" that is more musical than the primary song but still nasal. Calls include a soft gulping hiccup, "*d'jul,*" and a single "*pip.*"

PERTINENT PARTICULARS: As noted, Western Wood-Pewee seems more inclined to shift perches while hunting than Eastern Wood-Pewee (which almost always returns to the original perch after a sally). This tendency may relate to habitat more than behavioral programming. Eastern woodlands are generally more confining, and open perches in shorter supply. Western woodlands are more open, inviting longer sallies and greater perch options.

Western Wood-Pewee is often confused with Willow Flycatcher, particularly during migration, when birds may be out of habitat.

Eastern Wood-Pewee, *Contopus virens*

Flycatcher on a String

STATUS: Fairly common and widespread eastern woodland breeder. **DISTRIBUTION:** Eastern half of the U.S. and s. Canada—from N.S. across s. Que., s. Ont., and s. Man. to se. Sask. Western border cuts through cen. Dakotas, e. Neb., cen. Kans., w. Okla., and cen. Tex. Absent as a breeder in s. Tex. and all but n. Fla. **HABITAT:** Breeds in various forest habitats, including deciduous and coniferous (preferred in the South) woodlands and riparian and dry forests. Prefers edges, forest clearings, and, in the forest interior, places with more open canopy. **COHABITANTS:** Least Flycatcher, Great Crested Flycatcher, Summer Tanager, Scarlet Tanager. **MOVEMENT/MIGRATION:** Wholly migratory; winters in South America. A late-spring migrant, Eastern Wood-Pewee reaches the Gulf Coast in early April and migrates through May and early June. Fall migration is from early August through October. VI: 2.

DESCRIPTION: A dark, compact, small to medium-sized flycatcher (larger than *Empidonax* flycatchers; smaller than Eastern Phoebe) that haunts the middle forest canopy and habitually returns to the same perch. Sturdier and more angular than Eastern Phoebe, with a large, somewhat shaggy and peaked head, a large bill with a mostly yellow-orange lower mandible, a wide body, a short straight tail, and primaries that extend half the length of the tail.

Overall gray with a slight greenish cast to the back and no distinct contrast between the head and back. Wings of adults have obvious white wing-bars (buffy in immatures). The breast is covered by a smooth shadowy vest that is closed at the throat. The black eye, set on the pale gray face, is easily noted and is accentuated by a narrow half eye-ring encircling the rear of the eye.

The tail is twitchless.

BEHAVIOR: Repeat: the tail is twitchless, and the wings flickless. (Phoebes wag their tails; many empids flick their wings and jerk their tails.) The bird sits on a conspicuous perch, at middle heights or at the top of a tree, as still and untranquil as a rubber band held at full draw. Darts out on short triangular wings to snap up insects, and commonly snaps back to the same perch as though drawn by an elastic leash. During migration, occasionally hunts from low brushy edges of fields (behavior more typical of phoebes). Generally solitary; on territory, aggressive toward other species (not to mention other pewees), and in migration keeps to itself.

FLIGHT: Long, slender, and big-headed, with short triangular wings and a curiously flat, almost two-dimensional profile. Flight is quick, direct, slightly jerky, and sometimes slightly wandering or tacking, with rapid, steady, but slightly irregular wingbeats.

VOCALIZATIONS: Song is a slow, measured, two-part lament—a question followed by an answer. The first part is a plaintive two- or three-note question, "*Pee-whEE?*" or "*Pee-urr-EE?*" (ending on a high note), followed by a long pause and then a reply: "*PEE-urr.*" Sounds to some as though the bird's song is inhaled then exhaled. Late in the season, birds often dispense with the second part. Call is a flat "*pit.*"

Yellow-bellied Flycatcher, *Empidonax flaviventris*

The Easy Eastern Empid (Relatively Speaking)

STATUS: Uncommon to fairly common northern breeder; uncommon migrant. **DISTRIBUTION:** Breeds throughout the boreal forests of Canada from Nfld., N.S., and Lab. west to the cen. Yukon and extreme w. B.C. and possibly extreme e.-cen. Alaska (absent in s. B.C., s. Alta., s. Sask., and s. Man.). In the U.S., breeds in n. Minn., n. Wisc., the Upper Peninsula of Mich., the Catskill and Adirondack Mts. of N.Y., n. Vt., n. N.H., and all but s. Maine (also breeds, sparingly, in n. Pa.). Winters in s. Mexico and Central America. Migrates from the eastern prairies east, but avoids the Southeast and Fla. **HABITAT:** Wet, flat, poorly drained boreal forests, particularly bogs and muskeg. Found in both coniferous and mixed forests (with Black Spruce commonly the dominant tree type), but especially those that are well stratified with a canopy, a shrub layer, and an abundance of moss. Also, but less typically, occupies other northern forest habitats, such as forested talus slopes and jack pines. In migration, found in riparian woodlands, flood plains, forest understory, and thickets. **COHABITANTS:** Boreal Chickadee, Palm Warbler, White-throated Sparrow. **MOVEMENT/MIGRATION:** Migrates late in spring, early in fall, with fairly abbreviated migratory periods. Spring migration from late April or early May to early June, with a peak in mid to late May; fall from late July to early October, with a peak in late August and early September. VI: 2.

DESCRIPTION: It's the greenish-backed, somewhat pudgy empid with yellow on the throat seen east of the prairies. Smallish and compact (slightly larger than Least; smaller than Acadian), with a big round head, a short wide bill, a portly body, and a short tail. Least Flycatcher is more elfin-featured; Acadian Flycatcher is more robust, with a longer, heavier bill, a more peaked head, and a longer tail; and Alder and Willow Flycatchers also have a more peaked head and are overall gaunt.

Overall olive-toned above, with contrastingly slightly blackish wings. Shabbily olive-vested underparts show a distinct yellowish cast, particularly on the throat. The lower bill is rich yellow. The eye-ring is complete, conspicuous, and somewhat yellow, but normally not as crisply defined as Least. Wing-bars (white in adults, yellowish on immatures) are very conspicuous; in fact, they fairly pop against the backdrop of the bird's blackish wings. *Note:* In fall, some worn adults may show only traces of yellow, which in poor light will be difficult to see.

BEHAVIOR: Likes to forage where it's thick—in the lower, thicker portions of conifers and shrubs; in thickets; and in lush edge and openings in dense forests during migration. Favors wetter areas and woodlands or thickets adjacent to water. Feeds in tight places, often just below a vegetative canopy where it lunges for insects in the leaves above. Hovers frequently to pluck insects and commonly captures prey on the ground. Wing flicks and tail flicks with an abrupt upward jerk.

Usually solitary, but in migration may move with foraging migrant flocks made up of warblers, vireos, and other flycatchers. Responds apathetically to pishing, or may simply be loath to break into the open.

FLIGHT: Dark stocky profile; the darkish breast and greenish back are usually apparent. Flight is direct, with regularly spaced bouncy undulations.

VOCALIZATIONS: The two-noted "*Che-bunk*" song is similar to Least Flycatcher's "*Chebeck*," but softer, less emphatic, and more casually paced. Call is a plaintive, up-sliding, whistled "*chew-ee*" that recalls Eastern Wood-Pewee. Also makes a sharp, abrupt, sneezy, descending "*peeu.*"

Acadian Flycatcher, *Empidonax virescens*

Deep Dark Forest Flycatcher

STATUS: Fairly common breeder in mature closed-canopy eastern forests. **DISTRIBUTION:** Breeds north to s. Mass., s. N.Y., extreme s. Ont., cen. Mich., and e. Minn., west to e. Iowa, se. Neb., e. Kans., e. Okla., and e. Tex. (does not breed in coastal Tex., La., or s. Fla.). Winters in Central and South America. **HABITAT:** Prefers large tracts of mature deciduous forest, but also found in conifers, including tamarack swamps, white pine plantings, and hemlocks. Especially partial to wet areas or areas adjacent to streams. The nest, anchored to the end of a forked branch above the shrub layer and beneath the forest canopy, is often situated over a stream, road, or trail. In migration, usually stays within the forest interior. **COHABITANTS:** Red-eyed Vireo, Blue-gray Gnatcatcher, Louisiana Waterthrush, Summer and Scarlet Tanagers. **MOVEMENT/MIGRATION:** Spring migrants appear on the Gulf Coast in late March and reach northern nesting territories by mid-May. Fall migrants occur from late July through early October, with peak movements in late August and early September. VI: 1.

DESCRIPTION: A big sturdy empid (built like a light heavyweight) with a large head, a solid but well-trimmed body, long wings, and a long wide tail ("large," "solid," "long," and "wide" for an *Empidonax* flycatcher anyway). The bill is long, broad, and sturdy—the longest of any empid; the lower mandible is all yellow. The eye is distinctly large and prominent. The changeable head shape is sometimes domed, sometimes peaked with a flat forehead.

In all plumages, greenish brown-gray upperparts. Adults have grayish white underparts with a shadowy suggestion of a breast-band and little or no yellow on the belly and undertail coverts. The eye-ring is narrow, the wing-bars narrow and white. Immatures sometimes appear slightly scaly-backed and have more yellow on the underparts (sometimes even on the throat) and a more prominent olive breast-band. Wing-bars are broader and buffy.

BEHAVIOR: A bird of the forest interior—unlike most other *Empidonax* flycatchers, not prone to be found on woodland edge. Generally forages amid the leaves of the lower canopy and understory. (Acadian Worm-plucker would be a more apt name for this species.) Also darts out and hovers to pluck prey and, in true flycatcher fashion, snatches insects from the air. Usually does not return to the same perch after a successful sally, and in fact, when hunting, changes perches frequently, working up through the vegetative layers to the lower canopy.

When singing, males are often quite low (6–20 ft. from the ground) and in plain view near the trunk of a tree or an exposed branch above a watercourse. Early in the season, commonly remains on the same perch for several minutes, calling on average every 20–30 secs. Later in the season, vocalizes while moving and feeding.

Not as nervous as some *Empidonax* flycatchers, Acadian jerks its tail upward but does not commonly wing-flick.

FLIGHT: Flight is quick, direct, darting, and usually short (less than 10 ft.). When birds are covering greater distances, flight is strong, with shallow undulations. Acadian hovers with consummate ease on very rapid wingbeats and can even fly backwards for short distances.

VOCALIZATIONS: Males belt out their song—a loud, emphatic, slightly slurred "*WEE-SEE'T!*" (accent on both syllables). Females also give this call, but more softly. Classic call is a loud emphatic "*Wheet!*" (similar to the first note of the song). Also makes a loud, whistled, slightly slurred "*zWeee*"; a softer, hurried, slurping protest, "*p-t'lurp*"; and a softly uttered, excited "*pip-pip-pip-pippippip*" that is compressed and rushed at the end.

PERTINENT PARTICULARS: Key to the identification of this bird is its large size, overall broad proportions, large eye, greenish-tinged upperparts, and location in the forest interior. Like most empids, Acadian does not respond well to pishing, but may investigate if a mobbing action is incited.

Alder Flycatcher, *Empidonax alnorum*

Alder Bushwhacker

STATUS: Common and widespread northern breeder; common migrant east of the prairies. **DISTRIBUTION:** Breeds throughout most of Alaska (except extreme western and northern portions) and in subarctic Canada (except w. and s. B.C., s. Alta., s. Sask., and sw. Man.). Also widespread in the ne. U.S. Found in all states bordering Canada west to N.D., as well as in ne. Ind., n. Ohio, w. Mass., n. N.J., all but se. Pa., w. Md., e. W. Va., sw. Va., and nw. N.C. **HABITAT:** Wet deciduous thickets, primarily willow or alder thickets, but also deciduous (maple, birch, poplar) forest in early successional, dry fields with brushy edge or other brushy component. Likes to be near or over water, but not invariably so. In migration, prefers low wet or dry brushy edge or open fields with a brush component; also occurs in phragmites marsh. **COHABITANTS:** Yellow Warbler, Wilson's Warbler, Lincoln's Sparrow. **MOVEMENT/MIGRATION:** Late-spring and early-fall migrant. Spring migration from early- or mid-May to mid-June, with a peak in late May; fall from early August to late September, with a peak at the end of August. VI: 2.

DESCRIPTION: A lanky, long-faced, greenish-tinged flycatcher that maneuvers through brush like a frantic warbler. A large empid but a small songbird (larger than Wilson's Warbler; smaller than Yellow Warbler; the same size as the very similar Willow Flycatcher), with a long narrowish bill, a slender head, a nominally slender body, and a long straight-sided tail. The head is sometimes gently rounded, sometimes peaked toward the rear, *but always slender*; in combination with the very long bill, the slender head makes this species appear very long-faced. Despite the yellow lower mandible, and except from below, the bill usually appears dark.

Another one of the shade-splitting "ishish" birds. Overall grayish brown above, with a slightly greenish tinge; wings show narrow buffy or grayish wing-bars. Underparts show a whitish throat, a faint grayish wash across the breast, and a whitish or slightly yellowish belly. Shows a complete, well-defined, exceedingly narrow white eye-ring with no kick or tear at the rear.

BEHAVIOR: The bird thrives in low brush and tight vegetative confines (usually within 10 ft. of the ground). On territory, it flies at high speed into alder thickets and reappears somewhere else. It snatches insects from interior leaves and branches with fluttering leaps, hovers and plucks from outer branches, and engages in short, wild, wing-fluttering gyrations among the branches (before returning, in many cases, to the same perch). Alder looks like a flycatcher but in many ways hunts like a hyperkinetic foliage-gleaning warbler. Also sallies out to catch insects in the air, but secures most prey by gleaning.

Despite its sometimes energetic feeding habits, Alder is not particularly high-strung. Males vocalize from exposed perches—usually from the highest point around and often on a dead branch—and birds commonly sit for extended periods before relocating (or being chased off by another thicket-loving species). Alder twitches its tail in measure with its functionally two-noted call, but otherwise (and except for jerks of the head) sits calmly. Fairly shy. Does not respond to pishing.

FLIGHT: Slender, big-headed, long-tailed, fairly long-winged profile. Flight is straight, hurried, almost frantic, often low, and characterized by an upward jerkiness. Wingbeats are quick, strong, stiff, and mostly continuous, but punctuated by terse intermittent skips and pauses.

VOCALIZATIONS: Song is a short, rough, nasal, two-and-a-half-note rebuke: "*vr'r'ree Beer'r.*" The first part drags; the accent falls on the second part. The call is often phonetically rendered "*fee-bee-o*" (three syllables), but at any distance it attenuates to two notes (or one and a compressed two), and the rough burryness of the initial phrase seems less "*fee*" than a snarling "*v'r'ree.*" Call is a soft "*pip*" or "*pi/up*" that recalls water dripping from a faucet and is more musical and full-bodied than the terse dry "*whit*" call of Willow Flycatcher.

PERTINENT PARTICULARS: The combination of its fairly large size, lanky appearance, long face/bill, narrow white eye-ring, and generally brownish plumage distinguishes this species from all other empids except for its near twin, Willow Flycatcher. Alder is easily distinguished from Willow by song (Willow's burry two-noted "*Fitz-bew*" is accented on the first syllable, and Alder's "*vr'r'ree-Beer'r*" is accented on the second) and call (a soft mellow "*pip*" for Alder, a muttered spit "*whit*" for Willow), but away from breeding territories, the two species are very difficult to separate in the field. In the West, differences in upperparts (more greenish in Alder, more brownish in Willow) and wing-bars (whitish in Alder, buffy in Willow) are helpful. But in the East, where Willow is more like Alder in all respects, the most helpful characteristic is often humility (on the part of the observer).

Willow Flycatcher, *Empidonax traillii*

Plain Brownish Flycatcher

STATUS: Common and widespread, primarily northern breeder; widespread migrant except in the coastal Southeast and Florida. **DISTRIBUTION:** Unlike the northern-breeding Alder Flycatcher, Willow Flycatcher breeds primarily in the n. U.S. and border regions of Canada, nesting in s. B.C., s. Alta., s. Sask., s. Man., se. Ont., extreme s. Que., N.B., and P.E.I. In the U.S., found in Wash., Ore., scattered locations in e. and s. Calif., Idaho, w. Mont., e. Nev., Utah, all but s. Ariz., Wyo., w. and cen. Colo., all but s. and e. N.M., N.D., much of S.D., n. and e. Neb., s. and w. Minn., Iowa, Mo., s. Wisc., Ill., Ind., much of Ky., parts of Tenn., Pa., W. Va., w. Va., N.Y., Vt., s. N.H., coastal Maine, Mass., Conn., N.J., much of Md., and Del. Winters in Mexico and Central America. **HABITAT:** Brushy habitat, often adjacent to streams, marshes, or dry fields. Typical habitats include willow thickets, successional dry fields, shrub islands, and brushy edge adjacent to tidal and freshwater marsh, early grown clear-cuts, and woodland edge. Prefers areas with low shrubs and few tall trees and seems less wedded to wet areas than Alder Flycatcher. **COHABITANTS:** Yellow Warbler, Song Sparrow. **MOVEMENT/MIGRATION:** Late-spring migrant and early fall. Spring migration from early May to mid-June; fall from early July to late September. VI: 2.

DESCRIPTION: A lanky, plain, brownish, peweelike flycatcher that flies nimbly through low brush, not woodlands. A small flycatcher but a large empid (larger than Least Flycatcher; smaller than a pewee; about the same size as Alder and Dusky Flycatchers), with a large (both long and broad) bill, a peaked head with a flattish forehead, a long body, and a moderately long sturdy tail. Often indistinguishable from Alder, except by vocalizations and partially by range, but compared to Alder seems slightly bigger-billed, more peak-headed, and slightly shorter-tailed. Western birds in particular may show more yellow on the lower bill when viewed from the side.

Overall drab. Upperparts are distinctly brownish; eastern birds have a touch of olive on the back that is so subtle it's probably your imagination. Underparts show a whitish throat, a pale, brownish, often broken wash on the breast, and dirty white underparts sometimes slightly tinged with yellow. The eye-ring is inconspicuous or lacking. Wing-bars are narrow and dusky or buffy. The expression is plain. Willow is the plainest-faced empid; Alder is almost akin, and Gray Flycatcher is close behind.

BEHAVIOR: A brush-hugging flycatcher that hunts low, sometimes in fairly open habitats, but more often in heavy vegetation, including willow thickets, cattails, and phragmites. Hunts horizontally, making short flights between perches, but is not at all shy about making longer flights across open areas (100 yds. or more). Seems equally adept at snapping insects out of the air and at plucking prey from branches (where Alders may spend more time bushwhacking than flycatching). Males commonly vocalize from an exposed perch at the top of a bush (often the highest perch around) and usually remain in the same location for lengthy periods. Willow vocalizes at all times of day, including at night. When agitated, jerks its tail aggressively.

Fairly tame. Responds poorly to pishing.

FLIGHT: Appears slender, long-faced, and long-tailed. Flight is direct, rapid, darting, and slightly jerky, on wingbeats that are quick, down-pushing, and, though broken with frequent intermittent skips and pauses, nearly steady.

VOCALIZATIONS: Song is a low, rough, burry, two-note sneeze: "*r'r'rEE-yew.*" Also emits a low, rough, ascending, rippling "*r'lEEP*" or "*r'r'rEEP.*" Call is a low, terse, muttered spit, "*whit*" or "*pit.*"

PERTINENT PARTICULARS: Both Willow and Alder Flycatchers have broad-based and parallel-sided tails (as does Acadian). Other small flycatchers in this group have tails that are more narrow at the base and slightly to modestly flared along their length.

Least Flycatcher, *Empidonax minimus*

Subcanopy Flycatcher

STATUS: Common breeder in northern forests; common migrant east of the Rockies. Very rare migrant in the far West. A few winter in Florida.

DISTRIBUTION: Breeds across subtundra Canada (absent coastally in B.C.) from s. Alta. east to s. Nfld. In the U.S., breeds in ne. Wash. (a few in ne. Ore.), Mont., n. Wyo., N.D., n. and e. S.D., ne. Neb., Minn., w. Iowa, Wisc., n. Ill., Mich., n. Ind., nw. and ne. Ohio, most of Pa., and N.Y. (except Long Island). Also found along the Appalachians south to n. Ga. **HABITAT:** Dry mature or second-growth deciduous forests and mixed coniferous-deciduous forests that are fairly open or have closed canopies and open understories. In other words, this bird wants room to maneuver. In migration, makes greater use of forest edge. **COHABITANTS:** Yellow-bellied Sapsucker, American Redstart, Rose-breasted Grosbeak. **MOVEMENT/MIGRATION:** Spring migration from mid to late April to late May; fall from mid-July to mid-October. VI: 3.

DESCRIPTION: A small, compact, brownish gray empid with a bold, white, round eye-ring and a penchant for hanging out just below the canopy. Except for Buff-breasted Flycatcher, the smallest empid (half an inch smaller than Acadian, Willow, and Alder Flycatchers; a quarter-inch smaller than Hammond's Flycatcher; about the same size as Black-capped Chickadee and American Redstart), with a moderately sized, mostly dark bill, a large roundish head, a fairly slender body, and short wings that barely project beyond the base of the average to shortish tail. Yellow-bellied Flycatcher is bigger-headed and more portly; Acadian is larger, bigger-billed, and more robust; Alder and Willow are lankier and longer-billed; and Hammond's is longer-winged and narrower-billed. If looking at the head of this bird gives you a momentary impression of Gray-cheeked Thrush, you're picking up on the roundness of the head and the relative length of the bill.

Upperparts are grayish to olive brown (immatures), with white to buffy wing-bars; underparts show a whitish throat, a dingy chest, and a whitish or slightly yellow belly. *Note:* The eye-ring is white, complete, well defined, and shows a white kick or tear behind.

BEHAVIOR: On territory, likes to hang out just below the canopy (in migration, often lower), perched on a dead snag or twig that is not necessarily the most conspicuous perch in town. Active and noisy. Changes perches often; calls almost incessantly. Forages by flycatching and by plucking insects from foliage after short flight. Movements are quick, calculated, direct. Seems equally adept at both flycatching and gleaning, and conditions directly related to insect activity (and indirectly to habitat and temperature) probably dictate which method the bird employs. Also hovers and occasionally parachutes (on open wings) to a lower perch. Usually hunts in the lower part of the canopy or subcanopy, but in migration may hunt even closer to the ground and in brushier habitat at woodland edges. Avoids open unwooded areas.

Aggressive and territorial toward other Least Flycatchers and other forest species. A desultory wing-flicker, Least does jerk its tail when singing. Fairly tame; only moderately responsive to pishing.

FLIGHT: Much like other *Empidonax* flycatchers in overall shape and manner of flight, but small size, shorter proportions, and uncontrasting dinginess may help distinguish it.

VOCALIZATIONS: Song is an incessant series of dry emphatic hiccups: "*chibek chibek chibek chibek*" (accent on the second syllable). The sound has both a sharp quality and a nasal undertone, but the pattern and persistence distinguish it from all flycatchers except Yellow-bellied. The song of Yellow-bellied, which breeds in thick spruce wetlands and likes to stay low, has the same pattern as Least but is less emphatic and slightly lower-pitched. Least's call is a hard, often rapidly repeated "*PIT.*"

PERTINENT PARTICULARS: Least Flycatcher has often been associated with forest edges and clearings; although birds of this species certainly occur there, they generally avoid being in the open.

Hammond's Flycatcher, *Empidonax hammondii*

Kingletlike Flycatcher

STATUS: Common western breeder. **DISTRIBUTION:** Breeds from cen. Alaska south through the Yukon, B.C., sw. Alta., Wash., Ore., n. Calif., Idaho, w. Mont., w. Wyo., e. Utah, w. and cen. Colo., and n. N.M. Migrates mostly west of the Rockies. Winter range in Mexico reaches north just into se. Ariz. **HABITAT:** Closed-canopy, northern, and high-elevation forests of hemlocks, spruce, Douglas fir, and aspens. (Dusky Flycatcher prefers more open canopy.) **COHABITANTS:** Swainson's Thrush, Yellow-rumped Warbler, Red Squirrel. **MOVEMENT/MIGRATION:** Spring migration from early April to early June; fall from early August to early November. VI: 2.

DESCRIPTION: A small compact forest flycatcher that sings from (or near) the tops of conifers, forages in deciduous trees, and has a head and face that recall Ruby-crowned Kinglet. A small songbird (larger than a kinglet; smaller than a crossbill; about the same size as Yellow-rumped Warbler) and a medium-small empid (slightly larger than Least Flycatcher; slightly smaller than Dusky and Alder), with a small mostly blackish nub of a bill, a large rounded head (sometimes slightly peaked just back of center), long wings whose tips sometimes dangle from the bird's side and when fully folded extend beyond the undertail coverts, and a tail that looks short in no small part because of the length of the wings.

Grayish above (particularly gray-headed), with whitish wing-bars and a somewhat vague eye-ring that's fullest behind the eye. The ring is obvious enough, but often not as sharply defined as on some other empids. The whitish throat and slightly yellow-tinged belly are separated by a pale gray vest. In fall, birds are olive-tinged above, with a contrastingly gray head, and more yellowish below.

With its round head, petite bill, and vague eye-ring, no other flycatcher so quickly conjures the image of Ruby-crowned Kinglet.

BEHAVIOR: Although associated with coniferous forests, in mixed forest this species appears to spend as much time—if not more—foraging in deciduous trees, making quick short (often less than 1 ft.) sallying flights among leaves and branches, plucking insects en passe; also hovers and flycatches. Singing males commonly command the very highest point of a tall spruce or fir—often a dead branch—and vocalize for long periods. Also forages mostly in the canopy, although sometimes hunts low, in the understory, flying from low branch to low branch (somewhat in the manner of a foraging Swainson's Thrush).

Often a wing-flicker, but not a tail-jerker of histrionic proportions (adding to the bird's kingletlike quality). Generally unresponsive to pishing.

FLIGHT: Silhouette is like other small flycatchers: shows a projecting head, a long tail, and long wings. Flight is direct, quick, and furtively hurried and jerky, with near-steady wingbeats punctuated at short intervals by a skip/pause.

VOCALIZATIONS: The three-part song is similar to Dusky Flycatcher, but squeakier and somewhat more tentative, as if the bird is uncertain about the pronunciation: "*seeBIK; chur'rLE'E; SEEbur.*" The first phrase is quick, high, sharp, and rising (recalls Least Flycatcher's "*chebek,*" but the notes are higher and clearer). The second phrase is muttered and descending, and the third phrase is rougher, burry, slurred, high, and rising. The pattern and sequence are fluid. Birds may change the order and sometimes drop the muttered second phrase entirely, saying: "*seeBIK; chur'rLE'E.*" Call is a sharp but mellow "*pip!*" that recalls Alder Flycatcher. Also makes a low, mellow, whistled "*peur.*"

PERTINENT PARTICULARS: Hammond's Flycatcher molts before migrating in the fall. Dusky does not. The result is that in the fall Hammond's appears darker, crisper, and cleaner; Dusky is paler and shabbier.

Gray Flycatcher, *Empidonax wrightii*

The Phoebelike Empid

STATUS: Fairly common breeder of the intermountain West. Rare migrant along the Pacific Coast. Casual east of its normal range. **DISTRIBUTION:** Breeds from extreme s.-cen. B.C. into cen. and e. Ore., extreme s. Idaho, Nev. (except extreme s. Nev.), and bordering portions of Calif., Utah, sw. Wyo., w. and sw. Colo., n. and e. Ariz., and n. and w. N.M. Winters in Mexico, though some birds are found in se. Ariz. and near Big Bend, Tex. **HABITAT:** Dry woodlands and brush lands, including sagebrush, pinyon-juniper, and ponderosa pine. **COHABITANTS:** Dusky Flycatcher, Western Scrub-Jay, Pinyon Jay, Juniper Titmouse, Black-throated Gray Warbler. **MOVEMENT/MIGRATION:** Entire populations vacate breeding areas for the winter. Spring migration from late March to late May; fall from mid-August to late October. Migrants regularly occur as far west as southern California and east to trans-Pecos Texas. VI: 2.

DESCRIPTION: In a field distinguished by hair-splitting sameness, this empid is refreshingly easy to identify. Really. Overall large and long (the largest empid, Gray is a full quarter-inch larger than Dusky or Alder!), with a round head, a distinctly long straight bill, and a long tail. Other empids are described as "large" and "round-headed" and "long-tailed," but this one is the reference standard. Trust your eyes. It's an empid that reminds you of a phoebe, and if structural similarities were not enough, Gray Flycatcher even gives its tail a phoebelike wag.

In summer, on territory, overall gray top to bottom—no green, no buff, no yellow-gray. Uniformly pale gray above, paler gray below, with just the suggestion of a vest (open right down to the belly). Grayish wing-bars are almost nonexistent. With indistinct pale grayish spectacles/eye-rings, Gray Flycatcher has a face almost as plain as Willow Flycatcher. The expression is of wide-eyed innocence.

In fall migration and in fresh plumage, birds show stronger wing-bars, are washed with olive above, and show a tinge of yellow below. In other words, they look much more like Dusky at precisely the time when the two species may be seen together. Gray's greater bill size and tail-wagging behavior are the characteristics on which to focus.

BEHAVIOR: Likes to perch on dead twigs. Hunts near the tops of trees (seems very partial to certain perches) and in heavy brush; forages low (often near the ground) in the understory, which it hunts by changing perches frequently. Sallies for insects, sometimes flying nimbly around brush to intercept; also regularly captures insects on the ground. Commonly sits on the shady side of a bush. Habitually wags its tail when perched—*down, then up*, the opposite of other empids. When nervous, flicks both wings and tail. Fairly tame. Responds fairly well to pishing.

FLIGHT: For long distances, flight is straight, strong, fast, and slightly jerky, with stiff rapid wingbeats. For shorter distances (in brush), flight is nimble and erratic.

VOCALIZATIONS: Song is in two parts: first a low, rough, emphatic, two-noted "*chr'leep!*" or "*chr'wik!*" (often repeated several times), followed by a quick, soft, higher-pitched, whistled "*chir'EEp?*" (sounds like a baby bird begging for food). Call is a liquid softly spoken "*wheet.*" Also makes a high descending "*wheer?*"

Dusky Flycatcher, *Empidonax oberholseri*

The Intermediate Flycatcher

STATUS: Common breeder in western mountain woodlands. **DISTRIBUTION:** From the extreme s. Yukon to s. Alta. and w. Mont. south to the middle portions of N.M., Ariz., and s. Colo. (also disjunct populations in n. Baja and the northeast corner of Wyo. and w. S.D. [Black Hills]). Absent in the Central Valley of Calif. Winters just south of the U.S. border, with a few found in se. Ariz. **HABITAT:** A bird of brushy areas and dry open woodland habitat, including aspen groves, open coniferous forest (typical trees include ponderosa pine and Douglas fir), forests where logging has thinned or left openings, mountain chaparral, and willow thickets. **COHABITANTS:** (Unfortunately) Hammond's and Gray Flycatchers, which it closely resembles; also Pygmy Nuthatch. **MOVEMENT/MIGRATION:** Migrates through the intermountain West, with few records east of breeding range. Spring migration primarily from late April through May; fall in August and September. VI: 1.

DESCRIPTION: A medium-sized *Empidonax* flycatcher—intermediate in size, shape, and plumage between the slightly smaller Hammond's Flycatcher and the slightly larger, paler Gray Flycatcher. Dusky might even be the most average-looking empid. Appears round-headed, even flat-headed, and long-tailed. The bill is fairly thin and fairly long, seems flat on top, even ever so slightly upturned, and is intermediate in length between stubby-billed Hammond's and longer-billed Gray. The lower mandible is pale and dark at the tip.

Upperparts are gray (some individuals show a hint of green). Wings and tail are also gray, but with a brownish cast. Underparts are paler gray, with a whitish throat and (usually) a hint of yellow on the belly. Eye-rings may be conspicuous, and lores may be pale—giving the bird a "spectacled" look—but some individuals show indistinct eye-rings. Wing-bars are typically narrow and whitish. Some birds look like they are wearing an open vest; some have uniformly pale underparts.

Immatures are generally greener-tinged above, yellower below, with contrasting gray heads. Worn adults, in late summer, can be very grayish (more akin to the larger, longer-billed, downward-tail-wagging Gray Flycatcher).

BEHAVIOR: Likes to perch conspicuously—often on dead branches, often close to the trunk. Sallies out to capture insects, generally not returning to the same perch. Also works upward through foliage, gleaning insects from leaves and branches by sally flights and hovering. Occasionally flies and lands on the ground to grab prey. Likes open space; avoids vegetative clutter, but is fond of willow patches in forest openings.

On average, wing-flicks and tail-jerks less than Hammond's, but when agitated, becomes as jerky as the next empid.

FLIGHT: Like other empids. Light, direct, slightly bouncy or jerky.

VOCALIZATIONS: The male's variable three-part song is similar in pattern to Hammond's, but is clearer, cleaner, higher-pitched, and more emphatic or authoritative; some utterances are piercingly high. The standard sequence begins with a sharp two-note hiccup, "*SEEbik*," followed by a rough sliding "*bure'e'e*," followed by a piercing two-noted "*pee'HEE*" (for example, "*SEEbik bure'e'e pee'EE!*" or "*SEEbik pee'EE bure'e'e*"). Also emits a slower, more mellow, more halting or reflective-sounding, three-part whistle: "*Her. Herhee. Her.*" Call is a soft muttered "*pwit*."

Pacific-slope Flycatcher, *Empidonax difficilis*

"P'ooeet"

STATUS: Common summer breeder along the West Coast. **DISTRIBUTION:** Breeds, primarily coastally, from southern coastal Alaska through s. B.C. and extreme sw. Alta., south into w. and cen. Wash., Idaho, ne. Ore., and Calif. to the Mexican border. The eastern range is defined by the limits of the n. Rockies in Alta. and the Cascade and n. Sierra Nevada ranges. **HABITAT:** Wet and humid coniferous forests (mature and old-growth Douglas fir and hemlocks); also found in oak-pine woodlands, riparian canyon woodlands, alders, cottonwoods, live-oak woodlands, and some second-growth woodlands. Usually found close to streams, shaded ravines, and the open space below the tall canopy. Seems very attracted to man-made structures, frequently nesting in barns and carports and under decks. **COHABITANTS:** Chestnut-backed Chickadee, Winter Wren, Warbling Vireo, Cassin's Vireo, Swainson's Thrush, Hermit Warbler, Spotted Towhee. **MOVEMENT/MIGRATION:** Because this species wanders east in migration (primarily in fall) into Nevada and Arizona en route to wintering areas in Mexico, it is more likely to be seen than Cordilleran through lowland Arizona. Spring migration from March to June; fall from early August to mid-October. VI: 2.

DESCRIPTION: A small but robust yellowish green forest *Empidonax* flycatcher. While fairly easily distinguished from other *Empidonax* flycatchers breeding within its range, Pacific-slope is virtually identical to the Cordilleran Flycatcher of the Rockies and the Great Basin, whose range overlaps with Pacific-slope's in eastern Alberta, eastern Washington, and perhaps eastern Oregon. The species are best distinguished by range, habits, and calls.

Pacific-slope is a classically built empid: not long, not short, with a peaked, sometimes raggedy-crested head and a long bullet-shaped bill with a wholly bright yellow-orange lower mandible. Upperparts are olive brown (slightly duller than Cordilleran and, especially, Yellow-bellied Flycatchers), and underparts are distinctly yellowish. (Some birds, at a distance, seem ochre-colored about the head.) The bold white eye-ring has a tear or kick to the rear.

BEHAVIOR: Thrives in the open spaces just beneath the tall forest canopy (in the summer, prefers shaded perches). Males, vocalizing usually from exposed perches, shift perches after several minutes. Early in the season, males sing throughout the day; well into the nesting season, songs are usually limited to early morning. Forages by making a flycatching sortie from one perch and often landing on another. Also makes short flights to pluck insects from leaves and branches. Most flights are short. When agitated, pumps its tail up somewhat sluggishly and wing-flicks vigorously. Responds poorly to pishing, but may be attracted to imitations of Northern Pygmy-Owl.

FLIGHT: Flight is straight, fast, usually short, and slightly jerky, with rapid wingbeats punctuated by intermittent hesitations or skips.

VOCALIZATIONS: The whistled song consists of three spaced, very high, very thin, distinct phrases often sung in sequence: first a sharp, rising, high-pitched, two-noted "*p'sEE*," followed by a terse hiccup, "*p'ch*," and then a very short high "*hee!*" often uttered after a slight hesitation (for example: "*p'sEE p'ch hee!*"). Male's call is a high, sliding, compressed whistle that has the cadence and pattern of a human hailing whistle—"*WHEooEEE!!*" or "*P'ooeet*" or "*su-weet*"—but is a more compressed and sliding single note. (The call of Cordilleran Flycatcher is more distinctly two-noted.) The male's call is sometimes incorporated into the bird's song (for example: "*P'sEE WHEooEEE p'ch*"). Females make a high "*see*" that is similar to Cordilleran.

Cordilleran Flycatcher, *Empidonax occidentalis*

"W'r see!"

STATUS: Fairly common Rocky Mountain and Great Basin breeder. **DISTRIBUTION:** Breeds from (perhaps as far north as) se. B.C. and sw. Alta. to se. Wash., e. Ore., s. Idaho, w. Mont., w. and s. Wyo., Nev., Utah, w. Colo., and n., se., and w. Ariz. Also found in disjunct pockets in extreme ne. Calif., ne. Wyo., w. S.D., nw. Neb., s. Nev., and trans-Pecos Tex. **HABITAT:** Cool arid coniferous or mixed coniferous-deciduous forest, often in stream-cut canyons that offer the requisite canopy for shade and open areas for foraging. Likes to hang around cabins and outbuildings and often nests on them. In migration, like Pacific-slope and Acadian Flycatchers, gravitates to canopy woodlands and shade. **COHABITANTS:** Canyon Wren, Brown Creeper, Hermit Thrush, Western Tanager. **MOVEMENT/MIGRATION:** Entire population retreats into Mexico in winter. Spring migration from late March to early June; fall from mid-August to mid-October. VI: 1.

DESCRIPTION: A small but robust richly colored empid that likes to stay in the shadows. Average-sized (the same size as Pacific-slope) but overall broadly proportioned, with a large, peaked, slightly shaggy head, a large wide-based bill with a wholly yellow-orange lower mandible, a burly chest, and a long narrow tail.

Overall darker and more richly colored than most other western empids, with olive brown upperparts, olive yellow underparts, a bold whitish eye-ring with a distinctive teardrop kick to the rear, and broad dingy white wing-bars. Buff-breasted Flycatcher is more warm buff overall; Pacific-slope is virtually identical, but Cordilleran seems richer and more color-saturated.

BEHAVIOR: Likes high shady perches just below the canopy. Sallies out to snap up flying insects or plucks insects from leaves or the ground. Commonly changes perches after a sally. Males sing from a high perch. Birds respond mildly to pishing but usually remain just below the canopy. When agitated, Cordilleran flicks its tail and wings rapidly upwards.

FLIGHT: Like Pacific-slope, flight is direct, fast, somewhat jerky, and usually short.

VOCALIZATIONS: Song is similar to Pacific-slope, but slightly slower, richer, and more mellow, with the middle note a muttered trilled chirp as opposed to a terse hiccup: "*PeeEE pdl'lu ee!*" The male's call is a loud, simple, emphatic, two-note phrase, "*W'r Seet!*" that is louder and more robust than the thin, airy, up-slurred single-note rendering of Pacific-slope. Also sometimes makes a high thin "*see*" that is similar to Pacific-slope and a low muttered "*d d'jr'j.*"

Buff-breasted Flycatcher, *Empidonax fulvifrons*

Cherubic Flycatcher

STATUS: Very uncommon and geographically restricted. **DISTRIBUTION:** In the U.S., limited to the extreme southeast corner of Ariz. Found only at higher elevations (6,000–9,000 ft.), primarily in the Huachuca Mts. but also in the Chiricahua, Santa Rita, and Santa Catalina Mts., the Animas Mts. in sw. N.M., and the Davis Mts. in w. Tex. Ranges south through cen. Mexico and parts of Central America. **HABITAT:** Open-canopy pine and pine-oak forests with open, often grassy, understory. Often found in canyons (and campgrounds); frequently occurs near riparian areas and/or burned areas dotted with stands of live trees. **COHABITANTS:** Greater Pewee, White-breasted Nuthatch, Grace's Warbler, Hepatic Tanager. **MOVEMENT/MIGRATION:** Reaches Arizona from late March to mid-April; departs in September and early October. VI: 0.

DESCRIPTION: Tiny, cherubic, perch-shifting flycatcher with a wealth of identifying traits. Small, chickadee-sized, and plump, with an oversized rounded head, a rotund little body, and a short tail that is splayed and notched at the tip. The bill is small and straight, with a bright yellow-orange lower mandible.

Unlike other small flycatchers, Buff-breasted is overall brownish. The head and breast are washed with warm buff. The pale eye-ring is conspicuous but not crisply defined. Buffy wing-bars are *really* buffy, not the equivocal hair-splitting buffy attributed to other members of the genus. Immatures, with orangy underparts, are even more conspicuous. Late-summer adults have paler underparts but are still buffy.

BEHAVIOR: A perch-hunting and gleaning specialist—something of a vireo/flycatcher. Takes a strategic perch within the open lower interior of a pine or oak (or sometimes low in the tops of ground-covering shrubs like manzanita). Flies and plucks prey from the undersides of branches and needles in one smooth motion, or may hover. Returns to another perch. Sighting no prey, changes perches frequently (no prey, no gain). Also sits higher in the outer canopy and flycatches (in typical fashion) and lands on open ground and snatches prey there.

Vocalizes frequently, even in the middle of the day. The tail shivers when the bird sings and when it lands, and jerks (or at least twitches) when it calls.

FLIGHT: Flight is quick, nimble, bouncy, and usually short.

VOCALIZATIONS: Song is a pert, cheery, rapid, chirpy stutter: "*chid'd!*" Call is a sharp high-pitched spit, "*pit!*" and sometimes a more liquid "*pwit.*" Also makes a low rolling purr.

Black Phoebe, *Sayornis nigricans*

Aqua Phoebe

STATUS: Common resident species, but restricted to the West Coast and the Southwest. Very rare north and east of normal range. **DISTRIBUTION:** Breeds—and for the most part winters—west of the Sierras from extreme s. Ore. to Baja and from the Calif.–Ariz.–Nev. border across w. and s. Ariz. to N.M. and sw. Colo. south into sw. Tex. to the Edwards Plateau (and south into South America). In winter, some northern birds in Ariz. and N.M. vacate breeding areas. **HABITAT:** Almost any open habitat adjacent to water in some shape or form, including riverbanks, seacoast cliffs, lakeshores, seasonal ponds, cattle tanks, and swimming pools. Frequently found near human dwellings and structures, which it uses to support its mud-based nests. Forages over open water, grasslands, agricultural lands, and parks. **COHABITANTS:** Varied, but includes Killdeer, Song Sparrow, House Finch, humans. **MOVEMENT/MIGRATION:** Essentially nonmigratory. VI: 1.

DESCRIPTION: A distinctive, medium-sized, elegantly contoured and proportioned flycatcher of open areas that is unmistakably black and white—the only black-and-white flycatcher in North America. The head is fairly large and peaked, the bill short and straight, the tail long, narrow at the base, and flared toward the tip. Overall contours are crisp, trim, even somewhat angular.

The upperparts and breast are flat black. The belly and undertail are starkly white. Both the pattern and contrast are unmistakable.

BEHAVIOR: An active aerialist. Usually hunts from a fairly low perch, often near water or human dwellings, but may perch very high when not actively hunting. Takes prey in the air as well as on the ground, but generally does not hop. A patient hunter; usually sedentary while perched (no shifting around). Changes perches infrequently, but does not commonly return to the same perch after making a sortie for prey. Movements are quicker and more nimble and abrupt than Say's Phoebe. The tail wags nervously and is frequently fanned. Very vocal; calls frequently and sings occasionally in winter.

FLIGHT: Slender long-tailed profile, with short rounded wings. Flight is fast and direct with no undulations; movements are quick, jerky, and erratic. Flight is overall less leisurely than Say's Phoebe and lacks Say's floating quality.

VOCALIZATIONS: Song consists of two two-note whistled phrases whose pattern recalls Eastern Phoebe: "*F'brr*" (trilled, descending ending) and "*f'bee*" (ascending). Slightly higher-pitched than Eastern Phoebe. The pattern may be reversed—an ascending "*f'bee*" followed by a trilled descending "*f'brr.*" Call is a sharp, clear, whistled "*chee.*"

Eastern Phoebe, *Sayornis phoebe*

The Bridge Pewee

STATUS: Common to fairly common breeder. **DISTRIBUTION:** Found across most of the eastern half of

North America as well as in forested portions of n. and w. Canada. In the U.S., found west to cen. and s. N.D., on the e. Wyo. border, and in cen. Neb., se. Colo., w. Okla., and cen. Tex. Does not breed in coastal N.C. or S.C., s. Ga., Fla., sw. Ala., most of Miss., La., and se. Tex. In Canada, found from N.B. across s. Que., s. Ont., s. Man., all but sw. and ne. Sask., and most of Alta. to ne. B.C. north into the s. Northwest Territories. In winter, retreats to the s. U.S. south and east of a line drawn from Del. to e. Md., s. W. Va., e. Ky., n. Tenn., n. Ark., s. Okla., cen. Tex., se. N.M. south into Mexico. **HABITAT:** Breeds in a variety of woodland habitats, but prefers open woodlands or woodland edge, preferably near water. Has adapted well to humans and often chooses to use human structures—most notably bridges and buildings—to secure its nests. In migration, also favors woodlands and woodland edge; in spring, frequently found around lakeshores and ponds. In winter, also partial to wooded swamps. **COHABITANTS:** Barn Swallow, Cliff Swallow, Louisiana Waterthrush. **MOVEMENT/MIGRATION:** The earliest returning (and latest departing) flycatcher. In spring, birds may arrive before water is free of ice. Spring migrants appear as early March, but peak movements occur in late March and April, with numbers trailing off in early May. Fall migrants appear in late September, peak in October, and may continue until the first hard freeze. December sightings, north of the wintering range, are not uncommon. Occurs fairly regularly west, particularly southwest, of its normal range. VI: 3.

DESCRIPTION: A slender, conservatively cloaked, medium-sized flycatcher with a jaunty tail. The head is large and rounded, the all-black bill is fairly small, and the tail is long and animate—overall a slim, trim, nicely proportioned bird that seems longer, less angular, and more smoothly contoured than a wood-pewee.

In all plumages, plain dusky gray above, with a contrastingly darker face/crown and tail. Underparts are whitish with a pale broken shadow of a vest. In fall, immature birds are commonly washed with yellow below.

BEHAVIOR: Sits nearly erect on a conspicuous perch or on the forest edge—generally less than 20 ft. high and often close to buildings (when breeding). The bird signs its name with its tail, moving it with a measured down-pumping action—sometimes closed, sometimes slightly flared, often raised with a sideways swish. Sallies out to snatch insects on the wing. Sometimes hovers to pick insects from leaves, and also lands to grab insects from the ground.

Tame, often allowing close approach. Generally solitary, but tolerant of other species that also use man-made structures (such as Barn Swallow and American Robin).

FLIGHT: Overall slender and long-tailed. Flight is quick, direct, and not undulating, but slightly jaunty, bouncy, or buoyant.

VOCALIZATIONS: Song is a brisk, somewhat burry, assertive, ascending, two-note, whistled rendition of its name, "*ZweebEE*," followed after a brief pause by an abbreviated, descending, and more casually uttered "*z-b'r*": "*SweebEE* (pause) *z-b'r*."

Call is distinctive—a sharp prolonged "*chhip*" that drags or hesitates at the start and ends sharply and emphatically. Sometimes has a plaintive, tentative, or questioning quality.

PERTINENT PARTICULARS: Owing to overlapping ranges, Eastern Phoebe is most often confused with Eastern (or Western) Wood-Pewee. Pewees are overall more compact and angular, with longer bills (yellow on the lower mandible), larger peaked heads, and twitchless tails. Pewees usually inhabit deeper woodlands, range higher in the tree (often in the canopy), and are shyer around humans, more high-strung in their movements, and more aggressive toward other species. But when all else fails, the presence of white or buffy wing-bars on pewees and their all but absence on Eastern Phoebe is affirming (or disaffirming).

Say's Phoebe, *Sayornis saya*

Say's Ground-Tyrant

STATUS: Common western flycatcher with an extensive (and disjunct) longitudinal range. Casual vagrant in the East, mostly in fall. **DISTRIBUTION:** The northern breeding range extends across most of the interior of Alaska, the Yukon, the w. Northwest Territories, and n. B.C. The southern breeding range extends from se. B.C., s. Alta., and s. Man. through the Great Basin, the Rockies, and the Southwest, as well as s.-cen. Calif. (does not breed on the immediate coast north of Baja), south into Mexico. The eastern border cuts across e. N.D., e. S.D., e. Neb., cen. Kans., and n. and w. Tex. to the Edwards Plateau. In winter, vacates northern breeding areas. Winters across the Southwest from cen. to s. Calif., s. Nev., w. and s. Ariz., s. N.M., and s. Tex. almost to the La. border. **HABITAT:** At all seasons, prefers open, dry, sparsely vegetated country, including prairie, ranches, sagebrush plains, desert, tundra, rocky hillsides, agricultural land, orchards, parks, vineyards (also the area around abandoned and occupied buildings), and grassy cliffs adjacent to beaches. Avoids watercourses, but is attracted to boulder-strewn flood plains. **COHABITANTS:** In winter, feeds in the same habitat as Black Phoebe, Mountain and Western Bluebirds, Loggerhead Shrike. **MOVEMENT/MIGRATION:** Very early-spring migrant, returning before most other flycatcher species to the Great Plains in March or April and to Alaska in mid-May. Departs northern portions of the range in late August, with most migration noted between September and October. VI: 3.

DESCRIPTION: A slender medium-sized flycatcher (slightly larger than Eastern and Black Phoebe), with muted uncontrasting plumage and a tail that flares as it pumps. Looks somewhat like a slim pale American Robin. Overall fairly tubular in shape—head, body, and tail don't differentiate themselves. (Other phoebes are more classically contoured.) In all plumages, garbed in earth tones. Except for the contrastingly black tail and salmon wash on the belly and undertail coverts, overall gray-brown—the color of dry bare soil. Obvious color differences notwithstanding, the salmon-colored belly is the same tone as the rest of the bird and imparts a pale, cryptic, eye-defeating uniformity to birds seen at a distance.

BEHAVIOR: Likes it dry; likes to forage low. Sits conspicuously but low—usually within six feet of the ground—on branches, boulders, fences, cattle droppings, weeds, and short vegetation. Sallies out to flycatch, frequently landing and taking prey on the ground, and sometimes hunts from the ground. Also hovers with frequency and finesse; sometimes hops on the ground, but prefers to fly. In size, shape, behavior, and demeanor, suggests and may be mistaken for Mountain Bluebird. Changes perches frequently—generally more frequently than other phoebes. Often after taking a new perch, pumps its tail with an emphatic (but not jerky) stroke, often fanning the tail as it does so.

Generally solitary outside of breeding season, but sometimes migrates in small groups. Fairly vocal outside breeding season, but calls less than Black Phoebe.

FLIGHT: Overall slender and pale profile except for contrasting blackish tail. Flight is direct and buoyant, with a somewhat floating dreamlike quality and a slight jerkiness or bumpy rise on the down stroke. Usually flies low to the ground. In longer flights, sometimes jinks left and right, compromising the usual directness. Wingbeats are light, playful, and generally regular (short series interspaced not with glides so much as with rhythm-breaking pauses). Wings are broad and paddlelike, and the pale flight feathers sometimes glow with transfused light. At times, hovers briefly.

VOCALIZATIONS: Two-part whistled song consists of a series of spaced, low, mellow, mostly clear whistles followed by an ascending, trilled, urgent one: "*p'peurr p'peurr p'peurr p're'e'ep!*"

Most commonly heard call is a soft plaintive "*peeurr.*" Also gives a soft, nasal, downsliding trill; a high-pitched, up-sliding, trilled "*ch'wheep*" (like the trilled note in the bird's song); and muttered pipings uttered in association with the bird's mate.

Vermilion Flycatcher, *Pyrocephalus rubinus*

Ember-colored Flycatcher

STATUS: Common southwestern breeder; common to uncommon and more widely scattered winter resident. **DISTRIBUTION:** Common breeder in

s. Ariz., s. N.M., and s. Tex. south into Mexico; local and less common in se. Calif. and s. Nev. In winter, ranges tighten, hugging the Mexican border and ranging east to the lower Tex. coast, but some birds may be encountered as far east as Fla. **HABITAT:** Breeds in open riparian woodlands associated with arid and desert communities. Typical trees include cottonwood, willow, mesquite, and sycamore. Avoids dense woodlands and brush too dense for flycatcher foraging. In winter, often found in the same habitat, but also on open farmland, pastures, golf courses, savanna, and park habitat. **COHABITANTS:** During breeding, White-winged Dove, Black Phoebe, Yellow-breasted Chat, Blue Grosbeak; in Texas, Bell's Vireo, Painted Bunting. **MOVEMENT/MIGRATION:** Some birds appear to be year-round residents. Some northern birds migrate. Spring migration seems to occur primarily in March and April; fall movements between late August and late October. *Note:* These dates are somewhat speculative. VI: 3.

DESCRIPTION: Description of vermillion and brown adult males is almost unnecessary in light of the arresting beauty of the bird. Females and immatures are more cryptic but nevertheless distinctive. Fairly small (almost *Empidonax*-sized) adult males are overall squarish or stout, with a large peaked (sometimes shaggy-crested) head, a straight, thin, petite bill, a plump body, and a short, wide, animate tail. In structure, recalls somewhat an Old World chat or a blocky empid.

Females and immatures are slightly slimmer and more empidlike; gray-brown above and pale below, they have a head pattern that is a shadow rendering of the adult male. Most observers focus on the contrastingly white eye-line and white throat, which are so starkly unlike the pastel shadings of most small flycatchers. The bellies and undertails of immature females are washed with yellow; adult females and immature males have a distinctive rosy or salmon blush.

Confusion between female Vermilion Flycatcher and Say's Phoebe is possible, but the contrasting underparts of Vermilion (white breast, rosy belly) versus the cryptic all-dark gray breast blending to salmon underparts of Say's is distinctive and telling.

BEHAVIOR: A consummate but fairly casual perch-hunter. Perches in the open, often high at the top of a bush; low, atop a stiff weed or fence line; or on a branch below an open canopy. Moves theatrically on its perch, turning its entire body, inclining its head, fanning and bobbing its tail with a flourish. Sallies for insects are fairly slow and floating. Usually takes prey in the air (sometimes after a spirited chase), but also makes forays to the ground where it sometimes hovers above prey or settles with wings raised in a V. Very often returns to the same perch after a sortie.

Fairly confiding. Not social—most commonly found alone but often in close proximity to other breeding (or wintering) Vermilion Flycatchers.

FLIGHT: Silhouette is overall compact—short, blunt, almost square wings and a wide short wedge of a tail. Flight is fairly slow and direct, buoyant and floating, with tail fanned and wings wide. Movements seem sluggish for a flycatcher and recall somewhat a more nimble perch-hunting bluebird.

VOCALIZATIONS: Fairly quiet most of the year. Song is an excited rising-and-falling police whistle–like trill that begins with a stutter, "*p'pi'piTe'e'e'e,*" and is often sung in the dark, mostly at predawn but also at other times during the night. Call is a sharp high "*pee.*"

Dusky-capped Flycatcher, *Myiarchus tuberculifer*

Plaintive Myiarchus

STATUS: Common but geographically restricted. **DISTRIBUTION:** Tropical species whose breeding range extends into se. Ariz. and extreme sw. N.M. Also rare a breeder and visitor to w. Tex. Casual visitor to Calif., mostly in winter. **HABITAT:** In the United States, breeds in riparian corridors dominated by sycamore and evergreen live oak and in pine-oak woodlands. **COHABITANTS:** Arizona Woodpecker, Cordilleran Flycatcher, Sulphur-bellied Flycatcher, Hutton's Vireo, Bridled Titmouse. **MOVEMENT/MIGRATION:** In the United States, retreats south in winter; absent from late September to early May. VI: 2.

DESCRIPTION: A slender, trim, angular, pint-sized *Myiarchus* with a full-sized head. Small (about phoebe-sized), with a large peaked head, a long slender bill, a very slender, almost gaunt body, and a long narrow tail. Brown-crested and Ash-throated Flycatchers are larger, more robust, heavier, and thicker-billed. Proportionally, Dusky-capped has the longest bill for its size.

The plumage pattern is like other *Myiarchus* flycatchers — grayish brown above, a pale gray breast, and a yellowish lower breast and belly — but differs in some respects. The top of the head (cap) is slightly darker and more defined; the tail and folded wings show little or no rufous; the belly is bright yellow, and the contrast between the gray chest and the yellow belly is pronounced (more akin to Great Crested than Ash-throated or Brown-crested).

Brown-crested, another forest *Myiarchus* found in the same habitat as Dusky-capped, is larger, bulkier, and broader-billed and has obvious rufous highlights in the wings and tail. Ash-throated Flycatcher, an open-country *Myiarchus* that may occur in Dusky-capped's habitat — where arid side-canyon habitat meets lower riparian corridor — is overall paler and less contrastingly plumaged, with underparts showing only a faint yellow wash but wings and a tail that have distinct reddish tones.

BEHAVIOR: A nimble *Myiarchus*. Likes to hunt in confines that are tight — places so tight that only a tiny *Myiarchus* can weave a path — and low — lower than the canopy-haunting Brown-crested Flycatcher. As a rule, does not sit conspicuously high and does not forage above the canopy. Moves through the latticework of branches by hops and short flights, pausing at each relocation, turning its head in a visual search for prey. Usually hunts in pairs or in family groups. Paired adults vocalize to each other frequently. Single birds are also vocal.

Fairly tame; allows close approach. Responds quickly to pishing, especially to imitations of Northern Pygmy-Owl.

FLIGHT: Overall slender in flight and conspicuously small. The observer's first impression is of something smaller than a *Myiarchus*. Flight is quick, direct, and slightly jerky, with a tendency to rise and fall (not undulate) over longer flights. Wingbeats are given in a quick series punctuated by a short hesitation. Does not glide per se; strong downstroking wingbeats serve as a glide. Flight is more nimble and less flowing than other *Myiarchus*.

VOCALIZATIONS: Most commonly heard vocalization is a plaintive, smoothly descending, whistled "*wheeer?*" Sometimes rolls its r's and gives the call a compressed two-note quality: "*wheer'r'r'r.*" Also gives a strident "*deer deer deer deer*" and a raspy "*weedeer.*" Call is a sharp "*whit!*" Also emits a liquid, descending, three-noted strum: "*dree'dr lick.*"

Ash-throated Flycatcher, *Myiarchus cinerascens*

The Pale Myiarchus

STATUS: Common and widespread western breeder. Some birds winter in southwestern deserts. Rare but annual vagrant in the East. **DISTRIBUTION:** Breeds from s.-cen. Wash. to s. and cen. Ore. through almost all of Calif., Nev., s. Idaho, sw. Wyo., Utah, Ariz., N.M., w. and s.-cen. Colo., w. Okla., and the western two-thirds of Tex., south into Mexico. In the U.S., some birds winter in extreme se. Calif. and sw. Ariz., and small numbers winter regularly from s. Tex. around the Gulf Coast to Fla. **HABITAT:** Prefers to nest in open and dry scrub and open woodlands; also found in riparian woodlands and washes. Dominant vegetation includes Joshua trees, yuccas, saguaro cactus, assorted mesquites, pinyon-juniper woodlands, cottonwoods, and oak woodlands. Requirements include some large cavity-bearing trees for nesting. Avoids dense and wet woodlands, but in migration may be found in a variety of habitat types, including urban parks, woodlots, and windrows. **COHABITANTS:** Bell's Vireo, Juniper Titmouse, Black-throated Gray Warbler, Spotted Towhee, Curve-billed Thrasher, Cactus Wren, Scott's Oriole. **MOVEMENT/MIGRATION:** Except in parts of the desert Southwest, most Ash-throateds retreat to Mexico and Central America for winter. Spring migration from March through June, with a peak from late March through May; fall from mid-July to November, with a peak in August and early September. Strays as far as the Atlantic Coast, where it often appears in late October and November. VI: 3.

DESCRIPTION: A large, slender, brush-loving flycatcher—similar in structure and plumage to Brown-crested and Great Crested Flycatchers, but slightly smaller-billed and overall paler (particularly on the belly). Less robust than the other large *Myiarchus*, with a more slender build, a smaller bill, and a smaller head. The larger *Myiarchus* have heads and bills that appear oversized for the bird. Ash-throated has more balanced proportions; its bill in particular does not seem overly heavy.

Light brown upperparts with reddish highlights in the wing and tail are similar to other large *Myiarchus*, but the head is ashy, not as dark-capped, and the tail seems more uniformly reddish. Underparts are plainer and paler; in fact, in the harsh light that typifies open western habitats, they often appear uniformly pale, with little contrast between the pale grayish breast and the yellowish belly.

BEHAVIOR: The bird likes to forage low—sometimes in the middle heights, more often close to the ground. In dry sparsely vegetated habitat, it may sally frequently. Where brush is thicker, it becomes a stalker, moving by short flights through foliage, snaking through vegetation in a linear hunting pattern. Pauses, often on springy growth, are lengthy as the bird leans forward, sometimes crouching, moving its head in quick but not abrupt jerks, searching for prey, calculating its next move. Gleans from foliage. Sallies to the ground to grab prey, then returns to a perch. Generally does not flycatch. Pattern of movement is eminently predatory.

FLIGHT: Slender long-tailed profile with a projecting head and wide open wings. Classic *Myiarchus* flight, with mostly steady wingbeats that have the enjoined quality of both a flap and a glide. Overall flight is straight (no undulations) and buoyant but also slightly jerky, imparting a sense of urgency. Usually flares its tail when landing, flashing chestnut.

VOCALIZATIONS: Song is a repeated series of mostly two-note "*ka-brick*" phrases that may be hurried, trilled, or gurgled—"*chibri'r'r*"—or more emphatic—"*chbik*"—and sometimes include the odd, low, muttered "*wrk.*" The order changes, but the same phrases are used over and over. The delivery seems unenthused, matter-of-fact, and less dramatic or strident than Brown-crested. Also gives a low, descending, warbled whistle, "*cheer'r'r*" or "*p'cher'r'r,*" as well as a higher, louder, shriller "*wher'r'r'r*" that sounds like a police whistle. Call is a low mellow "*pip*" and a rough "*prrt.*"

Great Crested Flycatcher,
Myiarchus crinitus
Woodland "WHEEPer"

STATUS: Common breeder in eastern forests. DISTRIBUTION: Found throughout the e. U.S. and s. Canada west to the prairies, with the western border falling through cen. Alta. and cen. and se. Sask., cen. N.D., sw. S.D., w. Neb. (also the northeast corner of Colo.), w. Kans., w. Okla., and cen. Tex. south to about Corpus Christi. Winters in cen. and nw. South America (also in s. Fla.). HABITAT: Generally moist, open, and broken deciduous or mixed deciduous-coniferous woodlands, including forest edge. Does well in urban parks, cemeteries, and suburban areas rich in shade trees and orchards. Avoids boreal forest. In migration, also uses thicker brushy borders. COHABITANTS: Eastern Wood-Pewee, Red-eyed Vireo, Blue Jay, Tufted Titmouse. MOVEMENT/MIGRATION: Except in extreme southern Florida, wholly migratory. Spring migration from mid-April to late May; brief fall migration from late August to early October. VI:2.

DESCRIPTION: Large, richly colored, and noisy flycatcher. Except in Texas, the only large, gray-breasted, yellow-bellied, ruddy-tailed bird that says "*WHEEP!*" in the East. Large (larger than Eastern Wood-Pewee; smaller than Blue Jay) and overall lanky. The head is robust, proportionally oversized, and slightly peaked (or bushy-crested when piqued), with a long sturdy bill; the body and tail are long and incongruously slight. The head and bill are not quite as robust as Brown-crested Flycatcher, and the body not as slight as Ash-throated.

Olive brown above, with unmistakable reddish highlights in the folded wing and tail. Underparts show a well-defined pattern: a dark gray breast bordering bright yellow underparts and a mostly rufous tail. Overall darker and more crisply patterned and richly colored than the similarly sized Brown-crested Flycatcher, with a distinctly darker gray face, a considerably grayer and more sharply defined bibbed breast, brighter yellow underparts, and a redder tail.

But even in migration, the ranges of Brown-crested and Great Crested overlap only in south Texas.

BEHAVIOR: A shy, frequently sedentary, but noisy denizen of the forest canopy. Often perches high on dead branches just beneath the canopy, where it admonishes the universe with a rich, loud, rolling cacophony. Also hunts mostly in or just below the canopy, where it perches quietly for long periods (usually on an exposed perch, close to vegetation), moving its head with slow predatory movements, then sallying out (or up) to snatch prey from leaves and branches. Frequently hovers to pluck prey. Also flycatches and tenaciously pursues flying insects. Whether successful or not, does not commonly return to the launch perch.

Highly aggressive toward other birds and small mammals in its territory. Fairly indifferent toward humans, Great Crested has become very suburbanized in some places. Responds quickly to pishing. When aroused, raises its crest, leans forward, and scolds.

FLIGHT: Long-tailed, with fairly rounded wings and a long projecting head. The reddish highlights in the wings and tail are plainly visible. Flight is direct and buoyant. Occasionally glides between trees without flapping.

VOCALIZATIONS: Most often heard is a loud, monotonously repeated, two-part incantation: an ascending rolling "*oo-Wheep*," followed by a pause, then a descending more trilled "*Brrrrrp*" or "*berrrr-ep*" (recalls a police whistle). Also makes a loud, flat, repeated "*breep breep breep*" and a sliding two-note "*bee-erp.*" When piqued, unleashes a loud, rolling, stuttering harangue that begins with the repeated "*breep*" notes and degenerates from there. Call is a single loud "*wheep.*"

PERTINENT PARTICULARS: Great Crested Flycatcher is the only *Myiarchus* flycatcher commonly found in the East, but the smaller and much paler Ash-throated is a regular vagrant. In fall (particularly after September), any *Myiarchus* flycatcher—particularly one that hunts low, in open fields or along brushy edge—should be scrutinized.

Brown-crested Flycatcher, *Myiarchus tyrannulus*

Canopy Myiarchus

STATUS: Common southwestern breeder, but with restricted distribution. **DISTRIBUTION:** Found very locally in se. Calif., s. Nev., sw. Utah, and across s. Ariz. into sw. N.M., south into w. Mexico and Central America. Also found in s. Tex. from Del Rio north to San Angelo, then south and east to Corpus Christi (also at Big Bend in Rio Grande Village) and south into Mexico. **HABITAT:** Breeds in two distinctly different habitats. For most of its U.S. range, found in riparian woodlands dominated by cottonwood, mesquite, sycamore, and willow; in Texas, by mesquite, hackberry, and ash; and in suburban habitat and towns, by large shade trees. Also found in habitat dominated by saguaro cactus. Requires large trees and cactus for natural and excavated nesting cavities. **COHABITANTS:** In riparian areas, Western Wood-Pewee, Cassin's Kingbird, Bridled Titmouse, Summer Tanager, Hooded Oriole. In columnar cactus, Gilded Flicker, Gila Woodpecker. **MOVEMENT/MIGRATION:** All North American breeders retreat south for the winter. Spring migration from late March to late May; fall from early or mid-August to mid-October. VI: 2.

DESCRIPTION: A big, bulky, heavy-billed *Myiarchus* that likes to forage high. Overall large (slightly larger than Ash-throated Flycatcher; similar in size to Couch's Kingbird; identical in size to Great Crested Flycatcher), with a large peaked head, a very large, heavy, dark bill, a fairly hefty body, and a long tail. Longer and heavier-billed than Ash-throated and, to a lesser degree, many eastern Great Crested Flycatchers, but otherwise similarly proportioned.

Grayish brown above, with a distinct rusty wedge on the lower edge of the folded wing and rusty tones in the tail; a pale grayish throat and breast meld with bright yellow underparts. Ash-throated Flycatcher is overall slightly paler below, especially on the belly (it shows, at best, a yellow wash); Great Crested (whose breeding range overlaps slightly with Brown-crested in south Texas) is overall darker, showing a richer contrast between an olive-toned back and a grayer face/breast and more definition between the dark gray breast and bright yellow belly.

BEHAVIOR: Likes to stay in the canopy. Likes a bit of space to maneuver. Likes big trees. Forages mainly by sallying from perches to snatch prey from a more distant patch of foliage. Also flycatches more commonly than other western *Myiarchus*. Hunts more methodically or sluggishly than either Ash-throated or Dusky-capped.

A noisy bird, Brown-crested often calls attention to itself by lusty calling. Even when sedentary in the heat of the day, birds keep up a conversation with themselves.

FLIGHT: Long profile, with fairly short, broad, rounded wings. Flight is direct and slightly more sluggish or flowing, less jerky than Ash-throated. Wingbeats are continuous and slightly irregular.

VOCALIZATIONS: Raucous. Many sounds are similar to Great Crested Flycatcher. The dawn song, sung repeatedly, is a loud "*whit-will-do.*" Sometimes engages in loud harangues consisting of alternating whistled phrases that blend stuttering yelps and ringing trills, interspaced with the odd low mutter: "*wheer'wh'wher'r'r'r; wh'wh'wher'r'r'r; EEwh'we'r'r'r (brrp) wheer'wh'whe'r'r'r.*" Calls include a loud rolling "*whi'ut*" or "*wrreup*"; a loud, rolling, trilled "*bree bree bree*"; and a soft "*wrk wrk wrk wrk*" that recalls a clucking robin. Also emits a classic *Myiarchus* "*whip*" that is loud and low. Compared to the loud, rich, rolling vocal array of Brown-crested, Dusky-capped sounds weak and plaintive, and Ash-throated flatter and less flamboyant.

Great Kiskadee, *Pitangus sulphuratus*

One Big, Noisy, Distinctive Flycatcher

STATUS: Common resident, but restricted to south Texas. **DISTRIBUTION:** Tropical and subtropical species whose northern range is defined by a line drawn from just north of Corpus Christi to just north of Laredo, then northwest, along the Rio Grande, to just north of Del Rio. **HABITAT:** Thorn forest, riverside woodlands, and marsh edges. Avoids unbroken woodlands; in such habitat, found only along lakes, river oxbows, and clearings. Also increasingly common in well-vegetated suburban areas. **COHABITANTS:** Golden-fronted Woodpecker, Couch's Kingbird, Green Jay, Altamira Oriole. **MOVEMENT/MIGRATION:** Permanent resident. VI: 1.

DESCRIPTION: One big, noisy, robust, and boldly marked flycatcher. Large (larger than any kingbird) and stocky, with a big squarish head, a heavy bill, a portly body, and a short tail. Appears somewhat top-heavy and kingfisher-like.

Very boldly and distinctly patterned, with rufous brown upperparts, bright yellow underparts, and a white head branded with a black mask and cap. (If you are fortunate, you may see the adult's golden crown, and if you are blessed, you may see it raised.) Ages and sexes are similar (except that young birds lack the golden crown).

No other regular U.S. flycatcher looks like Great Kiskadee, but other tropical flycatchers, most notably the smaller Social Flycatcher, are similar.

BEHAVIOR: An eclectic feeder that is at home in the open and in tight vegetative confines. Most often seen sitting in the open (most often at the edge of lakes, rivers, marshes) about 10 ft. high, calling lustily. Forages from the open treetops, making flycatcher-like sallies for flying insects. Also forages by gleaning, hopping branch to branch in heavy cover (like a *Myiarchus* flycatcher and very unlike a kingbird); also hops on the ground (like a jay). Hovers over water (like a kingfisher) and dives, plucking prey from the water surface, but does not submerge. In suburban areas, easily attracted to fruit, and pilfers dog and cat food from bowls (explaining perhaps the evolutionary foundation of the bird's black mask). Also scavenges morsels of bait left on fishing docks, arriving at dawn, just before the Laughing Gulls and Boat-tailed Grackles. Commonly seen in pairs or small groups of three to five individuals. Quite tame, easily approached, and usually noisy, often vocalizing in chorus with several individuals.

FLIGHT: Overall stocky, with a big head, a short tail, and broad blunt wings; the bird is shaped more like a meadowlark or a kingfisher than a flycatcher. Flies much like a kingfisher too. Wingbeats are stiff, rapid, and strong, and given in a short series punctuated with terse set-winged glides (a wingbeat frozen for a fraction of a second). Flight is mostly straight but somewhat bouncy, characterized by frequent, abrupt, upward jerks (once again, like a kingfisher). Hovers well.

VOCALIZATIONS: The bird's signature (and namesake) call is a loud, strident, often repeated, and usually three-syllable chant: "*eh/t-wil-he/ar*" or "*Reh-kil'dare*," which someone with a tin ear transcribed as "*kiskadee*." Emphasis is on the first syllable. The sound seems like something between a laugh and a nasal whine. Birds also emit a loud, rising-and-descending, protesting whine: "*eeyaa!*" The dawn song, sung in two parts, "*wur; dihl dah*" or "*wur; it dilh dah*," is less squeaky or strident than the "*kiskadee*" call, but the quality is similar.

Sulphur-bellied Flycatcher, *Myiodynastes luteiventris*

Squeaky, Streaky, Rubber-ducky Flycatcher

STATUS: Fairly common but restricted in terms of habitat and U.S. range. **DISTRIBUTION:** Mexican and Central American species. In the U.S., found regularly only in the southeastern corner of Ariz. Casual visitor to Calif. and N.M. **HABITAT:** Riparian corridors in mountain canyons dominated by Arizona sycamore and walnut. **COHABITANTS:** Elegant Trogon, Arizona Woodpecker, Dusky-capped Flycatcher, Painted Redstart. **MOVEMENT/MIGRATION:** Absent in Arizona from late September to early May. VI: 1.

DESCRIPTION: A heavyset, large-headed, menacing-looking flycatcher with a squeaky voice. Medium-sized (like Western Kingbird), but distinctly more thickset, with a large flat head, a heavy, long, hook-tipped bill, a robust body, and a shortish tail.

In all plumages, a heavily streaked charcoal-and-whitish bird with a rufous tail. Despite the boldness of the pattern, the plumage is overall cryptic—a blend of low-definition dark grays and dirty whites (a pattern that replicates the dappled pattern of sunlight peaking through the canopy). Head/face is whitish with bold blackish stripes; the back is dappled charcoal and gray, and the underparts creamy with coarse blackish streaks on the breast and flanks. The "sulphur" belly is just a patch—it hardly qualifies as a field mark. The reddish tail is usually conspicuous, but no less so than the bold facial pattern. The flat head, heavy hook-tipped bill, and dark mask through the eye make the bird look ferocious and predatory.

BEHAVIOR: A predator that perch-hunts beneath the canopy. Works through and among the branches of the understory or the lower canopy as though stalking. Takes a perch. Sits for long periods with its head moving in falconlike (or troganlike) jerks. Leaps or sallies and snatches prey from limbs, foliage, or the air, or flies to another perch. Usually hunts alone or, if late in the breeding season, in family groups. Noisy. Usually calls attention to itself by its distinctive call. Fairly tame, but does not commonly sit high or in the open or stray far from the riparian corridors where it nests.

FLIGHT: Overall compact with short, wide, pointy wings. When hunting or changing perches, flight is straight, buoyant, and bluebirdlike; wingbeats are slow and flowing, with a deep lofty down stroke. When birds are vexed or traveling greater distances, flight is fast, direct, and slightly bouncy, with a rapid series of wingbeats interrupted by a terse closed-wing glide.

VOCALIZATIONS: Call is a high, squeaky, two-part squeal aptly described as the sound of a baby's squeak toy or a rubber ducky: "*squee-che/uh*" (often repeated). Other calls also have a squeaky and nasal quality.

Tropical Kingbird, *Tyrannus melancholicus*

The Trilling Kingbird

STATUS: Widespread in the Neotropics. Regular fall vagrant along the West Coast, with a few wintering. Casual along the Gulf Coast. **DISTRIBUTION:** In the U.S., uncommon se. Ariz. breeder; uncommon permanent resident in extreme s. Tex. Range. In Ariz., found in a square defined by a line drawn from just west of Nogales north to Tucson and east to almost the N.M. border. In Tex., found along the s. Rio Grande. **HABITAT:** Open and semi-open country, often in association with human habitation or altered habitat. In Arizona, closely associated with tall trees (primarily cottonwoods) surrounding ponds. In Texas, found on golf courses, football fields, and electric field stations. **COHABITANTS:** In Arizona, Gray Hawk, Western Wood-Pewee, Vermilion Flycatcher, Cassin's Kingbird. In Texas, White-winged Dove, Couch's Kingbird, Great Kiskadee. **MOVEMENT/MIGRATION:** In Arizona, arrives later than other *Tyrannus* flycatchers, April to early May; departs late July to mid-September. VI: 3.

DESCRIPTION: A slender, trim-lined, elegant kingbird. Medium-large (slightly larger and lankier than Western, which it resembles; about the same size as Couch's, which it closely resembles) and overall trim, even somewhat angular—a sense projected by the long bill, large flattish head, and fairly long, slightly flared, and (usually) conspicuously notched tail. Compared to all other yellow-bellied kingbirds (except Couch's), the bill is longer and narrower, giving Tropical Kingbird a slender long-faced appearance.

Overall pale. Pale bluish gray on the head and back, with a slaty mask (patch) through the eye and brownish gray wings and tail (except for the tail, more like Western than Cassin's). Underparts are bright but pale yellow. The very pale grayish olive wash across the breast often cannot be seen in bright light, so the bird appears all yellow below, giving way to white at the throat (no breast-band is discernible). Immatures are like adults, but have grayish olive backs, an olive wash on the breast, and pale edgings to the wing.

Because of the bird's overall slender profile, upright stance, gray paleness, and slender bill, a quick glance at a distance might suggest kinship with Northern Mockingbird.

BEHAVIOR: A flycatcher with finesse. Sallies, often from the crowns of trees, are lengthy, swooping, graceful, and somewhat leisurely. Birds frequently return to the same perch. While aerial sallies are the norm, the birds also perch-hunt from the middle heights to lower branches (including fences), snatching prey from vegetation, and on open ground, where it momentarily lands. They also sally low over ponds and pluck prey from the surface (but don't land!).

Fairly quiet and inconspicuous, vocalizing less than other kingbirds. Although intolerant of larger predatory birds near its nest (Gray Hawk, for example), seems much at ease with other kingbird species in its territory and even nests in the same trees.

FLIGHT: Flight profile is longish (long bill, longish tail) but short-winged. Wingbeats are rapid and shallow; flight seems more floating and buoyant than other kingbirds. Gliding sallies are more fluid and nimble; Tropical turns with more floating finesse and can even sally tight effortless turns around vegetation.

VOCALIZATIONS: Simple and straightforward call: a rapid, thin, high-pitched, rising twitter, "*breedeedeedeedeedeeda*," sometimes prefaced with one or more sharp introductory notes. Apparently does not make a single-note utterance.

PERTINENT PARTICULARS: This species can be quickly distinguished from Western and Cassin's Kingbirds, with which it most commonly associates in Arizona. Western and Cassin's have black or blackish tails that are darker than the wings; Tropical's tail has a brownish cast and is the same color as the wings. Its tail also has a flared tip very unlike the straight-edged square-cut tails of Western and Cassin's.

Couch's Kingbird, *Tyrannus couchii*

Texas Kingbird

STATUS: Fairly common, mostly resident species with a limited U.S. range. Casual visitor east along the Gulf Coast. **DISTRIBUTION:** Found across s. Tex. south of a line drawn from Del Rio to Galveston (but more common farther south, from just north of Laredo to just north of Corpus Christi). Ranges south across e. Mexico to the Yucatán Peninsula. **HABITAT:** Scrubby and thorny brush, agricultural land going into succession, riparian corridors, suburbia, and brushy clearings. In winter, often found along brushy edges adjacent to open habitat. Seems less tied to water than very similar Tropical Kingbird. **COHABITANTS:** Green Jay, Altamira Oriole, Golden-fronted Woodpecker, Bronzed Cowbird. In winter, may be found in loose flocks with other flycatchers, including Eastern Phoebe, Vermilion Flycatcher. **MOVEMENT/MIGRATION:** Permanent resident. VI: 2.

DESCRIPTION: A brightly colored tyrant flycatcher that most closely resembles Tropical Kingbird in plumage and shape. Largish kingbird (larger than Western or Cassin's; about the same size as Tropical), but less lanky and slightly more robust, with a shorter, heavier bill and a less deeply notched tail (two signature traits of Tropical).

Perhaps more brightly or richly colored than other tyrant flycatchers (including even Tropical). Upperparts are three-toned—pale gray head, greenish back, and brown-tinged wings and tail. Underparts are mostly bright yellow, with a dingy band of yellow or olive yellow on the upper breast ending at the whitish throat. Immatures are like adults.

BEHAVIOR: A consummate "flycatcher" in that it takes most of its prey from and in the air, but like Tropical (and particularly in winter), also lands and takes prey on the ground. Usually hunts over tall brush from a perch that offers superior height, but also uses taller treetops, roadside wires, and hedges (in other words, almost any elevated perch). Sallies are often skyward, and birds may climb more than 100 ft. before returning to their perches with a series of wing-braking glides. In winter in Texas, they sometimes gather in small groups of half a dozen and forage together in more open areas, including tall stands of trees (away from brush), which are more characteristic of Tropical Kingbird. Fairly tame; allows close approach.

FLIGHT: Long slender profile; fast strong flier. Flight is generally straight, with some yawing motion; wingbeats are given in a quick irregular series followed by a terse pause. Recalls a nimble buoyant American Robin. When hunting or returning to its perch, may be more casual in flight, floating with wings spread wide and slowed wingbeats that show a strong down stroke followed by a terse open-winged pause/glide.

VOCALIZATIONS: Song is a series of high, breezy, strident "squeals" that begins haltingly and rushes to a conclusion: "*w'wh; wa'waREEcheedoo.*" The pattern may remind you of Cassin's Kingbird, whose song is shorter, lower-pitched, and less strident. Couch's commonly heard call is a sharp high-pitched "*pip* (often repeated), followed by a breezy, rising-and-falling, trilled "*brE'e'e'r*" or "*b'b'breee*" or "*br'reeoo.*"

Cassin's Kingbird, *Tyrannus vociferans*

Dark Western Kingbird

STATUS: Fairly common, but more geographically restricted; in many places, less common than Western Kingbird. **DISTRIBUTION:** Distribution is somewhat fragmented. Breeds in se. Mont., e. Wyo., and along the western borders of S.D. (very local) and Neb. Also found in coastal Calif. from San Francisco to n. Baja. Primarily breeds from s. Nev. across s. Utah into w. Colo., south into n. and e. Ariz. and much of w. and cen. N.M. south into Mexico; also found in portions of trans-Pecos Tex. and the Big Bend region of w. Tex. Permanent resident locally in coastal s. Calif. **HABITAT:** Woodlands adjacent to open foraging areas, including riparian corridors dominated by cottonwood, pinyon-pine-juniper, and oak savanna. In California, also eucalyptus groves. Most partial to riverside areas, canyons, and higher elevations. Less likely than Western Kingbird to be found in open, flat, tree-poor habitat. **COHABITANTS:** Western Kingbird, Ash-throated Flycatcher, Hooded Oriole, Bullock's Oriole, Scott's Oriole. **MOVEMENT/MIGRATION:** Except for southern California, winters in Mexico. Spring arrival from March through May; departure late July through early November. Far less frequently recorded as a migrant east of its range than Western Kingbird. VI: 2.

DESCRIPTION: A stocky darkish kingbird wearing a dark gray vest and a white bow tie. Overall huskier, more compact, rounder-headed, darker, and more contrastingly plumaged than Western Kingbird, with a distinctly dark slate gray head and a fairly well-defined dark gray wash across the breast—a contrasting backdrop that accentuates the small white patch of a chin and upper throat (sometimes looks like a white mustache). Overall shape comes closer to Eastern Kingbird than Western, and though it is no shorter-billed than Western Kingbird, it sometimes appears small-billed.

Western Kingbird has the same pattern as Cassin's—gray head, whiter throat, grayish breast giving way to yellow belly—but Western is shades lighter, and its contrasts are weak. The white throat doesn't pop; the gray on the breast is a wash, not a dark band; and in harsh light, underparts may not show gray at all. Unlike Western, Cassin's often shows dusky pale edgings to the wing feathers. If you have to second-guess, it's a Western.

From behind, Cassin's blackish tail contrasts with the gray back, but not as much as on the paler-backed Western. The outer tail feathers lack Western's obvious white borders, but the tip of the tail shows an indistinct pale edge.

BEHAVIOR: Large, conspicuous, and noisy. Likes to sit high, atop a dead branch of a live tree or a handy utility line. Sallies out to snatch insects in the air and also excels at gliding to the ground and plucking prey, which it then carries to an elevated perch. Vocalizes early, often at first light.

FLIGHT: Appears stocky in flight, with little head projection; the tail is short, and the wings compact and rounded. Flight is direct, with classic stiff, open-winged, batty wingbeats that may or may not be interspaced with long, floating, open-winged glides. May glide all or most of the way to the ground to secure prey sighted there. When pursing insect prey, flies more nimbly and aggressively. Frequently sallies up rather than out to snatch prey. In leisurely flight from tree to tree, often (but not always) leaves its landing gear down (dangles its legs).

VOCALIZATIONS: Classic call is a loud, rolling, distinctive "*Ker-beer*" or "*Ch-beer*," with the first note terse, twangy, and ascending and the second note burry, dragging, and descending. Also sometimes does a truncated version, a rough nasal "*breer*." Also emits an excited "*khew khew khew khew khew*," with all notes evenly spaced and on the same pitch. The two-part dawn song consists of low rough notes and high nasal ones: several warm-up phrases consisting of paired low and high notes are followed by a low, rough, somewhat froglike, three-note chant: "*wha'reer; wha-wha'reer; wrk-Week-wer.*" Pattern and delivery are less frantic than in Western Kingbird; sounds like a Western with a bad cold.

Thick-billed Kingbird,
Tyrannus crassirostris
The Burliest Kingbird

STATUS: Very uncommon and geographically restricted U.S. breeder. Casual vagrant north of its range. **DISTRIBUTION:** In the U.S., breeds only in se. Ariz. and the extreme southwest corner of N.M. Range extends south along the west coast of Mexico. **HABITAT:** In the United States, riparian woodlands dominated by cottonwood and sycamore. **COHABITANTS:** Gray Hawk, Cassin's Kingbird, Rose-throated Becard. **MOVEMENT/MIGRATION:** Arrives in Arizona in May; departs by September. VI: 1.

DESCRIPTION: A *big* robust mesomorph of a flycatcher (the largest of the tyrant flycatchers) that likes to sit on a commanding post. Large not just in overall length but in bulk. Distinctly wide-bodied, with a big bull head (slightly peaked) and a short wide tail. The bill is thickly heavy, but not unusually long.

Dingy or brownish gray above, with a blackish mask through the eye. Whitish below; some adults, and especially immatures, show a pale yellowish wash on the belly. The demarcation between the dark face and the white throat and chest is sharp. Bright white underparts are conspicuous at great distances. At first glance (or even second), and its broad-beamed proportions notwithstanding, the bird looks much like a great big fat Eastern Kingbird.

BEHAVIOR: A perch-hunter. Sits high, commonly atop a dead snag on a living tree, and sallies out for insect prey, which it takes in flight. Usually returns to the same perch. Does not usually stray far from the nest, which may be in the same tree or one nearby. Generally intolerant of other birds near its nest, but allows Cassin's Kingbird to nest in the same tree (on the opposite side). Not impossible to see perched closer to the ground—or even at eye level—but except when bathing, almost never seen on the ground.

FLIGHT: Overall stocky profile. The short and very broad tail makes the bird looks like one big gliding surface. Excels in the art of short floating sallies. Short series of quick heavy wingbeats are followed by swooping glides and graceful banking turns. Longer flights are direct and steady, with regular wingbeats—more ponderous and less hurried and choppy than smaller kingbirds.

VOCALIZATIONS: Sounds screechier than other kingbirds. Most frequently heard vocalization is "*kiter'ree*" said softly, or more stridently, "*kity'b'ree!*" Also emits a shrill "*Skree,*" followed by a descending chatter, "*ch'ch'ch'ch'ch,*" and a rising, musical, goldfinchlike whine: "*t'tr'ee.*"

Western Kingbird, *Tyrannus verticalis*

Pale Western Kingbird

STATUS: Common and widespread breeder in the western United States and adjacent portions of Canada; winters in western Mexico and western Central America; also in south Florida. Rare but regular fall wanderer to the East Coast. **DISTRIBUTION:** Occurs from the prairies west. Eastern limit extends to cen. Minn., w. Iowa, w. Mo., e. Okla., and e. Tex. Northern limit extends to s. Man., s. Sask., s. Alta., and se. B.C. Generally absent as a breeder near the West Coast. A few pairs may be found east to the Mississippi River Valley. **HABITAT:** A mix of open, dry, and lower-altitude habitats, including prairie grasslands, farmland, rangeland, desert shrub, ranches, and towns. Key features of any habitat are a dotting of trees, shrubs, or man-made structures (such as utility lines) suitable for nesting or proximity to a woodland or riparian edge. **COHABITANTS:** Swainson's Hawk, Eastern Kingbird, Loggerhead Shrike, Bullock's Oriole, Western Meadowlark. **MOVEMENT/MIGRATION:** Entire population relocates south of the U.S. border in winter except for small numbers that winter in south Florida. Spring migration is short, from late April through May; fall is more protracted, from late July to November. VI: 3.

DESCRIPTION: The default yellow-bellied kingbird of the dry open western places. Medium-sized (slightly smaller and lankier than Eastern Kingbird; less compact than Cassin's), with nicely balanced proportions but (for a tyrant flycatcher) a smallish, fairly narrow bill.

Overall generally pale and uncontrasting—a sun-bleached prairie flycatcher with tri-toned upperparts: a pale gray head melting into a pale grayish green back, darker brownish gray wings, and a black tail bracketed with narrow white outer tail feathers. Also has tri-toned underparts: a white throat blending into a pale grayish breast blending into yellow underparts. Overall paleness and lack of strong contrast help distinguish this species from the very similar but overall darker and more contrasting Cassin's Kingbird.

The dark mask across the eyes is faint—a smudgy line. The white outer tail feathers are often impossible to see when birds are perched, and difficult to see even when the tail is fanned in flight. Also, on worn birds, those whitish edges may be missing.

BEHAVIOR: A perch-hunting flycatcher. Sits high, often just below the topmost branches of trees, and low, on fenceposts, thistles, and even cow droppings. Sallies out to snap insects out of the air, but hovers in front of foliage, plucking insects from branches and leaves, and lands on the ground to seize prey. Aggressively intolerant of hawks, corvids, and other Western Kingbirds in its territory, but where ranges overlap may be genially nonplused by the proximity of Eastern Kingbird. In migration, often seen hunting in small but spaced flocks.

FLIGHT: Straight, buoyant, and somewhat casual. Wingbeats are mostly steady and fluid, with a deep down stroke — less hurried and batty than Eastern Kingbird. Wingbeats are punctuated, however, by brief, intermittent, mostly closed-wing glides, and when flying great distances, birds sometimes jink or list. The pattern recalls American Robin, but overall flight is slower, more leisurely, more buoyant, and less pushing and twisting.

VOCALIZATIONS: Song is a hurried (almost strident), high, squeaky, testosterone-deprived "*ch'weet'will-do.*" Calls include a low "*pip*" and a low, liquid, muttered "*whi/ut?*" (the bird sounds like it's puzzled or asking a question). Also makes a variety of low muttered twittering or stuttering sounds that often precede the core phrases of the song.

Eastern Kingbird, *Tyrannus tyrannus*
White-tipped Tyrant

STATUS: Common and (despite the name) widespread North American breeder. Very rare migrant along West Coast states. **DISTRIBUTION:** Found across most of the U.S. and s. and w. Canada. In Canada, breeds from N.S. across s. Que., s. Ont., cen. Man., all but ne. Sask., Alta., and e. and s. B.C., north into the w. Northwest Territories and extreme se. Yukon. In the U.S., found everywhere except w. Tex., w. and cen. N.M., Ariz., s. and w. Utah, most of Nev., w. Wash., w. Ore., and Calif. Winters in South America. **HABITAT:** For nesting, selects virtually any open or semi-open habitat with a scattering of shrubs or trees in which to place its nest, but likes to be near water. In spring, also favors open habitats, including those lacking in perches (such as tilled agricultural land). In fall, becomes more arboreal, foraging in flocks on berries along forest edges and in the canopy. **COHABITANTS:** Western Kingbird, Warbling Vireo, Yellow Warbler, Orchard Oriole. **MOVEMENT/MIGRATION:** Relatively brief spring migration occurs primarily from mid-April to early June. Fall migration is equally abbreviated: most birds mass in premigratory flocks in August and make a mass departure by mid-September; lingering birds have been noted into early October. VI: 3.

DESCRIPTION: A dapper medium-small flycatcher distinguished by all-blackish upperparts, all whitish underparts, a small bill, and a distinctive white band on the tip of the tail. Big-headed, wide-chested, and overall compact, with a mercurial head shape — sometimes slightly peaked, sometimes roundly domed. The posture also varies — sometimes upright, sometimes more horizontal.

At a distance, the combination of the compact shape, the typically conspicuous perch (or the crown of a leafy tree), and the marked contrast between the black cap and white throat or black tail with a white tip is idiosyncratic.

BEHAVIOR: During the breeding season, adults sit conspicuously atop trees, telephone lines, fenceposts, tall grass, and flowers and also on the ground. Eastern perch-hunts for flying insects, sallies out to snap up prey, and banks and sails back to its perch on spread wings and tail. If multiple perches are handy, frequently shifts perches. Also hovers into the wind and forages for insects on the ground by flying short distances or hopping. In late summer and fall, when Eastern's diet shifts mostly to fruit, it moves from open habitats to forest edges and the canopies of fruit-bearing trees.

Aggressive toward other species — in fact, the bird's persistent and energetic harassment of crows and hawks (which may include perching on the retreating intruder's back and pulling feathers) is a signature behavior. Other species, such as crows and grackles, are frequently driven to the ground by Eastern Kingbird's frenzied onslaught.

Vocal except during incubation. Males vocalize frequently to define their territories; females and males communicate often; fledged young and adults constitute a filial cacophony.

In spring, flocks of 6–20 birds are common. In fall, more numerous premigratory concentrations, sometimes numbering hundreds or thousands of birds, gather in some eastern areas. During migration, birds are usually silent except at the roost.

FLIGHT: Very distinctive. Appears fairly compact, with short, pointy, triangular, and slightly angled-back wings and a blunt head; contrastingly dark above and white below. Flight is generally direct and strong, but also seems very effortful or perhaps timorous: there's not much forward movement for the vast amount of energy expended. Wingbeats are rapid, fluttery, and all down-stroke, giving the impression of a bird quivering in flight. Often projects a lighter-than-air quality, as if suspended, while flapping rapidly, from an invisible wire that controls and impedes its forward momentum. During aggressive interactions, however, capable of rapid acceleration characterized by wings moving at blurring speeds, swoops, banking descents, and climbs made at seemingly reckless speeds.

VOCALIZATIONS: A loud, rapid, strident scold begins with several short sharp notes and degenerates into a sputtering harangue: "*P'jeer, p'jeer, p'jeer,-jer,jer,jer,jer,jee.*" Also gives a shorter, more conversational "*P'jer-p'jee-ee.*"

PERTINENT PARTICULARS: Fork-tailed Flycatcher, a South American species that resembles Eastern Kingbird and has a long forked tail (reminiscent of Scissor-tailed Flycatcher), is a very rare vagrant to the United States, mostly in fall. When seen in spring, often keeps company with Eastern Kingbird.

Gray Kingbird, *Tyrannus dominicensis*

El Pitirre

STATUS: Common but restricted summer breeder in the southern United States. Casual farther north. **DISTRIBUTION:** Found across the West Indies and ne. South America. In the U.S., most common in coastal areas along the Florida Peninsula, especially in the Keys, but breeds as far north as Ga. and west to coastal Miss. **HABITAT:** Dry upland habitat near water. Often associated with mangroves, but also common in agricultural habitat, near fields, and in towns and suburban habitats (for example, the habitat that dominates the Florida Keys these days). Often seen on utility lines and TV antennas. **COHABITANTS:** Magnificent Frigatebird, Mangrove Cuckoo, Northern Mockingbird. **MIGRATION/MOVEMENT:** Breeders in the United States vacate from late September to late November; return as early as late March, continuing into June, with a peak in April and May. Wanders widely outside its range, particularly in fall, with mostly coastal sightings occurring north to New England and west to Texas. VI: 2.

DESCRIPTION: A noisy, rangy, long-billed, pale gray flycatcher that superficially resembles Eastern Kingbird but is built more like Tropical. Overall lankier and more angular than Eastern, with a rounder, flatter (not peaked) head and a distinctly longer, heavier bill. The tail is deeply notched when perched and straight-cut across the tip when fanned in flight. (Eastern's tail is conspicuously rounded.)

Gray Kingbird is an all-gray bird that is paler and overall less contrasting than Eastern Kingbird. Upperparts are mockingbird gray; underparts are paler but gray-washed (not bright white like Eastern). The blackish mask through the eyes is often hard to see at a distance; however, the lack of contrast between a blackish head and a gray back and white throat, which is characteristic of Eastern Kingbird, is just as defining.

BEHAVIOR: Sits high and conspicuously. Sallies out to snatch insects in the air or from the surface of the water. Often returns to the same perch. Sallies are often lengthy and assume the properties of a sortie, with the bird coursing back and forth after prey. Birds also plunge-pluck prey from leaves. Often found near houses. Like any self-respecting tyrant flycatcher, Gray is intolerant of crows, hawks, dogs, and cats that get too close to its nest. In migration, often seen flying and foraging with Eastern Kingbird. Very vocal.

FLIGHT: Lankier overall than Eastern, but with much wider and blunter wings. Lacks the white tip to the tail, but the contrast between the pale gray back and the darker tail is apparent—as is the straight-cut trailing edge to the tail when fanned. Gray's flight is more floating, flowing, and buoyant than Eastern Kingbird, and it flycatches with smoother finesse and less frantic action. Wingbeats are stiffer and less fluttery than Eastern's, and glides are long and swooping.

VOCALIZATIONS: Commonly (and often) heard call is a strident, high-pitched, two-part "*P'cher'r'r'r'r.*" The first part is short and clear (sometimes buzzy), followed by a descending trill. Sometimes sounds like "*peecheer'r'r'ri.*" Gray Kingbird is more musical than Eastern Kingbird, and less angry, stuttering, and shrill; sounds most like Tropical Kingbird. The colloquial name for the species in Cuba is "El Pitirre," an obvious phonetic rendering of the bird's call. Also has a chatter call that is often heard at dawn.

Scissor-tailed Flycatcher, *Tyrannus forficatus*

Kingbird on a Stick

STATUS: Common breeder in the southern plains and winter resident in extreme southern Florida; rare but widespread vagrant throughout the United States and Canada. **DISTRIBUTION:** Breeds in most of Kans., se. Colo., se. N.M., Okla., Tex., w. Mo., w. Ark., and w. La. A few pairs breed farther east. Winters in s. Mexico, Central America, and extreme s. Florida. **HABITAT:** Open grasslands dotted with trees and shrubs. Also found in orchards, pastures, golf courses, airports, and agricultural lands. **COHABITANTS:** Swainson's Hawk, Western Kingbird, Loggerhead Shrike, Eastern and Western Meadowlarks. **MOVEMENT/MIGRATION:** Entire breeding population relocates for the winter. Spring migration from mid-March to late May, with a peak in April; fall from late July to November, with a peak in September and October. VI: 3.

DESCRIPTION: An elegant, pale-bodied, dark-winged kingbird affixed to a long narrow stick of a tail. Medium-sized (the body is about the same size as Eastern or Western Kingbird), with a shortish (Eastern Kingbird–like) bill, a slender body, and an exceedingly long, narrow, tined tail (usually closed when perched). Tail length varies from twice as long as the body (adults) to just as long as the body (immatures). Whatever its length (unless tines are broken or molted), the tail always appears too long and narrow for the bird.

Very pale-bodied, with a whitish head contrasting with dark wings and a dark tail showing breaks or flashes of white. No other flycatcher is so starkly white-headed. Adults show salmon-blushed flanks and salmon underwings that bloom when birds take flight—and being flycatchers, they often do. Immatures are like adults but have shorter tails and a yellow blush on the flanks.

BEHAVIOR: Most commonly seen perched on the top strand of a barbed-wire fence in open country, but also uses utility lines, bushes, and trees and even sits on the ground. Forages mostly by sallies, snapping prey directly from vegetation and the ground or out of the air. Also, but less commonly, walks on the ground. Usually found alone or in pairs in summer; found in small flocks before and during migration. In migration, often flies very high in small spaced flocks. Fairly territorial, Scissor-tailed attacks birds ranging in size from Red-tailed Hawk to Lark Sparrow.

Fairly tame; usually easily approached. Most often noticed when it goes aloft, making nimble, acrobatic, and buoyant aerial sorties that display its long two-tined tail.

FLIGHT: Distinctive long-tailed silhouette (unless maneuvering, tails are usually closed). In flight, tail is mostly stiff and unbending (in contrast to the wavy tail of Fork-tailed Flycatcher); the first impression is of a swallow towing a string. Wings are short, pointy, and slightly cocked back—very swallowlike. When birds are flycatching, the splayed tail, with its shifting black-and-white pattern, is unmistakable; salmon underwing linings and flanks (which are yellow and hard to see on immatures) are also hard to overlook.

Flight is mostly direct and fast, with slight shifts in elevation — rises and falls that are too irregular to be called undulations. Wingbeats are quick and stiff, given in quick series punctuated by terse skip/pauses; flight has a slight jerkiness. You get the feeling that the long tail imparts a stability to the flight in the same way that a tail stabilizes a paper kite.

VOCALIZATIONS: Song, which recalls Western Kingbird, is a series of high squeaky "*pip*" notes that frequently become increasingly more excited and squeally. Calls include a soft low "*pik*" or "*kip*" and a soft, more nasal "*chuh*."

Fork-tailed Flycatcher, *Tyrannus savana*

Fork-tailed Kingbird

STATUS AND DISTRIBUTION: Tropical species. Very rare but widespread and annual vagrant across the U.S. and s. Canada, with most records concentrated in the East Coast (also Tex.). **HABITAT:** Open brushy or brush-lined fields, grassy areas, agricultural fields, and utility lines. **COHABITANTS:** Sometimes associates and migrates with Eastern Kingbirds, but in the United States is often alone. **MOVEMENT/MIGRATION:** Spring migration usually from late April and May; fall from September through November (with more records for the fall). VI: 3.

DESCRIPTION: An Eastern Kingbird with a long, thin, forked tail. Not including the tail, Fork-tailed is smaller than but shaped like Eastern Kingbird and Scissor-tailed Flycatcher; tail included, adult Fork-tailed is overall longer than Scissor-tailed. Unlike the stiff-tailed Scissor-tailed, the tail of Fork-tailed Flycatcher commonly droops or looks wilted at the tip when birds are perched.

Very similar to Eastern Kingbird, showing pale grayish upperparts, a blackish cap, and white underparts. Fork-tailed, however, has a long narrow tail that appears either twice as long as the body (adults) or slightly longer than the body (immatures) and has no white tip. (Eastern Kingbird's tail is short and white-tipped.) Scissor-tailed Flycatcher has an all-pale head and does not resemble Eastern Kingbird.

BEHAVIOR: Forages in the open, often close to the ground and sometimes on the ground. Active. Shifts perches frequently. Often sits on the highest point, but that point may be quite low — a fencepost or wire, a shrub, a low tree, a utility line. During the bird's long, energetic, sometimes acrobatic sallies in pursuit of insects, the long tail undulates, like a crazed flagellum. (In flight, Scissor-tailed's tail is surprisingly stiff.)

FLIGHT: Distinctive profile shows a small body, short triangular wings, and a long fluttering streamer tail. Wingbeats are rapid and mostly continuous. Short-distance flight is more undulating than Eastern Kingbird, but straighter and more direct when flying high or traveling greater distances.

VOCALIZATIONS: Unlike Eastern Kingbird and Scissor-tailed Flycatcher, a mostly quiet bird, but reportedly utters a weak but sharp "*tik*" or "*sik*" note and occasionally twitters or chatters.

PERTINENT PARTICULARS: Because of its slender profile, short triangular wings, and white underparts, a Fork-tailed flying high overhead may be mistaken for a cuckoo. This species has also been mistaken for an escaped Pin-tailed Wydah.

Rose-throated Becard, *Pachyramphus aglaiae*

The Chunky Vireo-Flycatcher

STATUS: Tropical species; rare and geographically limited breeder in the United States. **DISTRIBUTION:** In the U.S., found only along the border regions of se. Ariz. and the s. Rio Grande Valley of Tex. (where it is rare and often found in winter). **HABITAT:** In the United States, riparian woodlands dominated by sycamore and cottonwood. **COHABITANTS:** Northern Beardless-Tyrannulet, Brown-crested Flycatcher, Thick-billed Kingbird (Arizona), Couch's Kingbird (Texas). **MOVEMENT/MIGRATION:** Arrives in Arizona in May; departs by September. VI: 0.

DESCRIPTION: A stocky, squarish, flat-headed, often immobile bird that little resembles a flycatcher — and only tacitly behaves like one. About the size and shape of Downy Woodpecker (or a salt shaker), but with a large neckless head, a thick pencil stub for a bill, and a shortish narrow tail. The blackish crown is slightly shaggy—a crewcut-topped bird in need of a trim. The thickish bill is set at nearly a right angle to the blunt forehead.

Males are all gray with a black crown; the "rose"-colored throat patch is sometimes difficult to see. Black-capped females are ruddy brown above and buffy below, but in the shadowy light under which most observations occur, they look like darker-backed, paler-bellied birds with dark crowns, and Sulphur-bellied Flycatcher red tails. Immatures are like females but overall ruddier.

Idiosyncratic particulars aside, the overall shape of the bird is unique. Nothing else looks like a becard.

BEHAVIOR: Sits a lot, often high and partially or wholly obscured in a leafy tree overhanging water or a desert wash. Rose-throated Becard's presence is heralded by (1) its very distinctive but easily overlooked call and (2) its big, pendulous, soccer ball–sized nest enveloping a down-hanging branch.

Hunts like a vireo/flycatcher cross. Perches vertically—even clinging sideways to a branch, it remains upright—and moves its head in slow exploring fashion. Sallies by nearly throwing itself at an overhanging cluster of leaves, plucking off an insect, then returning to a perch. Usually the sally involves a short (less than 10 ft.) flight and hovering.

Perches are most commonly just below the canopy and frequently very close to the nest.

FLIGHT: Somewhat titmouselike (or nuthatchlike) in flight. Quick, with rapid wingbeats and undulating flight.

VOCALIZATIONS: Call is a high, thin, whiny, descending whistle: "*cheeuuuu.*" Sometimes vocalizations begin with a two-note stutter, "*Pita pit cheuuu,*" or a chatter.

PERTINENT PARTICULARS: As with many species, the peak activity period for this bird is early morning and evening. Territorial males vocalize before sunrise. Coupled with the bird's sedentary nature, this activity pattern deals a losing hand to observers who hope to see this bird at midday—even if you can see the nest! Plan your visit accordingly.

SHRIKES

Loggerhead Shrike, *Lanius ludovicianus*

Snub-billed Shrike

STATUS: Widespread breeder and wintering species; fairly common across most of the West; uncommon to very rare in the East. **DISTRIBUTION:** Breeds widely across North America, including Mexico and the prairie regions of Alta., Sask., and Man. Also absent across the Northeast, parts of the Midwest, and much of n. Idaho, w. Wash., w. Ore., and nw. Calif. In winter, northern birds retreat south of a line drawn from Md. south and west to n. Tenn., Mo., Kans., Colo., Utah, Nev., and s. Ore. **HABITAT:** Breeds and winters in expansive open areas that are dominated by short grass and have hunting perches, such as pastures, agricultural land, cemeteries, parks, golf courses, and mowed roadsides. Favored perches include isolated trees of moderate height (less than 30 ft.), hedgerows, utility lines, and wire fences. In winter, some birds also hunt in broken or open woodland habitat. **CO-HABITANTS:** American Kestrel, Western Kingbird, Eastern Kingbird, meadowlarks. **MOVEMENT/MIGRATION:** Over most of its range, a permanent resident. Spring migration for northern birds is early, late January through March; fall from mid-July to November. VI: 2.

DESCRIPTION: A large-headed, tassel-tailed, compact, mockingbird-like bird—but a predatory mockingbird. The conspicuously large head with a stubby bill looks well tailored and compact. (Northern Shrike is overall more gangly and long in the face.) Facing into the wind, assumes a tense crouch. With no wind, slouches. (Mockingbirds commonly perch upright and seem alert but not tense.) Also, shrikes hunker down on a perch, appearing legless. (Mockingbirds almost always show shank.)

Plumage is handsome and distinctive, a study in gray, black, and white. Overall darker than Northern Shrike and more crisply and contrastingly patterned than Northern Shrike or Northern Mockingbird. (Next to Loggerhead Shrike, mockingbirds look bleached out or muted.) The black mask is wider, more dominating, and more intimidating on Loggerhead—but you may have trouble noting the extent or width of the mask because the mask gets lost against Loggerhead's darker head. This contrast between the dark head/mask and the white throat/breast is particularly stark and striking. Also lost in the blackness of the mask are Loggerhead Shrike's eyes. On Northern Shrike, the narrow mask and (often) narrow white eye-ring make the eyes stand out.

BEHAVIOR: Its conspicuousness may be the bird's signature characteristic. Few other pale-breasted birds, if any, sit so brazenly in the open. Loggerhead almost always perches in the open, away from cover, and often at the top of the only vantage point around—a stout weed, a wind-whipped sapling, or a utility line (its favorite perch). Tail is animate. In no wind, moves in measured bobs. In a crosswind, bends sharply at a right angle to the bird.

Solitary. Hunts alone, from a perch, concentrating on the ground close to the perch. Drops onto prey with a rapid flurry of wings and a recklessly fast pounce. Also sallies out, flycatcher fashion, to snatch large insects out of the air. Carries prey in both its bill and feet and consumes it on a perch. Hops on the ground, with a very upright stance; sometimes opens its wings, exposing white patches, apparently to startle prey.

FLIGHT: Flight is direct and fast, with wingbeats too fast to count and the head projecting conspicuously ahead of the bird. Drops quickly from a perch. Flies low to the ground. Lands with a rapid climb to the perch. Longer flights are evenly undulating

VOCALIZATIONS: Silent for the most part, but gifted with an array of interesting, mostly two- or three-noted, often serially repeated incantations, including a muttered "*chur-lee*" or "*tur-lee*"; a Rusty Blackbird–like "*take-leak*"; a wrenlike thumb-running-down-the-comb trill; a catbirdlike "*naah, naah, naah*" scold; a harsh muttered "*d'gee'r*"; and a somewhat musical three-note stutter.

Northern Shrike, *Lanius excubitor*

Palid Shrike

STATUS: Uncommon northern breeder; uncommon to rare winter resident across southern Canada and the northern United States. **DISTRIBUTION:** Breeds throughout Alaska (except the North Slope), the Yukon, n. B.C., the Northwest Territories, s. Nunavut, n. Que., and n. and w. Lab. Winters from se. Alaska east through B.C., s. Alta., s. Sask., s. Man., s. Ont., and e. Que.; the southern limit cuts across ne. Calif., n. Nev., s. Utah, n. N.M., w. Okla., w. Kans., se. Neb., s. Iowa, n. Ill., n. Ind., n. Ohio, n. Pa., n. N.J., and Long Island, N.Y. **HABITAT:** Breeds in sparsely vegetated boreal (spruce) forest and creekside willow and alder thickets on open tundra. Winters in a variety of open habitats that incorporate at least some elevated hunting perches, but prefers a mix of open fields and brushy edge, such as agricultural land, barrier islands, game management properties, and shrub-pocked prairies. **COHABITANTS:** During breeding, Northern Hawk Owl, Boreal Chickadee, Gray Jay, Yellow Warbler, American Tree Sparrow. In winter, Northern Harrier, American Kestrel, Black-billed Magpie, American Tree Sparrow. **MOVEMENT/MIGRATION:** Except in southeast Alaska and the northern Yukon, all birds vacate breeding areas in winter. Spring migration from March to April; fall from October to December. VI: 3.

DESCRIPTION: A large somewhat gangly shrike with a pale and muted pattern. Medium-sized songbird (slightly larger than Loggerhead Shrike; slightly smaller than American Kestrel or Northern Mockingbird), with a large round head, a heavy hooked bill, a lean body, and a long tail. The profile and posture may recall Townsend's Solitaire. Longer-billed, less compact, and overall more menacing than Loggerhead Shrike. Northern's stiffer tail is less prone to swivel in the wind.

Adults are overall pale gray, with blackish wings and tail, scant contrast between upper- and lower parts, and a slight warmish (buffy) cast to the breast. Loggerhead Shrike is overall darker and more contrastingly patterned gray, white, and black. The black mask through Northern's eye is often indistinct—it looks more like a heavy line than a mask and does not hide the eye. (On Loggerhead, the mask envelopes the eye, making it difficult to see.) Underparts show very fine pale brown vermiculations that are visible at close range and at a distance impart a warm buffy cast to underparts. Immatures are overall browner, with obvious brownish barring below and a very indistinct (sometimes almost invisible) face mask. (A pale brownish shrike is automatically a Northern Shrike.) At close range, often shows a pale forehead and a pale base to the bill.

BEHAVIOR: A solitary hunter most commonly seen perched on a prominent point overlooking an open or semi-open area. Shifts perches frequently (every 3–5 mins.), and at times may perch lower in the partially concealed interior of a bush (unlike Loggerhead). Hunts with speed and stealth. Launches itself from perches, drops low, and approaches prey swiftly and directly. When changing perches, also drops, flies low, and rises abruptly. Aerially adept. Able to float to the ground, hover, and flycatch. Appears high-strung and intent. Follows the motion of small birds intently, twitching its tail nervously. Generally shy, and does not tolerate close approach. Usually silent, but males may vocalize in late winter.

FLIGHT: Long profile—showing long fairly tapered wings and a long tail—and contrastingly patterned, showing pale gray upperparts, black wings and tail, and white flashes in the wing and (sometimes) the sides of the tail. Flight is rapid and direct, usually low to the ground, and deeply and evenly undulating, with wingbeats given in a rapid series followed by a lengthy closed-wing glide. Wingbeats are slightly slower than Loggerhead; with Northern, you can usually discern individual wingbeats.

VOCALIZATIONS: Song is a series of short, musical, sometimes ringing, two-note whistles repeated several times (like a mockingbird). Calls include a low raspy scold, "*raaah*," and a more rapid, higher-pitched, protesting series whose pattern and quality recall a soft (or distant) peregrine: "*nah nah nah nah*." Usually silent away from breeding grounds, but sometimes become vocal in March.

VIREOS

White-eyed Vireo, *Vireo griseus*

Vireo with Attitude

STATUS: Common and widespread eastern breeder; more restricted in winter. **DISTRIBUTION:** Breeds across most of the e. U.S., north to s. Mass., se. N.Y., n. Pa., s. Mich., n. Ill., and n. and cen. Iowa; western border cuts across se. Neb., e. Kans., cen. Okla., and cen. Tex. (west to about the Pecos R.) south into Mexico. In winter, found on the Atlantic and Gulf coastal plains from N.C. south into Mexico and Central America. **HABITAT:** Deciduous brush and scrub (often with briars and a scattering of taller understory trees), dense forest understory in more open woodlands, overgrown successional fields, woodland edge, low coastal woodlands, streamside thickets, and (in Florida) mangroves. **COHABITANTS:** Carolina Wren, Gray Catbird, Brown Thrasher. **MOVEMENT/MIGRATION:** Spring migration ranges from March to mid-May, with a peak in late April and early May. Generally arrives a week or two ahead of the host of returning breeding bird species. Fall migration is fairly abbreviated, from late August to late October. Very rare vagrant west of its breeding range. VI: 3.

DESCRIPTION: A small compact vireo that likes to be heard but not seen. The bill is stout, the head large and neckless, the body portly, and the tail short and narrow. Looks (and behaves) like a plump warbler.

The grayish head and hindneck contrast with the greenish upperparts and yellowish underparts. (*Note:* The throat and belly are paler—grayer or whiter—than the flanks, but because the bird is often at eye level, the belly usually can't be seen.) Standout features are the bright yellow "spectacles" surrounding the beady white eyes and the dark-shouldered wings etched with narrow, crisp, white wing-bars. The expression is fierce.

Immatures have the overall compactness and plumage pattern of adults, but also have muted white, not yellow, spectacles and whiter underparts (the yellow is more washed out and restricted to the flanks). The eye is dark (not white) and large, giving the bird a gentle expression.

BEHAVIOR: Fairly tame; often allows close approach, but is also not inclined to leave its tangled fortress. Very active and warblerlike. Moves from branch to branch with hops and short flights. Often forages at eye level and below, rarely ascending more than 20 ft. above the ground. Feeds both vertically and horizontally, plucking prey from leaves, and often flies to snatch prey en passe. In winter, also forages on berries. Posture is generally erect, not horizontal. Singing birds seem to slouch.

A persistent singer. Sings into the heat of the day and late into the season (late September)—long after most species have stopped.

FLIGHT: Flight is quick, direct, strong, undulating, and usually short.

VOCALIZATIONS: Varied but nevertheless distinctive. Song is a short happy-go-lucky ditty comprising chips, whistles, whines, and harsh notes. Usually begins with a distinct "*chip*" or a low rough note and ends with another distinct "*chip*"—for example, "*chip!-ch'chur-e-ur-chip!*" or "*ch'-sp'l-eeur-chip!*" A classic rendering that captures the rhythm and spirit of the song is "*Spit. And see if I care. Spit.*" Individual utterances recall the sounds made by Yellow-breasted Chat but are higher-pitched, livelier, and faster-paced and notes are not repeated.

Repeats the same song for a time, then switches to another in the repertoire (which may number a dozen or more). In good habitat, it is not uncommon to hear several White-eyed Vireos singing different songs simultaneously.

Scold recalls a raspy Carolina Wren: "*eh-red-red-red-red-red.*"

Bell's Vireo, *Vireo bellii*

The Vireo That Says "$%#"&!!!"

STATUS: Uncommon to fairly common breeder in the central and southwestern United States. Casual migrant east of its normal range. **DISTRIBUTION:** From cen. N.D. and portions of S.D. across n. and e. Neb., all but n. Iowa, Ill., all but se. Ind., e. Colo., Kans., Mo., Okla., n. and cen. Ark., all but nw. and e. Tex., s. Calif., extreme s. Nev., w. and s. Ariz., and s. N.M., south across n. Mexico. Winters on the west coast of Mexico; a few winter in Fla. **HABITAT:** Thickets and dense shrubby growth, including regenerating forest, successional fields, mesquite woodlands, scrub oak, and tamarisk and willows in riparian areas. Absent in dense forest, sparsely vegetated desert, open grasslands, and agricultural land. Prefers brushy habitat near water. **COHABITANTS:** Brown Thrasher, Curve-billed Thrasher, Yellow Warbler, Eastern Towhee, Blue Grosbeak. **MOVEMENT/MIGRATION:** Spring from early March to early May, with a peak in late March through April; fall from early August to November, with a peak from September to mid-October. VI: 3.

DESCRIPTION: A slight, spry, long-tailed, pallid, effervescent, brush-hugging gremlin in vireo's clothing. Small (excluding the longish tail, tiny), with a small head, a large eye, and a longish slender, pale and somewhat flesh-colored bill. Its most distinctive feature is the long animate tail.

Overall pale and indistinctly marked. Western birds are pale gray above, paler gray below, with a frail eye-line and wing-bars. Eastern birds have grayish heads with poorly defined whitish "spectacles" and greenish upperparts; underparts are yellow-washed.

In size, plumage, and manners, Western birds resemble a long-tailed Lucy's Warbler. Eastern birds are more akin to small, longer-tailed, washed-out Blue-headed Vireo.

BEHAVIOR: Fairly active but methodical vireo that flicks its wings, jerks its tail, and hops from branch to branch gleaning insect prey. You will be lucky to see any of this. The bird forages within the brushy interior and is an ace at staying hidden. Males vocalize incessantly while feeding, often after other birds have quit for the day. Flights are short, usually ending in the next sizable bush. Birds sometimes hover, and are almost never seen on the ground. The tail vibrates as the bird sings.

A shy bird that responds to pishing but loses interest quickly.

FLIGHT: Overall pale and plain profile, with a long tail. Flight is quick, furtive, and undulating.

VOCALIZATIONS: Song is a raspy, hurried, rough-spoken, saucy admonishment: "*rch'eechr'rchee'-E'chur*" (ends with a low "*chur*" or "*chac*") or "*rch'eech'rcheecurEH*" (ends with a higher-pitched "*EH*" or "*EE*"). Song sounds like a vireo song on fast-forward. Call is a raspy "*naah*" uttered singly or sequentially (not as nasty or angry-sounding as White-eyed Vireo). Also utters a shorter raspy expletive: "*pe*" or "*whe*" or "*wr.*"

PERTINENT PARTICULARS: The eastern Bell's Vireo is similar enough to the usually more richly colored immature White-eyed Vireo to make identification troublesome. The paler, grayer western Bell's is sometimes confused with the longer-tailed Gray Vireo.

Black-capped Vireo, *Vireo atricapillus*

White-goggled Vireo

STATUS: Uncommon to rare geographically restricted breeder. **DISTRIBUTION:** Found only in cen. Tex., and locally in sw. Tex. (for example, in Big Bend) to the Mexican border (continuing south into Mexico) and in cen. Okla. **HABITAT:** Low, broken, brushy habitat of varying heights commonly situated on dry hillsides, gullies, and ravines. Prefers habitats dominated by deciduous trees (primarily oaks) with a scattering of junipers and thickets and vegetation that extends to ground level. Often associated with the successional growth following fires. **COHABITANTS:** Blue-gray Gnatcatcher, Golden-cheeked Warbler, Canyon Towhee. **MOVEMENT/MIGRATION:** Spring migration in April; most birds depart by the end of September. VI: 1.

DESCRIPTION: A stunning bird, a persistent singer, and a trial to see. Tiny (1 in. shorter than Blue-headed Vireo) but stocky, not in the least petite, with a large head and short tail. These features, combined with the striking plumage, which draws all eyes to the bird's head, can make the bird seem larger than it really is. The bill is longish, straight, and thin, and the face pointy.

Plumage is distinctive. A greenish-backed bird with a black head and bright white goggles, surrounding a red eye, that are bigger and more prominent than the "spectacles" of Blue-headed Vireo. Underparts on adults show a contrastingly white throat and yellow on the flanks. Immatures are like adults, but the contrasting pattern is muted—the bluish gray head and greenish back are akin to a small Blue-headed Vireo, *except* for the more prominent goggles and thinner, pointier bill.

BEHAVIOR: Likes to be where the brush is thick. Vocalizing males may be perched near the top of a bush but commonly cloak themselves in a leafy cluster—then sing persistently, even in the heat of the day.

Forages by moving in quick nimble hops; Black-capped is a more active feeder than most vireos, and certainly more active than the slow methodical Blue-headed Vireo. Forages mostly by gleaning from leaves; occasionally hovers. Flies little and not far. Responds well to pishing.

FLIGHT: An unglamorous, straightforward, undulating flight, with short series of quick wingbeats interrupted by terse closed-wing glides. Wings seem short and wide. On adult males and females, the black head contrasting with the greenish upperparts is obvious in flight no matter how brief the look.

VOCALIZATIONS: A lively, entertaining, and varied repertoire of smartly enunciated chips, husky trills, rising trills, rolling trills, sharp whistles, squeaks, and more complex and undecipherable utterances. Many notes are given sequentially and are sometimes repeated later in a song—for example, "*chchch grzreeee t't'tree ch pr'r'r'ree.*" Call is a harsh rising "*zreee.*" Also makes a short, muttered, two-noted stutter.

Gray Vireo, *Vireo vicinior*

Shrikelike Vireo

STATUS: Uncommon and habitat-specific western breeder and very local wintering species. **DISTRIBUTION:** Breeds in s. Nev., s. and se. Utah, w. Colo., n. and e. Ariz., and w. N.M.; also found in pockets in se. Calif., se. Colo., se. N.M., and the Big Bend region of Tex. Winters in sw. Ariz., trans-Pecos Tex., nw. Mexico, and very locally east of San

Diego. **HABITAT:** Hot dry slopes with scattered brushy vegetation, primarily pinyon-juniper, oak scrub, and (chamise) chaparral. Habitat may also include cactus and thorn scrub. **COHABITANTS:** Rock Wren, Verdin, Rufous-crowned, and Black-chinned Sparrows, Scott's Oriole, Varied Bunting. **MOVEMENT/MIGRATION:** Except for birds breeding in Texas, all Gray Vireos vacate breeding territory and retreat short distances to southwestern Arizona, eastern San Diego County, Calif., and Mexico. Spring migration from late March to early May, with a peak in April; fall from late August to mid-October, with a peak in mid-September. VI: 1.

DESCRIPTION: A large, lanky, gray vireo that thrives where few other species care to try. Overall long—the lankiest vireo—but solid (not gnatcatcher wispy), with a round head, a short, thick, gray nub of a bill, a compact body, and a long narrow tail. In appearance and posture, looks much like a miniature shrike.

Overall plain pale gray. Upperparts are darker and sometimes slightly brownish-tinged; pale underparts may seem whitish. The facial pattern and single, narrow, grayish wing-bar are indistinct bordering on inconsequential. In structure and plumage, Gray Vireo is most like the smaller and paler-billed Bell's Vireo (which will be down in the desert wash below the Gray Vireo belt).

BEHAVIOR: More active (or perhaps more far-ranging) than most vireos. Forages in brush, usually low (within 10 ft. of the ground), and most often among the interior branches (in other words, it's hidden). An accomplished and versatile hunter, Gray "stalks" prey among the branches with slow movements followed by a long hop or a flutter-dash intercept. Also flies readily from one new perch to another. Hovers in front of vegetation; also flycatches and sallies or pounces on prey on the ground. The long tail is often very animate—not just jerked but swished from side to side. Gray's movements and shape make it seem very gnatcatcher-like.

The male often sings from a conspicuous perch (such as atop a juniper) as well as in flight; though most vocal in the mornings, sings throughout the day. May fly great distances (up to a quarter-mile!) to find a new singing perch—often in the next gulch, that is, the one you are not standing in.

FLIGHT: Flight is strong, direct, and slightly jerky and undulating. Wingbeats are mostly constant, with very brief and regular skips. The bird's overall grayness, absence of any discernible plumage field marks, and long tail are apparent. Recalls a thin titmouse.

VOCALIZATIONS: A chopped-off-phrase singer (like Red-eyed, Plumbeous, and Philadelphia). Song is most like Plumbeous: a mix of hoarse and slurred phrases and a few sweeter ones, but with a quicker delivery, a more limited repertoire, and frequently repeated phrases; for example, "*chiree cheelr tu chiree chibu tu.*" Call is a dry "*charr,*" like Bewick's Wren. Also makes a harsh, descending, chatter scold.

Yellow-throated Vireo, *Vireo flavifrons*

Vireo of Bare Branches and Forest Border

STATUS: Fairly common but, where ranges overlap, almost always less common than Red-eyed Vireo. Casual visitor in the West. **DISTRIBUTION:** Breeds across most of the e. U.S. and se. Canada (absent in all but s. Maine, n. N.H., ne. Vt., the southern half of Fla., and coastal La. and Tex.). Western border extends to cen. N.D., cen. S.D., e. Neb., extreme e. Kans. and Okla., and e. and s.-cen. Tex. almost to the Mexican border. In Canada, found in se. Sask., s. Man., se. Ont., and s. Que. **HABITAT:** Mature, usually large or extensive deciduous and mixed deciduous-coniferous forests; avoids pure conifer woodlands and has a particular love of oaks. Most common at the forest edge and in places with more open understory. In migration, uses a variety of woodland habitats. **COHABITANTS:** Eastern Wood-Pewee, Red-eyed Vireo, Tufted Titmouse. **MOVEMENT/MIGRATION:** Spring migration from early April to early June; fall from mid-August to mid-November. VI: 3.

DESCRIPTION: A big beefy canopy vireo with yellow "spectacles," a yellow chest, and a white belly. Large (smaller but more robust than Red-eyed; about the same size and shape as Blue-headed), with a heavy vireo bill, a big round head, a heavy body, and a disproportionately (almost comically) short and narrow tail.

Upperparts are smoothly yellow-green; underparts are distinctly bisected with a bright yellow throat and chest and a bright white belly. The yellow spectacles around the eyes and broad white wing-bars are conspicuous. In size and shape, might be confused with a particularly bright Pine Warbler. Pine Warbler, however, is streaky and shabby-looking compared to the smooth, even, unblemished plumage of Yellow-throated Vireo, and even the sluggish Pine Warbler looks positively frenetic next to Yellow-throated Vireo.

BEHAVIOR: It's a bird that likes high canopy, sturdy bare (very often dead) branches, and long leaps. Forages mainly in the higher branches, often in the more open interior portions of the tree, moving in the stop-and-go fashion of vireos and avoiding slender springy perches. Unlike most arboreal vireos, Yellow-throated has a dinner table reach and specializes in locating and seizing more distant prey, often targeting insects on branches or trunks. (Other vireos mostly target leaves.) Leaps are longer, quicker, and springier than Red-eyed Vireo, and pauses in search of prey are lengthy. Short flights to seize prey on distant branches are common, and this bird is not at all shy about flying long distances as well, traveling over open areas or above the canopy to reach another tree. Males commonly sing as they forage. Responds well to pishing, but may not descend from the canopy.

FLIGHT: Profile shows a wide body with wide pointy wings, a projecting head, and a short, narrow, almost insignificant tail. The contrasting plumage pattern holds up. The greenish yellow head and upperparts and sharply defined yellow-and-white underparts are distinctive. Flight is mostly direct, with a slight jerkiness; wingbeats are deep, rapid, strong, given in a short series, and punctuated by a terse skip/pause.

VOCALIZATIONS: Lazy whistled song is a series of alternately higher then lower two- (sometimes three-)note phrases separated by a pause: "*twee'ree*" (rising); "*he'yew*" (falling); "*twee'l'ree*" (rising); or "*he'yew.*" Sounds somewhat as if the bird is saying: "*See me? Hear you.*" Some notes are pure in tone, but the overall quality is hoarse and slurred—a vireo with a sore throat. By comparison, Blue-headed Vireo's song is higher, clearer, and sweeter-sounding. Yellow-throated also makes a low, rough, descending, chatlike-sounding scold, "*cheh cheh cheh cheh.*"

Plumbeous Vireo, *Vireo plumbeus*

Lead-colored Spectacled Vireo

STATUS: Common vireo of western interior forests. **DISTRIBUTION:** Breeds in the cen. Rockies and the Great Basin. Found in extreme s. Idaho, cen. and e. Nev., Utah, the western half of Colo., n. and e. Ariz., all but e. and s. N.M., and s. and cen. Mexico; also found in parts of se. Mont., nw. and s. Wyo., the Black Hills of Wyo. and S.D., extreme nw. Neb., parts of e. Calif., and trans-Pecos Tex. Casual to e. Tex. and s.-cen. Alaska. **HABITAT:** Dry, mountain coniferous forests—particularly those dominated by ponderosa pine—or mixed forests that include pinyon-juniper-oak; also occurs in riparian woodlands in the Upper Sonoran region of Arizona and New Mexico. **COHABITANTS:** Mountain Chickadee, Pygmy Nuthatch, Grace's Warbler, Western Tanager. **MOVEMENT/MIGRATION:** Entire population vacates breeding range and winters primarily in western Mexico, but some birds are found in southern California and southern Arizona. Spring migration from late April to early June, with a peak in mid-May; fall from August to October, with a peak in September. VI: 2.

DESCRIPTION: A robust lead-gray vireo with stark white spectacles. Medium-sized and stocky with a large round head, a short thick bill, and a short tail. The bill is slightly but not appreciably longer than on Cassin's Vireo. Slightly longer overall than Cassin's and Blue-headed, but this is difficult to appreciate in the field.

The gray extreme of the "spectacled" vireo trinity. Plumage is a distinguished, unblemished, uncontestable gray—bluish or slate gray above, grayish and white below. The broad white spectacles, white lores, and white wing-bars pop against this crisp conservative backdrop. The slight yellowish or olive tinge on the flanks is barely discernible (and probably your imagination). Cassin's Vireo is comparatively muted, dull, and touched with green, particularly in the wings and flanks. Gray Vireo is longer-tailed and lacks the bold white spectacles and shows a single, faint white wing-bar.

BEHAVIOR: A patient and methodical feeder. Moves in short hops through the foliage, pausing, searching, plucking prey, moving on to the next perch, favoring all arboreal possibilities—outer leaves and twigs as well as inner branches (everything except the trunk and surrounding ground). Nimble but not acrobatic. Likes perches that bear its weight comfortably; rarely dangles. Feeding pattern is eclectic and random. Works trees methodically but not exhaustively. Generally flies to the next closest tree, then sometimes doubles back, working the same tree again.

Males vocalize as they feed. Fairly tame. Responds well to pishing.

FLIGHT: Appears stocky and distinctly gray. Wingbeats are quick, regular, and perfunctory. Undulating flight seems heavy or staid.

VOCALIZATIONS: Song is a series of rough, slurred, burry, sometimes pleasing, two-, three-, and three-and-a-half-note phrases: "*chee-oo-ree* (pause) *chur-ee* (pause) *chur-ee-oo*." Song is similar to Cassin's, not as lively, lighthearted, and musical as Blue-headed (the pauses between phrases are conspicuously longer), and not as hoarse and burry as Yellow-throated (not quite anyway). Most of the phrases sound both rough and slurred; r's are rolled, and few phrases are clearly enunciated. Plumbeous's song sounds like a drunk mixing up and melding synonyms—compressing two words into one (hence the three-and-a-half-syllable phrases). Call is a descending raspy-noted scold, like Blue-headed Vireo.

Cassin's Vireo, *Vireo cassinii*

The Intermediate Spectacled Vireo

STATUS: Fairly common West Coast representative of the "spectacled" vireo tribe. Regular fall migrant through the western Great Plains. **DISTRIBUTION:** Breeds from s. B.C. and extreme sw. Alta. to Wash., n. Idaho, w. Mont., n. and e. Ore., and Calif. except for the Central Valley and se. portions of the state; also found in portions of Baja. Winters in Mexico, with a few in s. Calif. and s. Ariz. **HABITAT:** Coniferous and mixed coniferous-deciduous forests, oak woodlands, and canyons. Partial to open forests dominated by deciduous trees. **COHABITANTS:** Band-tailed Pigeon, Steller's Jay, Swainson's Thrush, Townsend's Warbler. **MOVEMENT/MIGRATION:** Entire population vacates breeding range and retreats into Mexico (some birds to southern California and southern Arizona). Spring migration in April and May; fall from August to mid-October. VI: 2.

DESCRIPTION: A stocky white-"spectacled" vireo whose plumage homogenizes the richly patterned Blue-headed and the starkly gray Plumbeous. Medium-sized (the same size as Blue-headed; slightly smaller than Plumbeous) and overall compact, with a large round head, a stubby bill, a plump body, and a short tail, Cassin's is structurally similar to Plumbeous and Blue-headed.

Cassin's is more greenish than Plumbeous Vireo, and less richly and contrastingly patterned than Blue-Headed. Cassin's grayish head shades into a grayish green back with little change in tone. (The head of Blue-headed is contrastingly darker and bluer than the back.) Cassin's folded wings are touched with green, and the flanks washed with grayish yellow or olive yellow. (Plumbeous Vireo is uniformly gray, and the touch of color on the flanks is easily overlooked.) Cassin's face pattern melts into the white throat and is not sharply defined. (On Blue-headed, the contrast between the dark cheeks and white throat is crisp and stark.)

BEHAVIOR: A fairly active energetic vireo. Changes perches quickly. Frequently makes long vaulting hops to distant limbs and adjacent trees, with extravagant use of its wings. Has a distinct horizontal hunt pattern. Likes to work through the canopy. Hovers, lurches to seize prey above, and also flycatches, snatching prey in the air. More commonly, however, Cassin's uses the hop-stop-search-seize pattern of other vireos, though this species does seem to have a longer reach and a more expansive view of the hunting territory.

Fairly tame. Responds well to pishing.

FLIGHT: Direct, swift, no-frills, slightly undulating.

VOCALIZATIONS: Song, as with Blue-headed and Plumbeous, is a measured ensemble of two- to four-note, sweet, burry, and slurred phrases. Not as hurried, sweet, and pure-sounding as Blue-headed; not as rough and slurred as Plumbeous, but overall more like Plumbeous. Seems more prone to mix sweet phrases with rough ones and to use more consonants and fewer vowels than Plumbeous. For example: "*cl chl'ree* (pause) *cher-ee* (pause) *chlur.*" Favored song pattern uses two lazy phrases followed by a hurried compressed one. Call is a descending raspy-noted scold like Blue-headed Vireo.

Blue-headed Vireo, *Vireo solitarius*

Blue-headed Spectacled Vireo

STATUS: Common breeder across boreal Canada and New England and along the spine of the Appalachians. Very rare visitor to the western United States. **DISTRIBUTION:** Southwestern Nfld. and N.S. across s. Que., all but n. Ont., cen. Man., cen. and nw. Sask., all but s. Alta., ne. B.C., and the sw. Northwest Territories. In the U.S., found throughout New England, N.Y., n. N.J., most of Pa., ne. Ohio, n. Mich., and n. Minn.; also occurs in w. Md., extreme w. Va., most of W. Va., e. Ky., e. Tenn., w. and cen. N.C., extreme n. Ga., and ne. Ala. In winter, found on the coastal plain and lower Piedmont from se. Va. to cen. Tex., south into Mexico and Central America. **HABITAT:** Prefers mature or near-mature coniferous forest (spruce, hemlock, pine), but readily breeds in mixed woodlands—especially alder, birch, and willow—as well as in pure deciduous forest, particularly in the southern Appalachians. More important than forest species composition is the species' preference for large mature forest with a closed canopy and a rich but not dense shrub layer. In winter, also uses swamps, dense woodlands, and thickets. In migration, found in all manner of woodlands. **COHABITANTS:** In summer, Yellow-bellied Sapsucker, Red-breasted Nuthatch, Yellow-rumped Warbler, Blackburnian Warbler, Purple Finch. In winter, may join winter flocks with chickadees, nuthatches, kinglets, Yellow-rumped Warbler. **MOVEMENT/MIGRATION:** Among the earliest returning insectivores in the spring (and the earliest returning vireo in spring by one or two weeks); trans-Gulf migrants appear on the coast by mid-March, and northern breeders commonly reach territories before mid-May. Fall migration is protracted: some northern birds move south by early September, and birds linger north of their wintering areas well into November. Across much of the eastern United States, late September and October see peak southbound movements. VI: 3.

DESCRIPTION: A stocky slow-moving vireo with a dark helmeted head and white goggles. Medium-sized (larger than Philadelphia or White-eyed; smaller and stouter than Red-eyed; the same size as Warbling and Cassin's), with a large round head, a stout hooked bill, and a plump body. The tail seems too short and too narrow for so hefty a bird.

Plumage is a tasteful amalgam and is more contrasting and colorful than the other white-"spectacled" vireos (Cassin's and Plumbeous) but still coordinated and conservative (compared to many warblers). With its dark bluish gray head being distinctly darker and grayer than its greenish-washed back, Blue-headed has a helmeted and white-goggled appearance. Also, the border between the dark face and white throat is starkly defined. (On Cassins's, it's muted and blurred.) White underparts are washed with yellow-green on the sides and flanks.

Compared to other eastern vireos, which have creamy, grayish, or yellowish underparts, Blue-headed's are bright white (after allowing for the splash of yellow along the flanks). The expression is studied and gentle.

BEHAVIOR: A slow methodical feeder that likes to stay on limbs and sturdier branches, rarely venturing out onto the springy tips. Forages at heights in the middle portions of trees, where there is lots of room for the leap and grab or short sally flights it employs to secure prey from leaves and branches. Sometimes makes quick, nervous, easily overlooked wing flicks, but generally makes spare use of its wings (unless flying to secure prey) and moves by hopping from branch to branch.

Does not flock with other Blue-headed Vireos, a trait that harkens back to its former name, Solitary Vireo. Frequently forages with mixed flocks of warblers and year-round forest species (like chickadees and titmice) during migration and in winter. Fairly tame and also very responsive to pishing.

FLIGHT: Appears big, blunt-headed, and somewhat humpbacked, with a short tail. (The front of the bird reminds you of the head of a small Sperm Whale.) The contrast between the blue head, paler back, yellowish flanks, and white underparts holds at a distance. Flight is fairly slow, heavy, and slightly undulating. Wingbeats are given in a short series—the deep pushing wing beats give the impression that the bird is resting on the palms of its hands—followed by a terse skip/pause. Flight is slightly wandering or tacking.

VOCALIZATIONS: Song is a pleasing series of short two- and three-note phrases—some sweet and tonally pure, and some slurred—that can be phonetically rendered as "*see you*" (pause) "*here I am*" (pause) "*see me*" (pause) "*up here*" (pause) "*in the tree.*" Song recalls Red-eyed Vireo, but is sweeter, higher-pitched, and slower. (Actually, some phrases are more hurried, but the pauses are longer.) Call is a raspy, scolding, descending series: "*cheh, cheh, cheh, cheh, chech.*"

Hutton's Vireo, *Vireo huttoni*
Kingletlike Vireo

STATUS: Common but somewhat geographically restricted. **DISTRIBUTION:** There are two distinct breeding populations. The coastal population is found within 100 mi. of the West Coast from s. B.C. to n. Baja. The second, Mexican population reaches the U.S. in se. Ariz. and sw. N.M. and the Big Bend region of Tex. **HABITAT:** Mixed evergreen forest, particularly those rich in live oak. **COHABITANTS:** The coastal population associates with Cassin's Vireo, Chestnut-backed Chickadee, and Oak Titmouse, and the Mexican population with Plumbeous Vireo, Bridled Titmouse, and Painted Redstart. **MOVEMENT/MIGRATION:** Permanent resident and essentially nonmigratory; a few birds, however, tend to wander slightly outside their range in spring and fall. VI: 0.

DESCRIPTION: A small, plain, generic-looking vireo that passes easily for a Ruby-crowned Kinglet or an *Empidonax* flycatcher. Tiny and chickadee-sized, with a big round head, a small, dark, pointy bill (thick when viewed from the side, but thin from below), a plump body, and a short narrow tail. The eye is unusually large and bulging (should you get close enough to notice).

Overall dingy brownish gray (putty-colored) or dingy greenish gray with paler underparts. The whitish eye-ring and lores are obvious but not crisply defined. Two whitish wing-bars are evident (but know that Ruby-crowned Kinglets can also show two wing-bars).

BEHAVIOR: It's a vireo that likes to work a tree slowly. Moves in methodical and calculated hops between branches, usually in the interior and lower to middle portions of a tree (most commonly a live oak). Searches with terse movements of the head before moving to the next perch. Works a tree thoroughly and tenaciously, often returning to work it again later. Likes to be in places where branches are thick. Movements are overall looser and not as abrupt and jerky as many other vireos.

Seems reluctant to use its wings. Pauses longer when flight to another branch or tree is necessary. Generally plucks prey, usually from the underside of leaves and branches, while perched, but also leaps and makes fluttering sallies and also hovers. Hutton's is nimble enough to dangle beneath a leaf or twig, in (sluggish) chickadee fashion. Movements are not as frantic and nervous as kinglets; Hutton's does not usually flick its wings unless agitated.

In summer, commonly seen foraging in pairs. In winter, joins mixed flocks of chickadees, kinglets, and other flocking woodland species. A tame bird that responds well to pishing.

FLIGHT: Profile is short and stocky, with short rounded wings. Flight is weak, fluttery, and bouncy. When landing (especially downhill), brakes with a series of half-opened wing glides.

VOCALIZATIONS: Song is a monotonous repetition of the same simple whistled phrase: "*chu-wee chu-wee chu-wee.*" After a lengthy period, the bird switches to another phrase. Call is a low raspy spit, "*swit,*" or a low muttered "*jrt*" (often repeated) that pairs may use to communicate with each other while foraging.

Warbling Vireo, *Vireo gilvus*

The Plain-faced Riverside Vireo

STATUS: Common and widespread North American breeder and migrant. **DISTRIBUTION:** An extensive breeding range encompasses most of the U.S. and much of w. and s. Canada. Breeds across se. Alaska, the s. Yukon, the sw. Northwest Territories, most of B.C., all but ne. Alta., s. Sask., s. Man., sw. and extreme se. Ont., s. Que., and scattered portions of the Maritimes. In the U.S., very widespread and absent only in se. Wash., n. Ore., the Central Valley of Calif., se. Calif., w. and s. Ariz., s. and e. N.M., e. Colo., most of Tex. (except trans-Pecos and ne. Panhandle). Also absent as a breeder in the South from a line drawn from cen. N.J., n. Del., cen. Md., n. Va., cen. Ky., cen. Tenn., nw. Miss., s. Ark., and s. Okla. **HABITAT:** Deciduous and mixed deciduous woodlands, particularly those portions adjacent to streams, rivers, and lakes. Prefers mature trees. Woodlands need not be extensive; riparian corridors, tree-lined riverbanks, open city parks, and cemeteries are all acceptable. Seems particularly partial to cottonwoods. Also found in conifers in the western mountains. **COHABITANTS:** Eastern and Western Kingbirds, Baltimore, Bullock's and Orchard Orioles. **MOVEMENT/MIGRATION:** Entire population relocates to Mexico and Central America in winter. Spring migrants reach the southern United States by mid-March, but heaviest movements are noted in April through May. Early-fall migration: first southbound birds are recorded in early August, and peak movements occur in late August to mid-September. Sightings of the species in October are rare. VI: 2.

DESCRIPTION: Small, pale, plain vireo best known for its zesty and persistent singing. Overall compact but nicely proportioned, with a round head, a thin un-vireo-like bill, and a tail that is longer than shorter. In shape and behavior, seems warblerlike (most closely resembles Tennessee Warbler).

Plumage is overall pale, muted, and plain. Greenish gray upperparts and dingy white underparts lack clear contrast or definition. The large dark eye, set against a pale face or within a pale white halo, is prominent. The crown is grayer than the back, but not distinctly darker. The pale yellow wash on the underparts is limited to the flanks and sides of the breast (not the throat or center of the breast), and may be absent on some individuals.

The expression is benign or blank, but not as gentle as Philadelphia Vireo.

BEHAVIOR: A fairly active vireo, but calm in repose and not nervous; engages in no wing- or tail-flicking. Quick-acting, however, and given to abrupt movements and active wing-fluttering sallies to pluck insects from the undersides of leaves. Forages mostly on leaves in the canopy. Hovers easily and occasionally flycatches. Likes to work a tree thoroughly, then move to another (often distant or remote) tree. Partial to starting low and working up through the canopy. Males vocalize habitually, often pausing on the same branch for several renderings, then moving on to another branch or tree.

FLIGHT: Flight is moderately fast, undulating, and sometimes wandering. Executed with a quick series of wingbeats punctuated by a terse closed-wing skip/pause.

VOCALIZATIONS: Song is a rapid, conversational, saucy-sounding, highly variable warble. Some notes are sweet, while others are trilled or rough; all are somewhat slurred and rounded into the next: "*Ree rur ruh Ree rur rur rur REE.*" The last note usually (but not always) is louder, higher, and more emphatic than the rest. In tone and structure, sounds like a happy drunk making a conversational point at a party. More melodious and not as clipped, emphatic, or angry-sounding as Bell's Vireo. Scold call is a harsh, raspy, descending "*whe-hhh*" or "*raaaaah*" that has the pattern and something of the quality of the last part of Red-winged Blackbird's song: "*(tur-a) liiing.*" Also emits a soft, low, terse, and irregular spit, "*chut . . . chut,*" that is somewhat like the Indigo Bunting spit note.

Philadelphia Vireo, *Vireo philadelphicus*

Baleful-looking Vireo

STATUS: Widespread northern breeder, but generally uncommon and overlooked. **DISTRIBUTION:** The northernmost breeding vireo. Breeds across most of Canada, from ne. B.C. across n. and cen. Alta., cen. Sask., cen. Man., most of Ont., s. Que., N.B., s. Nfld., and (occasionally) other Maritime provinces. In the U.S., breeds in n.-cen. N.D., n. Minn., the Upper Peninsula of Mich., n. New England, and the Adirondacks of N.Y. Casual visitor in the w. U.S. **HABITAT:** Breeds in young second-growth deciduous woodlands, and less commonly in mixed coniferous-deciduous woodlands. Most partial to young aspen, birch, alder, and ash; does not commonly breed in towns. In migration, also favors younger second-growth trees and shrubs, particularly those bordered by water. **COHABITANTS:** Ruffed Grouse, Least Flycatcher, Red-eyed Vireo, Magnolia Warbler, American Redstart. **MOVEMENT/MIGRATION:** Entire population relocates to Central America for the northern winter. Migratory period is very brief. Spring migration from early May to early June; fall from late August to early October. VI: 2.

DESCRIPTION: Slightly yellowish, somewhat pudgy vireo with a snub bill and a gentle expression. Small and overall compact (larger than Northern Parula; smaller than Red-eyed or Warbling Vireo), with a large round head, a short stubby bill, a plump body, and a short narrow tail. Conspicuously pudgier, more rotund, and shorter-billed than the rangy long-snouted Red-eyed Vireo; very slightly smaller, plumper, stubbier-billed, and shorter-tailed than Warbling Vireo.

Plumage is mostly grayish above and slightly greenish-tinged and pastel-shaded, but lacking in pattern or clear contrasts except for a slightly darker cap and a smudgy dark line through the eye (particularly the lores). Sometimes shows a single faint wing-bar. Underparts are usually washed yellow, but in spring may be mostly white, showing only the faintest yellow wash on the upper breast.

Most closely resembles the longer-billed Warbling Vireo, but note Philadelphia's slightly darker, more contrasting face pattern and especially the dark eye-line that extends to the base of the bill. Warbling does not show yellow on the throat or center of the breast. Philadelphia's expression is gentle and serene, whereas Warbling looks blank and sometimes startled. Red-eyed Vireo has an intense borderline crazed expression and a head pattern that, in comparison to Philadelphia, makes the bird look like an over-painted dowager.

BEHAVIOR: A versatile treetop species that moves in hops and short flights. Brings most of its attention to bear on leaves and captures prey by plucking, leaping, and sometimes hovering—all in all more active and quicker than Red-eyed Vireo. In migration, often found among flocks of Red-eyed Vireos foraging in the lower strata of the forest edge (as well as in the canopy). Fairly tame. Responds well to pishing.

FLIGHT: Profile is stocky and distinctly snub-billed (seems to have a pushed-in face), with short roundish wings. The overall yellowish cast, slight contrast between the upper- and lower parts, and lack of any grabbing plumage marks are telling. Flight is less jerky than similar warbler species (especially Tennessee Warbler) and similar to Red-eyed Vireo, but bouncier and less smoothly undulating.

VOCALIZATIONS: Song is similar to Red-eyed Vireo—a varied series of hurried two- to four-note phrases. Sounds on average slightly higher-pitched, clearer, and sweeter than Red-eyed, with more distinct, less run-on phrases. Call is a series of descending nasal notes whose pattern recalls the scold of a Gray Squirrel: "*r'reh reh reh reh.*"

Red-eyed Vireo, *Vireo olivaceus*

Vociferous Leaf Wraith

STATUS: Common and widespread northern and eastern woodland breeder. **DISTRIBUTION:** In summer, found across most of Canada south of the tree line, from s. Nfld. to the se. Yukon and much of B.C. In the U.S., breeds everywhere except w. and s. Ore., Calif., Nev., Utah, Ariz., w. Colo., N.M., w.-cen. and coastal Tex.; also absent as a breeder in coastal La. and s. Fla. Winters in South America. **HABITAT:** A species of the canopy and subcanopy, Red-eyed Vireo is most commonly found in mature deciduous woodlands but also occurs in mixed pine-hardwood forest; prefers micro habitats dominated by deciduous trees. In migration, habitat choice is more eclectic, but the bird's partiality to deciduous trees is abiding. **COHABITANTS:** Hairy Woodpecker, Black-and-white Warbler, American Redstart. **MOVEMENT/MIGRATION:** Spring migration from April to early June, with a peak in mid-April through May; fall from mid-August to early November, with a peak from late August to early October. VI: 3.

DESCRIPTION: A large, sturdy, but slender-profiled vireo distinguished by its overaccented facial pattern and tenacious singing. Somewhat angular and overall compact, Red-eyed Vireo is particularly long-faced—the head is flat-topped or sloping (when agitated, a slight crest becomes raised), and the bill (seen in profile) is distinctly long, straight, and uniform along its length, ending with a vireo hook. (*Note:* The bill looks very different from below: wide at the base, acutely tapered to a pointy tip, and very flycatcher-like). The tail seems too short for the bird.

In all plumages, upperparts are plain, unadorned, olive green and grayish; underparts are mostly dull white with a yellow wash on the sides and the undertail. The distinctive head pattern—a gray crown separated from a white eyebrow by a mascara-fine black line and another abbreviated white line below—is the bird's signature feature. Except for Black-capped, all other vireos seem less made up, and their features more pastel-blended than makeup-room crisp. Red-eyed's expression is serious, no-nonsense.

BEHAVIOR: On territory, a persistent, almost incessant singer—continues to sing into the heat of the afternoon and late in the summer, after most other forest species have long since called it a season. A leaf specialist, Red-eyed forages during the breeding season among the leaves of the deciduous canopy and subcanopy. Methodical but fairly active feeder. Moves in a series of short hops (with no wing movement and each hop followed by a pause), punctuated by short relocation flights (usually in the same tree). Gleans caterpillars (and other insects) from vegetation, but may also hover and pluck prey from outer leaves.

In fall migration and in winter, switches largely to berries. In migration, occurs alone or in small loose groups of 3–10 individuals that may also include other vireo species and woodland warblers.

Responds well to pishing. Can be coaxed down from the canopy, and its alarm call incites other more reticent species.

FLIGHT: Body is angled up. The long bill is angled slightly but held conspicuously down. The contrast between the dark upperparts and the creamy whitish underparts (not bright white!) is obvious but blended, muted, not sharp or crisply defined. The bird often seems hooded, but the hood is shadowy and diffuse. Flight is fast, strong, more undulating, and less bouncy than warblers. Wingbeats show a deep pushing down stroke; appears to be resting momentarily on its hands.

VOCALIZATIONS: Makes a measured and monotonous (but not displeasing) series of short, slurred, and varied three-note phrases, each phrase punctuated by a pause and phonetically rendered as "*Here I am. See me? I'm up here. In the tree.*" Call is a nasal whine. When agitated, the whine is more strident and up-sliding—very reminiscent of a scolding catbird but more nasal and single-noted (Red-eyed Vireo: "*wheeeeeh*"; Catbird: "*Whe-yeah*").

PERTINENT PARTICULARS: Eye color is red in adults, brown in immatures, and not easy to see in either case. Do not consider it a crucial field mark. The crisply defined facial pattern is key to distinguishing Red-eyed Vireo from two other long-in-the-face (but geographically restricted) vireos — Yellow-green Vireo and Black-whiskered Vireo — and from several other vireo species characterized by rounder heads, stubbier bills, and plainer and more gentle facial expressions. Seen from below, the bird's overall compactness is suggestive, and contrast between the pale yellow wash on the vent and the otherwise immaculate underparts is, in most places, indicative.

Yellow-green Vireo, *Vireo flavoviridis*

The Perplexing Vireo

STATUS AND DISTRIBUTION: Breeds in coastal lowlands of Mexico (very close to the U.S. border) and Central America. Very rare summer visitor to the Rio Grande Valley of s. Tex. and occurs as a rare vagrant in s. Calif. (in fall), s. Fla., and elsewhere. **HABITAT:** Deciduous forest; typically riparian woodlands and scrubby forest edge. In the United States, most often recorded in the woodlands along the Rio Grande between Bentsen–Rio Grande Valley State Park and Brownsville, Texas. **COHABITANTS:** Rose-throated Becard, Olive Sparrow. **MOVEMENT/MIGRATION:** Migratory species that retreats from its breeding range into Amazonia in winter. Breeds between April and September. VI: 2.

DESCRIPTION: A brighter, plainer-faced Red-eyed Vireo with yellow-green upperparts and a pale to bright yellow wash on the sides, flanks, and undertail (and often the face). Larger-billed than Red-eyed, but otherwise similar. The face pattern is weak, shadowy, poorly defined, and more akin to Philadelphia Vireo except for the distinctly gray cap.

BEHAVIOR: A woodland species that prefers to forage in the middle and upper elevations and often in heavy cover, but generally avoids deep penetration into dense woods. Forages by gleaning in the foliage; occasionally hovers to pluck prey.

VOCALIZATIONS: Song is like that of Red-eyed Vireo — a series of short one- and two-note phrases — but more hurried and broken, less musical, and "chirpier" — sounds like a run-on House Sparrow. Other calls are like Red-eyed Vireo.

PERTINENT PARTICULARS: Until you see a true Yellow-green Vireo, you may conclude that the differences between this species and Red-eyed Vireo are subtle. If you are familiar with Red-eyed, your reaction to a first encounter with Yellow-green will probably be, "What's that?" not, "Oh, another Red-eyed Vireo." It's just too bright and may remind you of a warbler. Or its plainer face may put you in mind of Philadelphia Vireo. But if you are in the Rio Grande Valley in summer or coastal California in fall, you might want to think again.

Black-whiskered Vireo, *Vireo altiloquus*

Mangrove Vireo

STATUS: Uncommon to common but geographically restricted breeder. **DISTRIBUTION:** Found throughout the Caribbean. In the U.S., restricted to coastal Fla. from the Keys north along both coasts to Pasco Co. on the Gulf and New Smyrna Beach on the Atlantic. **HABITAT:** Primarily coastal mangrove, but also found in hardwood hammocks and woodlands surrounding coastal suburban communities; avoids pines. **COHABITANTS:** Mangrove Cuckoo, Yellow Warbler, Prairie Warbler. **MOVEMENT/MIGRATION:** Reaches Florida in late March or early April; departs in mid-September, with some birds lingering into October. In spring, this species has occurred in many locations along the Gulf of Mexico west to Texas; very rare along the Southeast coast. VI: 1.

DESCRIPTION: A duller, browner, pastel-shaded Red-eyed Vireo hiding in the top of a mangrove. Large (very slightly larger than Red-eyed), with a big bill, a pointy wedge-shaped face, a fairly slender body, and a shortish tail.

Plumage is much like Red-eyed Vireo, but the upperparts are browner and not quite so green, and the head pattern is muted, more monotoned, overall less contrasting (suggests a worm or bleached-out Red-eyed). The narrow dark whisker is usually distinct when set against the bland backdrop of the face pattern, but on some individuals it is more subtle and, given the clandestine nature of this species, not easy to see. Usually birds distinguish themselves by the muted face pattern and brownish upperparts. The whisker mark is used for confirmation.

BEHAVIOR: A canopy bugaboo. Males like to sit on a favorite perch in the upper canopy of a mangrove or broad-leafed trees where they will sing for hours, throughout the day. Also forages in the upper canopy, making only infrequent forays into the lower strata. Feeds mostly by hopping slowly and methodically and gleaning insects (and fruit) from among the leaves.

All these traits, of course, keep the bird, which is already not given to drawing attention to itself, mostly hidden from view. It gets worse. The bird seems immune to pishing and does not react to species that are more easily provoked. But it is a persistent singer.

FLIGHT: Like Red-eyed Vireo, but usually brief.

VOCALIZATIONS: Song is similar to Red-eyed Vireo: a series of deliberate two- to three-note phrases that have been phonetically likened to the recitation of names: "*whip-Tom-Kelly; sweet-John; John-to-whit.*" Songs are sometimes given in pairs—for example, "*ch'ree ch'lee; chl'r chl'er.*" Phrases range from high and churpy (not chirpy) to low, throaty, and slurred, but are overall more clipped and precise than the sliding whistled phrases of Red-eyed Vireo. If through some audio alchemy the song of a Red-eyed Vireo could be crossed with a House Finch, Black-whispered's "*churp*" would be the result. Bird also makes a nasal Red-eyed Vireo–like mew: "*yeaaaa.*"

JAYS, CROWS, AND RAVENS

Gray Jay, *Perisoreus canadensis*

Spruce Ghost

STATUS: Common widespread but not necessarily easy-to-find resident of northern and subalpine forests. **DISTRIBUTION:** Found across forested Canada and Alaska, from Nfld. and e. Lab. to w. Alaska (excluding the tundra, the northern prairies, and se. Ont.). In the U.S., found in the coastal areas and Cascade and Sierra Mts. of Wash., Ore. (also w. Ore.), and nw. Calif.; also found in the Rockies of Idaho, w. Mont., Wyo., Utah, Colo., n. N.M., the White Mts. of Ariz., and the Black Hills of Wyo. and S.D., as well as most of Maine, n. N.H., Vt., and the Adirondacks of N.Y. Very casually found south of its normal range. **HABITAT:** Northern coniferous forests that almost invariably are dominated by spruce or at least have a spruce component—black, white, Sitka, or Engelmann spruce. Also occurs in mixed forest on the Pacific Coast. Less commonly found in aspen and birch and, on Alaska's North Slope, mature dense willow thickets. **COHABITANTS:** Spruce Grouse, Northern Hawk Owl, Boreal Chickadee. **MOVEMENT/MIGRATION:** Permanent resident. VI: 1.

DESCRIPTION: An overgrown all-gray chickadee with a pale face and a dark cap. A large songbird and a medium-sized jay (slightly larger than Blue Jay; slightly smaller than Clark's Nutcracker; about the same size as Sharp-shinned Hawk, Merlin, and Steller's Jay), with a large round head, a small black bill, a robust body, and a disproportionately longish tail. Perches upright atop spruce trees and, except for the bill, seems very raptorlike.

Adults are wholly gray except for a pale head and blackish cap. At a distance, perched birds appear uniformly dark except for a pale face and collar. Darker immatures are uniformly dark charcoal gray except for a pale white mustache mark.

BEHAVIOR: Ghostlike in its ability to appear suddenly and silently on the limb just above your head. Incredibly curious and bold, Gray Jay is drawn to humans and even the sounds of human presence—a gunshot, the opening of a cooler, a crackling campfire. Moderately social. Found alone, in pairs, or in small family groups. Birds hunt for food visually, moving through the forest in short flights punctuated by short pauses. Most food is secured with the bill; larger items may be carried with the feet.

Unusually silent, especially for a jay. Often perches high atop spruces but where food is available, quickly descends to the ground (or the picnic table . . . or the gut pile . . .), where it hops nimbly. Curiously, does not respond quickly to pishing (at least, does not show itself). Seems most responsive to squeal calls (see "Pertinent Particulars") and responds vocally to imitations of Northern Saw-whet Owl.

FLIGHT: Profile shows a blunt head, a long and slightly fanned tail, and short round wings. The narrow pale tip of the tail is not easy to see (or necessary for identification). Flight is slow and buoyant, consisting of a short hurried series of wingbeats followed by a lengthy, slow-motion, open-winged glide. Flight is spooky—in its silence, its floating, time-suspended buoyancy, and in the bird's ability to glide through branches with smokelike ease.

VOCALIZATIONS: Varied, consisting of squeaky and mellow whistles, soft mewing notes, and harsh, low, and shrill chortles and chatters. Examples include a chatting "*cheh cheh cheh cheh*" (like a shrill Steller's Jay), a squealed "where?" a "*whe/a'r . . . whe/a'r*," and a soft "*why?*"

PERTINENT PARTICULARS: Gray Jays occur in the same range and habitat as Grizzly Bears and Black Bears, which also respond to pishing in general and squealing in particular. You might want to consider this before trying to imitate a wounded bird or animal in vegetation so dense that it prevents you from seeing what might be approaching (or is already there).

Steller's Jay, *Cyanocitta stelleri*

Dark Blue Crested Jay

STATUS: Common and widespread jay of western forests. **DISTRIBUTION:** Permanent resident from the Kenai Peninsula in Alaska south along the coast to s. B.C. and w. Alta. into the coastal areas and Cascade and Sierra ranges of Wash., Ore., Calif., and w. Nev. (also some montane forests of s. Calif.), through Rockies south into Mexico. **HABITAT:** Coniferous and mixed coniferous-deciduous forests, from the humid coastal forests of the Northwest to the arid oak-pine forests of the Southwest. Also found, in winter, in orchards, more open woodlands, suburban habitat, and, at all seasons, campgrounds. **COHABITANTS:** Mountain Chickadee, Chestnut-backed Chickadee, Gray Jay, Clark's Nutcracker, Varied Thrush. **MOVEMENT/MIGRATION:** Nonmigratory, but birds breeding at higher elevations irregularly descend, often relocating to favored habitats. Occasionally irruptive; moderate numbers have been recorded moving through areas where they are not commonly found (including desert areas and the western plains). VI: 1.

DESCRIPTION: North America's only all-dark crested jay—no white in the wings or tail. Slightly larger than the eastern Blue Jay but overall more robust. A conspicuously peaked shaggy-looking crest cocks forward when fully raised and billows tassel-like in the wind when slightly relaxed, making the bird appear anvil-headed when flattened.

At a distance, just an all-dark crested bird sitting at the top of a conifer; at close range, bird is overall dark blue with a blackish or charcoal brown cowl extending to the shoulders and back. The thin bluish or whitish scratch marks on the bird's forehead (found on some races) hardly add or detract from the identification process. Immatures are like adults but paler.

BEHAVIOR: Loud and boisterous, but in many places amazingly shy. Often remains high in the canopy of tall conifers, offering momentary views before plunging, vertically, into the canopy. Climbs with rapid reflexive hops along the interior branches of trees, seeming to run up the trunk in the spiraling manner of a squirrel. Also commonly descends to the ground to collect acorns and other nuts, many of which are cached in arboreal nooks and crannies. Comes readily to feeders, where it favors larger seeds like peanuts and sunflower seeds, as well as suet. Also feeds on the eggs and young of other forest birds.

Highly social. Pairs rarely separate. In fall and winter, commonly associates with neighboring jays to form small flocks.

FLIGHT: Appears all dark and remarkably stocky in flight (very unlike the longer, lankier profile of a scrub-jay or Blue Jay). Wings are short, wide, and roundly blunt; the tail is short and wide. Wingtips are slightly but conspicuously pale, as is the rump. Flight is buoyant and flowing—a series of deep wingbeats followed by short or lengthy, straight or undulating glides. A bird descending from a great height may maneuver solely by alternating open- and closed-wing glides (no wingbeats) that give it a flowing, grandly undulating flight. Landings are often flamboyantly fast, with flared wings used for braking.

VOCALIZATIONS: Many and varied. Common calls include a loud, harsh, braying scream, "*RAAAH!*" that loses force and volume at the end (sounds like a 150-pound cat protesting the inopportune planting of a foot upon its tail); a shorter multiple-note version, "*RAAH RAAH RAAH*"; a descending chortle, "*ch'lay ch'lay ch'lay ch'lay ch'lay*"; a muffled wooden-sounding rattle; and assorted soft mutterings and murmurings and clicks. A fair mimic of other birds, particularly hawks.

Blue Jay, *Cyanocitta cristata*

The Noisy Coxcomb

STATUS: The common and (except where Florida Scrub-Jay occurs) *only* eastern jay. **DISTRIBUTION:** Permanent resident across the e. and cen. U.S. and s. Canada. In Canada, found from Nfld. and the Maritimes across s. Que., s. Ont., s. Man., cen. Sask., cen. Alta., and portions of e. B.C. In the U.S., breeds primarily east of the Rockies in w. and s. Tex., but also found, in pockets, in all states except Calif., Nev., Ariz., and all but n. Utah. In winter, shows some expansion westward and southward. **HABITAT:** Most common in deciduous (especially oak) woodlands, but also found in coniferous and mixed habitat. The bird's westward expansion is partly explained by how very much at home it is in tree-rich towns and suburbs; also uses shelterbelts and riparian woodlands in the prairies. In winter, a standout at winter bird-feeding stations where peanuts and sunflower seeds are offered. **COHABITANTS:** Many and varied. **MOVEMENT/MIGRATION:** Although mostly a permanent resident, Blue Jay is the only true migratory jay, since some portion of the population does migrate, particularly when mast crop (like acorns) fails. Flocks may be sizable, numbering 3–100 birds. The migratory period is short—spring migration from late April to late May; fall from late September to October. VI: 3.

DESCRIPTION: A bright blue, white-trimmed, crested jay. Underparts are pale gray and garnished with a blue-black necklace. Handsome, rakish, unmistakable. In flight, appears not so crested, but shows a distinctive trailing white trim on wings and tail. Florida Scrub-Jay (limited to central Florida) has a pale gray back, no white in wings and tail, and no crest. Western Steller's Jay is wholly dark and darker blue with no white anywhere (except for an inconspicuous streak or two about the eye and forehead on some interior adults).

BEHAVIOR: During the nesting season, somewhat secretive and furtive. At other times of year, noisy, boisterous, assertive, and social, gathering in small loose flocks that move through woodland territories. Does not generally associate with other species, but responds quickly to (and is often the instigator of) mobbing actions, whether directed toward hawks, owls, cats, snakes, foxes, humans . . . and sometimes nothing at all! Forages frequently on the ground (also in trees of all heights), and likes to sit conspicuously high. Loves acorns and other large seeds, which it frequently carries and caches. Diet also includes large insects, eggs, and baby birds. Hops, with casual ease, on the ground. Bounds from limb to limb in trees, often "climbing" vertically. When threatened, executes a (usually) vertical (and always vocal) crash-dive into the nearest foliage. Very responsive to pishing and imitations of assorted owls.

FLIGHT: A curiously shaped bird that looks like it was assembled from leftover parts. The head is long and wedge-shaped, wings are short and paddlelike, and the tail is long, narrow, ovate (with a white trim), and slightly wilted. Overall pale, with translucent wingtips that flash when birds are high overhead; at a distance, suggests flickering candle flames. Flight is generally straight and somewhat hesitant or jerky, given with a mostly steady series of hurried wingbeats broken at short intervals with terse irregular skip/pauses. When landing or making short flights, sometimes undulates and glides or brakes on open wings. When crossing open areas, seems furtive or anxious (crest is lowered), but lands with a flourish—a rapid, floating, undulating, showoffish approach to the perch. Flocks are loose and strung out, and longer than they are wider.

VOCALIZATIONS: Wonderfully varied. Classic call is a loud brassy "*Jay Jay Jay*" that is often repeated and sometimes held: "*Jaaaay.*" Calls also include a "squeaky door" call, low soft murmurings, and a low flat rattle. Also an accomplished mimic, specializing in the calls of Red-shouldered and Red-tailed Hawks.

Green Jay, *Cyanocorax yncas*

Clearly Not a Blue Jay

STATUS AND DISTRIBUTION: Common, but in the U.S. found only in s. Tex. north and west to Laredo and north and east to just north of Corpus Christi. **HABITAT:** Woodlands, primarily second-growth and dense mesquite thickets. Generally not found in open suburban areas. **COHABITANTS:** Plain Chachalaca, Golden-fronted Woodpecker, Couch's Kingbird, Long-billed Thrasher. **MOVEMENT/MIGRATION:** Permanent resident. Occasionally reported as far north as San Antonio. VI: 0.

DESCRIPTION: Scrub-Jay is called a Blue Jay, and Steller's Jay is called a Blue Jay, but this bird, no matter how uninitiated the observer, cannot possibly be described as anything but a Green Jay. A medium-sized songbird (about the same size as mockingbird) but a smallish jay, Green Jay is slightly smaller than Blue Jay and more overall petite—not to mention crestless (though the head is not without distinction).

The bird is a visual knockout with its all-green body, skull-like black-and-blue-patterned head, and bright yellow outer tail feathers that flash in flight.

BEHAVIOR: Forages year-round in small, often well-spaced family groups. Almost never found far from brushy woodlands, and generally does not make long flights over open areas. *Very* nimble, *very* active, *very* furtive, hopping easily through brush dense enough to daunt a squirrel. Forages from the ground to treetops, starting low and working up, but spends much of its foraging time on or near the ground (particularly in winter).

Highly vocal, with an array of calls (an understandable trait for a social bird). More furtive than shy. Easily attracted to feeding stations (near brush), where it becomes habituated to people and is very partial to orange halves.

FLIGHT: Very slender long-tailed silhouette; being somewhat hunchbacked or droop-tailed, the bird looks uncomfortable or furtive. The pale green body makes the blackish blue head prominent, and the flashing yellow outer tail makes the bird unmistakable. Flight is jaunty, bouncy (more bouncy and nimble than Blue Jay), furtively fast, and usually short. Wingbeats are rapid, given in quick series, and punctuated by frequent glides on open wings and tail.

VOCALIZATIONS: Too numerous to recount, but similar to Blue Jay calls, except that Green Jay calls are faster, higher-pitched, and harsher. Most calls are brays or phrases given in triplets. "*Jay*" translates into "*cha,*" so that birds seem to be saying, "*cha cha cha,*" sometimes rapidly and with a scratchy quality; sometimes more musically, with a compressed two-note quality, "*ch'l ch'l ch'l; ch'l ch'l ch'l*"; and sometimes more slowly, roughly, and at a lower pitch: "*j/rah, j/rah, j/rah*" (recalling Steller's Jay). Also makes low murmurings, soft rattles, plaintive whines, and wheezy snores: "*ah'zz'ah.*"

Brown Jay, *Cyanocorax morio*

Buster Brown

STATUS AND DISTRIBUTION: Mexican species. Rare, somewhat irregular, and highly restricted in the U.S.; found exclusively along the Rio Grande between Rio Grande City and San Ygnacio, most often at Salineno and Chapeno. **HABITAT:** The open and dense woodlands flanking the Rio Grande. May frequent feeders, where they offer good viewing. **COHABITANTS:** Plain Chachalaca, White-tipped Dover, Audubon's Oriole, Gale and Pat DeWind. **MOVEMENT/MIGRATION:** Semipermanent resident. Sometimes absent from regularly frequented locations. VI: 0.

DESCRIPTION: A big, strapping, brown, chachalaca-like jay. Large! (dwarfs Green Jay; smaller than Plain Chachalaca; about the same size as Black-billed

Magpie), with a head like a crow, a slender tubular body, and a very long tail. Size alone distinguishes Brown Jay, and your first thought may well be: *It's a chachalaca.* But look at the bill, which is straight and crowlike, not snubbed and chickenlike.

Compared to other jays, a dull-looking bird. Upperparts are smoky brown (head and breast are slightly blacker), and the belly whitish. Immatures are like adults but have yellow bills. Stance is erect or slightly slouched. There is nothing delicate about this bird.

BEHAVIOR: A pack animal; family groups of six to ten birds move through woodlands in a staggered advance and a series of short flights—first one bird, then two, then more. Sometimes stealthy; often raucous and noisy. Birds forage at all levels, both hopping and gliding between perches; also forage in the leaf litter on the ground.

Fairly shy, even where habituated to people, and often difficult to see, particularly in dense riverside brush. Generally reluctant to cross open areas and seems to do so only after much deliberation.

FLIGHT: Very long-bodied and long-tailed profile, with long, broad, round-tipped wings. In size, shape, and plumage, recalls a chachalaca, but chachalacas have pale-tipped tails, unlike Brown Jay, and when chachalacas land, they raise and lower their tails (jays don't). Flight is straight, steady, slightly undulating, and lofting (more flowing and less frantic than chachalaca). Wingbeats are slow, deep, flowing, and loose; glides buoyantly with open wings and a partially fanned tail.

VOCALIZATIONS: A loud and raucous jay. Calls include a loud rough braying, "*raaough raaough raaough*"; a raspy, descending, whistled "*ee'u ee'u*" similar to Red-shouldered Hawk's whistle; a muffled "p'ow p'ow p'ow," whose first note suggests a popping sound; and a squeaky jaylike yelp, "*e'y'h.*"

Florida Scrub-Jay, *Aphelocoma coerulescens*

Long-tailed Scrub-Jay

STATUS: Locally uncommon to common; restricted in habitat and range. **DISTRIBUTION:** Found only in cen. Fla. **HABITAT:** Lives only in oak-scrub habitat—a fire-dependent habitat characterized by evergreen oaks with a scattering of pines and fairly sparse ground cover dominated by palmettos. Other key components are dry sandy soil and open unvegetated patches. Also found on roadsides, where it sits conspicuously on utility lines, and in orchards. **COHABITANTS:** Great Crested Flycatcher, White-eyed Vireo, Eastern Towhee; in winter, Palm Warbler. **MOVEMENT/MIGRATION:** Highly sedentary and site-tenacious. No migratory behavior. VI: 0.

DESCRIPTION: A large passerine but an average-sized jay (about the same length as Blue Jay), Florida Scrub-Jay is characterized by an overlarge, crestless, flat-topped head, a shortish bill, a conspicuously long tail, and a plumage pattern that seems bleached by the Florida sun. Overall slender, even gaunt (less robust than Western Scrub-Jay). Perched on a roadside utility line, it suggests a big-headed mockingbird.

Plumage is a blend of pale blues and grays, with no trace of white in the wings and tail but a fairly conspicuous gray-violet patch or saddle on the back. The facial pattern is weak, and the head is mostly blue, with a whitish eyebrow and throat. The expression is disdainful, disapproving.

BEHAVIOR: Highly social. Most birds are part of family groups consisting of an alpha pair and up to half a dozen prebreeding clansmen. Spends much of its time foraging on the ground or close to it in low shrubs. Moves by hopping nimbly on the ground and weaving a path through vegetation (far more adroit at this than Blue Jay). Visually hunts small vertebrate and invertebrate prey. Also gathers acorns and buries many of them in open sandy patches.

Tame, particularly where habituated to people. When perched, often sits conspicuously. Very responsive to pishing. When nervous, bobs and jerks its tail and flicks its wings. Generally less boisterous and vocal than Blue Jay.

FLIGHT: Uninspired flier. Flights are generally short—from one low tree to another. Silhouette has short roundish wings and a long tail. Flight is direct, comprising quick flapping series of wingbeats punctuated by sustained, buoyant, open-winged glides. Flight seems perfunctory, businesslike, but birds maneuver nimbly through brush.

VOCALIZATIONS: Call is a rough, grating, low-pitched bray, "*r'r'ra/ek,*" sometimes given as a multiple, "*r'r'ra/ek; ra/eck, ra/eck, ra/eck*"—like Western Scrub-Jay, but not as harsh, ringing, upsliding, or shrill. Has a number of other utterings, including a soft, muffled, raspy "*rh*" or "*jrr,*" but vocalizations are generally not as musical or comically improvisational as Blue Jay.

PERTINENT PARTICULARS: The ranges of Florida Scrub-Jay and Western and Island Scrub-Jays ensure that the birds will never be confused. Blue Jay and Florida Scrub-Jay do overlap, and the birds are often found in close proximity. In addition to the many plumage and structural differences, these birds distinguish themselves by habitat. Scrub-Jay, as the name attests, occurs in short dry scrub, generally close to the ground. Blue Jays like forest and tall trees and spend much of their time high up in the canopy. Also, Blue Jays commonly fly high, in the open, for long distances; scrub-jay rarely does so.

Island Scrub-Jay, *Aphelocoma insularis*

Santa Cruz Jay

STATUS AND DISTRIBUTION: Common but isolated and geographically restricted—in fact, isolated: Island Scrub-Jay is found *only* on the Calif. island of Santa Cruz (one of the Channel Is. 25 mi. off Santa Barbara). **HABITAT:** Primarily chaparral-oak and oak woodlands; also found in residual pine woodlands and scrub and eucalyptus groves. Not commonly found in grasslands. **COHABITANTS:** Northern Flicker, Orange-crowned Warbler. **MOVEMENT/MIGRATION:** Permanent resident. VI: Less than 0!

DESCRIPTION: Looks like a large, more richly colored, big-billed Western Scrub-Jay, *and it's the only jay on the island.* Western Scrub-Jay, common on the mainland, has never been recorded on Santa Cruz Island.

BEHAVIOR: Except for its communal breeding habits, behaves much like Western Scrub-Jay. Nonterritorial adults are often found in flocks. Like many birds on isolated islands, Island Scrub-Jay is almost fearless and seems as eager to engage you as you it. Responds quickly to pishing.

FLIGHT: Similar to Western Scrub-Jay.

VOCALIZATIONS: Similar to Western Scrub-Jay.

PERTINENT PARTICULARS: Finding this bird is no more difficult than taking a boat trip to Santa Cruz Island. Regular service is provided by Island Packers, Ventura, Calif. For information, call: 805-642-7688.

Western Scrub-Jay, *Aphelocoma californica*

Noisome Scrub Bandit

STATUS: Common noisy omnipresent western species. **DISTRIBUTION:** Resident of the West Coast, the s. and cen. Great Basin, and the Southwest, from w. Wash. and w. Ore. south through most of Calif. to Baja; also most of Nev., Utah, sw. Wyo., s. Idaho, w. and cen. Colo., n. and e. Ariz., most of N.M., and w. and cen. Tex. south into Mexico. **HABITAT:** A bird of dry scrub and lower elevations; particularly partial to oak woodlands, oak-pine, pinyon-juniper, and chaparral, but also found in other habitat types, including desert riparian woodlands, pine forests, orchards, suburbia, golf courses, and urban parks. **COHABITANTS:** Acorn Woodpecker, Bewick's Wren, Bushtit. Range and habitat overlap with Steller's Jay, Mexican Jay, Pinyon Jay. **MOVEMENT/MIGRATION:** Permanent resident, but may disperse in winter to lower elevations, deserts, and the western Great Plains, particularly in years with poor acorn production. VI: 2.

DESCRIPTION: A crestless blue jay with bright blue upperparts, a dark bandit's mask across the face, a brown (interior gray) saddle on the back, and a white throat bordered by a blue necklace (washed out on interior birds), contrasting with pale gray underparts. Perched birds seem big-headed, heavy-billed, no-necked, and somewhat top-heavy. With head couched and tail down, the body language says sullen or slouched. The expression is stern. Interior birds are overall paler and less crisply patterned but still distinctly marked.

BEHAVIOR: Noisy and social; calls attention to itself with its loud raucous cries. Usually found in small family troops. Often sits high and exposed in the treetops. Dives vertically (and noisily) for cover when threatened. At home on the ground, where it hops. Ascends trees in rapid bounds that look as though the bird is running up the tree.

FLIGHT: Hurried, furtive, silent, and direct. Flies as though it is uncomfortable being in the open. Overall slim and gangly in flight (but still big-headed). The body seems too thin to support the head, the wide round wings, and the long tail. Flight is a series of stiff rapid wingbeats punctuated by floating open-winged glides on wings held at a stiff right angle to the body (slightly angled down). Sometimes, particularly when flying short distances or approaching a perch, executes deep histrionic undulations; may be accompanied by exaggerated tail pumping.

VOCALIZATIONS: Classic call is a loud, harsh, scolding "*shrenk?!*" Recalls a parrot. Often repeated in a short series. Also makes a series of sputtering clucks and a rapid tapping that sounds like a cross between a murmur and a trill or a rattling exhalation.

PERTINENT PARTICULARS: The ranges of both Pinyon Jay and Mexican Jay overlap with the interior form of Western Scrub-Jay. Mexican Jay is overall more robust, and Pinyon Jay shorter-tailed. Neither species has the blue-bordered white throat of Western Scrub-Jay.

Mexican Jay, *Aphelocoma ultramarina*

Gray-breasted Jay

STATUS: A Mexican species that enjoys a limited distribution in the United States, where it is common. **DISTRIBUTION:** Southeast Ariz., sw. N.M., and (rarely) the Big Bend region of Tex. **HABITAT:** In Arizona and New Mexico, found in the "sky islands," where oak and mixed oak-pine-juniper woodlands dominate. Also found in riparian habitats dominated by sycamores (but close to where oaks flourish) and, in Texas, in slightly brushier habitat. Commonly frequents feeders. **COHABITANTS:** Bridled Titmouse, Painted Redstart, Hepatic Tanager. **MOVEMENT/MIGRATION:** Exhibits extraordinary degree of site fidelity. Family groups occupy their territory year-round. VI: 0.

DESCRIPTION: A big, sturdy, pastel-shaded jay (bigger and burlier than a scrub-jay), with a raucous demeanor and the pack instincts of a timber wolf. Overall stocky, with a big head, a somewhat short, contoured, but heavy bill, a portly body, and a medium-length tail. Perched birds, at rest, look particularly chubby, with heads drawn down, bodies puffed, and tails hanging nearly vertical.

Overall pale, showing a muted contrast between the turquoise blue upperparts and the soft gray underparts. The absence of eye-catching plumage characteristics (dark saddle, white eyebrow, white throat/streaky neckless/streaky throat) distinguishes this species from Western Scrub-Jay.

BEHAVIOR: A pack animal. Roving bands of Mexican Jays course through their territories, flying across open areas with the staggered coordination of a SWAT team. One rarely hears a single Mexican Jay; a single vocalizing bird invokes an echoing chorus. For most of the year, forages on acorns on the ground, moving across the forest floor with high springy hops. Pokes through the leaf litter and excavates for buried acorns. Also searches the canopy for insects and acorns as well as pinyon nuts (another important food item), and when agaves are in bloom, visits the blossoms for nectar and insects.

Very tame around campsites and easily attracted by handouts of seeds and nuts. Responds well to imitations of Northern Pygmy-Owl, but often not quickly.

FLIGHT: Flight is direct and straight, covering long distances, without undulations. Wings are short, broad, and paddlelike. Wingbeats are rowing and slightly irregular—a series of wingbeats are broken by a rhythm-breaking stutter-flap or, when birds are landing or descending from higher altitudes, a swooping, deeply undulating, open-winged glide. Wingbeats are sometimes very audible (perhaps intentionally?). At close range, wings sound like canvas rippling in the wind.

VOCALIZATIONS: Noisy. Most commonly heard call is a harsh ringing "*renk! renk!*" (similar to the "*shrenk*" call of a scrub-jay, but softer, more abrupt, and not as uncouth, though just as vehement). Also makes a low raspy "*w'rt*" and, when birds are feeding together, a very soft conversational "*wink . . . wink*" that recalls the "*bink*" flight call note of a bobolink.

Pinyon Jay, *Gymnorhinus cyanocephalus*

The Pinyon Pack Jay

STATUS: Common resident of the Great Basin but highly mobile, often inconspicuous, and habitat-specific. **DISTRIBUTION:** Central and s.-cen. Ore., e. Calif. (also n. Baja), most of Nev., se. Idaho, most of Utah, n. and cen. Ariz., s.-cen. Mont., much of Wyo., sw. S.D., w. and s. Colo., and all but e. N.M. **HABITAT:** Most commonly found at mid-elevation on hillsides dominated by pinyon pine(or pinyon-juniper). May also be found in sagebrush, chaparral, Joshua tree forest, and scrub oaks; also regularly found in other pine forests (particularly ponderosa pine and less commonly Jeffrey pine); also, more rarely, forages in fields. In winter, somewhat attracted to feeding stations. **COHABITANTS:** Western Scrub-Jay, Juniper Titmouse, Mountain Chickadee. **MOVEMENT/MIGRATION:** Usually nonmigratory, demonstrating year-round fidelity to their often large territories (25 sq. mi. or more!). Regional or more widespread failure of the pinyon pine crop results in periodic but irregular irruptions, at which times birds may appear up to hundreds of miles outside their normal range. VI: 2.

DESCRIPTION: A compact, spike-billed, short-tailed, *all-blue*, crestless jay perched in a pinyon pine as the rest of the pack forages below. Smallish (slightly smaller than Western Scrub-Jay; considerably smaller than Clark's Nutcracker), Pinyon Jay is shaped much like a slender crow with a long, straight (or slightly downturned), pointy bill, a large rounded head, a slender body, and a short tail.

Except for a slightly whitish throat, *uniformly pale blue or blue-gray*. At a distance, may appear capped or slightly darker-headed. Immatures are like adults but grayer.

BEHAVIOR: Almost *never* found alone. Forages and roosts year-round in large flocks (of 20–100 individuals on average, though flocks may include as many as 1,000 or more). Flocks move around their territory during the day, making lengthy flights to favorite foraging sites, food caches, and roost sites. Forages in trees, but also spends considerable time on the ground (where in pinyon-juniper habitat they become almost invisible). Sentinels perch high on the tops of trees, where often only their heads are visible. Unlike other jays, walks crowlike when foraging on the ground and also hops. In the open, flocks feed by leapfrogging over the pack, like blackbirds. Usually shy, but where acclimated to humans, can be very tame. Not very responsive to pishing. Often vocal in flight, but not necessarily when feeding.

FLIGHT: Overall stocky, but shows long broad wings (looks like a small crow). Very strong flier—almost always seen in tightly packed, oval-shaped, noisy flocks; habitually flies great distances. (Other jays, except migrating Blue Jays, fly short distances.) Wingbeats are quick, strong, regular, and somewhat rowing; flight is direct. Commonly takes off and descends near-vertically. (Flocks pitching in recall flocks of blackbirds descending into reeds to roost.) Tight-packed flocks are extremely maneuverable and recall the synchronized shifting of shorebirds.

VOCALIZATIONS: Like all jays, enjoys a verbal array, but is less brassy or harsh than other jays. Makes a soft nasal "*aar aar aar*" (the same pattern as a crow's "*caw caw caw*" but much softer) and a louder, harsher "*re'ah!*"

PERTINENT PARTICULARS: Distinguishing the bird isn't the problem. Western Scrub-Jay (which is also found in pinyon-juniper) has a conspicuously long tail; Steller's Jay is conspicuously crested; the larger Clark's Nutcracker is boldly patterned black-white-gray and is commonly found at higher altitudes. The problem is *finding* the bird. A good strategy is to position yourself at a high point within good habitat and scan and listen for a shifting flock. Or try driving along roadways that cut through pinyon-pine and ponderosa pine habitat.

Clark's Nutcracker, *Nucifraga columbiana*

Pied Jay

STATUS: Common resident of higher elevations in western mountains. **DISTRIBUTION:** Found from northern interior B.C. to sw. Alta., w. Wash., nw. Ore., and along the Cascades, Sierras, and scattered ranges throughout Calif., w. and cen. Mont., all but sw. Idaho, cen. and e. Nev., Utah, all but e. Wyo., all but e. Colo., ne. Ariz., and nw. and cen. N.M. A few birds are found in the Black Hills of S.D. Casual wanderer east to the Great Plains and beyond. **HABITAT:** Breeds in assorted open coniferous forests, especially ponderosa and lodgepole pine and pinyon-juniper. In summer (beginning in early June), birds move to the tree line and forage both on open rocky slopes and in timber; birds are often found at scenic overlooks along alpine highways, where they accept handouts of food. In fall and winter, descends to lower elevations, favoring pinyon-pine, white pine, bristlecone pine, and ponderosa pine. **COHABITANTS:** During breeding, Gray Jay, Mountain Chickadee, Pygmy Nuthatch, Western Tanager. Postbreeding, Common Raven, Violet-green Swallow, White-crowned Sparrow. **MOVEMENT/MIGRATION:** Nonmigratory, but engages in annual altitudinal population shifts (higher elevations in summer, lower in winter) and, in response to cone-crop failures, irruptive movements (noted between September and November) involving flocks of up to 50 individuals. VI: 3.

DESCRIPTION: A stocky, crestless, and conspicuously patterned black-white-and-gray jay. Overall large and robust, with a large head, a distinctly long, straight, spikelike bill, and a too-short tail. Seems too compact to be a jay; in fact, was initially described as a woodpecker.

Overall flat pale gray with black wings and tail—a pattern shared only with the smaller, more slender and hook-billed shrikes. The feathers around the eye are often paler than the rest of the bird. The white patches on the folded wing and white undertail coverts are usually visible but hardly necessary for identification. Ages and sexes are similar; young birds are tinged tan-brown.

BEHAVIOR: An active and social bird. Likes to sit conspicuously on the tops of trees (often on dead branches) as well as on rocks and rocky outcroppings. Forages in the canopy, where it uses its spikelike bill to extract seeds from cones. Also detaches unripened cones and carries them to a favored place (an "anvil") where they can be more easily manipulated. In summer, forages on the ground and on rocky slopes, hopping and sallying to grab insects on the ground and also hawking them in the air. Almost always found in small groups. Almost never silent for long. Tame. Responds well to pishing.

FLIGHT: Broad-winged and compact, with a flashy gray-white-and-black pattern and an aerial finesse that seems more flourish than flight. The unjaylike silhouette has long broad wings, a projecting head, and a short tail (shape is surprisingly similar to Lewis's Woodpecker). The black wings and black-centered white tail erupting from an otherwise plain gray bird make it seem magical. (You probably won't even notice the white patch on the trailing edge of the wing or be able to distinguish it from the flashy tail pattern.)

Direct flight is crowlike in cadence and execution, with strong, mostly steady, rowing wingbeats. Viewed from below, wings seem to flash out from the side (with no apparent up-and-down movement). When flying short distances, sallying, or hawking insects, birds fly with a slow buoyant series of wingbeats and brief or long open-winged glides. Clark's Nutcracker does not undulate except when flying downhill, and uses a series of open wings for braking. Flight is buoyant and nimble; birds execute abrupt banking turns with wings nearly vertical and use the advantages afforded by gravity, updrafts, and wind currents as a matter of course. Movements thus seem exaggerated or theatrical—executed with a flourish.

VOCALIZATIONS: Common call is a long, ascending, harsh purr or a trilled snarl: "*rrrrrah.*" Recalls the braying "scream" of Steller's Jay, but is longer and less guttural and rises at the end. Also gives a low ravenlike chortle, "*rawk rawk rawk*"; a more plaintive jaylike mew "*eyrrr*"; and a kingfisher-like rattle, "*ch'k'k'k'k'k*."

Black-billed Magpie, *Pica hudsonia*
Long-tailed Jay

STATUS: Common widespread western resident. **DISTRIBUTION:** Found from s. and cen. Alaska to B.C. and all but n. Alta., east across the southern half of Sask. In the continental U.S., found in e. Wash., the eastern two-thirds of Ore., ne. Calif., Idaho, Mont., Wyo., all but se. N.D., n.-cen. Minn., w. S.D., all but e. Neb., all but s. Nev., Utah, extreme ne. Ariz., Colo., n. N.M., w. Okla., and w. Kans. **HABITAT:** Generally a bird of open areas. Breeds in riparian thickets adjacent to open grasslands, meadows, and sage flats. Winters in a variety of open habitats, including city parks, open agricultural land, and feedlots close to vegetative cover. Does well where humans are present; avoids dense woodlands, but may be common in open forest campgrounds. **COHABITANTS:** Red-tailed Hawk, meadowlarks. **MOVEMENT/MIGRATION:** Permanent resident and essentially nonmigratory; though post-breeding dispersal may be extensive, the species does not commonly wander far outside its range. VI: 2.

DESCRIPTION: Unmistakable. A large, showy, black-and-white corvid (larger than Cooper's Hawk), with the head of a crow, the body of a jay, and an extra-extra-long tail that is all magpie. Overall long and robustly slender, with a heavy, black, crowlike bill, a large head, and a tail that is longer than the rest of the bird.

Boldly patterned black-and-white-bellied bird with iridescent blues in the wings and greens in the tail.

BEHAVIOR: Noisy, social, and generally not found far from people. Frequently seen sitting at the very tops of trees. Forages mostly on the ground and in open areas, eating a variety of items. Forages on grain, consumes carrion, steals food from other birds (including raptors), flips cow dung, eats dog feces, plucks insects from the grills and undersides of cars, works picnic areas for handouts or pilfering opportunities, and robs nests; also known to kill small mammals. Struts more than walks on the ground, with a head-bobbing action, but also hops and canters. Also "climbs" through the limbs of trees almost as quickly and nimbly as more strictly arboreal jays.

Bold, almost indifferent, where it is habituated to people. Its raptor avoidance strategy is to not stray far from cover. During breeding season, commonly harassed by smaller birds.

FLIGHT: The white-tipped paddle-shaped wings, in combination with the improbably long tail, render the bird almost unmistakable at any distance. Flight is slow, buoyant, unhurried, and straight, with regular rowing wingbeats and a shifting cadence—long series of wingbeats giving way to shorter, more hurried series. Glides with a swooping flourish upon landing, then salutes with a theatrically raised tail. (In landing, a white U-shaped pattern on the back is very prominent.) Often descends by using an undulating series of swooping, open-winged, braking glides (no flapping).

VOCALIZATIONS: Call is a loud, harsh, rising "*aaaay*" (a protracted "jay" scream without the "j" and with more whine, less brass). Also makes a lower, coarser, raspier, shorter "*r'chach*," often given in series.

Yellow-billed Magpie, *Pica nuttalli*
California Magpie

STATUS: Common resident within its limited range. **DISTRIBUTION:** A Calif. endemic found primarily in the Central Valley, southern coastal ranges, and the western foothills of the Sierra Nevadas. **HABITAT:** Oak savanna where large expanses of dry grassy habitat are dotted with oaks or other tall trees. Feeds in pastures, orchards, cultivated fields, and open grassland. Needs a stream or other water source within its territory. **COHABITANTS:** Acorn Woodpecker, Loggerhead Shrike, Western Scrub-Jay, European Starling, Brewer's Blackbird. **MOVEMENT/MIGRATION:** Permanent resident. VI: 0.

DESCRIPTION: The name says it all. The large size, distinctive long-tailed profile, bold black-and-white plumage pattern, and bright yellow bill (and skin on the face) make the bird difficult to overlook, much less misidentify. Slightly smaller than Black-billed Magpie (which occurs on the eastern slope of the Sierras in northeastern California), but the ranges of the two species do not overlap.

BEHAVIOR: Social and gregarious. Roosts colonially in oak groves. Forages in small flocks (up to 100 individuals) on the ground in open habitat dotted with trees. Usually walks but also hops, rapidly, in pursuit of fleeing prey. Plucks large insects, reptiles, or small mammals from the ground; also digs with its bill, flips over cow dung, and perch-hunts (like a raptor, but grasps prey with its bill, not its feet). Sometimes seen with starlings or blackbirds; almost never feeds with crows. Acorn Woodpeckers, intent on protecting their stock of acorns, are often antagonistic.

FLIGHT: Long-tailed silhouette, pied body, and flashing white wing patches are unmistakable. Flight is straight, buoyant, and somewhat leisurely, with steady rowing wingbeats given with both a flourish and a sense of nonchalance. The tail, stiffer than Black-billed Magpie, does not undulate or flow behind the bird.

VOCALIZATIONS: A raucous grating "*raah raah raah*"; similar to Black-billed Magpie.

American Crow, *Corvus brachyrhynchos*

"If people wore feathers and wings, very few would be clever enough to be crows."
—Henry Ward Beecher

STATUS: Common, widespread, and, in all probability, the large black corvid species most North American residents vilify after a "crow" has gotten into their curbside garbage. **DISTRIBUTION:** Breeds across most of the U.S. and Canada, from Nfld. and Lab. across s. Que., most of Ont., all but n. Man. and n. Sask., Alta., and the sw. Northwest Territories to B.C. and extreme se. Alaska. In the U.S., occurs everywhere except nw. Wash. (where Northwestern Crow occurs), se. California, cen. and s. Nev., cen. and s. Utah, most of Ariz., s. N.M., and w. and s. Tex. In winter, most Canadian birds far from coasts and the U.S. border retreat deeper into the U.S. **HABITAT:** Almost any open place that offers a few trees to perch in and a source of food, such as city parks, agricultural cropland adjacent to forest or woodlands, mall parking lots, cemeteries (*Note:* Eats carrion, but has not been documented to exhume human remains.), campgrounds, landfills, and the shorelines of seacoasts, rivers, and lakes. Avoids unbroken dense forest or high altitudes (where Common Raven dominates), as well as deserts and tree-poor grasslands (where Chihuahuan Raven thrives). **COHABITANTS:** Us. **MOVEMENT/MIGRATION:** Spring migration begins very early—most birds depart in March, but some leave as early as mid-February. Fall migration occurs from September to December, with a peak in October and early November. VI: 2.

DESCRIPTION: A large, sturdy, all-black bird with a straight heavy bill and a strut for a walk. Almost identical to Northwestern Crow of the Pacific Northwest (safely separable only by range). Differences between American Crow and Fish Crow are subtle, making range and vocal characteristics (not plumage or shape) the most reliable means of distinguishing the two species where their ranges overlap across much of the South and East.

American Crow is slightly larger than other crows and more sturdily built. Its bill is long and heavy (particularly at the base), its legs longish, its tail short and slightly rounded, and its plumage looser than Northwestern Crow and not quite as glossy and compact—particularly around the leggings, which may appear very shaggy, giving it a paunchy look. The head shape is somewhat angular and also variable: some birds have a pronounced forehead, and on others it's slight and sloping. Occasionally, in the right light, the head of American Crow takes on a slight brownish cast and is not as glossy as the folded wings.

All in all, American Crow looks more ravenlike than the slimmer, longer-tailed Fish Crow, which is more gracklelike.

BEHAVIOR: Highly and complexly social; where there is one crow, there is usually at least one more (very probably in a nearby tree serving sentry duty), and often many more. Large roosts may hold many thousands of birds. Generally forages on the ground, walking with a strut or a sailor's gait, picking up food items with its bill, and hopping for greater speed if necessary. Frequently caches food; crows carrying food in flight is a common sight outside the breeding season.

A resourceful and clever forager that can dig for food with its bill, tear open garbage bags, and pry open lids. Studies the movements of other birds to find nests and their young. Consumes carrion. Flies down migration-weakened passerine migrants and captures them with its bill. Pirates food, catches fish, eats from dog dishes, and plucks fruit from trees.

The bane of hawks and owls. Seems obsessed with driving away hawks (particularly Red-tailed Hawk) and letting the world know the location of every roosting Great Horned Owl.

Very vocal, but may also sit silently for long periods when, for any number of self-serving reasons, it does not want to draw attention to itself.

FLIGHT: As the proverbial crow flies—straight, direct, with little variation in altitude or course. Both the head and tail project well ahead and behind the broad, generally blunt wings. Wingbeats are steady, deep, and rowing when birds are covering distances. When accelerating (often while being pursued by another crow), wingbeats become quicker and deeper and the bird's body may twist and turn in flight. When landing, often glides short distances with a pronounced dihedral. Also wing and tail flicks upon landing.

Also soars, but only for brief periods. (Ravens soar habitually.) When going to roost during nonbreeding season, birds commonly travel in a narrow corridor. In migration, flocks are linear, well strung out, and unified but not compact.

VOCALIZATIONS: Many and varied—most are variations in pitch and pattern of the familiar "*Hah, Hah, Hah*" or "*Caw Caw Caw.*" Also gives an angry, descending, throaty growl, "*ahrrrrrr,*" that recalls Common Raven. Immature's call is higher-pitched and more nasal—similar to Fish Crow's.

PERTINENT PARTICULARS: American Crow is found in coastal areas and along rivers, but where both Fish and American Crows occur, American is most commonly found in upland areas, away from water.

Northwestern Crow, *Corvus caurinus*

Dwarf Marine Crow

STATUS: Common resident of the coastal Northwest. **DISTRIBUTION:** Kodiak I. and the Kenai Peninsula south to Puget Sound, Wash. Mostly coastal, but ranges inland (less than 100 mi.) along river courses and to elevations approaching 5,000 ft. **HABITAT:** Forest edge close to seacoasts and a variety of tidal habitats, including beaches and mud and sand flats. Also found in harbors, towns, dumps, picnic areas, tilled farmland, and campgrounds and around seabird colonies. **COHABITANTS:** Bald Eagle, Mew Gull, Glaucous-winged Gull, Black-billed Magpie, Steller's Jay, Chestnut-backed Chickadee (at nest sites). **MOVEMENT/MIGRATION:** Nonmigratory. VI: 0.

DESCRIPTION: A small, somewhat dwarfishly proportioned coastal crow. Often distinctly smaller than the widespread and familiar American Crow, but unfortunately, only very slightly smaller than the American Crows breeding in coastal Washington. Overall chunkier and more compact than the typical American Crow, with a stubbier bill, a larger and somewhat neckless head, a shorter tail, and shorter legs with smaller feet.

Overall dull black with only a slight purple gloss.

BEHAVIOR: Most commonly seen foraging below the high-tide line on coastal beaches (often among gulls); avoids grassy or vegetated substrates. Seems to lean back on its legs to walk, waddling rather than strutting. Stops frequently to pull at kelp or flip stones or debris to see what lies beneath. Like gulls, carries shellfish aloft and tries to drop it onto a hard surface. Fairly aquatic. Wades up to its breast to forage, and even dunks its head. Readily flies over open water, making crossings in excess of several miles. Fairly social, outside of breeding season, and generally vocal, often calling while feeding.

FLIGHT: Stockier than American Crow and short-tailed. Wings are fairly short and wide and show little taper. Flight is straight; wingbeats are quicker, choppier, and not as rowing as American Crow. Uses the wind adroitly to gain lift and shift positions.

VOCALIZATIONS: Varied and much like American Crow, but many utterances are slightly lower-pitched, hoarser, and less hurried. Classic caw is rendered "*raah*"; in series, often given with a slight hesitation: "*raah raah* (half-pause) *raah*."

Tamaulipas Crow, *Corvus imparatus*

Mexican Crow

STATUS AND DISTRIBUTION: Resident of ne. Mexico. Occurs irregularly in winter in Brownsville, Tex. Very rare in summer. **HABITAT:** Open or lightly vegetated areas, towns, agricultural lands, and (especially) landfills. **COHABITANTS:** Cattle Egret, Laughing Gull, Great-tailed Grackle, City of Brownsville municipal employees. **MOVEMENT/MIGRATION:** Much less regular in the U.S. in summer (but has nested). VI: 0.

DESCRIPTION: A small, small-billed, glossy blue crow—the only all-dark corvid species to be found in Brownsville except the much larger Chihuahuan Raven. Small corvid (dwarfed by Chihuahuan Raven; more stockily built but overall smaller than male Great-tailed Grackle), with a smallish bill, a large head with a high forehead and a bump on the noggin, a slender body, and a long tail. Overall more slender and not as blocky as American Crow.

The entire body is sleek, glossy blue-black. Immatures are similar but not so glossy. Chihuahuan Raven is larger, shaggier, and much heavier-billed. Male Great-tailed Grackle has a white eye and a keeled tail as long as the body.

BEHAVIOR: Tamaulipas Crow is as social as other crows, but expect to see individual birds at the northern limit of the range. Opportunistic and tame, this species forages, for the most part, on the ground.

FLIGHT: Overall small and slender (with flatish underparts), showing a short-billed face and a longish tail. Flies with head slightly elevated, back somewhat humped, and wings curled back at the hands. Flight is mostly straight, flowing, and very gracklelike except for a slight bounciness. Wingbeats are continuous, with a deep down stroke, but are slightly hesitant or irregular—not the steady rowing wingbeats of American Crow.

VOCALIZATIONS: Caw is a low, hoarse, and nasal "*rraah*," often given sequentially.

PERTINENT PARTICULARS: Places in the Brownsville area to intercept this bird include the NOAA weather station, the port of Brownsville, and the landfill.

Fish Crow, *Corvus ossifragus*

The Southern Crow with the New England Inflection

STATUS: A common and permanent resident within its fairly restricted range, except for birds in the northern interior part of the range. **DISTRIBUTION:** Found from coastal s. Maine to Tex., but south of Conn. expands inland, encompassing the coastal plains and lower Piedmont regions of the mid-Atlantic and southern states; also north, along the Mississippi and Missouri drainages to w. Ky., s. Ill., sw. and cen. Mo., e. Okla., and se.-e. Kans. A few are found in s. Maine, N.H., Vt., and upstate N.Y. **HABITAT:** Most often associated with water—seacoasts, tidal wetlands, lakes, and rivers—but this species is becoming increasingly urbanized and now joins American Crow in city parks, golf courses, landfills, and even agricultural habitats. **COHABITANTS:** Laughing Gull, Ring-billed Gull, American Crow; roosts with herons and egrets. **MOVEMENT/MIGRATION:** Some northern populations are migratory, and many birds relocate to places that offer greater food opportunities after the breeding season, such as landfills and mall parking lots. Spring migration from late February to April; fall from September to December. VI: 1.

DESCRIPTION: A small sociable crow that differs from American Crow in such small degrees that direct comparison is often needed to secure a confident visual identification. Voice constitutes the easiest and surest way to separate the species. Barring this, observers are advised to assess a number of characteristics to build a case for one species or the other.

Fish Crow is overall smaller, sleeker, slightly longer-tailed, and more smoothly contoured than American Crow, and not as bulgy or angular. Fish's head is smaller and rounder, and the bill heavy, shorter, and somewhat snubbed. The body is lean and cleaner lined; the feathers are smooth, tight, and compact. Americans seem shaggy; in the wind, their looser feathers billow. Also, the feathers of American's leggings (thighs) and vent seem particularly shaggy and often impart to the underparts a triangular paunch. The belly and vent of Fish Crow are usually smoother and more evenly contoured. In direct comparison, it is very apparent that Fish's legs are shorter and thinner than American's.

Overall black, Fish Crow's plumage is shinier and has a purple-blue gloss that is evident in good light. American Crow is overall duller.

When all the parts are compounded, the slighter, slimmer, shinier Fish Crow is an obvious crow, but a crow with grackle sympathies. The larger, bulkier, shaggier, heavier-billed, and heavier-legged American Crow is more ravenlike. So ask yourself when regarding an unidentified corvid: "Is there any way I could confuse this bird with a raven?" If the answer is a qualified "I guess," chances are you're looking at an American Crow. If the answer is "Not a chance," it's worth considering the possibility of Fish Crow.

BEHAVIOR: A very social crow, usually found in flocks; often nests in loose colonies (sometimes within heronries). Spends more time perched and less time on the ground than American Crow and is more nimble, perching on springy twigs and utility lines. (American Crow prefers sturdier limbs.)

Fish Crow is appreciably tamer than American Crow—allowing closer approach—and shows greater tolerance of prolonged scrutiny. (*Note:* All crows in parks and urban areas, however, are habituated to people.) Flocks are typically compact, globular, and tighter packed and less strung out than American Crow; even several birds tend to cluster.

FLIGHT: It is often easier to distinguish Fish and American Crows in flight. Fish Crow seems shorter and blunter-headed, pointier-winged, and longer-tailed than American Crow. Point-to-point flight is straight and direct; wingbeats are steady and, while rowing, stiffer, quicker, and not as fluid as American Crow. Fish Crow is also more nimble and buoyant in the air than American Crow; hovering, stalling, making tighter and more frequent turns, it seems more playful, less plodding; often sails and soars.

VOCALIZATIONS: Sounds like an American Crow with a cold (or a New England accent). "*Caw*" becomes "*kah*" or "*wah*" or "*aah*" and may be a single utterance or given in series: "*aah, aah, aah.*" Signature call is a nasal doubled denial: "*Uh, uh.*" Although immature American Crow's call has a nasal quality (similar to Fish Crow), it is usually higher-pitched, more drawn out, and more strangled-sounding and single-noted. "*Uh, uh*" is unique to Fish Crow.

PERTINENT PARTICULARS: Both Fish and American Crows have variable characteristics. Tail lengths and bill lengths differ. Fish Crow usually shows a round head with a distinct forehead, but American Crow's forehead can be either long and sloping or jutting upward (like Fish Crow). In other words, some individuals are easily distinguished and others are not. It's best to wait and listen.

Chihuahuan Raven, *Corvus cryptoleucus*
The Flatlands Raven

STATUS: Common to fairly common, but restricted in its range and habitat preference. **DISTRIBUTION:** In the U.S., found in se. Colo., sw. Kans., w. Okla., s. and e. Ariz., N.M., and w. and s. Tex. south into Mexico. **HABITAT:** Open, arid, and typically flat country, including desert grasslands, desert scrub, irrigated agricultural land, and open landfills. Also found in open foothills to the edge of the pinyon-juniper belt, Dumpsters, fast-food outlets, and mall parking lots. Higher and forested elevations are the domain of Common Raven. **MOVEMENT/MIGRATION:** Although permanent residents, in fall birds frequently vacate portions of their breeding range and cluster in large flocks orbiting around both agricultural and urban areas (and landfills) that promise ample food. In some locations, flocks may number in the thousands. Spring dispersal back to breeding areas occurs in February and March over most of the range; most winter flocking begins in September and October. VI: 0.

DESCRIPTION: A small gaunt raven—in size and proportions somewhere between American Crow and Common Raven (but favoring the raven). Raven-large, with a thick, heavy, but distinctly short blunt "Roman nose" of a bill. The top of the bill has a bristly sheath that mantles half to two-thirds of the length of the bill (visible at close range). Common Raven's bill is longer, only partially sheathed, and overall more daggerlike.

At close range and with an obliging wind, the billowing neck feathers disclose a white base. If catching such a glimpse seems unlikely, remember that since winds go hand in hand with open areas and loose feathering is an adaptation an all-black bird might need in a desert environment, seeing the white neck of the "white-necked" raven is not out of the question.

BEHAVIOR: Frequents and forages in dry open desert areas, unlike the raven of forests and mountains, the Common Raven, particularly during breeding season. In winter, however, Common Raven may range into open croplands and desert areas and forage with Chihuahuan Raven.

Commonly perches on telephone poles, windmills, and fenceposts and on flat open ground. (Unless foraging, Common Raven tends not to linger on the ground.) Also haunts busy highways (including interstates), where it patrols for roadkill (as does Common Raven), both from telephone pole perches and from the air. Although shy and not easily approached, attracted to human habitations as well as highways. Will fly to investigate humans entering its territory, but flybys are rarely close.

A social bird usually found in pairs, in groups, or in flocks. Calls frequently and usually gains notice because of its vocalizations. Like soaring hawks, usually most active later in the morning when thermals begin to perk. Does not harass hawks with the gusto shown by crows, but commonly mobs eagles.

FLIGHT: Body is slightly thinner and gaunter than Common Raven; lacks the larger species' full-bodied curvaceousness. The smaller, shorter, thicker bill may not be obvious in flight, but the straight-sided round-tipped tail differs from the diamond- or wedge-shaped tail of Common. Chihuahuan's wings are more crowlike—wider, less sinuously tapered, blunter at the tips—and lack the narrow turned-back tips that are so obvious on Common. Wingbeats are slow, shallow, and regular, but not as fluid or sinuous as Common Raven.

Typically does not fly high when foraging, but does soar to great heights to sport with eagles, play with other ravens, and do whatever it is that prompts ravens to soar to great heights.

Chihuahuan is a raven with crowlike qualities. While it is almost impossible to turn a Common Raven into a crow, there are times when Chihuahuan can seem very crowlike, and why not? Over most of its range there are no crows.

VOCALIZATIONS: Vocalizations are similar to Common Raven, but flatter, more parched, less musical, slightly higher-pitched, and generally less interesting and varied. Most commonly heard call is a harsh "*raah, raah, raah*"—a raven's croak that borders on Fish Crow nasal (the sound you might expect from a baby raven).

Common Raven, *Corvus corax*

The Mountain Raven

STATUS: Uncommon to common species; resident across much of North America. **DISTRIBUTION:** Extensive (ranging from Greenland to the Bering Sea to Central America), but not universally apportioned; absent from much of the U.S. In Alaska and Canada, absent only in the southern prairie regions of Alta., Sask., and Man. and from extreme se. Ont. In the continental U.S., found primarily from the Rockies west, but also in sw. Tex., n. Minn., n. Wisc., n. Mich., Maine, Vt., N.H., the Adirondack and Catskill Mts. of N.Y., and south along the Appalachians from n. N.J. and s. N.Y. to n. Ga. **HABITAT:** Highly variable—from arctic tundra to boreal forest, deciduous forest, prairies, seacoasts, cities, isolated settlements, agricultural lands, landfills, and deserts. Prefers wilderness and terrain at odds with the horizon. Outside these habitats, seems more closely associated with human-modified environments, such as agricultural land, highways, native villages, and small towns. **COHABITANTS:** Bald Eagle, Golden Eagle, Peregrine Falcon. **MOVEMENT/MIGRATION:** Permanent resident. VI: 1.

DESCRIPTION: A big, black, loosely feathered crow that is larger than most hawks (longer in length and wingspan than Red-tailed Hawk) and overall much longer, rangier, shaggier, and rougher-looking than crows (one-third again bigger than American Crow). The bill is long, menacing, and daggerlike, the throat bristly, the underparts loosely feathered, and the wings and tail long. The legs projecting from a shaggy skirt of feathers seem sturdy and short.

The entire bird is black and, in good light, sometimes shows a slight green or purple gloss.

Crows are overall sleeker, more compact, shorter-winged, and longer-legged than Common Raven. Chihuahuan Raven is raven-shaped but more nearly crow-sized, with a short thick "Roman nose" of a bill.

BEHAVIOR: Most commonly seen alone or in pairs, perched or aloft. Commonly seen perching on rocky ledges, bare tree limbs, utility poles, landfills, ski lodge and visitor center parking lots, and (sometimes) occupied dwellings. An eclectic and opportunistic scavenger-hunter. Although often found near carcasses or attending eagles and mammalian predators at recent kills, ravens also catch and kill a range of prey—from reptiles and amphibians to birds and small mammals—in addition to consuming large quantities of large insects and grain. Generally tolerant of crows, hawks, and owls, but very aggressive toward Golden Eagle.

In wilderness areas and places where regulations offer protection from humans, can be tame bordering on bold (for example, learning to open campers' tents and food containers). In other places, may be very shy and unapproachable. In winter in some places where food is concentrated, birds gather in large roosts and can be found in flocks exceeding several hundred.

Walks with a strutting gait. When taking off from the ground, commonly makes one or two hops. (Crows usually leap into the air.)

FLIGHT: A raven specialty. Profile is overall rangier and more sinuously lined than a crow—longer-headed, longer-tailed, and much longer-winged. Common Raven seems slimmer, but not slender. When gliding or half-soaring, wingtips taper acutely and turn back (a configuration that is visible at amazing distances). Fanned tails look distinctly pointy-tipped or wedge-shaped; partially closed, they look like singe-dip ice cream cones. Tails of crows are comparatively blunt, and Chihuahuan Raven's tail is more evenly rounded—somewhere between crow-blunt and raven-wedged. When soaring at a great distance or high altitude, the head disappears and the bird becomes all wide tail and acutely tapered wings.

Flight is buoyant, fluid, and graceful, never stiff. Wingbeats are slow, shallow, casual, and elastic, never mechanical. Glides and soars considerably more than crows, and seems more playful in the air, engaging in acrobatics for no other reason than to play. Often vocalizes in flight (most commonly when it has sighted you).

VOCALIZATIONS: Short, loud, varied, and comical. Some vocalizations are loud and harsh; others are low, gurgled, or muttered. All have a rough musical finesse that makes crows sound amateurish. The classic croak, "*raah raah raah*," is more baritone, less nasal, more hawking, than Chihuahuan. Also makes an ascending gargle, "*a'raaaaH*," and a vocal knocking that sounds like a woodpecker working (or a tapping on a door).

LARKS

Sky Lark, *Alauda arvensis*

Sky Canary

STATUS AND DISTRIBUTION: Resident species introduced to s. Vancouver I., B.C. Also occurs as a rare spring and fall vagrant in w. Alaska. **HABITAT:** Open dry areas devoid of trees and shrubs.

Common habitats include pastures, fallow fields, agricultural land, beaches, airports, golf courses, and grassy fields. In winter (particularly after blanketing snow), makes greater use of grain fields, agricultural land, and habitats bordering trees. COHABITANTS: In winter, Horned Lark, American Pipit, House Finch, Snow Bunting. MOVEMENT/MIGRATION: Permanent resident on south Vancouver Island. VI: 0.

DESCRIPTION: A cryptic, pinch-faced, blunt-crested, streaky-bibbed lark that is found on Vancouver Island (and in western Alaska) and is clearly not a Horned Lark. A small songbird (larger than American Pipit; smaller than European Starling; about the same size as Horned Lark and Snow Bunting), with a pointy larklike bill, a small round-topped (with crest flattened) or crested head, a long plump body, and a short tail. Usually crouches close to the ground.

Cryptically warm brownish or grayish brown above; whitish below with a well-defined streaky bib that is washed with buff and may trail down the sides and flanks of the bird. Overall more finely patterned than Horned Lark and lacks that species' bold black face pattern and narrow neck-band.

BEHAVIOR: Solitary and territorial during breeding season (but several breeding birds may occur in the same general area). Outside breeding season, forages in flocks on dry soil or short grass. Sometimes walks (or runs) with an upright posture, but when foraging often crouches. Crouches and freezes when approached, then launches itself, flies a short distance, and lands. Birds often perch on rocks, fenceposts, or runway signage. Males are celebrated for their protracted but largely unspectacular aerial vocal displays: they ascend on quivering wings, hover in place when they are almost too high to be seen, then descend after 2–3 mins. in slow spirals that end in a sudden and silent drop to the ground on folded wings. Song flights begin before dawn and continue until dusk.

FLIGHT: Overall stocky, with wide triangular wings that appear to sit high on the bird, like the wings of a single-engine Piper Cub. With its short slightly forked tail, from below the bird seems all wing and tail. The breast-band is plainly evident against the pale underparts. The partially fanned black tail is edged in white. In display flight, wingtips move so quickly that they blur. Direct flight is flowing, weaving, and undulating, with a broken skipping rhythm that recalls Snow Bunting. Wingbeats are loose and fluttery, and interrupted at intervals by brief glides.

VOCALIZATIONS: Song is a high, reedy, varied, and run-on outpouring of trilled and warbled notes (some borrowed from other birds). Sounds like a hoarse, hurried, and somewhat frantic canary. Calls include a two-noted "*peChee*," a liquid "*cheerup*," and a finger-being-run-down-a-comb trill that recalls Carolina Wren but is higher-pitched.

Horned Lark, *Eremophila alpestris*

Bird of Earth and Sky

STATUS: Common and widespread breeder and wintering species in North America. DISTRIBUTION: Breeds all across North America from the Canadian arctic and Alaska to the Mexican deserts. Absent as a breeder only from the forested interior of Canada, parts of New England, the n. Midwest, the West Coast from se. Alaska to n. Calif., and the Southeast coastal plain. Winters from s. Canada across the U.S. except the Florida Peninsula. HABITAT: A barren-ground specialist—at all seasons, prefers extensively open, dry, bare ground with little or scant weedy vegetation, such as tundra, desert, overgrazed land, arid prairies, bare fields (particularly where manure is spread), or high beach. Also favors very short grass, such as is found at airports and sod farms and on turf-depleted playing fields. COHABITANTS: In winter, longspurs, Snow Bunting. MOVEMENT/MIGRATION: Across extreme southern Canada and most of the United States, a permanent resident. Northern birds withdraw as winter advances, leaving northern breeding grounds as early as August and reaching southern range limits as late as December. Spring migration is likewise protracted: some birds depart in February and do not arrive on extreme northern territories until late May. VI: 4.

DESCRIPTION: Resembles a large, grayish brown, ground-hugging sparrow with a rakish facial pattern. Medium-small songbird (larger than a longspur or Snow Bunting; smaller than a starling), with a longish, somewhat flattened and ovular body (as if the bird has been stepped on), a pointy wedge-shaped face, longish wings, and a longish tail. Except when alert, the posture is generally horizontal and crouched, with the body angled toward the tail.

Adult males are overall pale-and-pastel-shaded grayish brown above, with a warm infusion of pink or peach on the hindneck, shoulders, breast, and (sometimes) flanks; whitish below. (*Note:* The hindneck and sides of the breast in some California birds are darker and more orange than the peach above.) The bold black face/breast pattern set against the yellow (or white) face and the overall cryptic backdrop of the bird is hard to overlook. The feather tuft "horns" are sometimes visible, sometimes not, but hardly necessary for the sake of identification. Always hornless females are colder-toned and streakier above (much more sparrowlike), with a washed-out face pattern and a shadow rendering of the breast-band. Often the most obvious feature is a large buffy or whitish eyebrow. A subdued juvenile plumage that somewhat resembles Sprague's Pipit is held for only a short time, and young birds are usually in the company of adults.

BEHAVIOR: Hardy, social, and often fairly tame. Often crouches when approached and flushes reluctantly. Forages on the ground, often on bare sand or soil, with bill turned down. Feet move with a mincing shuffling gait; bird seems to waddle. Raises its head frequently to look around, and runs forward in short spurts in a head-lowered crouch. Otherwise, seems to slink as it moves.

In winter, often found in flocks—sometimes numbering several hundred birds—with other ground-foraging birds (such as Snow Buntings and longspurs). Flocks are loose aggregations; feeding birds break into subflocks that forage off in their own direction for a time, then wander back. Approached too close, some or most members of the flock may fly while others stay. Flushed birds often fly low and land nearby, or they may climb, circle, and return. Calls when flushed.

FLIGHT: Slender, long-winged, and (except for the contrasting black tail), overall pale. The white outer tail feathers are not easily seen. Flight is bouncy, nimble, light, buoyant, and flowing. Two or three quick wingbeats are followed by a closed-wing glide. Flocks are fluid and shifting—fairly compressed at first, then more strung out and less cohesive as they gain altitude and travel. Low-flying flocks move in a flowing ground-hugging sheet, with individual birds undulating slightly and pairs veering off in brief tail-chases from the main body of the flock before rejoining it. Birds land by turning 180° (that is, facing the direction from which they came).

VOCALIZATIONS: Song is an ethereal and somewhat ventriloquial jingle. Two to three breathy chirps are followed by a rushed and compressed string of weak, clear, high-pitched notes that climb the scale as they accelerate. Song is often given in a hovering display and repeated after a brief pause. Call is high, thin, sometimes squeaky, and generally two-noted: "*chee-dee?*" or "*teedl.*" Birds sometimes insert a lisping "*chlip*" at the beginning.

SWALLOWS

Purple Martin, *Progne subis*

House Martin

STATUS: Uncommon to common widespread, primarily eastern breeder with a patchy distribution in the North and West. **DISTRIBUTION:** Breeds across much of the e. U.S. and s. Canada from N.B. to extreme s. Que., s. Ont., s. Man., s. Sask., and e.-cen. Alta. In the U.S., found mostly east of a line drawn across the cen. Dakotas, cen. Neb., w. Kans., w. Okla., and cen. Tex. (largely absent through interior portions of New England and along the spine of the s. Appalachians). In the West, very locally found in Wash., Ore., Calif., n. Utah, w. and cen. Colo., Ariz., N.M., and n. and cen. Mexico. Winters in South America. **HABITAT:** Open sky, often at great altitudes. In the East, owing to a near-dependence on multichambered Purple Martin "houses" and suspended hollow gourds, found largely near cities, towns, and, in rural areas, human dwellings. Through much of the West, found in mountain and coastal forests and in burned and logged forest (where woodpecker-excavated snags can be found). Prime habitat also includes cavity-rich sycamores and woodpecker-worked aspen woodlands adjacent to beaver ponds. In the desert Southwest, nests in saguaro cactus. **COHABITANTS:** Humans, Northern Flicker, Gila Woodpecker. **MOVEMENT/MIGRATION:** Possibly the earliest Neotropical migrants to return in spring, Purple Martins reach the U.S. Southeast as early as mid-January (Florida) and mid-February (Texas) and occupy all of their breeding range by early May. Western birds arrive much later. Fall migration is likewise early. Birds leave nest sites as soon as young are fledged (as early as June, but more commonly July through August). They concentrate in insect-rich areas and at sunset gather into large roosts in trees, in stands of tall marsh grass, and beneath bridges. These roosts may number in the tens, even hundreds, of thousands, and departures are often abrupt. By early October, almost all Purple Martins have left North America. VI: 2.

DESCRIPTION: A big, sturdy, broad-chested, big-headed swallow. Uniformly blue-black adult males are unmistakable (all other swallows have paler underparts). Females are also distinctive: bluish gray above, dirty gray below, with a pale collar and a darker chest. Although females might be confused with other small dark-backed pale-bellied swallows, the head on Purple Martin is distinctive: other swallows have round heads, whereas Purple Martin's head is peaked—topped by a flat-topped crewcut that slopes from crown to forehead.

BEHAVIOR: In the East, highly social and almost always found with other martins in the vicinity of colonies or late summer roosts. Colonies are active and noisy, particularly when galvanized by a predatory intruder. Purple Martin often sits with other swallows on wires where its larger size is immediately apparent. In spring and early summer, males go aloft before first light and vocalize. Birds rarely land on the ground except to secure nesting material or consume grit.

Commonly feeds at greater heights than most other swallows and consumes larger prey (such as beetles and dragonflies). After breeding, birds stage in insect-rich areas, roosting nightly in phragmites stands (where present), often with Tree Swallows; they also roost under bridges. These roosts commonly number in the tens of thousands of birds.

FLIGHT: Larger and more broadly proportioned than other swallows, showing long, broad-based, pointy wings, a projecting head, and a slightly notched tail—all in all a very falconlike (more specifically, merlinlike) silhouette. Adult males are unmistakably dark blackish. Females and immatures are grayish and show pale underparts that contrast with the dark underwings and breast. In active flight, appears harshly angular. When soaring or gliding, the leading edge of the wing curves back, giving the bird softer contours. The tail is somewhat short, distinctly notched, and closed; flares outward at the tip.

Flight is direct and powerful, displaying more stability and finesse than smaller swallows. Purple Martin is a telemark skier of a swallow in the company of slalom and mogul skiers. Wingbeats are deep and somewhat floppy or elastic. When hunting, often at high altitudes, Purple Martin is mostly a glider—setting its wings and covering ground (or sometimes holding into the wind and hardly covering any ground at all). Only Violet-green and Cliff Swallows seems to glide as much as the martin. When hunting, Purple Martin's wingbeats, given in sputtering series, seem not so much a primary means of propulsion as a way to provide energy for a glide. Birds soar often and well, often in multiple circles (unlike most other swallows).

VOCALIZATIONS: The sound from a colony is a bubbling mix of exclamatory chirps and throaty churps, liquid trills, gurgles, burbles, and periodic bleats—all with a pleasing musical quality. In flight, makes a low "*chur*" (often repeated), which has the quality of a stone skipped on ice, and a short musical trill. Conversational call is a soft "*chuh, chuh.*"

Tree Swallow, *Tachycineta bicolor*

Emerald Swallow

STATUS: Common, at times abundant, and widespread breeder and winter resident. **DISTRIBUTION:** Breeds across all but the northernmost regions of Canada and Alaska and throughout much of the continental U.S. Breeding does not occur across the South (absent in e. N.C., most of S.C., Ga., Ala., La., s. Mo., and Tex.), the central prairies (absent, for example, in most of Okla.), or the s. Great Basin (absent in se. Ore., sw. Idaho, most of Nev., and e. Calif.). Does breed in scattered locations in Ariz. and N.M. In winter, found coastally from Va. and N.C. to Central America (including all of Fla.) and in s. Calif. and sw. Ariz. **HABITAT:** At all times, prefers open habitat—fields, marshes, coastal estuaries, lakes, and swamps with standing dead timber. Nests in tree cavities (particularly old woodpecker holes), often in small colonies if a number of cavity-rich trees are clustered. Also quickly adopts man-made bluebird boxes. During migration (particularly) and in winter, gathers in tremendous roosts numbering hundreds of thousands, often in tall marsh vegetation such as phragmites. Before going to roost, gathers on utility lines. **COHABITANTS:** Often nests in association with Violet-green Swallow as well as Eastern, Western, and Mountain Bluebirds. In migration, often feeds and roosts with other swallows. **MOVEMENT/MIGRATION:** A very early migrant, Tree Swallow returns to the southern portions of the nesting range as early as late February; most northernmost nesters arrive before May. Fall migration begins the first week of July (just after first clutches fledge), peaks in September, and may continue (in coastal areas) well into November. In spring, a diurnal migrant that moves in small (sometimes mixed) swallow flocks. In fall in the East, often gathers in great single-species flocks. VI: 4.

DESCRIPTION: A green-backed swallow with brilliant white underparts. Large (larger than Bank Swallow or Violet-green Swallow; smaller than Purple Martin), compact, and classically proportioned, Tree Swallow shows a tail that is more notched than forked, and shorter than the wings when perched.

Adults are iridescent bluish green above, with darker, less iridescent wings and tail. The brilliant white underparts begin just below the eye (giving birds a dark face and white throat and making the black eye difficult to see) and extend the length of the bird. Young are plumper, more roundly proportioned, brown- to gray-brown-backed, and not so white below. Immatures also show a shadowy gray wash across the breast that suggests the dark breast-band of Bank Swallow, and an intruding white wedge pinching the base of the tail that recalls, but is less prominent than, the white saddlebags of Violet-green Swallow. Some second-year birds, retaining vestiges of their immature plumage, are duskier than adults.

BEHAVIOR: A very social gregarious swallow. Birds sit tightly packed on utility lines (also in massed numbers on tall rank weed and, less commonly, in packed flocks on the ground), roost in tremendous numbers in the East, and gather into massed balls of birds when threatened by a bird of prey. In winter this species forages for bayberries, and when cold weather shuts down insect activity, they also forage by plucking insects from the surface of the water.

A cavity-nesting bird, Tree Swallow quickly adopts bluebird boxes. Sometimes carries (usually white) feathers aloft and, releasing them, pursues them in the air.

FLIGHT: In flight in spring, adults are trim and angular, and the wings triangular. In late summer and early fall, flight feathers are molting and the bird looks disheveled and anything but trim. Immatures appear plumper and more roundly proportioned than adults and are overall duller and grayer, showing only traces of iridescence on the upperparts. Except where Violet-green Swallows occur, the wholly brilliant white breast and belly distinguishes them from other swallows. (Even Bank Swallows are plain white, not brilliant white.) *Note:* Immature Tree Swallow usually shows a dusky narrow band across the chest, a shadowy pattern that resembles the darker band of Bank Swallow, particularly at a distance, where contrasts tighten.

Flight is somewhat variable. Tree Swallow seems to incorporate and homogenize the flight characteristics of all the swallows. In general, flight is active, energetic, somewhat playful, level, and even-keeled; more than other swallows it recalls Purple Martin. Wingbeats are rapid, deep, and somewhat fluttery; the bird looks as though it's putting a great deal of energy into the wingbeats but the performance is lagging. Wingbeats are given in a long series without pause, followed by a glide. Often birds climb as they flap.

VOCALIZATIONS: Song is a series of high, clear, mostly two-note whistled phrases. Calls include a variety of liquid chirping and twittering. When agitated, birds chatter angrily. In flight, they make soft, conversational, clear or trilled chirps.

Violet-green Swallow, *Tachycineta thalassina*

The Swiftlike Swallow

STATUS: Common western breeder. **DISTRIBUTION:** Breeds from interior Alaska south into Mexico. The eastern border runs through the cen. Yukon and ne. B.C., then diagonally across Alta. to the southwest corner of Sask., extreme w. N.D. and S.D., e. Colo., w. Neb., e. N.M., and sw. Tex. Winters in s. Calif., Mexico, and Central America. Casual visitor in the East. **HABITAT:** Breeds in montane and canyon forests, particularly coniferous forests (also mixed); seems particularly fond of rocky cliff faces. **COHABITANTS:** White-throated Swift, Common Raven. **MOVEMENT/MIGRATION:** After breeding season (beginning in late July), birds leave nest areas and gather near watercourses. Spring migration from February to early May; fall from late July to early October. VI: 2.

DESCRIPTION: A small-bodied, long-winged, crisply patterned swallow whose shape and flight put it somewhere between a swallow and a swift. The body is short and slight, and the wings long and narrow (extending well beyond the tail when the bird is perched). In overall profile, Violet-green Swallow falls somewhere between Tree Swallow and Barn Swallow, but is smaller than either bird. The notched tail is difficult to see because of the bracketing folded wings. In flight, the tail is narrow at the base and slightly flared and notched—except during hard banking, when it's square-cut or rounded. Overall Violet-green is smaller, slighter, and more angular and elfin-featured than Tree Swallow.

Adult males and many females are distinctly patterned and colored. Their bright emerald green backs are much greener (and more eye-catching) than the blue-green backs of Tree Swallow (which intermittently flash dark green at the right angle). The exceedingly white underparts (they seem even whiter than Tree Swallow) distinguish these birds in a mixed-species flock. The black eye stands out against the all-white face of males and many females. (The eye is invisible on Tree Swallow.) The white rump sides wink as the bird twists and banks in flight; the narrow dark bisecting line disappears at any distance, making the rump appear just white. *Note:* The white saddlebags behind the wings shown by Tree Swallow (particularly immature Tree Swallow) are never so prominent that the bird seems truly white-rumped. If you have to second-guess, it's not a Violet-green.

Immatures (and to a lesser degree some females) are overall dusky above, dingy white below, with a dusky face that shows a shadow of the male's pattern. The eye remains easier to see than the eye of Tree Swallow.

BEHAVIOR: Quite gregarious. While solitary nesting is not atypical, birds often nest in colonies (most commonly in old woodpecker nest holes found in dead trees, but also on cliffs, sometimes with White-throated Swift and Cliff Swallow) numbering from a few individuals to more than 50 pairs. Forages and migrates in loose flocks that may number several hundred individuals, and while it commonly forages at greater heights than other swallow species, often mixes with other swallows and swifts, particularly White-throated Swift.

FLIGHT: Combines the maneuverability of a swallow with the speed and erratic stiffness of a swift—in particular the White-throated Swift, with which it is often found. Wings are longish and slender, although acutely tapered (not slender throughout their length like swifts); combined with the short body, the wings make Violet-green Swallow seem more swiftlike in its proportions than other swallows. Wingbeats are quicker, stiffer, and more jittery than Tree Swallow. Birds glide frequently and for extended periods, executing multiple circles or protracted set-winged coursing descents along the faces of cliffs. Because it glides with wings angled slightly down, but wingtips turned slightly up, Violet-green seems to wobble and to be slightly unsteady in flight. Tree Swallow seems always in control.

VOCALIZATIONS: Call is reminiscent of Tree Swallow, but higher-pitched, drier, and faster: "*chih chih chih*" or "*cheecheecheechee*"—sounds like electrical impulses stuttering down a wire.

Northern Rough-winged Swallow, *Stelgidopteryx serripennis*

Die Fledermaus (Flying Mouse)

STATUS: Fairly common and widespread. **DISTRIBUTION:** Breeds across s. Canada, from s. N.B. and se. Que. to coastal B.C., and virtually all of the U.S. (except n. Maine and sw. Ariz.). Some birds winter in s. Fla., along the Gulf Coast, and along the Mexican border, but most retreat into Mexico and Central America before the onset of cold. **HABITAT:** Nests in burrows dug into steep banks (principally, if not wholly, burrows excavated by other birds or mammals) and assorted other nooks and crannies, including steel and cement bridges, drainpipes, and cracks in concrete structures. Seems most attracted to nest sites over or in close proximity to water, but this is not an absolute requirement. **COHABITANTS:** Frequently nests at the edge of Bank Swallow colonies. Mixes with other swallows in migration and when feeding. **MOVEMENT/MIGRATION:** In spring, arrives between late February and April. Fall migration begins early, in mid-July, and continues into early November; most migration, however, occurs in July and August. VI: 2.

DESCRIPTION: Small, overall drab and stubby, and somewhat roundly proportioned, this mouse of a swallow is distinguished by its lack of distinction. When perched, it is slightly larger and pudgier than Bank Swallow, slightly smaller than Tree Swallow.

Plumage seems somewhat disheveled or shabby—rough! (not sleek and trim like Bank Swallow). Most of the bird is dull brown—from the plain featureless face to the dusky throat and upper breast to the wing and tail. The belly is dirty (not bright) white. Except for a somewhat darker tail and rufous wing-bars (found on immature birds), the bird shows no pattern or contrast. It's just a stubby bran-colored swallow with a dirty chest and dingy white underparts.

BEHAVIOR: As a breeder, often solitary, although several pairs may occupy the same bank or structure. In migration, usually found with other swallows; also generally outnumbered by other species.

Habitually feeds over water, often very low, and seems more inclined to forage over areas that other swallows find too obstructed or confining. Except at nest sites, less vocal than other swallows.

FLIGHT: Overall fairly compact and roundly contoured. Wings are proportionally broader and blunter than other swallows. The tail is broad and blunt when closed, short and rounded when fanned. Plain dingy brown above, whitish below, with a dirty chest. Shows no iridescence or paleness on the rump.

Flight is a little slower and more casual or tentative than other swallows; Northern Rough-winged may also be more deliberate and less erratic and wandering—a swallow that takes care of business without much panache. Wingbeats are slower, more shallow, more fluttery, less crisp, and more constant than Bank Swallow.

VOCALIZATIONS: Makes a low, short, burry snort, "*breet . . . breet,*" with little variation in pitch or tempo.

Bank Swallow, *Riparia riparia*

Ringed Swallow

STATUS: Uncommon to common widespread breeder. **DISTRIBUTION:** A Holarctic species, in North America Bank Swallow (called Sand Martin in the Old World) breeds across most of Canada but primarily in the n. U.S. Found from Nfld. and Lab. west to Alaska and north to the Arctic. The southern border runs north of a line drawn from Va. to n. Ark., n. Okla., n. N.M., Utah, n. Nev., and cen. Calif. Also breeds in cen. and s. Tex., but is absent as a breeder along the West Coast (w. Wash., w. Ore., s. Calif.). Winters south of the U.S. **HABITAT:** True to its name, Bank Swallow nests in tunnels that it excavates in vertical sand or dirt banks associated with riverbanks, seacoasts, and sand, soil, and gravel mining operations. Colonies may be found at altitudes higher than 5,000 ft., but most are in lowland areas. Because of the mercurial nature of Bank Swallow's nesting habitat, established colonies may disappear, but prime stable locations are commonly used for many years. **COHABITANTS:** May nest in association with Belted Kingfisher, Northern Rough-winged Swallow. In migration, found among mixed swallow flocks. **MOVEMENT/MIGRATION:** In spring, arrives in the southern parts of the range as early as mid-March and in arctic regions as late as June. Bank Swallows are early-fall migrants: adults and immatures begin moving south in July. The bulk of the population is gone by the end of September across most of North America, but lingering birds have been recorded into November. Early-fall migrants may be found in loose homogeneous groups, but Bank Swallows migrating later often join mixed swallow flocks. VI: 2.

DESCRIPTION: A small, slender, trim swallow distinguished by its brown upperparts and wide dark chest band bisecting white underparts. Except for Barn Swallow, Bank Swallow is the slimmest of the swallows, with a body/tail profile that tapers acutely. The tail appears disproportionately long and narrow and, when closed, shows a distinct notch (almost a fork) that is unlike the shorter, blunter tail of Northern Rough-winged Swallow.

Casually referred to as one of the two brown-backed swallows (the other is Northern Rough-winged), Bank Swallow is not as brown as some field guides portray it. Overall it's grayer-toned, paler, and more crisply and contrastingly plumaged than Northern Rough-winged. When perched, appears darker-winged and dark-masked. Note that the dark breast-band widens in the middle.

BEHAVIOR: A very social species most commonly seen in large buzzing groups over nesting colonies or foraging or traveling with other swallows in migration. Hunts flying insects by aerial pursuit (also picks prey from water surfaces), and generally forages low, within 50 ft. of the ground. Prefers to forage over open unforested habitat, but during migration and in northern spruce forests, may hunt above the canopy. Unless gathering nesting material, almost never seen on the ground. Like all swallows, perches on wires and also clings to nest-banks Chimney Swift fashion. Enters nest-burrow without landing and almost at full throttle.

FLIGHT: In flight, wings are long, narrow, pointy, and angled sharply back. Lines are generally straight and trim—more straight-cut and acutely angled along the leading edge and more curvaceous along the trailing edge. The body, behind the wing, seems to taper (almost) to a point.

In flight, the paler upperparts contrast with the darker wings. (Rough-winged is uniformly colored above.) The dark, narrow, crisply defined chest band bisecting the white underparts easily distinguishes the species. At some angles and at great distances, however, the white throat, curling up onto the head and contouring behind the dark ear patch, is more eye-catching; it makes the bird look white-faced (even somewhat white-headed) and can create the impression of a pale collar separating the head and back. Northern Rough-winged Swallow is dark-headed and uniformly brown above, in addition to being overall pudgier, stuffier, and more roundly contoured.

Bank Swallow's flight is fast, almost frantic—nimble, darting, and somewhat twisty-turny. When migrating or traveling to and from colonies, flight is generally direct, and wingbeats are measured and fairly regular. Hunting Bank Swallows are exceedingly energetic and three-dimensionally minded. Wingbeats are crisp without being stiff, rapid, deeply down-flicking, and near-constant—slower and more shallow in normal pursuit, rapid and deep as birds climb steeply to intercept prey. Horizontal glides are brief and infrequent, but short, descending, set-wing swoops (following a climb) are commonplace.

VOCALIZATIONS: Call is a buzzy, unmusical, not particularly loud "*brrrrr*" or "*breee*" that is sometimes doubled, "*brr-brrrt*," or, when excited, repeated: "*brr-brr-brrt.*" Bank Swallow's call is more excited, higher-pitched, and more varied in pitch and delivery than Northern Rough-winged Swallow's. The quality recalls the sound of electrical impulses running through a high-tension power line and may remind some of a scolding wren. Also makes a higher-pitched trill that recalls a finger being run down the teeth of a comb.

PERTINENT PARTICULARS: Although most commonly compared to Northern Rough-winged Swallow, Bank Swallow is probably most commonly confused with young Tree Swallow, which has gray-brown upperparts (with a slight greenish tinge), white underparts, and a shadowy dark collar (particularly evident at a distance). Tree Swallow also has a white collar projecting behind the face (like Bank Swallow).

Bank Swallow, however, is overall slender; Tree Swallow, particularly young Tree Swallow, is broad-winged, stubby, and shaped more like Northern Rough-winged Swallow. Also, Tree Swallow has white flank patches that intrude onto the dark back at the base of the tail. Bank Swallow does not. Finally, Bank Swallow, a very early-fall migrant, is rare across much of North America after September. Any October (or November or December) "Bank Swallow" likely is not.

In mixed swallow groups, when birds are perched along wires, feeding, or migrating, identifying Banks Swallow is often a simple matter of noting the size. In direct comparison, Bank Swallow is noticeably smaller than either Tree Swallow or Northern Rough-winged Swallow.

Cliff Swallow, *Petrochelidon pyrrhonota*

Overpass Swallow

STATUS: Uncommon and local across much of the East; common and widespread in the West. With the exception of Barn Swallow, has the most extensive breeding range of all North American swallows. **DISTRIBUTION:** Occurs across most of North America south of the tree line to almost Central America. Absent as a breeder across most of the coastal Southeast, essentially south and east of a line drawn from coastal Mass. to nw. Conn., n. N.J., se. Pa., w. Va., w. N.C., w. S.C., nw. Ga., n. Ala., n. Miss., cen. Ark., and e. Tex. Also does not breed in w. Ohio, parts of Tenn. and Ky., s. Tex., sw. Ariz., and arid southern interior portions of Calif. **HABITAT:** Historically, nested on vertical cliffs that offered protective overhangs. Now colonies also thrive on assorted man-made structures, including highway overpasses, office buildings, water culverts (beneath roadways), dams, barns, and other outbuildings situated in suitable habitat (grasslands, marshes, lakes, reservoirs, cities, towns, and riparian edge). Does not breed in deserts, at high altitudes, or in unbroken forest. **COHABITANTS:** Occasionally nests with Barn Swallow and, in Texas, Cave Swallow. Feeds in mixed swallow (and swift) flocks. **MOVEMENT/MIGRATION:** Entire population retreats to South America well before the onset of winter. Spring migration is very protracted across North America, owing to the species' vast latitudinal range. Earliest arrivals reach southern California in early or mid-February, while those nesting in the Alaskan interior may not arrive until mid-May (or later). Fall migration is comparatively brief; some birds depart in early July, and most have arrived south of the U.S. border by the end of September. VI: 4.

DESCRIPTION: A small, darkish, contrastingly patterned (and beautiful) swallow that is obliquely but readily distinguished from most other swallow species by the absence of bright white. With the pale spot on its forehead, somewhat resembles a husky crop-tailed Barn Swallow wearing a miner's lamp. Overall stocky, with a short blunt head, a short, wide, square-cut tail, and (in flight) wide, roundly contoured wings. Basically dark above, buffy gray below, with a pale (buffy) rump that contrasts with the dark upperparts and a sharply defined dark throat that contrasts with the paler underparts.

Immatures have the same basic plumage pattern as adults but are duller, lack the adult's rich colors, and have mostly dark foreheads.

BEHAVIOR: Highly colonial. Colonies may contain hundreds, even thousands, of distinctive, gourd-shaped nests constructed wholly of beaded mud. Birds forage collectively, feeding together in large groups where insects are concentrated (for example, along the lee side of cliffs in high winds, or over warmer lake surfaces when air temperatures are low). Operates freely in a three-dimensional world, seeking insects flying above, then climbing to intercept them.

FLIGHT: Overall compact, with a stubby body, broad wings, and distinctly rounded contours (shaped most like Northern Rough-winged Swallow). The tail is conspicuously square-tipped when closed, and very round when fanned. From below, the dark underwings and tail contrast with the pale body, and the pale rump patch is very apparent. The pale miner's lamp on the forehead is slightly less apparent at a distance.

A strong flier that beats its wings with a rapid fluttery motion. Not quite as nimble or acrobatic as some swallow species, but fast and more vertically oriented; climbs to intercept prey as a matter of course. Also glides more frequently than other swallows (except Purple Martin and Violet-green).

VOCALIZATIONS: Makes a low, short, fluttering "*chur*" or trilled "*pur tr'r'r'r*" that is not as buzzy or high-pitched as Bank Swallow. Also makes a low, muffled, plaintively mewed "*reeh*" or "*er'reeh*" when anxious or alarmed (sounds somewhat like a low catbird note or may recall a distant parakeet).

PERTINENT PARTICULARS: A very easily identified swallow. To distinguish this bird in flight, concentrate on the buffy rump, which is much easier to see than the pale forehead. To distinguish this species from Cave Swallow, rely on the sharply defined dark (in most lights, black!) throat.

Despite some plumage similarities, you cannot mistake this species for a Barn Swallow.

Cave Swallow, *Petrochelidon fulva*

Washed-out Cliff Swallow

STATUS: Locally common but in the United States geographically restricted breeder and winter resident. **DISTRIBUTION:** Breeds in se. N.M. and s. and cen. Tex. to the La. border and in extreme s. Fla. Western birds winter from s. Tex. into n. Mexico. A regular vagrant in late fall along the Atlantic Coast and the e. Great Lakes; casual in spring on the e. Great Lakes as well as the Great Plains. **HABITAT:** A swallow of warmer climates. Affixes its Barn Swallow–like mud nests to concrete highway bridges, wooden bridges, water culverts, and buildings (as well as sinkholes and caves). **COHABITANTS:** Breeds with Cliff and Barn Swallows. Feeds with other swallows and swifts. **MOVEMENT/MIGRATION:** Spring migration from mid-January to mid-May; fall from August to late November. VI: 3.

DESCRIPTION: Small (slightly smaller than Cliff Swallow), somewhat short-headed, roundly contoured swallow that resembles a washed-out Cliff Swallow. The pattern of a buffy throat patch and darker forehead is just the opposite of Cliff Swallow. The rump patch (not commonly visible when birds are perched) is slightly darker, more ruddy or cinnamon-tinged, and less contrasting than on Cliff. In some lights, the wings have a brownish cast (not seen on Cliff).

Florida Cave Swallows have richer, darker colors below but no white forehead, and Cliff Swallow, though a widespread nesting species in North America, doesn't nest in south Florida.

BEHAVIOR: Forages for insects on the wing, often in flocks (also with other swallows). Prefers to forage high, but particularly on cloudy or rainy days, will forage lower, over water.

FLIGHT: Silhouette shares the rounded contours of Cliff Swallow, but the plumage is overall muted, less contrasting, particularly between the only slightly darker underwings and the pale body. Like Cliff, glides a good deal. Flight is more delicate than Cliff, and wingbeats more fluttery. Glides are not quite so sweeping or lengthy.

VOCALIZATIONS: Song recalls a muffled Barn Swallow—an ensemble of sharp notes, warbles, trills, and nasal utterings that are sweeter than Cliff Swallow. Call is a low mellow "*wheet*" (sometimes repeated). Alarm call is a low, rough, descending protest: "*nyah.*"

PERTINENT PARTICULARS: This once rare North American resident is expanding its numbers and range primarily because of our species' love affair with poured concrete. The Texas subspecies is celebrated for its penchant for turning up thousands of miles away from its breeding range; it now annually shows up in November along the Atlantic seaboard and on the eastern Great Lakes and occurs as far north as Nova Scotia in spring. Any "Cliff Swallow" sighted in North America in November deserves close scrutiny.

An out-of-range Cave Swallow is usually deductively identified as follows: The first impression given by the overall shape, the pattern, and (particularly) the pale rump is that it's a Cliff Swallow. Then the overall paleness and lack of contrast undermine this conclusion. Then you search for the gold-standard field marks: a pale throat and an absence of white on the forehead.

Barn Swallow, *Hirundo rustica*

Tine-tailed Swallow

STATUS: Common and widespread; the most cosmopolitan of the swallows, with populations breeding on every continent except Antarctica. **DISTRIBUTION:** Breeds across most of subarctic North America, from s. Nfld. to se. Alaska. Absent across Lab., n. Que., much of n. Ont., n. Man., ne. Sask., most of the Northwest Territories, and Alaska. Also absent as a breeder in e. Calif., s. Nev., portions of Ariz., and the Florida Peninsula. Winters south of the U.S. border, but some birds turn up in December and January in Pacific states. **HABITAT:** Found almost everywhere, but is attracted to open agricultural regions, towns, and highway overpasses. In some places, notably coastal islands, persists in placing its cup-shaped mud-based nests in caves. Everywhere else uses vertical human structures with some overhead protection to anchor and protect nests, including porches, garages, docks, ships, bridges, and, of course, barns. In migration, most common over marshes (particularly coastal marshes). Frequently feeds over lakes and ponds; roosts in tall tightly packed marsh vegetation (such as cattails or phragmites). **COHABITANTS:** Other swallows, cows, sheep, farmers cutting hay. . . . Occasionally nests in association with Cliff Swallow. In migration, often associates with other swallows. **MOVEMENT/MIGRATION:** Owing to the bird's great distribution and general hardiness, migration is extremely protracted. In the southern United States, breeding birds may arrive beginning in late January and early February and reach northern points in April and May. Fall migration begins in July and ends in November (although December sightings, particularly in coastal areas, are possible). VI: 2.

DESCRIPTION: Seems almost unnecessary. Almost everyone is familiar with this elegant, medium-sized, tine-tailed charmer of a bird. Barn Swallow is exceedingly long and slender, with narrow pointed wings and a deeply forked tail that is very long in adult males, fairly long in adult females, and distinctly shorter and blunter in immatures. Except when the bird is soaring or banking, the tail is commonly closed, and at greater distances, the wire-thin tines disappear.

Except for male Purple Martin, Barn Swallow is the overall darkest of the swallows. In all plumages, it is uniformly blue-black above with iridescent highlights. The forehead and throat are red-orange (just dark at a distance); underparts vary from orange to buff to nearly white (young birds). Birds with orange underparts look particularly all dark at a distance; paler birds are distinguished by the uniquely slender shape and manifest contrast between the dark upper- and pale underparts, which in most birds still project a warmer or buffier tone than swallow species with white underparts. Young birds are duskier above, and less iridescent than adults.

At close range, when the tail is splayed, two narrow white windows bracket the tip of the tail, but this charming effect is hardly needed for identification.

BEHAVIOR: Social and communal, birds may nest singly but often form colonies involving dozens of birds (or more). These lively noisy birds spend a great deal of time on the wing and often begin vocalizing at first light, or even before. They tend to forage low, often just above the ground or vegetation, and to concentrate where livestock are grazing or hay is being mown (snapping up flushed insects). Barn Swallow also feeds at extremely high altitudes, gliding there more than it does close to the ground. Easily identified by its slim profile.

After the nesting season, Barn Swallows gather in large flocks with other swallows, roosting in tall marsh grass or hay fields. Unlike most swallows, Barn Swallow is very comfortable on the ground. Groups may gather near nest sites and during migration on beaches, parking lots, lakeshores, and other flat open places. They have also been known to capture insects on the ground.

In migration, Barn Swallows fly in mixed flocks with other swallows or in small homogeneous groups generally numbering less than a dozen.

FLIGHT: The most slender and elegant of swallows, Barn Swallow shows exceedingly long wings and a long narrow (sometimes forked, mostly closed) tail. Barn Swallow movements appear more fluid, less effortful, and more polished than other swallows—a professional skater among amateurs. Wingbeats are most commonly stiff, fairly slow or measured, and shallow. When the bird is accelerating, wingbeats become deep and pushing. Nonhunting flight is generally direct and often conducted within 30 ft. of the ground. Hunting birds are nimble, acrobatic, and capable of altering, even reversing, their course almost faster than the human eye can perceive. When hunting just over the ground, Barn Swallows dart left and right, twisting in flight, favoring one side and then the other, and varying their speed considerably—flying slowly, then racing forward, often to snap up prey. At high altitudes, they are masters of the fast arching glide, with wings angled sharply back. Slower banking turns are conducted with wings perpendicular to the body. Unless braking or banking, tails are closed (but the two tines are usually visible).

VOCALIZATIONS: On territory, very vocal. Song is an ensemble of musical mutterings, soft churrings, high-pitched squeaks, chitterings, and the occasional staccato rattle or buzz. It sounds as though the bird is nattering to itself. Alarm call is a sharp scolding "*Chee-Chee*" or "*Pt-Chee! Pt-Chee!*" (usually doubled).

CHICKADEES, TITMICE, VERDIN, AND BUSHTIT

Carolina Chickadee, *Poecile carolinensis*

Lesser Chickadee

STATUS: Common and widespread year-round resident. **DISTRIBUTION:** Across most of the southeastern quarter of the U.S. (except at higher altitudes), the only chickadee species found. Breeds south of a line drawn from cen. N.J. to se. Pa., across cen. Ohio, cen. Ind., and cen. Ill., across s. Kans. to n. Tex., then south and east, encompassing most of the eastern half of Tex. Absent in extreme s. Tex., the southern half of Fla., and w. Md. and at higher elevations of the Appalachians south to N.C. and Tenn. (where Black-capped Chickadee replaces it). **HABITAT:** Through much of its range, found in an assortment of deciduous, mixed, and coniferous forests and woodlands. Also found in suburban habitats; attracted to feeding stations. **COHABITANTS:** Blue Jay, Brown-headed Nuthatch, Tufted Titmouse, Blue-gray Gnatcatcher. **MOVEMENT/MIGRATION:** Nonmigratory; almost never wanders outside of its range. The bird breeds within sight of New York City (in New Jersey), but has never been recorded there. VI: 0.

DESCRIPTION: The black-capped chickadee of the Southeast, Carolina Chickadee is unlikely to be confused with its more northern congener because *its range and the range of the more northern Black-capped Chickadee abut but do not overlap* (except slightly at the borders). Even in the Appalachians, the two species are segregated by altitude—Black-capped occurs high, and Carolina is found at elevations below 3,500 ft. (the apparent break point). No other chickadee species is found in Carolina's range.

Carolina is smaller than Black-capped by almost half an inch, and not as portly and big-headed. Plumage differences are subtle (see "Black-capped Chickadee"), but in sum, Carolina is neater, trimmer, crisper, less contrasting, plainer, and perhaps slightly warmer-toned, overall more muted, and less gray than Black-capped. Black-capped looks like a scruffy ruffian of a chickadee; Carolina looks like a nice, well-groomed, well-bred chickadee.

BEHAVIOR: Little distinction between Black-capped and Carolina. Both forage mostly in low and middle heights. Both are agile, acrobatic, and adept at finding prey at the very tips of twigs and weeds, and both often hang upside down in their search efforts. Both gather in small flocks (5–10 individuals) in fall and winter. Both associate with other wintering species (titmice, nuthatches, kinglets, creepers). Both are attracted to bird feeders and favor sunflower seed and suet. Both hop between branches when foraging, with lavish use of wings to navigate even short distances.

Carolinas are slightly quicker than Black-capped and tend not to alight on the ground—not even for a moment, and not even to pick up a fallen seed.

FLIGHT: Flight is quick, fairly weak, short (generally no farther than the nearest cover), and evenly bouncy and jaunty.

VOCALIZATIONS: Song is the surest way to separate Black-capped and Carolina. Carolina's song is a two-note downscale whistle repeated quickly and twice: "*peeter, peeter.*" Black-capped says the same thing, but more slowly, more wistfully, and only once: "*Peee-ter.*" Carolina's classic chickadee call is quicker, higher-pitched, more strident, and with more notes (five to seven) than Black-capped: "*chickadededededede.*" Very vocal when foraging; its conversational high-pitched mutterings may be single-noted, involve multiple notes, or incorporate different sounds. For example, the ethereal high-pitched "*see . . . see*" is often followed by a rougher, more nasal "*ch'j'ur,*" which is almost a mock "*peeter.*" A stuttering call is used when pish-piqued.

Black-capped Chickadee, *Poecile atricapillus*

Greater Chickadee

STATUS: Common widespread permanent resident throughout its extensive range—the common black-capped chickadee across northern North America. **DISTRIBUTION:** Spans the continent from Nfld. and the Maritimes west across Canada south of the tree line to the s. Yukon and across s. and cen. Alaska. The southern border extends from n. N.J. across cen. Pa. and cen. Ohio; also extends south along the Appalachians to N.C., Tenn., cen. Ind., cen. Ill., n. Mo., cen. Kans., n. N.M., ne. Utah, n. Nev., s. Ore., and extreme nw. Calif. **HABITAT:** Found in deciduous and mixed forests and woodlands (avoids pure conifers); also occurs in riparian woodlands and willow thickets. Particularly in winter, frequents parks, suburban yards, disturbed brushy areas, and weedy fields. Seems most attracted to woodland edges, but also occurs in forest interiors. A perennial at feeding stations, where it favors sunflower seed. **COHABITANTS:** Black-cappeds are the core of foraging winter flocks, whose ranks also include woodpeckers, nuthatches, kinglets, creepers, and vireos, but if you see one in the company of a Brown-headed Nuthatch, be assured it is not a Black-capped Chickadee. **MOVEMENT/MIGRATION:** Nonmigratory, but irregularly irruptive; primarily northern chickadees move south in small groups to the established southern limits of the range (only occasionally intruding into the northern range of Carolina Chickadee). VI: 1.

DESCRIPTION: An obvious chickadee—a gray-backed pale-bellied parid with a black cap and triangular black bib. Slightly larger, larger-headed, and more portly and robust than near-look-alike Carolina Chickadee. Although plumage characteristics overlap, Black-capped is grayer-backed and has a brighter and more extensive white cheek patch, a more ragged-edged bib and cap, a whitish shoulder (wing coverts), more extensive white streaking in the folded wings, and a richer, more buffy wash localized more on the sides and flanks (does not cover the entire belly). Black-capped is a larger, more contrasting, more disheveled-looking chickadee than Carolina. Both chickadee species respond readily to pishing.

BEHAVIOR: Like Carolina, but less shy about landing on the ground.

FLIGHT: Also like Carolina (and other chickadees)—bouncy, with an abrupt upward bounce and lots of up-and-down movement in the tail.

VOCALIZATIONS: Song is a mellow whistled "*Peter*"—slower and more plaintive than Carolina, and not repeated. Black-capped's chickadee call is slower and lower-pitched than Carolina's. The "*Chicka*" part is sharp and strident, the "*Dee, dee, dee*" more raspy and slurred. When excited, calls become rapid and more strident.

PERTINENT PARTICULARS: Over most of their ranges, confusion between Black-capped and Carolina is unlikely. A glance at the range map suffices to distinguish them. In the narrow geographic lane where the two species meet, interbreeding does occur, plumage differences seem less distinct, and both species repeat the other's song. Some hybridization has been documented. Many individual birds are simply unassignable to one species or the other.

Generally very responsive to pishing, but in areas north of the range of screech-owls, may not respond to screech-owl calls. Try Northern Saw-whet (or Barred).

Mountain Chickadee, *Poecile gambeli*

The Nefarious Tit

STATUS: Common year-round resident of montane western forests. **DISTRIBUTION:** From extreme s. Yukon south through the Rocky Mt. regions of B.C. and w. Alta., along the Cascades in Wash. and Ore. and the Sierras of Calif. (also at higher elevations in s. Calif. and n. Baja), through the Rockies in Idaho, Mont., Wyo., Utah, w. Colo., n. and e. Ariz., and all but extreme e. and s. N.M. Except in n. Calif., not found in coastal forests and generally absent in the n. Great Basin. **HABITAT:** Breeds in montane pine forests, particularly forests dominated by pine, spruce fir, and pinyon-juniper. In winter, some individuals move to lower elevations, where they may be found in other habitat types, most notably riparian cottonwoods and plantings in residential areas. **COHABITANTS:** Black-capped Chickadee, Chestnut-backed Chickadee, Pygmy Nuthatch, Red-breasted Nuthatch. **MOVEMENT/MIGRATION:** Most birds and most populations are nonmigratory, but some shifting is noted within the bird's established breeding range, and immature birds often seek lower elevations in winter. Some years may witness more substantial movements that bring birds to desert areas and the western Great Plains. What detectable movement there is occurs in March and early April, and again in September and October. VI: 2.

DESCRIPTION: A small songbird and a medium-sized chickadee with a dark slash through its eye, resulting in a demonic expression. About the same size and shape as Black-capped Chickadee (ranges overlap slightly from Alberta to northern New Mexico), but overall grayer and scruffier, with a dingier, less contrasting face, grayer underparts, and little or (more commonly) no buffy wash along the flanks. Mountain Chickadee's most distinguishing characteristic is the white eyebrow that bisects the black cap, giving the bird the semblance of a black slash or mask running through the eye and an expression that seems downright nefarious. Immature birds are like adults, but the white eyebrow is limited and broken.

Bridled Titmouse has a facial pattern that recalls Mountain Chickadee, but is easily distinguished by its peaked crest (as well as by vocalization).

BEHAVIOR: Classic chickadee. A nimble acrobatic bundle of energy that feeds by picking and gleaning from foliage. In most portions of its range, spends most of its time foraging on the outer branches in the middle or upper canopy of conifers, but in other places (especially in pinyon-pines) spends much of its time at or near the ground. Where its range overlaps with Black-capped, Mountain Chickadee forages and nests in conifers, leaving the broad-leaf vegetation to Black-capped. Where range overlaps with Chestnut-backed Chickadee, Mountain concentrates its activity in pines, while Chestnut-backed favors Douglas fir.

A tame bird that is easily piqued by pishing or pygmy-owl imitations.

FLIGHT: Like other chickadees. Weak, quick, bouncy, somewhat wandering, with lots of up-and-down movement of the tail.

VOCALIZATIONS: Song is variable and similar to Black-capped, but breathier, raspier, and usually distinctly two-noted at the end: "*Fee-bay'bay.*" Whistled calls are hoarser, drier, lower-pitched, and slower (particularly drags on the last two notes) than Black-capped's "*Chicka day day.*" Also emits a more plaintive variation of the song "*p'bee bee bee*" that in pattern and quality recalls the last several notes of White-throated Sparrow's "*Sam Peabody*" song. Also makes a rapid excited "*chickachicka*" and high-pitched muttered "t"-sounding notes.

Mexican Chickadee, *Poecile sclateri*

Bearded Chickadee

STATUS: Not uncommon but highly restricted resident species. **DISTRIBUTION:** A Mexican species. In the U.S., found only in the Chiricahua Mts. of se. Ariz. and in the higher peaks of the Animas Mts. in sw. N.M. **HABITAT:** Found year-round at higher elevations in conifer forests that are ponderosa pine at lower elevations, spruce–Douglas fir at higher elevations. In winter, some birds descend to elevations that support pine-oak communities, sycamore, and alligator juniper. **COHABITANTS:** Steller's Jay, Red-breasted and Pygmy Nuthatches, Olive Warbler, Grace's Warbler. **MOVEMENT/MIGRATION:** Permanent resident; no evidence of wandering. Some seasonal altitudinal shifts to lower elevations in winter. VI: 0.

DESCRIPTION: Identification by geographic default. Mexican Chickadee is the only chickadee species found in the Chiricahua and Animas Mountains. If you are there, stop here. The only other "tit" in the area is the crested Bridled Titmouse, which may overlap with Mexican Chickadee in winter at lower elevations.

A small bird (the same size and proportions as Mountain Chickadee), Mexican Chickadee is mostly gray—the darkest and grayest of the chickadees—but has all the requisite trimmings: black cap, white cheeks, black bib. The cap is smaller, the bib is larger (makes the bird look bearded), and the white cheeks really stand out—the only all-white spot on an otherwise dark bird. Underparts are mostly gray with a pale line down the belly (giving underparts an open-vested appearance). The black eye peeks below the cap. On other chickadees (though not Mountain), the eye is flush with the lower border of the cap.

Mountain Chickadee, the chickadee species most likely to wander into Mexican Chickadee's limited U.S. range, is distinguished by the white eyebrow and usually faintly buff washed sides—but the underparts of even the grayest-flanked Mountain Chickadee will not appear darkly vested.

BEHAVIOR: Like other chickadees. Outside breeding season, forages in small flocks, often forming the nucleus of a mixed-species flock. Forages mostly in pines, at all altitudes, on the tree's periphery as well as the interior, targeting needles and twigs. Active, nimble, and animate, Mexican Chickadee hangs upside down, hovers, flycatches, and opens seeds with hammering blows of its bill.

Tame. Responds quickly and tenaciously to pishing, particularly imitations of Northern Pygmy-Owl. Seems less inclined than some other parids to leave cover and fly across open areas.

FLIGHT: Fairly slow, light, buoyant, and bouncy, consisting of a rapid, terse, sputtery series of wingbeats and a terse closed-wing glide.

VOCALIZATIONS: Knows the entire chickadee repertoire, which it renders with a hoarse, thin, breathy or wheezy, and generally rapid delivery. "Chickadee" is rendered "*sheeka deez.*" The sound has some of the hoarseness of Mountain Chickadee, but the delivery is quicker.

Chestnut-backed Chickadee, *Poecile rufescens*

Ruddy Chickadee

STATUS: Common resident species restricted largely to Pacific Coast locations. **DISTRIBUTION:** Found from the Kenai Peninsula in Alaska, south along the coast to the central coast of Calif. and Santa Barbara Co. (also along the Sierras to Mariposa Co.), inland to n. Idaho and extreme nw. Mont. **HABITAT:** Moist, dense, and largely coastal coniferous forests; also found in riparian habitats. Birds are particularly attracted to mature Douglas fir but will forage in black oak, western hemlock, and red cedar. In the southern part of its range, also very partial to willow and alder. Where ranges overlap with Black-capped and Mountain Chickadees, Chestnut-backed usually favors denser and more closed canopy and forages higher. **COHABITANTS:** Red-breasted Nuthatch, Brown Creeper, Golden-crowned Kinglet, Townsend's Warbler, Hermit Warbler, Wilson's Warbler. **MOVEMENT/MIGRATION:** Permanent resident, but does disperse to higher elevations after breeding. VI: 0.

DESCRIPTION: A small, dark, and distinctive chickadee (slightly smaller than Black-capped or Mexican Chickadee; slightly larger than Red-breasted Nuthatch); with a large round head, a plump body, and a shortish tail.

The brownish-black cap, black bib, and white cheek patch identify it as a chickadee. The rich chestnut back, rump, and flanks distinguish it from all other chickadees. (*Note:* Along the California coast, individuals have grayish-washed sides but retain chestnut backs.) The plumage of Boreal Chickadee (whose range overlaps with Chestnut-backed in Canada and Alaska) is most like Chestnut-backed, but Boreal is overall larger, more robust, and paler, has a brownish gray (not chestnut-colored) back, and is most commonly found in spruce forest, where they frequently forage low.

Overall, Chestnut-backed Chickadee is fairly dark and dingy with low contrast. The white cheeks are the bird's most prominent feature.

BEHAVIOR: Likes heights. Whether singly, in pairs, or in small troops, Chestnut-backed forages most commonly in the canopy of tall coastal conifers, focusing most of its attention on twigs and needles. Tends to be a thorough feeder, working a tree completely, starting with the lower branches, before moving on to the next one. In winter, Chestnut-backed forms the nucleus for winter flocks that may include nuthatches, kinglets, and creepers.

A noisy bird that maintains a lively and near-constant banter. Fairly nervous, Chestnut-backed hops with little pause between stops and often wing-flicks compulsively. Although tame, Chestnut-backed is usually either too high or too preoccupied to bother with you, so is only moderately responsive to pishing.

FLIGHT: As with other chickadees, flight is light, nimble, bouncy, straight, and usually short (just to the next tree). Profile also seems short—particularly short-tailed.

VOCALIZATIONS: Has no song but makes varied use of the "chickadee" call pattern, which, though similar to other chickadees, tends toward the more hurried, wheezy, higher end of the scale. The first part, "*Cheeka,*" is raspy and rapid; the second part, "*dee dee,*" is weak and wheezy. Sometimes the call begins with a descending whistled stutter, "*ch'ch'ch'ch'ch,*" followed by a thin wheezy "*dee dee.*" Has other variations, but the song quality is similar.

Boreal Chickadee, *Poecile hudsonicus*

Spruce Chickadee

STATUS: Fairly common resident of boreal forests. **DISTRIBUTION:** In appropriate habitat, found throughout most of subtundra Canada and Alaska; also occurs in Maine, n. Vt., n. N.H., the Adirondacks of N.Y., the Upper Peninsula of Mich., n. Wisc., n. Minn., extreme nw. Mont. and n. Idaho, and the mountains of n. Wash. Absent along the Pacific Coast, where Chestnut-backed Chickadee occurs. **HABITAT:** Boreal forests; partial, in most places, to spruce (or balsam fir) forest. Also occurs in mixed coniferous-deciduous forest, but rarely in pure deciduous forest. **COHABITANTS:** Spruce Grouse, Gray Jay, Black-capped Chickadee, Red-breasted Nuthatch, Pine Grosbeak. **MOVEMENT/MIGRATION:** Generally nonmigratory, but engages in occasional irruptive movements (at intervals of three to eight years) during which birds may join flocks of Black-capped Chickadees and move as far south as New Jersey, Pennsylvania, and states bordering Canada. VI: 2.

DESCRIPTION: A plump cryptically brownish chickadee showing very little white. A small songbird (only slightly larger than Black-capped Chickadee), with a large head that sometimes appears exaggeratedly domed or maned, a petite bill, a plump body, and a deeply notched tail. Overall slightly more robust or rotund than Black-capped.

Mostly grayish brown and cryptic, with tones (and even a pattern) recalling a hoodless Dark-eyed Junco. The extensive cap is distinctly brown, and the black bib small but well defined. The sides (and sometimes most of the underparts) are washed with a pinkish or ruddy buff that stops short of rust (like the flanks of Chestnut-backed Chickadee).

There is no white in the grayish wings. The breast is dingy. The pale cheek patch, almost the signature characteristic of a chickadee, is small and almost vestigial, reduced in some cases to little more than a mustache and often dingy or grayish (not white). Black-capped Chickadee and the more vibrantly colored Chestnut-backed Chickadee show bright white cheeks that contrast with dark caps and bibs.

BEHAVIOR: Fairly social; found for most of the year in small flocks (four to ten birds) and freely associates with Black-capped Chickadees, which usually outnumbers it. Also found with Chestnut-backed, Mountain, and Gray-headed Chickadees (to which it is clearly subordinate at feeding stations). Generally more timid and clandestine than Black-capped. Moves through woodlands by making short flights between trees, often worming its way into the protective interior of a tree rather than lingering on the edge. Like other chickadees, active and nimble, but seems to pause for longer periods and to be less inclined to hover. Not as responsive to pishing as other chickadees, but responds when other birds are aroused.

FLIGHT: Small with a blunt face, a long tail, and rounded wings. Overall dark and cryptically patternless, but shows warm tones (particularly below). Flight is like other chickadees—light and bouncy, with rapid fluttery wingbeats interrupted by terse skip/pauses.

VOCALIZATIONS: Song is slower, weaker, raspier, and more desultory than Black-capped's assertive "*chicka-dee dee.*" Boreal's song is rendered "*ch'ja d/jee d/jee,*" with the first notes sharp and the balance lazy, flat, and buzzy.

Gray-headed Chickadee, *Poecile cincta*

Arctic Willow Tit

STATUS: Rare arctic and subarctic resident. **DISTRIBUTION:** In North America, found primarily in a narrow geographic band bracketing the Arctic Circle and extending from about the eastern ends of Kotzebue and Norton Sounds east across the n. Yukon to the northwest corner of the Northwest Territories. Range may be more extensive but remains speculative. **HABITAT:** In North America, inhabits large tracts of mature willows along northern rivers; also reportedly inhabits stunted spruce at the tree line. **COHABITANTS:** Gray-cheeked Thrush, Common and Hoary Redpolls, Grizzly Bear. **MOVEMENT/MIGRATION:** Nonmigratory. Some birds (young?) may wander south, where they reportedly occupy aspen and spruce forests; also sometimes attracted to feeding stations in villages and other populated areas. VI: 0.

DESCRIPTION: A pale grayer Boreal-like Chickadee that breeds in mature willows, not the extensive spruce or mixed spruce-balsam forests favored by Boreal Chickadee. Large (slightly larger than Boreal and Black-capped), with a small bill, a big head, a fluffy and robust body, and a long tail.

Resembles Boreal Chickadee but overall is slightly grayer and paler, showing a much brighter and more extensively white face and cheeks and a scraggly, less crisply defined black bib.

BEHAVIOR: More important to identification than plumage is the fact that Gray-headed is the only brownish chickadee species living in a mature willow thicket in the arctic. Forages like other chickadees. Little regarding its social proclivities is known, but apparently the bird does not flock with other species in the winter (as is common with other chickadees), largely because few birds choose to remain in the arctic in the winter.

FLIGHT: Quick and nimble like other chickadees, with rapid wingbeats, but flight is straighter, more floating or lofting, and not as bouncy.

VOCALIZATIONS: Apparently has no signature song but rather pieces together a vocal ensemble of calls, including trills, gargles, and "*chick-a-dee*"–like renderings. Vocalizations are recognizably chickadee-like and include a low nasal "*yah, yah, yah*" ("*dee dee dee?*") whose quality sounds like a vireo annoyance call; also makes higher musical twitterings.

Bridled Titmouse, *Baeolophus wollweberi*

Crested Chickadee

STATUS AND DISTRIBUTION: Common but limited in its U.S. range; restricted to se. Ariz., extreme sw. N.M., and the mountains of w. Mexico. **HABITAT:** Most common in oak woodlands; also occurs in oak-pine woodlands (at higher elevations) and riparian woodlands (especially in winter); also found at feeders. **COHABITANTS:** Arizona Woodpecker, White-breasted Nuthatch, Painted Redstart. **MOVEMENT/MIGRATION:** Nonmigratory; permanent resident. VI: 0.

DESCRIPTION: A small, compact, somewhat plump, short-tailed, very chickadee-like titmouse. With a good look, the bird's distinctively "bridled" head pattern is unmistakable. The smallest titmouse (and one of the smallest parids—only Carolina Chickadee is smaller), Bridled Titmouse is most commonly confused with the similarly sized and somewhat similarly patterned Mountain Chickadee. Both birds have a broken head pattern, and both have an eye-catching black bib; moreover, the broken pattern of Bridled Titmouse's head often makes the crest difficult to see amid the branches and broken light of the tree canopy. The bird's expression is inconsequential: if you are close enough to gauge an expression, you are close enough to see the diagnostic head pattern and the bird's crest.

BEHAVIOR: An active, nimble, chickadee-like bird that clings to twigs and moves quickly and acrobatically in the canopy. During the nonbreeding season, frequently moves in groups, often accompanied by other winter species (such as Hutton's Vireo, Ruby-crowned Kinglet, Black-throated Gray Warbler, and White-breasted Nuthatch). Most foraging is done below the canopy and at midheight; forays to the ground are rare. Unlike most parids, responds poorly to pishing—in fact, flocks frequently cluster some distance away and turn silent, one of the few times this species is quiet, because otherwise . . .

. . . it's vocal to the point of being chateriferous!

FLIGHT: Flight is direct, light, buoyant, bouncy, and slightly wandering (much like other parids).

VOCALIZATIONS: Very chickadee-like in the variety of its utterings. Song is a rapid, rolling, compressed, musical stutter: "*peterpeterpeter.*" Also makes a rapid, high-pitched, chickadee-like stutter, "*ch'ch'c'h'h'h'h'h'h*"; a hurried, impatient, high-pitched "*ch'chee, chee, chee*" or "*ch, ch, ch, ch'rrrrr*" (with a raspy and stuttering ending); and a high-pitched "*tee*" or a lower-pitched "*t'chee.*"

Oak Titmouse, *Baeolophus inornatus*

Plain Titmouse with a Brownish Wash

STATUS: Common but geographically restricted woodland resident. **DISTRIBUTION:** Virtually endemic to Calif. Found from s. Ore to n. Baja. Absent in the wet northwestern forests of Calif., the Sierras, the San Joaquin Valley, and the arid desert regions of se. Calif. Just south of the Calif.–Ore. border the range of this species and the very similar Juniper Titmouse abut. **HABITAT:** Dry oak, oak-riparian, and pine-oak woodlands at lower and middle elevations. At times found in habitats devoid of oaks (junipers, for example). Common in suburban situations (where there are oak woodlands). **COHABITANTS:** Acorn Woodpecker, Nuttall's Woodpecker, Western Scrub-Jay, White-breasted Nuthatch, Bushtit. **MOVEMENT/MIGRATION:** Permanent resident. VI: 0.

DESCRIPTION: Except for extreme eastern portions of California (where Juniper Titmouse occurs), Oak Titmouse is the only small, active, plain gray, conspicuously crested bird to be encountered. Upperparts are slightly browner than underparts. The crest is sometimes raised into a sharp peak, sometimes lowered into a pointy sweptback ponytail or cowlick. Not easily confused with any other species except Juniper Titmouse, which inhabits dry juniper and pinyon-juniper woodlands and whose range, except in the vicinity of Lava Beds National Monument in extreme northeastern California, does not overlap. The expression is plain.

BEHAVIOR: A chickadee-quick arboreal specialist that forages on the branches, trunks, and foliage of trees (primarily oaks). Active much of the day, hopping among branches, flying between trees, exploring nooks and crannies for insect larvae. Frequently hangs upside down, and sometimes searches for food on the ground. Readily attracted to backyard bird-feeding stations.

Generally found in pairs; not very social. Tends to mind its own affairs. Highly vocal, and responds fairly well to pishing.

FLIGHT: Decidedly bouncy and generally short. Wingbeats are rapid and sputtery.

VOCALIZATIONS: Songs are varied, but the pattern is similar. Each song consists of a series of whistled (mostly) two-note utterances (a high note followed by a low note), which are repeated with a measured tempo three to seven times. For example: "*jeebur jeebur jeebur.*" Call is a somewhat chickadee-like "*see see see jrr*" or "*see see; jrr jrr*" (there are other variations as well), with the first notes chickadee-high and the trailing notes lower and raspy. Also makes a quick descending stutter: "*dedededededede.*"

Juniper Titmouse, *Baeolophus ridgwayi*

Plain Gray Titmouse

STATUS: Fairly common, fairly widespread resident of the intermountain West. **DISTRIBUTION:** Permanent resident from se. Ore. and se. Idaho across most of Nev. and adjacent portions of extreme e. Calif., Utah, extreme s. Idaho, sw. Wyoming, sw. Colo., most of n. and se. Ariz., and n. and w. N.M.; also found in the Guadalupe Mts. of Tex. **HABITAT:** Mixed pinyon and juniper woodlands and, in extreme northwestern portions of its range, juniper and sagebrush. Found at higher altitudes than the very closely related Oak Titmouse, which is associated with oak and oak-pine woodlands. **COHABITANTS:** Gray Flycatcher, Western Scrub-Jay, Pinyon Jay, Black-throated Gray Warbler. **MOVEMENT/MIGRATION:** Permanent resident. VI: 0.

DESCRIPTION: A small, plain, gray, crested "tit." Across most of its range, the only titmouse to be found (see "Pertinent Particulars").

Plumage is uniformly dull gray—and often somewhat darker than illustrations show—not washed with a brownish tinge like Oak Titmouse.

BEHAVIOR: Classic tit—active, nimble, moving through low shrubs and trees with hurried intensity, flying to the next tree with a fluttery bouncy flight. Hangs upside down, pries into bark, hammers to open seeds. Vocalizes occasionally, but is not as vociferous as many parids. Somewhat reticent—or perhaps simply easily missed in the uniform sameness of its habitat. Only moderately responsive to pishing. Often remains at a distance or partially hidden making restrained vocalizations.

FLIGHT: Flight is bouncy and undulating with rapid sputtery wingbeats given in a short rapid series followed by a closed-wing glide.

VOCALIZATIONS: Varied song is much like Oak Titmouse—a series of single notes or short (two- to three-syllable) phrases, repeated rapidly three to six times. For example: "*twee twee twee twee twee*" or "*seebol seebol seebol.*" Some songs seem warbled (like Northern Cardinal), some have a percussive quality (like a junco), and they all seem slightly faster on average than Oak Titmouse. Call is a somewhat chickadee-like "*chee chee chee ch'r,*" with the first notes chickadee-high and the last note descending and raspy.

PERTINENT PARTICULARS: Except for southeast Arizona, where the chickadee-faced Bridled Titmouse occurs, and the vicinity of Lava Beds National Monument on the California–Oregon border, where Oak Titmouse occurs, Juniper Titmouse is the only titmouse in its range.

Tufted Titmouse, *Baeolophus bicolor*

Understatement with a Crest

STATUS: Common and widespread resident of eastern woodlands. With the exception of the Black-crested Titmouse of Texas and eastern Mexico, the only titmouse found in the East. **DISTRIBUTION:** The e. U.S. and extreme se. Ont. The northern limit reaches s. Maine, N.H., Vt., n. N.Y., Mich. (not the Upper Peninsula), s. Wisc., and se. Minn. The western border cuts across cen. and s. Iowa, se. Neb., cen. Kans., w. Okla., and e. Texas. Absent in s. Fla. **HABITAT:** Deciduous forests, particularly forests with tall trees and a well-developed canopy; also found in mixed deciduous-coniferous forest, parks, suburbia, cemeteries, and orchards. **COHABITANTS:** Red-bellied Woodpecker, Carolina and

Black-capped Chickadees, White-breasted Nuthatch, Blue-gray Gnatcatcher. In Texas, abuts the range of Black-crested Titmouse west of a line drawn approximately from Fort Worth to Matagora Bay. **MOVEMENT/MIGRATION:** Permanent resident. Although nonmigratory, a few birds may wander slightly. VI: 1.

DESCRIPTION: An active, amiable, altogether enchanting bird distinguished by its pertly peaked head, black forehead, and large balefully black eyes. The expression is winsome. Larger than a chickadee. Gray above, whitish below, with a salmon wash along the flanks. It cannot be confused with any species except Black-crested Titmouse.

BEHAVIOR: Tame, somewhat single-minded in its foraging, and nonplused about human observers. During warmer months, most often forages higher in the canopy of deciduous trees. In winter, works lower, often in thickets, on the lower trunks of trees, on the forest understory, or in open areas close to trees. Searches nooks and crannies for insect larvae; hangs (in the fashion of tits) upside down and examines branches and clusters of leaves for edibles; opens seeds and acorns by stabilizing them with both feet and whacking them open with a hammering bill. Comes readily to feeders, where it prizes sunflower seed and shelled peanuts.

Generally found in pairs or alone. Sometimes forages in small titmice flocks or may join wandering mixed-species flocks in winter. Easily piqued by pishing, during which its scolding notes commonly draw other, more pish-resistant species.

FLIGHT: Flight is direct and fluttery. For short hops, generally undulates; longer flight across open areas is more direct.

VOCALIZATIONS: Very vocal. Sings all year, vocalizes frequently, and in fact is never quiet for long. Song is a clear, mellow, whistled "*peterpeterpeter.*" Calls, often very suggestive of chickadees, are varied and include a piercing ascending whistle with an abrupt down ending given somewhat conversationally, "*see-ch,*" and a raspy, nasal, scolding "*ch'yaah, yaah, yaah, yaah.*" When birds are piqued, the scolding call increases in volume, duration, and vehemence.

Black-crested Titmouse, *Baeolophus atricristatus*

Tit of Texas

STATUS AND DISTRIBUTION: Common resident of cen. Tex., with range extending north into the cen. Texas Panhandle (also sw. Okla.), west into the higher forested portions of trans-Pecos Tex., and south into e. Mexico. The eastern border is defined by a line running from Fort Worth to Matagora Bay. **HABITAT:** Assorted dense and open but commonly dry woodlands, including oak-juniper in the Edwards Plateau, coastal forest, and scrub and thornbush along the Rio Grande. **COHABITANTS:** Varied, despite its limited range; includes White-eyed Vireo, Green Jay, Western Scrub-Jay. **MOVEMENT/MIGRATION:** Permanent resident. VI: 0.

DESCRIPTION: A Tufted Titmouse with a rakish black crest and a white forehead—features that easily distinguish it from the slightly larger but otherwise similar eastern species and give this species a very expressive face. *Note:* Except where the ranges of Tufted and Black-crested abut, Black-crested is the only crested tit within its limited range.

BEHAVIOR AND FLIGHT: Like Tufted Titmouse.

VOCALIZATIONS: Very similar in pattern, quality, and repertoire to Tufted Titmouse (many are identical), but at times less phonetically precise. For example, whistled "*peterpeterpeter*" becomes "*he/ur he/ur he/ur.*" In south Texas, vocalizations may be somewhat slower, lower-pitched, more timid, and less assertive than Tufted Titmouse (or more northern Black-crested).

PERTINENT PARTICULARS: Where the ranges of Tufted and Black-crested Titmice meet, the species interbreed, and hybrids show head patterns that are usually midway between the parents (for example, a slightly darker crest).

Verdin, *Auriparus flaviceps*

The Bush Sprite

STATUS: Common desert species. **DISTRIBUTION:** Found from s. and cen. Tex. west across s. N.M. and s. and w. Ariz., north into the southeastern corner of Utah, s. Nev., and west into se. Calif. In Mexico, found in Baja and south to the middle of the country. **HABITAT:** Found in thorny desert scrub associated with desert washes and riparian habitat. Representative vegetation includes mesquite, acacia, palo verde, tamarisk, and chaparral. Also comes to feeders (particularly hummingbird feeders). Avoids open desert and dense forest. **COHABITANTS:** Cactus Wren, Black-tailed Gnatcatcher, Lucy's Warbler, Black-throated Sparrow. **MOVEMENT/MIGRATION:** Nonmigratory; permanent resident. VI: 0.

DESCRIPTION: Tiny, stumpy, furtive, active, and *plain.* Suggests a goldfinch and behaves like a chickadee. Overall small (smaller than a chickadee) and compact. The round overlarge head and a too short tail make the bird oddly and compactly balanced. Overall plain pale stealth gray, except for the yellow face of adults, which varies in brightness and intensity. (Immatures show no yellow at all.) The rusty shoulder patch is often difficult to see.

BEHAVIOR: Nimble, quick, active. Except in late summer, when family groups are still together, usually solitary or in pairs. Moves with abrupt chickadee-like antics through leafy and flowered branches and favors the exposed outer portions of the tree. Moving the body as much as the head, searches the branches by swinging side to side. Hops and flutters to the next perch, clinging chickadee-like and exploring for insects with a tiny wedge-shaped bill. Sometimes uses one foot for anchorage and the other to maneuver foliage for inspection. Usually forages in the middle heights, sometimes high, almost never on the ground.

Vocal. Calls frequently as it feeds. Sings from an exposed perch, often at the top of a tree or bush. Does not respond well to pishing.

FLIGHT: Usually of short duration, flight is buoyant, airy, with lots of wing flutter. Occasionally makes longer flights, passing over what may appear to be perfect opportunities for foraging, before landing at the end of a line of vegetation some distance away.

VOCALIZATIONS: Song is a high, clear, whistled "*pee-tee*" or "*pee-tee, tee.*" Also makes a single high-pitched "*chee*" or a compressed two-note "*chee'uh,*" which are repeated at irregular intervals. Calls are extremely varied and include a terse muttered "*chit*" and a nattering "*ch, ch, ch, ch, ch.*"

Bushtit, *Psaltriparus minimus*
Long-tailed Packtit

STATUS: Common and fairly widespread western resident. **DISTRIBUTION:** Found coastally from s. B.C. south to Baja and east into cen. and s. Ore., n. and e. Calif., extreme s. Idaho, Nev., Utah, w. and cen. Colo., n. and e. Ariz., w. and cen. N.M., w. Okla., and portions of Tex. (not e. Tex.), then south into Mexico. Very rare east to w. Kans. **HABITAT:** A generalist; occupies a variety of habitats but is most partial to open and mixed woodlands with an evergreen component (such as pine-oak or pinyon-juniper). Also found in mountain forests, drier habitats (coastal chaparral, for example), and riparian corridors. Avoids deserts but is common in vegetated suburban habitats adjacent to deserts. **COHABITANTS:** Year-round, Oak and Juniper Titmice, Verdin; in winter, kinglets. Often serves as the base for mixed-species flocks. **MOVEMENT/MIGRATION:** Permanent resident, but in places nomadic as birds at higher elevations move to lower elevations in winter. VI: 1.

DESCRIPTION: A tiny, pale, long-tailed, effervescent, beady-eyed wisp of a bird that would easily go unnoticed if it led a solitary life—which it does not. Tiny (kinglet-sized), with a hunched neckless head, a black nub of a bill, a plump body, and an extremely long tail. Looks like a long-tailed kinglet or a plump snub-billed gnatcatcher.

Overall pale and plain stealth gray with a muted wash of brown. Some individuals have a brownish cap or a blackish face. The expression is mean or frowning.

BEHAVIOR: A pack animal; except during breeding season, virtually never found alone and often found in groups of 6–40 birds. Almost never still. Flocks move like an animate cloud around and through trees and bushes. Individuals move from branch to twig in short hops or, more often, short flights (using their wings even when traveling a few inches). Jerks its head left and right then moves to the next perch. Despite the number of individuals involved and the obvious movement, foraging birds can be amazingly furtive. Movement is utilitarian, limited to perch changes, head search, and gleaning. Unlike gnatcatchers or many warbler species, does not side-swish its tail.

Very nimble; able to perch on the frailest twigs. Frequently dangles upside down (like a chickadee), sometimes with one foot. Forages at all heights, even in weedy patches beneath trees. Very tame, and almost indifferent to people. Moderately pish-piqued—less so than many other tits. When agitated, wing-flicks vigorously.

FLIGHT: Tiny, fairly pale, long-tailed, and short-winged. Flight is slow, sputtering, and slightly jerky or bouncy, with rapid steady wingbeats interrupted at regular intervals by an almost imperceptible skip. When moving short distances (between trees just a few yards apart), individuals go in staggered flights (first one, then the next). When traveling longer distances over open space, flocks move in a single animate cloud.

VOCALIZATIONS: Despite their numbers and frequent chatter, this species can be amazingly difficult to hear. Flock communications are soft, not meant to travel far, and so matter-of-fact that they just blend into the backdrop.

Flock sounds might loosely be described as soft, high-pitched, constant sputterings—"*chih chih chih chihs*" mixed in with "*tic tic tic tics*" and an occasional "*spik*" or buzz. Also makes an excited chickadee-like "*deedeedeedeedee.*"

PERTINENT PARTICULARS: If you are unfamiliar with the species, you may be concerned about distinguishing Bushtit from Verdin or Wrentit. Relax. Verdin is a compact plump little bird. Wrentit is a larger, longer-tailed, shy, skulking, ground-cover-hugging bird. You'll be lucky to get a look at it. Neither Verdin nor Wrentit travels in flocks.

NUTHATCHES

Red-breasted Nuthatch, *Sitta canadensis*

The Nuthatch That Says "Yank"

STATUS: Variously common to uncommon across its extensive breeding and wintering range. **DISTRIBUTION:** Breeds across forested Canada from Nfld. to the s. Yukon west into cen. Alaska. In the continental U.S., breeds in w. Wash., Ore., Calif., n. and s.-cen. Idaho (and adjacent portions of Wash. and Ore.), all but e. Mont., Wyo., Nev. (and the mountains in s. Calif.), Utah, all but e. Colo., n. and e. Ariz., and n. and w. N.M. In the East, found in n. Minn., n. Wisc., n. Mich., and New England south to n. N.J. and along the Appalachians to Tenn. In winter, may be found anywhere but extreme s. Tex. and the Florida Peninsula. **HABITAT:** Conifers! In migration and in winter, may be found in deciduous trees, but at all times much prefers conifers (spruce, fir, pine, hemlock); particularly in breeding season, is most partial to mature pines. Also comes readily to backyard feeding stations, where it favors sunflower seed. **COHABITANTS:** Varies greatly across such a vast range. During breeding, rarely found out of earshot of Yellow-rumped Warbler; also associates with Golden-crowned Kinglet and Red Crossbill. In winter, may be found with chickadees, titmice, kinglets. **MOVEMENT/MIGRATION:** Migratory and irruptive. Northern breeders appear to migrate every year; southern populations are largely resident. Southern movements vary in intensity and degree, with few birds one year and many more the following year (or perhaps the next). Spring migration from mid-March to early June; fall from late July to mid-November. VI: 2.

DESCRIPTION: A nuthatch of few words—in fact, for the most part, just one. A medium-small (1 in. smaller than White-breasted Nuthatch; slightly larger than Pygmy or Brown-headed), chunky, neckless tree-creeper of a bird. More compact than White-breasted, Red-breasted is a "toy" version with a shorter bill and tail.

Overall darker than White-breasted, with a slightly darker blue-gray back and ruddy underparts that range from rusty red to a barely discernible pale wash of color. Most distinguishing and reliable characteristic is the wide black eye stripe, which gives the bird its cross, mean expression.

BEHAVIOR: Active and animate feeder; more high-strung than White-breasted Nuthatch. Forages up and down trunks; frequently straddles a branch and hangs upside down to peer beneath it. Almost always found in conifers. Even in migration, seeks out evergreens, but may resort to hardwoods where evergreens are scarce. Forages alone, in pairs, or in small groups, and also joins winter flocks.

Tame, and almost indifferent to humans. Responds fairly quickly to pishing—particularly imitations of almost any small owl—but the response may initially be only a softly rendered call. Sometimes it takes a bit of coaxing to get a Red-breasted Nuthatch to leave its conifer fortress.

FLIGHT: Overall small and compact (virtually all wing), with short, broad, rounded wings, a pointy bill, and a short triangular tail. Warm (ruddy) underparts are often more distinctive in flight. Acutely bouncy, undulating flight is produced by a rapid sputtery series of wingbeats punctuated by a momentary frozen open-winged glide. In migration, flies alone or in small very loose groups.

VOCALIZATIONS: Song is a soft, nasal, monotonous "*yank, yank, yank*" that is more nasal and less abrupt than White-breasted Nuthatch.

White-breasted Nuthatch, *Sitta carolinensis*

Hardwood Nuthatch

STATUS: Common and widespread resident of primarily mature deciduous woodlands. **DISTRIBUTION:** Found across most of s. Canada and throughout much of the U.S. into Mexico (absent along the Florida Peninsula, much of the Gulf Coast, the southwestern coastal plain, the short-grass prairies, much of the Great Basin, and the Sonoran and Mojave Deserts). **HABITAT:** Primarily mature deciduous woodlands, but also occurs in mixed oak-pine woodland (particularly in the West) and suburban areas rich in mature trees; particularly favors woodlands with oaks. Nests in cavities situated near forest edges or openings (such as along streams). Easily attracted to backyard feeding stations, where it favors sunflower seed, peanuts, and suet. **COHABITANTS:** Chickadees, titmice. **MOVEMENT/MIGRATION:** Mostly nonmigratory; a permanent resident. Some fall and winter movement occurs throughout the bird's range, varying in scope from year to year. VI: 1.

DESCRIPTION: A chunky, neckless, child's fist-sized wind-up toy of a bird whose ability to hop headfirst down tree trunks distinguishes it as a nuthatch. The stark white face and breast contrast with the blackish cap and gunmetal blue back and also make the black eye a standout feature. (On other nuthatches, the eye is indistinct, even invisible, at any distance.) With its slightly upturned bill, the bird appears to be smiling. Does not use its short stubby tail to prop itself against the tree, as woodpeckers do.

BEHAVIOR: Perpetual motion in any direction. Moves in hitching headfirst jerks up and down tree trunks, and on top and beneath heavy limbs. Movements are quick but measured, and slightly slower than other nuthatches. Does not commonly land or forage on twigs. Pries into nooks and crannies and flakes off bark in search of insects. Wedges acorns, nuts, and seeds into crevices and hammers them open with its bill. Forages at any height, and sometimes searches for nuts on the ground.

Does not flock; found singly or in pairs. In winter, however, often joins winter flocks of chickadees and titmice as they move through its territory, though the nuthatch stops at the border. A vocal bird that often calls as it feeds. Responds very readily to pishing and owl calls.

FLIGHT: Flight is direct and jerky or bouncy—but less bouncy and more smoothly undulating than Red-breasted Nuthatch. A short series of flaps is followed by a descending closed-wing glide. At close range, wingbeats are audible.

VOCALIZATIONS: Song is a rapid yammering—one nasal note repeated in a series: "*whawhawhawha-whawhawha.*" Suggests a scolding admonishment. Foraging birds utter a series of soft muttered "*yink*" calls in an intermittent pattern that recalls Morse code. Classic call is a loud, nasal, and somewhat muffled "*yenk . . . yenk . . . yenk*" that is slower, more pleasing, and less nasal than Red-breasted. Interior western birds' call is similar but is held longer and is more trilled or burry.

Pygmy Nuthatch, *Sitta pygmaea*

Western Pack Nuthatch

STATUS: Common widespread western resident of pine forests. **DISTRIBUTION:** Distributed widely across much of the West between the front range of the Rockies and the Pacific Coast. The northern limits lie in s. B.C. and w. Mont. The eastern boundary falls across w. S.D., w. Neb., cen. Colo., cen. N.M., and trans-Pecos Tex. south into Mexico. **HABITAT:** Primarily open longleaf pine forests at higher elevations, including ponderosa and Jeffrey pines. Found in mature unlogged forests because of its dependence on dead trees or live trees with dead snags for roosting and breeding cavities. **COHABITANTS:** Steller's Jay, Mountain Chickadee, Red-breasted Nuthatch, Yellow-rumped Warbler, Red Crossbill. **MOVEMENT/MIGRATION:** Permanent resident; some irregular down-slope dispersal in winter; also wanders into the western Great Plains. VI: 2.

DESCRIPTION: A diminutive, pale, chatteriferous, hyperactive nuthatch that travels in packs. A tiny (smaller than Red-breasted Nuthatch), compact, tailless nub of a bird. The straight, thin, disproportionately long bill makes the bird seem all head.

Pale blue-gray above, with a grayish tan cap. Underparts are dirty white, often with a slight buffy wash. A dark shadowy eye-line separates the cap and white throat. The pale "bald spot" on the nape is often small and inconspicuous. Overall paler and less contrastingly plumaged than Red-breasted or White-breasted Nuthatch, whose ranges it overlaps.

BEHAVIOR: A nimble, active, high-strung nuthatch (makes Red-breasted Nuthatch look sluggish). Moves up, down, and along limbs, outer branches, and trunks with rapid jerky hops, pivoting frequently. Never uses its tail for a prop, and never still for long. Frequently hunts with a horizontal search pattern, moving across the tree and from limb to trunk to limb. Likes to hunt high and to hunt outer branches. Pries into bark, probes pine needles, and explores pinecones for insects. Hammers seeds open and flies in pursuit of flushed insects; also hovers.

Uncommonly social; almost never found alone. During the breeding season paired adults and several adult nest-helpers attend nest and young. Outside breeding season, family groups join in super-flocks or team up with other wintering species. Roosts communally in tree cavities, with reportedly over 100 nuthatches using the same cavity.

Uncommonly vocal. Members of the flock or family group communicate constantly with piping calls. Bold. Responds well to pishing and owl calls.

FLIGHT: Overall tiny and compact. With its stub of a body and short round wings, Pygmy looks like a head on wings. Flight is quick, jerky, and usually brief, with rapid sputtery wingbeats and deep bouncy undulations.

VOCALIZATIONS: Flocks surround themselves in a peeping fog of sound. Piped notes have a high, soft, clear, twittery, conversational quality: "*pih pih pih*," for example, or "*peeheehee*." Also utters a short two-note expletive, "*ta-tee, ta-tee*," repeated endlessly, and a high-pitched peeping twitter: "*ch'ch'cheep ch'ch'ch'ch cheep*." When pecking at seeds, the taps sound like a light-tapping woodpecker.

PERTINENT PARTICULARS: Size, overall paleness, and strength of numbers easily distinguish Pygmy from Red-breasted and White-breasted Nuthatches. Its range never overlaps with the similar and more eastern Brown-headed Nuthatch. Were they to occur together, Brown-headed's cap is unmistakably brown, whereas Pygmy's cap is brown-tinged but mostly gray, with little contrast in tone and color from the back. Also, the black eye-line is much more prominent on Pygmy, the bald spot is less conspicuous, and the vocalizations are piping and pleasing, not sharp, squeaky, and nattering.

Brown-headed Nuthatch, *Sitta pusilla*

Eastern Pack Nuthatch

STATUS: Uncommon to common permanent resident of southeastern pine forests. **DISTRIBUTION:** S. Del. and e. Va. to all but extreme s. Fla., west through all but w. N.C., w. S.C., n. Ga., n. Ala., and n. Miss.; also found in s. Ark., se. Okla., and e. Tex.

HABITAT: Mature southern coastal plain pine forests, particularly loblolly-shortleaf (in northern range) and longleaf-slash (in more southern portions of the range). **COHABITANTS:** Red-cockaded Woodpecker, Great Crested Flycatcher, Pine Warbler, Bachman's Sparrow. **MOVEMENT/MIGRATION:** Nonmigratory; permanent resident. VI: 0.

DESCRIPTION: A tiny, pale, highly social, brown-capped nuthatch. Decidedly smaller and overall dingier than White-breasted Nuthatch. Slightly smaller, paler, and less crisply and conspicuously patterned than Red-breasted. Overall compact—even by nuthatch standards. The head is disproportionately large, the bill disproportionately long, and the tail comically snubbed. Often the entire bird seems like little more than a long-faced head with wings.

Upperparts are pale brownish gray, and the crown pale brown. (Many illustrations show it as too dark.) Underparts are dingy—not white, not red. Overall subdued, although the brown cap and whitish nape are easily noted at close range; fortunately, the antics of nuthatches present opportunities to study the dorsal side of this treetop bird. The expression is sly and shrewd.

BEHAVIOR: Hyperactive; always moving in quick jerky hops up, down, and over the sides of trunks and branches and flying to the next branch or tree. Usually forages high in the canopy section of pines—sometimes on the trunk, sometimes on the limbs, sometimes on the outer branches, where it pries into cones and pokes into pine needle clusters. Occasionally flies to snap up flushed insects. Moves quickly as it forages, changing limbs frequently.

Highly social. Forages year-round in packs. In winter, these packs may associate with other wintering species, most notably and particularly Carolina Chickadee and Pine Warbler; birds are not necessarily in the same tree, but are in close proximity. Vocal!—members of the flock communicate constantly—but inconspicuous: birds typically forage high, and vocalizations are softly uttered and very hard to hear. Responsive to pishing and owl calls. Responding birds will cluster, sometimes clumping side by side on a branch.

FLIGHT: Flight is slow, struggling, somewhat like a bouncy bumblebee. Seems all head with short blunt wings. Wingbeats are given in a quick sputtery series, followed by an extra-terse closed-wing glide. Flight is bouncy and, given any appreciable distance, slightly erratic or wandering. Short flights between trees are the norm. Flights in excess of 30 yds. are the exception, but birds may fly several hundred yards across open habitat to reach a distant grove of trees.

VOCALIZATIONS: Foraging birds wrap themselves in an audible fog of soft, squeaky, chirpy, buzzy, wheezy natterings: "*chew, chew, chew; chi, chi, chi,*" or "*jee, jee, jee, jee d'jew, d'jew, d'jew,*" and a variety of single-note utterances: "*tip . . . dip . . . pic.*" The most often described call is the two-note, high-pitched, squeaky, wheezy, rubber ducky call: "*chee-d'h*" or "*wee-ja.*" Also makes a murmured "*peep, peep, peep.*"

CREEPERS AND WRENS

Brown Creeper, *Certhia americana*

Legless Perpetual-motion Bark Wren

STATUS: Common widespread but uncommonly cryptic breeder and winter resident of mature forests and woodlands. **DISTRIBUTION:** Distribution in North America is complex and somewhat disjunct. Breeds across most of s. Canada from s. Nfld. and the Maritimes to B.C., with the northern limit cutting across cen. Ont., Man., Sask., Alta., and nw. B.C. and the southeast coast of Alaska (except absent as a breeder in the southern prairies). In the e. U.S., breeds throughout New England, south to Md. and n. Va., west across Pa., e. W. Va., ne. Ohio, Mich., Wisc., ne. Ill., and n. Minn.; also found locally well south to Ala. and Miss. and west to the Dakotas. In the West, found coastally in Wash., Ore., and n. Calif., then south, in the mountains, to n. Baja. In the western mountains, breeds in Idaho, w. Mont., ne. Ore., w. and cen. Wyo., ne. Nev., Utah, w. and cen. Colo., e. and cen. Ariz., and n. and w. N.M., south into Mexico. In winter, vacates some of the northernmost reaches of its Canadian breeding range and moves south across s. Canada and virtually all of the U.S. except the Florida Peninsula and the Sonoran/Mojave region of the southwestern U.S. **HABITAT:** Primarily a breeder in mature northern coniferous forests, particularly wet or swampy woods; also occurs in mixed and deciduous habitat. More eclectic in its choice of wintering and migratory habitats; occupies all manner of woodlands, including riparian woodlands, early successional woodlands, city parks, and suburban areas well dotted with trees. **COHABITANTS:** In winter, woodpeckers, chickadees, titmice, nuthatches, Golden-crowned Kinglet. **MOVEMENT/MIGRATION:** Permanent resident across much of its range, but winter range is much more extensive. Spring migration from late March to early May; fall from mid-September to November. VI: 1.

DESCRIPTION: North America's only creeper. A tiny, slender, barklike shard of a bird that resembles an emaciated wren and hitches itself up tree trunks like a hyperactive jerky woodpecker. The whitish-flecked-and-brown or grayish brown upperparts replicate sunlight-dappled bark. The whitish underparts are usually plastered against the trunk of a tree. The bill is nut-pick-thin and curved, and the stiff tail (used to prop the bird against the trunk of trees) is slender, pointy, and scraggly-tipped. Legs? Seems not to have them. Does have feet, however, which it uses. . . .

BEHAVIOR: . . . to hitch itself up the trunks of trees like a wind-up woodpecker. Starts near the base of the tree, ascends in a jerky spiraling climb, and flies to the base of a nearby tree when branches begin to impede its climb. Probes into nooks and crannies with its bill. Most often targets mature trees with rough grooved bark, but also forages on heavy limbs and occasionally slender trees and branches (particularly dead ones with flaky bark). Furtive and generally quiet. Difficult to detect even when moving (which is almost always the case!). Freezes when disturbed, becoming part of the tree. Ascends headfirst (never moves head down the trunk), but can back down a trunk tail first.

Usually alone, although a mate may not be far away. In migration, occasionally found with small loose flocks that rarely number more than six. In winter, often associates with foraging flocks of titmice and chickadees. Fairly tame—in fact, seems almost oblivious to humans. Responds to screech-owl and Northern Pygmy-Owl imitations.

FLIGHT: Usually of short duration (only to the next large tree). Flight is quick and direct, with sputtery wingbeats and a jerky undulating carriage. Seems frail or feather-light. Also, when descending steeply (to the same trunk or a nearby tree), parachutes in a manner that replicates a falling leaf.

VOCALIZATIONS: Song is a simple whistled phrase that begins with one or two high, thin, held notes, then descends in a richer tumbling series: "*see-sooysooysoo.*" The song's pattern (not its tonal quality!) has given the bird the nickname "Meadowlark of the Woods." Call is a thin, extremely high-pitched trill that sounds like a shrill airy police whistle: "*sireee.*" Recalls Golden-crowned Kinglet but is single-noted and held longer. Vocalizations often prompt a response from another creeper (the mate?).

PERTINENT PARTICULARS: Never forages on the ground.

Cactus Wren, *Campylorhynchus brunneicapillus*

Thrasherlike Wren

STATUS: Common (and usually manifest) desert species. **DISTRIBUTION:** In the U.S., a permanent resident of the Chihuahuan, Mojave, and Sonoran Deserts. Found in s. and se. Calif. (also Baja), s. Nev., w. and s. Ariz., s. and cen. N.M., and w. and s. Tex., south into Mexico. **HABITAT:** Low-lying desert, including Joshua tree, cactus and mesquite flats (occasionally creosote-dominated flats), and cholla cactus communities; in some places, found in riparian habitat; not uncommon in suburban areas that border desert and are landscaped in a way that provides attractive habitat. In southern coastal California, local populations occur in coastal sage-scrub with prickly pear cactus. **COHABITANTS:** White-winged Dove, Black-throated Sparrow, Pyrrhuloxia, Scott's Oriole. **MOVEMENT/MIGRATION:** Permanent resident. VI: 0.

DESCRIPTION: A large thrasherlike wren (or a miniature thrasher) living in the desert. Big, starling-sized; the largest North American wren. Big-headed with a long, heavy, downturned bill; long-necked (for a wren); robust (starlinglike) body, and an uncommonly long tail that is never cocked above the body like in other wrens. In sum looks like a small, somewhat stocky thrasher.

Plumage is more patterned than a thrasher, but overall cryptic. Grayish brown upperparts are sprinkled with white flecks; creamy underparts are heavily black-spotted. The white eyebrow and clustered black spots forming a bib on the breast stand out. Barring on the tail does not stand out; the tail usually just looks blackish. The expression is dour.

BEHAVIOR: Not a skulking wren. Often stands atop vegetation and forages in the open by hopping, loping, walking, and running, primarily on the ground with head elevated. Stops at the edge of bushes, trees, or rocks to investigate possibilities, flipping over leaves and debris with its bill or peeking at the undersides of low-lying vegetation. Also gleans insects from shrubs by hopping among branches or clinging to the trunks of rough-barked trees. Sometimes leaps or lunges. Prey is grasped by the bill and consumed on the spot. Generally forages alone, but also goes in pairs. Presence is usually heralded by its very distinctive song or the presence of one (or more) round grassy nests in a cactus or bush.

Fairly tolerant. Responds poorly to pishing.

FLIGHT: Flight profile is also like a stocky thrasher, showing a projecting head, wide round-tipped wings held at a right angle to the body, and a long fanned tail (showing conspicuous white-tipped trim). Flight is low, slow, and straight. A rapid and prolonged flurry of fluttery wingbeats is followed by an equally long flat-winged glide that may be followed (if the bird's objective has not been gained) by shorter alternating series of wingbeats and glides.

VOCALIZATIONS: Song is a lengthy series of similar, harsh, low, croaking notes that gain in tempo. Has been aptly likened to the sound of an old car trying to start on a cold morning: "*rah, rah, rah, rahrahrahrahrahrahrah.*" Scold is similar to the song. Also makes a raspy nasal trill, "*wreeeh.*"

Rock Wren, *Salpinctes obsoletus*

Pale Scree-creeper

STATUS: Fairly common but habitat-specific western breeder and resident. **DISTRIBUTION:** Found across most of the mountainous West. Breeds from s.-cen. B.C. and sw. Alta. across s. Sask. south into Mexico. The eastern border falls across western portions of the Dakotas, Neb., Kans., w. Okla., and Tex. In the West, absent as a breeder in w. Wash. and Ore. and coastal Calif. south to San Francisco. Winters (and breeds) from n. Calif. across w. and s. Nev., extreme sw. Utah, ne. Ariz., n. N.M., and w. Okla. **HABITAT:** Dry and arid cliffs or rocky areas. Most commonly found in canyons, gullies, steep or mild slopes with fractured cliffs, rocky outcroppings, talus, and scree, and dry gravelly desert washes; also adopts human-modified habitats, including highway cuts, elevated roadbeds, and construction sites. **COHABITANTS:** White-throated Swift, Ash-throated Flycatcher, Gray Vireo, Canyon Wren, Black-throated Sparrow. **MOVEMENT/MIGRATION:** Northern birds withdraw into southern portions of the regular range. Spring migration from late March to early May; fall from late August through October. VI: 2.

DESCRIPTION: A pale, cryptic, sassy, rock-frolicking wren. Medium-sized—similar to Canyon Wren in size and shape, but overall a more proportionally balanced wren with a large head, shorter bill, oval body, and somewhat (perhaps proportionally) shortish tail.

Overall uniquely and cryptically pale. Gray or gray-brown with fine speckling above; whitish with a pale buffy wash below. Canyon Wren is overall ruddy with a grayish cap and prominent white throat and upper breast.

BEHAVIOR: Nimble and hyperactive. Forages in the open on rocky slopes, rock faces, and pebbly desert, through and over bushes, and in and out of cracks and crevices. Moves quickly in a rapid series of hops and fluttering flights. Sometimes pauses, turning its head left and right, gauging its next move. When disturbed, bobs its body or stands upright; otherwise crouched. Clambers up, down, and sideways across rock faces. Leaps for insects. Opens its wings and parachutes to lower sites. Rarely still for long. Movements and mannerisms are very similar to Canyon Wren.

Tame, and almost indifferent to people. Responds well to pishing.

FLIGHT: Flight silhouette is dominated by broad rounded wings and a widely fanned tail that shows a conspicuous buffy terminal band—the bird's only adornment. The cinnamon rump is easily overlooked. Flight is weak and fluttering, with rapid steady wingbeats; when dropping to a lower altitude, employs lots of open-winged parachuting glides.

VOCALIZATIONS: Song is a varied series of multiple notes or phrases, many high-pitched and slightly trilled, some warbled, others rapid and run-on, others spaced; each phrase is uttered two to eight times: "*cheerycheercheerycheerycherry* (pause) *reeh* (pause) *reeh* (pause) *reeh* (pause) *jurryjurryjurry jurry*." Another song is more run-on and sounds like a squeaky amateurish Winter Wren. Call is a short, loud, shrill, slightly trilled "*P'chee*."

Canyon Wren, *Catherpes mexicanus*
Cliff Nuthatch

STATUS: Fairly common and widespread permanent resident, but habitat-specific. **DISTRIBUTION:** From s.-cen. B.C. across the eastern two-thirds of Wash., all but w. Ore., most of Idaho, portions of Mont., w. S.D., most of Wyo., Nev., Utah, western two-thirds of Colo., most of Calif., Ariz., N.M., extreme w. Okla., and the western half of Tex. **HABITAT:** Specific to cliffs and steep rocky slopes, most commonly in arid and semi-arid areas. Classic habitat is rocky butte or cliff face or a stream-bearing canyon with steep rock-strewn slopes combined with some vertical cliffs. The nature of the sparse vegetation that accompanies such terrain seems inconsequential. Birds are also sometimes found locally in rocky areas with fairly dense riparian vegetation. **COHABITANTS:** Prairie Falcon, White-throated Swift, Rock Wren. **MOVEMENT/MIGRATION:** None. VI: 0.

DESCRIPTION: A nimble oddly proportioned wren that handles cliffs the way nuthatches navigate trees. Somewhat bottom-heavy wren with a wide but flattish body that looks like it was stepped on. Medium-sized, with a long, slender, needlelike, somewhat down-curved bill, a narrow wedge-shaped face, and a long and usually slightly fanned chestnut tail. When foraging, posture sometimes appears concave, with tail and bill raised.

Overall dark, with a contrastingly white throat and breast. At close range, appears ruddy brown, but at a distance, only the tail appears reddish, with narrow blackish bands (top and bottom) visible.

BEHAVIOR: An active animate wren that moves jerkily but nimbly up, down, over, and into cliffs—often disappearing into one crevice and appearing somewhere else. Moves in short and long hops (sometimes with wings fanned for balance) and in short and lengthy fluttering sallies. Climbs up, down, and across rock faces with nuthatchlike hops. Also clambers up or flies into low bushes and trees.

Movements often have a theatrical air. The bird pauses before a jump, as if posing, and turns its head (sometimes the body) left, then right. Sometimes pauses for extended periods with body flattened and tail up. Also crouches, with bill stiffly raised, and does "pushups" (bobs its body).

Not shy and not retiring. The bird is difficult to see only because of the nature of the habitat and its foraging habits.

FLIGHT: Wide rounded wings dominate the flight silhouette. Like a nuthatch, the bird seems like a head with wings—all gliding surface. Wingbeats are slow, constant, and fluttery; flight is weak, effortful, and not technically precise—somewhat baby-bird clumsy. Gives the impression that if it just touches down on the mountain, that will be fine. Descents are mostly parachuting glides with frequent changes of angle of descent (even direction) before landing.

VOCALIZATIONS: Harmonizing with the steep terrain, the song is a loud, descending, heart-gladdening tumble of notes that grow slower and more two-noted near the end: "*d'jeer'jew djew djew djew. . . . jewee jewee*"; sometimes ends with a low raspy "*chah*" or "*chah . . . chah . . . chah.*" Call is a flat, buzzy, descending "*breehr*" or "*jrrr*" (with a nasal twang) that is given frequently as birds forage.

Carolina Wren, *Thryothorus ludovicianus*
Ruddy Wren

STATUS: Common resident across much of the eastern United States. **DISTRIBUTION:** Found across the se. U.S., ranging north to Mass. and Conn., s. N.Y., extreme se. Canada, s. Mich., n. Ill., and se. Iowa. The western limit cuts across se. Neb., cen. Kans., cen. Okla., and cen. Tex., south into ne. Mexico. A few are found casually in fall and winter in New England and the Great Lakes region. **HABITAT:** Almost any woodlands with good tangled undergrowth. Particularly favors riparian woodlands, brushy edge, swampy areas, overgrown farmland (this species gives extra points for dilapidated buildings), and suburban (particularly unkempt) yards with a surfeit of dense shrubs and trees. May be attracted to feeders, where it favors suet. **COHABITANTS:** Tufted Titmouse, Northern Mockingbird, Gray Catbird, Eastern Towhee. **MOVEMENT/MIGRATION:** Mostly nonmigratory permanent resident whose numbers in northern portions of the range rise and fall depending on the severity of the winter. VI: 1/2.

DESCRIPTION: A portly, potbellied, humpbacked, medium-sized wren (largest eastern wren) with a big head, a sturdy, down-curved bill, and a fairly short tail. Upperparts are generally plain and unpatterned but distinctly ruddy brown; except for the whitish throat, underparts are uniformly warm buff. The prominent narrow white or buffy eye stripe is arrestingly long and often kicks up at the rear. No other wren looks like this one. The combination of overall plumpness, overall ruddiness, and the prominent whitish eye stripe eliminates all other possibilities. Smaller Bewick's Wren is overall thinner, rangier, and more high-strung, and has a long tail that it frequently twitches side to side. Carolina's expression is serious.

BEHAVIOR: Industrious but deliberate feeder that spends a great deal of time on the ground or in low-lying vegetation. Moves in jerky hops from the overturned flowerpot to the creeping ivy to the top of the fence to the trunk of a tree to the suet feeder to the inside of the shed with the door ajar. Can climb like a creeper and hang upside down like a nuthatch. Likes tight confines and eagerly investigates nooks, crannies, holes, buildings, junked cars. . . . Also likes to stay hidden, but at times remains stationary, in the open, for several seconds, jerking its breast up and down and perhaps contemplating its next series of moves.

Generally solitary or found in pairs. Vigorously defends its territory from neighboring Carolina Wrens, which are generally not far away.

FLIGHT: Flight is rapid, brief, and slightly undulating, with rapid wingbeats. Not a strong flier, but uses its wings frequently to assist its foraging and is capable of flying vertically from the base of a tree to the top in a single wing-assisted bound.

VOCALIZATIONS: Sings year-round, at any point during daylight hours and in all but the most atrocious weather. Song is usually sung in triplets. Whatever it says, it says three times (and sometimes four or five). Singing birds commonly have their tails down. (Foraging birds' tails are up-cocked.) Classic song, which has many variations, is a loud, rich, rolling "*tea-kettle, tea-kettle, tea-kettle*" or "*chur-ee, chur-ee, chur-ee.*" Scold is a loud, rapid, raspy snarl, "*chaaa . . . chaaa, chaaa,*" repeated in both a regular and an irregular sequence. Also makes a short musical muttering—a cross between a chip and a short trill—that, given in a lengthy trilled sequence, recalls the sound of a finger being run down a comb.

PERTINENT PARTICULARS: The song of Carolina Wren is similar to (and sometimes mistaken for) Kentucky Warbler—a bird of forest and dense understory. The pattern is similar, but the quality is different: the warbler's song is richer, usually more hurried and more ringing.

Bewick's Wren, *Thryomanes bewickii*
The Gnatcatcher-like Wren

STATUS: Rare and local in the East. Common in the West. **DISTRIBUTION:** More or less a permanent resident along the West Coast, from extreme s. B.C. across much of Wash., Ore., and all except se. Calif. Also found in nw. and extreme s. Nev., most of Utah, s. Wyo., all but sw. Ariz., w. and s. Colo., Ariz., Kans., Okla., w. Mo., w. Ark., and almost all of Tex. south into Mexico. Also found in restricted and diminishing pockets east of the Mississippi R. (primarily in the s. Appalachians). **HABITAT:** Brushy habitat, including thickets, riparian woodlands, desert scrub, chaparral, mesquite, well-vegetated suburban areas, and brushy borders to agricultural land and farm buildings. In the East, seems particularly drawn to unkempt yards festooned with junked cars and appliances. **COHABITANTS:** Many and varied but include Bell's Vireo, Bushtit, House Wren, Wrentit, Spotted Towhee, Song Sparrow. **MOVEMENT/MIGRATION:** For the most part a permanent resident, but individuals vacate extreme northern portions of the range, and wintering birds expand very slightly east and west of the breeding range. VI: 1.

DESCRIPTION: A medium-sized (between House Wren and Carolina Wren in size), rangy, rawboned wren whose shape and antics resemble a gnatcatcher. Overall long and trim, with a narrow, relatively straight bill and a long, thin, animate, gnatcatcher-long tail. Pacific Coast birds are plain dull brown above, with gray tails and pale grayish underparts. Eastern birds are more rufous brown above. Southwestern birds are overall grayer. In both plumages, the sheer plain uniformity of the plumage makes the long, narrow, white eyebrow a standout feature. The white tips to the outer tail feathers are not particularly evident (and hardly necessary for identification).

BEHAVIOR: An arboreal wren that forages actively in brush and branches (infrequently forages on the ground). Nimble, acrobatic, and high-strung (seems perpetually irritated or piqued), Bewick's moves in short hops and quick jerks and is never still for long. Clings to trunks. Hangs upside down. The long tail is gnatcatcher-animate, switching angrily from side to side. Likes to be where it is thick, but not shy about coming into the open or ranging high in trees. Responds well to pishing, and often scolds without prompting (just your presence is enough). Generally solitary, although sometimes joins foraging flocks.

FLIGHT: Rapid, direct, straight, and usually short. Sometimes rises and falls, but does not undulate.

VOCALIZATIONS: Song is highly variable—somewhat towheelike or Song Sparrow–like. Short introductory notes are followed by a warbled whistled trill. Often more elaborate and enhanced by buzzy notes. If you are in Bewick's Wren territory and hear a song that you just can't place, chances are it's Bewick's. Scold is a terse raspy "*ch'cha, ch'cha, cha, cha*" or "*chh, chh, chh, chh*" that is drier and higher-pitched than Marsh Wren. Also makes a low, flat, sequentially repeated "*b'jee*" or "*b'jrr*" that twangs like a mouth harp. Also emits a low raspy growl.

PERTINENT PARTICULARS: Among wrens, this species is most likely to be confused with Carolina Wren, whose range overlaps Bewick's in the eastern portions of the latter's range and whose size (greater than House Wren) and white eyebrow suggest Bewick's to some. Carolina is overall plump, stocky, and decidedly shorter-tailed and overall warmer-toned below—just the opposite of Bewick's. Be aware that light transfusing through the outer tail feathers of Carolina may give the impression that the bird has a white-trimmed tail. Ignore it. Concentrate on shape and Carolina's relative sluggishness.

House Wren, *Troglodytes aedon*

The Summer Wren

STATUS: Common and widespread, and familiar to rural, suburban, and small-town residents alike. **DISTRIBUTION:** The "Jenny Wren" has the most extensive north-south range of any native New World passerine—a distribution that extends from s. Canada to the tip of South America. In North America, except in coastal Va. and Calif., breeding and wintering populations are disjunct. Breeds from sw. N.B., e. Maine, and se. Que. across s. Ont., s. Man., s. Sask., much of Alta., and s. B.C. The southern limit of its breeding range reaches the N.C. Piedmont, n. S.C., n. Ga., n. Ala., n. Ark.,

n. Okla., cen. N.M., se. Ariz., e. and cen. Nev., and all of Calif. except the southeastern deserts. Winters across the s. U.S. from the Delmarva Peninsula to Fla., Ala., Miss., s. Ark., La., all but cen. Tex., extreme s. N.M., s. and w. Ariz., and s. and coastal Calif. north to San Francisco and south into Mexico. **HABITAT:** Traditionally a bird of woodland edge or open coniferous or mixed woodlands, House Wren has benefited greatly from the clearing of North America's forests and flourishes in all manner of human habitats—from city parks to well-vegetated suburban gardens to overgrown junk-strewn yards. A cavity-nesting species that is quick to occupy almost any natural or man-made nook or cranny, House Wren (true to its name) is readily attracted to bird boxes set about 4 ft. above the ground. In winter, favors thickets, brushy edge, riparian corridors, orchards, and (once again) well-vegetated parks and yards. **COHABITANTS:** You! Also Northern Mockingbird, Yellow Warbler, Common Yellowthroat, Song Sparrow, Northern Cardinal, towhees. **MOVEMENT/MIGRATION:** A fairly early-spring arrival, reaching breeding grounds just as leaves begin to emerge on low-lying shrubs; birds reach the southern portions of their range by mid-March and northern portions by early May. Fall migration begins in September and is mostly concluded by late October. During migration, after flying primarily if not exclusively at night, the birds occupy the same habitat they do during winter. VI: 1.

DESCRIPTION: A small, thin, elongated wren best known for its cheerful (and almost incessant) singing and undistinguished plumage. The bill is long, thin, and slightly drooped, making the bird appear long in the face. The medium-long, narrow, round-tipped tail may be very slightly raised, held in line with the body, or (most often) slightly drooped, giving a perched bird the suggestion of a slouch. Except in flight, the tail is always closed.

The plainest of the wrens, in all plumages House Wren is fairly monotoned—gray-brown above and below, although western birds (particularly those breeding in the Southwest, such as the "Brown-throated" Wren) tend to be slightly browner or warmer-toned. The front half of the bird is patently plain and almost devoid of plumage characteristics; the back half (wings, tail, undertail coverts, and, in some cases, flanks) is faintly barred. If you look real hard, you might find the suggestion of a narrow pale eyebrow. If you look even harder, you might find an indistinct half eye-line or an ear patch. The expression is not amused or mildly annoyed.

BEHAVIOR: An active, nervous, furtive bird that is never still for long—unless it is singing (usually from a sturdy exposed perch as high as 40–50 ft.). When perched, legs are comfortably spaced, not splayed. The bird moves quickly and dartingly through vegetation using short hops and brief direct flights. When not vocalizing, stays for the most part within 10 ft. of the ground (sometimes foraging on the ground). Movements often show a pronounced vertical inclination. Most birds move horizontally; House Wren likes to move up and down, pausing after every jump, exploring nooks and crannies, raising and lowering its body with movements too quick to see. Foraging Winter Wren, by comparison, hugs the ground.

Fairly tame and allowing close approach, these birds respond well to pishing during breeding season but are more reticent in winter.

FLIGHT: Direct and fast, usually just above the ground, with constant (or near-constant) rapid wingbeats. Birds also fly, almost vertically, to a perch-and-dive (also vertically) into protective cover. When changing perches, singing males may either brake with a flutter or "stick the landing" with more drama and energy than the mere act of perching requires. Unlike Marsh Wren, does not engage in aerial display.

VOCALIZATIONS: Song is a loud, rapid, rising-and-falling cascade of notes that begins with a repeated-note warm-up (an avian "*me-me-me-me meee*"), followed by a spirited and pleasing roller coaster of a song that rises and falls and drops off at the end. Calls are a low, atonal "*chur*"; a low creaking (like a door nudged by the wind); and an angry stutter, "*ch-ch-ch-ch-ch.*"

Winter Wren, *Troglodytes troglodytes*

Nub-tailed Wren

STATUS: Fairly common, widespread, noisome but secretive woodland wren. **DISTRIBUTION:** Breeds from Nfld. and ne. Que. across e. Canada to se. and cen. Man., cen. Sask., and much of Alta. and B.C., then north, along the southern coast of Alaska, to the Aleutian and Pribilof Is. In the U.S., breeds across most of New England, N.Y., n. N.J., and n. and w. Pa., along the Appalachians to Ga., and in Mich., Wisc., ne. Minn., w. Mont., n. Idaho, Wash., Ore., and n. and cen. Calif. (and a few pairs are found south in the Rockies). In winter, found across much of the e. U.S. from s. New England to n. Fla., west (skirting the s. Great Lakes) to se. Neb., cen. Kans., cen. Okla., and e. and cen. Tex. to just south of the Brazos R. Also found (sparingly) across the s. Great Basin, most of the Southwest, and the prairies north to n. Colo. and s. Neb. **HABITAT:** Breeds primarily in mature coniferous forest (particularly near streams, bogs, and lakes), but also inhabits barren sea cliffs, heath, deciduous forest, and riparian woodlands. In winter, inhabits a variety of woodlands, especially wet woodlands, and is most often found around fallen timber, stumps, brush piles, wet (log-strewn) edges of ponds, swamps, streams, and rivers. **COHABITANTS:** During breeding, Swainson's Thrush, Hermit Thrush, Varied Thrush, White-throated Sparrow. In winter, Rusty Blackbird (East). **MOVEMENT/MIGRATION:** Most eastern and some interior western populations are migratory. Spring migration from March to early May; fall from late September to December, with a peak in October and November. VI: 2.

DESCRIPTION: A tiny, brown, ground-hugging Ping-Pong ball with an abbreviated nub of a tail. Unmistakably tiny (kinglet-sized), but plump, almost rotund, with a petite, dark, needlelike bill and a short up-cocked tail. Uniformly dark ruddy brown, Winter Wren is darker and (except for House Wren) plainer than all other wrens, with slight contrast between upper- and lower parts; belly and flanks are scored by narrow and tight dark-on-dark barring. Against this dark backdrop, the only thing comparatively light about the bird is its narrow, slightly paler eye-line.

The only bird this species might be mistaken for is the pale-billed (not dark-billed) House Wren, which is overall paler, rangier, and (particularly) longer-tailed. Everything about Winter Wren says "stubby." Trust your eyes.

BEHAVIOR: "A feathered mouse" that moves, close to or on the ground, in short rapid hops from brush pile to uprooted tree stump to discarded refrigerator, investigating nooks and crannies, disappearing here and reappearing there. Sometimes pauses briefly atop a stump or log, bobbing nervously. Almost never seen foraging in bushes or trees, but males sing from elevated perches. When agitated, may flutter its up-cocked tail. Often navigates streambanks and is generally found close to water (swamps, floodplains, rivers, and streams). Unlike many wrens, not shy about being in the open, but any such unobstructed view is short-lived. The bird is perpetually in transit. Responds to pishing and owl calls (often by first calling).

FLIGHT: Tiny and overall compact. Flight is rapid, direct, and normally very short, on very rapid wings (recalls a small brown bumblebee) and commonly just above the ground.

VOCALIZATIONS: Song is a hurried, cascading, and prolonged ensemble of rising-and-falling warbles and trills—some musical, some more brittle, all incredibly loud to be generated by a bird so small. Songs often begin with a terse two-note stutter. Sharp single-note and three-note series intrude between the warbled and trilled segments of the song. Call of eastern birds is a distinctive and (usually) double-noted "*chimp-chimp*" that recalls a similar call of Song Sparrow; suggests a flat pluck of a banjo string. Call of western birds is a high, drier, sharper "*cheh-cheh*" reminiscent of the single-note call of Wilson's Warbler. When agitated, makes a high, dry, stuttering trill.

PERTINENT PARTICULARS: Aptly named, Winter Wren is the only wren found as far north as Alaska. In winter, ranges farther north than any other wren species. Unlike most of the Winter Wrens in North America, birds found along the Bering Sea inhabit open unforested sea cliffs and tundra—habitat very much in the eclectic character of Winter Wren (or simply "Wren") of northern Europe.

Sedge Wren, *Cistothorus platensis*
The Sedge Wraith

STATUS: Fairly uncommon to common, with shifting populations, a secretive species. **DISTRIBUTION:** Principal breeding range encompasses e.-cen. Alta., cen. and s. Sask., s. Man., s. Ont., and extreme s. Que.; in the U.S., n. and e. N.D., e. S.D., e. Neb., ne. Kans., Minn., Iowa, n. Mo., Mich., Ill., n. Ind., n. Ohio, and n. N.Y. (local). Also nests very sparingly and sporadically throughout New England, the mid-Atlantic, and the s. Midwest. Winters in the southeast coastal plain from s. N.J. to n. Mexico; also in cen. Tex., se. Okla., s. Ark., and w. Tenn. **HABITAT:** At all seasons, seeks out tall grass and sedges in wet or seasonal meadows; also the narrow ecotone lying between marshes and ponds and dry uplands. Generally not found in cattail marsh or areas with deeper standing water. Often found in wet grassy expanses with some shrubs. In winter, also occupies dry weedy fields, brackish and tidal wetlands, and phragmites bordering tidal marsh. **COHABITANTS:** Common Yellowthroat, Le Conte's Sparrow, Song Sparrow, Red-winged Blackbird. **MOVEMENT/MIGRATION:** In spring, migrates from mid-April to June. Fall movement is more protracted; some birds leave breeding territories in August, and others linger until driven out by freezing conditions in November. Migratory standards notwithstanding, this species is mercurial and nomadic. Birds may suddenly appear in their proper habitat throughout the summer. VI: 2.

DESCRIPTION: A small, buffy, short-billed wren with a distinctive chattery voice and a phobia about being seen. Smaller, paler, and more compact than Marsh Wren, with a distinctly shorter bill and a fairly short, somewhat ragged-tipped tail. Overall buffy and warm-toned, with pleasingly patterned but somewhat homogenized upperparts. The large dark eye stands out in a plain face. The streaked and white-frosted crown and back show little contrast (unlike Marsh Wren, which has a toupeelike brown crown and a disjunct streaked back). The expression is alert and innocent.

BEHAVIOR: The bird is a mouse—more often heard than seen. Feeds by scurrying about on the ground. Flushed once, it lands, runs, and is reluctant to flush again. Can be coaxed into the open by squeaking or persistent pishing, but is generally unresponsive. When nervous, flicks its tail downward with a rapid angry motion and flicks its wings so quickly that the motion is often overlooked. In winter, sometimes found with Marsh Wrens or other Sedge Wrens. Often vocalizes (calls) just at dawn and again at dusk.

FLIGHT: Overall small, compact, pale. Flight is weak, bouncy, and short; wingbeats are rapid, constant, and somewhat fluttery.

VOCALIZATIONS: Distinctive two-part song: an opening series of two to four sharp dry chips with a stone-skipped-on-ice quality is followed by a staccato chatter, "*ch,h' ch'h, ch'h-chachachachacha.*" Often confused with the rattle of Common Yellowthroat, whose call is slower, lower-pitched, and less brittle. Call suggests or replicates the notes of the opening chip series of the bird's song: "*chuh*" or "*chrr*" or "*chap*" (may be muffled or loud).

PERTINENT PARTICULARS: The bird is not difficult to identify. It is only difficult to see.

Marsh Wren, *Cistothorus palustris*
Clamorous Reed Wren

STATUS AND DISTRIBUTION: Common and widespread breeder across much of s. and cen. Canada, the n. U.S., and along all coasts. In winter, northern and interior birds withdraw primarily to the coastal regions and marshes well south of the freeze line (including to most of Mexico). **HABITAT:** Found almost exclusively in fresh- and saltwater marshes characterized by tall rank vegetation, particularly cattail, bulrush, phragmites, and tall stands of salt marsh cord grass. In winter, also found in *Salicornia* marshes. **COHABITANTS:** Virginia Rail, Common Yellowthroat, Swamp Sparrow, Red-winged Blackbird, Yellow-headed Blackbird. **MOVEMENT/MIGRATION:** In migration, Marsh Wren continues to seek out grassy wetland habitat. Spring migration from late March through May; fall from August to late October. VI: 2.

DESCRIPTION: A small, spry, somewhat portly wren with the agility of a gymnast and the vocal quality of an old gramophone. The head is large, the body plump, the bill long, thin, and slightly down-curved, and the tail fairly long and narrow and almost always up-cocked (if not vertical!). Unlike House and Winter Wrens, Marsh Wren's head and body seem anatomically distinct. Proportionally, most closely resembles a smaller, less portly Carolina Wren.

The upperparts are an amalgam showing a mostly brown back, a blackish brown crown, and ruddy highlights in the wings and tail. Underparts are pale with a broad warm ruddy or grayish brown wash along the sides. Marsh Wren is overall a fairly dark wren whose plumage accentuates a distinct white eyebrow and fine white streaking on the back. The expression is somewhat matter-of-fact.

BEHAVIOR: A skulker, even by wren standards—more often heard than seen. It clambers through the rank vegetation by hopping stalk to stalk or hitching itself up and down vertical reeds; also hops on the ground. Even counting aerial displays, 99.9% of a Marsh Wren's life occurs within 8 ft. of the ground. Singing males, with legs splayed, may anchor themselves midstalk, but often climb (or fly) to the springy tips of reeds, swinging jauntily (sometimes dangling) as they sing. One of the bird's hallmark gymnastic routines is a variation on the "iron cross"—poised triumphantly between two stalks with one foot on one stalk and the other foot on the other.

Singing males often execute an aerial display—a lofting arch during which they parachute to earth on fluttering wings, heads and tails cocked up.

In winter, birds bury themselves deep in marshes and can be very difficult to detect at all, much less see. At dawn and dusk, birds often concentrate in thick vegetated stands (usually with a southern or southeastern exposure) and call for several minutes, making a soft low "*cheh.*"

FLIGHT: Low, fast, and mostly direct, on rapid fluttering wingbeats. Flight seems effortful and somewhat wandering or uncontrolled, as though the bird applied too much English or didn't allow for windage and has to veer a bit just before landing. Flying birds recall large insects or perhaps bumblebees.

VOCALIZATIONS: When breeding, a very vocal bird that sings multiple variations of its song day and night. In structure, the song is somewhat like Sedge Wren, but more percussive, consisting of a two- to four-note introduction and a one- or two-part staccato stutter that drops off on the last note: "*chek, chek, Cheh-ch-ch-ch-ch-ch-ch-ch-cheh.*" Recalls a rapidly ticking Geiger counter, the clicking of an old rotary phone dial, or a cassette tape on fast rewind.

Rarely does the bird render the same version of the basic song twice. In every set, variations in duration, pitch, volume, speed, and quality are the norm, but the pattern remains the same.

Call is a muttered "*cheh*" or "*chek.*" Also makes a low grating growl.

PERTINENT PARTICULARS: Although House and Marsh Wrens may both occur where their habitats overlap (for instance, where standing dead trees encroach on flooded marsh), they are easily distinguished by color (pale versus dark), pattern (virtually none versus some), and posture (hunched versus alert with tail up-cocked). Sedge Wren is more likely to be confused with Marsh Wren; though Sedge prefers wet short-grass meadows, both species can be found in the same tall rank habitat in winter and during migration, particularly when birds go to roost.

Sedge Wren is overall stubbier, paler, buffier, and streakier, with a prominent outsized eye.

In coastal Georgia and South Carolina, a race of Marsh Wren occurs that is cold gray, lacking in reddish tones but otherwise similar in pattern and structure. Although it doesn't look exactly like its ruddy kin, it looks even less like Sedge Wren (or any other wren that might be found there).

Marsh Wrens respond truculently to pishing. Best results are had by persistently imitating the staccato portion of their song ("*Chuh-chuh-chuh-chuh-chuh-chuh*") or doing a slow, measured, softly uttered pish sequence ("*pssh, pssh, pssh; pssh, pssh, pssh*").

DIPPER AND BULBUL

American Dipper, *Cinclus mexicanus*

A Crop-tailed Aquatic Catbird (with a Wink and a Bob)

STATUS: Variously uncommon to common and widespread, but habitat-restricted. **DISTRIBUTION:** A western species with a contiguous range from cen. Alaska through all but ne. B.C., sw. Alta., n. and w. Wash., and Ore. to nw. and cen. Calif., most of Idaho, w. and cen. Mont., Wyo., and n. and w. Utah. Also ranges throughout most of Colo. and in disjunct pockets in the states already mentioned, as well as in s. Calif., Nev., Ariz., N.M., w. S.D., and parts of Mexico and Central America. **HABITAT:** Breeds on clear clean mountain streams; also found (in Alaska) on coastal streams. Key components are rock or sandy bottom, exposed rocks, fallen logs, and overhangs. Seems very attracted to underside of bridges for nesting. In winter, requires ice-free areas for foraging. Occasionally found on lakeshores. **COHABITANTS:** Harlequin Duck, Spotted Sandpiper (summer), Black Phoebe. **MOVEMENT/MIGRATION:** Migratory behavior varies. Where streams remain marginally ice-free all year, birds are nonmigratory. Where streams freeze, birds either retreat to lower elevations or move to different drainages. Winter range extends very little beyond breeding range. Spring migration from early February to late April; fall from October through December. VI: 1.

DESCRIPTION: A plump, all-gray, stream-loving bird that bobs and winks. Chunky, starling-sized, and fist-shaped, with a straight thrushlike bill, no neck, a plump body, and a cropped (baby bird) tail.

Catbird gray with pale pinkish gray legs and a white eyelid (nictitating membrane) that flashes across the eye.

BEHAVIOR: Almost invariably seen standing in water, standing beside water, or flying over water—usually swift-flowing turbulent water. Forages alone by walking along the shore or among exposed rocks and picking, swimming, or walking with head submerged, diving from the surface, or plunge-diving from rocks and banks into the water, where it may remain submerged for 15 secs. or more.

"Dips" its entire body with a springy jerk, usually blinking its white eyelid simultaneously. Dipping frequency increases when bird is agitated. Swims buoyantly, dives rapidly, and takes off from the water with whirring wings and pushing feet. Often perches on rocks in streams, beneath overhanging streambanks, and, where available, beneath bridges.

No other songbird is so wedded to, or uniquely adapted for, an aquatic environment. Fairly tame; often forages right past streamside observers.

FLIGHT: Comically compact and short-winged—alcidlike in shape and flight. Wingbeats are whirring and constant. Flight path follows the contours of the stream (usually less than 6 ft. above the surface and commonly just above the surface); almost never crosses land. Flight seems recklessly fast, tipsy, and somewhat out of control.

VOCALIZATIONS: Commonly silent. Song is a lively series of buzzy whistles, warbles, and trills, usually repeated. Recalls a thrasher or mockingbird. Call is a distinctive, loud, metallic "*kzeet*" that is often repeated, particularly when birds take flight.

Red-whiskered Bulbul, *Pycnonotus jocosus*

Cow-lick Crested Bulbul

STATUS/DISTRIBUTION: Introduced species. Fairly uncommon and local in Dade Co. (Miami area), Fl. Also a rare and local resident of Los Angeles Co., Calif. **HABITAT:** Parks, gardens, suburban neighborhoods boasting lush exotic plantings (in California, also citrus orchards). **COHABITANTS:** Yellow-chevroned Parakeets, Hill Myna, Spot-breasted Oriole, House Sparrow (and Northern Mockingbirds and Us). **MOVEMENTS/MIGRATIONS:** Nonmigratory.

DESCRIPTION: A distinctive and easily identified bird, unlikely to be confused with any other crested species (it's the cowlick crested bird perched atop the TV antenna, or support cable, near Miami). Smaller than European Starling; slightly smaller than Cedar Waxwing; larger than House Sparrow, with a conspicuous spiky crest that turns forward;

slender body and long tail. Blackish head and brown upperparts contrast with whitish underparts. Dark "spur" extending down the side of the breast is conspicuous; red spot on the face is often not. Immatures are like adults but lack the red "whisker."

BEHAVIOR: Found in pairs, trios, or small groups. Most often seen perched on exposed perches (including TV aerials, utility lines, and anchoring guide-cables, tree and bush tops). Captures insects by making short, sometimes vertical sorties. Also forages on berries while perched, gleans insects from tree trunks, and pursues insects on the ground. Feeding is largely confined to mornings and evenings. During most of the day, the birds retire to non-native vegetation at which point they are very hard to find.

FLIGHT: Appears round-winged and long-tailed (waxwings are pointy-winged and short-tailed). Uniformly brown above (except for touches of white on the tip of the tail). Black "spur" shows well against the white breast. Overall flight is fairly slow, undulating, jerky, and usually of short duration.

VOCALIZATIONS: Song is loud and musical whistle-chatter. Call is a loud lively *"Peter-grew"* (sometimes introduced with a "tick" or "kick" note).

KINGLETS

Golden-crowned Kinglet, *Regulus satrapa*

Conifer Kinglet

STATUS: Common and widespread northern and montane breeder and widespread winter resident. **DISTRIBUTION:** Breeds across Canada and in northern and mountainous parts of the U.S., from Nfld. and the Maritimes to s. Que., most of Ont., e. and cen. Man., cen. Sask., most of Alta., B.C., the s. Yukon, and s.-cen. Alaska. Also found in northern and interior New England, N.Y., n. and w. Pa., n. N.J., w. Md., e. W. Va., and the Tenn.–N.C. border; in the West, found in w. Mont., n. Idaho, w. Wyo., Utah, n. Ariz., w. and cen. Colo., and n. N.M.; also widespread in Wash., Ore., and (primarily) n. and cen. Calif. Winters coastally from Nfld. and Kodiak I. south. Widespread across all of the U.S. (except the deserts of se. Calif., s. Nev., s. Ariz., n. Minn., n. Wisc., n. Mich., and the Florida Peninsula). Also winters in s. B.C., s. Alta., se. Ont., se. Que., N.S., and N.B. **HABITAT:** Breeds primarily in spruce-fir forests, but also found in other conifer and mixed open or closed-canopy forests and, in the West, cottonwoods. Has expanded its breeding range south as a result of maturing spruce plantings in subboreal regions. In winter, occupies a variety of habitat, including hardwood forests and woodlands, swamps, and suburban parks, but prefers conifers and mature trees. In migration, often forages low in shrubs and also on the ground. **COHABITANTS:** Sharp-shinned Hawk, Red-breasted Nuthatch, Ruby-crowned Kinglet, Red Squirrel; in winter, joins assorted chickadees in mixed flocks. **MOVEMENT/MIGRATION:** In winter, most Canadian breeders retreat into the United States. Spring migration from late February to late May; fall from late August to late December. VI: 1.

DESCRIPTION: A tiny, effervescent, and crisply patterned pixie of a bird. Despite the diminutive size (smaller than any warbler), Golden-crowned is overall neckless and plump—a dumpling with a petite pointy bill, beady black eyes, and a short, narrow, deeply notched tail. In all plumages, a greenish gray bird with an eye-catching heavy-mascara face pattern, a racy white-and-yellowish wing pattern, and, in adults, a golden crown (orange-tinged in males). Ruby-crowned Kinglet is slightly larger, uniformly plainer, and less distinctly patterned.

BEHAVIOR: More social than Ruby-crowned Kinglet. In winter, often found in small kinglet groups, often in association with chickadees, titmice, and creepers. Also more reticent than Ruby-crowned, forages higher in the canopy or beneath the outer vegetation, and is more site-tenacious—forages longer in a tree before moving on. Also not as responsive to pishing—in fact, Golden-crowned is usually the last bird in a winter flock to arrive.

Hyperactive feeder, hopping quickly between branches and wing-flicking often (although less than Ruby-crowned). Forages by picking or hovering; is particularly adept at dangling chickadee-style from the tips of branches.

FLIGHT: Tiny size and overall compactness are distinctive. Overall pale. Flight is darting, but weak.

VOCALIZATIONS: Song sounds like the laughter of pixies. Begins with several exceedingly high thin notes that accelerate and degenerate into high squeaky laughter: "*see see seeseeseeseechchchhihih'll.*" To those unfamiliar with the laughter of pixies, the last part of the song may suggest the super-high-pitched stutterings of chickadees. Call is a merry, very high-pitched, often three-note twitter, "*chee tee tee*" or "*tee-tee-tee-tee-tee*"; the notes are sometimes pure-toned, and sometimes slightly trilled, just brushing the tympanic membrane.

Ruby-crowned Kinglet, *Regulus calendula*

A Bird That Moves Like Spit on a Skillet

STATUS: Common and widespread northern and montane breeder; common and widespread winter resident. **DISTRIBUTION:** Breeds in northern forests from Nfld., Lab., and the Maritimes across virtually all of Canada (except extreme se. Ont., s. Man., Sask., and se. Alta.) to cen. Alaska. Also found in n. New England, the Adirondacks, n. Mich., n. Wisc., n. Minn., and at higher altitudes in the Rockies and other major western mountain ranges south locally to the Mexican border. In winter, found south of a line drawn from Cape Cod, n. N.J., cen. Pa., s. Ohio, s. Ind., s. Ill., s. Mo., s. Kans., n. Tex., cen. N.M., n. Utah, s. Idaho, and s. B.C. south into Mexico. **HABITAT:** Breeding birds are almost always associated with conifers, particularly black spruce forest, spruce and tamarack muskegs, spruce-fir forests, some pines, and mixed forest. In migration, may be found in an array of habitats, including conifer and deciduous forests; tree-dotted parks and suburban neighborhoods, overgrown fields, and overgrown lots. Winters in all manner of deciduous and coniferous forest, woodland, and low vegetation, including tall reeds and phragmites marsh. **COHABITANTS:** Red-breasted Nuthatch, Golden-crowned Kinglet, Blackpoll Warbler, Palm Warbler, Rusty Blackbird, Red Crossbill. In winter, often found in mixed-species flocks with chickadees and Yellow-rumped Warbler. **MOVEMENT/MIGRATION:** Migration is slightly later in the spring and earlier in the fall than Golden-crowned Kinglet. Spring migration in April and May; fall from mid-September to mid-November, with a peak in October. VI: 1.

DESCRIPTION: A tiny, compact, hyperactive, and undistinguished bird that draws attention to itself by its perpetual motions and habitual wing-flicking. Overall diminutive (smaller than the smallest warbler; barely larger than Golden-crowned Kinglet), Ruby-crowned's head and body are plump and neckless, and its bill thin and petite. Generally plain and unpatterned—gray-green above, paler (dingy olive buff) below. Set against this nondescript backdrop, the faded eye-ring encircling the overlarge eye and (most often) a single pale wing-bar stand out. *Note:* Many illustrations show two wing-bars, but often only one is evident. The bird's expression is peeved.

The ruby crown is usually hidden and often subtle, but when birds are agitated, a shaggy crest may be raised and prominent.

BEHAVIOR: Constantly in motion; often flicking its wings. Forages horizontally, hopping along branches with reflexes that seem half again faster than the average warbler. May forage at any level, but frequently forages close to the ground—even in reeds and bushes. Frequently hovers near the tips of branches, but unlike Golden-crowned, does not usually cling to branches. Darts out to snatch insects out of the air. More active and mobile but less acrobatic than Golden-crowned.

In migration and winter, may travel with mixed-species flocks (particularly Golden-crowned Kinglet), but is often alone. In winter, a single kinglet, particularly a kinglet away from forest and woodlands, is usually this species.

FLIGHT: Overall tiny, plain, and pale. Flight is swift, jerky, and erratic, with lots of tacking movement.

VOCALIZATIONS: Song is a hurried, high-pitched, protracted jumble that begins with several low clear warm-up notes, degenerates into a bubbling tirade, and includes, somewhere in the mix, a clearly enunciated, chickadee-like, very high "*dee, dee, dee*" sequence before degenerating again. Slower, lower, more musical, and less strident-sounding than Golden-crowned. Call is a low, quick, staccato, two-note (sometimes three) stutter, "*j'jt*" or "*j'jt'jt.*"

When excited, calls sequentially and rapidly, recalling Morse code–like dots and dashes.

OLD WORLD WARBLERS AND GNATCATCHERS

Arctic Warbler, *Phylloscopus borealis*

Alaskan Willow Warbler

STATUS: Fairly common but geographically restricted Alaskan breeder. **DISTRIBUTION:** Found throughout much of w. and cen. Alaska (from Icy Cape to Bristol Bay), east through the northern and central mountains to the Dalton Highway; east again through the Alaska Range (passing between Anchorage and Fairbanks) to just short of Tok. Migrates across the Bering Sea to and from Alaska. **HABITAT:** In North America, found mostly in willows, most commonly along streams and rivers. Also found in spruce and shrubs. Avoids forest. **COHABITANTS:** Gray-cheeked Thrush, Yellow Warbler, Wilson's Warbler, Golden-crowned Sparrow, Common and Hoary Redpolls. **MOVEMENT/MIGRATION:** Arrives from early June to late June; departs between early August and mid-September. VI: 1.

DESCRIPTION: Small, dull, ever so slightly greenish "warbler" that makes persistent, ringing, rattling warbles from a willow thicket. Small songbird (larger than a redpoll; smaller than Golden-crowned Sparrow; approximately the same size as Yellow Warbler), with a stout warblerlike bill, a fairly hefty body, and a shortish tail. All in all (and except for the bill), somewhat vireo-like. When perched, seems to slouch. When it sings (which is often), the mouth blooms bright yellow.

Overall plain—ranging from greenish to olive brown above, dingy yellow or whitish below. This uninspired backdrop makes the long, narrow, (usually) pale, upward-kicking eyebrow stand out. Even the pale eyelash-fine wing-bar is usually apparent. The expression is mean.

BEHAVIOR: A methodical unglamorous generalist. Feeds just below the upper branches, plucking prey from the undersides of leaves. Sometimes flycatches; sometimes hovers; *most of the time* just moves casually through its leafy labyrinth in a series of hops and short fluttering flights. Flicks its wings as it forages, but without enthusiasm.

Males sing persistently from the tops of willows (often seeking out bare branches). After a lengthy bout of song from one perch, the bird flies to another, often distant perch (sometimes crossing streams in the process). Not shy. Responds to pishing, but not always enthusiastically.

FLIGHT: Smallish body, but broad wings. Appears uniformly bland above, pale or dingy below. Flight varies—undulating when flying short distances, more shallow undulations on longer flights. Wingbeats are strong but heavy or casual, given in an irregular but usually lengthy series punctuated by a brief closed-wing glide. Seems to slide through the air more than bound or bounce; flight may recall a thrush.

VOCALIZATIONS: Song is a dry, even, protracted, unmusical, rattling, metallic-sounding rattle, "*jih jih jih jih jih jih*," that's too slow to be called a trill. Call is a terse, hard, metallic "*d'zrt*"; also emits a sputtery "*tjjj-t*."

Blue-gray Gnatcatcher, *Polioptila caerulea*

The Twig Fairy

STATUS: Common widespread breeder in the United States and Mexico; winter resident in southern coastal areas and along the Mexican border. **DISTRIBUTION:** Breeds across most of the e. U.S. (as well as se. Ont.) and the Southwest. The borders of the eastern population cut across s. Maine, n. Vt. and N.H., extreme se. Que., se. Ont., cen. Mich., cen. Wisc., s. Minn., e. Neb., cen. Kans., w. Okla., and cen. Tex. south into e. Mexico. The borders of the western population fall across the Calif.–Ore. border, n. Nev., n. Utah, s. Idaho, sw. Wyo., w. Colo., e. N.M., and (in reduced numbers) trans-Pecos Tex. Winters coastally from N.C. and San Francisco Bay south to Central America; also in se. Calif., s. Ariz., and s. Tex. **HABITAT:** In the East, prefers deciduous woodlands—wet or dry, but favors wet. In the West, also uses pinyon-juniper, brush, chaparral, and riparian corridors. Wintering birds prefer woodlands associated with water, including mangrove swamp and willow thickets, but in the Southwest birds utilize mesquite and palo verde. **COHABITANTS:** Blue Jay or Western Scrub-Jay; Black-and-white Warbler or Black-throated Gray Warbler. In winter, often found in mixed-species flocks that include kinglets and (in the West) Bushtits. **MOVEMENT/MIGRATION:** Migrates early in spring (February to May), often arriving on breeding territories when trees are still in bud. In fall, departs as early as late July—soon after young have fledged—with peak migration in August and September; some birds linger into November. Migrates in small groups that are more like loose aggregations than flocks. VI: 3.

DESCRIPTION: A tiny pale sliver of a bird that seems to dance more than forage through the canopy and understory—a twig fairy. More slender than warblers, gnatcatchers call attention to themselves by the hyperactive twitching and sideways swishing of their long tails (half the overall length of the bird) and their mewing calls.

Overall pale blue-gray (underparts are paler), Blue-gray Gnatcatcher is distinguished by a narrow, complete, conspicuous white eye-ring and a black tail bracketed by white outer tail feathers. The eye-ring makes the bird look startled. With the tail folded, the white outer tail feathers make the entire underside of the tail appear white. The short bill is thin, straight, pointed, and dark, but not black. Eastern adult males are distinguished by a bluer crown and a black forehead and eyebrow. Females and immatures are overall plainer and more gray-blue than blue-gray; western birds are overall grayer.

BEHAVIOR: Almost always in motion. Its hyperactive behavior, coupled with its jerky sideways swishing of the tail, small size, slender proportions, and overall paleness, distinguishes the bird over all but the southwestern portion of its range, where it overlaps with Black-tailed and California Gnatcatchers. In woodlands, prefers to forage in the outer canopy and near the vegetative edge. Forages by hopping along branches, in a rapid jerky fashion. Frequently hovers to pick insects from a branch, and darts from the foliage to "gnatsnatch" insects it flushes. Adept at foraging low, in willows and understory. Does not commonly forage on the ground, but will land, briefly, to snatch prey.

Often solitary, but joins with other birds in winter flocks. Very responsive to pishing; leaves the canopy and often approaches too close for binoculars to focus.

FLIGHT: In flight, flutters when distances are short. Sustained flight is direct, energetic, and jerky, with a sputtery series of wingbeats that are too rapid to count followed by a terse closed-wing pause during which the bird seems to hang up or pause in flight. In flight, the tail is closed.

VOCALIZATIONS: Call is a distinctive high-pitched wheeze, most often given as a single drawn-out note, "*tseeeee*," that grows frail and ethereal at the end. Suggests air hissing from a tire. Other calls are doubled and trebled, but the wheezy quality remains: "*tseee-see*" or "*tseee-see-see*." Song is an ensemble of peevish natterings whose elements include a high-pitched series of chips and accelerated goldfinchlike phrases mixed in with the occasional classic "*zeeee*" or "*tsse, tsse*." Sounds like the bird is making it up as it goes along, but the mix usually includes enough classic wheezes to clue in the listener. Vocalizations of western birds are slightly lower-pitched, less thin and wheezy, and more raspy. Calls have a slight rising-and-falling quality, "*reeeur*" or "*reeuh*," and sound somewhat like the call of California Gnatcatcher.

California Gnatcatcher, *Polioptila californica*

Scrub Gnatcatcher

STATUS: Permanent resident; in places not uncommon (despite its endangered species status), but very habitat-restricted and geographically restricted. **DISTRIBUTION:** Primarily Baja. In the U.S., found only in coastal sections of San Diego, Orange, s. Los Angeles, and Riverside Counties below 1,500 ft. **HABITAT:** In the United States, almost exclusively low coastal sage-scrub. **COHABITANTS:** Wrentit, California Thrasher, California Towhee, California voters. **MOVEMENT/MIGRATION:** Nonmigratory. VI: 0.

DESCRIPTION: California Gnatcatcher is a Black-tailed Gnatcatcher with gray (not whitish) underparts hopping in coastal scrub. Very slightly smaller than, but structurally similar to, Black-tailed Gnatcatcher; shorter-billed and slightly more compact than Blue-gray Gnatcatcher.

Dusky, almost catbird-gray overall, showing little contrast between gray upperparts and gray underparts. The undertail is virtually all black, showing only faint traces of white (no spots like Black-tailed). The eye-ring is frail, thin, and sometimes broken or partial. Females and immatures show a distinct brownish wash on flanks, wings, and undertail.

Blue-gray Gnatcatcher (which winters in California Gnatcatcher territory) is overall paler, with whiter underparts, a distinct white eye-ring, and flashing white outer tail feathers. Black-tailed Gnatcatcher, whose range overlaps with California Gnatcatcher in the eastern edge of the latter's range, has paler underparts and white spots (sometimes shows as a white band) on the black undertail.

BEHAVIOR: Forages in coastal scrub by hopping, gleaning, sometimes hovering, and flycatching (much like other gnatcatchers). May be slightly more sluggish than other gnatcatchers, and forages lower (after all, the habitat is only waist-high). Flies little and low, and usually not more than a few feet. Most often seen in pairs, and tends to be vocal.

FLIGHT: Slender frail-looking bird—all gray with a blackish tail. Flight is slow and jerky.

VOCALIZATIONS: Makes a high-pitched, rising-and-falling, kitten- or catbirdlike mew, "*a/eeeah . . . a/eeeah,*" that is very unlike the harsh, multinote, wrenlike chattering of Black-tailed Gnatcatcher. To anglers, the sound of California Gnatcatcher may recall the sound of line being stripped off a spinning reel whose drag is set too low. California also makes a weak insectlike whine and wrenlike scoldings: "*raah, raah.*"

Black-tailed Gnatcatcher, *Polioptila melanura*

Whip-tailed Desert Scrub-gleaner

STATUS: Common resident species, but limited to southwestern deserts. **DISTRIBUTION:** Found year-round in the Sonoran, Mojave, Colorado, and Chihuahuan Deserts of the U.S. from se. Calif., s. Nev., and sw. Utah across w. and s. Ariz. and s. N.M., along the Tex.–Mexico border to McAllen, Tex. **HABITAT:** Arid and semi-arid desert scrub and open desert dominated by saguaro cactus, also creosote bush–lined desert washes. Not found in suburban areas where non-native vegetation dominates; usually not found in riparian areas and forested foothills, although may occur at the edge of such habitats where they abut more typical desert scrub areas. **COHABITANTS:** Ash-throated Flycatcher, Cactus Wren, Verdin, Black-throated Sparrow. **MOVEMENT/MIGRATION:** Mostly sedentary. VI: 0.

DESCRIPTION: A pale gray desert gnatcatcher with an almost all-black tail—top and bottom. A tiny long-tailed wisp of a bird with a small neckless head, a tiny blackish bill, a slight body, and a long, narrow, animate tail.

Pale, slightly brownish gray above (particularly on the wings); bluish tones are found only about the head. Underparts are paler grayish white. The black cap of breeding males is distinct; the tiny black dash of an eyebrow in nonbreeding plumage is usually indistinct. In all plumages the narrow white eye-ring is hard to see, often broken, and much less conspicuous than on Blue-gray Gnatcatcher of the wide-eyed stare.

The uppertail is all black, with a very narrow white outer edge. Folded, the undertail is black with obvious white spotting near the tip. The undertail of Blue-gray Gnatcatcher is all white; California Gnatcatcher's undertail is essentially all black with hard-to-see broken white hatch marks near the tip.

BEHAVIOR: An effervescent little twig fairy that makes short but studied and methodical hops among the tightly packed lower branches of desert brush and trees. In morning and evening, works the outsides; in the heat of the day, forages in the shady interior. Occasionally forages on the ground. Very infrequently "gnatsnatches" (flycatches), a technique favored by Blue-gray Gnatcatcher.

Most commonly forages singly or in pairs, vocalizing frequently, nervously jerking its tail up and down and swishing it side to side, but rarely fans its tail while foraging.

FLIGHT: Small, frail, long-tailed silhouette that above shows white tips on an otherwise black tail. Flight is weak, jerky, halting, and undulating. Wingbeats are sputtery and given in a rapid halting series that seems punctuated by a stall but no glide. Flies like a bird whose engine is running on fumes.

VOCALIZATIONS: Song is a terse, low, raspy, atonal buzz, "*chh chh chh chh*" or "*che che che*," repeated five to ten times. Also makes Verdinlike chips and a raspy wheeze, "*we/az*," that is lower-pitched than Blue-gray Gnatcatcher.

PERTINENT PARTICULARS: Black-capped Gnatcatcher, *Polioptila nigriceps* (a bird of western Mexico, VI: 1), is a rare visitor in extreme southeast Arizona (and has bred there as well). It closely resembles Black-tailed, but has an all-white undertail (like Blue-gray Gnatcatcher) and a longer bill than Black-tailed. The more extensive cap of breeding males extends below the eye, and all birds show no eye-ring (just a fine white tracing below the eye), so the eye is very difficult to see. The female has a browner back. Found in brushy thickets and along streams, Black-capped may team up with Black-tailed and Blue-gray Gnatcatchers in winter in mixed-species flocks that include Bridled Titmouse.

THRUSHES AND WRENTIT

Bluethroat, *Luscinia svecica*

Rainbow in a Plain Brown Wrapper

STATUS: Uncommon to locally common, but geographically restricted. **DISTRIBUTION:** In North America, found only in n. and w. Alaska from the Seward Peninsula east to n. Yukon (also an isolated population at Cape Romanzoff on the Yukon-Kuskokwim R. delta). Migrates across the Bering Sea to and from Alaska. **HABITAT:** Willow thickets (usually low) in foothills and flanking rivers and streams in lower tundra. **COHABITANTS:** Arctic Warbler, Eastern Yellow Wagtail, American Tree Sparrow, Hoary Redpoll. **MOVEMENT/MIGRATION:** Arrives in late May; retreats in August and very early September. VI: 0.

DESCRIPTION: A small, vocal, bran-colored, thicket-haunting chat with rust in the tail and a rainbow on its breast. Small stocky songbird (larger than a redpoll; slightly smaller than Northern Wheatear or a tree sparrow), with a head too large, a tail too short, and legs too long for the rest of the bird. The bill is short, narrow, and warblerlike; the tail is often held horizontally or slightly up-cocked.

A Janus-plumaged passerine—bland on one side, stunning on the other—that shows plain gray-brown upperparts and paler underparts emblazoned with a bright blue–rust-black-and-white bib. (The pattern is variable, but blue and rust tones dominate.) Females show a black beaded outline of the male's colorful bib (pattern is like Canada Warbler), and some show touches of blue and rust. Also, both males and females show a prominent pale eyebrow. Immatures resemble adults but have a duller throat pattern. Touches of the rusty patches that flank the base of the tail are evident on perched birds and conspicuous when birds fly, display, or flash their tails when nervous or disturbed.

BEHAVIOR: Unless perched high and singing or engaged in an aerial display, Bluethroat is a furtive species, spending most of its time foraging in dense willow and alder. A nervous bird with jerky quicksilver movements. Hops and runs. Forages on the ground in low vegetation; also flycatches. Flights between patches of cover are low and hurried (and reveal rusty tail patches). Compulsive singers, males sit conspicuously high when vocalizing and engage in towering displays that carry them to other nearby perches. Although a desultory wing-flicker and tail-wagger, the bird usually wing-flicks when it lands.

Found solitary or in pairs. In most places, not particularly shy (just loath to be seen).

FLIGHT: Small and compact, with short rounded wings. Rufous tail patches are manifest; otherwise, birds are just plain and dull. Flight is straight, fast, and jerky or skippy; wingbeats are given in a terse series followed by an equally terse closed-wing glide. Abrupt drops into cover appear borderline reckless.

VOCALIZATIONS: Sounds much like a flat canary. Song is a complex and varied cascade of chips, musical notes, trills, squeaks, sputters, and wheezes (many of which are difficult to render phonetically). An accomplished mimic. Often weaves parts of the calls and songs of other birds into the musical medley. Usually warms up to songs by uttering an accelerating series of low harmonic "*zrr*" notes. Call is a sharp dry "*chak*" and a higher whistled "*wheet!*"

Northern Wheatear, *Oenanthe oenanthe*

Stone Thrush

STATUS: Uncommon to locally common breeder, but restricted in both distribution and seasonal occurrence (lingers in North America for less than four months). **DISTRIBUTION:** There are two North American breeding populations. The first is found across w. and cen. Alaska (excluding southeast and southern coastal areas and the North Slope), all but se. Yukon, and the ne. Northwest Territories. Migrates across the Bering Strait. The second population occupies portions of Elsmere and Baffin Is., extreme n. Que., n. Lab., and coastal Greenland. Migratory routes to sub-Saharan Africa bypass most of North America, making the species a rare vagrant across much of the U.S. and Canada. **HABITAT:** Breeds in hilly or mountainous habitat so dry open, rocky, and barren that even a very accomplished Mountain Goat would starve. Also found in dry open tundra with intruding stone fields. In migration (as a vagrant), has turned up on construction sites, parking lots, rubble-strewn vacant lots, seawalls, short grassy areas, and garbage dumps. **COHABITANTS:** Rocks, lichen, Rock Ptarmigan, American Pipit, caribou. **MOVEMENT/MIGRATION:** Spring arrival from mid-May to mid-June; departure between early August and mid-September. Vagrants move to southern Canada and the continental United States mostly in September and October (less commonly in May). VI: 3.

DESCRIPTION: A bird that blends the colors of slate, obsidian, and sandstone with the shape of a bluebird and the jerky movements of a wren. Small, compact, thrushlike bird (larger than a redpoll; shorter than American Pipit), with a short straight (thrushlike) bill, an undersized flattish head, a robust potbellied body, a short tail, and disproportionately long legs. Posture is erect and, when alert, nearly vertical.

Breeding males are pale blue-gray above, creamy with a buffy wash below—the perfect pale backdrop to set off the bird's rakish black face mask, black wings, and black tail. Breeding females are similarly patterned but overall more muted and lack the full black face mask. In fall, adults and immatures are more monotoned, more pale brown above and buffy below. Males retain an identifiable measure of their summer plumage pattern, but females and immatures are pale warm brown and plain (except for the very bold and obvious black-and-white tail pattern that is flashed when they fly and forage). In size, shape, and posture, the bird closely resembles a bluebird (but bluebirds don't have white rumps).

BEHAVIOR: This solitary bird is almost defined by its breeding habitat. Nothing else seems willing or able to survive (much less thrive) in the open, rocky, barren places this bird calls home. Calls attention to itself by active feeding or sitting conspicuously atop prominent points—usually rocks. Hops and runs and stops, thrush fashion, to snatch prey or root in the earth. Also catches insects on the run (like a pipit), perch-hunts (like a bluebird), and sallies (like a flycatcher); often hovers (again, like a bluebird) and leaps into the air after prey. Very quick and agile. Able to spin 180° and pluck insects behind.

Navigates in a three-dimensional world, shifting positions vertically (to elevated outcroppings) or distant points. Seems to glory in being the only bird perched on a boulder the size of a dump truck. Flights may be short (a few feet) or several hundred feet (to the next ridge). Movements are quick, but not nervous. Northern Wheatear pauses and postures frequently, almost theatrically, with movements that seem quick, studied, and precise. Flicks its wings to the side (often after landing). Fairly tame, but circumspect near its nest.

FLIGHT: Short-bodied and wide-winged—but you'll probably fail to note these traits. The bright white of the bird's rump and the base of its tail is arresting. At close range, you can note the black T-shaped tip, but it's not necessary. No other North American thrushlike bird flashes white in the tail. Flight is quick, undulating, sometimes bouncy, and also somewhat buoyant and casual (or perhaps executed with finesse). Wingbeats are fairly quick, mostly steady, somewhat sweeping, and punctuated by intermittent skips or terse closed-wing glides.

VOCALIZATIONS: A short, hurried, mostly harsh-sounding warbled ensemble of chirps, squeaks, and whistles, often given in flight. Some notes have a harsh, screechy, blackbirdlike quality; others recall the more mellow murmured twitterings of Barn Swallow. One ditty is followed quickly by another. Calls include a raspy low "*chak*" as well as a soft, high, whistled "*wheet.*"

Eastern Bluebird, *Sialia sialis*

Eastern Wire-Thrush

STATUS: Common widespread and beloved eastern breeder and winter resident; in the words of the noted naturalist John Burroughs: "The bird that carriers the sky on its back and the earth on its breast." **DISTRIBUTION:** Breeds across the eastern two-thirds of the U.S. and s. Canada (except s. Fla.). The northern range extends to cen. N.S., s. Que., s. Ont. (broken distribution), s. Man., and s. Sask. The western border extends into e. Mont., e. Wyo., e. Colo., and w. Tex. (A disjunct resident population extends from se. Ariz. into Mexico.) Rare in Alta. and w. Colo. In winter, birds retreat south of a line drawn across s. Conn., cen. N.Y., n. Pa., cen. Ohio, cen. Ind., cen. Ill., n. Mo., s. Kans., and n. Tex. and expand west into w. Tex. and e. N.M. **HABITAT:** Frequents a variety of open and semi-open habitats—plowed fields, pastures, roadsides, open ridge tops, open pine forest, parkland habitat. Nests in natural or excavated tree cavities and is readily attracted to nest boxes built to bluebird specs. In winter, particularly in colder regions, birds also concentrate along river bottoms, where standing dead trees offer a surfeit of roosting cavities. Often seen perched on wire fences and utility lines. **COHABITANTS:** In breeding season, Tree Swallow; in winter, in the South, Pine Warbler. **MOVEMENT/MIGRATION:** An early-spring migrant (March to early May), and late-fall migrant (mid-September to late November), Eastern Bluebirds fly in small (6–30 birds) loose flocks with a horizontal spread. VI: 1.

DESCRIPTION: Medium-sized compact thrush (larger than Hermit Thrush; smaller than Gray-cheeked; about the same size as Veery and Western Bluebird), with a large head, a stubby bill, and a short tail. Overall stocky, even portly.

Adult males, with their bright blue upperparts, rust-colored breast (extending to the throat and sides of the neck), and white belly, are distinct; the tonal qualities impart an obvious contrasting pattern. Females are similarly patterned but overall paler; mostly gray above, with the tail and curved outline of the wing tinged with turquoise; orange-breasted, white-throated, and white-bellied below. Juveniles are overall gray and pale spotted.

The posture is hunched, almost slouched. The expression is peeved or cross. (In silhouette, the bird looks like Winston Churchill leaning on a cane. Really!)

BEHAVIOR: A perch-hunting songbird that hunts like a raptor or a shrike by stooping on insects in a fluttering glide, consuming them on the ground, then returning to the same or another perch. Crows and jays may hunt this way as well, and sometimes kingbirds. But no other small perching eastern bird is such a dedicated perch-hunter. Eastern Bluebird is immobile and intense when perched—no wing-flicking, no tail-wagging. The head moves with a jerk when something catches its eye. Overall its stance and movements seem precise, rehearsed, almost theatrical—a beautiful bird that knows all eyes are upon it.

Sometimes plucks prey from the ground without landing and snatches insects from the air in the fashion of a flycatcher. Also eats berries, which it plucks from branches while perched or snatches in flight. Does not hover-hunt to the degree that Western and (especially) Mountain Bluebirds do.

Almost never solitary, except during breeding season, when females are incubating. In fall and winter, forages in small flocks. Birds apportion themselves along utility lines, fenceposts, corn stubble, trees, and hedgerows. When surprised, they may collect in a single tree or on a utility line, then depart.

Quite vocal, particularly in flight. Observers are most often alerted to the bird's presence during migration by its very distinctive call.

FLIGHT: Mostly compact, with longish rounded wings and a short tail. From below, birds appear pale (and males dark-breasted). Flight is undulating, with a buoyant dreamlike quality. Seems to float when it comes in for a landing. Wingbeats are given in a quick series followed by a closed-wing glide.

VOCALIZATIONS: Song is a soft, mellow, somewhat hurried and run-on series of variable whistled or "churring" phrases: "*ch' ch'chrchrchcheer.*" Sounds as though the bird is singing softly so as not to wake anyone. Call is a short, whistled, three-note mantra, "*oo-EEE-ooo,*" that has a soulful dreamy quality. Not loud, the sound nevertheless carries great distances. Has a shorter single- or three-note call that sounds like a muffled muttered version of the common call, and also occasionally makes a nattering series of notes that sound like a Hermit Thrush played at high speed.

PERTINENT PARTICULARS: This bird loves telephone wires—in fact, in proper habitat, utility lines must be accorded favorite perch status. Another point easily overlooked but worth noting is that Eastern Bluebird is an odd, almost idiosyncratic size for an open-country bird. In the East, in open habitat, there are many small species that sometimes sit on open perches, including swallows and sparrows, and some medium-large species, such as starlings and meadowlarks, but very few species that fall in between, as bluebird does. Particularly in winter, the combination of uncommon size and a penchant for utility lines is enough to suggest (if not certify) an identification.

Western Bluebird, *Sialia mexicana*

Hooded Bluebird

STATUS: Common and fairly widespread breeder and winter resident west of the Great Plains. DISTRIBUTION: Breeds in s. B.C., parts of Wash., Ore., Idaho, w. Mont., Calif., w. Nev., s. and e. Utah, w. Colo., n. and e. Ariz., w. and cen. N.M., and the Texas Panhandle (also Mexico). In winter, retreats somewhat from northern and interior portions of its range, concentrating in coastal Ore., Calif., s. Utah, s. Colo., Ariz., w. and cen. N.M., and w. Tex. HABITAT: Breeds in open forests. Preferred breeding habitat is open parklike deciduous and coniferous forest, particularly ponderosa pine, pinyon-juniper, open oak woodlands, and oak-savanna. Also occurs at forest edges, in thinned/managed forests, on golf courses, and in open farmland where suitable nest sites are found. Avoids the large open meadows favored by Mountain Bluebird. In winter, found in open woodlands at lower elevations, as well as in riparian woodlands and agricultural lands. Particularly prevalent where mistletoe and juniper berries are abundant. COHABITANTS: Lewis's Woodpecker, Violet-green Swallow, Western Scrub-Jay, Pygmy Nuthatch, Mountain Chickadee. In winter, often associates with Mountain Bluebird and Yellow-rumped Warbler. MOVEMENT/MIGRATION: Less migratory than the other bluebird species; most birds make altitudinal rather than latitudinal shifts. Arrival dates in vacated areas range from late February to mid-April; noticeable movements are noted from mid-August into November. VI: 1.

DESCRIPTION: A smallish portly thrush—proportionally similar to Eastern Bluebird. The head is large and round, and the body plump, somewhat humpbacked, and potbellied. The bill is slightly thinner and not quite as stout as Eastern Bluebird.

The colors of adult males are richer and darker than Eastern Bluebird, and overall plumage less contrasting. From a distance, the dark rich blue of the head, throat, and back melds with the rust-colored breast and gray belly, making the bird look overall dark. Adult male Eastern Bluebirds never look all dark. Also, the wings and upper back of adult Western males are usually suffused with rusty feathers—a trait not found on Eastern. Western females are muted pastel renderings of the male. Immatures are even paler. Both have gray throats, necks, and bellies (Eastern Bluebirds have white throats, orange collars, and white bellies), and like adult males, are overall less contrasting and more uniformly tailored.

Western Bluebird's expression, like Eastern Bluebird's, is peeved.

BEHAVIOR: In summer, primarily a perch-hunter that drops on and glides toward prey sighted on the ground. Also snatches prey from the air flycatcher fashion and hover-hunts over open fields (much like, but somewhat less than, Mountain Bluebird). In winter, switches primarily to fruits and berries that it gleans from trees while perched.

Perhaps more nervous than Eastern Bluebird. Fidgets when perched, with wings and tail flicking. In winter, found in flocks sometimes with Mountain Bluebird, and where wintering ranges overlap, in Texas and New Mexico, with Eastern Bluebird. Often sit conspicuously, frequently on utility lines. Like other bluebirds, calls frequently.

FLIGHT: Like Eastern Bluebird.

VOCALIZATIONS: Song is a broken halting series of throaty, chirpy "chur" notes, "*chur churchur chur chur*," that are spaced, not run-on, and somewhat harsher and less pleasing than Eastern Bluebird. Call is a soft, plaintive, whistled, single-note "*purr*" or a compressed two-noted "*peurr*" that is lower-pitched and more muffled than Eastern Bluebird. Also has a rapid chattering scold: "*chachachachacha*."

Mountain Bluebird, *Sialia currucoides*

Prairie Bluebird

STATUS: Common species of open habitat across much of western North America. **DISTRIBUTION:** Breeds from cen. Alaska across the s. Yukon, interior B.C., all but n. Alta., cen. and s. Sask., and s. Man.; in the U.S., breeds in all but coastal Wash. and Ore., montane and ne. Calif., Idaho, Mont., Nev., Utah, Wyo., w. N.D., w. S.D., w. Neb., w. and cen. Colo., n. Ariz., n. and cen. N.M., and trans-Pecos Tex. In winter, retreats from Alaska, Canada, and the northern states; found in Ore., Calif., w. and s. Nev., s. Utah, e. Colo., sw. Neb., w. Kans., Ariz., N.M., w. Okla., and w. Tex. south into Mexico. **HABITAT:** Breeds in prairie-forest ecotone that offers a mix of shrubs, trees, and grasses, in burned and clear-cut areas, and in the transition zone between forest and alpine tundra. Like other bluebirds, nests in tree cavities and readily adopts nest boxes. Winters in parkland habitat, open pinyon-juniper woodlands, prairies and grasslands, and agricultural lands (including vineyards). Avoids drier desert areas and favors more open, less forested habitat than Western Bluebird. **COHABITANTS:** Breeding: Lewis's Woodpecker, Northern Flicker, Olive-sided Flycatcher, Tree Swallow, Townsend's Solitaire, Vesper Sparrow. In winter, Say's Phoebe, Western Bluebird, Western Meadowlark. **MOVEMENT/MIGRATION:** Spring migration from late February to late May; fall from late August to late November. VI: 3.

DESCRIPTION: Adult male is unmistakable—a pale blue shrouded bird. Female is paler, grayer, and, except for paler blue highlights in the folded wing, less contrasting than Western Bluebird; *however*, some females and immatures, particularly early in the fall, may have rufous-tinged breasts like female Western and Eastern Bluebirds.

Overall slightly longer, slimmer, and more angular than Western or Eastern Bluebirds, and stands more stiffly erect. (The profiles of the other two bluebirds appear stocky and slouched.) The head on Western seems flatter and proportionally smaller, and the bill longer and slimmer, giving the face a sharper appearance (accentuated by a pencil-fine line that runs from the bill to the eye). The back is generally straight (not hunched) and may even be slightly concave or swaybacked. The wings and tail are longer than Western and Eastern Bluebirds, but appear well proportioned in relation to the overall lankiness of the bird. The expression is benign.

BEHAVIOR: Like other bluebirds, a perch-hunter that searches visually for prey from an elevated vantage point. More than other bluebirds, uses weedy stalks, even ground perches. More than other bluebirds, hovers in the fashion of American Kestrel in search of prey. Highly social; in winter often found in large flocks (more than 100 birds), and commonly mixes with other bluebird species. Less nervous than Western Bluebird, with little wing- and tail-flicking. Also generally less vocal.

FLIGHT: Silhouette seems slender, small-headed, long-winged, and slightly hunchbacked. Flight is direct and casual, with a dreamlike quality. A fluttering series of wingbeats is punctuated by an open-winged glide on slightly down-angled wings.

VOCALIZATIONS: Song is a series of unevenly spaced, slurred, slightly vibrating "chur" notes: "*chr'r chr'r ch' cher ch'l.*" Sounds like a hoarse (and slightly drunk) Purple Martin. Call is a soft "*chur*" also uttered with a slight vibrancy. Also makes a plaintive, low, descending, single-note whistle, "*few*"; a low mellow trill; and a loud scuffing "*chuh*" or "*chuh/chuh.*"

Townsend's Solitaire, *Myadestes townsendi*

The Juniper Bluebird

STATUS: Common western bird of mountain forests and, in winter, lower elevations. Some more southern birds are permanent residents. **DISTRIBUTION:** Breeds in mountainous regions from cen. Alaska and the Yukon to cen. Mexico. In winter, vacates breeding areas north of s. Alta. and s. B.C. and expands east into the western prairies and south across more of Mexico. Rarely wanders farther east or to the West Coast. **HABITAT:** In summer, coniferous forest dominated by pines, up to the tree line. Also (less commonly) found on rocky outcroppings on alpine tundra. In winter, retreats to lower elevations where junipers dominate. Although less common outside the juniper belt, where found, the bird is always associated with an abundant berry crop (or, in summer, insects). **COHABITANTS:** Olive-sided Flycatcher, Mountain Bluebird; in winter, Pinyon Jay. **MOVEMENT/MIGRATION:** Spring migration in March and April; fall from September to November. VI: 3.

DESCRIPTION: An elongated, gray, snub-billed bluebird that perches high and doesn't like company. A large thrush (more than 1 in. larger than Mountain Bluebird; the same size as Spotted Towhee), Townsend's Solitaire has an overall slender and tubular shape—the head, body, and overwide tail run together with little in the way of distinguishing contours. The head is smallish (the thickset neck exaggerates the size), and the bill short, straight, and black—an Eastern Bluebird's bill fitted to a stretched-out Mountain Bluebird's body. Also, you cannot fail to notice the length of the tail; long and incongruously wide, the tail sometimes has a deeply notched tip, and at other times it tapers to a bulletlike point.

The posture is ridged, straight-backed, and nearly erect, with the bill slightly elevated. Townsend's Solitaire is often likened to a mockingbird (which it recalls in color and relative proportions), but solitaire stands at attention on its perch and does not slouch, as mockingbirds often do.

Overall medium gray—grayer than Northern Mockingbird, not as dark as Gray Catbird. The bold white eye-ring is evident at any reasonable distance. Salmon-colored slashes near the tip of the folded wing may be conspicuous or concealed. The white outer tail feathers are often evident only in flight. The expression is one of sympathetic concern.

BEHAVIOR: During the breeding season, furtive. A phantomlike bird that moves silently through foliage and lands with a balancing flutter. Seems very trogonlike in both shape and manner. In winter, a sedentary sentinel. A solitary bluebird that likes to sit conspicuous and high—in breeding season atop a pine, in winter preferably atop a juniper, but a telephone pole, utility line, or any high conspicuous perch will do. Forages by flycatching, both above and below the canopy. In winter, forages on juniper berries by plucking drupes while perched or by landing on the ground for fallen fruit. Also flutters and hovers to pluck fruit on the exterior of the tree and moves vertically within the tree by employing vigorous exaggerated flapping (the shape and manner recalling a feeding Elegant Trogon).

In winter, a loner. Aggressively defends its patch of junipers from other solitaires, of which, in proper habitat, there may be a number. Flights are usually short, but birds also make longer flights, often at considerable height, sometimes disappearing from view.

Responds readily to screech-owl and Northern Pygmy-Owl imitations.

FLIGHT: In flight, recalls an elongated snub-nosed bluebird, with a buff-colored wing-strut running the length of the underwing. Silhouette is long, slender, and tubular, with disproportionately short candle flame–shaped wings. For long flights, flies somewhat like American Robin. Flight is direct, slightly twisty-turny, sometimes jerky or skippy, and slightly wandering, but always more buoyant than a robin. Wingbeats are irregular, given in a terse series followed by an abbreviated closed-wing glide; down strokes are deep and exaggerated, with wings held close to the body. For short flights, wings and tail are broadly flared, and the flight bouncy and jerky (recalling a towhee). The white outer tail feathers are standout features.

VOCALIZATIONS: Song is a lengthy, hurried, unpatterned warble comprising clear whistles, low chirps, and musical trills—some high, some low. For example: "*chur ee ch'chr cheecheechee chur che churl chchch chee eeeee.*" Sounds a bit like a hurried, high-pitched, long-playing American Robin. In winter, gives a series of rough, short, vireo-like phrases. Call is a short, clear, somewhat ethereal, whistled "*HEE*" that recalls a squeaky bicycle wheel. One vocalizing bird often triggers a choral response from others.

Veery, *Catharus fuscescens*

The Thrush That "Veers"

STATUS: Common forest breeder in eastern Canada and the northeastern United States; less common in the northern Rockies and interior Pacific Northwest, the western limits of its range. **DISTRIBUTION:** Breeds from s. Nfld. and the Maritimes across s. Que., s. Ont., s. Man., s. and cen. Sask., s. and cen. Alta., and s. B.C. In the U.S., breeds throughout New England, n. N.J., Pa., and n. Ohio, along the Appalachians to n. Ga., and in Mich., all but s. Wisc., Minn., n. and e. N.D., Mont., Idaho, e. Wash., e. Ore., n. Utah, w. and s. Wyo., n. and cen. Colo., and cen. N.M. **HABITAT:** Breeds in thick, damp, deciduous woodlands. Prefers forest in early successional growth or, in more mature forest, disturbed areas or areas near streams where vegetation is thick. In some parts of its range, also breeds in mixed deciduous-coniferous woodlands and conifers. **COHABITANTS:** Red-eyed Vireo, Wood Thrush, Canada Warbler, Willow and Alder Flycatchers. **MOVEMENT/MIGRATION:** Entire population winters in South America. During migration, selects heavily vegetated portions of forests and forest edges—essentially the same habitat in which it chooses to breed. Spring migration is brief. Birds reach the southern United States by mid-April and arrive on breeding grounds by mid-May. In fall, migrates generally a week or two earlier than the other spot-breasted thrushes; the first birds appear in mid-August, and peak movements last from late August to mid-September. VI: 2.

DESCRIPTION: A medium-sized classically proportioned thrush—larger and more elongated than Hermit Thrush; smaller and more slender than Wood Thrush; similar in size to Swainson's and Gray-cheeked, but perhaps longer-faced. Overall the plainest, least patterned of the spot-breasted thrushes and, except for Wood Thrush, the ruddiest.

Upperparts—head, back, wings, tail—are wholly plain ruddy brown (the color is unique to this species); underparts are whitish (slightly browner in the West). The face pattern is weak and indistinct. The breast is lightly spotted, and the spots lose themselves in a warm blush the same color, like the spots, as the bird's upperparts. On the other thrushes, the spots are larger and darker (often blacker).

If a bird is slender and ruddy above, and the breast looks like the spots bled when the bird was put in the wash, it's Veery.

BEHAVIOR: Shy. Spends most of its time on the ground, but also forages in the understory—particularly when bushes are fruiting. Hops and turns over leaves. Posture is generally horizontal. Movements seem stiff, poised, almost theatrical, as if the bird knows it's standing before an appreciative viewer. Flies frequently. Pauses for long periods, studying the surroundings. Also perch-hunts from branches or logs, searching for prey, dropping to the ground to seize it. Sometimes makes short jumps to snatch berries. Flicks its wings and has an abbreviated tail wag that is less vigorous than Hermit Thrush.

FLIGHT: Strong, fast, direct, and short—generally to the nearest cover.

VOCALIZATIONS: A thrush that calls its name. Song is a descending series of somewhat slurred, somewhat ethereal notes. The first two notes sound as if they are inhaled; the rest are exhaled: "*veee-Ver-v'r-v'r-v'r-v'r.*" Call is a soft "*vrrr,*" sometimes uttered with a two-note quality. Also emits a higher-pitched "*vee-r-ry*" or "*v'ree.*" Flight call is similar; though it varies in pitch and enunciation, it remains mercifully unchanged in character. The bird says, "*Veer.*"

PERTINENT PARTICULARS: On the West Coast, birders are likely to encounter the "Russet-backed" form of Swainson's Thrush, a bird whose warmer upperparts, weaker face pattern, and weaker spotting make it a candidate for confusion with Veery, which is very rarely encountered along the West Coast.

Gray-cheeked Thrush, *Catharus minimus*
Thicket Thrush

STATUS: Common northern breeder, but in migration, less commonly encountered than Swainson's Thrush. **DISTRIBUTION:** Like the taiga and bordering tundra habitat that sustain it, Gray-cheeked's breeding range spans n. North America from Alaska (excluding the south coast), across most of the Yukon, southern portions of the Northwest Territories, n. Man., n. Ont., n. Que., Lab., and Nfld. Casual in the w. U.S. **HABITAT:** Dense shrubs of moderate height, both within a closed canopy and without. May prefer spruce forest, forest edge, or isolated stands of spruce (as well as other coniferous or deciduous trees), but readily occupies willow and alder thickets in open tundra and regenerating forest (without canopy). In migration, also seeks out dense woodland understory and avoids woodlands with open forest floor. **COHABITANTS:** Willow Ptarmigan, Spruce Grouse, Blackpoll Warbler, Fox Sparrow, American Tree Sparrow. In migration, Swainson's Thrush. **MOVEMENT/MIGRATION:** Passes east of the Rockies. Spring migration from late April to early June; fall from late August to mid-October. VI: 2.

DESCRIPTION: A gray-faced spot-breasted thrush. Medium-sized (slightly larger than Hermit or Bicknell's; smaller than Wood Thrush; about the same size as Swainson's Thrush and Veery), with a short stout bill, a round head, a somewhat plump or sagging body, and a short tail. Overall slightly more robust than Bicknell's (and possibly Swainson's).

Basically plain; grayer and colder-toned than other spot-breasted thrushes. Upperparts hesitate between grayish and olive brown. Underparts are whitish with a grayish wash on the sides and flanks and blackish spotting on the breast that varies in extent and degree (some birds are lightly spotted, some heavily). Key to the identification of this bird is the face, which is plain and cold-toned, if not downright gray (or flecked with gray), with a pale grayish white crescent behind the eye. Shows no trace of warm buff on the cheeks or well-defined buffy "spectacles" surrounding the eye.

Given a good look, the tail of Gray-cheeked may appear slightly warmer and browner than the balance of the upperparts, but Bicknell's tail is ruddier, and Hermit's tail is chestnut.

BEHAVIOR: In migration, a thicket skulker—a shadow-colored thrush moving through shadows. On territory, the bird is positively flamboyant. Males sit conspicuously high, in the open, singing lustily. Feeds primarily on the ground in the halting stop-and-go fashion of a robin, but hops (doesn't run). Also forages on berries (particularly in fall) and foot-patters in the leaf litter of the forest to activate insects.

In migration, vocalizes at night (most commonly at and just before dawn) and sings and calls during the day. Fairly shy, but responds to pishing.

FLIGHT: Stocky body and long broad wings. Like all spot-breasted thrushes, Gray-cheeked has underwings that show a pale wing-strut. Unlike Swainson's, shows a slightly ruddy tail. Flight is strong, fast, often weaving (around obstacles), and slightly halting or hesitant. Wingbeats are given in a near-continuous series, but punctuated by frequent, intermittent, terse skip/pauses.

VOCALIZATIONS: Song is a hurried, two-part, descending series of nasal wheezes: "*Ree'e Ree'e; reer re'r'r.*" Variable flight call is a descending, nasal, often somewhat two-noted *"whee'er!" or "quee'er"* that is similar to Veery's "*veer,*" but is harsher, shriller, more yelped and urgent sounding.

Bicknell's Thrush, *Catharus bicknelli*
The Not Quite Gray-cheeked Hermit-tailed Thrush

STATUS: Uncommon northeastern breeder; rare eastern migrant. **DISTRIBUTION:** Breeds primarily along the Gulf of St. Lawrence in e. Que. and nw. N.B. (also Cape Breton I., N.S.). In the ne. U.S., found in widely scattered locations (usually at higher altitudes and in stunted vegetation) in w. and cen. Maine, the White Mts. of N.H., the Green Mts. of Vt., and the Adirondack and Catskill Mts. of N.Y. Migrates to and from the Greater Antilles east of the Appalachians. **HABITAT:** Montane forests (at elevations exceeding 3,000 ft. in the United States; at lower elevations in much of Canada) dominated by young balsam fir with some spruce, white birch, and mountain ash.

Most often found in disturbed, stunted, regenerating, or transition habitats — traditionally in areas of high wind, severe winter ice pack, or fire, but more recently in areas disturbed by logging, road or power-line cuts, and ski resort trails. In migration, found in shady dense woodlands and shrub and scrub forests in coastal areas. **COHABITANTS:** Swainson's Thrush, Blackpoll Warbler, White-throated Sparrow, Dark-eyed Junco, Red Squirrel. **MOVEMENT/MIGRATION:** Spring migration from early May to early June; fall from early September to early November. VI: 1.

DESCRIPTION: Small eastern spot-breasted thrush with the face of Gray-cheeked and a ruddy tail more reminiscent of Hermit Thrush, *but usually very difficult to distinguish from Gray-cheeked.* Small (slightly smaller than Gray-cheeked and Swainson's; the same size as Hermit Thrush), with a short stout bill (extensively yellow on the lower mandible), a somewhat slender body, and a short tail. Overall more compact than Gray-cheeked. Appears shorter-tailed than Hermit Thrush.

Olive brown above, dingy white with a heavily spotted breast below. Closely resembles Gray-cheeked, but the back is slightly warmer brown, the face is a duller brown and usually plainer (less flaked or cracked), and the tail and lower edge to the wings are dull reddish (recalling a slightly dull Hermit Thrush).

BEHAVIOR: A bird that likes to stay hidden. On territory, found in pairs; in migration, usually solitary, foraging on or near the ground in thick cover. Reported to be fairly nimble and a rapid feeder — mixing short bouts of springy hops with short flights (like Swainson's Thrush?). Also leaf-gleans, hovers, and pursues flying insects. Males sing from exposed perches on both live and dead trees, but the song period is brief — late May to early July. During flight songs at dawn and dusk, birds rise abruptly, circle, and drop abruptly upon completion of the song. Does not habitually raise and lower its tail like Hermit Thrush, the other rusty-tailed thrush.

FLIGHT: Like Gray-cheeked, but shorter-winged and showing a more contrastingly reddish tail.

VOCALIZATIONS: Song is like Gray-cheeked — a series of descending phrases — but is thinner, wiry, higher-pitched, and more complex and usually rises at the end: "*wh whr ch'ch'cher cher ch'cherEE.*" Descending call suggests a very high nasal Veery and has a compressed two-noted quality: "*pee' urr.*" Flight call is like Gray-cheeked's descending nasal yelp, but is often terser, higher, flatter; more peeved than urgent-sounding.

PERTINENT PARTICULARS: What commonly happens when you encounter a Bicknell's Thrush is that you see a bird that you think is Gray-cheeked, but you can't seem to get that sanctifying and confidence-building sense of no-nonsense gray about the bird's face. Then you look at the tail and see that it is really pretty red. You think, *Maybe Hermit?* Then you notice that the tail is not animate. When an identification is vacillating between Gray-cheeked and Hermit, consider Bicknell's.

Swainson's Thrush, *Catharus ustulatus*

Spectacled Thrush

STATUS: Common northern and western breeder. Widespread migrant — the olive-backed thrush most likely to be seen even where Gray-cheekeds migrate. **DISTRIBUTION:** Breeds throughout the boreal forest, from the cen. sub–Brooks Range in Alaska across Canada (except s. Alta., s. Sask., s. Man., and extreme se. Ont.). In the U.S., breeds widely in Wash., Ore., coastal Calif. (also the Sierras), Idaho, ne. Nev., w. and cen. Mont., Wyo., n. and cen. Utah, w. and cen. Colo.; also found in scattered locations in n. Ariz., n. N.M., and sw. S.D. **HABITAT:** Varied, almost eclectic. Primarily a bird of mature tall coniferous forest with a preference for old growth and spruce-fir. In some places, the bird occupies deciduous riparian woodlands and in other places uses younger regenerating stands of conifers as well as mixed coniferous-deciduous woodlands. Seems most partial to habitats that offer a tall-tree component and a well-developed, fairly diverse understory. It also seems most likely to thrive at altitudes or forest strata away from the competition (Gray-cheeked, Bicknell's, Hermit, Varied). In migration, this flexibility translates into a more generalized use of habitat. In migration, uses forests with open understory, willow thickets, swamps, parks, suburban wood lots—many of which are more open and accessible to birders, explaining at least in part why Swainson's Thrush is more often encountered than brush-loving Gray-cheeked where their migratory ranges overlap. **COHABITANTS:** Across much of its breeding range, Boreal Chickadee, Gray Jay, Hermit Thrush, red squirrel; in coastal California, Wilson's and Yellow Warblers. In migration, Gray-cheeked Thrush, Veery. **MOVEMENT/MIGRATION:** Spring migration from late April to mid-June, with a peak from mid to late May; fall from early August to mid-October, with a peak in late August and early September. VI: 2.

DESCRIPTION: The bespectacled olive-backed thrush. Medium-sized (slightly larger than Hermit or Bicknell's; smaller than Wood; about the same size as Gray-cheeked and Veery), with a short stout bill, a rounded head, a well-proportioned body, and a short tail. Overall not as slender as Veery, but perhaps not quite as roundly compact as Gray-cheeked.

Upperparts are olive gray or olive drab over most of the bird's range, with warmer rufous cast in the Pacific Coast population. Underparts are grayish white with blackish spotting on the breast and an olive gray wash at least on the flanks (and sometimes the sides). Very similar to Gray-cheeked except for the face, which is overall warmer, buffier, and paler, particularly on the lores, which on Gray-cheeked are always gray. At close range, the pale lores and eye-ring make Swainson's look like it is wearing buffy spectacles. At a distance, the face of Swainson's just looks paler than the back, whereas the face of Gray-cheeked shows little contrast.

BEHAVIOR: More than its cousins, an arboreal hunter, spending a greater percentage of its foraging time in the midforest layer with its feet off the forest floor (but also hunts on the ground). Often hunts by flying from perch to perch—like Northern Goshawk, but the thrush moves every 10–15 secs., and the goshawk every 10 mins. Searches the ground from elevated perches and stoops or lunges for prey. Also flycatches and hover-plucks. Even when working the forest floor, Swainson's stays on the move and likes to keep its feet out of the dirt. It moves with springy hops, flies to a branch or stump just a few inches off the ground, sizes up the turf, then lunges for prey or flies to the ground for another series of hops or to another low perch. Because it stays on the move and hunts actively, it seems more at home in open or semi-open habitats than most other spot-breasted thrushes (with the notable exception of Hermit). In fall, often perches in low fruiting bushes, plucking or lunging for berries.

Males sing from high exposed perches, often for lengthy periods, and make an assortment of vocalizations between song sets off and on during the day. Shy and generally unobtrusive. Responds mildly to pishing and owl calls, but remains well back in the understory. Vocalizes at night, during migration, and during the day, particularly before sunrise.

FLIGHT: Compact body with long wide wings. Uniformly dark above, pale below, with a darker chest. The warm or pale face is often apparent. Flight is strong, fast, and mostly direct, with steady wingbeats executed with intermittent slight hesitations that result in a flight that is not exactly halting but not exactly seamless and smooth either.

VOCALIZATIONS: Song is wonderful. A hurried ascending tumble of notes that climb the scale in serial bounds: "*wr'wr'ooeeooOoEeOOEE.*" More ethereal, less harsh and nasal, and more seamless than Gray-cheeked, whose song is distinctly two-parted and runs down the scale. Swainson's song runs up the scale with no break in the sequence or pattern. Calls include a soft, mellow, liquid "*whit.*" Flight call is a short, plaintive, mellow "*peeh?*" that recalls the call of the common eastern spring peeper (a frog). So if it's May or August/September and you hear a spring peeper calling from somewhere in the vicinity of Cassiopeia, it's probably not a frog.

Hermit Thrush, *Catharus guttatus*

The Winter Thrush

STATUS: Common widespread northern and western breeder occupying a variety of forest habitats. The only spot-breasted thrush that winters in the United States (and southern Canada). **DISTRIBUTION:** Breeds from Nfld., the Maritimes, and se. Lab. across s. Que., Ont., and all but sw. Man., s. Sask., and se. Alberta; also breeds in B.C., s. Yukon, and southern and interior western portions of Alaska. In the e. U.S., breeds across New England, N.Y., n. N.J., most of Pa., and south in the Appalachians to N.C., n. Mich., n. Wisc., and n. Minn. In the w. U.S., breeds in w. and cen. Mont., all but sw. Idaho, portions of Wash., Ore., n. and cen. Calif., Nev., Utah, w. and s. Wyo., w. and cen. Colo., n. N.M., and Ariz. In winter, found along both coasts from s. New England and s. B.C. south. In the interior U.S., winters south of a line drawn from s. N.Y., n. Pa., cen. Ohio, s. Ind., s. Ill., cen. Mo., cen. Okla., extreme n. Tex., cen. N.M., cen. Ariz., s. Nev., and cen. Calif. south into Central America. **HABITAT:** Breeds in mature coniferous and hardwood forest, mixed forest, open northern taiga, and riparian canyons in the Southwest (among other woodland and quasi-woodland habitats). Forests may be wet or dry, but when breeding, more than other thrushes, Hermit Thrush prefers drier, more open understory and favors edge over forest interior. In winter, partial to wet, heavily vegetated, and berry-rich thickets and woodlands (often riparian or near springs or streams as well as well-vegetated residential areas). **COHABITANTS:** During breeding, Swainson's Thrush, Yellow-rumped Warbler, Dark-eyed Junco. In winter, Winter Wren, Gray Catbird, Fox Sparrow. **MOVEMENT/MIGRATION:** Migrates earlier in spring and later in fall than the bulk of the spot-breasted thrushes. Spring migration from April to mid-May; fall from mid-September to late November. In migration, may be found in the habitats described above as well as in the more vegetated portions of city parks and suburban yards. VI: 2.

DESCRIPTION: A compact spot-breasted thrush characterized by its small size and upright tail-cocked posture. If you have seen an unidentified spot-breasted thrush in North America between and including November and March, *stop now*—the bird is a Hermit Thrush. Otherwise, continue.

Small (about the same size or very slightly larger than Fox Sparrow) and compact, not quite so long and lean as the other brown- or gray-brown-backed spot-breasted thrushes (Swainson's, Gray-cheeked, Bicknell's, and Veery). Gray-brown upperparts contrast with the rustier lower edge to the folded wing and (particularly) the rusty tail. Bold black breast spots stand out against a whitish chest (washed with buff in other spot-breasted thrushes), and in very poor light, the white chest stands out.

BEHAVIOR: The bird signs its name with its tail. Signature characteristic is a quick upward jerk of the tail, which is then slowly lowered. Rapid wing-flicks accompany the single or multiple tail-jerks. Other thrushes sometimes jerk their tails (and wings); Hermit Thrush does it habitually.

Forages on the forest floor and in the understory. Particularly early in the morning (dawn), may feed alongside paved roads. On the ground, moves with a short series of springy energetic hops followed by a pause—like a hopping robin, only faster and with an air of impatience. Often flies to capture prey. Sometimes runs or walks. Gleans insects and berries from branches (generally within 10 ft. of the ground). Moves by hopping branch to branch; sometimes flutter-hops. Energetic. Usually punctuates each relocation with a jerk and a lowering of the tail.

FLIGHT: Compared to other spot-breasted thrushes, fairly compact, with shorter wings. Contrasting ruddy tail is fairly obvious; often shows a narrow pale bar on the *upperwing*. Typical flights are short, taking the bird to the nearest cover or a point less than 30 ft. away. Short flights are undulating, given with a hurried series of wingbeats punctuated by terse, open-winged, floating or braking glides. Longer flights across open areas are strong, direct, and fast.

VOCALIZATIONS: Varied, ethereal-sounding, two-part song begins with a held clear note followed by a high spiraling warble that sounds like it is ascending; for example: "*eeee chireel'ee* (pause) *urrr chee'eee'eee.*" The distinctive opening notes always vary in pitch, from low and mellow to high and shrill. Successive songs differ. Calls include a low admonishing "*chuck*" or "*churp*" that may be repeated at long intervals or given in a more rapid series of two or three. Most frequently at dawn, makes a high-pitched, assertive, rising whine, "*eeeeeh!*" that recalls Spotted Towhee's call. Flight call is a short, clear, plaintive whistle, "*peeur*" that is halfway between the truncated "*peep*" of Swainson's Thrush and the descending "*Veer*" of Veery.

PERTINENT PARTICULARS: In size and plumage, Hermit Thrush is most like Bicknell's Thrush—a northeastern spruce forest nester. The eastern form of Hermit has warm buffy flanks (gray in Bicknell's), a white eye-ring (grayish in Bicknell's), and black spots on a white chest (black over a buffy wash on Gray-cheeked).

Hermit is very responsive to pishing, responding quickly to imitations of a screech-owl, Northern Pygmy-Owl, or Northern Saw-whet Owl. Listen for the low "*chuck*" note.

Wood Thrush, *Hylocichla mustelina*

The Really Spotted Thrush

STATUS: Common breeder in eastern forests. **DISTRIBUTION:** Breeds across se. Canada from s. N.S. across s. Que. and se. Ont. to the eastern border of Lake Superior. In the U.S., breeds east of a line drawn from cen. Minn. to se. S.D., e. Neb., e. Kans., cen. Okla., and e. Tex. Does not breed in coastal Tex. or La. or on the Florida Peninsula. **HABITAT:** Found on the edge and interior of moist woodlands with a healthy but not necessarily dense understory and open forest floor with ample leaf litter. Partial to woodlands with streams or swamps; also occupies suburban woodlots. **COHABITANTS:** Acadian Flycatcher, Red-eyed Vireo, American Redstart, Hooded Warbler, Kentucky Warbler. **MOVEMENT/MIGRATION:** Completely vacates breeding territory in winter, migrating to southern Mexico and Central America. Spring migration from April to May; fall from late August to mid-October. In migration, seeks out a variety of woodland habitats (particularly those with berries), but also found in fairly open habitats with a forest component (like city parks). VI: 2.

DESCRIPTION: A large robust forest thrush—a standout among the spot-breasted thrushes. The body is unmistakably plump, the profile long-legged and stocky, and the head and bill large. Easily distinguished by its heavily spotted underparts and uniquely orange- or russet-toned head and neck, which, in deep shadows and compared to the bland gray-brown uniformity of most spot-breasted thrushes, seems almost to have a Day-Glo quality. Spots are larger, blacker, more extensive, more widely spaced, and, set against the bird's pure white underparts, more prominent than on other thrushes.

BEHAVIOR: A denizen of the forest floor, the understory, and shadows. Spends most of its time hopping through the leaf litter, flipping over leaves, probing for prey. Moves in the stop-and-go manner of a robin, but hops. Also hops when foraging among tree branches. Even when vocalizing, males typically remain in the understory, shunning the canopy. A persistently vocal dawn singer, often singing late into the morning and occasionally during the day. Also vocal at dusk, when shadows cement the trees.

FLIGHT: Direct and strong, somewhat robinlike, with a perfunctory businesslike quality.

VOCALIZATIONS: A celebrated songster. Songs are often, and not very accurately, described as flutelike. The pattern and tempo might be reproducible on a flute, but the quality of the sound, while no less pure-toned, doesn't sound like a flute. Varied song consists of an ensemble of rising-and-falling short phrases that have a dreamlike surreal quality. Song often begins with two harmonic stutters, then breaks into a short varied yodel that often ends on a held high note; for example: "*Eh-Eh oodle-oodle-eeee* (pause) *Er-Er pee-o-Ree* (pause) *Er-Er oddle-oodle-oodle EE.*" Common call is a low, descending, amphibian-like stutter, "*urh, urh, urh-urh-urh,*" that sounds as if the bird is clearing its throat. Also makes a descending "*pup-pup-pup*" (unflatteringly referred to as the "wet fart call"). Flight call, "*juee,*" is similar to several other thrushes, but more abrupt, vibrating, and harmonic.

Clay-colored Robin, *Turdus grayi*

American Robin in a Plain Brown Wrapper

STATUS AND DISTRIBUTION: Generally uncommon and very restricted, mostly winter visitor from Mexico, but has bred in the U.S. and a small resident population probably exists. In the U.S., found almost exclusively in the lower Rio Grande Valley, primarily from Bentsen–Rio Grande Valley State Park east to Weslaco, although in years of above-average influx, birds have wintered in the suburbs of McAllen and Brownsville, and a few have been found upriver as far as Zapata. **HABITAT:** In the United States, found in low dense thickets and riparian forest. In Central America (and occasionally the United States), forages on lawns (in much the same fashion as American Robin). **COHABITANTS:** White-tipped Dove, Olive Sparrow, needy listers. **MOVEMENT/MIGRATION:** Some Mexican (?) breeders disperse north in winter, with greatest numbers found in the Rio Grande Valley between December and March. VI: 0.

DESCRIPTION: Literally, a clay-colored robin. A large thrush (the same size and shape as American Robin, though perhaps slightly plumper) that is all brown—olive brown above, slightly paler brown and salmon-tinged below. The bill is dull olive.

BEHAVIOR: Robinlike, feeding mostly alone on the ground. Unlike American Robin, is most partial to foraging in the leaf litter beneath a thick understory. Also ventures into the open, hopping (in the halting stop-and-go fashion of American Robin) along edges and trails. Spends considerable time in trees, feeding on fruit in the canopy or perched, unmoving (except for an occasional jerk of its tail when it calls), on sturdy exposed limbs. Fairly tame, but more retiring where persistently pursued for a better view.

FLIGHT: In size, shape, and manner, like American Robin. Shows warm ruddy tones on the underwings. Flight is straight, strong, and very slightly bouncy, with deep regular wingbeats. In woodlands, flights are often just beneath the canopy and lengthy.

VOCALIZATIONS: Song sounds like American Robin imitating a thrasher. Calls include a quavering, slurred, compressed, three-note whistle: "*ee/ur'urh*." Also makes a low, muffled, robinlike cluck, "*churk churk churk*," and a "*skee*" reminiscent of American Robin's flight call.

American Robin, *Turdus migratorius*

Lawn Plover

STATUS: Common, widespread, almost ubiquitous. Although not North America's most abundant bird, almost certainly its most familiar. **DISTRIBUTION:** Breeds from the Bering Sea to s. Mexico and everywhere in Canada and the U.S. except for the Mojave Desert, s. N.M., s. and coastal Tex. and La., and most of the Florida Peninsula. Winters across much of s. Canada—s. Nfld., N.S., the south coast of N.B., se. Ont., s. Alta., and s. and coastal B.C. north, along the coast, to Kodiak I.—and everywhere in the U.S. except n. Maine. **HABITAT:** Breeds in a variety of habitats, including deciduous woodlands in the Northeast, plantation pine forests in the South, ponderosa pine forests in the northern Rockies and Idaho, spruce forests in northern Canada, and willow thickets on the North Slope of the Brooks Range. However, American Robin seems most habituated to, and is most often associated with, urban and suburban gardens, parks, and yards. In virtually all habitats, seeks at least some woody vegetation (for nesting and perching) and open areas that permit ground foraging (such as plowed fields, suburban lawns, or open understory). Also readily nests on human structures, such as porches and windowsills. In winter, gathers in large flocks that roost and forage in woodlands at lower elevations, including holly forests, orchards, and riparian woodlands. Below the snow belt, also forages on lawns and plowed pastures. **COHABITANTS:** Varied. In winter, often found with Cedar Waxwing. **MOVEMENT/MIGRATION:** Migrates extremely early in the spring—returning as soon as the ground partially thaws. Movements occur as early as February; migrants reach Alaska in April. The bulk of the fall migration occurs in October and November. VI: 4.

DESCRIPTION: A big, dark, chesty thrush with slate gray upperparts and a red breast. The female's breast is paler, more orange than red, and shows a slightly duller or paler head. The juvenile's breast is orange-washed and overlaid with measleslike dark spots.

Perches with bill elevated slightly. Demeanor is haughty.

BEHAVIOR: Forages in the open, on the ground. Runs and stops, runs and stops, with head erect; moves with dignity and poise. Wingtips often droop below the tail. Turns its head to peer down. Jabs forward with the bill and often extracts an earthworm. Young birds frequently hop, as do adults where the ground is covered by leaves. In summer, generally tame, allowing close approach. In winter, when gathered in large foraging flocks, less approachable. Also forages in trees for fruit, which it plucks from the tree while perched or by leaping or flying and snatching from branches. Very vocal. In summer, sings at first light, even at night (particularly in areas illuminated by artificial light). In winter, flocks are noisy, vocalizing an array of calls that are not complex and generally similar in quality.

FLIGHT: Short-distance flight is perfunctory and direct. Lands with braking flaps. Long-distance flight is strong, direct, and fast, sometimes twisting, more often even-keeled; though strong, also seems slightly jerky or halting, as though the bird is struggling somewhat from either towing a load or flying into the wind. Or perhaps running out of gas (and the engine is sputtering on fumes). Wingbeats are crisp, almost violent, and given in a rapid, flickering, slightly irregular or halting series punctuated by a brief half-closed-wing glide.

An all-dark bird with an evenly contoured, potbellied profile and a decidedly pale undertail. From below, the wide round wings dominate the silhouette.

Flocks are loose (much looser than blackbirds) and often have a wide horizontal spread. Birds are widely spaced and generally hold their place in the flock; there is surprisingly little shifting despite the lack of regimentation.

VOCALIZATIONS: Almost as recognizable as the sound of a sprinkler, the robin's song is a series of short two-note musical phrases that swing up . . . swing down . . . swing up . . . swing down . . . followed by a savoring pause. The second set follows. Then a third . . . and a fourth . . .

Calls include single cluck notes; a clucking "*beep*" followed by a three-note chuckle; and a loud, emphatic, protesting "*jeep! jeep!*" Flight call is a thin descending shriek with an emphatic ending: "*zzzek.*"

PERTINENT PARTICULARS: Rufous-backed Robin, *Turdus rufopalliatus,* is a rare visitor from western Mexico to the southwest United States. Slightly smaller than American Robin, with a rich yellow bill and more slender face, Rufous-backed wears a rufous saddle on the back and shows more limited rufous below and no white about the eyes. Shy, it commonly forages in woods and brush, but in the United States it may forage with flocks of robins on open lawns. Flight is deeply undulating (somewhat cardinal-like), with swooping open-winged glides.

Varied Thrush, *Turdus naevius*

Collared Robin-Flicker

STATUS: Common northwestern breeder; West Coast winter resident. **DISTRIBUTION:** Breeds throughout most of forested Alaska, the Yukon, w. Northwest Territories, B.C., sw. Alta., Wash., n. Idaho, w. Mont., w. Ore., and nw. Calif. In winter, birds breeding in the interior of Alaska, Canada, Mont., and Idaho withdraw to more temperate areas. Winter distribution ranges coastally from the s. Kenai Peninsula of Alaska to s. Calif. and inland across s. B.C., most of Wash., w. Ore., and much of Calif. west of the Sierras; also winters, irregularly, to w. Mont., e. Ore., w. Nev., the Ariz. border, and n. Baja. Very rare but annual winter visitor to the East, often at feeders. **HABITAT:** Classically a bird of mature, dense, unfragmented, moist coniferous forests but also Alder-thickets and poplar groves in the far north. In winter, more habitat-expansive, occupying parks, gardens, oak woodlands, and riparian areas as well as classic forest habitat. **COHABITANTS:** Gray Jay, Steller's Jay, Chestnut-backed Chickadee, Boreal Chickadee, Winter Wren, American Robin, Red Squirrel. **MOVEMENT/MIGRATION:** Spring migration from mid-March to early May; fall from late August to late November. VI: 3.

DESCRIPTION: An arrestingly plumaged, treetop-singing, ground-foraging, robinlike thrush — with flicker sympathies. A large songbird that is slightly smaller and shaped much like American Robin, but with a longer, thinner bill, a narrower head, a longer neck, a heavy body, and a shorter tail. Shape falls somewhere between American Robin and Northern Flicker.

Plumage is distinctive—a bold mosaic of black, blue gray, and orange makes Varied Thrush look like a robin that's been decorated for Halloween. The orange eyebrow and wide black collar bisecting bright orange underparts are obvious at almost any distance. Underparts are much brighter orange than on American Robin. The black collar might easily pass for the V on the chest of a flicker. The orange wing-bars and etched lines in the wings are not easy to see at a distance, but hardly necessary for identification. Females and immatures are like adult males, but less colorful and often showing only a shadow impression of a collar.

BEHAVIOR: A penchant for sitting high atop tall live trees in closed-canopy forest reduces singing males to little more than eerie disembodied notes. Birds commonly vocalize from the same perch for long periods, moving only to throw back their heads to vocalize. Varied forages on the ground and in low bushes. Moves in the stop-and-go fashion of a robin, but hops (robins commonly run) and maintains a more horizontal profile, with head and tail elevated to the same height (robins stand erect when they pause). Feeds by tugging on ground litter, hopping back, tossing vegetation aside, and surveying the exposed ground for edibles. In winter, spends more time in trees foraging for fruits and berries. In summer, found in territorial pairs. In winter, may be solitary, may gather in small flocks, or may join flocks of robins (in orchards or lawns). Fairly tame; responds somewhat reluctantly to pishing.

FLIGHT: Stockier than American Robin, with a pointier face and shorter tail. The broad orange stripe running the length of the wing is visible above and below. Flight is robinlike—straight and fast, with a rapid series of slightly irregular wingbeats punctuated by terse skip/pauses. For short distances, or when approaching a perch, flight becomes undulating and lofting, with open-winged braking glides. Males perched high atop trees can make near-vertical headfirst descents into the forest.

VOCALIZATIONS: Song is haunting, ethereal, unique—a single held note on an even pitch that sounds like someone blowing notes through a reedy harmonica; for example: "*eeeeeeee* (pause) *errrrrrrrr*." Given sequentially, notes vary in pitch and quality. Some are high and clear, some are low and buzzy, and some are slightly trilled. Pauses between notes are usually lengthy (8–15 secs.), and birds may vocalize for minutes on end. At times, songs are speeded up, held notes are abbreviated, or pauses are shortened (4–5 secs.). Call is a low "*churk*" or "*chup*." When agitated, makes a quick descending stutter: "*ch't't't*."

Wrentit, *Chamaea fasciata*

Skulkiferous Babbler-Tit

STATUS: Common West Coast resident. **DISTRIBUTION:** Found from the Columbia R. in Ore. south into n. Baja. Primarily coastal, but found inland to the western slopes of the Sierra and Cascade Mts. (but absent in the Central Valley of Calif.). **HABITAT:** Found in a variety of low, dense, brushy habitats. In coastal areas, prefers coastal scrub. Away from coasts, found in chaparral shrub understory in oak and pine forest; second-growth and riparian woodlands; blackberry and poison oak thickets and California grape; and brushy tangles and hedges in suburban yards and parks. **COHABITANTS:** Bewick's Wren, California Thrasher, Orange-crowned Warbler, California Towhee, Song Sparrow. **MOVEMENT/MIGRATION:** Permanent resident; virtually sedentary. VI: 0.

DESCRIPTION: A plain, skulking, hyperactive, somewhat incongruous bird that seems part wren and part tit, but is officially classed with the Old World babblers. Small (slightly larger than Song Sparrow) and oddly shaped, with a neckless plum-shaped head/body and an extremely long scraggly-tipped tail. The bill is stubby and dark (somewhat Bushtit-like, but slightly drooped). The pale beady eyes seem conspicuous in part because they are highlighted by a frail white eyebrow, and in part because the rest of the bird is so utterly and overall . . .

Plain. Plumage varies from plain pale brown to plain olive brown to plain ruddy brown to gray (with some warm blushes of salmon). Upperparts are usually slightly darker than underparts, and plumage sometimes appears somewhat furry or shaggy or blown-dry.

All in all, a rather homely-looking bird with a mean little expression.

BEHAVIOR: Maddeningly reticent. Even singing males (and females) commonly vocalize from the inner confines of a bush. Forages by making short hops and serial hops between branches. Incomparably quick; seems capable of moving on to the next perch before its feet touch the first. Not even kinglets move with such nimble swiftness. Commonly holds its tail at a stiff right angle to the body. When pausing on a perch, often pivots its entire body with jerky quickness. Plucks prey from nearby twigs and branches. If prey is out of reach, hops to intercept, then returns quickly to the original perch. Also makes startling long vertical leaps. Eats mostly insects but also some berries.

Pairs remain in the same small territory year-round but commonly forage alone. Wrentit forages at all levels in brush but almost never touches the ground. Flies with some reluctance. Responds somewhat to pishing, but that doesn't guarantee that it will appear in the open.

FLIGHT: Short-winged and extremely long-tailed. Flight is quick, weak, low, and short, with fluttery wingbeats and a floppy tail.

VOCALIZATIONS: Song consists of a series of low, measured, cardinal-like whistles, "*leep leep leep leep leep*," that accelerate into a warbling trill, "*leep leep leep leepleepleepleepleep*" (another "bouncing ball" call). Also makes a short, low, ticking trill, "*drrrt*," that sounds somewhat like the flutter of cards being shuffled.

PERTINENT PARTICULARS: Your glimpse of this bird might be momentary. Two species that frequent the same habitat are Song Sparrow and Bewick's Wren. Both are more overtly responsive to pishing than Wrentit.

Those not familiar with Wrentit or Bushtit commonly wonder whether the two species might be confused. No, they look very different. Bushtits travel in vocal flocks (except during breeding season) and almost always fly between perches, whatever the distance. Wrentits are solitary, mostly silent, and hop between perches—whatever the distance.

MIMIDS—CATBIRDS, MOCKINGBIRDS, AND THRASHERS

Gray Catbird, *Dumetella carolinensis*

Gray Thicket-Mimic

STATUS: A common widespread and familiar breeder; a coastal and southeastern winter resident. **DISTRIBUTION:** Breeds across s. Canada and much of the U.S.; found from the Maritimes across s. Que., s. Ont., s. Man., the southern half of Sask. and Alta., and s. B.C. In the U.S., breeds east of a line drawn through e. Wash., ne. Ore., sw. Idaho, ne. Nev., n. and e. Utah, ne. Ariz., n. N.M., sw. Kans., cen. Okla., and e. Tex. to the Brazos R. Does not breed on the immediate Gulf Coast or the Florida Peninsula. In winter, found along the coastal plain from se. Mass. to Central America. Casual visitor to the west coast of North America, from Calif. to Alaska. **HABITAT:** Catbirds like to be where the vegetative substrate is tangled and thick. If there are thorns involved, so much the better. In winter, catbirds show a marked preference for dense berry-rich thickets, especially those adjacent to a water source. Avoids unbroken woodlands, especially pine woodlands. **COHABITANTS:** Breeding: White-eyed Vireo, Spotted and Eastern Towhees, Brown Thrasher. In winter, Hermit Thrush, Brown Thrasher. **MOVEMENT/MIGRATION:** Spring migration from March through May; fall from late August to early November. VI: 2.

DESCRIPTION: A slight, slender, and dark gray bird bracketed by a black skullcap and a long black tail and garnished with rusty undertail coverts—all useful features when you get no more than a foliage-cloaked glimpse of the head or hindquarters of a bird tucked in the shadows.

Perched birds are more horizontal than erect and often have a slouched, almost casual way of sitting. The expression is one of wide-eyed child-like wonder.

BEHAVIOR: A skulker, Gray Catbird worms its way into tangles that would daunt Br'er Rabbit, moving from branches in short hops or brief flights. On the ground, moves speedily, half running, half hopping, with head erect and tail held just above horizontal. When foraging on fruit trees, easily lands on springy twigs to pluck fruit. When singing, males frequently sit high and fully exposed, often on the highest point of a tree. Birds sometimes flycatch, making short sorties from exposed perches.

Generally energetic and, except when singing, almost never still for long; not particularly nervous, however, or high-strung. Usually tolerant, and easily approached. In good habitat, several breeding pairs may nest in close proximity. In winter, if berries are in good supply, multiple birds may occupy the same thicket. In migration, multiple birds may be found foraging on fruit-bearing vines or trees.

FLIGHT: Fast, direct, generally close to the ground, and somewhat furtive—Gray Catbird is a bird of the thickets that is uncomfortable out in the open. Wingbeats are rapid and mostly continuous, but often interrupted in longer flights over open territory by a brief irregular skip/pause. The tail is partially fanned. The head is slightly raised above the body, making the bird look anxious or hyperalert. Although flight is straight, the bird seems to be struggling to keep itself from rising.

VOCALIZATIONS: Song is a casual careless babble consisting of clear whistled notes, squeaks, mewing sounds, and whines mingled with notes stolen from other bird songs. Gray Catbird seems to be simply trying out different sounds to see how they work, rarely repeating notes or phrases. Song may be continuous or have pauses between several note sets. Call is a descending petulant cross between a whine and a cat's meow: "*we-Ehhhh*" or "*Ehhhh.*" Also makes a muffled "*whurt.*" Alarm is a harsh, hurried, scratchy rattle: "*Eh-Eh-Eh*" or "*Eh-eh-Eh-Eh-Eh-ehek!*" Young birds make a high "*peep.*"

PERTINENT PARTICULARS: Gray Catbird responds readily to pishing and, despite its proclivity for dense vegetation, can be coaxed into the open.

Northern Mockingbird, *Mimus polyglottos*

Chimney Mimic

STATUS: Generally common (rare in northern portions of its range), but very widespread and familiar resident. **DISTRIBUTION:** Found across border portions of s. Canada, from N.S., cen. N.B., and se. Que. (rare) to Ore. and throughout the remainder of the U.S. and Mexico. Rare in B.C., Wash., Ore., Mont., the Dakotas, Neb., Minn., and Wisc.; casual farther north in Canada and Alaska. **HABITAT:** Preferred habitat is bushes (particularly fruit-bearing), hedges, thickets, and thorny vegetation. Throw in a bit of open habitat (like a well-trimmed lawn) and a chimney or television antenna to perch on, and it's perfect. **COHABITANTS:** Northern Cardinal, Us. **MOVEMENT/MIGRATION:** Some of the northernmost mockingbirds remain year-round, but others migrate or relocate. The mechanics, timing, and routes are poorly understood, but fall movements have been noted in September. VI: 2.

DESCRIPTION: A long, lithe, grayish grenadier of gardens and lawns. Larger than Gray Catbird and slender (even by the elongated standards of mimics), with a small head; a short, thin, black, slightly curved bill; and a long, narrow, animate tail that droops when the bird is perched or is held slightly up-cocked when it forages on the ground. Pale gray above, with two narrow, beaded, white wingbars; paler gray or somewhat buffy below with a whitish throat; a pencil-fine dark line connects the bill and the eye. Juveniles are like adults, but the breast is faintly spotted. The expression is mildly interested or mildly bored.

In flight (which it does frequently while foraging) and when engaged in "wing-flashing display," wings sprout large white patches (visible above and below). The tail has a blackish center and conspicuous white sides.

BEHAVIOR: Outside of breeding season, found solitary or in pairs. The terms "flock" and "mockingbird" are mutually exclusive. Birds are most commonly seen (and heard) sitting on your now-useless rooftop television antenna but are equally at home on utility lines and poles and in treetops or tall shrubs. Likes to sit high, and likes to sit conspicuously.

Often forages by walking, running, or hopping on lawns. A visual hunter that spies prey, runs to engage, stops, and jabs down with the bill. Also perch-hunts. In winter, forages heavily on fruit and berries that it plucks while perched. Despite its obvious love of open spaces, moves nimbly in thick cover, where it is often blocked from view.

"Wing-flashing display," a trademark characteristic, is performed with near-parade-ground pomp. The bird walks boldly erect, stops, opens its wings slightly, then partially, then fully, with wings held over the body. Folds wings. Marches again. The function of the display is a matter of debate.

An aggressive defender of its territory from other mockingbirds, dogs, cats, and, when guarding its food supply, easily intimidated fruit-eating birds such as waxwings, robins, and starlings.

Very vocal. Sings much or all of the year, and during breeding season, may sing at all hours of the day and night—particularly on moonlit nights or where territories abut well-lit areas.

FLIGHT: Generally straight, slow, buoyant, and somewhat exaggerated, with wings and tail more flared than efficient locomotion demands—as if the bird is consciously trying to display its showy white wing patches and white-edged tail.

VOCALIZATIONS: A loud persistent singer and consummate mimic. Song is a structured ensemble of phrases, most of which are snatches of songs from other bird species. Each phrase is repeated two to six times, followed by a savoring pause, followed by a new series. It sounds as though the bird is trying out everyone else's song, looking for one it likes.

Besides bird songs, this accomplished mimic integrates other sounds into the ensemble, including car horns, doorbell chimes, and snatches of popular songs.

Call is a dry, scratchy, scolding "*chak, chak, chak, chak*" (recalls a stick being dragged over a segmented sidewalk).

PERTINENT PARTICULARS: There are very few species that might be confused with Mockingbird. Loggerhead and Northern Shrikes have some plumage similarities and also like to sit conspicuously high in semi-open country. But shrikes have big heads, short heavy bills, and a dark mask across the face. In flight, a shrike's wingbeats are too rapid to count, and finally, shrikes are committed perch-hunters—they may not even know how to maneuver on the ground.

Townsend's Solitaire is uniformly gray (top and bottom) and shaped like a tubular bluebird (not as gaunt or rangy as a mockingbird) and stands upright. The smaller Sage Thrasher might be confused with a juvenile mockingbird, but young mockingbirds have spotted, not streaked, breasts and show all-white sides to the tail. Sage Thrasher's tail shows white corners.

Bahama Mockingbird, *Mimus gundlachii*, a very rare visitor to Florida, is more robust and browner, has streaks along the flanks, and lacks the prominent white wing patches seen on Northern Mockingbird.

Sage Thrasher, *Oreoscoptes montanus*
Sage Mockingbird

STATUS: Common but habitat-restricted interior western breeder; winters in the Southwest (mostly in Mexico). **DISTRIBUTION:** Breeds in the intermountain West from extreme s. B.C. (rare), se. Alta., and sw. Sask. south into e. Wash., Ore., e. Calif. (along the Nev. border), s. Idaho, s. and cen. Mont., extreme sw. S.D. (very rare), all but s. Nev., Utah, Wyo., w. and s. Colo., n. Ariz., and n. N.M. Winters in se. Calif. (scarce and local), s. Nev., s. Ariz., s. N.M., and w. Tex. north to Lubbock and east to Fort Worth and McAllen, then south into n. Mexico. Very rare along the Pacific Coast; casual visitor (mostly in the fall) in the East. **HABITAT:** Breeds almost wholly in open sagebrush flats. Winters in dry open country dominated by low brush and sage. **COHABITANTS:** Common Nighthawk, Horned Lark, Sage Sparrow, Brewer's Sparrow, Vesper Sparrow. **MOVEMENT/MIGRATION:** Entire breeding population shifts south for winter. Spring migration from late February to early May, with a peak in late March and early April; fall from late August to late October. VI: 3.

DESCRIPTION: Resembles a compact streaky-breasted mockingbird (or perhaps a mockingbird/pipit cross). A medium-sized songbird but a small thrasher, Sage Thrasher is the same size as Gray Catbird and Cactus Wren. Its slender body is less gangly and more compact than other thrashers, with a short, mostly straight bill and a medium-long, stiff, straight tail (not wilted like other thrashers). Face looks pipit-pointy.

Overall pale, overall streaky, and shabbily unkempt. Pale gray or brownish gray above, and pale whitish below (sometimes with a buffy or olive wash on the flanks), with a beaded pattern of fine dark streaks. (*Note:* In late summer, streaking may be very indistinct.) The face pattern—dark cheeks and a pale border—seems somewhat pipitlike. The narrow white wing-bars are apparent but don't stand out. The bright yellow eye (dusky in immatures) is very tiny and gives the bird a dim expression.

BEHAVIOR: Not particularly wary. Likes to perch high (that is, at knee height) on tops of sagebrush both to sing and to scrutinize observers. Spends most of its time on the ground, foraging with the halting run-stop-pick pattern of a robin. Runs more than walks, with head lowered. Stands almost fully erect with tail up-cocked when it stops (and lowers it after a few seconds). Raises and lowers its tail when disturbed.

Generally solitary; not a flocking bird. When pursued, commonly runs for cover. If perched, flies to a more distant perch rather than disappear into the closest cover.

FLIGHT: Quick, nimble, stiff, generally low (just over the tops of the sage), and slightly erratic. Overall slender, with moderately long broad wings held at a right angle to the body. Wingbeats are given in a lengthy series punctuated by a brief closed-wing glide. Flight undulates slightly. When approaching a landing point, the bird switches to a series of lengthier, more open-winged glides. The sooty gray tail is darker than the wings or body. Except on worn individuals, when tails are flared for a landing, the narrow white outer corners are conspicuous.

VOCALIZATIONS: A complex, long-winded, but comically musical song that integrates soft bluebirdlike churs and short tanager phrases with an assortment of warbling trills. Rather softly spoken, with little change in pitch or overall pattern and no harsh or squeaky notes. Call is a low "*chuck*."

PERTINENT PARTICULARS: In late summer, very worn birds may appear superficially similar to Bendire's Thrasher. Juvenile mockingbird is also a candidate for confusion.

Brown Thrasher, *Toxostoma rufum*

Eastern Roadrunner

STATUS: Uncommon to common widespread, primarily eastern breeder and southeastern winter resident. Except for extreme southwest portions of its breeding range, Brown Thrasher is the only thrasher likely to be encountered in the East. **DISTRIBUTION:** Breeds throughout the e. U.S. (east of the Rockies) and across s. Canada, from N.B. across s. Que., s. Ont., s. Man., s.-cen. Sask., and se. Alta. The western border cuts across cen. Mont., cen. Wyo., e. Colo., the Oklahoma Panhandle, and extreme n. Tex. southeast to the Brazos R. In winter, found coastally from s. Conn. to se. Va., then south of a line drawn across cen. N.C., nw. S.C., n. Ga., e. Tenn., s. Ill., s. Mo., e. Okla., the eastern half of Tex., south to Corpus Christi. A few are found west to se. N.M.; very rare farther west. **HABITAT:** At all times and seasons, prefers the low tangled thickets and dense understory associated with open-canopy forest, pine barrens, forest clearings, burns, and forest edge. In open country, uses hedgerows and woody draws; also inhabits parks and suburbia, particularly during migration and where low, dense, woody vegetation occurs. **COHABITANTS:** Gray Catbird, Yellow-breasted Chat, Eastern Towhee, Song Sparrow; in winter, Hermit Thrush, Fox Sparrow. **MOVEMENT/MIGRATION:** Early-spring migrants, Brown Thrashers return to vacated southern territories in early March and occupy northern portions of their range by early May. In fall, birds begin to depart in late August, peak in late September and early October, and continue into November. VI: 3.

DESCRIPTION: A large elongated will-o'-the-tangle characterized by a slender head, a heavy, fairly straight (slightly downturned) bill, and an extremely long narrow tail. Rich rufous upperparts and pale underparts are overlaid by charcoal brown streaks that look like stands of beaded teardrops. Long-billed Thrasher of south Texas is overall colder-toned, more cinnamon than rufous above, and whiter with blacker streaks below, and has a bill that is, as the name suggests, slightly longer, thinner, blacker, and more down-curved. Brown Thrasher's expression is nefarious.

BEHAVIOR: Likes to stay low, and likes to stay hidden. In summer, forages for the most part on the ground, sweeping leaves aside with its bill, then grasping prey or probing the earth for prey. Walks and runs with its long tail held level behind (recalling the figure of a roadrunner). In winter, spends more time foraging in vegetation for berries. Hops through tangles.

Singing males are frequently perched high, often at the very tops of trees in plain view. The stance is nearly vertical, with the tail down-cocked.

Generally solitary. In winter, if there is sufficient food, several individuals may occupy the same sizable thicket. At all seasons, often found in the same habitat with Gray Catbird, and in winter with Hermit Thrush.

FLIGHT: Two distinctive flights. Short flights (used when dropping or ascending to a perch) are jerky, almost theatrical, with the tail fanned and pumping. Longer flights are more direct, with little or slight undulation and tails mostly closed; wingbeats are steady and hurried, but overall momentum is fairly slow. Stays close to the ground, usually just above the vegetation, with flight following vegetative contours. Gives the distinct impression that it's uncomfortable being in the open and wants only to get to cover.

VOCALIZATIONS: Song is a rich, musical, and varied ensemble of two- and three-note phrases—each one uttered twice, then savored with a pause. The screechier, scratchier-sounding Gray Catbird utters phrases once; Northern Mockingbird, whose repertoire is stolen from the mouths of other birds, commonly says everything three or four or more times. Brown Thrasher's calls are many and varied and include a short, low, descending growl or asthmatic wheeze (often uttered just at dawn), "*rrrrr*"; an admonishing "*Tcheh*" that is loud and raspy at the beginning, high-pitched (almost shrill) at the end, and recalls a loud smacking kiss; a fast, descending, thrushlike "*eee-ur*"; a harsh angry "*kakaka*"; and a scratchy scrapping sound that recalls a stick being dragged across concrete (often heard when birds are flushed).

PERTINENT PARTICULARS: Thrashers don't like to be seen. They respond to pishing more reluctantly than many other thicket species, and often appear suddenly (in part or in full), then disappear just as suddenly. Persistence is key. Learning the bird's vocalizations is immensely helpful.

Wood Thrush shares the thrasher's breeding range and is also a rich rufous-onto-orange above and heavily marked below. Although Wood Thrush is more typically a bird of wet mature woodlands, sometimes the thrush and thrasher habitats are tangent or overlap. Note, however, the thrush's overall plump stocky shape, the heads-up, breast-up (not horizontal) stance, and the spotted (not beaded-into-streaks) breast. Note also that Wood Thrush has a large, somber, dark eye; Brown Thrasher has a beady, pale, malevolent eye.

Long-billed Thrasher, *Toxostoma longirostre*

South Texas Thrasher

STATUS AND DISTRIBUTION: Common resident, but the range is limited to s. Texas and ne. Mexico. The northern limit of the range is defined by a bulging line drawn through Del Rio, San Antonio, and Rockport. **HABITAT:** Dense dry scrub woodlands and mesquite thickets; riparian woodlands. **COHABITANTS:** Common Pauraque, Couch's Kingbird, Great Kiskadee, Green Jay. **MOVEMENT/MIGRATION:** Permanent resident. VI: 1.

DESCRIPTION: The brown-backed, gray-faced, spotty-breasted thrasher of south Texas. A large songbird (robin-sized) and a large thrasher, Long-billed is very similar in size and shape to the more widespread eastern Brown Thrasher. Shows long lean thrasher lines, with a long, somewhat down-curved bill, a narrow body, and a long tail. The bill is usually (though not always) slightly longer and/or more curved than Brown Thrasher, but it's also blacker and stands apart from the face.

Overall colder brown above and more contrastingly patterned below than Brown Thrasher. The gray face and (from some angles) head and nape are easily noted, given a reasonable look. The darker eye (yellow on Brown Thrasher; orange on Long-billed) is harder to see, and so Long-billed seems particularly beady-eyed. With their whiter color and blacker overlying spots and streaks, the underparts are more stark and contrasting in pattern.

BEHAVIOR: Most often seen singing on exposed perches above its thornbush fortress. Males sing from March to mid-summer, and may sing snatches of song the balance of the year. At other times, birds forage most commonly in the leaf litter of the forest floor, but also feed in low brush (particularly in winter).

Not particularly wary—perhaps because it's not easily approached. Can be coaxed to come to feeding stations.

FLIGHT: Long-tailed silhouette, with fairly short rounded wings. Flights are usually short and jerky, with exaggerated undulations.

VOCALIZATIONS: Musical song is similar to Brown Thrasher—a rambling run-on series of whistles, chips, churps, squeaks, trills, and nasal mewings, most of them doubled (but not all). The bird's incorporation of single, trebled, and quadrupled notes gives the song a more whimsical quality.

Song of Curve-billed Thrasher, whose range and habitat overlap with Long-billed Thrasher, is faster, less musical, and less complex. Long-billed calls are similar to Brown Thrasher.

PERTINENT PARTICULARS: The separation of Long-billed and Brown Thrashers is largely academic—in summer. Brown Thrasher does not breed in south Texas. In winter, some Brown Thrashers occur in the northern portion of Long-billed Thrasher's range.

Bendire's Thrasher, *Toxostoma bendirew*
Curve-billed Light

STATUS: Uncommon to fairly common breeder and resident of the Southwest, but distribution is spotty. Where range overlaps with Curve-billed Thrasher, generally less common. **DISTRIBUTION:** Breeds from extreme se. Calif. across southernmost Nev., s. Utah, most of Ariz., and the western half of N.M.; also south along the west coast of Mexico to the limits of the Sonoran Desert community. **HABITAT:** Generally dry open desert showing a mix of barren (typically sandy) ground, grassy areas, cactus, and widely spaced shrubs. In California and Nevada, partial to Joshua tree woodlands. Also sagebrush, at higher elevations, with scattered junipers. Avoids riparian woodlands and the dense mesquite associated with desert washes (the favored habitat of Curve-billed Thrasher), but seems very disposed to desert homes with surrounding trees and shrubs and to farmlands where shelterbelts are found. **COHABITANTS:** Scaled Quail, Cactus Wren, Curve-billed Thrasher. **MOVEMENT/ MIGRATION:** Birds from the northern half of the species' range retreat into the southern portion of the breeding range in winter. The fact that they do not expand beyond this range attests to the bird's habitat requirements. Spring migration from March to May, with a peak from mid-March to early May; fall from late June to November, with a peak in July and August. VI: 1.

DESCRIPTION: An abbreviated Curve-billed Thrasher. Overall smaller, slighter, slimmer, and slightly flatter, with a more peaked head and showing a shorter, straighter bill (slightly curved on the upper mandible, mostly straight lower mandible). Overall size, head shape, bill, and relative proportions might, at first glance, lead an observer to mistake a high-perched Bendire's for a stout mockingbird—not that mockingbird is a likely candidate for confusion or that such a misidentification would not soon be corrected. The point is that few birders would look at the silhouette of a Curve-billed Thrasher—a big, burly, stout-billed, dome-headed linebacker of a thrasher—and conjure the image of a mockingbird, but the more abbreviated lines and slighter proportions of Bendire's do seem mockingbird-esque.

Overall grayish brown (on average perhaps slightly browner or warmer-toned than Curve-billed) and more uniform, less contrasting or patterned than Curve-billed, especially on the breast. Breast of Bendire's is more faintly, finely, tightly, uniformly flecked or patterned than Curve-billed. Breast markings may be pencil-fine lines or tiny arrowheads, but the net effect is the same: more uniform underparts that are less mottled or patchy or patterned or dabbled or coarse.

The bill is also slightly paler (especially at the base), so that the contrast between the grayish brown face and the black bill, which is very pronounced on Curve-billed, is not so apparent on Bendire's. Also, the eye of Bendire's is yellow. Some Curve-billeds have a yellow eye, but on most it's orange.

Finally, Bendire's is generally less vocal than Curve-billed. If the bird whistles a loud "*Whit Wheet!*" it's not Bendire's.

BEHAVIOR: Much like Curve-billed. Forages mostly on the ground by picking, probing, and excavating in soil. Walks, runs, hops, and lopes. Also forages for fruit and insects by hopping through vegetation and gleaning from leaves and branches. Is not at all shy about being in the open and frequently perches high and in plain sight of observers. Can be attracted to backyard feeding stations.

FLIGHT: A series of hurried wingbeats interspaced with glides that are crisper and more buoyant than its larger cousin. Also seems more inclined to make longer flights.

VOCALIZATIONS: A hurried run-on ensemble of mostly high two- to five-note whistled phrases—some clear, and others high, squeaky, screechy, or squeally. Bendire's sounds more urgent and more complex, less slavishly double-noted, than Curve-billed. Call is a low, terse, raspy "*ch*" or "*ch/ch.*"

Curve-billed Thrasher, *Toxostoma curvirostre*

The Default Desert Thrasher—"Whit-Wheet!"

STATUS: Common and, across most of its southwestern range, the most common desert thrasher. **DISTRIBUTION:** Across the American Southwest from coastal Tex. to w. Ariz. and south into s. Mexico. The limits of the northern range are defined by a line drawn from just south of Galveston Bay north and west to the junction of Okla., Kans., and Colo., then south and west across N.M., then west across s. and cen. Ariz. Casual visitor to se. Calif. and in states north of its normal range. **HABITAT:** Less habitat-specific than other desert thrashers. Found in dry desert brush lands, grasslands dotted with cactus (especially cholla cactus), woodland edge, pinyon-oak, landscapes dominated by creosote bush, and dry desert hillsides dominated by cactus. Adapts well to human habitation and becomes suburbanized where sufficient desert vegetation is left standing and feeding stations are found. **COHABITANTS:** Harris's Hawk, Greater Roadrunner, White-winged Dove, Cactus Wren. **MOVEMENT/MIGRATION:** Essentially nonmigratory. VI: 2.

DESCRIPTION: A large sturdy desert thrasher best known for its loud, assertive, two-note, whistled call. Compact by thrasher standards, with a large domed head, a robust body, and a moderately long tail (but fairly short by thrasher standards). The namesake and signature characteristic is the curved bill. It is long (almost as long as the head is wide) and tapered, but still heavy throughout its length; though distinctly down-curved, it does not show the exaggerated bow found on several other species.

Plumage is overall dingy brownish gray above and somewhat mottled below. The modestly patterned underparts can be described as: (a) slightly paler underparts spattered with smudgy grayish spots or (b) brownish gray underparts dappled with pale sun-bleached areas. Very similar Bendire's Thrasher has dark pencil-fine streaks or tiny arrowheads on its breast and a shorter, straighter bill. Curve-billed's eye color ranges from golden to orange. Set against the bland backdrop of its face, the eye is often arresting (but by no means idiosyncratic). The expression is dour, angry, mean.

BEHAVIOR: Generally solitary or found in pairs. Sits high when vocalizing, and typically is not shy about being in the open (may fly considerable distances to the next perch). Feeds on the ground, again, often in the open, using its bill to probe the substrate, excavate promising areas, and sweep concealing materials aside. Usually walks quickly or runs. Occasionally hops. Also clambers and skillfully snakes its way through tangled brush; when not feeding, may sit for extended periods on the same perch. An early riser, Curve-billed often makes its distinctive call at first light.

FLIGHT: Usually low, hurried, direct, and short, with steady wingbeats and a braking down-pumping action with the wings and tail just before landing.

VOCALIZATIONS: Song is a run-on series of mostly two-note phrases, many of them harsh and buzzy and fairly low-pitched. Overall less complex and less musical than Bendire's—and perhaps less interesting. Call is a loud, distinctive, rolling "*whit-wheet!?*" or "*wh'whee'wheet!*" The bird sounds as if it's trying to get your attention.

California Thrasher, *Toxostoma redivivum*

Masked Thrasher

STATUS: Common resident with limited distribution. **DISTRIBUTION:** Endemic to Calif. and n. Baja, primarily in the coastal ranges, but also on the western slope of the Sierras and both sides of the Central Valley. **HABITAT:** Hillside chaparral, but also sagebrush, riparian and oak woodlands with brushy understory, and suburban yards and parks where brushy vegetation is allowed to flourish. **COHABITANTS:** California Gnatcatcher, Wrentit, Spotted Towhee, California Towhee. **MOVEMENT/MIGRATION:** Permanent resident. VI: 0.

DESCRIPTION: A large, sturdy, dark, dirt-colored thrasher with a triangular face mask. Large and robust (very slightly larger than Crissal or Le Conte's Thrasher and Western Scrub Jay), with a very long, heavy, and distinctly downturned bill, an oval and somewhat humpbacked body, and a long slightly wilted tail that it angles up when it runs.

Grayish brown above, slightly paler and warmer brown below, with a cinnamon-tinged belly and undertail. Except for the whitish throat, all dark. Except for the streaky triangular face mask (bracketed by a dark eye-line and dark malar stripe), unpatterned.

BEHAVIOR: A ground-foraging brush-loving bird most often seen vocalizing from the tops of brush or foraging, singly or in pairs, in the dirt at the edge of cover. Noisily searches for prey by tossing aside leaf litter with its bill or excavating holes with sideways sweeps of the bill. Also flips over twigs and cow pies. Primarily in late summer and winter, also eats fruit and berries (including domestic grapes,) and is attracted to bird-feeding stations that offer fruits and suet.

Vocalizing males (heard year-round) commonly use the highest point available—a tall shrub, a tree, a rooftop. California Thrasher usually runs but also hops. Flies reluctantly, usually running for cover with head down, tail slightly raised, and wings often partially opened.

FLIGHT: Profile is long, particularly long-tailed, with short, broad, rounded wings. Flight is straight, usually low to the ground, and short. Wingbeats are given in a hurried series followed by an open-winged glide with wings held at a right angle to the body.

VOCALIZATIONS: Song is an unhurried series of phrases (each repeated two to three times) consisting of whistled, trilled, buzzy, hash-sounding notes. Notes seem more distinct, more laid-back, and less formulaic than the run-on songs of Bendire's and Le Conte's Thrashers. Overall song has a squeaky quality and a cadence that recall Gray Catbird. Call is a low "*chuck*" or "*cherik.*"

PERTINENT PARTICULARS: Over most of its range, the only thrasher species. Overlaps slightly with the much grayer and paler Le Conte's Thrasher in the Mojave Desert and the southwest Joaquin Valley—where California Thrasher occurs in the more elevated brushy slopes, and Le Conte's in the drier, flatter, more open desert. Also *almost* abuts the range of the more similar but pale-eyed Crissal Thrasher in southeast California. Note the contrasting chestnut undertail of Crissal.

The bird most likely to be confused with California Thrasher is the similarly plumaged but smaller and smaller-billed California Towhee. *Note:* The thrasher prefers to run and hops only occasionally; the towhee prefers to hop and walks or runs only occasionally.

Crissal Thrasher, *Toxostoma crissale*

Bow-billed Thrasher of Desert Wash

STATUS: Permanent resident. Generally less common than other desert thrashers and somewhat more restricted in its habitat preferences. **DISTRIBUTION:** Limited to the Southwest, from se. Calif., s. Nev., and extreme sw. Utah south and east across the southern half of Ariz. and N.M. into w. Tex., and south into Mexico. **HABITAT:** Over most of its range, this "obligate riparian" species is closely linked with mesquite thickets and the taller riparian brush bordering dry desert streambeds. Also found in more open desert brush and on grassy pinyon-juniper hillsides. **COHABITANTS:** In desert washes, found with Verdin, Black-tailed Gnatcatcher, Curve-billed Thrasher, Abert's Towhee. In pinyon-juniper, found with Spotted Towhee. **MOVEMENT/MIGRATION:** Permanent resident. VI: 0.

DESCRIPTION: Large, dark, brownish gray, overall plain thrasher, primarily of brushy desert areas. In silhouette, best identified by its size (larger than Curve-billed Thrasher) and long, decidedly slender, and curved-onto-bowed bill. Where proximity and light conditions are favorable, the dark chestnut-colored undertail stands out. (In fact, it may peek out around the base of the tail when the bird presents a dorsal view.) The eye is small, beady, and difficult to see. (The eye of Curve-billed stands out.) The dark malar stripe also helps distinguish Crissal from the more common and habitat-eclectic Curve-billed.

BEHAVIOR: Crissal is found where the trees get slightly taller and the brush denser (for instance, around arroyos, abandoned structures, or corrals surrounded by shrubby thickets)—but is not necessarily found there. The birds are shy, cryptic, and ground-foraging. Singing males, poised high in the branches of a mesquite or juniper, give most birders their only view of the species. When approached, Crissal typically dives or branch-hops for the ground, where it makes its escape on foot. Runs, walks, and hops. Forages by digging in the leaf litter, probing at the base of grasses and shrubs. Its foraging efforts may take the bird into the open, but it is never far from cover.

FLIGHT: Straight, low, reluctant, effortful, and brief. Wings move with a fluttering steadiness, and tail is slightly spread. Sets its wings and glides before landing.

VOCALIZATIONS: Typical thrasher—a series of doubled and tripled notes and phrases, but uttered softly, almost murmured—as if the bird is trying not to be heard. Many of the phrases recall the song of American Robin. Also makes soft musical muttering: "*whrt.*"

PERTINENT PARTICULARS: Crissal is structurally similar to the slightly browner, more salmon-washed California Thrasher, but their ranges do not overlap. Le Conte's Thrasher is also similar in size and proportions, but is ghostly pale. Le Conte's is found in dryer open desert characterized by short stunted shrubs, but its habitat does overlap with Crissal's in a few places.

Le Conte's Thrasher, *Toxostoma lecontei*

Ghost Thrasher

STATUS: Uncommon, geographically restricted, habitat-specific resident. **DISTRIBUTION:** Desert species found in s. (primarily se.) Calif. (mostly in the Mojave Desert and San Joaquin Valley), s. Nev., and w. Ariz., south into nw. Mexico. **HABITAT:** Sparsely vegetated sandy desert flats or gently rolling hills. Much prefers areas dominated by saltbrush and creosote bush. Crissal and California Thrashers are found in the denser, taller, more riparian and closed-canopy habitats adjacent to Le Conte's habitat. **COHABITANTS:** Precious few. Loggerhead Shrike, Sage Sparrow, Brewer's Sparrow. **MOVEMENT/MIGRATION:** Nonmigratory; permanent resident. VI: 0.

DESCRIPTION: A plain, ghostly pale, slender thrasher of dry open desert. Medium-sized (slightly larger than Northern Mockingbird; slightly smaller than Crissal Thrasher), with a long thin, distinctly down-curved bill and a long narrow tail (often up-cocked as the bird runs). *Note:* Sage Sparrow, which occupies the same habitat, also runs with its tail up-cocked.

Overall plain and pale—distinctly paler and grayer than other thrashers, with no spotting, streaking, or clear facial pattern to speak of, just a dark smudge in front of the eye. The narrow and contrastingly darker tail looks browner above, blackish below. The cinnamon undertail coverts (darker than many illustrations show) are easy to see as the bird skulks away. The beady black eye gives the bird a not-too-bright expression.

BEHAVIOR: A shy skulking desert species most frequently seen running between widely scattered bits of vegetative cover (and in so doing, resembles a pale roadrunner). Forages, singly or in pairs, on the ground, moving in short stuttering steps, probing and excavating pits in bare earth. Hops when moving through bushes; does not hop on the ground. Also flips objects and chases grasshoppers.

Males vocalize from slightly elevated perches. All birds commonly escape by running swiftly instead of flying.

FLIGHT: Profile shows very short rounded wings and a long slightly drooped tail. Flight is direct, weak, reluctant, and usually short—a series of rapid, somewhat sputtery wingbeats followed by a protracted open-winged glide with wings held at a stiff right angle to the body.

VOCALIZATIONS: Song is like other thrashers—a varied, hurried, run-on series of doubled (sometimes tripled) phrases. Somewhat more musical and clear-noted than other thrashers (and not as screechy as Bendire's). If the song of California Thrasher is somewhat catbirdlike, Le Conte's leans more toward mockingbird. Call is a quizzical and distinctive "*soowheap?*" or "*wheeruk?*" that might recall the call of Mountain Quail.

STARLINGS AND MYNAS

European Starling, *Sturnus vulgaris*

Horatio Alger in Feathers (an American Success Story)

STATUS: Abundant, widespread, year-round resident. **DISTRIBUTION:** A real American. Established itself in the city. Moved into the suburbs. Kicked out the natives (cavity nesters like Red-headed Woodpecker, Purple Martin, and bluebirds). Now found virtually everywhere in North America (absent only across arctic and some boreal regions of Canada and much of Alaska). **HABITAT:** Most common where we are—cities, towns, suburbia, cemeteries, parking lots, garbage pits, cattle feedlots—but also occupies a variety of native, primarily open habitats, including agricultural and grazing land, parkland habitat, and even tidal marshes. Avoids large unbroken forests, chaparral, and deserts. Loves short-clipped grass as much as we do. Considers utility lines and television antennae prime perches. **COHABITANTS:** Besides us, mixes freely with many other flocking species, particularly blackbirds. **MOVEMENT/MIGRATION:** Once established, generally a resident; however, a number of individuals also migrate—early in spring (February and March), with somewhat more protracted movement in fall (August through December). VI: 4.

DESCRIPTION: Looks like a small dumpy blackbird with an acutely pointy face and a stub for a tail. Breeding adults are black with purple and green highlights that shimmer like gasoline on water. The bill is bright yellow. In winter, the dull black-and-brown body is festooned with tiny arrowpoint freckles and just a touch of iridescence on the wings; the bill is dark. Juveniles are overall dull brownish gray, but have the pointy-faced look and mannerisms of adults. The expression is nefarious, dull, cunning.

BEHAVIOR: A pack animal. Forages in groups throughout the year, gathering in flocks of 10–100 or where food is abundant (for example, cattle feedlots) in flocks that may number in the thousands at night roosts. Often mixes with blackbirds (Red-winged, grackles, cowbirds), as well as with American Robins on lawns, shorebirds on mud flats, finches at feeders.... Get the picture? Seems willing to try anything for a meal and able to adopt the feeding techniques of a number of species.

Usually forages on the ground, using its bill to explore lawns and soil for worms and grubs. Waddles with an oafish gait. Stands upright to gallop at a canter. Generally indifferent to other birds, edging them out of the way with its single-minded devotion to feeding. Also eats fruit from trees, with flocks festooning themselves over a tree and stripping it. Common at bird feeders, where it consumes everything.

Nests in tree cavities or any odd nook or cranny in almost any man-made structure. At night, may gather into large communal roosts on buildings, under bridges, in orchards, and (along with assorted blackbirds), in rank weeds bracketing rivers or covering marshes.

FLIGHT: With their pointy faces, short triangular wings, and stubby squared-off tails, birds resemble tiny arrowheads in flight. Flight is straightforward and single-minded, basically straight, and generally within 200 ft. of the ground. Rises and falls in a manner that seems more altitudinal correction than undulation. Wingbeats are rapid and sputtery, and glides usually short and abrupt. On approaching the point of landing, however, the bird may set its wings and execute a lengthy glide that seems impossibly long for such short wings, its body wobbling like a commercial jet landing in a high wind. Lands ungracefully, often with a hurried flurry of wingbeats.

Flocks are often strung out, with birds spaced. When flushed or when mobbing a hawk, flocks coalesce into a tightly packed oval cloud, every bird moving as one. Because underwings are paler than the rest of the bird, wheeling flocks flash dark and light. At a distance, a flock of mobbing starlings distinguishes itself by this now-you-see-it-now-you-don't character.

VOCALIZATIONS: Song is a somewhat muttered but colorful and eclectic stream of sounds—squeaks, screeches, whistles, murmurs, and gurgles—all intermingled with other bird vocalizations and accompanied by a snapping bill. A signature phrase, often included in any rendition, is a rising-and-falling whistle (sounds as though the bird is imitating a single-note wolf whistle or trying to mimic a wind gust on a Hollywood movie set). Alarm call is a metallic "*cheh!*" The most commonly heard flock vocalization is a short, low, protesting, gurgled "*raah*." The food-begging call of young birds is a strangled, higher-pitched, and longer-held rendition of the "*raah*" call. Also makes a stuttering staccato protest: "*reh'h'h'h'h*."

PERTINENT PARTICULARS: Mobbing starlings are indicators of a hawk in the air. Train your binoculars on the flock. The bird of prey, too distant to be seen by the unaided eye, is under the flock.

In flight, starling flocks are often confused with flocks of waxwings, which also fly in tight clusters. Waxwings are more buoyant, they often land with a series of undulations, and individuals within flocks are more shifting than starlings.

Another introduced species related to starlings is the Common Myna, *Acridotheres tristis*, a bird that has become very common in suburban areas and towns in south Florida. It resembles a large sooty-hooded juvenile (brown) starling with a bright yellow bill, a yellow teardrop over the eye, and bright yellow legs. In flight, it shows large white patches on the upper surface of the wings that are very conspicuous. These patches, combined with the bird's varied array of whistles, squawks, and chatter, is probably what drew your attention to the bird in the first place. Another introduced, suburbanized, but far less common myna, Hill Myna, *Gracula religiosa*, also occurs in Florida as well as California. Mostly black, it resembles a miniature crow with a heavy orange bill and a bright yellow wattle set against a black face. This species also shows white wing patches that are visible from above and below in flight, and like other mynas, Hill Myna is a noisy bird.

WAGTAILS AND PIPITS

Eastern Yellow Wagtail, *Motacilla tschutschensis*

Tundra Scold

STATUS AND DISTRIBUTION: Fairly common breeder on the coastal plains of n. Alaska (also extreme ne. Yukon west of the Mackenzie R. delta) and w. Alaska south to (but not including) the Alaska Peninsula. Regular migrant, sometimes in flocks, across the Bering Sea islands to and from Asia. Casual along the Pacific Coast south of the Aleutians. **HABITAT:** Open tundra with a grassy and shrubby or willow thicket component. Particularly fond of moist or wetter areas along creeks, roadsides, bluffs, and ditches. **COHABITANTS:** Bluethroat, Savannah Sparrow, Hoary Redpoll, Lapland Longspur. **MOVEMENT/MIGRATION:** Entire population is presumed to retreat to the southwest Pacific in winter. Spring migration from mid-May to mid-June; departs in August and early September. VI: 2.

DESCRIPTION: A slender yellow-bellied gnatcatcher-like bird hovering over your head and screaming, "*spee!*" Slender medium-sized songbird (larger than Lapland Longspur; smaller than White Wagtail; about the same size as American Pipit), with a thin pointy bill (contributing to a thin face), a slender body, and a distinctly long, narrow, animate tail—longer-tailed than a pipit but considerably shorter-tailed than White Wagtail.

Adults are greenish-backed with a gray head, black face mask, and blackish wings and tail showing white outer tail feathers—*which are obvious even when closed*—and bright yellow underparts. Immatures are mostly plain and drab, showing dull olive gray above, dingy with very limited pale yellow below, and a whitish throat. Might be confused with immature White Wagtail, but White is more distinctly gray, black, and white and more richly and contrastingly patterned, showing conspicuous wing-bars, a broad white eyebrow, and a black breast-band. Plainer Yellow Wagtail has a sliver of a white eyebrow, no breast-band, and no wing-bars.

BEHAVIOR: Active, conspicuous, vocal, omnipresent tundra species. On territory, sits conspicuously high (even if the highest point is only a few inches above the ground), calling loudly. Males engage in spirited aerial displays in which they climb at a steep angle and parachute to the ground on stiff fluttering wings, leveling off just before landing. The bird forages on the ground by walking and running; also hops in foliage and flycatches actively, adroitly, and frequently. Moves with a quick stiff nervousness. Flies frequently. Wags its tail infrequently, most often upon landing.

Usually solitary or found in pairs during breeding season, but gathers in small flocks (often in willow thickets) prior to and during migration. When nesting, fairly tame, but seems less perturbed about landing near you than it is about you approaching it. Scolds trespassers by hovering and bounding overhead, calling loudly. In migration, very skittish; does not allow close approach.

FLIGHT: A long slender profile shows short triangular wings and a long narrow tail (suggests an overgrown gnatcatcher). From below, the adult's yellow underparts contrast with the white-sided black tail. Wings appear translucent. Flight is distinctly jerky and bouncy. Wingbeats—which are given in a sputtery series so rapid that they leave the blurred impression of a triangular outline in the air—are followed by a lengthy closed-wing glide. Often flies considerable distances, calling frequently.

VOCALIZATIONS: Song is a high, hurried, somewhat ringing, even chattery warble of varying length; for example: "*cheehehehehehehehe.*" Song sometimes drops off at the end, with lower-pitched notes. Also makes a lower-pitched, churring chatter, "*chr'r'r'r'r'r'r,*" and has a two-part song (often given in flight) consisting of a high "*spee!*" followed by a lower, muttered "*zr'ip*"; for example: "*spee spee; zr'ip zr'ip.*" Call is a loud, sharp (almost shrill), upsliding whistle, "*speee!*" or "*sk/ree,*" that may sound slurred or may be given with a slight vibrancy.

White Wagtail, *Motacilla alba (combines Black-backed Wagtail,* Motacilla lugens, *and White Wagtail,* M. alba*)*

The Black-and-white (or White-and-black) Wagtail

STATUS: Uncommon breeder and geographically restricted casual vagrant along the West Coast. **DISTRIBUTION:** Widespread Old World species. In the U.S., found only in extreme w. Alaska (St. Lawrence I., outer coastal portions of the Seward Peninsula, and Capes Thompson and Lisburne). **HABITAT:** Coastal villages and sea cliffs. In North America, most commonly associated with human habitation. **COHABITANTS:** Snow Bunting, Lapland Longspur, Siberian Yupik people. **MOVEMENT/MIGRATION:** Arrives in late May; departs late August to mid-September. VI: 2.

DESCRIPTION: Distinctive black-and-white sliver of a bird with jerky head movements and a nervous tail. Medium-sized (larger and longer-tailed than Yellow Wagtail; larger and much more slender than

Snow Bunting), with a thin warblerlike bill, a slender body, and a very long, narrow, and animate tail.

Distinctly black and white with a variable pattern. All birds are light gray above, whitish or grayish white below, with a dark cap, a white face creased with a black eye-line, a black bib or breast-band, and varying amounts of white on the wings, ranging from all-whitish wings to white shoulders to two white wing-bars. Some birds have gray backs, some black. Immatures are like adults, but plumage is muted and less starkly gray, black, and white, and sometimes it is tinged slightly with light brown above, with a pale yellow wash to the face.

BEHAVIOR: A busy, jerky, methodical bird that spends most of its time walking along the short grassy edges of pools, along beaches, and within human settlements (also forages about garbage and marine mammal carcasses). Moves with perpetual nervous head-jerking movements and, particularly when perched or after rushing prey, wags its tail vigorously. Rushes forward with head down when it sights prey. Also flycatches (sometimes over water), and may pursue prey tenaciously and hover. Very tame, but when it flies, may travel long distances before landing (often on a human structure). In summer, found solitary, in pairs, or in family groups.

FLIGHT: Very slender with short triangular wings and a very long narrow tail. Conspicuously black, white, and gray; although variable in pattern, a black tail with white outer tail feathers is a common denominator. Flight is light, nimble, and evenly undulating in still conditions, jerky and bouncy when windy. Snappy wingbeats are given in a quick series followed by a lengthy, descending, closed-wing glide. Snow Bunting, which is also black and white and distinctly patterned, is plump (anything but slender) with black-tipped white wings and a bouncy flight with a lengthy series of wingbeats interrupted by terse *open-winged* glides.

VOCALIZATIONS: Simple song consists of a few twittering notes followed by a pause, followed by another bout of twittering. Call is a loud, hard, harsh, two-noted "*chizik*"; also makes a whistled "*chee wee.*"

PERTINENT PARTICULARS: The black-backed and white forms of this species have been regarded at different times as both single and separate species (currently regarded as separate species but perhaps not for long). The two forms appear to pair freely, supporting a determination that they are in fact a single species.

Red-throated Pipit, *Anthus cervinus*

Streak-backed Pipit

STATUS AND DISTRIBUTION: Old World species. Uncommon breeder on Alaska's Seward Peninsula, St. Lawrence I., and Little Diomede Is. Regular migrant on the Bering Sea islands. Very rare but regular fall migrant along the Pacific Coast (most records are for California). **HABITAT:** Breeds in low, wet, hummocky meadows and drier rocky areas adjacent to wetlands. In migration, found on estuaries, wet fields, pastures, and sod farms. **COHABITANTS:** During breeding, Pacific Golden-Plover, Rock Sandpiper, Lapland Longspur. In migration, often found in flocks of American Pipit. **MOVEMENT/MIGRATION:** Reaches Alaska in May and early June; departs during late August and early September; most records in California occur from late September to late October. VI: 2.

DESCRIPTION: An American Pipit with a boldly streaked back and a salmon-colored face and throat. Slightly smaller and slightly shorter-tailed than American Pipit.

Adults are brownish gray above, buffy white below, with crisp black-and-white streaking on the back, not the nape (similar to the pattern of Savannah Sparrow), and bold black streaking on the breast and sides. Backs of American Pipit are very plain, showing little or no dark streaking and, except for narrow wing-bars, no hint of white. Streaking on Red-throated's underparts is heavy but blurry. Breeding adults show a bright reddish-orange face and throat. Faces and throats of breeding females and fall males are slightly duller. Immatures are overall buffier and warmer-toned below, showing no trace of red, but still show boldly streaked backs and underparts. At all ages, birds show bright fleshy legs.

BEHAVIOR: Like American Pipit, but prefers slightly more cover. Calls regularly in flight.

FLIGHT: Similar to American Pipit, showing white outer tail feathers but a slightly shorter tail. Crisp bold streaking above and below sets it apart in a flock of American Pipits.

VOCALIZATIONS: A rapid run-on series of repeated notes, some distinct, some trilled, all rather sharp and ringing; for example: "*cheecheecheecheechch-chchchuchuchuchuweeweeweeweep'shee p'shee p'shee p'shee.*" Sounds like a hurried breathy canary. Call is a high, fine, drawn-out "*psssee*" that recalls Eastern Yellow Wagtail but is slightly higher, thinner, and less piercing.

American Pipit, *Anthus rubescens*

A Busy, Jerky, Single-minded Little Bird

STATUS: Common arctic and alpine breeder and wintering species well south of the snow line. **DISTRIBUTION:** Breeds in arctic and subarctic regions from Nfld. to Lab. and across n. Canada; all of Alaska (except for the southeast coast) and south through the mountainous portions of B.C. and Alta.; scattered at high elevations throughout the West. Winters from s. N.J. south and west throughout the coastal plain to Okla. (also s. Kans.), N.M., Ariz., much of the s. Great Basin, and Calif., as well as coastal Wash. and Ore. Outside the U.S., occurs south through Mexico to Central America. **HABITAT:** Breeds in dry tundra, in alpine meadows dominated by dwarf vegetation, and along stream courses. In winter, frequents sparsely vegetated (often wet) habitat or open areas covered by short-cropped plants, such as plowed fields, stubble fields, lawns, athletic fields, sod farms, beaches, mud flats, snow-free river edges, or ice-scoured *Spartina* wetlands. **COHABITANTS:** Horned Lark, rosy-finches, Lapland Longspur. **MOVEMENT/MIGRATION:** Highly migratory. Northern birds vacate breeding areas. Western birds descend to lower altitudes. Spring migration from March to early June; fall from early September to December. VI: 4.

DESCRIPTION: Slender, cryptic, ground-foraging bird that seems able to melt into the earth where it lands. Shaped like a warbler (in fact, in many ways recalls a Palm Warbler), American Pipit has a pointy face, a slightly humped back, a longish tail, and long legs. The overall shape is longer and more slender than a lark. When foraging, the tail is slightly up-cocked.

In all plumages, fairly muted, with a pattern that slights both color and contrast. Breeding birds are grayish-backed with buff or orangy buff on the face and underparts that may be streaked. In winter, birds are grayish brown above, with blurred dark streaks, and creamy or pale buff below, with streaks (sometimes heavy) that form a necklacelike pattern on the breast.

The face is bracketed by a pale eye-line and a pale mustache. The eye-line kicks up in the back, almost encircling the plain darker cap. Legs are dusky. The expression is preoccupied.

BEHAVIOR: Forages on the ground by walking in a wandering, virtually nonstop fashion. Also forages, at high altitudes in summer, on snowfields, where insects are understandably sluggish. The long strides, persistent head-bobbing, and frequent directional changes give the bird's movements a jerkiness that recalls a flickering old-time movie. Sparrows are more site-tenacious and more stop-and-go; unlike pipits, which are focused slightly ahead, larks are more plodding and concerned about what's at their feet. Standing pipits often bob their tails, sometimes vigorously. When foraging, tails are generally unmoving and straight or slightly up-cocked. Males vocalize using an aerial display.

Feeding flocks (numbering a dozen to several hundred birds) are generally silent and fairly loose and uncoordinated; birds move in the same general direction, but independently. When flushed, birds form more cohesive and linear flocks and become vocal.

Tame, frequently allowing observers to get quite close (but to little avail). Usually flush before being detected.

FLIGHT: Fairly nimble and bouncy, but not particularly fast. A rapid series of wingbeats is followed by a closed-wing glide. When landing, often brakes with a series of open down-angled wings that gives the bounce a floating quality. Sticks its landings with a vertical leg-throwing pounce, still facing the direction of approach.

VOCALIZATIONS: Says its name. Call is a high-pitched "*sip-it*" (more emphatic than the call of Horned Lark); when flushed, emits a single "*chee*" or "*chee, chee, chee,*" and sometimes a whispered "*chih.*" Song, usually sung in flight, is a variable and lengthy series of repeated identical notes or phrases often involving sequences of different-sounding notes; for example, "*ching ching ching ching ching*" (sounds like a metallic pinging) is followed by a high-pitched, somewhat raspy "*cheewee cheewee cheewee.*" Sequences often accelerate near the end.

Sprague's Pipit, *Anthus spragueii*

The Bug-eyed Short-grass Pipit

STATUS: Uncommon, geographically restricted, habitat-specific, skulking, and really hard to find. **DISTRIBUTION:** A prairie species that breeds in se. Alta., s. Sask., sw. Man., cen. and e. Mont., N.D., extreme nw. Minn. (rare), and nw. S.D. Winters in se. Ariz., extreme s. N.M., Tex. (except for the panhandle), s. Okla., cen. and s. Ark., all but se. La., and n. Mexico. **HABITAT:** At all seasons, expansive, short, dry, open grasslands that are tall enough to hide in and devoid of shrubs. Often found near or at the edge of bare or partially denuded patches of earth surrounded by grass, but does not favor the large, open, nongrassy areas preferred by American Pipit. Sprague's prefers native prairie grasses, but in winter and migration is also found in mowed fields of wheat stubble, alfalfa, and Bermuda grass. **COHABITANTS:** Savannah Sparrow, Chestnut-collared Longspur, Western Meadowlark. **MOVEMENT/MIGRATION:** Entire population shifts between breeding and wintering areas. Spring migration from late March to early May; fall from mid-September to late November. VI: 3.

DESCRIPTION: A pale, pointy-faced, bug-eyed pipit—head/face suggests an immature Horned Lark or (believe it or not) Upland Sandpiper. Small (smaller than Horned Lark; larger than Savannah Sparrow; about the same size as, or slightly larger than, American Pipit), with a pointy larkish face, a trim, somewhat chesty body, and a fairly short tail (for a pipit). A large dark eye set against the plain grayish to brownish gray face is prominent, giving the bird a dull uncomprehending expression. Sprague's Pipit looks bug-eyed; American Pipit looks beady-eyed.

Plumage recalls a pale streaky sparrow. Overall paler, buffier, and more finely patterned than winter American Pipit. Has a finely streaked (not all-dark) crown. The pale legs are generally not visible, but the bill flashes flesh tones (bill of American Pipit is dark).

BEHAVIOR: A "feathered mouse," weaving somewhat drunkenly through grass just tall enough for concealment, occasionally raising its head and showing all you need to identify it—a pointy-faced bug-eyed bird. Walks and runs with its head down, and does not wag its tail (not that you could see it doing so if it did). Flushes reluctantly, frequently calls when flushed, and usually ascends, circling high overhead, before executing a near-vertical descent; almost always returns to the same field even when flushed repeatedly. Often calls when flushed, and often raises its head to study the situation or assess your approach shortly before and after landing. Towering flight displays, conducted 50–100 yds. over the prairie, often last for over half an hour.

A solitary feeder, Sprague's is *not* a flocking species like American Pipit, and does not associate with American Pipit flocks; multiple birds may occupy the same field, however, and birds may form loose aggregations during migration. Frequently found in the short grass surrounding bare patches of earth and, in winter, in the same habitat occupied by Savannah Sparrow, Baird's Sparrow, and Chestnut-collared Longspur.

FLIGHT: Pointy-faced, pointy-winged, and fairly short-tailed profile. White outer tail feathers are conspicuous when birds are flushed, but not necessarily when birds are circling overhead; high-flying birds seem dirty-chested. Wingbeats are given in a short rapid series followed by a lengthy, descending, closed-wing glide. Flight is undulating, with a bouncy or jerky quality. Descending birds execute near-vertical drops followed by several braking undulations close to the ground, punctuated by an abrupt drop landing.

VOCALIZATIONS: Flight song is a prolonged descending series of high thin "*see*" notes given in two- to three-second bursts and then repeated. Call, often given several times when flushed, is a sharp squeaky "*squick*" that is often doubled.

PERTINENT PARTICULARS: This species shows a high degree of site fidelity. Finding the bird is often no more challenging than finding its somewhat specific habitat: extensive, short (2–4 in.), dry grasslands showing patches of bare earth. A good strategy is to walk the perimeter of such bare spots, about 8 yds. out from the edge. Sprague's Pipit usually climbs when flushed; sparrows line out low (and often leave the field).

WAXWINGS AND PHAINOPEPLA

Bohemian Waxwing, *Bombycilla garrulus*

Rufous-undertailed Waxwing

STATUS: Common breeder in Alaska and western and central Canada; more irregular breeder farther east (possibly all the way to Labrador); common but irregular winter resident in southwest Canada and in the northern Rocky Mountain states and northern prairies, becoming less common and even more irregular farther east all the way to Newfoundland (where it is almost annual). Irruptive. **DISTRIBUTION:** Breeds from Alaska to the Yukon, the Northwest Territories (south of the tundra), n. B.C., n. and cen. Alta., n. Sask., n. Man., and nw. Ont. During non-irruptive winters, vacates northern breeding areas, relocating to the southeast coast of Alaska, across all but the western portions of B.C., cen. and s. Alta., cen. and s. Sask., cen. and s. Man., s. Ont., s. Que., and the Maritimes. In the U.S., particularly during years showing a modest irruption, winters in e. Wash., e. Ore., Idaho, Mont., Wyo., n. Utah, nw. Colo., N.D., n. and w. S.D., n. and cen. Minn., Wisc., Mich., n. N.Y., and New England. During major irruptions, may wander as far south as northern portions of the states bordering Mexico, the central prairies, the Midwest, and the mid-Atlantic states. **HABITAT:** Breeds in taiga forests; also in open coniferous and coniferous-deciduous forests, most particularly areas that have been recently burned or that occur adjacent to water (beaver dams, rivers, swamps). In winter, shuns conifers and chooses to be near fruit-bearing bushes and trees. **COHABITANTS:** Breeding Mew Gull, Boreal Chickadee, Gray-cheeked Thrush, Yellow-rumped Warbler. In winter, American Robin, (less commonly) European Starling, and (especially) Cedar Waxwing. **MOVEMENT/MIGRATION:** Normal spring migration from March to late April; fall from September through November. At the eastern limits of the bird's range, may not arrive in numbers until December or even February. Within the winter range, some shifting of populations occurs all winter. VI: 3.

DESCRIPTION: A portly grayish waxwing with white in the wings and a cinnamon undertail. Medium-sized (larger and more robust than Cedar Waxwing; smaller than American Robin; about the same size as European Starling), with a large wedge-shaped face, a rakish backswept crest, a plump body, and a short wide tail. Overall more robust than Cedar Waxwing, as well as more humpbacked, more potbellied, bigger-headed, and shorter-tailed.

Overall grayer and slightly darker than Cedar Waxwing, with a brownish gray body and a slightly warmer brown head. No butterscotch on the breast and no yellow on the belly, but does share with its smaller cousin a conspicuously yellow-tipped tail. In a mixed flock, the greater size, dark undertail, and colder tones of Bohemian distinguish it. Alone or (more likely) among other Bohemians, the touches of white in the wing identify it. Don't depend on seeing the cinnamon undertail coverts on perched birds; they are most conspicuous in flight.

BEHAVIOR: In summer, pairs (sometimes in association with other pairs) perch atop spruces (and other tall perches) and sally out, snapping up flying insects. At other times, the bird forages almost wholly on fruit, which it plucks from branches while perched or on the wing; also hovers and plucks. Occasionally secures food by hopping on the ground.

For most of the year, highly social, flying in flocks of 50–300 birds (and larger flocks are not uncommon). Bohemians commonly mix with Cedar Waxwings and American Robins. Outside their normal range, the odd Bohemian is almost invariably associated with Cedar Waxwings, although Bohemian has also been known to feed with starlings and to forage alone.

Quite tame. Responds to pishing. Also vocal. When birds take flight, they commonly fly great distances (usually out of sight), and like Cedar Waxwing, they fly in tight-packed groups.

FLIGHT: Silhouette is much like European Starling—an elliptical body (pointed fore and aft), with short triangular wings set about midpoint. Overall uniformly dark, except for the bright white comma on the upper surface of the wing and the large bright reddish undertail that seems to encompass the whole trailing half of the bird. This is no small field mark. If it's there, you won't miss it. Bohemian is a strong flier. Flight is evenly undulating with a lengthy rapid series of wingbeats punctuated by short closed-wing glides or (particularly when flying short distances or landing) more bounding open-winged glides. Flight, particularly when banking, is more sluggish, more staid, and not as reckless as Cedar Waxwing.

VOCALIZATIONS: Commonly heard call is a low ripping trill that in length and pattern recalls the high ethereal sigh of Cedar Waxwing but is lower-pitched, more rippling, and slower; thus, you can more easily pick out individual notes than you can with Cedar Waxwing.

Cedar Waxwing, *Bombycilla cedrorum*

Crested Sigh

STATUS: Common and widespread breeder and winter resident. **DISTRIBUTION:** Breeds across boreal Canada and the n. U.S.; winters across southern border regions of Canada, from s. N.S. across extreme s. Que. and s. Ont. west to Vancouver I., south across all of the U.S. (except n. Maine) and Mexico into Central America. **HABITAT:** Nests in various open coniferous, deciduous, and mixed woodlands and successional fields dotted with trees and shrubs. Also found in riparian corridors, orchards, tree farms, and suburban habitats. In winter, found almost anywhere fruiting trees are found—open woodlands, forest edge, successional fields with shrubs and small trees, golf courses, hedgerows, shelterbelts, city parks, and suburban woodlots—almost everywhere but the forest interior. Loves cedars. **COHABITANTS:** In winter, American Robin, Yellow-rumped Warbler, and, occasionally, Bohemian Waxwing. **MOVEMENT/MIGRATION:** Cedar Waxwing is nomadic, and its movements mercurial, contingent upon the abundance or paucity of sugary fruits. The bird's migratory pattern is complex, with two distinct peaks in spring in central and eastern North America: a minor movement that occurs in February and March may correspond to the bird's search for food as winter stocks become depleted; the other, larger movement in May and June is made up of returning breeders. A late breeder, Cedar Waxwing does not begin nest-building in some areas until late June or even July. Fall migration begins in late August and continues into early December. VI: 4.

DESCRIPTION: Distinctive and distinguished. Can be confused only with the larger, portlier, grayer, and more northern Bohemian Waxwing (and in flight, European Starling). Named for the red tips on the secondary flight feathers, which suggest beaded wax, but the fact is that the entire body of Cedar Waxwing seems poured from wax. From the sweptback rakish crest to the yellow-tipped tail, every feather on the adult seems seamlessly set in place.

Adults are overall warm brown—pale butterscotch-colored—with a pale yellowish belly and white undertail coverts. The party-mask black facial pattern is apparent on perched birds, and the yellow-tipped tail is evident in flight. The folded wings, except for a pale line separating the wing from the back of the bird and the tiny red beads, are unmarked. (Bohemian Waxwing shows one or two touches of white on the folded wing.)

Immatures, seen throughout the fall, seem like a skeletal rendering of the adult—overall grayish brown, blurry but distinctly streaked below, with a facial pattern that seems a macabre death's head rendering of the adult's distinctive party mask. Like adults, the young have a crest and the yellow band on the tail and are virtually never seen alone.

BEHAVIOR: Extremely social. Found in flocks from several to a hundred birds throughout most of the year. Even during breeding season, it is common to see trios (although nest helpers are not documented for this species). Feeds on fruits and berries. Forages by perching on the outer branches of trees and plucking fruit, usually from an upright (sometimes dangling) position. Also hovers and sallies out to catch insects on the wing. Likes to sit high. Arriving flocks habitually claim the highest branches in live (and frequently) dead trees.

Festooning flocks, often packed in clusters, can be surprisingly cryptic. From a distance, the plumage is eye-defeating, and movement among the branches is minimized. Observers are usually alerted to the presence of the birds by their unique call (which increases in volume and urgency just before flocks take flight) or by the sight of the tightly packed highly coordinated flock lifting off from a tree.

FLIGHT: In flight, the birds look extremely pointy—pointy-faced and pointy-winged. This trait coupled with their short square-cut tails makes the birds look like arrowheads or a croissant with short triangular wings. Birds seem two-toned: dark in front and pale behind, or, at closer range, a pale butterscotch-colored bird with a pale vent.

Flight is undulating, fast, jaunty, and sometimes reckless given the close proximity of other birds in tightly packed flocks. Wingbeats are given in a long rapid series followed by a short closed-wing glide. Flocks are shifting—often tightly packed, sometimes spread out and loose, with individual birds shifting position. The effect is mesmerizing.

Other flocking birds that might be confused with Cedar Waxwing are starling and, in fall, Bobolink (See "Pertinent Particulars").

VOCALIZATIONS: No song, but highly vocal. Call is a high-pitched ethereal sigh—sometimes haunting, sometimes serene—that barely brushes the upper end of the human register. Sometimes the call is lower-pitched and trilled, and sometimes it's breathy and shrill—a vapor-thin scream. Birds vocalize constantly in flight, and even though the call seems to lack substance, it is audible for great distances.

PERTINENT PARTICULARS: At a distance, the tightly packed flock configuration of Cedar Waxwing is similar to that of both European Starling and Bobolink. Starlings, however, are more coordinated—individual birds don't shift position, and the flight is straight, not bouncy. Bobolink flocks are less tightly packed, and the birds often spread horizontally. Cedar Waxwing flocks are often stacked up vertically.

Phainopepla, *Phainopepla nitens*
Black Silky-Flycatcher

STATUS: Common but restricted resident and breeder of the Southwest. **DISTRIBUTION:** In winter, found in the Sonoran Desert of Ariz., Mexico, and s. Calif.; in late spring and summer, relocates northward to coastal and cen. Calif., s. Nev., extreme sw. Utah, cen. and se. Ariz., sw. N.M., and the western edge of trans-Pecos Tex. south to Big Bend. In Mexico, found throughout Baja and interior areas. **HABITAT:** In winter, desert riparian areas and washes dominated by acacia and mesquite where desert mistletoe thrives; also, found in less arid areas, amid oaks and cottonwoods where mistletoe flourishes. In summer, occurs in semi-arid and riparian woodlands dominated by live oak, sycamores, and Joshua trees, as well as the surrounding brushy areas. **COHABITANTS:** Cactus Wren, Pyrrhuloxia, House Finch, Lucy's Warbler, Yellow-breasted Chat, Blue Grosbeak. **MOVEMENT/MIGRATION:** Annual movement pattern is complex. Between March and May, birds breed in the Sonoran Desert and eastern Mojave. Between mid-April and early June, they relocate to adjacent regions, where they presumably breed again before returning to the core desert breeding areas in August through November. VI: 1.

DESCRIPTION: A slender, medium-sized, wispy-crested silky-flycatcher. Males, with their sparse cowlick crests, suggest a slender, gaunt, black cardinal. Females and immatures are similarly shaped, sooty gray, and likewise sparsely but conspicuously crested.

BEHAVIOR: Commonly seen perched atop a desert shrub or in transit, just over the tops of vegetation, to find another perch. Forages on both berries and insects. Clings to outer twigs of vegetation to pluck berries and (occasionally) insects. Also sallies out, hovers, and pursues insects with buoyant acrobatics. In winter (first breeding), usually defends a mistletoe berry–rich territory from other Phainopeplas (and fruit-eating birds). At this time of year, birds are often seen in loose groups. In summer (second breeding), birds are more social, even communal, nesting in small colonies; birds moving to and from roosts sometimes number 100 or more.

FLIGHT: Light, buoyant, nimble, direct, and somewhat jaylike; from below, the bird seems to be throwing its wings to the side rather than moving them up and down. The body is slender, and the wings broad and paddlelike; those of adult males show conspicuous white oval patches, which are visible both above and below. Females show a less conspicuous but still distinctive pattern. Wingbeats are quick but not rushed and given in an irregular series. Flight does not undulate but is sometimes twisty-turny or slightly wandering—birds jink left or right or show a little English, as though curving to compensate for windage.

VOCALIZATIONS: Song is a series of short, spaced, and softly uttered phrases—some musical and gurgling, others harsh and grating. Some phrases are recycled in the song, and a few are doubled. Interspaced between the sets of phrases is a signature rough, descending, purring trill. Song recalls the mindless babble of starlings but is less run-on and more softly rendered. Calls include a soft, plaintive, humanlike whistle, "*too-eee*"; a rough purring trill, "*trrrrr*"; and a burble, "*ch'r'leep*" or "*t'chr'leep*," that recalls Red-winged Blackbird.

WOOD-WARBLERS

Olive Warbler, *Peucedramus taeniatus*

Copper-hooded Warbler

STATUS: Fairly common but restricted by geography and habitat. **DISTRIBUTION:** Central and se. Ariz. and sw. N.M. south to Central America. **HABITAT:** High-altitude open-canopy pine forests. Particularly partial to ponderosa pine, but also found in sugar pine, Douglas fir, and true firs. **COHABITANTS:** Pygmy Nuthatch, Mountain Chickadee, Mexican Chickadee, Grace's and "Audubon's" Warblers. **MOVEMENT/MIGRATION:** Although a permanent resident throughout most of its breeding range, in the United States some Olive Warblers—perhaps most of them—vacate breeding areas, withdrawing south presumably. Does not commonly seek lower elevations in winter, but may move into canyons during the worst winter weather. VI: 0.

DESCRIPTION: A handsome, sturdy, long-billed warbler of tall pines and high places. A medium-sized warbler about the same size as Grace's Warbler, but more robust (somewhat Prothonotary Warbler–shaped), with a large head, a thin longish bill, and a fairly short tail that is flared, deeply notched and, in shape rather than length, recalls Cassin's Finch.

Adult males, with their burnt orange–hooded heads contrasting with grayish bodies, are unmistakable. Yellowish-headed females are similarly patterned but not as contrasting; the greatest contrast between the yellow hood and the rest of the bird occurs not on the back but at the bend of the folded wing. The dark party mask, which is prominent on males, is muted but still apparent on females and also serves to distinguish the more muted immatures.

On immatures, the hood is replaced by a yellow wash on the cheeks and sides of the neck. At a distance, this wash gives the face a yellowish cast and makes the shadowy mask more prominent. Note also the absence of any streaking on immatures and the uniform gray underparts.

BEHAVIOR: Loves pines; loves the canopy. Generally glimpsed foraging high along the inner branches and on the outer needles of long-needled pines. A slow deliberate feeder that moves in short hops along limbs, pausing to probe deep into needle clusters, sometimes hanging upside down. The pattern and cadence resemble the Pygmy Nuthatches often found in the same habitat—up to a point! Olive Warbler also hovers, sallies out after flushed prey, and flies to new perches with a regularity that would be unseemly for a nuthatch. While foraging, spends lavish amounts of time in the same tree. Birds shifting locations may fly to a nearby tree, but frequently they travel great distances instead.

Males vocalize while feeding. The bird does not respond well to pishing but will investigate the ruckus raised by birds that do. Don't expect Olive Warbler to leave the canopy.

FLIGHT: Seems robust in flight too. The white outer tail feathers are usually apparent. Flight is fast, strong, and generally just above the canopy, and often covers long distances. Flight is also direct and undulating, with moderately slow wingbeats given in series interrupted by terse skips. You can actually make out the shape of the wing—it's not all a blur!

VOCALIZATIONS: Song is a simple rendering of a compressed two-note phrase repeated three to five times; for example: "*d'jr d'jr d'jr d'jr d'jr*" or "*t'rjee t'rjee, t'rjee.*" Variations include a three-note phrase, "*her jee j'ree,*" and a two-part song comprising two different repeated phrases. The overall quality of the song is low and flat and has a slight twang on the second syllable. In cadence, pattern, and tonal quality, resembles a titmouse. Call is a short, soft, descending whistle, "*few,*" that recalls Western Bluebird; also makes a hard "*pit.*"

Blue-winged Warbler, *Vermivora pinus*

Mascara-masked Warbler

STATUS: Common and fairly widespread eastern breeder. **DISTRIBUTION:** Breeds across most of s. and cen. New England, the mid-Atlantic states, the interior Southeast, and the Midwest north to s. Maine, s. Ont., s. Mich., s. Wisc., and se. Minn., west to cen. Iowa, w. Mo., ne. Okla., and south to n. Ark., n. Ala., and n. Ga. Except where noted, absent as a breeder in the South. Casual vagrant in the West. **HABITAT:** A brush and shrub specialist. Found in second-growth deciduous forest, successional fields, brushy woodland swamps, and forest edge. In migration, also prefers to forage in dense, brushy, deciduous woodland habitat. **COHABITANTS:** Brown Thrasher, White-eyed Vireo, Golden-winged Warbler, Yellow Warbler, Yellow-breasted Chat, Eastern Towhee, Field Sparrow. **MOVEMENT/MIGRATION:** Wholly migratory; winters in southern Mexico and Central America. Spring migration from early April to early June, with a peak from mid-April to mid-May; fall from mid-July to early October, with a peak from mid-August to mid-September. VI: 3.

DESCRIPTION: A lemon yellow warbler with blue-gray upperparts (i.e., closed wings), bold white wing-bars, and a mascara-fine black line etched through the eye. Except for the dark eye-line, resembles a miniature Orchard Oriole. Small and lithe (slightly smaller than Yellow Warbler; about the same size as Wilson's Warbler), with a short pointy bill, a smallish pointy face/head, a fairly plump body, and a fairly short tail.

The very bright yellow head and underparts and dingier yellow or greenish back are complemented by bluish gray wings and tail. The white wing-bars are broad and conspicuous on adults, narrow on immatures. The thin black line cutting from the bill to the eye is easy to see. The white undertail coverts and white outer tail feathers are often conspicuous because of the bird's chickadee-like feeding antics. Ages and sexes are similar. The expression is impishly mean.

BEHAVIOR: A nimble acrobatic feeder that likes to work in low tight places (such as brushy understory). In spring and summer, often feeds actively, jumping frequently to new perches, hovering, focusing its attention on the outer and lower portions of bushes and trees. In fall, usually more methodical, clinging to vegetative tangles and dead leaf clusters (like Golden-winged and Worm-eating Warblers) and prying into the interior with its pointy nut pick–like bill. At all seasons, commonly hangs upside down (like chickadees). This behavior doesn't distinguish it from Golden-winged Warbler, but when views are quick or obscured by vegetation, it helps distinguish it from other mostly yellow warblers.

Males sing from exposed perches, either on the top of a bush or just below the canopy of a taller tree at the edge of the territory.

More responsive to pishing during breeding season. In migration, often reluctant to move into the open.

FLIGHT: Slight wisp of a warbler. Shows a slender profile with a long straight-sided tail. The lemon yellow underparts contrast with the white undertail. The white wing-bars are often easily seen, as are the white sides to the tail, particularly when the bird fans its tail. Flight is light, buoyant, and bouncy. Wingbeats are mostly steady, punctuated by very brief skip/pauses. At close range, the narrow black eye-line imparts an angry expression (even in flight).

VOCALIZATIONS: Simple easily recognized song consists of two prolonged, even, buzzy notes ("*beee buzzzz*"). The first note is higher and often has a zinging electric quality; the second is flatter and sometimes has a flatulent quality. Sounds like an asthmatic bird inhaling and exhaling. Also makes a single buzz followed or bracketed by excited stuttering notes (alternate song). Call is a weak dry "*chih.*" Flight call is a thin "*swee.*"

Golden-winged Warbler, *Vermivora chrysoptera*

Chickadee-bibbed Warbler

STATUS: Generally uncommon and geographically restricted breeder; eastern migrant. DISTRIBUTION: Found primarily the ne. and n.-cen. U.S. and se. and s.-cen. Canada. Breeds in s. Man., se. Ont., and s. Que., and in n. Minn., all but s. Wisc., portions of Ill., Mich., nw. Ind., e. Ky., e. and cen. Tenn., W. Va., w. N.C., w. Va., w. Md., all but se. Pa., nw. N.J., N.Y., nw. Conn., w. Mass., and Vt. Casual vagrant in the West. HABITAT: Open brushy and very commonly wet fields or marsh edge bordering woodlands; also occurs in sapling-dominated woodland swamps. Prime habitat includes fields undergoing succession and power-line cuts with brush. Favors wetter and younger successional habitats than Blue-winged Warbler. COHABITANTS: Willow and Alder Flycatchers, Blue-winged Warbler, Common Yellowthroat, Red-winged Blackbird, and Black-capped Chickadee, which it also resembles. MOVEMENT/MIGRATION: Fairly early- to mid-spring migrant (a little later than Blue-winged) and early-fall migrant. Spring migration from mid-April to late May, with a peak in early May; fall from late July to mid-October, with a peak in late August and early September. VI: 3.

DESCRIPTION: A trim pale gray warbler with a face pattern that shouts and gilded highlights that flash. Smallish (smaller than Yellow Warbler or Common Yellowthroat; the same size as Blue-winged), with a pointy face, a spiky bill, and a fairly short tail. Structurally similar to Blue-winged Warbler.

The adult male's white framed black face mask and chickadee-like bib are arresting. The bright yellow crown and shoulder seem incandescent against the plain, mostly gray body. Adult females and immatures show the same pattern, but the mask and bib are dark gray. It doesn't matter. The bordering white frame makes the dark mask stand out; the yellow crown and shoulders are duller, but still very apparent.

BEHAVIOR: A brush-loving bird that likes to work at low and middle levels—sometimes in the tangled depths of shrubs, sometimes on the outer branches of taller saplings. Fairly clandestine but also nimble and quick. Forages by hopping along branches, investigating leaves, and tenaciously probing clusters of live and dead vegetation; often dangles, chickadee fashion. Males make long flights across open areas, often to take a perch at or near the top of a sapling to vocalize. Females, as a general rule, forage closer to the ground.

Fairly bold. Responds well to pishing, but may be reluctant to leave protective cover, electing to show only its face.

FLIGHT: Small and slightly proportioned. Overall pale, whitish, and clean-cut. The dark face pattern is very conspicuous on adult males and fairly distinctive on females and immatures; the broad white sides of the tail are very obvious when birds fan their tails. Flight is direct, with a slight rising-and-falling action. Wingbeats are mostly continuous, given in a lengthy series punctuated by a terse skip/pause.

VOCALIZATIONS: Two-part song consists of a series of dry raspy buzzes—the first is long, thin, and higher-pitched; the second set of two to four notes (usually three) is hurried and huskier: "*jeee jer jer jer.*" Overall higher-pitched and more hurried than Blue-winged's two-note song. Sounds as though the bird is inhaling the first note and choking on the sequence. Has an alternate song similar to Blue-winged song. Call is identical to Blue-winged.

PERTINENT PARTICULARS: Golden-winged and Blue-winged Warblers hybridize freely, producing fertile young that mate and breed with either parent's species as well as other hybrids. Hybrids form the Brewster's Warbler group; these birds show the basic plumage pattern of Blue-winged Warbler but have yellowish, not white, wing-bars and lack the complete bright yellow underparts of pure Blue-winged. The young of paired hybrids show a mix of traits, of which the most stunning combination is the Lawrence's Warbler, which combines the face pattern of Golden-winged with the bright yellow head and underparts of Blue-winged.

Hybridization is as much the product of overlapping habitats as it is of genetic similarity. In general, Blue-wingeds favor drier and more densely wooded habitats; Golden-wingeds are found in wetter, more open places and younger successional growth, often at higher altitudes in the southern part of their range and where they overlap with Blue-winged.

Also, be aware that some Golden-winged Warblers sing songs more reminiscent of Blue-winged (and vice versa).

Tennessee Warbler, *Vermivora peregrina*
The New World's Old World Warbler

STATUS: Common breeder in northern forests; common migrant east of the high plains but not common on the East Coast. **DISTRIBUTION:** Breeds across much of subtundra Canada from the nw. Northwest Territories, se. Yukon, and w. B.C. to Nfld. and the Maritimes. (Absent in sw. B.C., se. Alta., sw. Man., s. Sask., and se. Ont.) In the U.S., breeds in n. Maine, n. N.H., n. Vt., the Adirondacks, and ne. Minn. Rare vagrant in the West. **HABITAT:** A variety of northern woodlands, including spruce forest, coniferous and mixed deciduous-coniferous habitat, young stands of deciduous trees, alder and poplar groves, and tamarack bogs. Seems to require in all habitats a dense deciduous shrub layer or low dense stands of early succession conifers. In spring migration, often forages high in the canopy of flowering trees (primarily oaks in the East). In fall, occurs in almost any woody habitat; birds very often forage low and along edges. **COHABITANTS:** Mixed in summer, but includes Swainson's Thrush, Nashville Warbler, Magnolia Warbler, Yellow-rumped Warbler, Dark-eyed Junco. In migration, joins mixed-species flocks. **MOVEMENT/MIGRATION:** Spring migration from mid-April to early June; fall from late July to early November. VI: 3.

DESCRIPTION: A small compact woodland subtlety. The plumage resembles one of the duller Old World warblers. Smallish warbler (smaller than Yellow-rumped; slightly smaller than Orange-crowned; the same size as Nashville), with a thin pointy bill, a neckless head, a compact body, and a shortish tail. Structurally similar to Orange-crowned Warbler, but longer-winged and distinctly shorter-tailed.

Males in spring are conservatively garbed but distinctive in their subtlety: no wing-bars, no tail spots, no bibs, caps, streaks, masks, spectacles. . . . In spring, a grayish head contrasts with olive green upperparts and whitish underparts. The bird is so plain that a frail grayish line through the eye and a narrow white eyebrow actually stand out.

Spring females are more yellow-tinged on the breast, showing less gray about the head (usually limited to the cap) and a yellow wash on the throat and breast. In fall, all birds (adults and immatures) are *really* plain—yellow-green above and yellowish below, *except* for contrastingly white (at least whitish) undertail coverts, and still showing a suggestion of a shadowy eye-line and a narrow pale eyebrow.

Most important, the plumage is smooth. Tennessee Warbler has no shadowy streaks on the underparts. That's Orange-crowned (which also shows a darker eye-line, a narrow, broken, white eye-ring, often a pale yellow edge at the bend of the wing, and yellow, not white, undertail coverts).

BEHAVIOR: In migration and summer, males are very vocal, often singing as they forage. On territory, singing males perch near the tops of trees (but generally not on the top) and may remain for some time on the same perch, where they can be maddeningly hard to find. A nimble active species that feeds mostly by gleaning off branches, leaves, and needles. Moves in short studied hops mixed with short searching pauses at all heights (from the shrub layer to the tallest outer branches). Hovers infrequently, but does hang from branches and pluck. In spring migration, seems most often to forage on the outer branches, often at the tops of trees. In fall, frequently forages low along wooded edge. On territory, usually works the tree's interior. In migration, often joins mixed-species flocks. Not shy. Responds well to pishing.

FLIGHT: Overall compact but disproportionately long-winged, short-tailed, and overall contrasting—greenish above, pale or yellowish below. In fall, the yellowish underparts contrasting with the pale undertail distinguish it from female Black-throated Blue (and Nashville). Flight is quick and slightly undulating; a lengthy series of wingbeats is followed by a terse closed-wing glide.

VOCALIZATIONS: Song is a series of two to four chattering note sequences, all hurried and given on the same pitch. First notes often sound hesitant or tentative. A sequence commonly begins with rapid chatter and ends with a staccato trill; for example: "*chi chi chi chi chi churchurchurch'h'h'h'h'h'h'h'h*" or "*chitsi chitsi chitsi see see see chchchchchchch.*" Most notes sound sharp, brittle, and crisply enunciated. Eastern Nashville Warbler's song is similar but slower, less emphatic, and usually more distinctly two-parted and more musical. Call is a soft weak "*chih.*"

PERTINENT PARTICULARS: Tennessee Warbler's muted plumage pattern is reminiscent of several vireo species, most notably Warbling, Philadelphia, and Red-eyed. Its smaller size, finer bill, and more active feeding behavior help distinguish it.

Orange-crowned Warbler, *Vermivora celata*

Knee-top (to Waist-high) Warbler

STATUS: Common northern and western breeder; common southwestern winter resident and migrant; much less common east of the Appalachians. **DISTRIBUTION:** Breeds from s. Lab. and ne. Que. across cen. Que., n. and cen. Ont., Man., all but s. Sask., all but s. Alta., B.C., most of the Northwest Territories, the Yukon, and all but n. Alaska; also found across much or most of Wash., Ore., Calif., Idaho, Nev., Utah, w. Mont., w. and cen. Wyo., w. Colo., e.-cen. Ariz., w. N.M., and trans-Pecos Tex. Winters along the coastal plain from se. Va. to La. and s. Ark.; also all of Tex. (except the panhandle), s. and w. Ariz., and much of Calif. west of the Sierras (a few are found in the Owen Valley and around Reno, Nev.), and coastally in w. Ore. and Wash.; also south into Mexico and Central America. **HABITAT:** Breeds in a variety of habitats, but is most partial to streamside thickets, riparian habitats, or woodlands rich in shrubs and low-lying vegetation. In winter, forages in all manner of deciduous thickets, woodlands, weedy fields, chaparral, parks, gardens, orchards, and exotic plantings. In spring, often forages high (especially in flowering oaks). In fall, most commonly found in weedy fields and bordering brushy edge—commonly below waist height. **COHABITANTS:** During breeding, Yellow Warbler, Wilson's Warbler, Lincoln's Sparrow. In winter, joins winter flocks with chickadees, kinglets, gnatcatchers, assorted warbler species. **MOVEMENT/MIGRATION:** Orange-crowned follows a complex pattern involving several subspecies, but over most of North America spring migration is from mid-March to mid-May, fall from mid-September to early November, with a peak in October. In California and along the West Coast, migration is very protracted, with spring movement from late February to early May, and fall from early July to late October. VI: 3.

DESCRIPTION: A plain-looking, mostly grayish or yellowish warbler with blurry or shadowy-streaked underparts; Orange-crowned is so blandly garbed that minuscule traits stand out. Medium-sized warbler and overall compact (larger than Tennessee; smaller than Yellow-rumped; about the same size as Yellow Warbler), with a short, spiky, slightly down-turned bill, a large head, a compact body, and a modestly long tail.

In all plumages, dull, bland (often dingy), and devoid of contrasts. Plumage ranges from wholly medium yellow (West Coast) to mostly gray with a touch of yellow on the belly and a yellow undertail. Except for West Coast birds, most individuals show a dull yellowish gray-green body and a grayer head/face. One candidate for confusion is Tennessee Warbler—or perhaps a very drab immature Yellow or Nashville Warbler.

In all but a very few very plain cases, the underparts of Orange-crowned Warbler show blurry or shadowy streaks. (Tennessee Warbler has no streaks.) In all plumages, the eyes are accented by tiny pale eye-arcs or broken spectacles and an eyelash-fine dark eye-line. (Yellow Warbler's face is plain; Tennessee's face shows a pale eyebrow, but the bird doesn't appear to be wearing spectacles.) In all plumages, the undertail coverts are yellow (white on Tennessee, except for a wash of pale yellow on a very few fall immatures). In most cases, a narrow yellow spot or line traces the bend of the folded wing (which Tennessee lacks). The orange crown is rarely visible.

BEHAVIOR: A very active, nimble, busy, even high-strung warbler that forages at all heights but is usually within 30 ft. of the ground, and often much closer than that; in fact, in winter, may forage among leaves on the ground. Something of a dead-leaf specialist, Orange-crowned is more commonly found amid the foliage, working a tree or bush thoroughly but quickly, making short hops and rapid head-jerking inspections. Stretches its neck and flutters its wings to reach prey just out of reach. Hangs upside down, clings to trunks, hovers, and sallies out to snap insects out of the air. Territorial (and migrating) males sing while foraging, and in winter, birds frequently call. Like several other species, Orange-crowned is drawn to sapsucker wells. Unlike many warbler species, seems much at home feeding in weedy growth (particularly in fall). In winter, may join mixed flocks or be found in small Orange-crowned groups. So bold as to be almost oblivious to observers at times, and easily piqued by pishing.

FLIGHT: Overall stocky and plain (except for the yellow on the undertail on gray individuals). Wingbeats are given in a rapid series punctuated by a terse skip/pause. Flight is quick and undulating.

VOCALIZATIONS: Song is a quavering warble—a low, flat, warbling trill that varies in delivery but remains consistent in its pattern—a hurried, rising-and-falling, slurred, but chattery warble or trill that seems to lose oomph and to drop (or sometimes rise) in pitch at the end; for example: "*chchchCHchChchchree.*" Call is a distinctive, deep, musical, sparrowlike "*t'chuh*" or sharper "*t'cheh*" (much like Field Sparrow). Neither call has a hard ending.

PERTINENT PARTICULARS: In winter, this species is very common in California, Arizona, Texas, and parts of the South. In migration, it is very common in early spring and late in the fall from the western Gulf Coast, Texas, and the Great Plains westward.

Nashville Warbler, *Vermivora ruficapilla*

False Connecticut Warbler

STATUS: Uncommon to fairly common northern breeder; widespread migrant with disjunct eastern and western breeding populations; uncommon and restricted winter resident. **DISTRIBUTION:** The eastern population breeds from the Maritimes and e. Que. across s. Que., s. and cen. Ont., and cen. Sask. In the U.S., found throughout New England, N.Y., nw. N.J., n. and w. Pa., n. Mich., n. Wisc., and n. Minn. The western population is found in s. B.C., sw. Alta., nw. Mont., w. and cen. Wash., e. and nw. Ore., n. Calif., and south along the Sierras to the s. Calif. mountains (local). Winters primarily in Mexico, but also found in very small numbers along the Calif. coast from Eureka to n. Baja and in coastal Tex. from the Brazos R. to Rio Grande City. **HABITAT:** Both populations favor deciduous or mixed second-growth forest or forests with open canopies and a dense shrub layer. Specific habitats include tamarack and spruce bogs, forest edge, burned-over jack pines in early stages of succession, and regenerating clear-cut areas. In migration, occupies a variety of habitats (often with mixed warbler flocks), including the tops and midstory sections of deciduous trees (particularly flowering oaks), the brushy edges of fields and streamsides or ponds, and drier mountain forest habitat in the West. In fall in the East, birds may forage in weedy fields. In winter in coastal California, often found in eucalyptus. **COHABITANTS:** In summer, Lincoln's Sparrow, Prairie Warbler, Mourning Warbler (eastern population), MacGillivray's Warbler (western population), White-throated Sparrow. In winter, joins mixed-species flocks, especially flocks containing Blue-gray Gnatcatcher. **MOVEMENT/MIGRATION:** Spring migration is fairly brief, spanning the continent from April to May. Fall is more protracted, extending from early August into November, with a peak in late August through October. VI: 2.

DESCRIPTION: A small active warbler that likes to forage on springy vegetative tips. Large and round-headed, short-tailed, and overall stubby (in fact, somewhat plump), Nashville Warbler has a petite bill that is short, sharp, and sometimes straight, sometimes slightly down-curved. The gray head, set against the bright yellow underparts and unmarked greenish back, is incongruously eye-catching. The general lack of patterning makes the completely white eye-ring stand out.

The combination of plain upperparts, a gray head, a bold white eye-ring, and, perhaps, some wishful thinking prompts some observers to confuse this species with Connecticut Warbler—a larger, more sluggish troglodyte of a warbler that never hangs upside down. This confusion is especially likely with female and immature Nashvilles, which show grayish throats that impart a hooded appearance at a distance (as does Connecticut). The throats of male Nashvilles are yellow.

BEHAVIOR: Very active. Likes to work the outer springy tips of foliage and likes to stay low; especially in fall, often forages near the ground. Western birds frequently bob their tails; eastern birds do so less consistently. East and west, birds actively swish tails left, right. Singing males may vocalize from the highest points of trees, and in spring birds frequently forage high in the canopy.

FLIGHT: Looks small (particularly short-tailed), plump, and plain—olive above, with a brighter rump, and bright yellow from the chin to the vent below. Some show a whiter belly. Shorter, rounder wings help separate Nashville from Tennessee. Often flies quite low, hugging the vegetation. Flight mixes rapid wingbeats with terse closed-wing glides and is both undulating and tacking.

VOCALIZATIONS: Variable musical song is given in two or three parts. The first part is slow and measured, and the balance (warbled or trilled) is faster: "*Seepit seepit (seepit) seeseeseesee,*" or "*see see see sirsirsir seeda seeda*" or "*see see see see chachachachacha.*" More musical, and less hurried and staccato than the three-part song of Tennessee Warbler. Call is a sharp, ringing, metallic "*tick*" or "*spik*" (similar to Virginia's Warbler). Flight call is a soft rising "*sip*" that is similar to Tennessee and Orange-crowned.

Virginia's Warbler, *Vermivora virginiae*

Tastefully Gray Warbler

STATUS: Fairly common but localized and habitat-specific southwestern breeder. **DISTRIBUTION:** Breeds in the s. Rockies and the mountains of the Great Basin in parts of s. Idaho, s. Wyo., sw. S.D., much of Nev., extreme e. Calif., Utah, w. and cen. Colo., portions of Ariz., N.M., and w. Tex. **HABITAT:** Found on dry steep slopes (above 4,000 ft.) in low brush adjacent to pinyon-juniper and dry montane coniferous forest, often with an understory of oak shrub. In fall migration, sometimes occurs in weedy-brushy fields. **COHABITANTS:** Black-throated Gray Warbler, Spotted Towhee. **MOVEMENT/MIGRATION:** Arrives late in the spring, from late April into late May. By late July, becomes scarce in breeding areas, but migration is generally later, from late August through September (and a few in early October). The migratory path is through Arizona, New Mexico, and west Texas. Rare on the West Coast, and casual east of west Texas. VI: 2.

DESCRIPTION: A small, slender, delicately proportioned, pale gray sprite with a staring white eye-ring. Shaped like a stretched-out Nashville with a slightly down-curved and more robust bill and a noticeably longer tail—a characteristic that enhances the bird's slenderness.

Overall a tastefully plumaged gray bird bracketed by splashes of yellow fore and aft—on the breast (though not all of it) and especially on the undertail coverts; the yellowish rump is hidden when wings are closed, but evident when the bird flies. A complete narrow white eye-ring gives the bird a staring expression. The adult male's (and sometimes the female's) rusty cap can be surprisingly conspicuous, and the crown may seem disheveled and shaggy. Adults and immatures most closely resemble immature Nashville Warbler. Virginia's is all gray above, however, whereas Nashville shows a grayish head contrasting with greenish wings.

BEHAVIOR: A hyperactive twig specialist that is rarely still for long. Moves with short rapid hops through low bushes and small trees, favoring the smaller branches within a tree's interior and giving each bush a thorough examination before flying the short distance to the next. Always moving—especially the tail, which the bird pumps with a downward flip and, when excited, twirls with a circular swish. Usually plucks prey from leaves and branches without resort to leaps or hovering—the advantage of foraging in tight confines. Not particularly shy, Virginia's is more aptly described as indifferent; nevertheless, its penchant for staying within the tree's interior makes it difficult to see.

FLIGHT: Light, nimble, weak, slightly undulating and jerky, and commonly short (no farther than the next bush). Rapid and deep wingbeats are given in terse bursts interrupted by terse skip/pauses.

VOCALIZATIONS: Song is high, sweet, and patterned like Nashville, but more anxious and impatient and less mellow-sounding; the ending is often hurried or abrupt. Song is in two (sometimes three) parts—the first, consisting of several measured evenly spaced introductory notes is followed by a new note pattern with a faster delivery. Alternatively, the second part may be a warbled trill; for example: "*t'lew t'lew t'lew t'lew t'lew zwee zwee.*" The notes in the first set usually have a two-noted quality. Call is a sharp, breathy, high-pitched, and protracted "*spee*" or "*spink*" or "*spik*" that seems to skip across the tympanic membrane. Sometimes uttered with a slight vibrancy.

Colima Warbler, *Vermivora crissalis*

Boot Spring Warbler

STATUS: For a rare bird, surprisingly common at precisely the right place at the right time. Otherwise nonexistent. **DISTRIBUTION:** In the U.S., found only in the Chisos Mts. of Big Bend National Park in Tex. Range extends into n. Mexico. **HABITAT:** Dry forested hillsides and wooded canyons dominated by oaks and junipers above 6,000 ft. Found on the trail head for the Pinacle Trail to Boot Spring and the Laguna Meadows Trail. Take either trail—or connect via the Colima Trail and do both. Climb two hours. As soon as you get to within a quarter-mile of the crest of the ridge, you are in the bird's breeding range. Bring water. **COHABITANTS:** Hutton's Vireo, Mexican Jay, Black-crested Titmouse. **MOVEMENT/MIGRATION:** Arrives as early as late March (usually not before April); begins to leave by mid-July; makes a late departure by mid-September. VI: 0.

DESCRIPTION: Hardly necessary. Within its extremely limited range, it's the only breeding warbler, although Virginia's occurs during migration. Largish (nearly 1 in. larger than Virginia's, which it most closely resembles), with a robust body, a longish straight bill, short wings, and a fairly long tail. Looks very much like a large, robust, browner Virginia's.

Fairly plain. Overall dingy brownish gray with an even grayer head, no yellow on the breast, but a rich golden (not pale yellow) wash on the rump and undertail. The white eye-ring is apparent but does not necessarily stand out. The rufous crown seems sparsely applied.

BEHAVIOR: A fairly sluggish warbler. Forages in the canopy and middle portions of the tree. Apportions its time equally between the outer foliage and more interior branches. Works a tree thoroughly before moving to the next (but not necessarily closest) tree. Feeds by gleaning, hopping nimbly but not energetically, plucking insects most commonly from the undersides of overhead branches and leaves. Also works its way into leafy clusters and occasionally sallies out in pursuit of flying insects.

Males sing while foraging, but also perch for extended periods, usually in the open and very often on an exposed dead branch at the very top of a tree. Unlike Virginia's and Nashville, does not flick or wave or wag its tail. Response to pishing is very subdued.

FLIGHT: Overall plain, darkish, slender-looking bird in flight, although the golden wash, particularly on the rump, is often apparent as the bird maneuvers in good sunlight. Flight is fairly slow but strong, with an even undulating pattern. Wingbeats are given in a moderately long series punctuated by a terse closed-wing glide.

VOCALIZATIONS: Song is a warbling chatter that often starts with one or two slower, more distinct notes, accelerates, then falters at the end; for example: "*cha chchchchchchchch chu chu (chu).*" Recalls a brittle Orange-crowned Warbler. Call is a "*chih*" or "*spee*" that is sometimes loud, sometimes soft, and has more body than Virginia's.

Lucy's Warbler, *Vermivora luciae*

Pale Mesquite Warbler

STATUS: Common but habitat- and range-restricted southwestern breeder. **DISTRIBUTION:** Along the border of Utah and Ariz., from the Four Corners to se. Nev. then south along the Colorado R. and (except for the eastern Mojave) east across s. Ariz., sw. N.M., and south into n. Mexico (local breeder in se. Calif., and a few pairs occur in extreme sw. Colo.). Winters along the west coast of Mexico. Very rare vagrant on the California coast; casual elsewhere north and east of its normal range. **HABITAT:** Riparian thickets in dry desert areas, especially mesquite but also willows and, at slightly higher elevations, cottonwood-sycamore–live oak woodlands. In fall migration, also found in weedy-brushy fields. **COHABITANTS:** Bell's Vireo, Verdin, Black-tailed Gnatcatcher, Phainopepla. **MOVEMENT/MIGRATION:** Arrives in March and April; departs early, primarily in July and August, with some birds lingering into September. VI: 1.

DESCRIPTION: A tiny gray nub of a warbler gifted with a charming pale plainness that in and of itself is almost diagnostic. Small (almost Ruby-crowned Kinglet–sized!), compact, and roundly chubby in its proportions, Lucy's shows a large round head, a short, thick, pointy bill, a stocky body, and a short tail.

Overall plain—pale gray above, paler gray to whitish below (sometimes shows a warm buff wash along the flanks and faintly on the breast). Set against a backdrop of vegetation, the paleness of the bird makes it seem almost translucent. The pale unpatterned face makes the large black eye and black stub of a bill stand out. A sparsely applied rust-colored cap is usually absent on immatures; a rust-colored rump patch is visible in snatches as birds feed. For juveniles (seen as early as June), rely on the bird's overall plainness, paleness, and dark-eyed balefulness—although the immature's buffy wingbars are often apparent, and the buffy rump patch against an all-gray backdrop is diagnostic.

The fact that the bird is also almost invariably seen, in breeding season, in mesquite along a desert wash should not be overlooked. The expression is innocent.

BEHAVIOR: A nimble, active, acrobatic foliage gleaner that specializes in foliage and tasseled flowers. Likes to forage high during breeding season, at the tops of mesquite trees and in the middle levels of taller cottonwoods. May drop to lower branches, and occasionally leaves the mesquite fortress along riverbanks to explore an orphaned shrub on adjacent flats.

Quick but not hyperactive, Lucy's is calculating in its movements. Hops or flutters to a likely twig, pauses, then searches left, then right. Hops to a hanging flowery tassel. Clings sideways, nuthatch fashion, and plucks prey. More constantly on the move, and dangles less, than Bridled Titmouse, but feeding behavior is similar.

Jerks its tail down as it feeds, and flicks its wings rapidly, most often just after landing. When its tail jerks, sometimes a bit of the rusty rump patch winks through.

FLIGHT: A tiny, overall plain and pale bird with a rusty or pale spot on the rump. Flight is weak, loose, and bouncy. Fairly lengthy and fluttery series of wingbeats are punctuated by intermittent skip/pauses.

VOCALIZATIONS: The song, a two- to three-part warbled trill, is like Nashville and Virginia's, but sharper and more hurried than Nashville and lacking in the two-noted quality of Virginia's; for example: "*sew sew sew sew ch'chchch*" or "*ch'chchchchchchch see see see see'sew.*" Call is a high, metallic, but softly uttered "*tink*" that is somewhat like Virginia's and Nashville.

PERTINENT PARTICULARS: Perhaps most likely to be confused with the also very gray Bell's Vireo, with which it shares the same habitat. Bell's has a very long and very active tail, its bill and face are the same noncontrasting gray as on Lucy's, and it also has faint wing-bars, a greenish back, and pale yellow on the sides and flanks. Where found together, the species often segregate out by elevation: Bell's forages lower, and Lucy's higher in the leafier vegetation.

Northern Parula, *Parula americana*

Plump and Petite

STATUS: Common to fairly common and fairly widespread breeder across most of eastern North America. **DISTRIBUTION:** There are two distinct breeding populations. The northern population extends from the Maritimes across s. Que., s. Ont., and se. Man.; also found in Maine, N.H., Vt., the Adirondacks, n. Mich., and n. Wisc. The more extensive southern population encompasses most of the se. U.S. (except the southern tip of Fla.), with the northern and western border cutting through s. N.Y., n. N.J., nw. Pa., ne. and sw. Ohio, cen. Ind., n. Ill., cen. Iowa, e. Kans., e. Okla., and e. Tex. to the Brazos R. and the Edwards Plateau. Winters in the West Indies, s. Mexico, and the eastern slope of Central America (uncommon in s. Fla.; rare in s. Tex.). In migration, common and widespread across the East; very rare west of its breeding range. **HABITAT:** Breeds in mature hardwood and/or coniferous forests in proximity to water—lakes, rivers, swamps, bogs—especially those that offer the requisite nesting material: Spanish moss (in the South) and old-man's-beard (in the North). In migration, occupies all manner of woodlands. **COHABITANTS:** Golden-crowned Kinglet, Blue-gray Gnatcatcher, Yellow-throated Warbler, Black-throated Green Warbler. In winter and migration, joins mixed-species flocks. **MOVEMENT/MIGRATION:** Spring migration from early March to early June; fall from late July to early November. VI: 3.

DESCRIPTION: A tiny, compact, plump-bodied, petite-billed, short-tailed mite of a warbler—so small it might pass for a kinglet. A well-marked bird, in all plumages the upperparts are blue or bluish with a telltale splash or blush of yellow-green on the mantle; the underparts are yellow in the front and white in the back. The breast of adult males and females is adorned with a blue/black-and-russet necklace. The throats and breasts of immatures are sometimes distinctly yellow and sometimes just a yellow wash.

Two short, thick, white wing-bars and the white tail spots are prominent. The broken white eye-ring, set against the dark face, is apparent at a considerable distance and helps distinguish this species from the all-dark-faced Tropical Parula of south Texas.

BEHAVIOR: Active nimble feeder that moves quickly, flying frequently, about the leaves, twigs, and outer branches of trees. Most often found in the canopy and subcanopy, although also forages in the understory, particularly during migration. Prefers to forage on the tree's exterior, often at the very tips of branches. Frequently hops while feeding and makes rapid picks at the foliage; sometimes hangs, chickadee fashion, or flutters to maintain its balance. Also stretches to reach prey, hovers, and flycatches. In spring, males pause frequently to sing.

In migration, associates with mixed-species flocks.

FLIGHT: The stocky proportions (plump rounded chest, short neck) and small (tiny!) size are very apparent; the tail is short and triangular and sometimes has a swollen notched tip. Generally pale—blue or greenish above (the yellow wash blends with the bluish back, making green) and whitish below (white underwings add to the general whiteness). The head often appears hooded. The white wing-bars, showing as two white spots, are often distinctive. Wingbeats are deep, rapid, and sputtery, given in a short regular series followed by a brief pause. Flight is slow, steady, and bouncy, with rising synchronized to the beat of the wings. Looks like a kinglet. Calls frequently in flight, and often turns its head from side to side, as if searching for prey.

VOCALIZATIONS: Song is a rapid, buzzy, ascending series of insectlike notes; the last one is held and then descends. Sounds like the song is climbing a ladder, hesitates at the top, then falls. Call is a sharp "*cheeep*" that drags at the beginning (and may recall Chimney Swift). Flight call is a short, high-pitched, steeply descending "*tswf*" given frequently and repetitively.

Tropical Parula, *Parula pitiayumi*

Masked Parula

STATUS AND DISTRIBUTION: Widespread tropical and subtropical species, but in the U.S. generally uncommon, highly restricted, and found only in s. Texas. Accounted locally common (as a breeder) in Kennedy Co. and rare in the lower Rio Grande Valley between Brownsville and Bentsen–Rio Grande Valley State Park. There are a few records for the s. Edwards Plateau and west to the Davis Mts. Casual north to La. and west to Ariz. **HABITAT:** Found in live-oak woodlands (Kennedy County) and deciduous riparian forest, usually in very tall trees like Texas ebony, sugar hackberry, and cedar elm. In both habitats, occurs only where there are epiphytes (such as Spanish moss) to use for nesting. **COHABITANTS:** In the lower Rio Grande Valley, Northern Beardless-Tyrannulet, Black-crested Titmouse, Altamira Oriole. In winter, forages in mixed flocks with Black-crested Titmouse, Blue-gray Gnatcatcher, assorted wintering warbler species. **MOVEMENT/MIGRATION:** Largely a permanent resident. VI: 1.

DESCRIPTION: A small, compact, dark blue and bright yellow, eyeless parula. Tiny warbler (smaller than Orange-crowned; slightly smaller than Nashville; slightly larger than Ruby-crowned Kinglet; in size and proportions, virtually identical to Northern Parula), with a thin pointy bill, a plump compact body, and a short narrow tail.

Male Tropical Parulas (and to a lesser degree, females and immatures) are slightly darker blue above than Northern Parula and brighter, more extensively yellow below, with yellow extending higher onto the face and lower onto the belly than on Northern. The face on males (and to a degree, females) looks blackish and dark-masked, making it difficult to see the bird's black eyes. You'll have no such problem with Northern Parula, whose eyes are outlined top and bottom with white half-circles.

Also, the breast of male Tropical is a rich orange that will remind you of Blackburnian Warbler's throat patch; females are orange-washed. The yellow breasts of male and female Northerns are bisected by a distinct dark band or necklace that is absent on immature female Northerns and all Tropical Parulas.

BEHAVIOR: Specific to its habitat. Almost always seen foraging at the top of the canopy, where it moves nimbly, but not necessarily quickly, on twigs and branches, on both the inner and outer branches. Favors the upper surfaces of leaves, and works thoroughly, remaining in the same tree for extended periods.

Males vocalize frequently and tenaciously during breeding season, singing while feeding and often vocalizing well into the morning. Birds begin singing as early as February in Texas, before migrating Northern Parulas arrive. *Note:* A few Northern Parulas winter in extreme south Texas.

FLIGHT: Overall tiny and short-tailed, with a bright yellow breast and a no-hooded appearance. Flight is quick, direct, bouncy, and mostly from canopy to canopy; wingbeats are given in a rapid series punctuated by a skip/pause or, when flying short distances, an open-winged braking.

VOCALIZATIONS: Song is very similar to Northern Parula—a rapid ascending series of scratchy insectlike notes that go flat or fall off at the end: "*ju' j'g'g'G'G'GEEuh.*" Also has a variation that rises at the end (like Cerulean): "*. . . g'g'GEE!*" Call is a high thin "*see*" (similar to Northern Parula).

PERTINENT PARTICULARS: If you get a glimpse of Tropical Parula, but not enough to see distinguishing field marks, what will probably stick in your mind is the surprising overall richness of the bird's colors. So if you find yourself thinking, *Gee, that sure was a brightly plumaged parula,* think Tropical—providing, of course, you are in south Texas.

Yellow Warbler, *Dendroica petechia*

A Low-down, Yellow, Beady-eyed Bushwhacker of a Warbler

STATUS: Common and widespread breeder across much of North America. **DISTRIBUTION:** As a breeder, found across virtually all of Canada and Alaska and the n. U.S. Absent across much of the s. U.S. The southern border of its breeding range cuts across se. Va., cen. N.C., nw. S.C., n. Ga., n. Ala., n. Ark., cen. Okla., n. N.M., n. Ariz., Nev., and Calif. Also found in se. and cen. Ariz. and sw. N.M. south into Mexico and along the Calif. coast to Baja. In winter, except for a nonmigratory subspecies native to extreme s. Fla. (the Golden Warbler) and extreme s. Tex. (the Mangrove Warbler) and a handful of birds that winter in s. Calif., Yellow Warbler is absent in the U.S. and Canada, wintering from s. Baja and the west coast of Mexico to s. Mexico, Central America, and n. South America. **HABITAT:** Breeds in willow thickets, cottonwoods, woodland edge, and young second-growth deciduous growth and brush, often near water. Golden and Mangrove Warblers are found in mangroves. In migration, Yellow Warbler frequents a variety of habitats, including city parks, orchards, and rank weedy growth, but favors the same habitat used when breeding. **COHABITANTS:** Varied, but in summer includes Alder and Willow Flycatchers, Bell's Vireo, Wilson's Warbler, Blue-winged Warbler, Chestnut-sided Warbler, Yellow-breasted Chat, Song and Lincoln's Sparrows. **MOVEMENT/MIGRATION:** Owing to the bird's broad latitudinal breeding range, spring migration is protracted, beginning in southern portions of the United States in early April and continuing into mid-June, with a peak from late April to mid-May across most of the United States and southern Canada. Among the earliest fall migrants, the first southbound Yellow Warblers appear in early July, peaking in early to mid-August (early to mid-September in the West), and continuing into early November. VI: 4.

DESCRIPTION: Basically an all-yellow warbler. Medium-sized, overall compact, and sharply featured—there's nothing round or cuddly about this warbler. Adults are distinguished by an overall *bright* yellow face (often both the face and head) and underparts. Upperparts are bright yellow to yellow-green, blending into somewhat soot-stained wings and tail on perched birds. Underparts on adult males (except southwestern birds) are conspicuously raked with narrow red streaks. Adult females frequently show a shadowy streak pattern. Most immatures are overall yellowish, but some individuals are washed-out pale gray with only a hint of yellow. *Hint:* If you are looking at a warbler that is utterly plain and devoid of any distinguishing field marks, you are probably looking at an immature Yellow.

In all plumages, the head and face are *plain.* With no conspicuous markings and only a shadowy cap, the bird's beady black eye and fairly hefty black bill stand out. In all plumages, fanned tails are bracketed by bright yellow panels that are unique to this species. The expression is placid.

BEHAVIOR: An active, often low-foraging warbler that likes to be in the open but is also adept at worming into tangles and grabbing insects in the clinches. Generally stays within 10 ft. of the ground, although may forage much higher, particularly along western watercourses. Moves in short hops, often with lots of wing motion and tail-flicking. Feeds by picking worms and insects from both sides of leaves and small branches. Also sallies to chase insects and habitually hovers to pluck prey. In sum, it's one of the most active and conspicuous of warblers.

Very responsive to pishing, and can be coaxed to approach repeatedly.

FLIGHT: Overall compact, pointy, and angular (not softly contoured), with a big, pointy, projecting head and a fairly short, broad, square-cut, pointy-cornered tail. Plumage is usually bright yellow; dingier individuals have a mustard-colored uniformity. Bright yellow tail panels are apparent when the tail is fanned; otherwise, the tail appears no brighter than the rest of the bird. Fairly strong direct flier, but jerky or bouncy, with wingbeats given in a short series followed by a terse closed-wing glide.

VOCALIZATIONS: Song is a short, simple, assertive ditty, comprising three identical pure notes followed by a hurried mumble and an emphatic end note, that is phonetically rendered: "*Sweet, Sweet, Sweet, Oh-so-Sweet!*" Loud, sweet, emphatic, slightly muted call says "*Chip!*" Flight call is a short "*zweet*" that has a buzzy vibrancy.

PERTINENT PARTICULARS: Yellow Warbler's most likely candidates for confusion include any yellower-toned Orange-crowned Warbler, female Wilson's Warbler, or female Common Yellowthroat. Orange-crowned is overall dingier with (usually) shadowy streaking below and a more complexly patterned face showing a pencil-fine dark eye-line and pale broken spectacles. Wilson's has a pug (not pointy) face, a longer tail, and (at the very least) a shadowy impression of a dark cap. Common Yellowthroat is a feathered mouse of a bird with a different shape and a tendency to skulk. And of course none of these other species have yellow tail panels.

Chestnut-sided Warbler, *Dendroica pensylvanica*

Lime-backed Robotic Warbler

STATUS: Fairly common, primarily northeastern breeder and eastern migrant. **DISTRIBUTION:** Breeds from the Maritimes and e. Que. across s. Canada to cen. Alta. (absent in s. Sask.). In the U.S., found throughout N.Y., New England, n. N.J., most of Pa., and south along the Appalachians to extreme n. Ga., Mich., n. Ohio, n. Ind., n. Ill., n. and extreme se. Wisc., and n. Minn. (Isolated populations also occur in s. Ind. and nw. Ark.) In migration, passes for the most part east of the prairies. Very rare in the West. **HABITAT:** Breeds in young second-growth deciduous habitat and forest edge. In migration, found in a variety of forest habitats. **COHABITANTS:** Gray Catbird, Blue-winged Warbler, Yellow Warbler, American Redstart. **MOVEMENT/MIGRATION:** Entire population relocates to Central America for the northern winter. Spring migration from early April to early June; fall from mid-August to mid-October. VI: 3.

DESCRIPTION: A stiff-moving, conspicuously plumaged, brush-hugging warbler. Medium-sized (larger than Blue-winged; smaller than Bay-breasted; the same size as Yellow Warbler), with a fairly sleek streamlined body. Distinguished by a stout but pointy bill, a tail cocked conspicuously above the body, and (often) wingtips jutting below the tail, imparting to the bird a stiff robotic posture (and movements to match).

Breeding males and females are easily distinguished by their bold pattern, whose most obvious elements include a bright yellow crown, a blackish face pattern, and a distinctive chestnut patch or stripe along the sides of the white breast (and flanks on males). In fall, heads and backs of adults and immatures show an unmistakable, almost Day-Glo bright lime green plumage; underparts are pale gray except for chestnut sides, which may be extensive, restricted, or (in the case of immature females) absent. It doesn't matter. The lime green back is unique and diagnostic in and of itself. Also unique are pale yellow wing-bars. Secondary but very suggestive characteristics on fall birds include a staring white eye-ring and pale grayish underparts. Don't forget (or overlook) the up-cocked tail.

BEHAVIOR: Hops stiffly and quickly through the understory, changing perches frequently. Movements seem stiff, tense, and deliberate (robotic!).

Favors small branches, twigs, and leaves; doesn't cling to trunks, doesn't wag its tail, and doesn't engage in dangling acrobatics. Usually plucks prey directly and adroitly from the undersides of leaves, although birds sometimes hover and fly after flushed prey. Usually forages alone, but in migration may be found among mixed warbler flocks, even into the canopy. Nevertheless, clearly prefers to forage from the lowest branches to about mid-height. Tame, and responds quickly and tenaciously to pishing.

FLIGHT: Slender but sturdy profile, with a long slim tail. Bicolored! The greenish back and plain, not-quite-white underparts are an odd combination that catches the eye. (*Note:* Underparts seem whiter in spring, partly because of the greater contrast with the darker back.) In fall, in early morning light, the golden light and gray underparts combine to give birds a purplish cast. Flight is bouncy and nimble, but stiff.

VOCALIZATIONS: Song is a quick, loud, clear, whistled greeting: "*S'swee S'swee S'swee z'WEECHU!*" phonetically rendered, "*Please Please Please to MEET YOU.*" Yellow Warbler also sings this song with this pattern, but Chestnut-sided is louder and more assertive, singing introductory notes that are always more emphatic. Song is variable, and the "*meet you*" is occasionally dropped. Call is a hardy, low, flat "*Chi/upf*" that may remind you of the chirp of House Sparrow. Flight call is an Indigo Bunting–like descending buzz, but higher-pitched.

Magnolia Warbler, *Dendroica magnolia*

A Bird of Short Phrases

STATUS: Common breeder in northern coniferous forests across all but westernmost Canada and in the northeastern United States. **DISTRIBUTION:** Breeds from s. Nfld., the Maritimes, and ne. Que. across cen. Que., n. Ont., cen. Man., cen. and nw. Sask., across cen. and n. Alta., ne. B.C., se. Yukon, and sw. Northwest Territories. In the U.S., breeds in Maine, N.H., Vt., w. Mass., N.Y., n. N.J., all but s. Pa., w. Md., e. W. Va., n. Ohio, n. Mich., n. Wisc., and ne. Minn. Winters in the West Indies (a few), s. Mexico, and Central America south to Panama. In migration, common across the eastern half of the U.S.; very rare but regular vagrant from the high plains west. Rarely winters in Fla. **HABITAT:** Nests in low, dense, coniferous growth, usually in stands of spruce in the North, hemlock in the South, but also found in mixed hardwood-coniferous forest in places with thick coniferous understory. In migration, also favors low, thick, or dense vegetation, often along forest edges, but may also be found in taller trees as well as in stands of tall grass. **COHABITANTS:** In summer, Boreal and Black-capped Chickadees, Tennessee Warbler, Yellow-rumped Warbler, Black-throated Green Warbler, Blackburnian Warbler. In migration, found in mixed warbler flocks. **MOVEMENT/MIGRATION:** Fairly brief migratory period. Spring from late April to early June; fall from late August to mid-October. VI: 3.

DESCRIPTION: A medium-sized classically proportioned warbler distinguished by a short straight bill and a moderately long narrow tail. Breeding-plumage males are strikingly patterned, and females only slightly less so. Black and gray above, yellow with bold black streaking below. Black streaks coalesce into a band or necklace across the chest. Broad white wing patches (sometimes wing-bars on first-spring females) are distinctive. Breeding-plumage females are grayer and more muted, but similar in pattern. Overall duller, immatures have gray heads, gray-green backs, and yellow underparts that may show shadowy dark streaks along the flanks. The white eye-ring set against the backdrop of a plain gray face gives the bird a startled expression. A pale washed-out necklace (or a tie-dyed chest) separates the throat from the chest.

In all plumages, the rump is yellow, and the upper surface of the fanned tail is bisected by a white band or paired windows that the bird sometimes flashes, American Redstart fashion. In all plumages, from below, the bird appears yellow in the front, white in the back. The closed tail is half white (basal half), half black.

BEHAVIOR: An under-leaf specialist that likes to forage low in the understory and on the outer tips of leafed-out branches. Hops and plucks. Frequently hovers. Sometimes sallies out to snap up flushed insects. Sometimes fans its tail, like American Redstart and Canada Warbler, displaying a broken white tail band. Especially in spring, may forage higher in the lower canopy with other migrating warblers; in fall also forages in tall grass or weeds (such as goldenrod).

FLIGHT: Shows a distinctly plump round head and body mated to a slender long tail and a tricolored pattern to the underparts—yellow in front, white behind, with a black tip to the tail. Owing to the white undertail coverts and white basal portion of the tail, the tail and body often seem unattached. On immatures, the separation between the gray face and yellow throat seems diffuse, not sharply delineated. Birds in flight often fan their tails, offering glimpses of the white tail panels. Flight is tacking (like American Redstart).

VOCALIZATIONS: Song is simple and distinctive—a short, clear, whistled "*wheeta, wheeta, wheet-oo*" (or something akin to that) that is repeated after a pause and is similar to American Redstart, but generally lower-pitched and not quite so strident. Call is an odd, husky, somewhat bisyllabic, unwarblerlike "*schliv*" that is given sparingly. Flight call is a short, soft, lazy "*g'e'e*" uttered with a trilled vibrancy.

PERTINENT PARTICULARS: In fall, immature Magnolia Warblers and Prairie Warblers are sometimes confused, and both may be found in low weedy growth. You can use plumage differences to distinguish them, or you can take the shortcut. Prairie Warbler habitually wags its tail; Magnolia does not.

Cape May Warbler, *Dendroica tigrina*
Tiger of the Treetops

STATUS: Uncommon northern breeder. **DISTRIBUTION:** Breeds across boreal Canada and the n. U.S. from the Maritimes and ne. Que. across s. Que. (absent in extreme s. Que.), cen. Ont., cen. Man., cen. and nw. Sask., n. Alta., into the sw. Northwest Territories, se. Yukon, and ne. B.C. In the U.S., breeds in cen. and n. Maine, n. N.H., the n. Adirondacks, n. Mich., n. Wisc., and n. Minn. Winters primarily in the West Indies (also coastal Yucatán and coastal Central America). A few birds may winter in s. Fla. Migrates mostly east of the Mississippi R. Casual in the western U.S. **HABITAT:** Breeds in mature or fairly mature coniferous forest (particularly spruce). In migration, found in a variety of habitats, including all manner of woodlands, thickets, orchards, residential gardens, and overgrown grassy areas, but once again is particularly partial to conifers, and in particular spruce. **COHABITANTS:** Tennessee Warbler, Blackburnian Warbler, Bay-breasted Warbler. **MOVEMENT/MIGRATION:** In spring, migration is most concentrated through Florida and east of the Mississippi River Valley; in fall, larger numbers migrate through the Appalachians and down the Atlantic Coast. Spring migration from mid-April to early June; fall from mid-August to late October. VI: 3.

DESCRIPTION: A beautiful treetop warbler with an ethereal voice. Medium-sized and overall compact and trim, Cape May Warbler has a pointy bill that is petite and very slightly drooped—a nut pick of a bill. The tail is short and slightly up-cocked.

Breeding males are stunning! The chestnut cheek patch and encircling bright yellow face, coupled with the white wing patch and black-and-yellow tiger-striped pattern below, render the bird unmistakable. Females and immature males are plainer, but easily distinguished by the yellow to yellow-green face, the shadow pattern of the male's cheek patch, and crisply streaked yellowish underparts and a pale yellow-washed (or greenish yellow) rump. A yellow or pale semicollar extends from the lower cheek toward the nape; though it doesn't encircle the neck, it is often visible as two pale points bracketing the neck when the bird is facing away.

Immature females are very nondescript—smooth, green-gray above, grayish with diffuse narrow streaking below. However drab, the birds always look neat, crisp, and well groomed. By comparison, immature Yellow-rumped Warbler has an end-of-the-workday look.

In separating immature female Cape May from Yellow-rumped, remember that the former has two frail wing-bars that bracket the pale greenish-edged wing coverts. At close range, you can see the etched lines; at a distance, the lines coalesce into a faint patch reminiscent of the adult male's bold white shoulder patch. Yellow-rumped just shows wing-bars. Also, immature females usually have a touch of greenish yellow behind the ear that may be very restricted—a gold blush that is barely feather-tip deep.

Don't forget the shape and the bill. Cape May Warbler is compact, trim, and shorter-tailed, with a thin slightly downturned bill. Yellow-rumped is heavier and straighter-billed.

BEHAVIOR: Even among the ranks of warblers, Cape May is conspicuously nimble. In spring and on territory, look for this bird in the treetops, particularly in short-needle conifers, where it forages in the open on the outer branches. In fall, it is more eclectic, foraging at all levels, and seems more inclined to forage behind the outer branches. Hops along limbs, picking up insects en route. Also hovers, flycatches, and, in migration, forages by hopping on the ground. Fruit, including grapes and berries, figures in the diet, and birds seem addicted to the sap oozing from sapsucker borings and sap-wells. Also likes nectar from flowers and sugar water from hummingbird feeders. Doggedly defends sap-wells and nectar sources from other Cape May Warblers.

FLIGHT: Overall short and stocky, angular, and all corners in flight, with a pointy face, an angular chest, sharp pointy wings, and sharp corners to the tail. (Distant birds appear no-headed.) Adult male's white wing patch is distinctive. The wing-bars and pale neck patch are visible on all but immature females, which appear overall dingy and dark. (If they show any color at all, it will be on the rump, which shows yellowish green.) Wingbeats are stiff, spare, and shallow. Flight is bouncy, straight, and fairly strong.

VOCALIZATIONS: Song is a short repeated sequence of high thin ethereal notes: "*seet, seet, seet, seet, seet.*" Occasionally notes have a compressed two-noted quality (more typical of Bay-breasted). Expect improvisation. Call is a terse, high, thin, soft "*chee*" or "*tee,*" sometimes uttered with a slight vibrancy; might recall a squeaky bicycle wheel. Flight call is a very high breathy "*zwee,*" which, at close range, also has a slight vibrancy.

Black-throated Blue Warbler, *Dendroica caerulescens*

The Dapper Dendroica

STATUS: Fairly common but fairly restricted breeder of eastern North America. **DISTRIBUTION:** Breeds in a belt from the Maritimes west across s. Que., s. Ont., n. Mich., n. Wisc., and ne. Minn., south through noncoastal New England, N.Y., n. N.J., n. and w. Pa., and along the Appalachians to n. Ga. Winters primarily in the Greater Antilles as well as on the east coast of the Yucatán Peninsula to and including northern Honduras (also extreme s. Fla.). **HABITAT:** Breeds in interior sections of large tracts of deciduous and mixed deciduous-coniferous forest, often at higher elevations or in steep terrain. Nests in dense understory, often in broadleaf evergreens such as mountain laurel. In migration, occupies a variety of woodland habitats, but prefers to remain in the forest interior. **COHABITANTS:** Blue-headed Vireo, Veery, Ovenbird, Canada Warbler. **MOVEMENT/MIGRATION:** Migrates along the Atlantic Coast, generally not west of the Appalachians except around the eastern Great Lakes. Both spring and fall migration are fairly compressed, with spring from mid-April through May; fall from August through October, with a peak in mid- to late September. VI: 3.

DESCRIPTION: A compact, medium-sized, yet wide-bodied warbler that differs so markedly in male and female plumage that early ornithologists considered the sexes different species. With their blue backs, black faces and throats, and contrastingly brilliant white underparts, males are stunning and unmistakable. Cryptic females, wrapped in olive and ash, are hardly less distinctive. Even the surprisingly obvious facial pattern—a dark cheek flanked by pale borders—and the white wedge bisecting the lower edge of the folded wing are hardly necessary for identification. The overall smooth grayish olive color of females alone is unique. Adults and immatures are essentially similar, although immature females often lack the white wing patch.

BEHAVIOR: Breeds and forages in the understory and lower canopy—generally 5–30 ft. high, but does sometimes feed on the forest floor and in the upper canopy. Seems particularly at home in dark damp places. A careful methodical feeder—slower and more halting than most of the arboreal warbler tribe. Feeds by hopping or flying to perches, pausing, then jerking its head to study the surrounding foliage. Specializes in snatching prey from the undersides of leaves by gleaning from a fixed perch, hovering, or snatching en passe.

Generally solitary, but in migration may be found in small flocks of Black-throated Blues or with other warbler and vireo species. Responds well to pishing.

FLIGHT: Fairly compact and somewhat angular in shape, with short wings and a medium-long, straight, untapered tail. Males, with their starkly white underparts and white underwings—when backlit, a longitudinal white stripe or crescent burns through the wing—are distinctive. Females are overall plain and dull olive, with slightly more yellow below. Usually the dark cheek stands out. Sometimes the pale eyebrow and white wedge on the wing project as pale points on the otherwise drab backdrop.

Flight is fairly straight, not erratic or wandering but slightly bouncy. The bird often rocks in flight, moving like a distance ice skater, favoring one side then the other. Wingbeats are quick and deep and generally given in a somewhat prolonged series followed by a terse closed-wing glide. Black-throated turns its head frequently to study its surroundings and calls frequently in flight (often multiple times). In migration, occasionally seen in small groups.

VOCALIZATIONS: Song is a slow, slurred, nasal, somnambulant, scale-climbing dirge; the last note is pushed, drawn out, and rising: "*zoo, zoo, zee, zee, zeeE*" ("*I am so lay zee*") or a more truncated "*zoo, zoo zee,*" which may recall Cerulean Warbler. Call is a short hard "*cht*" that recalls a junco. Flight call is a soft spit, "*tik*" or "*zwit*" (often repeated several times), that sounds like the muttered conversational call notes of Northern Cardinal.

Yellow-rumped Warbler, *Dendroica coronota*

The Swarm Warbler

STATUS: Common, widespread, and hardy. In winter and during the early portions of spring migration and the later parts of fall migration, the most common and most likely warbler to be found across most of the continental United States. Two similar groups are readily distinguishable: the more widespread (northern and eastern) "Myrtle" form and the "Audubon's" form, which breeds in the mountain West. **DISTRIBUTION:** This very widespread species breeds across Canada and the ne. U.S. from Nfld., Lab., and the Maritimes, across boreal subarctic Canada into Alaska (absent only in portions of s. Sask. and se. Alta.). In the U.S., breeds across most of New England, much of N.Y., nw. N.J., n. Pa., w. Md., ne. W. Va., n. Mich., n. Wisc., ne. Minn., and in every state west of the Rockies, including the Black Hills of S.D. and trans-Pecos Tex., at higher forested elevations south to Mexico. Winters south of a line drawn from e. Mass. and coastal Conn. across n. N.J., Pa., Ohio, extreme se. Ont., cen. Ill., n. Mo., Kans., s. Colo., s. Utah, s. Nev., and along the West Coast from extreme sw. B.C. south into Baja and encompassing most of cen. and s. Calif. Along the West Coast, in winter, "Audubon's" usually far outnumbers "Myrtle" (except locally). **HABITAT:** Breeds in coniferous forests as well as mixed forests. Although Yellow-rumped has very eclectic taste in tree species and forest type and composition, it's usually found on the edge or in patchy forest; avoids dense unbroken interior forests. In migration, found almost anywhere—from treetops to orchards to brush piles to lawns. In winter, likes open or broken habitat that offers a mix of bushes, trees, and open space, including riparian woodlands, overgrown borders, pine forest, orchards, residential areas with trees and shrubs, and dunes. Along the Atlantic Coast, heavily concentrated in thickets of bayberry and wax myrtle, which the birds consume. **COHABITANTS:** In summer, Ruby-crowned Kinglet and Black-throated Green, Magnolia, and Blackburnian Warblers. In winter and migration, joins mixed-species flocks with chickadees, titmice, nuthatches, kinglets, bluebirds. **MOVEMENT/MIGRATION:** (Brief) spring migration in April and May; (protracted) fall migration from late August to early December. VI: 4.

DESCRIPTION: A large, sturdy, athletically proportioned warbler that forages in numbers and signs its name with a distinctive "*check*" call (in the East), or a "*chih*" call ("Audubon's" in the West). The bill is short and heavy, the body lean, and the tail long and generally twitchless. Breeding adults are handsomely patterned—black on dark gray with white wing-bars above (and in males a golden crown); bright white below, with a black necklace (or chest band), black beaded streaks extending down the sides, and a distinctive yellow patch tucked below the bend of the wing.

"Myrtle" Warbler has a black face mask bracketed by a white eyebrow and white throat. "Audubon's" has an all-dark face and a distinctive slightly more restricted yellow throat. In all plumages, birds have a well-defined bright yellow rump patch that often peeks through the folded wings when birds are perched and is always visible when birds are in flight.

Nonbreeding plumage is a shadow rendering of the breeding plumage—including a fairly prominent touch or blush of yellow below the wing and, always, a distinctive rump patch. Females and immatures are overall dingy brown (adult males retain a grayish cast), and the plumage is somewhat dirty or diffuse, with smudgy streaks on the breast. "Myrtle" Warbler commonly retains a semblance of a face mask pattern (and a whitish eyebrow), although in some young birds it may be indistinct. The face of "Audubon's" is always plainer and usually grayer than "Myrtle," making the broken white eye-ring stand out.

Overall, in winter, "Myrtle" Warbler is browner and more overall patterned or contrasting than "Audubon's." The throats of many "Audubon's" Warblers show a hint of yellow. On pale-throated birds, note that the throat patch is more restricted and less crisply defined.

In flight, in both subspecies, the tail shows three bold points of light—the yellow rump and the white corners on the tail. Palm Warbler's outer tail shows more limited white, and the rump is dull and diffuse, not a bright yellow spot. Also, the white in the tail of "Audubon's" is more extensive, giving birds a tail pattern more reminiscent of Magnolia Warbler.

BEHAVIOR: Highly social. Migrates and forages in groups of several individuals to several score. Frequently perches in the open for extended periods, often on dead branches. Even in the understory, prefers sturdy perches that offer room to maneuver and see. Likes to forage at lower levels, even close to and frequently on the ground.

Foraging behavior is a blend of active and sedentary. Sits for long periods, searching visually for prey, then hops or flies to grab prey (or berries), often hovering. Flycatches with frequency and finesse. Changes trees, even locations, more frequently than many warblers, often flying considerable distances. Also forages by hopping along limbs, clinging to vertical tree trunks, stooping upon prey from a perch, hopping in short grass . . . in short, does a bit of everything. A jack-of-all-warblers.

When changing locations, particularly when crossing an open area, generally calls.

Also eats fruits and berries—in winter extensively. Favored food includes bayberry and wax myrtle as well as juniper berries, poison ivy, and greenbrier. Comes to feeders for suet, peanut butter, and fruit.

FLIGHT: Strong and fairly swift, with a coursing or tacking quality and a distinctive skippiness or jauntiness that seems cavalier more than slipshod aeronautics. The silhouette of this big warbler seems blunt-headed, humpbacked, and long-tailed (pinched at the base, flared toward the tip), with long, wide, blunt wings—overall sturdier than Palm Warbler (which migrates at the same time) and burlier and less pointy and angular than

Cape May. Overall color (in fall) is dingy, with touches of yellow below that make the white undertail stand out. Wingbeats rise to just above horizontal, and birds appear to make momentary glides on slightly down-pressed wings. Calls compulsively in flight.

VOCALIZATIONS: Fairly short, pleasing, warbled song is complex and mercurial; seems never the same twice. A two-parted song: the first part is a series of even repeated notes; the second part, given on a different, usually higher pitch, is more hurried. "Myrtle's" call is a low perfunctory "*chep*" or "*check*"—a cross between a chip and a cough. "Audubon's" call is a husky rising "*whik*"—more scuff than chip. Flight call is a somewhat whimsical rising "*swee*" or "*sweee*" and is often mixed with chip notes.

Black-throated Gray Warbler, *Dendroica nigrescens*

The Black-White-and-Gray Warbler

STATUS: Fairly common and widespread western breeder. **DISTRIBUTION:** Generally occurs west of the Rockies. Breeds from s. B.C. across w. Wash., throughout all but ne. Ore., s. Idaho, most of Calif. (except the southern deserts and the Central Valley), Nev., Utah, sw. and cen. Wyo., w. Colo., all but sw. Ariz., and all but e. N.M. Winters in Mexico and in very small numbers in s. Calif., s. Ariz., and (particularly) s. Tex. Casual visitor in the East. **HABITAT:** Brushy undergrowth in open pine and mixed pine-oak forests as well as pinyon-juniper and oak and big-leaf maple woodlands. **COHABITANTS:** Ash-throated Flycatcher, Oak, Juniper, and Bridled Titmice, Pinyon Jay, Western Scrub-Jay, Mexican Jay, Virginia's Warbler. **MOVEMENT/MIGRATION:** Early-spring migration from late March to early May; fall from mid-August to late October. VI: 3.

DESCRIPTION: A solidly proportioned medium-sized warbler that looks only superficially like a Black-and-white Warbler. Overall robust and fairly large-headed, with a stout short bill and a fairly long tail. Unlike Black-and-white, never seems long in the face.

Distinctively plumaged. The bright yellow dot on the lores notwithstanding, no other warbler is so utterly black, white, and gray; in fact, it is the silvery gray upperparts as much as the distinctive facial pattern that distinguishes adults of this species from Black-and-white and from breeding male Blackpoll. Immatures have the same distinctive pattern as adults but are basically gray and off-white, with little or no contrasting black on the face and a wash of buffy gray along the flanks.

BEHAVIOR: A slow, methodical, calculating feeder. Hops. Lands. Remains stationary. Moves its head in quick jerks, first left, then right, then hops to the next branch, from which it may pluck presighted prey. Forages in the middle heights of trees (in the Northwest, more commonly higher in the tree), favoring sturdy less springy branches for perches. It even includes the denuded branches of dead trees in its search pattern. Also engages in some hovering and flycatching.

Fairly tame. Allows close approach. Responds fairly well to pishing.

FLIGHT: Overall stocky with a rounded head (with a steep forehead), a slim abdomen, and a long slim tail. Flight is strong, steady, undulating, and slightly jerky, executed with a short rapid series of wingbeats followed by a terse closed-wing skip/pause. The tail is often slightly flared in flight, showing conspicuous and telltale white outer tail feathers.

VOCALIZATIONS: Song is usually short, simple, musical, and variable. The pattern is reminiscent of other members of the Black-throated Green complex, but more hurried and hoarse; for example: "*jer jer jer jeee!*" (last note pushed in pitch and volume) or "*zhu zhu zhur zer zee!*" Less wheezy than Townsend's, and not as lazy as Black-throated Green or Hermit. Call is a soft low-pitched "*t'ch*," with no hard ending. Recalls "Myrtle" Yellow-rumped Warbler, or even Common Yellowthroat, but is more robust and not as flat-sounding. Flight call is like Black-throated Green.

Golden-cheeked Warbler, *Dendroica chrysoparia*

Texas Golden-cheeked

STATUS AND DISTRIBUTION: Uncommon breeder within its very restricted and habitat-specific range (found only on or near the Edwards Plateau of cen. Tex.), and for this reason a federally listed endangered species. **HABITAT:** Old-growth oak woodlands suffused with Ashe junipers, whose supple bark provides nest material. Forest height of 20–30 ft. is adequate, but prefers areas with fairly dense or closed canopy. **COHABITANTS:** White-eyed Vireo, Black-capped Vireo, Western Scrub-Jay, Black-crested Titmouse, Blue-gray Gnatcatcher. **MOVEMENT/MIGRATION:** Wholly migratory. Arrives between the first week of March and late March; departs breeding areas from late June to the first week in September, though most birds are gone by August. VI: 3.

DESCRIPTION: The goldenest golden-cheeked. Medium-sized warbler (slightly larger than Carolina Chickadee; slightly smaller than Black-crested Titmouse; the same size and shape as other members of the Golden-cheeked Warbler complex), with a round head, a stout bill, a compact body, and a medium-length tail.

Strikingly patterned (especially adult males), showing black and blackish upperparts, a black throat, chest, and flanks, and grayish white underparts (no yellow). Wholly framed in black, the yellow cheeks pop! The distinctive pencil-fine black eye-line shows an upward or downward (sometimes upward *and* downward) kick just before merging with the black collar. The white wingbars are broad and prominent.

Adult females and immature males are patterned like adult males, but the dark cap and back are paler (not jet-black) and slightly greenish-tinged and show black streaking. Immature females are like adult females but lack the black throat and may show faint dark streaking on the back that, for field purposes, is invisible. Key to this species in all plumages is the combination of unmarked golden cheeks, the pencil-fine black eye-line, and the lack of yellow on the underparts. Black-throated Green Warbler, the only other golden-cheeked that occurs (as a migrant) through central Texas, has a much paler green crown and back, a smudgy spectacled mask across the face (not a crisp eye-line), a splash of yellow on the lower flanks and sides of the vent, and a pale, greenish, unstreaked back.

BEHAVIOR: A gleaning species that forages mainly in the upper and middle canopy of hardwoods (less often in junipers). Feeds rapidly, making short hops to nearby branches, moving its head in quick turns, then hopping again. Able to cling, chickadee fashion, but much prefers to work the upper side of leaves and branches and seems disinclined to hawk insects or hover.

Males sing while foraging as well as from exposed perches above the canopy. Shifts places of vocalization frequently, often flying considerable distances to a new perch. Fairly tame. Responds well to pishing but careful! Endangered species.

FLIGHT: Silhouette is like the other golden-cheekeds—overall compact, with a slightly long tail. Shows great contrast between the black or dark head/neck, golden cheeks, and whitish underparts. With wingbeats given in a quick series with intermittent skip/pauses, flight is distinctly bouncy.

VOCALIZATIONS: The pattern is variable, but the quality and some renderings are reminiscent of Black-throated Green Warbler. Common song is a buzzy, somnambulant, falling-and-rising, four-note chant, "*zoo zee z'dr zee*," or variations, including a three-note version: "*zee z'r zee*." Also makes a less complex "*z'r z'r z'r z'r see*," with the last note high, clear, and pushed. Call note is a low "*t'ch*" with no hard ending (similar to Black-throated Green). Flight call is like Black-throated Green.

Black-throated Green Warbler, *Dendroica virens*

Eastern Golden-cheeked

STATUS: Common northern and eastern breeder; common migrant east of the high prairies (farther east than Hermit and Townsend's). **DISTRIBUTION:** Breeds across boreal Canada from Nfld. and the Maritimes across s. Lab., s. Que., all but n. Ont., cen. Man., n.-cen. Sask., and cen. and n. Alta. to ne. B.C.. In the U.S., breeds throughout New England, N.Y., n. and s. N.J., all but se. Pa., ne. Ohio, most of W. Va., e. Ky., w. Va., e. Tenn., w. and e. N.C. (separate populations), south into coastal S.C., extreme ne. Ala., and n. Ga. Winters in the West Indies, e.

and s. Mexico, and Central America, and in small numbers in extreme s. Tex. and s. Fla. Casual migrant in the w. U.S. **HABITAT:** Boreal coniferous forest (partial to hemlocks and white pine) and mixed coniferous-deciduous woodlands with complex multistoried vegetative layers. Also found in red and white spruce, pitch pine, white cedar, and cypress swamps. In migration, may be found in almost any woody habitat—or even in nonwoody habitats, such as phragmites or the grasses bordering a forest edge. However, prefers foraging in trees (especially conifers). **COHABITANTS:** Blue-headed Vireo, Golden-crowned Kinglet, Blackburnian Warbler, Yellow-rumped Warbler. **MOVEMENT/MIGRATION:** Entire population relocates south of breeding areas for the winter. Spring migration from late March to early June; fall from late July to early November. *Note:* One of the latest of fall warbler migrants. VI: 3.

DESCRIPTION: Except in central Texas, the *only* member of the golden-cheeked warbler clan commonly found east of the Rockies. Medium-sized but robust (smaller than Yellow-rumped; larger than Tennessee), with a round head, a stout bill, a compact body, and a medium-length tail, which may seem short because of the long undertail coverts. Proportionally like Townsend's, Hermit, Golden-cheeked, or Black-throated Gray.

In all plumages, a boldly and distinctly marked bird, most conspicuously distinguished by its dark bordered, bright yellow face—the only eastern warbler to show this pattern—and thick white wing-bars (a very useful field mark in fall). Although the name is "Black-throated Green Warbler," the back is only somewhat greenish, and immature females may lack any trace of a black (or gray) throat. But every individual—from the most contrasting black-bibbed male to the palest immature female—has bright yellow cheeks that grab the eye. (The shadowy suggestion of a party mask across the eyes is easily dismissed.) The contrast between the unstreaked, greenish back and the bluish-gray wings distinguishes this species from western golden-cheekeds. Also, a touch or wash of yellow on the flanks/vent is often apparent.

BEHAVIOR: A fairly active species that forages most commonly on leaves and branches in the middle canopy forest layer and within the tree's interior, although, in migration, birds are commonly found much lower along brushy woodland edge and may even forage in weedy tangles close to the ground. By contrast, Townsend's and Hermit Warblers like the heights. Black-throated Green changes perches by hopping and flying frequently. Commonly hovers to pluck prey (or berries).

After breeding and during migration, often joins mixed flocks dominated by chickadees and other warblers. Fairly aggressive toward other warblers feeding in close proximity.

Not shy, and responds quickly to pishing. In spring and summer, males are tenacious singers.

FLIGHT: Profile appears broad at the shoulder, narrow at the hip, with a long straight-sided (and from below all-white!) tail. Appears yellow-headed at all angles; from below, shows a distinctive contrast between the yellow head, black collar, and white underparts. Flight is bouncy but borderline undulating; wingbeats are deep and stiff and given in a terse series punctuated by a skip/pause.

VOCALIZATIONS: Has a two-song repertoire. Both are simple, sonorous, slightly buzzy ditties consisting of five spaced and whistled notes, ending on an upscale note. The first rendering, "*zoo zoo zoo zuzeet!*" is mildly hurried; the second, "*soo see so so seet!*" is slower and dreamier. Call is a hard flat "*tchik*" (similar to Golden-cheeked, Townsend's, and Hermit, but perhaps more like Black-throated Gray). Flight call is a high-pitched "*wheet.*"

Townsend's Warbler, *Dendroica townsendi*

The Chickadee-like Golden-cheeked

STATUS: Common northwestern forest breeder; fairly common winter resident, primarily along the U.S. West Coast. **DISTRIBUTION:** Breeds from s. and e.-cen. Alaska south into the Yukon, w. and s. B.C., and sw. Alta., into Wash., n. Idaho, w. Mont., and cen. Ore. Winters coastally from Wash. to n. Baja; also in s. Ariz. and trans-Pecos Tex. and interior portions of Mexico into Central America. Regular migrant through the Rockies and the

Southwest, mostly in the fall. Casual in the East, mostly in the fall and early winter. **HABITAT:** Breeds in northern old-growth and mature coniferous and mixed coniferous forest, showing a marked preference for white spruce in the North, Douglas fir in the South. Forages primarily in the canopy, and in summer often forages in leafed-out deciduous trees (such as birch). In migration and winter, uses a variety of (primarily) wooded habitats, including riparian woodlands, orchards, junipers, pinyon-pines, oaks, chaparral, and exotic plantings. Seems partial to short-needled conifers, but also loves oaks. **COHABITANTS:** Gray Jay, Steller's Jay, Black-capped and Chestnut-backed Chickadees, Varied and Hermit Thrushes, Golden-crowned and Ruby-crowned Kinglets. In winter, often associates with chickadee flocks, kinglets, nuthatches, and Yellow-rumped Warbler; in migration, found in mixed warbler flocks. **MOVEMENT/MIGRATION:** Except in western Washington and Oregon, all Townsend's Warblers vacate breeding areas for winter. Spring migration from early April to late May; fall from early August through October. VI: 3.

DESCRIPTION: A plump, active, golden-cheeked warbler with a distinctive dark party-mask face pattern (looks a Black-throated Gray Warbler with a yellow head and breast). Medium-sized (larger than Red-breasted Nuthatch; smaller than Yellow-rumped; the same size and structure as other golden-cheeked warblers), with a short bill, a large head, a plump body, and a short tail.

Distinctly and contrastingly patterned. In all plumages, and at considerable distance, the bright yellow head and breast contrast with the dark party mask over the eyes and gray wings. From a distance, the bird looks yellow in front, and silvery gray and white to the rear. The white wing-bars are often narrow and inconspicuous on some immatures. The white outer tail feathers are not generally flashed as the bird feeds. No other member of the golden-cheeked tribe has yellow on the breast.

BEHAVIOR: As a feeder, somewhat more nimble, active, and acrobatic than Hermit Warbler. Hops, flutters, hovers, flycatches, and clings and picks chickadee fashion—in fact, behaves very much like a chickadee or kinglet. Likes to forage by hopping on top of tight-needled clusters—sometimes hanging partially down, sometimes reaching up to the undersides of overhanging branches. Hops, looks quickly left, right, then hops again. Wing-flicks and flutters its wings at distances where other warblers would simply hop to the next perch. Search pattern is often vertical; shows a preference for the outer tips of branches.

When breeding, forages primarily in taller canopy. Males commonly vocalize from the tallest perch, or they may sing while foraging. In migration and winter, commonly forages in much shorter trees (may even pursue insects to the ground). Fairly tame. Responds well to pishing.

FLIGHT: Overall stocky with a narrow tail. The golden head/breast is very distinctive. Quick wingbeats are given in a short series interrupted by a terse closed-wing skip/pause. Flight is slightly wandering, moderately fast, bouncy, and jerky more than undulating—seems very chickadee-like.

VOCALIZATIONS: Short simple song typically has two or three parts, and the final sequence is accelerated and higher-pitched; for example: "*jrrr jeee jer jeet*" or "*s'r see sir see sir see See See*" or "*der der der dir d'g'geeh.*" Notes are hurried and buzzy; sometimes the pattern may recall a chickadee. One song type is repeated at frequent intervals, then replaced with a new song. Call is similar to other members of the golden-cheeked clan—a soft "*tchik*" that recalls, somewhat, Common Yellowthroat but is distinctly sharper and higher-pitched. Flight call is a high rising "*zwee.*"

PERTINENT PARTICULARS: Townsend's and Hermit Warblers hybridize where their ranges overlap in Oregon and Washington (so, of course, they may occur farther south in fall and winter). Hybrids tend to more closely resemble Hermit by showing patternless cheeks, and some may show a shadow cheek patch reminiscent of Black-throated Green Warbler. Hybrid underparts are often more Townsend'slike.

Hermit Warbler, *Dendroica occidentalis*
Canopy Golden-head

STATUS: Uncommon to fairly common but restricted western breeder and fall migrant in the Southwest. **DISTRIBUTION:** Breeds from s. Puget Sound south along the western slopes of the Cascades to n. Calif., then south along the Calif. coast to just south of San Francisco and south along the Sierras to their southern limit (about the latitude of Bakersfield). A few pairs nest in the s. Calif. mountains. In winter, all but the southernmost coastal breeders withdraw. Some birds winter in coastal Calif. pine forests from just north of San Francisco to Santa Barbara, but most Hermit Warblers retreat to Mexico and Central America. In migration, generally found no farther east than the sw. N.M. mountains, but may be fairly common there in late summer and early fall. Accidental in the East. **HABITAT:** A canopy specialist. Breeds in both mature and young coniferous stands, but most populations appear to favor older, taller forests dominated by conifers, particularly Douglas fir in northern and coastal areas; in the Sierra Nevada, found in mixed conifers, including sugar, Jeffrey, ponderosa, and lodgepole pines and shows a partiality (not shown elsewhere in the bird's range) for broken or open canopy. In migration, still partial to conifers but more eclectic in spring, foraging (often in mixed groups) in cottonwoods, willows, live oaks, mesquites, orchards, suburban parks, and exotic plantings around human habitations. **COHABITANTS:** Red-breasted Sapsucker, Hammond's Flycatcher, Steller's Jay, Chestnut-backed Chickadee, Mountain Chickadee, Yellow-rumped Warbler. **MOVEMENT/MIGRATION:** Spring migration from early April to late May; fall from mid-July to October, with a peak in August and early September. VI: 2.

DESCRIPTION: A golden-headed treetop warbler. Like other members of the golden-cheeked complex, medium-sized, with a petite bill, a plump body, and a longish narrow tail. From below, head appears narrow and laterally compressed.

Has the simplest and most basic plumage of all members of the golden-cheeked clan—a proto-plumage. Blackish gray above, cleanly grayish white below. The adult male's bright yellow head is a standout feature, but the clean-edged black triangle on the throat (as contrastingly distinctive as the throat patch of Golden-winged Warbler) easily distinguishes this species from its streaky-flanked cousins. The adult female is like the male, with distinctively clean underparts and a contrasting dingy black throat patch, but has a golden face rather than a golden head.

Immatures have the same basic clean pattern of adults but are overall muted; they are somewhat dingier and less patterned and showy than other members of the golden-cheeked clan. Nevertheless, like adults, they have essentially no streaks on the flanks and less pattern on the face; in particular, they have no eye-line and only a shadowy face pattern. In the forest mountains of southeastern Arizona and southwestern New Mexico, confusion with female and immature Olive Warblers is possible.

BEHAVIOR: A bird that thrives in the canopies of the tallest conifers, infrequently foraging close to the ground. Hops along branches, emerging onto outer branches, then moving back into interior portions of the canopy. A fairly methodical no-frills feeder. Plucks prey from branches and explores needle clusters. Tends not to leap for prey, to sally, or to flycatch.

Males sing as they forage, working a tree for several minutes, then moving on to another (often not the nearest tree). Not particularly responsive to pishing (you're just too low to bother with), but flies quickly to confront other males in distant treetops.

FLIGHT: Profile is like other golden-cheeked warblers, but unaccountably whitish below. Flight is direct, moderately fast, and slightly jerky, with a rapid series of wingbeats and intermittent skips or missed beats.

VOCALIZATIONS: Song is highly variable; individual birds offer a mix. Some renderings are simple, such as "*weesee, weesee, weesee*" (recalls Black-and-white Warbler), and others are more complex. Many have a rising-and-falling pattern and buzzy quality that recall the short, lazy, sonorous song of

Black-throated Green Warbler, but Hermit's song is thinner: "*dee* (up) *dee* (down) *dee* (up) *dee* (down); *dee, dee dEEE* (pushed)" or "*ti ti t't't'ti!*" or "*ti-ti ti-ti ti* (up, down; up, down, up)." Still others have a more complex song that recalls Townsend's Warbler. Most songs end with a pushing high-pitched signature "*ziiig*" that sounds like the last note of Cerulean, or even Blackburnian. Call is a soft "*tchih*" that sounds much like others in the golden-cheeked clan.

Blackburnian Warbler, *Dendroica fusca*

Ember-throated Warbler

STATUS: Fairly common northern and eastern breeder. **DISTRIBUTION:** Nests from sw. Nfld. and N.S. west across boreal Canada to cen. and e. Sask. (and sparingly west into n.-cen. Alta.). In the U.S., breeds widely throughout New England and N.Y., south through Pa., e. Ohio, and nw. N.J., along the Appalachians to n. Ga., plus all but s. Mich., s. Wisc., and s. Minn. **HABITAT:** Breeds in cool, wet, mature coniferous forest and mixed coniferous-deciduous forest. In particular, in the East, favors hemlocks; also found in spruce and white pine. In migration, still favors conifers (where found, partial to cedars), but forages freely in deciduous trees—particularly flowering oaks in spring—and, in the absence of tall trees, bushes. **COHABITANTS:** Blue-headed Vireo, Golden-crowned Kinglet, Magnolia Warbler, Black-throated Green Warbler; in migration, other warblers and chickadee-based flocks. **MOVEMENT/MIGRATION:** Wholly migratory. Entire population relocates to southern Central America and northern South America. Migrates east of the Great Plains. Spring migration from mid-April to early June; fall from early August to mid-October, with a few departing in late November. Casual in the West. VI: 3.

DESCRIPTION: A stripy, contrasting, black-and-white (or gray-and-white) warbler with a flaming orange (or smoldering yellow) face/throat. Smallish and compact (smaller than Blackpoll; larger than Cerulean; about the same size as Black-throated Green), with a fairly stout bill and a moderately short tail.

The radiant orange face and throat and the boldly patterned black-and-white body of adult males make the bird unmistakable. Because of the size, compact shape, and stripy body, you may think you're looking at a Black-and-white Warbler (until you see the face). The adult female is more cryptic but shows a yellow-orange throat and the adult male pattern, right down to the dark face mask shaped like New York State. Immatures may be so boldly patterned as to resemble females, but many are more blurry. Immature Blackburnians have distinctly blackish upperparts that set off white wing-bars and narrow pale streaks on the back. Almost all show a wash of yellow on the throat and face and a shadowy face mask rimmed by a large pale eyebrow that connects to the pale sides of the neck.

BEHAVIOR: A canopy dancer. Flits and hops across the very tops of trees and the outer branches, often working its way right across the crown of a tree, then moving on to the next. When foraging, commonly droops its wingtips and cock its tail up slightly. Also works along branches in the tree's interior, looking up and down for prey. Likes to work in short-needle conifers. Sometimes hovers, but flycatches only infrequently.

Forages alone or with its mate. After breeding season, joins mixed-species chickadee-based flocks. Responds well to pishing.

FLIGHT: Looks like a petite Black-and-white. Not small but fairly slender, with a longish narrow head/neck, a pointy face, pointy wings, and a broad straight-sided tail (longer than Cerulean). The shape recall's Yellow Warbler, but the wings are set farther back. Breeding males are all field mark. (The orange throat and white wing patch are very apparent.) At all seasons, the black triangular cheek patch holds up. Tails shows lavish amounts of white.

VOCALIZATIONS: Song consists of run-on series of high even notes (initial pattern may recall Northern Parula) that are often hurried or stuttered at the end and close with an ethereal, superhigh, signature "*seee*" that barely brushes the registry of the human ear (recalls the high notes of

Golden-crowned Kinglet); for example: "*chewa-chewachewat't't'tsEEE.*" Alternate song is slower, more measured, and usually high-pitched, and it normally closes with the signature "*seee*"; for example: "*SEEpa SEEpa SEEa sEEE.*" Call is a loud "*chih*" that is surprisingly robust for such a small bird. Flight call is a buzzy "*seet*" that is similar to Yellow Warbler.

Yellow-throated Warbler, *Dendroica dominica*

Treetop Black-and-white Warbler

STATUS: Fairly common in its southeastern breeding range, but owing to its partiality to heights, often difficult to see. **DISTRIBUTION:** Breeds south of a line drawn from s. N.J. and se. Pa. west across n. Ohio, n. Ind., sw. Mich., n. Ill., and s. Wisc., then east of a line drawn from se. Iowa, w. Mo., e. Okla., and e. Tex. north of the Brazos R.; also the Edwards Plateau. (Does not breed in s. Fla., nor along the spine of the Appalachians from Pa. to n. Ga.) In winter, found coastally from S.C. to s. Tex. and the length of Fla., but most retreat to the Caribbean, the east coast of Mexico, and Central America. Casual in the West. **HABITAT:** Breeds in two distinct habitats: dry upland pine-oak forest (especially loblolly pine), and mature bottomland and wooded swamps including cypress, live oak, and (particularly in northern and western portions of its range) waterways with an abundance of sycamores. **COHABITANTS:** In summer, Pine Warbler, Northern Parula. In winter, may join mixed flocks with Carolina Chickadee and Tufted Titmouse (commonly only one Yellow-throated to a flock), but is most often found in palm trees and solitary. **MOVEMENT/MIGRATION:** An extremely early-spring migrant, Yellow-throated Warbler is among the first harbingers of spring. Birds appear in early March in southern portions of their range; northern birds arrive in April. Also departs early in the fall: some northern birds leave in late July, and most have departed by early September. VI: 3.

DESCRIPTION: A large, distinctly shaped, distinctly marked warbler that is best distinguished by its crisp, bold, black-white-and-yellow head pattern and long, thin, and (usually) down-curved bill. The head is big, the face long and pointy, and the body somewhat compact but not portly. Overlooking the brilliant yellow throat, in shape and also behavior, Yellow-throated recalls Black-and-white Warbler—a bird that is a true study in compactness.

The tail of Yellow-throated seems very narrow and blunt and occasionally shows a slightly notched tip. This would be a silly point to make if not for the fact that underside views are this bird's specialty, and along the East and Gulf Coasts, this species is most likely to be found with Pine Warbler—a bird that also has a yellow throat (and dull yellow breast) and whitish underparts (but a broad deeply notched tail).

BEHAVIOR: The bird loves treetops—it sings from treetops and forages deep in their leafy confines, emerging now and again to offer a tantalizing view and a snatch of song. A methodical feeder, Yellow-throated hops along limbs, submerging its head in pine needle cluster; hangs upside down on pinecones, probing with its bill; and clings to the trunk exploring crevices. It recalls a yellow-throated Black-and-white Warbler of the canopy or a nuthatch. Likes to work up a tree, starting low on a branch, working up and out, examining with equal intensity the surface of the branch beneath it and the underside of the branch above. Also hovers, briefly, in front of a hard-to-reach cluster of needles, leaves, or Spanish moss and flycatches.

In winter in south Florida, spends most of its time buried among dead palm fronds and fruit clusters at the center of palm trees (which are often not high). Birds may be located by listening for the telltale rustle in the leaves (or hearing chip notes). Elsewhere, in woodlands, may be found foraging in mixed flocks of chickadees, titmice, Blue-headed Vireos, and other warbler species.

FLIGHT: Moving from tree to tree, flight is weak, fluttery, and usually short. It somewhat recalls the wing-batting flight of Eastern Kingbird. Migratory flight is stronger. The long face and big bill are prominent, the facial pattern distinctive, and the tail appears long. Shows a clean gray back (not streaked like Blackburnian).

VOCALIZATIONS: The song is simple and sweet—easy to recognize and easy to remember. Song begins with a series of clear, sweet, seductively pitched, and theatrically spaced notes that catch listeners up in the rhythm, then fall and rise playfully at the end: "*Tee, tee, tee, tee, tee, tee, tee, tee, Too, Too, Sweet.*" Call is a full-bodied but not particularly loud "*chewf*" that falls off at the end. Flight call is similar to Northern Parula.

PERTINENT PARTICULARS: This species is most commonly confused with Common Yellowthroat—not because the birds look remotely alike, but because they have similar names.

Until seen, beginning birders often anguish about telling this species from the smaller Blackburnian Warbler. Don't worry. You won't confuse the two birds.

The challenge to Yellow-throated Warbler aspirants frequently comes down to sifting through the canopy separating this species from Pine Warbler, with which it shares habitat. In poor light or when backlit, Pine Warbler has a stout bill; Yellow-throated has a long thin bill. Pine Warbler has a broad deeply notched tail; Yellow-throated has a narrow blunt-tipped tail. Pine Warbler moves somewhat sluggishly, but like a warbler; Yellow-throated often moves like a nuthatch.

Grace's Warbler, *Dendroica graciae*

Crown Pixie

STATUS: Common breeder in the mountains of the Southwest. **DISTRIBUTION:** Southern Nev., sw. Utah, sw. Colo., all but sw. Ariz., and all but e. N.M.; also found in trans-Pecos Tex. (but not the Chisos Mts.). Casual vagrant in Calif. **HABITAT:** Breeds in dry montane pine and oak-pine forests. Prefers open parklike forests with broken canopy and is especially partial to forest dominated by ponderosa pine. In migration (rarely seen), prefers pines and other conifers; forages (uncommonly) in deciduous trees. **COHABITANTS:** Pygmy Nuthatch, Olive Warbler, Yellow-rumped Warbler, Western Tanager. **MOVEMENT/MIGRATION:** Entire U.S. population retreats into Mexico and Central America in winter, following a route that does not extend much beyond the breeding range. Spring migration from late March to mid-May; fall from early to mid-September to October. VI: 1.

DESCRIPTION: A small active warbler whose penchant for staying in the canopy is mitigated by its habit of feeding on the outermost twigs and needles. Overall compact and front-heavy, with a thin, pointy, petite bill, a round neckless head, and a fairly short tail that is very narrow at the base.

Shows bluish gray upperparts and whitish underparts, with some black streaks down the sides that are often not apparent owing to the bird's foraging height. Also, the gray facial pattern seems not to project at a distance. The bird's most conspicuous trait is a bright yellow face, throat, and (especially) breast. Although lacking in detail (the plumage as well as this description), the combination of a bright yellow face/breast on a gray-backed warbler foraging in the tops of pines is all the description needed to distinguish this species in its habitat and range. Female Olive Warbler has a dingy (not bright!) yellowish hood. (Immature Olive has little, if any, yellow on the throat and breast.) Immature Townsend's (which passes through Grace's Warbler's breeding range during fall migration) has an olive head and back (not gray) and is overall dingier. "Audubon's" Yellow-rumped Warbler is frequently found in Grace's Warbler's habitat (the birds may forage in the same tree), but the yellow on "Audubon's" is confined to a small well-defined patch on the throat.

The species that most closely resembles Grace's is the eastern Yellow-throated Warbler, whose range overlaps with Grace's only in winter. Yellow-throated's much longer bill and much more contrasting and crisply defined facial pattern distinguish it. Also, the feeding habits of the two species differ.

BEHAVIOR: A hyperactive warbler and a fine-foliage specialist that likes to be in pines and likes to forage in the canopy—hopping, forcefully, on narrow twigs and outer branches and burying its head into needle clusters. (Yellow-throated Warbler likes sturdier branches and trunks, works the interior, and seems comparatively sluggish.) Grace's is almost constantly moving in a coursing pattern that maps the canopy, sometimes making short hops, sometimes long fluttering jumps. When perched on more sturdy branches (as when males are singing), pivots its body left and right, as if releasing nervous energy.

Spends little time foraging in anything but pines. Does not hang upside down. Does sally out to catch insects on the wing.

FLIGHT: Direct, strong, and undulating, with rapid wingbeats.

VOCALIZATIONS: Song is usually in two (sometimes three) parts, each consisting of a series of repeated single notes; for example: "*zrr zrr zrr zrr zrr zreet zreet*" or "*chew chew chew chew chew t't't't't't* (warbled trill like a slow Chipping Sparrow) *chee chee chee.*" The end of the song is usually higher-pitched and faster than the beginning; overall, it has a hurried or impatient quality. The pattern and even some of the notes are much like Yellow-rumped's song, but at times the pattern is reminiscent of Wilson's Warbler; for example: "*chi chi chi chi chachacha.*" Call is a soft but rich "*chip.*"

Pine Warbler, *Dendroica pinus*

Mustard-colored Vireo-Warbler

STATUS: Common but somewhat localized eastern breeder and southeastern winter resident; curiously difficult to find in migration. **DISTRIBUTION:** Breeds for the most part from s. Canada (from s. N.B. across Maine to s. Que., s. Ont., and se. Man.) to the Gulf of Mexico (except s. Fla.) and east of the Mississippi, but the distribution of birds through much of the middle portion of its range (the mid-Atlantic states, the Midwest, the southern border states) is patchy and fragmented; a somewhat disjunct population breeds in e. Okla., w. Ark., e. Tex., and w. La. In winter, retreats south of a line drawn from s. N.J. (rare), across the Delmarva Peninsula, s. Va., w. N.C., s. Tenn., n. Ark., and e. Okla., with a few birds to s. Tex. A few birds wander north in fall and attempt to winter at suet feeders. Casual in the West. **HABITAT:** Almost always associated with pines, although also breeds in mixed oak-pine woodlands. In winter, in the southern United States, and during migration, may be found in deciduous woodlands and is especially fond of the insects drawn to flowering maples during spring migration. **COHABITANTS:** Where ranges overlap, often found with Yellow-throated Warbler; also Prairie Warbler. In migration and winter, joins mixed-species flocks with Carolina Chickadee, Brown-headed Nuthatch, kinglets, Eastern Bluebird, Yellow-rumped Warbler, Chipping Sparrow. **MOVEMENT/MIGRATION:** Permanent resident across much of the South. Northern birds vacate breeding areas between mid-September and mid-November; spring migration is very early, from late February to early May. VI: 3.

DESCRIPTION: A confusing fall warbler (even in the spring!). Large and robust (bordering on pudgy), bullheaded, and heavy-bodied (slightly larger than Blackpoll and Yellow-rumped Warbler; the same size as Yellow-throated Warbler), with a thick but still warbler-pointy bill, a long deeply notched tail, and an unabashed penchant for pines.

Plumage is roughed out and shabby—a watercolor field sketch of a warbler that the artist never returned to. Yellowish on the face and breast, olive or brownish above. The plumage pattern is overall muted, blended, and somewhat mustard-colored—even the generally wide white wing-bars seem not as crisp or contrasting as on most other warblers. Some adult males are wonderfully bright yellow below and show shadowy streaks on the sides of the breast. Duller females may be barely tinged with yellow; in fall, immature birds (particularly females) can be dingy grayish brown and generically patterned. In all plumages, birds have a "five o'clock shadow" on the face contrasting with a pale or yellow throat and paler sides to the neck. In all plumages, the back is plain and unstreaked. (In fall, the backs of Blackpoll and Bay-breasted Warblers show blurry streaks.) The white wing-bars are usually apparent, but with wings folded, only one may show. In flight, the outer tail feathers show white.

BEHAVIOR: Overall the bird's movements are methodical, studied, unhurried, almost lumbering—somewhere between warbler frenetic and vireo slothful. In fact, bright males might be mistaken for Yellow-throated Vireo. Pine Warbler does no wing-flicking, but does pump its tail infrequently. Likes to work a tree over well by prying deep into needle clusters—even hanging upside down and clinging to trunks—before moving to another tree. Sometimes drops to a lower limb and grabs a caterpillar as it falls; also hovers in front of prospective cover and flycatches insects flushed by its foraging. Visits suet feeders.

In breeding season, most commonly found in the middle and upper forest strata; in winter, often forages on the ground, in the open, and often with bluebirds and Chipping Sparrows.

FLIGHT: A large, long, sturdy warbler showing long blunt-tipped wings, a projecting head, and a long tail that appears narrow at the base and flared and notched at the tip. All birds show the suggestion of a pale collar, and most show whitish edges to the tail. A strong flier with spare irregular wingbeats that appear floppy or shallow.

VOCALIZATIONS: Song is a short, mostly musical, often lazy warble or trill characterized by great variation and improvisation. Notes are generally evenly spaced and similarly pitched, and more slurred than sharp. The song may increase in volume or cadence in the middle, split into two parts (for example, a fast trill breaking into a musical warble), or end with an improvised one- or two-note twist. Individual birds commonly repeat the same song pattern but may vary the vocalization in response to the song of a nearby rival. Call is an abrupt flat "*tcheh*" or "*chee*" that is often lacking in volume (and enthusiasm). Flight call is weak and inconspicuous.

PERTINENT PARTICULARS: In fall, away from pines, Pine Warbler often leaves an observer guessing. *Hint:* If you see a warbler and you can't decide what kind it is, chances are that it is either a very plain Pine Warbler or a very plain Yellow Warbler. If you see a small warblerlike bird and you aren't even certain that it *is* a warbler, consider Pine.

Kirtland's Warbler, *Dendroica kirtlandii*

The Jack Pine Warbler

STATUS AND DISTRIBUTION: Very rare and very restricted; the entire world breeding population of less than 2,000 adults is confined to 13 contiguous northern counties in lower Mich. (a few have recently been recorded in the Upper Peninsula) and adjacent Wisc., Ont., and Que. Winters exclusively in the Bahamas. **HABITAT:** Breeds in dense, regenerating, post-fire (or managed) jack pine–red pine forests with 6–15-year-old trees about 4–15 ft. tall. Typical nest sites are near breaks or openings covered by grasses and a mat of low bushes. In migration (where it is very rarely seen), the species is reported to be partial to low, open, broadleafed scrub. **COHABITANTS:** Breeds in association with Black-capped Chickadee, Hermit Thrush, Nashville Warbler, Dark-eyed Junco. **MOVEMENT/MIGRATION:** Spring migration from late April to late May; fall from early August to early October. The narrow route crosses Michigan, Ohio, West Virginia, western Virginia, western North and South Carolina, and Georgia, but very few birds are ever recorded. VI: 1.

DESCRIPTION: A large gray-and-yellow warbler singing from the top of a jack pine near Grayling or Mio, Michigan. Almost *huge* (closer in size to Hermit Thrush than Nashville Warbler), with a big, heavy, somewhat chatlike bill, a smallish head, a robust oval-shaped body, and a long, straight, sturdy tail.

Upperparts are gray with black streaks (same color and tone as Yellow-rumped Warbler), and underparts are bright pale yellow with black spotting and streaking down the sides. Some individuals show a streaky, usually broken band across the lower chest. Repeat: Kirtland's Warbler is *pale* yellow. Next to the rich yellow of so many other warbler species, this bird's underparts seem washed out. The white eye-arcs stand out; the frail wing-bars do not. In fall, upperparts of adults (and young) are somewhat browner, and young show fine spotting, not streaks, on the breast and sides.

Although Kirtland's shares some plumage traits with the considerably smaller Prairie Warbler and

Magnolia Warbler, its size, structure, gray streaked back, blackish face mask (males), and white eye-arcs are more likely to remind you of Yellow-rumped Warbler, particularly if you see the bird from behind.

BEHAVIOR: A vigorous and emphatic tail-wagger, Kirtland's is constantly moving its tail in a stiff down-pumping action when feeding (not commonly when singing). The tail-wagging action of Prairie and Palm Warblers seems nervous or jerky by comparison. Kirtland's is not at all shy. Singing males often sit conspicuously atop jack pines or dead branches and sing loudly and often, even late in the breeding season when young have fledged. Not only do birds allow close approach on territory, but curiosity prompts some individuals to sometimes approach you. Forages at all heights, from the ground to the crown, but most foraging seems to be from midheight to just above the ground, mostly on interior branches. Moves by hopping with quick but deliberate movements. Feeds by gleaning insects (in both pines and oaks); sometimes hovers to pluck insects from the tips of pine branches. In late summer, also feeds on ripened blueberries (which seem to dominate the shrub layer). Response to pishing? See "Pertinent Particulars."

FLIGHT: Large and long-tailed profile. The mostly plain grayish upperparts and pale yellow underparts recall Canada Warbler, but the outer corners of the tail of Kirtland's shows white. A strong flier whose wingbeats are given in a short series followed by a brief closed-wing glide. Short flights are bouncy; longer flights are slightly undulating.

VOCALIZATIONS: Variable song is low, loud, rich, and rising in volume and tempo; for example: "*(chip) ch ch ch chlw'r will tip tip*" or "*ch chuh chur ch'weer wheet! wheet!*" Recalls Northern Waterthrush—but what would a waterthrush be doing in a pine barrens? Alternate song sounds somewhat like House Wren, but whispered. Call is a low "*chup*" that sounds somewhat like Prairie Warbler.

PERTINENT PARTICULARS: This is an endangered species, and under federal law, you cannot harass an endangered species. But Kirtland's does respond readily and in fact ranks among the most pishable of warblers.

Prairie Warbler, *Dendroica discolor*

The Yellow Spectacled Warbler

STATUS: Common but somewhat local eastern breeder and Florida winter resident. **DISTRIBUTION:** The suggestion of its name notwithstanding, this species barely reaches the American prairies. Breeds in all the states east of the Mississippi R. except Minn. and Wisc., and absent in n. and cen. Maine, n. N.H., n. Vt., most of n. N.Y., and Mich. Breeders are also absent along the immediate Gulf Coast. West of the Mississippi, breeds in se. Iowa, cen. and sw. Mo., Ark., ne. Kans., e. Okla., w. La., and e. Tex. Winters across most of the Florida Peninsula and the Caribbean. **HABITAT:** Preferred habitat includes overgrown fields pocked with shrubs, vegetated dunes, second-growth woodlands, pine barrens, and, in Florida, mangroves. In migration, also favors overgrown fields and woodland edge. **COHABITANTS:** White-eyed Vireo, Gray Catbird, Pine Warbler, Yellow-breasted Chat, Eastern Towhee, Blue Grosbeak. **MOVEMENT/MIGRATION:** In spring, arrives fairly early; males commonly reach the northern portions of the range by late April or early May, continuing to late May. Fall migration is protracted: some birds depart in August and September, and others linger into November. VI: 3.

DESCRIPTION: A distinctly small warbler with a large round head, a well-proportioned body, and a long narrow tail that wags energetically. The bill is small, straight, and symmetrically tapered. Adult males and females are yellow-green above, bright yellow below. The bold black streaking on the sides is distinctive, and the wing-bars faint. The hallmark characteristic is the bird's distinctive "spectacled" facial pattern. The plumage of fall immatures is a paler greatly toned-down version of the adult pattern, but birds still show shadow streaks down the sides and a shadow spectacle pattern on the face. With its pale gray head, immature Prairie Warbler could be confused with immature Magnolia Warbler.

At a distance, birds are fairly undistinguished mustard-colored birds. In flight, the dark center of the tail contrasts with the pale back, but it's the wide white outer tail feathers that catch the eye.

BEHAVIOR: A warbler that likes to forage deep in the interior of bushes and likes to stay low, often feeding just above (less frequently on) the ground. Moves quickly but somewhat stiffly, preferring to make conservative short hops branch to branch. Also hovers and flycatches, but when foraging through brush, uses wings sparingly; when wings are needed to get the bird to the next branch, the movement is quick and cryptic and almost too rapid to perceive. More obvious, and virtually diagnostic, is the bird's persistent tail-wagging. While some other warbler species wag their tails, the action is periodic, not pervasive. Only the extremely rare Kirtland's and the larger, rangier, ground-loving Palm Warbler share Prairie Warbler's penchant for persistent tail-wagging.

Singing males are often very conspicuous, sitting at or near the tops of shrubs and young trees. The stance is slightly angled—more horizontal than erect. Birds spend most of their time less than 10–15 ft. off the ground, and some Prairie Warblers may go through their entire lives without ever seeing the canopy of a tree.

FLIGHT: Sleek and pointy-faced, with a slight body, short rounded wings, and a long thin tail. Appears greenish-backed and bright yellow below. The yellow in the face holds its tone (if not pattern). Flight is fast and direct, with shallow wingbeats and irregular undulations—some are short and deep, others long and shallow. Even more than many other birds with an undulating flight, suggests a stone skipped across water.

VOCALIZATIONS: Song is distinctive—a series of sonorous buzzy notes that start slow, gain momentum, and seem to climb the scale, or "climb the stairs": "*zoo, zoo, zoo, zo, zo zozoZHEET.*" Suggests a bird reciting a buzzy musical scale ("*Do, Re, Mi, Fa, Sol, La, Ti, DO*"). A variation descends the stairs, gaining momentum near the bottom, then leaps back up the scale on the last note: "*ZHEE, Zhuh, zhuh, zhuh zhuhzhuh zhEEEET!*" Another truncated variation is a somnambulant lazy "*Zoo, Zoo Zee*" or "*So Lay-Ze.*" Call is a soft "*chuh*" or "*cheh*" with the terminal "h" distinctly rendered.

Palm Warbler, *Dendroica palmarum*
The Wagtaillike Warbler

STATUS: Common breeder in forested regions of Canada and bordering northeastern United States. Common winter resident across the southern coastal plain; rare along the West Coast in fall and winter. **DISTRIBUTION:** Breeding range extends across boreal Canada from sw. Yukon and ne. B.C. to s. Lab. and the Maritimes (absent in the Canadian prairies). In the U.S., breeds in n. Minn., n. Wisc., the Upper Peninsula of Mich., and n. and cen. Maine. Winters from the Delmarva Peninsula (occasionally Long Island) south across coastal Ga., in all of Fla., and along the Gulf Coast. **HABITAT:** At all seasons, prefers open habitat with shrubs and a scattering of trees. Breeds in spruce or tamarack bogs and forest edge, usually near water. In migration, found in weedy fields, on forest edges, along hedgerows and fence rows (adjacent to plowed fields), on lawns, coastal dunes, and fairly open (unleafed) woodlands, and in woods bordering watercourses and lakeshores. In winter, often found in southeastern open pinewoods with brushy understory. **COHABITANTS:** In summer, Solitary Sandpiper, Yellow-bellied Flycatcher, Lincoln's Sparrow, Rusty Blackbird; in winter, Eastern Bluebird, Yellow-rumped Warbler, Pine Warbler, Chipping Sparrow. **MOVEMENT/MIGRATION:** Among the earliest (in the East) and latest of the warbler migrants. In spring, eastern ("Yellow") Palms appear as early as late March, and most are through by April—two to three weeks before the migration of most warblers. "Western" Palms are later migrants—from late April to mid-May. Fall migration is protracted, running from late August to November, with a peak in September and October, when migrations of most other warbler species have slowed or ended. VI: 3.

DESCRIPTION: A fairly large, plain, raw-boned warbler that looks and acts like a plain-backed pipit or wagtail. The face is plain and pointy, the tail longish and animate. Two easily identified subspecies both have plain dingy olive brown upperparts and (in spring) distinctive rust-capped crowns. "Western" Palm has pale underparts that are generally dirty white and diffusely streaked, with a splash of yellow on the throat and undertail. More distinctive "Yellow" Palm is uniformly dark yellow below, with diffuse reddish streaks.

Fall adults and immatures are overall dingy brown with a nondescript facial pattern distinguished only by a pale eyebrow. "Western" Palm retains a yellow undertail; "Yellow" Palm retains yellowish (but unstreaked or blurry-streaked) underparts. Rusty caps are absent.

In all plumages, the rump is extensively washed with yellow or yellow-green and the outer tail feathers are tipped with white. In flight, the retreating bird projects three pale points—a triangle defined by the pale rump and the white running lights on the corners of an all-dark tail.

BEHAVIOR: Behavior alone identifies this bird. Palm Warbler habitably forages on the ground, hopping or cantering more than running, often in open habitat more suited for sparrows; in fact, Palm Warbler sometimes forages with sparrows. Most other warblers confine their movements to trees and shrubs, rarely landing (much less foraging) on the ground. Also, Palm Warbler has a pipitlike habit of wagging its tail.

Palm Warbler also forages readily in brush, thick weedy growth, and (particularly in spring migration) treetops, often with flocks of Yellow-rumped Warblers. Moving actively and changing perches frequently, gleans insects by plucking, leaping vertically, and flycatching. May vigorously pursue flushed insects in tail-chase fashion, covering 20–30 ft.

In migration, often found in mixed flocks. In spring before trees have leafed out, may be found foraging in open woodlands on the forest floor.

FLIGHT: Long and slender profile (pipitlike!), with a pointy face, a flat crown, a hunched back, and a longish tail. Except for the yellow underparts (or yellow undertail coverts), overall dull and drab; in good light, brownish-backed (an odd color for a warbler). Flight is steady and even-paced, with relatively spare and shallow wingbeats and little or no gliding. The long, black, planklike tail shows white corners when fanned. Closed, from below, the tail appears to have a white tip.

VOCALIZATIONS: Song is a hurried trilled jumble of high and low notes. Notes sound scratchy. Song seems unrehearsed or amateurish by warbler standards. A trill that wants to be more than a trill. Not an enthusiastic singer. Song sets are well spaced. Call is a distinctive weak dry "*chech*" or "*check*" that is flatter and softer than Yellow-rumped Warbler. Flight call, "*seemp*," is protracted but ends abruptly.

PERTINENT PARTICULARS: In spring, most "Western" Palms pass west of the Appalachians through the Mississippi River Valley; nearly all "Yellow" Palms migrate east of the Appalachians. In fall, both subspecies occur in good numbers along the Atlantic Seaboard; however, "Western" Palms commonly outnumber "Yellow" Palms in the eastern Gulf states.

Bay-breasted Warbler, *Dendroica castanea*

Bay-flanked Warbler

STATUS: Numbers fluctuate, but generally an uncommon northern breeder and migrant except along the Gulf Coast in spring. **DISTRIBUTION:** Breeds from sw. Nfld. and the Maritimes across boreal Canada (south of Hudson Bay, north of the prairie regions) to ne. B.C., se. Yukon, and sw. Northwest Territories (absent in se. Ont.). In the U.S., breeds in Maine, n. N.H., n. Vt., the Adirondacks, n. Mich., and n. Minn. Casual migrant in the w. U.S. **HABITAT:** Breeds in thick mature boreal spruce-fir (less commonly mixed) forests, often near water; breeds less commonly in pines or hemlock. In migration, found in both coniferous and deciduous trees. Although still partial to spruce, in spring it also favors flowering oaks, where it forages high in the canopy. **COHABITANTS:** Tennessee Warbler, Cape May Warbler, Yellow-rumped Warbler. In migration, often found with Blackpoll and Cape May Warblers. **MOVEMENT/MIGRATION:** A trans-Gulf migrant, Bay-breasted travels generally east of the prairies in spring and fall. Spring migration from mid-April to early June; fall from mid-August to early November. VI: 3.

DESCRIPTION: A big, beefy, spruce-loving warbler with a splash of bay down the flanks. Large (larger than Black-throated Green; about the same size as, but stockier than, Yellow-rumped; the same size as Blackpoll, but overall bulkier), with a large head, a short, thickish, straight bill, a chunky body, and a shortish narrow tail.

Breeding males, with their black faces, pale nape patches, and bay-colored throats and flanks, are unmistakable; females and nonbreeding males bear a trace of bay down the sides and flanks. Immatures have mostly smooth mushroom yellow bodies that show a white belly, blackish wings, and broad white wing-bars—they resemble small compact Baltimore Orioles. Also, immature males usually show a pale peach-colored wash on the flanks and across the vent. By comparison, fall Blackpolls show blurry streaks on the back and breast and less contrast between the dark wing and the generally grayer-tinged back, and they are more distinctly yellow-breasted (not so extensively yellowish below). With its neat smooth plumage, Bay-breasted looks overall warm-toned and crisply contrasting; Blackpoll looks streaky, blurry, and yellow-breasted.

BEHAVIOR: A fairly sluggish methodical warbler that clomps as much as it hops through the foliage. On territory, tends to forage in a tree's interior at mid-height. In migration, favors treetops and outer twigs and branches, where it often occurs with Blackpoll.

During migration, fairly unresponsive to pishing (perhaps because it is feeding in the upper canopy), but more responsive in the fall.

FLIGHT: Profile is overall robust and somewhat sway-bellied, with long, broad, pointed wings. The wide white wing-bars on blackish wings are eye-catching; on immatures, the duller face and breast (warm buff or mushroom-toned) distinguish it from the yellow-fronted Blackpoll. Bay-breasted looks like a small stocky Baltimore Oriole or a bigger, buffier Blackpoll. Flight is strong and fast, with deep, stiff, fan-winged down strokes (looks as though the bird is caressing the air).

VOCALIZATIONS: Song is an ethereal series of high, thin, lisping notes with a slightly two-noted quality: "*t'see t'see t'see t'see*" or "*susie susie susie sue.*" Call is a classic "*chip.*" Flight call is a short, soft, high "*che'e,*" given with a slight vibrancy, that sounds like Blackpoll but is thinner and less buzzy.

PERTINENT PARTICULARS: On treetop-foraging males, the buffy ear/nape-patch is often easier to see than the bay breast and sides. In flight, in fall, compared to the similarly sized and shaped Blackpoll, Bay-breasted shows blacker wings with more prominent wing-bars and more contrast between the dingy (or rufous) flanks and the white belly. Bay-breasted never shows a bright yellow breast or bright white undertail coverts.

Blackpoll Warbler, *Dendroica striata*
The Chickadee Warbler

STATUS: Common northern breeder; variably common migrant east of the Rockies, with spring migrants more common in the Great Plains and the Mississippi River Valley than in the fall. **DISTRIBUTION:** Breeds from w. Alaska across boreal Canada from se. Yukon and ne. B.C. across n. Alta., nw. and cen. Sask., cen. and se. Man., all but se. Ont., Que., and the Maritimes. In the U.S., breeds in n. Minn., the Upper Peninsula of Mich., the Adirondacks of N.Y., n. Vt., n. N.H., nw. Mass., and Maine. Entire population retreats to South America in the winter on a course that carries birds first east to e. North America, then south, over open ocean, to the South American mainland, bypassing the w. U.S. and much of the s. U.S. In spring, migratory boundaries shift westward, encompassing the eastern two-thirds of the U.S.—that is, everywhere east of a line drawn from the Tex. coast to ne. Mont. Rare but regular vagrant in the w. U.S.

HABITAT: Breeds primarily in stunted or fragmented boreal spruce-fir forest and conifer bogs, but may also occur in mixed habitats (for example, in spruce-alder-willow thickets) where conifers are at least a component. In migration, remains partial to spruce, but may be found in a variety of forest and arboreal habitats. **COHABITANTS:** During breeding, Gray-cheeked Thrush, Palm Warbler, Boreal Chickadee, American Robin, American Tree Sparrow, "Red" Fox Sparrow. **MOVEMENT/MIGRATION:** A late-spring migrant that arrives in the southeastern United States in mid-April and continues north into mid-June—often the last lingering warbler of spring. In fall, birds begin to leave breeding grounds in mid-August. Peak movements along the Atlantic Coast occur from late September to late October, with diminished numbers continuing into early November—often in the company of Yellow-rumped Warblers, which remain in North America while Blackpolls still have a hemisphere to vault. VI: 3.

DESCRIPTION: A large, stocky, wide-bodied warbler. Black-and-white males are distinctive. Females and immatures are more generic but easily recognized. The head is large and longish, and the bill somewhat conical—wide at the base, pointy at the tip. Overall a pointy-faced bird. The body is plump, bordering on portly; the tail is comically short and often slightly up-cocked (like Chestnut-sided Warbler). The yellow legs (shanks) and feet are often difficult to note on fall birds.

As one of three black-and-white warblers, breeding male Blackpoll is distinguished by the combination of a black cap and a white face; note also the black whisker. The backs and breasts of females and immatures are greenish or yellow-green and streaked—adults have crisp black streaks that make them look pinstriped; young have blurry but discernible olive streaks. Two white wing-bars, set on dark wings, are standout features. In fall, the bright white undertail coverts contrast with the otherwise dingy or yellow-tinged body. Breeding adult males somewhat recall chickadees. Females and immatures resemble small compact female orioles.

BEHAVIOR: A sluggish, somewhat ungraceful warbler that likes to clomp around in the interior portion of trees, generally from midheight to the canopy; in fall, often forages lower. Sits for long periods, turning its head in jerks, plotting its next move. Hovers infrequently. Sometimes jerks its tail or flutters (more than flicks) its wings (often for balance) as it bumbles around like a warbler in a china shop. Seems somewhat indifferent to other warblers feeding around it—at times, in fall, seems fairly social with other migrating Blackpolls.

FLIGHT: In flight seems robust and chesty, with long, broad, pointy wings and a short heart-shaped tail. In fall, closely resembles Bay-breasted Warbler (also looks like a compact oriole), but shows more yellow on the front/breast and more white toward the tail; also, wing-bars are less showy. Flight is strong and somewhat tacking (but not as much as Yellow-rumped), with deep, stiff, caressing wingbeats and little or no gliding.

VOCALIZATIONS: Song is high, thin, and ethereal—a breathy "*See see see see see*" that starts with substance, grows frail, and disappears into thin air. Call is a classic "*chip*." Flight call is a short, high, buzzy trill, "*zeet*," similar to Yellow Warbler but usually higher and softer.

Cerulean Warbler, *Dendroica cerulea*

Canopy Blue (and Green)-backed Warbler

STATUS: Locally uncommon to fairly common, but by no means easy-to-see eastern breeder and migrant. **DISTRIBUTION:** Scattered throughout much of the e. U.S. and se. Canada (extreme s. Que. and se. Ont.), west to Minn., e. Neb., and e. Okla. (barely), south along the Appalachians to n. Ga. Shows extremely limited distribution in New England (absent in Maine, most of N.H., and all but nw. Vt.). Also absent in the Southeast coastal plain (from se. N.C. to Tex.), except for populations in nw. and ne. Miss. and s. Ark. Winters in South America. **HABITAT:** Prefers extensive, mature, broadleaf, deciduous forest, but may also occupy younger, second-growth, deciduous woodlands. Seems particularly partial to forest areas with a broken canopy or canopy with varied altitudinal strata. **COHABITANTS:** Red-eyed Vireo, Wood Thrush, Rose-breasted Grosbeak. **MOVEMENT/MIGRATION:** Spring migration from mid-April to late May, with a peak from late April to mid-May. Fall migration is very early: some birds depart in early July, and most birds are gone by September; peak migration is late July to late August. VI: 2.

DESCRIPTION: Methodical treetop specialist with whitish underparts and uniquely blue or greenish upperparts. Tiny and compact warbler (smaller than Blackburnian Warbler; larger than Blue-gray Gnatcatcher; slightly larger than Northern Parula), with a short thickish nub of a bill and a very short tail.

Adult males have distinctive pale blue upperparts, bright white underparts creased by a narrow dark breast-band, and streaked sides. On distant (high) birds, the breast-band may be difficult to see, but the head pattern showing a white face/throat and a sharply defined dark cape will be evident. Females are uniquely greenish blue above, whitish below, and, except for the absence of a narrow breast-band, patterned like males with two white wing-bars, streaked sides, and a more accentuated pale eyebrow that widens behind the eye. Immatures are like females, but more green-tinged above (bluish green rather than greenish blue) and yellowish, not creamy white, below. *Note:* Immature males show more blue above and are whiter below.

In shape and size, immatures most closely resemble the snub-billed Blackburnian Warbler, but Blackburnian is blackish and streaked above; young Cerulean is uniquely greenish and plain-backed. Also, the pale eyebrow of Blackburnian narrows behind the eye; Cerulean's eyebrow widens.

BEHAVIOR: For the most part, sticks to the leaves of the canopy and keeps its feet stuck to the branch. Forages by hopping in a deliberate halting fashion through the foliage (where it is frequently hidden), paying particular attention to the undersides of leaves. Males sing while foraging, often from the highest point in the tree or on an exposed dead twig. Cerulean works trees thoroughly, moving laterally to the tips of branches, then flies to the crown of another (and not necessarily the closest) tree. Although it does not hover or flycatch with any frequency, this species is not shy about making lengthy flights across open areas. On territory, appears to engage in more than its share of territorial disputes with other small canopy and subcanopy species (such as Blue-gray Gnatcatcher and American Redstart).

FLIGHT: Small and compact, with an extremely short tail and long pointy wings. (Profile shows more bird ahead of the wing than behind.) Males are very white below. In good light, the blue back of males and the incongruous green of females and immatures are apparent. In all plumages, the fanned tail (not commonly seen) shows a broken band of beaded white spots near the tip. Flight is rapid, surprisingly strong, and slightly bouncy, with a fairly lengthy series of wingbeats punctuated by a brief closed-wing glide.

VOCALIZATIONS: Song is a short, hurried, usually three-part series of short buzzy notes, with the last note louder and pushed: "*wr'r wr'r w'r ti'ti'ti tzEEE.*" Call is a sharp "*chip.*" Flight call is similar to Yellow and Blackpoll—a buzzy "*zeet.*"

PERTINENT PARTICULARS: While maintaining (and vigorously defending) individual territories, breeding Ceruleans appear to cluster, occupying some parts of the forest and avoiding others. Where you find one territorial male you are likely to find others within earshot.

Black-and-white Warbler, *Mniotilta varia*

Zebra Creeper

STATUS: Common breeding warbler of the northern and eastern forests of the United States and Canada. **DISTRIBUTION:** Breeds across boreal Canada from se. Yukon, sw. Northwest Territories, and ne. B.C. east across n. and cen. Alta., nw. and cen. Sask., cen. and s. Man., all but northernmost Ont., s. Que., the Maritimes, Nfld., and s. Lab. In the U.S., breeds in all states east of the Mississippi R., but absent on the southern coastal plain and in the agricultural belt of the Midwest. West of the Mississippi, breeds in e. Mont., n. and cen. N.D., n. and w. S.D., n. Minn., n. Neb., parts of Iowa, cen. and s. Mo., Ark., n. and cen. La., e. Okla., and ne. Tex. Winters coastally from s. N.C. to the Florida Panhandle (and all of the Florida Peninsula) and across se. Tex. south. Migrates primarily east of the Rockies. Rare but regular migrant in the w. U.S.

HABITAT: For breeding, favors mature and secondary deciduous woodlands, but occupies mixed coniferous-hardwood forests in the North and mature pine forests in parts of the South to south Texas. In migration, almost any woodland (or tree or bush) will serve. **COHABITANTS:** Hairy Woodpecker, Red-eyed Vireo, Blue Jay, Veery, American Redstart, Ovenbird. In migration, often found in mixed flocks. **MOVEMENT/MIGRATION:** Among the earlier returning spring warblers. Birds in the southern portion of the breeding range arrive in mid-March, and late migrants are still passing in early June; peak migration occurs in late April and May. In fall, some birds appear in late July, but the bulk of them move in late August through September; some birds linger into late October. VI: 3.

DESCRIPTION: A boldly striped black-and-white warbler that looks and moves like a nuthatch. The body is overall compact, the face long and pointy (creeperlike) with a long, thin, down-drooped bill, and the tail fairly short. Adult males are truly contrastingly black and white. Females and immatures are more subdued and more nearly black and gray and buff—but still distinctly patterned and still unmistakably Black-and-white Warblers.

BEHAVIOR: Active and animate. Clambers up and down tree trunks or along and around limbs with jerky, mechanical, nuthatchlike movements. Hops up and down limbs using a series of 180° investigative turns. Now you see the head . . . now the tail. Uses its long nut pick–shaped bill to pry in nooks and crannies. Generally restricts its feeding range to the middle levels, avoiding the forest floor and the canopy. Prefers stout limbs and branches, and forages most frequently in the interior portions of the tree. Joins other warblers on twigs and outer branches, however, and sometimes hovers and flycatches.

FLIGHT: Overall compact with short broad wings, but has a distinctly long pointy face that seems to be looking down. Very contrastingly patterned above (at a distance, melds into paler blue-gray uniformity), and very white below. The white tips on the outer tail feathers flash when the tail is fanned. Jerky or bouncy flight with a sputtery series of deep wingbeats (looks as though the bird is clapping its wings beneath the body) is followed by a terse skip/pause.

VOCALIZATIONS: Among the easiest warbler songs to recognize. Classic song is a leisurely series of monotonous, high, thin, squeaky, two-note phrases, "*weesee, weesee, weesee*," repeated five to eight times. On territory, also gives a more hurried and complex song, "*wee-seeseeseew'r-w'r-w'rchorychorycher*," that retains enough of the squeaky "*weesee*" quality and pattern to make it recognizable. Call is a flat, nasal, unmusical "*tchap*." Flight call is a short buzzy or hissing "*sss*."

PERTINENT PARTICULARS: Those unfamiliar with this bird who are focusing on plumage might conclude that Black-and-white Warbler can be confused with male Blackpoll or Black-throated Gray Warbler. Relax. Both Blackpolls and Black-throated Grays look and act like warblers—they have little pointy bills and long narrow tails, and they hop and flutter around on outer twigs. The bird that is more likely to be confused with Black-and-white Warbler is Yellow-throated Warbler, whose pseudo-nuthatch feeding habits, coupled with its somewhat pied plumage (and an appraisal that fails to take in the bright yellow throat), might lead an observer to the wrong conclusion.

Black-and-white Warbler is easily pish-piqued and is one of the species whose vocal reaction can be used to incite a mobbing action among more reticent warblers.

American Redstart, *Setophaga ruticilla*

Flash Dancer

STATUS: Common eastern and northern forest breeder; common eastern migrant; rare but regular vagrant along the West Coast. **DISTRIBUTION:** Breeds from se. Alaska across forested Canada, from the se. Yukon and n.-cen. and e. B.C. east to s. Lab., the Maritimes, and Nfld. In the U.S., breeds in every state east of the prairies (excluding much of the Atlantic coastal plain south of se. Va., all of Fla., and along the Gulf Coast). Also found in n. and e. Wash., parts of Ore. and Idaho, much of

Mont. and Wyo., ne. Utah, n.-cen. Colo., much of N.D., parts of S.D., and n. and e. Neb. Winters in the tropics, but a few birds are found in s. Fla. and coastal and e. Calif. **HABITAT:** Preferred nesting habitat is deciduous second growth or understory in mature forest, often near water and/or near the edge or in tree-fall openings. Also nests in mixed deciduous and coniferous woodlands, coniferous forest (in the West), bottomland swamps, riparian corridors, and willow and alder thickets. In migration, forages in almost all woodland and shrubby habitat. **COHABITANTS:** Red-eyed Vireo, Blue Jay, Veery, Wood Thrush, Yellow Warbler, Chestnut-sided Warbler, Black-and-white Warbler. **MOVEMENT/MIGRATION:** Both spring and fall migration are extremely protracted. Arrivals reach the southern United States in late March, with a peak in May. Second-year birds continue to migrate through early June. Fall migration begins in late July and continues into late October, with a peak in late August and September. VI: 3.

DESCRIPTION: A slender, active, acrobatic, and colorfully accented wisp of a warbler that dances through the forest understory. Medium-sized (the same size as Black-and-white Warbler, but radically different in shape), with a short, straight, wide bill, a small round head, a slender body, and a very long tail (which is often fanned).

Halloween-garbed (black and red-orange) adult males are unmistakable. Mostly grayish females and immatures show a splash of yellow (or orange) on the sides of the breast and conspicuous yellow panels in the tail when fanned (as it usually is). Folded wings and tail show a wink of yellow as well. In spring, second-year males show a spattering of black about the face and neck.

BEHAVIOR: In perpetual motion, American Redstart is active in the extreme, darting from branch to branch in the understory and on the woodland edge. Moves with wings drooped and half-open and the tail elevated and partially fanned, turning . . . turning . . . hopping to a higher branch, hoping to startle a moth, which it then pursues in an acrobatic sally (at times even tumbles down through the understory). Captures prey with an audible snap of the bill. Also hovers. Frequently forages with other redstarts (two birds flush better than one) and other warbler species in migration (as well as chickadees and titmice).

FLIGHT: Silhouette is distinctive. Overall very slender, with round wings and a long swollen-tipped tail. The adult male's black-and-orange plumage is distinctive. Females and immatures are gray above, pale below, and show a pale gray hood (most evident from below) and a black-tipped tail. The yellow panels in the tail are evident when birds bank; the yellow slash in the wing remains evident even when birds close their wings in flight. Flight is jerky and bouncy, tacking and yawing (twisty-turny). Wingbeats are given in a short series followed by a brief closed-wing glide.

VOCALIZATIONS: Highly variable song typically consists of a short series of sharp, high, sweet notes that change pattern at the end—for example, "*see see see see y'oo*" (the last note is down-slurped)—or may not vary at all: "*w'see w'see w'see w'see*" (no pattern change). Birds may repeat the same song or shift to a new one. Rather than fret about the variation, it's best to simply concentrate on the high clear quality, the simple pattern, the often down-slurped last note. The fact that the vocalist is singing from the understory is also helpful. Also helpful is that the song ends abruptly, almost emphatically. Call is a high, sharp, clear "*chip.*" Flight call is a breathy upsweeping "*zweet.*"

Prothonotary Warbler, *Protonotaria citrea*
Golden Swamp Warbler

STATUS: Common breeder. Winters in Central and South America. **DISTRIBUTION:** Breeds across much of the e. U.S. (also extreme se. Ont.). Breeds from Long Island to n. Fla. (absent in the Piedmont regions south to n. Ga. and along the immediate Gulf Coast). The northern limit of the bird's range extends to cen. Mich., s. Wisc., se. Minn., e. Iowa, w. Mo., se. Kans., cen. Okla., and e. and cen. Tex. Casual migrant in the West, New England, and the Maritimes. **HABITAT:** Hardwood swamps and wet woodlands without dense understory and with trees or nest boxes suitable for cavity-nesting. In migration, also occurs in dry woodlands, but prefers wet swamps or wooded shorelines. **COHABITANTS:** Wood Duck, Great Crested Flycatcher, Yellow-throated Warbler, Louisiana Waterthrush, Swainson's Warbler, Kentucky Warbler. **MOVEMENT/MIGRATION:** Entire population vacates the United States and Canada in winter. Early-spring and fall migrant. Spring migration from early March to mid-May, with a peak in March and April; fall from mid-July to October, with a peak in August and September. VI: 3.

DESCRIPTION: An animate mote of golden sunlight moving through dark swamps. Large (larger than Kentucky Warbler; smaller than Louisiana Waterthrush; the same size as Swainson's Warbler) and overall robust, almost nuthatch-shaped, with a large, neckless, wedge-shaped head, a portly body, and a short wide tail. Plumage is unmistakable. The male's head, neck, and body are rich saffron or orangy-yellow (females and immatures are bright yellow, with no orangy hues); bluish wings and tail, when fanned, show lavish amounts of white. Particularly on males, the long, straight, pointy, black bill and large black eye stand out against the plain all-yellow head. The white undertail coverts are distinctive but hardly necessary to distinguish this species.

BEHAVIOR: Forages by hopping on branches, stumps, trees, and the ground. Commonly forages low, over standing water, or along shorelines. Movements recall a ground-loving Black-and-white Warbler or a nuthatch. Not a contortionist. Shifts entire body when foraging (as opposed to leaning, reaching, or craning neck). Also forages (and vocalizes) high in the canopy. Nests in tree cavities (often in emergent stumps just above the water) as well as in bird boxes erected over standing (or vernal) water. Males vocalize while foraging, or they may sing from an open perch below the canopy. A vocal bird that responds well to pishing. Easily piqued by imitations of Barred Owl and screech-owls.

FLIGHT: Big, stocky, wide-bodied, big-headed, long-billed, broad-tailed warbler. Flight is strong, fast, and undulating, almost bounding. Wingbeats are heavy (for a warbler) and spare. In woodlands, moves nimbly and recklessly fast through vegetation. In contrast to the dark wings and back, the large golden head *glows*. In migration, seen from below, the short wide tail has a distinctive two-toned pattern—white at the base, dark at the tip.

VOCALIZATIONS: Song is a loud, clear, ringing series of like-sounding notes, "*tswee tSWEE tSWEE tSWEE tSWEE,*" with the first note in the series often softer or breathier. Most songs consist of four to eight notes, and sometimes up to a dozen. Call is a loud, emphatic, silver-toned "*tcheh*" that ends abruptly but without a hard consonant and projects the same sweet tonal quality as the notes of the bird's song. Flight call is like the softer opening note of the song.

PERTINENT PARTICULARS: This species is simply distinctive. If you see one, whether in its archetypical swampy habitat or out of habitat in migration, there is little chance of mistaking it. The robust size and shape, coupled with the all-yellow head and plain bluish wings, render it unmistakable.

Worm-eating Warbler, *Helmitheros verimorus*

Stripe-headed Taffy-colored Warbler

STATUS: Fairly common in its somewhat limited and fragmented range. **DISTRIBUTION:** This southern and eastern species breeds in cen. Mass., R.I., Conn., s. N.Y., N.J., much of Pa., Md., Del., n. and w. Va., W. Va., se. Ohio, s. Ind., s. Ill., much of Mo., much of Ark., n. La. (and extreme e. Tex.), coastal and w. N.C., w. S.C., n. Ga., the Florida Panhandle, Ky., much of Tenn., much of Ala., and Miss. Migrates east of its breeding range and also along the Tex. coast. Winters outside of the U.S. Casual in the West and the Maritimes. **HABITAT:** This ground-nesting species is partial to two distinct habitats. In the Appalachians, it is found in mountain laurel or rhododendron thickets on hillsides and steep slopes surrounded by mature deciduous forest. In mature coastal forests and mixed oak-pine barrens, whose terrain is decidedly flat, the bird selects dense understory. In migration, it likewise favors dense tangled interior understory.

COHABITANTS: White-eyed Vireo, Hooded Warbler, Swainson's Warbler. **MOVEMENT/MIGRATION:** Migrates fairly early in spring—by late March on the Gulf Coast, ahead of the great sweep of Canadian-zone breeders; most birds are on territory by early May. Fall migration begins very early in late July, peaks in August, and ends in mid-October. VI: 3.

DESCRIPTION: A conservatively garbed (by warbler standards) but distinctive warbler, Worm-eating Warbler combines the shape of a waterthrush, the acrobatic feeding habits of a chickadee, and a color that is almost unique. Medium-sized (smaller than Swainson's or Ovenbird; larger than White-eyed Vireo; the same size as Black-and-white Warbler and Hooded Warbler), with a stubby tail, prominent head, and large spikelike bill that make it appear pointy-faced, front-heavy, and somewhat like a waterthrush, except that Worm-eating Warbler has a flatter head and appears more neckless than a waterthrush.

The overall plain taffy-colored body (with warmer ochre tones on the head and breast) is virtually diagnostic. No other North American warbler boasts this combination. Swainson's Warbler is colder brown; female Black-throated Blue Warbler is overall olive. The bold black-and-buff-striped head pattern is an eye-catching standout feature. Males, females, and immatures are essentially similar.

BEHAVIOR: Worm-eating Warbler is a careful methodical feeder whose special talents include hanging upside down, chickadee-like, then probing deeply (often tenaciously) into and ripping apart dead-leaf clusters for the larval prey hidden within. On cold spring mornings, may forage in the canopy where sunlight strikes first—thrusting its long bill into last year's leaves or tasseled new growth. Moves by hopping along limbs, somewhat like Black-and-white Warbler. Uses its wings sparingly; for instance, while feeding, makes short hops instead of short or long flights.

Usually stays where it's thick and tangled—rarely moving higher than the understory and often foraging close to the ground (rarely on the ground). Often remains in the same small tangle for long periods without breaking into the open. Singing males generally vocalize from an exposed perch below the canopy, but may also sing while foraging.

In migration, Worm-eating Warblers are normally solitary. More indifferent than shy, birds grudgingly respond to persistent pishing, often after many other birds already incited to a mobbing action have begun to lose interest. Once activated, Worm-eating Warbler can often be coaxed to approach closely, particularly if the vegetation is thick, and may remain as long as you do, chipping tenaciously.

FLIGHT: Fairly robust and compact (hefty-looking), with a pointy face, a big chest, and a broad tail. The caramel color is idiosyncratic. Flight is direct, no-nonsense, and steady (not bouncy), with rapid wingbeats. When Worm-eating Warbler loses interest, it goes away. When it wants to be somewhere else, it takes a direct route.

VOCALIZATIONS: A rapid, brittle, almost angry trill that is faster and more insectlike than Chipping Sparrow. Songs begin and end abruptly and last about two seconds. Opening notes sometimes sound softer and breathier, and songs sometimes gain volume toward the end. Individual notes are distinct and clipped, but more compressed and less spaced or staccato than Chipping Sparrow. Call is a loud, flat, dry spit: "*chih.*" Flight call is a buzzy and doubled "*zeet/zeet.*"

PERTINENT PARTICULARS: The bird Worm-eating Warbler is most often confused with is Red-eyed Vireo, which is found in the same deciduous woodlands, is similarly sized and shaped, is somewhat similarly colored, and shares a distinctive striped head pattern. Nevertheless, these similarities between the two species are superficial. The head pattern is very different, and the color on the underparts (creamy on the vireo, taffy on the warbler) easily distinguishes them. But all you really need to know is how the two species apportion themselves in the forest. The vireo is a canopy species; the warbler likes to burrow in the understory. If you are looking at the bird's underparts, chances are it's the vireo.

Swainson's Warbler, *Limnothlypis swainsonii*

Plain Cane Warbler

STATUS: Uncommon southeastern breeder. **DISTRIBUTION:** Breeds in the coastal plain and Appalachians from se. Va., e. and w. N.C., s. w. W. Va., e. and extreme w. Ky., e. and w. Tenn., virtually all of S.C., Ga., the Florida Panhandle, Ala., Miss., La., Ark., extreme s. Mo., se. Okla., and e. Tex. **HABITAT:** In the coastal plain, associated with damp hardwood forest with dense understory (frequently cane) and an open forest floor characterized by heavy leaf litter but little to no herbaceous growth. In the Appalachians, associated with rhododendron and mountain laurel thickets. Also found in several other plant communities, but all share a similar habitat structure—canopy, dense understory, heavy leaf litter, and few plants cluttering the forest floor. **COHABITANTS:** Wood Thrush, Black-throated Blue Warbler (Appalachians), Worm-eating Warbler, Ovenbird, Hooded Warbler. **MOVEMENT/MIGRATION:** Entire breeding population vacates North America. Spring migration from mid-March to early May, with a peak in April; fall migration from mid-August to late October, with a peak from early September to mid-October. VI: 2.

DESCRIPTION: A drab but uniquely shaped warbler that boasts both the longest bill and the shortest tail. Medium-sized but overall stocky going on portly. The face is long and wedge-shaped—the product of a flat-sloping forehead and a uniquely long, straight, evenly tapered, and exceedingly pointy bill. The tail is short, barely a nub. Most closely resembles Worm-eating Warbler in shape, but is more compact, chesty, and distinctly front-heavy.

Plumage pattern most closely resembles Red-eyed Vireo—overall plain olive brown above, paler dingy olive, brownish gray below. Also has a simple vireo-like head pattern—ruddy brown crown, pale eyebrow, darkish eye-line. The bird is so drab that the pink legs, somewhat pinkish-based bill, and white undertail coverts become standout features. The expression is benevolent, kindly, benign.

Note: The ability to perceive the plumage subtleties of this totem to understatement depends on light conditions, which, given the dark dense habitat the birds favor, are rarely favorable. Rely on shape and voice.

BEHAVIOR: Virtually a disembodied voice. Males vocalize frequently, often from a high but unexposed perch or from deep within a brushy fortress. Often sits for extended periods singing from the same (hidden) perch, then shifts to another favored (hidden) perch as it makes the rounds within its large territory.

Forages low, most commonly on the ground or very low in the vegetation. Walks quickly or shuffles (doesn't exactly hop), using its long bill to root through the leaf litter. (Birds can sometimes be detected by the sound of their foraging.) The bird is erratic—changes direction often and quickly—and high-strung: the hind end seems to shiver as it forages. Responds poorly to most pishing, but will investigate chips.

FLIGHT: Looks like a pale waterthrush. Flight is direct, strong, fast, straight, and usually short. Wingbeats are rapid and mostly continuous, with a brief skip that gives the flight a slight jerkiness.

VOCALIZATIONS: Undeniably the bird's finest characteristic. Two-parted song is loud and assertive. Begins with two to four loud, ringing, slurred, down-sliding, well-spaced notes, "*SEER, Seer*" or "*SAY, SAY, Say*," followed by a loud, hurried, abbreviated, somewhat slurred "*SIST'rshrew*." Together, it's "*SEER, Seer, SIST'rshrew*." One colorful tour leader phonetically transcribes the song as "*oooh, oooh, stepped in pooh*."

The first part of the song recalls the beginning of Louisiana Waterthrush's song. The second part has also been phonetically rendered "*whip-poor-will*," a choice that approximates the cadence and pattern but not the ringing quality of the song. Call is a low, loud, somewhat squeaky "*chew*" or "*chi/ew*" that is more protracted than the terse spit of Worm-eating Warbler and not as flat.

Ovenbird, *Seiurus aurocapilla*

Spot-breasted Leaf Strutter

STATUS: Common and widespread breeder in northern and eastern forests. **DISTRIBUTION:** Range extends from Nfld. and the Maritimes to the se.

Yukon and e. B.C. (skirting the prairie regions). In the U.S., the range encompasses much of the forested land east of the Mississippi R. (except portions of the South below the Appalachians), extending west along the drainages of the Missouri and Mississippi Rivers. In winter, found in Fla. and the barrier islands of N.C. and along the Gulf Coast to s. Tex., but most birds retreat to the Caribbean, Mexico, and Central America. Very rare but regular migrant in the w. U.S. **HABITAT:** A bird of mature hardwood or mixed hardwood and coniferous forest, Ovenbird appears to require large forest tracts with an extensive unbroken forest canopy layer. Foraging and nesting almost completely on the forest floor, Ovenbird seems most attracted to drier upland tracts, particularly those with sparse undergrowth. During migration, also seeks out dry upland woodlands. **CO-HABITANTS:** Red-eyed Vireo, Blue Jay, American Redstart. **MOVEMENT/MIGRATION:** Spring migration from late March through May. In fall, some migrants arrive in coastal migrant traps as early as late July. Most birds migrate from early September to mid-October, continuing through late October. VI: 3.

DESCRIPTION: A plump, ground-hugging, dumpling-shaped warbler that is easily distinguished from waterthrushes and true thrushes (with which it shares only a superficial likeness) by shape, plumage, mannerisms, gait, and voice. Ovenbird, simply put, is very hard to misidentify.

Large warbler (larger than Worm-eating Warbler; smaller than any spot-breasted thrush; the same size as waterthrush), with a straight, somewhat thrushlike bill, a large round head, a portly body, long pink legs, and a short tail that is usually up-cocked, with wingtips drooped beside and below the tail. When perched, the bird seems to have its body angled down but the head raised and alert; when agitated, the head feathers are erect. When walking on the forest floor, the bird often leans forward in a crouch.

In all plumages, the upperparts are brown-olive (at close range, brownish with yellow highlights), and underparts are white overlaid with rows of beaded black streaks. The very large eye is accentuated by a bold white eye-ring. The bird seems to be staring, and the overall expression is startled or crazed. The handsome black-bordered orange crown patch is difficult to see if the bird is in the shadows—as it usually is.

BEHAVIOR: Very tame, and virtually a forest floor obligate, Ovenbird both feeds in and nests on the leaf litter of the forest floor. Walks slowly and steadily, with an upright posture and many a veer to the left and right; the head bobs, and the tail is moderately up-cocked and sometimes raised with an upward jerk. The movements seem stiff and somewhat robotic. Ovenbirds also walk along both logs and limbs, one foot in front of the other.

Despite their ground-hugging proclivities, singing males are often surprisingly high (over 20 ft.) and may remain on the same perch for extended periods. High or low, the preferred perch is a conspicuous branch, not far from the trunk, in a sapling or mature understory tree; Ovenbird does not sing from the canopy. Very tame, and often allows close approach—or may approach you.

FLIGHT: Round-headed, hunchbacked, and potbellied shape (rounder-looking than waterthrush), with the bill usually angled down. The bird's plain olive upperparts are almost idiosyncratic; the face, with its clownlike mask, shows a strong sense of pattern but at a distance birds seem pale-faced. The underparts show stark contrast between the striped chest and white belly (like immature Sharp-shinned Hawk). Flight is strong, direct, fast, and maneuverable in a perfunctory sort of way. Ovenbird may suggest a plump pale waterthrush but never seems as dark as a waterthrush.

VOCALIZATIONS: Song is a loud ringing (at close range, painfully so) report that rises in volume: "*Teacher, TEAcher, TEACHER, TEACHER! TEACHER!!!*" Also has a more complex flight song (often given at night) that sounds like a musical jumble but ends with "*teacher; teacher.*" Call is a loud, low, distinctive "*Chhup*" that drags a bit at the beginning and whose quality recalls a stone skipped across ice. When agitated, Ovenbird calls persistently. Notes are evenly spaced, but the volume, tone, and quality often vary.

PERTINENT PARTICULARS: Ovenbird is often slow to respond to pishing—in fact, it may be the last bird in a feeding flock to respond. But once activated, like the last guest to arrive at a party, they are persistent and the last to leave. When piqued, they commonly fly (often vertically) to a tree limb. Imitating the call of Ovenbird is often an effective way of rebooting a mobbing action after many or most birds have lost interest.

As mentioned, Ovenbird is sometimes confused with waterthrushes and true thrushes. Waterthrushes are long and pointy-faced, have a prominent white eye-line (not eye-ring), and bob their tails, not their heads. Thrushes are larger and more spotted than streaked, and they move very differently, hopping in short series and then stopping. Ovenbird just keeps walking along and very rarely hops.

Northern Waterthrush, *Seiurus noveboracensis*

Bog Waterthrush

STATUS: Common and widespread breeder across northern North America. For this and other reasons, the most likely of the two waterthrushes to be encountered. **DISTRIBUTION:** Breeds across both forested and barely forested portions of North America from w. and n. Alaska (south of the Brooks Range) east to Lab., Nfld., the Maritimes (absent along coastal Alaska except for the Seward Peninsula), B.C., and the Canadian prairies. In the U.S., breeds in ne. Wash., cen. Ore. (very locally), n. Idaho, w. Mont., nw. Wyo., n. N.D., n. Minn., n. Wisc., all but s. Mich., ne. Ohio, most of Pa., e. W. Va., nw. Va., w. Md., n. N.J., N.Y., and all of New England. Migrates primarily east of the Great Basin. Rare but regular farther west. In winter, retreats to the Neotropics but also occurs in s. Fla. **HABITAT:** As a breeder and migrant, favors dark wet areas with stagnant or slow-moving water and thick vegetation, including swamps, spruce bogs, lakeshores, and streams. In migration, also occurs in dry thickets and woodlands with dense understory, in well-vegetated suburban yards, and at the edge of cattail and phragmites stands adjacent to fresh and brackish water. **COHABITANTS:** Ruby-crowned Kinglet, Palm Warbler, Rusty Blackbird, Lincoln's Sparrow, White-throated Sparrow. **MOVEMENT/MIGRATION:** Migrates primarily in April and May; late July through October—in general, somewhat later than Louisiana Waterthrush. VI: 3.

DESCRIPTION: A streak-breasted, tannin-stained, brown shadow of a warbler that bobs its tail when it walks and "chips" incessantly when disturbed. Overall slender with a pointy face and a short tail. Closely resembles the similarly sized but bulkier, bigger-billed, and overall slightly colder-toned Louisiana Waterthrush.

Northern Waterthrush is uniformly charcoal brown above and buffy (looks tannin-stained) or whitish below. Underparts are overlaid with condensed rows of black beaded streaks. Louisiana Waterthrush is overall colder brown and more contrasting—gray-brown above, whiter and less densely streaked below. (Some adult Northern Waterthrushes, however, can be very white below.)

The bill of Northern Waterthrush is long and pointed but classically warbler-shaped. Louisiana's bill is larger and heavier at the base, seems more like an extension of the wedge-shaped face, and makes the head look something like a sloping doorstop.

Northern's pale eyebrow is finer, narrower, more evenly tapered fore and aft, and usually buffy. Louisiana's is whiter and crisper and widens behind the eye—sometimes showing an upturned kick or a blotch of white.

The throat of Northern Waterthrush is more streaked, but this mark is variable and difficult to see. Both birds may have a buffy wash on the flanks, but it is generally more contrasting and peachy on Louisiana. Both birds have pink legs, but Louisiana's are brighter.

BEHAVIOR: An active feeder that usually forages on wet ground near water but also forages in dry woodlands and low understory (unlike Louisiana Waterthrush, which rarely, if ever, does so). Walks (or runs) with the tail wagging up and down (not swinging side to side), picking up food from the ground or turning over leaves. Sometimes hops over obstacles and forages up and down fallen limbs. Occasionally flycatches. Despite its ground-hugging predilection, the male may sing from very high perches.

Responds quickly and aggressively to pishing and calls persistently. Not a flocking species, but it is not uncommon for several Northern Waterthrushes to occupy the same thicket or wet area during migration.

FLIGHT: Large, sturdy, dark, somewhat arc-backed or humpbacked warbler showing a pointing, projecting head, a broad, short, often slightly fanned tail, and broad pointy wings. Often appears to be looking down. Overall very dark (including dark underwings), showing paler (yellow-toned) underparts. Flight is heavy, strong, and direct or tacking, with deep, even, bouncy or bounding undulations. Fairly slow wingbeats are given in a short series (two to three beats) followed by a fairly protracted closed-wing glide. In migration, tends to be very aggressive, engaging and pursuing other warblers in flight.

VOCALIZATIONS: Loud, clear, whistled song begins with two or three choppy warm-up notes, then accelerates excitedly, drops in pitch, and ends abruptly; song is phonetically rendered: "*three three three twotwotwo oneone.*" Overall more emphatic and hurried than Louisiana. Call is a loud, ringing, assertive "*bweek*" or "*bwik*" with a micro-hesitation between the "b" and "w" and the hard "k"; slightly harder and sharper than Louisiana. Flight call is like Yellow Warbler—a breathy high-pitched "*tzeee*" or "*zwee*" but uttered with a rising inflection and held slightly longer.

PERTINENT PARTICULARS: Northern is the waterthrush more likely to be seen because of its wider range of habitats, its greater numbers, its greater distribution across North America, its longer migration period, and the fact that it migrates when birders are watching.

Louisiana Waterthrush, *Seiurus motacilla*

Stream Waterthrush

STATUS: Uncommon and restricted in both habitat (the limits of the eastern deciduous forest) and range (limited to the eastern half of the United States). **DISTRIBUTION:** Occurs in all states east of and bordering the Mississippi R. as well as e. Tex., e. and cen. Okla., e. Kans., e. Neb., and se. Ont. Absent in all but se. Minn., all but sw. Wisc., all but s. Mich., all but e. and se. Ohio, nw. N.Y., n. Vt., n. N.H., and all but s. Maine, and also absent along the southern coast from se. N.C. to Tex. In Fla., breeds only in the central panhandle. Winters mostly south of the U.S., but a few birds appear regularly in se. Ariz. Accidental farther west. **HABITAT:** Except during migration, Louisiana Waterthrush is almost always associated with clean flowing freshwater rivers and streams in deciduous woodlands. Late in the breeding season (when water flow may diminish) and in migration, also occurs in stagnant pools, along marshy edges, and in low wet areas in woodlands and suburban yards. **COHABITANTS:** Barred Owl, Acadian Flycatcher, Prothonotary Warbler. **MOVEMENT/MIGRATION:** Very early-spring migrant; some birds reach breeding areas in mid-March, and most birds arrive by late April. Fall migration is likewise early and short. Birds depart in late July, and most have gone by late August. (October encounters are very rare.) VI: 2.

DESCRIPTION: A large, stocky, conservatively garbed warbler that doesn't resemble a thrush, but doesn't exactly resemble most warblers either. The body is compact, beefy, and somewhat front-heavy owing in large part to the very prominent bill, the heavy chest, and the short wide tail (but also to the bird's low-in-the-front-high-in-the-rear posture). On some individuals (especially males), the bill and head recall a sloping doorstop.

Upperparts are overall cold brown; the head and back have a slightly grayish cast, and the wings and tail are browner. The white underparts are overlaid with neat, widely spaced, tastefully arranged, beaded, dark brown streaks that are not as densely packed as on Northern Waterthrush. The throat is generally unstreaked, and flanks are peachy buff. At a distance, the underparts are two-toned—white in the front, darker to the rear. (Northern Waterthrush seems more uniformly yellowish or whitish below.)

The bird's classic field mark is the broad white stripe over the eye that widens and whitens behind the eye, often flaring up, sometimes ending in a thickened blotch.

BEHAVIOR: The whole back end of the bird swings rhythmically as it walks—an idiosyncratic behavioral signature shared only by Northern Waterthrush (see "Pertinent Particulars"). Forages rapidly on the ground along streambanks or on river rocks, jabbing for insects and lifting leaves out of water. Generally walks, sometimes sidesteps, along branches. Almost never stationary. Even when not foraging, the bird bobs with the single-mindedness of a metronome.

Usually solitary, although mated pairs may forage together. Often sings from an exposed limb overhanging a stream (usually less than 25 ft. high). Also sings while foraging.

Overall fairly shy. Waterthrush ranges up and down the stream; in fact, this bird appears to limit its world and movements to the space between opposing streambanks. Often, birds are merely glimpsed in flight. Given the cloaking backdrop of sunlight on water, birds may not stand out, and observers are often notified of the bird's presence only by its distinctive call note.

FLIGHT: Big sturdy warbler. Shaped like Northern Waterthrush, but bigger-headed, bigger-billed, and chestier. Shows a more contrasting face pattern (the white eyebrow stands out) or just appears whiter in front. Underparts are not as contrasting or pale as Ovenbird, and not as overall dark as Northern Waterthrush. Flight is fast, powerful, and generally direct when close to the surface of the stream, more undulating when in the open.

VOCALIZATIONS: Song is loud, ringing, and theatrical. Begins with two to five high, clear, slurred, measured whistles, then collapses into a rapid muttered series (as if the bird has forgotten the rest of the song) or degenerates into a stuttering series of chips and sputters (as if the bird is angry that it forgot the rest of the song). One phonetic rendering is "*Chee Chee Chee cheetittiwee*" or "*Chee Chee Chee (one too) many.*" Call is a loud, brittle, emphatic "*t'ch*" or "*p'tch!*" that is not as sharp or hard-sounding as Northern. When irritated, birds call repeatedly.

PERTINENT PARTICULARS: Both Northern and Louisiana Waterthrushes wag or bob their tails. Northern Waterthrush is an enthusiastic up-and-down tail-wagger; Louisiana swings its posterior in an exaggerated circle, a movement that replicates the swinging posterior of a kneeling camel. Louisiana is also more leisurely, bobbing with a slower cadence than Northern.

Kentucky Warbler, *Oporornis formosus*
Masked Forest Warbler

STATUS: Fairly common, but more easily heard than seen. **DISTRIBUTION:** Breeds across the se. U.S. The western limits of its range fall across se. Neb., e. Kans., e. Okla., and ne. Tex.. The northern limits are defined by e. Iowa and s. Wisc., n. Ill., cen. Ind., n. Ohio, cen. Pa., n. N.J., extreme se. N.Y., and sw. Conn. (Absent in coastal S.C., Ga., the Florida Peninsula, and, except for portions of the Florida Panhandle, along the Gulf Coast.) Winters in the Neotropics. Migrates east of its breeding areas (but including the Gulf Coast). Casual in the West, New England, and the Maritimes. **HABITAT:** Mature, moist, unbroken hardwood forest with dense (especially herbaceous) understory. Proximity to a stream is the preferred habitat over much of the bird's breeding range, but streams are optional. Thick ground cover is mandatory to conceal the on- or near-ground nest. In places where Kentucky Warbler was once common, the denuding of the forest understory by an overabundant population of Whitetail Deer has had a great impact on their numbers. **COHABITANTS:** Barred Owl, Acadian Flycatcher, Ovenbird, Louisiana Waterthrush, Hooded Warbler. **MOVEMENT/MIGRATION:** Appears in Florida and Gulf Coast migrant traps as early as late March, but most birds return to breeding sites from April to mid-May (in the northern limits of their range). Departure begins in early August and is generally over by October. VI: 3.

DESCRIPTION: A striking and easily recognized warbler (once seen), Kentucky Warbler is only medium-sized (larger than Common Yellowthroat; smaller than Ovenbird; the same size as Black-and-white Warbler), but even by the standards of *Oporornis* warblers, this bird is stocky and heavily built. The body is chesty but not portly—Kentucky is a warbler built to compete in the decathlon. The head is prominent, the bill stout, and the tail short and wide. The undertail coverts are full and long, extending almost the length of the tail (and impart the heavy body look). The bird's legs (not commonly a standout feature on warblers) are long and pinkish.

Unadorned olive above and *bright* yellow below, the bird has a signature characteristic: the combination of distinct yellow "spectacles" and a rakish black mask. The mask of first-year birds is more diffuse, but the yellow spectacles are prominent.

BEHAVIOR: Although not a skulker, Kentucky Warbler lives where vegetation is thick and has no particular inclination to be seen. Even when singing, males tend not to move and often choose perches that offer partial concealment, such as branches at middle heights in trees. Wherever found, Kentucky Warbler likes to stay on or near the ground and rarely forages above the understory into the canopy. For most of the year, usually forages on the ground—moving in short hops followed by a brief pause, rummaging through the leaf litter with its bill, sometimes leaping straight up to pick a spider or caterpillar from a leaf or limb.

The birds also move by short calculated hops through the low understory—climbing and dropping with spare (usually no) use of wings. Although not unique, the bird's posture is distinctive: it leans forward, like a runner in the starting blocks or a swimmer on the mark. Often the posture makes the tail appear modestly up-cocked. Working through the foliage, balanced on its long legs, Kentucky often appears precariously angled, cantered like an arboreal waterthrush. When agitated, even wags its tail like a waterthrush—but not so constantly and not so dramatically.

Usually solitary, Kentucky Warbler sometimes associates with resident species and other warbler species during migration. Except during the breeding season, its association with other Kentucky Warblers is probably serendipitous.

FLIGHT: In forest confines, flights are usually short and undulating. During longer flights, Kentucky shows itself to be a strong, mostly direct flier.

VOCALIZATIONS: The song is simple, distinctive, and, even late in the breeding season, frequently given throughout the day—a loud, rolling, repetitive, upward-sliding chant, each two-syllable phrase identical: "*Tor-EE, Tor-EE, Tor-EE*" (usually rendered three to five times). The song recalls Carolina Wren, but is louder, more ringing, and less chirpy, and the phrases seem to roll one into the other; the phrases in Carolina Wren's song are more distinctly clipped. Also, the song of Kentucky Warbler is unvarying—each set of phrases is identical to the last. Carolina Wren habitually alters its songs after several sets; furthermore, Carolina Wren doesn't commonly sing from deep within hardwood forests, as Kentucky Warbler does.

The cadence and volume of Ovenbird's "*TEA-cher, TEA-cher, TEA-cher*" song may suggest Kentucky Warbler, but the quality is different, the volume of the Ovenbird song rises (Kentucky Warbler's song stays the same), and the emphasis in the Ovenbird's song falls on the first note, not the last.

The call note of Kentucky Warbler has something of the character of the song: a loud, rich, low-pitched, emphatic "*Churp*" or "*Chur*"; also makes a less frequently heard and higher-pitched "*Chit.*" When birds are highly agitated, they make a high-pitched "*tee.*" This "*tee*" call (heard when you have intruded too close to the bird's nest) may be repeated in sequence or spaced between the more familiar "*Churp.*"

PERTINENT PARTICULARS: Seen poorly, Kentucky Warbler might be confused with another *Oporornis* warbler, but one look at the face pattern changes that impression. A first-year male Common Yellowthroat has plumage characteristics that are similar to Kentucky Warbler (particularly an immature Kentucky Warbler), and during migration, both might occur in the same thick growth. But the yellowthroat is a slight, slim, round-featured wisp of a warbler; Kentucky is a big, stocky, full-sized warbler with all-yellow underparts. The yellow on Common Yellowthroat is restricted to the throat and the (much shorter) undertail coverts, and it doesn't extend to the distinctive yellow spectacles. Kentucky Warbler has them; Common Yellowthroats don't.

Connecticut Warbler, *Oporornis agilis*
The Anomalous Warbler

STATUS: Functionally uncommon, with a limited breeding range and a secretive nature, but in fact more common than common knowledge might have it. **DISTRIBUTION:** Breeds in a constricted band of boreal forest from e.-cen. B.C. across cen. Alta., cen. Sask., Man., s. Ont., ne. Minn., n. Wisc., the Upper Peninsula of Mich., and s. Ont. to the center of Lake Huron, then north into ne. Que. In spring, migrates to Fla. from South America via the West Indies and travels northwest from there to the Great Lakes (almost entirely west of the Appalachians). Rare along the East Coast and west of the Mississippi River Valley. In fall, the movement is more easterly and coastal (actually extra-coastal—birds fly out over the Atlantic), with most birds passing through a geographic window bracketed by the Great Lakes and New England to the north and Va. and coastal N.C. to the south. Casual fall migrant in the West. **HABITAT:** Over most of its breeding range, shows a marked preference for wet or moist woodlands, including spruce-tamarack forest and bogs, muskeg, and poplar thickets. Partial to more open and mature forest without dense understory. In migration, favors low, often dense, often wet or damp woodlots, thickets, briar patches, and weedy tangles adjacent to woods or islands of trees. Patches of proper habitat need not be large to be attractive; in proper habitat, this species regularly turns up in city parks and suburban yards. **COHABITANTS:** In summer, Yellow-bellied Flycatcher, Boreal Chickadee, Nashville Warbler. **MOVEMENT/MIGRATION:** Late migrant in both spring and fall. Running one to two weeks behind the flood of most warblers, Connecticut Warblers arrive in Florida from early to late May and reach breeding areas in late May to mid-June. Fall migration is from late August to mid-October, with a peak in late September. VI: 3.

DESCRIPTION: A big, bulky, funny-looking warbler. The head seems too slim and the tail too short for this potbellied *Oporornis.* The head especially seems ill cast—excruciatingly long, narrow, and pointy. The head coupled with the dark gray or brown hood and the distinctive, round, white, no-nonsense, complete, and *staring* eye-ring makes all other characteristics almost superfluous.

But worth relating anyway. The upperparts are dull green (appear olive in poor light) and the underparts green-yellow (appear dingy yellow in poor light). The yellow undertail coverts extend almost the length of the tail (enhancing the overall bulk of the bird). The bubblegum pink legs and overlarge feet add to the bird's improbability. All ages and sexes are fundamentally and recognizably similar. The expression is anxious, startled.

BEHAVIOR: A sluggish ground-hugging warbler that walks with a deliberate Ovenbird-like gait—a trait that easily distinguishes this species from the branch-hopping Mourning Warbler and MacGillivray's Warbler and the much smaller but similarly patterned Nashville Warbler (also immature Yellow Warbler). Bounces or tail-bobs as it walks, and often walks along branches, but this species also makes immobility an art form. Often sits motionless, sometimes for minutes on end, often astride heavy branches, not across the branch, as most birds sit. Does not wing-flick, but when nervous, moves its head in stiff scrutinizing jerks and reverses its position on the branch in abrupt 180° jumps.

Not social. Likes to forage alone in thick tangles on or close to the ground. Not very responsive to pishing, but can sometimes be coaxed into view by imitations of a low soft Eastern Screech-Owl.

FLIGHT: In flight, shucks much of its ungainliness. Overall large (between Blackpoll and a waterthrush), hang-bellied, and symmetrically proportioned, with a big head, long wings, and a short tail. Looks somewhat thrushlike. Flight is strong and somewhat undulating and weaving. Wingbeats are given in a quick series (two to three beats), followed by a closed-wing glide; much more casual than the multi-sputtering wingbeats of most warblers. The dark hood and yellow underparts stand out.

VOCALIZATIONS: Song is a loud rich series of phrases that gain in volume and tempo and have a halting or tripping delivery; for example: "*wr'cheetoo wr'cheetoo wr'cheetoo w'rcheet*" (phonetically: "*churpety churpety churpety chrup*"). *Note:* The notes are not trilled. Call, described as a nasal "*chimp*," is rare. Flight call (heard frequently) is a buzzy "*zeet*" similar to Blackpoll, Bay-breasted, and Yellow Warblers.

PERTINENT PARTICULARS: This is an aspired-to species. It is easy to turn other species into Connecticut; it is difficult to second-guess a real Connecticut once you see one.

Mourning Warbler, *Oporornis philadelphia*
Warbler with a Six-foot Ceiling

STATUS: Fairly common but by no means easy to see (especially in fall). **DISTRIBUTION:** Breeds from ne. B.C. to n. and cen. Alta., nw., cen., and se. Sask., s. Man., s. Ont., s. Que., se. Lab., Nfld., N.B., and N.S. In the U.S., found in extreme n. N.D., n. and e.-cen. Minn., most of Wisc., Mich., n. Ill., n. Ind., ne. Ohio, n. Pa., e. W. Va., w. Va., most of N.Y., Vt., n. N.H., and all but coastal Maine. Winters outside the U.S. and Canada. Casual fall migrant in the West. **HABITAT:** Dense thickets. Partial to young regenerating deciduous woodlands (burned or logged areas) less than 6 ft. in height, brushy clearings, streamside thickets, and raspberry thickets. In migration, found in dense weedy tangles along woodland edge as well as in brushy understory within woodlands. **COHABITANTS:** Yellow Warbler, Chestnut-sided Warbler, Common Yellowthroat. **MOVEMENT/MIGRATION:** A circum-Gulf migrant that arrives late in spring and departs early in fall. Spring migration from early May to mid-June; fall from late July to late September, with a peak in early September. Migration is conducted primarily west of the Appalachians. VI: 3.

DESCRIPTION: A robust-looking Common Yellowthroat–like bird that seems never to be more than a few feet from the ground. Medium-sized warbler (slightly larger than Common Yellowthroat; smaller than Connecticut; the same size as MacGillivray's), with a fairly hefty bill, a large head, a grayish hood (not a mask), a long body, and a short tail. Overall stocky and more classically warbler-shaped than Common Yellowthroat; Mourning shows a head, body, and tail, whereas the structure of Common Yellowthroat is generally less filled out and seems something of a larval form.

In all plumages, olive green above, bright yellow below. Adult males have a distinctive dark gray hood with blackish trim along the lower edge. Females show a crisp complete gray hood with no black trim. Immatures have a grayish head, a shadowy poorly defined bib with a broken border, and a yellow core. (By comparison, the throat of immature MacGillivray's is usually whitish or grayish.) Most Mourning adults show no eye-ring (as is typical of Connecticut) or bold white eye-arcs (like MacGillivray's). Immature Mournings show a narrow, barely broken, circular white eye-ring that is not as prominent as the eye-arcs of MacGillivray's but more crisply defined than the narrow pale eye-ring of young Common Yellowthroat.

Compared to immature Mourning, Common Yellowthroat is browner above, with less yellow below (limited to the throat and undertail, not the belly). Connecticut Warbler is bigger and more distinctly bibbed, with a complete eye-ring that makes it look like it's staring. For comparison with MacGillivray's Warbler, see "Description" under that species.

BEHAVIOR: Mourning Warbler really doesn't like to be in the open. It forages low, in dense cover, and sings from cover, starting low and working its way up to higher branches (which may sometimes be atop tall mature trees). Hops when it feeds, both on branches and on the ground. In fall migration, seems more likely to be found in rank weedy growth (often along woodland edge). Mourning is a solitary bird that does not flock with other migrants. Does not generally respond well to pishing (at the very least, doesn't usually show itself), but is more responsive on territory and in the fall.

FLIGHT: Appears long-winged and short-tailed. Flight is rapid, bouncy, normally short, and close to the ground, but it can also be very nimble, particularly when Mourning is chasing other birds on territory or in migration. Interestingly, Mourning appears not to engage in "morning flight"—a postdawn relocation (or reorientation) behavior evidenced by many migrating warblers (including Connecticut). Neither, for that matter, does Common Yellowthroat.

VOCALIZATIONS: Song is a loud, rich, ringing, rhythmic, two-part chant. The first part has three to four identical two-noted phrases; the final phrases are lower and more hurried or mumbled: "*chur'ry chur'ry chur'ry tori tor.*" The phrases are also often slightly trilled. Mourning's song is a dub-in favorite among TV and radio advertisers. If you have ever seen or heard ads for lawn care, gardening, etc., you have probably heard this bird's song in the background. Call is a terse dry spit, "*p'chit,*" with something of a wren or yellowthroat quality.

PERTINENT PARTICULARS: Until you see it, you might be concerned about separating Mourning from Connecticut. The reality is that you are more likely to confuse Mourning Warbler with female Common Yellowthroat.

MacGillivray's Warbler, *Oporornis tolmiei*
Eye-arcus Warbler

STATUS: Common western breeder and migrant. **DISTRIBUTION:** Breeds through the Rocky Mts. and other western ranges from extreme se. Yukon and se. Alaska south through all but ne. B.C., sw. Alta., Wash., Idaho, w. and cen. Mont., Ore., n. and cen. Calif., w. and n. Nev., Utah, Wyo. (including the Black Hills in sw. S.D.), and w. and cen. Colo. Also found in scattered portions of s. Calif., Ariz., and N.M. **HABITAT:** Breeds in a variety of densely vegetated mountain habitats at low and moderate elevations, including young stands of conifers in clear-cut areas, mixed coniferous-deciduous forest with a dense shrub base, riparian thickets and open forest clearings dominated by low shrubs, and regenerating clear-cuts and power-line right-of-ways. In migration, favors dense brushy and shaded areas, but seems able to accommodate almost any wooded or vegetated habitat, including agricultural land, suburban settings, and urban parks. **COHABITANTS:** Nashville Warbler, Wilson's Warbler, Lincoln's Sparrow. **MOVEMENT/MIGRATION:** Spring migration from late March to early June; fall from late July to mid-October. VI: 3.

DESCRIPTION: A brush-loving gray-hooded warbler with distinctive white arcs set above and below the eye. Medium-sized (slightly larger and more robust than Common Yellowthroat; smaller than Connecticut; the same size as Mourning Warbler), with a straight sturdy bill, a sturdy body, long pinkish legs, and a fairly longish tail (long by the standards of *Oporornis* warblers and, more importantly, longer than Mourning Warbler's). Overall seems a bit more classically warbler-shaped than Mourning. Because it often perches with the tail slightly elevated, its posture is reminiscent of Common Yellowthroat.

Like Mourning, a handsome bird, showing olive greenish upperparts, bright yellow underparts, and a distinctive gray hood. Adult males show more black on the face and less black on the lower part of the bib than Mourning. Adult males in particular, but also females and immatures, have short, thick, white arcs above and below the eye. The eye-ring on Connecticut is always full, with no breaks. The eye-ring on immature Mourning (and some adult females) is narrow and hard to see, and you usually have to look hard to determine whether the ring is full or broken at the corners. On MacGillivray's, there's no second-guessing. You see the arcs. The ends don't meet.

Immatures are more overall greenish yellow and typically show a pale but well-defined grayish bib with a touch of off-white bleeding through the throat. The bib on immature Mourning is poorly defined—a vague outline with lots of yellow (not white) showing through or, in some cases, just a narrow, washed-out, grayish band across the chest.

BEHAVIOR: A circumspect thicket-loving bird that likes to forage low but in summer also seems much at home in tall trees. Adult males commonly sing at or near the tops of tall conifers, and not infrequently birds forage well up, mostly in the interior portions of trees. A surprisingly active, almost aggressive bird. Hops from branch to branch, plucking insects from leaves and bark; also hovers and occasionally flycatches. Birds feed across a large area or vertically, through a tall tree, moving quickly and sometimes throwing their tails from side to side as they move.

Not a social bird. Tends to forage alone. Unlike Mourning and Connecticut, responds readily to pishing, and when provoked, often remains in view.

FLIGHT: Appears small but long-tailed. The bright yellow underparts and dark bib are conspicuous. Flight is rapid, direct, undulating or bouncy, and commonly low, with something of the furtive quality of Common Yellowthroat.

VOCALIZATIONS: Loud, rolling, two-part song is much like Mourning Warbler, but slightly higher, reedier, and more variable. The first part is usually high, rhythmic, ringing, and slightly trilled, and the second part is warbled or hurried; for example: "*chre'e'e chre'e'e chre'e'e jur jur jur jur*" or "*chre'e'e chre'e'e chre'e'e jeejeeju.*" Calls include a low scuff or scrape, "*t'cheh*" or "*chrch,*" that recalls Common Yellowthroat but is louder and less muffled. Also makes a sharper buntinglike "*spit.*"

Common Yellowthroat, *Geothlypis trichas*

Masked Wren-Warbler

STATUS: Common and widespread species across most of North America. **DISTRIBUTION:** As a breeder, absent only across n. Canada, Alaska (except for the southeast part of the state), and drier portions of the American Southwest and intermountain West. Winters from coastal Va. and Calif. to the U.S.–Mexico border south into Mexico, Central America, the Caribbean, and northern South America. In migration, in proper habitat, found almost anywhere. **HABITAT:** At all times, favors low dense habitat—most commonly reedy marshes or weedy fields, but also forages and nests in woodland edge, even dry pine or mixed pine-hardwood forest characterized by dense understory. **COHABITANTS:** Marsh Wren, Yellow Warbler, Song Sparrow, Red-winged Blackbird. **MOVEMENT/MIGRATION:** Spring migration (fairly early) from April to mid-May; fall (protracted) from late July to November, with some birds lingering into December. VI: 2.

DESCRIPTION: Furtive, wrenlike, wren-shaped warbler. Masked adult (and immature) males are distinctive; females are plain; the utter nondescriptness of immature females virtually distinguishes them. Medium-sized (smaller than Mourning or MacGillivray's; larger than Nashville; the same size as Yellow), slender, and smoothly, somewhat generically or amorphously contoured (a warbler in larval form). The bill is small and usually straight; sometimes slightly downturned, it can make the bird appear glum. The tail is long, narrow, and round at the tip. Because it is often slightly elevated, it makes the bird's upperparts appear concave.

Except for the arresting "bandit's mask" face pattern of the adult male, the plumage is plain, almost featureless. Upperparts are drab olive brown, and underparts are dingy gray-brown and white with a bracketing splash of yellow on the throat and undertail. Immature males have bright yellow throats that stand out against the olive brown upperparts and an identifiable mask. Immature females may have whitish throats (more commonly a faint yellow wash), but they always show some yellow on the undertail. *Note:* As can be expected with so widely distributed a species, there are a dozen or so subspecies; subspecies plumages do not deviate significantly from the standard pattern, however, with the exception of birds found in the Southwest, whose underparts show more yellow.

BEHAVIOR: The bird is a skulker but an active skulker that can be easily followed—if not actually viewed—by watching the quivering vegetation. Likes to stay low, generally feeding less than 5 ft. above the ground and very frequently on the ground. Hops, sometimes flutters, from branch to branch; also sidles up and down branches. Singing males sometimes perch high—10–20 ft. high—if perches are available. They also put on a vocal flight display—an energetic, wing-fluttering, tail-pumping flight that sometimes describes a looping arch or may be vertical.

Responds readily to pishing, usually climbing high enough to investigate that observers can get a good view. Retreat is usually straight down into the tangle. Not a flocking bird, but during migration it is uncommon to find more than one bird in proper habitat.

FLIGHT: Appears slight, slender, short-winged, and long-tailed. Flight is jerky and usually short (bird heads for the nearest cover). Seems very wrenlike. During migration, rarely leaves cover.

VOCALIZATIONS: Song is a loud, rich, measured chant: "*Witch-a-tee, Witch-a-tee, Witch-a-tee.*" Also has a flight song and a dry rattle that recalls, but is less brittle than, Sedge Wren. Call is a low-pitched, atonal, somewhat rough "*tcheh*"—sometimes "*chhh*" or "*tidge.*" Also makes a low muttered spit, "*sp,*" which may be repeated.

PERTINENT PARTICULARS: During migration, female Common Yellowthroat is sometimes misidentified as the slightly larger, bulkier, and skulking Mourning Warbler. Mourning is more classically shaped (that is, more warblerlike) and has bright yellow underparts, including the belly (which on Yellowthroat is pale).

Hooded Warbler, *Wilsonia citrina*
Forest Knight

STATUS: Fairly common to common in parts of the South, but somewhat habitat-restricted. **DISTRIBUTION:** The nesting range describes the boundaries of the old eastern deciduous forests—defined by a line from n. Conn. and R.I. to the southern shore of Lake Ontario, the north side of Lake Erie, s. Mich., s. Wisc., w. Ill., s. Mo., se. Kans., e. Okla., and e. Tex. to the mouth of the Brazos R. (The bird is absent in much of cen. and e. Ill., n. Ind., and w. Ohio, as well as on the Florida Peninsula.) Winters south of the U.S. A trans-Gulf migrant, Hooded Warbler passes east of the prairies. Very rare migrant in the West, n. New England, and the Maritimes. **HABITAT:** Inhabits the dense understory of mature deciduous woodlands, particularly riparian woodlands, ravines, and understory dominated by mountain laurel or rhododendron. Understory plant species composition is not critical, but density is. In some places, also occupies pine forests—again, with a dense understory. In migration, also seeks out low dense cover. **COHABITANTS:** Acadian Flycatcher, White-eyed Vireo, Worm-eating Warbler, Kentucky Warbler. **MOVEMENT/MIGRATION:** A trans-Gulf migrant, the first Hooded Warblers arrive on the coasts of Texas and Louisiana in late March and are established in the North by mid-May. April and early May are the peak migratory period, and at times, particularly at Gulf Coast migrant traps, Hooded Warbler ranks among the most common migrants. Fall migration has a more easterly component. Some birds depart breeding grounds by mid to late July, but peak movements usually occur in late August and September. By mid-October, most birds are south of the United States. VI: 3.

DESCRIPTION: An animate golden knight with a black helmet (and an open visor). A medium-sized (larger than Wilson's Warbler; smaller than Ovenbird; the same size as Canada or Kentucky Warbler) but overall husky bird, Hooded Warbler sometimes appears bigger than it is. Shows a large oval-shaped head with a long sloping forehead, a straight bill that looks large in combination with the bird's black lores, a plump body, and a long tail.

Olive above and bright yellow below. Males, with their yellow face peering through the visor of an all-black helmet, are unmistakable. Helmetless females and immatures wear an open-throated olive hood that frames a bright yellow almond-shaped face and sets off the large black eye. The smaller but similarly plumaged Wilson's Warbler also shows a yellow face and a large black eye, but Wilson's is capped, not hooded, has a more petite bill, and is overall much slighter and differently shaped.

In all plumages, Hooded Warbler displays prominent white outer tail feathers (absent in Wilson's) that are difficult to overlook since the bird will probably be flashing them.

BEHAVIOR: Often difficult to see in its leafy confines (even singing males prefer to stay at least partially concealed), Hooded Warbler offers birders two solid advantages. When feeding, it's active and animate. On territory, males are very vocal, singing well into the day. Hooded Warbler forages by making fairly short assertive hops through the often dense understory, with wings drooped and breast puffed (enhancing the sense of huskiness). Habitually fans and flashes its tail with a downward jerk, exposing bright white tail panels. Snaps up startled prey with a sally flight or may pursue prey on the ground. When foraging, stays close to the ground (within 15 ft.). Singing males more commonly choose perches in the canopy that offer some concealment. Responds quickly and loudly to pishing, but is often loath to break into the open.

FLIGHT: Appears surprisingly trim (but not small) and long-tailed in flight. The yellow face, yellow underparts, and white-sided tail are conspicuous. In woodlands, flight is undulating and short in distance and duration.

VOCALIZATIONS: Belts out its loud, clear, whistled song: "*t'Wee t'Wee t'WeeWEEchu!*"—or as it is often transcribed: "*weeta-weeta-wee-teo.*" Notes are ringing and slurred, and the delivery overloud, almost theatrical. Songs are repeated after a lengthy pause and delivered at any time of day. Birds also sing a more whimsical, less assertive, more softly whistled "*Chi-whee t'whew-wee*" or "*Ci-whew t'whew-wee-weeto.*" Call is also distinct—a loud, round-toned, and lingering "*Chip*" or "*Tink*" that drags a bit at the beginning and has the ringing quality of the bird's song. Also has a variation on this call that is flatter-sounding and not so loud. Also makes a soft high-pitched "*chi*" when approached near its nest (similar to the call made by Kentucky Warbler in similar circumstances).

PERTINENT PARTICULARS: Immature female Hooded Warbler might be confused with Wilson's Warbler, which also has yellow underparts, an olive back, and a distinctly capped appearance. Wilson's is a bird of streamside thickets (particularly willows) and is distinctly smaller and smaller-billed. Immature Yellow Warbler might also be confused with Hooded, but has a small beady eye and, unlike Hooded Warbler, a penchant for being in the open. And of course, neither Wilson's nor Yellow Warbler has distinctive white outer tail feathers, nor are the tails of these species habitually fanned or wagged.

Wilson's Warbler, *Wilsonia pusilla*

Golden Willow-Tit

STATUS: Common northern montane and West Coast breeder; common migrant in the West; uncommon in the East; winters primarily in Mexico and Central America, but also along the Gulf Coast and in coastal California. **DISTRIBUTION:** Except for extreme northern tundra regions, se. Ont.–Que., and the prairies, found across Canada and Alaska; also found in n. and e. Maine, n. Vt., n. N.H., the Adirondacks of N.Y., and, in the West, Wash., Ore., n. and cen. Calif., Idaho, w. Mont., and portions of Nev., Utah, Wyo., Colo., and n. N.M. **HABITAT:** Breeds in dense shrub thickets located primarily at the edge of water—streams, bogs, beaver ponds, and lakes. Also occurs in recently logged forest that is going into early stages of succession and, along the Pacific Coast, in well-shrubbed forest openings. Loves willows and alders. Does not breed in canopy woodlands, except where openings and clearings are found. In migration, also selects thickets, particularly those near water, but may be found in other vegetation, including weedy growth and forest edge. **COHABITANTS:** Yellow, Mourning, and MacGillivray's Warblers, Song and Lincoln's Sparrows. In migration, often found in mixed warbler and chickadee flocks. **MOVEMENT/MIGRATION:** Entire population vacates breeding areas. In the East, spring migration from late April to early June; fall from late August to mid-October, with a peak in October. In the West, birds are earlier, arriving in mid or late March and departing as early as late July. VI: 2.

DESCRIPTION: A small, active, yellow sliver of a warbler that distinguishes itself from other yellow warblers by size, shape, behavior, and field marks (or the lack of them). Overall fairly tiny (smaller than Yellow Warbler and Common Yellowthroat; slightly larger than Nashville Warbler) and slender, but with soft, rounded, almost amorphous contours. (Yellow Warbler, by comparison, looks sharp and pointy.) The head seems disproportionately large and neckless, and the pale pinkish bill is petite and stubby, a combination that gives the bird a pinched-face look. The eye is large, round, and particularly conspicuous set against the plain olive or yellowish backdrop of the bird's face. (Yellow Warblers usually appear more beady-eyed.)

Olive yellow above and bright to golden yellow below. No wing-bars, no tail spots, and no distinguishing adornment except for a very distinctive black skullcap on males and some females. On many females and immatures, the cap shows as a shadowy rendering. Less obviously capped individuals are more easily identified by the yellow eyebrow sandwiched between the olive cap and olive or dusky cheeks; the pattern may assume the proportions of a yellow party mask.

BEHAVIOR: Moves with the frenetic energy and many of the mannerisms of a Bushtit. Always active. Never sits on a perch for long. Turns its body left, then right, in rapid jerks. Hops, turns, leaps for prey, sallies out to snatch insects in the air, hovers, occasionally hangs chickadee fashion from branches, and flies across open areas to pluck prey from the branches of another tree. Even when the bird is stationary, the tail is usually animate, twitching up or switching sideways; more than one observer has likened the bird to a gnatcatcher.

Almost always found in dense thickets, often willow thickets, commonly within a few feet of the ground. Sometimes feeds in the canopy or on the ground, but maneuvering through low dense brush in hops and short flights is the bird's forte. Tame, almost indifferent to observers, and easily coaxed close by pishing.

FLIGHT: Profile features a distinctly long, dark, slim tail. Overall patternless, with bright yellow underparts and dingy yellow upperparts. Flight is fairly slow, often low to the ground, and undulating in a jerky fashion; appears somewhat gnatcatcher-like or like a stone being skipped across water. Wingbeats are quick and mostly steady, but with momentary intermittent hesitations.

VOCALIZATIONS: Song is a loud uncomplicated series of sharply enunciated chattering notes that break pattern near the end, dropping into a slightly lower, slower variation or accelerating and becoming higher-pitched; for example: "*ch ch ch ch ch ch chrchrchr.*" The delivery, which is somewhat variable in length and delivery, is too slow to be confused with a trill and not musical enough to be likened to a warble. Notes are somewhat husky or stuttered, never sweet, sharp, or clear. Whatever the variation, the basic quality and pattern of the song is consistent. Call is a sharp soft "*tch*" or "*tchimp*" similar to western Winter Wren.

Canada Warbler, *Wilsonia canadensis*

Necklaced Warbler

STATUS: Relatively uncommon northern breeder; uncommon migrant. **DISTRIBUTION:** Breeds across much of s. and cen. Canada and the ne. U.S. Breeds from ne. B.C., n.-cen. Alta., cen. Sask., cen. and e. Man., most of Ont., s. and cen. Que., and the Maritimes. In the U.S., breeds in ne. Minn., Wisc., n. Ill., n. and w. Mich., e. Ohio, most of New England, most of N.Y., n. and Appalachian Pa., and nw. N.J., along the Appalachians to n. Ga. Winters south of the U.S. Casual migrant in the West. **HABITAT:** Most often found in extensive, moist, mixed coniferous-deciduous forest with a dense understory. Often favors habitat with uneven terrain or slopes. Forest types include cedar swamps, alder and willow stands along streams, deciduous woodlands with a birch understory, rhododendron thickets, and mixed coniferous-deciduous woodlands, especially those with a cedar and maple component. May also elect to be near streams. **COHABITANTS:** Yellow-bellied Flycatcher, Blue-headed Vireo, Veery, Ovenbird. **MOVEMENT/MIGRATION:** Migrates mostly east of the Great Plains. Spring migration is late—from late April to early June; fall is early and brief—from late July to late September. VI: 3.

DESCRIPTION: An active warbler of the forest understory. Medium-sized chesty or athletic-looking warbler (larger than Wilson's Warbler; smaller than Northern Waterthrush; about the same size as American Redstart), with a straight heavy bill, a large head, and a chesty body that tapers acutely toward the long tail. When the bird is perched, the tail and head are often elevated, giving it a U-shaped, somewhat Ovenbird-like look.

Distinctive and distinguished-looking with unadorned plain bluish gray upperparts and bright yellow underparts. A complete white eye-ring accentuates the overlarge eye and gives the bird a staring expression. The yellow lores of adults merge over the bill of young birds, giving them a yellow forehead. The black-beaded necklace (shaped like a false eyelash) is distinctive on males, and less distinct but still obvious on females; it shows as a shadowy gray or tie-dyed pattern on immature females. Overall, young birds are paler, less contrasting, and slightly brown-tinged above. The undertail is white in all plumages—but you may be too fixated on the front of the bird to notice.

BEHAVIOR: A hyperactive bird that feeds almost always within shaded forest, from the lower canopy to the forest floor. Moves quickly, hopping along and among branches and making short flights. Often shows an up-cocked tail (sometimes flicked) and slightly drooped wings (also flicked). Sallies from perches to pluck prey. Hovers often, and flycatches with a finesse and enthusiasm that approaches American Redstart standards. Social in migration, and often found feeding in mixed warbler flocks. Responds quickly to pishing by approaching close, wing-flicking, and shifting its body back and forth on the branch. On territory, commonly sings from a low perch within vegetation, and in fall migration sometimes sings a whispered rendering of its song.

FLIGHT: Appears chesty and long-tailed. Distinctly plain—uniformly gray above, yellow with white undertail below. Flight is strong and bouncy.

VOCALIZATIONS: Song is a loud hurried jumble of sharp, clear, whistled notes that climb up and down the scale. May recall a frantic wordy Magnolia Warbler. Happily, the song is almost always preceded by an emphatic "*tchip*," followed by a slight hesitation, and ends loudly and emphatically. Call is a low rough "*chi/uh*" that drags at the beginning and ends abruptly; it sounds like the introductory note of the bird's song and recalls Lincoln's Sparrow.

Red-faced Warbler, *Cardellina rubrifrons*
Red-faced Chickadee-Warbler

STATUS: Common, but restricted by geography and habitat. **DISTRIBUTION:** In the U.S., found only cen. and se. Ariz. and sw. and s.-cen. N.M. Also occurs in the mountains of Mexico and Central America (winter only). Casual vagrant to s. Calif.; accidental elsewhere. **HABITAT:** Mountain fir, pine, and pine-oak forests above 6,600 ft. In many places, seems particularly drawn to steep, often narrow, maple-rich canyons surrounded by pines. **COHABITANTS:** Greater Pewee, Olive Warbler, Grace's Warbler, Western Tanager, Hepatic Tanager. **MOVEMENT/MIGRATION:** All birds withdraw south of the U.S.–Mexico border in winter. Spring migration from late February to late May; fall from early August to late September, with a peak from August to mid-September. VI: 1.

DESCRIPTION: Unmistakable. A chickadee-like bird with a red face and a black cap with ear flaps. Plump and neckless, with a topknot on the crown and a pushed-in face. Overall roundly contoured (somewhat like Wilson's Warbler), and usually appears soft and fluffy. The bill is tiny, black, and chickadee-like, and the tail is long and animate.

Gray above, pale grayish white below, with a distinctly patterned head—red-faced, black-capped, and pale-naped. Birds also have whitish rumps. But for the purposes of identification, if all you notice is a red face, the rest is superfluous. Except for the pale collar/nape and pale rump (visible both in flight and when perched), note the complete absence of white above. Males and females are alike; immatures show less red on the face.

BEHAVIOR: As much at home poking into needle clusters at the tops of mature pines as it is hopping among the branches of streamside oaks and maples or low scrub oak. Nests on the ground, but does not commonly forage there. An active warbler, but not nervous or frenzied. Moves in short nonchalant hops on and through the needles and foliage of outer branches, pivoting left and right as a matter of course, with its slightly raised tail waving from side to side. Usually plucks food from vegetation, often with a deft upward snatch; sometimes hovers. Actively pursues flying insects. Usually forages alone, but may move in mixed-species flocks, in which it remains the only Red-faced Warbler.

Tame, and easily approached. Responds well to pishing and pygmy-owl imitations.

FLIGHT: Flight profile is slender and long-tailed. Plain gray upperparts show two points of light—one on the nape, one on the rump. Flight is straight, light, nimble, slightly wandering, and bouncy or slightly jerky. A quick fluttery series of wingbeats is followed by a brief partially open-winged glide.

VOCALIZATIONS: Short, variable, enthusiastic song consists of a series of high, sweet, clear, whistled notes that ramble along the scale. Most notes or phrases are not repeated, and songs begin and end abruptly; sounds polished and slightly hurried with a quality reminiscent of Yellow Warbler. Call is somewhat variable—usually a low dry "*chrt*" (may recall Canada Warbler), sometimes a sharper "*tick.*"

Painted Redstart, *Myioborus pictus*

Red-bellied Whitestart

STATUS: Restricted and localized, but where they occur, fairly common to common. **DISTRIBUTION:** Breeds in the mountains and foothills of s. Ariz. (where it also sometimes winters), w. N.M., and trans-Pecos Tex., south into Mexico and Central America, where, except in northern portions of its Mexican range, it's a permanent resident. Casual in s. Calif.; accidental elsewhere. **HABITAT:** In the United States found in shaded canyons and associated riparian woodlands dominated by oak and pine and characterized by a mature canopy, a healthy understory, and (usually) permanent or seasonal water. In Mexico, also occupies arid woodlands. **COHABITANTS:** Dusky-capped Flycatcher, Mexican Jay, Bridled Titmouse. **MOVEMENT/MIGRATION:** Most birds in the northern portion of the breeding range retreat to central Mexico and Central America during winter; a few birds remain in the southeastern Arizona canyons. Found in the United States mostly between April and late September; some birds linger into early October. VI: 2.

DESCRIPTION: A large, active, arboreal warbler whose boldly patterned black-red-and-white plumage and flash-dance antics make identification no less challenging than detection. The conspicuous white flashes emanating from the fanned wings and tail catch the eye. One look at the brilliant red belly nails the identification. Don't fail in your appreciation to note the curiously peaked head and half-eye-ring (or crescent), which gives the bird a sleepy expression. Males and females are identical. Very young birds lack the red belly (the trait is acquired by fall), but are otherwise like adults.

BEHAVIOR: Active, nimble, almost theatrical. Birds dance with rapid quick jerks and turns along branches in the canopy and subcanopy. Opened wings and tails flash with startling whiteness—a device designed to startle and flush insect prey. Plucks sedentary prey; sometimes hovers. Pursues flushed prey and snaps it up in the air. Males (and females) frequently vocalize as they forage (usually the first clue to the bird's presence). Forages from the ground to the canopy, clinging to trunks, dancing along, investigating small twigs and leaves, but seems most active in the midstory. Shows a distinct preference for oaks. Also attracted to hummingbird feeders.

Hunting is active and concentrated. Birds spend great amounts of time working a small area deep in the canopy, rarely changing trees. No doubt this technique is an adaptation to help defeat the eyes of potential predators—but it also defeats the eyes of birders.

FLIGHT: Looks like a small black bird throwing twin white contrails. Flight is nimble, quick, and bouncy; wingbeats are given in a fluttery series interspaced with open-winged glides.

VOCALIZATIONS: Song is a short, pleasing, generally two-part, warbled whistle; for example: "*weetoo weetoo weetoo tew tew tew*" or "*weet'la weet'la weet'la weetawe.*" Notes and phrases are clear, run-on (without sounding hurried), and commonly doubled or trebled. Often songs seem to rise (sometimes rise and fall) in volume or tempo and, despite their basic pattern, to ramble a bit. Call is a short, sweet, musical, and frequently repeated two-note phrase, "*turl-lee,*" that is also slightly slurred and trilled.

Yellow-breasted Chat, *Icteria virens*

The Raucous Polyglot

STATUS: Fairly uncommon to common nesting species; widespread but also widely scattered. Regular but very uncommon wintering species along the Atlantic Coast. **DISTRIBUTION:** A southern warbler that breeds across most of the U.S. and s. Canada in extreme s. B.C., Alta., Sask., and north of Lake Erie. In the U.S., found regularly in every state except the six New England states; largely absent as a breeder across all but s. Mich., s. Wisc., Minn., and w. N.D.; also occurs on the Florida Peninsula and in s. Tex. Occurs in small numbers in fall and early winter in New England and the Maritimes. **HABITAT:** Prefers a mix of open space and low dense brush devoid of canopy. Also favors early successional fields moving toward deciduous or coniferous woodland. Prime habitats include power-line cuts, forest edge, overgrown and unkempt yards, hedgerows in managed game lands, marshland edge, riparian woodlands, and willow thickets. Preferred vegetation includes briars, multiflora rose, cedar, willows, cottonwoods, salt cedar, and mesquite. **COHABITANTS:** Willow Flycatcher, Yellow Warbler, Prairie Warbler, Blue Grosbeak, Lazuli Bunting, Indigo Bunting. **MOVEMENT/MIGRATION:** Except for small numbers of individuals that linger along the Atlantic Coast north to the Maritimes and survive at least into early winter, the entire population retreats into coastal and southern Mexico and Central America in winter. Spring migration from mid-April to early June; fall from mid-August to October. VI: 2.

DESCRIPTION: A heavyset, overgrown, and somewhat buffoonish-looking warbler (?) with white "spectacles," the bill of a tanager, and a conspicuously long tail. Overall large and robust (smaller than Summer Tanager; larger than Blue Grosbeak; about the same size as Brown-headed Cowbird), with a big head, a heavy and slightly downturned bill, a long thick neck, a sturdy body, and a long narrow tail. Upperparts are wholly grayish olive (slightly grayer on the head); underparts show a bright yellow throat and breast giving way to a white belly. The bird has no wing-bars or tail spots. Often enough, all you'll see is a white spectacled face peering at you from a shrubby fortress.

BEHAVIOR: Both a histrionic showoff and a shy skulker. During breeding season, males vocalize both day and night (particularly on moonlit nights), singing from a high perch and calling attention to themselves with a creative and raucous cacophony of sounds. Yellow-breasted Chat also lofts into the air and parachutes to a lower perch with an exaggerated rowing of wings and a pumping tail. Outside breeding season, chats are surreptitious, slipping stealthily through thick brush and occasionally feeding on the ground. Flies reluctantly across open areas.

Generally solitary. Seems not to flock with other species. Response to pishing varies; very probably birds do respond but remain partially or wholly concealed, showing only (if you are lucky) the head and face and a peak of yellow breast. A better response is often gained by imitating the bird's song. Territorial birds can even be incited to perform aerial displays. Except during breeding season, mostly silent, but does call (albeit surreptitiously) all year.

FLIGHT: Profile shows a projecting head, a heavy body, and a long tail. The yellow on the breast and underwings is particularly evident. Flight is direct, low to the ground, and brief—generally to the next patch of cover. Wingbeats are steady, hurried, and strong.

VOCALIZATIONS: Song is a loud, unhurried, persistent, and varied ensemble of repeated and single-note utterances separated by a savoring pause. Repertoire includes mellow whistles, high squeaks, slurping gurgles, raspy admonitions, liquid burps, and a sprinkling of other bird calls (American Crow is a favorite); for example: "*cha-cha-cha-cha-cha* (pause) *whirt!* (pause) *Wok Wok Wok Wok* (pause) *Ur* (pause) *Ur* (pause) *Cah Cah Cah Cah Cah* (pause) *toot!* (pause) *chachacha*." Calls include a low growl, "*chowl*"; a soft hollow "*tok*" (sometimes doubled) that sounds like a soft tapping on wood; and a loud, buzzy, descending "*bzeert*"—a contagious call to which other chats within earshot respond.

TANAGERS

Hepatic Tanager, *Piranga flava*

Masked Tanager

STATUS: Fairly common but geographically restricted breeder; rare winter resident. **DISTRIBUTION:** Extensive range in the tropics. In the U.S., found primarily in se. Ariz., N.M., and trans-Pecos Tex.; also occurs in the mountains of San Bernardino Co., Calif., and in extreme se. Colo. Casual on the Calif. coast, and accidental east of the normal range. **HABITAT:** Pine and pine-oak woodlands with an open configuration and limited understory. Particularly in dry years, also found in lower riparian canyons with sycamores and cottonwoods. **COHABITANTS:** Pygmy Nuthatch, Grace's Warbler, Western Tanager. **MOVEMENT/MIGRATION:** Most U.S. birds vacate breeding territories after nesting. Spring migration from late March to early June; fall from late August through October. VI: 1.

DESCRIPTION: An angular soot-smudged Summer Tanager. Large (slightly larger than Summer), with a large, flat-topped, almost rectangular head (head of Summer Tanager is slightly peaked), a robust, very slightly downturned bill, a long body, and a short tail. The bill shows a larger "tooth" that makes birds look formidable.

The adult male is reddish but dingier than Summer, with muted grayed-down back, wings, and flanks. Females and immatures are like a muted female Summer Tanager, with a yellowish head and breast and grayish upperparts and flanks. *Key point:* Both males and females have a sooty gray mask across the face. In combination with the flat head and the formidably heavy-"toothed" bill, it gives the birds a calculating predatory expression. Summer Tanagers, especially females, look calmly benign.

BEHAVIOR: Likes the heights. Forages almost exclusively in the canopy, moving in short flights between branches or in longer flights between trees. Hunts methodically, calculatingly. Perches and searches, often crouching low to peer under foliage. Most commonly hunts alone or with mate; later in the summer, forages in small groups and mixed-species flocks. Responds to other birds engaged in a mobbing action and to pygmy-owl calls.

FLIGHT: Lanky long-winged profile (despite the short tail). In combination with the plumage pattern, the profile may recall Pine Grosbeak. Flight is flowing and undulating. Wingbeats are regular and measured (you can see individual strokes), and given in a long series followed by a closed-wing glide.

VOCALIZATIONS: Often described as robinlike. Slower and less musical than Summer Tanager, with short phrases and distinct pauses. Call is a low soft "*chuh*" or "*chuk*" that is often interjected into the song.

Summer Tanager, *Piranga rubra*

Cardinal Tanager

STATUS: Common, mostly southern and southwestern breeder. **DISTRIBUTION:** Breeds generally south of a line drawn from s. N.J. to se. Neb. (absent along the Appalachians to ne. Tenn. and in s. Fla). Also found across e. Kans., e. Okla., all but n. Tex., s. and cen. portions of N.M., much of Ariz., extreme sw. Utah, s. Nev., se. Calif., and n. Mexico. Winters in Mexico and in Central and South America; a few winter in Fla., Calif., and s. Tex. Very rare but regular overshoot north and west of its regular range. **HABITAT:** In the East, found in open deciduous and mixed deciduous-coniferous (primarily oak and pine) forest, often near an edge (such as a roadway or power-line cut). Prefers moderately tall trees with open canopy and relatively little forest diversity. By comparison, Scarlet Tanager prefers taller trees and a closed canopy. In the Southwest, found in riparian woodlands dominated by willows and cottonwoods (also found in mesquite and salt cedar). Hepatic Tanager prefers higher elevations and mixed pine-oak forest; nesting Western Tanager prefers higher elevations and coniferous forest. **COHABITANTS:** In the East, Yellow-throated Vireo, Yellow-throated Warbler; in the West, Bell's Vireo, Blue Grosbeak. **MOVEMENT/MIGRATION:** Entire population vacates breeding areas for winter. Spring migration from early March to early June, with a peak in April and early May; fall from early August to late November, with a peak in September and October. VI: 3.

DESCRIPTION: A longish, peak-headed, long-billed tanager whose song has an oriole-like quality. Large (larger than Scarlet and Western; slightly smaller and less robust than Hepatic), with a distinctly long, heavy, pale (or yellow), and very slightly downturned bill, a slightly peaked crown, a long narrow body, and a long tail.

Summer Tanager's bill seems almost as long as the head is wide. The bills of Scarlet and Western Tanagers are heavy but snubbed by comparison; Hepatic's bill is medium-long but proportionally thicker and dark (not pale). Scarlet Tanager's head is round, and Hepatic's is squarish (not peaked).

The adult male is all red all year-round—more cardinal than scarlet—with dingy but not jet-black wings. Females and immatures are overall somewhat mustardy yellow, showing little contrast between the wings and the back; a few show touches or patches of orange. Similarly colored Scarlet Tanager females and immatures have darker wings contrasting with more greenish yellow bodies. The plumage of both male and female Hepatic Tanagers is muted by gray in the back, wings, and flanks, and, most conspicuously, on the face (forming a shadowy mask that Summer Tanager lacks).

BEHAVIOR: A canopy lover and something of a bee and wasp specialist—commonly attends honeybee hives. Sallies out to snap up insects on the wing; also hovers and plucks insects and fruit from branches and gleans from leaves and bark. Fairly active. When hunting, often flies between perches rather than hop to a nearer (and presumably less desirable) branch. Habitually frequents the edge of forest openings (a bee-rich environment perhaps?) and frequently flies across open areas.

Males are sedentary singers and like to vocalize from high exposed perches. Except early in the season, singing is most concentrated at the beginning and end of the day. Males and females call off and on all day—often in response to an avian or human intruder or to pishing (to which it responds reasonably well). At dusk, both males and females commonly end the day with bouts of calling.

FLIGHT: Longish and heavy-billed profile. Flight is undulating and flowing and is executed with a lengthy series of wingbeats followed by a closed-wing glide.

VOCALIZATIONS: Whistled song is a hurried enthusiastic series of two- to four-note "caroled" phrases whose pattern and quality recall an oriole. Common song consists of three to ten sets of phrases, followed by a pause, then by another set. Scarlet Tanager's song is, by comparison, slower, hoarser, and more recited than sung. Summer's distinctive call is a low, hurried, three-note hiccup: "*picatuk.*"

Scarlet Tanager, *Piranga olivacea*

The Black-winged Red Bird

STATUS: Common breeder of northeastern and Appalachian forests. **DISTRIBUTION:** Breeds from the s. Maritimes west across s. Que. and s. Ont. to s. Man.; south through New England and the mid-Atlantic states to e. Va., cen. N.C. and S.C., n. Ga., n. Ala., ne. Miss., and w. Tenn. The western border falls across e. N.D., e. S.D., e. Neb., e. Kans., and e. Okla. (absent in s. and e. Ark.). Casual in the West. **HABITAT:** Occupies a mix of habitat, including mixed deciduous-coniferous and even pure stands of balsam, aspen, and eastern hemlock, but much prefers mature broadleaf forest where the canopy is high and dense and oaks dominate. **COHABITANTS:** Hairy Woodpecker, Eastern Wood-Pewee, Acadian Flycatcher, Red-eyed Vireo. **MOVEMENT/**

MIGRATION: Entire population retires to South America for the winter. Spring migration from late March to late May, with a peak in late April to early May; fall from late August to early November, with a peak from mid-September to mid-October. VI: 3.

DESCRIPTION: A beautiful treetop tanager of mature eastern forests. Smallish and overall compact (smaller than Summer Tanager; slightly smaller than Brown-headed Cowbird; larger than Tufted Titmouse), with a heavy but not long bill, a rounded head, a stocky roundly contoured body, and a short tail. The bill is pale and considerably shorter than the head is wide.

Breeding males, with boot-black wings and tails set against a wholly scarlet body, are virtually unmistakable. Females, immatures, and nonbreeding-plumage males have the same pattern, but the bodies are greenish yellow and the wings blackish to grayish green. Scarlet Tanager is distinguished from female and immature Summer Tanagers by its more compact structure, rounder head, and smaller bill, the greenish cast to the plumage, and the greater contrast between the dark wing and pale body. Scarlet Tanager can be separated from Western Tanager by its lack of bold white or yellowish wing-bars. *Note:* Some immature Scarlet Tanagers show very thin faint wing-bars. Undecided? Western Tanagers have grayish backs; Scarlet Tanager backs are greenish.

BEHAVIOR: A canopy species that specializes in plucking and hover-gleaning insects from leaves, branches, and trunks. Does flycatch, but less often than Summer Tanager. Very commonly in fall migration, forages on fruit, which it plucks and hover-plucks. During breeding season, commonly forages alone or with a mate. In migration, forages and flies in small flocks with other tanagers.

Males sing frequently throughout the day, often from exposed perches high in the canopy. Despite the bird's near Day-Glo plumage, it can be amazingly difficult to spot beneath a shadowy canopy.

Responds readily to pishing and screech-owl calls, but does not necessarily leave the canopy. Most common response is to call (and often males and females call to each other).

FLIGHT: Profile is fairly stocky, with a big head, a wedge-shaped face, a short tail, and long, wide, pointy wings. Breeding males are unmistakable; females and immatures are mostly green. Flight is heavy and undulating, with a bounce; a short, quick, regular series of wingbeats is followed by a fairly prolonged closed-wing glide. The bird's abrupt rise during the flapping sequence imparts the bounciness to the flight.

VOCALIZATIONS: Song is a series of five hoarse, rising-and-falling, two-note phrases aptly described as sounding like "a hoarse robin," but the tanager's song is less musical and more slurred. Next to Summer Tanager, Scarlet sounds tired, bored, unenthused. Call is a low admonishing "*chik-bur*," and often just "*chik*."

Western Tanager, *Piranga ludoviciana*

Bar-winged Tanager

STATUS: Common and widespread western breeder. **DISTRIBUTION:** Breeds from se. Alaska, n. B.C., se. Yukon, and the southwest corner of the Northwest Territories south across w. Canada and east to cen. Sask. In the U.S., occurs west of a line drawn from e. Mont., w. S.D., w. Neb., the western two-thirds of Colo. and N.M., and trans-Pecos Tex. (Absent as a breeder across portions of the Great Basin from cen. Wash. to n. Nev., se. Calif., and sw. Ariz.) Winters in Mexico (some birds in coastal Calif.). Very rare visitor to the East, in migration and at winter feeders. **HABITAT:** Fairly generalized. Found in a variety of open coniferous and mixed coniferous-deciduous woodlands (also Gamble oak and pinyon-juniper). Forests may be wet or dry, mature or second-growth, at altitudes ranging from sea level to nearly 10,000 ft. In migration, may be found anywhere there is woody vegetation, including riparian woodlands, scrub, urban and suburban parks, and desert mesquite. **COHABITANTS:** Varies over the bird's range, but includes Hammond's Flycatcher, Steller's Jay, Black-capped Chickadee, Mountain Chickadee, Yellow-rumped Warbler. **MOVEMENT/MIGRATION:** Virtually all birds leave breeding areas and migrate south into Mexico for the winter. Spring migration from late

March to early June; fall from late June to early November. VI: 3.

DESCRIPTION: A bird so garishly colorful that it seems painted by a child. Medium-small forest bird (smaller than a starling; very slightly larger than Scarlet Tanager), with a round head, a thick, smallish, pale, somewhat conical bill, a long but plump body, and a shortish tail. Shape is very similar to the eastern Scarlet Tanager; more compact and roundly contoured than Summer or Hepatic Tanager.

The adult male is a visual knockout, with black upperparts, yellow underparts, and a bright red head; the bold yellow and white wing-bars alone would be a standout feature on any other bird. In winter, the red on the head is duller and limited to the face; otherwise, plumage is similar.

Females and immatures are more subdued. The head and underparts are yellowish (sometimes with a grayer or whitish belly), the wings and tail are olive, and the back is gray (at a distance, the bird seems to be wearing a saddle). Wings show two prominent wing-bars (the lower one may be faint). No other western tanager (except the extremely rare Flame-colored Tanager) shows wing-bars. A few Scarlet Tanagers may show faint narrowish wing-bars, but Scarlet is uniformly olive-backed (Western wears a gray saddle), and Scarlet has a gray or horn-colored bill. Western's bill is frequently orange-tinged, and often the face shows traces of orange too.

BEHAVIOR: A versatile but methodical and calculating arboreal feeder. Generally stays in the upper half of trees, but occasionally forages in the tops of lower shrubs (particularly in migration). Forages by slow stop-and-go hopping, hunting and gleaning through the foliage; favors the higher and outer branches. Also extremely adept at flycatching and consumes fruit and berries. In North America, feeds alone, with mate, or, in late summer, in family groups.

Responds well to pishing and pygmy-owl calls. Common response is to perch at the very top of a 20–40 ft. conifer and call. Often a pair of birds perch together.

FLIGHT: Shaped like Scarlet Tanager, but very slightly lankier or perhaps longer-tailed (?). Shows a yellow rump sandwiched between black wings, a dark back, and a black tail. Short flights are slightly bouncy, with series of stiff wingbeats. Lands with braking open-winged glides. Longer flights are flowing, undulating, and not bouncy, combining a short slightly arrhythmic series of wingbeats with lengthier closed-wing glides.

VOCALIZATIONS: Song is a hoarse rising-and-falling ensemble of two- and three-note phrases. Sounds much like Scarlet Tanager, but may be slightly more musical and hurried. Unlike Scarlet, Western punctuates its song sets with its very distinctive signature call. Call is a quick, ascending, slightly gurgled, three-note upscale strum on a xylophone: "*dlr'd'leet.*"

SEEDEATERS, TOWHEES, SPARROWS, JUNCOS, AND LONGSPURS

White-collared Seedeater, *Sporophila torqueola*

Swollen-billed Grass-Finch

STATUS AND DISTRIBUTION: Uncommon resident and one of the most geographically restricted of U.S. birds, found only in the Rio Grande Valley in Starr and Zapata Counties and most commonly in the riverside towns of Zapata and San Ygnacio. Range extends south to Panama. **HABITAT:** Open grassy areas (most often near water), thickets, and broken riparian woodlands and brush with a tall-grass component. Also found in riverside reed beds. **COHABITANTS:** Common Yellowthroat, Olive Sparrow, Lincoln's Sparrow (winter). **MOVEMENT/MIGRATION:** Permanent resident. VI: 0.

DESCRIPTION: A thumb-sized (and -shaped) swollen-billed sparrow—males resemble a browner male Lesser Goldfinch; females look like female Varied Bunting. A small and roundly contoured finch (even smaller than Verdin or Lesser Goldfinch), with a round and neckless head, a swollen nub of a bill, a somewhat elongated oval-shaped body, and a short, stiff, round-tipped tail. Frequently sits with body mostly erect and tail angled or pressed forward.

Adult male's plumage pattern recalls male Lesser Goldfinch—right down to the small white check on the folded wing—but shows whitish or buffy (not bright yellow) underparts, a buffy collar (or half-collar) across the nape, and a dark band or necklace on the breast. Black and buff is not a common plumage combination and pretty nearly distinguishes the bird. If hesitant, note the black helmet, narrow white wing-bars, and contrasting all-black tail. Immature males show a patchy but similar pattern.

Females are overall pale buffy brown with slightly darker wings and two frail wing-bars; they resemble tiny swollen-billed buntings.

BEHAVIOR: An active, energetic, furtive bird that moves with the gymnastic movements and quicksilver reflexes of a Lesser Goldfinch. When agitated, flicks its wings and jerks its tail up with a movement that is almost too rapid to perceive. Habitually clings to vertical perches, but holds its body erect. Hops short distances and flies (fast!) to perches more than a foot away. Feeds mostly on the seed-heads of tall grass, which it plucks while clinging to the stalk (almost never seen on the ground except where birdseed is scattered). Also feeds on acacia blossoms (in March) and occasionally hawks insects.

Males vocalize from high exposed perches and sing throughout the day. Social outside of breeding season, gathering in small flocks (fewer than 10 birds) or with other seed-eating birds. Often roosts in tall riverside reeds. Does not normally visit bird feeders. Not particularly responsive to pishing, but will come in to squeaking.

FLIGHT: Overall tiny and compact. Males show white flashes in the wings (sometimes hard to see) and black tails that contrast with pale underparts. Females are just plain plain. Wingbeats are snappy and rapid, approaching the speed of blur, and given in a quick series followed by an imperceptibly brief closed-wing glide. Flight is quick, nimble, and bouncy (but not as bouncy as Lesser Goldfinch).

VOCALIZATIONS: Variable song commonly has two to four parts. Usually begins with a soft, quick, abbreviated note (or two), followed by several louder, clear, whistled notes, followed by a stuttering trailing-off trill: "*wee wee chu chu chu chchch-chch.*" Sounds very buntinglike. Calls include a clear, plaintive, descending "*cheu*" (recalls Lesser Goldfinch or Red-winged Blackbird) and a low soft "*cheh*" (recalls House Sparrow's "churp").

Olive Sparrow, *Arremonops rufivirgatus*

Plump and Plain Sparrow

STATUS AND DISTRIBUTION: Common resident of s. Tex. north to just north of Rockport and Del Rio and just south of San Antonio. Also found in Mexico and Central America. **HABITAT:** Dense thorn-scrub woodlands, overgrown agricultural fields, riparian habitat with dense understory, and borders of fields and roadsides adjacent to brushy habitat. **COHABITANTS:** White-tipped Dove, Long-billed Thrasher, Lincoln's Sparrow (winter). **MOVEMENT/MIGRATION:** Permanent resident. VI: 0.

DESCRIPTION: A pudgy, cryptic, fairly undistinguished sparrow that forages slowly on the ground beneath a thorny fortress. Medium-sized (larger than Lincoln's Sparrow; smaller than Green-tailed Towhee; about the same size as Song Sparrow), with a large neckless head, a fairly long, narrow, pale bill, a plump body, and a fairly long tail that is slightly up-cocked when bird is feeding. The shape is almost unique among North American sparrows and is most akin to Rufous-crowned.

Overall bland with dull greenish-washed upperparts and a pale grayish head and underparts. The throat and breast are slightly paler but not distinctly white. The head pattern of pale brown stripes is surprisingly easy to see—largely because the rest of the bird is so plain. Plumage somewhat recalls a washed-out Green-tailed Towhee—another brush-loving species but a larger, more energetic bird (relatively speaking) with a distinctive all-rusty cap and contrasting white throat.

BEHAVIOR: A sluggish ground-feeding sparrow most often seen foraging (with little concern for you) on open ground near the edge of a thicket. Very commonly forages in pairs. Makes slow, short, single, and somewhat high hops. Never found more than a few yards from cover. When foraging in the open (along the edge of trails), hops more quickly and often in series (mostly to get back under cover). Seems addicted to bird-feeding stations, where it forages on seed spilled on the ground.

Males usually sing within the underbrush but occasionally take higher perches. Fairly unresponsive to pishing, but will vocalize, giving a high "*tip*" note.

FLIGHT: Slightly plump and longish-tailed silhouette; plumage is uniformly dull and fairly pale but in good light shows a greenish cast. Short flights are bouncy. Longer flights are straight, generally low to the ground, and surprisingly strong, with a moderately long series of rapid wingbeats punctuated by terse skip/pauses that impart a slight jerkiness to the flight (but nothing that could be called an undulation).

VOCALIZATIONS: Song is a distinctive accelerating series of chip notes all given on the same pitch: "*jh jh, ch, ch chch'h'h'h.*" In pattern, resembles Field Sparrow—but Olive Sparrow's notes are sharper—and Botteri's, a sparrow of open grasslands (not scrub woodlands). Olive's call is a soft lisping chip, "*cheh*" or "*t/cheh*" (recalls Orange-crowned Warbler) and a softer, more nasal "*jeeuh.*" Also makes a high, thin, zinging trill: "*gzzzz.*" When the bird is excited, call notes given in a series frequently change in volume, tone, and pitch, becoming a sharper and breathier "*chih!*"

Green-tailed Towhee, *Pipilo chlorurus*

The Gray-and-green Sparrow-Towhee

STATUS: Common breeder in the western United States; winter resident in the Southwest. **DISTRIBUTION:** Breeds throughout the Great Basin and flanking mountain ranges—se. Wash., s. and e. Ore., the mountains of Calif. (also n. Baja), s. Idaho, Nev., sw. Mont., Utah, n. Ariz., most of Wyo., w. and cen. Colo., much of N.M., and the Davis and Guadalupe Mts. of Tex. (rare). Winters in se. Calif., s. Nev., w. and s. Ariz., s. N.M., and sw. and cen. Tex. east to San Antonio and Corpus Christi. Very rare along the West Coast; casual in the East. **HABITAT:** In summer, favors thick brushy habitat, particularly big sagebrush. At lower elevations, breeds on dry flats and hillsides mixed with chaparral; at higher elevations, found in brush or in surrounding burned or disturbed areas in pine forests and meadow edges. Also breeds in assorted oak habitats and pinyon-juniper and in scrub-heavy portions of trans-Pecos Texas. In winter, almost any brushy area will serve, but birds are partial to mesquite and acacia-lined arroyos and places where fresh water is available. **COHABITANTS:** In summer, Sage Sparrow, Fox Sparrow; in winter, Brewer's Sparrow, Black-throated Sparrow, White-crowned Sparrow; at all times, Spotted Towhee. **MOVEMENT/MIGRATION:** Except for birds in central Arizona, southern New Mexico, and Texas, wholly migratory. Spring migration from March to early June; fall from late July to late December. VI: 3.

DESCRIPTION: A small towhee with an anomalous but commanding plumage combination: a gray body, a greenish back, a white throat, and a jaunty rust-colored cap. A large sparrow and small towhee (smaller than Spotted Towhee and Sage Thrasher; larger than Sage, White-crowned, and Olive Sparrows), Green-tailed's overall proportions (not size) recall White-crowned Sparrow. The conical bill seems small, the head often shows a peak to the rear (like an alert White-crowned), and the body is long and well proportioned, but the tail seems a full size short for the bird. The somewhat similarly plumaged Olive Sparrow, whose range overlaps with Green-tailed Towhee in the Rio Grande Valley, looks pudgy, round-headed, and heavy-billed by comparison.

Adults are an overall lovely gray but show difficult-to-ignore yellow-green highlights in the wing and tail. It's not a combination you see every day, and it *will* grab your attention. In combination with the rusty crown and bright white throat, the pattern is unique. Olive Sparrow is bland by comparison (you have to strain to see the greenish cast to the wings), with a dull reddish-striped crown and a throat that shows no white. Juvenile Green-taileds are streaked below and dull brownish gray above, but still show a greenish tinge to the wings.

BEHAVIOR: Forages mostly on the ground, almost always below brushy cover, so is difficult to see (although the double-shuffle scratch in the leaf litter is easy to hear). Hops quickly with tail slightly up-cocked; also runs. Usually found alone, but in winter may form loose associations with mixed-species sparrow flocks. Breeding males commonly vocalize from high perches (usually a bush, sometimes a rock). Hops nimbly to reach high perches, or may fly. If disturbed, usually descends vertically into heavy cover. Commonly avoids bird feeders, but may be attracted to seed spread on the ground.

During breeding season, responds moderately well to pishing. At other times, often one of the last members of a flock to show.

FLIGHT: Shaped like a longish well-proportioned sparrow. For short distances, flight is bouncy and sometimes arcing or bounding, with lots of exaggerated movement and conspicuous tail-pumping. For longer flights, the bird remains low but flight is straight and strong, with no tail-pumping and only slight undulations. Wingbeats are given in a rapid, fairly lengthy series followed by a momentary closed-wing glide; when landing, glides with open wings.

VOCALIZATIONS: Variable, short, run-on, four- to five-part ensemble of sharp notes, trills, and warbles (short notes are often repeated): "*cheep-chireecheet't't.*" Call is an easily recognized, somewhat querulous, catlike mew, "*weeeah,*" that is somewhat similar to Spotted Towhee but more two-noted and more pleasing. Also makes a high soft "*chee*" that is thinner, purer-toned, and less grating than the similar call of Spotted Towhee.

Spotted Towhee, *Pipilo maculatus*

Dot-winged Towhee

STATUS: Common and widespread western breeder, resident, and wintering species. **DISTRIBUTION:** In Canada and the U.S., breeds widely across the West, from s. B.C., Alta., and Man. south and west of a line drawn through the middle of the Dakotas, se. Wyo., cen. Colo., extreme e. N.M., and into Mexico. In winter, withdraws from northern interior portions of its range, vacating all but sw. B.C., most of Idaho, Mont., Wyo., the Dakotas, and northern portions of Colo. Wintering range extends south and east to include s. and e. Neb. and extreme w. Mo., all of Okla., and all but extreme e. Tex., south into n. Mexico. Casual in the East. **HABITAT:** A variety of low, brushy, and scrubby habitat, from riparian thickets to chaparral, blackberry thickets, manzanita thickets, and well-vegetated residential areas. In winter, uses the same types of habitats. Key characteristics, both in and out of breeding season, include a generally dry environment, a dense understory of broadleafed plants, some taller shading trees, and an ample litter layer for foraging. **COHABITANTS:** During breeding, California Quail, Bewick's Wren, Wilson's Warbler, Green-tailed and California Towhees, Song Sparrow; in winter, assorted sparrow species. **MOVEMENT/MIGRATIONS:** Different populations exhibit different migratory behavior—from full to partial to nonmigratory. Northern interior birds vacate breeding areas entirely; other populations are year-round residents or relocate only short distances. Spring migration (for northern birds) from late March to late May; fall from early September to late October. VI: 2.

DESCRIPTION: A big, colorful, long-tailed sparrow that sings from high perches but otherwise keeps a low profile in the brush. Has a large round or oval head, a thick neck, a conical bill, a plump body, and an extremely long tail that is held slightly elevated and almost doubles the overall length of the bird.

Strikingly patterned. Males and females have a blackish to dark grayish hood (and a dark red eye) and rusty sides; blackish upperparts (folded wings) are lavishly dappled with white spots, and the long black tail flashes white corners when the bird flies—just before disappearing into the brush. Females are not quite as richly colored as males. Compared to Eastern Towhee, Spotted Towhee's rusty sides seem to have more ragged edges, their tails seem longer, and, of course, their upperparts are spattered with white dots. (Eastern Towhee's upperparts are all black.) Also, Spotted's folded wing lacks the white check mark that is so obvious on Eastern's otherwise all-dark wing.

BEHAVIOR: Spends most its life foraging in thickets, or very close by, and is common at bird feeders. Moves with a springy hop, pausing after one or several hops to regard the ground or the vegetation around it. Also scratches leaf litter aside with a two-footed shuffle (birds can often be detected by the sound). Plucks food from the ground or reaches forward to glean from vegetation. Sometimes hops up the inclined base of a tree trunk to investigate bark; sometimes leaps to grasp prey.

Foraging patterns are not aimless. Birds course through vegetation, and when feeding in the open move from objective to objective (from tree to rock to bush to tree, etc.).

Fairly shy, but less so than Green-tailed Towhee. Where not habituated to people, bolts to the nearest cover when humans appear (which is usually at close quarters). The very vocal Spotted makes protesting noises from within its leafy labyrinth, confounding birders who are not familiar with its varied calls.

Despite their clandestine lifestyle, singing males are often perched very high in the tops of trees. In winter, Spotteds become fixtures at winter bird-feeding stations, but remain shyer than many other species.

FLIGHT: Flight silhouette is long, with a very long tail showing white outer corners and short rounded wings dappled with white. Flushed birds, flying short distances, explode from the ground and fly in a frenzied, wing-fluttering, tail-jerking display. For longer distances, wingbeats are rapid and given in a long series punctuated by short skip/pauses. Flight is slightly undulating and slightly jerky.

VOCALIZATIONS: Song varies depending on location. Those familiar with the "*drink-your-tea*" song of Eastern Towhee will note some quality and pattern similarities with Spotted Towhee. Over most of the bird's range, song begins with two or three (or more) similar introductory notes followed by a trill; for example: "*chr chr cheeeee*" or "*che che t'reeeeeee*" (phonetically rendered: "*drink, drink, y'r tee*"). Introductory notes are usually terse, flat, abbreviated whistles. Trills may be sharp and rattly or more musical and warbling. Spotted Towhee's song is not as musical or melodious as Eastern Towhee's. Spotteds in the Northwest dispense with the introductory notes and just trill. Call is like Eastern Towhee, but softer and more buzzy. Also emits a soft sharp "*chee*" (no hard ending) and (frequently) a nasty, whiny, rising-and-falling, catbirdlike whine/mew: "*eeeeEh'ha.*"

Eastern Towhee, *Pipilo erythrophthalmus*
Brush Robin

STATUS: Common and widespread in the East. **DISTRIBUTION:** Breeds across most of the e. U.S. and extreme se. Ont. and Que. (absent in n. N.H., cen. and n. Maine, the Upper Peninsula of Mich., and ne. Minn. The western limits reach se. Sask., cen. Minn., cen. Neb., e. Kans., w. Ark., e. Okla., and the Tex.–La. border. Vacates northern breeding areas; winters south of a line drawn from s. N.J. to cen. Ind., south and west to about cen. Okla. and south across e. Tex. to Corpus Christi. Casual farther west. **HABITAT:** A bird of dry, dense, brushy edge with deep ground litter. Habitat types include late successional fields, mixed pine-oak woodlands, oak-hickory woodlands, pine savanna, dune forest-scrub, riparian thickets, and small suburban woodlots. In winter, particularly at the northern limit of its range, concentrates in greenbrier tangles and multiflora rose, often near streams or open water. **COHABITANTS:** White-eyed Vireo, Brown Thrasher, Gray Catbird. In winter, White-throated Sparrow, thrashers, catbirds. **MOVEMENT/MIGRATION:** Spring migration from late March to mid-May; fall from mid-September to late November. VI: 2.

DESCRIPTION: A long-tailed sparrow garbed like an open-vested robin. Larger than any other sparrow (except other towhee species) and much smaller and leaner than American Robin, with a conical bill, a large head, a slender oval body, and a very long tail, which may be horizontal or slightly up-cocked.

The tricolored plumage pattern is bold and distinctive. Upperparts and hooded head/breast are jet-black (sooty brown on females), with broad rusty sides and a white belly (which may not be evident when birds are feeding on the ground). Most individuals have a white wedge at the lower edge of the folded wing, and some may show narrow white-etched lines in the wing; however, Eastern Towhee's upperparts are almost wholly black (or brown), not spotted or streaked, and the white in the wings does not form two beaded wing-bars (as on Spotted Towhee).

Except in the Southeast, where a white-eyed form of Eastern Towhee occurs, observers often note the angry red eye of this species.

BEHAVIOR: A ground-foraging species that is more easily heard than seen, sticks to the thickets, and, when seen in the open, is rarely far from cover. Feeds slowly and methodically, with short hops, long pauses, and a two-footed, back-scratching, leaf-scattering shuffle (easily heard when leaves are dry). Usually solitary, but may feed with its mate or, in migration, be found in small well-spaced groups with other sparrows. In winter, also associates with woodland sparrows and other species at backyard bird-feeding stations. In spring, when new growth is emerging, sometimes feeds in trees.

Fairly shy—never found far from cover—but also fairly vocal. In winter, may call anytime during the day, but is particularly vocal at dawn. Singing males usually sit conspicuously high on bushes and trees above or close to cover. It responds easily and well to pishing and screech-owl calls, usually calling quickly and hopping up, by degrees, through the foliage to expose itself. Perches upright, making white underparts easy to see.

FLIGHT: Profile is long, with short rounded wings that flash white near the tip and a long tail that shows white outer tips. Over long distances, flight is generally low, straight, and slightly undulating. Short flights (as when birds are ascending to a perch) are weak and floppy, with a pronounced open tail-pumping action.

VOCALIZATIONS: Song is a clear, ringing, easily recognized three-note whistle phonetically rendered "*Drink your tea,*" with the third note held and trilled ("*tee'e'e'e'e'e'e*"). The call—a loud two-noted whistle "*Ter'wrEEH!*" ("*Towhee!*")—is equally distinctive, but at a distance attenuates to a single-noted "*reeeh.*" Birds also make a harmonic buzz/trill and a broken "*jee'ee*" or almost three-noted "*jee'oo'whee*" that recalls the buzzy note of Swamp Sparrow.

Canyon Towhee, *Pipilo fuscus*

Buff-bibbed Towhee

STATUS: Common but restricted to the American Southwest and Mexico. **DISTRIBUTION:** In the U.S., found in cen. and w. Tex., extreme w. Okla., portions of se. Colo., N.M., and Ariz. Does not occur in Calif. **HABITAT:** For the most part, a bird of desert scrub, but found in a variety of habitats (and elevations) where dense brush and vegetation abut more open areas suitable for feeding. Specific habitats include riparian woodlands, rocky and sparsely vegetated desert hillsides, pinyon-juniper woodlands, and pine-oak forest. Found at higher elevations and in drier habitat than Abert's Towhee, but there is some overlap. Partial to suburbanized areas. **COHABITANTS:** Cactus Wren, Curve-billed Thrasher, Rufous-crowned Sparrow. **MOVEMENT/MIGRATION:** Nonmigratory; permanent resident. VI: 0.

DESCRIPTION: A modestly sized, somewhat raw-boned towhee whose overall plain brownish gray plumage sports several distinctive plumage traits. Slightly smaller, more slender, and more angular than Abert's Towhee. The head is particularly lumpy or angular, with a bump on the forehead and a peak to the rear (Abert's is more round-headed).

Plumage is fairly loose or shabby. Upperparts are brownish gray (grayer and paler than Abert's), and underparts are paler still. Against this bland backdrop, the rufous-tinged cap and dribble-stained buff-colored bib are standout features. (The bib makes the bird look somewhat double-chinned.) The expression is myopic, beady-eyed or confiding, unconcerned, nonplused.

BEHAVIOR: A fairly tame, sluggish, methodical, ground-feeding towhee that commonly forages in the open, often on rocky and somewhat barren habitat, but generally stays close to protective cover. Hops or shuffles on the ground. Runs when pressed. May forage by itself, in pairs, or at the edge of mixed flocks. Often found with Rufous-crowned Sparrow, which takes advantage of the towhee's leaf-scratching proclivities, but relative to most towhees, Canyon is a less-than-enthusiastic scratcher, preferring to search and pick.

FLIGHT: Low, slow, fluttery, and slightly bouncy, with wings wide and tail slightly fanned. Unlike Spotted or Green-tailed Towhee, this species is prone to long thrasherlike glides. The outer corners to the tail sometimes appear pale. Usually lands with a floating glide on short paddle-shaped wings.

VOCALIZATIONS: Song is a series of similar mellow musical notes whose quality and cadence recall Northern Cardinal: "*we, we, we, we, we.*" Call is a somewhat woodpecker-like trilled mew. Also makes a soft squeaky "*chuh*" or "*ch'ewh*" or "*che/uh.*"

California Towhee, *Pipilo crissalis*

Silver-toned Towhee

STATUS AND DISTRIBUTION: Common and widespread resident of Calif. west of the Sierras south through Baja (disjunct population found in sw. Ore.). **HABITAT:** Occupies an assortment of scrubby or shrubby habitat, from chaparral to riparian woodlands, oak woodlands, coastal scrub, city parks, orchards, gardens, and suburban yards (providing suitable cover exists nearby), but avoids desert habitat. In all habitats, seeks a mix of brushy cover and open areas suitable for foraging. **COHABITANTS:** California Quail, California Gnatcatcher, California Thrasher, Wrentit and Spotted Towhees; in winter, Golden-crowned and White-crowned Sparrows. **MOVEMENT/MIGRATION:** Nonmigratory; permanent resident. VI: 0.

DESCRIPTION: A classic towhee in a plain brown wrapper. Formerly and aptly named Brown Towhee. Split into two species, California and Canyon Towhee. The latter, a paler, more distinctly marked bird of more arid regions, shares no overlapping range with this species.

California Towhee resembles a large, puffy, long-tailed, grayish brown sparrow. The undertail is distinctly salmon-colored; the face, throat, and belly show traces of color. In good light, at close range, the entire bird seems suffused with warm tones, as if the cinnamon undercoating were showing through, but in fact California Towhee is just a gray-brown bird whose plumage, range, and short conical bill make it distinct and difficult to confuse with any other species.

BEHAVIOR: Slow-moving ground-foraging species. While tame and often willing to feed in the open, California Towhee rarely strays far from the protection of nearby brush. Usually hops and sometimes shuffles; feeds in the leaf litter by double-scratching with both feet. Feeds primarily on seed and readily comes to trays on the ground. In trees and shrubs, seems to clamber more than hop.

Often found in pairs. In winter, also associates with sparrows, particularly White-crowned and Golden-crowned.

FLIGHT: Wings seem very short. The long tail seems to hold the bird back. Flight is direct, usually short, and awkward or struggling, with lots of wing flutter for little ground gained. Sets wings, opens tail, and glides just before landing.

VOCALIZATIONS: Mellow plaintive song consists of three or four clear slightly metallic notes that accelerate into a trill: "*tee, tee, tee'tee t't't't't't't't.*" Call is a sharp silver-toned "*tee*" that recalls White-crowned Sparrow but has a purer hammer-tapped-on-metal quality. Also makes a lisping "*seet.*"

Abert's Towhee, *Pipilo aberti*

Masked Desert Towhee

STATUS: Fairly common but localized. **DISTRIBUTION:** Restricted to the American Southwest and border regions of Mexico. In the U.S., found in extreme se. Nev., sw. Utah, se. Calif., w. and s. Ariz., and extreme sw. N.M. **HABITAT:** Mesquite-scrub thickets and woodlands, especially riparian woodland in desert areas with low elevation (generally found at lower elevations than Canyon Towhee). Also occurs in well-vegetated suburban habitats, desert golf courses, and orchards. **COHABITANTS:** Gambel's Quail, White-winged Dove, Greater Roadrunner, Pyrrhuloxia. **MOVEMENT/MIGRATION:** Nonmigratory; permanent resident. VI: 0.

DESCRIPTION: A large, burly, all-brown towhee with a black mask and a pale bill. Slightly larger, bulkier, and longer-necked than Canyon Towhee, but fundamentally similar. Plumage is uniformly warm brown, with subtle infusions of pinkish or cinnamon in good light—a plain unpatterned backdrop that makes the blackish face and pale bill standout features. Ages and sexes are essentially similar. Canyon Towhee is overall browner, darker, and distinctly more patterned, particularly about the head, breast, and throat. The expression is mean or angry—the very opposite of Canyon.

BEHAVIOR: Forages on the ground, alone or in pairs, by scratching in the leaf litter or dry ground. Somewhat shy. Likes to stay within or at the edge of thick vegetation. Hops when foraging. Also runs with tail slightly up-cocked. More active than Canyon Towhee, and dominant where the two species occur together.

FLIGHT: Long profile with a long tail and short paddlelike wings. Except for the blackish tail (and ruddy undertail), plain grayish brown. Note the absence of pale tips to the corners of the tail. Flight is direct but bouncy, and almost always short (to the nearest cover). Wingbeats are floppy and amateurish. Flutters and glides with wings and tail open.

VOCALIZATIONS: Two-part song consists of a series of sharp high notes that accelerate into a low, harsh, scolding chatter: "*T' tee tee tee teeteetee-teechchchchcheechr.*" Calls are a high-pitched "*tee*" and a lower-pitched "*puhr*" (recalls a ground squirrel alarm call). Also makes a soft two-noted whistle, "*chew-ee,*" and a catlike mewing whine, "*ehh'h.*"

Rufous-winged Sparrow, *Aimophila carpalis*

Desert Chipping Sparrow

STATUS: Uncommon and very restricted in habitat and range. **DISTRIBUTION:** A Sonoran Desert species, in the U.S. found only in se. Ariz. **HABITAT:** Flat or hilly, grassy, desert thorn-scrub. Palo verdes and chollas are almost always part of the vegetative mix, and at least a partial covering of bunch grass is essential, except in drier desert areas, where the bird may be found in riparian washes. **COHABITANTS:** Verdin, Cassin's Sparrow, Botteri's Sparrow, Black-throated Sparrow. **MOVEMENT/MIGRATION:** Permanent resident. VI: 0.

DESCRIPTION: Slender, long-tailed, trimly proportioned and (often) bushy-crested sparrow of desert scrub. Medium-sized (smaller than Cassin's or Botteri's and Rufous-crowned Sparrows; slightly larger than Chipping Sparrow), with a small, slightly peaked, shaggy head; a small, conical, very slightly downturned bill; an average body; and a fairly long tail. Basically a slender classically proportioned sparrow that approximates Chipping or Field Sparrow in shape—very unlike the dumpy, large-headed, and long-tailed profile of Rufous-crowned Sparrow.

Mostly grayish brown above and pale gray below; more crisply patterned than Rufous-crowned Sparrow, with a sparse rusty cap and crisp blackish and rufous highlights on the back and wings. The gray face shows two distinct black whisker marks and a faint rusty line through the eye that flares to the rear (like American Tree Sparrow). The rusty wing patch is very difficult to see—don't let your identification hinge on it. The double mustache stripes are much more apparent.

BEHAVIOR: More arboreal than Rufous-crowned Sparrow, Rufous-winged sits out the heat of the day perched in low bushes, but feeds almost entirely on the ground, often on bare earth. Feeds by hopping and pecking around clumps of grass. Most easily observed when males vocalize from the tops of trees and low shrubs (also from the ground). Vocalizes most persistently after the onset of the monsoon rains (beginning in early July).

Usually found in pairs or family groups. Not a flocking sparrow, but in winter may forage with other local sparrows (most notably Black-throated). Usually elusive, shy, and difficult to find. Usually runs when pursued, but often lands in a small bush when flushed. Flushes close (singly or in pairs), and does not call in flight. Rare visitor to feeders.

FLIGHT: Shows a long-tailed silhouette. Flight is straight and slightly undulating, sometimes bouncy, executed with a stiff, hurried, fluttery series of wing-beats followed by a brief closed-wing glide.

VOCALIZATIONS: Whistled song is simple, distinctive, clear, and somewhat variable. Most commonly an accelerating musical trill: "*chip chp chpchpchpchpchp.*" A variation combines one or two sharp introductory notes with a blurred trill: "*chip chip chchchchchch.*" Introductory notes recall opening notes of Song Sparrow; trill sounds like a slow blurry Chipping Sparrow.

Cassin's Sparrow, *Aimophila cassinii*

The Lofting Sparrow with the Plaintive Voice

STATUS: Common to fairly common sparrow of western grasslands. **DISTRIBUTION:** Breeds from e. Colo. (rarely se. Wyo.), sw. Neb., and e. Kans. south into e. and s. N.M., se. Ariz., w. Okla., and all but e. Tex. south into Mexico. Birds from n. N.M. and n. Tex. northwards migrate south in winter. Casual visitor well west, north, and east of its normal range. **HABITAT:** Dry grasslands with scattered cactus (especially cholla), chaparral, mesquite, and oaks. Avoids overgrazed and agricultural fields. **COHABITANTS:** Ash-throated Flycatcher, Botteri's Sparrow, Vesper Sparrow, Black-throated Sparrow, Grasshopper Sparrow, Lark Bunting, Western and Eastern Meadowlark. **MOVEMENT/MIGRATION:** Northern breeders move south in winter. Spring migration from mid-April to late May; fall from early to mid-August to October. Populations are also nomadic and erratic. Birds may be found in numbers one year and be absent the next. In places where they always occur, numbers may be high one year and low the next. VI: 2.

DESCRIPTION: A medium-large, grayish, long-tailed grassland sparrow. Slightly smaller than Song Sparrow and about the same size as Botteri's, but less robust, with a more slender pinkish bill that contributes to a longer and thinner face, a slimmer body, and a long, wide, tongue depressor–shaped tail that may have a slight droop. Overall plain dull grayish brown (lacks ruddy and warm buffy tones) and slightly more patterned than Botteri's. The crown and neck have tiny blackish freckles. The back is dark-spotted; sometimes in fresh plumage, narrow white feather edges give it a frosted look. The throat is slightly paler than the grayish underparts, which have faint shadowy streaking on the breast and, most importantly, the flanks.

BEHAVIOR: Males are celebrated for their spirited and persistent aerial displays. Birds rise in a wing-fluttering lofting arc and parachute, on quivering wings. Displays may be perch to perch, perch to ground, or ground to perch. Also vocalizes from a perch—but not commonly from the highest point in the tree. (Botteri's commonly selects the highest, most conspicuous perch and only goes airborne very infrequently.)

Feeds on the ground; searches for seeds and insects by hopping and walking in relatively open areas (they still cannot be seen). Tail moves in concert with the song. The long trilled portion of the song sets the tail vibrating.

Flushes very close and flies very low and direct; drops quickly into cover and runs immediately. Does not call in flight, and usually does not flush into brush (prefers tall grass), but if it does land in brush, does so at the base of a shrub. Becomes increasingly more difficult to flush.

FLIGHT: Flies with its long tail drooped. In flight, fanned tails show pale grayish white corners.

VOCALIZATIONS: Song, most often given in flight, is short, plaintive, and unvarying. Begins with two brief tentative but pure-toned notes followed by a trilled whistle and a contemplative finale: "*tisi tseeeeeeee sooit* (lower) *seeit* (higher)." Call is a sharp "*chip.*"

Bachman's Sparrow, *Aimophila aestivalis*

The Pinewoods Sparrow

STATUS: Uncommon to fairly common woodland sparrow of the Southeast, but habitat-restricted and secretive; most often encountered when males are singing. **DISTRIBUTION:** Barely se. Va. to cen. Fla., west across S.C., Ga., Ala., Miss., all but coastal La., most of Ark., and e. Tex.; very local in portions of Tenn., Ky., and s. Mo. **HABITAT:** Open pinewoods with a dense ground cover comprising grasses, shrubs, and palmetto. Also occurs in grassy habitat in other woodland habitats and in clear-cut areas going through early succession, such as power-line cuts, clear-cut forest, and fallow pasture. **COHABITANTS:** Northern Bobwhite, Great Crested Flycatcher, Eastern Bluebird, Brown-headed Nuthatch, Pine Warbler, Eastern Towhee, Eastern Meadowlark. **MOVEMENT/MIGRATION:** Most birds (particularly southern birds) are nonmigratory. Northern birds seem to vacate breeding territory, and there is evidence of movement from late August to late October and of a return to northern areas between March and May. VI: 1.

DESCRIPTION: A medium-large, generally plain sparrow of grassy pine woodlands that is heard and then, with luck, seen. Resembles an unstreaked long-billed Song Sparrow. Overall well balanced and somewhat roundly proportioned, but with a long "Roman nose"–shaped bill, a smallish head with a sloping forehead, a slightly humped back, and a long, dark, round-tipped tail. Eastern birds are overall dingy grayish brown and low on contrast between the diffusely streaked upperparts and the plain grayish brown underparts—you have to look really close to see a rusty spattering on the neck. Western birds are overall ruddier, with rust and gray streaks above and a buffy wash surrounding a grayish oval belly below.

BEHAVIOR: Forages on the ground, where, given the density of the habitat, the bird will not be seen. Most sightings are of singing males that call attention to themselves by their distinctive song. Birds sing most commonly in the morning and evening and are tenacious, albeit sedentary, singers—singing from the same perch for long periods without shifting. Perches may be high, in the lower branches of pines, or low on a shrub or grass stalk (making detection challenging). Generally allows fairly close approach; when flushed, normally flies a short distance and disappears into ground cover. Walks, hops, and runs rather than flies when pursued. The tail moves in time with the song. Each short note in the series is reflected in a jerk of the tail.

FLIGHT: Generally short. Short flights are weak, fluttering, and bouncy, with tail pumped. Longer flights are fairly strong and undulating.

VOCALIZATIONS: Melodious song is in two or three parts and has a somewhat somnambulant quality. Frequently begins with a single clear whistle followed by a series of musical chips, trills, or warbles (or a combination of these sounds). Each successive song varies in composition; for example: "*seeee, chur, chur, chur, chur, chur; je'e'e'e chur, chur, chur, see, see, see, see, see.*" The introductory note is sometimes rough and buzzy instead of a clear whistle. The second part of the song often has the quality of the warble of a Pine Warbler, a bird found in the same habitat.

Botteri's Sparrow, *Aimophila botterii*

Clipped Notes and a Bouncing Ball

STATUS: Uncommon and localized; restricted range. **DISTRIBUTION:** A Mexican species whose range extends into extreme se. Ariz., sw. N.M., and se. Tex. **HABITAT:** In Arizona and New Mexico, arid, ungrazed, or lightly grazed grasslands dotted with oaks or mesquite. Most often found in areas dominated by sacaton grass. In Texas, birds inhabit coastal prairie with assorted shrubs, especially yuccas. **COHABITANTS:** In Arizona and New Mexico, Canyon Towhee, Rufous-winged Sparrow, Cassin's Sparrow, Black-throated Sparrow, meadowlarks. **MOVEMENT/MIGRATION:** Permanent resident in Mexico. In the United States, birds arrive during the second half of May and leave in September. VI: 0.

DESCRIPTION: A medium-large plain grassland sparrow with a song whose signature element is the sound of a bouncing ball. Large, sturdy, but sleek, with a large, thick, slightly drooped, grayish bill, a large flat head, a long body, and a long, narrow, round-tipped and often loosely folded or disarrayed tail. Overall slightly larger and more robust than the very similar Cassin's Sparrow, with a heavier (but not necessarily longer) bill.

Overall very plain dingy grayish brown with a slight ruddy tinge to the head and back; plain underparts are warmer and buffier. Except for a faint rusty eye-line that widens behind the eye, the face is almost patternless; underparts are smoothly uniform, with no streaks or shading. At close range, the crown and neck are suffused with fine ruddy freckling, and the back shows faint rusty streaks and fine dark lines (which meld into grayish brown uniformity at a distance and make Botteri's appear browner, warmer-toned, and overall less patterned above than Cassin's). Botteri's also lacks the dark streaks on the flanks that mark Cassin's (although these can be hard to see). Short-lived juvenile plumage has a streaked breast. The expression is innocent.

BEHAVIOR: Males perch conspicuously when vocalizing; otherwise, the bird spends most of its time foraging on the ground in open areas between clumps of grass. Moves quickly while foraging by walking, hopping, running, and fluttering — most often in pursuit of a grasshopper. Commonly drops into one place and takes off from another several yards away. Singing males take conspicuous perches at the tops of shrubs and trees, where they may remain for several minutes before flying to another perch. Botteri's sometimes seems clumsy, losing its balance, for instance, on too-small perches. Also has a moderately lofting, fluttering display flight, but much prefers to vocalize from a perch.

Breeds late, usually in July and August, to coincide with the monsoon rains. Found in loose colonies. Often several males can be heard from a single vantage point. Males and females commonly forage together. Tame and often allows close approach or lands near an observer and vocalizes. Flushes and flies much like Cassin's, that is, flushing from nearly underfoot and flying 30–60 ft., then dropping suddenly. When flushed, typically flies to the base of a nearby bush.

FLIGHT: Long-tailed and short-winged, with no eye-grabbing plumage points. Tail is uniformly dark, with no paler center and no white outer tips. Flight is fairly strong, straight, and ever so slightly bouncy or jerky, with steady, rapid, fluttery wingbeats. Does not glide. Does not commonly engage in a flight display.

VOCALIZATIONS: Variable song begins with a series of tentative, evenly spaced, somewhat clipped, whistled notes and ends in an accelerating trill, the pattern and cadence of which recall a Ping-Pong ball bouncing to a halt; for example: "*tip tse chibit swee treeee twit twittwit't't't't't't.*" Sometimes trills end with a high "*tee tee.*" A variation of the song has no trilled ending and is just a series of high, squeaky, slurred, clipped, trilled, sparrowlike notes and phrases rendered in a spaced vireo-like pattern; for example: "*zwee t'cur ch'lip chur zwee.*" Sounds as though the bird is reciting some arcane alphabet. Call is a breathy "*chip*"; also emits a high thin "*zwee.*"

Rufous-crowned Sparrow, *Aimophila ruficeps*

Rocky Hillside Sparrow

STATUS: Common resident of Mexico and the American Southwest. **DISTRIBUTION:** In the U.S., occurs in Calif. (excluding extreme n. Calif. and the Central Valley), extreme s. Nev., portions of Ariz., N.M., w. and cen. Tex., w. and cen. Okla., se. Colo., and sw. Kans.; isolated population in the Ozarks of Ark. **HABITAT:** Open to sparsely vegetated, arid, brushy or wooded, rocky and weedy hillsides from sea level to almost 9,000 ft. Prefers hillsides with a mix of shrub and tree species over monotypic habitats. **COHABITANTS:** Roadrunner, California, Canyon and Spotted Towhees, Rock and Canyon Wrens, Black-chinned Sparrow, Painted, Lazuli, and Varied Buntings. **MOVEMENT/MIGRATION:** Permanent resident. VI: 0.

DESCRIPTION: Plain, sluggish bordering on oafish, earth-hugging sparrow of inhospitable hillsides. Medium-large and overall robust, with a big fairly round head, a large eye, a large, pointy, conical bill, a plump body, and a distinctly long tail.

Overall grayish with rufous highlights above — a fairly distinct rufous crown that melds into the neck and diffuse rufous streaking on the back. Against this bland backdrop, the face seems expressive. The grayish face makes the white eyebrow, the white eye-ring, and the black-bordered white whisker stand out. Other rufous-crowned sparrows are smaller, more petite, and more classically proportioned (that is, they don't look plump and dumpy).

BEHAVIOR: Except for vocalizing males, generally not found in trees. Spends most of its time on the ground, slowly hopping in search of seeds and insects. Works its way into stands of grass and ground-hugging vegetation. Searches slowly and methodically. Usually found in pairs, but in summer and winter found in small family groups. Also may forage with other sparrows and towhees.

Not shy—in fact, fairly indifferent to observers—but does not like to be in the open. Runs across open areas. Forages close to cover. When approached, forages deeper into cover, and when pressed, flies with reluctance. Males sing from exposed perches—low shrubs (less than 6 ft. tall), rocks, and weed stalks. When flushed, flies low, eventually swooping up to perch on a bush or rock.

FLIGHT: Generally short, commonly downhill, straight, and slow, with deep, floppy, continuous wingbeats.

VOCALIZATIONS: Song is a short hurried chatter that incorporates sharp, mumbled, and scratchy notes. Recalls a chirpy House Wren, but more choppy and less fluid. Also gives sharp "*tik*" notes when excited. Call is a distinctive single note that has a somewhat mewing quality.

Five-striped Sparrow, *Aimophila quinquestriata*

California Gulch Sparrow

STATUS: Rare (fewer than 60–70 individuals in the United States), geographically isolated, and habitat-restricted. DISTRIBUTION: A Mexican species. In the U.S., found only along the border in the vicinity of Nogales (or more specifically, California Gulch, Sycamore Canyon, and less commonly Patagonia). HABITAT: Steep brush and cactus-covered hillsides (1,100–1,800 ft. in elevation) in narrow desert canyons blessed with year-round water. COHABITANTS: Black-tailed Gnatcatcher, Varied Bunting, damn little else. MOVEMENT/MIGRATION: Permanent resident. Difficult if not impossible to find in winter. VI: 0.

DESCRIPTION: A medium-large brown-and-gray sparrow with a distinctive black-and-white face pattern that is singing from the top of an ocotillo in California Gulch just after the onset of the rainy season. Large and robust (the same size as Rufous-crowned and Botteri's Sparrows), with a long neck, a big head, a long, heavy, slightly downturned bill, a long, plump, potbellied body, and a long tail.

All dark with brownish upperparts, gray underparts, a boldly striped head with bracketing black-and-white stripes, and a black breast spot. The white belly is often visible if birds are above you (which will almost certainly be the case). The expression is really, really cross.

BEHAVIOR: A ground- and shrub-foraging species that feeds slowly and methodically in short hops, moving from the ground to lower layers of shrubs, where it gleans somewhat like a vireo. Uses standing water to drink and bathe. You probably won't see any of this. What you'll see, between early July and late August (that is, after the onset of the rainy season, when birds start breeding), is a keyed-up male vocalizing from the top of what will probably be an ocotillo. A persistent and tenacious singer, Five-striped Sparrow commonly remains on the same perch for many minutes before shifting to another favored perch. Flights between perches may be long (for example, across the canyon).

FLIGHT: Robust profile with a long pointy face and a long tail. Flight is strong and straight, with a slight rise and fall (not really undulations). Wingbeats are given in a rapid near-continuous manner with the suggestion of a skip between otherwise lengthy and unbroken series.

VOCALIZATIONS: Song is a series of loud, short, simple phrases, repeated two or three times, followed by a pause, followed by another, different series. Series usually begin with an introductory note. Notes are short, sharp, brittle, emphatic, and stuttering. Calls include a low "*chuck*" and a higher "*seep.*"

American Tree Sparrow, *Spizella arborea*

The Winter Chippy

STATUS: Fairly common and widespread breeder across arctic and subarctic regions; common winter resident across southern Canada and the northern United States. DISTRIBUTION: Breeds from the west coast of Alaska across n. Canada to coastal Lab., with the southern limit of its breeding range cutting through n. B.C., s. Yukon, n. Sask., n. Man., n. Ont., cen. Que., and s. Lab. Entirely vacates breeding areas in winter. Winters from southernmost portions of Canada across much of the n. and cen. U.S., excluding w. Wash., Ore., most of Calif., and s. Nev.; also absent south of a line drawn through n. Ariz., cen. N.M., n. Tex., s. Ark., s. Tenn., and se. N.C. **HABITAT:** Nests in patches of stunted willows and spruce on open or boggy tundra. Winters entirely south of its breeding area in open weedy fields and marsh edges dotted with shrubs or hedges or adjacent to forest borders. Also readily attracted to backyard feeding stations. **COHABITANTS:** During breeding, Willow Ptarmigan, Northern Shrike, White-crowned Sparrow, Common and Hoary Redpolls. **MOVEMENT/MIGRATION:** Spring migration is very early, in March and April. In fall, birds depart in early September, but may not reach the United States until mid-October; they arrive at the southern reaches of their winter range by late November. VI: 2.

DESCRIPTION: An overall fairly pale but richly patterned rufous-capped sparrow whose musical lighthearted calls enliven the winter landscape. A medium-large sparrow but a large *Spizella* (larger than Chipping or Field Sparrow; about the same size as Song Sparrow), American Tree Sparrow is full-bodied without appearing pudgy or plump. The head is round or slightly peaked, with a fairly steep forehead and a bicolored bill (black above, bright yellow below) that is almost redpoll petite. The tail is long and narrow.

The back is distinctly streaked with rust, black, and gray. Underparts are very pale gray, sometimes with a blush of rust down the sides; almost always a larger patch or splash of rust appears on the breast near the bend of the wing, and birds are always adorned with a charcoal smudge of a breast spot on the otherwise unmarked chest. The all-gray head wonderfully sets off the ruddy cap (pushed jauntily forward) and rusty eye-line. At a distance, the gray face sandwiched between the ruddy cap and ruddy back, wings, and flanks is often very apparent.

BEHAVIOR: Active and nimble. Feeds on the ground in weed-rich places, moving actively about with rapid shuffling movements so quick that the legs seem to vibrate. Also leaps to snatch seeds from overlying grasses, forages in trees to a height of 30 ft. to feed on birch catkins, and reportedly flails seed-laden grasses with its wings, then picks the fallen seed from the ground. In winter, most commonly found in loose homogeneous flocks that move through an area as the birds feed (as opposed to staying in one location). In the lower reaches of the bird's winter range, individuals or several birds may mix with other flocking sparrows, including White-throated and Field. Responds readily to pishing—birds come in from great distances and often cluster in a bush or tree. Flocks are fairly vocal, but vocalizations are not as constant as some species. When flushed, may fly a considerable distance and quite high, landing high in a bush or small tree.

FLIGHT: Overall pale and long-tailed. Flight is quick, nimble, and undulating and shows more stability and control than the bird's smaller kin—perhaps because of its greater size, which in a mixed flock with Field and Savannah Sparrows is apparent. Wingbeats are given in a short series punctuated by a skip/pause.

VOCALIZATIONS: Song is a short, variable, several-parted ditty comprising run-on (and often repeated) high, clear, whistled notes and short warbles; for example: "*ee'EEweterterter tew tew tew.*" Song has a dreamy or lighthearted quality. Call is a soft, musical, lighthearted "*tee-dle*" or "*twee-dle.*" Also makes a soft, very high-pitched "*tink*" note.

Chipping Sparrow, *Spizella passerina*
Rusty-capped Lawn Sparrow

STATUS: Common; found at some point in the year across almost all of North America. **DISTRIBUTION:** Breeds widely throughout all of subarctic Canada, as well as in e.-cen. and se. Alaska. In the continental U.S., absent as a breeder only in portions of w. Wash., Ore., the Central Valley and parts of se. Calif., s. Ariz., and the central prairies from

e. Colo. and s. Neb. south through Tex. (excluding higher portions of trans-Pecos Tex. and the hill country), portions of the Gulf Coast, most of Fla. and se. Ga. In winter, found across the s. U.S. from s. Calif. to w. and s. Ariz., s. N.M., all but the Texas Panhandle, s. and cen. Okla., most of Ark., Tenn., Va., the Delmarva Peninsula, and s. N.J. **HABITAT:** Historically a bird of dry grassy forest (particularly coniferous) clearings, the "Chippy" has embraced civilization; in fact, this sparrow seems addicted to lawns—front lawns, corporate lawns, golf courses, highway rest stops, cemeteries, all are favored breeding habitats. For nesting and singing purposes, shows a preference for conifers. In winter, favors weedy fields, farmland, and brush. **COHABITANTS:** Mourning Dove, American Robin, House Finch. In winter, often joins mixed flocks that include Eastern and Western Bluebirds, Yellow-rumped and Pine Warblers, Clay-colored Sparrow and, especially, juncos. **MOVEMENT/MIGRATION:** Highly migratory; in winter, withdraws from most of its breeding range. Owing to the latitudinal scope of its breeding range, migration is protracted. In spring, southern birds arrive in March, northernmost birds in May. Fall migration begins in late August but is heaviest from late September to early November. VI: 2.

DESCRIPTION: A dapper sparrow that is small (smaller than American Tree and Song Sparrows or junco; the same size as Brewer's, Clay-colored, and Savannah Sparrows), slim, and well proportioned. The crown is round but sometimes appears flat-topped; the bill is small and somewhat narrow. The tail is narrow, slightly flared at the tip, and distinctly notched.

Breeding-plumage adults are crisply patterned, basically ruddy above, and clean gray below. The combination of the bright chestnut cap, prominent white eye stripe, and mascara-fine eye-line is distinctive and easily separates Chipping from all other rufous-capped sparrows. So don't sweat the details. You want to identify adult Chipping Sparrow? Just look at the face. The expression is smug.

In winter, adults retain the basic plumage pattern but are overall muted—brownish above and still grayish below, with some reddish highlights retained in the crown. Immatures are like winter adults but overall less distinctly marked. The crown is brown (not reddish) and narrowly streaked. The underparts and rump (which is almost impossible to see) are grayish and may help separate this species from the overall warmer-toned immature Clay-colored Sparrow.

BEHAVIOR: Almost indifferent to people—literally and figuratively. Over most of North America, the bird next door. In summer and winter, Chipping Sparrow forages on short-cut grass, roadsides, weedy patches, and hedgerows. Also flycatches—in fact, can be dogged in its pursuit of a flying insect. In winter, birds gather in flocks, often with Brewer's and Lark Sparrows in the Southwest.

Territorial males sing from elevated perches from water spigot height to treetops (particularly conifers); readily sing from man-made structures as well. An active sparrow, Chipping's movements are quick and perfunctory. Winter flocks typically flush as a group, calling on takeoff, and commonly land in a large or midsized tree.

FLIGHT: Light, nimble, buoyant, undulating, but also direct and no-nonsense. Appears overall pale, slender, nicely proportioned. Wingbeats are given in a quick hurried series broken with a terse skip/pause that imparts a light even bounciness or, when traveling longer distances, a brief closed-wing glide that makes the flight more evenly undulating.

VOCALIZATIONS: An enthusiastic and persistent singer, Chipping Sparrow sings all day. Song is most often a rapid emphatic trill or chatter of unvarying tempo and pitch, lasting 2–4 secs. After a short pause, the song is repeated. Sounds somewhat insectlike. A variation may be slower, with individual notes clearly enunciated. Call is a sharp but not particularly loud "*tcheh*" or "*tchee*" or "*tsit.*"

Clay-colored Sparrow, *Spizella pallida*

Hedge (and Edge) Sparrow

STATUS: Common Great Plains breeder; uncommon winter resident of south Texas. Common Great Plains migrant; uncommon in the Midwest.

Rare but regular and widespread migrant and vagrant elsewhere. **DISTRIBUTION:** Breeding range encompasses sw. Northwest Territories, e. B.C., Alta., w. and s. Sask., s. Man., s. Ont. (including southern portions of Hudson Bay), and s. Que. In the U.S., found in ne. Wash., Mont., n. Wyo., N.D., n. S.D., Minn., all but s. Wisc., all but s. Mich., w. Pa. (a few), and nw. N.Y. Winters in s. and w. Tex. and Mexico. **HABITAT:** Brush and brushy edge bordering open, dry, weedy or arid habitat and sagegrasslands; also favors willow and alder thickets, edges of bogs, and overgrown fields. **COHABITANTS:** During breeding, Alder and Willow Flycatchers, Golden-winged and Yellow Warblers, Lincoln's and Song Sparrows. In winter, associates with Chipping and Brewer's Sparrows. **MOVEMENT/MIGRATION:** Spring migration from late March to late May; fall from August to early November. VI: 3.

DESCRIPTION: Looks like a slender long-tailed female House Sparrow with a dark cheek patch or broadly striped head. A smallish sparrow (larger than Grasshopper Sparrow; smaller than Song Sparrow; about the same size as Brewer's and Chipping Sparrows) that resembles a very slightly plumper-bodied and slightly longer-tailed Chipping Sparrow.

Overall plain, pale, and innocuous—a backdrop that makes what would otherwise be a subtle head pattern more distinctive. Breeding adults are grayish brown and coarsely and sparingly streaked above; pale gray, with touches of buff on the breast, below. Immatures and winter adults are overall buffier, particularly on the breast.

The head shows an alternating pattern of dark and light stripes—a dark crown with a pale center stripe, and a dark face bracketed by a pale eyebrow and pale mustache. Describing the face as having a "cheek patch" is something of an overstatement, but it is set off a bit by slightly darker borders. The bracketing whitish (or buffy) eyebrow and mustache are more conspicuous and more determining. Although winter Chipping Sparrow has the same face pattern, it lacks the strong contrast. On Chipping, you won't notice a pale mustache (malar); on Clay-colored, you can't miss it.

Also in winter, and in combination with Clay-colored's overall warmer buffy plumage, the clean gray nape stands out. Overall grayer Chipping Sparrow also has a gray collar, but, again, it doesn't catch the eye. Brewer's Sparrow is just plain grayish brown and comparatively featureless, showing finely streaky upperparts (including the crown) but no defining pattern and little contrast overall.

BEHAVIOR: A quick, busy, nimble, somewhat nervous sparrow that loves brush. In summer, it hops through shrubs and thickets with warblerlike agility. After thoroughly working one bush, flies to the next (usually closest) bush. In winter, feeds more commonly in the open, often in dry sparsely vegetated terrain, but rarely strays far from brushy cover. Very social in winter and migration; usually found in flocks of 25–50 birds. Outside its normal range, usually teams up with Brewer's, Chipping, or Lark Sparrow.

Males sing persistently from a low perch, most commonly below the topmost branches. Tame and, for a sparrow, responds well to pishing. Usually flushes in flocks and flies into trees.

FLIGHT: Appears small, overall pale, and long-tailed. The brown rump helps distinguish it from gray-rumped Chipping Sparrow. Flight is quick and bouncy, with a quick fluttery series of wingbeats followed by a terse closed-wing glide (similar to Chipping and Brewer's Sparrows).

VOCALIZATIONS: Simple song is a series of three to six low, lazy, even, identical, buzzy notes: "*geeee geeee geeee.*" Sounds very insectlike and recalls Golden-winged Warbler. Call is a high weak "*chih*" that recalls Chipping Sparrow but is high, thinner, and not as sharp.

Brewer's Sparrow, *Spizela breweri*

The Sparrow of Understatement

STATUS: A common western species—at times and in places, the most common small sparrow. **DISTRIBUTION:** Breeds across the intermountain West from extreme w. N.D. north and west to se. Alta. and south to n. N.M. The western border of the breeding range crosses e. Wash., e. Ore., and e. Calif. south to the Mojave. Winters in the desert

Southwest (primarily the Sonoran and Chihuahuan Deserts) from s. Calif. and extreme s. Nev. across s. Ariz., s. N.M., and sw. Tex. south into Baja and cen. Mexico. A subspecies, Timberline Sparrow, *S. b. taverneri*, breeds at higher altitudes in the mountains of sw. Alta., B.C., the Yukon, and extreme e. Alaska. **HABITAT:** Breeds commonly in arid brush land dominated by sagebrush (a habitat as static as the bird is bland). Timberline Sparrow breeds mostly in stunted willow thickets. In winter, Brewer's is found in dry and desert habitat characterized by sagebrush, creosote bush, and sparse grass. In agricultural areas, common along vegetated roadsides. **COHABITANTS:** Sage Thrasher, Chipping Sparrow, Vesper Sparrow, Savannah Sparrow, Sage Sparrow, Green-tailed Towhee. **MOVEMENT/MIGRATION:** Spring migration from late February to early May, with a peak in March and April; fall from mid-August to early November, with a peak from September to mid-October. VI: 2.

DESCRIPTION: A pale, trim, neat, grayish brown wisp of a sparrow that is so patternless it makes other sparrows seem richly plumaged. Most closely resembles a pale, pencil-fine-patterned, immature Chipping Sparrow (or a slender female House Sparrow), but shows overall less contrast. Medium-small (slightly longer than Savannah or Clay-colored; smaller than Sage Sparrow; the same size as Field Sparrow), with a roundish nicely proportioned head that sometimes appears shaggy-crested, a short, almost goldfinch-fine, conical bill, a slender body, and a long, narrow, notched tail. Plumage is overall pale, plain, cryptic, and almost patternless. The head and back are grayish brown with crisp, fine, blackish streaks; underparts are slightly paler brownish gray. The complete narrow white eye-ring is often distinct. Both Chipping and Clay-colored Sparrows show greater contrast between darker upperparts and paler underparts and more distinctive and contrasting facial patterns.

BEHAVIOR: In summer, singing males sit prominently on some "high" point (sagebrush, barbed-wire fence), singing and changing perches frequently. Most likely to sing during mornings and evenings. Likes to stay in cover. In summer, forages for insects by hopping branch to branch within sagebrush. In winter, in small flocks, spends most of its time on the ground but also within concealing cover. Sometimes found with other sparrow species, most often Clay-colored, Savannah, and Vesper Sparrow and Lark Bunting. Responds quickly to pishing, usually by taking the highest closest perch. When flushed, commonly flies a short or moderate distance and perches prominently on a bush or shrub (that is, it does not bury itself in cover immediately). Often calls in flight and when flushed.

FLIGHT: Overall pale and gray with a long-tailed profile. Flight is straight, nimble, buoyant, bouncy, and usually just above the vegetation. Wingbeats are given in a short rapid series followed by a terse closed-wing glide.

VOCALIZATIONS: Song is a varied ensemble of buzzes and trills. Sounds like a dry sputtery canary; for example: "*jee jee jee; j'r'r'r'r'r; teeteetee; t'lr t'lr t'lr t't't't't'tee.*" Call is a short, thin, softly uttered "*chit*" or "*chee.*" Flight call is a lispy "*tseeet.*"

Field Sparrow, *Spizella pusilla*

Cherub in Sparrow's Clothing

STATUS: Common in its namesake habitat across the eastern half of the United States and southeastern Canada. **DISTRIBUTION:** Breeds east of a line that cuts across e. Mont., e. Wyo., e. Colo., the Tex.–Okla. border, and cen. Tex. south to the limit of the hill country. (Absent as a breeder across s. Tex., s. La., s. Miss., s. Ala., the Florida Peninsula, much of N.D., n. Minn., the Upper Peninsula of Mich., and n. Maine.) In Canada, breeds in se. Man., se. Ont., extreme s. Que., and s. Ont. In winter, found east and south of a line drawn from trans-Pecos Tex., se. N.M., the Texas Panhandle, cen. Kans., cen. and se. Mo., se. Iowa, cen. Ill., extreme s. Ont., s. N.Y., and se. Mass. Casual in the far West. **HABITAT:** Prefers successional weedy fields dotted, even half-overgrown, with brush. Particularly partial to fields dominated by broom sedge. For breeding, fields can be small (open ridge tops and woodland clearings). Wintering birds are more partial to larger open areas. **COHABITANTS:** American Kestrel, Northern Bobwhite, Blue-winged Warbler, Blue Grosbeak, Eastern Towhee, Song Sparrow. In winter, often joins mixed sparrow flocks. **MOVEMENT/MIGRATION:** Vacates northern portions of its range between September and early November; returning birds arrive on territory between late March and early May. VI: 2.

DESCRIPTION: A chubby rosy cherub of a sparrow. Small (larger than Chipping Sparrow; smaller than Tree or Song Sparrow) and distinguished by its pink bill, wide-eyed expression, and overall warm but muted plumage pattern. The head is large, and the body somewhat plump. The bill is stout, somewhat curved on the culmen, and bright pink. The tail is disproportionately long and narrow.

Field Sparrow is overall pale, with plumage that is not as crisply defined as on most sparrows. Males and females are identical, but some individual birds appear overall rosy or warm-toned and others more gray. The rusty pinkish crown, eye stripe, and buffy breast blur at the edges and meld with the gray face. The reddish and brown streaks on the back alchemize, with any appreciable distance, into rust. Against this homogenized backdrop, the bird's pink bill and crisp white eye-ring are striking. Pale wing-bars, while evident, are not standout features. At a distance, the buffy underparts often coalesce into a broad band across the chest.

BEHAVIOR: Owing to its penchant for weedy fields, Field Sparrow is easily overlooked, except in the nesting season, when males call attention to themselves with their sweetly melancholy song. Even then, Field Sparrow is not necessarily conspicuous, preferring to sit low in bushes where its outline doesn't break the horizon. Not skittish—in fact, almost tame—this species often allows observers to get within 20 ft. before flying to another (usually nearby) bush, where it will once again take an inconspicuous perch. The stance is erect. Forages by hopping on the ground for insects or seeds, but also clings to weed stalks, riding them to the ground.

Territories are small, so it is possible to encounter multiple males in close proximity. In winter, the birds may be solitary or found in mixed sparrow flocks, but they often gather in homogeneous flocks of up to 50 birds. Responds readily to pishing. Birds will approach and perch, often conspicuously, as a group on a strategic bush or hedge, and remain for several minutes. Usually flushes from the ground and lands in a prominent bush. Often calls in flight.

FLIGHT: Overall pale and tiny, with a plump body and a long narrow tail. Flight is bouncy, somewhat jaunty, particularly on short flights, and sometimes wandering or tacking on longer flights. Wingbeats are given in a short series followed by a skip/pause.

VOCALIZATIONS: A persistent singer. Sweet, clear, plaintive song is two to three descending whistles followed by a slow musical trill, all given at the same pitch or descending at the end: "*te-u, te-u, te-u, tew-tew-tew-tew-tew.*" The cadence has been likened to a Ping-Pong ball being dropped: "*bounce, bounce, bounce, bump, bump, bump, bump.*" Sometimes gives a two-part trill, the second usually higher-pitched and faster. Call is a high, clear, somewhat husky "*chip.*" In flight, utters a lispy drawn-out "*tsweeest.*"

PERTINENT PARTICULARS: While not particularly shy, Field Sparrow does not breed in close proximity to houses (as does Chipping Sparrow) and is not as common at feeders.

Black-chinned Sparrow, *Spizella atrogularis*

Juncolike Sparrow

STATUS: Fairly common southwestern breeder; more restricted and locally uncommon wintering species. **DISTRIBUTION:** In proper habitat, scattered throughout much of Calif., s. Nev., sw. Utah, Ariz. (excluding southwestern and northwestern portions), s. and cen. N.M., and parts of extreme sw. Tex. south into Mexico. In winter, found in s. Ariz., s. N.M., and w. Tex. south into Mexico. **HABITAT:** Breeds on arid mountain slopes with a mix of rock and brush. Particularly partial to habitats where brush is broken or clustered; avoids wholly overgrown hillsides. Vegetative community may be chaparral, manzanita, scrub oak and sagebrush-juniper, mesquite and grass dotted with ocotillo, and other cactus. In winter, found in rocky canyon washes, scrub, and dry forest edge, often in flocks of sparrows and juncos. **COHABITANTS:** Ash-throated Flycatcher, Rock Wren, California and Canyon Towhees, Black-throated Sparrow, Lazuli Bunting, Scott's Oriole. **MOVEMENT/MIGRATION:** Birds in California, Nevada, northern Arizona, and central New Mexico vacate in winter; where found year-round, birds descend from higher altitudes to lower canyons. Spring migration from mid-March to mid-May; fall from late July through October. VI: 0.

DESCRIPTION: A small brush-loving bird that looks like a cross between a junco and a sparrow. Overall slender with a small round head, a petite pink conical bill, and a disproportionately long and fairly wide tail.

The anomalous plumage is distinctive. Overall darkish and cryptic, Black-chinned has a smoothly plumaged gray head and underparts and a bright pink bill that tell you it's a junco, but the warm-toned brown wings say, "Uh, uh, sparrow!" Also lacks the white outer tail feathers of a junco. The male has a blackened face that is lacking in females, immatures, and most winter birds.

BEHAVIOR: In breeding season, males are commonly seen singing from conspicuous perches. Forages in summer, usually alone, by hopping between branches in the interior of low brush and on the ground. Does not appear in the open for long, but may fly considerable distances to and from its nest. In winter, feeds in small conspecific flocks or joins flocks of sparrows and juncos, including Canyon Towhee, Rufous-crowned Sparrow, Black-throated Sparrow, and Dark-eyed Junco. In winter, feeds mostly on seeds that it gleans on the ground or while perched in low twigs.

Fairly tame, but also very circumspect, even clandestine. Likes to stay hidden. Responds well to pishing in winter, but seems less responsive in summer. Flicks its wings when agitated.

FLIGHT: Flight silhouette is like a slender long-tailed sparrow or junco; the color is cryptically dark, with no white outer tail feathers and no sense of pattern or paleness. Flight is light and bouncy; a quick fluttering series of wingbeats is followed by a terse closed-wing glide.

VOCALIZATIONS: Distinctive. Several quick spaced introductory notes are followed by an accelerating and ascending musical trill. Sounds much like an accelerated Field Sparrow song, but with more "ping" in the notes and a faster acceleration. Call is a soft, high, frail "*see*"; also makes a high "*tip*."

Vesper Sparrow, *Pooecetes gramineus*

The Poor Man's Longspur

STATUS: Common in the Great Plains; uncommon and local across most of the East. **DISTRIBUTION:** Breeds across most of s. Canada (farther north, in the prairies, encompassing much of Man. and all of Alta.) and most of the northern half of the e. U.S. and northern two-thirds of the intermountain West (not in the Central Valley of Calif. and w. Ore. and Wash.). Winters across much of the s. U.S. and most of Mexico south of a line drawn from s. Va. to extreme s. Nev. to cen. Calif. **HABITAT:** Breeds in dry grasslands, often of poor quality or in early stages of succession, including grasslands dotted with shrubs such as scrubby oak and sage or bare earth, cropland going to weeds, recently reclaimed mine sites, grassy roadsides, prairie, pasture, or sagebrush steppe. Generally avoids taller grass and lush wet habitats and prefers areas that offer some elevated perches (fenceposts, shrubs, small trees) for vocalizing. In winter and migration, found in dry weedy fields bordering brush land, weedy roadsides and fencerows, cultivated farmland, grazed grasslands, and occasionally bare earth. **COHABITANTS:** American Kestrel, Killdeer, Brewer's Sparrow, meadowlarks. In winter, may be found in mixed flocks that include Brewer's, Savannah, and Grasshopper Sparrows. **MOVEMENT/MIGRATION:** Vacates almost its entire breeding range. Spring migration from late February to May, with a peak in late March and April; fall from late August through November, with a peak from September to early November. VI: 2.

DESCRIPTION: A large, robust, pale, and cryptically plumaged sparrow that signs its name with bracketing white outer tail feathers. Somewhat bulky, angular, and conical in shape, with a large robust head/chest and a longish body that tapers acutely to the tip of a tail that seems too short to those who use Song Sparrow as a reference. Recalls a longspur in shape and a shadowy, less crisply defined Savannah Sparrow in plumage.

Plumage is overall cryptic and uncontrasting. Pale dishwater gray-brown upperparts and dirty white underparts are overlaid with blurry streaks. Streaks on the chest form a broad-brush bib that usually coalesces into a dark central spot. Streaks along the flanks blur the lines between the upperparts and underparts—imparting a sense of streaky sameness to the bird.

The uniformity of the pattern, however, makes the pale markings about the face really stand out. These marks include pale lores, a pale eye-line, a pale mustache, and (especially) pale cheeks. The negative facial patterning, not the particulars, matters, because whereas the facial pattern of other sparrows is defined by dark markings on a pale surface, this one is more readily distinguished by a pale rendering on a bland surface. Also surprisingly evident is a narrow white eye-ring.

One thing you should *not* depend on is the chestnut shoulder patch. It is generally hidden on adults and absent on immatures. You can rely, however, on the bracketing bright white outer tail feathers that are evident when Vesper Sparrow flies and sometimes visible, as a narrow white edge, when the bird is foraging. When seen from below, the tail appears entirely white.

BEHAVIOR: Forages primarily on the ground. Hops, runs, and forages with a shuffling gait. Fairly eclectic feeder. Sometimes works a patch of ground thoroughly. Sometimes moves frequently to more promising locations. Not at all shy about standing in the open, and often forages, on dirt, far from taller vegetative cover. Crouches as it moves, but periodically stands tall, looking over the vegetation, particularly after shifting location.

A social bird, Vesper Sparrow is often found in winter and migration in small flocks and may mix with other species, most notably Savannah Sparrow, Brewer's Sparrow, and Lark Bunting. Fairly tame, allowing close approach. When flushed, may fly to a nearby vantage point and sit in the open to assess the situation. Singing males always take an elevated perch—sometimes inches from the ground, sometimes high atop a tree—but not necessarily the highest point around. Birds also sing in flight.

FLIGHT: Fairly stocky (somewhat like a longer-tailed Savannah) and overall pale except for a blackish tail bracketed by white outer tail feathers. Flight is strong, with swooping shallow undulations. A quick abbreviated series of rapid flickering wingbeats is followed by a somewhat protracted closed-wing glide (protracted for a sparrow anyway).

VOCALIZATIONS: Variable, usually three-parted song begins with two doubled or three clear, plaintive, spaced whistles (recalls Northern Cardinal), then accelerates into a run-on series of rich warbles and trills. The middle portion of the song suggests Dickcissel, especially at a distance; for example: "*tee tee tee t'chr ch'ch'ch sweesweesweetew.*" Many people hearing Vesper Sparrow are reminded of Song Sparrow. Calls include a high sharp "*spee*" or "*chee*" that recalls Savannah Sparrow.

PERTINENT PARTICULARS: Those who are unfamiliar with Vesper Sparrow often assume that they will have great difficulty distinguishing it from Song Sparrow. Actually, Song Sparrow, with its reddish plumage highlights, generally longish body, longer-appearing tail, and penchant for staying close to brush, is easily distinguished from the pale, grayish, overall stocky and wedge-shaped Vesper. It is the smaller, similarly patterned, and more similarly shaped—not to mention open country–loving—Savannah Sparrow that causes confusion. Several differences are key: Savannah usually has yellow about the face; Vesper, never. Savannah has an acutely short and narrow tail. Vesper has a tail that only seems short because the body is so long. The streaky plumage pattern on Savannah is neat, crisp, and fine. Vesper's plumage is broad-brush and blurred. The bill on most Savannahs is conical and petite. Vesper's bill is fairly hefty. And finally, those white (not just pale) outer tail feathers. They are the last word as the bird goes the other way.

Lark Sparrow, *Chondestes grammacus*

Harlequin Sparrow

STATUS: Common and widespread, primarily western breeder; common but more restricted winter resident. Uncommon and local breeder in the Midwest. Rare vagrant to the Atlantic Coast. **DISTRIBUTION:** Breeds in all states west of the Mississippi R. as well as se. B.C., se. Alta., s. Sask., s. Man., w. Wisc., most of Ill., and very locally in portions of Ind., Ohio, Ky., Tenn., and even N.C. In the West, absent as a breeder in the Cascades, Sierras, n. Rockies (Idaho), and desert areas in Nev., s. Calif., and w. Ariz. Winters in sw. Ore., coastal and interior portions of Calif., s. Idaho, nw. Utah, s. Ariz., s. N.M., and most of Tex. (excluding the panhandle). Also winters south into Mexico. **HABITAT:** At all seasons, prefers open habitat, including park grasslands, pinyon-juniper edges, pasture, assorted agricultural lands, lawns, and roadsides. Seems to prefer the ecotone between grassland and shrub-forest, open habitats dotted with shrubs and trees—such as oak savanna and sage-steppe—and disturbed, degraded, or successional habitats going to weedy succession. In winter, likes tilled agricultural land and oak savanna. **COHABITANTS:** Western Kingbird, Vesper Sparrow, Western Meadowlark, Bullock's Oriole. In winter, often occurs where you find bluebirds, Chipping Sparrow, and shrikes. **MOVEMENT/MIGRATION:** Vacates most of its breeding territory in winter, with some birds retreating as far as southern Mexico. Spring migration from mid-March through May; fall (extremely protracted) from mid-July to December. VI: 3.

DESCRIPTION: A large, distinctive, athletically proportioned sparrow distinguished by its bold facial pattern and long, round, white-rimmed tail. Overall large (larger than Song Sparrow or a junco; slightly smaller than White-crowned Sparrow), with a round head, a sloping forehead, a heavy conical bill, a long, robust, somewhat pot-bellied body, and a very long tail.

Overall fairly pale—pale brown with cryptic patterning above, distinctly whitish below. The overall pale backdrop makes the distinctive black-white-and-chestnut face pattern and black tail with the broad white border stand out. Other marks, like the black central breast spot, are almost superfluous. Immature birds are patterned like adults but the face pattern is brown, not chestnut.

BEHAVIOR: A bird that likes to forage on the ground as well as sit prominently at the tops of trees. Flushed birds frequently take the highest perch around and are partial to utility lines. Birds returning to the ground to feed come in quickly and like to stick the landing, often turning sideways or even 180° for better braking. A slow, methodical, deliberate feeder that walks with mincing steps; also hops, and sometimes runs. Wintering flocks commonly contain 40 or more individuals, but Lark Sparrows are also found in mixed flocks that include White-crowned, Savannah, and Vesper Sparrows as well as bluebirds and House Finch. An unusually quiet, calm, and not easily perturbed sparrow, but when it does take flight, it often flies great distances, commonly out of sight and commonly very high. Among all the sparrows, this one (and American Tree Sparrow) is perhaps the most likely to be seen overhead.

FLIGHT: Large broad-winged and long-tailed profile. Virtually the entire tail seems broadly edged in white (not just the corners, as reality would seem to have it). Flight is strong, straight, steady, and gracefully undulating, and often covers great distance or lasts a long time. More flowing than most other sparrows. Wingbeats are given in a lengthy series followed by a skip/pause.

VOCALIZATIONS: Variable song is a loud, rich, run-on ensemble of doubled and trebled notes, warbles, buzzes, and trills; for example: "*chee'he'he (buzz)chewchewchew ge'e'e'e wrtwrtwrt che'e'e'e chewchewchew.*" Song has a happy playful quality—the bird sounds like a whimsical happy-go-lucky canary—but also seems hurried. Four to eight parts consisting of doubled and trebled notes are interspaced with trills. Song is highly variable. Call is a short, sharp, high-pitched, pure-toned "*tee*" or "*teet,*" or sometimes squeaky "*tchee*"; often given in flight.

PERTINENT PARTICULARS: The broad white border on the tail is not always visible when birds are feeding or perched but is almost impossible to overlook in flight. Also easily seen, and virtually diagnostic, are the bird's distinct white underparts. No other sparrow appears so wholly white below. Front-facing Lark Sparrows perched high on treetops and utility lines almost gleam in the sunlight.

Black-throated Sparrow, *Amphispiza bilineata*

The Dapper Desert Sparrow

STATUS: Common sparrow of arid western lands. **DISTRIBUTION:** Breeds widely in the Great Basin and southwestern deserts. Contiguous range extends from se. Ore. across s. Idaho, south across most of Nev. and e. Calif., east across Utah, w. Colo., most of Ariz., s. and cen. N.M., and the southern and western half of Tex. south into Mexico. Also found irregularly in disjunct pockets in parts of e. Wash., ne. Ore., Calif., s. Colo., and extreme w. Okla. In winter, birds north of s. Calif. and in extreme s. Nev. and s. N.M. withdraw farther south. Very rare on the West Coast; casual vagrant farther north and east. **HABITAT:** Breeds in dry, open, and generally poorly vegetated areas (commonly slopes) dotted with low-lying shrubs or desert vegetation. Common plants include creosote bush, ocotillo, cholla (and other cactus), mesquite, sagebrush, and rabbit brush. At higher elevations, also found in pinyon-juniper and chaparral and on sparse grassy hillsides with scattered brush. In winter and migration, selects similar habitat. Comes to feeders. **COHABITANTS:** In some places, seemingly none, but does occur with Cactus Wren, Black-tailed Gnatcatcher, and Curve-billed Thrasher. In winter, Sage and Brewer's Sparrows. **MOVEMENT/MIGRATION:** Birds in the Great Basin and other northern areas move south in winter. Spring migration from early March to early May; fall from late August to early November. VI: 3.

DESCRIPTION: A sleek medium-sized sparrow (slightly smaller than Sage Sparrow; the same length as Brewer's Sparrow, but plumper) whose striking plumage weaves the soft colors of the desert into a smooth seamless fashion statement. Black-throated has an outsized head, a slender body, and a medium-short tail. The large head (sometimes peaked, sometimes roundly domed) and longish neck (retracted unless bird is feeding) make the rest of the bird seem gaunt. The conical bill seems disproportionately short.

Adult plumage is stunning—a near-seamless melding of brownish grays, grayish browns, and desert buff—the perfect backdrop to accentuate a bold black-and-white face pattern and, most importantly, a crisply defined triangular black bib—a pattern unique among North American sparrows (ignoring Harris's and House). The tail is blackish above and below, with white in the outer corners. (Sage Sparrow shows less white in the corners but shows a narrow white outer edge.)

Grayish immatures, lacking the black throat patch, are faintly streaked above and on the breast and so seem not so seamlessly smooth. They do, however, show a pale rendering of the adult's bold head pattern, including the broad white eyebrow, a feature that the overall scruffier, browner, more coarsely streaked Sage Sparrow does not show.

BEHAVIOR: Forages mostly on the ground, plucking seeds from low-lying plants or, during the breeding season, insects. Forages actively. Moves in short hops across the ground, with long, bouncy, wing-fluttering hops on vegetation, sometimes dancing over the tops of short vegetation—or even hovering! Also pursues insects in flight. Sometimes flicks wings quickly (and almost imperceptibly) above the body, perhaps to flush insects, and jerks its tail.

Males sit conspicuously atop a bush (or fence) and sing loudly; they also sing while foraging. Birds actively defend their territories, and boundary disputes, resulting in spirited tail-chases, are frequent. In winter, birds gather in flocks, often with other sparrow species. Black-throated responds quickly and repeatedly to pishing.

FLIGHT: A long black tail with white outer tips contrasts with an otherwise pale body. Flight is casual and loose, with a quick fluttering series of wingbeats interspaced with both closed- and open-winged braking glides. Flight is usually straight, but birds sometimes veer. Overall impression suggests a windblown object. Calls often in flight.

VOCALIZATIONS: Song is variable, but the pattern is fairly consistent: quick paired introductory notes are followed by one or more musical (or not so musical) trills or warbles that change pitch. Overall quality is sharp, brittle, tinkling. (Almost every description uses the word "tinkling" to describe the quality or elements of the bird's song.) Calls include a weak "*tink*"; a louder "bell-like version"; and a muffled but still sharp "*wheet wheet*" that suggests a distant Curve-billed Thrasher.

Sage Sparrow, *Amphispiza belli*

The Tell-tail Sparrow

STATUS: Common and fairly widespread western breeder; resident or wintering species. **DISTRIBUTION:** Breeds primarily in the Great Basin and Calif., from cen. Wash. to e. Ore. and e.-cen. and s. Calif. to Baja (absent through much of the center of the state), s. Idaho, w. and cen. Wyo., Nev., Utah, w. Colo., n. Ariz., and nw. N.M. Winters in e. and s. Calif., s. Nev., sw. Utah, Ariz., w. and s. N.M., w. Tex., and nw. Mexico. **HABITAT:** Dry open shrub-steppe. Specific habitats include sagebrush, coastal chaparral (Bell's subspecies), and desert scrub (saltbush, rabbit brush, etc.). In winter, found in dry washes and more desert areas with creosote bush, cactus, and mesquite. **COHABITANTS:** Sage Thrasher, Brewer's Sparrow, Black-throated Sparrow, Spotted Towhee. In winter, Le Conte's Thrasher, Black-throated, Vesper, and Brewer's Sparrows (but does not commonly mix with other sparrows). **MOVEMENT/MIGRATION:** Most northern birds and those breeding through much of the Great Basin are migratory; birds in eastern and southern California, southern Nevada, southwest Utah, and northern Arizona are resident. Spring migration from mid-February to early May; fall from late August to late November. VI: 1.

DESCRIPTION: A handsome, athletic, somewhat juncolike sparrow that raises its tail when running and jerks it when sitting. Medium-large (larger than Brewer's or Black-throated; smaller than White-crowned; slightly smaller than Song Sparrow), with a big, round, neckless head, a petite bill, a long body, and a long tail.

Tricolored and conspicuously marked with a grayish head, a plain brownish body, a white throat and belly (bracketed by streaky sides), and an all-black tail. (The outer tail feathers of interior birds have a narrow white edge.) The staring white eye-ring on the dark face is conspicuous. Seen facing you, the white throat and breast pop, making the small, dark, triangular patch on the center of the breast very conspicuous (as are the bracketed black whisker marks on the California Bell's subspecies). The black tail may appear brownish above (until the bird flies).

BEHAVIOR: Almost never found away from brushy confines, where it spends most of its time foraging on open ground. Hops when it feeds; runs when startled or when crossing open areas (with tail conspicuously raised—looking like a sturdy miniature thrasher). Moves quickly through brush with nimble hops. Males typically sing from the tops (or near the tops) of the highest shrubs in their territory. When perched or when foraging on the ground, habitually jerks or twitches its tail upward like a flycatcher.

In winter, forms small flocks, sometimes with other sparrow species. Generally quiet. Shy. Does not respond well to pishing—in fact, usually goes the other way.

FLIGHT: Shows great contrast between the brownish body and black tail (with or without white edges). Flight is bouncy, with a rapid, somewhat fluttery series of wingbeats followed by a closed-wing glide (similar to Black-throated). Tail is usually partially fanned. When flying short distances, pumps its tail.

VOCALIZATIONS: Three- to four-part song is variable but simple and closely approximates the stereotypic "*chirpity chirpity churp*"; for example: "*chrr ch'le chle'le chirp*" or "*chirpe'e'e chirpah.*" The pattern is slow and even, and the notes are low and rough; some are slightly trilled. Call is a soft, almost inaudible, bell-like "*tink*"; also has a weak, high, brittle "*tchee*" or "*tee.*" In confrontational situations, emits a low, muffled, blackbirdlike "*chrrp.*"

Lark Bunting, *Calamospiza melanocorys*

One-Bird Pet Shop

STATUS: Common breeder in the Great Plains; common to abundant wintering species in the Southwest. **DISTRIBUTION:** Breeds from s. Alta. and Sask. south across all but w. Mont. and w. Wyo., all but e. N.D. and S.D., w. Neb., nw. and e. Colo., w. Kans., e. N.M., the Texas Panhandle, and extreme w. Okla. Less common and periodic breeder 100–200 mi. east and west of this core range. Winters from s. and cen. Ariz. to s. and e. N.M. and the western two-thirds of Tex. south into Mexico. Casual in the far West and the East. **HABITAT:** Breeds in dry, open, short-grass prairie, grasslands, and agricultural lands; in winter, uses similar habitat, but also frequents desert communities, harvested agricultural fields, weedy roadsides, and cattle feedlots. **COHABITANTS:** Horned Lark, Vesper and Brewer's Sparrows, McCown's and Chestnut-collared Longspurs, Western Meadowlark. **MOVEMENT/MIGRATION:** All birds vacate breeding territories for the winter. Both spring and fall migration are protracted—spring from mid-February to late May; fall from early July to early November. Nomadic in winter, wandering in large flocks. Also regularly occurs as a vagrant across the United States and southern Canada. VI: 3.

DESCRIPTION: A large, stocky, gregarious prairie sparrow that is almost never found alone. Overtly stocky, even portly, and shaped most like House Sparrow, with a large round head, a heavy conical bill, a plump body, and a shortish tail. (But where House Sparrows seems to crouch, Lark Bunting holds itself erect.)

Breeding-plumage males are unmistakable—an all-black body with a bluish bill and a long white patch tracing the lower edge of the wing. Nonbreeding males, females, and immatures are dingy grayish brown above, pale and heavily streaked below; the streaks often coalesce into a prominent breast spot. Winter males retain some black mottling, especially on the throat and underparts. Overall color, tone, and diffuse pattern suggest female House Finch, except for the face, whose strong pattern recalls Purple Finch or Vesper Sparrow.

In all plumages, wings show a pale line that defines the lower edge of the folded wing. In flight, it blossoms into a large pale patch that dominates much of the inner half of the wing.

BEHAVIOR: Highly social and very gregarious. In summer, males cluster in their territories and engage in virtually nonstop aerial displays that show off their white wing patches to best effect. In winter and migration, birds gather into large vocal, nomadic flocks. Birds feed mostly on the ground—often near bare earth, and always away from overhanging cover—by hopping, walking, and loping. When disturbed, flocks fly a short distance and festoon themselves over fences and bushes before returning to feed, cryptically evaporating into the ground. Birds land with a fluttering braking flourish, and also flycatch.

Fairly tame. Responds moderately well to pishing.

FLIGHT: In flight, the bird's overall stockiness is enhanced. Wings are short and blunt; the tail is likewise short and, when fanned, exceedingly round. The large whitish wing patch stands out against the darker brown (or black) upperwing; the tip of the tail shows a narrow white trim (most evident when the tail is fanned).

Wingbeats are stiff, choppy, and even (as much rise above the body as drop). Standard flight is mostly direct and comprises a long series of wingbeats punctuated by intermittent, irregular, closed-wing glides (makes the flight slightly undulating and jerky). Also sometimes employs an open-winged bouncy glide (a sparrow that glides!) that is sometimes extensive and seems clumsy and unsteady, like the flight of House Sparrow or a baby bird.

VOCALIZATIONS: Song is wonderful. A quick pattern of repeated whistled notes, trills, and throaty pronouncements. The pattern recalls a mockingbird: notes repeated several times, then a shift to a new note. Many of the notes themselves are like those used by Northern Cardinal. (But cardinals repeat the same note ad nauseam; Lark Bunting uses the cardinal repertoire in one song.) Just a single singing Lark Bunting sounds like an entire pet shop — but where there is one male, there are usually more.

Call is a distinctive low "*woit*" or "*whoy*" that recalls the "*boink*" sound of a cartoon character being struck on the head — but softer.

Savannah Sparrow, *Passerculus sandwichensis*

Pollen-faced Sparrow

STATUS: Common and widespread grassland species; as a breeder, migrant, or wintering species, found everywhere in North America north of Central America. **DISTRIBUTION:** Breeds across the whole of mainland Canada (also Nfld.) and Alaska south to n. N.J., along the Appalachians to all but s. Ga., n. Ky., s. Ind., s. Ill., n. Iowa, n. Neb., w. Colo., nw. N.M., n. Ariz., Utah, Nev., and s. Calif. In winter, found south of a line extending from Cape Cod, cen. Md., w. Va., w. N.C., n. Ga., sw. Ky., cen. Mo., s. Kans., n. N.M., cen. and nw. Ariz., sw. Utah, s. Nev., and cen. Calif. Permanent resident (subspecies) in Calif. and Baja marshes. **HABITAT:** Prefers open, grassy, often sparsely vegetated habitats, most commonly pastures, cultivated fields, tundra, weedy fields, tidal wetlands, roadsides, dunes, and airports. **COHABITANTS:** Northern Harrier, American Kestrel, Short-eared Owl, Vesper Sparrow, meadowlarks. In winter, may mix with other grassland sparrow species. **MOVEMENT/MIGRATION:** Owing to the latitudinally extensive breeding range, both spring and fall migration are protracted. Spring migration from March to early June (when birds reach northernmost reaches of the range), with a peak in April; fall from early August to mid-November, with a peak in October. VI: 4.

DESCRIPTION: A small, compact, fairly streaky, and neatly trim sparrow with a very pointy bill and a short tail — suggests a neat, trim, crop-tailed Song Sparrow. Medium-small and compact (larger than Grasshopper Sparrow; smaller than Song Sparrow or Vesper; the same length as Brewer's or Clay-colored), with a small, narrow, pointy bill, a stocky body, a very short tail, and pale flesh-colored legs. The head is generally rounded and usually a little rough or shaggy on top.

Depending on the subspecies, individuals may be generally brownish or grayish, but except for two essentially coastal subspecies (see "Pertinent Particulars"), birds appear very neat, trim, and crisply patterned, with dark (usually black) streaks down the back and across the breast and sides; the central breast spot is usually conspicuous. Almost always shows a telltale splash (sometimes just a dash) of yellow on the lores, the eyebrow, and or the face. So it looks like a neat, trim, petite-billed, crop-tailed Song Sparrow with pollen on its face.

BEHAVIOR: A social sparrow; except when breeding, often found in small flocks, but does not commonly associate with other sparrows. Breeds in taller grasses and sedge; in winter and migration, also forages in dry sparsely vegetated habitat (such as grass-bordered asphalt roadsides, the edges of tilled fields, or grassy dunes). Usually tame, allowing close approach. When flushed, birds often fly to the nearest tree or cover (usually not far away) and commonly perch in the open, often for extended periods. If no high perches are available, Savannah Sparrow drops back into the grass. Sits easily on springy twigs and grasses, flicking its wings only when off balance. Usually forages on the ground, where it normally walks, sometimes runs and hops. Very vocal. Responds well to pishing. Flushed birds fly up (10–30 ft. high) and fly far before dropping to the ground. Often calls in flight.

FLIGHT: Overall compact with a very short, trim, triangular tail and proportionally long wings. Except for the pale outer tail edges (usually apparent) and the dusting of yellow about the face, appears uniformly streaky above; the streaky chest makes the bird appear bibbed. Flight is straight and strong, with slightly to markedly bouncy undulation. Wingbeats are rapid and mostly steady, but interrupted at regular intervals by a terse skip/pause.

VOCALIZATIONS: Song is high-pitched, buzzy, and weak. Begins with several short notes followed by a thin musical hiss that drops in pitch at the end: "*t-t-t-tseeee'tsrrr.*" Sounds much like Grasshopper Sparrow, but with an extra note at the end. Call and flight call are a high, thin, prolonged "*tseep*" or "*tsee*" that goes thin at the end. Also emits a ticking "*tip*" when agitated.

PERTINENT PARTICULARS: Two somewhat restricted regional forms of this sparrow are different enough to cause confusion. "Ipswich" Sparrow, *P.s. princeps*, breeds on Sable Island, Nova Scotia, and winters in coastal dunes and beaches from New England to Georgia. "Ipswich" Sparrow is distinctly larger and paler than other Savannah Sparrows, with a more muted pattern; though it sometimes associates with other Savannah Sparrows, it is just as likely to be found alone. The "Large-billed" Sparrow, *P.s. rostratus*, found in fall and winter in coastal southern California and the Salton Sea, is also considerably larger, paler, and less crisply marked than the typical Savannah Sparrow and is best distinguished by its more bulbous and pale-based bill, which gives it the head shape of Seaside Sparrow. Often forages on rocky shorelines devoid of vegetation.

Also, several subspecies residing in the *Salicornia* marshes of California are much darker than typical Savannahs.

Grasshopper Sparrow, *Ammodramus savannarum*

Buff-breasted Sparrow

STATUS: Variously common to uncommon but widespread breeder and southern winter resident. **DISTRIBUTION:** Breeds across much of the e. and cen. U.S. and s. Canada. Absent as a breeder from coastal N.C. to e. Tex. (except for isolated populations in sw. Ga. and s.-cen. Fla.) across much of the U.S. and Canada west of the Rockies but with populations in s. B.C., e. Wash., e. Ore., s. Idaho, sw. Wyo., n. Utah, coastal and cen. Calif., and se. Ariz. In winter, found across the s. U.S. from coastal N.C., all but northern portions of S.C., Ga., Ala., Miss., La., s. and cen. Tex., sw. N.M., s. Ariz., and Calif. south into Mexico. **HABITAT:** Expansive dry grassy fields with sparse, less vegetated places, bunched grass, or (more commonly in the prairies) grasslands with some intruding brush. In the East, commonly found in fields with broom sedge (also in alfalfa fields and airports); in the prairies, favors little bluestem. In winter, often occupies denser unbroken grasslands. **COHABITANTS:** Savannah and Baird's Sparrows, Chestnut-collared Longspur, Eastern and Western Meadowlarks. **MOVEMENT/MIGRATION:** Spring migration from mid-April to mid-June; fall from early August to mid-November. VI: 2.

DESCRIPTION: A small, flat-headed, spindly-tailed grassland sparrow with a buffy face and breast and a persistent insectlike voice. Small (slightly smaller than Savannah, Henslow's, and Baird's Sparrows; slightly larger than Le Conte's), with a wedge-shaped face (the combination of a flat head and a large, conical, somewhat bulbous bill), a short plump body, and a very short, exceedingly narrow, sparsely feathered tail (usually notched, sometimes splayed).

Overall pale and plain—particularly plain-faced. Some descriptions focus on the complexity of the pattern, but unless you are close, all features melds into cryptic uniformity. Adults are mostly grayish above, with pale edged feathers. (Distant birds have a frosted appearance.) Plain unstreaked underparts show a buffy breast-band and pale underparts. Because the head/face shows a dark crown with a pale central stripe and a small dark spot below and behind the eye on an otherwise plain buffy face, the dark eye stands out. At a distance, the narrow white eye-ring and touches of red in the upperparts are difficult to discern (the bird sometimes shows a reddish shoulder), but the gray-and-red collar becomes more contrasting and more evident. Juveniles are paler, overall warmer-toned, and somewhat scaly-backed, with a breast overlaid with a pattern of narrow beaded streaks.

BEHAVIOR: Except when males are singing, very hard to see. Feeds on the ground, often near the edge of more open grassy areas, hunting for insects (eats mostly seeds in winter). Walks and runs with a crouch. Almost always solitary, although, in winter, may feed with or close to Savannah and Vesper Sparrows.

Singing males vocalize from grassy stalks (also from fenceposts, wires, bushes, even utility lines), singing persistently for several minutes (at least) before flying 20–40 yds. to shift perches. Favors certain perches and returns continuously. Sings throughout the day and not uncommonly at night. When flushed, usually flies less than 20 yds. before landing on the ground (less commonly on a branch or fence wire). Responds poorly to pishing. Flushes close (5–10 ft.) and flies low, below eye level, for a short distance before dropping into the grass or vaporizing into a shrub.

FLIGHT: Overall small-onto-tiny, plain, and pale, a very short tail and short rounded wings. Flight is weak, somewhat buoyant, short, low, and direct or sometimes tacking (zigzagging). Wingbeats are rapid, constant, and quivering. Lands with a braking flutter and a gentle controlled crash.

VOCALIZATIONS: Basic song is insectlike—a high, quick, ticking hiccup followed by high, weak, buzzy trill: "*tic-up t/cheeeeeeeee.*" Call is a bisyllabic "*trillic*" (sometimes has a three-note quality). Also, on territory, renders a song that is a rising-and-falling jumbled cascade of notes; resembles a high, thin, hurried Winter Wren.

PERTINENT PARTICULARS: At close range, adults and young show a touch of bright yellow along the bend of the lower edge of the wing. It's apparent because it's so incongruous, and it helps distinguish juvenile Grasshopper Sparrow from the similarly sized and shaped Le Conte's.

Baird's Sparrow, *Ammodramus bairdii*
Dribble-bibbed Sparrow

STATUS: Uncommon, restricted by geography and habitat, and just plain hard to find. **DISTRIBUTION:** Breeds in the northern prairies from se. Alta., s. Sask., and sw. Man. to all but w. and s. Mont., all but extreme e. N.D., and n. S.D. Infrequently occurs in nw. Minn. Winters primarily in n.-cen. Mexico, but also found in se. Ariz., extreme sw. N.M., and southern reaches of trans-Pecos Tex. **HABITAT:** Most commonly found in dry, medium-tall, mixed-grass prairie with some shrubs or last season's clumped (and elevated) grass. Most numerous in, and restricted to, large tracts of native prairie, but also occurs in fields comprising non-native species, hay fields, and lightly grazed pasture. In winter, found in tall dense grasslands or grazed pasture. **COHABITANTS:** Sprague's Pipit, Savannah and Grasshopper Sparrows, Chestnut-collared Longspur. **MOVEMENT/MIGRATION:** Spring migration from mid-April to early June; fall from late August to late October. VI: 2.

DESCRIPTION: An ochre-faced Savannah Sparrow with a melodious voice. Small and fairly stocky (larger than Grasshopper; smaller than Vesper; the same size, shape, and proportions as Savannah Sparrow), with a medium-sized bill, a rounded head, a robust body, and a short but sturdy tail. Because of its size, relative proportions, and several shared plumage traits, you would do better to think of Baird's Sparrow as a slightly paler, more sparsely and contrastingly marked Savannah Sparrow than a streaky-breasted Grasshopper Sparrow.

Overall contrasting and well patterned, with traits that, again, resemble Savannah Sparrow: a well-patterned face, a sparse dribble-streaked breast, and a pale or pinkish bill. The back shows blackish streaks on a brown backdrop. The breast is marked with short, dark, sparse, and widely spaced streaks that look like chocolate dribbled down a shirt. The face shows a pale eyebrow and a double-striped throat (a dark mustache and a bold lateral throat stripe). The more complex pattern makes the face more akin to Savannah Sparrow than the plainer-faced Grasshopper Sparrow. The ochre- to buff-colored wash on the face stands out, in good light, but may be disquietingly indistinct on overcast days. Juvenile Baird's Sparrows are more scaly-backed but otherwise like adults.

Savannah Sparrow is more heavily and obviously streaked, particularly down the sides. Although the faces of some Savannah Sparrows can be lavishly washed with yellow (not ochre or buff!), on most individuals the yellow is limited to the lores. Juvenile Grasshopper Sparrow, like Baird's, shows a streaked breast and ochre-colored face, but is smaller, overall paler, flatter-headed, more bulbous-billed, more spindly-tailed, plainer-faced, and more faintly and finely streaked on the breast.

BEHAVIOR: In spring and summer, male Baird's perches on an elevated platform—a bush, old grass, a sturdy plant—that nevertheless is often below grass-top level. Sings persistently from this perch, changing perches periodically (usually flies less than 50 yds.). Rarely perches on wires or posts. Beyond this, there is little to say. The bird spends most of its life on the ground in the grass. The only other way it can be seen is when it flushes and drops back into the grass. Responds poorly to pishing.

FLIGHT: In size and shape, somewhat like Savannah Sparrow, but with rounder and broader wings. Overall pale, but the dark streaks on the back can be surprisingly distinctive. Outer tips to the tail may show whitish, but the bird lacks the more obviously pale outer tail feathers of Savannah, and also unlike Savannah, Baird's does not call in flight. Flushes and flies like Grasshopper Sparrow; not like Savannah. Flight is straight and fairly strong, and shows a more tempered rise and fall than the bouncier undulations of Savannah. Rapid wingbeats fall somewhere between the strong flap of Savannah and the fluttery flap of Grasshopper and are given in a lengthy series followed by a terse closed-wing glide.

VOCALIZATIONS: Melodious song (particularly in comparison to the buzzy reedy songs of most other prairie sparrows) begins with several sharp notes followed by a melodious lower-pitched trill. Calls reportedly include a whiny "*deerdeerdeer*," a sharp "*chip*," a lower "*chrp*," and a high "*tseet*" like Grasshopper Sparrow.

Henslow's Sparrow, *Ammodramus henslowii*

The Prairie Hiccup

STATUS: Uncommon, restricted, and localized within its range. **DISTRIBUTION:** Breeds very sparingly in the northeastern grasslands from cen. N.Y. south, with more birds in the Appalachians and Midwest from cen. Pa., w. Md., and w. W. Va. across Ohio, Ind., s. Mich., n. Ky., Ill., se. Minn., most of Iowa and Mo., se. Neb., e. Kans., and extreme ne. Okla. Isolated coastal populations occur in N.C. Winters in the se. U.S. from coastal N.C. south into all but extreme s. Fla. and west through s. Ga., Ala., Miss., s. Ark., all of La., and e. Tex. Very rare migrant along the Atlantic Coast. **HABITAT:** Breeds primarily in tall-grass prairie that has a dense ground litter and has not been burned for at least three years. Avoids smaller grasslands, grasslands adjacent to woodland edge, and grasslands with a lot of intruding shrubs. (But having a few shrubs to use for singing perches is okay.) Also found in weedy fields, planted reclaimed strip mines and, in coastal areas, the upper edge of coastal marshes (formerly?). In winter, most concentrated in open wet or dry pine savanna with an understory dominated by grasses (principally broom sedge and wiregrass) and sedges. **COHABITANTS:** Dickcissel, Grasshopper and Savannah Sparrows (which favor more open, less mature grasslands), Eastern Meadowlark, Bobolink. In winter, Bachman's Sparrow. **MOVEMENT/MIGRATION:** The entire population relocates south in winter, but birds are infrequently encountered. Spring migration from March to late May, with a peak in April; fall from early September to late November, with a peak in October. VI: 2.

DESCRIPTION: A small, fairly dark, secretive sparrow with an olive or greenish head, a reddish body, and an innocuous song. Overall compact and top-heavy (smaller than Savannah; the same size as Grasshopper Sparrow and Le Conte's), with a big head, a heavy, somewhat bulbous bill, a plum-shaped body, and a fairly short, narrow, and scraggly-tipped tail. The head is flattish but, perhaps because it's so large, less so than Le Conte's or the sharp-tailed sparrows. The tail is short, especially in flight, but when perched, this trait seems less obvious than on Grasshopper Sparrow.

Overall plumage is dark. Back, wings, and tail are reddish brown (bordering on maroon!) with black streaks; underparts are dingy white or grayish with a streaky chest band and streaks extending down buff-washed flanks. (*Note:* In winter, in fresh plumage, the pale feather edges give the backs of some individuals a frosted look.) The greenish or olive-colored head varies in intensity and color—from very green to yellowish. The color and intensity notwithstanding, the pale uniform backdrop makes the head and facial pattern pop and the pinkish bill fairly conspicuous. The dull reddish tail may be very conspicuous on backlit vocalizing males.

Juveniles are like adults, but less streaked on the breast and more reminiscent of the paler, buffier Grasshopper Sparrow; like adults, however, juvenile Henslow's shows rufous wings and tail and a greenish head.

BEHAVIOR: A ground-skulking species that walks and runs through matted grass, flushes with great reluctance, and grounds itself after a short weak flight. Vocalizing males, however, are often very conspicuous: perched on a grassy stem or bush, they project their insectlike call skyward with a backward snap of the head. Found in clusters, with several different males within voice range. Males generally remain exposed for several minutes before flying to another low perch that may be over 100 yds. away. A somewhat clumsy bird. Wing-flutters and jerks its tail when newly perched (mostly for positioning and balance), but remains motionless once stabilized, except for the head. Responds poorly to pishing.

Seems partial to certain perches. Frequently returns to the same blade of grass or bush as it maps out its territory with song. Flushes at 5–10 ft.; flies a short distance (10–20 ft.) before dropping. Difficult to flush again.

FLIGHT: Effortful and slightly unsteady, but not weak. Profile shows short, blunt, wide wings and an extremely short tail. Flights are generally straight, low, and short (less than 100 yds.), with periodic (and barely perceptible) bounces or jerks. Wingbeats are rapid, fluttery or batty, and mostly steady, but a fractionally frozen down stroke intermittently breaks the cadence with a skip or a hesitation.

VOCALIZATIONS: Song is a terse, squeaky, bi- or trisyllabic hiccup: "*t'sl'eet*" or "*t'les'sr'leep.*" In tonal quality (and even pattern), has a Horned Lark squeakiness ("*tweedle-ee*"). Second syllable is rising and slurred, and while it ends abruptly, it has no hard consonant sound (that is, there's no hard "k").

Le Conte's Sparrow, *Ammodramus leconteii*

Bobolinklet

STATUS: Fairly common, but secretive bordering on clandestine. **DISTRIBUTION:** Breeds primarily in the prairie regions of Canada and the U.S.; found across the s. Northwest Territories, ne. B.C., Alta., all but s. Sask., w. and s. Man., ne. Mont., cen. and e. N.D., ne. S.D., n. Minn., n. Wisc., and the Upper Peninsula of Mich. Also occurs in e. Mont. and scattered locations in w. Ont. and cen. Que. Winters in the s.-cen. U.S., including s. and cen. Mo., s. and sw. Ill., w. Ky., w. Tenn., e. and s.-cen. Okla., Ark., e. Tex. (south to Corpus Christi), La., Miss., Ala., the Florida Panhandle, and in small numbers up the Atlantic Seaboard to se. Va. Casual visitor in the w. U.S. and in the Northeast states. **HABITAT:** Over most of its range, associated with wet grassy and sedge meadows often adjacent to bogs, lakes, and marshes, but also found in dense, dry, grassy pastures, fallow fields, hay fields, and grain fields. In winter, occurs in an assortment of tall, dense, usually dry, weedy fields and grasslands. **COHABITANTS:** Sedge Wren, Savannah Sparrow, Grasshopper Sparrow, Bobolink. **MOVEMENT/MIGRATION:** Spring migration from late March to late May; fall from early September to early December. VI: 3.

DESCRIPTION: It resembles a tiny winter Bobolink. A dollop of a sparrow—in fact, the smallest sparrow (distinctly smaller than Savannah Sparrow; slightly smaller than Nelson's Sharp-tailed and Grasshopper Sparrows). Overall dumpling-shaped, with a flattish head, a small bill, and a short, thin, scraggly-tipped tail. Nelson's Sharp-tailed Sparrow has a pointier face and a pointier, thinner bill and is overall slightly more elongated—more pickle- than dumpling-shaped.

Brownish-backed and warmly washed below (you'll be lucky to see the white belly and dark streaks down the sides). Three things stand out: (1) the boldly striped head pattern—black-on-orange buff—that resembles a winter Bobolink; (2) the streaked mantle or shawl, with the streakiness extending onto the sides of the breast and contrasting with the plain nape; (3) the orange-buff wash on the face (and sometimes the breast). Three things do not stand out: (1) the fine pinkish or plum-colored beaded streaks on the neck (just looks plain at a distance); (2) the touches of white in the upperparts; (3) a triangular cheek patch. (What you'll note is the black Bobolink-like line extending behind the eye.) Juveniles resemble adults, including the Bobolink face pattern, but usually lack the contrasting orange on the face.

BEHAVIOR: Presumed to spend most of its life foraging on the ground, hidden in tall grass, from which it flushes (from almost underfoot), then flies a distance of 30–50 ft. before dropping back into its grassy fortress. In winter, solitary. In summer, possibly colonial.

More easily located (but not necessarily viewed) in spring and summer, when males vocalize on territory. Usually perches low, on a sturdy tuft of grass or weed but hidden below the surrounding tops of taller grass. Sometimes sings from taller open perches (until displaced by a Savannah Sparrow). Never perches on fences or utility lines. Sings persistently and often remains on the same perch for long periods. When nervous, flicks its wings high over the body and jerks its tail. Moderately tame. Responds poorly to pishing, but sometimes surprises you. Difficult to flush more than twice.

FLIGHT: Small, overall compact, and oval-shaped, with an abbreviated tail. Wingbeats are deep and fluttery, given in a rapid flurry followed by a skip/pause. Overall flight is weak, undulating (sometimes jerky or bouncy), and clearly not as strong and controlled as Savannah Sparrow.

VOCALIZATIONS: Short three-part song is mostly buzz. Begins with a barely audible, terse, crackling prefix, then a thin, even, flat buzz, followed by a closing note so abrupt that it is more accurately described as a punctuation point than a syllable: "(crackle) *e'e'e'e'e'e'e'e tp* or (.)." Savannah Sparrow's song is more musical and complex and ends with a flourish (a distinct single note), not a barely audible period. Nelson's Sharp-tailed sounds like a weak slurred hiss. Le Conte's Sparrow's call is a weak soft "*tih*" or "*t/chu.*"

Nelson's Sharp-tailed Sparrow, *Ammodramus nelsoni*

Orange-bibbed Sparrow

STATUS: Common but secretive. **DISTRIBUTION:** Breeds from ne. B.C. and sw. Northwest Territories across n. and e. Alta., s. and cen. Sask., s. Man., extreme ne. Mont., cen. and e. N.D., ne. S.D., and n.-cen. Minn. Also found coastally along the southern shore of Hudson and James Bays and the Gulf of St. Lawrence in e. Que. and coastal Nfld., the Maritimes, and coastal s. N.H. Winters along the Atlantic and Gulf Coasts from Long Island to the Mexican border (with prairie birds found primarily along the Gulf of Mexico). Also found rarely but annually in the marshes of Calif. **HABITAT:** Except for birds found in the Maritimes, Maine, New Hampshire, and James Bay, breeds in freshwater marshes dominated by tall rank grasses (such as cord grass, cattail, and bulrush) as well as in dry upland fields. Coastal birds are found in salt marshes and wet grasslands flanking rivers and lakes (less commonly in dry fields). In migration, found in both wet fields and marshes. In winter, found mostly in tidal salt or brackish marshes, where it favors the taller grasses flanking creeks and wetter areas. **COHABITANTS:** During breeding, Sora, Black Tern, Le Conte's Sparrow, Red-winged Blackbird. In winter, Saltmarsh Sharp-tailed Sparrow, Seaside Sparrow. **MOVEMENT/MIGRATION:** Spring migration from late April to early June; fall from early September to mid-November. VI: 2.

DESCRIPTION: A small, pointy-faced, short-tailed marsh sparrow with an undifferentiated orange face and breast and a penchant for staying hidden. Small (slightly smaller than Saltmarsh Sharp-tailed Sparrow; slightly larger than Le Conte's Sparrow or Grasshopper Sparrow), with a spiky bill, a round-topped head, a plump body, and a short tail. Slightly smaller-billed and not as acutely pointy-faced as Saltmarsh Sharp-tailed Sparrow.

Adults have grayish necks and backs and orangy brown wings. The face and breast show bright to dull orange with no contrast in color or tone. (The face of Saltmarsh Sharp-tailed is bright orange contrasting with a paler buff or orange wash on the breast.) The head appears dark-crowned with an orange-bordered gray face. (Head of Le Conte's looks striped.) The orange breast, overlaid with dull, faint, frail, or blurry streaks, is rich, full, and sharply truncated, giving the bird a bibbed appearance. Saltmarsh Sharp-taileds are so heavily streaked on the breast that it's hard to see the orange wash, and because many individuals have whitish throats and breasts, the underparts show an open pale orange vest as opposed to a complete bib. Coastal Nelson's Sharp-tailed is duller than interior birds, but the basic pattern remains the same — most notably, the orange tone of the face is the same tone as the breast.

BEHAVIOR: A somewhat nocturnal skulker. Males sing from grassy perches (and during aerial displays), but much singing is done at night and twilight. Forages on the ground. Flushes reluctantly, and then commonly flies only a short distance. In winter and migration, forages in the taller, denser, wetter vegetation along tidal creeks and shores, and at night and high tides gathers in tight stands of vegetation (such as phragmites) adjacent to feeding areas. Except when roosting, solitary and does not flock, although multiple birds may forage in the same habitat or area.

Fairly responsive to persistent squeaking; birds will take visible perches on tall grasses and reeds and sometimes approach.

FLIGHT: Appears short-tailed and up-angled. Flight is straight, low (no higher than the grass tops, like Le Conte's and Grasshopper), effortful, and usually short, on rapid, fluttery, uninterrupted wingbeats. Often lands with a crash or a fluttery hover.

VOCALIZATIONS: A frail sizzling hiss that drops off at the end, Nelson's Sharp-tailed's song has been aptly described as the sound of a hot poker thrust into water: "*p'shhhh-shr.*" Call is a hard sharp "*tik.*"

PERTINENT PARTICULARS: In late April and May along the Atlantic Coast, breeding Saltmarsh Sharp-tailed Sparrows vacate the taller, ranker, wetter, and low-lying salt marsh vegetation (cord grass) in favor of the shorter, higher, drier (*patens*) grasses where they breed. Nelson's Sharp-taileds linger in the cord grass.

Saltmarsh Sharp-tailed Sparrow, *Ammodramus caudacutus*

Spiky-billed Sparrow

STATUS: Common but restricted by geography and habitat. **DISTRIBUTION:** Breeds coastally from s. Maine to the Delmarva Peninsula. In winter, found coastally from Long Island to cen. Fla. **HABITAT:** Tidal marshes and upland edge, particularly marshes dominated by short *Spartina* patens grass (or salt hay); also favors cord grass and black grass. Frequently found in taller grass along tidal creeks; may roost in phragmites, particularly in winter and at high tide. **COHABITANTS:** Clapper Rail, Seaside Sparrow. **MOVEMENT/MIGRATION:** Spring migration from mid-April to late May; fall from late August to November. VI: 0.

DESCRIPTION: A pointy-faced salt marsh sparrow distinguished by an orange face contrasting with whitish underparts. Smallish (smaller than Seaside Sparrow; very slightly larger than Nelson's Sharp-tailed Sparrow) and wispy-looking, with a flat head, a spiky-billed long-faced appearance, a slender oval body, and a short scraggly tail.

Overall paler than Seaside Sparrow. Distinguished by a rich orange face pattern and heavily streaked but conspicuously paler whitish underparts. Streaks and a pale orange wash form a broken breast-band (pale in the middle) that separates the whitish throat from the white belly. The orange face contrasts markedly with the white throat.

Nelson's Sharp-taileds (both interior races and Atlantic) are shorter-billed and have a pointy orange face that shows essentially no contrast with the throat or breast. The breast-band on Nelson's is brighter and complete, not broken or vested, and the dark streaks on the underparts are less blurry and weaker, less dominating than the streaked breast and flanks of Saltmarsh (also less broken or distinctly spotted). Juvenile Saltmarshes are much like adults but more extensively and brighter orange below.

BEHAVIOR: Forages mostly on the mud among dense salt marsh grasses or along tidal creeks by walking or hopping. Sometimes chases or leaps for prey. Commonly runs when crossing open areas. Flushes reluctantly (at 5–15 ft.) and usually flies only a short distance (less than 50 yds.). Rarely found away from tidal areas.

In winter, may be found in small loose flocks or with Nelson's Sharp-tailed and Seaside Sparrows; also roosts with these species in taller marsh grasses, including phragmites. Responds somewhat to most pishing, but can be coaxed to hitch itself up on taller perches by squealing; if you are persistent, the bird may subsequently approach for a closer look.

FLIGHT: Distinctly short-tailed and up-angled posture. The orange face and white belly project at great distances. Wingbeats are rapid, sputtery, and mostly steady with intermittent skips. Flight is direct, low, without undulations; though its flight is effortful, Saltmarsh Sharp-tailed seems a stronger and more agile flier than Seaside Sparrow. During breeding season, some flights may be lengthy (several hundred yards) and frequent; otherwise flies low, short, and close to the ground—just above the marsh grass, and never higher than 10 ft. above the ground. Usually disappears when it lands (unless responding to pishing).

VOCALIZATIONS: Seldom-heard song is a slow, weak, thin, two-parted sizzle, "*'eee'rl e/raaah,*" that is more stumbling or slurred and less wheezy than Nelson's Sharp-tailed. The first part, higher, thinner, and more hissing/sizzling; the second part, lower, flatter, and more rasping. The pattern might recall Red-winged Blackbird's "*tur-a-ling*" but with the song played at high speed and the "*ling*" coming first.

PERTINENT PARTICULARS: Distinguishing this species from Seaside Sparrow is easy. If you see an all-dark sparrow with a conspicuous white throat, it's not Sharp-tailed. Yes, Sharp-tailed has a white throat, but at a distance, given the lack of contrast, it doesn't stand out like the throat of Seaside Sparrow. *Note:* Immature Seaside Sparrow has a white belly, but also shows dusky yellow lores on an all-dark face.

Distinguishing this species from the interior race of Nelson's Sharp-tailed is also fairly easy. If the bird's face and breast are uniformly bright orange (that is, there's no contrast between the richer orange face and the paler orange wash on the breast), it is not Saltmarsh Sharp-tailed. Separating Atlantic Nelson's Sharp-tailed Sparrow from Saltmarsh is more difficult. The Atlantic Nelson's is overall more washed-out, less contrasting, and uniformly bland, with a slightly smaller bill and very blurry streaking on the orange-washed breast. The streaking on Saltmarsh Sharp-tailed is crisp, dark, and dominating.

And of course, the throat of Saltmarsh Sharp-tailed Sparrow is white; the throat of Nelson's Sharp-tailed is orange.

Seaside Sparrow, *Ammodramus maritimus*

Sparrow Dressed for a Funeral

STATUS: Common in its limited range and specific habitat. **DISTRIBUTION AND HABITAT:** With small exceptions, Seaside Sparrow is restricted to the tidal *Spartina* marshes of the Atlantic and Gulf Coasts from coastal N.H. to Tex. (excluding n. New England and parts of the Florida Peninsula). One race, localized to a tiny portion of Fla.'s Everglades, the Cape Sable Sparrow, is resident in brackish and freshwater marshes dominated by saw-grass. From the mid-Atlantic south, a permanent resident within its range; in winter, however, populations shift somewhat, with many northern birds relocating south. A few birds wander north into Maine in late summer and fall; larger numbers occupy Fla. coastal marshes south of their breeding range. **COHABITANTS:** Clapper Rail, Forster's Tern, Saltmarsh Sharp-tailed Sparrow. **MOVEMENT/MIGRATION:** Spring movements in April and May; fall from late August to early November. VI: 1.

DESCRIPTION: A large sparrow easily identified by its habitat, size, shape, and distinctively undistinguished plumage—a sparrow dressed for a funeral. The head is big and flat. The long bill seems swollen—thick, long, and slightly downturned (sparrowlike only with an imaginative stretch). The body is heavy and somewhat potbellied (particularly in flight). The tail is short, and the tip jagged-edged or deeply notched. When the bird is singing, the tail is often fanned with individual tail feathers splayed.

Plumage is overall gray-brown and somewhat shabby, with diffuse dark streaking above and below and an infusion of brown in the wings. Yellow lores, a white mustache stripe, and (particularly) a yellow highlight at the bend of the wing don't project well over great distances or through heat waves. What does project is an all-dark sparrow with an obvious white throat. Because Saltmarsh Sharp-tailed Sparrow (which often breeds in the same general habitat) is overall paler, particularly on the breast, its white throat is not as distinct at a distance. Seaside's expression is unfriendly or sinister.

Gulf Coast birds are overall warmer, browner, and more heavily streaked. Cape Sable Sparrow is distinguished by an olive cast about the head and back (in good light, some Atlantic Coast birds also show this trait), but primarily by distinctive streaking on a paler breast.

BEHAVIOR: A sluggish clumsy sparrow. Seaside Sparrow forages by walking (or running!) on the marsh surface, and less frequently on exposed mud. Because its habitat is not easily navigated by humans, Seaside is most easily seen during the breeding season, April to July, when males sing from elevated perches on open marsh—although "elevated" might be something of a stretch. Often only the bird's head is visible in the marsh grass. The posture is erect, and the singing persistent. The bird shifts perches often.

In winter, birds forage singly. At night or at high tide, they gather to roost (often in stands of phragmites). Seaside prefers to forage in the taller cord grass found along tidal creeks than in shorter marshland grasses. Those hoping to find the birds outside the nesting season are most successful walking the tidal edges of creeks.

Although difficult to find and flush (flushes at 10–20 ft.), responds to pishing (especially squealing), and will perch, exposed, for a better vantage. When disturbed, wing-flicks with an exaggerated rowing motion.

FLIGHT: Flight is direct, straight, and hurried. Seaside rarely flies far (no more than 100 ft.) and rarely more than a foot or three above the marsh. Wingbeats are fast and fluttery. The head is elevated, the body inclined. In flight, note the short scraggly tail and distinctly unmarked, uniformly plumaged wings. Landings are a barely controlled crash that often submerges the bird in the grass. Singing (or curious) birds hitch themselves up stalks and sit with assumed equanimity. When breeding, also has a towering display flight.

VOCALIZATIONS: Distinctive but variable. Commonly in three parts. Begins with a two-part buzzy trill or a musical slurp, followed by a prolonged insectlike trill with a ringing quality and a hard ending: "*ching-ching-Chinnnnng*" or "*chur-a-liiiiiing.*" Song resembles a distant, buzzier, less musical Red-winged Blackbird and may be confused with Saltmarsh Sharp-tailed Sparrow. At a distance, the beginning of the song attenuates and becomes a single buzzy introductory note or phrase often overlaid with a clear note. Call is a soft, low, persistent "*chep*" or "*chup*" uttered with an irregular cadence; when agitated, utters a high-pitched, excited, repetitive "*tink, tink, tink.*"

PERTINENT PARTICULARS: Never mind the beginning of the song—just listen for the trailing ringing "*Chinnnng.*" And never mind the particulars—just look for a big all-dark sparrow with a white throat sitting one sparrow-length above the salt marsh.

"Red" Fox Sparrow, *Passerella iliaca*

Eastern Fox Sparrow

STATUS: Common northern breeder; common winter resident in the Southeast. **DISTRIBUTION:** Breeds across n. Canada from Nfld., Lab., Cape Breton I., N.S., and N.B. west across most of Que., n. Ont., n. Man., n. Sask., n. Alta., n. (but not coastal) B.C., w. Northwest Territories, Yukon, and interior and w. Alaska. Winters south and east of a line drawn from Cape Cod to s. Iowa, n. Mo., e. Kans., cen. Okla., and ne. Tex. (absent along the southern coastal plain between Tex. and S.C.). Migration passes east of the high plains. Very rare but regular in the w. U.S. **HABITAT:** Breeds in dense cover, commonly brushy woodland edge. Partial to spruce bogs, young regenerating growth, and low stands of spruce, willow, and alder thickets. Prefers areas with bare ground for foraging. In winter, found in brushy woodlands and woodland edge, greenbrier tangles, heavy brush along streams, and brushy suburban yards, where it comes readily to platform feeders and seed spilled on the ground. **COHABITANTS:** During breeding, Yellow Warbler, Wilson's Warbler, White-crowned Sparrow, redpolls. In winter, Hermit Thrush, Eastern Towhee, Northern Cardinal, White-throated Sparrow. **MOVEMENT/MIGRATION:** Spring migration from late February to April; fall from October to December. VI: 3.

DESCRIPTION: A large, robust, richly colored, and handsomely patterned sparrow—distinctly larger and bulkier than the other sparrows it commonly associates with, such as Song and White-throated Sparrows. Only Harris's Sparrow, whose winter range overlaps Fox Sparrow in the central Great Plains, and Golden-crowned Sparrow are as large or larger. Almost the size of Hermit Thrush, "Red" Fox Sparrow has a big domed head, a portly and slightly humpbacked body, and a long tail. The bill is narrowly conical, undersized for the bird, and except on top, bright yellow. In all plumages, the rusty upperparts suffused with gray on the head and rump are arresting, beautiful, and diagnostic. No other sparrow is so richly reddish above. Whitish underparts are lavishly overlaid with thick, beaded, rusty streaks that coalesce into a large central breast splotch (more than a breast spot). The expression (hardly necessary) is stern.

BEHAVIOR: Likes to stay low where it's wooded and thick; likes even more to be on the ground. Forages in leaf litter, scratching, towhee fashion, with both feet. Doesn't move much. Tenaciously digs in the same place, excavating holes in the leaf litter that may partially conceal it. A social feeder in small well-spaced flocks. Mixes freely with other woodland sparrow species—who often attend Fox Sparrow diggings for overlooked morsels.

Males are very vocal. Sings even in winter, particularly on warm days. Responds well to pishing and screech-owl calls, frequently after other species have responded, but will then remain tenaciously in view after other species have lost interest. When piqued, generally sits high—at eye level or slightly above. When flushed, often flies straight up to take a high perch in a tree.

FLIGHT: Heavy, direct, and slightly but evenly undulating, with a regular pattern consisting of a series of wingbeats followed by a short closed-wing glide.

VOCALIZATIONS: Whistled song is loud, rich, whimsical, varied, and generally three-parted. The first part consists of two notes, the first sliding up and the second sliding down; the middle part is several varying notes; the final part is three spaced notes: "*per-E Eur p'er it reh-ee-ee-ee.*" Sounds like a lazy or dreamy R2-D2 of *Star Wars* fame. Call is a dry, robust, somewhat raspy "*t'chak*" that recalls Brown Thrasher.

"Sooty" Fox Sparrow, *Passerella iliaca unalaschensis*

One-Color Fox Sparrow

STATUS: Common northwest coastal breeder, resident, and winter resident. **DISTRIBUTION:** Breeds coastally and within 150 mi. of the coast from the e. Aleutians to nw. Wash. Winters from coastal B.C. to n. Baja. **HABITAT:** A thicket sparrow—breeds in heavily vegetated habitats (deciduous or mixed) beside creeks, muskegs, ponds, and other openings. In winter, found in woodland understory and, in California, chaparral. **COHABITANTS:** During breeding, Varied Thrush, Hermit Thrush, Wilson's Warbler, Golden-crowned Sparrow; in winter, Hermit Thrush, Spotted Towhee. **MOVEMENT/MIGRATION:** Spring migration from late March to early May; fall from late August to early November, with a peak in early October. VI: 1.

DESCRIPTION: An overall dark, plain, and unpatterned Fox Sparrow showing no color contrast between the head and the body. Large and robust (slightly smaller but fatter than Golden-crowned Sparrow; about the same size as and structurally similar to other fox sparrows), "Sooty" Fox Sparrow is mostly one color, dark sooty brown, and shows no contrast between a grayer head (or face) and back and rustier wings and tail, as is typical of other fox sparrows. *Note:* Some birds may show a slightly more reddish rump or tail. Don't be daunted. Underparts are more heavily spotted than on other fox sparrows—so much so that "Sooty's" breast often appears all dark with white spotting (rather than appearing pale and heavily streaked with a dark central splotch).

BEHAVIOR: Like other fox sparrows. Spends most of its time foraging on the ground beneath thick brush. Males commonly vocalize from high perches (often tall spruce).

FLIGHT: Like other fox sparrows, but appears uniformly dark above and mostly dark below.

VOCALIZATIONS: Song is light, lively, quick, and buntinglike, with notes that are distinct and spaced (not sliding into each other like "Red" Fox Sparrow). Individuals commonly repeat the same song, or they may vary the ending; for example: "*ch pee'r chREE ch'da seeseeseeseeju,*" or ending with "*chchch*" or "*cheeteeteeteetee.*" Call is like "Red" Fox Sparrow.

"Slate-colored" Fox Sparrow, *Passerella iliaca schistacea*

Rocky Mountain Fox Sparrow

STATUS: Uncommon to fairly common western breeder and southwestern winter resident whose breeding range abuts "Red" Fox Sparrow in the North, "Sooty" Fox Sparrow in coastal British Columbia, and "Thick-billed" Sparrow in the northern Sierras. **DISTRIBUTION:** Fairly widespread breeder in the Rockies and other western ranges. Found in cen. and s. B.C. (breeds within a few miles of the coast), extreme sw. Alta., n. and e. Wash. (also the Cascades), e. and cen. Ore., Idaho, w. Mont., w. Wyo., n. and cen. Nev., n. Utah, and n.-cen. Colo. Winters in w. and s. Calif., s. Ariz., s. N.M., the extreme western portion of trans-Pecos Tex., and extreme n. Mexico. **HABITAT:** Breeds in dense riparian thickets beside streams or mountain meadows; also, above the tree line, in dense stands of stunted spruce and low vegetative tangles. In winter, favors montane chaparral and thickets along streams. **COHABITANTS:** During breeding, Wilson's Warbler, MacGillivray's Warbler (at lower elevations), White-crowned Sparrow (at higher elevations). In winter, Hermit Thrush, White-crowned Sparrow, California Towhee, Spotted Towhee. **MOVEMENT/MIGRATION:** Timing is somewhat uncertain, but "Slate-colored" Fox Sparrow is known to be an early migrant that arrives earlier than "Red" Fox Sparrow in the fall (as early as late August) and departs wintering areas as early as late March in the spring. VI: 1.

DESCRIPTION: A grayish fox sparrow with warm brown touches in the wings and tail and a small (at least not overtly large) bill. Large and robust (larger than Song Sparrow; smaller but fatter than Golden-crowned; about the same size as White-crowned), with a conical bill (whose proportions won't startle you), a large head, and a robust longish tail. May appear slightly smaller-billed and longer-tailed than the average fox sparrow.

Overall fairly dull, mostly grayish above, with rufous brown in the wings and tail. Underparts are variably streaked: some birds show heavy diffuse streaking, and some birds lighter spotting. Spotting may be blackish, blackish going onto brown on the flanks, or mostly reddish. Except for the (usually) smaller bill, looks much like "Thick-billed" Fox Sparrow.

BEHAVIOR: Like other fox sparrows. Likes to be in brush; in winter likes to associate with flocks of other sparrow species.

FLIGHT: Like other fox sparrows. The contrasting pattern (gray body and ruddy wings and tail) distinguishes it from mostly all-dark "Sooty" and mostly all-red "Red" Fox Sparrow.

VOCALIZATIONS: A short, whistled, happy-go-lucky ditty that blends run-on high notes and high and low notes—some lazy and sliding, some hurried and doubled, some warbled; for example: "*erEEcher ch'h weewewewh'r.*" Every male has several song variations and never repeats the same song twice. Call is a low "*chew*" or "*chuh.*" Also emits a soft high "*t'see.*"

"Thick-billed" Fox Sparrow, *Passerella megarhyncha*

Sierra Fox Sparrow

STATUS: Common but restricted breeder and wintering resident. **DISTRIBUTION:** Breeds from s. Wash. (a few) to cen. Ore. south through the Sierras; also locally in the mountains from Santa Barbara Co. south and east to San Bernardino and San Diego Cos. and extreme n. Baja. Winters from cen. and s. Calif. to n. Baja, but the status and distribution are not well known. **HABITAT:** Breeds in thickets (often near streams), mountain meadows, recently logged or disturbed areas in early succession, and young conifers at higher elevations. In winter, found in chaparral and brushy thickets. **COHABITANTS:** During breeding, Mountain Quail, Dusky Flycatcher, MacGillivray's Warbler, Wilson's Warbler, Green-tailed Towhee. **MOVEMENT/MIGRATION:** Entire population moves from higher elevations to lower habitat zones (chaparral and lower mountain slopes) in winter. Spring migration mostly in March and April; fall in August and September. VI: 0.

DESCRIPTION: A large, big-billed, long-tailed sparrow whose shape and behavior seem towheelike. Like other fox sparrows, but the bill on many individuals is large and bulbous, made all the more prominent by the blunt forehead (more tapered on other fox sparrows). Also, the tail is conspicuously long and, when foraging, slightly elevated (towheelike). Plumage is similar to "Slate-colored" Fox Sparrow—gray head and back, rufous wings and tail, but spotting on the breast and flanks (taking the form of tiny black arrow points) that is sharper, crisper, blacker, and more contrasting against the very white underparts. Ignoring for the moment the streaking down the sides, the contrasting breast pattern recalls Hermit Thrush.

BEHAVIOR: Behaves very much like a towhee—in fact, like California Towhee. Forages, mostly on the ground, by hopping and scratching. Somewhat sluggish and confiding, allowing close approach. Males sing from one of several low perches (often in a small conifer or thicket), and territories seem very confined. Lusty and persistent singer, often vocalizing during the day.

FLIGHT: Somewhat sluggish, fluttery, and effortful. Wingbeats are patterned—a series and then a halting pause. Flight is slightly undulating. At close range, wingbeats are surprisingly audible.

VOCALIZATIONS: Also sounds like a towhee (this time like a Green-tailed). Song is loud, musical, and happy-sounding, but not as comical as "Red" Fox Sparrow. Consists of a variable mix of high and low single and double notes interspaced with short warbles and trills: "*Tew wheet tew wheet chica chica wheeta wheeta tee tee tew.*" Males change the order and pattern of the song routinely. Call is a loud, sharp, metallic "*t'chee*" or "*t'chink*" that is similar to California Towhee.

Song Sparrow, *Melospiza melodia*

The Default Sparrow

STATUS: Common, widespread, and probably not far from where you are right now. **DISTRIBUTION:** Breeds across most of Canada south of the tree line, from s. Nfld., e. Que., and the Maritimes to the Pacific Coast and s. Yukon. Occurs in s.-cen. Alaska coastally out to the end of the Aleutian chain. In the U.S., breeds across most of the Northeast and the West; absent in the southeast coastal plain, the Deep South, the central plains (north to the Dakotas), and drier desert portions of the American Southwest. In winter, vacates interior portions of Canada and extreme northern portions of the bordering prairie states as well as n. Maine; spreads south across all of the U.S. into n. Mexico. **HABITAT:** At all times of year, found in brushy weedy edge habitat (often near water). Particularly fond of overgrown farmland, marsh edges, roadside thickets, weedy vacant lots, well-shrubbed yards, and parks. **COHABITANTS:** Too numerous to list. **MOVEMENT/MIGRATION:** Migratory, but a permanent resident across much of its range. Spring migration in March and April; fall from September to November. VI: 4.

DESCRIPTION: A long-tailed bumpkinish-looking sparrow with a grayish face, coarsely streaked underparts, and (most commonly) touches of red in the wings and tail. Medium-sized (larger than Savannah; smaller than White-crowned; about the same size as Vesper Sparrow and juncos), Song Sparrow is generally lanky, with a flattish head, a fairly long and heavy bill, a somewhat longish face, a pickle-shaped body, and a long tail.

Regional variations in plumage range from overall dark in the Aleutians to pale and rusty in the Southwest. Generally drab grayish brown above with touches (or lavish amounts) of chestnut in the wings and tail. Creamy underparts are heavily and lavishly streaked, with the streaking coalescing into a large blotch or breast spot. Sometimes the streaking is blackish, crisp, and fine. More often it is broad and blurry as if the streaks bled into the backdrop. In most birds, the grayish face contrasts with the bordering browner or ruddy brown face pattern.

On some birds, the overall plumage appears shabby or unkempt; on other birds, it's blurred, muted, homogenized, or not crisply defined. This bland uniformity highlights Song Sparrow's prominent dark muttonchop malar stripe, big dark breast splotch, and less patterned rufous-tinged wings and tail. The expression is generally unconcerned.

BEHAVIOR: Fairly deliberate and unhurried; almost always forages on the ground, picking seeds from the surface or searching through leaves using a two-footed double scratch. Not at all shy about being in the open (but usually stays close to cover). Moves by hopping. In winter, frequently found in small flocks or with other sparrows, but unlike many other sparrows, also frequently forages alone (often along roadsides) and late into dusk. Comes to feeders. Fairly tame. Responds quickly and tenaciously to pishing, usually sitting conspicuously high. When flushed (at 15–25 ft.), flies 50–100 ft. or into the nearest bush or hedge.

FLIGHT: Appears long-tailed and mostly uniform above. Undulating flight is direct, strong, bouncy, and buoyant but also somewhat effortful, often given with exaggerated tail-pumping and/or side-slipping and the tail is often partially fanned. Maybe the overly long tail is holding the bird back? Wingbeats are mostly constant but broken, with a regular skip/pause. Lands with buoyant open winged braking.

VOCALIZATIONS: Sings all year, though infrequently in winter. Individuals sometimes sing at night. Song is variable but simple in pattern and easily recognized. Usually begins with two to four (usually three) quick, clear, identical notes followed by a long trill and an ending that has several short notes: "*chee, chee, chee tureeeee, uree tee tee.*" Call is a distinctive, low-pitched, unmusical bleat, "*j'mp*" or a low "*churp,*" that recalls House Sparrow. Also makes a high-pitched longish "*seet?*" that is thinner and higher-pitched than White-throated Sparrow.

PERTINENT PARTICULARS: There are several other streaky-breasted sparrows. The smaller Savannah is a stocky, distinctly short-tailed grassland sparrow with neatly and finely streaked underparts. The streaks on Song Sparrow are most often coarse, blurry, and broad-brush. The smaller Lincoln's Sparrow inhabits wooded edge and thickets (often near water) and, like Song Sparrow, is gray and rufous above. Lincoln's is also crisply patterned, with streaking on buff-washed underparts that is black, broken, and penpoint-fine. Fox Sparrow is a bigger, more robust sparrow (anything but lanky) with dark, broad, coarse, spotted breast streaks that look like they were applied with a putty knife.

Lincoln's Sparrow, *Melospiza lincolnii*

The Gentrified Song Sparrow

STATUS: Common northern breeder; common western migrant and wintering species; uncommon and elusive migrant and winter resident in the East. **DISTRIBUTION:** Breeds in subarctic and subalpine regions from Nfld. to cen. Alaska, south to n. New England and n. N.Y., west through n. Mich., n. Minn., and most of s. Canada to B.C. (absent in the southern prairie provinces of Canada); also south through the Rockies to n.

N.M. and the Pacific Coast ranges to cen. Calif. Winters from coastal s. B.C. to Baja and across the s. U.S. from Ariz. to s. N.M., all but nw. Tex., Okla., se. Kans., s. Mo., Ark., La., and coastal Miss. and Ala., up the Atlantic Seaboard to se. Va. **HABITAT:** Prefers damp, if not wet, brushy areas. Breeds in boggy northern marshes and along streams with a brushy component (particularly thick stands of willows). In migration and winter, found in any brushy or shrubby habitat, forest edge, or weedy field near cover (wet or dry, although shows a partiality to wetter areas). A few come to feeders. **COHABITANTS:** During breeding, Alder and Willow Flycatchers; Yellow, Wilson's, Mourning, and MacGillivray's Warblers. **MOVEMENT/MIGRATION:** Spring migration from mid-April to early June; fall from early September to early November. VI: 2.

DESCRIPTION: A neater, trimmer, finer, more gentrified Song Sparrow with a buffy whisker and buffy wash across the chest. Small and compact (smaller than Song Sparrow; about the same size as Swamp Sparrow), with a larger, rounded, often shaggy-crested head, a small conical bill, and a proportionally long tail (though not as long as Song or Swamp Sparrow). Overall neater, trimmer, and less gangly-looking than Song Sparrow (which looks like a bumpkin by comparison) and more angular than Swamp Sparrow.

Darkish upperparts are a blend of gray and brown lavishly overlaid with fine black streaking. Plumage is grayer than Song Sparrow; streaking is neater, trimmer, and finer and makes the streaking on Song Sparrow look blurry and amateurishly applied. Basically whitish underparts are enriched by a pale buffy or orange-buff wash across the chest and down the sides, overlaid by crisp, black, pencil-fine streaking. The buffy wash also creeps onto Lincoln's Sparrow's face—showing as a buffy whisker on a gray cheek.

All these marks seem distinctive enough—and they are when birds are seen close and well. But at a distance, some dark alchemy blends this sparrow's traits into a bland uniformity that even discerning eyes skip right over; as a result, this species is often overlooked. There is no trick that could be offered here for making this identification. Only mindfulness will work.

BEHAVIOR: Begin by being mindful of habitat. Lincoln Sparrow likes to be in or near brush and near wet places. Doesn't like to be out in the open as much as Song Sparrow or (especially) Savannah Sparrow. Fairly sedentary; when feeding, likes to work an area thoroughly. Feeds on the ground by shuffling, sometimes hopping quickly. Responds slowly to pishing, but often flies to a high perch (below the canopy) to assess the situation before returning to feed. Sits with tail slightly upcocked.

In migration, sometimes found feeding with other sparrows. In winter, generally found in small single-species groups, alone, or occasionally with Song and Swamp Sparrows.

FLIGHT: Mostly straight, but slightly jerky or bouncy.

VOCALIZATIONS: A sparrow that sounds like a House Wren. Song is varied—a rollicking, carefree, but somewhat hurried series of repeated notes, trills, and warbles, usually given in three to five parts, with the opening notes recalling a towhee and the final utterances typically muttered or swallowed; for example: "*ching ching chur'd'd'd'd' chee'ing (d'l-weeder).*" Overall tone is lower-pitched than most sparrows and sounds as if the bird is singing with its mouth full. Commonly heard calls include a low scrapping "*chup*"; a sharp "*chip*"; and a weak, high-pitched, electric, buzzy trill, "*spee*" or "*speer.*"

PERTINENT PARTICULARS: While confusion with Song Sparrow is likely, Lincoln's Sparrow may also be confused with Swamp Sparrow, particularly immature Swamp Sparrow—a bird that, like Lincoln's, is overall darkish with a gray face and rusty brown wings. Swamp Sparrow is overall darker and distinctly rustier, and though it likes wet areas, it prefers rank weeds and fields over brush or willow thickets.

Swamp Sparrow, *Melospiza georgiana*
Dark Marsh Sparrow

STATUS: A common sparrow in its range and preferred habitat, which is, as the name implies, wet. **DISTRIBUTION:** Breeds in the ne. U.S. as well as in

most of subarctic Canada. Found from Nfld., s. Lab., and the Maritimes west to the w. Northwest Territories, e. B.C. and se. Yukon (absent in s. Alta. and s. Sask.). In the U.S., breeds throughout New England and the mid-Atlantic states as well as in n. Ohio, n. Ind., Mich., n. Ill., Wisc., Minn., n. Iowa, e. N.D., and e. S.D. Winters from s. New England and extreme se. Ont. across n. Ohio and s. Iowa south to the Gulf of Mexico and Mexico (excluding the southern half of Fla.) as well as in e. Kans., e. and cen. Okla., and all of Tex. (except the panhandle); found locally in s. N.M., se. Ariz., and (a few) along the West Coast from s. Wash. to s. Calif.
HABITAT: At all seasons, prefers freshwater marshes, streamsides, and beaver ponds characterized by rank vegetation or willows. In winter and migration, occupies brackish marshes, dense weedy cover in upland areas, and brushy edge. **COHABITANTS:** American Bittern, Virginia Rail, Sora, Marsh Wren, Red-winged Blackbird. In migration and winter, other sparrows. **MOVEMENT/MIGRATION:** Spring migration from late March to mid-May, with a peak in April and early May; fall from mid-September through November, with a peak from late September to early November. At times and in places in fall, Swamp Sparrow may be abundant, outnumbering all other sparrow species. VI: 3.

DESCRIPTION: An overall dark and coarsely patterned sparrow distinguished by its reddish brown body contrasting with a mostly grayish face. Small but nicely proportioned (slightly larger than Lincoln's Sparrow; smaller than Song Sparrow). The fairly flat head with little forehead and the long, pointy, conical bill (recalls somewhat Red-winged Blackbird) give Swamp Sparrow a pointy-faced look that suggests, but is nowhere as exaggerated as, Seaside Sparrow. The body is fairly compact, and the tail proportionately long, with a rounded tip. The shape recalls Lincoln's Sparrow, but Swamp Sparrow is slightly more amorphously proportioned and less angular (a more larval or proto-sparrow shape).

Overall distinctly darkish sparrow, with rusty wings that contrast with a grayer face and breast. (In summer, many birds show a ruddy cap.) Underparts are also dark—gray in the front, buffy brown (even rufous) on the flanks—showing little contrast with the dark upperparts.

Swamp is a somewhat shabby sparrow too. The patterning on the upperparts is coarse, and the underparts are smudgy, laced with shadowy streaks. Combined with the all-dark visage, the bird often looks disheveled. Winter adults and immatures are duller and lack red caps, but the overall dark plumage and reddish tones persist.

BEHAVIOR: A busy active sparrow. Not shy, but likes to stay where it's thick and likes to stay near the ground. Forages by walking or hopping along wet edges and clambering about on vegetation protruding from the water. Often flicks its wings as it moves with a motion almost too quick to see. Fairly clamorous. In summer, males sing day and night from exposed perches but rarely more than 10 ft. above the ground.

Not gregarious, and not a flocking species. In winter, particularly in wet habitat, solitary birds are the rule, but in upland areas occasionally joins with other sparrows. Does come to feeders, but not readily (and usually only in harsh weather). In migration, sometimes occurs in great numbers in rank vegetation, dry weedy fields, and seed-bearing crops. When flushed, returns quickly to feed. Responds well to pishing. Flushes at 15–20 ft., flies a fairly short distance (30–75 ft.), and drops into the marsh, the grass, or the base of a small shrub.

FLIGHT: Retains the look of a small, fairly compact, all-dark sparrow. Flight is direct, with little undulation, fairly strong, and usually short.

VOCALIZATIONS: Song is a slow, rich, even trill, "*tew tew tew tew tew*," or trilled notes with a compressed two-note quality, "*tewa, tewa tewa tewa.*" Slower and more pleasing than Chipping Sparrow, the song has a cadence and quality that may recall a hurried Northern Cardinal. Call is a loud, empathic, somewhat low-pitched "*chimp*" or "*chemp*" (might recall Eastern Phoebe). Other vocalizations include a short, low, buzzy "*chee*" (similar to Lincoln's Sparrow).

PERTINENT PARTICULARS: If you don't get lost in the details, this is an easy sparrow to identify: an overall small rufous-onto-dark sparrow plus wetlands equals Swamp. In size, shape, and habitat preference, the species most likely to be confused with Swamp Sparrow is Lincoln's Sparrow. While Lincoln's is not as dark, rusty, or scruffy-looking (in fact, Lincoln's looks crisp, clean, and gentrified!), the buffy wash across the chest overlaid with crisp fine black streaking easily distinguishes it.

White-throated Sparrow, *Zonotrichia albicollis*

Eastern Woodland Sparrow

STATUS: A common northern and eastern woodland breeder and winter resident (rare in the West) distinguished by two distinct plumages and a pure-toned song that goes right to the heart. **DISTRIBUTION:** Breeds across much of subarctic Canada, from Nfld., s. Lab., and the Maritimes west to se. Yukon and ne. B.C. (absent in s. Alta., s. Sask., s. Man., and extreme se. Ont.). In the U.S., breeds throughout New England, N.Y., n. Pa., all but s. Mich., n. Wisc., and n. Minn. In winter, vacates most of its breeding territory. Winters south of a line drawn from s. N.S. south across coastal Maine, s. New England, s. and cen. N.Y., se. Ont., s. Mich., s. Wisc., cen. Iowa, se. Neb., e. and s. Kans., w. Okla., the Texas Panhandle, cen. N.M., and se. Ariz. Also found (sparingly) in parts of Colo. and along the West Coast from sw. B.C. to s. Calif. **HABITAT:** Breeds in coniferous and mixed forest, selecting dense second-growth understory along the edges of ponds, clearings, and burned areas. In winter and migration, prefers dense brushy woodlands with an adjacent edge (particularly wet areas or woodlands flanking watercourses), but individuals and small flocks are content to occupy small thickets, hedgerows, city parks, overgrown fields, and vegetated suburban yards—*particularly* yards whose owners maintain bird feeders. **COHABITANTS:** At all seasons, Hermit Thrush, Yellow-rumped Warbler, Dark-eyed Junco. In winter, several other flocking sparrow species. In the West, usually found among White-crowned and Golden-crowned Sparrows. **MOVEMENT/MIGRATION:** Spring migration from early April to late May; fall from late September to late November. VI: 3.

DESCRIPTION: The common eastern flocking sparrow of woodland edge. Large and robust (larger than Song Sparrow; smaller than Fox Sparrow), with a fairly heavy bill, a large gently round-topped head, a portly body, and a long narrow tail. Seems plumper and less athletically lean than White-crowned Sparrow. Stance is fairly horizontal.

White-throated Sparrows come in two plumages—stunning and shabby. The crisply patterned white-striped form, with its boldly striped head, yellow lores, and black-bordered white throat contrasting with gray underparts, is distinctive. Birds with tan eye stripes range from distinctly patterned to borderline nondescript.

Most, if not all, tan birds have at least a shadow rendering of the head pattern of the white-striped form, a hint of yellow about the lores, and a dark-edged pale throat garnished with two faint malar stripes. Dingier individuals are overall warm brown or rufous above (particularly on the wings) and gray or gray-brown below. Underparts are frequently marked with shadowy streaks that may blur the line on heavily marked birds between upper- and underparts, homogenizing the plumage and making the bird look like a dark, diffusely patterned, nondescript sparrow—which is, in subtle fact, a distinguishing characteristic! With the exception of some Song Sparrows and Swamp Sparrows and immature Golden-crowned Sparrows, the plumage of sparrows is crisp and trim. Juvenile White-throateds are crisply streaked below.

BEHAVIOR: This is a sparrow of the woodland—a habitat not favored by most sparrows—and a flocking sparrow. Though it associates with other sparrows, over much of its winter range it's usually found in fairly sizable (10–100 individuals) homogeneous flocks. If you chance upon a poorly marked individual, chances are good that a very distinctly marked White-throated Sparrow is scratching in the leaf litter nearby to help guide you toward an identification.

White-throated frequently forages by hopping on the ground at the forest edge and often in the open (but generally not far from cover). Also forages in shrubs and trees and is particularly fond of fresh spring growth. Birds are also fairly vocal and will sing through the winter; they are particularly noisy at dawn and dusk as flocks settle in or leave the roost. Startled, they disappear into the closest wooded cover.

FLIGHT: Flights are usually of short duration—from the ground to cover and from a feeder. Flight is direct and undulates slightly; wingbeats are given in a rapid series broken by a skip/pause.

VOCALIZATIONS: The whistled song is simple, plaintive, pure-toned, and easily recognized. It has been described as a dirge ("*Old Sam Peabody, Peabody, Peabody*") or a lament ("*Oh, Sweet, Canada, Canada, Canada*"). Don't be dismayed if all you hear is a lower-pitched introductory whistle followed by five clear, higher-pitched, evenly spaced notes: "*Oh, Say, Say, Say, Say, Say.*" Or the "*say*" note may quaver: "*Oh Sa'ay, Sa'ay, Sa'ay, Sa'ay, Sa'ay.*"

Call note is a sharp explosive "*spink!*" given at dawn and dusk and when birds are excited. Also heard is a high thin flock location call, "*seEep,*" sometimes given with a slight trill. Similar to Song Sparrow's "*seet,*" the White-throated call is more abrupt and matter-of-fact and not so questioning or ethereal. Also makes an excited "*kll kll kll kll kll*" when flocks are disturbed.

PERTINENT PARTICULARS: White-throated Sparrow is easily attracted to winter feeding stations and is particularly fond of white proso millet and black-oil sunflower seed. Also responds quickly, dramatically, and en masse to almost any form of pishing.

Harris's Sparrow, *Zonotrichia querula*

The Burliest Sparrow

STATUS: Fairly common but restricted both on its entirely Canadian breeding grounds and wholly U.S. wintering territory. **DISTRIBUTION:** Breeds in the overlap zone between tundra and forest north of a line drawn from the Mackenzie R. delta in the Northwest Territories of Canada south and east to the region just south of Hudson Bay and n. Ont. Winters in the central prairies from se. S.D. south through e. and cen. Neb., w. Iowa, Kans., w. Mo., Okla., w. Ark., and cen. and e. Tex. Very rare or casual visitor east and west to both coasts. **HABITAT:** Breeds in mixed taiga forest habitat, from mostly forested regions to open tundra dotted with spruce. Most frequently found in woodlands adjacent to open tundra or clearings, particularly where some conifer component is found. In winter, prefers dense deciduous woodlands (often riparian) adjacent to open prairie, pasture, farmland, and roadsides; also found in hedgerows, shelterbelts, brush piles, and feeding stations. **COHABITANTS:** In winter, found in association with Fox, White-throated, and White-crowned Sparrows and Dark-eyed Junco, Osage orange trees. **MOVEMENT/MIGRATION:** Spring migration from March to early June, with a peak in April and May; in fall, birds reach prairie regions in September and October, but some birds (mostly juveniles) do not complete their migration until December. VI: 3.

DESCRIPTION: Next to towhees, our largest, burliest sparrow (larger than Fox Sparrow or House Sparrow; smaller than Red-winged Blackbird). Overall robust, with a distinctly large, often peaked head, a big conical pink bill, and a very long tail. Plumage is striking, contrasting, and distinctive—a brownish-backed bright white–bellied bird with a black crown, face, and throat contrasting with a paler, plain grayish, brownish, or ochre-tinged face/head. The grayish-brown back is overlaid with black streaks; whitish underparts are bedecked with a bold or sparse black bib, and heavy streaking extends along the sides and flanks. Immature birds are more sparingly marked but similarly shaped (the triangular bib is penciled in, not blackened in).

BEHAVIOR: A sparrow that likes to forage on the ground and sit high in trees. On territory, sings from the tops of trees (often the *only* tree). In winter, when flushed, generally heads for the safety of a thicket or flies to the top of a taller tree. Usually found in small flocks (6–12 birds) with the generally more numerous White-throated or White-crowned Sparrows. Moves by hopping. In winter, feeds on seeds on the ground, but also sits high in bushes and trees and nibbles on emergent growth in spring. Somewhat shy. Often the last bird in a flock to respond to pishing, and often sits with a branch between you and it.

FLIGHT: A strong flier that sometimes travels several hundred feet. Flight is mostly straight but occasionally describes shallow undulations. Wingbeats are strong, regular, and given in a lengthy series punctuated by a terse closed-wing glide.

VOCALIZATIONS: Song consists of one to four clear spaced notes that recall White-throated Sparrow but with no change in pitch. Call is a loud "*wenk*," like White-throated Sparrow, but not as sharp. Rally call is a low, rapid, somewhat metallic "*ch'weh, ch'weh, ch'weh*" that sounds like a bugle charge.

PERTINENT PARTICULARS: As distinctive as this species might be, under the catalytic influence of wishful thinking, Harris's Sparrow does bear a superficial likeness to House Sparrow—another brown-backed, gray-headed, black-bibbed, ground-foraging sparrowlike bird. House Sparrow is smaller, overall stocky, and much shorter-tailed than Harris's Sparrow and lacks a pink bill. House Sparrow is a bird of open, often barren areas that offer a bit of shrub and vegetation for cover. Harris's Sparrow is a bird of brushy woodland edges.

White-crowned Sparrow, *Zonotrichia leucophrys*

Dapper Hedge Sparrow

STATUS: Common and widespread—often the dominant sparrow in the West; less common in the East. **DISTRIBUTION:** Breeds primarily across arctic and subarctic Canada and Alaska as well as the western mountains of B.C. and n. and w. Alta., south to cen. Calif., cen. Nev., Utah, w. and cen. Colo., n.-cen. Ariz., and nw. N.M., and along the Pacific Coast from s. B.C. to cen. Calif. Winters widely across the U.S. (and s. B.C. and extreme se. Ont.) and is absent only in n. New England, all but se. N.Y., n. Pa., most of Mich., Wisc., Minn., n. Iowa, N.D., S.D., and e. Mont. **HABITAT:** Preferred breeding habitat includes a mosaic of grass and bare ground with a scattering of shrubs or low-lying conifers—a combination that defines and includes tundra scrub, broken taiga forest, alpine meadows and disturbed forest areas, and coastal scrub (among other habitats). In winter, prefers brushy weedy habitat and open habitats bordered by hedges, thickets, or trees. Often found along roadsides, where it is attracted to grit; on farms where livestock are fed grain; and, of course, in residential neighborhoods where bird-feeding stations are maintained. **COHABITANTS:** Varied. Breeding birds in arctic and alpine tundra are found with ptarmigans, Yellow and Wilson's Warblers, Fox and Golden-crowned Sparrows, redpolls. In winter, often found in mixed sparrow flocks (often constituting the dominant species), including Golden-crowned Sparrow, Brewer's Sparrow, Vesper Sparrow, House Sparrow. **MOVEMENT/MIGRATION:** Most subspecies are migratory (Pacific Coast birds are year-round residents). Migratory timetables differ, but most birds' migratory period falls between April and May in spring, September and November in fall. VI: 3.

DESCRIPTION: A large, lean, handsome, athletic-looking sparrow characterized by trim lines and crisp conservative plumage. The head is usually conical or peaked, the bill distinctly pink, orange, or yellow (unlike White-throated and Golden-crowned Sparrows) and tastefully petite. The body is sturdy without being portly or plump, and the tail is long. *Note:* The head shape is variable. While peak-headedness is the norm, feathers often flatten, giving the bird a flat-headed appearance.

The distinctive black-and-white-striped head pattern of adults stands out against the cryptically patterned brown upperparts and uniformly pale gray underparts. Immatures are patterned like adults, but overall plainer, browner, and less showy; the distinctive head pattern is similar but muted, with the black bracketing crown stripes replaced with a rich brown cap that is creased with a well-defined tawny center. The tawny center is somewhat like the crown pattern found on adult (less so immature) Golden-crowned Sparrow, but White-crowned Sparrow has a distinctive facial pattern—a fine eye-line and dark cheeks—whereas Golden-crowned Sparrow is plain-faced. Also, the yellow on the crown of immature

Golden-crowned is not well defined, is more restricted to the forehead, and often washes over onto the lores.

BEHAVIOR: A quick, nervous, active, alert, and noisy sparrow that forages in flocks and rarely strays far from cover — generally no more than 10 ft. away. Feeds in the open or at the edge of shrubs and cover. Picks and scratches for seed and shifts position frequently and nervously by rapid hopping or short flights. When disturbed, sits high, posture erect. When perched for extended periods, seems to slouch.

In winter, found in flocks that are fairly noisome and quarrelsome, with feeding birds frequently displaced. In migration and winter, occurs with other species, including Chipping Sparrow, Golden-crowned Sparrow, and Lark Sparrow. Does not generally forage or occur in woodlands (with White-throated Sparrow). Calls frequently. Sings throughout the year.

FLIGHT: Slender long-tailed profile. Overall pale and plain — brown and gray. Flight is strong and mostly direct, with regular, slightly bouncy undulations. Wingbeats are strong, rapid, and deep, and given in a quick series followed by a brief skip/pause. When birds are landing or flying short distances, flight is more bouncy, and open wings are used for braking.

VOCALIZATIONS: Variable three- to four-part song has a lazy dreamlike quality. Commonly begins with a clear whistled note, followed by a series of quick compressed notes, followed by a closing three-note sequence that sounds flat and buzzy. The last note is held and descending: "*Seee t'ut't tee tee turrrr.*" The bird has many variations on this theme, including one in which the clear opening note is followed by a hurried incongruous "*d'jt, d'jt,*" whose cadence and pattern recall the opening notes of the *William Tell Overture*. Coastal birds sound buzzier. Calls include a clear silver-toned "*chip*" or "*tink*" that has a hammer-on-anvil distinctiveness; an explosive "*Spink!*"; a quizzical high-pitched "*seep?*"; and a rapid challenging sputtering, "*w'hr, w'hr, w'hr, w'hr,*" which sometimes precedes the song.

Golden-crowned Sparrow, *Zonotrichia atricapilla*

Thickset Thicket Sparrow

STATUS: Common but somewhat restricted in its range. **DISTRIBUTION:** Breeds in shrubby transition zones along coasts and near the tree line in Alaska, the Yukon, B.C., and w. Alta. Winters from coastal and s. B.C. to n. Baja (and rarely in portions of the lower Colorado Valley). Very rare visitor in the interior West; casual in the East. **HABITAT:** Breeds in the transition zone between forest and tundra, often in alders, willows, or stunted spruce. In winter and migration, found in brush, chaparral, riparian thickets, suburban gardens, and well-vegetated parks. Unlike White-crowned Sparrow, prefers denser woodier areas, although much overlap occurs. **COHABITANTS:** Spruce Grouse, Willow Ptarmigan, Yellow and Wilson's Warblers, Dark-eyed Junco, White-crowned and Fox Sparrows, moose. In winter, White-crowned Sparrow, California Quail, California Towhee. **MOVEMENT/MIGRATION:** Spring migration in April and May; fall migration begins as early as late August, with a peak across most U.S. wintering grounds from mid-September through October. VI: 2.

DESCRIPTION: A big, burly, somewhat heavyset sparrow whose proportions are similar to the closely related White-crowned; slightly larger, however, and overall bulkier, plainer-faced, and more uniformly and less contrastingly patterned. The head is large and broad, and the bill heavy (not petite!) and mostly gray. The body is plumper and wider than White-crowned, and the tail heavier and longer. In the company of White-crowned, the greater size and bulk of Golden-crowned stand out.

Breeding adults are distinctive. It's hardly necessary to note the yellow to green-yellow crown framed within the black toupee on this large plain brown-gray sparrow. In winter, adults retain some of their head pattern. Young Golden-crowneds are similar to immature White-crowneds but overall plainer, and they show more drab gray-brown uniformity between the upperparts and underparts. Immature White-crowneds are overall browner and warmer-toned, with a distinctive contrast between the brown upperparts and the paler cleanly unpatterned underparts.

Where immature White-crowned shows a distinctive head pattern—a golden brown crown stripe flanked by crisply defined rich brown borders—immature Golden-crowned has a muted, poorly defined, slightly darker crown that usually shows only a restricted dab or dusting of yellow on the forehead and on the lores. (The bird looks as though it has pollen on its face.) The face of Golden-crowned is distinctly plain, almost dull. The face of White-crowned, with its well-defined cap and more distinctive eye-line, is more expressive. And of course, White-crowned has a yellow, orange, or pinkish bill; the bill of Golden-crowned is dull gray.

BEHAVIOR: Very much like White-crowned, with which it often associates while foraging on the ground by hopping, rarely far from protective cover; when disturbed, flies to cover. Generally quiet and less vocal than White-crowned.

FLIGHT: Direct, strong, slightly undulating. Wingbeats are rapid and mostly constant, but the series are interrupted or punctuated by brief, barely perceptible skips.

VOCALIZATIONS: Simple plaintive song consists of three lazily whistled descending notes: "*s'Zee'eee'ur.*" Call is a loud sharp "*pink.*"

Dark-eyed Junco, *Junco hyemalis*

A Brittle Twitter and a Flash of White

STATUS: Common Canadian breeder (also at higher elevations in the United States) and common wintering species in coastal and southern Canada and most of the United States. **DISTRIBUTION:** Breeds across virtually all of Canada and Alaska except northern tundra regions and the southern prairies. In the U.S., found throughout most of New England and south through the Appalachians to n. Ga.; also in n. Mich., n. Wisc., and n. Minn. In the w. U.S., breeds widely through the intermountain West south to the Guadalupe Mts. of Tex. and Baja and in the Pacific states. Winters across all of the U.S. and n. Mexico except the Florida Peninsula and s. La. Also in coastal Alaska, southern portions of the w. Canadian provinces, U.S. border regions, and throughout the Maritimes north to Nfld. **HABITAT:** Breeds predominantly in mature coniferous forest but also in late successional deciduous forest. Winters primarily in riparian areas and brushy edge adjacent to open weedy fields, open woodlands, roadsides, pines or pinyon-juniper, brushy or wooded suburban habitats, parks, and bird feeders. **COHABITANTS:** Varies widely, but includes White-throated, White-crowned, Harris's, and American Tree Sparrows. In winter, often found with Chipping Sparrow. **MOVEMENT/MIGRATION:** Most populations are migratory, but degree and distance vary greatly. The most widespread group, the Slate-colored form, migrates from early March to early May in spring; from late September to early December in fall. VI: 4.

DESCRIPTION: A mostly dark gray (or gray-headed and ruddy-backed and -sided) sparrow with a conspicuously pink bill and bright white edges to the tail. A medium-large sparrow (smaller than White-throated or White-crowned; larger than Field or Chipping; about the same size as or slightly smaller than Song Sparrow) and slightly plumpish but trimly contoured, with a large round or flat-topped head, a conically petite and pink bill, and a proportionally long tail.

Plumage varies between populations. The two most common types are the very widespread "Slate-colored" Junco—which is all slate gray or brownish gray except for a whitish belly—and the primarily western "Oregon" Junco, which, with its blackish or dark-gray-hooded head and rufous back and sides (contrasting with the white belly), resembles a miniature Spotted or Eastern Towhee. Overall paler "Pink-sided" Junco resembles a washed-out "Oregon" with dark lores and pinker sides. "White-winged" Junco is like "Slate-colored" but is larger and paler, has finely beaded white wing-bars, and shows more white in the tail.

Two other forms, "Gray-headed" and "Red-backed," are like pale "Slate-colored" but with a rufous mantle and dark lores; they very closely resemble Yellow-eyed Junco of southeastern Arizona except for the dark eyes and absence of reddish tones in the wings.

Note: In all cases, birds show dark upperparts contrasting with an oval white belly, and all except "Red-backed" have a bright pink bill. White outer tail feathers are often apparent on the folded tail.

The expression is angry, almost sullen (but you have to be very close to see this).

BEHAVIOR: In winter, highly social and vocal; forages on the ground in flocks not far from cover. Habitually seen on roadsides, along woodland edges, and at bird feeders. Often mixes with other sparrow species and bluebirds. An active, agile, but methodical feeder. Hops when it feeds. Birds are usually tame, but when startled, flocks explode into a swirling, twittering, scattering cloud that recalls leaves carried in a wind gust. The bright white outer tail feathers flash conspicuously and disappear as soon as birds land. Birds often retreat just to the edge of a thicket or to a high perch to assess the situation. Very responsive to pishing.

On territory (also in late winter), commonly sings from a conspicuous perch.

FLIGHT: Bright white outer tail feathers render all other characteristics secondary. Flight is strong, lively, nimble, and bouncy or slightly jerky—a short (or sometimes longer) series of wingbeats is followed by a terse skip or, occasionally, a brief closed-wing glide. Flares and pumps its tail in flight, exposing flashing glimpses of white outer tail feathers. When flushed, often climbs abruptly; when returning to the ground, really sticks its landing.

VOCALIZATIONS: Song is a rapid musical trill that varies in length and pitch and is more rapid than Pine Warbler, more musical than Chipping Sparrow. It sometimes ends with three notes: "*tew tew tew.*" Calls include a terse but not loud or emphatic "*tick*"; an abrupt spit "*sp*"; a high "*ti*"; and a high, rapid, brittle, three-note twitter, "*tw't't,*" often given when birds are flushed.

Yellow-eyed Junco, *Junco phaeonotus*

Shuffling Junco

STATUS: Common junco of Mexican mountains with an extremely limited range in the United States. **DISTRIBUTION:** In the U.S., found only in the "sky islands" of extreme se. Ariz., with a few pairs in sw. N.M. Occurs in Mexico south through the central mountains, into Central America. **HABITAT:** Breeds in drier pine and oak and pine woodlands at higher elevations; also at the edge of meadows. In winter, retreats to slightly lower elevations, where, in oak and scrub canyon woodlands, it mixes with assorted Dark-eyed Junco subspecies and comes readily to feeders. **COHABITANTS:** Arizona Woodpecker, Dusky-capped Flycatcher, Mexican Jay, Bridled Titmouse, Painted Redstart. **MOVEMENT/MIGRATION:** Nonmigratory, but moves to slightly lower elevations in winter. VI: 0.

DESCRIPTION: A largish, longish, and attractive junco species with a bicolored bill and staring yellow-orange eyes. Genetically distinct from the multi-plumaged Dark-eyed Junco complex. Slightly larger, bigger, rounder, sometimes more peak-headed, and curiously longer-necked than Dark-eyed Junco. (The length is apparent when feeding birds extend their necks to search for danger, as they often do.) The bicolored bill (dark upper mandible, yellow lower) is also longer.

Discounting the rusty wings and saddle, Yellow-eyed Junco is the overall palest (particularly on the head and breast) junco. In mixed flocks, the bird can be picked out by this paleness alone (coupled with the brandishing of the bird's underparts when it raises its very long neck). The namesake quality, the bright yellow (actually yellow-orange) eye, is very apparent. Other juncos, with their tiny black eyes couched in a shadowy mask or an overall dark hood, appear eyeless. Yellow-eyed's expression is deadpan, as though the bird is waiting for the other shoe to drop. Juveniles, which resemble shabby sparrows with white outer tail feathers, have dark eyes.

BEHAVIOR: Forages on the ground by walking, hopping, or shuffling in a fashion that is a combination of both. Other juncos just hop. The difference in the rhythm (not to mention the pattern) of movement is readily apparent in a mixed flock. Sometimes bobs its head, blackbird fashion, as it walks, and forages with wingtips drooped.

Fairly calm (but no more so than Dark-eyed Junco).

FLIGHT: Straight, quick, bouncy or undulating, and generally short.

VOCALIZATIONS: Variable, but basically a two- or three-part musical whistle: "*p'chee, chee, chuchu-chuchu, eeeeee.*" The trilled ending rises, pushing the scale in an electric buzz. Call is a sharp "*t'chew*" or "*chink*" that drags slightly at the beginning and is more musical and robust and held longer than Dark-eyed Junco.

McCown's Longspur, *Calcarius mccownii*

The Pallid Grinning Longspur

STATUS: Uncommon to common prairie breeder and wintering species. **DISTRIBUTION:** Breeds in the northern prairies from s. Alta. to (primarily) sw. Sask., all but w. and extreme se. Mont., w. N.D., n. and e. Wyo., extreme sw. Neb., and extreme n.-cen. Colo. Winters from se. Ariz. across s. N.M., extreme w. Okla., and w. and cen. Tex. south into n. Mexico. Casual to the West Coast and east of its normal range. **HABITAT:** Sparsely vegetated, dry, open prairie and overgrazed lands showing a mix of stunted grasses and cacti; also found in dirt agricultural fields. Prefers shorter grasses than Chestnut-collared Longspur. **COHABITANTS:** Mountain Plover, Horned Lark, Lark Bunting, Chestnut-collared Longspur. **MOVEMENT/MIGRATION:** Entire population vacates breeding areas in winter. Spring migration from late February to May; fall from mid-August to mid-November. VI: 2.

DESCRIPTION: A pale, grayish, compact, mouselike prairie bird. In winter, closely resembles a female House Sparrow. Overall stocky with a large flattish head, a large and slightly upturned lower mandible, a long neck, a plump oval-shaped body, and a very short tail. Overall more robust than Chestnut-collared Longspur, with a longer, heavier, and distinctly marked pink bill (dark in breeding males). The upper mandible slopes straight down from the forehead, and the bent lower mandible, angling up halfway along its length, makes the bird look like it is grinning.

In all plumages, paler and grayer above than other longspurs; in winter, plainer and paler, with chests that show a wash of color but no texture (no barring or streaking). Breeding males, with their black-slashed whitish faces, black chests, and chestnut epaulets, are distinctive; in winter, they retain an identifiable vestige of breeding plumage. Females have paler heads contrasting with grayer backs (may seem pale-hooded) and traces of reddish epaulets (often hard to see). Immatures are pale, distinctly plain-faced, and slightly buffier than adults, with a buffy unstreaked wash across the chest (a barely discernible shadow rendering of the breeding male's black chest).

BEHAVIOR: A versatile ground-foraging bird. Forages by "stalking" on the ground, either walking or running (doesn't hop). Pursues prey on the ground and (in summer) in the air, and frequently hawks flying insects (a tactic less commonly used by Chestnut-collared Longspur). In spirited display flights, males flutter aloft to a height of 20–30 ft., raising their back feathers and parachuting to the ground on set wings and a fanned tail, which shows off the bird's black-on-white T-shaped tail pattern. Often circles during the descent, and sometimes travels a considerable distance from the launch point.

Gregarious. In summer, despite their established and defended territories, one McCown's Longspur is never far from the next. In migration and winter, birds gather in medium-sized flocks. McCown's associates commonly with Horned Lark and less commonly with other longspurs (most often with Lapland, occasionally Chestnut-collared, very rarely Smith's).

FLIGHT: Overall compact in flight, with a conspicuously short tail but long pointy wings. Flight is deeply undulating, like other longspurs. Wingbeats are given in a quick flurry followed by a closed-wing glide.

VOCALIZATIONS: Song is a bubbly, loosely patterned, rising-and-falling ensemble of soft musical and tinkling phrases: "*ch'lee ch'ler ch'leedlee chlr chee ch'lr.*" Song begins with a pattern that wanders off script, although many initial phrases keep popping up throughout the song. Song is sometimes punctuated by a high thin "*p'eeee*" note, or sometimes this note is interjected in the song. In tone and pattern, the song recalls a long-winded Western Meadowlark, but many notes have the squeaky intonations of Horned Lark. Rattle call is a short, low, flat, rattling trill, "*tr'r'rt,*" like Lapland Longspur but shorter and more mellow. Calls also include a short whistled "*whip*"; a terse, low, muttered "*clup*"; a soft "*bink*" or "*chih*"; and a sharp high-pitched "*chee-dee.*"

Lapland Longspur, *Calcarius lapponicus*

Barren Grounds Bunting

STATUS: Very common arctic breeder; common, widespread, and at times abundant winter resident. **DISTRIBUTION:** Holarctic breeder. In North America, breeds from the Aleutians and coastal and some interior portions of Alaska across extreme n. Canada and the arctic archipelago to n. Lab. and coastal Greenland. The southern limits include coastal portions of Hudson Bay. In winter, found south of the standing snowpack across much of the U.S. and border regions of Canada—from coastal Nfld. across s. Canada to s. B.C., south to n. Calif. (less commonly to s. Calif.), s. Utah, n. N.M., and n. Tex. to the Gulf of Mexico and east to the mid-Atlantic. Most common in the prairies; uncommon and localized in the Great Basin and the Southeast; very rare in Ariz., w. and s. N.M., s. Tex., Florida Peninsula, and interior portions of Maine and N.H. The *only* longspur likely to be found wintering east of the Mississippi R. **HABITAT:** Breeds in wet and dry tundra, avoiding the rocky unvegetated terrain favored by Snow Bunting. In winter, found in open and commonly barren habitats, including prairie, weedy fields, upper beaches, tilled agricultural land, and fields with spread manure. In migration, in forested regions, occurs on lake and river shores. **COHABITANTS:** In winter, commonly associates with Horned Lark and Snow Bunting. **MOVEMENT/MIGRATION:** Complete migrant and somewhat nomadic, moving farther south when heavy snows blanket wintering areas. Spring migration from late February to early June; fall from late August to early December. VI: 3.

DESCRIPTION: Looks like a too heavily marked Savannah Sparrow or Vesper Sparrow with the ground-hugging posture of a shuffling supplicant. Overall plump, compact, and medium-sized bunting (larger than Savannah Sparrow; smaller than Snow Bunting or Horned Lark; about the same size as other longspurs), with a short thick bill, a large neckless head, a plump body, a short tail, and long wings that enhance the sense of short-tailedness. Feeding birds seem to crouch and appear slightly humpbacked; wingtips commonly droop below the tail.

Breeding males have a distinctive all-black face/throat, yellow bill, and chestnut nape. The female's pattern is a paler, broken, shadowy replication of the male but clearly shows a duller chestnut patch on the nape. Winter plumage is overall more streaky and contrasting than other longspurs, and field marks, though subtle, are also bountiful. Easily noted is the stirrup-shaped face pattern (recalling Savannah and Vesper Sparrows) contrasting with the warm or ochre-colored face and nape. The back shows neat narrow rows of black streaks, and the flanks show heavy blurry streaking. Most birds show a dark "five o'clock shadow" of a breast-band, which contrasts with the white about the throat and belly. At all seasons, the dull chestnut or rufous panel in the wings of males is diagnostic.

McCown's and Chestnut-collared Longspurs are overall plainer, paler, and grayer. Smith's Longspur is overall plainer, warmer-toned, or tawny.

BEHAVIOR: Highly gregarious. Forages in large single-species flocks or with Horned Larks and Snow Buntings. Actively feeds by walking or running (sometimes hopping) with a heads-up (almost House Sparrow–like) stance, picking at seeds exposed by melting snow. At other times, feeds more methodically, hunkering close to the ground, moving with a slow mincing shuffle and frequent pauses as it manipulates seeds plucked from the ground. Individuals in flocks feed fairly independently, and flocks commonly forage for long periods before relocating. (Snow Buntings, by comparison, are flighty and energetic feeders that like to stay in more compact, quickly shifting, and coordinated flocks.)

Often chooses to forage in places where even an accomplished Savannah Sparrow would starve—such as bare earth or sandy beach—but even then favors places with at least some sparse grass or vegetation and seems most comfortable working the denuded edge of a more densely vegetated place. Smith's Longspur (the other longspur with narrow white outer tail feathers rather than a mostly white tail) is a bird of thick short grass, not barren landscapes.

A tame bird, Lapland Longspur often doesn't flush until underfoot. Usually calls when it flies—a rough, short, rattly trill interspaced with a diagnostic "*tew.*"

FLIGHT: Short-bodied and blunt-headed, with long, pointy, triangular wings that appear wholly pale and translucent from below. Birds explode from the ground, climb quickly, and sometimes circle and land. Strong flier. A quick sputtery series of wingbeats is punctuated by a terse closed-wing glide. When landing, employs a series of open-winged swooping glides. Flight is undulating, jaunty, bouncy. Flocks are loose and strung out.

VOCALIZATIONS: Short repetitive song has the quality of Horned Lark and the pattern of Eastern Meadowlark: "*(tch) see-you.*" The "*chew*" or "*tew*" call and the dry rattle, "*pr'r'r't,*" are similar to Snow Bunting. Also makes a squeaky "*tewtewtew*" and a soft "*peedle,*" which are most often given on the breeding grounds.

PERTINENT PARTICULARS: When mixed with Horned Larks, Lapland Longspurs look shorter and plumper and are often chased by the larks.

Smith's Longspur, *Calcarius pictus*

The Tawny Longspur

STATUS: Uncommon, geographically restricted, and difficult to see. **DISTRIBUTION:** There are two distinct breeding populations. One is distributed along a narrow habitat-band between subarctic tundra and the tree line, from Alaska's Brooks Range to the northwest coast of James Bay, Ont. The second occupies the Wrangel Mts. from se. Alaska to the northwest corner of B.C. In winter, found exclusively in the central and eastern prairies from se. Kans., sw. Mo., e. Okla., w. Tenn., Ark., nw. Miss., nw. La., and ne. Tex. In spring migration, occurs east to Iowa, Ill., and Ind. Casual well east and west of its normal range. **HABITAT:** Breeds on open, flat, wet or moist tundra characterized by sedges and grasses interspaced with heath and a smattering of stunted trees. In winter and migration, also prefers short grass or prairie (either mowed or grazed) with wet patches or puddles. Also found at airports. **COHABITANTS:** During breeding, American Golden-Plover, American Tree Sparrow Lapland Longspur. In winter, Le Conte's Sparrow, Savannah Sparrow, Sprague's Pipit, meadowlarks. Generally not found with other longspurs. **MOVEMENT/MIGRATION:** In spring, birds begin leaving southern portions of their winter range by mid-February and are gone by April. Arrive on breeding grounds from late May into early June. In fall, birds begin departing the Arctic in late August and reach U.S. wintering areas from mid-October to early December. VI: 2.

DESCRIPTION: A long-winged, long-billed, longish-tailed longspur distinguished from all other longspurs by the overall tawny to ochre-colored cast to its plumage (and by habitat!). Breeding birds, with a bold black-and-white face pattern and ochre-colored underparts, are unmistakable. In winter, birds are more cryptic and generally hard to see; often only the heads of foraging birds are visible. Look for the fairly distinct facial pattern, the relatively long, narrow, acutely pointed bill, and the very narrow white eye-ring around a largish eye. The back appears somewhat white-flecked or frosted. Underparts (difficult to see) are uniformly tawny or buffy (no white on the belly!), with shadowy streaks on the breast and sides.

BEHAVIOR: Often forages in flocks or small groups that simply melt into the grass. Very habitat-specific. Habitually feeds in expansive, flat, short (ankle-high) grasslands with a mix of high and low areas and shallow standing puddles or wet areas. Typical prairie grasses include long spike stridens, dropseed, and (especially) three awn.

Waddles and worms its way through grasses by walking or running. Flushes reluctantly (almost underfoot), and birds in feeding flocks may not all flush at the same time. Almost always calls as it flushes and continues to call at spaced intervals in flight. Explodes from the grass. Climbs rapidly and often stays aloft for some time (30 secs.) before descending (usually within 200 yds. of its launch point).

FLIGHT: Fairly long-winged, pointy-winged, and long-faced (for a longspur). Overall warm or buffy color is easily noted. The white wing-bars may be prominent, particularly on winter males. The white on the tail is limited to the outsides of the tail. Flight is strong and undulating; a series of deep rapid wingbeats are followed by a closed-wing glide. When landing, birds generally drop the last foot into the grass.

VOCALIZATIONS: Song is a short, mostly high, clear, whistled warble: "*(e'eeu) See See See S'E'E Se' you.*" The opening notes are soft and muttered, and the penultimate "*see*" note is hoarse or stuttered. Rattle call is a low sputtery rattle, "*t'r'r'r't,*" that is like Lapland Longspur but the notes are slightly more spaced (and so less urgent-sounding). Does not utter a "*chew*" note like Lapland.

PERTINENT PARTICULARS: In winter, in its classic grassy habitat, this bird can be very hard to see. It frequently shifts fields and may fly great distances. Birders attentive to the rattle call may locate birds overhead. Otherwise, birds are best found by walking through appropriate habitat, paying particular attention to areas with shallow standing water. Birds almost always flush before being seen, but are likely to land nearby. Mark the spot well. Approach to within 40 ft. Scan with binoculars. Usually the birds are mindful of your approach and will be watching, with just their slightly darker heads protruding from the paler backdrop of grass.

Chestnut-collared Longspur, *Calcarius ornatus*

The Tall-grass Longspur

STATUS: Fairly common prairie nester and common winter resident. **DISTRIBUTION:** In summer, found from s. Man. and N.D. to extreme w. Minn., west to s. Alta., all but w. Mont., and ne. Wyo. Also found in pockets in se. Wyo., s. S.D., extreme w. Neb., and ne. Colo. In winter, found in s. Kans., all but e. Okla., n. and w.-cen. Tex., s. and e. N.M., and s. Ariz. south into cen. Mexico. Also very local in se. Calif. Casual in the East. **HABITAT:** Breeds in dry short-grass native prairie, showing a partiality for grazed and burned areas. Also occurs in mowed fields and grass-covered areas (such as airfields). Where found with McCown's Longspur, prefers denser, taller, often wetter areas. In winter, inhabits dry and desert grasslands; prefers taller grassland than McCown's. Also found in cut alfalfa and grassy farm fields. **COHABITANTS:** Ferruginous Hawk, Mountain Plover, Lark Bunting, Western Meadowlark. **MOVEMENT/MIGRATION:** Entire breeding population relocates in winter. Spring migration from late February to early May; fall from September to late November. VI: 3.

DESCRIPTION: The smallest longspur (just slightly larger and plumper than Savannah Sparrow). Overall compact and petite (particularly in comparison to McCown's Longspur—a veritable mesomorph of a longspur). The head is small and gently rounded, the bill small, conical, and compact, the body plump and oval-shaped, and the wings and tail short.

The yellow face, rufous nape, and jet-black underparts on breeding males are distinctive. Winter males commonly retain enough pattern (and enough rufous on the nape) to be readily identified. (But beware: Lapland Longspur also shows a reddish nape.) Breeding females are overall grayish brown, but commonly show a shadow rendering of the male's head pattern, including some rufous on the neck.

In winter, at its drabbest, Chestnut-collared Longspur is overall grayish brown—grayish with little pattern above, and grayish (often with shadowy streaks) below. Lapland Longspur is overall darker with lots of pattern, and with its contrasting earth tones—brown, buff, black, white, even some rufous—gray is at a premium on this bird. Smith's Longspur is overall warm ochre or buff. McCown's is even paler and plainer (particularly about the face) than Chestnut-collared, particularly the underparts, which may have some faint shadowy streaking on the breast but are primarily whitish.

Two traits are not particularly useful. The tail patterns of Chestnut-collared and McCown's both show lavish amounts of white and a similar inverted black T-shaped tail pattern and are difficult to distinguish in the field. Also, the bills of winter McCown's and Chestnut-collared are both pink; however, the bill on McCown's is larger and generally brighter pink, so it stands out.

More useful is the fact that McCown's is overall stockier and larger-headed in particular, with a protruding forehead and a very robust bill. Also, Chestnut-collared's expression is alert or stern; McCown's, principally because of the large and curiously contoured bill, always looks like it's smiling.

BEHAVIOR: When breeding, the male's aerial displays are arresting. At other times of the year, Chestnut-collared is very circumspect and nearly invisible in its vast grassy fortress. Forages like a supplicant, with its neckless head down and its back arched or bowed, moving along with shuffling mincing steps (doesn't hop). Feeding flocks are generally disorganized, with every bird facing in a different direction. The birds are site-tenacious when feeding and move very little (compounding the difficulty in finding them). Although they are usually found in taller grass than McCown's, they are most likely to feed in the shorter areas (such as grazed areas or areas close to roadsides). If flushed, or if just arriving to feed, birds often perch on fence wires, with a distinctly slouched posture.

FLIGHT: Bouncy, rollicking, and wandering, with rapid wingbeats and a terse closed-wing glide. When landing, brakes with several open-winged glides, then hovers, briefly, and descends almost vertically, sticking the landing. In flight, tails show considerable white (much more than Smith's, Lapland, and Vesper Sparrows), much like McCown's.

VOCALIZATIONS: Song is reminiscent of Western Meadowlark, with seven to eight mostly clear notes/phrases and the last one lower and slurred: "*s'See sir See sir See'es'lew.*" Flight call is distinctive and frequently given (almost always when birds take off); a soft, pleasing, slightly squeaky, compressed, two-note mewing, it sounds like a cross between a bluebird's "*churr*" and Horned Lark's squeaky "*squeedle*": "*kweedle,*" or sometimes "*kweedle-eu'u.*" Also occasionally makes a low, short, trilling chortle that recalls a small woodpecker's rattle call.

PERTINENT PARTICULARS: Distribution and habitat are salient clues where longspur identification is concerned. This particular species thrives in grass, particularly taller native grass (about a foot high), and though it may forage in close-cropped, even bare, overgrazed portions of grasslands, it is not a bird commonly found in barren or plowed fields or on beaches or mud flats.

In winter, when you stumble into a flock, *stop.* No matter how many birds leave, there will almost invariably be one (or a dozen or half a hundred) still on the ground. Search low sparsely vegetated areas until you find them creeping along like mice.

Snow Bunting, *Plectrophenax nivalis*

Spark of Winter

STATUS: Common arctic and tundra breeder; uncommon to common wintering species, with distribution and abundance apparently linked to snow accumulation. **DISTRIBUTION:** In North America, breeds from Greenland across arctic Canada to the arctic regions and higher elevations of Alaska as well as along the Aleutians. A few are found in the mountains of n. B.C. and the Yukon. In winter, except for coastal s. Alaska, where it is a year-round resident, winters south of its breeding range, from Nfld. to the s. Yukon, south to the mid-Atlantic states, west to n. Colo. and n. Utah, then north and east through the Great Basin to Wash. Occurs irregularly farther south. **HABITAT:** Breeds in rocky areas in proximity to tundra (wet or dry); also on rocky seacoast. Winters in a variety of open, mostly barren landscapes, including beaches, lakeshores, prairie, weedy fields, harvested cropland, and cleared roadsides after heavy snows. **COHABITANTS:** At all times of year, Horned Lark and Lapland Longspur. **MOVEMENT/MIGRATION:** Spring migration is very early and very protracted; some birds depart wintering areas in February (arriving in arctic areas in March) and some females do not arrive on territory until June. In fall, northernmost birds may leave in late August, and influxes of birds may continue into southern portions of the winter range until December. Peak fall movements occur from late October into November. VI: 3.

DESCRIPTION: A sturdy chubby-faced bunting of cold open places that melts into winter landscapes and explodes into white flight. Seems structurally incongruous—plump and somewhat hump-backed in front, slender and longish behind. Plumage appears soft and fluffed, but the feathers lie tight, not billowy.

Breeding males are starkly black and white. Nothing (except for the geographically isolated McKay's Bunting and albino finches and sparrows) approaches this plumage. Females and all winter-plumaged birds blend the colors of a snow-whipped weedy field or a cobble-strewn beach. Upperparts are brown and gray; underparts are dirty snow-colored. The large roundly contoured head and upper breast are touched with rusty buff and bedecked with a tiny black eye and a short yellow conical bill. The horizontal white slash (hinting of hidden magic) usually runs the length of the folded wing. The expression is serene.

BEHAVIOR: Moves like a penitent, and flies like a crazed snowflake. Forages in the open, on the ground, in windswept places—barren places that would starve a House Sparrow. Its tendency to huddle low makes it look stepped on. Shuffles as it walks, but also hops and sidles with a sideways bent. Forages where debris has accumulated—in small, sparse, weedy patches, or at the ecotone where beach and dunes meet. Sometimes leaps to grab seeds from overhanging grasses, and is also reported to chase the waves, Sanderling fashion. Although mostly terrestrial, also perches on snow fence, utility lines, treetops, and the sunny side of roofs. In bad weather or windy conditions, crouches in crevices.

Very social. Feeds in flocks that shift positions frequently and number anywhere from two or three to several hundred. Often associates with longspurs and Horned Lark.

FLIGHT: Arresting pattern of black, brown, and, especially, white in the wings and sides of the tail. Given the deeply undulating pattern, the rapid flutter and fixed-wing glide, and the "tumbling" turnover in tightly packed flocks (birds in the rear overfly the flock to get in front), observers often cannot conclude where the white is on the bird. It doesn't matter. No other songbird (unless wholly or partially albinistic) shows so much white.

Silhouette shows extremely long broad wings for so small a bird. Flight is fast, deeply undulating, and changeable, incorporating short hops and longer, more swooping glides. Often vocalizes in flight.

VOCALIZATIONS: Song is bright, cheery, and variable, mixing rough and sharp notes. Classic pattern consists of a single phrase repeated two or three times; for example: "*t'jee t'jew t'jee't'jew t'jee't'jew*" or "*p'chee p'cheer.*" Often adds an improvised ending, such as: "*w'rtch w'rtch wer chew.*" Sounds like a monotonous bunting. Two classic calls are both given in flight: a low, rippling, buzzy rattle that recalls Lapland Longspur (and Carolina Wren!—if you hear a Carolina Wren growl but it's overhead, think Snow Bunting), and a loud, clear, whistled "*chew!*" (similar to Lapland Longspur). Also makes a low flat "*t'cheh.*"

McKay's Bunting, *Plectrophenax hyperboreus*

Outermost Bunting

STATUS: Uncommon, very local, and remote. **DISTRIBUTION:** Breeds almost solely on two islands in the Bering Sea between Siberia and Alaska—Hall and St. Matthew—but also very irregularly on St. Lawrence I. as well as St. Paul. In winter, found on the Alaskan Bering Sea coast from Kotzebue south to Cold Bay and irregularly on the southern coast of Alaska and the Aleutians. Outside Alaska, records exist for B.C., Wash., and Ore. **HABITAT:** Rocky tundra. Winters in coastal marshes, beaches, and areas showing exposed bare ground or vegetation. **COHABITANTS:** In summer, Rock Sandpiper, Long-tailed Jaeger; in winter, Snow Bunting. **MOVEMENT/MIGRATION:** Vacates breeding grounds and winters on the coastal mainland of Alaska. Fall migration mostly in October; returns in May. The range of winter records falls between December and March. In late April to mid-May, apparently stages due east of breeding grounds at Hooper Bay. VI: 1.

DESCRIPTION: A whiter Snow Bunting. Slightly larger (apparent in direct comparison), but otherwise structurally similar to Snow Bunting. Breeding males are almost entirely white, with only the wing and tail tips showing touches of black. By comparison, male Snow Bunting has a black back. Breeding females show all white except for touches of gray on the forehead and a smattering of gray and black on the back. Female Snow Bunting is overall dark above, with a grayish wash on the head and a fully dark back.

Winter males are much like summer birds but show touches of rust on the head, face, and side of the breast. Winter females are like pale renderings of Snow Bunting but show more white on the back, wing, and tail and more restricted rust in the plumage.

BEHAVIOR: Much like Snow Bunting, with which they flock in fall and winter and by whom they are greatly outnumbered. Reportedly breeds in logs washed up on beaches, but more commonly in depressions and protected clefts of exposed rocky hillsides. In summer, on St. Paul Island, male McKay's occasionally pairs with Snow Bunting. Very tame.

FLIGHT: In almost all respects like Snow Bunting, but McKay's males are almost all white, showing only a touch of black on the wingtips and a spot of black near the tip of the tail. Females are patterned like Snow Bunting but are overall paler.

VOCALIZATIONS: Similar to Snow Bunting.

PERTINENT PARTICULARS: A small all-white bunting-like bird seen outside Alaska is far, *far* more likely to be an albino form of some more common resident species than a wayward McKay's Bunting. A trip to St. Matthew Island in summer is probably your best bet (but a trip to Nome or Bethel in winter is more logistically feasible).

CARDINALS, GROSBEAKS, BUNTINGS, AND DICKCISSEL

Northern Cardinal, *Cardinalis cardinalis*

The Red Bird

STATUS: A common widespread, mostly eastern and southwestern resident. **DISTRIBUTION:** Found across all of the e. U.S. east of the prairies (except absent in n. Minn., extreme n. Wisc., most of the Upper Peninsula of Mich., ne. Vt., n. N.H., and n. Maine). Also occurs in se. Ont., extreme se. Que., and sw. N.S. The western border cuts across se. S.D., cen. and sw. Neb., se. Colo., w. Kans., the Oklahoma Panhandle, w. Tex., s. N.M., and cen. and sw. Ariz. south into Mexico. **HABITAT:** A species of hedges and woodland edge. Northern Cardinal's only habitat requirement seems to be one woody-stemmed tangle big enough to conceal its nest and room to forage. It *thrives* in suburbia, but is also well established in small woodlands, riparian corridors, city parks, and mesquite-rimmed desert washes. Common at bird feeders. **COHABITANTS:** House Wren, Northern Mockingbird, Song Sparrow. **MOVEMENT/MIGRATION:** Permanent resident. VI: 2.

DESCRIPTION: Over much of North America, it's the crested red bird in your backyard. Note the big red-red conical bill and black border on the face. If it lacks a distinctly peaked crest and has no black on the face, and if you live in more southern portions of the cardinal's range, you may be looking at Summer Tanager. Note, too, that some adult male cardinals may be vibrant red and others may have darker, duller wings and tails.

The adult female is like the adult male, but more plumage-challenged—light brown with distinct reddish highlights in the wings and tail that are particularly evident in flight. Immatures are like females, but bills are gray, not red, and the crest is stunted. If you live in desert portions of southeast Arizona, southern New Mexico, or southwest Texas, be aware that you have one other crested bird with reddish highlights to contend with—the Pyrrhuloxia (see "Pertinent Particulars").

BEHAVIOR: Usually seen foraging on the ground, not far from bushes or cover. Feeds by hopping, and is sometimes found with other ground-feeding sparrows. A fairly active bird, but also a retiring bird. Even where it is habituated to people, it doesn't tolerate close approach and is often timid at bird-feeding stations. In winter, retreats to brushier tangles. Also a common (and aspired to) frequenter of backyard feeding stations, where it is most active at dawn and dusk. Prefers larger seed (like sunflower and safflower).

Both males and females sing—males from high (often very high) exposed perches, females from the nest. Males may sing all year, particularly in more southerly portions of their range; in the north, they sing mid-February through August.

FLIGHT: Shows a long tail and round wings. The male's color is distinctive; the female's hidden reddish hues bloom in flight. Flight is direct, unhurried, undulating (almost flippant), and generally short (from one close bush to another) and low to the ground. Wingbeats are given in a short series followed by a brief closed-wing slide.

VOCALIZATIONS: Song is a loud ensemble of clear pure-toned notes or phrases frequently repeated. The most universal patterns are phonetically rendered "*wheeta wheeta wheeta*," "*wheet whee wheet*," "*what Cheer! what Cheer! what Cheer!*" or (slow) "*tew, tew, tew, tew*" and (rapid) "*too-too-too-too too*." Call is distinctive and easily learned—a short sharp "*teet*" (often truncated, sounding more like "*t'k*"). Also makes a soft high-pitched twitter.

PERTINENT PARTICULARS: The only bird that Northern Cardinal is likely to be confused with is Pyrrhuloxia—a likewise crested and closely related species that is restricted to desert regions of the American Southwest. Male Pyrrhuloxia is mostly gray with limited amounts of red on the body. Females and (particularly) immature cardinals and Pyrrhuloxia are very similar and can be best separated by looking at bill shape—short, blunt, and rounded (like a miniature parrot's bill) on Pyrrhuloxia; longer, conical, and sharp on the cardinal. Also, the bill of adult female Pyrrhuloxia is yellow, whereas the bill on female cardinal is bright red-red. Note, too, that the crest of Pyrrhuloxia is long and spiky; the cardinal's crest is typically fuller and bushier.

Pyrrhuloxia, *Cardinalis sinuatus*

Desert Cardinal

STATUS: Uncommon to common permanent resident of dry and desert areas in the Southwest. **DISTRIBUTION:** From se. Ariz. across s. N.M. and w. and s. Tex. to just south of Houston and south into the northern half of Mexico as well as s. Baja. Casual visitor north and west of its normal range. **HABITAT:** Generally dry sparsely to moderately vegetated areas, including classic Sonoran Desert habitat, desert scrub, mesquite grasslands, riparian washes, and suburban areas where native vegetation is left to flourish. **COHABITANTS:** Greater Roadrunner, Cactus Wren, Curve-billed Thrasher, Northern Cardinal. **MOVEMENT/MIGRATION:** Permanent resident that occupies the same habitat year-round over much of its range; however, a shift in winter in some places to riparian habitats has been noted. VI: 2.

DESCRIPTION: A gray-and-red-trimmed cardinal with a spiky crest, a pushed-in face, and a not-red bill. Slightly smaller, slighter, and more roundly proportioned than Northern Cardinal. While both species are acutely crested, Pyrrhuloxia's crest is long, narrow, and pointy—looks like a punk cut. Northern Cardinal's crest is wide, wedge-shaped, and shaggier—looks like the bird is having a bad feather day. The shape of the faces also differs. Pyrrhuloxia is very round-faced—a curvature that begins with the contouring forehead and curls around the blunt round yellow bill. Northern Cardinal's face is pointy and wedge-shaped—starting at the raised crest and running, in a straight line, to the tip of the conical orange-red bill.

The differences in bill color cannot be overemphasized. Adult Pyrrhuloxia's bill is yellow; adult male and female cardinals have red bills. Immature Pyrrhuloxia has a gray bill, and young cardinals a black bill.

Adult male Pyrrhuloxia's resemblance to female Northern Cardinal is superficial. The overall contrast between the gray body and crisply defined bright red feathers of Pyrrhuloxia is very unlike the brownish body and pastel-shaded reddish areas of the female cardinal. Female Pyrrhuloxia and young birds are more like Northern Cardinal, but Pyrrhuloxia is always grayer, and the reddish crest, wing edge, and tail edge more restricted and more crisply defined. When in doubt, just look at the face. Round with a blunt button of a bill — Pyrrhuloxia. Pointy-faced with a conical bill — Northern Cardinal. Pyrrhuloxia's expression is out-of-sorts.

BEHAVIOR: Forages by hopping on the ground and moving branch to branch (or cactus blossom to cactus blossom). Favors drier, more open, and sparsely to moderately vegetated habitat than cardinal, but where feeding stations are well maintained, both birds occur. In winter, often found in small groups that orbit around more heavily vegetated or riparian areas. Subordinate to cardinal at a feeding station. Although fairly bold, birds that remain sedentary for any length of time often work their way into the protective confines of a tree or shrub, where they may be plainly visible but afforded some protection from predators.

FLIGHT: Straight, somewhat leisurely, generally low, and slightly undulating, with the hint of a skip or bounce. Wingbeats are given in a quick, fluttery, and audible series, followed by a brief closed-wing glide. The head and body rise as the wings beat, then flatten when the bird glides. When landing or descending, brakes with bounding, swooping, open-winged glides.

VOCALIZATIONS: Many of Pyrrhuloxia's vocalizations are similar to Northern Cardinal but slightly softer and not as full-bodied or vehement. Song is a whistled "*ch'chew, chew, chew, chew.*" A "*teet*" call is like Northern Cardinal, but, again, not as full-bodied. Also makes a sharp, descending, and accelerating series of notes that sound like a cross between a spit and a chip: "*spik . . . spik . . . spik-spikspikspikspik.*"

Rose-breasted Grosbeak, *Pheucticus ludovicianus*

Grosbeak Shot Through the Heart

STATUS: Common northeastern and western Canadian breeder; common migrant east of the Rockies. **DISTRIBUTION:** In Canada, breeds from se. Yukon and sw. Northwest Territories to ne. B.C., all but s. Alta., cen. and se. Sask., s. Man., s. Ont., s. Que., throughout the Maritimes, and w. Nfld. In the U.S., breeds west to n. and e. N.D. to e. S.D., e. Neb., and e. Kans., south to cen. Mo., s. Ill., s. Ind., s. Ohio, w. Md. (and south along the Appalachians to n. Ga.), and n. N.J. Found throughout New England. Winters south of the U.S. Very rare but regular visitor in the w. U.S. **HABITAT:** Deciduous and mixed woodlands, especially younger second-growth woodlands and moist woodlands. Very partial to woodland edges of streams, roads, pastures, suburban yards, parks, and orchards. In migration, found in a variety of woodland habitats and at feeders in spring (and winter). **COHABITANTS:** Red-eyed Vireo, Blue Jay, Veery, Scarlet Tanager. **MOVEMENT/MIGRATION:** Wholly migratory. Spring migration from late March to early June; fall from late July to early November. VI: 3.

DESCRIPTION: A robust canopy-loving forest grosbeak. Crimson-bibbed black-and-white males look like they've been shot through the heart. Medium-sized songbird (larger than Scarlet Tanager; smaller than Baltimore Oriole; about the same size and shape as the western Black-headed Grosbeak), with a huge, pale pinkish, somewhat parrotlike bill, a large egg-shaped head, an oval body, and a shortish tail. Overall looks like the head is too large and the tail too short.

Males are distinctive. Grayish brown females look like big, plump, pink-billed, treetop-frequenting sparrows with a dark mask through the eye and crisp, narrow, dark streaks etched completely across white underparts (recalls a giant Purple Finch). By comparison, adult female Black-headed Grosbeak is somewhat buffy or butterscotch-breasted, with very fine streaking (if any) limited to the sides and a grayish bicolored bill — dark on top, paler on the bottom. Immature male Rose-breasted resembles the female but shows a blush or touch of red on the breast. Immature has warm, somewhat buffy underparts (like female Black-headed Grosbeak), but shows a heavier and more lavishly streaked breast (like female Rose-breasted).

BEHAVIOR: Despite the male's distinctive plumage, not necessarily a conspicuous bird. Spends much of its time in the deep leafy canopy, where it moves little and feeds deliberately, plucking insects and fruit from leaves, branches, and trunks. In spring, in migration along the Gulf Coast, may feed in low brush or even on the ground. Also hovers and, despite its apparent stockiness, occasionally fly-catches. Particularly in spring, feeds on tree flowers and buds and is attracted to feeders offering large seeds (such as sunflower). In migration, sometimes found in small flocks. Males sing from elevated perches (usually in the canopy and frequently hidden). Fairly tame. Responds well to pishing.

FLIGHT: Overall robust and chesty with a big blunt head and a short, wide, squared-off tail (which is frequently flared). The hands of the long wide wings often seem to get stuck or momentarily frozen in the down position. Wings of males show silver dollar–sized white patches, and underwing linings flash pinkish red (bright yellow on Black-headed). Underwings of female Rose-breasted are dull yellow (bright yellow on Black-headed). Flight is undulating, more flowing than a tanager, but also somewhat indecisive. Deep wingbeats are given in a lengthy irregular series followed by a short closed-wing glide.

VOCALIZATIONS: A sweet, slow, flowing, robinlike series of whistled notes, but overall somewhat softer and more timid. Call is a sharp harsh squeak, "*EEK*," that has been well described as the sound of a sneaker skuffing a gym floor; squeakier than the similar call of Black-headed Grosbeak.

Black-headed Grosbeak, *Pheucticus melanocephalus*

Western Parrot-bill

STATUS: Common breeder across much of the West. **DISTRIBUTION:** Breeds from s. B.C., s. Alta., extreme sw. Sask., and w. N.D. south. Range includes all of Wash., Ore., Idaho, Wyo., Nev., Utah, and Colo., as well as n. and coastal s. Calif., n. and e. Ariz., w. and cen. N.M., trans-Pecos Tex., nw. Kans., the western two-thirds of Neb., and the western half of S.D. Breeding range extends south into Mexico. Casual visitor in the East. **HABITAT:** A habitat generalist, commonly found in some form of woodland. Typical habitats include riparian woodlands, mature pine and pine-oak forest, pinyon-juniper, deciduous groves in mountain canyons, and oak savanna. Does not occur in dense coniferous forest, desert areas away from riparian woodlands, and pure grasslands. Also attracted to feeders. **COHABITANTS:** Western Wood-Pewee, Western Tanager, Bullock's Oriole. **MOVEMENT/MIGRATION:** Entire population withdraws into Mexico during winter. Spring migration in April and May; fall from early July to early October, with a peak in August and early September. Occurs regularly across west Texas as a migrant; less commonly to the coast. VI: 3.

DESCRIPTION: A medium-sized, massive-headed, gargantuan-billed songbird (dwarfs Blue Grosbeak; larger than Western Tanager; same size as Rose-breasted Grosbeak). Large egg-shaped head is fitted with a supersized finchlike bill that is dark and down-curving on the upper mandible. The body is tubular and robust, and the tail a bit too short for the bird. The structural package is somewhat parrotlike.

Breeding adult males are unmistakable with their burnt orange body, blackened head, wings, and tail, and white slashes on the wing. Non-breeding males, females, and immatures show a shadow rendering of the breeding male pattern, but upperparts are grayish brown and streaky, and underparts range from buff to butterscotch, with pencil-fine dark streaks etched down the sides and flanks (and sometimes partially across the breast). The head pattern is bold: the bird looks like it's wearing dark goggles highlighted by a very obvious broad buffy eyebrow.

Female and immature Black-headed Grosbeaks are identical in structure and similar in plumage to the eastern Rose-breasted Grosbeak. Black-headed Grosbeak is overall buffier or warmer-toned and has buffy eyebrows (white in Rose-breasted), fine streaking or none at all on the breast (coarse and/or wide and blurry streaking on Rose-breasted), and a blackish upper mandible and a grayish or pinkish lower mandible (Rose-breasted has a pink or horn-colored bill, with little contrast between the upper and lower mandibles).

Also, the underwing linings of Black-headed Grosbeak are bright yellow (dingy yellow or ochre in Rose-breasted). When birds are fully exposed in flight, sometimes a fluffy glimpse of the yellow underwing linings peeks through at the bend of the wing.

BEHAVIOR: A slow slothful feeder. Perched high on branches or low bushes, moves with slow deliberate hops, long pauses, and slow appraising turns of the head (one eye up, one down), then reaches out for morsels (insects for fruit) with a long neck up . . . out . . . or down. Sometimes hangs head down (parrotlike). Consumes whatever it grabs on the spot, and takes its time about it. In places, feeds habitually in short brush or the ground (something Rose-breasted Grosbeak does much less commonly). Also drawn to sugar-water and backyard feeding stations, where it favors large seeds.

Tame, but very circumspect when near the nest. Comes readily to pishing. Males sing frequently even late in the breeding season.

FLIGHT: Overall stocky profile—big head, chunky body, short tail. The bold colors of adult males stand out in flight. Added to this are the prominent white oval patches that flash near the tips of each wing. (The white-tipped corners to the tail are easily overlooked.) Females and immatures are a very bland uniform brown above, *but* the underwing linings are bright yellow. (Linings of Rose-breasted are dirty yellow or mustard-colored.)

Flight is strong, powerful, effortful, and fast. Wingbeats are rapid and fluttery, and given in a lengthy series followed by a brief closed-wing glide. Flight is smoothly undulating, but birds drop quickly on the glide.

VOCALIZATIONS: Song is a light carefree ensemble of whistled two-note phrases that alternate: one high, next low, next high (some slightly trilled). Sounds very much like Rose-breasted Grosbeak, but more hurried and less lyrical, with choppier phrases. (With Rose-breasted, phrases flow into the next.) Call is a short sharp "*ick*." Young birds make a plaintive whistled "*peeuu*."

Blue Grosbeak, *Guiraca caerulea*

Mesobunting

STATUS: Common summer resident. **DISTRIBUTION:** Widespread across the s. U.S. (except on the Gulf Coast and in s. Fla.), with a range that extends north to n. N.J., se. Pa., the Va.–W. Va. border, s. W. Va., s. Ohio, s. Ind., s. Ill., cen. Mo., s. N.D., most of S.D., Neb., Kans., e. and portions of sw. Colo., n. N.M., s. Utah, Nev., and s. and cen. Calif. Range extends south into Mexico and Central America. Very rare north of its normal range. **HABITAT:** Primarily weedy and overgrown fields with shrubs, hedgerows, islands of trees, or woodland edge; also found in riparian woodlands, mesquite, salt cedar, southern pine forests, and desert washes. In migration, may be found in open grassy areas, pastures, croplands, and lawns (but rarely backyard feeding stations). **COHABITANTS:** Northern Mockingbird, Yellow-breasted Chat, Lazuli and Indigo Buntings. **MOVEMENT/MIGRATION:** Spring migration from mid-April to early June; fall from late August to mid-November. VI: 3.

DESCRIPTION: A big burly Indigo Bunting–like bird with rust-colored slashes on the wing. A small grosbeak (larger, stockier, more robust than Indigo Bunting; smaller than Yellow-breasted Chat or Brown-headed Cowbird; smaller than Rose-breasted or Black-headed Grosbeak), with a large conical bill that dominates the face, a large, often peaked head, a chunky body, and a short tail. When foraging, the tail is commonly fanned, showing a broad bulbous tip. Indigo Bunting is smaller, less angular, more finely proportioned, and particularly smaller-billed. Also, Blue Grosbeak commonly perches tensely erect. Indigo Bunting appears more relaxed.

Males are overall deep blue with blue-black wings and tail. The silver bill stands out, and broad rusty wing-bars are also conspicuous. Females and immatures are warm brown to gray-brown and overall plain except for one or two prominent cinnamon-tinged wing-bars. The heads often appear warmer-toned than the backs.

BEHAVIOR: Moves like it loses points if it touches the ground. Forages by clinging to, hopping through, or flying to the next bit of low vegetation, be it grass stalk, weed, or bush. Sometimes hovers or pursues prey in the air; also hops on the ground, but seems awkward doing so. A persistent singer that often sings during the middle of the day; males sing from high perches, usually the highest perch, and very commonly utility lines. Seems more inclined to sing from perches in open fields, whereas Indigo Bunting is more inclined to perch along the woodland edge. When foraging, Blue Grosbeak frequently fans and pumps its tail with a jaunty flourish. When anxious, raises its crown into a peaked crest.

Somewhat shy; generally flies when approached. In migration, may be found in small flocks and among Indigo and other buntings.

FLIGHT: Overall stocky, big-headed, and short-tailed. Adult males often show a blackish tail and brown on the wings; females and immatures seem blandly uniform or may show a ruddier back. Flight is direct and slightly undulating or jerky, often covers great distance, and is executed without much finesse—this is a meat-and-potatoes flier. Stiff wingbeats are given in a short irregular series punctuated by a terse closed-wing glide. Almost invariably fans and pumps its tail upon landing.

VOCALIZATIONS: Song is a rich, rough, somewhat garbled and hurried warble consisting of two-note phrases that rise and fall in pitch. Sounds like a hoarse House Finch or a finch singing with its mouth full. Distinctive call is a loud, emphatic, metallic "*chink*" or "*spink.*" Also gives a buzzy "*grrr*" that sounds like a light Dickcissel's flight call but is buzzier and less flatulent.

PERTINENT PARTICULARS: If you are color-sensitive, Blue Grosbeak and Indigo Bunting can be separated by color alone. Male Blue Grosbeak is blue bordering on purple; Indigo Bunting is a paler, purer blue. Female Blue Grosbeak is warm-toned; female Indigo Bunting is comparatively colder and grayer. But perhaps the best approach is to focus on overall shape. Blue Grosbeak looks like a stocky heavy-headed grosbeak; Indigo Bunting seems more akin to a finch.

Lazuli Bunting, *Passerina amoena*

Bunting with Wing-bars

STATUS: Common and widespread western breeder; restricted winter resident. **DISTRIBUTION:** Breeds across most of the w. U.S. and s. Canada. Found in s. B.C., extreme s. Alta., and Sask.; in the U.S., in all but w. Wash. and Ore., Calif. except southeast regions, Idaho, Mont., w. Dakotas, w. Neb., Nev., Utah, n. Ariz., all but e. Colo., sw. Kans., n. N.M., and the Oklahoma Panhandle. Winters in w. Mexico; some wintering birds are found in se. Ariz. **HABITAT:** A variety of open brushy habitats to an altitude of nearly 10,000 ft., but shows a partiality for riparian corridors in arid regions. Also found on brushy hillsides, aspen thickets, sagebrush, chaparral with grassy patches, regenerating disturbed areas, suburban gardens, and sometimes feeders. Winters along the borders of weedy fields. **COHABITANTS:** Willow Flycatcher, Yellow Warbler, Blue Grosbeak, Black-billed Magpie. **MOVEMENT/MIGRATION:** Entire population withdraws to Mexico and southeast Arizona in winter. Spring migration from mid-March to early June; fall migration occurs in two stages, with an initial post-breeding relocation to southern California, southern Arizona, and southwest New Mexico in August. VI: 2.

DESCRIPTION: A beautiful vocal bunting that carries the western sky on its back and a sandstone blush across its breast. Small finch (slightly smaller than House Finch; the same size as Indigo Bunting), with a peaked head, a longish conical bill, and a short slightly notched tail. Like Indigo Bunting, but slightly leaner, more angular, and stiffer.

Breeding males are distinctive, with sky blue upperparts, a rusty wash across the breast, and bright white underparts. Winter males are shabbier and browner but retain enough of the breeding pattern to be identified. If slim profile and conical bill are insufficient to separate the bird from the larger, plumper Western Bluebird, the bunting's broad white wing-bars will eliminate confusion.

Females and immatures are plain pale brown but have characteristics that harken back to the male's pattern. Female and immature Lazulis have crisp, narrow, broken, whitish wing-bars that stand out against the blue-grayish-tinged brown wings. The wing-bars on other female and immature buntings are indistinct. Also, females and (particularly) immature Lazulis have a warm blush across the chest (a shadow of the male's rusty breast) and a paler belly. On other buntings, upperparts and underparts are more uniformly bland. *Note:* Some immature Lazulis may show faint streaking that is more characteristic of young Indigo Bunting. Lazuli's expression is mean.

BEHAVIOR: Males are tenacious singers—singing well into summer after most other breeders have called it a season. Usually sings from a high conspicuous perch—most commonly from the very top of a tree or bush—and favors certain perches. Frequently sings from the same perch for many minutes, but also patrols the border of its territory.

Forages most commonly on the ground or on reedy stalks using short hops or flights. In summer, also forages in leafy trees, where it gleans insects and fruit and occasionally flycatches or even hovers. Easily attracted to bird feeders, where, like most finches, it favors millet.

Tame where habituated to people. Often more circumspect and difficult to approach when breeding.

FLIGHT: Perhaps slightly lankier and longer-profiled in flight than Indigo Bunting and (particularly) Varied Bunting. Flight is direct, with slightly bouncy shallow undulations. Wingbeats are rapid and given in a short irregular series of deep open-winged flaps punctuated by intermittent skips.

VOCALIZATIONS: Song is a quick, loud, lively ensemble of repeated, clear, bright, two- and three-note phrases; for example: "*ch'weh ch'weh che che chchch ch'wee'r.*" The pattern often stumbles or drops off at the end and is similar to Indigo Bunting but more chatty and more varied. Call is a dry spit: "*sp'k.*"

PERTINENT PARTICULARS: A word of caution. Female Blue Grosbeak—with buffy wing-bars, a bluish tinge in the wing and tail, and a warm blush across the breast—might be confused with immature Lazuli Bunting. The Blue Grosbeak is larger, bulkier, and shorter-tailed, has a peaked head and a very large bill, and is usually overall warm-toned, particularly about the head.

Indigo Bunting, *Passerina cyanea*

The Blue Canary

STATUS: Common and widespread—the blue bunting found east of the prairies. **DISTRIBUTION:** Breeds across most of the e. and s. U.S. and se. Canada. In Canada, found in s. Man., s. Ont., and s. Que. In the U.S., breeds in all states east of the prairies except n. Maine, the Florida Peninsula, and coastal La. The western border cuts across e. N.D., cen. S.D., w. Neb., ne. and sw. Colo., Utah, and w. Ariz. Rare in the West Coast states and w. Canada; rare but regular in the Maritimes. **HABITAT:** Breeding birds like disturbed overgrown habitat and brushy woodland edge bordering fields, such as power-line cuts, pasture bordering wood lots, roadsides, and abandoned fields going into succession, particularly those offering taller trees or utility lines for singing males. In migration, forages in fields, lawns, bushes, and treetops. In the West, often found near water. **COHABITANTS:** Eastern Kingbird, Field Sparrow, Blue Grosbeak. **MOVEMENT/MIGRATION:** Except for south Florida and south Texas, winters entirely outside the United States. Migration is fairly brief; spring migration from mid-April to early June; fall from late August to early November. In April, on the Gulf Coast, great numbers of these "blue canaries" may pile up on the lawns of residents. VI: 3.

DESCRIPTION: A small, robust, but nicely proportioned bunting that is a regular gender-linked

Jekyll and Hyde: breeding males are just blue (slightly more purple on the head), and females and immatures are just brown. Both plumages are distinctive. Indigo Bunting is a small sparrowlike bird (considerably smaller than Blue Grosbeak; larger than Yellow Warbler; the same size as Chipping Sparrow), with a slightly peaked head, a small conical bill, a slender body, and a narrow shortish tail.

Breeding males are wholly blue, but paler than the deep blue-onto-black plumage of male Blue Grosbeak. The conical bill is silver-blue and glints like gunmetal but appears short and sparrowlike (bunting abbreviated, not grosbeak gross). Winter and first-spring males are a mixture of brown and blue.

Females and immatures are just plain pale or warm brown and almost devoid of distinguishing characteristics, which is, by default, the distinguishing characteristic. They do, however, have some other distinguishing traits. Breasts show shadowy streaking (other female buntings are typically smooth-breasted with no hint of streaking). Also, female and immature Indigos show two faint, pale (or buffy), equivocal wing-bars. (Lazulis show broad and buffy or crisper, whiter, but still no-nonsense wing-bars; Varied and Painted Buntings have no discernible wing-bars at all.) Also, the throat of female and immature Indigos is slightly whiter/paler than the breast, and the tail is darker than the rest of the body (sometimes with a bluish cast).

BEHAVIOR: Forages for seeds and insects on the ground, in low bushes, and in treetop vegetation. In migration, sometimes seen in small flocks (particularly in spring on the Gulf Coast), and sometimes with other ground-feeding species (most notably other bunting species). Hops on the ground. Clings acrobatically to reeds. When disturbed, often flies to the top of a bush, crest erect, and swishes its tail side to side. Males on territory are enthusiastic singers, often singing all day from a high perch along the woodland edge; they are less inclined than male Blue Grosbeak to sing in open brushy fields.

FLIGHT: Plump round-featured profile with a distinctly short, rounded, often slightly open tail. Flight is fairly strong and direct, with frequent, brief, shallow undulations. Aerial display is a long fluttering glide with the head elevated.

VOCALIZATIONS: The song is a bright, cheery, lighthearted incantation. Whistled notes are sharp, clear, high-pitched, and usually repeated: "*ch'wee, ch'wee, ch'wee, ch, ch, see't, see't.*" As one rendering has it: "*what! what! where? where? see it! see it.*" Variations are many, but birds often repeat the same song pattern. Call is an avian spit: "*sp*" or "*sp'k.*" Flight call is a short, flat, buzzy, insectlike trill.

PERTINENT PARTICULARS: Female and immature buntings are difficult to separate. Indigo Bunting is uniformly warm brown with shadowy streaking below. Lazuli has more distinctive wing-bars and a normally unstreaked breast that is warmer-toned than the rest of the bird. Varied Bunting is uniformly brown (warm brown in winter, cold brown in summer) with an abbreviated, somewhat pushed-in bill. Painted Bunting is distinctly greenish above, drab or yellow below.

Varied Bunting, *Passerina versicolor*

Dark Chubby Bunting

STATUS: Uncommon breeder with a very limited range in the United States. **DISTRIBUTION:** Primarily a Mexican species; in the U.S., breeds only in se. Ariz., se. and sw. N.M., trans-Pecos Tex., and along the Rio Grande. **HABITAT:** Steep desert hillsides covered in thornbush and assorted cactus and riparian thickets in desert canyons and along desert washes dominated by acacia and willows. Favored territories bracket both hillsides and boast flowing or seasonal watercourses. In Texas, also found in thickets of acacia, juniper, and scrub oak. **COHABITANTS:** Ash-throated Flycatcher, Cactus Wren, Bewick's Wren, Bell's Vireo, Black-tailed Gnatcatcher, Lucy's Warbler, Rufous-crowned Sparrow. **MOVEMENT/MIGRATION:** Birds in the United States and northern Mexico migrate. Spring migration from mid-March to late May; fall from mid-September to October. VI: 0.

DESCRIPTION: A dark roundly contoured desert bunting that blooms in sunlight. A smallish sparrowlike bird (slightly smaller than House Finch; the same size as Indigo and Painted Buntings), but more compact and rotund (chubby!), with a round head, an abbreviated, roundly contoured nub of a bill, and a moderately long tail. The bill is key. The bills on other buntings are longer, and the upper culmen is never so distinctly curved as the bill of Varied Bunting, which makes the bird look comparatively pug-faced. If you look at a female or immature bunting and you have to keep pushing the image of White-collared Seedeater out of your mind, you're looking at Varied Bunting.

Adult males are unmistakable—under good conditions. In poor light, they just look black, and distant birds might easily be mistaken for a miniature Bronzed Cowbird. In fair light or at distances, males look maroon (the reds and blues in the plumage meld). In good light and at close range, the bird blooms.

Females and immatures are uniformly plain pale brown (caramel-colored or putty-colored) with a slight bluish tinge to the wings and tail. They have no wing-bars, no shadowy streaking on the breast, no white on the throat, and no buff tones. Just a uniformly plain bunting with a pug face.

BEHAVIOR: Not shy, and when feeding, very preoccupied. Commonly forage in pairs (sometimes alone), on or near the ground, moving in a linear search pattern that cuts across the face of a hillside. Not hyperactive but more or less constantly on the move, hopping along the ground or flying to the next shrub or tuft of grass. Moves by hopping and flutter-flying. Likes to stay close to cover. While foraging, flicks its wings and moves its tail with a downward jerk.

Gleans insects from foliage, hawks flying insects while perched, and occasionally hovers. Males vocalize from a high perch (given the nature of the vegetation, this means less than 10 ft.) as well as from the ground. Responds fairly well to pishing.

FLIGHT: Quick, light, nimble, and bounding, flight incorporates a rapid terse series of wingbeats punctuated by a brief pause or para-glide.

VOCALIZATIONS: Song is similar to other buntings—a quick, lively, complex ensemble of short run-on notes and phrases. Song of Varied Bunting is lower-pitched, hoarser, hurried, mumbled, more even-pitched and "chirpy," and generally less glittery and exciting than the song of Indigo Bunting. Varied does not repeat phrases as much as other buntings, but notes like "*chip*" and "*chr*" and "*chee*" crop up frequently. Often the song loses force and volume at the end, as if the bird is losing interest. The call, a "*spit*," may be slightly lower-pitched than Indigo. Other calls include a "*spee*" that sounds slightly more musical than Indigo and a low descending buzz.

Painted Bunting, *Passerina ciris*
Thicket Bunting

STATUS: Uncommon to fairly common summer resident and wintering species in the Southeast. **DISTRIBUTION:** There are two distinct breeding populations. The coastal population is found from coastal N.C. to n. Fla. The second, primarily inland population occurs in w. Miss., s. Mo., sw. Tenn., Ark., s. and e. Kans., most of Okla., most of Tex., and (barely) extreme s. and ne. N.M. Winters in s. Fla., Mexico, and Central America. Rare in se. Ariz.; casual to the West Coast and north of the regular range (often at feeders). **HABITAT:** Basic requirements seem to be brushy edge adjacent to open fields or brushy woodlands with some open components. Atlantic coastal population clearly favors woodland edge adjacent to coastal wetlands. Interior birds are partial to streamside brush. Has adapted to roadsides, fields going into early succession, and weedy lots with scattered trees and shrubs. Comes readily to suburban feeding stations. **COHABITANTS:** Eastern Kingbird, Orchard Oriole, Northern Cardinal. **MOVEMENT/MIGRATION:** Entire population migrates. Spring migration from April to mid-May; fall from late July to October. VI: 3.

DESCRIPTION: Both males and females, while very different, are unmistakable but some early-fall immatures may require a second look. Painted Bunting is a small sparrowlike bird that is similar in size and shape to other buntings except for a slightly longer, thicker, more bluntly contoured (bullet-shaped) bill that is not as cleanly conical as Indigo and Lazuli bills and longer than the snubbed bill of Varied. Painted may also appear slightly rounder-headed than Indigo and Lazuli, and perhaps slightly shorter-tailed.

The adult male, with colors that make the bird look like it was painted by a crayon-wielding three-year-old, is unique. Adult females and immature males, with lime green backs and greenish yellow underparts, are less eye-catching but no less distinctive. Other female buntings are brown—some warm brown, some grayish brown, but brown, not greenish. Female Scarlet and Summer Tanagers, which are greenish or yellow-green, are larger birds of forest and treetops.

Immature female Painted Bunting in early fall is overall grayish (brighter by mid-fall), but the upperparts (and especially the rump) are infused with greenish hues that distinguish it from other buntings. The expression on females and immatures is somewhat baby-faced.

BEHAVIOR: Despite its plumage (or because of it), this species is shy, often difficult to find, particularly in Florida in winter. Forages on the ground or in weedy fields or grassy marsh; also in low shrubs. Hops and plucks seed from vegetation, sometimes leaping to grasp seed heads. Works an area thoroughly; doesn't move around a great deal. Does come readily to feeders, where it may remain for long periods, but when foraging is completed, it buries itself in a nearby shrub and just stays there.

Males sing from exposed perches that may be very high. Response to pishing varies. In summer, some birds respond very well. In winter, birds seem less responsive.

FLIGHT: Like other buntings. Fairly slow, direct, bouncy flight.

VOCALIZATIONS: Song suggests Indigo Bunting but is more warbling, more hurried, more complex, and more carefree or carelessly varied (doesn't double its notes). Suggests Warbling Vireo to some. Rough or slightly trilled notes sometimes crop up in the flood of high and low notes. Each song varies slightly in structure, duration, and range of notes (whereas Indigo frequently repeats the same song over and over again). Call is a soft "*pip*" or "*whip*" that is softer and less emphatic than the "*spit*" call of Indigo Bunting.

Dickcissel, *Spiza americana*

Raspberry (Calling) Sparrow

STATUS: Common breeder in its core range; irregular in the wider periphery. **DISTRIBUTION:** Core range includes se. S.D., e. Neb., all but w. Kans., n. and cen. Okla., n.-cen. Tex., cen. and s. Iowa, n. and w. Mo., w. Ill., e. Ark., and cen. and s. Tex. Peripheral breeding area encompasses the Dakotas, s.-cen. Neb., e. Colo., e. N.M., w. Okla., the Texas Panhandle, s. Minn., all but n. Wisc., Mich., Ill., se. Mo., Ark., all but coastal La., Ind., Ohio, Ky., Tenn., n. Ala., n. Miss., s.-cen. Pa., and w. Md. Irregular breeder in s. Canadian prairies, west to the Rocky Mts., and east to s. N.Y., to N.J., and to (but not including) the Atlantic and Gulf coastal plain. Winters mainly outside the U.S., but a few individuals may join flocks of House Sparrows in the e. U.S. Casual migrant in the far West. **HABITAT:** A variety of open, dense, tall grasslands and agricultural lands, including native prairie, hay fields, unmowed roadsides, fencerows, fallow agricultural fields (rice and sorghum), lightly grazed pasture, and fields in early stages of succession. In migration, uses marshes for roosting. In winter, found at bird feeders frequented by House Sparrows. **COHABITANTS:** During breeding, Upland Sandpiper, Grasshopper Sparrow, Bobolink, Eastern and Western Meadowlarks; nonbreeding, House Sparrow. **MOVEMENT/MIGRATION:** Spring migration from early April to early June; fall from late July to early November, with a peak in August and September. Somewhat erratic, even nomadic, in distribution—common in an area one year, absent the next. VI: 3.

DESCRIPTION: A large-billed House Sparrow in a tall weedy field. Medium-sized sparrowish songbird (slightly larger than House Sparrow; smaller than Bobolink), with a heavy, swollen, conical bill, large oval head, a long, fairly lean body, and a shortish tail. Resembles a rangy big-billed House Sparrow.

Breeding males, with their grayish bodies, rusty wings, and boldly patterned heads and breasts, are unmistakable. At a distance, the gray head contrasts with the rusty upperparts (folded wings). Females resemble female House Sparrows, but with a broader, paler, more prominently yellow-tinged eyebrow, a slight yellow wash across the chest and face, and a touch of red on the shoulders. Immatures are like immature House Sparrow but usually show a trace of yellow on the face or sides of the throat and blurry streaking on the breast, which may also show a trace of mustard yellow. (House Sparrow is slightly warmer-toned and clean-breasted and has a smaller, pinker bill.)

BEHAVIOR: Territorial males sing from exposed perches (fenceposts, barbed wire, tall weed, shrub) from dawn to dusk and often fly great distances when changing perches. Outside breeding season, highly gregarious; found in large flocks. Usually forages in dense vegetation, either on the ground or from seed heads, but may join House Sparrows and feed in the open. Flies in large, tight, writhing flocks that look like smoke on the horizon. Birds flying overhead often announce themselves with their distinctive embarrassing-sounding call, which can usually be heard before birds are in view. Somewhat responsive to pishing.

FLIGHT: Shaped like a slender big-headed Bobolink showing a large, pointy, projecting head, an oval body, wide but pointy-tipped wings, and a short notched tail. Flight is straight, strong, and slightly rising and falling (occasionally more bouncy). Wingbeats are given in a short series followed by a closed-wing glide. In tall grass, lands with a controlled wing-fluttering crash.

VOCALIZATIONS: Song consists of an abrupt stutter followed by sharp, dry, whistled notes; for example: "*chip chip* (hesitation) *cheedle cheedle* (*cheedle*)" ("*Dick Dick cissel cissel* [*cissel*]") or "*ch* (hesitation) *chr chr chr*." Call is a low dry "*chek*." Flight call is a low, short, flatulent raspberry or Bronx cheer from on high.

ICTERIDS—BLACKBIRDS AND ORIOLES

Bobolink, *Dolichonyx oryzivorus*

Upside-Down Icterid

STATUS: Common breeder in southern Canada and the northern United States. **DISTRIBUTION:** In Canada, found from sw. Nfld., e. Que., and the Maritimes across s. Que., s. Ont. (also sw. James Bay), s. Man., s. Sask., se. and cen. Alta., and s. B.C. In the U.S., breeds throughout New England, N.Y., n. N.J., w. Md., the Va.–W. Va. border, n. and cen. Ohio, n. and cen. Ind., n. and cen. Ill., Mich., Wisc., Minn., Iowa, n. Mo., N.D., S.D., most of Neb., parts of Kans., most of Mont., much of Wyo. and Colo., parts of Idaho, n. Utah, n. and e. Wash., e. Ore., and ne. Nev. Winters in South America. Migrates primarily east of the Rockies, with great concentrations occurring along the Atlantic Coast in fall. **HABITAT:** Breeds in pasture and hay fields (formerly more numerous in native prairie grasslands). In migration, may occupy the same habitats, but often roosts in freshwater marshes, particularly marshes rich in wild rice; also found on short cut grass (such as airport runways), particularly grass rich in dandelions. **COHABITANTS:** Upland Sandpiper, Savannah Sparrow, Eastern and Western Meadowlarks, Red-winged Blackbird. **MOVEMENT/MIGRATION:** In spring, sexes largely segregate, with males preceding females by about a week. Birds arrive in the United States in late April; peak movements occur in May. In fall, southbound birds (usually males) may appear in coastal locations as early as late June, but most birds pass from late August into mid-September, with some birds continuing into early November. Rare but regular fall migrant along the Pacific Coast. VI: 3.

DESCRIPTION: Small for a blackbird (which it is but hardly resembles), but large for a sparrow (which females and immatures very closely resemble). Bobolink is slightly smaller than Brown-headed Cowbird and has a distinctive shape and posture: a large head, a shortish conical bill, a long sturdy body, and a short tail with an uneven, ragged or pointy, sometimes forked tip. Sits very erect but often seems to slouch.

Males in breeding plumage are striking and, compared to most birds, inversely patterned—black below, brightly patterned with a saffron-colored nape, and white rump and scapulars above. Breeding females are dull olive above overlaid with dark streaks on the back and flank, and buffy below (in a word, very sparrowlike). In fall, all ages and sexes are like females, overall tawny with dark streaks, and similar (in plumage) to the much smaller Le Conte's Sparrow (except that Bobolink has a plain unstreaked nape).

BEHAVIOR: Forages for grass seeds and insects, for the most part on the ground where it walks. Also clings to grass stalks, often flapping its wings to keep its balance. Very social, foraging and flying in flocks of 10–100 and frequently in flocks numbering in the many hundreds or even thousands. Displaying males may sing from perches (frequently the highest point around) but more frequently fly just above the grass, head raised, tail down, wings fluttering rapidly. The white rump is fully exposed when birds are in full song.

Fairly tame. Upon landing, migrating birds often perch on reeds in the open for extended periods. When disturbed, birds depart with a near-vertical climb. In migration, very vocal. Observers are usually alerted to the presence of passing flocks (even flocks flying beyond the limit of the unaided eye) by their distinctive call. Flocks are generally loose, strung out horizontally, and somewhat shifting. The tawny yellowish-colored bodies catch the light, making flocks seem like pale sparks high overheard. In migration, flocks are generally homogeneous.

FLIGHT: In flight, shows a long projecting head, wide pointy wings, and a short scraggly tail. Flight is strong, fast, and direct, with a rapid sputtery series of wingbeats followed by a brief closed-wing glide; there is a frozen moment between the wingbeats and the glide when the very pointy wing shape is evident. Flight is characterized by intermittent bounciness more than even undulations.

VOCALIZATIONS: Song of male is lively, bubbling, comical, and complex. Begins with a few pinging or metallic-sounding notes or two-note phrases that accelerate in tempo and volume into a binking, gurgling musical frenzy. One bird can sound like a whole pet shop full of caged birds in full chorus. Flight call is a short silver-toned "*bink*" or "*ink*" that seems somewhat ethereal or muffled by distance. In large flocks, multiple vocalizations recall the sound of distant ceramic wind chimes—some notes higher-pitched, some lower.

Red-winged Blackbird, *Agelaius phoeniceus*

Ember-winged Spring Herald

STATUS AND DISTRIBUTION: Common (at times abundant) and widespread, this familiar and aptly named species breeds across most of North America (absent only in extreme tundra regions) and winters throughout most of the continental U.S. (excluding some colder northern states), coastal Canada, and se. Alaska. For much or all of the year, most of us are never far from a Red-winged Blackbird. **HABITAT:** Preferred nesting habit is shallow freshwater marsh surrounded by rank growth, particularly cattails (also tule, phragmites, willows, shrubs, and trees), and also nests in brackish or saltwater wetlands, dry upland fields, pastures, and thickets bordering ponds and watercourses in city parks. Beginning in midsummer, Red-winged Blackbirds gather with other blackbirds in large roosts (most commonly in heavily vegetated wetlands). In southern portions of the bird's range, these roosts may hold through the winter and are most common in grain-producing regions. In winter, may be found in marshes, fields, and other breeding habitats, but the heaviest concentrations are found in grain fields and feedlots. **COHABITANTS:** Mallard, Northern Harrier, Marsh Wren, Song Sparrow, meadowlarks. In winter, other blackbirds. **MOVEMENT/MIGRATION:** Fall migration is two-tiered: some birds migrate (or redistribute) in August and September; another, generally larger push occurs in October and November. In spring, males precede females and in many places are the first returning bird of spring, arriving in February and March. Flocks of females follow about two weeks behind males. VI: 4.

DESCRIPTION: An overall stocky medium-sized blackbird (larger than Brown-headed Cowbird or any sparrow; smaller than Common Grackle or Yellow-headed Blackbird), with a fairly conical bill and a short tail. Birds usually perch upright with tails down-cocked and often show a hunched or humpbacked silhouette.

The adult male is as distinctive as the female is cryptic. Males are glossy (but not iridescent!) black with rakish, usually yellow-trimmed red epaulets on the shoulders. Why labor the obvious—except to say that the red is often concealed on perched birds (and sometimes the yellow trim too). The very sparrowlike, mostly grayish brown females seem woven out of grass and are conspicuously streaky, both above and below. Upperparts are usually suffused with warm even reddish tones; underparts are boldly streaked, with as much white on brown as brown on white. Viewed closely, the streaking is distinctive and crisp from breast to tail; seen at a distance, females are counter-shaded—dark above, paler below. Immature males are like females, but sport small raggedy epaulets. Most immature birds and females have an orange- or salmon-colored wash on the throat and face that helps distinguish them from the overall darker, colder-toned (and white-throated) Tricolored Blackbird (found only on the West Coast). Females and immature males usually have a prominent, pale, sometimes white eyebrow stripe that distinguishes them from most other blackbirds.

BEHAVIOR: Displaying males sit conspicuously, spreading their wings, flaring their tails, and arching their backs so as to show off their namesake characteristic to best effect. Females (sometimes a dozen or more in a male's territory) perch lower. These nimble birds can perch on grass and twigs and often sidle along branches and down stalks (almost like primates). Spends a great deal of time foraging on the ground, particularly during the nonbreeding season, and often walks among other blackbirds on open ground.

Very tame, displaying males often allow approach to less than 20 ft. Both males and females aggressively defend nests (and young) from intruders by hovering overhead and calling loudly.

FLIGHT: Flight profile is compact for a blackbird—short-billed and short-tailed. Flight is generally direct, fairly slow, and somewhat casual, undulating, or bouncy, with some tacking left and right. (This lateral movement seems related more to indifference than design.) In migration, flocks numbering from a dozen to several hundred birds are long, linear or globular, loosely packed, and hardly synchronized (individual birds rise and fall independently). Leaving the roost, birds may fly in parallel formation, forming a bouncing line or horizontal sheet moving across the sky. When flushed by a predator or pursued, flocks pack tight and move as one.

VOCALIZATIONS: The male's classic song is a loud musically gurgled "*tur-a-leee.*" Variations are many but usually have three or four distinctive parts. Western bird's song quality often seems harsher, less musical, and more two-noted—a complex gurgled opening followed by a two-note call that falls off at the end: "(gurgle) *Gee-erh.*" At a distance, you don't hear the gurgle. Calls include a banjolike twang; a harsh low "*Chack*" call often given in flight; a plaintive often repeated "*cheep . . . cheep . . . cheep*"; and a high piercing whistle. Alarm call (usually given while hovering) is an angry irregular "*chak, chak, chak*" interspaced with periodic lower-pitched "*chuck*" calls and often punctuated by a high-pitched descending "*pee-er*" whistle.

PERTINENT PARTICULARS: In California, Red-winged Blackbird identification is complicated by the addition of one subspecies, the "Bicolored" Blackbird, and another distinct species, the Tricolored Blackbird. Male "Bicolors" have all-red epaulets with no yellow border. Females are overall darker than Red-wingeds (streaked underparts are generally fused black on the belly) but are still somewhat warm-toned, particularly about the throat and face.

Tricolored males are glossier overall and have a broad white border on the epaulets. Females are cold charcoal gray and more overall dark than streaked. The throat and face lack the orange and salmon tones of Red-winged (and to a lesser extent, "Bicolored").

Over most of Red-winged's range, however, females are most often confused with sparrows. The distinctive streaking and somewhat conical bill, coupled with the bird's habit of foraging on the ground (underneath feeders) and migrating without the company of adults, is a perfect formula for confusion.

Psst! Look at the salmon wash on the face and throat. There aren't many 8-in. sparrows like that.

Tricolored Blackbird, *Agelaius tricolor*

White-trimmed Blackbird

STATUS: Fairly common but geographically restricted resident species with a patchy distribution. Generally less common than Red-winged (and "Bicolored" Red-winged Blackbird), but where they are found, often nests in huge colonies and forages in large flocks. **DISTRIBUTION:** Primarily w. Calif. (west of the Sierras and deserts) from Humbolt and Shasta Cos. south to Baja. Also occurs in small isolated pockets in Ore., s. Wash., and possibly w. Nev. **HABITAT:** For breeding and roosting, prefers large freshwater cattail and bulrush marshes, but also nests in an assortment of upland and agricultural habitats, including wheat fields, hay fields, willows, blackberry, thistle, tamarisk, and safflower. Forages in pastures, grazed rangeland, mowed fields, park lawns, cattle feedlots, dairies, orchards, vineyards, and other mostly open habitats frequented by blackbirds. **COHABITANTS:** In winter, commonly roosts and forages with Red-winged and Brewer's Blackbirds, Brown-headed Cowbird, European Starling. **MOVEMENT/MIGRATION:** Shows large-scale seasonal shifts. After breeding, some birds concentrate in the Sacramento Valley of California from August to October, then relocate to the Sacramento–San Joaquin River delta and northern San Joaquin Valley until March and April, when they return to breeding marshes. VI: 0.

DESCRIPTION: A Red-winged Blackbird whose epaulets show a white border (not yellow). (Females and immatures are more problematic.) Medium-sized blackbird in size and shape, and except for a thinner, more oriole-like, less conically finchlike bill, differs little from Red-winged Blackbird.

Plumage of adult males is also like Red-winged Blackbird, except for the glossier blue-black color and the narrow border on the rich red shoulder patch, which is buffy white from late summer to early winter and otherwise bright white. Male Red-winged Blackbird has a yellow border; male "Bicolored" (resident in California) has no border (just red epaulets).

Females and immatures are overall darker, colder-toned, and less streaky or conspicuously patterned than female Red-winged. Dull blackish plumage shows narrow leaden streaks or a grayish infusion about the throat and upper breast; some individuals seem to have a grayish sheen about the breast and head, like a leaden cowl or bib. Unlike Red-winged, the upperparts show no rufous tones, and the throat and face have no rose or orange blush.

BEHAVIOR: A colonial nester that gathers in dense noisy colonies that commonly number in the thousands. Territorial males defend only the area immediately surrounding nests, and nests may be only a few yards apart. For this reason, Tricolored males do not engage in aerial displays. Instead, they drape themselves over marsh grass and agricultural plantings (such as wheat and hay fields). After breeding, birds continue to roost and feed in monotypic flocks but are also (and commonly) found in mixed-species flocks, where direct comparison with similar species is possible.

Moves and behaves much like Red-winged Blackbird.

FLIGHT: Profile is like Red-winged Blackbird, but more pointy-winged. Often flies in compact flocks. Flight is somewhat less buoyant than Red-winged.

VOCALIZATIONS: Song is a rough nasal gurgle, "*wr'ah wree'u,*" that recalls a snarling cat. Also gives a squeaky, nasal, parrotlike "*chlah, chlah, chlah,*" frequently in series, and a raspy "*w'raah*" that sounds like a domestic chicken. Call is like Red-winged but lower-pitched and more nasal.

Eastern Meadowlark, *Sturnella magna*

The Meadowlark That Doesn't Say "Churk"

STATUS: Common eastern and southwestern resident and northeastern breeder. **DISTRIBUTION:** Breeds throughout most of the e. U.S. (absent in nw. Maine) and much of s. Canada, including sw. and se. Ont., s. Que., and parts of N.B. The western border extends across cen. Minn., e. and s. Neb., along the S.D. border, cen. Kans., s. Okla., south across cen. Tex. (and south into Mexico). In winter, birds in most of Canada, Minn., n. Wisc., Mich., cen. Pa., n. N.Y., and n. New England retreat south. The Lilian's subspecies (see "Pertinent Particulars") is resident in trans-Pecos Tex., e. and s. N.M., cen. and e. Ariz., and bordering Mexico. **HABITAT:** Grasslands, pastures, prairie, pine savanna, airports, overgrown fields, and the upper drier portions of tidal wetlands. In winter and migration, found in orchards, roadsides, and plowed or harvested fields. In winter in drier areas, favors lawns. Except for "Lilian's," where Eastern and Western Meadowlarks overlap, Eastern selects the wetter, denser habitat and Western the sparser, drier, more disturbed areas, the shorter-grass areas, and agricultural fields (although there is overlap in winter). **COHABITANTS:** Upland Sandpiper, Grasshopper Sparrow, Savannah Sparrow, Bobolink. **MOVEMENT/MIGRATION:** Northern birds (above the permanent snow belt) are migratory. Spring migration from late February to early May, with a peak in mid-March and April; fall from late August to late November, with a peak in October. VI: 1.

DESCRIPTION: A boldly patterned fencepost singer as synonymous with eastern farmland as the robin is synonymous with suburban lawns. Medium-sized songbird (slightly larger than but shaped much like European Starling; similar in size and shape to Western Meadowlark), with a long pointy bill, a slender head, a plump body, a short tail, and long legs.

Upperparts are grayish brown and complexly patterned; the bold head pattern shows black stripes through the eye and crown. Underparts are mostly yellow with bold black streaking down paler sides and a signature black V (sometimes U) on the chest. In winter, many individuals are overall paler and dingier but comparably patterned.

Eastern Meadowlark is almost identical to Western Meadowlark, except that Eastern shows bolder, blacker lines on the head in both breeding and nonbreeding plumage. Also, Eastern's face pattern is tricolored—gray, whitish, and yellow (the gray cheeks and yellow throat are knit by a pale band). On Western, the cheeks show gray bordering on yellow (with no intervening pale band). Viewed head on, the more restricted yellow shows as a patch (or a yellow goatee) on the throat of Eastern. On Western, the whole throat appears yellow. Also, Eastern's flanks are dingy and darkly streaked; Western show brighter, whiter flanks, with less prominent spotting or streaking.

Western is overall slightly paler and colder-toned than Eastern, and this difference may be useful in winter in mixed flocks.

BEHAVIOR: Forages primarily on the ground, where it walks with a slow, preoccupied, waddling gait. Very territorial in summer, singing from the ground or an elevated perch; seems born to fence-posts, but will generally select the highest sturdy perch around, including fences and utility wires. Also sings in flight. In winter and migration, often gathers in large flocks (some in excess of 100 birds) that may include Western Meadowlark. Individuals are generally well spaced, but normally forage in the same direction.

Fairly shy. When approached, crouches and tries to slink away. When pressed, explodes into flight.

FLIGHT: A stocky, potbellied, short-tailed, pointy-faced profile set on short triangular wings. May be slightly more slender-winged than Western, but like Western, seems hardly more aerodynamic than a bumblebee. The flared tail pattern showing a brown center and bracketing white sides is distinctive. The white sides and dark center are about equal widths in Eastern; in Western, the white sides are slightly narrower than the dark center. Birds explode into flight with frenzied sputtery wing-beats that are given in a rapid series punctuated by short open-winged glides that may become long glides as birds descend to land. (Western's wingbeat is slower and heavier, and the bird's takeoffs are more lofting than exploding—it's just plain slower off the line.) Eastern's flight is direct, straight, and, despite the sputter-glide pattern, remarkably steady and even-keeled. The bird turns its head from side to side, as if looking for a good place to land.

VOCALIZATIONS: Song is a high, plaintive, and somewhat wistfully whistled "*WE'he SEE'ee you*," with the first notes clear and strong, and the last slightly slurred and descending. Common calls include a low, rough, nasal mutter, "*dzert*" (most commonly given when made anxious by a human approach), and a dry scratchy rattle, "*ch'h'h'h'h*," that is sometimes followed by a sweeter "*t'le/ur*." Rattle call is higher-pitched than Western's. Also gives a high, thin, somewhat wheezy "*szwee . . . . szwee*," most commonly in flight. *Note:* Eastern Meadowlark doesn't say "*churk*." That's Western.

PERTINENT PARTICULARS: The "Lilian's" Eastern Meadowlark (southwestern subspecies) is as pale as Western Meadowlark, whose range it overlaps, partially in breeding and wholly in winter. Where found together, "Lilian's" prefers taller grasslands, and Western prefers sparser disturbed habitats, such as agricultural fields and roadsides (although there is much overlap and the birds are often found together). "Lilian's" is particularly easy to distinguish in flight. The tail is mostly white, showing only a narrower dark center. Western's tail is mostly dark, with narrower white sides.

Western Meadowlark, *Sturnella neglecta*

The Meadowlark That Says "Churk"

STATUS: Common and widespread, mostly western resident; northern populations are migratory. **DISTRIBUTION:** In Canada, breeds from s. B.C. to cen. and s. Alta., cen. and s. Sask., s. Man., and extreme sw. Ont. (rare) to Lake Ontario. In the U.S., breeds in nw. Ohio, n. Ind., Mich., and west of a line drawn across n. Ill., n. Mo., e. Kans., cen. Okla., and cen. Tex. In winter, birds in Canada and northern states subject to permanent snow cover withdraw into southern portions of the breeding range and expand south into Mexico and east to about the Mississippi R. The winter range lies south of a line drawn from extreme s. B.C. across ne. Wash., ne. Ore., sw. Idaho, n. Utah, n. Colo., cen. Neb., n. Iowa, and s. Wisc. **HABITAT:** Particularly common in native grasslands and rangeland, but also found in all manner of open habitat, including agricultural fields, orchards, vineyards, weedy fields, roadsides, and desert grasslands. Across most of its range, prefers drier well-drained areas. **COHABITANTS:** Killdeer, Upland Sandpiper, Horned Lark, Lark Bunting; in winter, European Starling, Savannah Sparrow. **MOVEMENT/MIGRATION:** Northern birds are migratory. Spring migration from late February to early May, with a peak from mid-March through April; fall from late August to late November. VI: 2.

DESCRIPTION: A well-marked and familiar grassland bird with the manner of a quail and a voice that is almost synonymous with wide-open western spaces. In structure and plumage, almost identical to Eastern Meadowlark, from which it is most easily distinguished by voice and, where the species do not overlap, range. A medium-sized songbird (slightly larger than European Starling; the same size as Eastern Meadowlark), Western is overall stocky, sturdy, potbellied, and chesty, with a long pointy face, a comically short tail, and long legs (resembles a large, sturdy, long-legged starling).

Upperparts are pale grayish brown with tight rows of darker streaking on the head and back. Underparts are bright yellow except for a conspicuous black V etched on the chest and grayish white flanks sparsely overlaid with dark spots and streaks. The head is conspicuously striped; the dark tail flashes white sides in flight.

Overall slightly paler than Eastern Meadowlark (except for the "Lilian's" subspecies). Shows paler stripes on the head (dark gray, not blackish), more extensive yellow on the throat, paler and grayer sides, and flanks usually overlaid with less extensive dark spotting. (See "Eastern Meadowlark" for more discussion.)

BEHAVIOR: In winter, feeding birds, commonly in flocks of a dozen or more, are often well concealed and visible only in snatches or when flushed. When breeding, singing males sit conspicuously atop elevated perches (from cow pies to fenceposts to treetops), where they and males in adjacent territories sing variations of their short comical song at short intervals and for long hours.

Waddles more than walks. Stops at intervals, raises its head, and looks around, then returns to feeding. When approached, "*churks*" nervously and waddles or runs, but when pressed, lofts into flight, often with an initial near-vertical ascent. Moderately tame. When perched atop a utility line, may let observers stand below; when approached on the ground, usually flushes before its finer plumage points can be noted with binoculars.

FLIGHT: Distinctive profile—flat on top, potbellied, short-tailed (very starlinglike). Flies (and glides) with the body angled slightly up, giving the impression that it's struggling for altitude. The white sides to the tail are only slightly less prominent than on most Eastern Meadowlarks but show less white than "Lilian's." (Western has a dark tail with white sides; "Lilian's" has a white tail with a dark center.) Lofts into the air and then, after a slight hesitation (as if the engine has stalled or the bird is checking its bearings), lines out. Flight appears sluggish and effortful. Wingbeats are hurried, steady, fluttery, down-pumping, and mostly continuous, punctuated by a long, somewhat unsteady, set-winged glide just before landing. Eastern Meadowlark explodes into flight (like quail!), appears more high-strung and jittery in the air, and flies with a series of short, rapid, sputtery wingbeats interrupted by frequent short glides.

VOCALIZATIONS: Song is a short, loud, rich, somewhat bubbly ditty of five to eight notes. Sounds as if the bird has a good one- or two-note start, loses its timing, and rushes to a jumbled finish. Songs are variable (individual birds have a repertoire of eight to twelve songs), but all renditions have the same good start, rushed-ending pattern, and careless happy-go-lucky quality. Call is a low, descending, blackbirdlike "*churk*"—a musical cross between a chirp and a burp. Also makes a dry rattle that is similar to but lower-pitched than the rattle of Eastern Meadowlark. Flight call is a high ethereal wheeze—like Eastern's, but slightly lower-pitched.

Yellow-headed Blackbird, *Xanthocephalus xanthocephalus*

Saffron-hooded Cacophony

STATUS: Common, mostly western breeder; winter resident in the Southwest; fairly regular wanderer outside its normal ranges. **DISTRIBUTION:** In Canada, breeds in e. and cen. B.C., all but n. Alta., all but n. Sask., w. and s. Man., and sw. Ont. In the U.S., breeds in e. Wash., Ore., e. and cen. portions of Calif., Idaho, Mont., N.D., S.D., all but ne. Minn., Wisc., Nev., Utah, Wyo., Colo., Neb., w. and n. Kans., all but e. Iowa, nw. Mo., n. Ill., extreme w. and e. Ariz., N.M., and n. Tex.; also found in isolated pockets in Ind., Ohio, and extreme s. Ont. Winters in s. Calif., s. Ariz., N.M., and Tex. from the trans-Pecos region east along the Rio Grande Valley to the Gulf of Mexico, south into Mexico. **HABITAT:** Breeds primarily in deeper freshwater marshes in tall, lush, emergent stands of aquatic vegetation (primarily cattail, bulrush, and phragmites); also nests in wetlands near meadows and aspen stands and in larger lakes in mixed forest habitat. In winter, forages in grasslands, croplands, and cattle pens; returns to roost at night in marshes. **COHABITANTS:** American Coot, Marsh Wren, Black Tern, Red-winged Blackbird. **MOVEMENT/MIGRATION:** Except in central and southern California and western Arizona, all birds vacate breeding areas in winter. Spring migration from late March to mid-June (males migrate one to two weeks earlier than females); fall from mid-August to early November. VI: 3.

DESCRIPTION: A large, sturdy, noisome yellow-headed blackbird. Males are considerably larger and more robust than females, and females are only slightly larger than Red-winged Blackbird. Overall modestly sturdier, longer, and more tubular-shaped than Red-winged Blackbird; also slightly larger-headed, larger-billed, and broader-tailed. Males in particular seem very large-headed.

Plumage of adult males is distinctive—jet-black with a saffron hood extending over the breast and a white edge to the wing. Females and immatures are uniformly sooty brown with no streaking above or below and varying degrees of yellow or mustard-colored pattern on the face, throat, and breast. Yellow on females and immatures is never absent, but it is sometimes easy to overlook.

BEHAVIOR: Nests in noisy colonies in reed beds over water. Males sit conspicuously atop reeds or engage in flight displays that show their golden heads and white wing patches to effect. Red-winged Blackbirds may nest in proximity but generally in more marginal landward areas.

In summer, forages by clinging to reed stalks and plucking insects near the water, or may hunt grassy upland areas. In winter, flocks forage in harvested or plowed fields and at feedlots, often engaging in the roll-over maneuver employed by other blackbirds: birds in the back of the flock leapfrog over the flock to advance the leading edge. Males and females tend to segregate; males remain farther north and so are more likely to be encountered in the United States in winter.

At times, unaccountably shy. When disturbed, often flies to the tops of tall trees. Outside of its range, forages with other blackbirds, most commonly Red-wingeds and grackles.

FLIGHT: In flight, has a longer and sturdier profile than Red-winged, with a large head, a heavy bill, a compact body, and broad, pointy, longer wings. All in all, looks more like an oversized longer-billed cowbird. The male's white wing patch is conspicuous (immature males have just a trace of white). Flight is undulating, flowing, and not as bouncy as Red-winged. Wingbeats are stiff, quick (less hurried and more regular and perfunctory than Red-winged), and given in a long series followed by a close-wing glide. In their daily activities, Yellow-headeds frequently glide on open wings, perhaps more often and with more finesse than Red-wingeds.

VOCALIZATIONS: Song is an amazing cacophony of disparate sounds—old car horns, creaky rusty gates, whistles, and clackings (and males rarely sing alone). Calls include a loud nasal "*raaah*" that has been described as "the nastiest sound in nature" and a low loud "*t'churk*" or "*chur'rk*" (has a compressed two-note quality).

Rusty Blackbird, *Euphagus carolinus*
Swamp Blackbird

STATUS: Fairly uncommon but widespread northern breeder; winters across the southeastern and central United States. **DISTRIBUTION:** Breeds across all but n. Canada and Alaska, from Nfld. to w. Alaska. Also breeds in n. New England, the Adirondacks, the Upper Peninsula of Mich., and extreme n. Minn. Winters from coastal Mass. south to cen. Fla. and west to e. S.D., cen. Neb., Kans., Okla., and Tex. Casual in fall and winter in the western states. **HABITAT:** Breeds in wet coniferous and mixed northern forest and in willow thickets. Partial to beaver ponds, muskegs, ponds, and swampy lakeshores. In winter and migration, found in wet woodlands and on the edges of wooded swamps; also occasionally forages with other blackbirds in plowed fields and on short grass. **COHABITANTS:** During breeding, Solitary Sandpiper, Gray-cheeked Thrush, White-crowned Sparrow; in winter, Winter Wren, Palm Warbler, Red-winged Blackbird, Common Grackle. **MOVEMENT/MIGRATION:** Entire population vacates breeding areas. Spring migration from early March to mid-May; fall from mid-September to early December. VI: 3.

DESCRIPTION: A compact blackbird with a somewhat gracklelike bill and a penchant for keeping its feet wet. Slightly larger than Red-winged Blackbird (and considerably smaller than Common Grackle), with a longish, slightly downturned, gracklelike bill and a shortish tail. Most like Brewer's Blackbird in shape, but with a thinner, less conical bill and a shorter tail.

Breeding males are overall blackish with subtle purple highlights in the head and greenish highlights on the body. Females are overall grayish brown with warm highlights on the head and upperparts and a pale eyebrow. In winter, all plumages usually show conspicuous rusty highlights on the face, back, wings, and breast. Most birds also show a distinctive dark triangular mask or eye patch on a warm brownish face that contrasts with a pale eye.

BEHAVIOR: Commonly forages in wet woodlands, where it frequently stands in shallow water—a habitat shunned by most blackbirds. Noisy and gregarious. Birds commonly walk somewhat sluggishly (but sometimes run), feeding methodically as they search through leaf litter and flip submerged leaves. While foraging, sometimes moves its tail in a slight upward jerk. Joins other blackbirds in mixed-species flocks in fields. When not feeding, perches at all heights in surrounding trees.

Response to pishing is variable, but when attracted, often sits for extended views.

FLIGHT: Slightly larger than Red-winged Blackbird, but with a distinctly longer, pointier face, a pointier wing, and a more bulbous-tipped or lobed tail. In winter, the pale brown head often contrasts with darker wings and tail. Flight is smoothly undulating (more gracklelike and not as bouncy as Red-winged Blackbird); a short or lengthy series of wingbeats is followed by a short pause. When flushed, flight may be slightly twisty-turny. Commonly calls in flight.

VOCALIZATIONS: Song is a series of whistled squeaks and soft gurgles. Most often heard phrase is phonetically and ribaldly rendered "*Take a leak!*" (sometimes truncated to "*leak*"). The quality of the call (also given in flight) suggests Common Grackle, but Rusty Blackbird is higher and squeakier. Also in flight, makes a low "chuck" call that is lower-pitched than Red-winged or Brewer's.

Brewer's Blackbird, *Euphagus cyanocephalus*

The Archetypical Icterid

STATUS: Common and widespread but mostly western species. **DISTRIBUTION:** In Canada, breeds from s. and cen. B.C. across most of Alta., s. and cen. Sask., s. Man., and s. Ont., almost to Que. In the U.S., breeds west of a line drawn across w. S.D., w. Neb., e. Colo., and n. and e. N.M. (absent across most of N.M., s. and w. Ariz., s. Nev., and se. Calif.). Breeding range also extends east across most of N.D., all but s. Minn., and Wisc. (and locally to ne. Ill.). In winter, vacates all of Canada except for B.C. and much of the n. U.S., expanding south and east. In the U.S., winters west and south of a line drawn from w. Mont., w. Wyo., n. Colo., n. and e. Kans., n. Ark., w. and s. Ky. and Tenn., and cen. Ga. A few small groups are found in the mid-Atlantic and the Southeast. Winters south into Mexico. Absent, except for vagrants, in e. Canada, New England, and along much of the Eastern Seaboard. **HABITAT:** At all seasons, prefers open and semi-open dry, often human-modified habitats such as agricultural croplands, orchards, golf courses, residential and corporate lawns, feedlots, city parks, shopping malls and eateries, and roadways. Also found in burned-over or logged areas, forest edges, mountain meadows, marshy edges, lakeshores, pastures, grassy areas, beaches, and rocky coasts. **COHABITANTS:** Varied. In winter, often associates with other blackbirds, particularly Red-winged; also House Finch. **MOVEMENT/MIGRATION:** Northern and eastern populations vacate their breeding range. The balance of the population undergoes seasonal shifts from higher to lower elevations—as well as to areas with abundant food resources, such as cattle feedlots—but continues as permanent residents. Spring migration from late February to late April; fall from late September to early November. VI: 3.

DESCRIPTION: A handsome nicely proportioned blackbird distinguished by its heads-up carriage and the male's two-toned glossy iridescence—the quintessential icterid. The head is distinctly round, the forehead prominent, and the bill fairly long, straight, and conical (giving the bird a pointy-faced appearance). The body is plump, and the tail is tastefully long without being unwieldy and somewhat flared or swollen at the tip. Red-winged Blackbird is overall stockier, with a shorter, thicker bill and a shorter tail; grackles have heavier down-curved bills and long keel-shaped tails. All cowbird species are flat-headed and have heavy bills that seem like extensions of the face. Rusty Blackbird has a longer, slightly more downcurved bill and a more bulbous-tipped tail.

The plumage of adult males is distinctive. Overall black with purple iridescence on the head and green iridescence shot throughout the body. (In winter, a few males have rust-tipped feathers on the head and breast but not on the wings, as does male Rusty Blackbird.) Females are unequivocally plain, virtually devoid of plumage pattern, and cold brown-gray, the color of lead (although in good light and at close range they may show some green highlights in the wings). The eye is dark, tucked into the shadowy lores and almost indiscernible. Winter Rusty Blackbird has malevolent bright yellow eyes and a bolder face pattern.

BEHAVIOR: Highly social, often flying or foraging in flocks with other blackbird species and starlings. Very much at home on bare ground or short grass, Brewer's forages by walking a few steps and pausing to search for prey. Sometimes turns over stones or cattle droppings to inspect for prey. Seems very focused and preoccupied. With its distinctly erect and heads-up stance, regular and high-stepping gait, and head bobbing forward and backward as it walks, Brewer's gives a theatrical parade-ground performance. Red-winged Blackbird walks with a more rolling gait (a sailor's gait), keeps its head down, hops more, and flies frequently (even seems disinclined to walk).

Frequently perches on utility lines, with bill slightly elevated. Fairly tolerant of humans, and in places where they have become habituated to people (such as city parks and outdoor eateries), may be virtually underfoot.

FLIGHT: Profile is overall longer and slimmer than Red-winged Blackbird, with distinctly pointier wings. Flight is direct, straight, only slightly undulating, and more energetic but less buoyant and bouncy than Red-winged Blackbird. Wingbeats are given in a fast, hard, stroking series followed by a pause (but no glide). Calls frequently in flight.

VOCALIZATIONS: Song is a weak, wheezy, uprising gurgle: "*queeek!*" Call is a short scrapping "*chuh*" or "*cheh*" (an audible scuff). Also emits a liquid "*wh'rt*" that recalls the sound of a water drop from a leaky faucet.

PERTINENT PARTICULARS: There are structural and plumage characteristics that help distinguish Brewer's and Rusty Blackbirds, but two fundamental considerations make confusion unlikely. The first is range. Along the Atlantic Seaboard, Brewer's is rarely seen; west of the Great Plains and south of Canada, Rusty is rarely encountered. The second consideration is habitat. Brewer's is a bird of dry, open, often sparsely vegetated habitat; Rusty forages in wet wooded swamps. Neither species is likely to be found in the habitat of the other, although the rare Rusty in the West may join a flock of Brewer's.

Common Grackle, *Quiscalus quiscula*
The Purple Crackle

STATUS: Common and at times abundant east of the Rockies, Common Grackle ranks among suburban North America's most familiar birds, even if many people mistakenly label them "crows" and those half in the know call them "purple crackles." **DISTRIBUTION:** Breeds across most of subarctic Canada from nw. B.C. and cen. Alta. east across most of Sask., all but n. Man., most of Ont., s. and e. Que., the Maritimes, and sw. Nfld. In the U.S., breeds east of a line drawn across w. Mont., se. Idaho, n. Utah, se. Colo., w. N.M., and nw. Tex. south and east to Corpus Christi. In winter, retreats from most of Canada and the n. and w. U.S. Found east of a line drawn across se. S.D., cen. Neb., w. Kans., the Oklahoma Panhandle, w. and s. Tex., and se. N.M., and south of a line extending across s. Minn., s. Mich., extreme se. Ont., cen. N.Y., s. Vt., s. N.H., and s. Maine. **HABITAT:** Needs trees to nest (deciduous or coniferous), but is fundamentally a bird of more open country—pastures, feedlots, cornfields, plowed fields, suburban lawns, city parks, and marshy areas (even sandy beaches). Also readily occupies open woodlands (including boreal forest) and swamps, but avoids the interiors of extensive, dense, mature woodlands. Nests in small loose colonies and roosts and sometimes feeds in tremendous aggregations, often with other blackbirds. **COHABITANTS:** Breeds among many woodland species; if a place can support Gray, Fox, or Red Squirrel, it can probably support Common Grackle. In winter, may associate with other blackbirds. **MOVEMENT/MIGRATION:** Migratory in the northern and western portions of its range, this is the species that sheaths suburban yards in a foraging carpet of birds in fall and whose long linear flocks are reminiscent of smoke on the horizon. Spring migrants return early, in March and early April, and leave late, in October and November. VI: 3.

DESCRIPTION: A large black bird (much larger than other blackbirds; smaller than other grackles; about the same size as Mourning Dove) distinguished by a long formidable bill, a long, straight, stiff, keel-shaped tail (less so in fall and winter), an eye-catchingly malevolent yellow eye, and a plumage shot through with iridescence about the head and neck that may be green, blue, or purple, depending on the viewing angle and regional plumage variations. The wings and belly are brownish or bronzed, but in poor light and at a distance, the entire bird just seems black. Females are like males but smaller and less iridescent. Young birds are overall sooty brown and lack iridescence.

The head is flat and lacks a forehead; the large, pointed, barely down-curved bill is particularly heavy at the base and seems an extension of the face. When the bird is perched, its bill is commonly angled up. In flight, the tail is often creased in a sharp V but appears round or lobed when the bird is sitting or walking. The expression is nefarious.

BEHAVIOR: Gregarious; usually found with other grackles or other blackbirds. At feeders, generally dominant over other birds. Stalks more than walks, and moves with a haughty demeanor. Perched birds sit more horizontally than most icterids, often with the head and bill angled conspicuously up.

Birds begin collecting in large communal roosts right after the nesting season (as early as late June) and disband in March. Birds leaving (and returning to) roosts may form linear flocks that are miles long. Birds leave at first light and return around sunset.

FLIGHT: Longish blackbird with a horizontal profile and a distinctive silhouette showing a pointy face, relatively short pointy wings, and a very long, (mostly) straight, narrow tail that is sometimes blunt, sometimes acutely pointed. The leading edge of the wing is straight; the trailing edge is curved and tapers to a point (recalls the wing on the old British *Spitfire* fighter of Battle of Britain fame). From below, the tail appears long, straight, and, while narrow, somewhat curiously contoured: it's narrower along the proximal half and slighter wider distally. Viewed from behind, it is often bent in a sharply creased keel-like V. Wingbeats are stiff, quick, regular, and perfunctory, and given in a short series followed by a brief closed-wing pause. Flight is direct and shows a very slight rise and fall. By comparison, Boat-tailed and Great-tailed Grackles are larger, blunter-winged, with wingbeats that are more fluttery. Common Grackle's flight is more jerky and less flowing than Boat-tailed or Great-tailed but not as bouncy as Red-winged Blackbird.

Flocking birds form dense linear aggregations. Although not exactly an orderly formation, birds generally hold their place in the flock. Sometimes entire flocks turn or wheel en masse — a synchronized display that seems too perfect for words.

VOCALIZATIONS: Noisy! Song is loud and harsh — a guttural protest followed by a breathy strangled screech. Call is a flat "*chack*" or "*kaak.*" In flight, utters a more truncated "*kek*" (often repeated).

Boat-tailed Grackle, *Quiscalus major*

Coastal Great-tailed

STATUS AND DISTRIBUTION: Common, conspicuous, and coastal; a year-round resident from Long Island south to about mid-coastal Tex. Except in Fla., where it occurs along inland watercourses, Boat-tailed Grackle is almost never far from salt water. Casual in s. New England. **HABITAT:** Virtually a salt marsh obligate; usually found in tidal marshes, but also occurs on beaches and in coastal towns. During snowstorms, may visit feeders away from marshes. **COHABITANTS:** Clapper Rail, Forster's Tern, Seaside Sparrow. Does not commonly flock with other species, although breeding birds may be found in close proximity to Red-winged Blackbird, Common Grackle, and, in coastal Texas and Louisiana, Great-tailed Grackle. **MOVEMENT/MIGRATION:** Some northern birds migrate in October, returning in March, but for the most part the bird is a permanent resident throughout its range. VI: 1.

DESCRIPTION: Large, conspicuous, and not at all shy, a male Boat-tailed Grackle is distinguished by an outlandishly long tail and noisy displays that border on buffoonery. The head shape is variable, but generally round (not flat-topped) on both males (especially) and females. Males often display a pronounced forehead, made all the more conspicuous by how thin the bill seems for a grackle this size. The tail, which is as long as the body (and in flight appears longer), is heavy and fan-shaped and curves up in a pronounced U (not a sharp V) shape. More flexible than Common Grackle's tail, Boat-tailed's flutters in the breeze.

In the grackle tradition, adult males are blue-black with greenish highlights (all black at a distance and in poor light). Females and immatures are *conspicuously smaller* (about Common Grackle size), but warm brown below and on head, not black, with no iridescent highlights (except for subtle touches in the blackish wings). The face shows a dark eye-line or a shadowy triangle (somewhat akin to nonbreeding Rusty Blackbird). Eye color varies. Atlantic Coast birds have yellow eyes; the eyes of Florida and Gulf Coast birds are brown.

BEHAVIOR: Sits conspicuously and forages by walking on the ground; also clambers deliberately among branches. When flushed, generally flies (unhurriedly) a short distance to take another nearby perch. Not as gregarious as Common Grackle, but roosts in colonies at all times of the year. Several males may display in close proximity, and nesting females form colonies. Birds going to roost fly en masse from a pre-roost gathering place in a long low-flying flock. Does not commonly mix with other species.

FLIGHT: Mostly direct, sometimes with a very shallow rise and fall. Seems labored, effortful, in slow motion. The wings seem undersized, and the wingbeats hurried, rubbery, or elastic. The trailing edge of the wing appears to ripple or flutter. Overall more curvaceous or contoured than the sharp-edged pointy-lined Common Grackle. The head is somewhat raised, and the underparts show an S-shaped curvature. Migrating flocks and flocks heading for or leaving roost sites are linear—they seem to flow over the landscape.

VOCALIZATIONS: Perhaps this bird's most distinctive feature. Males are a virtual one-bird cacophony. Vocalizations may recall a loose fan belt catching ("*ehhhhhh!*") or recall a screeching parrot ("*Kreeeeh, Kreeeh, KREhhhh*"). Other calls suggest a high-scoring pinball machine ("*Eeh-Eeh-Eeh-Eeh-Eeh*") or book pages being flipped. Call is a "*chuck*" that is lower-pitched and not as angry-sounding as Common Grackle's "*chack.*"

PERTINENT PARTICULARS: Although Common Grackle can occur in Boat-tailed habitat, in those coastal areas where Boat-tailed occurs, there should be little chance of confusion. Male Boat-tailed's greater size, bluer color, and distinctive tail easily distinguish it. The brown female and immature Boat-taileds are likewise distinctive (as are groups made up of both black and brown grackles).

The greater challenge lies in distinguishing Boat-tailed from Great-tailed Grackle where their ranges overlap. See "Great-tailed Grackle" for more discussion.

Great-tailed Grackle, *Quiscalus mexicanus*

Western Great-tailed

STATUS: Common and widespread, primarily southwestern and Great Plains resident. **DISTRIBUTION:** In the U.S., found in southern half of Calif., (north to San Francisco), Nev., Utah, Ariz., N.M., most of Colo., s. S.D., s. Neb., Kans., n. and w. Mo., w. Iowa, Tex. (except inland e. Tex.) and coastal La. Also isolated pockets are found in se. Ore. and n.-cen. Iowa. In winter, birds in most northern portions of Utah, Colo., S.D., Neb., Kans., and Mo. retreat south, but some remain there locally. Casual vagrant farther north and east. **HABITAT:** Open and usually dry habitats with some trees and proximity to water, including agricultural lands, scrub prairie, pastures, broken chaparral, freshwater marshes, citrus groves, golf courses, feedlots, airports, cemeteries, urban and suburban settings, and landfills. Along the Gulf Coast, found in tall reeds along shallow lakeshores and islands of standing woody vegetation in salt marsh. Avoids dry or desert habitat without water and unbroken forest. **COHABITANTS:** Cattle Egret, Red-winged Blackbird, cowbirds, cattle, humans. **MOVEMENT/MIGRATION:** In winter, birds in northern populations migrate or centralize around optimum habitat, but little is known about timing or routes. VI: 2.

DESCRIPTION: A large, noisy, flamboyant blackbird known by most people for the male's grotesquely large tail and, among birders along the Gulf Coast (from Louisiana to about Rockport, Texas), for the female's maddening resemblance to female Boat-tailed Grackle. Conspicuously large grackle (much larger than Common Grackle; slightly longer than Boat-tailed), with a slightly curved heavy-based bill, a slender flat-topped head, a long neck, and a slender body. The adult male's long, lobe-tipped, and often slightly drooping tail seems much too large for the bird. The female's tail is more modest and the overall shape more akin to a heavy-billed Common Grackle.

Adult males are wholly black with purple and blue iridescence (no brown, no green) and a bright yellow eye. (Common Grackle has a purple head and brown body, and male Boat-tailed's body shows flashes of green.) Females, which also have bright yellow eyes, have brownish heads and underparts, blackish wings, and a tail showing an iridescent gloss. The face shows a dark blackish mask across the eyes that is more prominent than on Boat-tailed. Overall tone is colder, but many females show warm brown underparts like female Boat-tailed. Immatures are like female but show less iridescence and have brown eyes (not bright yellow).

BEHAVIOR: Very social, roosting in trees and feeding communally across a variety of open habitats, including shallow standing water. Birds are spaced (not packed like Common Grackles) and forage with a strutting walk. Commutes daily between roost sites in flocks of varying size, with flocks most commonly spread horizontally. Males are best known for their noisy histrionic displays (often conducted by several tightly packed males) in which they raise their bills high (an avian swordsman's salute), fan their gargantuan tails, and flap their wings in a fluttering frenzy—all the while vocalizing. Fairly tame. Associates with other blackbirds.

FLIGHT: Both males and females are slender, almost sinuously contoured, with a long slender head/bill/neck, a slender body, and (when flying long distances) a long, straight, ruler-shaped tail with a slight notch at the tip. From the side, the male's tail rises at the base, then flattens or droops—it looks as though it was broken and mended poorly. The tail shape on Great-tailed is something of a chimera. When birds are taking off, it is often fanned and flat (or very slightly upturned), showing a broad wedge shape and a pointy tip that recall a raven. Tails of males may also bend into an acute V shape (U-shaped in Boat-tailed), and the tail may even be compressed laterally, so that it looks like a partially open fan. Often the tails on males appear distinctly lobed.

Flight is hypnotically flowing. Somewhat stiff (not fluttery) wingbeats may be steady or given in a varying series followed by a brief closed-wing glide. The female's flight is rising and falling more than undulating. The male's flight is usually steady and direct but sometimes shows some rise and fall. Great-tailed often flies very high (Boat-tailed likes to hug the ground), and when descending, may glide for several hundred yards. When gliding, wings are angled slightly down, but wingtips are turned up.

VOCALIZATIONS: Males sound like an arcade, uttering a variety (and geographically variable) array of up-sliding whistles, castanet rattles, strangling rasps and gasps, low chortling chatters, piercing chatters, nasal scrapes, rasps, growls, and low shrill trills—most with a mockingbird repetitiveness followed by a pause between sets. The male's call is a loud gruff "*chak*," and the female's a softer guttural "*churk*."

PERTINENT PARTICULARS: Great-tailed versus Boat-tailed: not a problem over most of the bird's range, but certainly a challenge along the Gulf Coast in Louisiana and Texas. Adult males are not that difficult to separate. Great-taileds are larger and have larger, more bulbous and billowing tails (when folded), pointier, more wedge-shaped tails (in flight), flatter heads, and heavier bills; they show purple-blue iridescence (not green) and have bright yellow (not dark) eyes.

Females are a problem; differences are measured in tendencies, not absolutes. Great-tailed Grackle females tend to be heavier-billed, flatter-headed, and overall colder-toned and to show less overall contrast between blackish upperparts and brownish gray underparts—but they do show a more contrasting facial pattern. But (indicative of how difficult it is to tell the species apart) where they overlap along the Gulf Coast of Louisiana and Texas, the normally yellow-eyed Boat-tailed Grackle has brown eyes.

What does this suggest to you?

Immature Great-taileds, with dark eyes carried into the fall, are more problematic.

Shiny Cowbird, *Molothrus bonariensis*

The New Cowbird on the Block

STATUS: Uncommon, restricted, and recently established tropical species. **DISTRIBUTION:** Widespread in South America. An expanding population along the Antilles reached the Florida Keys in 1985. Now a mostly spring and summer resident in s. Fla. and the Keys (a few may winter). Rare but regular in spring north to Ala. and in summer north to S.C.; accidental elsewhere. **HABITAT:** Any open habitat, particularly one dotted with shrubs or trees. In the United States, highly suburbanized—found on lawns, in gardens, at feeders in agricultural areas, and along roadsides. Forages on the ground as well as the outer canopy of trees. **COHABITANTS:** Brown-headed Cowbird. **MOVEMENT/MIGRATION:** Permanent resident throughout much of its tropical range. A few birds remain year-round in the United States, but the population is augmented by a new wave of colonizing birds every spring. As expected in a colonizing species, Shiny Cowbird has turned up a considerable distance away from its established range—as far north as Nova Scotia and as far west as Texas. So far, the species is more likely to turn up in coastal areas than in the interior. VI: 2.

DESCRIPTION: A small pointy-faced blackbird. Recalls a chubby Brewer's or a trimmer, more angular Brown-headed Cowbird. Shiny Cowbird is the same size as Brown-headed Cowbird, *with which it commonly associates*, but its bill is longer, pointier, and more symmetrically conical. Brown-headed Cowbird's bill is stouter and more House Finch–like, right down to the curved upper culmen. The head on Shiny is flatter and peaked toward the rear. (The head of Brown-headed is round.) In sum, Shiny Cowbird is pointy-faced, and Brown-headed is not. Also, Shiny has a slightly humped back (flat on Brown-headed) and a slightly longer tail.

Adult males are overall black (no brown head) with an attractive iridescent purple gloss on the head and body and a slight greenish cast to the wings and tail. Plumage overall is much shinier and more purple than Brown-headed Cowbird.

Overall pale brown, female Shiny is very similar to female Brown-headed, but warmer brown, slightly shabbier, and more contrastingly patterned. The facial pattern is more distinct, the black bill shows greater contrast with the face, the upper- and underparts are imbued with subtle diffuse streaking, and there is more contrast between the darker upperparts and underparts. By comparison, Brown-headed's plumage is grayer, uniformly undistinguished, and overall smoother. The face of Brown-headed is particularly plain.

Shiny's expression is benign. Female Brown-headed seems beady-eyed, as though the bird has something to hide.

BEHAVIOR: Similar to Brown-headed Cowbird. Commonly forages on the ground, working an area well. Moves with short mincing steps and the bill turned down. Also forages in trees, working the outer leaves and branches. Shiny is a tame and social bird that mixes with other blackbirds and allows close approach.

FLIGHT: Like Brown-headed Cowbird in shape, but not so pointy-winged and more angular; more Brewer's Blackbird. Flight is nimble, buoyant, and slightly undulating.

VOCALIZATIONS: Song has been described as three to four "bubbling" purr notes followed by an ascending series of short notes. Flight call is like Brown-headed's whistled "*pee-ee*," but softer and held slightly longer. Call is a soft "*chrk, chrk*" that is somewhat like Brewer's but more muffled.

Bronzed Cowbird, *Molothrus aeneus*
Megaheaded Cowbird

STATUS: Fairly common semitropical species with a restricted range in the United States. Where found, much less numerous than Brown-headed Cowbird. **DISTRIBUTION:** In the U.S., breeds along the Calif.–Ariz. border, s. Ariz., the southwestern portion of N.M., s. Tex. from Big Bend to just north of Corpus Christi, and extreme s. La. In winter, found only in s.-cen. Ariz., extreme se. Tex., and s. La. Mostly permanent resident in Mexico and Central America. Accidental or casual north of its normal range. **HABITAT:** Forages in all manner of open areas with a scattering of brush, including pastures, rangeland, agricultural land, lawns, golf courses, and backyard bird-feeding stations. Roosts communally in rank marsh reed and thickets, frequently with other blackbirds. In winter, found primarily around feedlots and golf courses. **COHABITANTS:** Often forages with Brown-headed Cowbirds; roosts with other icterids; parasitizes the nests of birds as large as Green Jay, Northern Mockingbird, Hooded Oriole, and Long-billed Thrasher and down to the size of Song Sparrow and Yellow-green Vireo. **MOVEMENT/MIGRATION:** Not clearly understood. Most birds in the United States vacate breeding areas in fall; movements in August have been noted, as well as the return of birds in March and April. VI: 2.

DESCRIPTION: A solidly built, bulky-headed, menacing-looking blackbird—larger and bulkier than Brown-headed Cowbird, with which it often associates. The large head slopes down to a heavy bill (no forehead), giving the bird a flat-headed or anvil-headed appearance (a sense enhanced by the male's thick ruff or mane). Legs are disproportionately long and heavy.

Males are inky blue onto black, with a malevolent red eye. In good light, the head may show a slight brownish cast, but unlike Brown-headed, it usually appears just blackbird black. Eastern Bronzed Cowbird is all blackish and duller than western males. Western female Bronzed is dark brownish gray with bronzy greenish highlights in the distinctly darker wings. Female Brown-headed is overall paler, uniformly dull gray brown color, and shows no contrast except for the blackish bill set against the pale face (a contrast that is less pronounced in Bronzed female). Like the male, female Bronzed has a red eye and looks just as menacing. The expression is malevolent.

BEHAVIOR: Forages on the ground along with other ground-feeding birds, often in association with cattle. Its long legs seem to stomp as the bird walks, usually in a coursing pattern. Also known to pick ticks off the backs of cattle. The male's advertising display to females is arresting: the bird fluffs its manes, crouches, and flutters its wings rapidly before going aloft and hovering 2–3 ft. above the female.

FLIGHT: Direct, strong, unwavering, but perfunctory, on very pointed wings that seem to move with regular mechanical indifference. Not very nimble (or perhaps just casual), birds altering their course execute wide sweeping turns.

VOCALIZATIONS: Song is a short ensemble of thin strangled notes. Recalls some of the notes of starling.

Brown-headed Cowbird, *Molothrus ater*
The Parasitic Blackbird

STATUS: Common and widespread, this onetime prairie species and full-time pariah of the bird world has expanded its range to incorporate most of North America except the northern portions of Canada and Alaska. **DISTRIBUTION:** Breeds across most of the U.S. (excluding only the southernmost half of Fla.) and s. Canada. The northern limit reaches cen. Lab., N.B., s. Que., s. Ont., s. Man., cen. and nw. Sask., all but ne. Alta., sw. Northwest Territories, and ne. and sw. B.C. In winter (except for extreme se. Ont. and N.S.), retreats south of Canada and in the U.S. retreats from much of the interior West (found coastally from Wash. to Baja) and the northern prairie states as well as all but s. Wisc. and s. Mich., n. N.Y., n. Vt., n. N.H., and n. and w. Maine. Winter range extends south into Mexico and s. Fla. **HABITAT:** Breeds in and occupies almost any open (or semi-open) upland habitat (including suburbia) that is dotted with trees and has brushy edge and an ecotone between woodlands and open areas—that is, areas populated by the wide variety of host species in whose nests the female cowbird deposits her eggs. Only the interiors of mature, dense, unfragmented forest and woodlands seem impervious to cowbirds. Winters in open grazed or short-grass areas, particularly feedlots, agricultural fields, and pastures. **COHABITANTS:** Found with other icterids, particularly Red-winged Blackbird and Brewer's Blackbird; also starlings. **MOVEMENT/MIGRATION:** Spring migration (early) from early in March and April; fall migration (extremely protracted) begins as early as July, with peak migrations in late October and November. VI: 2.

DESCRIPTION: A small, almost finch-sized and -shaped blackbird (larger than House Finch; smaller than a starling or Red-winged Blackbird; the same size as Shiny Cowbird) that is noticeably smaller than all other icterids in mixed flocks. The head is smallish, flat-topped, and neckless, the face short and pointy, the body overall compact, and the tail short (and moderately flared or rounded at the tip). The bill is stubby, conical, and curved on the upper culmen—in a word, finchlike.

The plumage of adult males—black with green iridescences topped off with a dull brown hood—is distinctive. At a distance, birds appear truly black—even blacker than most icterids. Females are overall dull brownish gray. So undistinguished is the bird that its black eye, dark bill, slightly paler throat, and very subtle streaking on the breast—characteristics that would be dismissed in a less field mark–challenged species—stand out.

With their finely scaled backs and more distinct streaking below, immatures appear even more House Finch–like than adult females. *Note:* House Finch hops; cowbirds walk.

BEHAVIOR: For much of the year, the gregarious Brown-headed Cowbird is one of the flock: in warmer temperatures, it forages among flocks of hoofed mammals for the insects they raise (a trait that harkens back to its days of following bison herds across the prairies); in winter, it mixes with other blackbirds in search of grain. In mixed flocks, Brown-headed is distinguished as the smallest, and female Brown-headeds are the palest. Cowbirds also signal their presence in a flock with their raised tails (other icterids commonly hold their tails more horizontally). Also found in homogeneous flocks in which birds "roll over": birds on the back side of the flock leapfrog over those in the front as the flock advances.

In breeding season, males and females collect (often at the very tops of tall trees) and display, with heads tilted upward. The male's display involves fanning its wings and tail as it appears to feign falling from the branch.

FLIGHT: Overall compact, with softer angles than other icterids but distinctly pointy wings. Males seem overall black, and the bodies of females two-toned—slightly paler in front. Flight is direct but nimble, and more buoyant and less jerky and bounding than Red-winged Blackbird, but at the same time perfunctory and joyless.

VOCALIZATIONS: Song is a gulp or gurgle followed by a shrill musical whistle (or two falling water drops, "*plop, plop,*" followed by the whistle). The female's call is a loud "*pee-seee.*" Also makes a musical rattle and a dry rattle.

PERTINENT PARTICULARS: Easily distinguished in the East; more easily confused with Brewer's Blackbird or Bronzed Cowbird in the West.

Orchard Oriole, *Icterus spurius*

Elfin Oriole

STATUS: Common breeder across much of the United States east of the Rockies; less common in the Northeast. **DISTRIBUTION:** In the U.S., breeds from the Atlantic Coast to the foothills of the

Rockies, with the western border falling across e. Mont., e. Wyo., e. Colo., and e. N.M. The northern range just edges into se. Sask. and extreme s. Man. Does not breed in n. New England, upper N.Y., n. Wisc., n. Minn., s. Fla., or the s. Rio Grande Valley. Rare visitor in the far West. **HABITAT:** Open, usually grassy habitat studded with or bordered by stands of both young and mature deciduous trees. Typical habitats include orchards, suburban areas, woodland edge bordering grassland or pasture, overgrown fields, and highway rest areas. **COHABITANTS:** Eastern and Western Kingbirds, Warbling Vireo, Yellow-breasted Chat, Indigo Bunting, Baltimore and Bullock's Orioles. **MOVEMENT/MIGRATION:** Entire population vacates the United States and Canada after breeding. Spring migration from late March and early April to late May; fall from early July to early September. VI: 3.

DESCRIPTION: An elfin oriole of open places that might be mistaken for an oversized warbler. Small and slender (slightly smaller than Yellow-breasted Chat; closer to Louisiana Waterthrush in size than to Baltimore Oriole!), with a short, mostly straight, blackbirdlike bill, a slender body, and a shortish tail. Seen hopping through low foliage or in flight, an observer's first impression is frequently "warbler." No other oriole makes this first impression.

Adult males are unique. A black hood, back, wings, and tail are mated to a burnt orange body. Females and immatures have overall greenish yellow bodies and grayish wings creased by two well-defined whitish wing-bars. The overall greenish tinge distinguishes them from all but Hooded and Scott's Orioles, whose ranges overlap with Orchard along the Rio Grande Valley of Texas. (Other female and immature orioles have orange or bright yellow bodies.) The slightly larger Hooded Oriole is distinctly longer-tailed and longer-billed (in other words, it doesn't resemble a warbler), and there is nothing elfin about the much larger Scott's, a big solid oriole.

The second-year male Orchard resemble females with a black face and throat. Insofar as second-year males are most likely to push the limits of the bird's range and will be calling attention to itself by singing from the tops of trees or saplings, it is a plumage pattern worth knowing outside the bird's established range.

BEHAVIOR: Active and nimble. Forages in treetops and understory as well as in weedy stands and tall grass. Like other orioles, acrobatic, moving fluidly among branches, using gravity-defying moves to change perches or reach for prey that might engender envy in the most accomplished rock climbers. Pumps its tail habitually (another warblerlike quality).

Males are vocal, singing well into the day, usually from an exposed perch high atop a tree. Tame, allowing close approach (even when foraging near the ground). Responds readily to pishing.

FLIGHT: Profile is very slender, with a long extremely narrow tail. Flight is stiff, direct, and quick. Wingbeats are steady and rapid, and given in a long series punctuated by a regular skip/pause. Flight does not undulate but rises and falls slightly.

VOCALIZATIONS: Song is a lively, varied, whimsical mix of slower whistled phrases and quick stuttering double or trebled notes, usually hurried at the end; for example: "*r'ee tu t't't chee tu'lee.*" Song is overall less patterned and musical and more varied and higher-pitched than the song of Baltimore Oriole, with notes that rise and fall across a wide range and have a distinct Tin Pan Alley quality. If Baltimore Oriole sings at the Met, Orchard croons at a saloon. Call is a low "*chuk.*"

Bullock's Oriole, *Icterus bullockii*

The Cottonwood Oriole

STATUS: Common and widespread western breeder. **DISTRIBUTION:** Found across most of the West, east to the high plains. The northern border falls across s. B.C., s. Alta., and extreme sw. Sask.; the eastern border extends to the western border of the Dakotas, w. Neb., w. Kans., w. Okla., and the western half of Tex. Winters in Mexico (a few in California). Casual in the East. **HABITAT:** Primarily riparian woodlands dominated by mature cottonwoods; also found in orchards, open and semi-open oak or mesquite woodlands, cottonwoods, ranch yards, and any stands of mature trees surrounded by open areas. Prefers taller, more mature trees than Orchard Oriole. **COHABITANTS:** Warbling Vireo, Yellow Warbler, Lesser Goldfinch. **MOVEMENT/MIGRATION:** Almost the entire breeding population relocates to Mexico in winter. Spring migration from early March to late May; fall from early July to early November. VI: 3.

DESCRIPTION: A feathered dichotomy: adults are readily distinguished from Baltimore Oriole, but immatures (particularly immature females) must be separated only with care. Medium-sized, compact, robust oriole with a short, straight, moderately thick (blackbirdlike) bill and a short tail. Slightly larger but structurally identical to Baltimore.

Brightly plumaged orange-black-and-white adult males are easily distinguished by their large (in fact, dominating) white wing patch (other orioles have wing-bars) and distinctive bohemian head pattern—an orange face bedecked with a black beret, black wraparound eyeglasses, and a black goatee. Adult females (and immatures) are overall paler and less coarsely and contrastingly patterned than adult female Baltimore. Upperparts are grayish, and underparts are partitioned—yellow face, throat, and upper breast, grayish white belly, yellow or orange-yellow undertail. *Note:* There is little or no hint of orange on immature and female Bullock's Orioles, except perhaps the tail, which shows a *Myiarchus* flycatcher glow of orange when backlit.

The problem is that the plumage of immature female Baltimore is very similar to the plumage of adult and immature female Bullock's.

Go back to the adult males: females and immatures of both species harken back to their well-marked male counterparts. The face on adult male Bullock's is bright orange with a dark cap and a bold black line through the eye. The face on females and immatures is pale yellow with a slightly darker cap and a pencil-fine line through the eye. And the goatee? Immature males develop a black goatee by October of their hatching year. What's more, some adult females (presumably older adults) also show a sparse black goatee.

By comparison, the head of immature female Baltimore is olive drab and dark, like adult male Baltimore. It's too dark to show an eye-line.

Also look at the wing-bars. On Baltimore, the two wing-bars are well defined and distinct. On immature Bullock's, the wing-bars are linked by pale feather edges that blur the distinction and, at a distance, fuse to give the suggestion of a wing panel (just like the adult male). At close range, the upper dominant wing-bar is jagged or serrated. The wing-bar of Baltimore Oriole is a narrow clean-edged bar.

BEHAVIOR: An active oriole that is nimble enough to work the exterior leaves and twigs and that shows its versatility, particularly later in breeding season, by foraging low on bushes, grasses, and sagebrush and even lands on the ground. Moves branch to branch with acrobatic hops and flutters. Often hangs down, stretching its neck (in defiance of balance and gravity) to pluck prey. Also eats fruit, drinks nectar from agaves, and can be attracted to sugar-water.

Usually forages alone. In migration and winter, gathers in small flocks. Fairly tame, and responds well to pishing.

FLIGHT: Like Baltimore Oriole. Straight, with some horizontal shifting, on strong, mostly steady rapid wingbeats.

VOCALIZATIONS: Like Baltimore Oriole, song is a short, loud, one- or two-part ditty that incorporates chattering and musical notes; for example: "*ch chchurch will-do!*" or "*ch chchurch will-do! we'ch'we'chew*" (first part mostly stuttered; last part whistled). Overall more choppy and chattery and less musical than Baltimore. Song is often punctuated by oriole chatter that is duller and drier than Baltimore. Chatter call is slightly rougher or raspier than similar call of Baltimore. Call is a plucked note with a compressed two-note quality: "*t'wik!*"

Hooded Oriole, *Icterus cucullatus*

The Palm Leaf Oriole

STATUS: Common but, in the United States, restricted to the southwestern border states.

DISTRIBUTION: Western and s. Calif., s. Nev., sw. Utah, w. and s. Ariz., sw. N.M., and the Rio Grande Valley region of Tex. from Big Bend to the coast, north to Corpus Christi. Very rare or casual north and east of the normal range. **HABITAT:** Historically a bird attracted to the stands of tall trees associated with desert oases, especially palms but also riparian cottonwoods, willows, and sycamores. In Texas, also traditionally associated with stands of Texas ebony and mesquite. Now habituated to the artificial "oases" that typify suburban and urban areas (particularly those with a fan-leafed palm tree component) and shade tree–rich desert ranch houses. **COHABITANTS:** Orchard and Bullock's Orioles, Cassin's Kingbird, Bronzed Cowbird. **MOVEMENT/MIGRATION:** Entire population retreats to Mexico in winter. VI: 3.

DESCRIPTION: An Americanized Neotropical beauty living in suburbia. A smallish oriole (slightly larger than Orchard; smaller than Bullock's), but structured like a larger oriole species, with a longish distinctly down-curved (gracklelike) bill and a slender disproportionately long tail. Other smaller orioles have shorter and more conical bills (more blackbirdlike). Only the larger Audubon's Oriole (south Texas) has a tail so long.

Adult males, with their orange to golden yellow bodies, black faces and bibs, and white wing-bars, are unmistakable. Females have grayish upperparts, with no streaking, and yellowish underparts (the same yellow as female Summer Tanager), and immature males, which are like females but have a black bib, closely resemble female and immature male Orchard Oriole. To distinguish, remember, first, that except in Texas and eastern New Mexico, the ranges of Hooded and Orchard do not overlap. And second, remember that the elfin-featured Orchard Oriole appears warblerlike (especially in flight). With its longer down-curved bill and long tail, Hooded Oriole never seems warblerlike.

BEHAVIOR: Forages in vegetation, at all heights, but rarely alights on the ground (as Orchard does commonly). Maneuvers through the canopy and outer leaves of more open understory. Clings mostly to leaves and leaf clusters and reaches for prey (sometimes head down, sometimes perpendicular to the branch, and sometimes upside down). Movements are nimble and acrobatic, but fairly sluggish, even slothful. Also feeds on nectar and is attracted to hummingbird feeders.

FLIGHT: Flight profile is slender, with a long and resplendent tail. Wingbeats are stiff and deep, neither hurried nor leisurely, and given in a prolonged series. Flight shows typical oriole directness but is slightly twisty-turny and tends somewhat to rise and fall.

VOCALIZATIONS: Song is variable but has a warbled foundation with imitations and improvisations thrown in. Most commonly heard call is a soft mellow query, "*w'eyk*" or "*wh'reh?*" (in plainer phonetic English: "*wink?*" or "*reep?*"). Recalls the "*bink*" flight call of Bobolink, and sometimes notes combine the Bobolink "*bink*" with a House Finch–like whine. Also makes a rapid "*wha'her?*" that recalls the flock call of California Quail and is sometimes given with a stuttering beginning: "*ch'cheh wha her?*" Also has a classic oriole rattle.

PERTINENT PARTICULARS: Despite their brilliant plumage, males perched within the canopy foliage can be difficult to see. When not feeding, and when perched close to the nest, they may remain sedentary for long periods. Be patient—if a Bronzed Cowbird appears during the breeding season, an oriole will betray itself by ousting the interloper.

A Mexican species, Streak-backed Oriole, *Icterus pustulatus*, is a very rare visitor to Arizona and southern California. It most closely resembles Hooded Oriole in plumage and size (only slightly larger), but is overall more robust, with a shorter, straighter bill and a short widish tail. The most notable differences relating to plumage include the black-streaked or beaded backs (solid black or gray in Hooded) and the bold white double wing-bars on wings that are also shot through with white streaking along their folded length. The more uniformly dark wings of Hooded Oriole show a single wide upperwing-bar and a narrow trace of a lower wing-bar.

Spot-breasted Oriole, *Icterus pectoralis*
Florida Orange-headed Oriole

STATUS AND DISTRIBUTION: Introduced from Central America. Uncommon and localized; found exclusively in se. Fla., principally in the Fort Lauderdale/Miami area. **HABITAT:** Urban and suburban areas where mature flowering trees and bushes abound. **COHABITANTS:** Yellow-chevroned Parakeet, Eurasian Collared-Dove, Common Myna, Mockingbird. **MOVEMENT/MIGRATION:** Permanent resident. VI: 0.

DESCRIPTION: An Altamira-like oriole seen in Miami. Large (larger than Baltimore Oriole; slightly smaller than Altamira, whose range it overlaps in Central America but not in the United States), with a heavy, pointy, slightly downturned bill (less robust than Altamira Oriole), a nominally large head, a slender body, a black spattered breast, and a shortish round-tipped tail.

Adults are mostly bright rich orange, the only orange-headed oriole species in Florida (very similar to Altamira Oriole and somewhat similar to Hooded Oriole, which do not normally occur east of Texas). The face, throat, breast, wings, and tail are jet-black, and the sides of the breast are spattered with black spots. Wings show an orange shoulder patch and two white patches, one large and one small, on the feathers of the folded wing.

Second-year birds are much like adults, but are paler yellow-orange and show no black spotting. Immatures are like two-year-olds, but greener-backed and overall golden; lacking any black on the face or throat, they show a wholly bright orange head, setting off a beady black eye. Adult male Baltimore Oriole has a black head and conspicuous wing-bars, which young Spot-breasted lacks.

BEHAVIOR: Methodical and not particularly social—usually found in pairs or family groups (although in the tropics forages with Altamira and Streak-backed Orioles). Forages on nectaring flowers (also sugar-water); investigates bundles of dead leaves and searches for insects among leaves and branches in the canopy. Males sing all year, usually from an exposed perch (including utility lines). Tame and not shy around humans (as habitat preference attests).

VOCALIZATIONS: Song is a series of loud, slow, rich, melodious, sometimes slurred whistled notes and phrases. Sounds to some like a human just learning to whistle. Calls include a nasal "*yenk*."

Altamira Oriole, *Icterus gularis*
Grosbeak Oriole

STATUS AND DISTRIBUTION: Common resident but extremely restricted in the U.S.; found only in the lower Rio Grande Valley of Tex. Range extends along the eastern and southwestern coasts of Mexico into Central America. **HABITAT:** Open and dense riparian woodlands. **COHABITANTS:** Plain Chachalaca, White-tipped Dove, Long-billed Thrasher, Olive Sparrow. **MOVEMENT/MIGRATION:** Permanent resident. VI: 0.

DESCRIPTION: A large, heavy-bodied, heavy-billed, bright orange tropical oriole. Large (largest oriole species in the United States—about the size of Green Jay) and overall robust bordering on bulky, with a straight heavy-based bill, a large head, a husky body, and a shortish tail—in other words, just the opposite in shape of the similarly plumaged adult male Hooded Oriole. If you look at the bird and think tanager or grosbeak instead of oriole, that's very understandable. There is little about this bird that seems oriole-lithe.

Altamira is strikingly patterned bright orange and black, with black lores and throat and an orange wedge cutting across the shoulder. (Adult male Hooded Oriole shows a white wedge.) Adults are identical; immatures are similar, but not so brilliantly orange, and their tails are drab olive, not black.

Despite its striking plumage, and because of its subarboreal habits, observers are most often alerted to the bird's presence by its very simple and oriole-like song.

BEHAVIOR: A sluggish arboreal oriole that excels in tight places. Often seen perched high atop trees. When it leaves its perch, it very commonly heads straight down into a leafy fortress. Forages at all levels in trees and forests (except the ground), hopping (sometimes walking) slowly and methodically from branch to branch or flying short distances. Consumes insects and fruit, but can be attracted to sugar-water and on occasions comes to feeders for seed. Often jerks its tail up as it moves.

Usually found in pairs, but sometimes in trios or small groups in winter. Also seems to associate with Green Jays (or perhaps it's the other way around). Not particularly wary or shy. Adults commonly call to each other while feeding.

FLIGHT: Stocky oriole showing a bulky-headed, short-tailed, somewhat potbellied profile—but the stunning orange-and-black plumage pattern is arresting. For short distances, flight is jerky and theatrical; birds approach nearby limbs with exaggerated descending swoops. Over greater distances, flight is more direct and seems somewhat heavy and jaylike, with wingbeats that describe a deep down stroke. When flying across open areas, Altamira jerks its head left and right (looking for danger?) and often approaches landing points with braking, lofting, open-winged glides that give the flight a slight jerkiness.

VOCALIZATIONS: Simple, measured, unhurried, whistled song consists of two or three notes repeated in a rising-and-falling sequence, "*tew tew* (rising) *tee;* (hurried) *tootootoo* (rising) *tee* (falling) *tew.*" The pattern is sophomoric—sounds like a talented six-year-old learning to play basic song patterns on a flute—but has an eminently oriole-like quality. Also gives a whistled "*p'tew*"; a softly murmured "*e/ah*"; and a harsh nasal "*rah*" (sometimes repeated).

Audubon's Oriole, *Icterus graduacauda*
Thicket Whistler

STATUS AND DISTRIBUTION: Uncommon and geographically restricted resident. In the U.S., found only along the lower Rio Grande Valley of Tex. (accidental north to San Antonio) west to Laredo. Also breeds on the east and west coasts of Mexico. **HABITAT:** A forest and forest edge species. In the United States, found most commonly in riparian woodlands, thorn-scrub woodlands, and live-oak forest. In thorn-scrub habitat, usually most commonly associated with the taller, denser vegetation along arroyos. Also attracted to backyard feeding stations. **COHABITANTS:** White-tipped Dove, Olive Sparrow, Altamira Oriole. **MOVEMENT/MIGRATION:** Permanent resident. VI: 0.

DESCRIPTION: A distinctive, black-headed, yellow-bodied oriole seen close to the Rio Grande. Large and slender (larger than Orchard or Hooded; smaller and slimmer than Altamira), with a short fairly straight bill (shorter, heavier-based, and straighter than Scott's Oriole), but a long tail.

The pattern of adults is unique. There is no other bright yellow–bodied, wholly black–headed oriole with a yellow (actually greenish yellow) back. The somewhat similar Scott's Oriole has a black, not yellow, back (and other distinguishing characteristics, not the least of which is a different range). Immatures, showing little or no black on the head, are dull yellow below and grayish above, but still boast greenish yellow backs. This characteristic, combined with their larger size and restricted range, distinguishes them.

BEHAVIOR: Not cryptic, but certainly surreptitious. Adept at moving through thick brush and walking and hopping, quickly and nimbly, along branches, gleaning insects, flying little (its behavior and habitat the exact opposite of Scott's). Somewhat less vocal than Altamira (adding to its elusive quality), and though both sexes sing, they often do so from concealed perches. Despite its woodland predilections, Audubon's seems drawn to edges (like most orioles) and is attracted to backyard feeding stations, where it favors sunflower seed, oranges, peanut butter–lard–corn meal mix, and sugar-water.

Usually found in pairs. Vocalizes throughout the day.

FLIGHT: Very slender long-winged profile and distinctive contrasting black-and-yellow pattern. May seem humpbacked, with a somewhat drooped tail. Flight is straight and quick, but very jerky and undulating. Short series of quick wingbeats are punctuated by regular skip/pauses. Wingbeats are audible at close range.

VOCALIZATIONS: Simple, plaintive, somewhat unenthused whistled song consists of mellow notes and slurred phrases (many repeated): "*ee-ur ee-ur ee* *yer-e-yer ee-ur'eeur.*" Pairs communicate with a whistled "*pew.*" Also makes a nasal "*yike.*"

Baltimore Oriole, *Icterus galbula*

The Eastern Arboreal Oriole

STATUS: Common breeder across most of the eastern United States and southern and central Canada; uncommon wintering species in the Southeast. **DISTRIBUTION:** Breeds from N.S. across N.B., extreme s. Que., extreme s. Ont., s. Man., s. and cen. Sask., s. and cen. Alta., and the e.-cen. border of B.C. In the U.S., breeds across most of the East and Midwest (east of the short-grass prairies), with populations reaching n. Mont., w. N.D., w. S.D., w. Neb., ne. Colo., w. Kans., w. Okla., and ne. Tex. Absent as a breeder on the coastal plain from s. Va. to Tex. Winters primarily in s. Mexico, Central America, and nw. South America, but also in Fla., Cuba, and (in small numbers) along the Gulf and Atlantic Coasts north to s. N.J. Rare but regular in the far West. **HABITAT:** Breeds in open woodlands, particularly mature deciduous woodlands, woodland edge, riparian woodlands, shade trees in suburban neighborhoods, and city parks. **COHABITANTS:** Eastern Kingbird, Warbling Vireo, American Robin. **MOVEMENT/MIGRATION:** Wholly migratory. Spring migration from early April to late May; fall from early July to mid-October. VI: 3.

DESCRIPTION: A colorful, slender, orange-toned, treetop vocalist. Medium-sized (larger and more robust than Orchard Oriole; about the same size as Bullock's Oriole and Eastern Kingbird) but compact for an oriole, with a smallish head, a long pointy bill/face, a nominally slender body, and a fairly short but slender tail. Larger and more robust and blackbird-shaped (less warblerlike) than Orchard Oriole.

Seen sitting in the treetops, adult males are unmistakable—a bright orange bird with a black hood and upperparts, with one orange and one white wing-bar and wispy white fringing in the wings. Adult male Orchard Oriole is brick red below, and male Bullock's Oriole has a black-and-orange-patterned head and a broad white patch on the wings. Female and immature Baltimores are dingy gray above and orange-washed below, with two prominent white wing-bars. Female and immature Orchards are greenish yellow, with no hint of orange tones. See "Description" under "Bullock's Oriole" on separating Bullock's from Baltimore.

BEHAVIOR: A bird whose life is wedded to dense leafy canopy. Males and females sing, sometimes from exposed perches, sometimes from within the leafy confines of a tall deciduous tree. Even in migration, commonly lands on the tallest point of a tree and stands alertly erect, searching for danger, before immersing itself in the foliage. Rarely lands on the ground, but forages in low shrubs and tall grass.

An active feeder that makes short hops and longer wing-assisted leaps, pokes through leaves for caterpillars, and sometimes hangs upside down to probe or seize prey or fruit or drink nectar. In winter, sometimes roosts communally (in the Northeast, often in groves of cedars). In migration, travels in small flocks (3–10 birds), flying in a loosely strung-out linear formation. (Sometimes seen flying with Eastern Kingbirds or Scarlet Tanagers.) Fairly tame. Responds fairly well to pishing or even poor imitations of its song. Can be attracted with sugar-water or fruit.

FLIGHT: Slender but not particularly long; seems pointy-faced and short-tailed. The orange-yellow tail contrasting with the dark gray wings is often very apparent. Flight is direct and straight, fluid and unhurried, and slightly jerky or executed with a bumpy rise and fall but no undulations. Wingbeats are quick, crisp, and deep, given in a short irregular series punctuated by a skip/pause; the pattern and cadence recall American Robin (but the overall flight is more flowing and nimble and less theatrical than the robin).

VOCALIZATIONS: Song is a short, clear, pleasing, somewhat hurried and variable series of whistled notes, "*turee tu tu tu,*" followed by a savoring and occasionally interspaced with a low, angry, chattering rattle: "*chahahahaha*" (the bird's call).

Scott's Oriole, *Icterus parisorum*

Yucca Oriole

STATUS: Uncommon but fairly widespread southwestern breeder. **DISTRIBUTION:** Breeds in scattered locations across the southern half of Calif., Nev., Utah, w. Colo., sw. Wyo. (barely), Ariz., w. and s. N.M., and trans-Pecos Tex., extending east to cen. Tex., then south into Mexico. Winters in Mexico (a few in s. Calif.). Casual north and east of its normal range. **HABITAT:** An arid-lands oriole; almost always found where yuccas flourish. Fairly specific to dry arid slopes, the pinyon-juniper belt of desert foothills, live oak savanna, and the arid yucca-rich plains between mountain ranges. Avoids true desert regions where cactus dominates. **COHABITANTS:** Ash-throated Flycatcher, Juniper Titmouse, Black-tailed Gnatcatcher, Crissal Thrasher, Rufous-crowned Sparrow. **MOVEMENT/MIGRATION:** Almost all U.S. breeders retreat into Mexico after the breeding season. A few are regular in winter in desert edge in southern California. Spring migration from late March to early May; fall from early August to early October. VI: 1.

DESCRIPTION: A slender but solid-looking hooded oriole of yucca-studded hillsides. Medium-sized (same size as Bullock's Oriole), with a large head, a longish, heavy-based, slightly down-curved bill, a slender body, and a short wide tail. Not as long-billed or rangy-looking as Hooded Oriole; not as compact as Bullock's.

Adult males are unmistakable. A bright, bright yellow–bodied bird wearing a jet–black-hooded cowl that covers the head, breast, and back. Black wings show yellow epaulets and a broken white wing-bar. The boldly patterned tail is yellow at the base; the balance is black.

No other male oriole is so black, so yellow, so boldly and creatively patterned.

Brighter females and overall dingier immatures show at least a muted shadow pattern of the adult male (and females and advanced immature males show black on the face and throat). The body is yellow to olive, and the upperparts, wing, and distal half of the tail are dingy gray. A key point is that the head, back, and breast are hooded like adult male, but gray on females and grayish olive on immatures. It's not striking, contrasting, or crisply defined, but it is complete. Other yellow-green-type orioles have yellow on the face, breast, and nape. Scott's is truly hooded.

BEHAVIOR: Not generally an arboreal oriole. Feeds low, in shrubs, but primarily on yuccas and junipers, where it drinks nectar and gleans insects attracted to the blossoms. Feeds alone, with a mate, or, in mid to late summer, in family groups. Frequently covers great distances while foraging, moving from tree to tree (or yucca to yucca). Not as nimble as many orioles, but does hang upside down. Also not as tame as other orioles—or perhaps has more ground to cover and less time to waste.

Singing males sit conspicuously high on a bush or—you guessed it—a yucca.

FLIGHT: Oriole-shaped, but with a shorter tail. The adult male, with its black head, wings, and tail and yellow belly, upperwing coverts, and base of the tail, are distinctive. Females and immatures are overall dingy with olive bodies and grayish wings (similar to many other non-adult male orioles), but have diffuse yellow sides to otherwise darkish tails (once again, shades of the male's pattern!).

Flight is heavy but lofting, and smoothly undulating—more undulating than other orioles. Wingbeats are stiffer, slightly slower, and more mechanical than smaller orioles, and given in a long steady series followed by a short but discernible closed-wing glide. Flights frequently cover great distances.

VOCALIZATIONS: The song, which sounds like an oriole imitating Western Meadowlark, is a short, simple, pure-hearted, whistled pattern that repeats after a pause; for example: "*ter lew dee ter lew de t'r t'lee.*" Call is a rough low "*churk.*" Also, in flight, makes a low mewing "*eeeh.*"

FINCHES

Gray-crowned Rosy-Finch, *Leucosticte tephrocotis*

Gray-crowned Alpine-Finch

STATUS: Fairly common and widespread western alpine breeder and winter resident. **DISTRIBUTION:**

Breeds (mostly at higher altitudes) across most of Alaska (including the Aleutians but excluding the North Slope), through the n. Rockies in the Yukon and extreme w. Northwest Territories, B.C., w. Mont., ne. Idaho, n.-cen. Ore., and in higher portions of the Cascades and Sierras south to cen. Calif. In winter, vacates most of Alaska (except for the Aleutians, the Pribilofs, the Alaska Peninsula, and Kodiak I.), the Yukon, and most of B.C. Found in s. B.C., s. Alta., se. Sask., e. and cen. Wash., e. and cen. Ore., ne. Calif., Idaho, Mont., most of Nev., Utah, Wyo., sw. S.D., w. and cen. Colo., and n. N.M. Rare in w. Neb. Casual farther east. **HABITAT:** Breeds above the tree line, often near snow or glaciers, almost invariably on talus cliffs or rock piles. In winter, found in open rocky country as high as snow levels allow (with some open ground for foraging), but descends to lower elevations, where it may be found in meadows, along roadsides, and in cultivated fields. Also comes to ski resorts and backyard feeding stations. Sometimes found in mixed flocks with other rosy-finches. **COHABITANTS:** During breeding, White-tailed Ptarmigan, Horned Lark, American Pipit, Pika. **MOVEMENT/MIGRATION:** The extent and timing of migration differs geographically and is somewhat weather-dependent. Fall movement seems heaviest in late October and early November. In spring, departs by April and early May. VI: 2.

DESCRIPTION: A big, fairly sleek, gray-capped, russet-colored finch that thrives where topography is at odds with the horizon and rock and ice vie for supremacy. Medium-sized (smaller than Snow Bunting or White-crowned Sparrow; about the same size as American Pipit,) with a round head, a short (abbreviated) conical bill, and a long slender body that tapers acutely toward the long folded wings and tail. Seems front-heavy. Usually forages by crouching.

Overall dark russet brown, with a black face/throat, a silver-gray crown (or head), and rosy touches in the folded wing, flank, and belly. It can be confused only with Brown-capped or Black Rosy-Finch.

BEHAVIOR: Forages almost exclusively on the ground, more in the fashion of a Horned Lark than a finch. Hugs the ground, ambling or waddling more than walking, foraging among the rocks, at the edge of snowpack and short alpine vegetation. Prefers fairly level substrate for feeding. Occasionally makes short flights, often landing on exposed rock. In summer, usually found alone or in pairs (occasionally several individuals). In winter, highly gregarious; often forages in large flocks (sometimes exceeding 1,000 individuals) with other rosy-finches as well as Horned Lark.

On the ground, in summer, not particularly vocal and exceptionally tame. Allows close approach (to within several feet). Because it commonly flies great distances, observers often see it overhead, en route to someplace else. Birds descend rapidly from higher elevations. Flocks in transit are very vocal and very obvious, commonly making one or more wide sweeping circles over a potential landing site before setting down. Flock movements are synchronously coordinated, recalling flocks of shorebirds.

FLIGHT: A fluid-lined bird with a fluid flamboyant flight. Overall slender-bodied with a blunt head and a slender and notched (or partially fanned) tail. The dominant characteristic is the bird's conspicuously long, wide, pointy wings, which are straight-cut along the leading edge and curved along the trailing edge.

Overall incongruously dark! You will be perplexed by the sheer absence of patterning—no wing-bars, no tail spots, no contrast top to bottom. At close range, however, note that the all-gray tail and gray crown and, from below, the translucent wings (flight feathers) glow in contrast to the rest of the bird.

Flight is both playful and dashing, often sweeping and nimble, and occasionally reckless in its disregard for descents at high speed. Undulations are buoyant, sweeping, sometimes bounding. Wingbeats are given in a strong, deep, usually short and hurried series followed by a lengthy open and closed-wing glide. Commonly calls in flight.

VOCALIZATIONS: In flight and on the ground, makes a low flat "*chr'r*" or "*t'chr'r.*" Also sometimes makes a shorter multiple-call note, "*jj jj jj,*" a liquid gurgle, and a high twittering.

Black Rosy-Finch, *Leucosticte atrata*

Gray-crowned Dark Morph

STATUS: Common but habitat-restricted Rocky Mountain and Great Basin breeder; more widespread winter resident. **DISTRIBUTION:** In scattered locations, at elevations above the tree line, breeds in se. Ore., n. Nev., Idaho, w. Mont., w. Wyo., and Utah. In winter, moves down-slope and expands in a southerly direction, occurring in se. Ore., all but s. and nw. Nev., s. Idaho, Utah, sw. Mont., all but e. Wyo. and e. Colo., and n.-cen. N.M. Casual to n. Ariz. and e. Calif. **HABITAT:** Nests above the tree line in cliffs and rock fields close to tundra, glaciers, and snowfields where birds forage. In winter, commonly remains above the tree line in years with reduced snowfall, retreating down-slope and into sheltered canyons during storms. During down-slope movements, birds usually remain above the snow line, concentrating along roadsides, in feedlots, and frequently at ski resorts and backyard bird-feeding stations. **COHABITANTS:** In summer, American Pipit, Horned Lark, Clark's Nutcracker, big-horn sheep; in winter, associates with Horned Lark, Gray-crowned and Brown-capped Rosy-Finches. **MOVEMENT/MIGRATION:** Somewhat weather-dependent; spring migration from mid-March to late May; fall from early October to early December. VI: 1.

DESCRIPTION: A Gray-crowned Rosy-Finch with an all–blackish brown body and touches of rose in the wings, undertail, and rump (not the back). Females and immatures are paler and overall cold grayish brown, showing little or no rose tones, but like adult males, they have a contrasting gray swath running from just above the eye to the nape (reminiscent of a winged Mercury).

The only bird this species is likely to be confused with is the overall paler, ruddier, and more northerly breeding Gray-crowned Rosy-Finch, whose breeding range overlaps with that of Black Rosy-Finch on the Idaho–Montana border and widely in winter. Gray-crowned shows either more extensive rose tones throughout the body or a more extensive gray cap (essentially an all-gray head). Brown-capped Rosy-Finch, a third species found only in extreme southern Wyoming and Colorado and in northern New Mexico in winter, is overall plainer and more uniformly colored—all rich rosy brown (males), dull grayish brown (females), or just plain gray (young). Some percentage of Brown-capped males show gray crowns, but owing to the rich ruddy tones on the back, they are more likely to be confused with Gray-crowned than Black Rosy-Finch.

BEHAVIOR: Like other rosy-finches. Highly social. In winter, found in flocks that may number several hundred individuals. In spring, found in smaller flocks. In summer, several pairs may feed together or birds may be in pairs. Highly mobile. In summer, commonly forages at the wet edge of melting snowpack and the shorelines of thawing lakes; in winter, forages on exposed tundra. In winter, after heavy snows, flocks are commonly found along roadsides or perched on utility lines (see "Behavior" under "Gray-crowned Rosy-Finch").

Very tame, allowing close approach.

FLIGHT: Like Gray-crowned Rosy-Finch, but overall darker or grayer.

VOCALIZATIONS: Like Gray-crowned Rosy-Finch.

Brown-capped Rosy-Finch, *Leucosticte australis*

Plain Rosy-Finch

STATUS AND DISTRIBUTION: Common, but the most geographically restricted of the three rosy-finches, breeding primarily in the higher mountains of w. and cen. Colo.; also in the Medicine Bow Mts. of s.-cen. Wyo. and the Sangre de Cristo Mts. of N.M. In winter, more widespread at lower elevations, but remains limited to Colo. and n.-cen. N.M. Appears not to wander. **HABITAT:** Like other rosy-finches, an alpine specialist. **COHABITANTS:** White-tailed Ptarmigan, American Pipit, Yellow-bellied Marmot. **MOVEMENT/MIGRATION:** Spring migration from mid-March to late May; fall from late September to early December. VI: 0.

DESCRIPTION: A geographically restricted rosy-finch wrapped in a plain rose, or gray, wrapper. Males are overall a rich rose brown and generally show more rose tones throughout the plumage than other male rose-finches but overall less pattern and (usually) no gray on the head. Females are plainer and lead-colored (like a male with most of the color leached out); immatures are like females but paler gray.

For the most part, the uniformity in plumage and absence of a contrasting gray crown distinguish this species. Be aware that some male Brown-cappeds apparently do show a gray stripe over and behind the eye *like that found on some Gray-crowned Rosy-Finches.* Summer presents no identification problems, since ranges do not overlap. In winter in Colorado and northern New Mexico, however, there's a big problem. Brown-capped has a slightly smaller bill. But some Brown-capped Rosy-Finches brandishing narrow gray Mercury wings on the side of the head (not all gray or mostly gray heads) may be impossible to identify with certainty.

BEHAVIOR/FLIGHT/VOCALIZATIONS: Like other rosy-finches.

Pine Grosbeak, *Pinicola enucleator*

Club-billed Mesofinch

STATUS: Fairly common resident of taiga and high-elevation coniferous forests; moderately migratory east of the Rockies and slightly irruptive in the East. **DISTRIBUTION:** Breeds across northern taiga regions from cen. Alaska across n. Canada to the Maritimes and Nfld. The southern limit cuts across cen. Yukon, ne. Alta., n. Sask., n. Man., all but sw. and se. Ont., all but extreme s. Que., and n. Maine. Also breeds in coastal Alaska from Kodiak I. and the s. Alaska Peninsula to Queen Charlotte I. (Canada). Also breeds in the Cascades and Rockies from n. B.C. to Wash., Idaho, and w. Mont. and in isolated pockets in ne. Ore., the Sierras in cen. Calif., portions of Wyo., Utah, Colo., n. Ariz., and the White Mts. of Ariz. In winter, occurs regularly south to s. Alta., s. Sask., ne. N.D., n. Minn., se. Ont., n. N.Y., n. Vt., N.H., and cen. Maine. During irruptions (which occur only east of the Rockies), birds may reach n. Neb., cen. Iowa, n. Ill., n. Ind., n. Ohio, cen. Pa., and n. N.J. **HABITAT:** Name notwithstanding, over much of its range Pine Grosbeak is found in open spruce and tamarack forest habitat found at the northern edge of the boreal forest and at tree lines (but seems to adapt to areas that have been opened by partial or selective logging). In parts of the Rockies, associated with fir and lodgepole pine. Avoids dense coniferous forest. In winter, often found in habitats dominated by willow, birch, ash, maple, and crabapple, and also comes to feeders. Seems attracted to roadsides, where it gets grit. **COHABITANTS:** Gray Jay, Boreal Chickadee, White-winged Crossbill; during breeding, Blackpoll Warbler, White-crowned Sparrow. **MOVEMENT/MIGRATION:** Many birds remain within the breeding range throughout the winter. Regular movements to more southern locations and to lower elevations occur from late October to late December, and from early February to late April with birds east of the Rockies. VI: 3.

DESCRIPTION: A large long-tailed finch with the neckless head of a snapping turtle. Large songbird (considerably larger than any other finches; smaller than Gray Jay; only slightly smaller than American Robin), with a jelly bean–shaped bill (shows a hook at close range), a big flat-topped neckless head, a long cylindrical body, a conspicuously long tail, and short legs.

Males are distinctive — a lead gray body with an all–pinkish red head, breast, and rump (sometimes shows a reddish back). White wing-bars and dark gray wings are usually easy to see. Females and immatures are just lead gray with a blush of yellow (or orange) on the head and rump. The yellow rump patch is usually conspicuous in flight, but the yellow on the head is often hard to see. Both sexes often seem to have a smudgy shadow across the eye.

BEHAVIOR: Individuals are normally encountered perched at the top of a spruce or flying overhead, bound for someplace else. Groups (numbering 5–10) are often seen feeding on the buds of trees or hopping beside roadways where the snow has been cleared. Forages a great deal on the ground, moving with short shuffling steps (very unfinch-like), but also hops when it's in a hurry. Feeds in trees at all heights, feasting on buds (also seeding dandelions along roadsides), and is reported to flycatch (albeit clumsily) and hover.

Fairly social (but less so than other finches). In summer, found in pairs, and in winter in small flocks numbering fewer than 10 birds. In summer, quite retiring and difficult to find. In winter, more obliging and very tame, allowing close approach.

FLIGHT: Overall elongated profile shows long round-tipped wings and a long tail; you might think robin because nothing else fits the size or relative proportions. Overall plumage is mostly dark and plain, but the red on males is conspicuous, and the female's yellow rump patch vaults distance.

Flight is graceful, flowing, and evenly undulating, with wingbeats given in a series of three to four (slow enough to count), followed by a long, buoyant, closed-wing glide. In migration (at least along ridges), birds often fly in a string.

VOCALIZATIONS: Song is a soft varied ensemble of whistled mutterings often given in series; many of the notes sound vaguely like they could be made by Black-headed Grosbeak, but many others sound more like an exuberant parakeet. Call, often given in flight by eastern birds, is a loud "*tew tew tew*" that sounds like Greater Yellowlegs. Birds in the West give a different call.

PERTINENT PARTICULARS: When you look at this bird, your mind may go blank. It's too big for a finch. The bill's too blunt (unless it's deformed!), and the body's too long for a grosbeak. It's the size of a robin, but the shape's wrong. One choice left.

Purple Finch, *Carpodacus purpureus*

Robust and Raspberry Red Finch

STATUS: Uncommon to common breeder in northern forests and at higher elevations; widespread and irregularly irruptive winter resident in the eastern United States and along the Pacific Coast. **DISTRIBUTION:** Breeds across s. Canada from Nfld. and the Maritimes to B.C. and north into the s. Yukon and sw. Northwest Territories (absent in the southern prairie regions of Sask., Alta., and se. B.C.). Permanent resident across extreme s. Que., s. Ont., s. Man., and sw. B.C., as well as in New England, N.Y., n. N.J., most of Pa., w. Md., e. W. Va., ne. Ohio, n. Mich., n. Wisc., n. Minn., and western portions of Wash., Ore., and along the coast and in the Sierras of Calif. In winter, found anywhere east of a line drawn between w. N.D. and cen. Tex. (but absent in s. Tex. and the Florida Peninsula). During irruptive years, may appear in w. and s. Ariz. and N.M. Casual on the western plains. **HABITAT:** Prefers to nest in moist, cool, coniferous forest, particularly at the edge of clearings associated with bogs, but also occurs in mixed coniferous-deciduous forests, deciduous forest, suburban areas, orchards, and cemeteries. In winter, also occupies weedy fields, hedgerows, young second-growth woodlands, and (especially) backyard feeding stations. **COHABITANTS:** Red-breasted Nuthatch, Yellow-rumped Warbler; in winter, often flocks with goldfinches, House Finch, Pine Siskin. **MOVEMENT/MIGRATION:** Northern birds vacate breeding areas every year, but in some years more extensive and widespread irruptions occur. Spring migration from February to late May; fall from September through December. VI: 2.

DESCRIPTION: The robust *Carpodacus.* The same size as House Finch, but plumper (and very slightly smaller than Cassin's Finch), with a heavy, conical, almost grosbeaklike bill, a big round or squarish head, a plump body, and a short deeply notched tail. Overall bigger-headed, heavier-billed, and more portly than the more generically contoured House Finch or the sharply featured spiky-billed Cassin's Finch.

Adult males appear extensively but diffusely wine-stained on the head, breast, flanks, and even bleeding through the brown back and wings. On House Finch, brighter red to orange tones are limited to the head, face, throat, and breast. House Finch looks like it had wine thrown in its face; Purple Finch was dipped in it. On Cassin's Finch, the patch of red on the peaked crown is conspicuously brighter and richer (somewhat redpoll-like).

On adult male Purple Finch, the very white belly and flanks show no brownish streaks (unlike House Finch), and the face shows a shadow of the female's dark cheeks.

Females and immatures are grayish brown above and white with broad blurry streaking below. The dark cheek patch and bordering pale eyebrow easily distinguish Purple from the very plain-faced female House Finch. In keeping with its more siskinlike features, female Cassin's Finch shows narrow, crisp, siskinlike streaking (but no yellow in the wings).

BEHAVIOR: A tree-loving finch that likes to forage on the higher and outer branches, feeding on flowers, buds, and (especially) seeds. Sometimes feeds on the ground, but less regularly than House Finch. Vocalizing males commonly sing from the highest point on a tree. Outside breeding season, commonly forages in flocks (often mixing with other finches), and is easily attracted to backyard feeding stations (where it is particularly partial to black-oil sunflower seed).

Fairly tame. Responds readily to pishing, but when attracted, commonly perches at the very tops of trees. Calls frequently in flight.

FLIGHT: Overall stockier, heavier, and more humpbacked than House Finch. The too-short tail is slightly flared and deeply notched. Wingbeats are given in a crisp regular series followed by a closed-wing glide. Overall flight is less jerky and more evenly undulating than House Finch.

VOCALIZATIONS: Song is a hurried, mumbled, run-on warble that is sweeter and less histrionic than House Finch. Calls are a low flat "*pik*," often given in flight, and a whistled vireo-like "*chuee.*"

Cassin's Finch, *Carpodacus cassinii*

The Siskinlike Purple Finch

STATUS: Common and widespread western breeder, resident, and wintering species. **DISTRIBUTION:** Breeds from se. B.C. and extreme sw. Alta. through most of the Rockies, e. Cascades, and Sierras, including e. Wash., w. and s. Ore., and n. and e. Calif. (and isolated locations at higher elevations farther south), Idaho, w. Mont., n. and cen. Nev., Utah, w. and cen. Wyo., w. and cen. Colo., n. Ariz., and n. N.M. Winter range expands to include the mountains of cen. and s. Calif., e. Ariz., and all but e. N.M. south into cen. Mexico. Also sometimes wanders farther east into the western prairies. **HABITAT:** Breeding and nonbreeding, open coniferous mountain forests. Also reported to breed occasionally in open pinyon-juniper. **COHABITANTS:** Yellow-rumped Warbler, Evening Grosbeak, Pine Siskin. **MOVEMENT/MIGRATION:** Most (but not all) birds appear to move to lower elevations in fall. Spring migration from mid-March to late May; fall from early September to late November. There is evidence of some post-breeding movement to higher elevations. VI: 2.

DESCRIPTION: The *Carpodacus* finch with the features of a siskin and plumage closer to Purple Finch. A small songbird, but the largest of the *Carpodacus* trio (slightly larger than Purple and House Finches; about the same length as Red Crossbill). Overall more trim, angular, and pointy than Purple Finch, with a peaked head, a slightly longer, thinner, pointier, more siskinlike bill, and a slightly longer and more splayed and forked tail.

Plumage is like Purple Finch, but crisper and more contrastingly patterned. The reddish head, back, and breast are more rose red than wine red. The forehead and forecrown are conspicuously richer red than the rest of the head, the breast is merely blushed with color (not stained), and the dark lines on the back are distinct (not blurred). Females and immatures are overall colder gray than Purple Finch, sometimes have a faint olive wash, and are more finely and crisply streaked both above and (particularly) below—much more siskinlike. Some birds even show a siskinlike touch of yellow on the cheeks.

BEHAVIOR: Forages on buds and seeds in the canopy; also frequently forages on the ground. Highly social. Breeds semicolonially; for the balance of the year, found in flocks with other Cassin's Finches, but also occurs in mixed flocks with other finches (including siskins). Mostly arboreal, but commonly forages for grit along roadways (both summer and winter). Perches upright; not as acrobatic as siskins. Easily attracted to feeders. Frequently seen flying over forest canopy.

FLIGHT: Profile appears longer-winged than Purple or House Finch (recalls a rosy-finch). Flight is quicker, more bounding, and less bouncy or jerky than House Finch. Pattern is a series of rapid wingbeats followed by a closed-wing glide.

VOCALIZATIONS: Song is a rapid gush of rising-and-falling notes with some musical, some rough and burry, and many doubled or trebled; song is longer and more complex than Purple Finch and does not end with an emphatic "*p'chew*" like House Finch. Sounds like a wandering warbler song or a hurried verbose House Finch. Call is a squeaky querulous "*chweup.*" Flocks also make soft conversational churs, purrs, and squeaks.

House Finch, *Carpodacus mexicanus*

The Everywhere Finch

STATUS: Common resident. Originally a bird of western habitats, but now, in the wake of a very successful period of accidental introduction and colonization, the default finch in urban and suburban habitats across most of the United States. **DISTRIBUTION:** Found almost everywhere in the U.S. and bordering portions of Canada (extending from N.S. across s. N.B., extreme s. Que., s. Ont., extreme s. Man., Sask., and Alta. into s. and cen. B.C. In the U.S., absent only in n. Maine, the cen. and s. Florida Peninsula, parts of s. La., and portions of the n. Rockies. **HABITAT:** In the West, breeds in disturbed and arid regions near water; also at lower elevations of open coniferous forests and riparian corridors, and very common in orchards, residential areas, parks, and wine country. In the East, most common in settled habitats around houses and lawns, particularly those whose landscaping includes conifers. In rural or forested regions, found exclusively near homes and buildings. **COHABITANTS:** Mourning Dove, Rock Pigeon, other finches. **MOVEMENT/MIGRATION:** Nonmigratory in the West; somewhat migratory in the East, with some northern birds retreating to more southern portions of the breeding range in winter. Spring migration presumably occurs between February and April; fall from October to early December. VI: 1.

DESCRIPTION: A slim, small-headed, blandly contoured, somewhat pedestrian-looking finch. Head is elongated and flatly curved (not peaked) on top. The finch bill is short, somewhat swollen or bulbous or blunt, and not acutely pointy or conical. (On some individuals, a curved upper mandible and down-curved culmen make the bird look predatory.) The slender slightly notched tail seems long because the wings are so short.

Males are overall dingy grayish brown with blurry streaking on dirty white underparts. The reddish areas on males vary in extent, intensity, and even color—red is common, orange uncommon, and yellow rare—but are generally more confined than on Purple Finch (relegated to the top of the head, face, throat, and, especially, breast) and seem brighter or perhaps overapplied on a bird so otherwise drab. Also distinguishing male House Finches are the broad dark streaks on the lower breast and flanks (absent in Purple; fine and indiscernible on Cassin's). Females are overall drab with a plain face and blurry streaks that contrast little with the dingy (not white!) underparts—all in all, a fairly undistinguished bird. The tail is less notched than on the other two species. The expression is cross, stupid, mean.

BEHAVIOR: Highly social. Outside breeding season, forages in flocks, often associating with House Sparrow. Active but not particularly quick or agile; in fact, seems downright clumsy next to the other birds—chickadees, titmice, siskins, redpolls—coming to feeders. Spends much time foraging on the ground. (Cassin's and Purple Finches are more arboreal.) Perches on the highest points available; flocks commonly align themselves along utility lines.

FLIGHT: Shorter, rounder-winged, and longer-tailed than other finches. Flight is bouncy or bounding, but effortful. The bird always seems like it's flying into the wind. Doesn't undulate so much as escalate. Long fluttering series of wingbeats are punctuated by a pause, causing birds to climb rapidly but making them appear to be bounding up the stairs. Flocks are fairly cohesive but not coordinated—more a mob than a flock.

VOCALIZATIONS: Song sounds like a spirited descending monologue (in rapid conversational tones), ending with an exclamation, "*P'chee, p'che/p'chee p'chee, p'chu/choo p'choo* (half pause) *P'Choo!*" Call is a soft muttered "*ch'dip*" (often repeated).

Red Crossbill, *Loxia curvirostra*

Eclectic Parrot-Finch

STATUS: Common resident of northern and higher-altitude coniferous forests, but nomadic, irruptive, and unpredictable. In some locations, may rank among the dominant species, yet in other parts of its "normal" range, may be rare. **DISTRIBUTION:** Regularly breeds in the boreal forests of s. Canada—from Nfld. to N.S., e. N.B., s. Que., s. Ont., s. and cen. Man., most of Sask., forested portions of Alta., B.C., sw. Northwest Territories, s. Yukon, se. and s. coastal Alaska. Also resident in e. Maine, s. N.H. and Vt., n. N.Y., n. Mich., n. Wisc., n. Minn., cen. and w. Neb., w. N.D., e. Mont., and w. S.D. Also breeds in montane forests in w. Va. and N.C., as well as in the Sierras, the Cascades, s. Calif. and Baja, the Rockies, and cen. Mexico. Irregular breeder in coniferous forests elsewhere. In fall and winter, particularly during years when cone crops fail within the bird's established winter range, birds may wander extensively outside of normal breeding areas, some as far south as northern portions of the Gulf Coast states, cen. Tex., and nw. Mexico. **HABITAT:** Coniferous forests with a seasonally bountiful cone crop. Specifically targets spruce, hemlock, and most pine species (much less spruce-dependent than White-winged Crossbill). **COHABITANTS:** Red-breasted Nuthatch, Yellow-rumped Warbler, Pine Siskin, White-winged Crossbill, Red Squirrel. **MOVEMENT/MIGRATION:** Wanders year-round throughout its normal breeding range in search of food, but seems most prone to travel from May to June and from September to November. Migrates by day, presumably to be able to assess cone availability. During irruption years, fall movements begin earlier, in August, and are often preceded by, or coincide with, irruptions of Red-breasted Nuthatch. VI: 3.

DESCRIPTION: A squat top-heavy finch—about the same size and shape as a cardboard toilet paper tube. Medium-sized (smaller than Evening Grosbeak; larger than Pine Siskin; only slightly larger than Purple or Cassin's Finch), with a large dome-topped head, a "crossed" bill, a stocky body, and a narrow deeply forked tail that appears much too small for the bird. The bill varies greatly in size (depending on subspecies), and unless viewed closely, the crossed mandibles are often hard to note. Most birds show a heavy, down-hooked, parrotlike bill, or the bill may appear short, heavy, and blunt, adding to the bird's overall stockiness.

Males are overall dull red with brownish wings and tails (recalls a dingy Vermilion Flycatcher). Females are dull yellow or grayish with yellow highlights and darker wings and tails (may recall a dirty fall male Scarlet Tanager). Both males and females often seem to have dirty faces. Juveniles also have hooked bills and are streaked like siskins. Some birds (subadults?) are dull orange—halfway between adult female and adult male. *Note:* In all plumages, there are no bright white wing-bars. That's the other crossbill.

BEHAVIOR: Very social. Commonly found in flocks numbering several individuals to a dozen birds (occasionally more). Flocks work favored trees thoroughly, frequently returning until cones are stripped. Very parrotlike in its behavior, but nimble. Moves incrementally through a tree, hopping branch to branch, using its bill (as well as its feet) to maneuver. Flocks may be quiet when feeding or noisy (particularly just before flying). When not feeding, Red Crossbill tends to sit at the very top of conifers (as do singing males). When feeding, has an amazing capacity to worm its way into the interior branches of trees. Combined with its sluggish feeding habits, this trait often makes the bird difficult to view.

Usually tame. Although generally unresponsive to pishing, very responsive to pygmy-owl imitation in the West.

FLIGHT: Overall compact, but with a blunt head, an abbreviated (but distinctly notched) tail, and very long, broad, pointy wings that are straight-cut along the leading edge and curved along the trailing edge. Flight is strong, almost angry, and fast, direct, and often bounding, with a rapid rise when the bird is flapping but at other times merely undulating. Rapid flexing wingbeats are executed with a strong down stroke and given in a lengthy series followed by a lengthy closed-wing glide. Birds are often seen flying great distances, above the canopy. Flocks are well spaced and noisy.

VOCALIZATIONS: Simple but varied song is a two- to four-part series of repeated "*kip*," "*tink*," "*churp*," and wheezy notes—some slow, some hurried. Call is a short, loud, somewhat musical and distinctive "*kip*" that is usually doubled or given in a series: "*kip kip*" or "*kip kip kip*." Call, which varies between subspecies, is similar to White-winged Crossbill but more hollow-sounding or full-bodied and not as sharp.

PERTINENT PARTICULARS: Eight different Red Crossbill "types" have been identified. They differ primarily in bill size, call notes, and distribution. Although their classification is uncertain, their basic similarity in the field is a matter of fact.

White-winged Crossbill, *Loxia leucoptera*

Spruce Parrot-Finch

STATUS: Irregular but widespread, primarily northern resident—sometimes common, sometimes absent. **DISTRIBUTION:** Boreal forest region of Canada and the U.S.—from Alaska (south of the Brooks Range and western coastal region) across virtually all subarctic Canada (excluding coastal B.C., s. Alta., s. Sask., sw. Man., and extreme se. Ont.). In the U.S., occurs in n. and e. Wash., cen. Ore., n. and e. Idaho, w. Mont., w. Wyo., ne. Utah, n. Minn., n. Wisc., n. Mich., n. N.Y., n. N.H. and Vt., and n. and cen. Maine (also in irregular pockets in w. Colo.). Wanders irregularly south to n. Nev., cen. Utah, n. N.M., cen. Neb., s. Mo., s. Ill., cen. Ky., w. Va., Md., Del., and N.J. **HABITAT:** Spruce (also tamarack) forest with bountiful cone crops. **COHABITANTS:** Gray Jay, Boreal Chickadee, Pine Grosbeak, Red Squirrel. Also associates with Red Crossbill and Pine Siskin. **MOVEMENT/MIGRATION:** Nomadic and periodically irruptive. Populations shift in response to the success or failure of spruce cone crops and engage in three (sometimes more) movements per year. One broad-based shift occurs in mid to late May (when cones begin to develop); another in October and November when cones ripen; and a third in February, when regional food stocks become depleted. Widespread irruptions occur primarily during the late fall movements. VI: 3.

DESCRIPTION: A stocky, top-heavy, heavy-billed, slight-of-tail finch that sits at the top of spruces flashing conspicuous white wing-bars. Medium-sized (larger than Pine Siskin or Dark-eyed Junco; smaller than Pine Grosbeak; the same size as Red Crossbill), with a bulky crossed bill, a large, round (or slightly peaked), neckless head, a robust body, and a short, narrow, deeply notched tail. Appears conspicuously big-headed and robust with a ridiculously undersized tail.

Adult males are rosy red (slightly brighter than Red Crossbill) with blackish wings and tail and some dark smudgy markings about the face. Females are dull grayish, with touches of yellow on the breast and rump and a suggestion of blurry streaking. Overall gray and heavily streaked, streaky juveniles look like bulky big-billed siskins.

In all plumages, shows two conspicuous bright white patches (or bars) in the wing. At all ages, the heavy bill is crossed (like you'd cross your fingers) at the tip, *although often only the downturned tip is visible, so the bill looks parrotlike*—looks not at all like other finches.

BEHAVIOR: Highly social. At all times of year, found in flocks, but singing males are commonly solitary. Forages almost exclusively in spruce, where it maneuvers with slow awkward steps and fluttering wings (and often the assist of their bills) among the interior and exterior portions of the top of the tree. Seems very parrotlike. Also forages on cones on the ground, maneuvering, in finch fashion, with short hops. Often silent when feeding and difficult to see (birds just become part of the tree). Singing males, however, are usually conspicuously perched at the very top of (or just below the top of) a spring spruce (showing some preference for bare dead perches). When perched and feeding, movements seem sluggish, jerky, and somewhat amateurish. Movements that other birds make with their heads and necks the neckless crossbill must accomplish by shifting its entire body.

Flocks signal their departure from trees by growing suddenly vocal and lifting off simultaneously. Vocal in flight, flocks usually travel great distances about 100 ft. above the canopy.

FLIGHT: Overall stocky with long, broad, pointy wings and a short, narrow, conspicuously notched tail. Seems hang-bellied or portly, and wing-bars are conspicuous at great distances. Except for the wing-bars, the shape and pattern may recall Scarlet Tanager. Flight is strong, straight, and undulating. Wingbeats are given in a hurried series followed by a lengthy closed-wing glide. Also engages in more nearly level flight with two wingbeats followed by an abbreviated closed-wing glide.

VOCALIZATIONS: Song (given from a perch and in flight) is a lengthy run-on series of chattering trills and rattles, each given with a shifting pitch; for example: "*re'e'e'e'e'e'e'e'ch'h'h'h'h'h'h'h'checheche-chechecheche twi'i'i'i'i'i'i'i'i.*" Some note sequences are almost musical, and occasionally two-note utterances are larded in. Sounds much like a hurried brittle canary. Flight call, a sharp "*kip*," is like Red Crossbill, but weaker, higher-pitched, and scratchier or more brittle-sounding. Also makes a low dry stutter, "*chuh/chuh*" or "*chuhchuhchuh*," that sounds much like a redpoll.

Common Redpoll, *Carduelis flammea*

Catkin Finch

STATUS: Common arctic and subarctic breeder; common winter resident across northern North America and farther south during periodic irruptions. **DISTRIBUTION:** Circumpolar. In North America, breeds throughout Alaska, the Yukon, the Northwest Territories, nw. B.C., n. Sask., n. Man., n. Ont., n. and cen. Que., Lab., and Nfld.; also in Greenland. In winter, most birds withdraw from northern portions of their range, extending the range south across s. Canada and the northern border states (also found in ne. Ore., Wyo., and Mass.). During irruption years, birds regularly reach Utah, Colo., n. Okla., n. Mo., n. Tenn., and n. N.C. **HABITAT:** Breeds in a variety of northern habitats, including the edges of spruce forest, stunted spruce, willow thickets, stands of birch, and low tundra scrub in sheltered pockets. Avoids dense forest, but is common in urban and suburban areas where trees and woody vegetation flourish. In winter, found in open woodland, brushy edge, weedy fields, and backyard feeding stations (especially thistle). Loves birch catkins—simply loves them. **COHABITANTS:** In summer, Wilson's Warbler, Tree Sparrow, White-crowned Sparrow, Hoary Redpoll. In winter, Pine Siskin, American Goldfinch. **MOVEMENT/MIGRATION:** During normal years, fall migration extends from late August to early December; spring return is from late February to early June. During irruption years, fall migration is protracted, with nomadic flocks (and in some cases wholesale invasions) arriving in February and March. VI: 3.

DESCRIPTION: Streaky, stubby, effervescent pipsqueak of a finch with a small red beret and a black goatee. Small (smaller than House or Purple Finch; slightly larger than American Goldfinch or Pine Siskin), with a small goldfinchlike bill, a round head, a compact body, and a moderately long tail with a slightly flared deeply notched tip. Overall plumper and less pointy-faced and angular than Pine Siskin.

Grayish brown above, grayish or dingy white below, and overall somewhat darkly and diffusely streaked, *but the streaks on the sides and flanks of the paler underparts are conspicuously dark, thick, and well defined.* Breeding males sport a rose-colored blotch on the breast. In all plumages, Common Redpoll shows a small well-defined red cap on the noggin and a small black splotch surrounding the bill (absent on immatures). Other finches show extensive and often diffuse amounts of red on the head, face, and breast and no black on the face or throat.

BEHAVIOR: A nimble, vocal, highly social bundle of nervous energy. Rarely still for long. Often seen (and heard) flying high overhead (singly or in pairs, trios, small groups, or large flocks that in migration may number in the thousands). A nimble outer branch specialist, often seen dancing on the upper surface of spruce needle clusters or dangling, chickadee fashion, as it savages seed from a birch catkin. In summer, birds vying for mates halt their bounding aerial displays only long enough to play hide-and-seek with prospects.

Forages at all heights—from treetops to grassy seed heads—but seems loath to let its feet touch the ground (except at feeders). Very tame, allowing close approach (or circles and lands right where you are standing). Responds well to pishing.

FLIGHT: Small and overall compact, showing short triangular wings and a narrow tail that is notched and flared at the tip. Indistinctly marked and fairly pale overall, except that the back is darker, and the underparts are lighter with some streaking. The rose-colored patch on the breast of breeding males stands out. The heavily streaked paler rump does not stand out and can be seen only with some effort.

Flight is bouncy, bounding, and energetic—the bird flies like it's having fun and has energy to burn. Rapid wingbeats are given in a short (sometimes lengthier) burst followed by a closed-wing glide. Climbs steeply and circles often; flight sometimes wanders. Calls constantly in flight.

VOCALIZATIONS: Repertoire consists of two basic vocalizations—buzzy fluttering trills and short flat notes given in series. These may be given separately or in sequence; for example: "*che'e'e'e'e chr chr chr*" or "*chee chee chee chu'r'r'r'r'r'r'r.*" Most sounds have a flat, hollow, or tinny quality. Also makes sounds that recall some of the vocalizations of American Goldfinch and a hard even flight call that has something of the stone-skipped-on-ice quality of a crossbill: "*cheh, cheh, cheh.*"

PERTINENT PARTICULARS: The plumage of Common Redpoll is variable, with adult males ranging paler than females and immatures. Although readily identifiable as redpolls, these pale males approach and probably overlap with female and immature Hoary Redpolls in overall color and pattern. Differences and similarities are discussed under "Hoary Redpoll."

Hoary Redpoll, *Carduelis hornemanni*

Feathered Snowball

STATUS: Common, primarily arctic breeder; regular winter resident in northern regions, but numbers vary year to year. Rare straggler to the northern United States and western Canada during redpoll irruptions. **DISTRIBUTION:** Breeds along the west and north coast of Alaska and in extreme n. Canada, to the west side of Hudson Bay; also in n. Que. and coastal Greenland. Winters regularly through w. and cen. Alaska, cen. Yukon, the Northwest Territories, ne. Alta., most of Sask., Man., all but se. Ont., Que., and n. Nfld.; also in border regions of N.D., Minn., Wisc., and the Upper Peninsula of Mich. During irruptions, may occur as far south as ne. Wash., n. Idaho, n. Wyo., cen. Neb., cen. Iowa, Ill., Ind., Ohio, w. Md., and N.J. **HABITAT:** Willow thickets, stunted spruce, and low mats of vegetation on open tundra. Generally breeds deeper into the tundra than Common Redpoll, but still requires pockets of woody vegetation (however low and stunted). In winter, uses the same habitat, but some birds join Common Redpoll farther south along wooded edges, fields, towns, feeders, etc. **COHABITANTS:** In summer, Yellow Wagtail, Lapland Longspur, American Tree Sparrow, Common Redpoll. In winter, Common Redpoll. **MOVEMENT/MIGRATION:** Amazingly, some birds remain in northern breeding areas year-round, but numbers fluctuate. Others move south between late August and early December and return north from March to early June. VI: 2.

DESCRIPTION: A paler, grayer, generally plumper redpoll. A small dwarfishly proportioned finch (slightly larger than Common Redpoll), with a short nub of a bill, a large dome-shaped head showing a pushed-in face and sometimes a puffy mane, a plump fluffy body, and a moderately long deeply notched tail. The bill is key. Common Redpoll has an average-looking finch bill—conical but longer than it is wide at the base. Most Hoaries have a petite, abbreviated, pushed-in nub of a bill that's only as long as the base is wide. In addition, many Hoaries show a protruding puff of feathers on the forehead, making the bill even more diminutive and obscure-looking. If you look at the face of a redpoll and it looks like a normal finch, it's Common. If it looks like it was hit in the face with a potato-filled sock, it's Hoary.

Overall pale—*hoary*—with grayer upperparts and paler whitish underparts distinguished by a general paucity of streaking. Adult males are ghostly—snowballs with a red cap, a black chin, and a dark line through the eye. The flanks, undertail coverts, and uppertail coverts are white, with just a few faint pencil-fine lines. In breeding plumage, underparts are blushed with a pale suggestion of pink (and no obvious blotch, as on Common). Females and immatures are darker and more heavily streaked than adult male Hoary and approach some paler adult male Commons in overall tone and pattern. Hoary will usually appear paler and grayer above and whiter below, with streaks on the flanks and on the rump that are narrowly and crisply defined.

BEHAVIOR: Like Common Redpoll—except how lingering residents survive in the darkness of arctic winters is a behavioral enigma.

FLIGHT: The overall paleness projects in flight. The whiter underparts are conspicuous; the pale rump on adult males is blatant, and conspicuous on females and immatures. On Common Redpoll, with rumps overlaid with dark streaking, you have to study the bird to discern a paler rump. On Hoary Redpoll, the pale rump is obvious.

VOCALIZATIONS: Like Common Redpoll.

Pine Siskin, *Carduelis pinus*

The Bratty, Streaky, Little Pipsqueak at the Thistle Feeder

STATUS: Common breeder in northern regions and mountain forests; irruptive winter movements are frequent and widespread. **DISTRIBUTION:** Breeds from s.-cen. Alaska across boreal Canada to Lab. and N.B. (absent in southern portions of the prairie provinces). In the U.S., found year-round in the western mountain ranges from Wash. to cen. Calif. (also in isolated pockets farther south), Idaho, w. Mont., locally in the Dakotas, Nev., Utah, w. and se. Wyo. (also sw. S.D.), w. and cen. Colo., n. and e. Ariz., and n. and w. N.M., south through the mountains of Mexico. Also occurs in n. Minn., n. Wisc., n. Mich., n. N.Y., Vt., N.H., and all but s. Maine. In winter, birds commonly forage over the northern two-thirds of North America, and during major irruptions can be found anywhere but the Florida Peninsula. **HABITAT:** Breeds in open coniferous forests, including artificial settings (arboretums, cemeteries, Christmas tree plantations); forages in pines as well as shrubs, thickets, deciduous trees, grasslands, lawns, and backyard feeding stations, where it favors thistle. **COHABITANTS:** Red-breasted Nuthatch, Cassin's Finch, Evening Grosbeak, Red Squirrels; in winter, commonly associates with flocks of American Goldfinch and Common Redpoll. **MOVEMENT/MIGRATION:** Birds from interior Alaska and across northern portions of their Canadian range move south in winter. Also engages in widespread (but not necessarily continental) irruptions that propel large numbers well south of the normal winter range. Spring migration from late March to late May; fall from mid-September to late November. VI: 3.

DESCRIPTION: A quick, nervous, dark, drab, ultra-streaky pipsqueak of a bird that makes up in sass and belligerence what it lacks in size. A small, acutely angular, top-heavy, and pointy-faced bird that resembles a gaunt streaky goldfinch, but has a distinctly longer, thinner, pointier bill, a narrow chest, and a short thin tail.

Overall dull and streaky, above and below, with subtle or blatant infusions of yellow about the face, wings, flanks, tail, and (commonly) the bolder lower wing-bar. The face is more distinctly patterned than American or female Lesser Goldfinch and House Finch, with a dark eye-line that curls around behind the eye and a pale eyebrow. Note the absence of red anywhere on the plumage. The expression is myopic, beady-eyed.

BEHAVIOR: Extremely social species that feeds in flocks even during breeding season and in winter commonly associates with goldfinches, redpolls, and other finch. Active, acrobatic, and nimble—siskins seem to do everything just a little bit quicker than other species and make goldfinches seem sluggish and House Finch seem tethered. Feeds in the canopy, often dangling upside down when feeding on buds or extracting seeds. Also feeds in shrub, on grasses along roadsides, on the ground, and (particularly) at backyard feeding stations, where it favors (and may swarm all over) the thistle feeder. On the ground, moves in short shuffling hops. Somewhat fearless; when feeding flocks are flushed, Pine Siskin is usually the last bird out and the first one back.

Vocally quarrelsome and belligerent, constantly challenging occupants for their perches or standing its ground against larger birds (up to the size of Evening Grosbeak).

FLIGHT: Overall tiny, pointy-winged, and short-tailed. The bold yellow wing stripes are conspicuous. Flight is quick, bouncy, and jerky, often wandering or tacking; wingbeats are given in a hurried series followed by a descending closed-wing glide, followed by a rapid fluttering ascent. Vocalizes frequently in flight.

VOCALIZATIONS: Song is a rapid compressed series of stuttery, wheezy, raspy, chatty, and querulous call notes. Call notes include a rising "*eeeh?*" (like a goldfinch), a "*Chee*" (a cross between a chip and a snarl), and a more buzzy and ascending "*zzzzeh!*" (the celebrated "zipper" call). Also gives a "*chew ch ch ch reeeh?*" in flight. Many of the sounds are similar to a goldfinch but harsher and squeakier.

Lesser Goldfinch, *Carduelis psaltria*

The Diminutive Finch

STATUS: Common in California and south-central Texas; somewhat less common and numerically mercurial across the balance of its range. **DISTRIBUTION:** Breeds in w. North America—generally west of a line drawn from the w. Ore.–Wash. border, across s. Idaho, ne. Nev., Utah, ne. Colo., sw. Neb., ne. N.M., and cen. Tex. (less common but does breed somewhat north of this line to sw. S.D.). Where found, primarily a year-round resident; does withdraw, however, from some northern portions of its range (such as Colo., most of Utah, Neb., and Nev.) and establishes wintering populations in extreme se. Calif., coastal Mexico, the lower Rio Grande Valley of Tex., and ne. Mexico, where it does not breed. **HABITAT:** Uses a wide variety of habitats, including oak woodlands, oak-conifer forests, pine groves, riparian corridors, chaparral, desert oases, brushy gullies, suburban yards, and weedy patches. Readily attracted to backyard feeding stations. **COHABITANTS:** Varied. Feeds with Lawrence's and American Goldfinches; also House Finch. **MOVEMENT/MIGRATION:** Little is known about the bird's migratory patterns. In areas where birds are absent in winter, they arrive in April and May and depart in October. VI: 1.

DESCRIPTION: A tiny, active, acrobatic, social, and somewhat canary-like finch. Distinctly smaller, somewhat slimmer, more compact, more angular, and less pudgy than American Goldfinch, with which it often associates. The bill is pointier and more petite, and the head often seems neckless. Lesser is distinguished from American by both size and plumage, especially in winter, when Lesser is more extensively yellow, particularly on the undertail coverts.

Adult males are yellow below and all or mostly dark above—all black in Texas, and mostly black (except for the dark green back) over the rest of the bird's range. Breeding male American Goldfinches are overall bright yellow with black restricted to the wings, tail, and a patch on the forehead. Lesser, by contrast, has bright white T-shaped wing-bars.

Female and immature Lesser Goldfinches are distinctly greenish to olive above, yellow below. The backs and underparts of nonbreeding male, winter female, and immature American Goldfinches are brown or putty-colored, with no olive and no yellow except for a subtle wash on the head and throat.

Breeding female American Goldfinch (March through October) is most like Lesser Goldfinch, but on American, the dominant wing-bar is buff, not white, and the bill is yellow-orange, whereas it's black on Lesser.

BEHAVIOR: Like other goldfinches. Social outside of breeding season; forages in flocks on or near the ground, clinging to weed stalks or feeders, frequently hanging upside down, extracting and husking seeds. Lesser's movements are quicker, and its activity more frenetic, than larger finches. At feeders, hovers more and sometimes jerks its tail nervously. (American Goldfinch feeds more quietly.)

Lesser Goldfinch is subordinate to and easily displaced by American Goldfinch, Pine Siskin, and House Finch. Often feeds on the ground or waits while other birds command perches.

FLIGHT: Quick, nimble, and bounding. The large white wing patches on the outer wing flash prominently.

VOCALIZATIONS: Song is a lighthearted but somewhat unenthused ensemble of single, doubled, and tripled notes interspaced with wheezes, canary-like trills, stutters, and clear notes—all given from a high perch. Imitates other birds' sounds as well. Call is a wheezy goldfinch's "*wheeee?*" Also makes a clear, plaintive, descending "*pee'r.*" In flight, makes a descending three-note "*chink chink chink.*" Winter flocks often engage in conversational twittering and nattering.

Lawrence's Goldfinch, *Carduelis lawrencei*

California Grayfinch

STATUS: Uncommon, erratic, localized, and geographically restricted. **DISTRIBUTION:** Breeds entirely in Calif. (and n. Baja) west of the Sierras and the Cascades. Winters in sw. Calif. and Baja and wanders somewhat regularly in fall and winter to s. Ariz. (and border portions of se. Calif.) and casually to Nev., N.M., and extreme w. Tex. A few birds may remain in northern breeding areas. **HABITAT:** An arid-land specialist. Breeds in open and riparian woodlands adjacent to chaparral and brush, grasslands, and weedy fields. A water source (stream, pond, cattle trough) is requisite. Prefers oaks, but also nests in cottonwoods. In winter, concentrates in riparian habitats in arid areas. Forages in grasses, fields, roadsides, mesquite, cultivated fields, desert oases, open washes, orchards, gardens, parks, and suburban yards. Often nests in conifers. In late summer, some birds apparently move to higher elevations, concentrating in streamside thickets bordering grasslands. **COHABITANTS:** Habitually found with Lesser Goldfinch. In winter, often found in flocks with bluebirds, Lark Sparrow, House Finch, Lesser Goldfinch. **MOVEMENT/MIGRATION:** Erratic, irruptive, and somewhat nomadic. Birds are often common one year and absent the next. Departs breeding areas from early July to mid-September; departs northern California in September; returns between early March and early April. VI: 1.

DESCRIPTION: A pale gray goldfinch whose life vacillates between the interior of oaks and grassy meadows and wet streamsides. A tiny finch (slightly smaller than American Goldfinch and Pine Siskin; slightly larger than Lesser Goldfinch) that is shaped like Lesser Goldfinch but has a more pushed-in face, a stubbier, more House Finch–like bill, and a slightly longer tail.

Adults are overall pale and gray. The yellow breast, the golden highlights in the wings, and especially the black face of adult males are standout features. Females and immatures are plain-faced—no shading, no patterning, no hint of yellow—and uniformly colored (winter females and immatures are brownish gray). The wing pattern is confused: wing-bars are not well defined, but obvious yellow streaking bleeds through the folded wing—a touch of color that stands out against an otherwise bland backdrop. If you're looking at a goldfinch whose face seems House Finch–plain and whose wings are yellow-gilded like a drab Pine Siskin, you're looking at Lawrence's.

A splash of yellow on the breast of adult females is usually apparent. The pale bill color is often no paler than Lesser Goldfinch.

BEHAVIOR: Highly social; occurs in flocks even during nesting season. Often found in flocks of Lesser Goldfinches and, in winter in open country, in flocks with bluebirds and House Finches. In the dry season, almost always found near water. Sits in streamside thickets or in the interior of streamside trees and throughout the day flies down to drink or goes off to forage in weedy patches on hillsides, beside watercourses or roadways, or along fencerows.

More finicky about its diet than Lesser Goldfinch. In spring, feeds primarily on fiddleneck, and in winter on chamise. Also attracted to salt from blocks set out for cattle.

Tame. Not as responsive to pishing as Lesser Goldfinch.

FLIGHT: Males are distinctive. Females and immatures are overall pale and can be picked out of flocks with Lesser Goldfinches by their pale underwing linings, the white band bisecting the tip of the tail, and (usually) the splash of yellow in the wings. Flight is like other goldfinches—light, nimble, and bouncy. Climbs on rapid wingbeats and descends on a closed-wing glide.

VOCALIZATIONS: Song is like other goldfinches—a lengthy and complex blend of rising-and-falling, tinkling, trilling, warbling notes that are not repeated. Generally higher-pitched but softer, breezier, and more musical than Lesser and American. Calls include a nasal "*teeyer*" and a high silver bell–like "*tee*" or "*teenk*." In flight, makes a distinctive bell-like "*tink-ooo*" that is very different from the three-note stutter of Lesser Goldfinch.

PERTINENT PARTICULARS: During the dry season, the key to finding this species is finding standing water and the flocks of Lesser Goldfinches that will almost certainly be in the area—feeding on the water's edge, hidden in willows, or calling from within the canopy of nearby oaks. Pishing is generally not productive. The better strategy is to wait until the birds come down to the water's edge to drink, as they do frequently throughout the day.

American Goldfinch, *Carduelis tristis*

Says "Potato Chip," Flies with a Dip

STATUS: Common, widespread, and familiar—known to a broad cross-section of North American residents as the "wild canary." **DISTRIBUTION:** Breeds across s. Canada from s. B.C. across s. and cen. Alta. (except western portions), s. and cen. Sask., s. Man., s. Ont., s. Que., the Maritimes, and sw. Nfld. In the U.S., breeds north of a line drawn from s. Calif., cen. Nev., cen. Utah, nw. and se. Colo., w. Okla., extreme ne. Tex., cen. La., cen. Miss., cen. Ala., cen. Ga., cen. S.C., and se. N.C. In winter, except for extreme se. B.C., se. Ont., extreme s. Que., N.B., and s. N.S., absent in Canada. In winter in the U.S., found everywhere except for n. Mont. and e. Wyo. Casual north to Alaska. **HABITAT:** In all seasons, prefers weedy fields (particularly fields rich in thistle), forest edge (particularly forests laced with seed-bearing trees, such as birch and sweet gum), and suburban yards (particularly those sporting commercial thistle feeders). **COHABITANTS:** During breeding, varied. In winter, often associates with House Finch, Pine Siskin, other goldfinches. **MOVEMENT/MIGRATION:** Northern birds are migratory; southern breeders are year-round residents. Spring migration (late for finches) from May to June; fall (late) from October to January. In addition, populations shift in response to food availability throughout the winter. VI: 1.

DESCRIPTION: A small, tame, acrobatic, vocal, and charming bird. Canary-like right down to the conical finchlike bill, but more compact, with a plumper body and a short, narrow, deeply notched tail that is too petite for the rest of the bird. Bright yellow males, with jet-black caps, wings, and tail, are unmistakable. Except for blackish wings and tails, winter males, females, and immatures are overall plain and putty-colored (with no hint of green on the back), except for a wash of yellow on the face, two almost equal-sized, bold, broad, buffy or creamy (not white!) wing-bars, and white undertail coverts. The expression is sleepy.

In winter in the West, confusion with the smaller, slimmer, and more angular and compact Lesser Goldfinch is possible. Shape and size differences notwithstanding (and they're easily seen in direct comparison), Lesser Goldfinch is overall greener above, yellower below (especially the undertail coverts), with plumage that is less homogenized overall.

BEHAVIOR: In summer, most often seen dangling jauntily from a thistle plant—the seeds are an important food source, and the thistledown a primary component of the bird's cup-shaped nest. In winter (also in summer), most often seen swarming over somebody's thistle feeder—or trying to.

American Goldfinch is generally subordinate to other finches, including Pine Siskin and House Finch, but dominant over Lesser.

A highly social flocking bird found in groups of less than 12 to more than 200. Forages by clinging to weed stalks just above the ground or to the highest outermost branches of seed-bearing trees. Nimble, acrobatic, and somewhat chickadee-like, American Goldfinch has snappier movements—indeed, its every movement seems more accelerated than most other birds. Dangles chickadee-like and flaps its wings rapidly to keep its balance. Lands on the ground to drink, typically not to feed—in fact, seems somewhat terraphobic.

Flocks are active but uncoordinated, with birds shifting perches at whim. Often calls attention to itself with a twittering banter. Often found where other birds (robins, Yellow-rumped Warblers, Cedar Waxwings) are feeding, but does not seem to associate with these species. Does mix with other goldfinch species and with Pine Siskin.

FLIGHT: Jaunty and bouncy in the extreme. Highly exaggerated U-shaped undulations are marked by a rapid upward jerk when the bird begins a series of wingbeats, following a closed-wing plunge. Undulates sharply even when flying a few feet (such as from branch to branch). Often calls in flight. Flocks are basically globular and amorphous; individuals are widely and randomly spaced because of (and perhaps in deference to) each individual's reckless bouncy flight. Individuals swirl around within the flock, shifting locations constantly, so that it seems more an exercise in quantum mechanics than a flock.

In flight, except for breeding males, individuals seem surprisingly cryptic or pale. Combined with the wide spacing between birds, this trait make flocks difficult to detect against the sky. In flight, dark underwings contrast with overall pale bodies—but wingbeats are so rapid that the black blurs to gray.

VOCALIZATIONS: Song is a somewhat bubbly (and incessant) series of repeated phrases punctuated by "*jee*" and "*zwee*" notes. Classic call is a whiny, ascending, two-note query: "*jur-EEEE?*" Flight call is a lighthearted descending skip down the scales, "*J'je,j'jur,*" often phonetically rendered by an earlier generation of birders as "*Per-chic-o-ree,*" but now expressed as "*Potato Chip.*"

Evening Grosbeak, *Coccothraustes vespertinus*

Golden Mesofinch

STATUS: Fairly common breeder in northern and western forests; widespread irruptive migrant across most of the United States. **DISTRIBUTION:** Breeds (and often remains) across most of s. Canada, including all but nw. B.C., w. and cen. Alta., cen. Sask., cen. and se. Man., s. Ont. (north to James Bay and excluding extreme se. Ont.), s. Que., the Maritimes, and s. Nfld. In the U.S., occurs in much of Wash., Ore., n. Calif., Idaho, w. Mont., w. and s.-cen. Wyo., n. and sw. Utah, n. and e. Ariz., w. and cen. Colo., n. N.M., n. Minn., n. Wisc., n. Mich., n. N.Y., all but southern portions of Vt. and N.H., and most of Maine. In winter, ranges south (and northwest into the s. Yukon); birds sometimes reach s. Calif., the Mexican border, the Gulf states, and the Southeast. **HABITAT:** Does best in mature, open-canopy, mixed coniferous woodlands, including spruce, fir, and pines. Also found in mixed coniferous-deciduous habitats and some northern hardwood forests, but in the United States seems to avoid pine-oak. In winter, more eclectic—found in pines, oak-pines, pinyon-juniper, and small wood lots in urban and suburban areas where ample amounts of sunflower seed are being offered. **COHABITANTS:** Purple Finch, Cassin's Finch, Red Crossbill, Pine Siskin. **MOVEMENT/MIGRATION:** Irruptive and irregular. Fall migration begins in August, peaks in October and November, and extends into December; in some years, however, a second movement, presumably linked to depleted food supplies, occurs in late winter. Spring movements occur between early March and early June, with a peak from late April to mid-May. VI: 3.

DESCRIPTION: A big meso-billed finch that calls attention to itself. Large, robust, and starling-sized (absolutely dwarfs a goldfinch), with a massive pale greenish or yellowish bill, a large head/neck, a portly body, and a short tail.

Males are unmistakable—these yellow mustard-colored birds have a blackish hood, wings, and tail and a white saddle on the lower back. In breeding plumage, the huge lime green bill is too anomalous and imposing to ignore. The yellow undertail seems to glow. The yellow eyebrows make birds look fierce or cross. Females and immatures are mostly gray, lightly washed with yellow, and trimmed with black wings and tails that show touches of white. They look somewhat like grossly oversized female Lawrence's Goldfinch—but goldfinches are petite, and Evening Grosbeaks look like sumo wrestlers. You will have no more trouble identifying this bird than you had finding it because...

BEHAVIOR: ... Evening Grosbeak is flamboyant and noisy (except in breeding season). Moves in flocks that announce themselves by explosive call notes and rippling trills and distinguish themselves by their energetic roller-coaster flight and large white patches on the wings. You usually hear Evening Grosbeaks before you see them, and you might have to look very high. A social bird in winter and migration, often found in flocks of 10–40 birds (at times as many as 300). In spring, small groups are common, and in summer, sightings of individual birds are the norm.

Commonly seen foraging at the tops and outer branches of trees. Also moves freely on the ground and is commonly seen along roadsides picking up grit (particularly where snowplows have cleared). Hops (high) when feeding. Birds arriving at feeding stations commonly perch high for a short period before descending to the feeder. Fairly tame. Fairly tolerant of other species at feeding stations. Responds to pishing.

FLIGHT: Big, stocky, front-heavy profile, with a big projecting head, a short tail, and wide wings. Males appear black and white in flight—mostly dark with flashing white patches on the inner wing. Females and immatures are plainer but show flashes of white in the outer wing (and, at close range, a pale tip to the tail). Flight is strong and bounding—finchlike but more ponderous. Wingbeats are given in a fairly lengthy series followed by a closed-wing glide.

VOCALIZATIONS: The same vocalizations are used when perched, in flight, and (apparently) on territory. Basic call is a loud, shrill, explosive, whistled report, "*PHEEU!*" or "*PEEER!*" that is often repeated and given by multiple individuals in a flock. Also makes a somewhat musical rippling trill.

OLD WORLD SPARROWS

House Sparrow, *Passer domesticus*

Sidewalk Sparrow

STATUS: Common and widespread, but particularly common in urban areas and around farm buildings. **DISTRIBUTION:** Found throughout the U.S. (except Alaska) and Mexico and across most of subtundra Canada. Casual in se. Alaska. Almost always found in habitat modified for humans. **HABITAT:** Prime habitats include zoos, parks, amusement pier boardwalks, city parks, outdoor cafés (even enclosed shopping malls!), and hedges near feedlots. Does not occur in forests, native grasslands, or deserts except where human dwellings intrude. **COHABITANTS:** Rock Pigeon, starlings, grain-fed horses, humans. **MOVEMENT/MIGRATION:** Permanent resident and essentially nonmigratory, except that fall movement has been noted at some East Coast locations. VI: 4.

DESCRIPTION: A stocky, large-headed, short-legged, street-tough finch with a robust bullet-shaped bill and a small black beady eye. Females are overall more pointy-faced than males. About the same size as House Finch, but distinctly broader-beamed and more portly. Breeding males (some are in plumage by February), with their gray crown, white cheeks, bright chestnut nape and upperparts, grayish underparts, and black bill, throat, and breast, are dapper and distinctive. In nonbreeding plumage, the breast is gray (the black under the bill is limited to a narrow goatee), and the bill is pale or yellowish; otherwise, similar to breeding plumage.

Females and immatures are just plain drab—grayish brown with an unstreaked crown and a somewhat streaked back above, brownish gray with no streaks below. Against this bland backdrop, the broad buffy eyebrow and pale bill are standout features.

BEHAVIOR: Noisy, quarrelsome, and highly social bird. Except in breeding season, always found in flocks. Forages primarily on bare ground (or pavement), moving with quick jaunty hops or flying directly to snap up morsels. When not feeding, individuals and flocks frequently bury themselves in dense bushes or vegetation—where their near-constant chips and other vocalizations betray them.

Acrobatic, in a clumsy sort of way. Lands on exterior twigs and clings tenaciously as the branch wilts and the bird hangs upside down. Able to land sideways on branches (and signs and brickwork). Also flycatches and pursues slow-moving insects with clumsy determination. Likes worming into tight confines; builds its bulky nests in the nooks and crannies of man-made structures (as well as in bird boxes erected for other bird species).

FLIGHT: Overall stocky, short-winged, and short-tailed. Flight is fast and direct, with constant, rapid, near-whirring wingbeats, yet it also has an odd flowing quality—birds rise and fall rather than undulate.

VOCALIZATIONS: Chirpy and chatteriferous species with an array of chirping calls (and nothing that really deserves to be called a song). Call is a vaguely musical "*ch'eur*" that sounds like a cross between a chirp and a slurp. Also makes a more muffled and twangy "*ch'uhr*," which is doubled and shortened as an alarm call, "*chr/chr*," and sometimes rendered in a chattering series: "*chr-chr-chr-chr-chr.*"

PERTINENT PARTICULARS: Males and females are almost always found together (aiding greatly in the identification of females). In the East, flocks should be checked for the presence of Dickcissel (which sometimes join House Sparrow flocks), which are more slender, heavier-billed, and more anvil-headed and show a pale wash of yellow about the face and breast.

Eurasian Tree Sparrow, *Passer montanus*

St. Louis Sparrow

STATUS: Common to fairly common resident, but geographically restricted. DISTRIBUTION: Found only in se. Iowa, ne. Mo., and w.-cen. Ill. HABITAT: Urban, residential, and rural areas, often in association with ponds, lakes, and surrounding trees. In winter, concentrates around grain storage areas and farms. COHABITANTS: House Sparrow, Cardinals' fans. MOVEMENT/MIGRATION: A permanent resident that shows some modest relocation between breeding and wintering areas. VI: 1.

DESCRIPTION: A better-looking House Sparrow. In direct comparison, overall roundly compact but conspicuously smaller and less robust, with a smaller head, a smaller bill, and a shorter tail.

Brown with traces of gray above, gray with buffy sides below. In all plumages, distinguished from House Sparrow by an all–dull rufous brown cap (House Sparrow has a gray forehead and a redder brown cap), a black cheek spot, and a trim black goatee (House Sparrow has only a shabby black bib). Unlike House Sparrow, all ages and sexes of Eurasian Tree Sparrow are similar.

BEHAVIOR: During breeding season, often found in wooded parks, farms, and rural areas (being out-competed for nest sites by the larger House Sparrow in more urban areas). At other times, hangs around with House Sparrows in trees and shrubs, on utility wires and dusty street sides, and, of course, at bird-feeding stations.

In winter, gathers in flocks of up to 500 individuals near grain storage areas. Feeds primarily by hopping on the ground, often employing the roll-over technique seen in some blackbird species (birds in the rear overfly the flock to get to the front). Occasionally gleans from vegetation and leaps for passing insects. At night, roosts in dense protective vegetation (frequently planted conifers) or in barns and tree cavities. All in all, behaves much like its more widespread relative.

FLIGHT: Like House Sparrow. Wingbeats are constant and whirring; flight is mostly straight, but with a buoyant rising-and-falling quality.

VOCALIZATIONS: Chirpy, like House Sparrow, but slightly higher-pitched and somewhat bisyllabic. In flight, makes a sharp "*teck*."

PERTINENT PARTICULARS: Not a difficult species to find (when you're in its range). If you're having trouble finding the bird in urban or suburban areas, home in on a plot of trees near a lake, where Eurasian Tree Sparrow is more likely to be than House Sparrow. Like House Sparrow, likes to dust-bathe, so drive down roadsides looking for curb-side clusters.

Index

Numbers in **bold** indicate main entries.